HANDBOOK OF ENVIRONMENTAL DATA ON ORGANIC CHEMICALS

HANDBOOK OF ENVIRONMENTAL DATA ON ORGANIC CHEMICALS

SECOND EDITION

Karel Verschueren

Environmental Advisor
Heidemij/Adviesbureau
and
Department of Public Health and Tropical Hygiene
Agricultural University of Wageningen
Netherlands

VNR VAN NOSTRAND REINHOLD COMPANY
NEW YORK   CINCINNATI   TORONTO   LONDON   MELBOURNE

Copyright © 1983 by Van Nostrand Reinhold Company Inc.

Library of Congress Catalog Card Number: **82-10994**
ISBN: 0-442-28802-6

All rights reserved. Certain portions of this work copyright © 1977 by Van Nostrand Reinhold Company, Inc. No part of this work covered by the copyright hereon may be reproduced or used in any form or by any means—graphic, electronic, or mechanical, including photocopying, recording, taping, or information storage and retrieval systems—without permission of the publisher.

Manufactured in the United States of America

Published by Van Nostrand Reinhold Company Inc.
135 West 50th Street, New York, N.Y. 10020

Van Nostrand Reinhold
480 Latrobe Street
Melbourne, Victoria 3000, Australia

Van Nostrand Reinhold Company Limited
Molly Millars Lane
Wokingham, Berkshire, England

15 14 13 12 11 10 9 8 7 6 5 4 3 2 1

Library of Congress Cataloging in Publication Data
Verschueren, Karel.
 Handbook of environmental data on organic chemicals.

 Bibliography: p.
 1. Organic compounds—Environmental aspects—Handbooks, manuals, etc. I. Title.
 TD196.073V47   1983   363.7'384   82-10994
 ISBN 0-442-28802-6

CONTENTS

I **INTRODUCTION** 1

II **ARRANGEMENT OF CATEGORIES** 1

    A. Properties 1
    B. Air pollution factors 1
    C. Water pollution factors 2
    D. Biological effects 2

III **ARRANGEMENT OF CHEMICALS** 2

IV **EXPLANATORY NOTES** 2

    A. Properties 3
        1. Boiling points 4
        2. Vapor pressure 5
        3. Vapor density 6
        4. Water solubility 8
        5. Octanol/water partition coefficient 14
        6. Specific Gravity 36
    B. Air Pollution Factors **40**
        1. Conversion between volume and mass units of concentration 40
        2. Odor 42
        3. Atmospheric degradation 59
        4. Natural sources 62
        5. Manmade sources 63
        6. Emission control methods and efficiency 63
        7. Sampling and analysis: methods and limits 63
    C. Water Pollution Factors 65
        1. Biodegradation 65
        2. Oxidation parameters 68
        3. Impact on biodegradation processes 72

CONTENTS

      4. Waste water treatment 75
      5. Alteration and degradation processes 88
  D. Biological Effects 96
      1. Arrangement of data 96
      2. Classification list 96
      3. Organisms used in experimental work with polluting substances or in environmental surveys 99
      4. Discussion of biological effects tests 109
  E. Glossary 124
  F. Abbreviations 132

ENVIRONMENTAL DATA 137

FORMULA INDEX 1202

BIBLIOGRAPHY 1229

Errata Sheet for

HANDBOOK OF ENVIRONMENTAL DATA ON
ORGANIC CHEMICALS
Second Edition

by Karel Verschueren
ISBN 0-442-28802-6

Pages 882 and 892 unfortunately were interchanged in printing. There are no missing pages.

HANDBOOK OF ENVIRONMENTAL DATA ON ORGANIC CHEMICALS

I INTRODUCTION

Since the publication of the first edition of this handbook in 1977, much more information has become available about the presence and fate of new and existing organic chemicals in the environment. These data, when given a wide distribution, will no doubt reduce the misuse of dangerous chemicals and hence their impact on the environment. The "Handbook of Environmental Data on Organic Chemicals" has now been updated and covers not only individual substances but also mixtures and preparations.

II ARRANGEMENT OF CATEGORIES

The information in the categories listed below is given for each product in the sequence indicated; where entries are incomplete, it may be assumed that no reliable data were provided by the references utilized.

Name: the commonly accepted name is the key entry.
Synonym: alternate names, as well as trivial names and identifiers, are indicated.
    Obsolete and slang names have been eliminated as far as possible
Formula: the molecular and structural formulas are given.

A. PROPERTIES The chemical and physical properties typically given are: physical appearance; molecular weight (m.w.); melting point (m.p.); boiling point (b.p.) at 760 mm Hg unless otherwise stated; vapor pressure (v.p.) at different temperatures; vapor density (air = 1); saturation concentration in air at different temperatures (sat. conc.); the maximum solubility in water at various temperatures (solub.); the specific gravity; the logarithm of the octanol/water partition coefficient (log P_{oct})

B. AIR POLLUTION FACTORS The following data are given: conversion factors (between volume and mass units of concentration); odor threshold values and characteristics; atmospheric reactions; natural sources (and background concentrations); manmade sources (and ground level concentrations caused by such sources); emission control methods (and results); methods of sampling and analysis.

C. WATER POLLUTION FACTORS   Analogous to the previous category, the following data are listed: biodegradation rate and mechanisms; oxidation parameters, such as BOD, COD, $KMnO_4$ value, TOC, ThOD; impact on treatment processes and on the BOD test; reduction of amenities through taste, odor, and color of the water or aquatic organisms; the quality of surface water and underground water and sediment; natural sources, manmade sources, waste water treatment methods and results; methods of sampling and analysis.

D. BIOLOGICAL EFFECTS   Residual concentrations, bioaccumulation values and toxicological effects of exposing the products to: ecosystems, bacteria, plants, algae, protozoans, worms, molluscs, insects, crustaceans, fishes, amphibians, birds, mammals and man.

The "explanatory notes" give a more detailed description of the compiled data, explanations of the definitions and abbreviations used throughout the book, and indicates how the data can be used to prevent or reduce environmental pollution.

III   ARRANGEMENT OF CHEMICALS

The chemicals are listed in strict alphabetical order; those comprised of two or more words are alphabetized as if they were a single word.   The many prefixes used in organic chemistry and disregarded in alphabetizing since they are not considered an integral part of the name; these include *ortho-*, *meta-*, *para-*, *alpha-*, *beta-*, *gamma-*, *sec.*, *tert.*, *sym-*, *as-*, *uns.-*, *cis-*, *trans-*, *d-*, *l-*, *dl-*, *n.*, N-, as well as all numerals denoting structure.   However, there are certain prefixes that are an integral part of the names (iso-, di-, tri-, tetra-, cyclo-, bio-, neo-, pseudo-), and in these cases the name is placed in its normal alphabetical position, e.g., dimethylamine under D and isobutane under I.

IV   EXPLANATORY NOTES

The reader should consult the appropriate sections of this chapter if he is not acquainted with the definitions and abbreviations used throughout the book. The data are given in the following sequence (each item will be discussed in detail):

A. PROPERTIES
   1. formula
   2. physical appearance
   3. molecular weight (m.w.)
   4. melting point (m.p.)
   5. boiling point (b.p.)
   6. vapor pressure (v.p.)
   7. vapor density (v.d.)
   8. saturation concentration (sat. conc.)
   9. solubility (solub.)
   10. specific gravity (sp. gr.)
   11. logarithm of the octanol/water distribution coefficient (log P_{oct})

B. AIR POLLUTION FACTORS
  12. conversion factors
  13. odor
  14. atmospheric reactions
  15. natural sources
  16. manmade sources
  17. control methods
  18. sampling and analysis
C. WATER POLLUTION FACTORS
  19. biodegradation
  20. oxidation parameters
  21. impact on biodegradation processes
  22. odor and taste thresholds
  23. water and sediment quality
  24. natural sources
  25. manmade sources
  26. waste water treatment
  27. sampling and analysis
D. BIOLOGICAL EFFECTS
    —residual concentrations
    —bioaccumulation values
    —toxicological effects
  28. ecosystems
  29. bacteria
  30. algae
  31. plants
  32. protozoans
  33. worms
  34. molluscs
  35. insects
  36. crustaceans
  37. fishes
  38. amphibians
  39. birds
  40. mammals
  41. man

A. PROPERTIES

Only the most relevant chemical and physical properties are given. Flash points, flammability limits, autoignition temperature etc. have been omitted because they are not of direct concern to the environmentalist. These and other dangerous properties of chemicals can be found in "*Dangerous Properties of Industrial Materials*" by I. Sax.

Chemicals are never 100% pure, but the nature and quantity of the impurities can have a significant impact on most environmental qualities. The following parameters are very sensitive to the presence of impurities:

water solubility
odor characteristic and threshold values
BOD
toxicity

The following data illustrate this point (from Shell's Chemical Guide):

product: *diethylene glycol* (Shell)
$O(CH_2CH_2OH)_2$

	"normal grade"	*"special grade"*
distillation range:	240–255°C	242–250°C
acidity (as CH_3COOH):	max. 0.2 wt%	max. 0.002 wt%
ash content:	max. 0.05 wt%	max. 0.002 wt%
BOD_5:	0.12	0.05
COD:	1.49	1.51
goldfish LD_{50}(24 hr):	5000 mg/l	5000 mg/l

product: *ethyleneglycol*
$HOCH_2-CH_2OH$

	"normal grade"	*"special grade"*
distillation range:	194–205°C	max. 2°C, incl. 197.6°C
ash content:	max. 0.002 wt%	max. 0.001 wt%
BOD_5:	0.47	0.15
BOD_5 after adaptation:	0.81	0.67
COD:	1.24	1.29
goldfish LD_{50}(24 hr):	5000 mg/l	5000 mg/l

When no data are available, the distillation range can give a first indication on the presence of impurities. Therefore, in this work, whenever a distillation range (boiling range) is given, the environmental data should be interpreted carefully.

product: *triethanolamine*
$N(CH_2CH_2OH)_3$

	"normal grade"	*"85%"*
triethanolamine content	min. 80 wt%	min. 85 wt%
BOD_5	0.02	0.03
BOD_5 after adaptation	0.17	0.90
COD	1.50	1.50

After adaptation of the culture, the "85%" grade is much more biodegradable than the less pure "commercial" grade.

1. Boiling Points. The boiling points of the members of a given homologous series increase with increasing molecular weight. The boiling points rise in a uniform manner as shown in Figs. 1 and 2.

If a hydrogen atom of one of the paraffin hydrocarbons is replaced by another atom or a group, an elevation of the boiling point results. Thus alkyl halides, alcohols,

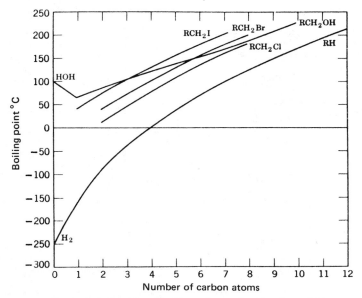

Fig. 1. Relationship of boiling points and molecular weight.

aldehydes, ketones, acids, etc., boil at higher temperatures than the hydrocarbons with the same carbon skeleton.

If the group introduced is of such a nature that it promotes association, a very marked rise in boiling point occurs. This effect is especially pronounced in the alcohols and acids, since hydrogen bonding can occur.

2. Vapor Pressure. The vapor pressure of a liquid or solid is the pressure of the gas in equilibrium with the liquid or solid at a given temperature. Volatilization, the evaporative loss of a chemical, depends upon the vapor pressure of the chemical and on environmental conditions which influence diffusion from the evaporative surface. Volatilization is an important source of material for airborne transport and may lead to the distribution of a chemical over wide areas and into bodies of water (e.g., in rainfall) far from the site of release. Vapor pressure values provide indications of the tendency of pure substances to vaporize in an unperturbed situation, and thus provide a method for ranking the relative volatilities of chemicals. Vapor pressure data combined with solubility data permit calculations of rates of evaporation of dissolved organics from water using Henry's Law constants, as discussed by MacKay and Leinonen (1943) and Dilling (1944).

Chemicals with relatively low vapor pressures, high adsorptivity onto solids, or high solubility in water are less likely to vaporize and become airborne than chemicals with high vapor pressures or with less affinity for solution in water or adsorption to solids and sediments. In addition, chemicals that are likely to be gases at ambient temperatures and which have low water solubility and low adsorptive tendencies are less

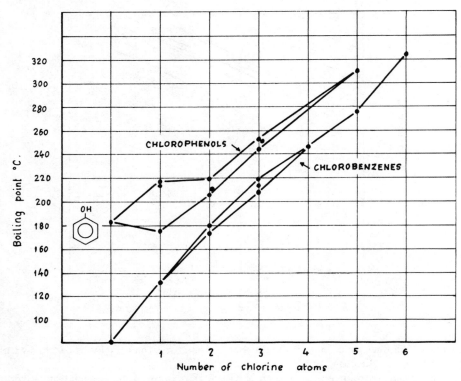

Fig. 2. Relationship of boiling points and molecular weights for chlorinated benzenes and phenols.

likely to transport and persist in soils and water. Such chemicals are less likely to biodegrade or hydrolyze, but are prime candidates for photolysis and for involvement in adverse atmospheric effects (e.g., smog formation, or stratospheric alterations). On the other hand, nonvolatile chemicals are less frequently involved in significant atmospheric transport, so concerns regarding them should focus on soils and water.

Vapor pressures are expressed either in mm Hg (abbreviated mm) or in atmospheres (atm).

If vapor pressure data for certain compounds are not available, they can be derived graphically from their boiling points and the boiling point/vapor pressure relationship for homologous series. An example is shown in Fig. 3.

3. Vapor Density. The density of a gas indicates whether it will be transported along the ground, possibly subjecting surrounding populations to high exposure, or whether it will disperse rapidly.

The concentration term *vapor density* is often used in discussion of vapor phase systems. Vapor density is related to equilibrium vapor pressure through the equation of state for a gas:

$$PV = nRT$$

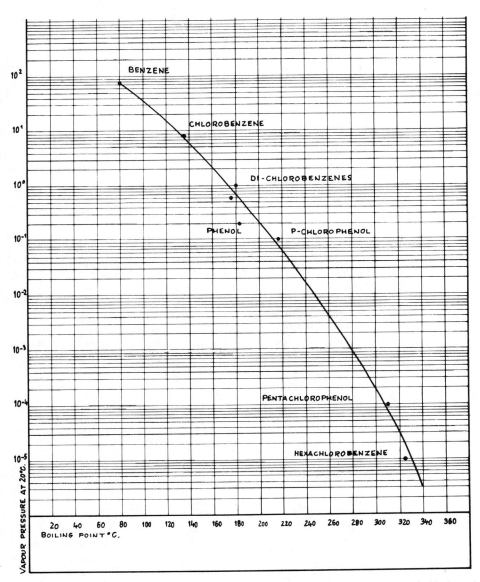

Fig. 3. Relation between boiling point and vapor pressure for homologous series of chlorinated benzenes and phenols.

or if the mass of the substance and the gram molecular weight are substituted for the number of moles n, the following equation is obtained:

$$\text{vapor density (Vd)} = \frac{PM}{RT}$$

where P is the equilibrium vapor pressure in atmospheres and
$R = 0.082$ liter atmospheres/mol/K
M = gram molecular weight
T = absolute temperature in degrees Kelvin.

In this book the relative vapor density (air = 1) is given because it indicates how the gas will behave upon release.

4. Water Solubility

4.1. Objectives. The water solubility of a chemical is an important characteristic for establishing that chemical's potential environmental movement and distribution. In general, highly soluble chemicals are more likely to be distributed by the hydrological cycle than poorly soluble chemicals.

Water solubility can also affect adsorption and desorption on soils and volatility from aquatic systems. Substances which are more soluble are more likely to desorb from soils and less likely to volatilize from water. Water solubility can also affect possible transformation by hydrolysis, photolysis, oxidation, reduction, and biodegradation in water. Finally, the design of most chemical tests and many ecological and health tests requires precise knowledge of the water solubility of chemicals. Water solubility is an important parameter for assessment of all solid or liquid chemicals. Water solubility is usually not useful for gases because their solubility in water is measured when the gas above the water is at a partial pressure of one atmosphere. Thus, solubility of gases does not generally apply to environmental assessment because the actual partial pressure of a gas in the environment is extremely low.

4.2. Interpretation of Data. It is not unusual to find in the literature a wide range of solubilities for the same product. The oldest literature will generally yield the highest solubility values. The reasons are twofold: first, in the years before and immediately after the Second World War, products were not as pure as they are today; secondly, recent determinations are based upon specific methods of analysis, such as gas chromatography. Nonspecific determinations do not distinguish between the dissolved product and the dissolved impurities; the latter, when much more soluble than the original product, will move to the aqueous phase and be recorded as "dissolved product". Nonspecific methods include turbidity measurement and TOD.

The measurement of aqueous solubility does not usually impose excessive demands on chemical techniques, but the measurement of the solubility of very sparingly soluble compounds requires specialized procedures. This problem is well illustrated by the variability in the values quoted in the literature for products such as DDT and PCBs. This situation happens to be of some consequence, in that many of those compounds that are known to be significant environmental contaminants, such as DDT and PCBs, are those that have very low water solubilities.

4.3. Influence of the Composition of Natural Waters. The composition of natural waters can vary greatly. Environmental variables such as pH, water hardness, cations, anions, naturally occurring organic substances (e.g., humic and fulvic acids, hemicelluloses) and organic pollutants all affect the solubility of chemicals in water. Some bodies of water contain enough organic and inorganic impurities to significantly alter the solubility of poorly soluble chemicals.

The solubility of lower *n*-paraffins in salt water compared with fresh, distilled water is higher by about one order of magnitude, this difference decreasing with and increase in the molecular weight of the hydrocarbon. The increased solubility in seawater is due to simultaneous physical and chemical factors. The solubility of several higher *n*-paraffins (C_{10} and higher) has been determined in both distilled water and seawater. In all cases, the paraffins were less soluble in seawater than in distilled water. The magnitude of the salting out effect increases with increasing molar volume of the paraffins in accordance with the McDevit-Long theory. This theory of salt effects attributes salting in or salting out to the effect of electrolytes on the structure of water. Since the data in the literature indicate that the lower paraffins (below C_{10}) are more soluble, and that the higher *n*-paraffins (C_{10} and higher) are less soluble in seawater than in distilled water, it is possible to speculate upon the geochemical fate of dissolved normal paraffins entering the ocean from rivers. If fresh water is saturated or near saturated with respect to normal paraffins (e.g., because of pollution), salting out of the higher paraffins will occur in the estuary. The salted out molecules might either adsorb on suspended minerals and on particulate organic matter or rise to the surface as slicks. In either case, they will follow a different biochemical pathway than if they had been dissolved. The salting out of dissolved organic molecules in estuaries applies not only to *n*-paraffins, but to all natural or pollutant organic molecules whose solubilities are decreased by addition of electrolytes. Thus, it is possible that regardless of the levels of dissolved organic pollutants in river water, only given amounts will enter the ocean in dissolved form owing to salting out effects of estuaries. Estuaries may act to limit the amount of dissolved organic carbon entering the ocean, but may increase the amount of particulate organic carbon entering the marine environment.

4.4. Molecular Structure–Solubility Relationship. Since water is a polar compound it is a poor solvent for hydrocarbons. Olefinic and acetylenic linkages or benzenoid structures do not affect the polarity greatly. Hence, unsaturated or aromatic hydrocarbons are not very different from paraffins in their water solubility. The introduction of halogen atoms does not alter the polarity appreciably. It does increase the molecular weight, and for this reason the water solubility always falls off. On the other hand, salts are extremely polar. Other compounds lie between these two extremes. Here are found the alcohols, esters, ethers, acids, amines, nitriles, amides, ketones and aldehydes—to mention a few of the classes of frequent occurrence.

As might be expected, acids and amines generally are more soluble than neutral compounds. The amines probably owe their abnormally high solubility to their tendency to form hydrogen-bonded complexes with water molecules. This theory is in harmony with the fact that the solubility of amines diminishes as the basicity decreases. It also explains the observation that many tertiary amines are more soluble in cold than in hot water. Apparently at lower temperatures the solubility of the hydrate is involved, whereas at higher temperatures the hydrate is unstable and the solubility measured is that of the free amine.

Monofunctional ethers, esters, ketones, aldehydes, alcohols, nitriles, amides, acids, and amines may be considered together with respect to water solubility. As a homologous series is ascended, the hydrocarbon (nonpolar) part of the molecule continually increases while the polar function remains essentially unchanged. There follows, then, a trend toward a decrease in the solubility in polar solvents such as water.

In general an increase in molecular weight leads to an increase in intermolecular forces in a solid. Polymers and other compounds of high molecular weight generally exhibit low solubilities in water and ether. Thus formaldehyde is readily soluble in water, whereas paraformaldehyde is insoluble:

$$CH_2O \rightarrow HO(CH_2O-)_xH$$
water soluble ..... water insoluble

Methyl acrylate is soluble in water, but its polymer is insoluble:

$$CH_2=CHCOOCH_3 \rightarrow (-CH_2\underset{\underset{COOCH_3}{|}}{CH}-)_x$$
water soluble ..... water insoluble

Glucose is soluble in water, but its polymers—starch, glycogen, and cellulose—are insoluble. Many amino acids are soluble in water, but their condensation polymers, the proteins, are insoluble.

Lindenberg (1803) proposed a relationship between the logarithm of the solubility of hydrocarbons in water and the molar volume of the hydrocarbons. If the logarithm of the solubilities of the hydrocarbons in water is plotted against the molar volume of the hydrocarbons, a straight line is obtained. This relationship has been worked out further by C. McAuliffe, and solubilities as a function of their molar volumes for a number of homologous series of hydrocarbons have been presented graphically (242).

From the given correlation between molecular structure and solubility the following conclusions may be drawn:

Branching increases water solubility for paraffin, olefin and acetylene hydrocarbons, but not for cycloparaffins, cyclo-olefins, and aromatic hydrocarbons.
For a given carbon number, *ring formation* increases water solubility.
Double bond addition to the molecule, ring or chain increases water solubility. The addition of a second and third double bond to a hydrocarbon of given carbon number proportionately increases water solubility (Table 2).
A *triple bond* in a chain molecule increases water solubility to a greater extent than two double bonds.

Cary T. Chiou et al. (382) found a good correlation between solubilities of organic compounds and their octanol/water partition coefficient. Furthermore functional groups such as chlorine atoms, methyl groups, hydroxyl groups, benzene rings, etc. showed additive effects on the logarithm of the octanol/water partition coefficient ($\log P_{oct}$) of the parent molecule.

This allowed the calculation of $\log P_{oct}$ values for many organic compounds based on the $\log P_{oct}$ value for the parent compound and the additive effects of the functional groups. Because of the correlation between solubilities of organic compounds

TABLE 1. Influence of Functional Groups on Solubility of benzene derivatives.

	Functional group	$S_{mg/l}$ solubility mg/l (temp. °C)		$\log S_{mg/l}$	$\Delta \log S_{mg/l}$ $\log S_{C_6H_5X} - \log S_{C_6H_6}$
aniline	$-NH_2$	34,000	(20°)	4.53	1.28
phenol	$-OH$	82,000	(15°)	4.91	1.66
benzaldehyde	$-COH$	3,300		3.52	0.27
benzoic acid	$-COOH$	2,900		3.46	0.21
nitrobenzene	$-NO_2$	1,900		3.28	0.03
benzene	–	1,780		3.25	0.00
fluorobenzene	$-F$	1,540	(30°)	3.19	-0.06
thiophenol	$-SH$	470	(15°)	2.67	-0.58
toluene	$-CH_3$	515		2.71	-0.54
chlorobenzene	$-Cl$	448	(30°)	2.65	-0.60
bromobenzene	$-Br$	446	(30°)	2.65	-0.60
iodobenzene	$-I$	340	(30°)	2.53	-0.72
diphenylether	O-⌬	21	(25°)	1.32	-1.93
diphenyl	-⌬	7.5	(25°)	0.88	-2.37

and $\log P_{oct}$, it is not surprising to find the same additive effects of functional groups on their water solubility. Table 1 shows this influence of functional groups on the solubility of benzene derivatives. Solubilities of homologous series of organic compounds are plotted in Figs. 4 to 9.

Effects which cannot be taken into account by this additive-constitutive character of the solubility are:

- steric effects which cause shielding of an active function
- intra- and intermolecular hydrogen bonding (see trihydroxyphenols)
- branching
- inductive effects of one substituent on another
- conformational effects, e.g., "balling up" of an aliphatic chain.

TABLE 2. Influence of Double Bonds on Aqueous Solubility of Cyclic Hydrocarbons (at room temperature) (242).

Hydrocarbon	Solubility mg/l
cyclohexane	55
cyclohexene	213
1,4-cyclohexadiene	700
benzene	1,780

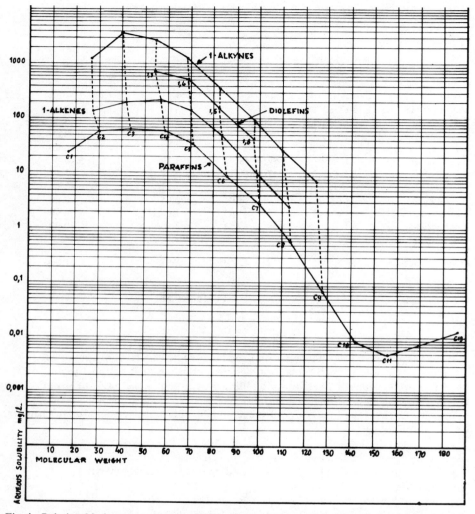

Fig. 4. Relationship between aqueous solubility and molecular weight for saturated and unsaturated straight-chain hydrocarbons.

4.5 Solubility of Mixtures. Mixtures of compounds, whether they are natural such as oil or formulations such as many pesticides, behave differently from the single compounds when brought into contact with water. Indeed each component of the mixture will partition between the aqueous phase and the mixture.

Components with a high aqueous solubility will tend to move toward the aqueous phase while the "unsoluble" components will remain in the other phase. From this follows that the fractional composition of the "water soluble fraction" (WSF) will

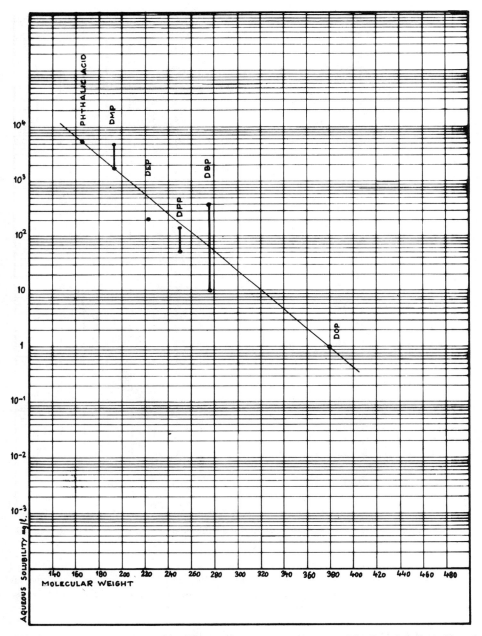

Fig. 5. Relationship between aqueous solubility and molecular weight for phthalic acid and phthalates.

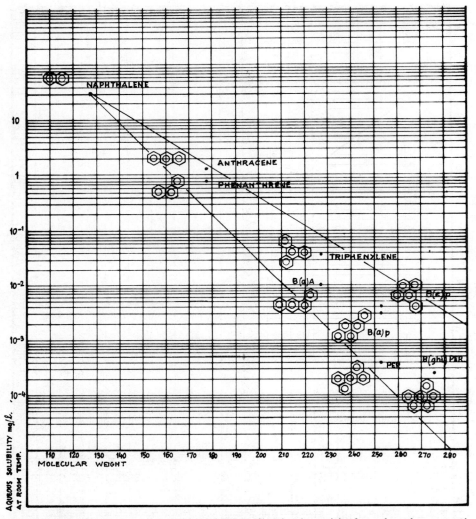

Fig. 6. Relationship between aqueous solubility and molecular weight for polynuclear aromatic hydrocarbons.

differ from the original composition of the mixture and that concentrations of the components in the WSF are generally lower than the maximum solubilities for the individual components. Examples are shown in tables 3 and 4.

5. Octanol/Water Partition Coefficient. The ability of some chemicals to move through the food chain resulting in higher and higher concentrations at each trophic level has been termed *biomagnification* or *bioconcentration*. The widespread distribu-

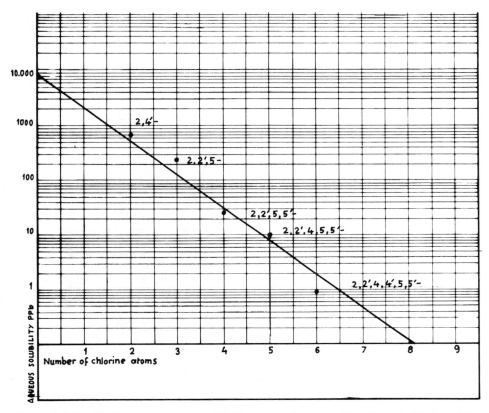

Fig. 7. Relationship between aqueous solubility and molecular weights for polychlorinated biphenyls (PCB's).

tions of DDT and the polychlorinated biphenyls (PCBs) have become classic examples of such movement.

From an environmental point of view this phenomenon becomes important when the acute toxicity of the agent is low and the physiological effects go unnoticed until the chronic effects become evident. For this reason prior knowledge of the bioconcentration potential of new or existing chemicals is desired. However, the determination of the bioconcentration factor of a chemical on a number of animals or in a food chain is expensive and time consuming. If a simple relationship could be established between physicochemical properties of a chemical and its ability to bioconcentrate, it would be of great benefit in planning the future direction of any development work on a new chemical and in directing research efforts to determine the distribution and ultimate fate of a limited number of selected chemicals.

5.1. Definition. The partition coefficient P is defined as the ratio of the equil-

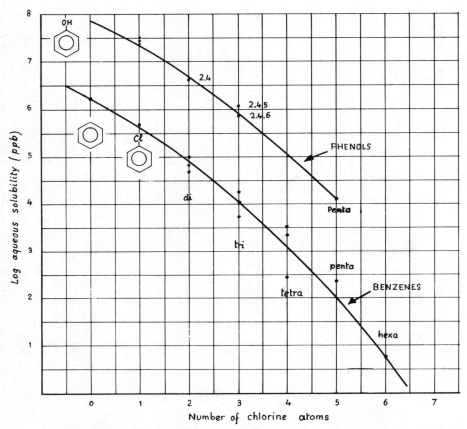

Fig. 8. Relationship between aqueous solubility and molecular weights for homologous series of chlorinated benzenes and phenols.

ibrium concentration C of a dissolved substance in a two-phase system consisting of two largely immiscible solvents, in this case n-octanol and water:

$$P = \frac{C_{\text{octanol}}}{C_{\text{water}}}$$

In addition to the above, the partition coefficient is ideally dependent only upon temperature and pressure. The partition coefficient P is the quotient of two concentrations and is a constant without dimensions. It is usually given in the form of its logarithm to base ten (log P).

The n-octanol/water partition coefficient has proved useful as a means to predict soil adsorption (419), biological uptake (416), lipophilic storage (415), and biomagnification (417, 418, 339, 193).

The bioconcentration of several chemicals in trout muscle was found to follow a

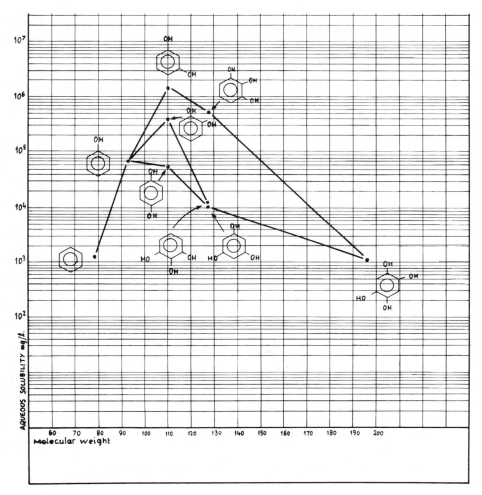

Fig. 9. Relationship between aqueous solubility and molecular weights for hydroxylated benzene derivatives.

straight-line relationship with *n*-octanol/water partition coefficient (193). Bioconcentration in this work was defined as the ratio of concentration of the chemical between trout muscles and the exposure water measured at equilibrium. The relationship was established by measuring the bioconcentration in trout of a variety of chemicals over a wide range of partition coefficients. An equation of the straight line of best fit was determined and used to predict the bioconcentration of other chemicals from their *n*-octanol/water partition coefficients. The predicted values agreed with the experimental values in the literature. Values are expressed as their decimal logarithmus.

The linear relationship between bioconcentration factor and partition coefficient is

TABLE 3. Comparison of Aqueous Solubility of Some PCB Isomers in the Water Soluble Phase of Aroclor 1242 with Maximum Solubility of Individual Isomers (1909).

Isomer	Solubility µg/l in WSF of Aroclor 1242	Max. solubility for individual compound
4-	15.1	2,000 (calculated)
2,2'-	21.2	900
2,4'-	138.9	637
2,5,2'-	61.4	248
2,5,2',5'-	22.3	26.5

The same is true for many mineral oils and petroleum products of which the WSF consists mainly of the more soluble aromatic compounds benzene, toluene, xylene and their alkyl homologues.

given by

$$\log B_f = 0.542 \log P_{oct} + 0.124 \qquad (1)$$

where B_f = bioconcentration factor
P_{oct} = octanol/water partition coefficient.

The relationship is shown in Fig. 10.

The largest compilation of n-octanol/water partition coefficients has been made by Albert Leo et al. (1457).

By far the most extensive and useful partition coefficient data were obtained by the classical way of shaking a solute with two immiscible solvents and then analyzing the solute concentration in one or both phases.

Examples of physico-chemical determinations which may be appropriate are:

- photometric methods
- gas chromatography
- HPLC
- back-extraction of the aqueous phase and subsequent gas chromatography.

TABLE 4. Composition of Aroclor 1242 and Its Water Soluble Fraction (WSF) (1909).

	Aroclor wt %	WSF wt %	WSF/Aroclor ratio of wt %
monochlorobiphenyls	3	19.4	6.5
dichlorobiphenyls	13	31.8	2.4
trichlorobiphenyls	28	31.3	1.1
tetrachlorobiphenyls	30	16.5	0.55
pentachlorobiphenyls	22	0.9	0.04
hexachlorobiphenyls	4	–	< 0.02
	100	100	

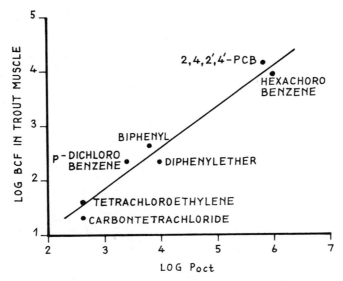

Fig. 10. Relationship between octanol/water partition (P_{oct}) coefficient and bioaccumulation factor (BCF) in trout muscle (1458).

5.2. Calculation of Partition Coefficients. Since partition coefficients are equilibrium constants, it should not be surprising that one finds extrathermodynamic relationships between values in different solvent systems. This relationship can be expressed by the general equation:

$$\log P_2 = a \log P_1 + b \qquad (2)$$

e.g.,

$$\log P_{toluene} = 1.135 \log P_{octanol} - 1.777 \qquad (3)$$
$$(n = 22; r = 0.980; s = 0.194)$$

$$\log P_{cyclohexanone} = 1.035 \log P_{octanol} + 0.896 \qquad (4)$$
$$(n = 10; r = 0.972; s = 0.340)$$

Many $\log P_{oct}$ partition coefficients mentioned in this book have been calculated by A. Leo et al. (1457) using the above and other equations. Furthermore it was found that the $\log P_{oct}$ of a compound could well be calculated from the $\log P_{oct}$ of another compound of the same homologous series by adding or subtracting a number of times a constant value.

Fig. 11 shows that the $\log P_{oct}$ in a homologous series increases by approximately 0.5 per CH_2. The additive effect on $\log P_{oct}$ of functional groups such as chlorine atoms, methyl groups, or benzene rings is clearly demonstrated on Figs. 12 and 13 as well as in Table 5.

Additivity was first established for a wide variety of groups in a study of the substituent constant, π, defined by the following equation:

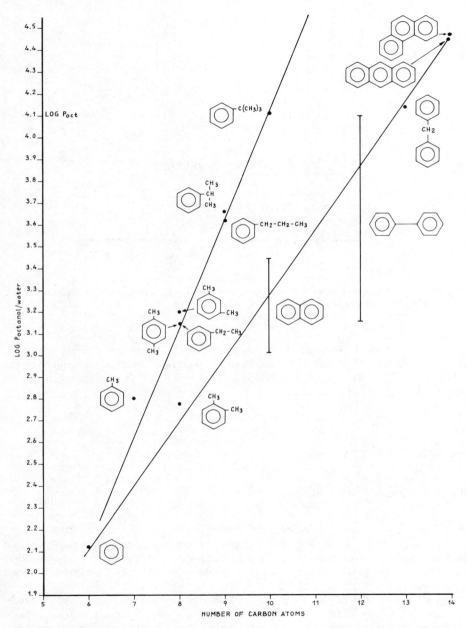

Fig. 11. Additive effect of CH_2, CH_3 and benzene rings on the log P_{oct} partition coefficient.

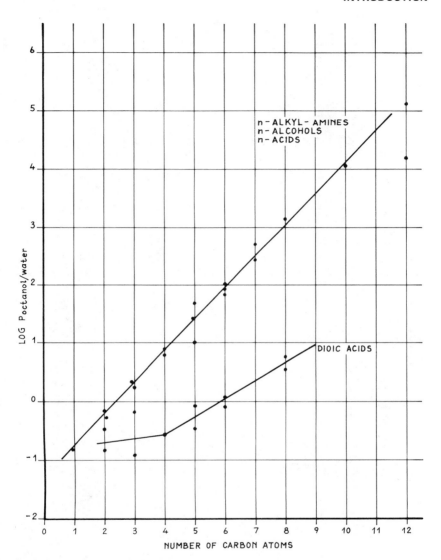

Fig. 12. Relationship between log P_{oct} and homologous series of *n*-acids, *n*-alcohols, *n*-alkylamines and dioic acids.

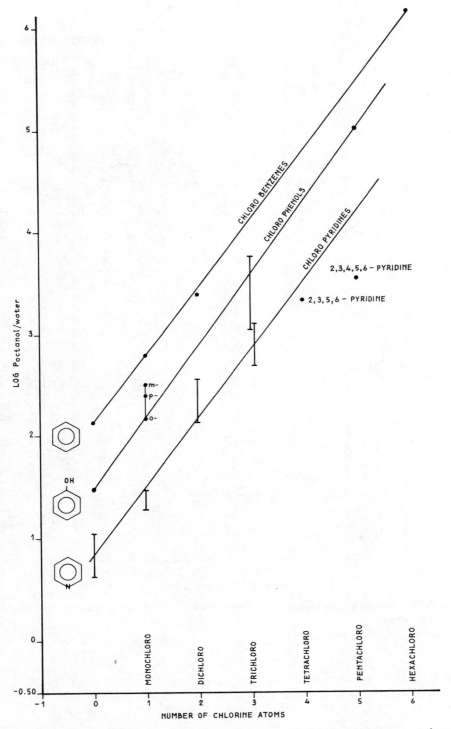

Fig. 13. Additive effect of chlorine atoms on the log P_{oct} partition coefficient of benzene, phenol, and aniline.

TABLE 5. Influence of Functional Groups on n-Octanol/Water Partition Coefficient of Benzene Derivatives.

Product	Functional Group	$\log P_{oct}$	$\Delta \log P_{oct}$ $\log P_{C_6H_5X} - \log P_{C_6H_6}$
benzenesulfonic acid	$-SO_3H$	-2.25	-4.38
benzenesulfonamide	$-SO_2NH$	0.31	-2.44
aniline	$-NH_2$	0.90	-1.23
phenol	$-OH$	1.46	-0.67
benzaldehyde	$-COH$	1.48	-0.65
benzonitrile	$-CN$	1.56	-0.57
benzoic acid	$-COOH$	1.87	-0.28
nitrobenzene	$-NO_2$	1.85/1.88	-0.28
nitrosobenzene	$-NO$	2.00	-0.13
benzene	–	2.13	–
fluorobenzene	$-F$	2.27	+0.14
thiophenol	$-SH$	2.52	+0.39
toluene	$-CH_3$	2.80	+0.67
chlorobenzene	$-Cl$	2.84	+0.71
bromobenzene	$-Br$	2.99	+0.86
iodobenzene	$-I$	3.25	+1.12
diphenyl	$-C_6H_5$	3.6	+1.47
diphenylether	$-O-C_6H_5$	4.21	+2.08

$$\pi_X = \log P_X - \log P_H \qquad (5)$$

when P_X is the derivative of a parent molecule P_H and thus π is the logarithm of the partition coefficient of the function X. For example π_{Cl} could be obtained as follows:

$$\pi_{Cl} = \log P_{chlorobenzene} - \log P_{benzene} \qquad (6)$$

It has been found that π values are relatively constant from one system to another as long as there are no special steric or electronic interactions of the substituents not contained in the reference system. π values for aliphatic and aromatic positions are shown in table 6. Other effects which must be taken into account in the additive-constitutive character of log P are:

- steric effects, which can cause shielding of an active function by inert groups
- inductive effects of one substituent on another
- intra- and intermolecular hydrogenbonding
- branching
- conformational effects, e.g., "balling up" of an aliphatic chain

Because of the difficulties of estimating the effect on $\log P_{oct}$ of steric, inductive and conformational effects, calculated $\log P_{oct}$ values of complex molecules can only be approximate and can be wrong by 1 to 2 orders of magnitude. However, for most simple molecules calculated values are correct within 1 order of magnitude.

TABLE 6. Comparison of Aromatic and Aliphatic π Values.

Function	Aromatic π $\log P_{C_6H_5} - \log P_{C_6H_6}$	Aliphatic π $\log P_{RX} - \log P_{RH}$
NH_2	-1.23	-1.19
I	1.12	1.00
$S-CH_3$	0.61	0.45
$COCH_3$	-0.55	-0.71
$CONH_2$	-1.49	-1.71
$COOCH_3$	-0.01	-0.27
Br	0.86	0.60
CN	-0.57	-0.84
F	0.14	-0.17
Cl	0.71	0.39
COOH	-0.28	-0.67
OCH_3	-0.02	-0.47
OC_6H_5	2.08	1.61
$N(CH_3)_2$	0.18	-0.30
OH	-0.67	-1.16
NO_2	-0.28	-0.85
CH_2	0.50	0.50

5.3. Relationship between Aqueous Solubility and Octanol/Water Partition Coefficient. Unfortunately, the partition coefficients of many components of environmental significance are not always available despite a recent extensive compilation (1457), or cannot be easily calculated from parent molecules. Assessment of partition coefficients from a more readily available physical parameter would therefore be useful. By definition, the partition coefficient expresses the equilibrium concentration ratio of an organic chemical partitioned between an organic liquid (e.g., *n*-octanol) and water. This partitioning is, in essence, equivalent to partitioning an organic chemical between itself and water. Consequently, one would suspect that a correlation might exist between the partition coefficient and the aqueous solubility. Based on experimental values of aqueous solubility and *n*-octanol/water partition coefficient for various types of chemicals the following regression equation was found (382):

$$\log P_{oct} = 5.00 - 0.670 \log S \tag{7}$$

where S is the aqueous solubility in μmol/l. Or, if the solubility is expressed in mg/l, Eq. (7) becomes

$$\log P_{oct} = 4.5 - 0.75 \log S \text{ (mg/l)}$$

or

$$\log P_{oct} = 7.5 - 0.75 \log S \text{ (ppb)} \tag{8}$$

This equation has been obtained empirically from Figs. 14 and 15. This correlation as can be seen on figure 14 covers many classes of chemicals from hydrocarbons and organic halides to aromatic acids, pesticides, and PCBs. It also spans chemicals of different polarities (from nonpolar or polar) and of different molecular states (both liquid and solid).

Cary T. Chiou et al. (382) found for Eq. (7) a correlation coefficient of 0.970, which allows an estimation within one order of magnitude of the partition coefficient of a given compound from its aqueous solubility. However, when more data points are

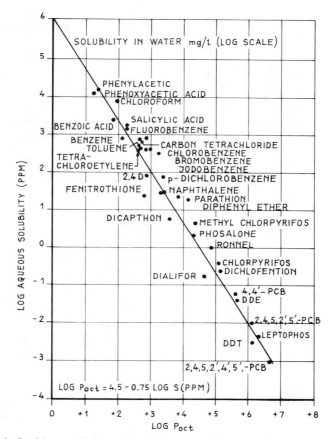

Fig. 14. Partition coefficients and aqueous solubilities of various organic chemicals.

added (Fig. 15) the scatter increases considerably for solubilities >100 mg/l. A few products deviate even considerably from the regression equations (7) and (8), as shown in the following data (see also Fig. 15):

pentachlorophenol
aqueous solubility   14 mg/l at 20°C
$\log P_{oct}$: experimental   5.01
     calculated eq. (7)   3.84
          eq. (8)   3.76

l-tyrosine
aqueous solubility   480 mg/l at 25°C
$\log P_{oct}$: experimental   −2.26
     calculated eq. (7)   +2.7
          eq. (8)   +2.5

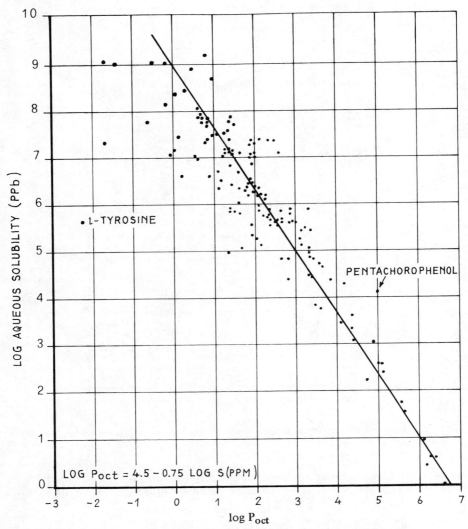

Fig. 15. Partition coefficients and aqueous solubility of various organic chemicals.

The regression equation, however, remains the same, although calculated P_{oct} values from its aqueous solubility may be wrong by more than 1 order of magnitude.

Obviously, it would be doubtful that Eqs. (7) and (8) would apply for salts, strong acids, and bases, since the activities of these solutes in this case cannot be approximated by their concentrations. As moreover with materials such as aliphatic acids or bases, the partition coefficient can vary drastically with changes in pH.

As previously stated, the partition coefficient is related to physical adsorption on solids, biomagnification, and lipophilic storage. Equation (9) below would extend these correlations to cover compounds using their aqueous solubilities without requiring the partition coefficient data. Based on reported biomagnification data of some selected organic chemicals in rainbow trout (*Salmo gairdneri*) the following regression equation was calculated:

$$\text{Log (BCF)} = 3.41 - 0.508 \log S \qquad (9)$$

where BCF is the bioconcentration factor in rainbow trout and S is aqueous solubility in μmol/l. If the aqueous solubility is expressed in mg/l, Eq. (9) becomes

$$\log \text{(BCF)} = 3.04 - 0.568 \log S_{mg/l} \qquad (10)$$

where $S_{mg/l}$ is the aqueous solubility in mg/l.

5.4. Ecological Magnification (E.M.). The process of bioaccumulation involves a number of fundamental events:

1. partitioning of the foreign molecule under consideration between the environment and some surface of the organism
2. diffusional transport of these molecules across cell membranes
3. transport mediated by body fluids, such as exchange between blood vessels and serum lipoproteins
4. concentration of the foreign molecule in various tissues depending upon its affinity for certain biomolecules, such as nerve lipids
5. biodegradation of the foreign material

The bioaccumulation process is thus seen to be a result of both kinetic (diffusional transport and biodegradation) and equilibrium (partitioning) processes. The relative importance of these processes is at present undecided. Intuition dictates that a molecule will not bioaccumulate in an organism if its degradation rate is greater than its accumulation rate. Experience with DDT may be considered a massive experiment from which it may be concluded that degradation occurred too slowly compared to the transport and partitioning of DDT into the higher levels of the food chain, thus permitting toxic levels to result.

R. L. Metcalf and coworkers (1643) have correlated the ecological magnification values for a number of organic compounds (pentachlorobiphenyl, tetrachlorobiphenyl, trichlorobiphenyl, DDE, chlorobenzene, benzoic acid, anisole, nitrobenzene, aniline) from the fish of model ecosystems with both water solubility and with the octanol/water partition value. For the limited number of compounds included, the correlation between physical properties and biomagnification is excellent. The regression equations were

$$\log \text{E.M.} = 4{,}48 - 0{,}47 \log \text{solub. (ppb)}$$
$$\log \text{E.M.} = 0{,}75 + 1{,}16 \log P_{oct} \qquad (11)$$

where solub. = aqueous solubility
P_{oct} = octanol/water partition coefficient.

W. H. Könemann (1833) reported the occurrence of a nonlinear relation between log P_{oct} and log BCF in the fat of guppies exposed to the chlorinated benzenes. The bio-

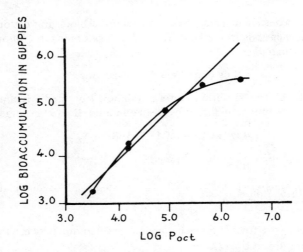

Fig. 16. Relation between log bioaccumulation and log P_{oct} for six chlorinated benzenes.

accumulation of chlorobenzenes increased with log P_{oct} until reaching an optimum value at log P_{oct} = 6.5 (Fig. 16). This reduction was caused by a sharp decrease in the magnitude of the uptake rate constant beyond the optimum value at log P_{oct} = 5.4 (Fig. 17). The correlations between ecological magnification and water solubility or octanol/water partition coefficient, as described above, are only valid for compounds which do not exhibit significant biodegradation.

Kapoor et al. (1937; 1938; 1939) studied in a model ecosystem the behavior of 8 DDT analogues covering a wide range of biodegradability. The basic methodology involved systematic study of the DDT molecule by replacing the environmentally stable C—Cl bonds with other groups of suitable size, shape, and polarity that could

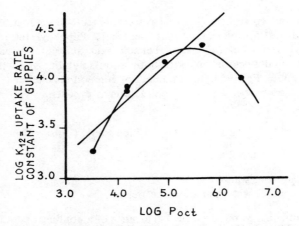

Fig. 17. Relation between log K_{12} and log P_{oct} for six chlorinated benzenes.

serve as degradophores by acting as substrates for the mixed function oxidase enzymes widely distributed in living organisms. The action of the enzymes was shown to result in substantial changes in the polarity of the molecule, so that the degradation products were excreted rather than stored in lipids as was DDT and its chief degradation product DDE.

A summary of model ecosystem data for a number of DDT analogues with degradaphores incorporated into aromatic and aliphatic moieties of the molecule is presented in Table 7 and Figs. 18 and 19. Figures 18 and 19 show that the difference between octanol/water partition coefficient and ecological magnification for the DDT analogues can be largely explained by the biodegradability index. The higher the biodegradability index, the larger is the difference between octanol/water partition coefficient (as a predictive measure of ecological magnification) and ecological magnification itself.

Most of the correlation equations between water solubility (or octanol/water partition coefficients) and ecological magnification have indeed been established on biorefractive compounds or on homologous series of compounds with comparable biodegradability characteristics. Therefore the equations are not universally applicable as shown by Fig. 20, representing the relationship between water solubility of DDT, DDE, DDD, and the DDT analogues and ecological magnification in mosquito fish of a terrestrial-aquatic model ecosystem. More than 30 pesticides were studied by Metcalf and Sanborn (1881), who found a highly significant correlation between log E.M. and log water insolubility (Fig. 21). Also for phenol and chlorinated derivatives a similar correlation exists (Fig. 22).

Since pesticides are now being "engineered" to be less "persistent," correlations between physico-chemical properties of compounds and ecological magnification will become less meaningful for such compounds unless degradation velocity is taken into account. For this reason the following equation is proposed which predicts the upper limit for ecological magnification based on lipid solubility only (water insolubility):

$$\log \text{E.M.} = 6 - 0.66 \log \text{solub. (ppb)} \qquad (12)$$

As this correlation is only very approximate, preference has been given to an easy to remember equation. The equation is therefore not based on a regression analysis.

Figure 21 shows at the same time that the real E.M.'s are 1 to 2 orders of magnitude smaller than the E.M.'s predicted from Eq. (12). Indeed, of 43 compounds 17 true E.M. values are smaller by less than 1 order of magnitude than the predicted values, 18 E.M. values differ by more than one but less than 2 orders of magnitude; and 6 E.M. values differ by more than 2 orders of magnitude.

5.5. Adsorption. In considering the adsorption of a compound from solution onto a solid, the following factors will influence the extent to which the adsorption process occurs:

- physical and chemical characteristics of the adsorbent (adsorbing surface)
- the actual surface area of the solid
- the nature of the binding sites on these surfaces and the actual distribution of these adsorption sites

An index of the tendency to adsorb onto a solid is the solubility of the compound.

TABLE 7. Biodegradability and Ecological Magnification in Fish of DDT Analogues.

Number of DDT analogue	R_1	R_2	R_3	$\log P_{oct}$[a]	log E.M.	$\log P_{oct}$ minus log E.M.	$\log (100 \times \text{B.I.})$	$\log H_2O$ solub. ppb[b]
1	Cl	Cl	CCl_3	6.1	4.93	1.17	0.18	0.9
2	Cl	Cl	$HCCl_2$	5.4	4.92	0.48	0.73	1.8
3	CH_3O	CH_3O	CCl_3	4.7	3.19	1.51	1.97	2.7
4	CH_3	CH_3	CCl_3	6.0	2.15	3.85	2.85	1.0
5	CH_3S	CH_3S	CCl_3	5.9	0.74	5.16	3.67	1.1
6	Cl	CH_3	CCl_3	6.05	3.15	2.9	2.53	0.9
7	CH_3	C_2H_5O	CCl_3	5.9	2.60	3.3	2.08	1.1
8	CH_3O	CH_3S	CCl_3	5.3	2.49	2.81	2.44	1.9
9	CH_3O	CH_3O	$C(CH_3)_3$	4.7	3.21	1.49	1.81	2.7
10	Cl	Cl	$HC(CH_3)NO_2$	4.4	2.05	2.35	2.51	3.1

Calculated from data from Metcalf et al. (1940). Kapoor et al. (1937), Hirwe et al. (1941).
B.I. = Biodegradability Index = ratio of polar compounds to non polar compounds.
[a]Calculated.
[b]Derived from $\log P_{oct}$.

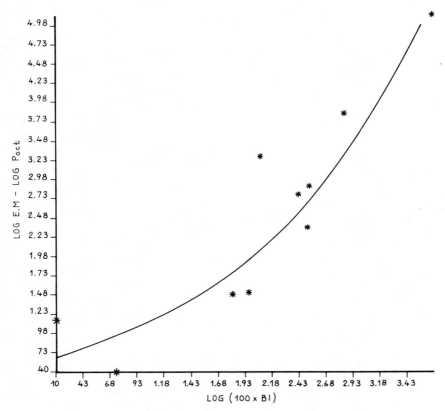

Fig. 18. Influence of biodegradability of DDT analogues on experimental and predicted ecological magnification.

For homologous series of compounds, decreasing solubility can be interpreted as an increasing tendency to leave the water. Also the n-octanol/water partition coefficient ($\log P_{oct}$) has proved useful as a means to predict soil adsorption, as shown in Fig. 23.

5.6. *Structure–Toxicity Correlations.* Kopperman and coworkers (1832) noted that structure–toxicity correlations are possible only if the compounds examined have an identical mode of action. However, if toxic mechanisms of similar compounds are identical then the internal concentration of toxicant required to elicit a specific biological response should be consistent. Since the internal concentration equals the external concentration multiplied by the partition coefficient, the external concentration is proportional to the reciprocal of the partition coefficient, and a plot of log external concentration vs log partition coefficient should have a slope of − 1.

T. Wayne Schultz and coworkers (1662) have observed excellent correlation between toxicity to the ciliate *Tetrahymena pyriformis* and partition coefficient within the following series of organic contaminants: pyridines, anilines, phenols, quinolines and

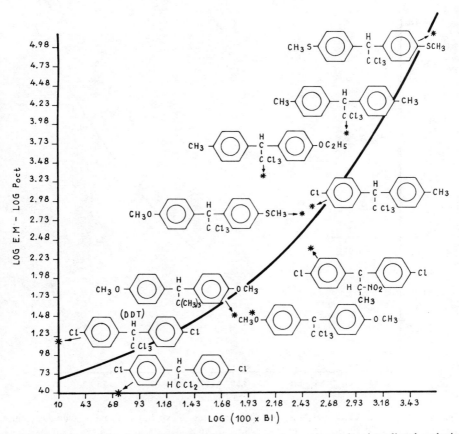

Fig. 19. Influence of biodegradability of DDT analogues on experimental and predicted ecological magnification.

benzenes. However no significant correlation between toxicity and partition coefficient was observed when data from all the tested contaminants were combined. Generally, an increase in alkyl substitution increases toxicity and decreases solubility. Furthermore, hetero-atom substitution into or onto the ring severely alters both the toxicity and solubility of the compound (Fig. 24). Kunio Kobayashi (1850) found that an increase of the Cl-atom number in the chlorophenols promoted an accumulation of the chlorophenols by fish and lead their concentrations in the fish to a lethal level even when guppies were exposed to rather low concentrations, and consequently increased the fish-toxicity of chlorophenols. Kopperman et al. (1832) found approximately the same correlation for goldfish (Fig. 25). The data are summarised in Table 8.

Because all the correlations obtained are only valid within homologous series and for the test-organism concerned, their predictive power is limited. Moreover concentration values have mostly been expressed in moles/liter. Although this is certainly more

INTRODUCTION 33

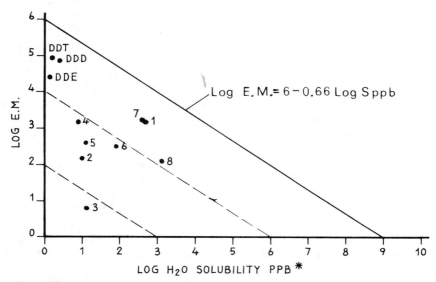

Fig. 20. Relationship between water solubility of DDT, DDE, DDD, and DDT analogues and ecological magnification in mosquitofish of terrestrial-aquatic model ecosystem.

*Calculated for DDT analogues (see Table 7).

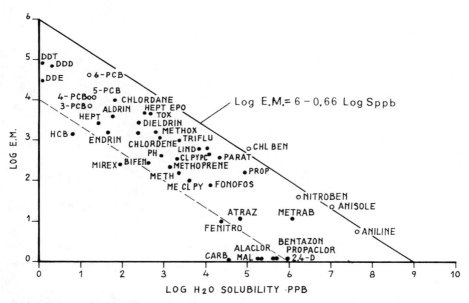

Fig. 21. Relationship between water solubility of pesticides and ecological magnification in mosquito fish of terrestrial-aquatic model ecosystem; after Metcalf and Sanborn (1881).

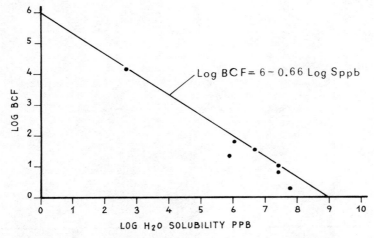

Fig. 22. BCF versus solubility for phenols and chlorinated phenols for goldfish (after Ref. 1850).

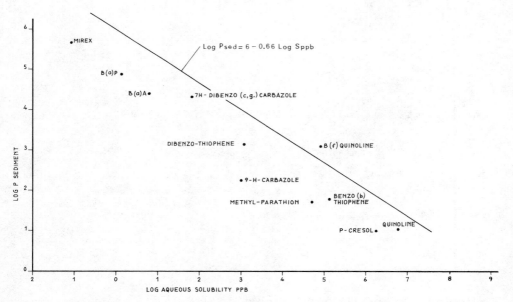

Fig. 23. Plot of log water solubility and log partition coefficient (log $P_{sediment}$) for organic compounds adsorbed on natural sediment collected in Coyote Creek, California (from EPA-60017-78-074, May 1978).

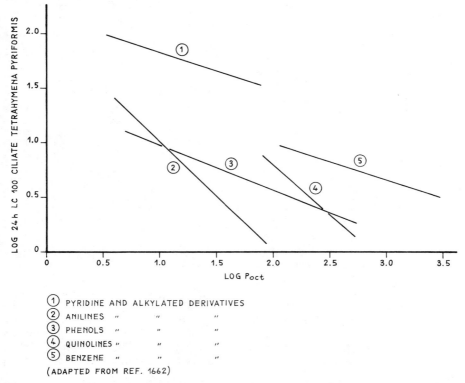

① PYRIDINE AND ALKYLATED DERIVATIVES
② ANILINES "   "   "
③ PHENOLS   "   "   "
④ QUINOLINES "   "   "
⑤ BENZENE   "   "   "
(ADAPTED FROM REF. 1662)

Fig. 24. A least squares linear regression of log 24h LC_{100} (mMole/l) vs. log partition coefficient in octanol-water system for the ciliate *Tetrahymena pyriformis*.

correct from a scientific standpoint of view than using mg/l, it is advised to express concentrations in µg/l or mg/l in order to make the use of the correlations more practicable.

Figure 26 shows that the correlation between the aqueous solubility expressed in ppm (mg/l) and LD_{50} for goldfish are still good enough. Figure 25 illustrates that solubilities can be used instead of octanol/water partition coefficients to derive acute toxicity of compounds, within one order of magnitude if in a homologous series the toxicity is known for two or more compounds.

5.7. Prediction of Skin-Adsorptive Properties of Chemicals. In his book *Nonelectrolytes* N. V. Lazarev demonstrated correlations between narcotic and lethal concentrations of nonelectrolytes in the blood, on the one hand, and their oil/water partition coefficients and solubilities in water, on the other. Rumiantsev and Norvikov (1942) have demonstrated the possibility of calculating LD_{50} values for substances absorbed through the skin of animals from the octanol/water partition coefficient (P_{oct}). They derived the equation:

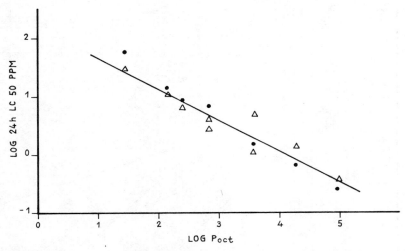

Fig. 25. Toxicity versus octanol/water distribution coefficient for goldfish ● and guppies △ (after Refs. 1833 and 1850) exposed to phenol and chlorinated phenols.

$$\log LD_{50} \text{ (rabbits via skin)} = 1.25 + 0.199\, P_{oct}$$

$$(n = 72,\ t = 0{,}633,\ \overline{S} = \pm 0.376)$$

Moreover, they have correlated skin absorption LD_{50} values for rats and rabbits with intragastric (ig) LD_{50} values for rats (4-hr contact for rats and 24-hr contact for rabbits):

$$\log LD_{50} \text{ (rats via skin)} = 0.79 \log LD_{50} \text{ (rats ig)} + 0.77$$

$$\log LD_{50} \text{ (rabbits via skin)} = 0.77 \log LD_{50} \text{ (rats ig)} + 0.87$$

They have also derived a multiple regression equation relating the LD_{50} skin for rabbits to the octanol/water partition coefficient P_{oct} and to the LD_{50} for rats with intragastric administration:

$$\log LD_{50} \text{ (rabbits via skin)} = 0.645 + 0.54 \log LD_{50} \text{ (rats ig)} + 0.09\, P_{oct}$$

6. Specific Gravity. *Hydrocarbons* are usually lighter than water. As a given homologous series of hydrocarbons is ascended the specific gravity of the members increases, but the increment per methylene radical gradually diminishes. Curves I, II, and III in Fig. 27 show the change in density for the alkanes, 1-alkenes, and 1-alkynes. It will be noted that the specific gravity of the acetylenic hydrocarbon is greater than that of the corresponding olefin, which in turn is more dense than the paraffin hydrocarbon with the same number of carbon atoms.

The replacement of one atom by another of higher atomic weight usually increases the density. Thus Curve IV Fig. 27, which represents the specific gravities of the normal alkyl chlorides, lies above the curves of the hydrocarbons. It will be noted that the alkyl chlorides are lighter than water and that the specific gravities *decrease* as the number of carbon atoms is increased.

TABLE 8. Phenol and Chlorinated Phenols: Structure–Toxicity Data.

	24h LC_{50} (ppm)[a]		solub. ppm	log P_{oct} exp.	log p_{oct} calculated	BCF goldfish [b]
	goldfish (1850)	guppies (1833)				
phenol	60	30	67.000 (10°C)	1.47	1.55	1.9
o-cnlorophenol	16	11	28.500 (20°C)	2.17	2.27	6.4
m-chlorophenol	–	6.5	–	2.48	2.27	–
p-chlorophenol	9.0	–	27.100 (20°C)	2.41	2.27	10.0
2,4-dichlorophenol	7.8	4.2	4.600 (20°C)	–	2.93	34
3,5-dichlorophenol	–	2.7	–	–	2.93	–
2,4,6-trichlorophenol	10.0	–	800 (24°C)	3.37	3.69	20
2,4,5-trichlorophenol	1.7	–	1.190 (25°C)	3.72	3.69	62
2,3,5-trichlorophenol	–	1.6	–	–	3.69	–
2,3,6-trichlorophenol	–	5.1	–	–	3.69	–
3,4,5-trichlorophenol	–	1.1	–	–	3.69	–
2,3,4,6-tetrachlorophenol	0.75	–	–	–	4.42	93
2,3,4,5-tetrachlorophenol	–	0.77	–	–	4.42	–
2,3,5,6-tetrachlorophenol	–	1.37	–	–	4.42	–
pentachlorophenol	0.27	0.38	14 (20°C)	5.01	5.19	475

[a] pH 7.3.
[b] Values were obtained in goldfish which died in the concentration of each chlorophenol most close to 24h LC_{50}.

38 HANDBOOK OF ENVIRONMENTAL DATA ON ORGANIC CHEMICALS

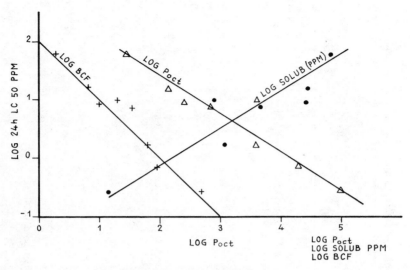

Fig. 26. Toxicity and accumultion of phenols in goldfish compared to the aqueous solubility and octanol/water partition coefficients (after Ref. 1850).

Curves VI and VII in Fig. 27 show that the specific gravity of the primary alkyl bromides and iodides is greater than 1.0 and that in these homologous series the specific gravity decreases as the number of carbon atoms is increased. The downward slope of Curves IV, VI, and VII is due to the fact that the halogen atom constitutes a smaller and smaller percentage of the molecule as the molecular weight is increased by increments of methylene radicals. The relative position of the curves in Fig. 27 shows that the specific gravity increases in the order

$$RH < RF < RCl < RBr < RI$$

provided that comparisons are made on alkyl halides with the same carbon skeleton and of the same class. Similar relationships are exhibited by secondary and tertiary chlorides, bromides, and iodides.

The specific gravities of aryl halides also arrange themselves in the order of increasing weight of the substituent (Table 9).

An increase in the number of halogen atoms present in the molecule increases the specific gravity. Compounds containing two or more chlorine atoms or one chlorine atom together with an oxygen atom or an aryl group will generally have a specific gravity greater than unity.

The introduction of functional groups containing oxygen causes an increase in the specific gravity. The curves in Fig. 28 represent the change in specific gravity of some of the common types of compounds.

The ethers (Curve VIII) are the lightest of all the organic oxygen compounds. The aliphatic alcohols (Curve IX) are heavier than the ethers but lighter than water. The specific gravity of the alcohols becomes greater than 1.0 if a chlorine atom (ethylene chlorohydrin), a second hydroxyl (ethylene glycol), or an aromatic nucleus (benzyl alcohol) is introduced.

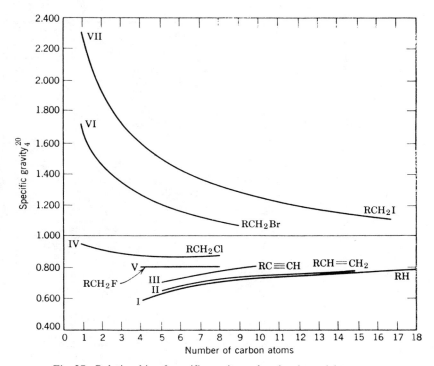

Fig. 27. Relationship of specific gravity and molecular weight.

The amines (Curve X) are not as dense as the alcohols and are less associated. Association also causes the specific gravity of formic acid and acetic acid to be greater than unity; the higher liquid fatty acids are lighter than water (Curve XI).

The simple esters (Curve XII) are lighter than water, whereas esters of polybasic acids (Curve XIII), halogenated, keto, or hydroxy esters are heavier than water. Introduction of the aromatic ring also may cause esters to be heavier than water. Since the hydrocarbons are lighter than water, it is to be expected that esters containing long hydrocarbon chains will show a correspondingly diminished specific gravity.

In general, compounds containing several functional groups—especially those groups that promote association—will have a specific gravity greater than 1.0. Merely noting

TABLE 9. Specific Gravities of Aryl Halides.

Compound	B.p., °C	Sp. gr.$_4^{20}$
Benzene	79.6	0.878
Fluorobenzene	86	1.024
Chlorobenzene	132	1.107
Bromobenzene	156	1.497
Iodobenzene	188	1.832

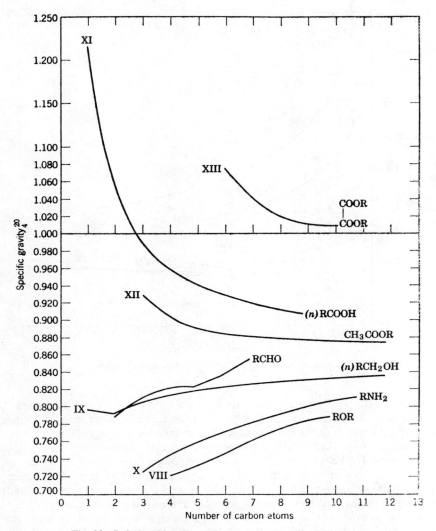

Fig. 28. Relationship of specific gravity and moelcular weight.

wether a compound is lighter or heavier than water gives some idea of its complexity. This is of considerable value in the case of neutral liquids.

B. AIR POLLUTION FACTORS

1. Conversion between Volume and Mass Units of Concentration. The physical state of gaseous air pollutants at atmospheric concentrations may be described generally by the ideal gas law:

$$pv = nRT \tag{1}$$

where p = absolute pressure of gas
v = volume of gas
n = number of moles of gas
T = absolute temperature (°K)
R = universal gas constant.

The number of moles (n) may be calculated from the weight of pollutant (W) and its molecular weight (m) by:

$$n = \frac{W}{m} \tag{2}$$

Substituting equation (2) into equation (1) and rearranging yields:

$$v = \frac{WRT}{pm} \tag{3}$$

Parts per million refers to the volume of pollutant (v) per million volumes of air (V):

$$\text{ppm} = \frac{v}{10^6 V} \tag{4}$$

Substituting equation (3) into equation (4) yields:

$$\text{ppm} = \frac{W}{V} \frac{RT}{pm\,10^6} \tag{5}$$

Using the appropriate values for variables in equation (5), a conversion from volume to mass units of concentration for methane may be derived as shown below:

$$T = 293.16 \text{ K } (20°\text{C})$$

$$p = 1 \text{ atm}$$

$$m = 16 \text{ g/mol}$$

$$R = 0.08205 \text{ l-atm/mol K}$$

$$\text{ppm} = \frac{W(\text{g}) \times 10^3 \text{ (mg/g)}}{V(\text{l}) \times 10^{-3} \text{ (m}^3/\text{l)}} \times \frac{0.08205 \text{ (l-atm/mol K)} \times 2.93.16 \text{ (K)}}{1 \text{ (atm)} \times 16 \text{ (g/mol)} \times 10^6}$$

$$1 \text{ ppm} = 0.665 \text{ mg/m}^3$$

$$1 \text{ mg/m}^3 = 1.504 \text{ ppm}$$

Whenever conversion factors are used, reference must be made to the pressure and temperature at which the conversion factors have been calculated. Unfortunately, the conditions of "Air at Normal Conditions" or "Standard Air" or "Air at STP" (STP = Standard Temperature and Pressure) often vary from country to country. ASTM D 1356-60 defines "Air at Normal Condition" as: "Air at 50 per cent relative humidity, 70°F (21°C), and 29.92 inches of mercury (760 millimeters of mercury). These conditions are chosen in recognition of the data which have been accumulated on air-handling equipment. They are sufficiently near the 25°C and 760 millimeters of mercury commonly used for indoor air contamination work that no conversion or correction ordinarily need be applied."

For outdoor air (ambient air) pollution control, it is important to use the correct conversion factors. "Normal conditions" vary from 0°C and 760mm Hg(dry) for Canada over 20°C and 760mm Hg for Australia to 25°C and 760mm Hg for Brasil.

Calculations of minimum chimney heights, in order to ensure sufficient dispersion of the waste gases before they reach ground level, are often based on the difference between Maximum Immission Concentrations (M.I.C.) and the existing pollutant concentration at ground level. Significant differences in chimney heights can be obtained by not using the appropriate conversion factors.

Because of the lack of "standardization" of "standard" conditions worldwide, and because outdoor and indoor conditions are different, different conversion factors will continue to be used. The conversion factors on the data sheets should, therefore, be regarded only as approximate values. Knowing the molecular weight of the compound, the correct values at 0°C and 20°C can be found in Tables 10 and 11.

2. Odor

Threshold Odor Concentration (T.O.C.). A starting point in relation to quantification of odors seems to be the definition of a threshold odor concentration. There are at least three different odor thresholds that have been determined: 1. the absolute perception threshold, 2. the recognition threshold, and 3. the objectionability threshold.

At the perception threshold concentration one is barely certain that an odor is detected, but is too faint to identify further. Furthermore, the sense of smell results must be a statistical average because of biological variability. The thresholds normally used are those for 50% and for 100% of the odor panel. When the T.O.C. is given without any qualification, it is usually the 50% recognition threshold.

For the sake of clarity, a number of definitions are listed here:

Hedonic tone: the pleasure or displeasure that the odor judge associated with the odor quality being observed.

Absolute Odor Threshold: the concentration at which 50% of the odor panel detected the odor.

50% Recognition Threshold: the concentration at which 50% of the odor panel defined the odor as being representative of the odorant being studied.

100% Recognition Threshold: the concentration at which 100% of the odor panel defined the odor as being representative of the odorant being studied.

PPT_{50} (*Population Perception Threshold*): the concentration at which 50% of the people who have a capable sense of smell are able to detect an odor.

PIT_{50} (*Population Identification Threshold*): that concentration at which 50% of the population can identify and describe the odor, or at least compare its quality with another odor.

TABLE 10. Conversion Factors at 0°C and 760 mm Hg.

1 ppm = m mg/m³	1 mg/m³ = ppm	1 ppm = m mg/m³	1 mg/m³ = ppm	1 ppm = m mg/m³	1 mg/m³ = ppm	1 ppm = m mg/m³	1 mg/m³ = ppm
16 .714	1.401	51 2.277	0.439	86 3.839	0.260	121 5.401	0.185
17 .759	1.318	52 2.321	0.431	87 3.884	0.257	122 5.446	0.184
18 .804	1.244	53 2.366	0.423	88 3.929	0.255	123 5.491	0.182
19 .848	1.179	54 2.411	0.415	89 3.973	0.252	124 5.536	0.181
20 .892	1.121	55 2.455	0.407	90 4.018	0.249	125 5.580	0.179
21 .937	1.067	56 2.500	0.400	91 4.062	0.246	126 5.625	0.178
22 .982	1.018	57 2.545	0.393	92 4.107	0.243	127 5.670	0.176
23 1.027	0.974	58 2.589	0.386	93 4.152	0.241	128 5.714	0.175
24 1.071	0.934	59 2.634	0.380	94 4.196	0.238	129 5.759	0.174
25 1.116	0.896	60 2.679	0.373	95 4.241	0.236	130 5.804	0.172
26 1.161	0.861	61 2.723	0.367	96 4.286	0.233	131 5.848	0.171
27 1.205	0.830	62 2.768	0.361	97 4.330	0.231	132 5.893	0.170
28 1.250	0.800	63 2.812	0.356	98 4.375	0.229	133 5.937	0.168
29 1.295	0.772	64 2.857	0.350	99 4.420	0.226	134 5.982	0.167
30 1.339	0.747	65 2.902	0.344	100 4.464	0.244	135 6.027	0.166
31 1.384	0.722	66 2.946	0.339	101 4.509	0.222	136 6.071	0.165
32 1.429	0.700	67 2.991	0.334	102 4.554	0.220	137 6.116	0.164
33 1.473	0.679	68 3.036	0.329	103 4.598	0.217	138 6.161	0.163
34 1.518	0.659	69 3.080	0.325	104 4.643	0.215	139 6.205	0.161
35 1.562	0.640	70 3.125	0.320	105 4.687	0.213	140 6.250	0.160
36 1.607	0.622	71 3.170	0.315	106 4.732	0.211	141 6.295	0.159
37 1.652	0.605	72 3.214	0.311	107 4.777	0.209	142 6.339	0.158
38 1.696	0.590	73 3.259	0.307	108 4.821	0.207	143 6.384	0.157
39 1.741	0.574	74 3.304	0.303	109 4.866	0.206	144 6.429	0.156
40 1.786	0.560	75 3.348	0.299	110 4.911	0.204	145 6.473	0.154
41 1.830	0.546	76 3.393	0.295	111 4.955	0.202	146 6.518	0.153
42 1.875	0.533	77 3.437	0.291	112 5.000	0.200	147 6.562	0.152
43 1.920	0.521	78 3.482	0.287	113 5.045	0.198	148 6.607	0.151
44 1.964	0.509	79 3.527	0.284	114 5.089	0.197	149 6.652	0.150
45 2.009	0.498	80 3.571	0.280	115 5.134	0.195	150 6.696	0.149
46 2.054	0.487	81 3.616	0.277	116 5.179	0.193	151 6.741	0.148
47 2.098	0.477	82 3.661	0.273	117 5.223	0.192	152 6.786	0.147
48 2.143	0.467	83 3.705	0.270	118 5.268	0.190	153 6.830	0.146
49 2.187	0.457	84 3.750	0.267	119 5.312	0.188	154 6.875	0.145
50 2.232	0.448	85 3.795	0.264	120 5.357	0.187	155 6.920	0.145

IPT (*Individual Perception Threshold*): the lowest concentration of a particular odor at which a subject gave both an initial positive response and a repeated response when the same stimulus was given a second time.

T.O.N. (*Threshold Odor Number*): the number of times a given volume of the gas sample has to be diluted with clean, odorless air to bring it to the threshold level (detected by 50% of a panel of observers). The T.O.N. is thus the value of the intensity of an odor expressed in odor units.

TABLE 11. Conversion Factors at 20°C and 760 mm Hg.

1 ppm = m mg/m³	1 mg/m³ = ppm	1 ppm = m mg/m³	1 mg/m³ = ppm	1 ppm = m mg/m³	1 mg/m³ = ppm	1 ppm = m mg/m³	1 mg/m³ = ppm
15 0.624	1.603	51 2.120	0.472	87 3.617	0.276	123 5.113	0.196
16 0.665	1.504	52 2.162	0.463	88 3.658	0.273	124 5.155	0.194
17 0.707	1.414	53 2.203	0.454	89 3.700	0.270	125 5.196	0.192
18 0.748	1.337	54 2.245	0.445	90 3.741	0.267	126 5.238	0.191
19 0.790	1.266	55 2.286	0.437	91 3.783	0.264	127 5.280	0.189
20 0.831	1.203	56 2.328	0.430	92 3.824	0.261	128 5.321	0.188
21 0.873	1.145	57 2.369	0.422	93 3.866	0.259	129 5.363	0.186
22 0.915	1.093	58 2.411	0.415	94 3.908	0.256	130 5.404	0.185
23 0.956	1.046	59 2.453	0.408	95 3.949	0.253	131 5.446	0.184
24 0.998	1.002	60 2.494	0.401	96 3.991	0.251	132 5.488	0.182
25 1.039	0.962	61 2.536	0.394	97 4.032	0.248	133 5.529	0.181
26 1.081	0.925	62 2.577	0.388	98 4.074	0.245	134 5.570	0.180
27 1.122	0.891	63 2.619	0.382	99 4.115	0.243	135 5.612	0.178
28 1.164	0.859	64 2.660	0.376	100 4.157	0.241	136 5.654	0.177
29 1.206	0.829	65 2.702	0.370	101 4.199	0.238	137 5.695	0.176
30 1.247	0.802	66 2.744	0.364	102 4.240	0.236	138 5.737	0.174
31 1.289	0.776	67 2.785	0.359	103 4.282	0.233	139 5.778	0.173
32 1.330	0.752	68 2.827	0.354	104 4.323	0.231	140 5.820	0.172
33 1.372	0.729	69 2.868	0.349	105 4.365	0.229	141 5.861	0.171
34 1.413	0.708	70 2.910	0.344	106 4.406	0.227	142 5.903	0.169
35 1.455	0.687	71 2.951	0.339	107 4.448	0.225	143 5.945	0.168
36 1.487	0.668	72 2.993	0.334	108 4.490	0.223	144 5.986	0.167
37 1.538	0.650	73 3.035	0.329	109 4.531	0.221	145 6.028	0.166
38 1.580	0.633	74 3.076	0.325	110 4.573	0.219	146 6.070	0.165
39 1.621	0.617	75 3.118	0.321	111 4.614	0.217	147 6.111	0.164
40 1.663	0.601	76 3.159	0.317	112 4.656	0.215	148 6.152	0.163
41 1.704	0.587	77 3.201	0.312	113 4.697	0.213	149 6.194	0.161
42 1.746	0.572	78 3.242	0.308	114 4.739	0.211	150 6.236	0.160
43 1.788	0.559	79 3.284	0.305	115 4.780	0.209	151 6.277	0.159
44 1.829	0.547	80 3.326	0.301	116 4.822	0.207	152 6.319	0.158
45 1.871	0.534	81 3.367	0.297	117 4.864	0.206	153 6.360	0.157
46 1.912	0.523	82 3.409	0.293	118 4.905	0.204	154 6.402	0.156
47 1.954	0.512	83 3.450	0.290	119 4.947	0.202	155 6.443	0.155
48 1.995	0.501	84 3.492	0.286	120 4.988	0.200	156 6.485	0.154
49 2.037	0.491	85 3.533	0.283	121 5.030	0.199	157 6.526	0.153
50 2.079	0.481	86 3.575	0.280	122 5.072	0.197	158 6.568	0.152

O.I. (*Odor Index*): a dimensionless term which is based upon vapor pressure and odor recognition threshold (100%) as follows:

$$\text{O.I.} = \frac{\text{vapor pressure (ppm)}}{\text{odor recognition threshold (100\%) (ppm)}}$$

where 1 atm = 1,000,000 ppm

INTRODUCTION   45

TABLE 12. Threshold Odor Concentrations and Odor Index of Organic Chemicals.

Product	Odor Index
acetaldehyde	4,300,000
acetic acid	15,000
acetic anhydride	12,800
acetone	720
acetonitrile	2,400
acetophenone	660
acrolein	19,300
acrylic acid	4,210
acrylonitrile	2,630
allylalcohol	13,800
allylchloride	17,900
allylisothiocyanate	901,000
ammonia	167,300
amylacetate	2,500
anethole	4,400
aniline	400
benzaldehyde	22,000
benzene	300
p-benzoquinone	790
benzylchloride	28,000
bornylacetate	1,700
1,3-butadiene	2,530
n-butane	480
n-butanol	120
sec. butanol	400
tert. butanol	55,900
n-butylacetate	1,200
n-butylamine	395,000
α-butylene	43,480,000
β-butylene	3,330,000
1,2-butyleneoxide	260,563
butylglycolether	1,650
n-butylchloride	6,300
n-butylether	13,400
n-butylformate	2,300
n-butylmercaptan	49,340,000
n-butylsulfide	658,000
butyraldehyde	2,395,000
n-butyric acid	50,000
camphor	41
caproic acid	43,900
ε-caprolactam	20
caprylic acid	164,500
carbon disulfide	44,430
carbon tetrachloride	540
carvacrol	6,600
chloral	980,000
α-chloroacetophenone	330
chlorobenzene	52,600
chlorobromomethane	350
chloroform	70
chloropicrin	22,200
chloroprene	2,392,000
cinnamaldehyde	53,000
o-cresol	60
m-cresol	80
p-cresol	260
crotonaldehyde	125,000

HANDBOOK OF ENVIRONMENTAL DATA ON ORGANIC CHEMICALS

TABLE 12. (Continued)

Product	Odor Index
cyanogenchloride	1,316,000
cyclohexane	203,000
cyclohexanol	26,300
cyclohexanone	21,900
1-decanol	31,000
1-decene	3,900,000
diacetonealcohol	774
di-N-butylamine	5,500
o dichlorobenzene	26
p-dichlorobenzene	26
1,1-dichloroethylene	1,300
2,2-dichloroethylether	60
1,2-dichloropropane	1,100
α-dicyclopentadiene	26,000
diethylamine	880,000
2-diethylaminoethanol	46,000
diethylselenide	3,200,000
diethylsuccinate	800
diethylsulfide	14,400,000
diethylketone	1,900
diglycol	120
diglycolacetate	250
diisobutylcarbinol	2,500
diisobutylketone	45
diisopropylamine	108,000
diisopropylether	3,227,000
N,N-dimethylacetamide	37
dimethylamine	280,000
dimethylethanolamine	292,000
N,N-dimethylformamide	35
1,1-dimethylhydraxine	41,300
dimethylsulfide	2,760,000
1,4-dioxane	230
1,3-dioxolane	720
diphenylether	130
diphenylsulfide	14,000
diphosgene	11,960
di-N-propylamine	395,000
1-dodecanol	1,800
enanthic acid	7,900
epichlorohydrin	160
ethane	25,300
ethanol	11
ethanolamine	130
2-ethoxy-3,4-dihydropyran	10,900
ethylacetate	1,900
ethylacrylate	38,160,000
ethylamine	1,445,000
ethylhexanoate	760,000
ethylidenenorbornene	75,600
ethylisoamylketone	660
ethyl-isovalerate	88,000
2-ethyl-1-butanol	3,100
ethylbutyrate	1,982,000
ethylene	57,100
ethylenediamine	1,100
ethylenebromide	550
ethylenedichloride	410
ethyleneglycol	3

INTRODUCTION 47

TABLE 12. (Continued)

Product	Odor Index
ethylglycol	200
ethylglycolacetate	6,300
ethyleneimine	105,300
ethyleneoxide	2,100
ethylether	1,939,000
2-ethyl-1-hexanol	480
2-ethylhexylacrylate	7,300
ethylmercaptan	289,500,000
N-ethylmorpholine	32,100
ethylpelargonate	109,000
ethylsilicate	16
ethyl-n-valerate	178,000
eugenol	37,600
formaldehyde	5,000,000
formic acid	2,200
furfural	5,260
furfurylalcohol	66
glycoldiacetate	1,700
heptane	200
1-heptanol	23,100
2 heptanone	171,000
hexachlorocyclopentadiene	700
n-hexanol	14,300
sec.hexylacetate	12,500
hexyleneglycol	2
hydrazine	5,300
hydrocinnamic alcohol	253,000
hydrogen cyanide	163,000
hydrogen sulfide	17,000,000
α-ionone	1,050,000
isoamylacetate	526,000
isoamylalcohol	300
isoamylisovalerate	1,050,000
isoamylsulfide	1,640,000
isobutane	3,000,000
isobutanol	320
isobutene	4,640,000
isobutylacetate	3,300
isobutylacrylate	525,000
isobutylcellosolve	34,400
isobutyraldehyde	948,000
isodecanol	300
isopentanoic acid	9,600
isophorone	900
isopropanol	60
isopropylacetate	2,100
isopropylamine	637,000
isopropylbenzene	89,600
isopropylmercaptan	1,052,000
isovaleric acid	365,500
linalylacetate	66,000
maleic anhydride	0.2
menthol	100
mesityloxide	460
methacrylonotrile	72,000
methanol	22
methylacetate	1,100
methylamine	940,000
methylamylalcohol	12,650

48 HANDBOOK OF ENVIRONMENTAL DATA ON ORGANIC CHEMICALS

TABLE 12. (*Continued*)

Product	Odor Index	1 ppt	10 ppt	100 ppt	1 ppb	10 ppb	100 ppb	1 ppm	10 ppm	100 ppm	1000 ppm	1 %
methylanthranilate	101,000				o							
2-methyl-2-butanol	1,700					o—o—o						
methylbutyrate	19,000,000					o						
methylchloride	200,000								oo			
4-methylcyclohexanol	3,700									o		
methylenchloride	2,100								o—ooo			
N-methylethanolamine	400							o—o				
methylethylketone	3,800							ooooo	oo			
2-methyl-5-ethylpyridine	137,000				ooo							
methylformate	300											o
methylglycol	140						o—oo		—o			
methylglycolacetate	14,400						o—o					
methylisoamylalcohol	6,600						o—o					
methylisoamylketone	75,100					o—oo						
methylisobutylketone	1,000						o—ooo—	—oo				
methylisopropenylketone	184,000						o					
methylmercaptan	53,300,000	oo		o—o—oo—oo								
methylmethacrylate	108,000						o—ooo					
2-methylpentaldehyde	131,500					o—o						
2-methyl-1-pentanol	24,000					o—o						
methylsalicylate	113,400			o								
α-methylstyrene	19,400					o—o—o						
monochloroacetic acid	1,460					o						
morpholine	75,200					o—od						
naphthalene	2,400		o—|—o									
nitrobenzene	200		o—o				—o					
nitromethane	460								o			
1-nitropropane	40									o		
2-nitropropane	85									o		
n-nonane	9,800						o					
n-octane	100					o				—o		
1-octanol	33,000				o							
2-octanol	506,000			o								
2-octanone	4									o		
n-octylacetate	2,800						oo					
pelargonic acid	164,000			o								
n-pentane	570							o		—o		
2,4-pentanedione	384,100				ooo							
1-pentanol	368					o—o		o—o				
2-pentanone	2,000								o			
1-pentene	376,000,000				o							
phenol	16					oo—o—o—o—oo		—o				
phenylisocyanide	3,950,000					o						
phenylmercaptan	94		oo									
phosgene	1,600,000							o—oo—o				
α-picoline	228,800				ooo							
α-pinene	469,000				o							
piperonal	6,100				o							
propane	425											o
n-propanol	480					o—o—oo—		—oo				
propeneoxide	15,000							o—o				
propionaldehyde	3,865,000					o o—o						
propionic acid	112,300				o—ooo							
n-propylacetate	1,600					o—o		—o				
propylene	14,700								o—o			
propylenediamine	184,600				o—o—o							
n-propylmercaptan	263,000,000	o—o—o										
pyridine	2,390		o—	—o		ooooo—o—o						
safrole	34,000				o							

TABLE 12. (Continued)

Product	Odor Index	1 ppt	10 ppt	100 ppt	1 ppb	10 ppb	100 ppb	1 ppm	10 ppm	100 ppm	1000 ppm	1 %
skatole	30,000	o—	—o									
styrene	263				o—o	—ooo—	—ooo—	—o				
styrene oxide	1,000					o—	—o					
1,1,2,2-tetrachloroethane	3,300							o				
1,1,2,2-tetrachloroethylene	370								oo—	—oo		
tetrahydrofuran	5,800								o			
1,2,3,5-tetramethylbenzene	136,000				o							
thymol	155,000			o								
toluene	720						ooo—	—ooo—	—o			
toluene-2,4-diisocyanate	6					oo—	—oo—	—o				
o-tolylmercaptan	39,000		o—	—o								
1,1,1-trichloroethane	330									oo		
trichloroethylene	300								oo—o—	—o		
trichlorofluoromethane	4,300							o—		—oo		
1,1,2-trichloro,1,2,2,-trifluoroethane	2,631								o—o—	—o		
triethylamine	235,000						o—o					
trimethylamine	493,500	o—					—o—	—o				
n-undecane	8,400							o				
valeric acid	256,300				oo							
vanilline	822,000	o										
vinylacetate	198,600							o—ooo				
vinylchloride	100											o—o
vinyl-2-pyridine	6,600							o				
o-xylene	300				o—	—o—o	ooo—o—	—o				
m-xylene	2,100							o				
p-xylene	18,200						o					
xylidine	82,000				o							

The O.I. is, in essence, a ratio of the driving force to introduce an odorant into the air versus the ability of an odorant to create a recognized response. It is a concept that provides information pertaining to the potential of a particular odorant to cause odor problems under evaporative conditions. The O.I. was first proposed by T. M. Hellman and F. H. Small in 1973 as a tool to predict whether complaints are likely to arise under certain evaporative conditions. Examples of evaporative conditions are spills, leaks, and solvent evaporation processes. The O.I. takes into account the vapor pressure of a compound, which is a qualitative measure of the potential of an odorant to get into the air, as well as the odor recognition threshold, which is a measure of the detectability of an odorant in the air. The values of odor index listed in Tables 12 and 13 range from a high of 1,052,000,000 for isopropylmercaptan to a low of 0.2 for maleic anhydride.

In its present form, the O.I. does not differentiate between "good" and "bad" odor qualities. It could incorporate a quality factor which would reflect a consideration of the odor quality, e.g., a "bad" odor might have a higher quality factor than a "good" odor. Since the O.I. is proposed as a general indicator of odor pollution, it would be reasonable to utilize categories of odor index values for comparison, rather than comparing individual values. These values can be separated into three categories: Category I—O.I. higher than 1,000,000 (high odor potential); Category II—O.I. between

TABLE 13. 100% Odor Recognition Concentration, Odor Index of Chemicals, Arranged by Chemical Class.

Chemical	Formula	Molecular Weight	Odor Index	100% Odor Recogn. Concentr.
BTX aromatics				
benzene	C_6H_6	78	300	300 ppm
toluene	$C_6H_5CH_3$	92	720	40 ppm
xylenes	$C_6H_4(CH_3)_2$	106	360–18,200	0.4–20 ppm
1,2,3,5-tetramethylbenzene	$C_6H_2(CH_3)_4$	134	136,000	2 ppb
isopropylbenzene	$C_6H_5CH(CH_3)_2$	120	89,600	40 ppb
ethylesters				
ethylacetate	$CH_3COOC_2H_5$	88	1,900	50 ppm
ethylbutyrate	$C_3H_7COOC_2H_5$	116	1,982,000	7 ppb
ethyl n-valerate	$C_4H_9COOC_2H_5$	132	178,000	60 ppb
ethylhexanoate	$C_5H_{11}COOC_2H_5$	144	760,000	4 ppb
ethylpelargonate	$C_8H_{17}COOC_2H_5$	186	109,000	1 ppb
ethyldecanate	$C_9H_{19}COOC_2H_5$	200		0.17 ppb
ethylundecylate	$C_{10}H_{21}COOC_2H_5$	214		0.56 ppb
methylesters				
methylformate	$HCOOCH_3$	60	300	2000 ppm
methylacetate	CH_3COOCH_3	74	1,100	200 ppm
methylbutyrate	$C_3H_7COOCH_3$	102	19,000,000	2 ppb
ketones				
acetone	CH_3COCH_3	58	720	300 ppm
methylethylketone	$CH_3COC_2H_5$	72	3,800	30 ppm
diethylketone	$C_2H_5COC_2H_5$	86	1,900	9 ppm
methylisobutylketone	$CH_3COCH_2CH(CH_3)_2$	100	1,000	8 ppm
methylisoamylketone	$CH_3COCH_2CH_2CH(CH_3)_2$	114	75,100	70 ppb
ethylisoamylketone	$C_2H_5COCH_2CH_2CH(CH_3)_2$	128	660	4 ppb
2 pentanone	$CH_3-CO-CH_2-CH_2-CH_3$	86	2,000	8 ppm
2-heptanone	$CH_3-CO-CH_2CH_2-CH_2CH_2CH_3$	114	171,000	20 ppb
2-octanone	$CH_3CO(CH_2)_5CH_3$	128	4	250 ppm
diisobutylketone	$[(CH_3)_2CHCH_2]_2CO$	142	45	50 ppm
mercaptans				
isoamylmercaptan	$(CH_3)_2CHCH_2CH_2SH$	104.2		0.2 ppb
methylmercaptan	CH_3SH	48	53,300,000	35 ppb
ethylmercaptan	CH_3CH_2SH	62	289,500,000	2 ppb
propylmercaptan	$CH_3CH_2CH_2SH$	76	263,000,000	0.7 ppb
isopropylmercaptan	$(CH_3)_2CHSH$	76	1,052,000,000	0.2 ppb
butylmercaptan	$CH_3CH_2CH_2CH_2SH$	90	49,000,000	0.8 ppb
isobutylmercaptan	$(CH_3)_2CHCHSH$	90.2		0.83 ppb
t.butylmercaptan	$(CH_3)_3CSH$	90.2		0.81 ppb
phenylmercaptan	C_6H_5SH	110	940,000	0.2 ppb
o-tolylmercaptan	$C_6H_5(CH_3)SH$	125	39,000	2 ppb
sulfides				
hydrogen sulfide	H_2S	34	17,000,000	1 ppm
methylsulfide	$(CH_3)_2S$	62	2,760,000	0.1 ppm
ethylsulfide	$(CH_3-CH_2)_2S$	90	14,400,000	4 ppb
propylsulfide	$(C_3H_7)_2S$	118		19 ppb
butylsulfide	$(CH_3CH_2CH_2CH_2)_2S$	146	658,000	2 ppb
isoamylsulfide	$[(CH_3)_2CHCH_2CH_2]_2S$	174	1,640,000	0.4 ppb
phenylsulfide	$(C_6H_5)_2S$	186	14,000	4 ppb
pentylsulfide	$(C_5H_{11})_2S$	174		0.028 ppb
di-isopropylsulfide	$[(CH_3)_2CH]_2S$	118		3.2 ppb

TABLE 13. (Continued)

Chemical	Formula	Molecular Weight	Odor Index	100% Odor Recogn. Concentr.
acrylates				
ethylacrylate	$CH_2=CH-COOC_2H_5$	100	38,160,000	1 ppb
isobutylacrylate	$CH_2=CH-COOCH_2-CH(CH_3)_2$	128	525,000	20 ppb
ethylhexylacrylate	$CH_2=CH-COOC_8H_{17}$	184	7,300	150 ppb
butyrates				
methylbutyrate	$CH_3-CH_2-CH_2-COOCH_3$	102	11,000,000	3 ppb
ethylbutyrate	$CH_3-CH_2-CH_2-COOC_2H_5$	116	1,982,000	7 ppb
amines				
ammonia	NH_3	17	167,300	55 ppm
methylamine	CH_3NH_2	31	940,000	3 ppm
ethylamine	$CH_3CH_2NH_2$	45	1,445,000	0.8 ppm
isopropylamine	$(CH_3)_2CHNH_2$	59	637,000	1 ppm
butylamine	$CH_3(CH_2)_3NH_2$	73	395,000	0.3 ppm
dimethylamine	$(CH_3)_2NH$	45	280,000	6 ppm
diethylamine	$(C_2H_5)_2NH$	73	880,000	0.3 ppm
dipropylamine	$(C_3H_7)_2NH$	101	395,000	0.1 ppm
diisopropylamine	$[(CH_3)_2CH]_2NH$	101	108,000	0.8 ppm
dibutylamine	$(C_4H_9)_2NH$	129	5,500	0.5 ppm
trimethylamine	$(CH_3)_3N$	59	493,500	4 ppm
tri-ethylamine	$(C_2H_5)_3N$	101	235,000	0.3 ppm
ethanolamine	$HOCH_2CH_2NH_2$	61	130	5 ppm
methylethanolamine	$HOCH_2CH_2\!\!>\!\!NH$ $\;\;\;\;\;\;\;\;\;\;\;\;CH_3$	75	400	3 ppm
dimethylethanolamine	$HOCH_2CH_2\!\!-\!\!N(CH_3)_2$	89	292,000	40 ppb
alkanes				
ethane	CH_3CH_3	30	25,300	1500 ppm
propane	C_3H_8	44	425	11000 ppm
butane	C_4H_{10}	58	480	5000 ppm
isobutane	C_4H_{10}	58	3,000,000	2 ppm
pentane	C_5H_{12}	72	570	900 ppm
heptane	C_7H_{16}	100	200	200 ppm
octane	C_8H_{18}	114	100	200 ppm
nonane	C_9H_{20}	128	9,800	0.4 ppm
undecane	$C_{11}H_{24}$	156	8,400	0.2 ppm
alkenes				
ethene	$CH_2=CH_2$	28	57,100	800 ppm
propene	$CH_3CH=CH_2$	42	14,700	80 ppm
1-butene	$CH_3CH_2CH=CH_2$	56	43,480,000	0.07 ppm
2-butene	$CH_3CH=CHCH_3$	56	3,330,000	0.6 ppm
isobutene	$(CH_3)_2C=CH_2$	56	4,640,000	0.6 ppm
1-pentene	$CH_3CH_2CH_2CH=CH_2$	70	376,000,000	2 ppb
1-decene	$CH_3(CH_2)_7CH=CH_2$	140	3,900,000	20 ppb
ethers				
ethylether	$CH_3CH_2OCH_2CH_3$	74	1,939,000	0.3 ppm
isopropylether	$(CH_3)_2CHOCH(CH_3)_2$	100	3,227,000	0.06 ppm

TABLE 13. (Continued)

Chemical	Formula	Molecular Weight	Odor Index	100% Odor Recogn. Concentr.
butylether	$C_4H_9OC_4H_9$	130	13,400	0.5 ppm
phenylether	$C_6H_5OC_6H_5$	170	130	0.1 ppm
aldehydes				
formaldehyde	H CHO	30	5,000,000	1 ppm
acetaldehyde	CH_3CHO	44	4,300,000	0.3 ppm
propionaldehyde	CH_3CH_2CHO	58	3,865,000	0.08 ppm
acrylaldehyde (acroleine)	$CH_2=CHCHO$	56	19,300	20 ppm
butyraldehyde	$CH_3CH_2CH_2CHO$	72	2,395,000	40 ppb
isobutyraldehyde	$(CH_3)_2CHCHO$	72	948,000	300 ppb
crotonaldehyde	$CH_3CH=CHCHO$	70	125,000	0.2 ppm
methylpentaldehyde	$C_5H_{12}CHO$	101	131,500	0.15 ppm
furfuraldehyde	C_4H_3OCHO	96	5,260	0.2 ppm
benzaldehyde	C_6H_5CHO	106	22,000	5 ppb
cinnamaldehyde	$C_6H_5CH=CHCHO$	132	131,500	2 ppb
acids				
formic acid	HCOOH	46	2,200	20 ppm
acetic acid	CH_3COOH	60	15,000	2 ppm
propionic acid	CH_3CH_2COOH	74	112,300	40 ppb
butyric acid	$CH_3(CH_2)_2COOH$	88	50,000	20 ppb
valeric acid	$CH_3(CH_2)_3COOH$	102	256,300	0.8 ppb
caproic acid	$CH_3(CH_2)_4COOH$	116	43,900	6 ppb
enanthic acid	$CH_3(CH_2)_5COOH$	130	7,900	20 ppb
caprylic acid	$CH_3(CH_2)_6COOH$	144	104,500	8 ppb
pelargonic acid	$CH_3(CH_2)_7COOH$	158	164,000	0.7 ppb
capric acid	$CH_3(CH_2)_8COOH$	172		1.96 ppb
lauric acid	$CH_3(CH_2)_{10}COOH$	(200.3)		3.4 ppb
alcohols				
methanol	CH_3OH	32	22	6000 ppm
ethanol	CH_3CH_2OH	46	11	6000 ppm
propanol	$CH_3CH_2CH_2OH$	60	480	45 ppm
butanol	$CH_3(CH_2)_2CH_2OH$	74	120	5000 ppm
pentanol	$CH_3(CH_2)_3CH_2OH$	88	368	10 ppm
hexanol	$CH_3(CH_2)_4CH_2OH$	102	14,300	0.09 ppm
heptanol	$CH_3(CH_2)_5CH_2OH$	116	23,100	0.06 ppm
octanol	$CH_3(CH_2)_6CH_2OH$	130	33,000	2 ppb
decanol	$CH_3(CH_2)_8CH_2OH$	158	31,000	6 ppb
dodecanol	$CH_3(CH_2)_{10}CH_2OH$	186	1,800	7 ppb
phenolics				
phenol	C_6H_5OH	94	16	20 ppm
cresols	$(CH_3)C_6H_4OH$	108	60–260	0.2–0.7 ppm
acetates				
methylacetate	CH_3COOCH_3	74	1,100	200 ppm
ethylacetate	$CH_3COOC_2H_5$	88	1,900	50 ppm
propylacetate	$CH_3COOC_3H_7$	102	1,600	20 ppm
isopropylacetate	$CH_3COO(iC_3H_7)$	102	2,100	30 ppm
butylacetate	$CH_3COOC_4H_9$	116	1,200	15 ppm
isobutylacetate	$CH_3COO(iC_4H_9)$	116	3,300	4 ppm
amylacetate	$CH_3COOC_5H_{11}$	130	2,500	20 ppm
isoamylacetate	$CH_3COO(iC_5H_{11})$	130	526,000	20 ppb
sec. hexylacetate	$CH_3COO(sec.C_6H_{13})$	144	12,500	0.4 ppm
octylacetate	$CH_3COOC_8H_{17}$	172	2,800	300 ppb

INTRODUCTION 53

100,000 and 1,000,000 (medium odor potential); Category III—O.I. lower than 100,000 (low odor potential). The odor indexes calculated by the author are based on the highest 100% recognition levels mentioned in the literature, including those mentioned by Hellman and Small.

Table 12 gives an alphabetic list of some 260 organic compounds with their odor threshold limits (perception as well as recognition levels) and the O.I. based on the

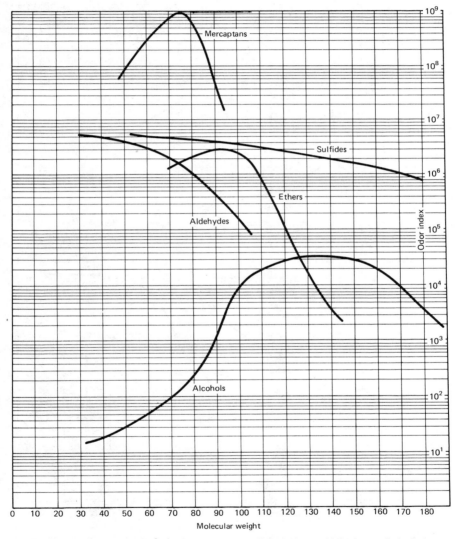

Fig. 29. Odor index (at 20°C) of mercaptans, sulfides, ethers, aldehydes, and alcohols.

highest odor threshold. The compounds have been arranged per chemical class in Table 13.

The threshold odor concentrations for 100% recognition and the associated odor indexes are shown graphically in Figs. 29 through 32. Per chemical class, we observe a smooth evolution of the O.I. and of the 100% recognition levels as a function of the molecular weight of the compounds. The correlations between threshold odor concen-

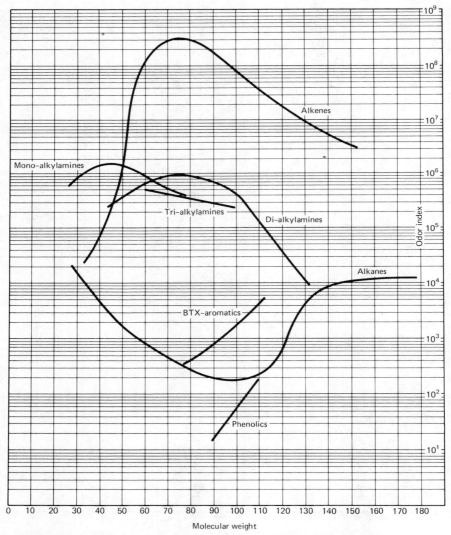

Fig. 30. Odor index (at 20°C) of alkanes, mono-, di-, tri-alkylamines, BTX-aromatics, alkanes, and phenolics.

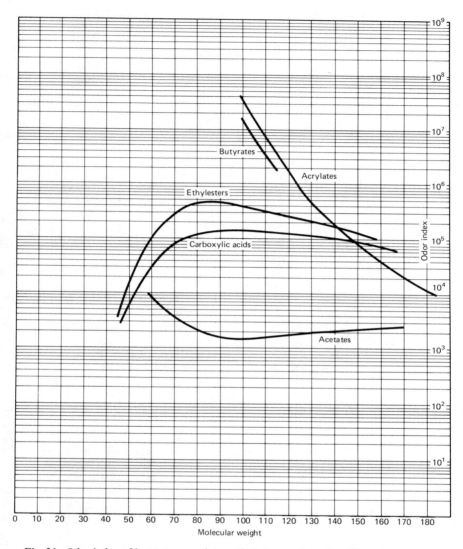

Fig. 31. Odor index of butyrates, acrylates, ethylesters, carboxylic acids, and acetates.

trations for straight-chain hydrocarbons with a terminal functional group and the molecular weight are represented in Fig. 32 as smooth curves. The odor index, which is a function of the vapor pressure of the product, must be calculated at the temperature of the evaporating product. The O.I. mentioned in the foregoing tables have all been calculated at 20°C. It is clear that certain products, in practice, will have a higher odor index than mentioned in these lists because they are handled at higher temperatures.

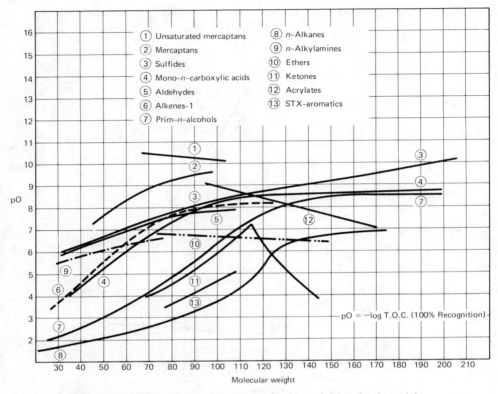

Fig. 32. Threshold odor concentrations as function of the molecular weight.

The 100% recognition level has been taken as the basis for the calculation of the O.I. The 100% recognition level is the concentration at which all members of an odor panel recognize the odor. It shows less variation than the absolute perception level, which is much lower. For this reason, the highest recognition level mentioned in the literature has been taken as the basis for these calculations.

In general, we can say that straight-chain aliphatic molecules have the highest recognition level, and that the level decreases with increasing molecular weight. For nearly every class of straight-chain molecules, the threshold odor concentration decreases with increasing molecular weight. The influence of functional groups on the 100% recognition levels and on the odor indexes is given in Tables 14 and 15. The functional groups of the small molecules have dominating influence on their threshold, as shown in Table 16 for molecules with one carbon atom. Branched chains often show different results, probably because of steric effects. The functional groups can intensify each other's effects on the threshold odor concentration, but can also reduce the effect depending upon their position in the molecule. In general, a double bond reduces the threshold odor concentration. This is the case in aliphatic compounds, mercaptans, and ketones, but not in aldehydes, as shown in Table 17. The merit

INTRODUCTION 57

TABLE 14. Classification of Chemical Classes According to Their 100% Recognition Threshold.

Highest Odor Threshold – alkanes

to ↓
    BTX-aromatics
    alcohols
    chlorinated alkanes
    amines: tri-alkylamines
             di-alkylamines
             mono-alkylamines
    carboxylic acids
    alkenes
    aldehydes
    sulfides
    disulfides
    trisulfides
    mercaptans

Lowest Odor Threshold – unsaturated mercaptans

TABLE 15. Classification of Chemical Classes, According to Their Odor Index (at 20°C).

$O.I. > 10^6$: mercaptans
               alkenes
               sulfides

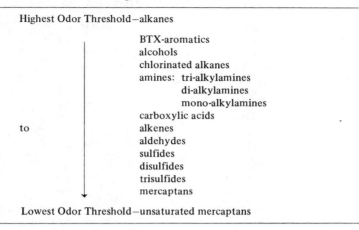

butyrates ⎫
acrylates |
aldehydes ⎬ of low molecular weight
ethers |
alkylamines ⎭

$O.I.$ between 10^4 and 10^6: di-alkylamines
                                  tri-alkylamines
                                  higher ethylesters
                                  carboxylic acids

aldehydes ⎫
ethers   ⎬ of high molecular weight
alcohols ⎭

$O.I. < 10^4$: alkanes
               acetates
               BTX-aromatics
               lower alcohols
               phenolics

TABLE 16. Influence of Functional Groups on T.O.C. (100% Recognition).

C_1	T.O.C. (100% recogn.)
CH_4	±10,000 ppm
CH_3OH	6,000 ppm
$CHCl_3$	300 ppm
CH_2Cl_2	214 ppm
CCl_3F	209 ppm
CCl_4	200 ppm
$CH_2(OH)_2$	60 ppm
CH_3Cl	30 ppm
$HC(=O)OH$	20 ppm
$HC\equiv N$	5 ppm
CH_3NH_2	3 ppm
CCl_3NO_2	1 ppm
$HC(=O)H$	1 ppm
$ClC\equiv N$	1 ppm
CH_3SH	0.035 ppm
CHI_3	0.00037 ppm

TABLE 17. Influence of Double Bonds on the Odor Recognition Threshold.

Product	Formula	100% Recognition Threshold
propionaldehyde	CH_3CH_2CHO	0.08 ppm
acroleine	$CH_2=CHCHO$	20 ppm
Butyraldehyde	$CH_3CH_2CH_2CHO$	0.04 ppm
Crotonaldehyde	$CH_3CH=CHCHO$	0.2 ppm
butane	$CH_3CH_2CH_2CH_3$	5,000 ppm
1-butene	$CH_2=CHCH_2CH_3$	0.07 ppm
pentane	$CH_3CH_2CH_2CH_2CH_3$	900 ppm
1-pentene	$CH_2=CHCH_2CH_2CH_3$	0.002 ppm
propylmercaptan	$CH_3CH_2CH_2SH$	0.7 ppb
allylmercaptan	$CH_2=CHCH_2SH$	0.05 ppb
butylmercaptan	$CH_3CH_2CH_2CH_2SH$	0.8 ppb
crotylmercaptan	$CH_2CH=CHCH_2SH$	0.055 ppb

of the previous analysis is that by extrapolation, the T.O.C. of products for which only the formula is known can be estimated.

Comparison of Techniques for Organoleptic Odor-Intensity Assessment.

1. Odor room Odor threshold can be determined by using an *odor room*. A known volume of odorous air is admitted to the room of volume V until the volume S of odorous air is found which just, and only just, allows the odor to be detected within the room. The ratio V/S at this point is the threshold dilution.

2. The syringe method In the *ASTM syringe method* (ASTM D 1391-57), variable volumes of odorous sample air are made up to 100 cu m with clean air in a series of syringes. Variable volume samples may then be taken from these syringes and similarly made up to achieve yet higher dilutions and so on until odor threshold is reached. Assessment for odor is by injection into the nostril of panel members, one syringe being supplied to each person for each dilution tested. Hypodermic needles were sealed to produce end caps or joined together to produce transfer tubes in order to retain the sample within the syringes until required and to effect the dilution operations respectively.

3. The dynamic dilution method The general concept of *dynamic dilution* involves the mixing of sample air supplied at a known flowrate with dilution air at known flowrates. After each change in dilution flowrate, panel members to whom the issuing mixture is presented register the presence or absence of odor, and in this manner threshold dilutions are determined. In order to achieve a sufficiently wide range of dilutions, it is normal to use a two-stage system in which a sample of diluted odorous air from the first dilution operation is introduced to a second mixing chamber and is there diluted with flowing clean air in the second stage.

The first apparatus used in this work for dynamic dilution, the *Mk1*, was constructed entirely of glass and was provided with one assessment port situated in a small box to which the observer introduced his head to assess odor. Rotameters were used to measure flowrates. The maximum dilution flow in each of the dilution stages was 30 l/min. Sample air entered tangentially to provide more efficient mixing with the axially flowing dilution air in the mixing chambers. The *Mk2* instrument was developed from experience gained with the *Mk1*. It was again a two-stage system but much larger than the earlier version, allowing for initial and secondary sample flows of 2–10 l/min and a maximum dilution rate in both stages of 2000 l/min. Samples were introduced by allowing air to enter the rigid container, thus pressuring the flexible bag at a controlled rate to achieve turbulent mixing with dilution air in the vertical mixing tube. Dilution air was supplied by a compressor via an active carbon trap. The entire apparatus, including orifice plates and valves, was constructed in Teflon and Teflon-coated mild steel. The mixing tubes of each dilution stage were provided with three observation ports so that three panel members could simultaneously record results.

3. Atmospheric Degradation. The atmospheric media provides an excellent matrix for photochemical or oxidative alterations of chemical contaminants. The energy provided by sunlight is able to break carbon-carbon and carbon-hydrogen bonds, cause the photodissociation of nitrogen dioxide to nitric oxide and atomic oxygen, and photolytically produce significant amounts of hydroxyl radicals. The relatively high concentration of oxygen (20.9 V/V) makes it one of the most important participants in various reactions with air pollutants since the rates are concentration dependent. Similar reasoning can be used for reactions with water vapor (0.1–5 V/V) and carbon dioxide (0.03–0.1% V/V).

Although hydrocarbons are essential to the formation of photochemical smog, not every hydrocarbon produces the manifestations such as eye irritation, plant damage, and visibility reduction that are usually associated with smog. The reason for this is the differences in reactivity and chemical structure of hydrocarbons.

We will define "reactivity" as the tendency of an atmospheric system containing organic products and nitrogen oxides to undergo a series of reactions in the presence of sunlight. These series of reactions lead to various smog manifestations which are principally: ozone (oxidant) production, eye irritation, hydrocarbon consumption, nitric oxide oxidation, aerosol formation, and plant damage. However, because of the variety of ways in which smog is manifested, a number of reactivity scales have been developed. These scales cover:

1. rates of hydrocarbon consumption (hydroc. cons.)
2. rates of photoxidation of NO to NO_2 (NO ox.)
3. rates and yields of ozone formation
4. eye irritation (EI)
5. chemical reactivity
6. OH rate constants

There appear to be at least three important reactions leading to hydrocarbon degradation:

1. H-atom abstraction by OH radicals
2. addition to olefins by OH radicals
3. addition to olefins by ozone

These same processes appear to dominate the life cycles of hydrocarbons in the background atmosphere and form part of the natural atmosphere's cleansing mechanism.

In the polluted atmosphere, the degradation products accumulate to the extent where they may be discernible by their effects on visibility, plant life, and human health.

The OH rate constant scale is based on the critical importance of OH radicals in photochemical smog formation through reactions (1) and (2) (1700). The OH rate constant is an objective quantitative measure of reactivity which can be accurately measured by a wide range of experimental techniques. A reactivity scale is illustrated in Table 18, together with the estimated lifetime for each hydrocarbon for its photochemical oxidation on a sunny summer day in southeast England. This calculation employed a computer simulation of the chemical reactions occurring in the atmosphere to obtain a typical OH concentration.

The O_3 and OH concentrations were 0.16 ppm and 0.75×10^7 molecules/cm^3, respectively. Ambient concentrations data were utilized where possible for C_1–C_5 hydrocarbons, NO_x, CO, and SO_2.

Relative reactivity scales were developed for the sake of comparison, using as a stan-

TABLE 18. Reactivity Scale Based on OH Rate Constants and Estimated Lifetimes under Photochemical Smog Conditions in SE England (1699).

Product	OH rate constant	Lifetime (hr)
methane	0.007	5291
acetylene	0.18	206
ethylene	5.2	7.2
propylene	33.3	1.1
butane	2.4	15
cis-2-butene	62.4	0.6
pentane	3.3	11
hexane	6.3	5.9
benzene	1.3	28
toluene	6.4	5.8
o-xylene	14.3	2.6
2,3-dimethyl-2-butene	111	0.33

dard a fast hydrocarbon, set at a value of 1–100. In the Glasson–Tuesday Scale, sometimes referred to as the Jackson scale, 2,3-dimethyl-2-butene is the reference hydrocarbon and it has a relative reactivity for NO oxidation (NO_2 formation) of 100. Table 19 presents an abbreviation of this scale.

The natural atmospheric environment is a complex, highly reactive system involving photochemical and secondary dark reactions to both organic and inorganic substances. To reproduce such a complex system in the laboratory, the following parameters must be controlled:

- spectral distribution and intensity of the light
- concentration of the reactants in the test chamber
- temperature
- surface reactions (major difference between results in test reactors and in the atmosphere)

Four general laboratory procedures have been used to study atmospheric degradation:

- long-path infrared (LPIR) cells
- plastic containers
- glass flask reactors
- smog chambers

With LPIR cells, the test chemical is placed into the cell and irradiated.

The rate of reaction is determined by following the disappearance of the test chemical and formation of some products by *infrared analysis*. Fluorescent lights are usually used for irradiation and various configurations and chamber materials (stainless steel, Teflon-coated interior, and Pyrex glass) have been used. Known amounts of nitric oxide and nitrogen dioxide can be added to simulate air pollution situations.

TABLE 19. Abbreviated Glasson-Tuesday (Jackson) Reactivity Scale (Basis: NO_2 formation).

Product	Relative Reactivity
methane	0
acetylene	0
ethylene	2.88
propylene	5.93
butane	1.27
1-butene	6.04
2-butene	15.5
pentane	1.58
1-pentene	4.63
2-pentene	11.05
hexane	1.58
hexene	3.39
benzene	0.56
toluene	2.20

A popular method for studying the atmospheric chemistry of both artificial and natural atmospheric samples is the use of plastic containers made of either FEP (fluorinated ethylene-propylene), Teflon, Tedlar, or Mylar. The sample is placed in the bag, irradiated with either natural or artificial sunlight, and analyzed periodically, usually by gas chromatography. Glass flask reactors are used in a similar way, but are usually smaller in volume than the plastic bags and are more commonly irradiated with artificial lights.

In order to study air pollution reactions, many researchers have resorted to large photochemical *smog chambers*. These chambers are extremely complicated and expensive to build and operate. Most chambers can be run in a dynamic or static mode, although for simplicity and precision, the static procedure is more predominantly used.

4. Natural Sources. Although the amount of manmade organic chemicals (excluding lubricants) which enter the environment may perhaps be as much as 20 million tons a year, this total is very small in comparison with the enormous tonnages of organic compounds naturally produced. Over the ages, degradation and emanation cycles have become established through which an equilibrium seems to be maintained. Although there is a great deal of knowledge about the detailed mechanisms involved, the way in which many cycles operate is still obscure. Some of them are on a massive scale; for instance, it has been estimated that swamps and other natural sources emit as much as 1600 million tons of methane into the atmosphere each year. Even cattle, which emit methane equivalent to 7% of their energy intake, must contribute a world total of several million tons. It is estimated that the world's atmosphere contains a total of 4800 million tons of methane and it is evident that a balance situation exists.

Terpene-type hydrocarbons, emitted into the air by forests and other vegetation, amount to an estimated 170 million tons a year. It is believed that these polymerize in the air, and presumably they are eventually eliminated from the air by rainfall or deposition in aerosols.

This heading includes not only information about the natural sources (emission rates) but also the background concentrations at ground level at remote places such as Pt. Barrow in Alaska, although these low concentrations might be mainly of anthropogenic origin.

5. Manmade Sources. Emission rates for various sources, such as diesel and gasoline engines, municipal waste incinerators, central heating furnaces, etc., as well as ground level concentrations in residential, urban, and industrial areas have been considered.

6. Emission Control Methods and Efficiency. Gaseous (or liquid or dissolved) molecules that adhere to the surface of a solid by physical or chemical attraction are said to be adsorbed. The surface is the adsorbent and the material adsorbed is the adsorbate. The amount of material adsorbed depends upon the number of available "active" sites on the adsorbent. Although there are several materials that adsorb, including activated carbon or charcoal, silicagel, alumina, clays, and powders, the only practical adsorbent for most air pollution work is activated carbon. Activated carbon is the only nonpolar adsorbent of note and therefore, the only one that can be used satisfactorily in the presence of water vapor.

Vapors with boiling points higher than $0°C$ are easily adsorbed. The amount of adsorbed material (percent by weight of adsorbent) retained after exposure to saturated air and flushing with clean air is the retentivity of the adsorbent.

7. Sampling and Analysis: Methods and Limits.

Gas Detection Devices with Test Tubes. The gas detection devices are based on the adsorption of the tested air pollutant on solid surfaces with the simultaneous occurrence of a color reaction. They can be used for monitoring the atmosphere at the work place, and in the environment near the source. The test tubes are made of glass. The polluted air is sucked through the tube by means of a handpump. The tubes as a rule initially contain a layer of reagents along the direction of the gas stream. The reagents serve to retain all the interfering substances.

In order to obtain semiquantitative values, a reagent is applied in a concentration over the carrier (solid surface) such that, with a certain number of strokes (e.g., 5 or 10), the length of the colored layer is a measure of the pollutant concentration in the tested air. The reproducibility of the quantitative results is stated by the manufacturers. The standard deviation of the results is about 6–8% in favorable and 10–20% in less favorable cases. The correctness of the results depends on the calibration, specificity of the detection reaction, and the elimination of interfering substances.

Spectrophotometry. Infrared, visible, and ultraviolet spectrophotometry have been utilized for analysis of gaseous constituents in flue gases and in ambient air. The principle involved is the selective absorption of light at a given wavelength by a particular substance. The degree of absorption is proportional to the concentration of the particular compound.

1. Infrared analyzers. Two different types of infrared instruments are available.

The *nondispersive* type relies on selective absorption of infrared light at a particular wavelength by a single compound. It makes a comparison in signals between a reference and a sample cell. It may be adapted for continuous monitoring by drawing gas through the sample cell. *Dispersive* infrared analyzers make a wavelength scan from one to 16 microns on a single sample. Variable sensitivity may be attained by changes in the path length. It is not adaptable for continuous monitoring.

2. *Ultraviolet analyzers.* Ultraviolet radiation may be used for analysis of gaseous constituents in a manner analogous to those employed with infrared techniques.

3. *Visible spectrophotometry.* The absorption of some organic compounds in the visible range has been developed as a continuous monitoring method for certain emissions.

Flame Ionization. Hydrogen flame ionization analyzers operate on the principle that a small electric current is generated when gases containing carbon atoms are oxidized to carbon dioxide in a hydrogen flame when a potential is applied across the flame. The flame ionization units do not respond to water, carbon monoxide, carbon dioxide, hydrogen sulfide, oxygen and nitrogen to any measurable degree.

Combustibles Analysis. *Combustibles* gas analyzers operate on the principle that the electric resistance of a wire placed in a flue gas changes as its temperature changes. The unit is first balanced for zero combustibles at the flue gas temperature and resulting changes in temperature of the wire are then caused by the presence of gases which become oxydized on contact with the wire. The combustible analyzers in use are nonspecific since they are sensitive to hydrogen, hydrogen sulfide, carbon monoxide, and organic compounds.

Differential Ultraviolet Photometry. Low concentrations of phenol can be selectively monitored by differential ultraviolet photometry using the characteristic wavelengths shift with ph. Phenol at pH of 12 shows a shift towards higher wavelength and a smaller increase in UV absorption versus an acidic phenol solution at pH of 5. This difference in absorption is linearly proportional to the phenol concentration and is specific since no other contaminants expected in water will show a change.

Photometry. Concentrations of many liquids, gases, and vapors can be reliably measured or controlled using a photometric analyzer. In general, all aromatics, carbonyls, and many inorganic salts absorb radiation from the choice of light sources available. Concentrations of such materials can be accurately measured, assuming that appropriate measuring and reference wavelengths can be selected. On the other hand straight chain hydrocarbons, inorganic gases, and the lower alcohols generally do not absorb visible or ultraviolet radiation. They act as transparent background materials. Air and water vapor are also transparent background materials at the wavelengths used.

The *minimum full scale ranges* (with accuracies of ±2% of full scale) are listed in parts per million (ppm) for vapor samples and 10^{-6} moles per liter for liquids. The minimum full scale listed is based on a one meter cell length. However, lower full scale ranges can often be provided for vapors by increasing the absolute pressure of the sample. The following products act as transparent background materials; acetylene, argon, butane, *i*-butanol, *n*-butanol, carbon dioxide, carbon monoxide, ethane, ethanol, ethylene, ethylene glycol, helium, hydrochloric acid, hydrogen, krypton, methane, methanol, neon, nitrogen, oxygen, propane, propylene, *i*-propanol, *n*-propanol, water, xenon.

Gas Chromatography. Gas chromatography provides for separation of constituents in a gaseous mixture based on their differences in relative affinity for a given packing material in the column. This continuous process of repeated absorption and desorption results in their passing through the column at different rates. They can be analyzed separately at the column exit by means of a suitable detector. It may be necessary to concentrate the samples prior to analysis to provide for a sufficient amount of material to be measured accurately. This may be done by freeze out, solid adsorption, or absorption in an inert solvent or a chemically reactive solution. Detectors frequently used include thermal conductivity electron capture, flame ionization, and thermistor. Thermal conductivity is theoretically sensitive to any gas but is limited in its sensitivity. Electron capture (EC) is useful for halogen containing compounds, oxides of nitrogen, oxides of sulfur, and certain other gases, but is adversely affected by water. Flame ionization (FID) is useful for organic materials, while flame photometry is useful for phosphorus and sulfur containing substances.

Gas Chromatography–Mass Spectrometer (GC/MS). The use of the mass spectrometer as a detector for the gas chromatograph was developed during the 1960's. The minicomputer was invented at this time and applied to GC/MS to utilize the information produced by the mass spectrometer. Computerized GC/MS quickly revolutionized the field of trace organic analysis, and made very significant contributions to research in medicine, biochemistry, odors, and organic geochemistry. One advantage of the technique is that it substantially increases the capacity of a staff to handle large numbers of environmental samples and to accurately identify specific organic compounds.

C. WATER POLLUTION FACTORS

1. Biodegradation

1.1. Objectives. The major use of data on biodegradation is for assessing the persistence of a chemical substance in a natural environment. The natural environment, for the purpose of such tests, is natural waters and various soils (including hydrosoils). If the compound does not persist, information is needed on whether the compound degrades to innocuous molecules or to relatively persistent and toxic intermediates. Secondary concerns for environmental persistence are the possibilities that (1) toxic substances may interfere with the normal operation of biological waste treatment units or (2) toxic substances not substantially degraded within a treatment plant may be released to the natural environment.

The test procedures for biodegradation give an estimate of the relative importance of biodegradability as a persistence factor. The data are used to evaluate biodegradation rates in comparison with standard reference compounds. Most tests provide opportunities for biodegradation with relatively dense microbial populations which have been allowed to adapt to the test compound. Those compounds which degrade rapidly (in comparison with reference compounds) and extensively (as judged by such evidence as CO_2 evolution and loss of dissolved organic carbon) are likely to biodegrade rapidly in a variety of environmental situations.

Such compounds may persist in specialized environments and under circumstances which are poorly represented by these preliminary tests. Compounds which produce little indication of biodegradation in these tests may be relatively persistent in a wide variety of environments.

Reliable conclusions about biodegradation are not generally possible on the basis of structure alone. Biodegradation is the most important degradative mechanism for organic compounds in nature, in terms of mass of material transformed and extent of degradation. Therefore, information on biodegradability is very important in any evaluation of persistence and is generally needed on organic compounds that can be solubilized or dispersed in or on water. Highly insoluble compounds are not testable at present without the use of radioisotopes or complex analytical measurements, nor are there methods to study biodegradation with very low substrate concentrations.

These types of tests do not differentiate all chemical compounds as relatively non-biodegradable or rapidly and extensively biodegradable. Results for many materials lead to intermediate conclusions. A more complete understanding of the biodegradability of such compounds would result from advanced tests, such as those employing radiolabelled substrates.

Anaerobic microbial degradation or organic compounds is an important mechanism for degrading waste materials in both the natural environment and in waste treatment plants. However, there are few relatively simple state-of-the-art methods at present for evaluating the potential for anaerobic biodegradation. The types of methods most frequently cited employ microcosms such as flooded soils in flasks and require the use of radiolabelled substrate. Methane from fermentation of organic substrates is the end of a food chain process involving a wide variety of anaerobic bacteria.

The anaerobic digestion test compares the production of methane and CO_2 by anaerobic bacteria in sludge samples with and without added test material.

A desirable goal of degradation testing is to obtain some estimate of the rate at which a compound will degrade in the environment. It is relatively easy to estimate reaction rates for such degradation processes as hydrolysis, photolysis and free-radical oxidations. Environmentally realistic reaction rates for biodegradation are much more difficult to obtain. Among the more important environmental variables which can effect the rate and the extent of biodegradation are (1) temperature, (2) pH, (3) salinity, (4) dissolved oxygen, (5) concentration of test substance, (6) concentration of viable microorganisms, (7) quantity and quality of nutrients (other than test substances), trace metals, and vitamins, (8) time, and (9) microbial species.

1.2. The Determination of Biodegradability. Biodegradability tests may be divided into two groups of tests: first, die-away tests in static systems; and second, tests in flow-through systems (continuous cultures).

In die-away tests, the concentration of the substance under investigation (the substrate) is determined analytically as a function of time. During the experiment, the substrate is contained in a fixed amount of test medium. In flow-through systems, a constant flow of the medium is fed into a completely mixed 'reactor', in which the

TABLE 20. Survey of Test Methods for the Determination of Biodegradability (1716).

Aspects	Static systems		Flow-through systems	
	Die-away tests in shake-flasks	Die-away tests in percolating filters	Activated sludge systems	Chemostats
1. Comparison with environmental conditions	discharge in a river	none	aerobic purification plant	natural waters
2. Mechanism of adaptation	no special mechanism (growth and dying)	growth and dying	growth not specially related to the substrate	sufficiently fast growth, other microorganisms are washed from the system
3. Duration of test	4–8 weeks	first substrate 4 weeks second, etc. substrates 1 week each	6–12 weeks	6 weeks
4. Properties of substrate	no restrictions	soluble in the medium non-volatile	soluble in the medium non-volatile	no solids
5. Media	no restrictions	no restrictions	artificial sewage	no restrictions
6. Analysis	measurement of mineralization specific analysis	measurement of mineralization specific analysis	specific analysis	measurement of mineralization specific analysis; bioassays
7. Equipment	laboratory glassware shaking machine	glass columns and pumps	perspex or glass system and accurate metering pump	chemostat, accurate metering pump, pH-control
8. Oxygen availability	aerobic and anaerobic	aerobic	aerobic	aerobic and anaerobic
9. Temperature	25°C or the temperature of the environment under investigation	20–25°C	20–25°C	25°C, or temperature of environment under investigation

medium volume is also kept constant. The degradation is calculated from the difference between the substrate concentrations in the inlet and effluent streams.

Four test methods for the determination of biodegradability are discussed (Table 20):

1. die-away tests in shake-flask cultures
2. die-away tests in percolating filters
3. tests with activated-sludge systems
4. tests with continuous cultures (chemostats)

The method chosen depends on a number of factors including the following: the equipment needed; substrate properties; concentration of the substrate; duration of the tests; adaptation of the inoculum; and biomass.

2. Oxidation Parameters. The conventional oxidation parameters, such as biological oxygen demand (BOD), chemical oxygen demand (COD) using potassium dichromate, permanganate value ($KMnO_4$), total organic carbon (TOC), total oxygen demand (TOD), and theoretical oxygen demand (ThOD) are mentioned here. The oxidation parameters are dimensionless (grams oxygen consumption per gram of product), unless stated otherwise. BOD values are assumed to be measured at 20°C, unless indicated otherwise.

Biochemical Oxygen Demand: BOD. When water containing organic matter is discharged into a river, lake, or sea, natural purification by biological action takes place. Thus, biochemical oxidation is brought about by naturally occurring microorganisms which use the organic matter as a source of carbon. Dissolved oxygen in the water sustains respiration. Naturally, this is a simplified picture of a very complex set of reactions, the rates of which depend on the temperature, the type of organic matter present, the type of microorganisms, the aeration, and the amount of light available. A number of years ago an attempt was made to produce a test which would match the rate of biochemical oxidation that would occur in a river into which organic containing water was discharged. The 5 day test has been generally adopted with the knowledge that this does not necessarily represent the time required for total oxidation of the organic matter present. In some cases, a test period of longer than 5 days is specified. It follows, therefore, that the BOD_5 test should always be considered in conjunction with other data and a knowledge of the system being studied. Application of the test to organic waste discharges allows calculation of the effect of the discharges on the oxygen resources of the receiving water. Data from BOD tests are used for the development of engineering criteria in the design of waste water treatment plants.

The BOD test is an empirical bioassay-type procedure which measures the dissolved oxygen consumed by microbial life while assimilating and oxidizing the organic matter present. The sample of waste, or an appropriate dilution, is incubated for 5 days at 20°C, in the dark. The reduction in dissolved oxygen concentration during the incubation period yields a measure of the biochemical oxygen demand. The standard dilution method prescribes the use of several dilutions because faulty results can be obtained by the toxicity of the sample (yielding a low BOD), or by depletion of the oxygen by a too high concentration of biodegradable organics. The bottles which after 5 days still contain approximately 50% of the original oxygen content are selected for the BOD calculation.

Seventy-seven analysts in fifty-three laboratories analyzed natural water samples plus an exact increment of biodegradable organic compounds. At a mean value of 194 mg/l BOD, the standard deviation was ±40 mg/l. There is no acceptable procedure for determining the accuracy of the BOD test.

BOD values sometimes show a wide range of variation. The main reason is that microorganisms have a tremendous capacity for adaption. Whenever the samples are inoculated with microorganisms from a very polluted river, we may suppose that these microorganisms are already partly adapted to the specific organics. When an inoculum is taken from a polluted river, or from an industrial waste water treatment plant, this should be mentioned.

Chemical Oxygen Demand (COD). Because the BOD method takes 5 days to carry out, other methods have been sought for measuring oxygen requirements of a sample. The methods should be simple, quick, and reliable. One popular method is the chemical oxygen demand test based on the oxygen consumed from boiling acid potassium dichromate solution. This is a severe test and a high degree of oxidation takes place. Interference from chloride ions can be a problem. Several procedures are available to overcome this problem. They are based mainly on the complexion of the chloride ion with mercuric sulphate. This procedure is recommended in many standard methods. It has been demonstrated that high COD results are achieved when chloride ions are present and true compensation is not made. Finally, it should be noted that the COD test is not a measure of organic carbon although the same reactions are involved.

In this procedure, organic substances are oxidized by potassium dichromate in 50% sulfuric acid solution at reflux temperature. Silver sulfate is used as a catalyst and, as mentioned, mercuric sulfate is added to remove chloride interference. The excess dichromate is titrated with standard ferrous ammonium sulfate, using orthophenanthroline ferrous complex as an indicator. Eighty-nine analysts in fifty-eight laboratories analyzed a distilled water solution containing oxidizable organic material equivalent to 270 mg/l COD. The standard deviation was ±27.5 mg/l COD and the mean recovery was 96% of the true value.

The ASTM Standard Method of Test for Chemical Oxygen Demand of Waste Water D1252-67, states in its scope (Section 1.1) that it "provides for addition of reagent so as to minimize loss of volatile organic materials". Shell Research (The Netherlands) follows a slightly modified procedure which seems more effective in avoiding such evaporation losses and which is also faster. In the first step of the ASTM procedure, 1 g of $HgSO_4$ and 5 ml of H_2SO_4 are added to a 50 ml sample (or aliquot diluted to 50 ml) "very slowly with swirling"; the method further prescribes "cool while mixing". This, however, takes several minutes, during which time an increase in temperature at least locally in the mixture cannot be avoided and light components may escape from the open conical flask. According to Shell's modified method, from a 25 ml dispenser, a Standard solution A, containing 5 ml of H_2SO_4, 1 g of $HgSO_4$, and 20 ml of water is added to a 30 ml sample. This takes only seconds, no heat is evolved, and the reaction flask can immediately be stoppered. The 25 ml of dichromate is then dispensed as prescribed. According to the ASTM method, 1 g of the catalyst Ag_2SO_4 and (slowly and through the open end of the condenser) 70 ml of H_2SO_4 are added separately. In the modified procedure, 70 ml of Stock Solution B is also added through the condenser. This Stock Solution B contains 1 g of Ag_2SO_4 in 70 ml of H_2SO_4. As with

TABLE 21. COD Measured on Chemical Compounds.

Compound	Trade name	Recovery ASTM procedure g/g	%	Modified procedure g/g	%
benzene		2.15	70	2.80	91
3-broco-1-propene	allyl bromide	0.72	68	0.93	88
3-chloro-2-methyl-1-propene	methallyl chloride	1.18	61	1.72	89
3-chloro-1-propene	allyl chloride	0.86	51	1.33	80
1,3,5-cycloheptatriene		2.30	74	2.78	89
1,5-cyclooctadiene		2.62	80	3.11	96
decene mixture	amylene dimer	1.68	49	2.34	68
1,3-dichloropropane		0.84	74	1.03	91
5-methyl-3-heptanone	ethyl amyl ketone	1.94	67	2.61	91
4-methyl-2-pentanone	methyl isobutyl ketone	2.16	79	2.46	90
nonene mixture	propylene trimer	1.64	48	2.12	62
trimethylpentenes mixture	diisobutylene	1.59	46	2.17	63

the addition of $HgSO_4$ in the first step, this dosage of Ag_2SO_4 already dissolved in H_2SO_4 saves time weighing separate 1 g portions.

In Table 21, the Chemical Oxygen Demand of a number of chemical compounds obtained with the two procedures are compared. The results are expressed as grams of oxygen uptake per gram of compound and as the percentage of the calculated theoretical complete oxidation to H_2O and CO_2 (assuming that the organic chlorine and bromine are converted to the respective halides). Likewise, in Table 22, COD results measured on petroleum spirits (solvents) with a boiling range of 80–110°C and of varying aromaticity are presented. The figures show that the degree of oxidation depends much on the aromatic content of the solvent.

Total Oxygen Demand (TOD). The Total Oxygen Demand method is based on the quantitative measurement of the amount of oxygen used to combust the impurities in a water sample.

TABLE 22. COD Measured on 80/110 Petroleum Spirits.

Aromaticity, %w	COD, g/g	
	ASTM procedure	Modified procedure
0	0.09	0.09
40	0.83	0.94
60	1.19	1.47
80	1.59	2.08

Ideally, to measure the ultimate oxygen demand, the complete combustion of all the oxidizable elements in the sample, is desired. Thus, it is desired that the following reactions take place since the end products listed represent the highest, stable oxidation states of these elements (or ions) in nature:

$$C_nH_{2m} + \left(n + \frac{m}{2}\right)O_2 \rightarrow nCO_2 + mH_2O$$

$$2H_2 + O_2 \rightarrow 2H_2O$$

$$2N^{-3} + 3O_2 \rightarrow 2NO_3^-$$

$$S^{--} + 2O_2 \rightarrow SO_4^{--}$$

$$SO_3^{--} + \tfrac{1}{2}O_2 \rightarrow SO_4^{--}$$

$$N_2 + O_2 \rightarrow \text{no reaction}$$

When interpreting TOD values, it should be kept in mind that not only hydrocarbons are oxidized but all oxidizable matter, as demonstrated by the foregoing reactions.

Total Organic Carbon (TOC). The instrument measuring the TOC of polluted water is very similar to the TOD analyzer, except for the last step. A microsample of the waste water to be analyzed is injected into a catalytic combustion tube which is enclosed by an electric furnace, thermostated at 950°C. The water is vaporized and the carbonaceous material is oxidized to carbon dioxide and steam. The carrier gas, CO_2, O_2, and water vapor enter an infrared analyzer sensitized to provide a measure of CO_2. The amount of CO_2 present is directly proportional to the concentration of carbonaceous material in the injected sample.

Theoretical Oxygen Demand (ThOD). The Theoretical Oxygen Demand is the amount of oxygen needed to oxidize hydrocarbons to carbon dioxide and water.

$$C_nH_{2m} + \left(n + \frac{m}{2}\right)O_2 \rightarrow nCO_2 + mH_2O$$

The ThOD of C_nH_{2m} equals

$$\frac{32\left(n + \frac{m}{2}\right)g}{(12n + 2m)g} = \frac{8(2n + m)}{6n + m}$$

When the organic molecule contains other elements, such as N, S, P, etc., the ThOD depends on the final oxidation stage of these elements. Most authors do not bother to define the ThOD which they mention in their publications; however, ThOD can easily be calculated.

Example: ammonium acetate: CH_3COONH_4; molecular weight: 77

a) if N remains as NH_4^+

$$CH_3COONH_4 + 2O_2 \rightarrow 2CO_2 + H_2O + NH_4^+ + OH^-$$

or

$$\text{ThOD} = \frac{64g}{77g} = 0.83$$

b) if N is oxidized to N_2

$$2CH_3COONH_4 + 5.5O_2 \rightarrow 4CO_2 + 7H_2O + N_2$$

or

$$\text{ThOD} = \frac{5.5 \times 32}{2 \times 77} = 1.14$$

c) if N is oxidized to NO_3^-

$$2CH_3COONH_4 + 7.5O_2 \rightarrow 4CO_2 + 5H_2O + 2NO_3^- + 2H_2^+$$

or

$$\text{ThOD} = \frac{7.5 \times 32}{2 \times 77} = 1.56$$

Expressing other parameters, such as BOD and COD, as a percentage of ThOD, can be highly misleading. In the standard BOD_5 day test, nitrification does not occur yet, and consequently the BOD_5 of compounds containing nitrogen can be expressed as a percentage of the ThOD, under the assumption that the nitrogen remains unchanged. In determining BOD with acclimated seed tested over a longer period, nitrification may occur. Hence, it alters the oxygen demand and also the ThOD, because N is oxidized to NO_3^-. When using ThOD values, the final oxidation stage of the elements other than C, H, and O should be mentioned.

3. Impact on Biodegradation Processes. Certain compounds are very toxic to microorganisms used in waste water treatment processes and in the BOD test, and inhibit, even at low concentrations, the biodegradation processes.

Microbial activities are associated with three major biogeochemical cycles (i.e., carbon, nitrogen, and sulfur cycling) and with decomposition of organic matter. Data obtained will provide preliminary indications of possible effects of the test chemical on the cycling of elements and nutrients in ecosystems. In addition, the tests will aid in the formation of hypotheses about the ecological effects of chemicals, hypotheses which can be used in the selection of higher level tests when appropriate.

Data on the effects of chemicals on microbial populations and functions can be obtained from laboratory studies employing nonradioisotopic analytical techniques. Studies of effects on microbial functions constitute a more direct approach and are preferred to studies of effects on populations. The activities to be observed are:

- organic matter (cellulose) decomposition by following CO_2 evolved from organically bound carbon
- nitrogen transformations by following the release of organically bound nitrogen (in urea) and the formation of ammonia

- sulfur transformation by following the reduction of sulfate to sulfide by *Desulfovibrio*.

3.1. Cellulose Decomposition. Cellulose is one of the most abundant organic materials in plants. Dead plant matter, consisting largely of cellulose, is degraded by a number of species of bacteria and fungi. This degradation is accomplished in a stepwise manner going from cellulose to cellobiose to glucose to organic acids and CO_2.

The degradation of cellulose is of particular interest to farmers. The fertility of farmland depends in part on the presence of organisms which degrade dead plant matter. The organisms which decompose the various forms of cellulose are some of the most important microorganisms contributing to humification processes in soils. The inhibition of, or interference with, cellulose degradation by toxic chemicals adversely impacts the recycling of carbon and soil fertility by retarding the breakdown of the vast amounts of cellulose that enter the soil.

There are two basic methods available for evaluating cellulose degradation by microorganisms. One method uses variations of soil burial tests in which a cellulose substrate is buried in soil containing a mixed culture that includes cellulolytic microorganisms. The other type employs a liquid culture of cellulolytic microorganisms and measures evolved CO_2.

One advantage of the liquid culture method over soil burial is time. Typical soil burial tests last for 7 to 8 weeks, while liquid culture requires only 7 days. In addition, test compounds in liquid culture are more apt to be in contact with the microorganisms than in soil burial tests because possible soil sorption of test substance is avoided. Pure culture studies in liquid medium also reduce the possibility of lowered toxicity due to biodegradation of the test material. Studies using pure cultures have the two advantages of simpler experimental procedures and greater reproducibility of data.

The major disadvantages of pure cultures are that they are a poor representation of the natural environment and that they fail to allow for synergistic and competing reactions. It could be argued, for example, that inhibition of, e.g., *Trichoderma longibrachiatum* should not be taken as evidence that the test compound will inhibit cellulose degradation since other organisms in the soil may be unaffected by the concentration of test material and may degrade cellulose and/or transform the test material. This argument has some validity and only additional testing with mixed cultures in natural soil samples would resolve that question.

3.2. Nitrogen Transformation. Almost all microorganisms, higher plants, and animals require combined nitrogen. In addition, the nitrogen cycle is a major biogeochemical cycle. The main aspects of this cycle are fixation of gaseous nitrogen, ammonification of organically bound nitrogen, nitrification, and denitrification.

Ammonification is a key initial step in the reintroduction of nitrogen from protein wastes into the soil and is one of the more readily measured reactions of the nitrogen cycle. As soon as an organism dies and its organic waste returns to the soil, biological decomposition begins and fixed nitrogen is released. The breakdown of proteins and other nitrogen-containing organics in soil and the production of ammonia are the work of widespread and varied microflora. The amino groups are split off to form ammonia in a series of reactions collectively called ammonification. Urea, a waste product found in urine, is also decomposed by numerous microorganisms with the formation of ammonia. This reaction can serve as a convenient assay method for ammonification activity. There is a strong correlation between an organism's ability to degrade urea and its capacity to degrade protein. Information from such testing would be used to assess the likelihood that the test chemical interferes with the normal conversion of organically bound nitrogen into ammonia.

Consideration of at least some aspects of the nitrogen cycle is essential for assessing the effects of substances on microorganisms. An easily measured feature of the nitrogen cycle is the oxidation of nitrite to nitrate by *Nitrobacter* bacteria. This focuses on a part of the nitrogen cycle less important than the critical step involving the conversion of organic nitrogen into ammonia. The method uses urea as a readily obtainable, reproducible nitrogenous organic compound. Some investigators have used pieces of liver or kidney tissue, vegetable meal, dried blood, and casein hydrolysate as nitrogen sources, but these substances are not standardizable. Percolation techniques may also be used for the same purposes in the study of soluble proteins, peptones, polypeptides, and amino acids in solution. Methodology using fertile soil and urea and following the evolution of ammonia nitrogen is relatively simple to conduct and will provide meaningful results. Urea is a suitable nitrogen source not only because it is readily available and relatively easy to handle, but in addition, the ability to degrade urea has been associated with general proteolytic capabilities.

Nitrification is one of the most sensitive conversions in the soil. Nitrification may be inhibited by concentrations of chemicals not inhibiting other important biochemical reactions, so there are reasons for choosing nitrification as the process which would give results most useful for assessment. One of the arguments against this is that ammonification is a more vital part of the nitrogen cycle than nitrification. In addition, the temporary inhibition of nitrate formation is often beneficial in that it slows down the loss of nitrogen from the root areas of plants by leaching and denitrification. Lastly, nitrification by *Nitrobacter* may be too sensitive to inhibition by chemicals in laboratory studies and might give results not representative of natural circumstances and environments. Although inhibition of nitrifying activity could lead to ammonium ion accumulation and serious problems such as root damage, ammonification is considered the more appropriate process to be examined as a first step in looking for effects of the nitrogen cycle.

3.3. Sulfur Transformation (Sulfate Reduction). The sulfur cycle is one of the major biogeochemical cycles. Sulfur is found in soil, waters, and the atmosphere in its most reduced form as a result of volcanic action and the decomposition of organic matter. Free sulfur occurs in some soils, and the oxidized form, sulfate, occurs in all soils. Both sulfur and hydrogen sulfide (H_2S) are oxidized by combustion and by microorganisms to sulfates which are used by plants. Sulfur is essential to all living organisms as a part of sulfur-containing amino acids. It is released from organic compounds (plant and animal wastes) by anaerobic decomposition, as H_2S, in its most reduced state. Sulfate-reducing bacteria are widely distributed in nature where anoxic conditions exist—as in sewage, sediments, muds, and in bovine rumina.

Two groups of bacteria are able to reduce sulfate to H_2S. The best known of these are the *Desulfovibrio* organisms. These organisms have been thoroughly studied because they are responsible for serious odor and corrosion problems associated with sulfate reduction. However, these problems tend to obscure the necessary and beneficial role played by *Desulfovibrio* in the sulfur cycle. *Desulfovibrio* plays an important role in the sulfur cycle and functions under conditions of low oxidation-reduction potential (anaerobic conditions) and provides some indication of potentially adverse effects under such conditions.

Standard methods for the determination of sulfate-reducer activity are described in Standard Methods (APHA, 1975) and by the American Society for Testing and Materials (1978).

4. Waste Water Treatment

4.1. Biological Oxidation. In order to describe the basic investigation procedures and results for a wide variety of biological test methods in a compact way, the information is presented in columns, using the following column headings: methods; temperature °C; days observed; feed mg/l; days acclimation; % removed; % theoretical oxidation. The methods column describes experimental procedures and their sequence. For example, acclimation may have been achieved in activated sludge with assimilation or oxidation measured by respirometer, BOD, or other methods. Temperature and observation time are given as precisely as possible. Two entries in the days of observation column indicate the time in an oxidation unit such as activated sludge and corresponding time for incubation of the BOD test. Feed concentration (mg/l) occasionally was indefinite. The reference may have indicated 500 mg/l feed concentration with performance tests by BOD technique. Obviously, the BOD test bottle did not always use the cited concentration and the percent of theoretical oxidation could be altered by concentration differences. A question mark indicates this situation. Days of acclimation entered include the time prior to test runs but not necessarily the minimum requirements. Performance estimates are entered in two columns. Percent of theoretical oxidation is included where oxygen utilization under specified conditions was measured. A difference between the influent and effluent is shown by percent removed. Percent of theoretical oxygen demand was obtained from a ratio of the observed oxygen demand divided by the calculated amount of oxygen required to convert the chemical to carbon dioxide, water, and ammonia (carbonaceous). Additional oxygen requirements to convert ammonia to nitrate were not used in the calculation of the theoretical oxygen demand unless chemical analysis of nitrogen (CAN) was indicated. Percent of theory is used to facilitate comparisons in oxidation of different compounds because each chemical has a different oxygen requirement for stabilization. More than one reference citation for a single entry indicates similar methods and results but not always precise replication. Significant differences are reported in separate entries.

Biological treatment performance is difficult to evaluate without information on both percent oxidation and percent removed. A characteristic balance of time, oxidation, and biosorption is necessary for effective continuous treatment. Percent oxidation shows the degradation within the test period. It rarely exceeds 60% of the influent oxygen demand within the detention period of an activated sludge or other biological treatment unit. Percent removal may be 90 or more because biosorption retains a significant fraction of the remaining load. The retained material is subject to further oxidation in process; hence, it is not likely to degrade the effluent.

4.2. Stabilization Ponds. The principle design parameter is the first-order BOD removal coefficient, K. The evaluation of K is the key to the whole design process. The following method has been proposed by Dhandapani Thirumurthi and is based on the following definitions:

Standard BOD removal coefficient Ks—a constant and standard value of Ks has been chosen that corresponds to an arbitrarily selected standard environment. Under these standard environmental conditions, a stabilization pond will perform with the BOD removal coefficient, Ks. The standard environment consists of a. a pond temperature of $20°C$, b. an organic load of 60 lb/day/acre (672 kg/day/ha), c. absence of industrial toxic chemicals, d. minimum (visible) solar energy at the rate of 100 langleys/day, and e. absence of benthal load.

Design BOD removal coefficient, K—design coefficient K corresponds to the actual environment surrounding the pond. Hence, the value of K will be used when a pond is being designed. When the critical environmental conditions deviate from one or more of the defined standard environmental conditions, suitable correction factors must be incorporated:

$$K = K_s C_{Te} C_o C_{Tox}$$

where C = correction factor
Te = correction for temperature
o = correction for organic load, and
Tox = correction for industrial toxic chemicals

In the absence of industrial wastes, the factor C_{Tox} will equal unity. Laboratory investigations indicated that, in the presence of certain industrial organic compounds, green algae could not synthesize chlorophyll pigment. Without chlorophyll production, photosynthesis cannot be sustained by algae and therefore, oxygen production will stop. The resultant decrease in dissolved oxygen (DO) concentration in the pond will result in a drastic reduction in BOD removal efficiency. Based on this observation, C_{Tox} values have been calculated for various concentrations of selected organic chemicals. Ponds designed to treat toxic industrial wastes will perform at lower efficiencies and, thus, C_{Tox} values must be determined by laboratory investigations or by previous field experience.

4.3. The Activity of Mutant Microorganisms. In the normal biological cycle adaption and mutation are constantly taking place, permitting the survival of the participating microorganisms. Many of the present day organic toxicants with which we are contaminating our environment contain a large number of carbon-chlorine bonds in addition to other substitution groups which, due to molecular configuration and complexity, do not permit the process of adaption to proceed at a normal rate. In the effort to overcome this obvious biochemical disability, it was decided to obtain soil and marine samples from various parts of the world to isolate and study the naturally occurring microorganisms. The numerous isolates were tested to determine their ability to degrade low levels of various synthetic organic substrates such as aryl halides, aryl and alkyl amines, halogenated phenols, inorganic and organic cyanides, halogenated insecticides, halogenated herbicides, and various synthetic surfactants. These various organic chemicals were selected for biodegradation studies because they are high on the list of toxic recalcitrant molecules with which we are defiling our environment. After determining the capability of the various microorganisms to degrade selected organic molecules, tests were run to determine the maximum toxicant concentration to which they would adapt. Increasing concentrations of the challenging organic molecules were added to the growing cultures over a period of 21 days to determine the maximum level of tolerance to which the biomass would adapt. Upon completion of the adaption process, the microorganisms were exposed to programmed radiation to develop mutants with advanced biochemical capabilities. From the several thousand mutants obtained, 397 were isolated that were outstanding in their ability to degrade various types of inorganic and organic compounds. These included: 180 pseudomonas, 45 nocardia, 102 streptomyces, 15 flavobacterium, 12 mycobacterium, 14 aerobacters, 14 achromobacters, 10 vibrio, and 10 micrococcus. Activated sludges were prepared from the various mutant microorganisms and the maximum tolerated dose of toxicant, as determined by previous tests, was added to determine the maximum velocity and degree of molecular dismutation. It has been fairly well established that none of the highly halogenated organic compounds are used by microorganisms as a source of metabolic carbon. Degradation or molecular change has been found to occur most readily at levels of maximum metabolic activity. It is quite apparent that the observed changes are enzymatic and relate more to detoxification than to the use of highly halogenated organic compounds in metabolic processes. It is possible that the biodegradation observed in microbial sludges is caused essentially by extracellular enzymes which would cause a desorption of the more soluble degradation intermediates.

4.4. Solvent Extraction: Distribution Coefficient. Water soluble organic compounds can be extracted from water by solvents which are much less soluble in water. The ratio of the concentration of the compound in the solvent to the concentration of the compound in the water, after a sufficient contact time, is called the *distribution coefficient*.

4.5. Stripping. Air-stripping has been demonstrated to be a feasible technique for removing a portion of the organics from wastewater. These operations are typically carried out in a cooling tower type device (induced or forced draft) or in an air sparged (bubble) vessel. Air stripping occurs to varying degrees in conventional waste water treatment techniques such as dissolved or dispersed air flotation and of course aerobic biodegradation. An analytical simulation model of desorption in aerated stabilization basins (1811) indicates that significant removals of selected industrial chemicals are occurring.

A study of ten common industrial chemicals in eleven full-scale aerated basins showed that 20–60% removal efficiencies were possible without biochemical oxidation. Detention times ranged from 1.7 to 14.2 days. Laboratory observation of surface agitator desorbers support these data.

Equations have been derived to predict the evaporation rates of hydrocarbons and chlorinated hydrocarbons. These compounds have high rates of evaporation even though the vapor pressure is low. Evaporation "half lives" of minutes and hours are due to the high activity coefficient displayed by these components in aqueous solution.

From the above it appears that a significant amount of volatile components are being stripped and/or desorbed in conventional secondary treatment operations involving the use of air or the presence of large air-water interfaces.

Fundamental desorption concepts. The volatile character of dissolved constituents in wastewater can be adequately quantified by the experimental determination of two parameters:

1. *Volatile fraction* (F_v). This measure denotes the maximum amount (in%) of the original organic pollutants in a water sample that can be removed by air contact. This is also the maximum efficiency of treatment that can be achieved by stripping the wastewater with air.

The organic pollutants in the wastewater can be expressed as BOD, COD, TOC and other gross pollutant measures and/or concentration of individual constituents.

2. *Relative volatilization rate* (K/a). This measure denotes the ratio of the rate of volatile removal by air contact to the rate water is evaporated in the same apparatus. If the experimental value of K/a is greater than unity, stripping with air may be a feasible treatment operation for this wastewater.

If K/a is unity, stripping will have no effect on removing volatile constituents from this wastewater and if K/a is less than unity or zero, stripping will result in an increase of this constituent in the wastewater. This parameter, like F_v, is dimensionless and both are determined from a single desorption experiment.

L. J. Thibodeaux (1812) used a desorption apparatus consisting of a packed column (Fig. 33) and developed a mathematical model which, if the concentration of water is much greater than the concentration of the pollutant, yields the following solution

$$\log X_t = \log X_0 - (K/a - 1) \log \frac{M_0}{M_t}$$

where X_t = concentration of pollutant in the desorption apparatus at sample time t (mg/l)
X_0 = concentration of pollutant in the desorption apparatus at start of experiment (mg/l)
M_t = quantity of water in the desorption apparatus at sample time t (g)
M_0 = quantity of water in the desorption apparatus at start of experiment (g)
K = the specific desorption rate of the organic component in the packed column
a = the specific desorption rate of water component in the packed column

The relative volatilization rate K/a can be obtained graphically be preparing a log-log plot of X_t and M_0/M_t. The slope of a straight line through these raw data points yields ($K/a - 1$)

The following relative volatilization rates were obtained for pure compounds at room temperature and initial concentrations of approx. 100–1000 ppm:

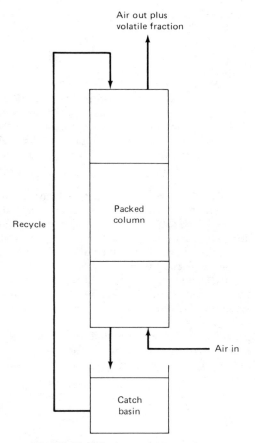

Fig. 33. Volatile desorption apparatus.

	Experimental K/a
methanol	7–13
i-propanol	3–15
n-propanol	12–23
n-butanol	10–18
formic acid	4–7
acetic acid	1–5
propionic acid	0–1.5
acetaldehyde	171–207
acetone	36–100

benzene	107
phenol	0.5–1.9
furfural	10
sucrose	0.26

From these experiments it appears that formic acid, acetic acid, propionic acid and phenol have K/a values which are not significantly different from 1 and therefore these compounds will not be removed significantly by air stripping. Acetaldehyde, acetone and benzene show high relative volatilization rates and can thus be removed by stripping. Sucrose on the contrary concentrates in the solution upon stripping, as can be deducted from its K/a value of 0.26.

4.6. Adsorption. The affinity of a chemical substance for particulate surfaces is an important factor affecting its environmental movement and ultimate fate. Chemicals that adsorb tightly may be less subject to environmental transport in the gaseous phase or in solution. However, chemicals that adsorb tightly to soil particles may accumulate in that compartment. Substances which are not tightly adsorbed can transport through soils, aquatic systems and the atmosphere.

Adsorption isotherms. The experimental information is usually expressed as an *adsorption isotherm*. The adsorption isotherm is the relationship, at a given temperature, between the amount of the substance adsorbed and its concentration in the surrounding solution.

In very dilute solutions, such as encountered in the environment or in wastewaters, a logarithmic isotherm plotting usually gives a straight line. In this connection, a useful formula is the Freundlich equation, which relates the amount of impurity in the solution to that adsorbed as follows:

$$\frac{x}{m} = KC^{1/n}$$

where x = amount of substance adsorbed
m = weight of the substrate
x/m = amount of substance adsorbed per unit weight of substrate K and n are constants
C = unadsorbed concentration of substance left in solution or, in logarithmic form,

$$\log x/m = \log k + 1/n \log C$$

in which $1/n$ represents the slope of the straight line isotherm (Fig. 34). Isotherm tests also afford a convenient means of studying the effects of different adsorbents and the effects of pH and temperature.

Isotherm tests are also important in water purification by activated carbon. From an isotherm test it can be determined whether or not a particular purification can be effected. It will also show the approximate capacity of the carbon for the application, and provide a rough estimate of the carbon dosage required.

The inherent adsorbability of a chemical in pure component tests does not necessarily predict its degree of removal from a dynamic, multicomponent mixture. How-

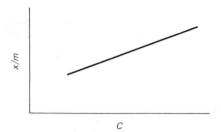

Fig. 34. Freundlich isotherm.

ever, pure component data do serve as a useful background for understanding why multicomponent interactions occur. Tests with individual solutions showed that aromatic compounds, even those containing polar substituent, were quite amenable to adsorption (region I in Fig. 35), while substituted aliphatic compounds (alcohols, amines, aldehydes, ketones, carboxylic acids) decreased sharply in amenability to adsorption with decreasing molecular weight (region II in Fig. 35). The lower envelope (region III in Fig. 35) shows the relatively poor adsorption of glycols and glycol ethers (ethylene glycol; di-, tri-, and tetraethylene glycol; various glycol adducts of low-molecular-weight alcohols) (1935).

The adsorbability of a chemical can be expressed as a fraction (or percentage) of the amount adsorbed or as the distribution coefficient between the substrate and the solution, as is illustrated in Fig. 35 and in Table 23.

The activated carbon/water distribution coefficient ($P_\text{activated carbon}$ or P_AC) is the ratio between the amount of substance adsorbed per unit of substrate (x/m) (e.g., mg substance /kg activated carbon) and the unadsorbed concentration of substance left in the solution (c):

$$P_\text{AC} = \frac{x/m}{c}$$

or

$$\log P_\text{AC} = \log \frac{x/m}{c}$$

Table 23 and Fig. 36 show clearly that as the solubility decreases, there is a corresponding increase in amenability to adsorption. Relationships between aqueous solubility and adsorbability on activated carbon is graphically presented in Figs. 37–39 for series of homologous compounds.

The magnitude of adsorption of a compound onto a substrate or distribution between two immiscible liquids or between a liquid and a gas tells something about the escaping tendency of the compound from its original solution. The aqueous solubility itself is also a measure of the escaping tendency of a compound from the aqueous phase. In Subsections 4 and 5 of Section A of these notes we have seen that solubility is determined in the first place by the molecular structure, but is affected by the characteristics of the aqueous phase such as temperature, pH, and composition (salt

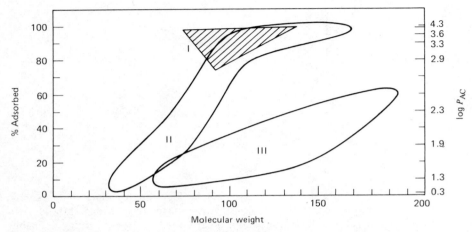

Fig. 35. Adsorption of chemicals on activated carbon. Region I, aromatic compounds. Region II, aliphatic O- and/or N-containing compounds. Region III, glycols and glycol ethers (polyglycols). Solutions of 1 g/l or at solubility limit if <1 g/l. Dosages of powdered activated carbon: 5 g/l. log P_{AC} = logarithm of distribution coefficient activated carbon/water.

content). The equation log P = 6 − 0.66 log S_{ppb} has been used extensively in Subsection 5 of Section A to indicate the relationship between solubility and maximum accumulation in sediments, in fish, or in an ecological system. The slope of the line is −0.66, which means that the logarithm of the distribution coefficient increases by 0.66 for each decrease of the logarithm of the solubility by 1. The slope of this equation gives the relative impact of solubility on the above-mentioned distribution processes.

Accumulation on a substrate such as activated carbon is not only affected by the composition of the solute but also by the characteristics of the carbon such as microstructure and carbon surface chemistry. Moreover the adsorbability is a function of the concentration of the substance adsorbed as expressed by the Freundlich isotherm. Relationships between adsorbabilities and physical properties of compounds are therefore less evident. On Fig. 36 an attempt has been made to show the relationship between the adsorbability on activated carbon and aqueous solubility for 165 organic compounds of most different chemical classes, as listed in Table 23 (from Refs. 32 and 1936). For most of the compounds a range of adsorbabilities is given, represented by the vertical lines. The higher adsorption values have most often been measured at the lowest residual substance concentration (i.e., at highest carbon dosage for the same initial substance concentration). Compounds which were completely miscible in water were for this exercise arbitrarily given a maximum solubility of 1 kg/l (or 10^9 ppm), neglecting herewith the specific gravity.

From Fig. 36 it can be seen that the line represented by the equation

$$\log P_{AC} = 8 - 0.66 \log S_{ppb}$$

gives a good fit between the solubilities and the lower adsorption values. From this

TABLE 23. Accumulation of Organic Compounds on Activated Carbon (32, 1936).

Compound	$\log P_{AC}$	pH	$\log S_{ppb}$
acetates and acrylates			
methylacetate	1.85	–	8.5
ethylacetate	2.31	–	7.9
propylacetate	2.78	–	7.3
butylacetate	3.04	–	6.8
prim. amylacetate	3.17	–	6.3
isopropylacetate	2.63	–	7.5
isobutylacetate	2.96	–	6.8
vinylacetate	2.56	–	7.4
ethyleneglycolmonoethylether acetate	2.59	–	8.4
ethylacrylate	2.84	–	7.3
butylacrylate	3.65	–	6.3
alcoholamines			
monoethanolamine	1.20	–	9
diethanolamine	1.90	–	8.9
triethanolamine	2.00	–	9
monoisopropanolamine	1.70	–	9
diisopropanolamine	2.22	–	8.9
alcohols			
methanol	0.86	–	9
ethanol	1.35	–	9
	0.9–2.4	5.3	
propanol	1.67	–	9
butanol	2.36	–	7.9
n-amylalcohol	2.74	–	7.2
n-hexanol	3.63	–	6.8
isopropanol	1.46	–	9
allylalcohol	1.48	–	9
isobutanol	2.16	–	7.9
t-butanol	1.92	–	9
2-ethylbutanol	3.07	–	6.6
2-hexylethanol	4.14	–	5.8
aldehydes			
formaldehyde	1.30	–	9
acetaldehyde	1.40	–	9
propionaldehyde	1.90	–	8.3
butyraldehyde	2.35	–	7.9
acrolein	1.94	–	8.3
	2.5–2.8	5.2	8.6
crotonaldehyde	2.23	–	8.2
benzaldehyde	3.50	–	6.5
paraldehyde	2.75	–	8.0

TABLE 23. (Continued.)

Compound	log P_{AC}	pH	log S_{ppb}
amines			
di-N-propylamine	2.94	–	9
butylamine	2.33	–	9
di-N-butylamine	3.13	–	9
allylamine	1.96	–	9
ethylenediamine	1.37	–	9
diethylenediamine	1.94	–	9
hexamethylenediamine	<2.0	–	–
aniline	2.78	–	7.5
o-anisidine	4.0–4.8	–	–
p-nitroaniline	4.5–6.7	7.0	–
4,4′-methylene-bis-(2-chloroaniline)	5.2–5.9	7.5	–
cyclohexylamine	<2.0	–	–
α-naphtylamine	4.6–6.1	7.0	–
β-naphtylamine	5.1–5.9	7.5	–
diphenylamine	4.4–6.1	7.0	–
benzidine dihydrochloride	4.8–6.2	–	–
3,3-dichlorobenzidine	5.3–7.2	7.2	–
N-nitroso-di-n-propylamine	3.5–4.5	5.3	7.0
N-nitrosodiphenylamine	4.8–6.3	7.0	–
aromatics			
benzene	3.3–3.7	5.3	6.2
	3.58	–	5.8
toluene	3.7–4.3	5.6	5.7
	2.88	–	5.7
ethylbenzene	4.3–4.7	7.3	5.2
	3.02	–	5.3
p-xylene	4.0–4.7	7.3	5.3
styrene	4.9–5.4	7.0	5.5
	3.19		
naphtalene	4.7–5.5	5.6	4.5
acenaphtene	4.9–6.2	5.3	3.5
acenaphtylene	5.3–6.9	5.3	3.6
fluorene	5.4–7.4	5.3	3.3
anthracene	6.2–6.8	5.3	3.0
phenanthrene	5.8–6.8	5.3	3.2
fluoranthene	6.5–7.0	5.3	2.4
3,4-benzofluoranthene	6.9–7.6	–	0.1
dibenzo(a,h)anthracene	5.1–5.7	7.1	–0.3
benzo(a)pyrene	6.5–7.0	–	0.5
benzo(ghi)perylene	6.3–7.0	–	–0.7
4-aminodiphenyl	4.6–6.7	–	–
4-nitrodiphenyl	5.5–6.7	7.0	–
polychlorinated biphenyl 1221	5.5–6.3	5.3	4.0
polychlorinated biphenyl 1232	5.9–6.9	5.3	3.0

TABLE 23. (Continued.)

Compound	log P_{AC}	pH	log S_{ppb}
chlorinated benzenes			
benzene	3.3–3.7	5.3	6.3
chlorobenzene	4.8–5.0	7.0	5.7
1,2-dichlorobenzene	4.6–5.2	5.5	5.2
1,3-dichlorobenzene	4.2–5.0	5.1	5.0
1,4-dichlorobenzene	4.9–5.4	5.1	4.9
1,2,4-trichlorobenzene	4.4–5.3	5.3	4.5
hexachlorobenzene	4.0–4.3	5.3	0.8
chlorinated pesticides			
α-endosulfan	5.7–8.9	5.3	2.7
β-endosulfan	5.8–6.3	5.3	2.4
endosulfan sulfate	6.0–6.6	5.3	–
endrin	5.9–6.3	5.3	2.4
aldrin	5.9–6.1	5.3	1.0
dieldrin	6.1–7.2	5.3	2.3
chlordane	5.8–7.3	5.3	1.7–3.3
heptachlor	6.0–6.4	5.3	1.7
heptachlorepoxide	6.4–7.0	5.3	2.5
DDE	7.7–9.0	5.3	1.0
DDT	7.4–8.2	5.3	0.3
ethers			
isopropylether	2.90	–	7.1
butylether	4.59	–	5.5
bis(2-chloroethyl)ether	3.1–3.6	5.3	7.0
bis(2-chloroisopropyl)ether	3.7–3.9	5.4	6.2
2-chloroethylvinylether	3.4–3.5	5.3	7.2
4-chlorophenylphenylether	5.2–5.5	5.3	4.8
4-bromophenylphenylether	5.1–5.4	5.3	4.6
glycols and glycolethers			
ethyleneglycol	1.16	–	9
diethyleneglycol	1.86	–	9
triethyleneglycol	2.34	–	9
tetraethyleneglycol	2.44	–	9
propyleneglycol	1.43	–	9
dipropyleneglycol	1.60	–	9
hexyleneglycol	2.50	–	9
ethyleneglycolmonomethylether	1.50	–	9
ethyleneglycolmonoethylether	1.95	–	9
ethyleneglycolmonobutylether	2.40	–	9
ethyleneglycolmonohexylether	3.13	–	7.0
diethyleneglycolmonoethylether	2.18	–	9
diethyleneglycolmonobutylether	2.98	–	9
ethoxytriglycol	2.63	–	9

TABLE 23. (Continued.)

Compound	log P_{AC}	pH	log S_{ppb}
halogenated paraffins and olefins			
chloroform	3.4–3.8	5.3	6.9
trichlorofluoromethane	3.4–4.7	5.3	–
carbon tetrachloride	4.2–4.4	5.3	5.9
dichlorobromomethane	3.9–4.7	5.3	–
dibromochloromethane	3.7–5.7	5.3	–
bromoform	4.4–5.1	5.3	6.1
methylenechloride	2.9–3.0	5.8	7.3
chloroethane	2.8	5.3	6.8
1,1-dichloroethane	3.4–4.0	5.3	6.7
1,2-dichloroethane	3.7–3.9	5.3	6.9
	2.94	–	6.9
1,1,1-trichloroethane	3.5–5.0	5.3	6.6
1,1,2-trichloroethane	4.1–4.6	5.3	6.7
1,1,2,2-tetrachloroethane	4.2–5.5	5.3	6.5
hexachloroethane	5.0–6.7	5.3	4.7
1,1-dichloroethene	3.8–4.6	5.3	5.6
1,2-*trans*-dichloroethene	3.5–4.6	6.7	5.8
trichloroethene	4.7–5.2	5.3	–
tetrachloroethene	5.0–5.7	5.3	5.2
dichloropropane	3.7–4.4	5.3	6.4
	3.41	–	6.5
1,2-dibromo-3-chloropropane	5.0–5.9	5.3	–
1,2-dichloropropene	4.0–4.9	5.3	–
ketones			
acetone	1.74	–	9
methylethylketone	2.25	–	8.4
methylpropylketone	2.66	–	7.6
methylbutylketone	2.92	–	?
methylisobutylketone	3.05	–	7.3
methylisoamylketone	3.06	–	6.7
diisobutylketone	>4.78	–	5.7
cyclohexanone	2.61	–	7.4
acetophenone	3.84	–	6.7
	4.4–5.1	–	6.7
isophorone	3.75	–	7.1
	3.8–5.0	5.5	7.1
morpholines			
N-methylmorpholine	2.17	–	9
N-ethylmorpholine	2.36	–	9
organic acids			
formic acid	1.79	–	9
acetic acid	1.80	–	9
propionic acid	1.98	–	9

TABLE 23. (Continued.)

Compound	log P_{AC}	pH	log S_{ppb}
butyric acid	2.47	–	9
valeric acid	2.98	–	7.4
caproic acid	3.81	–	7.0
acrylic acid	2.56	–	9
benzoic acid	3.31	–	6.5
	3.6–3.8	–	6.4
oxides			
propylene oxide	1.85	–	8.6
styrene oxide	3.61	–	6.5
phenols			
phenol	3.8–4.3	7.0	7.8
2-chlorophenol	4.1–4.6	5.8	7.5
2,4-dichlorophenol	4.1–6.5	5.3	6.7
2,4,6-trichlorophenol	4.5–5.2	6.0	5.9
pentachlorophenol	4.6–5.5	7.0	4.1
2,4-dimethylphenol	4.1–5.2	5.8	7.2
p-nonylphenol	5.1–6.5	7.0	–
4-chloro-3-methylphenol	4.2–5.6	5.5	6.6
4,6-dinitro-2-methylphenol	4.4–5.7	5.2	5.4
p-hydroxyphenol	4.3–5.2	3.0	–
	3.00	–	7.8
2-nitrophenol	4.2–5.5	5.5	6.3
4-nitrophenol	4.0–5.4	5.4	7.2
2,4-dinitrophenol	4.1–4.6	7.0	6.7
α-naphthol	4.6–5.2	7.0	–
β-naphthol	4.6–5.4	7.0	–
phthalates			
dimethylphthalate	4.5–5.7	7.0	6.6
diethylphthalate	4.1–5.0	5.4	6.0
di-n-butylphthalate	5.1–5.5	3.0	6.7
bis(2-ethylhexyl)phthalate	5.6–5.9	5.3	4.7
butylbenzylphthalate	5.6–6.1	5.3	–
pyridines			
pyridine	2.26	–	9
2-methyl-5-ethylpyridine	3.22	–	?
uracils			
uracil	3.8	7.0	–
5-fluorouracil	3.7–3.9	7.0	–
5-chlorouracil	4.0–4.3	7.0	–
5-bromouracil	4.2–4.8	7.0	–

TABLE 23. (Continued.)

Compound	log P_{AC}	pH	log S_{ppb}
miscellaneous			
2-acetylaminofluorene	4.6-6.8	–	–
acridine orange	4.7-6.2	–	–
acridine yellow	4.6-6.2	–	–
acrylonitrile	2.3-2.9	5.3	7.9
adenine	4.3-5.1	–	–
adipic acid	3.7-3.8	3.0	7.2
anethole	5.0-6.2	–	–
benzothiazole	4.6-6.1	–	–
α-BHC	5.9-7.4	5.4	3.3
β-BHC	6.0-7.0	5.4	2.3
γ-BHC	5.5-6.5	5.3	3.8
bis(2-chloroethoxymethane)	3.5-4.0	5.3	7.9
2-chloronaphthalene	5.2-6.2	5.5	3.8
1-chloro-2-nitrobenzene	4.7-5.6	7.0	–
cholinechloride	<2.0	–	–
cyclohexanone	3.5-3.8	7.3	–
cytosine	3.5-3.6	7.0	–
dimethylaminoazobenzene	5.9-7.2	7.0	–
N-dimethylnitrosamine	3.2-4.3	7.5	–
dimethylphenylcarbinol	4.8-6.5	7.0	–
2,4-dinitrotoluene	4.7-5.7	5.4	5.4
2,6-dinitrotoluene	4.5-5.7	5.4	–
1,1-diphenylhydrazine	4.3-5.3	7.5	–
1,2-diphenylhydrazine	5.9-6.6	5.3	5.3
ethylenediaminetetraacetic acid	3.4-3.8	7.0	–
guanine	4.5-5.4	7.0	–
hexachlorobutadiene	6.0-6.9	5.3	3.3
hexachlorocyclopentadiene	6.0-6.7	5.3	2.9
nitrobenzene	5.3-6.6 3.65	7.5	6.3
phenylmercuric acetate	5.2-6.3	7.0	–
1,2,3,4-tetrahydronaphthalene	4.6-5.0	7.36	–
thymine	4.0-4.4	7.0	–

analysis we conclude that the lower adsorption values are 2 orders of magnitude higher than could be predicted from accumulation through partitioning processes, i.e., from the equation $\log P = 6 - 0.66 \log S_{ppb}$. The slope of the line, however, has remained the same (-0.66), which means that the relative effect of solubility on this adsorption process has not been affected.

For compounds which are completely miscible with water, no correlation between solubility and adsorbability can be drawn. Molecular weight and functional groups will generally determine the adsorbability, as shown in Fig. 40.

5. *Alteration and Degradation Processes.* The alteration and degradation processes can be divided into three categories:

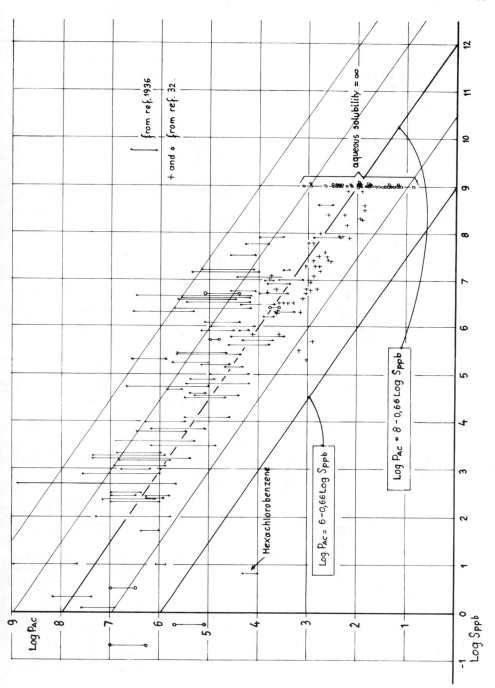

Fig. 36. Relationship between aqueous solubility and adsorbability on activated carbon for organic compounds.

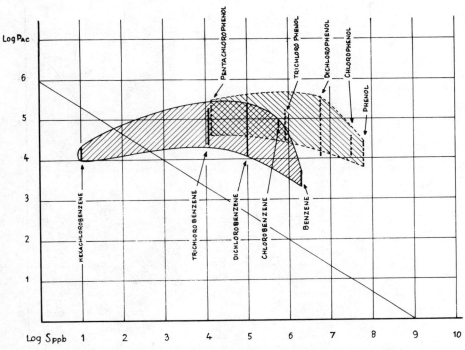

Fig. 37. Relationship between adsorption on activated carbon and aqueous solubility of phenol, benzene, and their chlorinated derivatives.

1. *biodegradation*, effected by living organisms
2. *photochemical degradation*, i.e., nonmetabolic degradation requiring light energy
3. degradation by chemical agents (*chemical degradation*), i.e., nonmetabolic degradation which does not require sunlight

Biodegradation of organic compounds appears to be the most desirable because it results generally in completely mineralized end-products. In contrast, photochemical and other nonmetabolic processes usually result in only slight modifications in the parent compound.

Hydrolysis of environmental compounds by chemical agents has been extensively studied and correlation between laboratory and field results is facilitated by the ease of measuring one of the more important rate-determining factors, pH, both in laboratory and in field.

Chemical processes in soil can be studied after sterilization of the soil by autoclaving, chemical treatment, or γ-irradiation. The sterilization processes often alter the soil to such an extent that any process observed could be artificial. Nevertheless, it should be noted that any biodegradation technique where the test chemical is incubated with a natural medium (e.g., soil or water) would allow transformations effected by chemical

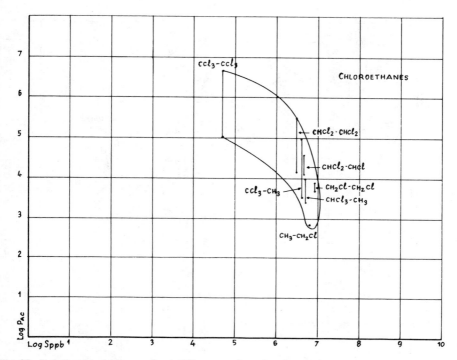

Fig. 38. Relationship between adsorbability on activated carbon and aqueous solubility for chloroethanes.

agents to take place; therefore, these processes would be partially considered during biodegradation testing.

5.1. Soil Degradation Studies. The low concentrations usually studied and the interferences introduced in many analytical procedures by naturally occurring materials in soil make radiotracer techniques almost a requirements.

The key factors affecting the degradation of synthetic chemicals in soil appear to be:

- soil type
- depth of soil
- test chemical concentration
- soil microorganisms and acclimation
- physical soil properties (including pH, temperature, O_2 availability, redox potential, and moisture content).

Many of these factors are similar to those previously mentioned for water.

Soil degradation of chemicals has been studied in a number of apparatuses. The two methods recommended by EPA are the *closed (biometer flask)* and *open systems*. In

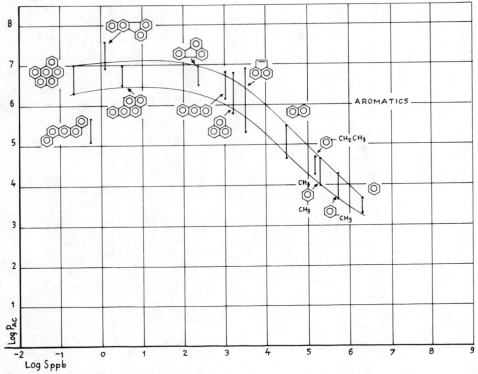

Fig. 39. Relationship between adsorbability on activated carbon and aqueous solubility for aromatics.

the open system, a multipurpose manifold assembly, which allows a constant level of oxygen to be maintained in the decomposition chambers, is used and traps for volatile products can be incorporated. In the biometer flask approach, soil and test chemical are incubated in a flask that has a NaOH side arm for collecting $^{14}CO_2$.

The soil is air dried and screened before being placed in the reaction flask. After treatment with the test chemical, the soil is brought to 75% of field capacity and incubated. Under aerobic conditions, the rate of degradation is followed by $^{14}CO_2$ evolution and the soil is extracted in a sequence of non-polar to polar solvents in order to characterize metabolic products.

To study breakdown under anaerobic conditions, the flask is evacuated and flushed with nitrogen several times and then flooded with a water barrier to protect against oxygen leakage. Under anaerobic conditions, methane gas, one of the major degradation products, can be extracted and analyzed, but it is much less convenient to follow than carbon dioxide.

The multipurpose manifold assembly has the advantage that the effect of different gaseous environments can be studied, but the drawback is that it is very complex for multiflask experiments. The biometer flask is simple and convenient for multiflask experiments.

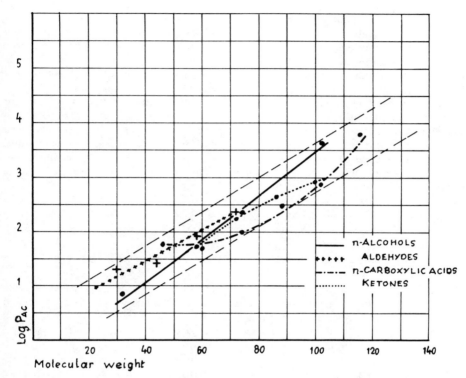

Fig. 40. Relationship between molecular weight and adsorbability on activated carbon for low-molecular-weight organic compounds.

As was the case with water degradation studies, the degradation pathways in soil are determined in two ways:

1. isolating stable intermediates in soil usually by use of the radiolabeled compounds
2. determining all the steps of the degradation using pure cultures and cell-free extracts.

The advantages and disadvantages of the two methods are basically the same as with the water systems.

5.2. Photochemical Degradation. Chemicals introduced into aqueous media in the environment can undergo transformation by direct photolysis in sunlight into new chemicals with different properties than their percursors. Data on direct photolysis rate constants and half-lives establish the importance of direct photolysis in sunlight as a dominant transformation of chemical in aqueous media.

Although numerous papers have been published on the photolysis of chemicals in solution, rate constants for direct photolysis of chemicals in water under environmental conditions (i.e., in sunlight) have emerged only in the last few years. Zepp and Cline

(1948) published a paper on photolysis in aquatic environments with equations for direct photolysis rates in sunlight. These equations translate readily obtained laboratory data into rate constants and half-lives for photolysis in sunlight. Rate constants and half-lives can be calculated as functions of season, latitude, time-of-day, depth in water bodies, and the ozone layer. Several published papers concerning the photolysis of chemicals in the presence of sunlight verify this method.

The soil and water media do not provide as good a matrix as the atmosphere for photochemical alterations because of the attenuation of the incident light. Nevertheless, photochemical processes on soil or vegetable surfaces and in water have been shown to be important with some compounds and, therefore, experimental procedures have been developed to study these processes. A number of different procedures can be used to study photochemical reactions. The selections that have to be made include:

- light source and apparatus for irradiation
- media of reaction—solution (choice of solvent), solid, adsorbed on another solid, thin film
- effect of environmental parameters—add sensitizer, effect of oxygen, pH

The use of sunlight as the irradiation source has been a common experimental technique. However, the inherent variation of sunlight in both wavelength distribution and intensity results in poor reproducibility, inconvenience, and lack of experimental control, and therefore many researchers have resorted to artificial sources (such as medium to low pressure mercury lamps, fluorescent lamps).

Photochemical energy reaches the earth from the sun in the form of photons with wave lengths covering the spectra from infrared to the far ultraviolet, including the visible. The principal energy sources of photons for environmental degradation reactions are those in the ultraviolet (UV) region. The UV radiation which reaches the earth's surface represents a wavelength range between 2900 and 4500 Å. Shorter wavelength UV, with high energy, is prevented from reaching the earth's surface by absorption in the upper atmosphere.

PNA's are prime candidates for photochemical degradation, since they have high absorptivities in the environmental UV range. Some typical data for molar absorptivities at principal bands are the following:

Compound	Maxima Å	Molar absorptivity	Maxima Å	Molar absorptivity
anthracene	3560	8.5×10^3	3750	8.6×10^3
1,2-dibenzanthracene	2875	9.0×10^4	3410	7.1×10^3
chrysene	3200	1.2×10^4	3610	6.9×10^2

Monoaromatics have much lower molar absorptivities in this range, generally in the region of 10^2-10^3 liters per mole-cm. Consequently, PNA's should more readily absorb energetic radiation to put the molecule into a higher energy state for oxidative and other environmental reactions.

Studies made on benzo(a)pyrene show it to be particularly sensitive to light. Under white fluorescent light, the photooxidation of benzo(a)pyrene absorbed onto calcite suspended in water followed a first-order rate. One would anticipate that in a natural aquatic system, photooxidation rates would be greatest in shallow systems or in the upper layers of deeper waters. A report of the Stanford Research Institute gives a photooxidation half life for benzo(a)pyrene in a lake environment of 7.5 hours. One would expect photooxidation to be low in sediments, however, because of a lack of penetrating radiation.

5.3. Chemical Transformations (hydrolysis). Chemicals introduced into aqueous media in the environment can undergo hydrolysis and be transformed into new chemicals with properties different from their precursors. In addition, processes other than nucleophilic displacement by water may occur (e.g., elimination or isomerization). The importance of these transformations of chemicals as dominant pathways in aqueous media can be determined quantitatively from data on hydrolysis rate constants and half-lives.

Hydrolysis data will generally be important in assessing risks from organic chemicals that have hydrolyzable functional groups (e.g., esters, amides, alkyl halides, epoxides, and phosphoric esters). Hydrolysis refers to the reaction of an organic chemical (RX) with water with the resultant net exchange of the group X for the OH group from the water at the reaction center. Therefore,

$$RX + HOH \longrightarrow ROH + HX$$

In the environment, hydrolysis of organic chemicals occurs in dilute solution. Under these conditions, water is present in a large excess and the concentration of water is essentially constant during hydrolysis. Hence, the kinetics of hydrolysis are pseudo-first-order at a fixed pH.

Processes other than nucleophillic displacement by water can take place. For example, X can be lost from RX via an elimination reaction. These elimination reactions exhibit kinetics behavior (i.e., pH independent or first-order acid or base dependent) similar to those *reactions where* OH substitution occurs.

The hydrolysis reaction can be catalyzed by acidic or basic species, including OH^- and H_3O^+ (H^+). The promotion of hydrolysis by H^+ or OH^- is known as *specific acid* or *specific base catalysis*, as contrasted to general acid or base catalysis encountered with other cationic or anionic species. So far, the published laboratory data (1945; 1446; 1953; 1952) indicate that hydrolysis rates are the same in sterile natural freshwaters and in buffered distilled water at the same temperature and pH. Thus, only specific acid or base catalysis together with neutral water reaction need be considered. Although other chemical species may catalyze hydrolysis reactions, the available concentrations of these species in the environment are usually too low to have an effect and are not expected to contribute significantly to the rate of hydrolysis (1954).

An extensive amount of information has been published on the hydrolysis of a wide variety of organic chemicals. However, most of the literature relating to environmental hydrolysis of chemicals pertains to pesticides. Much of this data is incomplete for the range of pH and temperature of environmental concern. Effects of buffer salts are often unrecognized.

D. BIOLOGICAL EFFECTS

1. Arrangement of Data. The following classification has been followed in arranging the data on residues bioaccumulation, and toxicity:

bacteria
algae
plants
protozoans (Phylum Protozoa)
worms (Phyla Platyhelminthes and Aschelminthes)
molluscs (Phylum Mollusca)
segmented worms (Phylum Annelida)
arthropods (Phylum Arthropoda)
   insects (Class Insecta)
   crustaceans (Class Crustacea)
   spiders and allies (Class Arachnidae)
chordates (Phylum Chordata)
   vertebrates (Subphylum Vertebrata)
   fishes (Class Pisces)
   amphibians (Class Amphibia)
   reptiles (Class Reptilia)
   birds (Class Aves)
   mammals (Class Mammalia)

A more detailed classification list is given on the following pages, as well as a list of major organisms used throughout the book with their scientific and common names.

2. Classification List

protozoans	(*Phylum Protozoa*)
	the flagellates (Class Mastigophora)
	rhizopods (Class Rhizopoda)
	amoebas (Order Amoebina)
	the spore-formers (Class Sporozoa)
	the ciliates (Class Ciliata)
mesozoans	(*Phylum Mesozoa*)
sponges	(*Phylum Parazoa*)
hydroids	*jellyfishes and corals* (*Phylum Cnidaria or Coelenterata*)
	sea firs (hydroids), hydras and siphonophores (Class Hydrozoa)
	jellyfishes (Class Scyphozoa)
	sea anemones and corals (Class Anthozoa)
flatworms	*flukes and tapeworms* (*Phylum Platyhelminthes*)
roundworms	*rotifers and allies* (*Phylum Aschelminthes*)
	roundworms (Class Nematoda)
	rotifers (Class Rotifera)
molluscs	(*Phylum Mollusca*)
	gastropods (Class Gastropoda)
	Subclass Pulmonata

	bivalves (Class Bivalvia = Pelecypoda)
	Subclass Lamellibranchia
	squids, cuttlefishes, octopuses (Class Cephalopoda)

segmented worms *(Phylum Annelida)*
  polychaetes (Class Polychaeta)
    nereids (Family Nereidae)
      Nereis diversicolor
  oligochaetes (Class Oligochaeta), including earthworms
    Family Tubificidae
      Tubifex tubifex
  leeches (Class Hirundinea)

arthropods *(Phylum Arthropoda)*
  millipedes (Class Diplopoda)
  centipedes (Class Chilopoda)
  insects (Class Insecta)
    exopterygote insects
      mayflies (Order Ephemeroptera)
      dragonflies (Order Odonata)
      stoneflies (Order Plecoptera)
      grasshoppers, crickets and allies (Order Orthoptera)
      cockroaches and praying mantids (Order Dictyoptera)
      termites (Order Isoptera)
    endopterygote insects
      butterflies and moths (Order Lepidoptera)
      flies (Order Diptera)
      fleas (Order Siphonaptera)
      ants, wasps, bees (Order Hymenoptera)
      beetles (Order Coleoptera)
  crustaceans (Class Crustacea)
    Subclass Branchiopoda
      Order Anostraca
        Artemia salina (brine shrimp)
      Order Notostraca
      Order Conchostraca
      waterfleas (Order Cladocera)
        Daphnia
    copepods (Subclass Copepoda)
      Order Calanoida
        Calanus
      Order Cyclopoida
        Cyclops
    malacostracans or higher crustaceans (Subclass Malacostraca)
      isopods, sowbugs, woodlice (Order Isopoda)
      amphipods, scuds, and sideswimmers (Order Amphipoda)
        Suborder Gammaridea
          Gammarus

      decapods (Order Decapoda)
       Suborder Natantia
        Section Caridea
         Crangon vulgaris
         Crangon crangon
       Suborder Reptantia
        Section Astacura
         Homarus americanus
         Homarus vulgaris
         Nephrops norvegicus
        Section Brachyura
         Carcinus maenas (green crab)
         Callinectus sapidus (blue crab)
   spiders and allies (Class Arachnida)
    scorpions (Order Scorpiones)
    harvestmen (Order Opiliones or Phalangida)
    ticks and mites (Order Acari)
    spiders (Order Araneae)
  starfishes, sea urchins, and allies (Phylum Echinodermata)
   starfishes (Class Asteroidea)
   sea urchins, sand dollars, and allies (Class Echinoidea)

chordates *(Phylum Chordata)*
  vertebrates (Subphylum Vertebrata)
   fishes (Classes Marsipobranchii, Selachii, Bradyodonti, and Pisces)
   amphibians (Class Amphibia)
    newts and salamanders (Order Caudata)
    frogs and toads (Order Salientia)
   reptiles (Class Reptilia)
    tortoises, terrapins, and turtles (Order Chelonia = Testudines)
    crocodilians (Order Loricata = Crocodylia)
    lizards and snakes (Order Squamata)
     snakes (Suborder Ophidia = Serpentes)
   birds (Class Aves)
   mammals (Class Mammalia)
    insect-eating mammals (Order Insectivora)
    bats (Order Chiroptera)
    the primates (Order Primates)
    ant-eaters, sloths, armadillos (Order Edentator)
    hares, rabbits, pikas (Order Lagomorpha)
    rodents (Order Rodentia)
    whales (Order Cetacea)
    flesh-eating animals (Order Carnivora)
     dogs, cats, weasels, bears (Suborder Fissipeda)
     seals, sea-lions, walruses (Suborder Pinnipedia)
    elephants (Order Proboscidea)

3. Organisms Used in Experimental Work with Polluting Substances or in Environmental Surveys

I.

English name	Latin name
abalone	*Haliotis* spp.
american char	*Salvelinus fontinalis* (Mitchell)
american toad	*Bufo americanus*
anchovy	*Stolephorus purpureus*
anole	*Anolis carolinensis*
armadillo	*Daysypus novemcinctus*
atlantic kelp	*Laminaria digitata*, L. *agardhii*
atlantic menhaden	*Brevoortia tyrannus*
atlantic ribbed mussel	*Volsella demissa*
atlantic salmon	*Salmo salar*
barnade	*Balanus* spp.
barn owl	*Tyto alba*
bay anchovy	*Anchoa* (*Anchoiella*) *mitchilly*
bay mussel	*Mytilus edulis*
bay scallop	*Aequipecten* (*Pecten*) *irradians*
big brown bat	*Eptesicus fuscus*
bigmouth buffalo	*Ictiobus cyprinellus*
black bullhead	*Ameiurus melas* or *Ictalurus melas*
bleak	*Alburnus alburnus* (L)
bloodworm	*Glycera dibranchiata*
blue crab	*Callinectes sapidus*
bluegill sunfish	*Lepomis macrochirus*
bobwhite quail	*Colinus virginianus*
box turtle	*Terrapene* sp.
brook trout	*Salvelinus fontinalis*
brown bullhead	*Ameiurus nebulosus* or *Ictalurus nebulosus*
brown shrimp	*Penaeus aztecus*
brown shrimp	*Crangon crangon*
brown trout	*Salmo trutta*
bullfrog	*Rana catesbiana*
bullheads	*Ameiurus* or *Ictalurus* (see also catfish)
bumble bee	*Bombus*
calico crab	*Ovalipes ocellatus*
calico scallop	*Aequipecten gibbus*
california ground squirrel	*Spermophilus beecheyi*
california sea mussel	*Mytilus californicus*
canada goose	*Branta canadensis*
carp	*Cyprinus carpio*
cat	*Felis domestica*
catfish	*Ictalurus*

English name	Latin name
catfish (American)	*Ameiurus nebulosus* (Le Sueur)
channel catfish	*Ictalurus punctatus*
chicken	*Gallus domesticus*
chinook salmon	*Oncorhynchus tschawytscha*
chub	*Squalius cephalus* (L.)
cod	*Gadus morrhua*
coho salmon	*Oncorhynchus kisutch* (Walbaum)
common cockroach	*Periplaneta americana*
common earthworm	*Lumbricus terrestris*
common frog	*Rana temporaria*
common toad	*Bufo bufo*
cone shells	*Conus* spp.
coalfish	*Gadus virens*
copepod	*Pseudocalanus minutus*
cotton rat	*Sigmodon hispidus*
cottontail	*Sylvilagus* sp.
crab	*Ranina serrata*
	Pertunus sanquinolentes
	Podophtalmus vigil
creek chub	*Semolitus atromaculatus*
cricket	*Gryllus* sp.
croaker	*Micropogon undulatus*
crucian carp (goldfish)	*Carassius carassius*
cutthroat trout	*Salmo clarki*
dace	*Leuciscus leuciscus* (L.)
deer mouse	*Peromyscus maniculatus*
dog	*Canis familiaris*
domestic chicken	*Gallus gallus*
domestic New Zealand white rabbit	*Oryctolagus cuniculus*
Dungeness crab	*Cancer magister*
earthworm	*Lumbricus terrestris*
eastern oyster	*Crassostrea virginica*
eastern chipmunk	*Tamias striatus*
edible frog	*Rana esculenta*
eel	*Anguilla anguilla* (L.)
	Anguilla vulgaris
egret	*Ardeola* sp.
English sole	*Parophrys vetulus*
European badger	*Meles meles*
European hares	*Lepus europaeus*
fathead minnow	*Pimephales promelas*
field cricket	*Gryllus pennsylvanicus*
flatworms	*Platyhelminthes*
Fowler's toad	*Bufo fowleri*
fruit fly	*Drosophila* sp.

English name	Latin name
garter snake	*Thamnophis sirtalis*
goatfish	*Mulloidichthys* spp.
goldfish	*Carassius auratus*
green crab	*Carcinides maenas*
green frog	*Rana clamitans*
green sunfish	*Lepomis cyanellus*
gulf menhaden	*Brevoortia patronus*
guinea pig	*Cavia*
guppy	*Lebistes reticulatus*
hard clam	*Mercenaria (Venus) mercenaria*
hares	*Leporidae*
herring	*Clupea harengus*
honey bee	*Apis melliferra*
horned lizard	*Phrynosoma cornutum*
housefly	*Musca domestica*
house sparrow	*Passer domesticus*
japanese quail	*Coturnix coturnix*
kangaroo rat	*Dipodomys* sp.
kelp	*Macrocystis pyrifera*
killifish	*Fundulus*
king crab	*Paralithoides camtschatica*
lake trout	*Salvelinus namaycush*
land snail	*Helix* sp.
largemouth bass	*Micropterus salmoides*
leopard frog	*Rana pipiens*
little neck clam	*Protothaca staminea*
lobster	*Panulirus japonicus: P. pencillatus*
longear sunfish	*Lepomis megalotis*
mallard	*Anas platyrhynchos*
marine pin perch	*Lagodon rhomboides*
meadow vole	*Microtus pennsylvanicus*
mealworm	*Tenebrio* sp.
Mexican axolotl	*Ambystoma mexicanum*
midge	*Chironomus plumosus*
milkfish	*Chanos chanos*
mink	*Mustela vison*
minnow	*Phoxinus phoxinus*
mosquito fish	*Gambusia affinis*
mosquito fly	*Culex*
	Aedes
	Anopheles
mountain bass	*Kuhlia sandvicensis*
mullet	*Mugil cephalus*
mummichog	*Fundulus heteroclitus*
northern lobster	*Homarus americanus*

English name	Latin name
northern pike	*Esox lucius*
nutria	*Myocaster coypus*
old world mouse	*Mus musculus*
old world rat	*Rattus norvegicus*
opossum	*Didelphis virginianus*
pacific oyster	*Crassostrea gigas*
pacific herring	*Clupea pallasii*
pacific sardine	*Sardinops caerula*
pack rat	*Neotoma lepida*
perch	*Perca fluviatilis*
pheasant	*Phasianus* sp.
pickerel frog	*Rana palustris*
pigeon	*Columba livia*
	Columba sp.
pinfish	*Lagodon rhomboides*
pink salmon	*Oncorhynchus gorbuscha*
pink shrimp	*Penaeus duorarum*
pismo clam	*Tivela stultorum*
plaice	*Pleusonectes platessa*
pollack	*Gadus pollachius*
pompano	*Trachinotus carolinus*
pompano, jack cravally	*Caranx* spp.
pumpkinseed	*Lepomis gibbosus*
purple sea urchin	*Stronglo centrotus purpuratus*
rabbits	*Sylvilagus*
rainbow trout	*Salmo gairdnerii*
rainwater killifish	*Lucania parva*
razor clam	*Siliqua patula*
rat snake	*Elaphe* sp.
red algae	*Porphyra* spp
red seaweed	*Gracilaria virrucosa, G. foliifera*
red snapper	*Lutianus campechanus*
rhesus monkey	*Macaca mulatta*
ribbed limpet	*Siphonaria normalis*
ringnecked pheasant	*Phasianus colchicus*
roach	*Rutilus rutilus* (L.)
round worms	*Aschelminthes*
sailfin molly	*Poecilia* (*Mollienisia*) *latipinna*
salmons	*Salmo* or *Oncorhynchus*
salmon (atlantic)	*Salmo salar* (L.)
saltwater limpet	*Helcioniscus exaratus H. argentatus*
sandworm	*Nereis virens, Nereis vexillosa*
sea anemone	*Anthopleura elegantissima*
sea lamprey	*Petromyzon marinus*
sea lettuce	*Ulva* spp.

English name	Latin name
sea urchin	*Arbacid puntulata; Lytechnius* spp *Echinometra* spp.
sailfin molly	*Poecilia (Mollienisia) latipinna*
sand shrimp	*Crangon septemspinosa*
segmented worms	Annelida
scup	*Stenotomus chrysops*
shore crabs	*Hemigiapsis* spp.
sheep	*Ovis* sp.
sheepshead minnow	*Cyprinodon variegatus*
shiner perch	*Cymatogaster aggregata*
shrimp	*Crangon* spp *Pandalus* spp
sprat	*Clupea sprattus*
smallmouth bass	*Micropterus dolomieui*
snapping shrimp	*Crangon* spp.
snapping turtle	*Chelydra serpentina*
soft shell clam	*Mya arenaria*
southern flounder	*Paralichthys lethostigma*
spiny lobster	*Panuliris argus*
spring peeper	*Hyla crucifer*
stable fly	*Stomoxys calcitrans*
staghorn sculpin	*Leptocottus armatus*
starling	*Sturnis vulgaris*
starry flounder	*Platichthys stellatus*
stickleback (12-spined)	*Pygosteus pingitius* (L.)
striped bass	*Roccus saxatilis* *Morone saxatilis*
striped mullet	*Mugil cephalus*
suckers	*Catostomus* or *Ictiobus*
summer flounder	*Paralichthys dentatus*
sunfish (common)	*Lepomis humilis*
surf clam	*Spirula solidissima*
surgeon fish	*Acanthurus* spp.
swallows	Hirudinidae
swine (miniature)	*Sus scrofa*
squirrel monkey	*Saimiri sciureus*
tench	*Tinca tinca* (L.)
sea moss	*Chondrus crispus*
threespine stickleback	*Gasterosteus aculeatus*
treefrog	*Hyla versicolor*
trout	*Salmo* or *Salvelinus*
turkey	*Meleagris*
walleye	*Stizostedion vitreum*
water flea	*Daphnia*
water shrimp	*Gammarus pulex*

English name	Latin name
western chipmunk	*Eutamias* sp.
white shrimp	*Penaeus setiferus*
white sucker	*Catostomus commersoni*
whiting	*Gadus merlangus*
winter flounder	*Pseudopleuronectes americanus*
woodfrog	*Rana sylvatica*
yellow bullhead	*Ictalurus netalis*

II.

Latin name	English name
Acanthurus spp	surgeon fish
Aedes	mosquito fly
Aequipecten gibbus	calico scallop
Aequipecten (Pecten) irradians	bay scallop
Alburnus alburnus	bleak
Ambystoma mexicanum	Mexican axolotl
Ameiurus melas	black bullhead
Ameiurus nebulosus	brown bullhead or American catfish
Anas platyrhynchos	mallard
Anchoa (Anchoiella) mitchilly	bay anchovy
Anguilla anguilla	eel
Anguilla vulgaris	eel
Annelida	segmented worms
Anolis carolinensis	anole
Anopheles	mosquito fly
Anthopleura elegantissima	sea anemone
Apis	honey bee
Apis melliferra	honey bee
Arbacid puntulata	sea urchin
Ardeola spp.	egret
Aschelminthes	round worms
Balanus spp	barnacle
Bombus	bumble bee
Branta canadensis	canada goose
Brevoortia patronus	gulf menhaden
Brevoortia tyrannus	atlantic menhaden
Bufo americanus	American toad
Bufo bufo	common toad
Bufo fowleri	Fowler's toad
Callinectes sapidus	blue crab
Cancer magister	Dungeness crab
Canis familiaris	dog
Caranx spp.	pompano, jack cravally
Carassius auratus	goldfish
Carassius carassius	crucian carp

Latin name	English name
Carcinides maenas	green crab
Catostomus	suckers
Catostomus commersoni	white sucker
Cavia	guinea pig
Chanos chanos	milkfish
Chelydra serpentina	snapping turtle
Chironomus plumosus	midge
Chandrus crispus	sea moss
Clupea herengus	herring
Clupea pallasii	pacific herring
Clupea sprattus	sprat
Colinus virginianus	bobwhite quail
Columba livia	pigeon
Conus spp.	cone shells
Coturnix coturnix	japanese quail
Crangon crangon	brown shrimp
Crangon septemspinosa	sand shrimp
Crangon spp.	snapping shrimp
Crassostrea gigas	pacific oyster
Crassostrea virginica	eastern oyster
Culex	mosquito fly
Cymatogaster aggregata	shiner perch
Cyprinodon variegatus	sheepshead minnow
Cyprinus carpio	carp
Daphnia	water flea
Daysypus novemcinctus	armadillo
Didelphis virginianus	opossum
Dipodomys sp.	kangaroo rat
Drosophila sp.	fruit fly
Echinometra spp.	sea urchin
Elaphe sp.	rat snake
Eptesicus fuscus	big brown bat
Esox lucius	northern pike
Eutamias sp.	western chipmunk
Felis domestica	cat
Fundulus	killifish
Fundulus heteroclitus	mummichog
Gadus merlangus	whiting
Gadus morrhua	cod
Gadus pollachius	pollack
Gadus virens	coalfish
Gallus gallus	domestic chicken
Gambusia affinis	mosquito fish
Gammarus pulex	water shrimp
Gasterosteus aculeatus	threespine stickleback
Glycera dibranchiata	bloodworm

Latin name	English name
Gracilaria verrucosa, G. foliifera	red seaweed
Gryllus pennsylvanicus	field cricket
Haliotis spp.	abalone
Helcioniscus exaratus, H. argentatus	saltwater limpet
Helix sp.	land snail
Hemigrapsis spp.	shore crabs
Hirudinidae	swallows
Homarus americanus	northern lobster
Hyla crucifer	spring peeper
Hyla versicolor	treefrog
Ictalurus	catfish
Ictalurus melas	black bullhead
Ictalurus natalis	yellow bullhead
Ictalurus nebulosus	brown bullhead
Ictalurus punctatus	channel catfish
Ictiobus	suckers
Ictiobus cyprinellus	bigmouth buffalo
Kuhlia sandvicensis	mountain bass
Lagodon rhomboides	marine pin perch
Laminaria digibacta, L. agardhii	atlantic kelp
Lebistes reticulatus	guppy
Lepomis cyanellus	green sunfish
Lepomis gibbosus	pumpkinseed
Lepomis humilis	common sunfish
Lepomis macrochirus	bluegill sunfish
Lepomis megalotis	longear sunfish
Leporidae	hares
Leptocottus armatus	staghorn sculpin
Lepus europaeus	european hares
Leuciscus leuciscus	dace
Lucania parva	rainwater killifish
Lumbricus terrestris	common earthworm
Lutianus campechanus	red snapper
Lytechnius spp.	sea urchin
Macaca mulatta	rhesus monkey
Meleagris	turkey
Macrocystis pyrifera	kelp
Meles meles	European badger
Mercenaria (Venus) mercenaria	hard clam
Micropogon undulatus	croaker
Micropterus dolomieui	smallmouth bass
Micropterus salmoides	largemouth bass
Microtus pennsylvanicus	meadow vole
Morone saxatilis	striped bass
Mugil cephalus	mullet

Latin name	English name
Mugil cephalus	striped mullet
Mulloidichthys spp.	goatfish
Musca domestica	housefly
Mus musculus	old world mouse
Mustela vison	mink
Mya arenaria	soft shell clam
Myocaster coypus	nutria
Mytilus californicus	California sea mussel
Mytilus edulis	bay mussel
Nereis virens Nereis vexillosa	sandworm
Neotoma lepida	pack rat
Oncorhynchus gorbuscha	pink salmon
Oncorhynchus kisuth	coho salmon
Oncorhynchus tschawytscha	chinook
Oryctolagus cuniculus	domestic New Zealand white rabbit
Ovalipes ocellatus	calico crab
Ovis sp.	sheep
Pandalus spp	shrimp
Panuliris argus	spiny lobster
Panulirus japonicus; P. pencillatus	lobster
Paralithoides camtschatica	king crab
Paralichthys dentatus	summer flounder
Paralichthys lethostigma	southern flounder
Parophrys vetulus	english sole
Passer domesticus	house sparrow
Penaeus aztecus	brown shrimp
Penaeus duorarum	pink shrimp
Penaeus setiferus	white shrimp
Perca fluviatilis	perch
Periplaneta americana	common cockroach
Peromyscus maniculatus	deer mouse
Petromyzon marinus	sea lamprey
Phasianus colchicus	ringnecked pheasant
Phoxinus phoxinus	minnow
Phrynosoma cornutum	horned lizard
Pimephales promelas	fathead minnow
Platichthys stellatus	starry flounder
Platyhelminthes	flatworms
Pleuronectes platessa	plaice
Podophthalmus vigil	crab
Poecilia (Mollienesia) latipinna	sailfin molley
Porphyra spp.	red algae
Portunus sanquinolentus	crab
Protothaca staminea	little neck clam
Pseudocalanus minutus	copepod

Latin name	English name
Pseudopleuronectes americanus	winter flounder
Pygosteus pungitius	stickleback (12-spined)
Rana catesbiana	bullfrog
Rana clamitans	green frog
Rana esculenta	edible frog
Rana palustris	pickerel frog
Rana pipiens	leopard frog
Rana sylvatica	woodfrog
Rana temporaria	common frog
Ranina serrata	crab
Rattus norvegicus	old world rat
Roccus saxatilis	striped bass
Rutilus rutilus	roach
Saimiri sciureus	squirrel monkey
Salmo	trout
Salmo clarki	cutthroat trout
Salmo gairdnerii	rainbow trout
Salmo salar	atlantic salmon
Salmo trutta	brown trout
Salvelinus	trout
Salvelinus fontinalis	American char or brook trou
Salvelinus namaycush	lake trout
Sardinops caerula	pacific sardine
Semolitus atromaculatus	creek chub
Sigmodon hispidus	cotton rat
Siliqua patula	razor clam
Siphonaria normalis	ribbed limpet
Spermophilus beecheyi	California ground squirrel
Squalius cephalus	chub
Spirula solidissima	surf clam
Stenotomus chrysops	scup
Stizostedion vitreum	walleye
Stolephorus purpureus	anchovy
Stomoxys calcitrans	stable fly
Stronglo centrotus purpuratus	purple sea urchin
Sturnis vulgaris	starling
Sus scrofa	swine, miniature
Sylvigalus	rabbits
Sylvilagus sp.	cottontail
Tamias striatus	eastern chipmunk
Tenebrio sp.	mealworm
Terrapene sp.	box turtle
Thamnophis sirtalis	garter snake
Tinca tinca	tench
Tivela stultorum	pismo clam

Latin name	English name
Trachinotus carolinus	pompano
Tyto alba	barn owl
Ulva spp.	sea lettuce
Volsella demissa	atlantic ribbed mussel

4. Discussion of Biological Effects Tests

4.1. Ecological Effects Tests. Releases of hazardous chemical substances into the environment during manufacturing, processing, distribution, use, or disposal, whether accidental or planned, can have an adverse impact on both natural and man-modified ecosystems and their components. The societal costs may include degradation of the environment; losses in sport and commercial fisheries, shellfish populations, and wildlife resources; losses in agriculture; losses in tourism and property values; and other adverse impacts.

Testing for such effects requires the selection of indicators (i.e., indicative parameters) that provide for wide taxonomic representation and include a range of biological processes. Ideally, testing at levels of ecological organization above the individual species would provide information more directly related to ecological consequences of the release of a hazardous chemical. However, the development and standardization of such tests is difficult due to the complexity of the species interactions that characterize ecosystems. A major thrust for the future therefore will be the development of test methods that address interactions such as those which occur between predator and prey, among competitors for habitat or food, and between disease and host organisms. As methods such as microcosm studies and other laboratory model systems are developed, they may help to address these ecological testing needs.

Laboratory testing below the level of the organism is also potentially useful. It is generally rapid and readily amenable to standardization, but most such testing has not yet been shown to be usable for ecological impact assessment. Since many cellular and subcellular functions are common to a wide range of organisms, they have the potential of being applicable to many sets of ecological circumstances.

However, most ecological effects tests currently in use employ single species test populations of vertebrates, invertebrates, or plants. Individual species represent an intermediate level of biological organization between subcellular functions and community/ecosystem interactions. Many single species tests are considered to be state-of the-art and have correlated well with actual ecological effects of chemicals.

The following criteria can be used to select tests:

- The test results are significant and useful for risk assessment.
- The test applies to a wide range of chemical substances or categories of chemical substances.
- The test is cost effective in terms of personnel, time, and facilities.
- The test is adequately sensitive for detection of the subject effect.

Confidence in extrapolation from simple tests to ecologically significant impacts depends not only on the appropriate kinds of tests but also on selection of appropriate organisms to be used in those tests. Organisms useful for assessment testing should have characteristics such as the following:

- The organism is representative of an ecologically important group (in terms of taxonomy, trophic level, or realized niche).
- The organism occupies a position within a food chain leading to man or other important species.
- The organism is widely available, is amenable to laboratory testing, easily maintained, and genetically stable so uniform populations can be tested.
- There is adequate background data on the organism (i.e., its physiology, genetics, taxonomy, and role in the natural environment are well understood) so data from these tests can be adequately interpreted in terms of actual environmental impacts.

4.2. Plant Effects Tests. All organisms require energy to perform vital functions and use radiant energy of sunlight as the ultimate source of energy. Green plants use this energy directly through the process of photosynthesis when suitable inorganic carbon and nutrients are present. The sun's energy is converted into chemical energy which is stored in plants in the form of sugars, starches and other organic chemicals to be used by the plants themselves or by other organisms as energy sources.

Photosynthesis is the source of virtually all atmospheric oxygen. Plants also synthesize vitamins, amino acids, and other metabolically active compounds essential to many organisms. In this context, the maintenance of the biosphere depends upon the normal functioning of green plants. Data from the tests in this section are expected to provide preliminary indications of the effects of chemical substances on the following groups of plants: Blue-green algae, diatoms, green algae, monocotyledons, and dicotyledons.

Blue-green algae are one of two groups of organisms capable of converting atmospheric nitrogen into forms which can be utilized by all living organisms. Diatoms and unicellular green algae are responsible for most of the world's photosynthesis and are the primary food energy base for most organisms inhabiting aquatic environments. They are, therefore, necessary for all aquatic life and for human food taken from fresh water and marine environments (e.g., fish).

Monocotyledons and dicotyledons are the two major groups of terrestrial flowering plants. The most highly evolved and widely distributed plants, they are the source of most human food and fiber, as well as shelter and important medicinals. These groups of plants are recognized as major converters of radiant energy to chemical energy in terrestrial environments.

Test 1. Algal inhibition. Testing for inhibition or stimulation of the growth of algae indicates the extent to which a test chemical can affect primary producers in lakes, streams, estuaries, and oceans. It can also generally indicate phytotoxicity or plant growth stimulation. Substances which drastically inhibit growth at or near concentrations expected in the environment may reduce aquatic productivity. However, even those substances which inhibit algal growth only partially, or which stimulate growth, at or near concentrations expected in the environment, might shift relative algal populations so that undesirable species could increase.

If diatoms or green algae grow less in the presence of the chemical than blue-green algae, for example, a bloom of the less desirable blue-green species could develop.

Algae are often selected to represent aquatic primary productivity because they constitute the major mechanism for fixation of energy in most aquatic locations. Techniques for culturing and measuring them are simpler and less expensive than for larger plants or for attached organisms (e.g., periphyton). The parameters recommended to measure potential effects on algae are inhibition of dry weight increase and changes in cell size. There are other potential effects which are not recommended. Lethality, for example, is commonly measured for other organisms, but it is difficult to determine for microscopic organisms and for nonmotile organisms. Inhibition of photosynthesis and/or respiration could be measured, but the balance between photosynthesis and respiration is accumulated as growth. In addition, growth represents the product which is important in the food chain and is therefore more directly relevant to assessment.

Uncertainties in using these data in assessment center upon whether the selected species are adequate indicators of the potential for stimulation or toxicity to nonselected algal species, whether benefits of algae to the food chain can be accurately predicted from changes in dry weight and cell size, and whether there are significant interactive effects (such as competition) which are affected at lower concentrations of the chemical substance than any individual species.

Test 2. Lemna Inhibition. Testing for inhibition of *Lemna* (duckweed) growth can serve as an indicator of the extent to which a chemical substance can interfere with aquatic higher plants. Chemicals which inhibit *Lemna* growth only partially at or near concentrations expected in the environment may shift relative populations of macrophytes and algal species.

Lemna was selected to represent aquatic macrophytes because it is well known genetically and physiologically and is one of the few aquatic macrophytes used extensively in culture. It is easy to culture and tests can be performed at less expense than would be possible with larger plants. Aquatic macrophytes possess many of the developmental/structural characteristics unique to higher plants. Roots are particularly important in this regard because they are frequently more sensitive than other plant parts, and their inclusion among the measurements made on *Lemna* serves as an important indicator of potential toxicity. Unfortunately, roots may also be especially sensitive to such environmental variables as nutrients and temperature. For that reason, and because foliage may also be sensitive, it is important to hold cultural conditions constant during testing. The fact that Lemna is not completely typical of higher plants introduces an uncertainty into its use as a test organism.

Of the two general approaches to culturing *Lemna*, the most commonly used is static (batch) culture in which nutrients and test substance are present at the beginning of the experiment and decline as the plant takes them up. A less commonly used technique is a flow-through system in which nutrients and test chemical are continuously replenished during the test. Whichever system is used, response can be measured several ways. Since roots are generally most sensitive, a measure of their growth often indicates maximum sensitivity of the species. However, roots are less useful in assessment than measures more directly related to productivity and standing crop of plants in the ecosystem. For these two purposes, dry weight and frond number are commonly used.

The major uncertainty in the use of *Lemna* data in assessment lies in extrapolation to a wide variety of aquatic plants. No single species can fully represent the diverse life forms and habitats of plants which grow in water.

4.3. Animal Effects Tests. The potential of a chemical to produce adverse ecological effects can be indicated by the results of preliminary testing of "representative" animals. Preliminary tests and test organisms should be selected on the basis of taxonomic, ecological, toxicological, and chemical exposure criteria. Test schemes should reflect those ecological hazards that a specific chemical substance may cause. Test responses—death, reproductive and/or behavioral dysfunction, and impairment of growth and development—are important factors for hazard assessment.

4.3.1. Invertebrates. Toxicity of a chemical substance to invertebrates is an important factor in preliminary assessment of impact on ecosystems. Invertebrates have broad ecological roles and show various sensitivities to chemicals.

(a) Aquatic invertebrates. A number of aquatic invertebrates for acute tests are listed below.

Marine and estuarine invertebrates
  copepods: *Acartia tonsa, Acartia clausi*
  shrimp: *Penaeus setiferus, P. duorarum, P. aztecus*
  grass shrimp: *Palaemonetes pugio, P. intermedius, P. vulgaris*
  sand shrimp: *Crangon septemspinosa*
  shrimp: *Pandalus jordani, P. danae*
  bay shrimp: *Crangon nigricauda*
  mysid shrimp: *Mysidopsis bahia*
  blue crab: *Callinectes sapidus*
  shore crab: *Hemigrapsus* sp., *Pachygrapsus* sp.
  green crab: *Carcinus maenas*
  fiddler crab: *Uca* sp.
  oyster: *Crassostrea virginica, C. gigas*
  polychaetes: *Capitella capitata*

Freshwater invertebrates
  daphnids: *Daphnia magna, D. pulex, D. pulicaria*
  amphipods: *Gammarus lacustris, G. fasciatus,* or *G. pseudolimnaeus*
  crayfish: *Oronectes* sp., *Cambarus* sp., *Procambarus* sp., or *Pacifasiacus leniusculus*
  stoneflies: *Pteronarcys* sp.
  mayflies: *Baetis* sp. or *Ephemerella* sp.
  mayflies: *Hexagenia limbata* or *H. bilinata*
  midges: *Chironomus* sp.
  snails: *Physa integra, P. heterostropha, Amnicola limosa*
  planaria: *Dugesia tigrina*

(b) Terrestrial invertebrates. The ecological role and suitability for toxicity testing of a number of terrestrial invertebrates is discussed below.

Phylum annelida
Class: Oligochaeta (earthworms). Family Lumbricidae: *Lumbricus terrestris* (common earthworm).
Ecological role: earthworms occur in upper soil levels and feed on decaying organic matter. They are particularly important as soil mixers, aerators and drainers and serve as food for many insectivores (robins, woodcock, mice, shrews).
Suitability for toxicity testing: The diversity and wide distribution of worms make them desirable test species. Earthworms are particularly valuable because of their role in soil ecosystem, their past use, and ease of maintenance.

Phylum Mollusca
Class: Gastropoda (pulmonate snails): *Helix aspersa*.
Ecological role: Terrestrial snails and slugs are primary consumers and eat a varied diet of plant materials. They are a food source for larger insectivores.
Suitability for toxicity testing: Helix sp. is a very widely distributed snail and abundant in certain moist habitats.

Phylum Arthropoda
Class Arachnida: members of this class are spiders, mites and ticks, scorpions and harvestmen.
Ecological role: Mites and ticks are parasitic on plants and animals, deriving their substance directly from the fluids of their hosts. Spiders are carnivorous invertebrates whose food consist entirely of small animals, primarily insects. All arachnids are potential food sources for insectivores.
Suitability for toxicity testing: Mites and ticks are easily maintained under controlled conditions. Spiders are excellent test subjects because they are predators on many insects, they are relatively easy to maintain, if not to breed, and their web-building provides a very useful experimental tool.
Class Insecta; order Orthoptera: this order includes many large and well known insects—crickets, grasshoppers, roaches, locusts, and praying mantids. Toxicological research has been done on wide-ranging and easily obtainable species: *Periplaneta americana*, the common cockroach; *Gryllus pennsylvanicus*, the field cricket; *Schistocera gregaria*; and *Locusta migratoria*, locusts; *Mantis* sp., *Stagmomantis* sp., and *Tenodera* sp., mantids.
Ecological role: Crickets and cockroaches are omnivorous insects and will feed on many kinds of organic matter. Locusts and grasshoppers are vegetarians and can occur in very large numbers, sometimes defoliating the countryside. Praying mantids are predators and feed primarily on insects. All of these species are possible food items for insectivorous invertebrates and vertebrates.
Suitability for toxicity testing: The insects in this order are easily maintained and very abundant. Praying mantids, being strictly carnivorous and relying heavily on insects for food, might accumulate certain chemicals or be more heavily exposed to target animals.

—Orders Hemiptera, Homoptera: hemipterans are true bugs; the homopterans are closely associated with the bugs.
Ecological role: Hemipterans and homopterans are feeders on organic fluids, primarily plant juices. They can be destructive agricultural pests. These insect groups are food for insectivorous invertebrates and vertebrates. Aphids or plantlice are a large group of Homopterans and frequent pests on vegetation. The herbivorous aphid species are good selections for use in studies of environmental and contaminants that may accumulate or deposit on vegetation.
—Order Coleoptera: Beetles that have been used in research are frequently pest species, though not exclusively. Included are ground beetles (*Harpalus, Agonum, Feronia*), lady beetles (*Hippodamia, Coleomegilla*), and flour beetles (*Tribolium*).
Ecological role: Some beetles are pests on agricultural crops and other are predacious ground dwelling species (e.g., *Harpalus*). Others feed on fungi and carrion. All beetles are potential food for insectivorous invertebrates and vertebrates.
—Orders Lepidoptera: Butterflies and moths are conspicuous and well known insects.
Ecological role: The larvae of butterflies and moths, often severe agricultural pests, are economically much more important than the adults, some of which never feed. They frequently supply food for insectivorous predators.
—Orders Diptera: The "flies" are a well known group of insects and one of the larger orders. Mosquitoes, stable flies, house flies, and blow flies are pests of man and other animals.
Ecological role: Many adult dipterans are vectors of disease and nuisance pests of other animals. However, they also can represent staple foods for insectivorous predators (i.e., bats, swallows, frogs). Aquatic larvae are frequently major food sources for fish in quiet waters. The double association of some forms (e.g., mosquitoes) to aquatic and terrestrial systems at different times during their life cycles may make them particularly suitable subjects in experiments in which land-water transfer of a substance is to be studied.
—Order Hymenoptera: The Hymenoptera contain ants, sawflies, ichneumons, chalcids, wasps, and bees.
Ecological role: Many Hymenoptera are important as pollinators and as parasites on other insects. They feed on pollen, plant juices, and many on other liquid foods.

Phylum arthropoda
Class Crustacea: Wen Yuh Lee (1717) recommends three laboratory-cultured crustaceans for use in marine pollution studies because they are characterized by wide distribution, a short life cycle, high reproductive potential, and are representative of the plankton and benthos in coastal waters and the intertidal zone. The recommended crustaceans are: *Acartia tonsa* (a planktonic copepod), *Sphaeroma quadridentatum* (an isopod), and *Amphitoc valida* (an amphipod).
The selection of specific organisms should further be based on vulnerability to marine pollutants in a critical life state (usually the larval or temporary plank-

tonic stage), commercial or biological value, availability and ease of collection, ease of rearing and maintaining in the laboratory, existing knowledge on ecological requirements (927). Laboratory-bred populations show several advantages in a long-term toxicity study. One of the most important is that the cultured population is able to grow and reproduce in the laboratory and is available whenever either a test is to be undertaken to determine the toxicity of products (e.g., oils) or to rank products such as dispersants in toxicity. The advantages and disadvantages of natural and laboratory cultured populations are summarized in Table 24.

Crustacean life cycle. Aquatic invertebrates are the most common food chain links between phytoplankton and desirable species of fish and shellfish. The extent to which chemical substances affect reproduction and growth of aquatic invertebrates is important since healthy stocks of fish and shellfish are dependent on adequate sources of food. A life cycle test is desirabe for assessment because it would give a better estimate of total hazard than, for example, an acute toxicity test.

Freshwater crustacean: Daphnia. One of the most widely performed and economical life cycle test uses Daphnia, a freshwater zooplankton. Daphnia have a planktonic existence, short life cycle, and can be easily cultured. They have been widely used in toxicological testing, and are sensitive to toxicants. Although no invertebrate life cycle tests have been completely standardized, the Daphnia life cycle test has been used extensively by many researchers. Reproduction and life-cycle tests on Daphnia begin with newborn daphnids,

TABLE 24. Comparison between Field- and Laboratory-Cultured Populations for Marine Pollution Studies (1717).

Field population	Laboratory population
A. Disadvantages	
1. Careful handling needed during collection and transportation	1. Genetic drift from wild conditions
2. Seasonal variation in mortality and in healthy condition	
3. Variations in size and life stages	
4. Seasonal changes in population abundance	
5. Various physiological states due to zonation	
B. Advantages	
1. Natural population realistic	1. Close to the normal physiological state, capable of growing and reproducing in captivity
	2. Ages known
	3. Available throughout the year
	4. Biochemical comparison possible between laboratory and suspected polluted populations
	5. Useful for chronic toxicity tests

which are exposed to a chemical substance in culture for approximately three or four weeks. Reproductive impact of the chemical substance is evaluated by comparing the number of young produced by the organisms exposed to a chemical substance with the number produced by controls. Chronic lethal effects are evaluated by observing survival of the daphnids initially exposed throughout their life cycle.

Marine crustacean: mysid shrimp. Mysidopsis bahia is an excellent species for life cycle test for *marine invertebrates* due to its sensitivity to known toxic chemicals, ease of culturing, short life cycle, and importance in near-shore marine food webs.

4.3.2. Vertebrates

(a) Fish toxicity tests. The term "toxicity test" covers a wide and increasing range of types of investigations, and with fish such tests can include such things as:

(a) The study of the toxic properties of fish (that is, fish toxicology).
(b) The use of fish for detecting the presence of, or determining concentrations of, e.g., metals, toxins, hormones, in solution (i.e., bioassay and tests for screening for the presence or absence of some defined response).
(c) Laboratory and field tests of selective piscicides.
(d) Basic toxicological research into the metabolism and detoxification of substances by this class of animal.
(e) Tests to compare the relative lethal toxicities of different substances under some fixed but arbitrary set of conditions, to one or more species of fish.
(f) Tests to compare the lethal toxicity of a given substance to a single species of fish under a range of test conditions (e.g., pH, dissolved oxygen, hardness) to determine the effects of environmental conditions on toxicity.
(g) Simple, but unscientifically based, laboratory tests made under fixed conditions to assess arbitrarily for some administrative convenience the acceptability of a substance or effluent.
(h) Laboratory tests of the effects of a substance on survival, growth, reproduction, and so on, in fish.
(i) Laboratory and field tests of the effects of a waste (or of a chemical for use in agriculture) on fishes, on fish populations, and, where these exist, on fisheries.
(j) The laboratory use of fish to monitor aqueous wastes for harmful effects.
(k) The laboratory use of fish to monitor waters being abstracted from rivers for drinking, food processing, irrigation, and so on, for harmful effects.
(l) The use of fish in cages to monitor a river water or aqueous domestic and industrial wastes for harmful effects.

Toxicity studies are conveniently classified on the basis of the duration of exposure (as shown in Table 25), which of course automatically reflects the concentration of the poison. In a full investigation these tests would typically follow each other in the order given, should the substance under effect and the nature of exposure warrant it. Special studies are made for carcinogenic, teratogenic, and mutagenic effects.

Rainbow trout (*Salmo gairdneri*) and bluegill sunfish (*Lepomis macrochirus*) are suggested as standard test species. These two fish, one a cold water species and the

TABLE 25. Broad Scheme of Toxicity Testing.

Type of test	Information sought
(a) Acute	The lowest concentration having effect within a few days of continuous exposure. The effect is related to the breakdown of physiological systems and typically death is the response sought.
(b) Subchronic	The highest concentration having no effect within perhaps one-tenth of the normal average life span. Used to determine the mode of action and functional and physiological changes.
(c) Chronic	The highest concentration having no effect over the life time of the animal.

other a warm water species, have generally been the most sensitive to most previously tested chemical substances. The choice of species may depend upon the geographical area of expected chemical release and available testing facilities. If salt water exposure is probable, testing a marine species is advised. Even though marine species generally are no more sensitive than fresh water species, toxic effects can be significantly modified by water chemistry.

Fish embryo juvenile test. Objectives: The objective of this test is to give preliminary indication of potential effects of a chemical on fish. Chemical substances can have significant chronic effects on individual fish at concentrations 2 to 500 times lower than LC_{50} values. When long-term exposure is probable, data on chronic effects contribute to assessment. Differences in sensitivities between a chronic and embryo-juvenile test are generally small or negligible, while differences in cost are great. The embryo-juvenile test is usually an excellent, cost effective substitute for a chronic fish study.

Test description: In an embryo-juvenile test, fish eggs and fry hatched from these eggs are exposed to a chemical at several concentrations for a few weeks. Effects on hatchability of eggs, and growth and survival of fry are determined by comparing responses at each concentration with the control.

Fish bioconcentration test. Objectives: Bioconcentration, the uptake of a compound from water into living tissue, affects the movement, distribution and toxicity of chemical substances in the environment. A substance that bioconcentrates may affect life far removed from the initial points of environmental release and may alter ecological processes at concentrations much lower than predicted from acute and subacute results. Bioconcentration is the first step in the process of food chain biomagnification. Results of bioconcentration studies are useful in assessing risk to the environment especially when the substance is highly lipid soluble (e.g., octanol/water partition coefficient is greater than 1000), poorly soluble in water, and does not undergo rapid chemical or biological transformation.

Test description: Fish are exposed to the chemical substance in water for 28 days or until their tissue concentration reaches steady state. After steady state or 28 days, fish are placed in uncontaminated water for 7 days. During the exposure (1–28 days or steady-state) and depuration (1–7 days) periods, fish are sacrificed periodically and the concentration of chemical substance in their tissues is measured. The bioconcentration factor, the relative uptake rate, and the depuration rate constants are estimated from these data. The most commonly tested species are fathead minnow (*Pimephales promelas*) and bluegill sunfish (*Lepomis macrochirus*).

(b) Amphibians and Reptiles: Anurana. The aquatic forms of frogs and toads, the tadpoles have been widely used to study developmental biology. Common species in North America and Europe are:

Rana catesbiana: bullfrog
Rana clamitans: green frog
Rana palustris: pickerel frog
Rana pipiens: leopard frog
Rana sylvatica: woodfrog
Hyla crucifer: spring peeper
Hyla versicolor: treefrog
Bufo americanus: american toad
Bufo fowleri: Fowler's toad
Rana esculenta: edible frog
Rana temporaria: common frog
Bufo bufo: common toad

Anurans are carnivorous animals that feed on a great variety of invertebrate species, particularly insects. Many larger predators utilize them as a food source. Birds, snakes, turtles, and mammals feed on the adults, and the tadpoles provide food for predators associated with aquatic habitats.

Chelonia. Turtles are basically omnivorous reptiles and are important elements in the aquatic systems. They are predators on all types of invertebrates and some species are avid consumers of aquatic vegetation.

(c) Birds. Birds are a fairly large vertebrate group. They are very important in the world ecosystem. They are primary and secondary consumers, feeding on plants, invertebrates and vertebrates alike. They in turn are food for mammalian predators (including man), a few amphibians and reptiles, and a few species of birds. Many avian species are good indicators of environmental quality.

A toxicity test recommended by EPA (1464) is the *Quail Dietary Test*. The objective of a quail dietary test is to give preliminary indication of possible effects of a chemical substance on terrestrial birds. This test is designed to determine the concentration of a substance in food that will be lethal to 50 percent of a test population as well as to observe behavioral, neurological, and physiological effects. The test is appropriate for a chemical which might be found in or on terrestrial bird food. The bobwhite quail (*Colinus virginianus*) is an appropriate test species since it is easily and economically reared, widely available, and generally more sensitive to many hazardous chemicals than other common test species.

(d) *Mammals*
   Lagomorpha. The lagomorphs are composed of two families, the rabbits and hares (Leporidae) and the pikas (Ochotonidae).

Ecological role: Rabbits (*Sylvilagus*) and hares (*Lepus*) are primarily herbivorous, utilizing many plant foods, but preferring grasses and herbaceous materials. They are typically openland-edge inhabitants and remain active year round.

Suitability for toxicity testing: Domestic rabbits are readily available and easy to use in toxicity testing.

Rodentia; family Muridea: lab rats and mice: The choice of research animals in the past has been based on availability and ease of maintenance. No group of animals was more available than the old world rats (*Rattus norvegicus*) and mice (*Mus musculus*) which have adapted so well to living with man. The white lab rats and mice of today are albino strains of these species that have been selectively bred for research purposes.

Ecological role: Lab rats and mice are bred for man's uses only and have no ecological role in the wild.

Suitability for toxicity testing: Laboratory bred rats and mice are readily available and easily maintained. However, because they have been selectively bred for laboratory work and long removed from the genetic influences of wild types, their relationship to natural fauna is indefinable, and the results of studies using them may not be extrapolable to populations in the wild.

Carnivora; subgroup Canidae, Felidae—Domestic species. This subgroup of carnivores includes the dog (*Canis familiaris*) and the cat (*Felis domestica*), both familiar animals in the laboratory. They are adapted to living with man, and are very suitable for lab work.

Suitability for toxicity testing: This group is available in large numbers and adapted to laboratory life. The animals are handled with ease and are most suitable subjects for experimental work. How suitable they are for toxicity testing is difficult to say because their gene pools have been manipulated by man and, in some important ways, have been free for many years from natural selection pressures.

Primates. Because of their close relationship to man, nonhuman primates have been used in many fields of research. Primates are omnivorous animals. The most studied primate species other than man is the rhesus monkey (*Macaca mulatta*).

Ecological role: Nonhuman primates are omnivorous animals of tropical and subtropical regions of the world. Their staple foods are fruits, leaves, roots, insects; they frequently consume larger animals. They are prey for large predators.

Suitability for toxicity testing: Because of the long life spans of primates they are invaluable in long-term research. Nonhuman primates are desirable as experimental subjects because of their close relationship to man.

4.4. Toxicity, Carcinogenicity, Mutagenicity.

Acute toxicity testing. Acute tests, the simplest toxicity tests, determine whether a single exposure to a given chemical can produce a critical effect in a test animal. Here we summarize the five most common acute toxicity tests.

1. Acute Oral. The most frequently performed toxicity test is the acute oral. In this procedure, the test animal, usually a rat, receives a single gastrointestinal dose of the test material. A 14-day observation period follows administration of the potion.

Acute oral testing typically involves 15 animals. These animals are caged in groups of five, and dosages are varied among the different groups in order to establish a concentration effect. Technicians administer the dose by introducing the test material directly into the stomach by means of a tube (intubation) or into the throat by means of a syringe (lavage).

Gross necropsies (the pathologist's term for postmortem examinations) are performed on all animals—those that die and those that are sacrificed at the end of the observation period. All tissues and organs are visually inspected to determine gross effects caused by the chemical.

Laboratory personnel then statistically analyze data from animals that die as a result of the acute oral test. The amount of test material required to kill 50 percent of the animals with a 95 percent degree of confidence under a stated set of conditions is called the LD_{50}, the lethal dose for 50 percent of the test population. This value is expressed in milligrams of chemical per kilogram of body weight of animal. Thus, in a direct extrapolation, an LD_{50} of 200 mg/kg in the rat would be equivalent to 14 grams (0.5 ounce) for a 70 kg (155 lb) man.

We must be cautious in interpreting acute oral LD_{50} values. A number of factors often influence the figure. The LD_{50} can vary with the kind of test animal and even with the subspecies within a given species. It can also be affected by the method of dosing as well as by the animal's age and habitat.

We must recognize the uncertainty implicit in extrapolating from one species to another, and we must compensate for it with appropriate safety factors when we project animal data to humans.

2. Acute Dermal. The acute dermal test is similar to the acute oral test except that the test animal most frequently used is the rabbit, and technicians apply the test material over the closely clipped hair of the animal. The laboratory staff prepare solutions of the test material in various concentrations and apply 0.5 ml portions on two areas of each animal—one on abraded skin and the other on intact skin. As a precautionary measure each test area is sealed under a patch for 24 hours. Again, the test includes a 14-day observation period, and examiners conduct gross pathological examinations on all animals. Acute dermal LD_{50} values are calculated in the same manner as acute oral LD_{50}'s.

3. Acute Inhalation. As the name suggests, the acute inhalation test involves exposure to vapors, gases, dusts and other materials that can be inhaled. In this case, the rat is the typical test animal; it is exposed to a particular concentration of test material for a certain period.

Animals that survive the exposure period are kept under observation for 14 days, and all undergo gross necropsies. Mortalities are reported in terms of LC_{50} or LCT_{50}— lethal concentration or lethal concentration with time, respectively, required to kill 50 percent of the animals at 95 percent confidence level. The units may be expressed in milligrams/liter, milligrams/cubic meter or parts per million (for vapors of pure liquids).

4. Primary Skin Irritation. Similar in many respects to the acute dermal test, the primary skin irritation test has a different purpose. It is designed to determine whether or not a material irritates, sensitizes or corrodes the skin.

In this test, technicians apply a 0.5 ml or 0.5 g portion of test material over the clipped hair of the test animal, usually a rabbit. Exposure time can vary from 4 hours to 24 hours, depending on the protocol (the plan of the experiment). Examiners perform no necropsies but inspect the condition of the skin. They note only topical effects and express the results numerically from 0 (to indicate no effect) to 8 (to indicate corrosive action or extreme damage to the skin tissue).

5. Primary Eye Irritation. The other acute test designed to be nonlethal is the primary eye irritation test. In this test technicians introduce small amounts of test material into the eye of the rabbit. Either 0.1 ml of liquid or 0.1 g of solid is applied with and without post-application irrigation. After 1 week, laboratory personnel evaluate the effects on the rabbit's eye. They inspect conjunctiva, iris and cornea and rate the degree of injury using a composite score range from 0 for no effect to 10 for corrosive action.

Subchronic tests. Most subchronic toxicity studies begin with information generated in the acute tests. Dosage information is a case in point. Fractional amounts of the acute oral LD_{50} are used in range-finding experiments to establish the long-term dosage level to be used in chronic tests. In these subchronic tests the laboratory group administers 20, 10, and 5 percent of the acute oral LD_{50} to the test animals repeatedly for test periods that can run from 14 to 90 days. Subchronic testing also helps to identify target organs in the body as well as early cumulative effects of toxic materials.

Tests like these that are administered periodically and at less than the acute LD_{50} dose are called subacute or subchronic tests, the latter because they are conducted for a period shorter than the animal's life span. The terms subacute and subchronic are often used interchangeably. Acute and subchronic testing differ in the fact that subchronic studies always include a nontreated group of animals. This group acts as a control and insures that the observed effects are treatment-related.

Chronic toxicity testing. The move from subchronic to chronic studies involves mainly a difference in length of exposure. Chronic gastrointestinal (oral), inhalation, and dermal studies analogous to the acute and subchronic series determine what happens after a lifetime of exposure.

Certain other tests are chronic by definition. Tumorigenic or oncogenic (cancer) studies, for example, are regularly carried out over the lifetime of the animal. Even though some materials are tumorigenic enough to cause tumors in acute and subchronic exposures, most tumorigenic information comes from chronic studies.

Chronic tests also involve large animal populations. An individual test usually requires 200 animals; these are separated into groups of 50 for each of three dosage levels plus a control group.

Reproductive tests. Reproductive studies include teratological experiments to provide information on embryonic effects, and three-generation protocols to determine long-term reproductive effects.

Teratological Studies. The objective of teratological studies is to determine the ability of a material to cause foetal malformations. Groups of pregnant animals, usually rabbits, are fed varying doses of a test material as part of their regular diet. Just before normal delivery, each foetus is removed by caesarean section and carefully examined for abnormalities.

Three-Generation Reproductive Studies. Three-generation studies address the possibility of finding abnormal reproductive effects in either males or females as the result of exposure to a chemical agent. Rats are the preferred species for these studies. Young rats of both sexes receive the test compound as part of their diet. When the rats reach maturity they are mated, and their progeny are similarly treated. The process is repeated through the third generation. Examiners keep records of all pertinent data such as frequency of pregnancy, litter size, pup size, and any abnormalities that appear. Three-generation studies can reveal changes that are less obvious than those found through teratological tests. Subtle problems that are evidenced by genetic changes can often be detected by this extended procedure.

Animal bioassays for carcinogenicity. Chemicals found to cause cancer in animal tests are generally considered capable of causing cancer in humans. This assumption rests on the following observations:

- Chemicals known to cause cancer in humans cause cancer in animals, although the cancers may be in different organs. A possible exception is arsenic, a human carcinogen not conclusively shown to cause cancer in animals.
- Chemicals that induce cancer in one mammalian species will also induce cancer in other mammals, although one species may be more susceptible than another. There is little evidence that any cancer-causing chemical affects only one kind of animal.
- The ways that cancers develop are similar in humans and animals. Most types of human cancer can be induced in laboratory animals.
- The action of carcinogen molecules on target tissue is found to be similar in several animal species and humans. At the cell level, there is no reason to expect different types of responses in different species.

The fact that high dosages of test chemicals are administered to test animals should not be interpreted to mean that only unrealistically massive doses would be considered hazardous to human health. Animal bioassays for carcinogenicity are meant to screen chemicals for cancer causing potential, not to predict the frequency at which cancers will appear in human populations.

Mutagenicity testing. The constitution of living organisms is determined by the type of information it carries in the nuclear DNA (called "genetic" information). This information determines whether the organism is uni- or multicellular, the type of protein and enzymes that are contained in the cells, the site and shape of the organs, the type of development and the appearance of the fully developed individual. This genetic information is transmitted from parents to progeny and is responsible for the continuation of the species.

Genetic patterns alter with the evolution of the species leading to a change in the characteristics of successive generations. Such alterations are usually beneficial but they are not invariably so. The alteration in the inheritable characteristics is called a *mutation.*

Certain chemicals also produce an alteration in the heritable characteristics of living cells and are therefore called *mutagenic.* Since it is not always possible to be certain that a mutation is harmless it is the usual practice to assume that such a change is untoward and undesirable and likely to be harmful. Every chemical found to induce mutations in lower animals and organisms is assumed to be potentially mutagenic to man.

The significance of mutagenic changes acquired greater relevance when it was found that most of the commonly acknowledged carcinogens for animals and man are mutagenic to bacteria and to mammalian cells in culture. It is now widely accepted that the demonstration of an unequivocal mutagenic change by a chemical raises serious questions about the safety of the chemical, and although such results are not usually accepted at face value, they are always given considerable attention. In current toxicological practice, mutagenicity tests are looked upon as an essential part of the toxicological tests carried out on a compound. However, a mutagenic result cannot be interpreted in isolation and in the absence of good toxicology and knowledge of the type and extent of any human exposure.

Commonly employed mutagenicity tests in vitro. Bacterial tests: The most widely used test is the one developed by Bruce Ames. It employs bacteria (*S. typhimurium*) which have been bred for several generations under special conditions of growth which have deprived them of the ability to synthesize the essential amino acid histidine. Mutagenic agents mutate back these specially bred organisms to the wild (i.e., natural) state so that they no longer require the addition of histidine to the growth medium for survival. Other organisms with a different types of defect have been developed by other workers, notably Prof. Bridges of Sussex University, who developed a type of *E. coli*. Approximately 75% of carcinogens have given positive results in the Ames or *E. coli* test and it is reckoned that the test has a good predictability value. It is not, however, without drawbacks, since false positive results have been recorded with compounds known to be free of carcinogenic activity (e.g., formaldehyde and nitrate) and false negative results have been given with some carcinogens (e.g., nickel carbonyl).

Carcinogenicity versus mutagenicity. Chemicals are tested for mutagenicity on Petri plates with several specially constructed mutants of *Salmonella typhimurium*. Homogenates of rat (or human) liver are added directly to the Petri plates, thus incorporating an important aspect of mammalian metabolism into the in vitro test. In this way, a wide variety of carcinogens requiring metabolic activation can be detected easily as mutagens.

For each chemical, quantitative data are presented as revertants per plate (histidine revertants on a Petri plate/number of micrograms tested). For mutagenic compounds, data are from the linear region of dose-response curves, and for nonmutagenic chemicals, data are presented as "less than" figures at the highest dose level tested. The number of revertants per nmol is an indication of the mutagenic potency of the chemical in the test system.

The relationship of the mutagenic potency of a mutagen on a particular strain to the overall mutagenic potential of the compound for DNA in general, and to carcinogenic potency, remains to be determined. In addition the standard assay represents a compromise between various factors. Thus, comparisons between potency of different chemicals must be undertaken with caution.

A number of chemicals are listed in Table 26 in increasing order of mutagenicity in the *Salmonella* test. Nonmutagens have been tested over a wide dose range both with and without the liver microsome activating system. The test system is not suitable for metals entering the bacteria because of the large amounts of Mg salts, citrates, and phosphates in the medium.

TABLE 26. Mutagenic Potency of Known Carcinogens in the *Salmonella* Test (1883).

Carcinogen	Revertants per nmol	
2,4-diaminotoluene	0.60	
benzo(c)pyrene	0.43	(weak carcinogen)
benzidine	1.4	
ethylene-imine	2.0	(without liver homogenate)
β-naphthylamine	8.5	
benz(a)anthracene	11	
4-nitrobiphenyl	11	(without liver homogenate)
benzo(a)pyrene	121	
dibenz(a,c)anthracene	175	
2-aminoanthracene	510	
aflatoxin B1	7057	

E. GLOSSARY

acaricide (miticide) a material used primarily in the control of plant feeding mites (acarids), especially spider mites.

acroosteolysis regression of bone by decalcification, especially in the extremities, such as the finger tips. Exposure to vinyl chloride monomer is one cause.

actinomycetes a group of branching filamentous bacteria, reproducing by terminal spores. They are common in the soil. Selected strains are used for production of certain antibiotics.

adenoma nonmalignant enlargement of glandular type, often called *benign tumor*.

adjuvant an ingredient which, when added to a formulation; aids the action of the toxicant. The term includes such materials as wetting agents, spreaders, emulsifiers, dispersing agents, foaming adjuvants, foam suppressants, penetrants and correctives.

algicide A chemical intended for the control of algae

alkaloid A physiologically active, usually naturally occurring nitrogenous compound alkaline in reaction. Many are characteristic of specific plants, i.e., nicotine in tobacco.

anaesthetic A chemical that induces insensibility to pain, such as chloroform or diethyl ether. Vinyl chloride is also an anaesthetic.

angiosarcoma (haemangioarcoma) Malignant growth on the inner linings of the blood vessels, often found in areas of high blood vessel concentrations, such as the liver or kidneys. Vinyl chloride monomer is known to cause angiosarcoma of the liver.

anthelmintic A material used for the control of internal worms (helminths) parasitic in man and animals.

antibiotic Any of certain chemical substances produced by microorganisms such as bacteria and fungi (molds) and having the capacity to inhibit the growth of or destroy bacteria and certain fungi causing animal and plant diseases.

anticoagulant rodenticide a rodenticide which kills rats by inducing uncontrolled internal bleeding, e.g., Warfarin.

approximate fatal concentration the geometric mean between the largest concentration allowing survival for 48 hr and the smallest concentration that was fatal in this time for practically all fish.

avicide a substance to control pest birds.
benthic referring to aquatic organisms growing in close growth with the substrate.
benthos (or **benthon**) aquatic microorganisms capable of growth in close association with the substrate.
bio-accumulation (synonym: **bioconcentration**) the process in and by which chemical substances are accumulated in living organisms.
- direct bio-accumulation
  1. The process in which a chemical substance accumulates in organisms by direct uptake from the ambient medium through oral, percutaneous, or respiratory routes.
  2. The increase in concentration of test material in or on test organisms (or specified tissues thereof) relative to the concentration of test material in the ambient environment (e.g., water) as a result of partitioning, sorption or binding processes.
- indirect bio-accumulation: the process in which a chemical substance accumulates in living organisms through uptake via the food chain.

bio-accumulation factor the ratio of the concentration of the test chemical in the test animal to the concentration in the test environment (e.g., water) at steady state (apparent plateau) or the ratio of the uptake rate constant (k_1) to the depuration rate constant (k_2).
bio-accumulation potential
  1. the ability of living organisms to concentrate a chemical substance either directly from the ambient medium or indirectly through the food chain.
  2. the property of a chemical substance to be accumulated due to inherent properties, such as high lipophilicity and sorptivity.

biodegradability the ability of an organic substance to undergo biodegradation.
biodegradation molecular degradation of an organic substance, resulting from the complex action of living organisms.
biomagnification a term describing a process in which chemicals in organisms at one trophic level are concentrated to a level higher than that in organisms at the preceding (lower) trophic level.
bird repellent a substance which drives away birds or discourages them from roosting.
bloom a concentrated growth or aggregation of plankton, sufficiently dense as to be ready visible.
blue green algae the group Myxophyceae, characterized by simplicity of structure and reproduction, with cells in a slimy matrix and containing no starch, nucleus, or plastids and with a blue pigment in addition to the green clorophyll
cancer a malignant tumor anywhere in the body of a person or animal; its origin is usually in the several types of epithelial tissue and it invades any of the surrounding structures. The characteristic of metastasis or seeding to other organs of the body is positive to this diagnosis. Leukemia can be regarded as a cancer of the blood.
carbamate insecticides carbamates are esters of carbamic acid and like the organophosphorus compounds they inhibit chlorinesterase. Carbamic acid: $H_2N-COOH$.
carcinogen a highly controversial term, applied generally to any substance that produces cancer, as well as to highly specific chemicals suspected of being the cause of cancer development in any one of many target organs or the body in test animals or human beings. The words "cancer suspect agent" have been used by U.S. authorities to cover this possibility. However, NIOSH has broadened this terminology

to include any agent reported in the literature to cause or be suspected of causing a *tumor* development, *malignant* or *benign* (see **oncogenic**), e.g., some mineral oils, vinyl chloride, benzene, beta-naphthylamine, hydrazine, nickel, etc.

chelating agents (chelates) chelating agents are readily soluble in water and have found wide use in many areas through their control of metal ions. These chelants (chelated metal ions) are used, e.g., in the fields of textiles, water treatment, industrial cleaners, photography, pulp and paper, agriculture, etc. In agriculture, both macronutrients and micronutrients are essential for proper plant growth. In some areas, the intensification of agricultural practices has resulted in depletion of available micronutrients. In order to achieve adequate agricultural production, it is necessary to add micronutrients to these soils. A chelated micronutrient is made by reacting a metallic salt with one of the chelating agents, forming a protecting glove around the metal and retarding the normal soil chemistry reactions that tie up that metal. Thus it is more available to the plants.

chemosterilants compounds which sterilize insects sexually to prevent reproduction.

chlorinated organic pesticides the organochlorine chemicals form one of the three principal families; including, e.g., aldrin, benzene hexachloride, chlordane, DDT, endosulfan, heptachlor, lindane, methoxychlor, toxaphene, etc.

cholinesterase a body enzyme necessary for proper nerve function that is destroyed or damaged by organic phosphates or carbamates taken into the body by any path of entry.

cholinesterase-inhibiting pesticides a class of pesticides having related pharmacological effects, including aldicarb, carbaryl, carbofuran, chlorpyrifos, parathion, etc.

cohort a group of individuals selected for scientific study of toxicology or epidemiology.

compatibility the ability of two or more substances to mix without objectionable changes in their physical or chemical properties.

contact herbicide a herbicide which kills primarily by contact with plant tissues rather than by translocation (systemic herbicides).

contact insecticide a chemical causing the death of an insect with which it comes in contact. Ingestion is not necessary.

controls the most important factor in any statistically meaningful study of animals, humans or bacteriological experiments. The nature, number and reproducibility of the controls determine the accuracy and significance of the conclusions from the experimental cohort results.

coupling agent a solvent that has the ability to solubilize or to increase the solubility of one material in another.

critical range the range of concentrations in mg/l below which all fish lived for 24 hr and above which all died. Mortality is given as a fraction indicating the death rate (e.g, $\frac{3}{4}$).

cyclodiene insecticides the cyclodienes include mainly aldrin, chlordane, dieldrin, endrin, heptachlor, strobane, endosulfan, and toxaphene. They are characterized by their endomethylene bridge structure.

defoliant a preparation intended for causing leaves to drop from crop plants such as cotton, soybeans, or tomatoes, usually to facilitate harvest.

dessicant a preparation intended for artificially speeding the drying of crop plant parts such as cotton leaves and potato vines.

disinfectant a substance which destroys harmful bacteria, viruses, etc., and makes them inactive.
disinfestant a substance which destroys infesting organisms such as insects, mites, rats, weeds, and other organisms multicellular in nature.
dispersant a material that reduces the cohesiveness of like particles, either solid or liquid.
encapsulated pesticides pesticides enclosed in tiny capsules of such material as to control release of the chemical and extend the period of diffusion
epidemiology attempts to evaluate the health of a defined human population and to determine cause-and-effect relationships for disease distribution. Factors such as age, sex, and ethnicity are its parameters. It may be defined as the study of the distribution and determinants of disease and injuries in human populations. Examples of epidemiological studies are those that associated a lower incidence of dental caries with fluoridated drinking water and a higher incidence of lung cancer with cigarette smoking.
eradicant fungicide a fungicide used to destroy ("burn out") fungi which have already developed and produced a disease condition.
fumigant a substance or mixture of substances which produce gas, vapor, fume or smoke intended to destroy insects, bacteria or rodents.
fungus (plural, **fungi**) all non-chlorophyll bearing thallophytes (i.e., all non-chlorophyll bearing plants of a lower order than mosses and liverworts) as, for example rusts, smuts, mildews, and molds. Many cause destructive plant diseases. The simpler forms are one-celled; the higher forms are branching filaments.
green algae organisms belonging to the class Chlorophyceae and characterized by photosynthetic pigments similar in color to those of the higher green plants. The storage food is starch.
haematology examination of the blood.
herbicide a chemical intended for killing plants or interrupting their normal growth. Herbicides are used in five general ways:
1. pre-planting: applied after the soil has been prepared but before seeding.
2. pre-emergence (contact): nonresidual dosages are used after seeding but before emergence of the crop seedlings.
3. pre-emergence (residual): applied at time of seeding or just prior to crop emergence; it kills weed seeds and germinating seedlings.
4. post-emergence: application after emergence of a crop.
5. sterilant (non-selective): used to effect a complete kill of all treated plant life.

histopathology microscopic examination of tissue.
inoculum the inoculum is a combination of microorganisms which, in degradability experiments, is added to the test system in order to obtain degradation of the compounds under investigation (substrate).
insecticides the various insecticides fall into six general categories according to the way in which they affect insects:
1. stomach: toxic quantities are ingested by the insect.
2. contact: kills upon contact with an external portion of the body.
3. residual contact: remains toxic to insects for long periods after application.
4. fumigant: possesses sufficient natural or induced vapor pressure to produce lethal concentrations.

5. repellent: does not kill but is distasteful enough to insects to keep them away from treated areas.
6. systemic: capable of being absorbed into the plant system where they make plant parts insecticidal.

Various classes of insecticides, according to their composition, and examples of each, include:
- inorganics: calcium and lead arsenates, sodium fluoride, sulfur, and cryolite.
- botanicals: pyrethrum, nicotine, rotenone.
- chlorinated hydrocarbons: DDT, BHC, lindane, methoxychlor, aldrin, dieldrin, heptachlor, toxaphene, endrin.
- organic phosphates: parathion, diazinon, malathion, ronnel.
- carbamates: sevin, zectran.

isomer a chemical the molecules of which contain the same number and kind of atoms as another chemical but arranged differently; i.e., normal (straight chain) octylalcohol and its isomer, isooctyl alcohol. Stereoisomers are those isomers in which the same number and kind of atoms are arranged in an identical manner except for their relative positions in space; e.g., endrin is a stereoismer of dieldrin.

juvenile hormone a hormone produced by an insect in the process of its immature development which maintains its nymphal or larval form.

larvicide a substance intended to kill especially the larvae of certain insect pests such as mosquitoes.

leaching downward movement of a material in solution through soil.

LC_{50} (lethal concentration fifty) a calculated concentration which, when administered by the respiratory route, is expected to kill 50% of the population of experimental animals. Ambient concentration is expressed in milligrams per liter.

LD_{50} (lethal dose fifty) a calculated dose of a chemical substance which is expected to kill 50% of a population of experimental animals exposed through a route other than respiration. Dose concentration is expressed in milligrams per kilogram of body weight.

median tolerance limit (TL_m) this term has been accepted by most biologists to designate the concentration of toxicant or substance at which 50% of the test organisms survive. In some cases and for certain special reasons, the TL_{10} or TL_{90} might be used. The TL_{90} might be requested by a conservation agency negotiating with an industry in an area where an important fishery exists, and where the agency wants to establish waste concentrations that will definitely not harm the fish. The TL_{10} might be requested by a conservation agency which is buying toxicants designed to remove undesirable species of fish from fishing lakes.

metabolism all natural and synthetic chemicals ingested or inhaled by the living body, either animal or vegetable, are continually subjected to chemical transformation in the organism into other products by myriad chemical reactions, such as synthesis and oxidative transformation in the organs of the body. Many of these primary and intermediate products find their way to body excretions through lung exhalation, urine, feces, or other expirations. The tracing of these routes are important for specific chemicals and possible relation to disease. Isotope-tagged materials are used for these research studies. These studies are often called *pharmacokinetic* and *metabolism research*.

metastasis in medicine the shifting of pathogenic cells of a disease, such as a malignant tumor, from one part or organ of the body to another unrelated to it.

mite mites are tiny organisms closely related to ticks in the group Acarina. They have eight legs as do spiders, except the newly hatched mites which have only six. Some mites, such as the chicken mite and the chigger, are parasitic on higher animals. A large family of mites is known as the spider mites from the habit of spinning a web on undersides of leaves where they feed.

mold any fungus exclusive of the bacteria and yeasts, which is of concern because of its growth on foods or other products used by man. Fungus with conspicious profuse or wooly growth (mycelium or spore masses). Occurs most commonly on damp or decaying matter and on surface of plant tissues.

molluscicide a compound used to control snails which are intermediate hosts of parasites of medical importance to man

mothproofer a substance which, when used to treat woolens and other materials liable to attack from fabric pests, protects the material from insect attack.

mutagenesis the alteration of the genetic material of a cell in such a manner that the alteration is transmitted to subsequent generations of cells. A particular case is where a genetic change can be passed from parent to offspring.

mutation a sudden variation in some *inheritable* characteristic of an individual animal or plant, as distinguished from a variation resulting from generations of gradual change. It is an effect attributed to an action prior to conception of the embryo. It has been correlated with increased incidence of chromosome breaks in the reproductive cells, male or female. Possibility of such an action by VCM has recently been noted in an epidemiology study on PVC production workers compared with PVC fabricators and rubber workers at the same plant site. Research in this field in recent years has focused on the development of tests for the effect of chemicals on rapidly reproducing species, such as bacteria and fruit flies. The Ames test, in particular, has been used to identify chemicals that accelerate reversion of specially developed mutant strains of a bacterium, *Salmonella*. Certain types will rapidly revert with chemicals which are known to be carcinogenic in animals and humans. This research is interesting, but does not mean that a chemical that is a mutagen will be proven to be a carcinogen. Mutation studies in animals involve two- or three-generation exposures—i.e., through at least one pregnancy and preferably two—to look for mutations in genetic material of the chromosomes.

necrosis destruction of cells.

nematode a member of a large group (phylum Nematoda) also known as threadworms, roundworms, etc. Some larger kinds are internal parasites of man and animals. Nematodes injurious to plants, sometimes called eelworms, are microscopic, slender, wormlike organisms in the soil, feeding on or within plant roots or even plant stems, leaves, and flowers.

nematocide a material, often a soil fumigant, used to control nematodes infesting roots of crop plants; e.g., ethylene dibromide.

no effect level implies that the animals remained in good condition. In most experiments, blood and urine tests were made. The urine tests included specific gravity, pH, reducing sugars, bilirubin, and protein. The blood tests included hemoglobin concentration, packed cell volume, mean corpuscular Hb content, white and differential cell counts, clotting function, and the concentration of urea, sodium, and potassium. Control tests for hematological examination were made on the group of animals before exposure. No effect level in this context means: no toxic signs, organs normal in autopsy, blood and urine tests normal (if made).

nymph the early stage in the development of insects which have no larval stage. It is the stage between egg and adult during which growth occurs in such insects as cockroaches, grasshoppers, aphids, and termites.

organochlorine insecticides the principal pesticides included under organochlorines are the *bis*-chlorophenyls (e.g., DDT) and the cyclodienes (aldrin, etc.) with 50% chlorine content or more. These insecticides are characterized by their persistence in the environment.

organophosphorus pesticides organophosphorus compounds are anticholinesterase chemicals which damage or destroy cholinesterase, the enzyme required for nerve functions in the animal body. Use of some of these pesticides may involve danger for the applicator. Examples of the leading series are as follows, where R represents some organic radical:

- phosphate: dicrotophos

$$\begin{array}{c} CH_3 \\ \diagdown \\ P \\ \diagup\diagdown \\ CH_3O-R \end{array} \quad \begin{array}{c} O \\ \diagup\!\!\!\diagup \\ \end{array}$$

- phosphorothioate: parathion

$$\begin{array}{c} C_2H_5O \\ \diagdown \\ P \\ \diagup\diagdown \\ C_2H_5OO-R \end{array} \quad \begin{array}{c} S \\ \diagup\!\!\!\diagup \\ \end{array}$$

- phosphorodithioate: phorate

$$\begin{array}{c} C_2H_5O \\ \diagdown \\ P \\ \diagup\diagdown \\ C_2H_5OS-R \end{array} \quad \begin{array}{c} S \\ \diagup\!\!\!\diagup \\ \end{array}$$

organotin fungicides several tin-based organic fungicides are commercially available, e.g., triphenyltin acetate, triphenyltin hydroxide, tricyclohexyltin hydroxide.

oxidation pond an enclosure for sewage designed to promote the intensive growth of algae. These organisms release oxygen that stimulates transformation of the wastes into inoffensive products.

pathogen any micro-organism which can cause disease. Most pathogens are parasites but there are a few exceptions.

photosynthesis process of manufacture by algae and other plants of sugar and other carbohydrates from organic raw materials with the aid of light and chlorophyll.

phytoplankton plant microorganisms, such as certain algae, living unattached in the water. Contrasting term: **zooplankton**.

phytotoxicity degree to which a material is injurious (poisonous) to vegetation. It is specific for particular kinds or types of plants.

plant growth regulator a preparation which in minute amounts alters the behavior of ornamental or crop plants or the produce thereof through physiological (hormone) rather than physical action. It may act to accelerate or retard growth, prolong or break a dormant condition, promote rooting, or in other ways. See **chlorophenoxy-propionic acid**; **gibberellic acid**.

post-emergence herbicide a chemical applied as a herbicide to the foliage of weeds after the crop has emerged from the soil.

protozoa unicellular animals, including the ciliates and nonchlorophyllous flagellates.

quaternary ammonium compounds a type of organic nitrogen compound in which the molecular structure includes a central nitrogen atom joined to four organic groups as well as an acid group of some sort. Nitrogen forms such pentavalent compounds, as is shown in the simplest example, ammonium chloride (NH_4Cl). When the hydrogen atoms are replaced by organic radicals, the compound is known as a quaternary ammonium compound, e.g., tetramethyl ammonium chloride. These compounds are in contrast to trivalent nitrogen compounds where the nitrogen combines with only three hydrogen atoms, as in ammonia, or these are replaced by one to three radicals (e.g., the carbamate structure).

red algae a class of algae (Rhodophyceae) most members of which are marine. They contain a red pigment in addition to chlorophyll.

reentry time the period of time immediately following the application of a pesticide to a field when unprotected workers should not enter.

rodent a member of the animal group (order Rodentia) to which rats, mice, gophers, and porcupines belong.

rodenticide a preparation intended for the control of rodents (rats, mice, etc.) and closely related animals (such as rabbits).

saprophytic the capability by some plants, including certain bacteria and molds, of utilizing dead organic matter as nutrients.

sensitization an increased reaction on the second or subsequent exposure to a compound due to an immunological mechanism.

spray drift the movement of airborne spray particles from the intended area of application.

surface-active agent (surfactant) a substance that reduces the interfacial tension of two boundary lines. These materials can be classed as nonionic, anionic, and cationic. Most emulsifying agents are of the nonionic type, i.e., they do not ionize. Wetting agents and detergents are primarily anionic, i.e., they become ionized in solution, the negative molecule exerting primary influence.

synergist a material which exhibits synergism. The joint action of different agents so that the total effect is greater than the sum of the independent effects.

systemic pesticide a pesticide that is translocated to other parts of a plant or animal than those to which the material is applied.

teratogenicity the observation of stillbirths, birth defects, or malformations which may be caused by toxic chemicals affecting the growing fetus transplacentally. These studies are growing in importance as the percentage of female workers in our chemical plants increases. Animal studies on this type of effect are becoming essential not only for food and drug chemicals, but also industrial chemicals.

toxicology is the science that attempts to determine the harmful effects of materials on human populations by testing animals. It may be defined as the science that prescribes limits of safety for chemical agents. For example, the study that reported tumor development in animals when they were fed saccharin at high dosage levels was a toxicological study. Toxicology, the prospective science, warns of the

potential danger to man of a given chemical substance, and epidemiology, the retrospective science, considers a given population exposed to the chemical and determines that indeed this is or is not a hazardous case.

trademark defined as "a word, letter, device, or symbol used in connection with merchandise and pointing distinctly to the origin or ownership of the article to which it is applied." A trade-name is actually a trademarked name.

translocation distribution of a chemical from the point of absorption (plant leaves or stems, sometimes roots) to other leaves, buds, and root tips. Translocation occurs also in animals treated with certain pesticides.

triazine herbicides these materials include atrazine, simazine, etc., based on a symmetrical triazine structure, where R_1, R_2, and R_3 are a variety of attached radicals:

<chemical structure: symmetrical triazine with R_1, R_2, R_3 substituents>

weed any plant which grows where not wanted.

wetting agent a substance which appreciably lowers the interfacial tension between a liquid and a solid, and increases the tendency of a liquid to make complete contact with the surface of a solid.

wood preservative there are three main classes of wood preservatives: toxic oils (e.g., creosote) that evaporate slowly and are relatively insoluble in water; salts that are injected as water solutions into the wood; and those consisting of a small percentage of highly toxic chemicals in a solvent or mixture of solvents other than water. The waterborne types are simple to apply but they are subject to leaching, are more or less poisonous to warm-blooded animals, and some are corrosive to iron.

zooplankton protozoa and other animal microorganisms living unattached in water.

F. ABBREVIATIONS

abs. perc. limit—absolute perception limit
A.C.—activated carbon
A.S.—bench scale activated sludge, fill and draw operations
ASC—activated sludge, continuous feed and effluent discharge
ASCF—activated sludge, fed slowly during aeration period
atm—atmosphere
avg—average
BCF—bioconcentration factor
BOD_5—biological oxygen demand after 5 days at $20°C$
b.p.—boiling point
CA—chemical analysis for the test material
cal—calorie
CAN—chemical analysis to indicate nitrogen transformation
cu ft—cubic foot
cu m—cubic meter (m^3 in equations)

°C—degree centigrade
CO_2—carbon dioxide used to follow oxidation results
COD—chemical oxygen demand
conc.—concentration
det. lim.—detection limit
DO—dissolved oxygen
dyn. dil.—dynamic dilution
EC—electron capture
EC—effective concentration
effl.—effluent
EIR—eye irritation reactivity
°F—degree Fahrenheit
F—flow-through bioassay
6 f. abs. app.—six-fold absorber apparatus
f.p.—freezing point or fusion point
FT—flow-through bioassay
g—gram
GC-EC—gas chromatography–electron capture
GC-FID—gas chromatography–flame ionization detection
geom.—geometric
glc's—ground level concentration
HC. cons.—hydrocarbon consumption
HC's—hydrocarbons
hr—hour
95%ile—95 percentile
I.D.—internal diameter
i.m.—intramuscular
infl.—influent
inh.—inhibitory or toxic action noted
inhal.—via inhalation
i.p.—intraperitoneal
I.R.—infrared
i.v.—intravenous
kcal—kilocalorie
kg—kilogram
km—kilometer
LC_{50}—lethal concentration fifty
LD_{50}—lethal dose fifty
liq.—liquid
m—meter
m—month
MATC—maximum acceptable toxicant concentration
max—maximum
mg—milligram
min—minute
min.—minimum
MLD—median lethal dose = LD_{50}

mm—mmHg
m.p.—melting point
mph—miles per hour
μ—micron
m.w.—molecular weight
n-normality
nat—natural acclimation in surface water
NEN—Nederlandse norm (Dutch standard test method)
NFG—nonflocculant growth activated sludge
NOD—nitrogenous oxygen demand
NO ox—nitric oxide oxidation
n.s.i.—no specific isomer—means that in the literature, no reference has been made to a specific isomer. It does not necessarily mean that the stated information is valid for all isomers.
O.I.—odor index
or.—oral
O.U.—odor units
P—plant treatment system for mixed wastes including the test chemical
p.m.—particulate matter
PMS—photoionization mass spectrometer
p.p.—pour point
ppm—parts per million
ppb—parts per billion
R—renewal bioassay
R.C.R.—relative chemical reactivity
RD_{50}—concentration associated with 50% decrease in respiratory rate
Resp—special respirometer
RW—river water oxidation substrate
S—static bioassay
sat. conc.—saturation concentration in air
sat. vap.—saturated vapor
S.C.—subcutaneous
scf—standard cubic foot
Sd—seed material
sel. strain—selected strain, pure culture of organisms
sew—municipal sewage
sew. dil.—sewage dilution oxidation substrate
sp. gr.—specific gravity
SPM—suspended particulate matter
std. dil. sewage—the standard dilution technique has been used with normal sewage as seed material
STP—standard temperature and pressure
solub.—maximum solubility in water
$t^{1/2}$—half life
TF—trickling filter
THCE—total hydrocarbon emmissions
theor.—theoretical

ThOD—theoretical oxygen demand
TL$_m$—median threshold limit
T.O.C.—threshold odor concentration
TOC—total organic carbon
TOD—total oxygen demand
T.O.N.—threshold odor number
TSP—total suspended particulates
UVS—spectrophotometry with ultra violet light
v.d.—relative vapor density-air = 1
VLS—spectrophotometry with visible light
v.p.—vapor pressure
vs—versus
W—Warburg respirometer
yr—year

A

abate (O,O-dimethylphosphorothioate-O,O-diester with 4,4′-thiodiphenol; O,O,O′,O′-tetramethyl-O,O-thiodi-*p*-phenylene phosphorothioate; temephos; abathion; difenthos; nimitox)

$$H_3C-O-\underset{\underset{H_3C-O}{|}}{\overset{\overset{S}{\|}}{P}}-O-\underset{}{\bigcirc}-S-\underset{}{\bigcirc}-O-\underset{\underset{O-CH_3}{|}}{\overset{\overset{S}{\|}}{P}}-O-CH_3$$

Use: mosquito larvicide.

D. BIOLOGICAL EFFECTS
  —Algae:
    Dunaliella euchlora: 1000 ppb (36% reduction in O_2 evolution)
    Dunaliella euchlora: 100 ppb (23% reduction in O_2 evolution)
    Phaeodactylum tricornutum: 1000 ppb (38% reduction in O_2 evolution)
    Phaeodactylum tricornutum: 100 ppb (28% reduction in O_2 evolution)
    Skeletonema costatum: 1000 ppb (55% reduction in O_2 evolution)
    Skeletonema costatum: 100 ppb (23% reduction in O_2 evolution)
    Cyclotella nana: 1000 ppb (80% reduction in O_2 evolution)
    (O_2 evolution measured by Winkler Light-and-Dark Bottle Technique 1 1 of culture incubated 20 hrs in pesticide soln. then placed in test bottles.)   (2353)
  —Crustacean:
    amphipod: *Gammarus lacustris*:   (96 hr, LC_{50}) 82 µg/l   (2124)
    amphipod: *Gammarus pulex*:   (1 hr, $LC_{90\text{-}95}$) >> 1.0 ppm   (1653)
    decapod: *Uca pugnax*:   (24 hr) 4.31 ppm
                                                  Killed or eliminated the escape response of 50% of the organisms tested.   (1151)
    decapod: *Metapenaeus monoceros*: (72 hr, LC_{50} - S) 45 µg/l
    decapod: *Penaeus monodon*:   (72 hr, LC_{50} - S) 45 µg/l   (1470)
    korean shrimp: *Palaemon macrodactylus*: (TL_{50}) 2550(994-6540)ppb
                                                   96 hr static lab bioassay
    korean shrimp: *Palaemon macrodactylus*: (TL_{50}) 249(72.5-853)ppb
                                                   96 hr intermittent flow lab bioassay   (2353)
  —Insects:
    ephemeroptera: *Baltis rhodani*: (1 hr, $LC_{90\text{-}95}$) 0.001–0.002 ppm

trichoptera: *Brachycentrus subnubilis*: (1 hr, LC_{90-95}) 0.1-0.2 ppm
               Hydropsyche pellucidula: (1 hr, LC_{90-95}) 0.5-1.0 ppm
odonata: *Agrion*: (1 hr, LC_{90-95}) 0.04-0.05 ppm
diptera: *Simulium ornatum*: (1 hr, LC_{90-95}) 0.2-0.5 ppm   (1653)
Pteronarcys californica: (96 hr, LC_{50}) 10 µg/l   (2128)
—Fish:
   Salmo gairdneri: (96 hr, LC_{50}) 158 µg/l   (2137)
   Chingatta (*Channa gachua*): (96 hr, LC_{50}) 217 mg/l (50% EC) - S   (1494)
   eel: *Anguilla japonica*: (72 hr, LC_{50} - S) 7500 µg/l
   fish: *Mugil cephalus*: (72 hr, LC_{50} - S) 600 µg/l
   fish: *Mugil carinatus*: (72 hr, LC_{50} - S) 23 µg/l   (1470)
—Mammals:
   rat: acute oral LD_{50} 8600 mg/kg
   rabbit: acute dermal LD_{50} >400 mg/kg   (1854)

abathion *see* abate;

abietic acid (abietinic acid; sylvic acid)
$C_{19}H_{29}COOH$ (phenanthrene ring system)
major active ingredient of rosin.
A. PROPERTIES: yellowish resinous powder m.w. 302.44; m.p. 172-175°C
D. BIOLOGICAL EFFECTS:
  —Fish:
   Salmo gairdneri: (96 hr, TL_m - S) 0,7 mg/l   (441)
   Coho salmon (*Oncorhynchus kisuth*) Juvenile: (96 hr, LC_{50}) 0,41 mg/l   (1495)

abietinic acid *see* abietic acid

abs *see* Teepol 715

acenaphthene (1,8-hydroacenaphthylene; ethylenenaphthalene; periethylenenaphthalene)

Manufacturing source: petroleum refining; shale oil processing; coal tar distilling.
    (347)
Users and formulation: dye mfg; plastics mfg; insecticide and fungicide mfg.
    (347)
Natural sources (water and air): coal tar.   (347)
Man caused sources (water and air): combustion of tobacco; constituent in asphalt; in soots generated by the combustion of aromatic fuels doped with pyridine.
    (347; 1723)
A. PROPERTIES: m.w. 154.21; m.p. 90-95°C; b.p. 279°C; d. 1.069
C. WATER POLLUTION FACTORS:
  —Aquatic reactions: adsorption on smectite clay particles from simulated seawater

at 25°C—experimental conditions: 100 µg acenaphthene/l, 50 mg smectite/l—
adsorption: nil. (1009)
—Odor thresholds: T.O.C. in water at room temp.: 0.08 ppm, range 0.02-0.22 ppm,
14 judges. (321)
   20% of population still able to detect odor at 0.026 ppm
   10% of population still able to detect odor at 0.014 ppm
   1% of population still able to detect odor at 0.0019 ppm
   0.1% of population still able to detect odor at 0.00021 ppm
D. BIOLOGICAL EFFECTS:
—Mutagenicity: acenaphthene induced significant mutation to 8-azaguanine resistance in *S. typhimurium* at concentrations as low as 1000 µM (1723)

1,2 acenaphthenedione *see* acenaphthenequinone

acenaphthenequinone (1,2 acenaphthenedione)

Use: dye synthesis.
A. PROPERTIES: yellow needles; m.w. 182.18; m.p. 249-252°C (dec.).
C. WATER POLLUTION FACTORS:
  —Water quality:
   in Eastern Ontario drinking waters (June-Oct. 1978): n.d.-1.1 ng/l (n = 12)
   in Eastern Ontario raw waters (June-Oct. 1978): 0.9-10.6 ng/l (n = 2) (1698)

acenaphthylene
  Manmade sources: in soots generated by the combustion of aromatic hydrocarbon
  fuels doped with pyridine. (1723)
A. PROPERTIES: m.w. 152.2; m.p. 80-83°C; b.p. 280°C; sp.gr. 0.899; solub. 3.93 mg/l
  at 25°C in distilled water.
C. WATER POLLUTION FACTORS:
  —Water quality:
   in Eastern Ontario drinking waters (June-Oct. 1978): 0.1-2.0 ng/l (n = 12)
   in Eastern Ontario raw waters (June-Oct. 1978): 0.1-0.5 ng/l (n = 2) (1698)
D. BIOLOGICAL EFFECTS:
  —Mutagenicity: did not induce significant mutation to 8-azaguanine resistance in
   S. typhimurium at concentrations up to 1300 µM, the limit of solubility under
   the assay condition. (1723)

acetaldehyde (ethanal; ethylaldehyde; aldehyde; acetic aldehyde)
CH_3CHO
  Manufacturing source: organic chemical mfg. (347)
  Users and formulation: organic chem. mfg.; perfumes, flavors, aniline dyes, plastics,
  synthetic rubbers mfg., silvering mirrors, hardening gelatin fibers. (347)
  Natural sources (*water and air*): metabolic intermediate in higher plants; alcohol

140 ACETALDEHYDE

fermentation; sugar decomposition in body; by-product of most hydrocarbon oxidations. (347)
Man caused sources (water and air): vehicle exhaust; open burning and incineration of gas, fuel oil and coal; evaporation of perfumes; lab use. (347)

A. PROPERTIES: colorless liquid or gas; m.w. 44.1; m.p. -123.5°C; b.p. 20.2°C; v.p. 740 mm at 20°C; v.d. 1.52; sat.conc. 1811 g/cu m at 20°C; THC 279 kcal/mole; solub. in all proportions, sp.gr. 0.783 at 18/4°C; log P_{oct} 0.43 (calc.).

B. AIR POLLUTION FACTORS: 1 mg/cu m = 0.55 ppm, 1 ppm = 1.831 mg/cu m
  —Odor: characteristic: quality: green sweet, apple ripener (73)
                               hedonic tone: pungent (129)

odor thresholds   mg/cu m

[odor threshold chart: 10^{-7} to 10^4 mg/cu m; detection points around 10^{-2} to 10^{-1}; recognition points around 10^{-2} to 10^{-1}; not specified points around 10^{-2} to 10^{-1} and near 10^2]

                     (54; 73; 291; 279; 307; 610; 666; 710; 788; 836; 842; 871)

  odor index: 5.000.000 (2)
  —Natural sources: glc's Pt Barrow Alaska, Sept. 1967: n.d. -0.3 ppb (101)
  —Manmade sources: in gasoline exhaust: 0.8-4.9 ppm (195) (1053)
                       7.2-14.3 vol. % of total exhaust
                       aldehydes · (394; 395; 396; 397)
                in diesel exhaust: 3.2 ppm (311)
  —Control methods: activated carbon: retentivity; 7 wt% of adsorbent (83)
     wet scrubber: water at pH 8.5: outlet: 1700 odor units/scf
                 $KMnO_4$:   outlet: 200 odor units/scf (115)
  —Sampling and analysis: PMS: det. lim. 3-7 ppm (200)
     second derivative spectroscopy: det. lim. 400 ppb (42)
     photometry: min.ful scale 2200 ppm (53)
     non dispersive I.R.: min. full scale 275 ppm (55)
     detector tubes: UNICO: det. lim. 40 ppm (59)

C. WATER POLLUTION FACTORS:
  —BOD_5: 1.3 std.dil.sew. (27)
           1.27 std.dil.sew. (41)
  ThOD: 1.82

odor thresholds   mg/kg water

[odor threshold chart: 10^{-7} to 10^4 mg/kg water; detection points near 10^{-2} to 10^{-1}; recognition points near 10^{-3}]

                                                       (129; 873; 889; 908)

—Waste water treatment: A.S. after   6 hr: 11.0% ThOD
                                    12 hr: 21.5% ThOD
                                    24 hr: 27.6% ThOD   (88)
   A.S.: BOD, 20°C, 1/3-5d. observed, 30 d acclim.: 93% removed   (93)
   A.C. adsorbability: 0.022 g/g carbon, 11.9% reduction, infl. 1000 mg/l, effl. 723 mg/l   (32)
   anaerobic lagoon: COD/day/1,000 cu ft   infl. mg/l   effl. mg/l
                              13   30   10
                              22   80   35
                              48   80   40   (37)
—Sampling and analysis: photometry: min. full scale 124×10^{-6} mole/l   (53)

D. BIOLOGICAL EFFECTS:
   —Protozoan: threshold conc. of cell multiplication inhibition of the protozoa *Uronema parduczi* Chatton-Lwoff: 57 mg/l   (1901)
   —Fish: pinperch: TLm(24 hr) 70 mg/l   (248)
          sunfish: TLm(96 hr) 53 mg/l   (226)
   —Mammalia: cat: inhalation: (1/1) 13,600 ppm, 1/4 hr
                               (0/1) 4,100 ppm, 3-5 hr
                               (0/1)   256 ppm, 5 hr   (104)
                               (6/6) 16,000 ppm, 4hr   (54)
              rat: inhalation: LC_{100} sat.vap., 2 min
                               LC_{50} 20,500 ppm, 30 min
                   single oral dose: LD_{50}: 1.93 g/kg
                   subcutane: LD_{50}: 0.64 g/kg   (54)
   —Man: eye irritation sensitive persons: 25 ppm, 15 min
         eye irritation: 50 ppm, 15 min
         irritation of respiratory tract: 134 ppm, 30 min
         irritation of nose and throat: 200 ppm, 15 min   (54)

acetaldehydecyanohydrin   *see* lactonitrile

acetaldol   *see* 3-hydroxybutanal

acetamide   (ethanamide; acetic acid amine)
CH_3CONH_2
   Uses: organic synthesis, general solvent, lacquers, explosives, wetting agent.
A. PROPERTIES: m.w. 59.07; m.p. 81°C; b.p. 222°C; THC 282.6 kcal/mole; solub. 975 g/l at 20°C, 1780 g/l at 60°C; sp.gr. 1.159 at 20/4°C; log P_{oct} −1.58/−1.26 (calculated)
B. AIR POLLUTION FACTORS:
   —Odor threshold: recognition: 140-160 mg/cu m   (610)
C. WATER POLLUTION FACTORS:
   —BOD_5: 0.745
    $KMnO_4$: 0.755
    ThOD: 1.08   (30)
   —At 100 mg/l no inhibition of NH_3 oxidation by *Nitrosomonas sp.*   (390)
   —Waste water treatment: A.S. after   6 hr: 1.2% ThOD
                                       12 hr: 3.3% ThOD
                                       24 hr: 12.0% ThOD   (89)

ACETAMIDOACETIC ACID

D. BIOLOGICAL EFFECTS:
— Carcinogenicity: weak
— Mutagenicity in the *Salmonella* test: neg, <0.0008 revertant colonies/nmol; <70 revertant colonies at 5000 µg/plate (1883)
— Toxicity threshold (cell multiplication inhibition test).

bacteria (*Pseudomonas putida*):	>10.000 mg/l	(1900)
algae (*Microcystis aeruginosa*):	6.200 mg/l	(329)
green algae (*Scenedesmus quadricauda*):	>10.000 mg/l	(1900)
protozoa (*Entosiphon sulcatum*):	99 mg/l	(1900)
protozoa (*Uronema parduczi* Chatton-Lwoff):	>10.000 mg/l	(1901)

— Fish:

Gambusia affinis:	TLm (72 hr): 15.5–20 g/l	(30)
mosquito fish:	TLm (24; 48; 96 hr): 26.3; 26.3; 13.3 g/l	(41)

acetamidoacetic acid *see* N-acetylglycine

acetanilide (N-phenylacetamide; antifebrin)

$$\text{C}_6\text{H}_5-\text{NH}-\overset{\text{O}}{\underset{\|}{\text{C}}}-\text{CH}_3$$

Uses: stabilizer for cellulose ester coatings

A. PROPERTIES: m.w. 135.16; m.p. 114°C; b.p. 305°C; THC 1010 kcal/mole; solub. 5.63 g/l at 25°C, 35.0 g/l at 80°C; sp.gr. 1.21 at 4/4°C; log P_{oct} 1.16

B. AIR POLLUTION FACTORS:
— Odor threshold: 270 mg/cum (735)

C. WATER POLLUTION FACTORS:
— BOD_{10}: 1.20 std. dil. sew. (256)
— Biodegradation rates: adapted A.S. at 20°C — product is sole carbon source: 94.5% COD removal at 14.7 mg COD/g dry inoculum/hr
— Waste water treatment:

methods	temp °C	days observed	feed mg/l	days acclim.	% removed
NFG, BOD	20	1–10	50	365 + P	51
NFG, BOD	20	1–10	600–1000	365 + P	inhibition
RW, BOD	20	2–10	50	12	78

(93)

D. BIOLOGICAL EFFECTS:
— Fish:
Lepomis macrochirus: static bioassay in fresh water at 23°C, mild aeration applied after 24 hr

material added ppm	% survival after				best fit 90h LC_{50} ppm
	24 hr	48 hr	72 hr	96 hr	
320	40	40	40	0	
180	100	100	100	40	100
100	100	100	100	50	
79	100	100	100	70	

(352)

Menidia beryllina: static bioassay in synthetic seawater at 23°C mild aeration applied after 24 hr

material added	% survival after				best fit 96h LC$_{50}$
ppm	24 hr	48 hr	72 hr	96 hr	ppm
320	50	50	0	–	
210	100	100	100	0	
180	100	100	100	100–20 at 120 hr	115

(352)

C. auratus: BCF 1.23 (1871)

—Mammals: rats, monkeys tolerated oral doses of 200–400 mg/kg for many weeks.

(211)

acetate, sodium salt
(a) $NaC_2H_3O_2$
(b) $NaC_2H_3O_2 \cdot 3H_2O$

Uses: dye and color intermediate; pharmaceuticals; cinnamic acid; soaps, photography; meat preservation; medicine; electroplating; tanning; buffer in foods

D. BIOLOGICAL EFFECTS:
—Fish: *Lepomis macrochirus*: TLm, 24 hr: 5.00 mg/l (1294)
—Insects: *Culex* sp. larvae: TLm, 24 hr: 7.500 mg/l
TLm, 48 hr: 7.425 mg/l (1294)

acetic acid (ethanoic acid; methanecarboxylic acid; glacial acetic acid; vinegar acid) CH_3COOH

Manufacturing source: beetsugar mfg.; winery; vinegar mfg.; textile mills; wood distillation plants. (347)

Users and formulation: food processing plants; organic chemical mfg.; nylon and fiber mfg.; dyestuff and pigments mfg.; vitamins, antibiotics, hormones mfg.; rubber mfg.; photographic chemicals mfg.; ester solvents mfg.; plastics mfg. (347)

Natural sources (water and air): both plants and animals as normal metabolite (347)

Man caused sources (water and air): domestic use of vinegar; photographic film developing, lab use. (347)

A. PROPERTIES: colorless liquid; m.w. 60.05; m.p. 16.7°C; b.p. 118.1°C; v.p. 11.4 mm at 20°C, 20 mm at 30°C; v.d. 2.07; sp.gr. 1.049 at 20/4°C; sat.conc. 38 g/cu m at 20°C, 63 g/cu m at 30°C; log P_{oct} -0.31/-0.17

B. AIR POLLUTION FACTORS: 1 mg/cu m = 0.401 ppm, 1 ppm = 2.494 mg/cu m
—Odor: characteristic: quality: sour; hedonic tone: pungent

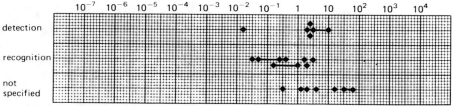

(57; 279; 610; 652; 670; 671; 679; 683; 709; 741; 753; 755; 778; 779; 825; 826; 835; 836)

ACETIC ACID

odor index: 15,000 (2)
—Control methods: thermal incineration: min. temp. for odor control: 743°C (94)
—Sampling and analysis:
gaswashing bottle: medium 200 ml water;
sampling rate: 0.12 cu ft/min; test conc. 520 ppm;
absorption efficiency: +99% (103)

C. WATER POLLUTION FACTORS
—BOD_5: 0.34; 0.45; 0.52; 0.53 (287)(284)(288)(281)
0.586; 0.62; 0.77; 0.85; 0.88 (282)(289)(260)(259)(285)
0.62 st.dil.acclim. (41)
76% bio.ox. in fresh water (23)
66% bio.ox. in sea water (23)
36% ThOD (220)
—BOD_{10}: 82% bio.ox. in fresh water
88% bio.ox. in sea water (23)
—BOD_{15}: 85% bio.ox. in fresh water
88% bio.ox. in sea water (23)
—BOD_{20}: 0.9 (30)
96% bio.ox. in fresh water
100% bio.ox. in sea water (23)
—BOD_{35}^{25}: 0.84 in seawater/inoculum: enrichment cultures of hydrocarbon oxidizing bacteria (521)

odor thresholds mg/kg

	10^{-7}	10^{-6}	10^{-5}	10^{-4}	10^{-3}	10^{-2}	10^{-1}	1	10	10^2	10^3	10^4
detection					•				•	•		
recognition												
not specified								•—•—•				

(97; 294; 295; 875; 896; 924)

—Manmade sources: in domestic sewages: 2.5 to 36 mg/l (85)
avg. content of secondary sewage effluent: 0.130 mg/l (86)
—Waste water treatment:
A.S.: 50% theor. oxidation of 500 ppm by phenol acclimated sludge after 12 hr aeration (26)
A.C.: adsorbability: 0.048 g/g carbon, 24% reduction; influent: 1,000 mg/l, effluent: 760 mg/l (32)
coagulation with 3 lb alum/1,000 gal: 8% BOD reduction (95)
A.S.: after 6 hr: 30% of ThOD
12 hr: 34.9% of ThOD
24 hr: 40.2% of ThOD (89)
stabilization pond design: toxicity correction factor: 1.6 at 270 mg/l in influent (179)

ACETIC ACID

methods	temp. °C	days observed	feed mg/l	days acclim.	theor. oxidation	% removed	
A.S., BOD	20	1-5	333	15	–	99	
A.S., Resp., BOD	20	1-5	716	<1	53	99	(93)

powdered carbon: at 100 mg/l sodium salt (pH 7.5)—carbon dosage 1000 mg/l: 1% adsorbed (520)
—Sampling and analysis: photometry: min. full scale: 430×10^{-6} mole/l (53)

D. BIOLOGICAL EFFECTS:
—Plants
—Phytotoxity:

	EC 50* mg/cu m	(95% conf.)
wheat	23.3	(4.7-48.0)
alfalfa	7.8	(5.8-10.2)
tobacco	41.2	(4.5-79.4)
soybean	20.1	(12.2-28.2)
corn	50.1	(35.1-65.4)

*EC 50, concentration required to cause visible injury in 50% of the leaves of the plant population exposed for a 2 hr fumigation period. (1831)
—Toxicity threshold (cell multiplication inhibition test)
  bacteria (*Pseudomonas putida*): 2850 mg/l (1900)
  algae (*Microcystis aeruginosa*): 90 mg/l (329)
  green algae (*Scenedesmus quadricauda*): 4000 mg/l (1900)
  protozoa (*Entosiphon sulcatum*): 78 mg/l (1900)
  protozoa (*Uronema parduczi* Chatton-Lwoff): 1350 mg/l (1901)
—Protozoa: *Vorticella campanula*: perturbation level: 12 mg/l (30)
—Arthropoda: brine shrimp: TLm (24-48 hr): 42-32 mg/l (23)
    D. magna: TLm (24 hr): 47 mg/l (26)
    Gammarus pulex: perturbation level: 6 mg/l (30)
—Mollusca: *Limnea ovata*: perturbation level: 15 mg/l (30)
—Fishes: bluegill: TLm (96 hr): 75 mg/l (23)
  Lepomis macrochirus: TLm (24 hr): 100-1000 mg/l (26)
  mosquito fish: TLm (24 hr-96 hr): 251 mg/l (41)
  goldfish: lethal at 423 mg/l, 20 hr (154)
  goldfish: period of survival at pH. 6.8: 48 hr to 4 days at 100 ppm
  goldfish: period of survival at pH 7.3: 4 days at 10 ppm (157)
  creek chub: LD_0: 100 mg/l, 24 hr; Detroit river (243)
  creek chub: LD_{100}: 200 mg/l 24 hr; Detroit river (243)
  mosquito fish: TLm (24 hr): 251 mg/l; turbid Oklahoma (244)
  goldfish: LD_0: 10 mg/l; long time exposure in hardwater (245)
  fathead minnows:
  static bioassay in lake superior water at 18 - 22°C:
    LC_{50}: (1; 24; 48; 72; 96 hrs): >315; 122; 92; 88; 88 mg/l
  static bioassay in reconstituted water at 18 - 22°C, pH ≤5.9
    LC_{50}: (1; 24; 48; 72; 96 hrs): 175; 106; 106; 79; 79 mg/l (350)

ACETIC ACID AMINE

—Insects: *Culex sp.* larvae: TLm (24-48): 1,500 mg/l (26)
—Mammals: not specified: single oral dose: LD_{50}: 3.3 g/kg
 guinea pigs, mice: inhalation: LC_{50}: 5,000 ppm, 1 hr (211)
—Man: severe toxic effects: 200 ppm = 500 mg/cu m, 60 min
 symptoms of illness: 40 ppm = 100 mg/cu m, 60 min
 unsatisfactory: 20 ppm = 50 mg/cu m, (185)
—Carcinogenicity: none
—Mutagenicity in the *Salmonella* test: none
 <0.004 revertant colonies/nmol
 <70 revertant colonies at 1000 µg/plate (1883)

acetic acid amine *see* acetamide

acetic aldehyde *see* acetaldehyde

acetic anhydride (acetic oxide; acetyl oxide; ethanoic anhydride) $(CH_3CO)_2O$
A. PROPERTIES: colorless liquid; m.w. 102.09; m.p. -68°C to -73°C; b.p. 139.9°C; v.p. 3.5 mm at 20°C, 5 mm at 25°C, 7 mm at 30°C; v.d. 3.52; sp.gr. 1.083 at 20/20°C; THC 431 kcal/mole, LHC 412 kcal/mole; sat.conc. 19 g/cu m at 20°C, 38 g/cu m at 30°C
B. AIR POLLUTION FACTORS: 1 mg/cu m = 0.236 ppm, 1 ppm = 4.24 mg/cu m
 —Odor: characteristic: quality: sour acid
 hedonic tone: neutral to pleasant
 T.O.C.: abs.perc.limit: 0.14 ppm
 50% recogn: 0.36 ppm
 100% recogn: 0.36 ppm
 O.I.: 14,611
 —Sampling and analysis: photometry: min. full scale: 60×10^{-6} mole/l (53)
D. BIOLOGICAL EFFECTS:
 —Algae: *Chlorella pyrenoidosa*: toxic at 360 mg/l (41)
 —Toxicity threshold (cell multiplication inhibition test)
 bacteria (*Pseudomonas putida*): 1150 mg/l
 green algae (*Scenedesmus quadricauda*): 3400 mg/l
 protozoa (*Entosiphon sulcatum*): 30 mg/l (1900)
 —Mammals: rat: single oral LD_{50}: 1.78 g/kg
 inhalation: death at 2,000 ppm, 4 hr (211)
 —Man: severe eye and skin irritant (211)

acetic ether *see* ethylacetate

acetic oxide *see* acetic anhydride

acetoacetic ester *see* ethylacetoacetate

acetoin (3-hydroxy-2-butanone; acetylmethylcarbinol; dimethylketol) $CH_3COCHOHCH_3$

ACETONE 147

Uses: preparation of flavors and essences.
A. PROPERTIES: slightly yellow liquid or crystalline solid; m.w. 88.10; m.p. 15°C; b.p. 148°C; sp.gr. 1.002 at 15/4°C; oxidizes gradually to diacetyl on exposure to air
C. WATER POLLUTION FACTORS:
—Waste water treatment: A.S.: after 6 hr: 2.6% ThOD
　　　　　　　　　　　　　　after 12 hr: 4.3% ThOD
　　　　　　　　　　　　　　after 24 hr: 13.8% ThOD   (88)

acetone (dimethylketone; 2-propanone; DMK)
$CH_3-CO-CH_3$
Users and formulation: mfg. smokeless powder; paints, varnishes, lacquers mfg.; organic chemical mfg.; pharmaceuticals mfg.; sealants and adhesives mfg.; solvents for celluloses acetate, nitrocellulose, acetylene.   (347)
Natural sources (water and air): normal microcomponent in blood and urine; minor constituent in pyroligneous acid; oxidation of alcohols and humic substances.
         (347)
A. PROPERTIES: clear colorless liquid; m.w. 58.08; m.p. -95°C; b.p. 56.2°C; v.p. 89 mm at 5°C, 400 mm at 39.5°C, 270 mm at 30°C; v.d. 2.00; sp.gr. 0.791 at 20°C; THC 431 kcal/mole, LHC 407 kcal/mole; sat.conc. 553 g/cu m at 20°C, 825 g/cu m at 30°C; $\log P_{oct}$ -0.24
B. AIR POLLUTION FACTORS: 1 mg/cu m = 0.415 ppm, 1 ppm = 2.411 mg/cu m
—Odor: characteristic: quality: sweet, fruity
　　　　　　　　　　　　hedonic tone: pleasant to neutral

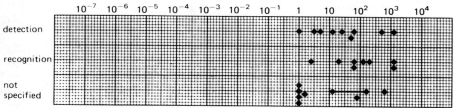

(19; 278; 279; 291; 606; 643; 655; 675; 676; 709; 721; 722; 741; 749; 790; 804; 829; 844)
　　odor index: 1740   (19)
　　threshold for unadapted persons: 0.03% in diluent
　　threshold after adaption with pure odorant: 5.0% in diluent   (204)
　　USSR: human odor perception: non perception: 0.8 mg/cu m
　　　　　　　　　　　　　　　　perception: 1.1 mg/cu m
　　　　animal chronic exposure: no effect: 0.5 mg/cu m
　　　　human reflex response: no response: 0.35 mg/cu m
　　　　　　　　　　　　　　　adverse response: 0.55 mg/cu m   (170)
—Natural sources: glc's: Pt Barrow Alaska, Sept 1967: 0.3 to 2.9 ppb   (101)
—Manmade sources: in cigarette smoke: 1100 ppm   (66)
　　　　　　　　　in gasoline exhaust: 2.3 to 14.0 ppm, (partly propionaldehyde)   (195; 1053)

ACETONE

—Sampling and analysis: photometry: min. full scale: 2,200 ppm (53)
I.R. spectrometry: det. lim.: 10 ppb (56)
non dispersive I.R.: min. full scale: 125 ppm (55)
detector tubes: UNICO: det. lim.: 500 ppm (59)
DRAGER: det. lim.: 100 ppm (58)
PMS: det. lim.: 5-9 ppm (200)

C. WATER POLLUTION FACTORS
- BOD_5: 0.81; 0.5; 1.0 (26)(27)(30)
  0.31-1.63 std.dil.acclim.sew. (41)
  0.544; 0.69; 1.39 std. dil. sew. (282)(285)(260)
  0.31 at 10 ppm, std. dil. sew. (269)
  0.05-0.72 at 1.7-20 ppm, std. dil. sew. (280)
  1.43 at 12 ppm, std. dil. sew. (268)
  0 at 440 ppm, Sierp, 10% sewage (280)
  1.76 (277)
  56% bio. ox. in fresh water
  38% bio. ox. in sea water (23)
  46% ThOD; 55% ThOD (220)(79)
- BOD_{10}: 1.20 std. dil. sew. (256)
  1.63 at 12 ppm, std. dil. sew. (268)
  67% bio. ox. in fresh water
  76% bio. ox. in sea water (23)
  72% ThOD (79)
- BOD_{15}: 83% bio. ox. in fresh water
  69% bio. ox. in sea water (23)
  78% ThOD (79)
- BOD_{20}: 1.78 (26)
  84% bio. ox. in fresh water
  76% bio. ox. in sea water (23)
  78% ThOD (79)
- BOD_{10}^{20}: 71.8% ThOD at 2.5 mg/l in mineralized dilution water with settled sewage seed (405)
- BOD_5^{20}: 0.81 at 10 mg/l, unadapted sew.; lag period: 2 days
- BOD_{20}^{20}: 1.78 at 10 mg/l, unadapted sew. (554)
- COD: 2.0; 1.12; 2.07; 1.63; 1.92 (23)(27)(36)(41)(277)
  100% ThOD (220)
- $KMnO_4$: 0.006 (30)
  acid 2% ThOD
  alkaline 6% ThOD (220)
- TOD: 2.20 (26)
- TOC: 100% ThOD (220)
- ThOD: 2.20
- Impact on biodegradation processes:

BOD: days at 20°C	BOD of original sample	BOD of sample + 50 ppm acetone
5 days	5 ppm	0 ppm
10 days	13 ppm	1 ppm

ACETONE 149

15 days	14 ppm	2 ppm	
20 days	15 ppm	4 ppm	
25 days	16 ppm	6 ppm	
30 days	16 ppm	6 ppm	(172)

digestion on sludge is inhibited from 4 g/l (30)
nitrification of activated sludge is decreased with 75% at 840 mg/l (30)
slight inhibition of microbial growth after 24 hr exposure at 5 ppm (523)
approximately 50% inhibition of ammonia oxidation in Nitrosomonas at 8100 mg/l (407)

odor thresholds   mg/kg water

10^{-7} 10^{-6} 10^{-5} 10^{-4} 10^{-3} 10^{-2} 10^{-1} 1   10 10^2 10^3 10^4

detection

recognition

not specified

(97; 294; 296, 894; 907; 908)

—Manmade sources:
in year old leachate of artificial sanitary landfill: 0.6 g/l (1720)
—Waste water treatment:
A.C.: adsorbability: 0.043 g/g carbon; 21.8% reduction infl.: 1,000 mg/l, eff.: 782 mg/l (32)
anaerobic lagoon:

lb COD/day/1,000 cu ft	infl. mg/l	effl. mg/l	
13	150	60	
22	150	80	
48	150	70	(37)

aeration by compressed air (stripping effect): 72% removal in 8 hr (30)
A.S. after 6 hr: 0.1% of ThOD
        12 hr: 0.0% of ThOD
        24 hr: toxic (88)

methods	temp °C	days observed	feed mg/l	days acclim.	% removed	
A.S., COD	20	1/3	333	30 +	86	
NFG, BOD	20	1–10	250–1,000	365 + P	47	(93)

—Sampling and analysis: photometry: min. full scale: 90×10^{-6} mole/l (53)

D. BIOLOGICAL EFFECTS:
—Toxicity threshold (cell multiplication inhibition test):
  bacteria (*Pseudomonas putida*):   1700 mg/l   (1900)
  algae (*Microcystic aeruginosa*):   530 mg/l   (329)
  green algae (*Scenedesmus quadricauda*):   7500 mg/l   (1900)
  protozoa (*Entosiphon sulcatum*):   28 mg/l   (1900)
  protozoa (*Uronema parduczi* Chatton-Lwoff): 1710 mg/l   (1901)
—Arthropoda: *Daphnia magna*:   TLm (24–48 hr): 10 mg/l   (26; 153)
              brine shrimp: TLm (24–48 hr): 2,100 mg/l   (23)
             Gammarus pulex: TLm: 5,500 mg/l   (30)

ACETONECYANOHYDRIN

—Fishes:
  mosquito fish: TLm (24; 48; 96 hr): 13,000 mg/l (244)
  goldfish: LD_{50} (24 hr): 5000 mg/l (277)
  bluegill sunfish (*Lepomis macrochirus*): LC_{50}, 96 hr: 8300 mg/l (593)
  fingerling trout: LC_{50}, 24 hr: 6100 mg/l (flow through system) (592)
  guppy (*Poecilia reticulata*): 14 d LC_{50}: 7032 ppm (1833)
—Amphibians:
  mexican axolotl (3-4 w after hatching): 48 hr LC_{50}: 20.000 mg/l
  clawed toad (3-4 w after hatching): 48 hr LC_{50}: 24.000 mg/l (1823)
—Mammals: mouse: inhalation: LC_{100}: 46,000 ppm, 1 hr
  rat: inhalation: LC_{100}: 126,000 ppm, 2 hr
  guinea pig: inhalation: LC_{100}: 50,000 ppm, $\frac{3}{4}$ hr
  cat: inhalation: LC_{100}: 21,000 ppm, 3 hr (54)
  rabbit: single oral lethal dose: 10 ml/kg
  dog: single oral lethal dose: 8 g/kg (211)
  rat: ingestion: single dose oral toxicity: LD_{50}: 9750 mg/kg (1546)
—Man: repeated exposure to 25–920 ppm: chronic conjunctivitis, pharyngitis, bronchitis, gastritis, gastroduodenitis (54)
  light irritation of the mucous membrane above 300 ppm (54)
  severe toxic effects: 4,000 ppm = 9650 mg/cu m, 60 min
  symptoms of illness: 800 ppm = 1930 mg/cu m, 60 min
  unsatisfactory: >400 ppm = 965 mg/cu m, 60 min (185)
  estimated minimum lethal dose by ingestion: 50 ml (277)
—Carcinogenicity: none
—Mutagenicity in the salmonella test: none
  <0.0004 revertant colonies/nmol
  <70 revertant colonies at 10^4 µg/plate (1883)

acetonecyanohydrin (α-hydroxyisobutyronitrile; isopropylcyanohydrin; 2-hydroxy-2-methylpropanenitrile)
$(CH_3)_2 C(OH)CN$
  Uses: insecticides; intermediate for organic synthesis especially methylmethacrylate.
A. PROPERTIES: colorless liquid; m.w. 58.10; m.p. -19°C; b.p. 81°C at 23 mm, 120°C decomposes; v.p. 0.8 mm at 20°C; v.d. 2.95; sp. gr. 0.93 at 19°C
B. AIR POLLUTION FACTORS: 1 mg/cu m = 0.28 ppm, 1 ppm = 3.54 mg/cu m
C. WATER POLLUTION FACTORS:
  —Waste water treatment: A.C.:

influent	carbon dosage	effluent	% removal
1,000 ppm	10 X	400 ppm	60
200 ppm	10 X	110 ppm	45
100 ppm	10 X	70 ppm	30 (192)

D. BIOLOGICAL EFFECTS:
  —Fishes:
    Lepomis macrochirus: static bioassay in fresh water at 23°C, mild aeration applied after 24 hr

material added	% survival after				best fit 96 hr LC_{50}
ppm	24 hr	48 hr	72 hr	96 hr	ppm
5.6	0 (1 hr)				
3.2	0 (1 hr)				
1.0	0 (4 hr)				
0.75	10	10	10	10	0.57
0.5	100	88	94	69	(352)

Menidia beryllina: static bioassay in synthetic seawater at 23°C: mild aeration applied after 24 hr

material added	% survival after				best fit 96 hr LC_{50}
ppm	24 hr	48 hr	72 hr	96 hr	ppm
0.75	0 (2 hr)				
0.50	50 (2 hr)	50	50	50	0.50
0.25	100	100	100	100	(352)

—Mammals: mice: acute oral LD_{50}: 15 mg/kg (211)

acetoneoxime *see* acetoxime

acetonitrile (ethanenitrile; methylcyanide)
CH_3CN

Use: solvent; manufacture of synthetic pharmaceuticals.

A. PROPERTIES: colorless liquid, m.w. 41.05; m.p. -44°C, -41°C; b.p. 82°C; v.p. 74 mm at 20°C, 115 mm at 30°C; v.d. 1.42; sp. gr. 0.79 at 20/4°C; LHC 296-302 kcal/mole; sat. conc. 163 g/cu m at 20°C, 249 g/cu m at 30°C log P_{oct} -0.34

B. AIR POLLUTION FACTORS: 1 mg/cu m = 0.586 ppm; 1 ppm = 1.706 mg/cu m
—Odor: T.O.C.: 68 mg/cu m = 39.8 ppm
  detection: 1950 mg/cu m (643)
—M.I.C.: P.E.L.: 160 ppm for 30 min
  80 ppm for 1 hr
  40 ppm for 2 hr (45)

C. WATER POLLUTION FACTORS:
—Waste water treatment: biodegradation by mutant microorganisms: 500 mg/l at 20°C% disruption: parent: 100% in 9 hr
  mutant: 100% in 1.5 hr (152)

methods	temp °C	days observed	feed mg/l	days acclim.	theor. oxidation	% removed
A.S., Resp, BOD	20	1-5	490	<1	nil	17
RW, CO_2, CAN	20	4	10	19	60	100
RW, CO_2, CAN	20	8	39	25+	60	100
RW, CO_2, CAN	5	25	39	25+ at 20°C	60	100
ASC, C N	22-25	28	139	28	70+	98+	(93)

—Impact on biodegradation processes: at 100 mg/l no inhibition of NH_3 oxidation by *Nitrosomonas* sp. (390)

3-(α-ACETONYLBENZENE)-4-HYDROXYCOUMARIN

D. BIOLOGICAL EFFECTS:
—Toxicity threshold (cell multiplication inhibition test)

bacteria (*Pseudomonas putida*):	680 mg/l	(1900)
algae (*Microcystis aeruginosa*):	520 mg/l	(329)
green algae (*Scenedesmus quadricauda*):	7300 mg/l	(1900)
protozoa (*Entosiphon sulcatum*):	1810 mg/l	(1900)
protozoa (*Uronema parduczi* Chatton-Lwoff):	5825 mg/l	(1901)

—Fishes: fathead minnows: TLm (96 hr): hard water: 1,020 mg/l   (41)
        fathead minnows: TLm (96 hr): soft water: 1,000 mg/l   (41)
        bluegill: TLm (96 hr): soft water: 1,850 mg/l   (252)
        guppies: TLm (96 hr): soft water: 1,650 mg/l   (41)

—Mammals: rats: acute oral LD_{50}: 1.7 to 8.5 g/kg
         guinea pig: acute oral LD_{50}: 0.18 g/kg
         rat: inhalation: LC_{50}: 7,500 ppm, 8 hr
         dog: inhalation: LC_0: 8,000 ppm, 4 hr
         rat: inhalation: no effect: 330 ppm, 90 days   (211)
—Man: no specific response: 160 ppm, 4 hr   (211)

3-(α-acetonylbenzene)-4-hydroxycoumarin   *see* warfarin

acetophenone   (methylphenylketone; hypnone; acetylbenzene; phenylmethylketone)

Manufacturing source: organic chemical industry; coal processing industry.   (347)
Users and formulation: perfume mfg.; solvent for synthesis of pharmaceuticals, rubber, chemicals, dyestuffs and corrosion inhibitors; plasticizer mfg.; tobacco flavorant; intermediate in synthesis of pharmaceuticals.   (347)
Natural sources (water and air): oils of castoreum and labdanum resin; buds of balsam poplar; heavy oil fraction of coal tar.   (347)

A. PROPERTIES: colorless liquid; m.w. 120.1; m.p. 19°C; b.p. 202°C; v.p. 1 mm at 15°C; v.d. 4.14; sp.gr. 1.03 at 20/4°C; solub. 5,500 mg/l; THC 989 kcal/mole; sat.conc. 1.96 g/cu m at 20°C, 3.80 g/cu m at 30°C; log P_{oct} 1.58

B. AIR POLLUTION FACTORS: 1 mg/cu m = 0.20 ppm, 1 ppm = 4.99 mg/cu m
  —Odor: characteristic: quality: sweet, almond
                 hedonic tone: pleasant
    T.O.C.: 0.01 mg/cu m = 2 ppb   (57)
         average: 0.17 ppm, range: 0.0039 to 2.02 ppm, 17 panellists   (210)
         abs. perc. limit: 0.30 ppm
         50% recogn.: 0.60 ppm
         100% recogn.: 0.60 ppm
         O.I.: 2,183   (19)
    USSR: human odor perception: 0.01 mg/cu m

human reflex response: no response: 0.003 mg/cu m
adverse response: 0.007 mg/cu m
animal chronic exposure: adverse effect: 0.07 mg/cu m (170)
—Manmade sources: in gasoline exhaust: < 0.1 to 0.4 ppm (195)
C. WATER POLLUTION FACTORS:
—BOD_5:   0.518; 0.5 (30)(27)
　　　　　 32% ThOD (220)
—BOD_{10}: 1.40 std. dil. sew. (256)
—COD:　 100% ThOD (220)
—$KMnO_4$: 0.020 (30)
　　　　　 acid 2% ThOD, alkal. 13% ThOD (220)
—ThOD:　 2.532 (30)
—Biodegradation:

$$\underset{\text{lactone-forming mono-oxygenases}}{\underset{\text{Ph-CO-CH}_3 \;\;\xrightarrow{\text{NADPH}_2,\; O_2}\;\; \text{Ph-O-CO-CH}_3}{}}$$

(426)

—Taste in fish: at 0,5 mg/l (41)
T.O.C.: average: 0.17 mg/l, range: 0.0039 to 2.02 mg/l (97)
　　　　0.17 mg/l (294)
　　　　0.065 mg/l (297)
T.O.C. in water: 68 ppm (326)
Water quality:
in Kanawha river: 13.11.1963: raw: 19 ppb
　　　　　　　　　　　　　　 sand filtered: not detectable
　　　　　　　　　 21.11.1963: raw: 29 ppb
　　　　　　　　　　　　　　 after aeration: 17 ppb
　　　　　　　　　　　　　　 carbon filtered: not detectable (159)
—Waste water treatment:
A.C.: adsorbability: 0.194 g/g carbon, 97.2% reduction, infl.: 1,000 mg/l;
　effl.: 28 mg/l (32)
Conventional municipal treatment: infl. 0.019 mg/l, effl. n.d. (404)
Conventional + A.C.:　　　　　　 infl. 0.021 mg/l, effl. n.d. (404)
NFG, BOD, 20°C, 1-10 days observed, feed: 100-1,000 mg/l, acclim.: 365 +
　P; 14% removed (93)
D. BIOLOGICAL EFFECTS:
—Fish: fathead minnows: static bioassay in Lake Superior water at 18-22°C:
　LC_{50} (1; 24; 48; 72; 96 hr): $> 200; > 200$; 103; 158; 155 mg/l (350)
—Mammals: not specified: single oral dose: 0.9-3 g/kg
　　　　　　　rat: repeated oral dose: no effect: 102 mg/kg/day, 30 days
　　　　　　　rat: inhalation: no deaths: sat. vap., 8 hr (211)

acetothioamide　*see* thioacetamide

acetoxime (2-propanoneoxime; acetoneoxime)
$(CH_3)_2 CNOH$
  Uses: organic synthesis (intermediate); solvent
A. PROPERTIES: m.w. 73.09; m.p. 61°C; b.p. 136.3°C; sp. gr. 0.97 at 20/20°C
C. WATER POLLUTION FACTORS:
  —BOD_5:   0.075
  —$KMnO_4$: 2.180
  —ThOD: 2.180   (30)
D. BIOLOGICAL EFFECTS:
  —Bacteria: *Pseudomonas*: toxic at 0.3 g/l   (30)

aceturic acid   see N-acetylglycine

acetylacetone   see 2,4-pentanedione

acetylaminoacetic acid   see N-acetylglycine

N-acetyl-2-aminoethanoic acid   see N-acetylglycine

acetylbenzene   see acetophenone

acetylene (ethine; ethyne)
CHCH
A. PROPERTIES: colorless gas; m.w. 26.04; m.p. −81.8°C; b.p. −84°C sublimes; v.d. 0.91; sp. gr. 0.62 (liquified); THC 312 kcal/mole, LHC 302 kcal/mole   (67)
B. AIR POLLUTION FACTORS: 1 mg/cu m = 0.92 ppm, 1 ppm = 1.08 mg/cu m
  —Odor threshold: detection: 240 mg/cu m   (637)
                   1300–2750 mg/cu m   (609)
  —Atmospheric reactions: reactivity; NO ox.: ranking: 0.1   (63)
  —Manmade sources:
     diesel engine: 14.1% of emitted hydrocarbons   (72)
     reciprocating gasoline engine: 12–17% of emitted HC's   (78;391;392;393)
     rotary gasoline engine: 3.3% of emitted HC's   (78)
     expected glc's in USA urban air: 15 to 250 ppb   (102)
  —Estimated lifetime under photochemical smog conditions in SE England: 206 hr   (1699; 1700)
  —Sampling and analysis:
     nondispersive I.R.: min. full scale: 125 ppm   (55)
     detector tubes: DRAGER: det.lim.: 500 ppm   (58)
                      AUER: det.lim.: 200 ppm   (57)
                      UNICO: det.lim.: 50 ppm   (59)
D. BIOLOGICAL EFFECTS:
  —Seed plants: sweet pea: declination in seedling: 250 ppm, 3 days
                tomato: epinasty in petiole: 50 ppm, 2 days   (109)
  —Fish: river trout: TLm 33 hr: 200 mg/l   (30)

acetylenedichloride   see 1,2-dichloroethyelene

acetylenetetrabromide (1,1,2,2-tetrabromoethane; *sym*-tetrabromoethane)
$CHBr_2-CHBr_2$
A. PROPERTIES: colorless to yellow liquid; m.w. 345.70; m.p. 0.1°C; b.p. 239-242°C; v.p. 0.1 mm at 20°C; v.d. 11.9; sp. gr. 2.964 at 20/4°C; solub. 651 mg/l at 30°C;
B. AIR POLLUTION FACTORS: 1 mg/cu m = 0.07 ppm, 1 ppm = 14.37 mg/cu m
  —Manmade sources: downtown Los Angeles: glc's: 0.068 to 0.234 ppm C
  East San Gabriel Valley: glc's: 0.050 to 0.103 ppm C (52)
D. BIOLOGICAL EFFECTS:
  —Mammals: rats, guinea pigs, monkey: growth depression in guinea pigs, all animals increase liver weight: 14 ppm, 7 hr/days, 5 days/w., 100 to 106 days
  rabbits, guinea pigs: oral LD_{50}: 0.4 g/kg
  rat: single oral dose: survived: 0.6 g/kg
  rat: single oral dose: succumbed: 1.6 g/kg (211)

acetylenetetrachloride *see* 1,1,2,2-tetrachloroethane

acetylenylcarbinol *see* propargylalcohol

N-acetylethanolamine (N-(2-hydroxyethyl)acetate)
$CH_3CONHC_2H_4OH$
A. PROPERTIES: m.w. 103.12; m.p. 63-65°C; b.p. 166°C at 8 mm; sp. gr. 1.108 at 25/4°C
C. WATER POLLUTION FACTORS
  —BOD_{10}: 1.10 std.dil.sew. (256)
  —Waste water treatment: NFG, BOD, 20°C, 1-10 days observed, feed: 200-1,000 mg/l; acclim.: 365 + P: 40% removed (93)

N-acetylglycine (N-acetyl-2-aminoethanoic acid: aceturic acid; acetamidoacetic acid; acetylaminoacetic acid)
$CH_3CONHCH_2COOH$
  Use: medicine.
A. PROPERTIES: m.w. 117.10; m.p. 206°C; solub. 21,700 mg/l at 15°C; log P_{oct} -1,80/-1,31 (calc.)
C. WATER POLLUTION FACTORS:
  —A.S.: after 6 hr: 9.3% ThOD
        12 hr: 10.0% ThOD
        24 hr: 18.5% ThOD (89)

acetylmethylcarbinol *see* acetoin

4-acetylmorpholine (N-acetylmorpholine)

156   ACETYL OXIDE

A. PROPERTIES: m.w. 129.16; m.p. 14°C; b.p. 152°C at 50 mm (decomposes); v.p. 0.02 mm at 20°C; v.d. 4.46; sp. gr. 1.12 at 20/20°C
C. WATER POLLUTION FACTORS
  —BOD_{10}: nil. std.dil.sew (256)
  —Waste water treatment: NFG, BOD, 20°C, 1-10 days observed, feed: 200–1,000 mg/l, acclim.: 365 + P: nil % removed (93)
D. BIOLOGICAL EFFECTS:
  —Mammals: rat: single oral LD_{50}: 6.13 g/kg
           rat: inhalation: no deaths: sat.vap., max 8 hr (211)

acetyl oxide *see* acetic anhydride

N-acetyl-p-phenylenediamine *see* p-aminoacetanilide

acid butylphosphate *see* butylphosphate

acraldehyde *see* acrolein

acridine

Manufacturing source: coal tar
Uses: manufacturing of dyes (1590)
A. PROPERTIES: small colorless needles; m.w. 179.21; m.p. 108°C, sublimes at 100°C; b.p. 346°C; sp. gr. 1.1 at 20/4°C; log P_{oct} 3.40
B. AIR POLLUTION FACTORS:
  —Manmade sources:
    in airborne coal tar emissions: 8.44 mg/g of sample or 297 μg/cu m of air
    in coke oven emissions: 32.4–172.8 μg/g of sample
    in coal tar: 9.18 mg/g of sample
    in wood preservative sludge: 3.47 g/l of raw sludge (993)
D. BIOLOGICAL EFFECTS:
  —Inhibition of photosynthesis of a freshwater, non-axenic unialgal culture of *Selenastrum capricornutum* at
    1% saturation: 92% carbon-14 fixation (vs. controls)
    10% saturation: 75% carbon-14 fixation (vs. controls)
    100% saturation: 1% carbon-14 fixation (vs. controls) (1690)
  —*Daphnia pulex*: bioaccumulation factor: 29.6 (initial conc. in water: 245 ppb)
    LC_{50}, 24 hr: 2.92 mg/l
    immobilization concentration: IC_{50}: 1.71 mg/l (1050)
  —Arthropoda: *Daphnia*: toxic at 0.7 mg/l (30)
  —Fish: toxic at 5 mg/l (30)
  —Bioaccumulation: fathead minnow (*Pimephales promelas*): bioaccumulation factor
    uptake from water: 125

uptake via interaction with contaminated sediment: 874
uptake via ingestion of contaminated zooplankton (*Daphnia pulex*): 30
uptake via ingestion of benthic invertebrates (*Chironomus tentans*)
living in contaminated sediments: 51
the calculated rates of uptake of acridine via ingestion of contaminated invertebrates (0.02 µg/g/h) and ingestion of sediment (0.01 µg/g/h) were negligable compared with direct uptake from water (1.40 µg/g/h) in a hypothetical system with all compartments in equilibrium. (1715)

acrolein (acraldehyde; acrylic aldehyde, allylaldehyde: 2-propenal; acrylaldehyde; aqualin)

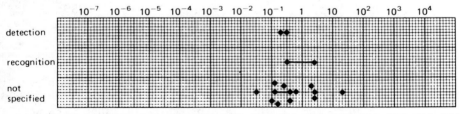

A. PROPERTIES: colorless to yellowish liquid; impurities of technical grade: 0.1% m (w) hydrochinon to prevent polymerisation; m.w. 56.1; m.p. $-87.7°C$; b.p. $52.5°C$; v.p. 220 mm at $20°C$, 330 mm at $30°C$; v.d. 1.94; sp. gr. 0.8427 at 20/20°C; THC 393 kcal/mole, LHC 379 kcal/mole; sat.. conc. 671 g/cu m at $20°C$, 974 g/cu m at $30°C$, solub. 20.8% at $20°C$.

B. AIR POLLUTION FACTORS: 1 mg/cu m = 0.43 ppm, 1 ppm = 2.328 mg/cu m
 —Odor: characteristic: quality: burnt sweet, hot fat, acrid
   hedonic tone: pungent

odor thresholds   mg/cu m

(2; 73; 151; 210; 241; 274; 298; 307; 625; 635; 710; 724; 741; 789; 813; 842)
 —USSR: human odor perception: 0.8 mg/cu m
   human reflex response: adverse response: 0.6 mg/cu m
   animal chronic exposure: adverse effect: 0.15 mg/cu m (178)
 —Manmade sources: in cigarette smoke: 150 ppm (66)
           in gasoline exhaust: 0.2 to 5.3 ppm (195)
           2.6–9.8 vol. % of total exhaust aldehydes
                        (394;395;396;397)
 —Control methods:
   activated carbon: retentivity: 15 wt% of adsorbent (83)
   wet scrubber: water at pH 8.5: outlet: 140,000 odor units/scf
           $KMnO_4$ at pH 8.5: outlet: 1 odor unit/scf (115)
   catalytic incineration over commercial Co_3O_4 (1.0-1.7 mm) granules, catalyst charge 13 cm³, space velocity 45000 h^{-1} at 100 ppm in inlet:
     80% decomposition at $190°C$

99% decomposition at 210°C
80% conversion to CO_2 at 190°C
95% conversion to CO_2 at 230°C
— Sampling and analysis:
photometry: min. full scale: 4,400 ppm (53)
sintered disc absorber 30 1 air/30 min: VLS: lower limit: 30 μ g/cu m/30 min
(208)

scrubbers: liquid lift type; 15 ml of liquid, 4.5 l/min gas flow 4 min scrubbing trap method:

	% removal
H_2O, 0°C	22
NH_2OH soln; 0°C	94
H_2SO_4 (conc.), 55°C	96
open tube, −80°C	13
ethanol, −80°C	91

C. WATER POLLUTION FACTORS
— BOD_5: 0.0 NEN 3235-5.4 (277)
— COD: 1.72 NEN 3235-5.3 (277)
— Odor threshold: 0.11 mg/kg (895)
— Experimental concentrations of 0.1 mg/l can significantly taint the flesh of rainbow trouts to make them unpalatable (1788)
— A.C.: adsorbability: 0.061 g/g carbon; 30.6% reduction, infl.: 1,000 mg/l, effl.: 694 mg/l (32)

methods	temp °C	days observed	feed	days acclim.	% theor. oxidation	
TF,Sd,BOD	20	10	?	?	33	(32)
RW,Sd,BOD	20	10	?	100	33	(93)

D. BIOLOGICAL EFFECTS:
— Protozoa (*Uronema parduczi Chatton-Lwoff*): inhibition of cell multiplication starts at 0.44 mg/l (1901)
— Bacteria: *Pseudomonas putida*: inhibition of cell multiplication starts at 0.21 mg/l (329)
— Algae: *Microcystis aeruginosa*: inhibition of cell multiplication starts at 0.04 mg/l (329)
— Insects: mayfly nymphs (*Ephemerella walkeri*): lowest observed avoidance conc. > 0.1 mg/l (1621)
— Fishes: goldfish: LD_{50} (24 hr) < 0.08 mg/l modified ASTM D 1345 (277)
  rainbow trout (*Salmo gairdneri*): lowest observed avoidance conc. 0.1 mg/l (1621)
  Lepomis macrochirus: 80 μg/l, 24 hr, LC_{50} (2106)
  Salmo trutta: 46 μg/l, 24 hr, LC_{50} (2108)
  Lepomis macrochirus: 79 μg/l, 24 hr, LC_{50} (2108)
— Fathead minnow: flow through bioassay:
  incipient TLm: 84 μg/l
  maximum acceptable toxicant concentration: 11.4 μg/l (443)

ACROLEIN 159

—Mammals:

Species	Conc. ppm	Duration	Inhalation Exposure Effect	
cat	870	2.5 hr	death during exposure	(963)
	650	2.25 hr	death within 18 hours	
rabbit	10.5	6 hr	$\frac{1}{2}$ died ⎫ histology: tracheobronchitis	(975)
guinea pig			$\frac{2}{4}$ died ⎬ with edema, consolidation,	
mice			$\frac{9}{20}$ died ⎭ congestion, and emphysema	
rat	8	4 hr	$\frac{1}{6}$ died	(984)
rat	0.66	60 days continuous	$\frac{7}{10}$ died	(985)
rat	8	4 hr	2, 3 or $\frac{4}{6}$	(986)
mouse	133–178	10 min.	60% mortality	(211)
	10	few min	lethal for most	
rat	131	30 min	LC$_{50}$ in 3 wk observation	(964)
guinea pig			mean lethal dose: 130 g-min/M^3	(961)
mouse			mean lethal dose: 70 g-min/M^3	
rabbit			mean lethal dose: 140 g-min/M^3	
man	153	10 min	death	(211)
cat	87	2.5 hr	severe pulmonary irritation	(963)
	17.5	4 hr	severe pulmonary irritation	
	11	up to 9 hr	salivation, lacrimation, nasal irritation, gradual light narcosis	
guinea pig	0.6	2 hr	increased pulmonary resistance and tidal volume, decreased respiratory	(987, 988)
guinea pig	17	1 hr	rate and minute volume	(972)
rats, dogs, guinea pigs	0.7 and 3.7	8 hr/day × 5 day/wk × 6 wk	0.7 ppm - all animals appeared normal; histologic sections showed inflammation and emphysema.	(985)
monkey			3.7 ppm - dogs and monkeys showed symptoms of respiratory irritation. All species showed abnormal weight gain; histologic sections showed morphologic changes in respiratory tract.	
	0.22	90 days	Severe respiratory irritation at 1.0 –	
	1.0	continuous	1.8 ppm; inflammatory changes in respiratory tract at all concentrations. At 1.8 ppm	
	1.8		squamous metaplasia and basal cell hyperplasia of the trachea in monkeys.	
rat	131	30 min	marked respiratory distress; pulmonary edema, hemorrhage, hyperemia	(964)
rat	0.55	11–21 days, continuous	upper respiratory tract irritation and increased susceptibility to airborne infections; adaptation after 4 weeks exposure	(989)
rabbit trachea	13.5–56	within 1 hr	ciliostasis	(965)
	0.25	5 min	moderate irritation	(211)
man	0.8	10 min	extremely irritating; only just tolerable	(976)
	1.2	5 min		(975)

1-2	5 min	87% of test panel reported irritation	(978)
1	5 min	82% of test panel reported irritation	
.5	5 min	35% of test panel reported irritation	
.5	5 min	19% of test panel reported irritation	
.5	12 min	91% of test panel reported irritation	
1.8	30 sec	odor	(990)
	1 min	slight eye irritation	
	2 min	slight nasal irritation; distinct eye irritation	
	3 min		
	4 min	profuse lacrimation; practically intolerable	
5.5	5 sec	odor, moderate nasal and eye irritation	
	20 sec	painful nasal and eye irritation	
	60 sec	marked lacrimation; practically intolerable	
21.8		intolerable	(211)
30.25	5 min	moderate irritation	
4	5 min	severely irritating to eyes	(977)
0.06		eye irritation .471 on scale 0-2	(991)
1.3-1.6		eye irritation 1.182 on scale 0-2	(992)
2.0-2.3		eye irritation 1.476 on scale 0-2	
150	10 min	fatal	(211)

acrylaldehyde *see* acrolein

acrylamide (propenamide; acrylic amide)
$CH_2=CHCONH_2$
    Uses: synthesis of dyes, etc.; polymers or copolymers as plastics, adhesives, soil conditioning agents; flocculants   (1590)
A. PROPERTIES: m.w. 71.08; m.p. 84-85°C; v.p. 2 mm at 87°C, 10 mm at 117°C; v.d. 2.46; solub. 2,050 g/l
B. AIR POLLUTION FACTORS: 1 mg/cu m = 0.34 ppm, 1 ppm = 2.95 mg/cu m
C. WATER POLLUTION FACTORS
  —BOD_5: 0.97 std. dil./acclimated   (41)
  —Manmade sources: in paper mill treated effluent: 0.47-1.2 μg/l
                      colliery: coal washing effluent: 1.8 μg/l
                             tailings lagoon: 39-42 μg/l   (214)
                         in sewage effluents: 0.280 mg/l   (231)
  —Waste water treatment:
  RW, COD, 20°C, 2 days observed, feed: 10 mg/l, 33 days acclim., 100% removed
  $KMnO_4$ oxidation: 1 mg/l at pH 5-8.5, 4 hr contact time, 100% removed
  MnO_2 column/pH 5-7, 0.1 hr contact time: 17-33% removed
  Ozone: 3 mg/l infl. at pH 7, 0.5 hr contact time: 100% removed
  Cl_2: 10 mg/l infl.: pH 1.0: contact time 4 hr, 100% removed

	10	5.0	4	64% removed
	10	8.5	4	0% removed
A.C.	8	5.0	0.5	13% removed	(214)

  —Sampling and analysis: GC-FID: det. lim.: 0.1 μg/l   (215)
D. BIOLOGICAL EFFECTS:
  —Fishes:

harlequin fish (*Rasbora heteromorpha*)

	24 hr	mg/l 48 hr	96 hr	3 m (extrap.)
$LC_{10}(F)$	390	220	103	
$LC_{50}(F)$	460	250	130	10

(331)
brown trout yearlings: 48 hr LC_{50}: 400 mg/l (static bioassay) (939)

acrylic acid (propenoic acid; ethylenecarboxylic acid)
$CH_2=CHCOOH$
   Use: monomer for polyacrylic and polymethacrylic acids and other acrylic polymers (1590)
A. PROPERTIES: m.w. 72.06; m.p. 12–14°C; b.p. 141°C; v.p. 3.2 mm at 20°C, 10 mm at 39°C; v.d. 2.50; sp. gr. 1.06 at 16°C; THC 327 kcal/mole; sat. conc. 12.6 g/cu m at 20°C, 22.8 g/cu at 30°C; log P_{oct} 0.31/0.43 (calc.)
B. AIR POLLUTION FACTORS: 1 mg/cu m = 0.33 ppm, 1 ppm = 3.00 mg/cu m
   –Odor: characteristic: quality: rancid, sweet
                       hedonic tone: unpleasant
   –T.O.C.: absolute: 0.094 ppm
            50% recognition: 1.04 ppm
            100% recognition: 1.04 ppm
   –O.I.: 105,700 (19)
   –Emission limits: if emission >0.1 kg/hr, than M.E.C. = 20 mg/cu m (178)
C. WATER POLLUTION FACTORS
   –COD: 100% ThOD
   –$KMnO_4$: acid: 55% ThOD
              alkaline: 42% ThOD
   –TOC: 95% ThOD (220)
   –Natural sources: produced by marine algae such as *Phaeocystis* and *Polysiphonia lanosa*; as a result of hydrolysis of dimethyl-β-propiothetin (514)
   –Waste water treatment:
     A.C.: adsorbability: 0.129 g/g carbon, 64.5% reduction, infl.: 1,000 mg/l, effl.: 355 mg/l (32)
     Sew P, CO_2, 26°C; 10 days observed, feed: 12 mg/l, 20 days acclimation, 35% theor. oxidation (93)
D. BIOLOGICAL EFFECTS:
   –Toxicity threshold (cell multiplication inhibition test)
     bacteria (*Pseudomonas putida*):   41 mg/l   (1900)
     algae (*Microcystis aeruginosa*):   0.15 mg/l   (329)
     green algae (*Scenedesmus quadricauda*):   18 mg/l   (1900)
     protozoa (*Entosiphon sulcatum*):   20 mg/l   (1900)
     protozoa (*Uronema parduczi* Chatton–Lwoff): 11 mg/l   (1901)
   –Mammals: rat: inhalation: no effect level: 80 ppm, 20 × 6 hr (65)
             rat: single oral LD_{50}: 2.5 g/kg (211)

acrylic aldehyde *see* acrolein

acrylic amide *see* acrylamide

acrylon *see* acrylonitrile

acrylonitrile (acrylon; carbacryl; cyanoethylene; fumigrain; 2-propenenitrile; VCN; ventox; vinylcyanide)
$CH_2=CHCN$
  Uses: the major use of acrylonitrile is in the production of acrylic and modacrylic fibers by copolymerization with methylacrylate, methylmethacrylate, vinylacetate, vinylchloride, or vinylidenechloride. Other major uses include the manufacture of acrylonitrile-butadiene-styrene (ABS) and styrene-acrylonitrile (SAN) resins. Acrylonitrile is also used as a fumigant.
  Formulations:
  acritet = 34% acrylonitrile, 60% CCl_4
  ventox = acritet
  carbacryl: equal volumes of acrylonitrile and CCl_4
  acrylofume: 3.95% acrylonitrile; 30% CCl_4; 30% chloroform; 0.5% chloropicrin
A. PROPERTIES: colorless liquid; m.w. 53.06; m.p. $-83°C$; b.p. $77.4°C$; v.p. 100 mm at $23°C$, 137 mm at $30°C$; v.d. 1.83; sp.gr. 0.8004 at $25°C$; sat. conc. 257 g/cu at $20°C$, 383 g/cu m at $30°C$; log P_{oct} -0.92
B. AIR POLLUTION FACTORS: 1 mg/cu m = 0.454 ppm, 1 ppm = 2.203 mg/cu m
  —Odor: characteristic: quality: onion, garlic
  hedonic tone: pungent
  —T.O.C.: recogn.: 3.72–51.0 mg/cu m
  1.7–23 ppm
  $PIT_{50\%}$: 21.4 ppm
  $PIT_{100\%}$: 21.4 ppm (2)
  average: 18.6 ppm
  number of panellists: 16 (210)
  41.9 mg/cu m = 19 ppm (279)
  45 mg/cu m = 20.4 ppm (291)
  detection: 3.4 mg/cu m (819)
  recognition: 47 mg/cu m (741)
C. WATER POLLUTION FACTORS
  —BOD_5: 0.72 std. dil./acclim. (41)
  nil (26)
  —BOD_{10}: 0.70 std. dil. sew. (256)
  BOD_5^{20}: nil at 10 mg/l, unadapted sew
  BOD_{30}^{20}: 1.21 at 10 mg/l, unadapted sew; lag period: 12 days (554)
  —COD: 1.39 (36)
  —ThOD: 3.17 (36)
  —Impact on biodegradation processes: BOD test is not influenced up to 1 g/l (30)
  at 100 mg/l no inhibition of NH_3 oxidation by *Nitrosomonas* sp. (390)
  —Reduction of amenities:
  T.O.C.: average: 18.6 mg/l, range: 0.0031 to 50.4 mg/l (97)
  29 mg/l (294)
  T.O.C. in water: 1.86 ppm (326)
  2.02 ppm
  3.9 ppb

Aquatic reactions: photooxidation by u.v. light in aqueous medium at 50°C:
  24.2% degradation to CO_2 after 24 hr (1628)
—Waste water treatment:
  biodegradation by mutant microorgansims: 500 mg/l at 20°C
  % disruption: parent: 84% in 24 hr
              mutant: 100% in 4.0 hr (152)

methods	temp °C	days observed	feed mg/l	days acclim.	% theor. oxidation	% removed
NFG, BOD	20	1–10	100–1,000	365+ P	–	25
RW, BOD	20	5–10	50	27	–	67
RW, COD	20	2	10	32	–	100
RW, CO_2, CAN	20	19	10	8	60	100
RW, CO_2 CAN	20	9	40	30+	60	100
RW, CO_2 CAN	5	33	40	30+ at 20°C	60	100
ASC, C N	22–25	28	89	21	70+	95+

(93)

—A.C.: influent; carbon; effluent; % reduction
  ppm   dosage   ppm
  1,000   10X   490   51
    100   10X   72   28 (192)

D. BIOLOGICAL EFFECTS
  —Bacteria: *Pseudomonas putida*: inhibition of cell multiplication starts at 53
     mg/l (329)
  —Arthropoda:
  Crangon crangon in sea water at 15°C:
    exposure time; EC_{50} after recovery period in unpolluted sea water (328)
                    0 min   24 hr
      1 min   >32,000 ppm   >10,000 ppm
      3 min   >32,000 ppm   <10,000 ppm
      9 min   15,000 ppm   ≪10,000 ppm
  Crangon crangon in sea water at 15°C:
    exposure time: LC_{50}, mg/l (328)
      3 min   >32,000
      9 min   ±15,000
     27 min   4,800
      1 hr   3,600
      3 hr   480
      6 hr   180
     24 hr   25
     48 hr   20
     72 hr   6
     96 hr   6
  —Fishes: fathead minnows: TLm (96 hr) hard water: 14.3 mg/l
           fathead minnows: TLm (96 hr) soft water: 18.1 mg/l

bluegill: TLm (96 hr) soft water: 11.8 mg/l
guppies: TLm (96 hr) soft water: 33.5 mg/l
pinperch: TLm (96 hr) soft water: 24.5 mg/l (41)
pinperch: TLm (24 hr) sea water: 24.5 mg/l (248)
bluegill sunfish: TLm (24 hr) soft water: 25 mg/l (252)
minnows: TLm 24 hrs: 37.4 mg/l (319)
TLm 48 hrs: 24.0 mg/l

Gobius minutus in sea water at $15°C$:
exposure time; LC_{50}, mg/l (328)

exposure time	LC_{50}, mg/l
3 min	±18,000
9 min	± 8,000
27 min	±18,000
1 hr	± 1,800
3 hr	435
6 hr	150
24 hr	20
48 hr	15
72 hr	14
96 hr	14

Gobius minutus in sea water at $15°C$: (328)
exposure time; EC_{50} after recovery period in unpolluted sea water

	0 min	24 hr
1 min	±18,000 ppm	>10,000 ppm
3 min	±18,000 ppm	3,200 ppm
9 min	± 8,000 ppm	> 3,200 ppm

Lagodon rhomboides: LD_0, 24 hr: 20 mg/l
LD_{50}, 24 hr: 24,5 mg/l
LD_{100}, 24 hr: 30 mg/l (439)

—Mammals: mice: acute oral LD_{50}: 35 mg/kg
rat: acute oral LD_{50}: 78 mg/kg
guinea pig: acute oral LD_{50} 90 mg/kg
?: repeated oral doses—no effect: 0.05%, 2 years (211)
rat inhalation: slight transitory effect: 129 ppm
rat inhalation: fatal: 636 ppm, 4 hr
rabbit inhalation: slight transitory effect: 97 ppm
rabbit inhalation: fatal: 258 ppm
cat inhalation: sometimes fatal: 152 ppm
guinea pig inhalation: slight transitory effect: 267 ppm
dog inhalation: very slight effects: 29 ppm
dog inhalation: $\frac{3}{4}$: 110 ppm (211)

rats: ingestion: 35 ppm (4 mg/kg body wt/day): mild signs of toxicity (decreased water and food consumption, decreased body wt).
100 ppm (10 mg/kg body wt/day) during 12 months: stomach papillomas (1 of 20 rats); central nervous system tumors (6 of 20 rats)
Zymbal gland carcinoma (2 of 20 rats) (481)
inhalation: 80 ppm (6 hr/d, 5 d/week, 1 year): 3 of 26 rats: central nervous

ADIPIC ACID DINITRILE 165

      system tumors. (481)
   animal carinogen: BRD 1977 (487)

actellic (pirimiphosmethyl; PP-511)

$$(CH_3O)_2-\overset{\overset{S}{\|}}{P}-O-\underset{N(C_2H_5)_2}{\text{pyrimidine}}$$

Use: insecticide and acaricide
D. BIOLOGICAL EFFECTS:
  —Fishes:
    carp (*Cyprinus carpio*): 48 hr LC_{50}; 0.005 ml/l (1199)
    rainbow trout (*Salmo gairdneri*): 48 hr LC_{50}; 0.001 ml/l
    guppy (*Poecilia reticulata*):   48 hr LC_{50}; 0.004 ml/l (1199)

adipic acid (hexanedioic acid; 1,4-butanedicarboxylic acid)
$COOH-(CH_2)_4-COOH$
A. PROPERTIES: m.w. 146.14; m.p. 151–153°C; b.p. 100 mm at 265°C; v.d. 5.04; sp.gr. 1.37; solub. 15,000 mg/l at 15°C; THC 669 kcal/mole; log P_{oct} 0.08
B. WATER POLLUTION FACTORS
  —BOD_5:   0.598 (30)
                  36% of ThOD (220)
  —BOD_{20} °C: 1.115 (3)
  —COD: 97% of ThOD (220)
  —$KMnO_4$ value: 0.003 (30)
                  acid 2% ThOD, alkal. 1% of ThOD (220)
  —ThOD: 1.423 (30)
  —Impact on biodegradation processes:
    A.S.: after 6 hr: 1.3% of ThOD
           12 hr: 1.3% of ThOD
           24 hr: 7.1% of ThOD (89)
  Aquatic reactions: photo-oxidation by u.v. in aqueous medium at 90–95°C; time for the formation of CO_2 (% of theoretical)   25%: 2.0 hr
                                                    50%: 5.0 hr
                                                    75%: 32.4 hr (1628)
D. BIOLOGICAL EFFECTS:
  —Bacteria; Algae, Protozoae: no effect: 100 mg/l (30)
  —Fishes: *L. macrochirus*: TLm (24 hr): <330 mg/l (153)
        fathead minnows: static bioassay in Lake Superior Water at 18–22°C
            LC_{50} (1; 24; 48; 72; 96 hr): >300; 172; 114; 97;
                97 mg/l (350)
  —Mammals: rat: inhalation: no effect level: 126 g/l, 15 × 6 hr (65)

adipic acid dinitrile *see* adiponitrile

adiponitrile (1,4-dicyanobutane; hexanedinitrile; adipic acid dinitrile; hexanedioic acid dinitrile; adipyldinitrile; tetramethylenedicyanide)
$CN-(CH_2)_4 CN$

A. PROPERTIES: colorless liquid; m.w. 108.15; m.p. 1°C; b.p. 295-306°C; v.p. 2 mm at 119°C; v.d. 3.73; sp. gr. 0.962 at 20/4°C
B. AIR POLLUTION FACTORS: 1 mg/cu m = 0.22 ppm, 1 ppm = 4.50 mg/cu m
C. WATER POLLUTION FACTORS
   —Waste water treatment:

methods	temp °C	days observed	feed mg/l	days acclim.	% theor. oxidation	% removed
RW,CO$_2$, CAN	20	13	10	8	60	100
RW,CO$_2$, CAN	20	9	40	30+	60	100

methods	temp °C	days observed	feed mg/l	days acclim.	% theor. oxidation	% removed
RW,CO$_2$, CAN	5	33	40	30+ at 20°C	60	100
ASC,CAN	22-25	28	120-160	28+	80+	98+ (93)

(41)

   —Oxidation parameters: COD: 1.9

D. BIOLOGICAL EFFECTS
   —Fishes: fathead minnows. TLm (96 hr): hard water: 820 mg/l
            fathead minnows: TLm (96 hr): soft water: 1,250 mg/l
            bluegill: TLm (96 hr): soft water: 720 mg/l
            guppies: TLm (96 hr): soft water: 775 mg/l   (41)
            bluegill sunfish: TLm (24 hr): soft water: 1,250 mg/l   (252)
   —Mammals: guinea pig: S.C. LD_{50}: 50 mg/kg   (211)

adipyldinitrile *see* adiponitrile

adronol *see* cyclohexanol

aerozine-50
   Composition: 50% hydrazine + 50% uns. dimethylhydrazine
   Use: rocket fuel
D. BIOLOGICAL EFFECTS:
   —Fish:
     guppy (Lebistes reticulatus): static bioassay:
       LC_{50}, 24 hr: 12.3 mg/l in hard water at 22-24.5°C
              48 hr: 4.37 mg/l in hard water at 22-24.5°C
              72 hr: 2.72 mg/l in hard water at 22-24.5°C
              96 hr: 2.25 mg/l in hard water at 22-24.5°C
       LC_{50}, 24 hr: 5.08 mg/l in soft water at 22-24.5°C
              48 hr: 2.86 mg/l in soft water at 22-24.5°C
              72 hr: 1.95 mg/l in soft water at 22-24.5°C
              96 hr: 1.17 mg/l in soft water at 22-24.5°C   (474)

comparison between previously exposed and unexposed guppies—pre-exposure period of 14 days at 1/25th 0.1 mg/l of toxicant concentration of 2.5 mg/l:
mean survival time of pre-exposed fish: 43.10 hr in soft water at 22–24.5°C
unexposed fish: 67.55 hr in soft water at 22–24.5°C
(474)

AGE *see* allylglycidylether

alamine 336 (aliphatic amines, C_{12}–C_{18})
Alamine is a trademark for a series of primary, secondary and tertiary aliphatic amines, organic substituted ammonia derivatives, chain length from C_{12} to C_{18}, with varying degrees of unsaturation. (1590)
Use: corrosion inhibitors, ore flotation agents, textile finishing agents, rubber compounding. (1590)
D. BIOLOGICAL EFFECTS:
—Fish: rainbow trout: 96 hr LC_{50}, S: 7.5–10 mg/l (1500)

***dl*-alanine** (*dl*-2-aminopropanoic acid; *dl*-α-aminopropanoic acid)
$CH_3CH(NH_2)COOH$
A. PROPERTIES: m.w. 89.09; m.p. 295°C; b.p. 200°C sublimes; solub. 166,000 mg/l at 25°C, 322,000 mg/l at 75°C; THC 388 kcal/mol
C. WATER POLLUTION FACTORS
—BOD_5:   0.942
             87% ThOD (274)
—COD: 77% ThOD (0.05 n $K_2Cr_2O_7$) (274)
—$KMnO_4$ value: 0.005
             0% ThOD (0.01 n $KMnO_4$) (274)
—ThOD: 1.080 (30)
—Natural sources: in soil: silty loam: 6 to 160 µg/kg soil
                  (n.s.i.): clay soils: 30 to 400 µg/kg soil (174)
—Manmade sources: excreted by man in urine: 0.55 mg/kg body wt/day (203)
                  in domestic sewage effluent: 5 µg/l (227)
—Waste water treatment:
A.S.:   α-DL-alanine   β-alanine
       after 6 hr: 11.7% of ThOD   1.7% ThOD
            12 hr: 27.0% of ThOD   6.9% ThOD
            24 hr: 43.0% of ThOD   16.0% ThOD   (89)

methods	temp °C	days observed	feed mg/l	days acclim.	% theor. oxidation	% removed
A.S., BOD	20	1–5	333	15	–	97
A.S., Resp, BOD	20	1–5	500	<1	39	96

(93)
powdered carbon: at 100 mg/l, carbon dosage 1000 mg/l: 1% absorbed (n.s.i.)
(520)

alar (succinic acid, 2,2-dimethylhydrazide
A plant growth regulator currently being used on a variety of food crops and fruit

trees. Since the chemical is applied as a foliar spray, direct soil contamination results. Also since alar is water soluble, it can be washed off the plants onto the soil. (1640)

C. WATER POLLUTION FACTORS:
— Biodegradation: the half-life of alar, on four soils under greenhouse conditions, was 3 to 4 days. Microbial degradation is the major route of dissipation. Major degradation product is CO_2, which is liberated in amounts of up to 84% in 14 days. (1640)

alcoholethoxylates *see* alfonic

prim-n-**alcohols** *see* dobanol

aldehydine *see* 2-methyl-5-ethylpyridine

aldicarb (Temik; 2-methyl-2(methylthio)propionaldehyde-O-(methylcarbamoyl)-oxime; Ambush)

$$CH_3-S-\underset{\underset{CH_3}{|}}{\overset{\overset{CH_3}{|}}{C}}-CH=NO\overset{\overset{O}{\|}}{C}NHCH_3$$

Use: systematic insecticide; acaricide and nematocide for soil use.

A. PROPERTIES: white crystalline solid, sp.gr. 1.195 at 25/20°C; v.p. 1×10^{-4} mm at 25°C solub. 0.6% at 25°C (technical grade)

C. WATER POLLUTION FACTORS:
— Degradation: aldicarb degrades quite rapidly in soils with the evolution of CO_2. Under field conditions a half-life of about 7 days in loam soil was found.
The following metabolites were identified: aldicarbsulfoxide [2-methyl-2-(methylsulfinyl)propionaldehyde-O-(methylcarbamoyl)oxime] and aldicarb-sulfone [2-methyl-2-(methylsulfonyl)-propionaldehyde-O-(methylcarbamoyl)-oxime]
— Aquatic reactions: leaching: leaching of aldicarb in Houston clay and Lufkin sandy loam is insignificant, but appeared to move more freely through columns of coarse sand (1611)

D. BIOLOGICAL EFFECTS:
— Mammals:
acute oral LD_{50} (rat): 0.9 mg/kg (technical aldicarb)
acute dermal LD_{50} (rabbit): >5,0 mg/kg (technical aldicarb)
acute oral LD_{50} (rat): approx. 7.0 mg/kg (temik 10G)
acute dermal LD_{50} (rat): dry, 2100–3970 mg/kg (temik 10G) (1854)

aldrin (1,2,3,4,10,10-hexachloro-1,4,4a,5,8,8a-hexahydro-1,4-*endo,exo*-5,8-dimetha-nonaphthalene; HHDN; aldrex; aldrite; aldrosol; drinox; octalene; seedrin liquid)

Use: insecticide, fumigant
A. PROPERTIES: brown and white crystalline solid; mp 104–105,5°C, v.p. 2,3 × 10^{-5} mm at 20°C (technical grade m.p. 49–60°C), solub. 0.01 mg/l
B. AIR POLLUTION FACTORS:
 —Atmospheric reactions (photochemical transformations):

Aldrin → Dieldrin
↓   ↓
Photo aldrin → Photo dieldrin

C. WATER POLLUTION FACTORS:
 —Odor threshold: 0.017 mg/kg water (326; 915)
 —Biodegradation: Metabolic pathway of aldrin and dieldrin under oceanic conditions: (1311)

Aldrin → Dieldrin → Photo dieldrin
     ↘   ↓
       Aldrin diol

 —Conversion of aldrin to dieldrin was 80% complete after 8 weeks in river water kept in a sealed jar under sunlight and artificial fluorescent light—initial conc. 10 µg/l (1309)

— Aqueous reactions: persistence in river water in a sealed glass jar under sunlight and artificial fluorescent light-initial conc. 10 µg/l

% of original compound found

after	1 hr	1 wk	2 wk	4 wk	8 wk	
	100	100	80	40	20	(1309)

photo-oxidation by u.v. light in aqueous medium at 90-95°C time for the formation of CO_2 (% of theoretical):

25%: 14.1 hr
50%: 28.2 hr
75%: 109.7 hr (1628)

— 75-100% disappearance from soils: 1-6 years (1815; 1816)
— calculated half-life in water at 25°C and 1 m depth, based on evaporation rate of $3{,}72 \times 10^{-3}$ m/hr: 185 hr (437)
— Water and sediment quality:
  in Northern Mississippi water: avg. 0.21 ng/l; range 0.01-0.49 ng/l (1082)
  Hawaï: sediment: 5.5-11.02 ppb (1174)

D. BIOLOGICAL EFFECTS:

— aquatic vascular plants: of lake Päijänne, Finland (1972, 1973)
  mean: 2 µg/kg dry weight n = 114
  S.D.: 5 µg/kg dry weight
  min.: 0 µg/kg dry weight
  max.: 36 µg/kg dry weight (1055)

— Arthropoda
  Crustaceans

	mg/l	LC_{50}	
Gammarus lacustris	9800	96	(2124)
Gammarus fasciatus	4300	96	(2125)
Palaemonetes kadiakensis	50	96	(2125)
Asellus brevicaudus	8	96	(2125)
Daphnia pulex	28	48	(2127)
Simocephalus serrulatus	23	48	(2127)

Korean shrimp (*Palaemon macrodactylus*): 96 hr static lab bioassay:
  0.74(0.51-1.08) ppb; TL_{50}, 96 hr (2352)
Korean shrimp (*Palaemon macrodactylus*): 96 hr intermittent flow lab bioassay:
  3(1.1-8.5) ppb; TL_{50}, 90 hr (2352)
Sand shrimp (*Crangon septemspinosa*): 96 hr static bioassay:
  8 ppb; LC_{50}, 96 hr (2327)
Grass shrimp (*Palaemonetes vulgaris*): 96 hr static bioassay:
  9 ppb; LC_{50}, 96 hr (2327)
Hermit crab (*Pagurus longicarpus*): 96 hr static bioassay:
  33 ppb; LC_{50}, 96 hr (2327)
Daphnia magna: LC_{50}, 24 hr: 30 µg/l
                        48 hr: 28 µg/l (1002; 1004)
isopod (*Asellus*): LC_{50}, 24 hr; 80 ppb (1681)

— Toxicity ratios of aldrin (A) to photo-aldrin (PA), calculated from respective 24 hr LC_{50} and LT_{50} values:

A/PA
Crustacea
  Daphnia pulex (water flea)   1.43

   Gammarus spp. (amphipod)   1.83
   Asellus spp. (isopod)   2.13 (2.0)*
   Insects
    Aedes aegypti larvae (mosquito) 5.68 (6.0)*
    Musca domestica (house fly) 2.13 (2.0)*
   Fish
    Lebistes reticulatus (guppy) 2.41
    Pimephalus promelas (bass) 1.42
    Lepomis macrochirus (blue gill) 3.60 (2.9)*
                   (1681*; 1684; 1685)
 shrimp (*Nephrops norvegicus*): avg 0.2 ppb fresh wt. ($n = 7$)
  from Central Mediterranean—1976/77); range 0.1–0.4 ppb fresh wt. (1774)
—Mollusca
toxicity:
Hard clam (*Mercenaria mercenaria*): 10 day two-cell stage fertilized, 500 ppb,
 37% survival   (2324)
 10 day eggs introduced into test media, 1000 ppb, 0% survival (2324)
 48 hr 50 percent of eggs develop normally, >1000 ppb, TLM (2324)
 12 day 50 percent of larvae survive, 410 ppb, TLM   (2324)
bioaccumulation: BCF: 5 aquatic molluscs: 350–4500   (1870)
residue: mussel (*Mytilus galloprovincialis*): avg 1.02 ppb fresh wt. (n = 4)
 (from Central Mediterranean—1976/77): range 0.4–1.7 ppb fresh wt. (1774)
—Insects:

          µg/l
Pteronarcys californica 1.3 96 hr LC_{50} (2128)
Pteronarcys californica 180 96 hr LC_{50} 2.5 µg/liter (30 day LC_{50}) (2118)
Acroneuria pacifica   200 96 hr LC_{50} 22 µg/liter (30 day LC_{50}) (2118)
 fourth instar larvae *Chironomus riparius*: 24 hr LC_{50}: 0.8 µg/l (1853)
 house fly (3 days old female *Musca*): LD_{50}: 14 µg/fly   (1681)
 mosquito (late 3rd instar *Aedes aegypti* larvae): 24 hr LC_{50}: 3 ppb (1681)
 Pteronarcys: LC_{50}, 48 hr: 43 µg/l   (1003)
 Hydropsyche larvae: significant modification of net construction after 48 hr
  exposure: 20 µg/l   (1006; 1007; 1008)
—Fish: Toxicity:
Mummichog (*Fundulus heteroclitus*): 96 hr static bioassay: 4 ppb; 96 hr, LC_{50}
                       (2328)
Mummichog (*Fundulus heteroclitus*): 96 hr static bioassay: 8 ppb; 96 hr, LC_{50}
                       (2329)
Striped killifish (*Fundulus majalis*): 96 hr static bioassay: 17 ppb; 96 hr, LC_{50}
                       (2329)
Atlantic silverside (*Menidia menidia*): 96 hr static bioassay: 13 ppb; 96 hr, LC_{50}
                       (2329)
Striped mullet (*Mugil cephalus*): 96 hr static bioassay: 100 ppb; 96 hr, LC_{50}
                       (2329)
Bluehead (*Thalassoma bifasciatum*): 96 hr static bioassay: 12 ppb; 96 hr, LC_{50}
                       (2329)
Northern puffer (*Sphaeroides maculatus*): 96 hr static bioassay: 36 ppb; 96 hr,
 LC_{50} (2329)

American eel (*Anguilla rostrata*): 96 hr static bioassay: 5 ppb; 96 hr, LC_{50}
(2329)
Threespine stickleback (*Gasterosteus aculeatus*): 96 hr static bioassay: 27.4 ppb; 96 hr, TLM (2333)
Shiner perch (*Cymatogaster aggregata*): 96 hr static bioassay: 7.4 ppb; 96 hr, TL_{50} (2354)
Shiner perch (*Cymatogaster aggregata*): 96 hr intermittent flow bioassay: 2.26 ppb (1.08–4.74); 96 hr, TL_{50} (2354)
Dwarf perch (*Micrometrus minimus*): 96 hr static lab bioassay: 18 ppb; 96 hr, TL_{50} (2354)
Dwarf perch (*Micrometrus minimus*): 96 hr intermittent flow bioassay: 2.03 ppb (1–4.2); 96 hr, TL_{50} (2354)
Pimephales promelas: 28 µg/l: 96 hr LC_{50} (2113)
Lepomis macrochirus: 13 µg/l: 96 hr LC_{50} (2113)
Salmo gairdneri: 17.7 µg/l: 96 hr LC_{50} (2119)
Oncorhynchus kisutch: 45.9 µg/l: 96 hr LC_{50} (2119)
Oncorhynchus tschawytscha: 7.5 µg/l: 96 hr LC_{50} (2119)
Striped bass (*Morone saxatilis*): 96-hr LC_{50},s: 0.010 mg/l
Banded killifish (*Fundulus diaphanus*): 96-hr LC_{50},s: 0.021 mg/l
Pumpkinseed (*Lepomis gibbosus*): 96-hr LC_{50},s: 0.02 mg/l
White perch (*Roccus americanus*): 96-hr LC_{50},s: 0.042 mg/l
American eel (*Anguilla rostrata*): 96-hr LC_{50},s: 0.016 mg/l
Carp: 96-hr LC_{50},s: 0.004 mg/l
Guppy: 96-hr LC_{50},s: 0.02 mg/l (1193)
Salmo gairdneri: 96 hr LC_{50}: 10 µg/l (1001)
bluegill: 24 hr LC_{50}: 260 ppb (1681)
bluegill: 96 hr LC_{50}: 0.013 ppm
rainbow trout: 96 hr LC_{50}: 0.036 ppm (1878)
susceptible mosquito fish: 48 hr LC_{50}: 36 ppb
resistant mosquito fish: 48 hr LC_{50}: 2735 ppb (1851)
Residues: in marine animals from the Central Mediterranean (1976-1977): fish:
anchovy (*Engraulis encrasicholus*): avg 0.26 ppb fresh wt. ($n = 12$)
range 0.1–0.8 ppb fresh wt.
striped mullet (*Mullus barbatus*): avg 0.59 ppb fresh wt. ($n = 10$)
range 0.2–1.7 ppb fresh wt.
tuna (*Thunnus thynnus thynnus*): avg 0.14 ppb fresh wt. ($n = 5$)
range 0.1–0.2 ppb fresh wt. (1774)
—Birds: mean concentration in gamebird muscle in upper Tennessee (USA); mg/kg fresh weight ± standard error

	grouse	quail	woodcock
Johnson county	0.29 ± 0.03 ($n = 12$)	0.28 ± 0.08 ($n = 6$)	0.18 ± 0.06 ($n = 6$)
Carter county	0.30 ± 0.04 ($n = 9$)	0.20 ± 0.06 ($n = 7$)	0.28 ± 0.08 ($n = 6$)
Washington county	0.59 ± 0.16 ($n = 10$)	0.23 ± 0.07 ($n = 11$)	0.32 ± 0.09 ($n = 6$)

Quail: acceptable LD_{50}: 4 mg/kg
FAO/WHO Residue tolerance limit: 0.03–0.3 mg/kg/day (1338)
—Mammals: Toxicity:
acute oral LD_{50} (rats) approximately 67 mg/kg
acute dermal LD_{50} (rats) 98–>200 mg/kg (1854; 1802)

ingestion: rats fed for two years at dietary level of 5 ppm suffered no ill effects but liver damage resulted at the 25 ppm level (1855)
residues
Hooded Seal, Greenland: Fat: 0.028 mg/g. (1180)

aldol *see* 3-hydroxybutanal

algerite alba *see* hydroquinone monobenzylether

algerite powder *see* N-phenyl-α-naphthylamine

aliphatic amines *see* alamine 336

alkylarylsulfonate *see* teepol 715

alkylbenzenes (*see also* dobane)

D. BIOLOGICAL EFFECTS:
—Bioaccumulation factors: In muscle of Coho salmon:

	weeks of exposure*				1 week of depuration
	2	3	5	6	(after 6 wks of exposure)
C_2-substituted benzenes	1.1	2.4	2	1	n.d.
C_3-substituted benzenes	10	30	50	10	n.d.
C_4–C_5-substituted benzenes	150	170	550	200	n.d.

In muscle of starry flounder (*Plathichthys stellatus*):

	weeks of exposure*		weeks of depuration (after 2 wks of exposure)	
	1	2	1	2
C_2-substituted benzenes	20	4	1	n.d.
C_3 substituted benzenes	500	70	6	10
C_4–C_5-substituted benzenes	9300	1700	980	2600

(n.d. = not detected) (1659)
*exposure to approx. 1 ppm of the water soluble fraction of Prudhow Bay crude oil

alkylbenzenesulfonate, linear (*see also* Teepol 715)
D. BIOLOGICAL EFFECTS:
—Fish: chain length C_{10-15}: 46.7% active material (supplied by Unilever, U.K.):

	48 hr LC_{50}	96 hr LC_{50}
Rasbora heteromorpha	0.9 mg/l	0.7 mg/l
Slamo trutta	0.2–0.4 mg/l	0.1–0.5 mg/l
Idus idus	0.8–0.9 mg/l	0.4–0.6 mg/l
Carassius auratus	1.2 mg/l	

(1905)
15.4% active material (supplied by Hüls, Germany):

ALKYLBENZENE SULFONIC ACID

	48 hr LC_{50}	96 hr LC_{50}	
Rasbora heteromorpha	7.6 mg/l	6.1 mg/l	
Salmo trutta	2.0–5.3 mg/l	0.9–4.6 mg/l	
Idus idus	2.1–2.9 mg/l	1.9–2.9 mg/l	
Carassius auratus	4.9 mg/l		(1905)

28.5% active material (supplied by Procter and Gamble, U.K.):

Rasbora heteromorpha	5.1 mg/l	4.6 mg/l	
Salmo trutta	0.7–2.3 mg/l	1.4 mg/l	
Idus idus	1.3–1.7 mg/l	1.2–1.3 mg/l	
Carassius auratus	2.4 mg/l		(1905)

alkylbenzene sulfonic acid *see* dobanic acid

alkylethoxysulfate
D. BIOLOGICAL EFFECTS:
—Fish: (chain length C_{12-15}):

	48 hr LC_{50}	96 hr LC_{50}	
Rasbora heteromorpha	3.9 mg/l	–	
Salmo trutta	1.4–2.6 mg/l	1.0–2.5 mg/l	
Idus idus	3.4–7.2 mg/l	3.3–6.2 mg/l	
Carassius auratus	5.7 mg/l		(1905)

alkylnaphthalenes
D. BIOLOGICAL EFFECTS:
—Fish: bioaccumulation: bcf in muscle of Coho salmon*

	weeks of exposure				1 week depuration
	2	3	5	6	after 6 wks of exp.
C_2-substituted naphthalenes	30	40	85	40	n.d.
C_3-substituted naphthalenes	50	30	140	80	n.d.

bcf in muscle of Starry flounder*

	weeks of exposure		weeks of depuration after 2 wks of exp.	
	1	2	1	2
C_2-substituted naphthalenes	2400	540	270	700
C_3-substituted naphthalenes	3400	1000	420	1600

*exposure to approx. 1 ppm of the water soluble fraction of Prudhoe Bay crude oil (1659)

alkylolefinsulfonate
D. BIOLOGICAL EFFECTS:
—Fish: chain length C_{16-18}:

	48 hr LC_{50}	96 hr LC_{50}	
Rasbora heteromorpha	0.9 mg/l	0.5 mg/l	
Salmo trutta	0.3–0.6 mg/l	0.5 mg/l	
Idus idus	1.0 mg/l	0.9 mg/l	
Carassius auratus	1.9 mg/l		(1905)

chain length C_{14-16}:

	48 hr LC_{50}	96 hr LC_{50}	
Rasbora heteromorpha	4.8 mg/l	3.3 mg/l	
Salmo trutta	2.5–5.0 mg/l	2.5–5.0 mg/l	
Idus idus	3.7–6.8 mg/l	3.4–4.9 mg/l	
Carassius auratus	5.7 mg/l		(1905)

alkylphenolethyleneoxide condensate see nonidet NP 50

sec-alkylsulfate see Teepol 610

d-trans-allethrin (see also pyrethroid) (dl-2-allyl-4-hydroxy-3-methyl-2-cyclopenten-1-one ester of d-trans-chrysanthemum monocarboxylic acid; allylhomologue of cinerin I; a pyrethroid)
Allethrin refers to a mixture of cis and trans allethrins

$$(CH_3)_2C{=}CH{-}CH\!\!\!\diagdown\!\!\!\diagup\!\!CH{-}\overset{O}{\overset{\|}{C}}{-}O{-}\!\!\!\!\overset{\overset{CH_3}{|}}{\bigcirc}\!\!\!\!{-}CH_2{-}CH{=}CH_2$$

Use: synthetic insecticide.
A. PROPERTIES: clear, amber-colored viscous liquid, bp approx. 160°C, sp. gr. 1.01 at 20/20°C.
D. BIOLOGICAL EFFECTS:
  —Crustaceans:

Gammarus lacustris: 11 µg/l: 96 hr LC_{50}	(2124)
Gammarus fasciatus: 8 µg/l: 96 hr LC_{50}	(2126)
Simocephalus serrulatus: 56 µg/l: 48 hr LC_{50}	(2127)
Daphnia pulex: 21 µg/l: 48 hr LC_{50}	(2127)

  —Insects:

Pteronarcys californica: 2.1 µg/l: 96 hr LC_{50}	(2128)
fourth instar larval Chironomus riparius:	
24 hr LC_{50}: 41.9 ppb	
0.4 ppb (+piperonyl butoxide)	(1853)

  —Fish:

Lepomis macrochirus: 56 µg/l: 96 hr LC_{50} (n.s.i.)	(2137)
Salmo gairdneri: 19 µg/l: 96 hr LC_{50} (n.s.i.)	(2137)

	96 hr LC_{50} (µg/l)		
	static test	flow through test	
fathead minnows	80.0	53.0	
channel catfish	–	14.6	
yellow perch	7.8	–	(1626)

	96 hr LC_{50} (µg/l)*	
	static test	flow through test
coho salmon	22.2	9.4
steelhead trout	17.5	9.7

channel catfish   >30.1   27.0
yellow perch   –   9.9
*Toxicity calculated for formulation with 90% of active ingredient   (1626; 444)
bluegill (*Lepomis macrochirus*) static bioassay:
96 hr TLm: 36–60 µg/l at 12–22°C   (444)
Atlantic salmon (*Salmo salar*): 48-hr Lethal threshold concentration LTC)-S: 16.5 µg/l   (1196)
—Mammals:
         rat: 920 mg/kg   (1855)
acute oral LD_{50}:   rat: 860 mg/kg (*d-trans-*)   (1854)
         mice: 480 mg/kg   (1855)
in diet: rats tolerated diets containing 2,000 ppm for one year without ill effects   (1855)

allylacetate (2-propenylethanoate)
$CH_3COOCH_2CH=CH_2$
A. PROPERTIES: m.w. 100.11; b.p. 103–104°C; sp. gr. 0.928
B. AIR POLLUTION FACTORS: 1 mg/cu m = 0.245 ppm, 1 ppm = 4.1 mg/cu m
—Emission limits:
wet scrubber: water at pH 8.5: outlet: 1,700 odor units/scf
         $KMnO_4$ at pH 8.5: outlet: 25 odor units/scf   (115)
D. BIOLOGICAL EFFECTS:
—Mammalia: rat: acute oral LD_{50}: 0.13 g/kg
         rat: inhalation: 3/6: 1,000 ppm; 1 hr   (211)

allylalcohol (propenylalcohol; 2-propen-1-ol; propenol-3; vinylcarbinol)
$CH_2=CH-CH_2OH$
Use: contact pesticide for weed seeds and certain fungi.
A. PROPERTIES: m.w. 58.1; m.p. -129°C; b.p. 96.9°C; v.p. 20 mm at 20°C, 32 mm at 30°C; v.d. 2.00; sp.gr. 0.8250 at 20/4°C; THC 442 kcal/mole; sat.conc. 57 g/cu m at 20°C, 98 g/cu m at 30°C; log P_{oct} 0.17
B. AIR POLLUTION FACTORS: 1 mg/cu m = 0.414 ppm, 1 ppm = 2.414 mg/cu m
—Odor: characteristic: quality: alcoholic
         hedonic tone: not unpleasant

odor thresholds   mg/cu m

	10^{-7}	10^{-6}	10^{-5}	10^{-4}	10^{-3}	10^{-2}	10^{-1}	1	10	10^2	10^3	10^4
detection									♦	♦		
recognition												
not specified								♦	♦♦♦	♦		

(57; 211; 307; 643; 649; 708; 710; 788)

C. WATER POLLUTION FACTORS:
—BOD_5:   0.2 std.dil.   (27)
         9.1% of ThOD   (220)
         1.66 NEN 3235–5.4   (277)

—BOD_{10}: 1.6 std.dil. (256)
　　　　　　55.0% of ThOD (220)
—BOD_{15}: 78.2% of ThOD (220)
—BOD_{20}: 81.8% of ThOD (220)
—COD:　　　2.10 NEN 3235-5.3 (277)
　　　　　　96% of ThOD (220)
—$KMnO_4$:　acid 76, alkal. 30 (220)
—TOC:　　　100% of ThOD (220)
—ThOD:　　2.2
—Impact on biodegradation processes: 75% inhibition of the nitrification process in activated sludge at 19.5 mg/l (43)
—Aqueous reactions
　photo-oxidation by u.v. light in aqueous medium at 50°C: 13.9% degradation of CO_2 after 24 hr (1628)
　faint odor: 0.017 mg/l water (129)
—Waste water treatment: A.C.: adsorbability: 0.024 g/g carbon, 21.9% reduction, infl.: 1,010 mg/l, effl.: 789 mg/l (32)
　NFG, BOD, 20°C, 1-10 days observed, feed: 200-1,000 mg/l, acclimation 365 + P, 57% removed (93)

D. BIOLOGICAL EFFECTS:
—Plants: highly phytocidal (1855)
—Fish: goldfish: LD_{50} (24 hr) = 1 mg/l: modified ASTM 1345 (277)
—Mammalia: mouse: single oral dose LD_{50} 96 mg/kg
　　　　　　rabbit: single oral dose LD_{50} 71 mg/kg
　　　　　　rat: single oral dose LD_{50} 64-105 mg/kg
　　　　　　rat: repeated ingestion no effect 4-12 mg/kg (211)
　　　　　　rat: inhalation LC_{50} 165 ppm, 4 hr
　　　　　　rat: inhalation LC_{50} 76 ppm, 8 hr
　　　　　　rat: inhalation LC_{10} 60 ppm, 60 × 7 hr
　　　　　　rat: inhalation no effect 20 ppm, 60 × 7 hr (211)
—Man: slight eye irritation: 6.25 ppm
　　　nasal irritation: <0.78 ppm
　　　pulmonary discomfort: >25 ppm (211)

allylaldehyde *see* acrolein

allylamine (2-propenylamine)
$CH_2=CHCH_2NH_2$
A. PROPERTIES: m.w. 57.07; b.p. 53°C; v.d. 2.0; sp.gr. 0.76 at 20/4°C
B. AIR POLLUTION FACTORS: 1 mg/cu m = 0.428 ppm, 1 ppm = 2.33 mg/cu m
　—Odor: characteristic: similar to ammonia, irritating
　　T.O.C. = 6.3 ppm; <6 mg/cu m; 14 mg/cu m (279; 688; 710)
C. WATER POLLUTION FACTORS:
　—Reduction of amenities: faint odor: 0.067 mg/l (129)
　—Waste water treatment:
　　A.C.: adsorbability; 0.063 g/g carbon, 31.4% reduction, infl.: 1,000 mg/l, effl.: 686 mg/l (32)

178   ALLYLBENZENE

    degradation by *Aerobacter*: 200 mg/l at 30°C
        parent: 78% in 93 hr
        mutant: 100% in 13 hr   (152)

D. BIOLOGICAL EFFECTS:
    —Toxicity threshold (cell multiplication inhibition test)
        bacteria (*Pseudomonas putida*): 700 mg/l   (1900)
        algae (*Microcystis aeruginosa*): 0.35 mg/l   (329)
        green algae (*Scenedesmus quadricauda*): 2.2 mg/l   (1900)
        protozoa (*Entosiphon sulcatum*): 23 mg/l   (1900)
        protozoa (*Uronema parduczi Chatton-Lwoff*): 3140 mg/l   (1901)
    —Amphibian:
        mexican axolotl (3-4 wk after hatching): 48 hr LC_{50}: 1.8 mg/l
        clawed toad (3-4 wk after hatching): 48 hr LC_{50}: 5.0 mg/l   (1823)
    —Mammalia: rat: acute oral LD_{50}: 106 mg/kg
                  rat: inhalation: LC_{50}: 286 ppm, 4 hr
                  rat: inhalation: LC_{50}: 177 ppm, 8 hr
                  rat: inhalation: deaths 40 ppm, 50 × 7 hr
                  rat: inhalation: reduced growth: 5 ppm, 50 × 7 hr   (211)

allylbenzene *see* 3-phenylpropene

allylchloride (3-chloro-1-propene; chloroallylene; 3-chloro-1-propylene)
$CH_2=CH-CH_2CL$
A. PROPERTIES: colorless to pale yellow; m.w. 76.53; m.p. -136°C, -134.5°C; b.p. 44-45.5°C; v.p. 340 mm at 20°C, 440 mm at 30°C; v.d. 2.64; sp.gr. 0.94 at 20/4°C; solub. 100 mg/l; sat. conc. 1229 g/cu m at 20°C, 1772 g/cu m at 30°C
B. AIR POLLUTION FACTORS: 1 mg/cu m = 0.314 ppm, 1 ppm = 3.18 mg/cu m
    —Odor: T.O.C.: 0.660 mg/cu m = 0.21 ppm   (307)
                1500 ppm   (279)
                    50% recognition: 0.21 ppm
                    100% recognition: 25 ppm   (211)
        PIT_{50}: 0.21 ppm
        PIT_{100}: 0.47 ppm   (2)
                1.5 mg/cu m = 0.47 ppm   (57)
        characteristic: garlic-onion pungency, green
C. WATER POLLUTION FACTORS:
    —COD: 0.86 = 51% ThOD—ASTM procedure
               1.33 = 80% ThOD—modified Shell procedure   (272)
    —BOD_5: 0.23 = 14% ThOD   (277)
    —Impact on biodegradation processes: 75% decrease of nitrification by activated sludge at 180 mg/l   (30)
    —Reduction of amenities: odor threshold: average: 14,700 mg/l
                                        range: 3,660 to 29,300 mg/l   (30)
    —Waste water treatment: evaporation rate from water at 25°C of 1 ppm solution:
        50% after 27 min
        90% after 89 min   (313)(289)

D. BIOLOGICAL EFFECTS
 —Toxicity threshold (cell multiplication inhibition test)
  bacteria (*Pseudomonas putida*): 115 mg/l (1900)
  green algae (*Scenedesmus quadricauda*): 6.3 mg/l (1900)
  protozoa (*Entosiphon sulcatum*): 8.4 mg/l (1900)
  protozoa (*Uronema parduczi Chatton-Lwoff*): >240 mg/l (1901)
 —Fishes:

Testfish	dilution water	(24 hr)	TLm (48 hr)	(96 hr)	
fatheads	soft	24.0	24.0	19.8	
fatheads	hard	25.9	24.0	24.0	
bluegills	soft	59.3	42.3	42.3	
goldfish	soft	26.6	20.9	20.9	
guppies	soft	57.7	53.5	51.8	(158)

  goldfish: LD_{50} (24 hr) 10 mg/l (277)
 —Mammalia: rat: max. exposure time concentrations: 290 ppm, 3 hr
  inhalation 2900 ppm, 1 hr
  29,300 ppm, 15 min
  rat, guinea pig, rabbit: liver injury: 8 ppm, 7 hr/day, 5 days/w, 28 days
  rat, guinea pig, rabbit: died after 127–134 exposures at 3 ppm
  rat: oral LD_{50}: 0.7 g/kg (211)
 —Man: very irritating to the eyes and upper respiratory tract
  eye irritating: 50–100 ppm (211)

allylene *see* methylacetylene

allyl-2,3-epoxypropylether *see* allylglycidylether

allylglycidylether (AGE; allyl 2,3-epoxypropylether; 1-allyloxy-2,3-epoxypropane; 1,2-epoxy-3-allyloxypropane; glycidylallylether; [(2-propenyloxy)methyl] oxirane)

$$CH_2=CHCH_2OCH_2\underset{O}{CH\diagdown\diagup CH_2}$$

Use: component of epoxy resin systems. The epoxy group of the glycidylether reacts during the curing process and glycidylethers are therefore generally no longer present in completely cured products.
A. PROPERTIES: colorless liquid; m.w. 114.15; m.p. −100°C forms glass; b.p. 153.9°C; v.p. 3.6 mm at 20°C, 5.8 mm at 30°C; v.d. 3.94; sp.gr. 0.969 at 20/4°C; solub. 141,000 mg/l; sat.conc. 22 g/cu m at 20°C, 35 g/cu m at 30°C
B. AIR POLLUTION FACTORS: 1 mg/cu m = 0.21 ppm, 1 ppm = 4.74 mg/cu m
 —Odor: threshold value: 47 mg/cu m (57)
C. WATER POLLUTION FACTORS
 —BOD_5: 0.06 NEN 3235-5-4 (277)
 —COD: 1.99 NEN 3235-5-3 (277)
D. BIOLOGICAL EFFECTS
 —Fish: goldfish: LD_{50} (24 hr): 78 mg/l
  LD_{50} (96 hr): 30 mg/l (277)

180 2-ALLYL-4-HYDROXY-3-METHYL-2-CYCLOPENTEN-1

—Mammalia: mouse: single oral LD_{50}: 0.39 g/kg
rat: single oral LD_{50}: 1.60 g/kg
mouse: inhalation: LC_{50}: 270 ppm, 4 hr
rat: inhalation: LC_{50}: 670 ppm, 8 hr (211)
rat: 400 mg/kg i.m. injections on days 1, 2, 8 and 9; animals sacrificed on day 12: focal necrosis of the testis in 1 of 2 of the 3 surviving rats, atrophy or loss of lymphoid tissue in 2 of 3 rats, decreased leukocyte count. (1406)

2-allyl-4-hydroxy-3-methyl-2-cyclopenten-1-one ester of chrysanthemum monocarboxylic acid see d-trans-allethrin

allylisosulfocyanate see allylisothiocyanate

allylisothiocyanate (mustard oil; 2-propenylisothiocyanate; allylisosulfocyanate)
$CH_2=CHCH_2NCS$
Use: fumigant; ointments and mustard plasters; military poison gas
A. PROPERTIES: m.w. 99.15; m.p. $-100°C$; b.p. $151°C$; v.p. 1 mm at $-2°C$, 10 mm at 38.3 °C, 40 mm at $67.4°C$; sp.gr. 1.015 at $15/4°C$; solub. 2,000 mg/l; v.d. 3.4; $\log P_{oct}$ 2.11 (calculated)
B. AIR POLLUTION FACTORS: 1 mg/cu m = 0.243 ppm, 1 ppm = 4.120 mg/cu m
—Odor: characteristic: mustard oil, irritant

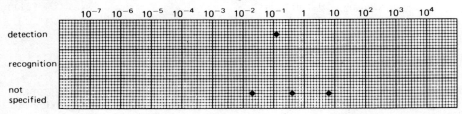

(279; 307; 602; 710; 827)
—Emission limits: wet scrubber: water: pH 8: outlet: 2,500 odor units/scf
$KMnO_4$: pH 8: outlet: 1 odor units/scf (115)
C. WATER POLLUTION FACTORS
—Impact on biodegradation processes: 75% inhibition of the nitrification process in activated sludge at 1.9 mg/l (43)
—Reduction of amenities: faint odor: 0.0017 mg/l (129)

1-allyloxy-2,3-epoxypropane see allylglycidylether

allylsulfocarbamide see allylthiourea

allylsulfourea see allylthiourea

allylthiourea (allylsulfocarbamide; thiosinamine; allylsulfourea)
$C_3H_5NHCSNH_2$
Use: medicine, corrosion inhibitor, organic synthesis

A. PROPERTIES: white crystalline solid; slight garlic odor; bitter taste sp.gr. 1.22; m.p. 78°C
C. WATER POLLUTION FACTORS:
   —Impact on biodegradation processes: approximately 50% inhibition of NH_3 oxidation by Nitrosomonas at 1.2 mg/l (407)
D. BIOLOGICAL EFFECTS:
   —Toxicity threshold (cell multiplication test):
   bacteria (*Pseudomonas putida*):   140 mg/l
   green algae (*Scenedesmus quadricauda*): 41 mg/l
   protozoa (*Entosiphon sulcatum*):   13 mg/l   (1900)

altosid-SR-10   (Isopropyl(2E-4E)-11-methoxy-3,7,11-trimethyl-dodeca-2,4-dienoate; methoprene; ZR-515):

$$CH_3O-\underset{\underset{CH_3}{|}}{\overset{\overset{CH_3}{|}}{C}}-CH_2CH_2CH_2-\overset{\overset{CH_3}{|}}{CH}-CH_2-CH=CH-\overset{\overset{CH_3}{|}}{C}=CH-\overset{\overset{O}{\|}}{C}-O-\overset{\overset{CH_3}{|}}{\underset{\underset{CH_3}{|}}{CH}}$$

Use: insect growth regulator: prevents adult emergence of mosquitoes; houseflies; stable-flies and blackflies by preventing metamorphosis of final instar larvae
A. PROPERTIES: amber liquid, sp.gr. 0.9261 at 20°C, v.p. 2.37×10^{-5} at 25°C; solub. 1.39 ppm; v.p. 1.6×10^{-4} at 40°C
D. BIOLOGICAL EFFECTS:
   —Arthropods
   amphipod: *Gammarus aequieaudu*:
      adult female: 96 hr LC_{50} 2150 μg/l
      adult male: 96 hr LC_{50} 1950 μg/l   (1128)
   decapod: *Rhithropanopeus harrissii*: adult: 1.30 mg/l, 12–15 days. Progressive inhibition of vitellogenesis and stimulation of spermatogenesis after 30–45 days. (1157)
   —Fish:
   juvenile rainbow trout: 96 hr LC_{50}: 106 mg/l
                                            95% conf. lim: 92–121 mg/l
   coho salmon: 96 hr LC_{50}: 86 mg/l
                                            95% conf. lim: 81–91 mg/l   (1059)
   blue gill: TL_{50}:   4.6   ppm (static)
   trout: TL_{50}:   4.39 ppm (static)
                                       106   ppm (static, when aerated)
   channel catfish: TL_{50}: >100   ppm (static)   (1855)
   —Mammals
   acute oral LD_{50} (rat):   >34,600 mg/kg
   acute dermal LD_{50} (rabbit): > 3,000 mg/kg
   acute inhalation LC_{50} (rat): >   210 mg/kg   (1855)
   teratogenicity: no effects on rats at 1,000 mg/kg   (1855)

ametryn   (6-ethylamino-4-isoproylamino-2-methylthio-1,3,5-triazine; 2-ethylamino-4-isopropylamino-6-methylmercapto-*s*-triazine; gesapax; evik; G 34162)

AMIDOL

$$\text{H}_3\text{C}-\text{S}-\underset{\underset{N}{\|}}{\overset{\overset{N}{\|}}{C}}\underset{\underset{}{}}{\overset{}{}}\text{NH}-\text{CH}(\text{CH}_3)_2 \quad \text{with } \text{NHC}_2\text{H}_5$$

Use: herbicide
A. PROPERTIES: colorless crystals, m.p. 84–85°C, solub. 185 ppm at 20°C, v.p. 8.4 × 10^{-7} mm at 20°C
D. BIOLOGICAL EFFECTS:
　—Algae:
　　Chlorococcum sp. (technical acid): 20 ppb; 50 percent decrease in O_2 evolution;
　　Chlorococcum sp.: 10 ppb; 50 percent decrease in growth; measured as ABS (525 mu) after 10 days
　　Dunaliella tertiolecta: 40 ppb; 50 percent decrease in O_2 evolution;
　　Dunaliella tertiolecta: 40 ppb; 50 percent decrease in growth; measured as ABS (525 mu) after 10 days
　　Isochrysis galbana: 10 ppb; 50 percent decrease in O_2 evolution;
　　Isochrysis galbana: 10 ppb; 50 percent decrease in growth; measured as ABS (525 mu) after 10 days
　　Phaeodactylum tricornutum: 10 ppb; 50 percent decrease in O_2 evolution;
　　Phaeodactylum tricornutum: 20 ppb; 50 percent decrease in growth; measured as ABS (525 mu) after 10 days　(2348)
　—Mammals:
　　acute oral LD_{50} (rat):　　　1,110 mg/kg (technical)　(1854)
　　acute dermal LD_{50} (rabbit): >8,160 mg/kg　(1855)
　　in diet: rats fed ninety days at 100 mg/kg/day were comparable to controls except for slight histological changes in the liver　(1855)

amidol　*see* 2,4-diaminophenol hydrochloride

p-aminoacetanilide　(N-acetyl-*p*-phenylenediamine)

$$\text{NH}_2-\text{C}_6\text{H}_4-\text{NH}-\text{C}(=O)-\text{CH}_3$$

Uses: intermediates; azo dyes
A. PROPERTIES: Colorless or reddish crystals, m.p. 162°C, b.p. 267°C
C. WATER POLLUTION FACTORS
　—Biodegradation rates: adapted A.S. at 20°C—product is sole carbon source: 93.0% COD removal at 11.3 mg COD/g dry inoculum/hr　(327)

aminoacetic acid (aminoethanoic acid; glycine; glycocoll)
H₂NCH₂COOH
  Use: stabilizing agent in photoprocessing
A. PROPERTIES: crystalline; m.w. 75.07; m.p. 233°C decomposes; b.p. 289/292°C decomposes; sp.gr. 1.601; solub. 253,000 mg/l at 25°C, 575,000 mg/l at 75°C; log P_{oct} -3.03/-1.7 (calculated)
C. WATER POLLUTION FACTORS
  —BOD₅: 0.385 (1828)
        0.548 (30)
        86% ThOD
  —COD: 64% ThOD (0.05 n Cr₂O₇)
        reflux COD: 99.2% recovery
        rapid COD: 36.1% recovery (1828)
  Poor recovery was exhibited by the rapid COD method. A possible interference was volatilization of the compound or compound oxidation products in the open digestion flask at the high test temperatures. (1828)
  —KMnO₄ value: 0.008 (30)
        0% ThOD (0.01 n KMnO₄)
  —ThOD: 0.720: 0.640 (30; 274)
  —Impact on conventional biological treatment systems: Unacclimated system at 500 mg/l: biodegradable (1828)
  —Natural sources: normal constituent of proteins (211)
  —Manmade sources: excreted by man in urine: 2.3 to 18.0 mg/kg body wt/day (203)
  —Waste water treatment:
    A.S.: after 6 hr: 4.1% of ThOD
        12 hr: 8.1% of ThOD
        24 hr: 16.9% of ThOD (89)

methods	temp °C	days observed	feed mg/l	days acclim.	% theor. oxidation	% removed
A.S., Resp, BOD	20	1–5	720	<1	57	87
A.S., BOD	20	1–5	333	15	–	93

        (93)
    powdered carbon: at 100 mg/l, carbon dosage 1000 mg/l: 1% absorbed (520)

***p*-aminoanisole** *see p*-anisidine

***p*-aminoazobenzene** *see p*-phenylazo-aniline

***p*-aminobenzenesulfonic acid** *see p*-anilinesulfonic acid

2-aminobenzimidazole (AB)

m-AMINOBENZOIC ACID

Manmade source: degradation product (hydrolysis) of the fungicides benomyl, carbendazim, and thiophanate methyl (1373, 1374)

Use: in photographic industry as an antifoggant

A. PROPERTIES: m.w. 133.15, m.p. 229-231°C

C. WATER POLLUTION FACTORS:
 —Biodegradation: total evolution of ^{14}C in CO_2 and remaining radio-activity in soil after 218 days of incubation at 25°C with 4 ppm ^{14}C-2 aminobenzimidazole*

	not inoculated	inoculated with adapted soil
^{14}C-evolution	48.2%	84.2-84.3%
^{14}C-soil residues	40.6%	10.7-18.0%
% recovery	88.8%	94.9-102.3%

(1714)

—Effect of AB on nitrification by *Nitrosomonas* sp.

conc. ppm	µg nitrite recovered/ml medium incubation days	
	6	15
0	3.8	108.0
10	3.7	146.4
100	2.8	3.9
	by *Nitrobacter agilis*	
0	289.5	0
10	352.0	0
100	622.5	661.3

m-aminobenzoic acid

A. PROPERTIES: m.w. 137.13; m.p. 174/179°C; b.p. sublimes; sp.gr. 1.511 at 20/4°C; solub. 5,900 mg/l at 15°C; log P_{oct} 0.14/0.27 (calculated)

C. WATER POLLUTION FACTORS
 —Biodegradation:
 decomposition by a soil microflora: >64 days (176)
 adapted A.S. at 20°C—product is sole carbon source: 97.5% COD removal at 7.0 mg COD/g dry inoculum/hr (327)

D. BIOLOGICAL EFFECTS
 —Mammalia: mouse: I.P.: LD_{50}: 250-500 mg/kg (211)

*structure of 2-aminobenzimidazole labeled with ^{14}C as indicated by an *:

***o*-aminobenzoic acid** *see* anthranilic acid

***p*-aminobenzoic acid**

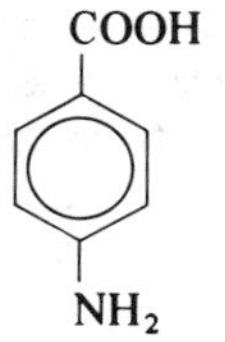

COOH
—
NH₂ (on benzene ring)

A. PROPERTIES: m.w. 137.13; m.p. 187°C; solub. 3,400 mg/l at 9.6°C; log P_{oct} 0.68
C. WATER POLLUTION FACTORS
—Biodegradation:
   decomposition by a soil microflora: 8 days (176)
   adapted A.S. at 20°C—product is sole carbon source: 96.2% COD removal at 12.5 mg COD/g dry inoculum/hr (327)
—Lag period for degradation of 16 mg/l by wastewater or by soil at pH 7.3 and 30°C: less than 1 day (1096)
—Impact on biodegradation processes: at 100 mg/l no inhibition of NH_3 oxidation by *Nitrosomonas* sp. (390)

aminobenztrifluoride
$CF_3C_6H_4NH_2$
Users and formulations: pharmaceutical intermediate.
A. PROPERTIES: colorless to oily yellow liquid; aniline like odor; m.w. 161.13; m.p. 3°C; b.p. 189°C; v.d. 5.56; sp.gr. 1.303 at 15.5/15.5°C
C. WATER POLLUTION FACTORS
—Water quality: in river Waal (Netherlands): average in 1973: 0.4 µg/l (342)

1-aminobutane *see n*-butylamine

2-aminobutane *see* sec. butylamine

2-aminobutanedioic acid *see dl*-asparatic acid

2-aminobutanedioic amide *see l*-asparagine

4-amino-6-*t*-butyl-3-methylthio-1,2,4-triazin-5(4H)-one (Metribuzin; Sencor; Bay 94337; Sencoral; Sencorer; 4-amino-6-(1,1-dimethylethyl)-3-(methylthio)-1,2,4-triazin-5(4H)-one)

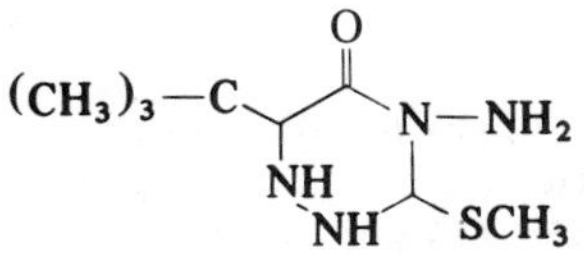

Use: herbicide, effective in the control of broad leafed and grassy weeds encountered in the growing of potatoes and tomatoes

A. PROPERTIES: white crystalline solid, m.p. 125–126.5°C; solub. 1200 ppm
C. WATER POLLUTION FACTORS:
—Aquatic reactions: Metribuzin has been shown to undergo nonbiological degradation in four Manitoba soils under dry conditions at 15°C, and the rate law describing this degradation has been shown to be somewhat less than first order. Calculated times for 50% loss (at a "normal" application rate of 1.8 ppm metribuzin) vary with soil type from approximately 90–115 days for Red River, Almasippi, and Stockton soils to three times this period for Newdale soil. At higher application rates, half lives are somewhat longer (1056)
D. BIOLOGICAL EFFECTS:
—Harlequin fish (*Rasbera heteromorpha*):

mg/l	24 hr	48 hr	96 hr	3 m (extrapolated)	
$LC_{10}(F)$	105	130	130		
$LC_{50}(F)$	145	140	140	100	(331)

—Mammals:
acute oral LD_{50} (rat): 1936–1986 mg/kg
dermal LD_{50}: 2000 mg/kg (1854)

dl-aminocaproic acid see *dl*-norleucine

aminocarb (metacil; 4-(dimethylamino)-3-methylphenyl-N-methylcarbamate(ester); 4-dimethylamino-*m*-tolylmethylcarbamate)

$$H_3C-HN-\overset{\overset{O}{\|}}{C}-O-\underset{CH_3}{\underset{|}{\bigcirc}}-N(CH_3)_2$$

Use: nonsystemic insecticide, molluscicide
A. PROPERTIES: white crystalline solid; m.p. 93–94°C
C. WATER POLLUTION FACTORS:
—aquatic reactions: persistence in riverwater in a sealed glass jar under sunlight and artificial—fluorescent light—initial conc. 10 μg/l

	% of original compound found				
after 1 hr	1 wk	2 wks	4 wks	8 wks	
100	60	10	0	0	(1309)

D. BIOLOGICAL EFFECTS:
—Crustacean: *Gammarus lacustris*: 12 μg/l; 96 hr LC_{50} (2124)
—Mussels: BCF in mussel (*Mytilus edulis*): 3.8–4.9 (on wet weight) (1864)
—Insects: fourth instar larval *Chironomus riparius*: 24 hr LC_{50}: 376.6 ppb (1853)

—Mammals:
acute oral LD_{50} (rats): approxim. 50 mg/kg
acute intraperitoneal LD_{50} (rats): 21 mg/kg
acute dermal LD_{50} (rats): 275 mg/kg (1885)
in diet: rats fed 5 mg/kg/day for twenty-eight days showed no symptoms of poisoning (1885)

***o*-aminochlorobenzene** *see o*-chloroaniline

***m*-aminochlorobenzene** *see m*-chloroaniline

***p*-aminochlorobenzene** *see p*-chloroaniline

aminocyclohexane *see* cyclohexylamine

3-amino-2,5,dichlorobenzoic acid (chloramben; amiben (ammonium salt); amilon-WP; vegiben)

$$\text{COOH} \atop \text{Cl} - \bigcirc - \text{Cl, NH}_2$$

Use: selective preemergence herbicide
A. PROPERTIES: white odorless crystalline solid, m.p. 200–201°C, v.p. ~7 × 10⁻³ mm at 100°C, solub. 700 mg/kg at 25°C
C. WATER POLLUTION FACTORS:
—Control methods:
A.C. type BL (Pittsburgh Chem. Co.): % absorbed by 10 mg A.C. from 10^{-4} M aqueous solution at pH 3.0: 50.5%
7.0: 12.4%
11.0: 5.2%   (1313)

D. BIOLOGICAL EFFECTS:
—Algae:

Technical acid	*Chlorococcum* sp.	1.15 × 10⁵ ppb	50 percent decrease in O₂ evolution	*f*
Technical acid	*Chlorococcum* sp.	5. × 10⁴ ppb	50 percent decrease in growth	Growth measured as ABS. (525 mu) after 10 days
Technical acid	*Dunaliella tertiolecta*	1.5 × 10⁵ ppb	50 percent decrease in O₂ evolution	*f*
Technical acid	*Dunaliella tertiolecta*	5. × 10⁴ ppb	50 percent decrease in growth	Growth measured as ABS. (525 mu) after 10 days
Technical acid	*Isochrysis galbana*	1 × 10⁵ ppb	50 percent decrease in O₂ evolution	*f*
Technical acid	*Isochrysis galbana*	1.5 × 10⁴ ppb	50 percent decrease in growth	Growth measured at ABS. (525 mu) after 10 days
Technical acid	*Phaeodactylum tricornutum*	1.0 × 10⁵ ppb	50 percent decrease in O₂ evolution	*f*
Technical acid	*Phaeodactylum tricornutum*	2.5 × 10⁴ ppb	50 percent decrease in growth	Growth measured at ABS. (525 mu) after 10 days
Ammonium salt	*Chlorococcum* sp.	2.225 × 10⁶ ppb	50 percent decrease in O₂ evolution	*f*
Ammonium salt	*Chlorococcum* sp.	4. × 10⁶ ppb	50 percent decrease in growth	Growth measured as ABS. (525 mu) after 10 days
Ammonium salt	*Dunaliella tertiolecta*	2.75 × 10⁶ ppb	50 percent decrease in O₂ evolution	*f*
Ammonium salt	*Dunaliella tertiolecta*	4. × 10⁶ ppb	50 percent decrease in growth	Growth measured as ABS. (525 mu) after 10 days
Ammonium salt	*Isochrysis galbana*	1.5 × 10⁶ ppb	50 percent decrease in O₂ evolution	*f*
Ammonium salt	*Isochrysis galbana*	3.5 × 10⁶ ppb	50 percent decrease in growth	Growth measured as ABS. (525 mu) after 10 days

(2348)

—Mammals:
acute oral LD_{50} (male albino rats): 5620 mg/kg (as free acid)   (1554)
acute dermal LD_{50} (albino rats): 3160 mg/kg   (1555)

2,2'-aminodiethanol *see* diethanolamine

p-aminodiethylanilinehydrochloride

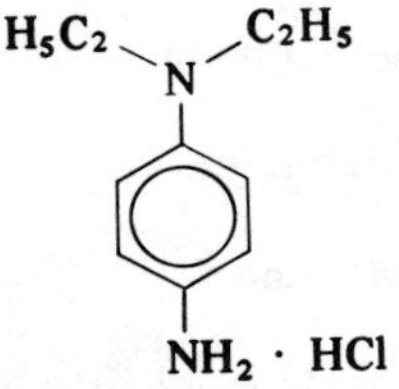

Use: color photography
A. PROPERTIES: b.p. 260-262°C
D. BIOLOGICAL EFFECTS:
—Green algae (*Microcystis aeruginosa*): LD_{100}: 1 ppm (1094)

2-amino-3,5-diiodobenzoic acid

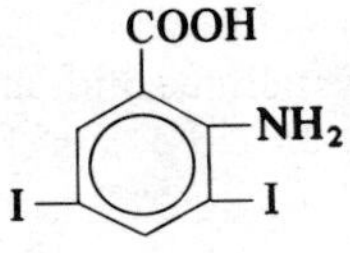

C. WATER POLLUTION FACTORS:
—Impact on biodegradation processes: at 100 mg/l no inhibition of NH_3 oxidation by *Nitrosomonas* sp. (390)

p-aminodimethylaniline (dimethylaminoaniline; dimethyl-p-phenylenediamine)

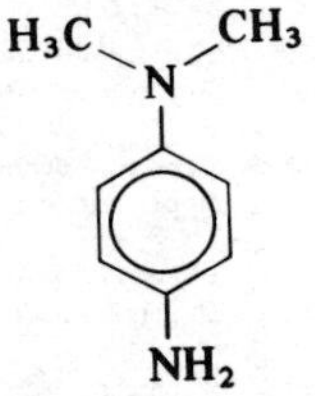

Use: base for production of methyleneblue; photodeveloper; reagent for cellulose
A. PROPERTIES: colorless, asbestos-like needles, m.p. 41°C, b.p. 257°C
D. BIOLOGICAL EFFECTS:
—Green algae (*Microcystis aeruginosa*): LD_{100}: 2 ppm (1094)

4-amino-3,5-dimethylphenol *see* 4-amino-3,5-xylenol

2-amino-4,6-dinitrophenol *see* picramic acid

1-aminododecane *see* n-dodecylamine

aminoethane *see* ethylamine

2-aminoethanol *see* ethanolamine

aminoform *see* hexamethylenetetramine

d-α-**aminoglutaramic acid** *see l*-glutamine

dl-α-**aminoglutaric acid** *see dl*-glutamic acid

2-amino-5-guanidopentanoic acid *see dl*-arginine

dl-α-**amino-δ-guanidovaleric acid** *see dl*-arginine

dl-**2-aminohexanoic acid** *see dl*-norleucine

2-amino-3-hydroxybutanoic acid *see* threonine

2-amino-4-hydroxybutanoic acid *see* homoserine

(2-amino-4-hydroxyphenyl)propanoic acid *see dl*-tyrosine

2-amino-3-hydroxypropanoic acid *see dl*-serine

α-**amino-β-imidazolepropionic acid** *see dl*-histidine

2-amino-3-indolylpropanoic acid *see* tryptophane

l-α-**aminoisocaproic acid** *see l*-leucine

dl-α-**aminoisovaleric acid** *see dl*-valine

l-**2-amino-3-mercaptopropanoic acid** *see l*-cysteine

aminomethane *see* methylamine

p-**aminomethoxybenzene** *see p*-anisidine

dl-**2-amino-3-methylbutanoic acid** *see dl*-valine

dl-α-**amino-γ-methylmercaptobutyric acid** *see dl*-methionine

d-**2-amino-3-methylpentanoic acid** *see l*-isoleucine

l-**2-amino-4-methylpentanoic acid** *see l*-leucine

p-**aminomethylphenylether** *see p*-anisidine

1-amino-2-methylpropane *see* isobutylamine

dl-**2-amino-4-methylthiobutanoic acid** *see dl*-methionine *d*-α-amino-β-methylvaleric acid   *see l*-isoleucine

1-amino-2-nitrobenzene   *see o*-nitroaniline

1-amino-3-nitrobenzene   *see m*-nitroaniline

1-amino-4-nitrobenzene   *see p*-nitroaniline

1-aminopentane   *see n*-amylamine

2-aminopentanedioic acid   *see dl*-glutamic acid

2-aminopentanedioicamide   *see l*-glutamine

o-aminophenol   (*o*-hydroxyaniline)

A. PROPERTIES: m.w. 109.12; m.p. 170/174°C; b.p. sublimes; solub. 17,000 mg/l at 0°C, log P_{oct} 0.52–0.62
C. WATER POLLUTION FACTORS
—Biodegradation:
decomposition period by a soil microflora: 4 days   (176)
adapted A.S. at 20°C—product is sole carbon source: 95.0% COD removal at 21.1 mg COD/g dry inoculum/hr   (327)
—Reduction of amenities: organoleptic limit: 0.01 mg/l USSR 1970   (181)
D. BIOLOGICAL EFFECTS
—Algae: *Chlorella pyrenoidosa*: 47 mg/l: toxic   (41)

m-aminophenol   (*m*-hydroxyaniline)

Use: dye intermediate
A. PROPERTIES: m.w. 109.12; m.p. 122/123°C; solub. 26,000 mg/l; log P_{oct} 0.15/0.17
C. WATER POLLUTION FACTORS:
—Biodegradation:
decomposition by a soil microflora: >64 days   (176)
adapted A.S. at 20°C—product is sole carbon source: 90.5% COD removal at 10.6 mg COD/g dry inoculum/hr   (327)
lag period for degradation of 16 mg/l by waste water or soil suspension at pH 7.3 and 30°C: >25 days   (1096)

—Impact on biodegradation processes:
at 0.6 mg/l, inhibition of degradation of glucose by *Pseudomonas fluorescens*
(293)
at 9 mg/l, inhibition of degradation of glucose by *E. coli*
D. BIOLOGICAL EFFECTS:
—Algae: *Chlorella pyrenoidosa*: toxic at 140 mg/l (41)

p-aminophenol (*p*-hydroxyaniline; rodinal)

A. PROPERTIES: m.w. 109.12; m.p. 184°C decomposes; b.p. sublimes; solub. 11,000 mg/l at 0°C; THC 760 kcal/mol; log P_{oct} 0.04
C. WATER POLLUTION FACTORS
—Biodegradation: adapted A.S. at 20°C—product is sole carbon source: 87.0% COD removal at 16.7 mg COD/g dry inoculum/hr (327)
D. BIOLOGICAL EFFECTS
—Bacteria: *Escherichia coli*: toxic: 8-10 mg/l
—Algae: *Chlorella pyrenoidosa*: toxic at 140 mg/l (41)
    Scenedesmus: toxic at 6 mg/l (30)
—Arthropoda: *Daphnia*: toxic: 0.6 mg/l (30)
—Fish: goldfish: approx. fatal conc.: 2.0 mg/l; 48 hr (226)
    fathead minnows: static bioassay in Lake Superior water at 18-22°C: LC_{50} (1; 24; 48; 72; 96 hr): 24; 24; 24; 24; 24 mg/l (350)
—Carcinogenicity: negative
—Mutagenicity in the *Salmonella* test: neg; <0.01 revertant colonies/nmol
    <70 revertant colonies at 500 µg/plate
(1883)

aminophenolsulfonic acid
C. WATER POLLUTION FACTORS:
—Biodegradation: adapted A.S. at 20°C—product is sole carbon source: 64.6% COD removal at 7.1 mg COD/g dry inoculum/hr (327)

2-amino-3-phenylpropionic acid *see* phenylalanine

dl-2-aminopropanoic acid *see dl*-alanine

1-amino-2-propanol *see* isopropanolamine

1-((4-amino-2-propyl-5-pyrimidinyl)methyl)-2-picoliniumchloride, hydrochloride *see* amprolium

2-aminopyridine (α-pyridylamine)

A. PROPERTIES: m.w. 94.11; m.p. 56°C; b.p. 204°C; v.d. 3.25; log P_{oct} −0.22
B. AIR POLLUTION FACTORS: 1 mg/cu m = 0.26 ppm; 1 ppm = 3.91 mg/cu m
C. WATER POLLUTION FACTORS
 —Biodegradation: adapted A.S. at 20°C—product is sole carbon source: 97.3% COD removal at 41.0 mg COD/g dry inoculum/hr (327)
D. BIOLOGICAL EFFECTS
 —Mammalia: rat: acute oral LD_{50}: 0.2 g/kg
  mouse: acute oral LD_{50}: 0.05 g/kg (211)
 —Man: readily absorbed through the skin
  transient symptoms: 5.2 ppm, 8 hr (211)

4-aminopyridine

Use: a frightening agent for protecting grain crops from blackbird; intermediate.
A. PROPERTIES: m.w. 94.12, m.p. 155–158°C, b.p. 273,5°C; log P_{oct} 0.28
C. WATER POLLUTION FACTORS:
 —degradation in soils: degradation of 4-aminopyridine-^{14}C to $^{14}CO_2$ was negligible in soils incubated up to 2 months under anaerobic conditions; under aerobic incubation, after 3 months at 30°C and 50% moisture, evolution of $^{14}CO_2$ ranged from 0.4% for a highly acidic loam (pH 4.1) to more than 50% for a lighter-textured, alkaline, loamy sand (pH 7.8) (1642)
D. BIOLOGICAL EFFECTS:
 —Fish:
  bluegill: static bioassay: 96 hr TLm: 2.82–7.56 mg/l at 12–22°C
  channel catfish: static bioassay: 96 hr TLm: 2.43–5.80 mg/l at 12–22°C (445)

l-α-aminosuccinamic acid see *l*-asparagine

dl-aminosuccinic acid see *dl*-aspartic acid

aminothiourea see thiosemicarbazide

o-aminotoluene see *o*-toluidine

4-amino-*m*-toluenesulfonic acid

A. PROPERTIES: m.w. 187.21
D. BIOLOGICAL EFFECTS:
  −Fish:
    mosquito fish: TLm (24 hr): 425 mg/l
                       (48 hr): 410 mg/l
                       (96 hr): 375 mg/l   (41)

3-amino-1,2,4-triazole (amitrol; amerol; amizol; cytrol; herbizole; weedazole)

Use: herbicide and plant growth regulator
A. PROPERTIES: m.w. 84.08, m.p. 153−156°C, solub. 280 g/l at 25°C
C. WATER POLLUTION FACTORS:
  −Impact on biodegradation processes: ~50% inhibition of NH_3 oxidation in *Nitrosomonas* at 70 mg/l (n.s.i.)   (407)
D. BIOLOGICAL EFFECTS:
  −Crustacean:

Gammarus fasciatus		no effect 100,000 µg/l 48 hr	(2125)
Daphnia magna	30,000 µg/l	48 hr LC_{50}	(2125)
Cypridopsis vidua	32,000 µg/l	48 hr LC_{50}	(2125)
Asellus brevicaudus		no effect 100,000 µg/l 48 hr	(2125)
Palaemonetes kadiakensis		no effect 100,000 µg/l 48 hr	(2125)
Orconectes nais		no effect 100,000 µg/l 48 hr	(2125)

  −Fish:

Lepomis macrochirus		no effect 100,000 µg/l 48 hr	(2125)
Oncorhyncus kisutch	325,000 µg/l	48 hr LC_{50}	(2106)
bluegill	100 ppm	48 hr LC_{50}	(1878)

  −Mammals:
    acute oral LD_{50} (rats):   1,100−2,500 mg/kg   (1885)
                (male albino rat):   25,000 mg/kg   (1854)
    acute dermal LD_{50} (rats):   >10,000 mg/kg   (1855)
    in diet: rats fed 50 ppm for 68 weeks suffered no effect on growth or food intake but the male rats developed an enlarged thyroid after thirteen weeks
      rats fed 500 ppm for 17 weeks and returned to normal diet 2 weeks before sacrifice appeared to have normal thyroids   (1856)
    cancer-suspect agent   (1857)

4-AMINO-3,5,6-TRICHLOROPICOLINIC ACID

    carcenogenicity: +
    mutagenicity in the *Salmonella* test: none, <0.001 revertant colonies/nmol
                                       <70 revertant colonies at 5000 µg/plate
                                                                                                              (1883)

4-amino-3,5,6-trichloropicolinic acid *see* picloram

2-amino-1,3,5-trimethylbenzene *see* mesidine

4-amino-3,5-xylenol (4-amino-3,5-dimethylphenol)

                           OH
                           |
            H_3C — ⬡ — CH_3
                         |
                         NH_2

D. BIOLOGICAL EFFECTS:
    —Fish:
        bluegill (*Lepomis macrochirus*): 96 hr LC_{50},S 0.32 mg/l (99% purity)   (1502)

amitrole *see* 3-amino-1,2,4-triazole

ammonia
NH_3
A. PROPERTIES: colorless gas—liquified by compression; m.w. 17.03; m.p. $-77.7°C$;
    b.p. $-33.4°C$; v.p. 10 atm. at $25.7°C$, 8.7 atm. at $20°C$; v.d. 0.6; sp.gr. 0.817 at
    $79°C$; solub. 895,000 mg/l at $0°C$, 531,000 mg/l at $20°C$, 440,000 mg/l at $28°C$;
    THC 91.5 kcal/mole, LHC 75.7 kcal/mole
B. AIR POLLUTION FACTORS: 1 mg/cu m = 1.414 ppm, 1 ppm = 0.707 mg/cu m
    —Odor: hedonic tone: extremely pungent

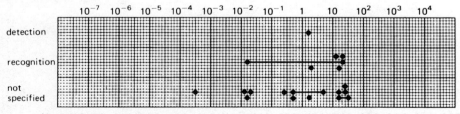

    (2; 5; 10; 73; 210; 279; 629; 652; 658; 664; 670; 674; 741; 800; 816; 821; 857)
    USSR: human odor perception: non perception: 0.4 mg/cu m
                                          perception: 0.5 mg/cu m
           human reflex response: no response: 0.22 mg/cu m
                                adverse response: 0.35 mg/cu m
       animal chronic exposure: no effect: 0.2 mg/cu m
                                 adverse effect: 2.0 mg/cu m   (170)

—Manmade sources: combustion sources:
                amount of emission
    coal   2 lb/ton
    fuel oil   1 lb/1,000 gal
    natural gas 0.3 to 0.56 lb/10^6 cu ft
    butane   1.7 lb/10^6 cu ft
    propane 1.3 lb/10^6 cu ft
    wood   2.4 lb/ton
    forest fires 0.3 lb/ton   (111)

ammonia discharged daily in metropolitan area of 100,000 persons using each heating system: domestic heating fuel:
    coal: 2,000 lb NH_3
    oil: 800 lb NH_3
    gas: 0.3 lb NH_3 (112)

downtown Tokyo: Jan.–May 1969: glc's
              20 to 152 µg/cu m Nessler's procedure
              4.0 to 25.8 µg/cu m pyridine-pyrazolone procedure   (205)

—Sampling and analysis:
second derivative spectroscopy: min.det.limit: 1 ppb   (42)
photometry: min. full scale: 1,800 ppm   (53)
I.R. spectrometry: det.lim.: 0.22 ppm   (56)
non dispersive I.R.: det.lim.: 250 ppm
detector tubes: UNICO: det.lim.: 20 ppm   (59)
                AUER: det.lim.: 5 ppm   (57)
                DRAGER: det.lim.: 5 ppm   (58)
impinger, 800 l air/30 min; VLS: det.lim.: 5 µg/cu m/30 min Nessler reagent   (208)

sampler: gaswashing bottle: medium: 200 ml water
                           sampling rate: 0.12 cu ft/min
                           test conc.: 162 ppm
                           absorption efficiency: 84%   (102)

C. WATER POLLUTION FACTORS
—Impact on biodegradation processes: 2.0 mg/l affects the selfpurification of water courses   (181)
—Reduction of amenities; faint odor: 0.037 mg/l   (129)
—Waste water treatment: ammonia is oxidized by ozone, the reaction is first order with respect to the conc. of ammonia and is catalyzed by OH^- over the pH range 7–9
—Sampling and analysis: Nessler reagent: det.lim.: 0.01 mg/l   (208)

D. BIOLOGICAL EFFECTS
—Oligosaprobic and β-mesosaprobic organisms: toxic; 0.08 to 0.4 mg/l   (30)
—Fish: fathead minnows: TLm (96 hr) hard water: 8.2 mg/l   (41)
        goldfish: TLm (24–96 hr): 2 to 2.5 mg/l   (154)
        coho salmon: 96-hr LC_{50}, F: 0.45 mg/l   (1505)
        guppy fry: 72-hr LC_{50}, S: 74 mg/l   (1503)
        guppy fry: 72-hr LC_{50}, S: 1.26 mg/l   (1503)
        cutthroat trout (*Salmo clarki*) fry: 96-hr LC_{50}, F: 0.5–0.8 mg/l   (1503)
                                 36-d LC_{50}, F: 0.56 mg/l   (1504)

rainbow trout: fertilized egg: 24 hr LC_{50}, S: >3.58 mg/l
alevins (0–50 days old): 24 hr LC_{50}, S: >3.58 mg/l
fry (85 days old): 24 hr LC_{50}, S: 0.068 mg/l
adults: 24 hr LC_{50}, S: 0.097 mg/l (951)
walking catfish: 48 hr LC_{50}: 0.28 mg/l (static bioassay) (948)
—Mammals:
mice: RD_{50} (respiratory rate): 303 ppm (1330)
—Man: lethal >1,700 ppm
severe toxic effect: 500 ppm, 1 min
symptoms of illness: 200 ppm
unsatisfactory: > 100 ppm (185)

ammoniumacetate
CH_3COONH_4
Uses: drugs; textile dyeing; foam rubbers; vinylplastics
A. PROPERTIES: m.p. 114°C, sp. gr. 1.07
C. WATER POLLUTION FACTORS:
—waste water treatment: A.S., Resp., BOD, 20°C, 1–5 days observed, feed: 1,000 mg/l;1 day acclimation: 79% theoretical oxidation (93)
D. BIOLOGICAL EFFECTS:
—Fish: mosquito fish: 24 hr, TLm: 238 mg/l
96 hr, TLm: 238 mg/l (41)

ammoniumcarbazotate see ammoniumpicrate

ammoniumchloride
NH_4Cl
Uses: dry batteries; soldering; manufacture of various ammonia compounds fertilizer; electroplating
A. PROPERTIES: white crystals, sublimes 350°C, sp. gr. 1.54
D. BIOLOGICAL EFFECTS:
—Fish:
Carassius carassius: 24 hr TLm: 640 mg/l (1295)
Daphnia magna: 24 hr TLm 202 mg/l
48 hr TLm: 161 mg/l
72 hr TLm: 67 mg/l
96 hr TLm: 50 mg/l (153)
100 hr TLm: 139 mg/l (1295)
Lepomis macrochirus: 24 hr–96 hr TLm: 725 mg/l (1295)
—Snail egg—Lymnacae sp.: 24 hr TLm: 241 mg/l
48 hr TLm: 173 mg/l
72 hr TLm: 73 mg/l
96 hr TLm: 70 mg/l (153)

ammoniumfluoride
NH_4F
Uses: fluorides; antiseptic in brewing; wood preservation
A. PROPERTIES: white crystals; sp.gr. 1.31; decomposed by heat

D. BIOLOGICAL EFFECTS:

	24 hr LC_{50}	48 hr LC_{50}	96 hr LC_{50}
fathead minnows (*Pimephales promelas*)	438 mg/l	417 mg/l	364 mg/l
grass shrimp (*Palaemonetes pugio*)	160 mg/l	93 mg/l	75 mg/l

(1904)

ammoniumoxalate

$$\begin{array}{c} COONH_4 \\ | \\ COONH_4 \cdot H_2O \end{array}$$

Uses: analytical chemistry, safety explosives
A. PROPERTIES: colorless crystals; soluble in water; sp.gr. 1.5 decomposed by heat
C. WATER POLLUTION FACTORS:
  —Biodegradation: adapted A.S.—product as sole carbon source—92.5% removal at 9.3 mg COD/g dry inoculum/hour   (327)

ammoniumpicrate (ammoniumcarbazotate; ammoniumpicronitrate)

$$O_2N-\underset{NO_2}{\overset{ONH_4}{\bigcirc}}-NO_2$$

Use: explosive, medicine
A. PROPERTIES: yellow crystals, sp.gr. 1.72, m.p. decomposes
B. BIOLOGICAL EFFECTS:
  —Fish:
  Lepomis macrochirus: static bioassay in fresh water at 23°C, mild aeration applied after 24 hr

material added	% survival after				best fit 96 hr LC_{50}
ppm	24 hr	48 hr	72 hr	96 hr	ppm
790	90	70	30	0	
560	90	70	30	0	220
320	90	80	50	30	
180	90	80	70	60	
100	100	100	100	92	

(352)

Menidia beryllina: static bioassay in synthetic seawater at 23°C, mild aeration applied after 24 hr

material added	% survival after				best fit 96 hr LC_{50}
ppm	24 hr	48 hr	72 hr	96 hr	ppm
180	0	—	—	—	
100	60	0	—	—	66
75	95	30	30	20	
56	100	100	100	100	

(352)

ammoniumpicronitrate *see* ammoniumpicrate

ammonium salts, quaternary *see* arquad

ammoniumsulfate
$(NH_4)_2SO_4$
Uses: fertilizers; fermentation; viscose rayon; tanning; food additive
A. PROPERTIES: brownish-gray to white crystals; sp.gr. 1.77; m.p. 513°C with decomposition
D. BIOLOGICAL EFFECTS:
—*Daphnia magna*: 25 hr TLm: 423 mg/l
  50 hr TLm: 433 mg/l
  100 hr TLm: 292 mg/l   (1295)

ammoniumsulfite
$(NH_4)_2SO_3 \cdot H_2O$
Uses: chemical intermediate; medicine; photography
A. PROPERTIES: colorless crystals; acrid; sulfurous taste; sublimes at 150°C with decomposition; soluble in water; sp.gr. 1.41
D. BIOLOGICAL EFFECTS:
—*Daphnia magna*: 25 hr TLm: 299 mg/l
  50 hr TLm: 273 mg/l
  100 hr TLm: 203 mg/l   (1295)

amprolium (1-[(4-amino-2-propyl-5-pyrimidinyl)methyl]-2-picoliniumchloride, hydrochloride)
Use: a coccidiostat
D. BIOLOGICAL EFFECTS:
—Fish:
  guppy (*Poecilia reticulata*): static bioassay: 48 hr TLm: 270 mg/l at 24°C
  rainbow trout: static bioassay: 48 hr TLm: 1550 mg/l at 15°C
                                                     (446)

prim-**amylacetate** (*n*-amylacetate; 1-pentanolacetate; amylacetic ester)
$CH_3COOC_5H_{11}$
A. PROPERTIES: m.w. 130.2; b.p. 148°C at 737 mm; sp.gr. 0.879 at 20/20°C; solub. 1,800 mg/l at 20°C; THC 1,042 kcal/mole
B. AIR POLLUTION FACTORS: 1 mg/cu m = 0.188 ppm, 1 ppm = 5.32 mg/cu m
—Odor: characteristic: quality: sweet, ester, banana
  hedonic tone: pleasant

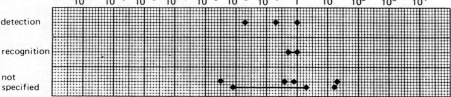

(19; 210; 307; 602; 636; 665; 671; 708; 786)

odor index: 25.047 (19)
threshold: unadapted panellists: 20 ppm
after adaption with pure odorant: 3,000 ppm (204)
USSR:
human odor perception: non perception: 0.5 mg/cu m
perception: 0.6 mg/cu m = 0.11 ppm
human reflex response: no response: 0.12 mg/cu m
adverse resp.: 0.30 mg/cu m (170)

C. WATER POLLUTION FACTORS:
- BOD_5: 0.9 at 440 mg/l, st.dil. (27)
0.31 at 1.7-20 ppm, std.dil.sew. (280)
0.88 at 440 ppm, Sierp, sew. (280)
- ThOD: 2.34
- Impact on biodegradation processes:
inhibition of degradation of glucose by *Pseudomonas fluorescens* at 350 mg/l
inhibition of degradation of glucose by *E.coli* at: >1000 mg/l (293)
- Reduction of amenities:
odor threshold: average: 0.08 mg/l (918; 889)
range: 0.0017 to 0.86 mg/l (97)(294)
- Waste water treatment:
A.C.: adsorbability: 0.175 g/g carbon, 88.0% reduction, infl.: 985 mg/l, effl.: 119 mg/l (32)

D. BIOLOGICAL EFFECTS:
- Bacteria: *Escherichia coli*: no effect level: 1 g/l (30)
- Toxicity threshold (cell multiplication inhibition test)
bacteria (*Pseudomonas putida*): 145 mg/l (1900)
algae (*Microcystis aeruginosa*): 63 mg/l (329)
green algae (*Scenedesmus quadricauda*): 80 mg/l (1900)
protozoa (*Entosiphon sulcatum*): 226 mg/l (1900)
protozoa (*Uronema parduczi Chatton-Lwoff*): 550 mg/l (1901)
- Crustacean:
Daphnia: 48 hr threshold toxic effect at 23°C: 440 ppm (356)
- Fish:
mosquito fish: 24-96 hr TLm: 65 mg/l (41)
creek chub: LD_0: 50 mg/l, 24 hr in Detroit river water
LD_{100}: 120 mg/l, 24 hr in Detroit river water (243)
creek chub: 24 hr critical range: 350-500 ppm (355)
goldfish: 96 hr TLm: 10 ppm (354)
Lepomis macrochirus: static bioassay in fresh water at 23°C, mild aeration applied after 24 hr:

material added	% survival after				best fit 96 hr LC_{50}
ppm	24 hr	48 hr	72 hr	96 hr	ppm
760	0 (<24 hr)				
560	narcosis	100	100	100	650
320	narcosis	100	100	100	

(352)

Menidia beryllina: static bioassay in synthetic seawater at 23°C: mild aeration applied after 24 hr:

material added ppm	% survival after				best fit 96 hr LC_{50} ppm
	24 hr	48 hr	72 hr	96 hr	
560	50	0	–	–	
320	90	30	10	0	180
180	100	80	70	50	
100	100	100	100	90	

(352)

amylacetic ester   *see prim*-amylacetate

***prim-n*-amylalcohol**   *see n*-pentanol

***sec*-act.amylalcohol**   (2-pentanol; methylpropylcarbinol; 1-methyl-1-butanol)
$CH_3(CH_2)_2CHOHCH_3$
A. PROPERTIES: m.w. 88.15; b.p. 119°C; v.d. 3.04; sp.gr. 0.809 at 20/4°C; solub. 53,000 mg/l at 30°C; log P_{oct} 1.34 (calculated)
C. WATER POLLUTION FACTORS:
—Waste water treatment: A.S., BOD, 20°C, 1–5 days observed, feed: 333 mg/l, 30 days acclimation, 91% removed   (93)
D. BIOLOGICAL EFFECTS
—Mammalia: rabbit: single oral LD_{50}: 4.25 ml/kg   (211)

***tert*-amylalcohol**   *see* 2-methyl-2-butanol

***n*-amylaldehyde**   *see n*-valeraldehyde

***n*-amylamine**   (pentylamine; 1-aminopentane)
$CH_3(CH_2)_4NH_2$
A. PROPERTIES: m.w. 87.16; m.p. -55°C; b.p. 104°C; v.p. 35 mm at 26°C; v.d. 3.01 sp.gr. 0.77 at 20/4°C; log P_{oct} 1.05 (calculated)
B. AIR POLLUTION FACTORS: 1 mg/cu m = 0.281 ppm, 1 ppm = 3.56 mg/cu m
C. WATER POLLUTION FACTORS:
—Waste water treatment:
degradation by *Aerobacter*: 200 mg/l at 30°C:
   parent: 100% in 25 hr
   mutant: 100% in 9 hr   (152)
D. BIOLOGICAL EFFECTS
—Fish: creek chub: LD_0:   30 mg/l, 24 hr in Detroit river water
   LD_{100}: 50 mg/l, 24 hr in Detroit river water   (243)

amylcarbinol   *see n*-hexanol

amylchloride   (1-chloropentane; pentylchloride)
$CH_3(CH_2)_3CH_2Cl$
A. PROPERTIES: m.w. 106.60; m.p. -99°C; b.p. 108.2°C; v.d. 3.67; sp.gr. 0.883 at 20/4°C
C. WATER POLLUTION FACTORS
—Waste water treatment: A.S.: after 6 hr: 1.5% of ThOD
   12 hr: 1.8% of ThOD
   24 hr: 2.8% of ThOD   (88)

n-amylcyanide: *see* hexanenitrile

α-*n*-amylene   *see* 1-pentene

β-*n*-amylene   *see* 2-pentene

amylene dimer
   Use: intermediate in the manufacture of alkylated phenols
   Composition: approx. 83% C_{10} olefin isomers, 10% C_{15} olefin isomers
A. PROPERTIES: sp.gr. (15/4°C): 0.783; boiling range; 150–260°C; solub. 15 mg/l at 20°C
C. WATER POLLUTION FACTORS:
   —BOD_5: 0.14 = 4% ThOD
      0.24 = 7% ThOD (after adaptation)   (277)
   —COD: 1.68 = 49% ThOD   (277)
D. BIOLOGICAL EFFECTS:
   —Fish: *Carassius aurates*: not noxic in saturated solution (16 mg/l)   (277)

amylenehydrate   *see* 2-methyl-2-butanol

amylmercaptan   *see* pentylmercaptan

n-**amylmethylketone**   *see* 2-heptanone

amylxanthate, potassium
   Use: flotation agent
A. PROPERTIES: solid, decomposes
D. BIOLOGICAL EFFECTS:
   —Fish:
      rainbow trout (*Salmo gairdneri*):
      96 hr LC_{50}: Cyanamid C 350: 32–56 mg/l (static test)
                      Dow Chemical Z 6: ~ 18 mg/l (static test)
         LC_{100}: 56 mg/l after <4 days exposure (static test)
                1.0 mg/l after 28 days exposure (flow through test)   (1087)
   —Mammals: ingestion: rat: single oral LD_{50}: 1000 to 2000 mg/kg   (1546)

anesthesin   *see* benzocain

aniline   (aminobenzene; phenylamine)

A. PROPERTIES: m.w. 93.1; m.p. -6°C; b.p. 184°C; v.p. 1 mm at 35°C, 0.3 mm at 20°C; v.d. 3.22; sp.gr. 1.02; solub. 34,000 mg/l; sat.conc. 1.5 g/cu m at 20°C, 3.4 g/cu m at 30°C; log P_{oct} 0.90/0.98
B. AIR POLLUTION FACTORS: 1 mg/cu m = 0.259 ppm, 1 ppm = 3.87 mg/cu m
   —Odor: characteristic: hedonic tone: pungent

ANILINE

odor thresholds   mg/cu m

[Chart showing odor thresholds on log scale from 10^{-7} to 10^4 mg/cu m, with rows for detection, recognition, and not specified]

(2; 73; 279; 291; 297; 610; 664; 695; 703; 741; 840; 845)

USSR: human odor perception: non perception: 0.34 mg/cu m
                                  perception: 0.37 mg/cu m
        human reflex response: adverse response: 0.07 mg/cu m
          animal chronic exposure: adverse effect: 0.05 mg/cu m   (170)
—Sampling and analysis: photometry: min. full scale: 33 ppm   (53)
                        Test tubes: DRAGER: min.det.lim.: 1 ppm   (58)

C. WATER POLLUTION FACTORS:
—BOD_5: 1.5 std.dil. at 1.5–3 mg/l   (27)
           1.49–2.26 std.dil.sewage   (41)
           1.76   (30)
           1.42   (36)
           62% ThOD   (274)
—BOD_5^{20}: 1.42 at 10 mg/l, unadapted sew.: lag period: 3 days
—BOD_{20}^{20}: 2.02 at 10 mg/l, unadapted sew.: lag period: 3 days   (554)
COD: 94% ThOD (0.05 n Cr_2O_7)   (274)
          2.34; 2.4   (36)(41)
—$KMnO_4$: 9.48   (30)
            88% ThOD (0.01 n $KMnO_4$)   (274)
—ThOD: 2.66; 3.09; 3.18; 2.41   (30)(27)(36)(274)
—Impact on biodegradation processes:
   75% inhibition of nitrification in the activated sludge process at 7.7 mg/l   (43)
   degree of inhibition of NH_3 oxidation by *Nitrosomonas sp*:
      at 100 mg/l,   86% inhibition
      at <1 mg/l, ~ 50% inhibition   (390)
   inhibition of photosynthesis of a freshwater non-axenic uni-algal culture of
     Selenastrum capricornutum:
     at   10 mg/l: 90% carbon-14 fixation (vs controls)
     at   100 mg/l: 34% carbon-14 fixation (vs controls)
     at 1000 mg/l: 3% carbon-14 fixation (vs controls)   (1690)

effect on BOD:

BOD 20°C days	BOD of original sample	BOD after addition of 100 ppm aniline
5	6	0
10	12	2
15	14	3
20	16	5
25	17	6
30	17	7

(172)

—Reduction of amenities: T.O.C.: 70 mg/l, range: 2.0 to 128 mg/l (294; 30)
    3000 mg/kg (907)
—Biodegradation:
  decomposition period by a soil microflora: 4 days (176)
  adapted A.S. at 20°C—product is sole carbon source: 94.5% COD removal at 19.0 mg COD/g dry inoculum/hr (327)
  Warburg app., activated sludge from mixed domestic/industrial treatment plant: 44–58% depletion at 20 mg/l after 6 hr at 25°C
—Aqueous reactions: photo-oxidation by u.v. light in aqueous medium at 50°C; 28.5% degradation to CO_2 after 24 hr (1628)
—Waste water treatment:
  A.C.: adsorbability: 0.150 g/g carbon; 74.9% reduction, infl.: 1,000 mg/l, effl.: 251 mg/l (32)
  adsorption on Amberlite X AD-7: retention efficiency: 100%, infl.: 4.0 ppm, effl.: 0 ppm (40)
  air stripping constant: k = 0.198 $days^{-1}$ at 100 mg/l (82)
  degradation by *Aerobacter*: 500 mg/l at 30°C:
          parent: 100% ring disruption in 54 hr
          mutant: 100% ring disruption in 12 hr (152)
—Sampling and analysis: photometry: min.full scale: 0.4×10^{-6} mole/l (53)
D. BIOLOGICAL EFFECTS:
—Bacteria: *E. coli*: no effect at 1 g/l (329)
—Algae:
  Scenedesmus: toxic: 10 mg/l
  Microcystis aeruginosa: inhibition of cell multiplication starts at 0.16 mg/l (329)
  Microcystis aeruginosa. LD_{50}: 20 ppm (1094)
—Protozoa: ciliate (*Tetrahymena pyriformis*): 24 hr LC_{100} = 21.5 m mole/l
    (1662)
—Arthropoda: *Daphnia*: toxic: 0.4 mg/l
—Toxicity threshold (cell multiplication inhibition test):
  bacteria (*Pseudomonas putida*):   130 mg/l
  green algae (*Scenedesmus quadricauda*):   8.3 mg/l
  protozoa (*Entosiphon sulcatum*):   24 mg/l (1900)
  protozoa (*Uronema parduczi Chatton-Lwoff*): 91 mg/l (1901)
—Amphibian: lethality and teratogenicity to early embryonic stages of South African clawed frog, Xenopus laevis:

conc. mg/l	day							
	1		2		3		4	
	A/S[a]	%	A/S	%	A/S	%	A/S	%
0	0/50	0	0/50	0	0/50	0	0/50	0
10	0/50	0	0/47	0	4/36	11	4/36	11
50	1/50	2	3/48	6	3/48	6	3/48	6

(1418)

[a] A/S = abnormals/survivors
  mexican axolotl (3–4 w after hatching): 48 hr LC_{50}: 440 mg/l
  clawed toad   (3–4 w after hatching): 48 hr LC_{50}: 560 mg/l (1823)
—Carcinogenicity:?

- Mutagenicity in the *Salmonella* test: neg., <0.005 revertant colonies/nmol; <70 revertant colonies at 1000 µg/plate (1883)
- Man: severe toxic effects: 80 ppm, 60 min
  symptoms of illness: 20 ppm
  unsatisfactory: >10 ppm (185)

anilinechloride *see* aniline hydrochloride

aniline hydrochloride (anilinechloride)

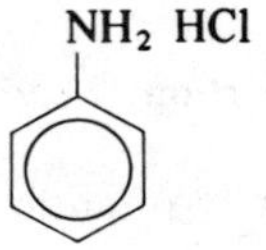

$$\text{C}_6\text{H}_5\text{NH}_2 \cdot \text{HCl}$$

A. PROPERTIES: white crystals, darkens in light and air; m.p. 189°C; b.p. 245°C; sp.gr. 1.2215
D. BIOLOGICAL EFFECTS:
   - Fish: goldfish. approx. fatal conc. 5.5 mg/l, 48 hr (226)

***o*-anilinesulfonic acid**

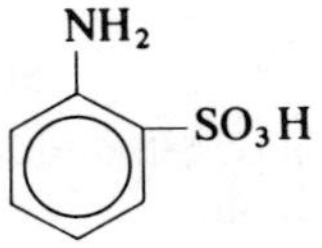

$$\text{C}_6\text{H}_4(\text{NH}_2)(\text{SO}_3\text{H}) \text{ (ortho)}$$

C. WATER POLLUTION FACTORS:
   - Waste water treatment: decomposition period by a soil microflora: >64 days (176)

***m*-anilinesulfonic acid**

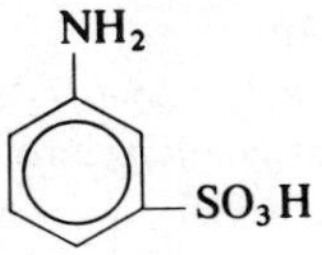

$$\text{C}_6\text{H}_4(\text{NH}_2)(\text{SO}_3\text{H}) \text{ (meta)}$$

C. WATER POLLUTION FACTORS:
   - Waste water treatment: decomposition period by a soil microflora: >64 days (176)

***p*-anilinesulfonic acid** (sulfanilic acid *p*-aminobenzenesulfonic acid)

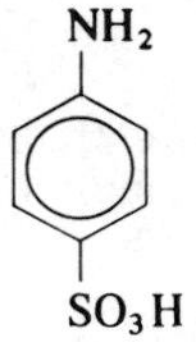

$$\text{C}_6\text{H}_4(\text{NH}_2)(\text{SO}_3\text{H}) \text{ (para)}$$

A. PROPERTIES: colorless crystals; m.w. 191.20; b.p. 288°C, decomposes; solub. 10.8 g/l at 20°C; 66.7 g/l at 100°C
C. WATER POLLUTION FACTORS:
—Biodegradation: decomposition by a soil microflora in >64 days (176) adapted A.S. at 20°C—product is sole carbon source: 95.0% COD removal at 4.0 mg COD/g dry inoculum/hr (327)
—Oxidation parameters:
BOD_5: 1.11 std.dil.sew. (282, 163)
$KMnO_4$ value: 4.79 (30)
D. BIOLOGICAL EFFECTS:
—Mammalia: mouse: acute oral LD_{50}: >3.2 g/kg (211)

aniline yellow *see* p-phenylazoaniline

o-anisic acid (2-methoxybenzoic acid; salicylic acid methylether)

A. PROPERTIES: m.w. 152.14; m.p. 101°C; b.p. 200°C; sp.gr. 1.180; solub. 5,000 mg/l at 30°C; log P_{oct} 0.80/2.93 (calculated)
C. WATER POLLUTION FACTORS:
—Waste water treatment: decomposition period by a soil microflora: 4 days (176)

m-anisic acid (3-methoxybenzoic acid)

A. PROPERTIES: colorless needles; m.w. 152.14; m.p. 107/110°C; b.p. 170/172°C at 10 mm Hg; log P_{oct} 2.02
C. WATER POLLUTION FACTORS:
—Waste water treatment: decomposition period by a soil microflora: 16 days
(176)

p-anisic acid (4-methoxybenzoic acid)

A. PROPERTIES: m.w. 152.14; m.p. 275/280°C; sp.gr. 1.385 at 4/4°C; solub. 400 mg/l at 18°C; log P_{oct} 1.96

C. WATER POLLUTION FACTORS:
—Waste water treatment: decomposition period by a soil microflora: 2 days
(176)

o-anisidine (2-methoxyaniline)

A. PROPERTIES: m.w. 125.15; m.p. 6.2°C; b.p. 224°C; v.p. <0.1 mm at 30°C; v.d. 4.25; sp.gr. 1.0923 at 20/4°C; log P_{oct} 0.95
B. AIR POLLUTION FACTORS: 1 mg/cu m = 0.20 ppm, 1 ppm = 5.12 mg/cu m
C. WATER POLLUTION FACTORS:
—Waste water treatment: degradation by *Aerobacter*: 500 mg/l at 30°C
                      ring disruption: parent: 92% in 120 hr
                                     mutant: 100% in 16 hr   (152)
—Decomposition period by a soil microflora: >64 days   (176)

m-anisidine (3-methoxyaniline)

A. PROPERTIES: m.w. 123.15; m.p. <−12°C; b.p. 251°C; sp.gr. 1.096 at 20/4°C; log P_{oct} 0.93
C. WATER POLLUTION FACTORS:
—Waste water treatment: degradation by *Aerobacter*: 500 mg/l at 30°C
                      ring disruption: parent: 80% in 120 hr
                                     mutant: 100% in 24 hr   (152)
—Decomposition period by a soil microflora: >64 days   (176)

p-anisidine (4-methoxyaniline; *p*-aminoanisole; *p*-aminomethoxybenzene; *p*-aminomethylphenylether)

A. PROPERTIES: m.w. 123.15; m.p. 57/59°C; b.p. 243°C; v.d. 4.25; sp.gr. 1.0605 at 67/4°C; log P_{oct} 0.95
B. AIR POLLUTION FACTORS: 1 mg/cu m = 0.20 ppm; 1 ppm = 5.12 mg/cu m

C. WATER POLLUTION FACTORS:
—Waste water treatment: degradation by *Aerobacter*: 500 mg/l at 30°C:
ring disruption: parent: 86% in 120 hr
mutant: 100% in 12 hr (152)
—Decomposition period by a soil microflora: 64 days (176)

anisole *see* methylphenylether

anol *see* cyclohexanol

anone *see* cyclohexanone

ansar *see* disodiummethanearsonic acid

anthanthrene (dibenzo (*cd, jh*) pyrene; dibenzo (*def-mno*) chrysene)

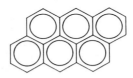

B. AIR POLLUTION FACTORS:
—Manmade sources:
in tailgases of gasoline engine: 2–73 µg/cu m (340)
in tailgases of gasoline engine: 0.017–0.026 mg/l gasoline consumed (1070)
in gasoline (high octane-number): 0.028–2.1 mg/l (1220; 380)
in bitumen: 0.04–0.30 ppm (500)
emission from space heating installation burning:
coal (underfeed stoker): 0.29 mg/10^6 Btu input
gas: 0.2 mg/10^6 Btu input (954)
—Ambient air quality: glc's in The Netherlands

(ng/cu m)		Delft	The Hague	Vlaardingen	Amsterdam	Rotterdam
summer	1968	n.d.	–	n.d.	–	n.d.
	1969	<1	<1	<1	n.d.	<1
	1970	–	<1	<1	<1	<1
	1971	1	<1	n.d.	n.d.	n.d.
winter	1968	n.d.	n.d.	4	2	2
	1969	1	1	3	1	3
	1970	<1	–	2	–	2
	1971	n.d.	n.d.	1	<1	1

(1277)

in the average American urban atmosphere, 1963:
2.3 µg/g airborne particulates, or
0.26 ng/g cu m air (1293)
glc's in Birkenes (Norway): Jan.–June '77: avg.: 0.03 ng/cu m
range: n.d.–0.25 ng/cu m
(*n* = 18)
glc's in Rørvik (Sweden): Dec. '76–April '77: avg.: 0.04 ng/cu m
range: n.d.–0.27 ng/cu m
(*n* = 21) (1236)

anthracene

Uses: dyes

A. PROPERTIES: sp.gr. 1.25; m.w. 178.23; m.p. 216.2 - 216.4°C;
b.p. 340°C; v.d. 6.15; log P_{oct} 4.45
solub: 1.29 mg/l at 25°C in distilled water
0.6 mg/l at 25°C in salt water
0.075 mg/l at 15°C

B. AIR POLLUTION FACTORS:
—Manmade sources:
in coke oven emissions: 46.4 - 942.8 µg/g of sample (960)
emissions from space heating installation burning:
coal (underfeed stoker): 0.85 mg/10^6 Btu input
gasoil: 3.9 mg/10^6 Btu input (954)
in gasoline: 1.55 mg/l
in exhaust condensate of gasoline engine: 0.53 - 0.64 mg/l gasoline consumed
(1070)

emissions from typical European gasoline engine 1608 cu cm—following European driving cycles—using leaded and unleaded commercial gasolines: 18.2–392.5 µg/l fuel burnt (1291)
in gasoline (high octane number): 2.59 mg/l (380)
in outlet of waterspray tower of asphalt hot-road-mix process: 1600 ng/cu m
in outlet of asphalt air-blowing process: 220,000 ng/cu m (1212)

—Ambient air quality:
glc's in Birkenes (Norway): Jan.–June '77: avg: 0.03 ng/cu m
range: n.d.–0.23 ng/cu m
(n = 18)
glc's in Rørvik (Sweden): Dec. '76–April '77: avg: 0.09 ng/cu m
range: n.d.–0.95 ng/cu m
(n = 21) (1236)
glc's in Budapest 1973: heavy traffic area: winter: 62.8 ng/cu m
(6–20 hr) summer: 55.1 ng/cu m
low traffic area: winter: 21.5 ng/cu m
summer: 49.9 ng/cu m (1259)

organic fraction of suspended matter:
Bolivia at 5280 m altitude (Sept.–Dec. 1975): 0.037–0.055 µg/1000 cu m (+ phenanthrene)
Belgium, residential area (Jan.–April 1976): 0.09–0.89 µg/1000 cu m (+ phenanthrene) (420)

—Sampling and analysis: photometry: min.ful scale: 3.8 ppm (53)

C. WATER POLLUTION FACTORS
—BOD_5: 2% of ThOD (220)
0% of ThOD (274)

- $BOD_{35}^{25°C}$: 1.18 in seawater/inoculum: enrichment cultures of hydrocarbon oxidizing bacteria (521)
- COD: 35% of ThOD (220)
  94% of ThOD (0.05 n Cr_2O_7) (274)
- $KMnO_4$: 0.632; acid 5% of ThOD (30; 220)
  alkal. 1% of ThOD
  0% of ThOD (0.01 n $KMnO_4$) (274)
- ThOD: 3.41 (521)

Anthracene → Anthracene *cis*-1,2-dihydrodiol →

1,2-Dihydroxy anthracene → *cis*-4-(2-hydroxynaphth-3-yl)-2-oxobut-3-enoic acid →

4-(2-Hydroxynaphth-8-yl)-2-oxo-4-hydroxybutyric acid → (Pyruvate) → 1-Hydroxy-2-naphthaldehyde → 1-Hydroxy-2-naphthoic acid →

→ Salycilate, etc.

- Biodegradation to CO_2:

Sampling site	concentration ($\mu g/l$)	month	incubation time (hr)	degradation rate ($\mu g/1/day$) × 10^3	turnover time (days)
Control station	15	–	–	0	∞
Near oil storage tanks	15	–	–	70 ± 30	290
Near Oil storage tanks	15	–	–	0	∞
Skidaway river	15	Oct.	24	0	∞
Skidaway river	15	Oct.	72	8 ± 2	2000

(381)

210 ANTHRANILIC ACID

—Aquatic reactions:

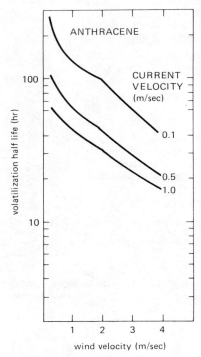

Variation in predicted volatilization rates of anthracene at 25°C under varying conditions of wind and current velocity in a stream 1.0 m in depth (1391)

Adsorption on smectite clay particles from simulated seawater at 25°C—experimental conditions: 100 μg anthracene/l; 50 mg smectite/l—adsorption: 0.90 μg/mg = 46% adsorbed (1009)
In estuarine waters: at 4 μg/l, 6% adsorbed on particles after 3 hr
at 15 μg/l, 22% adsorbed on particles after 3 hr (381)
After 3 hr incubation in natural seawater, 11% of 15 μg/l were taken up by suspended aggregates of dead phytoplankton cells and bacteria (957)
—Sediment quality: in sediments in Severn estuary (U.K.): 0.1–6.4 ppm dry wt. (includes phenanthrene) (1467)
—Sampling and analysis: photometry: min.full scale: 0.015×10^{-6} mole/l (53)

D. BIOLOGICAL EFFECTS:
—Algae: inhibition of photosynthesis of a freshwater, non axenic uni-algal culture of *Selenastrum capricornutum* at:
  1% saturation: 99% carbon-14 fixation (vs. controls)
  10% saturation: 104% carbon-14 fixation (vs. controls)
  100% saturation: 99% carbon-14 fixation (vs. controls) (1690)
—Crustacean:

Daphnia pulex: *bioaccumulation* factor: 760, Initial conc. in water: 0.02 ppb, equilibrium reached after 4 hr

excretion of ^{14}C—after 16 hr incubation with ^{14}C-anthracene and subsequent transfer to clean water resulted in a rapid release (1 hr) of about 30% of the total ^{14}C, a more slow elimination of roughly 60% with a half life of 3.3 hr and a tightly bound residue of 8%

metabolism: the observed rate of metabolite excretion during the first 24 hr of excretion was only 6% of the total ^{14}C outflux rate (1597)

—Molluscs: Uptake and depuration by oysters (*Crassostrea virginica*) from oil-treated enclosure:

time of exposure days	depuration time days	concentration oysters $\mu g/g$	water $\mu g/l$	accumulation factor oysters/water
2	–	5.6	13	430
8	–	2.5	1	2,500
2	7	1.2	–	–
8	7	0.4	–	–
8	23	0.1	–	–

half-life for depuration: 3 days (957)
—Fish: trout: no effect level: 5 mg/l, 24 hr (30)
—Carcinogenicity: negative
—Mutagenicity in the *Salmonella* test: negative
<0.01 revertant colonies/n mol
<70 revertant colonies at 1000 µg/plate (1893)

anthranilic acid (*o*-aminobenzoic acid)

$$\text{COOH} \atop \bigcirc\text{-NH}_2$$

Uses: dyes; drugs; perfumes; pharmaceuticals
A. PROPERTIES: m.w. 137.13; m.p. 145/147°C; b.p. sublimes; solub. 3,500 mg/l at 14°C; log P_{oct} 1.21
C. WATER POLLUTION FACTORS
—BOD$_5$: 1.32 st. dil. sewage (281)(16)
—Waste water treatment: acclimated activated sludge: 50% theor. oxidation, infl.: 250 mg/l, 30 min aeration (92)
—Biodegradation:
decomposition period by a soil microflora: 2 days (176)
adapted A.S. at 20°C—product is sole carbon source: 97.5% COD removal at 27.1 mg/COD/g dry inoculum/hr
lag period for degradation of 16 mg/l by waste water or by soil at pH 7.3 and 30°C: less than 1 day (1096)

impact on biodegradation processes: at 100 mg/l no inhibition of NH_3 oxidation by *Nitrosomonas* sp. (390)

anthraquinone (9,10-dihydro-9,10-diketoanthracene)

Uses: intermediate for dyes and organics, bird repellent for seeds

A. PROPERTIES: yellow green crystals; m.w. 208.20; m.p. 286°C sublimes; b.p. 379/381°C; sp.gr. 1.419 at 20/4°C

B. AIR POLLUTION FACTORS:
  —Ambient air quality:
   organic fraction of suspended matter:
    Bolivia at 5200 m altitude (Sept.–Dec. 1975): 0.064–0.065 μg/1000 cu m
    Belgium, residential area (Jan.–April 1976): 0.57–1.00 μg/1000 cu m (428)
   glc's in residential area (Belgium) Oct. 1976:
    in particulate sample: 1.59 ng/cu m
    in gasphase sample: 5.66 ng/cu m (1289)
  —Sampling and analysis: photometry: min. full scale: 1.4 ppm (53)

C. WATER POLLUTION FACTORS:
  —Impact on biodegradation processes: at 2.5 mg/l inhibition of the selfpuration activity of natural waters (30)
  —Inhibition of photosynthesis of a freshwater non-axenic uni-algal culture of *Selenastrum capricornutum*: at 1% saturation: 97% carbon-14 fixation (vs. controls)
   10% saturation: 91% carbon-14 fixation (vs. controls)
   100% saturation: 85% carbon-14 fixation (vs. controls) (1690)
  —Water quality:
   in Eastern Ontario: drinking waters (June–Oct.'78): 0.1–2.1 ng/l ($n = 12$)
   raw waters (June–Oct. '78): 0.9–4.7 ng/l ($n = 2$) (1698)
  —Sampling and analysis: photometry: min. full scale: 0.057×10^{-6} mole/l (53)

D. BIOLOGICAL EFFECTS:
  —Mammals: oral LD_{50} (mice) >5000 mg/kg (1855)

anthraquinone-α-sulfonic acid

C. WATER POLLUTION FACTORS:
  —BOD$_5$: 0 std. dil. sewage (n.s.i.) (281)(161)
D. BIOLOGICAL EFFECTS:
  —*Daphnia magna*: 24 hr TLm: 186 mg/l (sodium salt)
                    48 hr TLm: 186 mg/l (sodium salt)
                    72 hr TLm: 186 mg/l (sodium salt)
                    96 hr TLm: 50 mg/l (sodium salt)   (153)
                   100 hr TLm: 12 mg/l (sodium salt)   (1295)
  —Snail eggs, *Lymnaea* sp.: 24–96 hr TLm: 186 mg/l (sodium salt) (153)

antifebrin *see* acetanilide

aquathol K (7-oxabicyclo(2,2,1)heptane-2,3-dicarboxylic acid; 1,2-dicarboxy-3,6-endoxocyclohexane; 3,6-endoxohexahydrophthalic acid; endothall; accelerate; des-i-cate; hydout; hydrothol; ripenthol, triendothal)

Use: herbicide, defoliant, desiccant, growth regulator
A. PROPERTIES: mp. 144°C, solub. 100 g/kg at 20°C
D. BIOLOGICAL EFFECTS:
  —Algae:

		ppb		
Chlorococcum sp.	Technical acid	1 × 10^5	50 percent decrease in O$_2$ evolution	*f*
Chlorococcum sp.	Technical acid	5 × 10^4	50 percent decrease in growth	Measured as ABS (525 mu) after 10 days
Dunaliella tertiolecta	Technical acid	4.25 × 10^5	50 percent decrease in O$_2$ evolution	*f*
Dunaliella tertiolecta	Technical acid	5 × 10^4	50 percent decrease in growth	Measured as ABS (525 mu) after 10 days
Isochrysis galbana	Technical acid	6 × 10^4	50 percent decrease in O$_2$ evolution	*f*
Isochrysis galbana	Technical acid	2.5 × 10^4	50 percent decrease in growth	Measured as ABS (525 mu) after 10 days
Phaeodactylum tricornutum	Technical acid	7.5 × 10^4	50 percent decrease in O$_2$ evolution	*f*

Phaeodactylum tricornutum	Technical acid	1.5×10^4	50 percent decrease in growth	Measured as ABS (525 mu) after 10 days
Chlorococcum sp.	Amine salt	$>1 \times 10^6$	50 percent decrease in O_2 evolution	f
Chlorococcum sp.	Amine salt	3×10^5	50 percent decrease in growth	Measured as ABS (525 mu) after 10 days
Dunaliella tertiolecta	Amine salt	$>1 \times 10^6$	50 percent decrease in O_2 evolution	f
Dunaliella tertiolecta	Amine salt	4.5×10^4	50 percent decrease in growth	Measured as ABS (525 mu) after 10 days
Isochrysis galbana	Amine salt	$>1 \times 10^6$	50 percent decrease in O_2 evolution	f
Isochrysis galbana	Amine salt	2.25×10^4	50 percent decrease in growth	Measured as ABS (525 mu) after 10 days
Phaeodactylum tricornutum	Amine salt	$>1 \times 10^6$	50 percent decrease in O_2 evolution	f
Phaeodactylum tricornutum	Amine salt	2.5×10^4	50 percent decrease in growth	Measured as ABS (525 mu) after 10 days

(2348)

—Crustacean: *Gammarus lacustris*: no effect at 100,000 µl/96 hr (dipotassium salt)

(2124)

—Insects: lowest observed avoidance conc., mayfly nymphs (*Ephemerella walkeri*): >10 mg/l (dipotassium salt)

—Fish:
rainbow trout (*Salmo gairdneri*), lowest observed avoidance conc.: >10 mg/l (dipotassium salt) (1621)
Pimephales promelas: 320,000 µg/l LC_{50}, 96 hr (dipotassium salt) (2131)
Lepomis macrochirus: 160,000 µg/l LC_{50}, 96 hr (dipotassium salt) (2131)
Pimephales notatus: 110,000 µg/l LC_{50}, 96 hr (disodium salt) (2132)
Lepomis macrochirus: 125,000 µg/l LC_{50}, 96 hr (disodium salt) (2132)
Micropterus salmoides: 120,000 µg/l LC_{50}, 96 hr (disodium salt) (2132)
Notropis umbratilus: 95,000 µg/l LC_{50}, 96 hr (disocium salt) (2132)
Micropterus salmoides: 200,000 µg/l LC_{50}, 96 hr (disodium salt) (2132)
Oncorhynchus tschawtscha: 136,000 µg/l LC_{50}, 96 hr (disodium salt) (2132)

—Mammals:
acute oral LD_{50} (rats): 51 mg/kg (acid)
182–197 mg/kg (sodium salt)
in diet: endothall-Na, fed for 2 years to rats at 1000 ppm caused no ill effects

(1855)

arachidic acid *see* eicosanoic acid

dl-arginine (2-amino-5-guanidopentanoic acid; *dl*-α-amino-δ-quanidovaleric acid; *dl*-N-δ-guanylornithin)

$NH_2C(NH)NH(CH_2)_3CH(NH_2)COOH$
   Uses: biochemical research; medicine; pharmaceuticals; natural sources; widely found in animal and plant proteins
A. PROPERTIES: m.w. 174.21; m.p. 217/218°C decomposes; b.p. 238°C (*l*-arg.); solub. 150,000 mg/l (*l*-arg.); log P_{oct} −2.59 (calculated)
C. WATER POLLUTION FACTORS
   —Manmade sources: excreted by man:
      in urine: 0.34 to 0.5 mg/kg body wt/day
      in feces: 2.9 to 5 mg/kg body wt/day
      in sweat: 5.8 to 21.4 mg/100 ml (203)
   —Waste water treatment:
   A.S.: after 6 hr: 2.1% of ThOD
          12 hr: 7.7% of ThOD
          24 hr: 16.5% of ThOD (89)
   A.S., BOD, 20°C, 1–5 days observed, feed: 333 mg/l, 15 days acclimation, 94% removed (93)
   —Impact on biodegradation processes: approximately 50% inhibition of NH_3 oxidation in *Nitrosomonas* at 1.7 mg/l (*l*-isomer) (408)

aroclor 1016 (*see also* polychlorobiphenyls)
A polychlorobiphenyl containing 16% of chlorine
A. PROPERTIES: solub. 0.22–0.25 mg/l; 0.049 mg/l at 24°C composition of the water soluble fraction (WSF) (1666)

monochloro isomers	110.8 µg/l	or	12.2% of WSF
dichloro isomers	279.6 µg/l	or	30.9% of WSF
trichloro isomers	328.7 µg/l	or	36.3% of WSF
tetrachloro isomers	186.5 µg/l	or	20.6% of WSF
total WSF	905.5 µg/l		100% (1909)

C. WATER POLLUTION FACTORS: Partition coefficient to natural sediments
   physical-chemical characteristics of sediment

sediment	TOC (%)	pH	% sand	% salt	% clay	partition coeff.
USDA Pond	0.8	6.4	–	–	–	1370
Doe Run Pond	1.4	6.1	56.0	44.0	<1.0	1290
Hickory Hill Pond	2.4	6.3	55.0	45.0	<1.0	1300
Oconee River	0.4	6.5	93.0	6.0	2.0	620

D. BIOLOGICAL EFFECTS: Concentration of Aroclor 1016 in fish confined to a live-cage for 14 days near Rodger's Island in the Hudson river: average conc. of Aroclor 1016 in Hudson water: 0.17 µg/l

		bioaccumulation factor
creek chubsucker (*Ermyzon oblongus*):	2200 µg/kg (whole fish)	13000
yellow perch (*Perca flavescens*):	1800 µg/kg (whole fish)	10600
pumpkinseed (*Lepomis gibbosus*):	2500 µg/kg (whole fish)	14700
brown bullhead (*Ictalurus nebulosus*):	3800 µg/kg (whole fish)	22300
	2800 µg/kg (whole fish)	17000

(1434)

aroclor 1221 (*see also* polychlorobiphenyls) = 1 polychlorobiphenyl containing 21% of chlorine

AROCLOR 1232

A. PROPERTIES: solub. 0.59 mg/l at 24°C
 —Composition of the water soluble fraction (WSF)

monochloro isomers:	3240.7 µg/l	or	92.2% WSF
dichloro isomers:	232.9 µg/l	or	6.6% WSF
trichloro isomers:	31.4 µg/l	or	0.9% WSF
tetrachloro isomers:	11.5 µg/l	or	0.3% WSF
total WSF	3516.3 µg/l		100% (1909)

C. WATER POLLUTION FACTORS: biodegradation at 0.1 mg/l:

	normal sewage	adapted sewage	
after 24 hr:	0%	49%	
after 135 hr:	12%	96%	(997)

D. BIOLOGICAL EFFECTS:
 harlequin fish (Rasbora heteromorpha)

mg/l	24 hr	48 hr	96 hr	3m (extrapolated)	
LC_{10} (F)	1.1	1.05	0.98		
LC_{50} (F)	1.3	1.15	1.05	0.5	(331)

 cutthroat trout: 96 hr LC_{50}: 1.2 mg/l   (1617)

aroclor 1232
A polychlorobiphenyl containing 32% chlorine

D. BIOLOGICAL EFFECTS:
 —Fish:
 cutthroat trout: 96 hr LC_{50}: 2.5 mg/l
 harlequin fish (*Rasbora heteromorpha*): mg/l   (1617)

	mg/l				
	24 hr	48 hr	96 hr	3m (extrapolated)	
LC_{10} (F)	0.52	0.27	—		
LC_{50} (F)	0.9	0.56	0.32	0.03	(331)

aroclor 1242

Users and formulation: dielectric liquids; thermostatic fluids; swelling agents for transmission seals; additives or base for lubricants, oils and greases; plasticizers for cellulosics, vinyls and chlorinated rubbers.

A. PROPERTIES: v.p. 50 mm at 225°C; sp.gr. 1.41 at 65/15.5°C; solub. 0.10 mg/l at 24°C
 —Composition:

monochlorobiphyenyls:	3%	
dichlorobiphenyls:	13%	
trichlorobiphenyls:	28%	
tetrachlorobiphenyls:	30%	
pentachlorobiphenyls:	22%	
hexachlorobiphenyls:	4%	(1837)

 composition of the water soluble fraction (WSF)

monochloro isomers:	136.7 µg/l	or	19.4% of WSF
dichloro isomers:	223.6 µg/l		31.8%
trichloro isomers:	220.3 µg/l		31.3%
tetrachloro isomers:	116.2 µg/l		16.5%
pentachloro isomers:	6.4 µg/l		0.9%
total WSF:	702.7 µg/l		100% (1909)

AROCLOR 1242 217

C. WATER POLLUTION FACTORS:
  —Control methods: calculated half life time based on evaporative loss for a water depth of 1 m at 25°C: 12.1 hr (330)
  —Aquatic reactions: partition coefficient to natural sediments:
  physical-chemical characteristics of sediment

sediment	TOC (%)	pH	% sand	% salt	% clay	partition coeff.
Oconee River	0.4	6.5	93.0	6.0	2.0	540
USDA Pond	0.8	6.4	–	–	–	1210
Doe Run Pond	1.4	6.1	56.0	44.0	<1.0	1090
Hickory Hill Pond	2.4	6.3	55.0	45.0	<1.0	1250

(1068)

D. BIOLOGICAL EFFECTS:
  —Bioaccumulation: BCF's for fathead minnows (from water) after 8 months exposure: 32,000–274,000 (1838)
  —Toxicity:
  —Crustacean:
  amphipod (*Gammarus fasciatus*): 4 days (F)LC_{50}: 0.010 mg/l
  crayfish (*Orconectes nais*): 7 days (S)LC_{50}: 0.030 mg/l
  —Insects:
  naiad of damselfly (*Ischnura verticalis*): 4 days (F)LC_{50}: 0.40 mg/l
  naiad of dragonfly (*Macromia* sp.): 7 days (s)LC_{50}: 0.80 mg/l
  —Fish:
  rainbow trout:

days (f) LC_{50}:	mg/l
5	0.067
10	0.048
15	0.018
20	0.010
25	0.012

  bluegills:

days (F) LC_{50}:	mg/l
15	0.164
20	0.125
25	0.120
30	0.084

  channel catfish:

days (F)LC_{50}:	mg/l
15	0.219
20	0.150
25	0.132
30	0.087

(1617)

  harlequin fish (*Rasbora heteromorpha*):

	24 hr	48 hr	96 hr	3 m (extrapolated)	
		mg/l			
LC_{10} (F)	0.63	0.275	–		
LC_{50} (F)	0.96	0.6	0.37	0.05	(331)

  cutthroat trout: 96 hr (S) LC_{50}: 5.4 mg/l (1617)
  rainbow trout: acute oral toxicity: >1.5 g/kg

aroclor 1248
A polychlorobiphenyl, containing 48% chlorine.
Composition: dichlorobiphenyls: 2%
              trichlorobiphenyls: 18%
              tetrachlorobiphenyls: 40%
              pentachlorobiphenyls: 36%
              hexachlorobiphenyls: 4%   (1837)

C. WATER POLLUTION FACTORS:
  —Control methods: calculated half-life time based on evaporative loss for a water depth of 1 m at 25°C: 9.53 hr   (330)
  —Complete dechlorination of Aroclor 1248 was obtained with 69% Nickel on Kieselguhr in the presence of sodium hydroxide and 50 atm of hydrogen at 115°C for 6 hr

D. BIOLOGICAL EFFECTS:
  —Bioaccumulation:
    fathead minnow (*Pimephales promelas*): BCF: 1.2×10^5 (1919)
    BCF's for fathead minnows (from water) after 8 months exposure: 60,000–120,000   (1838)
    channel catfish: after 77 days exposure: bioaccumulation factor: 56,370
                                                                              (1435)
    channel catfish: bioaccumulation factor (conc. in water: 0.0058 mg/l)
                    after 7 days of exposure: 3500
                    after 14 days of exposure: 8000
                    after 28 days of exposure: 29000
                    after 56 days of exposure: 34000
                    after 77 days of exposure: 56370   (1617)

  —Toxicity:
  —Crustacean: amphipod (*Gammarus fasciatus*): 4 days (S) LC_{50}: 0.052 mg/l
                                                                              (1617)

  —Fish:
    rainbow trout:

days (F) LC_{50}	mg/l
5	0.054
10	0.038
15	0.016
20	0.0064
25	0.0034

    bluegills:

days (P) LC_{50}	mg/l
5	0.136
10	0.115
15	0.111
20	0.106
25	0.100
30	0.078

    fathead minnow (*Pimephales promelas*): 30 d LC_{50} = 4.7 g/l   (1919)
    channel catfish: 10 days (F) LC_{50}: 0.121 mg/l
                     15 days (F) LC_{50}: 0.121 mg/l

20 days (F) LC_{50}: 0.115 mg/l
25 days (F) LC_{50}: 0.104 mg/l
30 days (F) LC_{50}: 0.075 mg/l
cutthroat trout: 96 hr LC_{50}: 5.7 mg/l (1617)
rainbow trout: acute oral toxicity: >1.5 g/kg (1617)

aroclor 1254
A polychlorobiphenyl containing 54% chlorine
A. PROPERTIES: solub. 0.057 mg/l at 24°C (1666)
 —Composition: tetrachlorobiphenyls: 11%
 pentachlorobiphenyls: 49%
 hexachlorobiphenyls: 34%
 heptachlorobiphenyls: 6% (1837)
 —Composition of the water soluble fraction (WSF)

trichlorobiphenyls:	1.8 µg/l	or	2.6% of WSF
tetrachlorobiphenyls:	28.3 µg/l		40.2%
penta and hexabiphenyls:	40.2 µg/l		57.2%
total WSF	69.8 µg/l		100% (1909)

C. WATER POLLUTION FACTORS:
 —Control methods: calculated half-life based on evaporative loss for a water depth of 1 m at 25°C: 10.3 hr (330)
 Dechlorination was achieved in 2-propanol with 2.0 mmol $NiCl_2$ 60 mmol $NaBH_4$ and 0.3 mmol aroclor 1254. Biphenyl constituted 97% of the reaction products, and monochloro- and dichlorobiphenyl the remaining products (1551)
 —Biodegradation: biodegradation at 0.05 mg/l:

	normal sewage	adapted sew.
after 24 hr	0%	52%
after 135 hr	0%	43% (997)

 —Impact on biodegradation processes: effect on degradation of glucose by mixed culture derived from activated sludge

concentration (mg/l)	increase in lag period/hours	respiration rate (%)
1	0	100
10	0	110
100	0	135
1,000	>200	0 (997)

D. BIOLOGICAL EFFECTS:
 —Bioaccumulation: *bioaccumulation factors for aquatic invertebrates*:

	conc. in water	exposure period (days)				
species	µg/l	1	4	7	14	21
daphnid	1.1	2100	3800	—	—	—
amphipod	1.6	4400	5200	6000	6300	6200
crayfish	1.2	160	240	350	500	750
grass shrimp	1.3	1300	1900	2000	2200	2600
stonefly	2.8	640	710	740	750	740
dobsonfly	1.1	300	1000	1260	1460	1500
mosquito	1.5	2300	3400	3500	—	—
phantom midge	1.3	2400	2500	2600	2700	—

 channel catfish: bioaccumulation factor (conc. in water: 0.0024 mg/l):

AROCLOR 1254

after 7 days of exposure: 6,500
14 days 12,000
28 days 28,000
56 days 42,500
77 days 61,190 (1617)

—Fish:
brook trout: 118 d BCF 40,000–47,000 (F)
128 d LC_{50}, (F): 6.2 µg/l (1531)
channel catfish: after 77 days exposure: bioaccumulation factor: 61,190 (1435)
fathead minnows: after 244 days exposure: bioaccumulation factor: 109,000–238,000 (1436)
rainbow trout: after 30 days exposure: bioaccumulation factor: 34,000–46,000 (1437)
brook trout fry (*Salvelinus fontinalis*): BCF: 40,000–47,000 (1917)
BCF's for fathead minnows (from water) after 8 months exposure: 46,000–307,000 (1838)

—Toxicity:
—Algae: *Tetrahymena pyriformis*: 10 ppb, 13.30% decrease in population size, 96 hr static lab bioassay (2350)
—Insects:
naiad of damselfly (*Ischnura verticalis*): 4 days (F) LC_{50}: 0.20 mg/l
naiad of dragonfly (*Macromia* sp.): 7 days (S) LC_{50}: 1.0 mg/l (1617)

—Crustaceans:

Penaeus duorarum	Pink shrimp	0.94 ppb	51 percent mortality	15 day chronic exposure in flowing seawater
Penaeus duorarum	Pink shrimp	3.5 ppb	50 percent mortality	35 day chronic exposure in flowing seawater
Leiostomus xanthurus	Spot	5 ppb	50 percent mortality	18 day chronic exposure in flowing seawater
Lagodon rhomboides	Pinfish	5 ppb	50 percent mortality	12 day chronic exposure in flowing seawater

amphipod (*Gammarus fasciatus*): 4 days (S) LC_{50}: 2.4 mg/l
crayfish (*Orconectes nais*): 7 days (S) LC_{50}: 0.10 mg/l
grass shrimp (*Palaemonetes kadiakensis*): 7 days (F) LC_{50}: 0.003 mg/l
striped hermit crab (*Clibanarius vittatus*): exposure to 30 µg/l for 96 hr did not produce mortalities in static test (1573)

—Fish:
rainbow trout:

days (F) LC_{50}	mg/l
10	0.160
15	0.064
20	0.039
25	0.027

bluegills:

days (F) LC_{50}	mg/l
15	0.303
20	0.260
25	0.239
30	0.177

channel catfish:

days (F) LC_{50}	mg/l
10	0.303
15	0.286
20	0.293
25	0.181
30	0.139

harlequin fish (*Rasbora heteromorpha*):

mg/l	24 hr	48 hr	96 hr	3m extrapolated	
LC_{50} (F)	1.6	0.82	0.56		
LC_{50} (F)	6.2	1.45	1.1	0.1	(331)

cutthroat trout: 96 hr LC_{50}: 42 mg/l (1617)
rainbow trout: acute oral toxicity: >1.5 g/kg
22-day-old fry of deepwater ciscoes (*Coregonus*):
   at 10 mg/l for 96 hr: 20% mortality
   5 days LC_{50}: 3.2 mg/l (1861)

—Mammals:
rats: Dietary exposure to 5–800 ppm can inhibit the growth of at least one experimental tumor, the Walker 256 carcinosarcoma. (507)
Continuous feeding of aroclor 1254 in a diet produced varying degrees of dermatitis after 10 weeks. These skin lesions were found in 15 of 60 animals fed the PCB at 100 ppm, 4 of 60 at 30 ppm and 1 of 60 at 10 ppm for 10 to 20 weeks (1613)

—Carcinogenicity: It is concluded that under the conditions of this bioassay, Aroclor® 1254 was not carcinogenic in Fischer 344 rats; however, a high incidence of hepatocellular proliferative lesions in both male and female rats was related to administration of the chemical. In addition, the carcinomas of the gastrointestinal tract may be associated with administration of Aroclor 1254 in both males and females. (1702)

aroclor 1260

A polychlorobiphenyl containing 60% chlorine.

A. PROPERTIES: solub. 0.080 mg/l at 24°C (1666)

—Composition:
pentachlorobiphenyls:	12%
hexachlorobiphenyls:	38%
heptachlorobiphenyls:	41%
octachlorobiphenyls:	8%
nonachlorobiphenyls:	1%

(1837)

C. WATER POLLUTION FACTORS:
—Control methods: calculated half-life based on evaporative loss for a water depth of 1 m at 25°C: 10.2 hr (330)

D. BIOLOGICAL EFFECTS:
—Fish:

fathead minnow:	30 d LC_{50}:	3.3 µg/l	(F)	
	250 d BCF:	270,000	(F)	(1530)
rainbow trout:	30 d LC_{50}:	51 µg/l	(FT)	
bluegill:	30 d LC_{50}:	400 µg/l	(FT)	
channel catfish:	30 d LC_{50}:	433 µg/l	(FT)	
cutthroat trout:	96 hr LC_{50}:	61 mg/l	(S)	(1617)
rainbow trout:	acute oral toxicity:	>1.5 g/kg		

aroclor 1262
A polychlorobiphenyl containing 62% chlorine.
A. PROPERTIES: solub. 0.052 mg/l at 24°C (1666)
D. BIOLOGICAL EFFECTS: Harlequin fish: 96 hr LC_{10} (F): not toxic below 100 mg/l
(331)

arquad (ammonium salts, quarternary)
Trademark for a series of quarternary ammonium salts containing one or two alkyl groups ranging from C_8 to C_{18}.
Uses: corrosion inhibitors; emulsifiers; germicides and sanitizing agents; textile fabric softeners
C. WATER POLLUTION FACTORS: Biodegradation: at ± 18 mg/l, no degradation after 28 days exposure to nonadapted sewage (488)

asparacemic acid *see dl*-aspartic acid

l-**asparagine** (2-aminobutanedioic amide; *l*-α-aminosuccinamic acid; *l*-β-asparagine)
$NH_2COCH_2CH(NH_2)COOH$
Uses: biochemical research; preparation of culture media; medicine.
Natural sources: widely distributed in plants and animals, both free and combined with proteins.
A. PROPERTIES: m.w. 132.13; m.p. 236°C decomposes; b.p. 235°C decomposes, sp.gr. 1.543 at 15/4°C; solub. 24,600 mg/l at 25°C, 866,000 mg/l at 100°C
C. WATER POLLUTION FACTORS
—Waste water treatment:
A.S.: after 6 hr: 10.3% of ThOD
12 hr: 19.5% of ThOD
24 hr: 24.7% of ThOD (89)

dl-**aspartic acid** (2-aminobutanedioic acid; *dl*-aminosuccinic acid; asparacemic acid)
$COOHCH_2CH(NH_2)COOH$
Uses: biochemical and clinical studies
Natural source: a naturally occurring nonessential amino acid in young sugarcane and sugarbeet molasses
A. PROPERTIES: m.w. 133.10; m.p. 278/280°C decomposes; sp.gr. 1.663 at 12/12°C; solub. 8,200 mg/l at 25°C, 47,900 mg/l at 75°C
C. WATER POLLUTION FACTORS
—Manmade sources: excreted by man: in urine: 0.37 to 3.7 mg/kg body wt/day
(203)

—Waste water treatment:
   A.S.: (dl-asp.acid): after 6 hr: 8.9% of ThOD
                              12 hr: 16.2% of ThOD
                              24 hr: 28.8% of ThOD   (89)

asulam (asulox: active ingredient: 40% w/v methyl[(4-aminophenyl)sulphonyl]-carbamate (as Na salt); methylsulfanilylcarbamate; methyl-4-aminobenzenesulphonyl-carbamate)

$$H_2N-\underset{}{\bigcirc}-\underset{O}{\overset{O}{\underset{\|}{\overset{\|}{S}}}}-NH-\overset{O}{\underset{\|}{C}}-OCH_3$$

Use: herbicide
A. PROPERTIES: colorless crystals; m.p. 143–144°C; solub. 0.5%
C. WATER POLLUTION FACTORS:
   —Impact on biodegradation processes:
      cellulose decomposition, measured as weight loss of buried cotton cloth, was reduced by 8–38% in treated soil at 16 ppm, after incubation at 19°C for 8 weeks and by 0–60% in treated soil at 160 ppm   (1825)
      experiments using pure cultures of soil-inhabiting fungi and actinomycetes, some of which were cellulolytic, showed that asulam at 10 ppm had either no, or only a temporary, effect on growth   (1825)
D. BIOLOGICAL EFFECTS:
   —Fish:
      rainbow trout: 96 hr LC_{50} (S), >5,000 mg/l (active ingred.)
      channel catfish (*Ictalurus punctatus*):   96 hr LC_{50} (S), >5,000 mg/l
      goldfish (*Carassius auratus*):   96 hr LC_{50} (S), >5,000 mg/l
      bluegill (*Lepomis macrochirus*):   96 hr LC_{50} (S), >3,000 mg/l   (1108)
   —Mammals: acute oral LD_{50}:
      rat: >5,000 mg/kg (potassium salt)   (1854)
      mice: =5,000 mg/kg (potassium salt)
      rabbits: >2,000 mg/kg (potassium salt)
      chicken: >1,000 mg/kg (potassium salt)   (1855)

asulox *see* asulam

atrazine (2-chloro-4-ethylamino-6-isopropylamino-*s*-triazine)

$$C_2H_5NH-\underset{N}{\overset{\overset{Cl}{\underset{|}{\underset{N\nwarrow\;\nearrow N}{}}}}{\bigcirc}}-NH-\underset{CH_3}{\overset{CH_3}{\underset{|}{\overset{|}{CH}}}}$$

Uses: most widely used chemical for pre-emergence weed control in corn. In Hawaii it is important to the culture of sugarcane, pineapple, and macadamia nut.

224 ATRAZINE

A. PROPERTIES: colorless crystals, m.p. 173–175°C, v.p. 3×10^{-7} mm at 20°C, solub. 70 ppm at 25°C

C. WATER POLLUTION FACTORS:
  —Biodegradation and aqueous reactions:
   in submerged soils: in 90 days 0.005% of atrazine-^{14}C was recovered as $^{14}CO_2$ (from ring labeled atrizine);
      48% to 85% of atrazine was hydrolyzed in 30 days, depending upon soil type
   chemical hydrolysis of atrazine to hydroxyatrazine is the principal pathway of detoxication in soil. Biological dealkylation without dehalogination occurs simultaneously leading to 2-chloro-4-amino-6-isopropylamino-*s*-triazine:

$$\text{H}_5\text{C}_2-\text{NH}-\underset{\text{N}}{\overset{\text{OH}}{\text{triazine}}}-\text{NH}-\text{CH}(\text{CH}_3)_2 \xrightarrow{\text{microbial degradation}}$$

chemical hydrolysis ↑

$$\text{H}_5\text{C}_2-\text{NH}-\underset{\text{N}}{\overset{\text{Cl}}{\text{triazine}}}-\text{NH}-\text{CH}(\text{CH}_3)_2 \longrightarrow \text{H}_2\text{N}-\underset{\text{N}}{\overset{\text{Cl}}{\text{triazine}}}-\text{NH}-\text{CH}(\text{CH}_3)_2$$

      microbial dealkylation   (1307)
   75–100% disappearance from soils: 10 months   (1815)

D. BIOLOGICAL EFFECTS
  —Bacteria: *Pseudomonas putida*: inhibition of cell multiplication starts at >10 mg/l   (329)
  —Algae: *Microcystis aeruginosa*: inhibition of cell multiplication starts at 0.003 mg/l   (329)

Chlorococcum sp.	Technical acid	100 ppb	50 percent decrease in O$_2$ evolution	*f*
Chlorococcum sp.	Technical acid	100 ppb	50 percent decrease in growth	Measured as ABS (525 mu) after 10 days
Dunaliella tertiolecta	Technical acid	300 ppb	50 percent decrease in O$_2$ evolution	*f*
Dunaliella tertiolecta	Technical acid	300 ppb	50 percent decrease in growth	Measured as ABS (525 mu) after 10 days
Isochrysis galbana	Technical acid	100 ppb	50 percent decrease in O$_2$ evolution	*f*
Isochrysis galbana	Technical acid	100 ppb	50 percent decrease in growth	Measured as ABS (525 mu) after 10 days

Phaeodactylum tricornutum	Technical acid	100 ppb	50 percent decrease in O_2 evolution		f
Phaeodactylum tricornutum	Technical acid	200 ppb	50 percent decrease in growth	Measured as ABS (525 mu) after 10 days	

—Bioaccumulation:
 BCF: snails: 2-15
　　　 algae: 10-83
　　　 fish: 3-10 (1891; 1892; 1893)
—Fish:
 bluegill: flow through bioassay: 2 y TLm: 5.4-8.4 mg/l at 27°C
　　　　　　　　　　　　　　　 MATC: 0.09 mg/l at 27°C
 fathead minnow: flow through bioassay: 1 y TLm: 11-20 mg/l at 25°C
　　　　　　　　　　　　　　　　　　 MATC: 0.21 mg/l at 25°C
 brook trout: flow through bioassay: 1.5 y TLm: 4.0-6.0 mg/l at 9-16°C
　　　　　　　　　　　　　　　　　 MATC: 0.06 mg/l at 9-16°C
(448)
—Mammals:
 acute oral LD_{50} (rats): 3,080 mg/kg
　　　　　　　　　 (mice): 1,750 mg/kg (1855)
 acute dermal LD_{50} (rabbits): 7,500 mg/kg (1855)
 in diet: when fed for 2 years to rats at dietary levels of 100 and 1,000 ppm, no effect was observed. (1855)

avadex *see* diallate

azinphosethyl (O,O-diethyl-S-(4-oxo-3-H-1,2,3-benzotriazine-3-yl)-methyldithiophosphate; S-(3,4-dihydro-4-oxobenzo(*d*)-(1,2,3)-triazin-3-ylmethyl)diethylphosphorothiolothionate; triazotion; ethylguthion; gusathion A)

$$(C_2H_5O)_2PSS\ CH_2-\text{benzotriazinone}$$

Use: nonsystemic insecticide and acaricide
A. PROPERTIES: colorless crystals, m.p. 53°C, b.p. 111°C at 0.001 mm, v.p. 2.2 × 10^{-7} mm at 20°C, sp.gr. 1.284 at 20/4°C
D. BIOLOGICAL EFFECTS:
 —Crustaceans:
 Simocephalus serrulatus: 48 hr LC_{50}, 4 µg/l (2127)
 Daphnia pulex: 48 hr LC_{50}, 3.2 µg/l (2127)
 —Fish:
 Salmo gairdneri: 96 hr LC_{50}, 19 µg/l (2137)
 —Mammals:
 acute oral LD_{50} (male rats): 17.5 mg/kg
　　　　　　　　　 (female rats): 12.5 mg/kg

AZINPHOSMETHYL

acute dermal LD_{50} (rats): 250 mg/kg (2 hr exposure)
intraperitoneal LD_{50} (rats): <7.5 mg/kg (1855)

azinphosmethyl (guthion; gusation M; O,O-dimethyl-S-[(4-oxo-1,2,3-benzotriazin-3(4H)-yl)methyl] phosphorodithioate; S-(3,4-dihydro-4-oxobenzo(d)-(1,2,3)-triazin-3-ylmethyl)dimethylphosphorodithioate)

Uses: nonsystemic insecticide and acaricide of long persistence; cholinesterase inhibitor

A. PROPERTIES: brown waxy solid, m.p. ~73°C; solub. 29 ppm at 25°C; sp.gr. 1.44 at 20.4°C; m.w. 317.3

C. WATER POLLUTION FACTORS: Odor threshold: detection: 0.0002 mg/kg water
(915)

D. BIOLOGICAL EFFECTS:
—Molluscs

Crassostrea virginica	American oyster	Eggs	620 ppb	TLM 48 hr, s	48 hr static lab bioassay
Mercenaria mercenaria	Hard clam	Eggs	860 ppb	TLM 48 hr, s	48 hr static lab bioassay
Mercenaria mercenaria	Hard clam	Larvae	860 ppb	TLM 12 d, s	12 day static lab bioassay

(2324)

—Crustacean
Gammarus lacustris: 0.15 µg/l, 96 hr LC_{50} (2124)
Gammarus fasciatus: 0.10 µg/l, 96 hr LC_{50} (2126)
Gammarus pseudolimneaus: 0.10 µg/l, 30 day, no effect (2124)
Palaemonetes kadiakensis: 1.2 µg/l, 120 hr LC_{50}; 0.16 µg/l (20 day LC_{50})
(2126)
Asellus brevicaudus: 21.0 µg/l, 96 hr LC_{50} (2126)
—Insects:
Pteronarcys dorsata: 21.1 µg/l, 96 hr LC_{50}; 4.9 µg/l (30 day LC_{50}) (2134)
Pteronarcys californica: 1.5 µg/l, 96 hr LC_{50} (2128)
Acroneuria lycorias: 1.5 µg/l (30 day LC_{50}), 1.36 µg/l, 30 day, no effect (2134)
Ophiogomphus rupinsulensis: 12.0 µg/l, 96 hr LC_{50}; 2.2 µg/l (30 day LC_{50}), 1.73 µg/l, 30 day, no effect (2134)
Hydropsyche bettoni: 7.4 µg/l (30 day LC_{50}), 4.94 µg/l, 30 day, no effect
(2134)
Ephemerella subvaria: 4.5 µg/l (30 day LC_{50}), 2.50 µg/l, 30 day, no effect
(2134)
—Fish:
Pimephales promelas: 93 µg/l, 95 hr LC_{50} (2119)
Lepomis macrochirus: 5.2 µg/l, 96 hr LC_{50} (2119)
Lepomis microlophus: 52 µg/l, 96 hr LC_{50} (2121)

Micropterus salmoides: 5 µg/l, 96 hr LC$_{50}$ (2121)
Salmo gairdneri: 14 µg/l, 96 hr LC$_{50}$ (2121)
Salmo trutta: 4 µg/l, 96 hr LC$_{50}$ (2121)
Oncorhynchus kisutch: 17 µg/l, 96 hr LC$_{50}$ (2121)
Perca flavescens: 13 µg/l, 96 hr LC$_{50}$ (2121)
Ictalurus punctatus: 3290 µg/l, 96 hr LC$_{50}$ (2121)
Ictalurus melas: 3500 µg/l, 96 hr LC$_{50}$ (2121)
goldfish: 4.3 mg/l, 96 hr LC$_{50}$
minnow: 0.24 mg/l, 96 hr LC$_{50}$
carp: 0.70 mg/l, 96 hr LC$_{50}$
sunfish: 0.05 mg/l, 96 hr LC$_{50}$
bluegill: 0.02 mg/l, 96 hr LC$_{50}$
threespine stickleback (*Gasterosteus aculeatus*): 4.8 ppb, TLm, 96 hr (2333)
rainbow trout fingerlings: 96 hr LC$_{50}$ (S): 7.10 mg/l (1101)
fathead minnow: flow through bioassay: 96 hr TLm: 1.9 mg/l at 25°C
goldfish: flow through bioassay: 96 hr TLm: 2.37 mg/l at 25°C
fathead minnow: flow through bioassay: MATC: 0.51 µg/l at 23–25°C (455)
—Mammals:
acute oral LD$_{50}$ (male guinea pigs): 80 mg/kg (1855)
acute oral LD$_{50}$ (rats): 13–16.4 mg/kg
acute dermal LD$_{50}$ (rats): 220 mg/kg (1854)
in diet: no mortality occurred in rats fed 1 mg/kg/day for 60 days (1855)

azobenzene (diphenyldiimide; benzeneazobenzene)

$$\langle\bigcirc\rangle-N=N-\langle\bigcirc\rangle$$

Use: manufacture of dyes and rubber accerlators; fumigant, acaricide
A. PROPERTIES: yellow or orange crystals; m.w. 182.23; m.p. 68.3°C; b.p. 293°C; sp.gr. 1.09; log P$_{oct}$ 3.82
C. WATER POLLUTION FACTORS: Impact on biodegradation processes: at 100 mg/l no inhibition of NH$_3$ oxidation by *Nitrosomonas* sp. (390)
D. BIOLOGICAL EFFECTS:
—Mammals: in diet: dogs fed 63 days on diet containing 600 ppm, suffered high mortality and liver damage (1855)
—Carcinogenicity: ?
—Mutagenicity in the *Salmonella* test: +
1.4 revertant colonies/nmol
379 revertant colonies at 50 µg/plant (1883)

azodrin (dimethylphosphate of 3-hydroxy-N-methyl-*cis*-crotonamide-O,O-dimethyl-O-(2-methylcarbamoyl-1-methylvinyl)phosphate; monocrotophos)

$$(CH_3O)_2\overset{O}{\underset{\|}{P}}O\underset{H_3C}{\diagdown}C=C\underset{\underset{O}{\overset{\|}{C}}NHCH_3}{\diagup}{}^H$$

AZODRIN

Uses: systemic insecticide; acaricide

C. WATER POLLUTION FACTORS:
—aquatic reactions: persistence in riverwater in a sealed glass jar under sunlight and artificial fluorescent light—initial conc. = 10 μg/l:

% of original compound found after

1 hr	1 wk	2 wk	4 wk	8 wk	
100	100	100	100	100	(1309)

D. BIOLOGICAL EFFECTS:
—Crustacean: copepod (*Acartia tonsa*): 96 hr LC_{50} (S), 240 μg/l   (1129)
—Fish: harlequin fish (*Rasbora heteromorpha*):

	mg/l				
	24 hr	48 hr	96 hr	3m (extrapolated)	
LC_{10} (F)	580	580	280		
LC_{50} (F)	750	730	450	150	(331)

—Mammals:
acute oral LD_{50} (rat): 8–23 mg/kg
acute dermal LD_{50} (rabbit): 354 mg/kg   (1854)

B

B(a)A *see* benzo(*a*)anthracene

balan (N-butyl-N-ethyl-2,6-dinitro-4-trifluoromethylaniline; N-butyl-N-ethyl-α, α, α-trifluoro-2,6-dinitro-*p*-toluidine; benefin; benfluralin; bethrodine; quilan)

$$CH_3-CH_2-N-(CH_2)_3-CH_3$$

(structure: 2,6-dinitro-4-trifluoromethyl phenyl ring with N-butyl-N-ethyl amino substituent, O_2N and NO_2 groups, CF_3 group)

A. PROPERTIES: yellow-orange crystals; m.p. 65 - 66,5°C; v.p. 4×10^{-7} mm at 25°C; solub. 70 ppm at 25°C

B. BIOLOGICAL EFFECTS:
  —Crustacean: *Gammarus faciatus*: 1100 μg/l, 96 hr LC_{50} (2125)
  —Fish: harlequin fish (*Rasbora heteromorpha*):

mg/l	24 hr	48 hr	96 hr	3 m (extrapolated)	
LC_{10} (F)	1.0	0.95			
LC_{50} (F)	1.4	1.3	1.2	1.0	(331)

  —Mammals:
  acute oral LD_0 (male and female rats): >10,000 mg/kg
  (technical balan)(1854)
  acute oral LD_{50} (mice): >5,000 mg/kg (1855)
  acute oral LD_{50} (rabbits, dogs, chicken): >2,000 mg/kg (1855)
  in diet: in 3 month feeding tests, the "safe" level for rats was 1,250 ppm, for dogs 500 ppm (1855)

B(a)P *see* benzo(*a*)pyrene

baygon (2-(1-methylethoxy) phenol methylcarbamate; propoxur; arprocarb; Bay 39007; Blattanex; Suncide)

(structure: phenyl ring with $O-\overset{O}{\underset{\|}{C}}-NH-CH_3$ carbamate group and $O-CH(CH_3)_2$ isopropoxy group)

230 BAYTEX

Use: insecticide
A. PROPERTIES: white to tan crystalline solid; m.p. 91°C; solub. ~ 2000 ppm
C. WATER POLLUTION FACTORS:
—Aquatic reactions: persistence in river water in a sealed glass jar under sunlight and artificial fluorescent light—initial conc. 10 µg/l:

% of original compound found

after	1 hr	1 wk	2 wk	4 wk	8 wk
	100	50	30	10	5

(1309)

D. BIOLOGICAL EFFECTS:
—Algae:

Dunaliella euchlora	1000	ppb	25 percent reduction in O_2 evolution	O_2 evolution measured by Winkler Bottle technique 1 l. of culture incubated 20 hrs in pesticide solution, then placed in test bottles 4 hrs.
Dunaliella euchlora	100	ppb	32 percent reduction in O_2 evolution	
Dunaliella euchlora	10	ppb	27 percent reduction in O_2 evolution	
Phaeodactylum tricornutum	1000	ppb	23 percent reduction in O_2 evolution	
Phaeodactylum tricornutum	100	ppb	28 percent reduction in O_2 evolution	
Phaeodactylum tricornutum	10	ppb	40 percent reduction in O_2 evolution	
Skeletonema costatum	1000	ppb	30 percent reduction in O_2 evolution	
Skeletonema costatum	100	ppb	23 percent reduction in O_2 evolution	
Skeletonema costatum	10	ppb	29 percent reduction in O_2 evolution	
Cyclotella nana	1000	ppb	53 percent reduction in O_2 evolution	(2325)

—Crustaceans:

Gammarus lacustris	34 µg/l	96 hr, LC_{50}	(2124)
Gammarus fasciatus	50 µg/l	96 hr, LC_{50}	(2126)

—Insect:

Pteronarcys californica	13 µg/l	96 hr, LC_{50}	(2128)

fourth instar larval *Chironomus riparius*: 24 hr LC_{50}: 64.4 ppb   (1853)

—Mammals:
acute oral LD_{50} (rat): 95–104 mg/kg
dermal LD_{50} (rat): >1000 mg/kg   (1854)

baytex (baycid; entex; hebaycid; mercaptophos; tiguvon; O, O-dimethyl-O-(3-methyl-4-(methylthio)phenyl)phosphorothioate; fenthion)

$$CH_3O-\underset{CH_3O}{\overset{O}{\underset{\|}{P}}}-O-\text{C}_6\text{H}_3(CH_3)-S-CH_3$$

BAYTEX

Use: systemic and contact herbicide
A. PROPERTIES: b.p. 87°C at 0.01 mm; v.p. 3×10^{-5} mm at 20°C; sp.gr. 1.25 at 20/4°C; solub. 55 ppm at room temp.
C. WATER POLLUTION FACTORS:
  —Aquatic reactions: persistence in river water in a sealed glass jar under sunlight and artificial fluorescent light—initial conc. 10 µg/l:

after	1 hr	1 wk	2 wk	4 wk	8 wk	
% of original compound found	100	50	10	0	0	(1309)

D. BIOLOGICAL EFFECTS:
  —Algae:

Dunaliella euchlora	1000	ppb	27 percent reduction in O_2 evolution	O_2 evolution measured by Winkler Light-and-Dark Bottle Technique
Dunaliella euchlora	100	ppb	27 percent reduction in O_2 evolution	
Dunaliella euchlora	10	ppb	16 percent reduction in O_2 evolution	
Phaeodactylum tricornutum	1000	ppb	29 percent reduction in O_2 evolution	
Phaeodactylum tricornutum	100	ppb	29 percent reduction in O_2 evolution	1 l. of culture incubated 20 hrs in pesticide soln. then placed in test bottles.
Phaeodactylum tricornutum	10	ppb	35 percent reduction in O_2 evolution	
Skeletonema costatum	1000	ppb	19 percent reduction in O_2 reduction	
Skeletonema costatum	100	ppb	51 percent reduction in O_2 evolution	
Skeletonema costatum	10	ppb	26 percent reduction in O_2 evolution	
Cyclotella nana	1000	ppb	50 percent reduction in O_2 evolution	
Cyclotella nana	100	ppb	48 percent reduction in O_2 evolution	

(2348)

  —Insects:
  Pteronarcys californica   96 hr LC_{50}:   4.5 µg/l   (2128)
  ricefield spider: *Oedothorax insecticeps*: LD_{50}: 500 ppm   (1814)
  insect larvae (*Chaoborus*): LC_{50}, 48 hr: 0.008 ppm
    (*Cloeon*): LC_{50}, 48 hr: 0.012 ppm   (1323)
  —Molluscs: gastropoda (*Lymnea stagnalis*): 48 hr, LC_{50}: 6.4 ppm   (1323)
  —Crustaceans:

Gammarus lacustris	8.4	µg/l	96 hr, LC_{50}	(2124)
Gammarus fasciatus	110	µg/l	96 hr, LC_{50}	(2126)
Palaemonetes kadiakensis	5	µg/l	120 hr, LC_{50} 1.5 µg/l (20 day LC_{50})	(2126)
Orconectes nais	50	µg/l	96 hr, LC_{50}	(2126)
Asellus brevicaudus	1800	µg/l	96 hr, LC_{50}	(2126)

Simocephalus serrulatus	0.62 µg/l 48 hr, LC_{50}		(2127)
Daphnia pulex	0.80 µg/l 48 hr, LC_{50}		(2127)
Korean shrimp	(*Palaemon macrodactylus*): 5.3 (3.13–8.92) ppb TL_{50}, 96 hr, S		
Korean shrimp	(*Palaemon macrodactylus*): 3.0 (1.5–6.0) ppb TL_{50}, 96 hr, F		(2348)

Gammarus pulex: 48 hr LC_{50}: 0.014 ppm   (1323)

—Fish:

Pimephales promelas	2440 µg/l	96 hr LC_{50}	(2121)
Lepomis macrochirus	1380 µg/l	96 hr LC_{50}	(2121)
Lepomis microlophus	1880 µg/l	96 hr LC_{50}	(2121)
Micropterus salmoides	1540 µg/l	96 hr LC_{50}	(2121)
Salmo gairdneri	930 µg/l	96 hr LC_{50}	(2121)
Salmo trutta	1330 µg/l	96 hr LC_{50}	(2121)
Oncorhynchus kisutch	1320 µg/l	96 hr LC_{50}	(2121)
Perca flavesens	1650 µg/l	96 hr LC_{50}	(2121)
Ictalurus punctatus	1680 µg/l	96 hr LC_{50}	(2121)
Ictalurus melas	1620 µg/l	96 hr LC_{50}	(2121)

96 hr, LC_{50}, mg/l		96 hr, LC_{50}, mg/l		
catfish	1.7	bluegill	1.4	
bullhead	1.6	bass	1.5	
goldfish	3.4	rainbow	0.9	
minnow	2.4	brown	1.3	
carp	1.2	coho	1.3	
sunfish	1.9	perch	1.7	(1934)

—Mammals:
acute oral LD_{50} (rat): 215–300 mg/kg
acute dermal LD_{50} (rat): 320–330 mg/kg   (1855)

BBCP *see* 1,2-dibromo-3-chloropropane

B(*b*)F *see* benzo(*b*)fluoranthene

BBP *see* butylbenzylphthalate

BCME *see* dichloromethylether

behenic acid (docosanoic acid; *n*-docosoic acid)
$CH_3(CH_2)_{20}COOH$

Natural sources: a minor component of the oils of the type of peanut and rapeseed.
Use: cosmetics; waxes, plasticizers; stabilizers

A. PROPERTIES: m.w. 340.58; m.p. 80.2°C; b.p. 306°C at 60 mm; sp.gr. 0.8221 at 100/4°C; solub. 1,000 mg/l at 10°C

B. AIR POLLUTION FACTORS
—Manmade   sources:
glc'c: Detroit freeway interchange: Oct.–Nov. 1963: 61.6 µ g/1000 cu m air
New York high traffic location: 119.2 µ g/1000 cu m air   (100)

—Organic fraction of suspended matter:
Bolivia at 5200 m altitude (Sept.–Dec. 1975): 0.29–0.69 µg/1000 cu m
Belgium, residential area (Jan.–April 1976): 3.2–8.4 µg/1000 cu m (428)
glc's: Botrange (Belgium): woodland at 20–30 km from industrial area, June–July 1977: 3.2; 3.8 ng/cu m ($n = 2$)
Wilrijk (Belgium): residential area: Oct.–Dec. 1976: 10.3; 15.8 ng/cu m ($n = 2$) (1233)
glc's in residential area (Belgium) Oct. 1976: in particulate sample: 13.7 ng/cu m (1289)

benefin *see* balan

benomyl (methyl-1-(butylcarbamoyl)-2-benzimidazolecarbamate; benlate; tersan 1991)

$$O=C-NH-CH_2-CH_2-CH_2-CH_3$$
$$NH-COOCH_3$$

Use: in agriculture as a systemic fungicide for controlling a broad spectrum of phytopathogenic fungi including rice pathogens.

B. WATER POLLUTION FACTORS:
—Impact on biodegradation processes:
Effect of benomyl on nitrification by *Nitrosomonas* sp.

conc. ppm	µg nitrite recovered/ml medium incubation		(days)
	6	15	
0	3.8	108.0	
10	2.0	2.1	
100	1.7	2.0	
	by *Nitrobacter* agilis		
0	289.5	0	
10	444.2	513.2	
100	444.2	440.5	(1372)

—Aquatic reactions: Benomyl is transformed readily to methyl-2-benzimidazole carbamate (MBC) and 2-aminobenzimidazole (AB) in soils and in water (1373; 1374)

D. BIOLOGICAL EFFECTS:
—Mammals: acute oral LD_{50} (rat): >10,000 mg/kg (1854)

bensulide (N-(2-ethylthio)benzene sulphonamide-S,O,O-diisopropylphosphorodithioate; S-(O,O-diisopropylphosphorodithioate) of N-(2-mercaptoethyl)benzenesulfonamide; R 4461; Betasan; Prefar; Exporsan)

$$\begin{array}{c} O \\ \| \\ -S-NH-CH_2-CH_2-S-P \\ \| \\ O \end{array} \begin{array}{c} S \\ \| \\ O-CH-(CH_3)_2 \\ O-CH-(CH_3)_2 \end{array}$$

234 BENZ(a)ACRIDINE

Use: selective preemergence herbicide
A. PROPERTIES: amber-colored liquid; solub. 25 ppm; sp.gr. 1.23 at 20/20°C
D. BIOLOGICAL EFFECTS:
 —Fish: channel catfish (*Ictalurus punctatus*): 96 hr LC_{50}, S: 379 µg/l (1202)
 —Mammals:
acute oral LD_{50} (male rat):	1082 mg/kg	(1854)
acute oral LD_{50} (male albino rats):	339 mg/kg	(1855)
acute dermal LD_{50} (albino rats):	3950 mg/kg	

in diet: well tolerated by rats and dogs fed 90 days at dietary levels up to 250 ppm and 625 ppm, respectively (1855)

benz(*a*)acridine

A. PROPERTIES: m.w. 229; log P_{oct} 4.45 (calculated)
C. AIR POLLUTION FACTORS:
 —Ambient air concentrations: in the average American uraban atmosphere—1963: 2 µg/g airborne particulates or 0.2 ng/cu m air (1293)
 Cleveland (Ohio, USA): max. conc.: 31 ng/cu m (144 values) (n.s.i.) (1971/1972)
 annual geom. mean (all sites): 1.37 ng/cu m (n.s.i.)
 annual geom. mean of TSP (all sites): 12 ppm (wt) (n.s.i.) (556)
 glc's in residential area (Belgium)—Oct. 1976: in particulate sample: 0.85 ng/cu m (1239)

D. BIOLOGICAL EFFECTS:
 —Crustacean: *Daphnia pulex*: bioaccumulation factor: 352 (initial conc. in water: 18 ppb) 24 hr LC_{50}, 0.449 mg/l
 immobilization concentration IC_{50}: 0.362 mg/l (1050)

benz(*c*)acridine

C. AIR POLLUTION FACTORS:
 —Ambient air quality: in the average American urban atmosphere—1963: 4 µg/g airborne particulates or 0.6 ng/cu air (1293)

benzaldehyde (benzenecarbonal; oil of bitter almonds)
C_6H_5CHO

BENZALDEHYDE 235

A. PROPERTIES: m.w. 106.1; m.p. $-26°C$; b.p. $179°C$; v.p. 1 mm at $26°C$, 40 mm at $90°C$; v.d. 3.66; sp.gr. 1.05 at $15/4°C$; solub. 3,300 mg/l; log P_{oct} 1.48

B. AIR POLLUTION FACTORS: 1 mg/cu m = 0.227 ppm, 1 ppm = 4.410 mg/cu m
  —Odor: characteristic: quality: bitter almonds

odor thresholds   mg/cu m

[chart: detection, recognition, not specified across 10^{-7} to 10^{4}]

(279; 307; 610; 708; 710; 724; 739; 771; 788; 795)

  —Manmade sources:
    in gasoline exhaust: <0.1–13.5 ppm   (195; 1053)
                             3.2–8.5 vol % of total exhaust aldehydes
                                                                        (394; 395; 396; 397)
    in diesel exhaust: 0.3 ppm   (311)
  —Control methods:
    wet scrubber: water at pH 8.5: outlet: 80 odor units/scf
                      $KMnO_4$ at pH 8.5: outlet: 1 odor unit/scf   (115)
  —Sampling and analysis
    second derivative spectroscopy: det.lim.: 100 ppb   (42)
    photometry: min.full scale: 3.9 ppm   (53)

C. WATER POLLUTION FACTORS
  —BOD_5:   1.62   (36)
               36% of ThOD   (220)
  —BOD_5^{20}: 1.62 at 10 mg/l, unadapted sew.; no lag period   (554)
  —BOD_{20}^{20}: 1.78 at 10 mg/l, unadapted sew.   (554)
  —BOD_{10}: 1.50 std.dil.sew.   (256)
  —COD: 1.98   (36)
               94% of ThOD   (220)
  —$KMnO_4$: acid 20% of ThOD; alkaline: 2% of ThOD   (220)
  —ThOD: 2.42   (36)
  —Biodegradation: adapted A.S. at $20°C$—product is sole carbon source: 99.0%
    COD removal at 119.0 mg COD/g dry inoculum/hr   (327)
  —Impact on biodegradation processes:
    at 400 mg/l standard BOD_{10} test perturbated   (30)
  —Reduction of amenities: faint odor: 0.003 mg/l   (129)
    tentative T.O.C.: 0.002 mg/l; 0.44 ppb   (27)
    T.O.C. in water: 0.18 ppb; 3.0   ppb
                          4.29 ppb; 0.436 ppb
                          4.0 ppb; 3.0   ppb   (326)
    detection: 0.035 mg/kg   (882)
  —Waste water treatment:
  —A.C.: adsorbability: 0.188 g/g C, 94% reduction, infl.: 1,000 mg/l, effl.: 60 mg/l
                                                                                              (32)

—A.S. acclimated to the following aromatics:
(infl.: 250 mg/l benzaldehyde, 30 min aeration)
phenol: 38% theor. oxidation
benzylalcohol: 30% theor. oxidation
anthranilic acid: 35% theor. oxidation   (92)

methods	temp °C	days observed	feed mg/l	days acclim.	% removed	
NFG, BOD	20	1–10	200–400	365+ P	50	
NFG, BOD	20	1–10	600	365+ P	19	
NFG, BOD	20	1–10	800	365+ P	7	(93)

A.C.: influent ppm	carbon dosage	effluent ppm	% reduction	
1,000	10X	9	99	
500	10X	6	99	
100	10X	2	98	(192)

—Sampling and analysis: photometry: min.full scale: 0.16×10^{-6} mole/l   (53)

D. BIOLOGICAL EFFECTS
—Toxicity threshold (cell multiplication inhibition test):

bacteria (*Pseudomonas putida*):	132	mg/l	(1900)
algae (*Microcystis aeruginosa*):	20	mg/l	(329)
green algae (*Scenedesmus quadricauda*):	34	mg/l	(1900)
protozoa (*Entosiphon sulcatum*):	0.29	mg/l	(1900)
protozoa (*Uronema parduczi Chatton-Lwoff*):	22	mg/l	(901)

—Fish: minnows: stop eating: 17.1 mg/l of 85% solution   (226)
—Mammalia: rabit: subcutane LD_{50}: 5.0 g/kg   (211)

benzaminoacetic acid *see* hippuric acid

benzanthrone

Use: dyes
A. PROPERTIES: pale yellow needles; m.p. 170°C; m.w. 230.27
C. AIR POLLUTION FACTORS:
—Ambient air quality: glc at Botrange (Belgium): woodland at 20–30 km from industrial area—June 1977: 0.18 ng/cu m   (1233)

1-benzazine *see* quinoline

benzene

BENZENE   237

Manufacturing source: petroleum refinery; solvent recovery plant; coal tar distillation; coal processing; coal coking. (347)
Users and formulation: mfg. styrene, phenol, detergents, organic chemicals, pesticide, plastics and resins, synthetic rubber, aviation fuel, pharmaceuticals, dye, explosives, PCB, gasoline, tanning, flavors and perfumes, paints and coatings; nylon intermediates; food processing; photographic chemicals. (347)

A. PROPERTIES: colorless liquid; m.w. 78.11; m.p. 5.5°C; b.p. 80.1°C; v.p. 76 mm at 20°C, 60 mm at 15°C, 118 mm at 30°C; v.d. 2.77; sp.gr. 0.8786 at 20/4°C; solub. 1780 mg/l at 20°C; sat. conc. 319 g/cu m at 20°C, 485 g/cu m at 30°C; log P_{oct} 2.13 at 20°C

B. AIR POLLUTION FACTORS: 1 mg/cu m = 0.31 ppm, 1 ppm = 3.26 mg/cu m
  —Odor: T.O.C.: 0.516 mg/cu m = 0.160 ppm (307)
                43 mg/cu m = 13.3 ppm (307)
                1–300 ppm (9)
                4.68 ppm (3)
                1 ppm (5)
                60 ppm (298)(12)(210)
                100 ppm (4)
                320 ppm (13)
                180 mg/cu m = 60 ppm (278)
                100.7 mg/cu m = 31.0 ppm (279)
                2.8 mg/cu m = 0.86 ppm (291)
                  recognition: 10,5–210 mg/cu m (73)
                ED_{50}: 38 mg/cu m (307)
                $PIT_{50\%}$: 2.14 ppm
                $PIT_{100\%}$: 4.68 ppm (2)
                distinct odor: 310 mg/cu m = 90 ppm (278)
      USSR: human odor perception: 3.0 mg/cu m = 1 ppm
              animal chronic exposure: adverse effect: 3.2 mg/cu m (170)
  —Atmospheric reactions: R.C.R.: 0.276 (49)
    reactivity: HC cons.: ranking: 0.5
              NO ox.: ranking: 0.04–0.15 (63)
  —Natural sources: glc's Pt Barrow Alaska, Sept 1967: not detectable to 0.4 ppb (101)
  —Manmade sources:
    diesel engine: 2.4% of emitted HC.s (72)
    rotary gasoline engine: 1.3% of emitted HC.s
    reciprocating gasoline engine: 2.2% of emitted HC.s (78)
    expected glc's in USA urban air: range: 10 to 50 ppb (102)
    emitted by household central heating system on gasoil: approx. 20 ppm at CO_2 = 7%
    6 g/kg gasoil at CO_2 = 6%
    2.2 g/kg gasoil at CO_2 = 7% (182)
    in gasoline exhaust: 0.1 to 42.6 ppm (partly methylvinylketone) (195)
    in gasoline: 1.8–5 vol % (312; 34)
    in exhaust of gasoline engines: 62-car survey: 2.4 vol % of total exhaust HC's (391)
    exhaust emissions from 1975 GM models equipped with bead-type converters: %

238 BENZENE

of total hydrocarbons: avg. 2.86%, range 1–7% (1975 federal test procedure)
mg/mile: avg. 19.9, range 8–75
exhaust emissions from non catalyst cars (mostly 1974 GM models)
% of total hydrocarbons: 4.6% avg.
mg/mile: 113 avg. (1386)
for late model cars it has been estimated that over 90% of automotive benzene comes from exhaust and less than 10% from fuel evaporation; this does not include any benzene lost during tanker-to-station and station-to-car fuel transfers (1386)
evaporation from gasoline tank: 0.3–0.4 vol % of total evaporated HC's
evaporation from carburettor: 0.1–1.8 vol % of total evaporated HC's
(398; 399; 400; 401; 402)

—Ambient air quality:
glc's in Netherlands:
in tunnel Amsterdam–1973: avg. 6 ppb ($n = 3$)
in tunnel Rotterdam–1974.10.2: avg. 25 ppb ($n = 12$)
max. 33 ppb
The Hague–1974.10.11: avg. 13 ppb ($n = 12$)
max. 29 ppb
Roelofarendsveen–1974.9.11: avg. 2 ppb ($n = 12$)
max. 3 ppb (1231)
glc's in USA during winter 1977–1978:

	averages	range (8 hr average)
urban/suburban areas:	< 0.5–3.6 ppb	< 0.5–12.1 ppb
rural areas:	< 0.5–0.6 ppb	< 0.5–1.8 ppb
remote areas:	< 0.5–0.7 ppb	< 0.5–2.0 ppb

(1401)
glc's in Los Angeles 1966: avg.: 0.015 ppm ($n = 136$)
highest value: 0.057 ppm (1319)

—Photochemical reaction: estimated lifetime under photochemical smog conditions in SE England: 28 hr (1699; 1707)

—Control methods:
platinized ceramic honeycomb catalyst: ignition temp.: 180°C
inlet temp. for 90% conversion: 250–300°C (91)

—Sampling and analysis:
I.R.: det.lim.: 0.15 ppm (18)
second derivative spectroscopy: det.lim.: 25 ppb (42)
non dispersive I.R.: min.full scale: 40 ppm (55)
6 f.abs.app.; 40 l air/30 min; UVS: det.lim.: 5 mg/cu m/30 min (208)
photometry: min.ful scale: 210 ppm (53)
detector tubes: UNICO: det.lim.: 10 ppm (59)
AUER: det.lim.: 5 ppm (57)
DRAGER: det.lim.: 5 ppm (58)
effectiveness of scrubbers at 325 to 375°F: inlet: 15 ppm benzene:
1) $HgSO_4/H_2SO_4$: 89% absorbed
2) $PdSO_4/H_2SO_4$: 53% absorbed
3) 1) + 2): 91% absorbed (198)
PMS: det.lim.: 2–9 ppm (200)

BENZENE

C. WATER POLLUTION FACTORS:
- BOD_5: 0 std.dil.sew. (41, 27)
  - 10% ThOD (220)
  - 58% bio.-ox. (acclim.) (23)
  - 24% bio.-ox. (non acclim.) (23)
- BOD_{10}: 67% bio.-ox. (acclim.) (23)
  - 27% bio.-ox. (non acclim.) (23)
- BOD_{15}: 76% bio.-ox. (acclim.) (23)
  - 24% bio.-ox. (non acclim.) (23)
- BOD_{20}: 80% bio.-ox. (acclim.) (23)
  - 29% bio.-ox. (non acclim.) (23)
- BOD_5: 2.18 NEN 3235-5.4 (277)
  - 0% of ThOD (274)
- BOD_{10}: 1.20 std.dil.sew. (256)
- $BOD_{35}^{25°C}$: 1.88 (62)
- $BOD_{35}^{25°C}$: 1.58 in seawater/inoculum: enrichment cultures of hydrocarbon oxidizing bacteria (521)
- COD: 0.25; 1.40 (27)(23)
  - 19% ThOD (220)
  - 2.15 ASTM procedure (272)
  - 2.80 modified Shell procedure (272)
  - 33% ThOD (0.05 n $K_2Cr_2O_7$) (274)
- $KMnO_4$: 0.032 (30)
  - acid 0 (220)
  - alkaline 0 (220)
  - 0% ThOD (0.01 n $KMnO_4$) (274)
- TOC: 40% ThOD (220)
- ThOD: 3.10 (23)
- Biodegradation:
- Biodegradation to CO_2 in estuarine water:

Concentration (μg/l)	Month	Incubation time (hr)	Degradation rate (μg/l/day) $\times 10^3$	Turnover time (days)	
6	June	24	200 ± 10	30	
12	June	24	260 ± 25	46	
24	June	24	330 ± 30	75	(381)

incubation with natural flora in the groundwater—in presence of the other components of high-octane gasoline (100 μl/l)

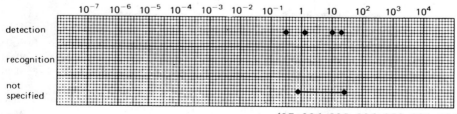

(97; 226; 295; 296; 875; 898; 924)

—Water quality:
in river Maas at Eysden (Netherlands) in 1976: median: 0.1 µg/l;
range: n.d. to 5.7 µg/l
Keizersveer (Netherlands) in 1976: median: 0.1 µg/l;
range: n.d. to 1 µg/l (1368)
in lake Zürich (Switzerland): at surface: 28 ppt; at 30 m: 22 ppt
in Zürich area: spring water: 18 ppt; ground water; 45 ppt; tap water: 36 ppt
(513)
—Aquatic reactions: evaporation:
Calculated half-life in water at 25°C and 1 m depth, based on evaporation rate of 0.144 m/hr: 4.81 hr;
based on evaporation rate of 0.137 m/hr: 5.03 hr (437)
soil adsorption: Freundlich constants for benzene sorption after 16 hr of incubation—concentrations 10–1000 ppb:

adsorbent	K	$1/n$
Hastings silty clay loam	2.4	0.89
Overton silty clay loam	1.8	0.94
Al-saturated montmorillonite	30.9	1.08
Ca-saturated montmorillonite	4.4	0.99	(1862)

—Waste water treatment:
A.S.: 33% theoretical oxidation of 500 ppm benzene by phenol acclimated activated sludge after 12 hr aeration (26)
A.C.: adsorbability: 0.08 g/g C, 95% reduction, infl.: 416 mg/l, effl.: 21 mg/l
(32)
anaerobic lagoon: 13 lb/day/1,000 cu ft: infl.: 10 mg/l, effl.: 5 mg/l (37)
ion exchange: adsorption on Amberlite X AD-2: retention efficiency: 100%, infl.: 100 ppm, effl.: 0 ppm (40)
air stripping constant: K = 1.71 days^{-1} at 100 mg/l (82)
A.S., Sd, BOD, 14 d acclimation: 2% of ThOD after 5 days at 20°C
A.S., W, 14 d acclimation: 3% of ThOD after $\frac{1}{4}$ days at 20°C, feed:
50–200 mg/l (93)
air flotation after chemical addition: 78% removal (173)

A.C.: influent ppm	carbon dosage	effluent ppm	% reduction	
500	10X	27	95	
250	10X	23	91	
50	10X	20	60	(192)

—Sampling and analysis: photometry: min.full scale: 8.5 × 10^{-6} mole/l (53)

D. BIOLOGICAL EFFECTS:
—Toxicity threshold (cell multiplication inhibition test):
bacteria (*Pseudomonas putida*): 92 mg/l (1900)
algae (*Microcystis aeruginosa*): >1400 mg/l (329)
green algae (*Scenedesmus quadricauda*): >1400 mg/l (1900)
protozoa (*Entosiphon sulcatum*): >700 mg/l (1900)
protozoa (*Uronema parduczi Chatton-Lwoff*): 486 mg/l (1901)
—Algae:
Chlorella vulgaris: 50% reduction of cell numbers vs controls, after 1 day incubation at 20°C: at 525 ppm (343)

inhibition of photosynthesis of a freshwater, non axenic unialgal culture of
Selenastrum capricornutum
at 10 mg/l: 95% carbon-14 fixation (vs. controls)
100 mg/l: 84% carbon-14 fixation (vs. controls)
1000 mg/l: 5% carbon-14 fixation (vs. controls) (1690)
ciliate (*Tetrahymena pyriformis*): 24 hr LC_{100}: 12.8 mmole/l (1662)
—Crustacean:
Grass shrimp (*Palaemonetes pugio*): 96 hr LC_{50}: 27 ppm (940)
Crab larvae—stage 1 (*Cancer magister*): 96 hr LC_{50}: 108 ppm (941)
Shrimp (*Crangon franciscorum*): 96 hr LC_{50}: 20 ppm (942)
brine shrimp: TLm (24, 48 hr): 66–21 mg/l (41)
—Fish:
minnows: min. lethal dose: 5–7 mg/l; 6 hr (226)
bluegill sunfish: LD_{50}: 24–48 hr: 20 mg/l
LD_{100}: 24 hr: 34 mg/l; 2 hr: 60 mg/l (226)
goldfish: LD_{50} (24 hr): 46 mg/l, modified ASTM D 1345 (277)
fatheads: soft water: TLm (24, 96 hr): 35.5 to 33.5 mg/l
fatheads: hard water: TLm (24, 96 hr): 24.4 to 32 mg/l
bluegills: soft water: TLm (24, 96 hr): 22.5 mg/l
goldfish: soft water: TLm (24, 96 hr): 34.4 mg/l
guppies: soft water: TLm (24, 96 hr): 36.6 mg/l (58)
mosquito fish: TLm (24, 96 hr): 395 mg/l
young Coho salmon: no significant mortalities up to 10 ppm after 96 hrs in artificial sea water at 8°C
: mortality: 12/20 at 50 ppm after 24 up to 96 hrs in artificial sea water at 8°C
: 30/30 at 100 ppm after 24 hrs in artificial sea water at 8°C (317)
Bass (*Morone saxatilis*): 96 hr LC_{50}: 5.8–10.9 ppm (942)
Herring and anchovy larvae (*Clupea pallasi; Engraulis mordex*):
35–45 ppm caused delay in development of eggs and produces abnormal larvae
10–35 ppm caused delay in development of larvae; decrease in feeding and growth, and increase in respiration (944)
Chinook salmon (*Onchorhynchus tschawytsche*) 5–10 ppm: initial increase in
Striped bass (*Morone saxatilis*) respiration (945)
pacific herring (*Clupea harengus pallasi*): BCF:
in eggs: 10.9
yolk-sac larvae: 6.9
feeding larvae: 3.9 (1918)
eel: infiltration ratio: flesh/water: 0.31; eel flesh 0.14 ng/g, water: 0.45 ng/g
(412)
eel (*Anguilla japonica*): BCF: 3.5; half-life: 0.5 days (1926)
guppy (*Poecilia reticulata*): 14 d, LC_{50}: 63 ppm (1833)
brown trout yearlings: 1 hr, LC_{50}: 12 mg/l (static bioassay) (939)
—Amphibian:
mexican axolotl (3–4 w after hatching): 48 hr LC_{50}: 370 mg/l
clawed toad (3–4 w after hatching): 48 hr LC_{50}: 190 mg/l (1823)

—Mammals: rat: LD_{50}: 5600–5700 mg/kg body weight (226)
embryotoxicity through inhalation: CF-1 mice and New Zealand white rabbits were exposed to 0 or 500 ppm of benzene for 7 hr per day from days 6 through 15 (mice) and 6 through 18 (rabbits) of gestation. Little evidence of maternal toxicity was seen in either species. Although some signs of embryonal toxicity were observed in both mice and rabbits, a teratogenic effect was not discerned in either species inhaling 500 ppm of benzene (1688)
—Man: EIR:1.0 (49)
    severe toxic effects: 1500 ppm, 60 min
    symptoms of illness: 500 ppm
    unsatisfactory: 50 ppm (185)

benzeneazobenzene see azobenzene

benzenecarbonal see benzaldehyde

benzenecarbonitrile see benzonitrile

benzenecarbonylchloride see benzoylchloride

1,2-benzenediamine see o-phenylenediamine

1,3-benzenediamine see m-phenylenediamine

1,4-benzenediamine see p-phenylenediamine

1,2-benzenedicarboxylic acid see o-phthalic acid

1,3-benzenedicarboxylic acid see isophthalic acid

1,4 benzenedicarboxylic acid see terephthalic acid

1,2-benzenediol see catechol

1,3-benzenediol see resorcinol

1,4-benzenediol see hydroquinone

m-benzenedisulfonic acid

C. WATER POLLUTION FACTORS:
—Biodegradation: adapted A.S. at 20°C—product is sole carbon source: 63.5% COD removal at 3.4 mg COD/g dry inoculum/hr (327)

benzenehexachloride *see* hexachlorocyclohexane

benzenesulfochloride (benzenesulfonylchloride; benzenesulfonchloride)

Users and formulation: dye mfg. (intermediate); accelerator in alkyl resin formation; phenol mfg. (intermediate); mfg. resorcinal (intermediate) (347)
A. PROPERTIES: m.w. 176.62; m.p. 14.5°C; b.p. 246°C decomposes; sp.gr. 1.378 at 23°C
D. BIOLOGICAL EFFECTS: brown trout yearlings: 48 hr LC_{50}: 3 mg/l (static bioassay) (939)

benzenesulfonic acid

A. PROPERTIES: m.w. 158.17; m.p. 1.5°C $+H_2O$, 43/44°C anh.; b.p. decomposes; $\log P_{oct}$ -2.25 (calculated)
C. WATER POLLUTION FACTORS
 —Biodegradation: decomposition by a soil microflora: 16 days (176)
  adapted A.S. at 20°C—product is sole carbon source: 98.5% COD removal at 10.6 mg COD/g dry inoculum/hr (327)
 —Waste water treatment:
  ion exchange: adsorption on Amberlite X AD-2: 31% retention effic. infl.: 3.0 ppm, effl.: 2.1 ppm (40)
  A.C.: adsorptive capacities for benzenesulfonate:

carbon 0.273 mm particle size	benzenesulfonate mole/g C at 30°C
Columbia LC	131
Dareo S 51	101
Fisher	120
Norit	92
Nuchar C-190	54

D. BIOLOGICAL EFFECTS
 —Mammalia: mouse: acute oral dose LD_{50}: 0.4 to 3.2 g/kg (211)
 —Man: severe skin irritation (211)

benzenethiol *see* thiophenol

1,2,3-benzenetriol *see* pyrogallol

benzenyltrichloride *see* benzotrichloride

benzethoniumchloride (hyamine 1622; a synthetic quaternary ammonium compound)
$C_{27}H_{42}ClNO_2 \cdot H_2O$
Use: antiseptic; cationic detergent
A. PROPERTIES: colorless, odorless plates, very bitter taste; m.p. 164–166°C; m.w. 466.1
C. WATER POLLUTION FACTORS:
  —Biodegradation: at 18 mg/l no degradation after 28 days by nonadapted sewage
  (488)
D. BIOLOGICAL EFFECTS:
  —Bacteria: *Staphylococcus aureus*: at 20 mg/l bacteriolytic action after 1 hr; bacteria isolated from Rhône (France) water: at 20 mg/l no significant reduction of growth rate (488)
  —Fish

Pimephales promelas	1600 µg/l	96 hr LC_{50}	(2131)
Lepomis macrochirus	1400 µg/l	96 hr LC_{50}	(2131)
Oncorhynchus kisutch	53000 µg/l	96 hr LC_{50}	(2100)

  —Man: highly toxic by ingestion; 1 gram may be fatal

benzidine (p,p'-bianiline; 4,4'-diaminobiphenyl)

$$H_2N-\text{\textlangle}\bigcirc\text{\textrangle}-\text{\textlangle}\bigcirc\text{\textrangle}-NH_2$$

Use: organic synthesis; manufacture of dyes
A. PROPERTIES: grayish-yellow, white or reddish gray crystalline powder; m.w. 184.23; m.p. 116/129°C; b.p. 402°C; v.d. 6.36; sp.gr. 1.250 at 20/4°C; solub. 400 mg/l at 12°C, 9,400 mg/l at 100°C
B. AIR POLLUTION FACTORS:
  —Method of analysis: air is drawn through a glass-fiber filter followed by a bed of silica gel to collect these substances as either particles or vapors. The compounds are extracted from the sampler and analyzed by HPLC with sensitivities in the range of 3 µg/cu m for 48 hr air samples (1686)
C. WATER POLLUTION FACTORS:
  —$KMnO_4$: 1.896 (30)
  —Biodegradation:
    possible bio-oxidation products scanned by GC/MS: N-hydroxybenzidine; 3-hydroxybenzidine; 4-amino-4'-nitrobiphenyl; N,N'-dihydroxybenzidine; 3,3'-dihydroxybenzidine; 4,4'-dinitrobiphenyl (419)
    W, A.S. from mixed domestic/industrial treatment plant: 85–93% depletion at 20 mg/l after 6 hr at 25°C (419)
D. BIOLOGICAL EFFECTS:
  —Carcinogenicity: +
  —Mutagenicity in the Salmonella test: +
    1.4 revertant colonies/nmol
    265 revertant colonies at 50 µg/plate (1883)

benzidinedihydrochloride

HCl·H₂N―⟨○⟩―⟨○⟩―NH₂·HCl

C. WATER POLLUTION FACTORS:
  —Control methods: concentration at various stages of water treatment works
    (+ chrysene):
    river intake: 0.090 µg/l
    after reservoir: 0.072 µg/l
    after filtration: 0.033 µg/l
    after chlorination: 0.012 µg/l   (434)
  —Impact on biodegradation: NH₃ oxidation by *Nitrosomonas* sp.:
    at 100 mg/l: 84% inhibition
    at 50 mg/l: 56% inhibition
    at 10 mg/l: 12% inhibition   (390)

benzo(a)anthracene   (1,2-benzanthracene; B(a)A)

Manmade sources:
  in gasoline: 0.232 mg/l; 0.04 mg/l; 0.272 mg/l   (380; 1052; 1070)
  in bitumen: 0.13–0.86 ppm   (506)
  in Kuwait crude oil: 2.3 ppm
  in South Louisiana crude oil: 1.7 ppm   (1015)
  in wood preservative sludge: 5.18 g/l of raw sludge   (960)
A. PROPERTIES: m.w. 228; solub. 0.010 mg/l; 0.044 mg/l at 24°C (practical grade)
B. AIR POLLUTION FACTORS:
  —Manmade emissions:
    emissions from typical European gasoline engine (1608 cu cm) - following Euro-
      pean driving cycles - using leaded and unleaded commercial gasolines: 7.3 -
      32.4 µg/l fuel burnt   (1291)
    in exhaust condensate of gasoline engine: 0.5 - 0.08 mg/l gasoline consumed
                                                                        (1070)
                               280 ppm   (1069)
    emissions from asphalt hot-mixing plant: 5–24 ng/cu m
      (in high volume particulate matter): avg. 11 ng/cu m   (491)(1379)
    in coke oven airborne emissions: 105–2740 µg/g of sample   (960)
    in cigarette smoke: 0.3 µg/100 cigarettes   (1298)
  —Ambient air quality: organic fraction of suspended matter:
    Bolivia at 5200 m altitude (Sept.–Dec. 1975): 0.040–0.005 µg/1000 cu m (+
      chrysene)

Belgium, residential area (Jan.–April 1976): 2.2–13 µg/1000 cu m (+ chrysene)
(428)

glc's in The Netherlands (ng/cu m):

		Delft	The Hague	Vlaardingen	Amsterdam	Rotterdam	
summer	1968	2	–	3	–	1	
	1969	1	<1	1	n.d.	1	
	1970	1	<1	3	1	2	
	1971	2	1	3	1	1	
winter	1968	16	6	29	20	17	
	1969	10	7	25	11	12	
	1970	8	4	19	7	6	
	1971	4	2	14	4	5	(1277)

glc's in Birkenes (Norway) Jan.–June 1977: avg: 0.34 ng/cu m
range: n.d.–1.86 ng/cu m ($n = 16$)
in Rørvik (Sweden) Dec. '76–April '77: avg: 0.62 ng/cu m
range: 0.02–4.60 ng/cu m ($n = 21$)
(1236)

glc's in Budapest 1973: heavy traffic area: winter: 2.2 ng/cu m
(6–20 hr) summer: 56.6 ng/cu m
low traffic area: winter: 11.2 ng/cu m
summer: 85.1 ng/cu m (1251)

glc's in residential area (Belgium) oct. 1976:
in particulate phase: 12.2 ng/cu m
in gas phase: 3.87 ng/cu m (+ chrysene) (1289)

glc's in the average American urban atmosphere–1963:
~ 30 µg/g airborne particulates or
~ 4 ng/cu m air (1293)

Cleveland (16 sites) 1971/72: 1.2 ng/cu m (556)
Los Angeles (4 sites) 1971/72: 1.2 ng/cu m (560)
Lyon (1 site) 1972: 0.33 ng/cu m (561)
Rome (1 site) 1970/71: 2.8 ng/cu m (562)
Budapest (1 site) 1971/72: 8.21 ng/cu m (563)

Cleveland (Ohio, USA): max. conc. 140 ng/cu m (448 values)
(1971/72) annual geom. mean (all sites): 1.37 ng/cu m
annual geom. mean of TSP (all sites): 11 ppm (wt)
(556)

C. WATER POLLUTION FACTORS:
—Biodegradation: was not observed during enrichment procedures (1882)
biodegradation to CO_2 in estuarine water:

concentration (μg/l)	incubation month	time (hr)	degradation rate (μg/l/day) $\times 10^3$	turnover time (days)	
10	Jan.	48	0	∞	
10	June	48	0	∞	(381)

—Aquatic reactions: adsorption: in estuarine water: at 10 µg/l, 53% adsorbed on
particles after 3 hr (381)
after 3 hr incubation in natural seawater, 59% of 3 µg/l were taken up by suspended aggregates of dead phytoplankton cells and bacteria (957)

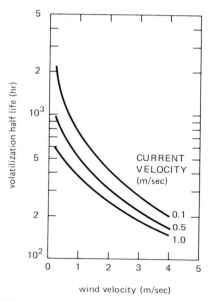

Variation in predicted volatilization rates of benzo (a) anthracene at 25°C under varying conditions of wind and current velocity in a stream 1.0 m in depth   (1391)

— Water and sediment quality: in sediment of wilderness Lake—Colin Scott—
  Ontario (1976): 7 ppb (dry wt)   (932)
  in rapid sand filter solids from Lake Constance water: 0.2–0.3 mg/kg in river
  water solids: river Rhine: 0.4 mg/kg
         river Aach at Stockach: 0.9–3.2 mg/kg
         river Argen: 0.4 mg/kg
         river Schussen: 2.6 mg/kg   (531)
  in river water solids:

river Gersprenz at Munster (W. Germany):	4.3–18.8 ng/l
river Danube at Ulm (W. Germany):	11–14 ng/l
river Main at Seligenstadt (W. Germany):	7.0–385 ng/l
river Aach at Stockach (W. Germany):	101–199 ng/l
river Schussen (W. Germany):	57 ng/l

       (530)

— Manmade sources:

in domestic effluent:	0.191–0.319 ppb
in sewage (high percentage industry):	0.343–1.36 ppb
in sewage during dry weather:	0.025 ppb
in sewage during heavy rain:	10.36 ppb

       (531)

— Control methods:

	Dec. 1965	May 1966
in raw sewage	31.4 ppb	1.87 ppb
after mechanical purification	6.0 ppb	0.28 ppb
after biological purification	0.05 ppb	0.06 ppb

       (545)

ozonation: after 1 min. contact time with ozone: residual amount: 4.5% (550)
chlorination: 6 mg/l chlorine for 6 hr:
  initial conc. 16.11 ppb: 53% reduction (549)

D. BIOLOGICAL EFFECTS:
—Molluscs: oysters: uptake and release

time of exposure days	conc. in oysters μg/g	water μg/l	depuration time days	conc. in oysters μg/g	water μg/l
2	4.1 (0.5)	3.8	14	0.2	0
4	1.7 (1.1)	1.1	14	0.4	0
9	1.6 (1.3)	0.3	14	0.4	0
15	0.8	0.1			

Figures between brackets refer to compartment surrounded by a 60 μm Nitex filter to filter out large particles (579)

Uptake and depuration by oysters (*Crassostrea Virginica*) from oil-treated enclosure

time of exposure days	depuration time days	concentration oysters μg/g	water μg/l	accumulation factor oysters/water
2	–	2.8	5.3	530
8	–	1.8	0.1	18,000
2	7	1.9	–	–
8	7	1.0	–	–
8	23	0.3	–	–

half-life for depuration: 9 days (957)

—Carcinogenicity: +
—Mutagenicity in the *Salmonella* test: +
  11 revertant colonies/nmol
  2640 revertant colonies at 57 μg/plate (1883)

benzocaine (ethyl-*p*-aminobenzoate; anesthesin)

$$\text{COO}-\text{CH}_2-\text{CH}_3$$

(benzene ring with NH$_2$ at para position)

Use: medicine; suntan preparations.

A. PROPERTIES: white, crystalline, odorless, tasteless powder; m.p. 88-90°C; m.w. 165.19

C. WATER POLLUTION FACTORS:
—Impact on biodegradation: NH_3 oxidation by *Nitrosomonas* sp.:
  at 100 mg/l: 30% inhibition
  at 50 mg/l: 27% inhibition
  at 10 mg/l: 0% inhibition (390)

benzo(a)chrysene (picene)

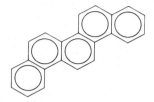

B. AIR POLLUTION FACTORS:
 —Manmade sources: in gasoline: 1 µg/l
 in exhaust condensate of gasoline engine: 1 µg/l gasoline
 consumed (1070)

benzo(ghi)cyclopenta(pqr)perylene *see* 1,12-methylenebenzo(ghi)perylene

benzo(b)fluoranthene (3,4-benzfluoranthene; B(b)F)

Manmade sources:
in Kuwait crude oil: <1 ppm
in South Louisiana crude oil: <0.5 ppm (1015)
in gasoline: low octane number: 0.16–0.49 mg/kg ($n = 14$) (385; 1070)
in gasoline: high octane number: 0.26–1.34 mg/kg ($n = 13$) (385)
in high octane gasoline: 3.9 mg/kg ($b + j + k$-isomers) (1220)
in fresh motor-oil: 0.08 mg/kg ($b + j + k$-isomers) (1220)
in used motor-oil after 18 European driving cycles: 2.8–3.3 mg/kg
in used motor-oil after 5,000 km: 45.2–82.2 mg/kg
in used motor-oil after 10,000 km: 55.6–141.0 mg/kg
in used motor-oil after 10,000 km and 18 European driving cycles:
 63.0–166.4 mg/kg ($b + j + k$-isomers) (1220)
in lubricating motor-oils: 0.04–0.30 ppm (6 samples) (379)
in bitumen: 0.40–1.60 ppm (506)
A. PROPERTIES: m.w. 252
B. AIR POLLUTION FACTORS:
 —Manmade sources:
  in cigarette smoke: 0.3 µg/100 cigarettes (1298)
  in exhaust condensate of gasoline engine:
   0.019–0.048 mg/l gasoline consumed (1070)

in exhaust condensate of gasoline engine: 162 µg/g (1069)
in tail gases of gasoline engine: 16–66 µg/cu m (n.s.i.) (340)
in gasoline: 243 µg/kg (n.s.i.) (340)
—Ambient air quality: in air in Ontario cities (Canada):

location	April–June '75		July–Sept. '75		Oct.–Dec. '75		Jan.–March '76	
	ng/1000 cu m air	µg/g p.m.	ng/1000 cu m air	µg/g p.m.	ng/1000 cu m air	µg/g p.m.	ng/1000 cu m air	µg/g p.m.
Toronto	866	9.7	798	8.4	1387	16.8	783	10.1
Toronto	890	11.8	693	10.6	1259	20.7	1829	13.4
Hamilton	813	5.6	2626	18.9	7841	113.1	2297	27.5
S. Sarnia	371	6.0	243	5.2	938	17.9	289	10.6
Sudbury	255	7.8	173	4.1	417	18.7	650	27.8

(999)

Cleveland (Ohio, USA): max. conc.: 170 ng/cu m (448 values)
(1971/1972) annual geom. mean (all sites): 3.62 ng/cu m
annual geom. mean of TSP (all sites): 26 ppm (wt) (556)
glc's in Birkenes (Norway)–Jan.–June '77: avg.: 0.62 ng/cu m
range: n.d.–5.85 ng/cu m ($n = 18$)
in Rørvik (Sweden)–Dec. '76–April '77: avg.: 1.24 ng/cu m
range: n.d.–9.68 ng/cu m ($n = 21$)
(1236)
in Botrange (Belgium): Woodland at 20 - 30 km from industrial area–June–July 1977: 1.62; 1.99 ng/cu m ($n = 2$) (+(b) isomer)
in Wilrijk (Belgium): residential area–Oct./Dec. 1976: 14.8; 67.1 ng/cu m ($n = 2$) (+(b) isomer) (1233)

C. WATER POLLUTION FACTORS:
—Water, sediment and soil quality:
in groundwater (W. Germany–1968): 0.6–5.7 µg/cu m ($n = 10$)
in tapwater (W. Germany–1968): 2.6–5.4 µg/cu m ($n = 6$) (955)
in dried forest soil samples:
South of Darmstadt (W. Germany): 30–110 ppb ((b) + (j) isomers)
near Lake Constance: 25– 35 ppb ((b) + (j) isomers) (528)
in sediment of Wilderness Lake, Colin Scott, Ontario (1976): 25 ppb (dry wt)
(932)

river Gersprenz at Munster (W. Germany): 10.4–13.2 ng/l
river Danube at Ulm (W. Germany): 23.9–24.2 ng/l
river Main at Seligenstadt (W. Germany): 12.2–362 ng/l
river Aach at Stockach (W. Germany): 76–332 ng/l
river Schussen (W. Germany): 41 ng/l (530)
in rapid sand filter solids from Lake Constance water: 0.4 mg/kg
in river water solids: river Aach at Stockach: 2.7–6.7 mg/kg
river Argen: 1.2 mg/kg
river Schussen: 4.9 mg/kg (531)
Control methods:
in domestic effluent: 0.036–0.202 ppb
in sewage (high percentage industry): 0.525–0.87 ppb
in sewage during dry weather: 0.039 ppb

in sewage during heavy rain: 9.91 ppb (531)
chlorination: 6 mg/l chlorine for 6 hr:
  initial conc. 5.18 ppb: 34% reduction (549)
concentration at various stages of water treatment works (+(j) and (k) isomers)
  river intake: 0.147 µg/l
  after reservoir: 0.132 µg/l
  after filtration: 0.039 µg/l
  after chlorination: 0.021 µg/l (434)

	Dec. 1965	May 1966	
in raw sewage	23.7 ppb	0.58 ppb	
after mechanical purification	4.8 ppb	0.25 ppb	
after biological purification	0.15 ppb	0.05 ppb	(545)

—Manmade sources:
primary and digested raw sewage sludge: 0.21–0.42 ppm
liquors from sewage sludge heat treatment plants: 0.03–0.55 ppm
sludge cake from heat treatment plants: 0.22–0.52 ppm
final effluent of sewage works: 0.03 ppb (1426)

benzo(*j*)fluoranthene (10,11-benzofluoranthene; b(*j*)F)

Manmade sources:
  in gasoline: 9 µg/l (1070)
  in Kuwait crude oil: <1 ppm
  in South Louisiana crude oil: <0.9 ppm (1015)
A. PROPERTIES: m.w. 252
B. AIR POLLUTION FACTORS:
  —Manmade sources:
    in cigarette smoke: 0.6 µg/100 cigarettes (1298)
    in airborne coal tar emissions: 0.12 mg/g of sample or 4 µg/cu m of air sampled
      in coke oven emissions: 18.7–285.3 µg/g of sample
    in coal tar: 0.73 mg/g of sample
    in wood preservative sludge: 0.31 g/l of raw sludge (993)
    in exhaust condensate of gasoline engine: 11–27 µg/l gasoline consumed (1070)
    in exhaust condensate of gasoline engine: 94 µg/g (1069)
    emissions from typical European gasoline engine (1608 cu cm)—following European driving cycles—using leaded and unleaded commercial gasolines: 0.3–9.2 µg/l fuel burnt (1291)
C. WATER POLLUTION FACTORS:
  —Water and sediment quality:

river Gersprenz at Munster (W. Germany): 4.6–14.7 ng/l
river Danube at Ulm (W. Germany): 10.1–23.4 ng/l
river Main at Seligenstadt (W. Germany): 14.2–337 ng/l
river Aach at Stockach (W. Germany): 144–420 ng/l
river Schussen (W. Germany): 53 ng/l (530)
in rapid sand filter solids from Lake Constance water: 0.5 mg/kg
in river water solids: river Aach at Stockach: 0.8–4.6 mg/kg
river Argen. 1.4 mg/kg
river Schussen: 6.2 mg/kg (531)
in sediment of Wilderness Lake, Colin Scott, Ontario (1976): 8 ppb (dry wt)
(932)

—Manmade sources:
in domestic effluent: 0.037–0.205 ppb
in sewage (high percentage industry): 1.1–1.74 ppb
in sewage during dry weather: 0.057 ppb
in sewage during heavy rain: 10.79 ppb (531)

—Control methods:

plant location	water source	river water conc. ng/l	drinking water conc. ng/l[*]	% removal/ transformation
Pittsburgh, Pa.	Monongahela river	36	0.3	99.2[**]
Huntington W. Va.	Ohio river	5	0.3	94.4
Philadelphia, Pa.	Delaware river	42.6	n.d.	100.0

[*]treatment provided: lime, ferric sulfate or chloride; A.C., chlorination and fluoridation
[**]two stages A.C.: powdered and granular carbon (958)

	Dec. 1965	May 1966	
in raw sewage	29.6 ppb	0.56 ppb	
after mechanical purification	11.6 ppb	0.14 ppb	
after biological purification	–	0.09 ppb	(545)

benzo(*k*)fluoranthene   (11,12-benzofluoranthene; B(*k*)F)

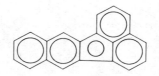

Manmade sources:
in bitumen: 0.34–1.41 ppm (506)
in gasoline: 9 μg/l (1070)
in Kuwait crude oil: <1 ppm
in South Louisiana crude oil: <1.3 ppm (1015)

A. PROPERTIES: m.w. 252
B. AIR POLLUTION FACTORS:
   —Manmade sources:

in airborne coal tar emissions: 9.96 mg/g of sample or 350 µg/cu m of air sampled
in coal tar: 32.5 mg/g of sample (993)
emissions from typical European car engine (1608 cu cm)—following European driving cycles—using leaded and unleaded commercial gasolines: 0.2–18.7 µg/l fuel burnt (1070; 1291)
—Ambient air quality:
 glc's in Birkenes (Norway): Jan.–June 1977: avg.: 0.48 ng/cu m
  (+ j-isomer) range: n.d.–3.99 ng/cu m ($n = 18$)
  in Rørvik (Sweden): Dec. '76–April '77: avg. 0.42 ng/cu m
  (+ j-isomer) range: n.d.–3.49 ng/cu m ($n = 21$) (1236)
 glc's in residential area (Belgium)—Oct. 1976:
  (+ b-isomer) in particulate sample: 23.1 ng/cu m
   in gas phase sample: 2.01 ng/cu m (1289)
 organic fraction of suspended matter:
 Bolivia at 5200 m altitude (Sept.–Dec. 1975):
  0.036–0.055 µg/1000 cu m (+benzo (b) fluoranthene)
 Belgium, residential area (Jan.–April 1976):
  2.9–30 µg/1000 cu m (+benzo (b) fluoranthene) (428)
 in air in Ontario cities (Canada):

location	April–June '75 ng/1000 cu m air	µg/g p.m.	July–Sept. 75 ng/1000 cu m air	µg/g p.m.	Oct.–Dec. '75 ng/1000 cu m air	µg/g p.m.	Jan.–March '76 ng/1000 cu m air	µg/g p.m.
Toronto	428	4.8	571	6.0	916	11.1	508	6.5
Toronto	328	4.3	285	4.4	597	9.8	519	3.8
Hamilton	419	2.9	1425	10.3	5145	74.2	443	5.3
S. Sarnia	81	1.3	70	1.5	439	8.4	104	3.8
Sudbury	57	1.3	74	1.8	197	8.8	271	11.6

(999)

C. WATER POLLUTION FACTORS:
 —Water and sediment quality:
  river Gersprenz at Munster (West Germany): 4.8–9.6 ng/l
  river Danube at Ulm (West Germany): 7.7–14.1 ng/l
  river Main at Seligenstadt (West Germany): 4.2–130 ng/l
  river Aach at Stockach (West Germany): 132–173 ng/l
  river Schussen (West Germany): 33 ng/l (530)
  in rapid sand filter solids from Lake Constance water: 0.5–1.0 mg/kg
  in river water solids:
   river Rhine: 0.6 mg/kg
   river Aach at Stockach: 1.3–2.4 mg/kg
   river Argen: 1.4 mg/kg
   river Schussen: 3.3 mg/kg (531)
  in Thames river water:
   at Kew Bridge: 80 ng/l
   at Albert Bridge: 40 ng/l
   at Tower Bridge: 120 ng/l (529)
  in sediment of Wilderness Lake, Colin Scott, Ontario (1976): 25 ppb (dry wt)
  (932)
  in groundwater (W. Germany—1968): 0.2–1.8 µg/cu m ($n = 10$)

BENZO(g,h,i)FLUORANTHENE

   in tapwater (W. Germany–1968):   1.0–3.4 µg/cu m ($n = 6$)   (955)
—Manmade sources:
primary and digested raw sewage sludge: 0.1–0.42 ppm
liquors from sewage sludge heat treatment plants: 0.03–0.45 ppm
sludge cake from heat treatment plants: 0.09–0.33 ppm
final effluent of sewage work: 0.03 ppb   (1426)
in domestic effluent: 0.031–0.193 ppb
in sewage (high percentage industry): 0.336–0.46 ppb
in sewage during dry weather: 0.022 ppb
in sewage during heavy rain: 4.18 ppb   (531)
—Control methods:

plant location	water source	river water conc. ng/l	drinking water conc. ng/l[*]	% removal/ transformation
Pittsburgh, Pa.	Monongahela river	19	0.2	99.0[**]
Huntington, W. Va.	Ohio river	3.6	0.2	94.4
Philadelphia, Pa.	Delaware river	33.0	n.d.	100

[*]treatment provided: lime, ferric salt, A.C., chlorination and fluoridation
[**]two stages A.C.: powdered and granular carbon   (958)

	Dec. 1965	May 1966
raw sewage	8.1 ppb	0.21 ppb
after mechanical purification	2.2 ppb	0.09 ppb
after biological purification	0.05 ppb	0.04 ppb (545)

chlorination: 6 mg/l chlorine for 6 hr: initial conc. 68.74 ppb: 56% reduction
   (549)

benzo(g,h,i)fluoranthene

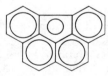

Manmade sources:
in lubricating motor-oils: 0.16–0.96 ppm (6 samples)   (379)
in bitumen: 0.11–0.43 ppm   (506)
in Kuwait crude oil: <1 ppm
in South Louisiana crude oil: 1 ppm   (1015)
in gasoline: 0.003 mg/l   (1070)

B. AIR POLLUTION FACTORS:
—Manmade sources:
in exhaust condensate of gasoline engine: 0.11–0.24 mg/l gasoline consumed
   (1070)
in airborne coal tar emissions: 3.29 mg/g of sample or 115 µg/cu m of air sampled
in coke oven emissions: 151–677 µg/g of sample
in coal tar: 4.40 mg/g of sample
in wood preservative sludge: 0.91 g/l of raw sludge   (993)

C. WATER POLLUTION FACTORS:
—Sediment quality: in sediment of Wilderness Lake, Colin Scott, Ontario (1976):
14 ppb (dry wt) (932)

benzo(a)fluorene (1,2-benzofluorene)

Manmade sources:
in bitumen: 0.02–0.10 ppm (506)
in coal tar: 20.13 mg/g of sample (n.s.i.) (933)
in gasoline: 1.5 mg/l (1070)
A. PROPERTIES: m.w. 216.28; m.p. 187–189°C; b.p. 407°C
B. AIR POLLUTION FACTORS:
—Manmade sources:
in coke oven emissions: 87.4–971.2 µg/g of sample (960)
in exhaust condensate of gasoline engine: 0.08–0.14 mg/l gasoline consumed
(1070)
in airborne coal tar emissions (n.s.i.): 24.5 mg/g of sample or 862 µg/cu m of
air sampled (993)
—Ambient air quality:
Cleveland (Ohio, USA): max. conc.: 7.5 ng/cu m (400 values)
(1971/1972) annual geom. mean (all sites): 0.25 ng/cu m
annual geom. mean of TSP (all sites): 2 ppm (wt)
(556)
organic fraction of suspended matter:
Bolivia at 5200 m altitude (Sept.–Dec. 1975): 0.012–0.024 µg/1000 cu m
(+ benzo(c)fluorene)
Belgium, residential area (Jan.–April 1976): 0.38–0.54 µg/1000 cu m (+
benzo(c)fluorene) (428)
glc's in residential area (Belgium)—Oct. 1976 (n.s.i.):
in particulate sample: 2.33 ng/cu m
in gasphase sample: 1.87 ng/cu m (1289)
in Birkenes (Norway), Jan.–June 1977: avg.: 0.16 ng/cu m
range: n.d.–1.45 ng/cu m
($n = 18$)
in Rørvik (Sweden), Dec. '76–April '77: avg.: 0.35 ng/cu m
range: n.d.–3.87 ng/cu m ($n = 21$)
(1236)

benzo(b)fluorene

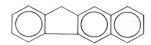

256 BENZO(c)FLUORENE

Manmade sources:
in gasoline: 1.4 mg/l (1070)
in bitumen: 0.02–0.09 ppm (506)
A. PROPERTIES: m.w. 216.28; m.p. 209–210.5°C
B. AIR POLLUTION FACTORS:
—Manmade sources:
in coke oven emissions: 16.7–109.4 µg/g of sample (960)
in exhaust condensate of gasoline engine: 0.065–0.112 mg/l gasoline consumed (1070)

emissions from typical European gasoline engine (1608 cu cm)—following European driving cycles—using leaded and unleaded commercial gasolines: 37.1–251.5 µg/l fuel burnt (1291)
—Ambient air quality:
glc's in Birkenes (Norway), Jan.–June 1977:
avg.: 0.04 ng/cu m
range: 0.02–0.34 ng/cu m (n =18)
in Rørvik (Sweden), Dec. '76–april 1977:
avg.: 0.2 ng/cu m
range: n.d.–2.26 ng/cu m (n = 21) (1236)
C. WATER POLLUTION FACTORS:
—Sediment quality: in sediments in Severn estuary (U.K.): 0.2–1.2 ppm dry wt (1467)

benzo(c)fluorene

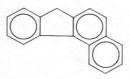

B. AIR POLLUTION FACTORS:
—Manmade sources: in coke oven emissions: 38.9–627.0 µg/g of sample (960)
C. WATER POLLUTION FACTORS:
—Control methods:
Concentration at various stages of water treatment works (+ (j) and (k)-isomers):
river intake: 0.147 µg/l
after reservoir: 0.132 µg/l
after filtration: 0.039 µg/l
after chlorination: 0.021 µg/l (434)

	Dec. 1965	May 1966
In raw sewage	23.7 ppb	0.58 ppb
After mechanical purification	4.8 ppb	0.25 ppb
After biological purification	0.15 ppb	0.05 ppb

(545)

In domestic effluent: 0.036–0.202 ppb
In sewage: high percentage industry: 0.525–0.87 ppb
during dry weather: 0.039 ppb
during heavy rain: 9.91 ppb (531)

Chlorination: 6 mg/l chlorine for 6 hr: initial conc. 5.18 ppb: 34% reduction
(549)

benzofuran (coumarone)

A. PROPERTIES: log P_{oct} 2.67
B. AIR POLLUTION FACTORS:
  —Manmade sources: in gasoline exhaust: <0.1–2.8 ppm   (195)

benzoic acid

Users and formulation: food preservative; pharmaceutical and cosmetic preparations; mfg. of alkyl resins; intermediate in the synthesis of dyestuffs and pharmaceuticals; production of phenol and caprolactam; plasticizer mfg. (to modify resins-PVC, PV acetate, phenol-formaldehyde)   (347)
Natural sources (water and air): cranberries, prunes, ripe cloves, bark of wild black cherry tree, scent glands of beavers, and oil of anise seeds.   (347)
A. PROPERTIES: white powder; m.w. 122.1; m.p. 121.7°C; b.p. 249°C; v.d. 4.21; sp.gr. 1.27; solub. 2,900 mg/l; 2.700 mg/l at 18°C; log P_{oct} 1.87 at 20°C
C. WATER POLLUTION FACTORS:
  —Biodegradation:
   decomposition period by a soil microflora: 1 day   (176)
   adapted A.S. at 20°C—product is sole carbon source: 99.0% COD removal at 88.5 mg COD/g dry inoculum/hr

BENZOIC ACID

- Bacterial degradation: pathway (1219)

  lag period for degradation of 16 mg/l by waste water or by soil at pH: 7.3 and 30°C: less than 1 day (1096)
- BOD_5: 1.65 (275)

  1.34 to 1.4 std.dil.sew. (281)(282)

  47% of ThOD (220)

  1.25 (30)

  1.45 (36)

  1.36 std.dil.acclimated (41)
- BOD_{10}: 1.40 std.dil.sewage (256)(166)
- BOD_{20}: 1.45 (30)
- BOD_5^{20}: 1.45 at 10 mg/l, unadapted sew.: no lag period
- BOD_{20}^{20}: 1.52 at 10 mg/l, unadapted sew.
- COD: 1.95 (36)

  100% of ThOD (220)
- COD: 1.88 (41)
- $KMnO_4$ value: 0.032 (30)

  acid 3% ThOD; alkaline: 1% ThOD (220)
- ThOD: 1.96 (36)
- Waste water treatment:

  A.C.: adsorbability: 0.183 g/g C, 91.1% reduction, infl.: 1,000 mg/l, effl.: 89 mg/l (32)

  A.C.(Pittsburgh Chemical Comp.–type BC): % adsorbed by 10 mg A.C. from 10^{-4} M aqueous solution: at pH 3.0: 49.7%

  at pH 7.0: 11.2%

  at pH 11.0: 2.5% (1313)

  coagulation with 3 lb alum/1,000 gal: 8% BOD reduction (95)

  ion exchange: adsorption on Amberlite X AD-2: 23% retention effic., infl.: 1.0 ppm, effl.: 0.8 ppm; at pH 3.2: 100% retention effic., infl.: 1.0 ppm, effl.: 0 ppm (40)

methods	temp °C	days observed	feed mg/l	days acclim.	% theor. oxidation	% removed
A.S., BOD	20	1–5	333	15		99
NFG, BOD	20	1–10	400–1,000	365 + P		46
Sel. Strain	30	1/12	2 mole/Warburg flask	none	66	

(93)

- Manmade sources: in primary domestic sewage plant effluent: 0.003 mg/l (517)

D. BIOLOGICAL EFFECTS:
   −Toxicity threshold (cell multiplication inhibition test):

bacteria (*Pseudomonas putida*):	480 mg/l	(1900)
algae (*Microcystis aeruginosa*):	55 mg/l	(329)
green algae (*Scenedesmus quadricauda*):	1630 mg/l	(1900)
protozoa (*Entosiphon sulcatum*):	218 mg/l	(1900)
protozoa (*Uronema parduczi Chatton-Lwoff*):	31 mg/l	(1901)

   −Arthropoda: *Daphnia magna*: immobilization at 146 mg/l; prolonged exposure (226)
   −Fish: mosquito fish: TLm (24 hr): 240 mg/l
   (48 hr): 255 mg/l
   (96 hr): 180 mg/l (41)
   goldfish: lethal in 7 ot 96 hr: 200 mg/l (226)
   orange spotted sunfish: lethal in 1 hr: 550 to 570 mg/l (226)
   −Mammalia: rat: acute oral dose LD_{50}: 1.7 g/kg (211)
   −Carcinogenicity: none
   −Mutagenicity in the *Salmonella* test: none
     <0.0009 revertant colonies/nmol
     <70 revertant colonies at 1000 µg/plate (1883)

benzoic trichloride *see* benzotrichloride

benzonitrile (Benzenecarbonitrile; phenylcyanide)

A. PROPERTIES: m.w. 103.12; m.p. −13°C; b.p. 190.7°C; sp.gr. 1.01 at 15/15°C; solub. 10,000 mg/l at 100°C; log P_{oct} 1.56
B. AIR POLLUTION FACTORS:
   −Sampling and analysis: photometry: min.full scale: 13.6 ppm (53)
C. WATER POLLUTION FACTORS
   −Waste water treatment:

methods	temp °C	days observed	feed mg/l	days acclim.	% theor. oxidation	% removed
RW, CO₂, CAN	20	4	10	28	60	100
RW, CO₂, CAN	20	7	51	30+	60	100
methods	temp °C	days observed	feed mg/l	days acclim.	theor. oxidation	% removed
RW, CO₂, CAN	5	18	51	30+ at 20°C	60	100
ASC, CAN	22−25	28	134−179	28	75+	99+

(93)

   −Sampling and analysis: photometry: min.full scale: 0.56×10^{-6} mole/l (53)

D. BIOLOGICAL EFFECTS:
—Toxicity threshold (cell multiplication inhibition test):

bacteria (*Pseudomonas putida*):	11 mg/l	(1900)
algae (*Microcystis aeruginosa*):	3.4 mg/l	(329)
green algae (*Scenedesmus quadricauda*):	75 mg/l	(1900)
protozoa (*Entosiphon sulcatum*):	30 mg/l	(1900)
protozoa (*Uronema parduczi Chatton-Lwoff*):	119 mg/l	(1901)

—Fish:

fathead minnow:	hard water:	96 hr TLm:	78 mg/l	(41)
fathead minnow:	soft water:	96 hr TLm:	135 mg/l	(41)
bluegill:	soft water:	96 hr TLm:	78 mg/l	(252)
guppies:	soft water:	96 hr TLm:	400 mg/l	(41)
adult bluegills:	no organoleptic influence at 35 mg/l			(226)

benzo(*ghi*)perylene (1,12-benzoperylene; B(*ghi*)P)

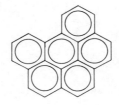

Manmade sources:
in fresh motor-oil: 0.12 mg/kg
in used motor-oil after 5,000 km: 108.8–207.6 mg/kg
in used motor-oil after 10,000 km: 153.0–289.4 mg/kg   (1220)
in Kuwait crude oil: <1 ppm
in South Louisiana crude oil: <1.6 ppm   (1015)
in bitumen: 1.37–5.50 ppm   (506)
in gasoline: low octane number: 0.32–1.24 mg/kg (n = 13)
        high octane number: 0.42–9 mg/kg (*n* = 15)   (380; 385; 1070; 1220)

A. PROPERTIES: m.w. 276; solub. 0.00026 mg/l at 25°C
B. AIR POLLUTION FACTORS:
—Manmade sources:
emission from space heating installation burning:
  coal (underfeed stoker): 4.5 mg/10^6 Btu input
  gasoil: 0.3 mg/10^6 Btu input
  gas: 1.8 mg/10^6 Btu input   (954)
emissions from typical European car engine (1608 cu cm)—following European driving cycles—using leaded and unleaded commercial gasolines: 2.7–111.8 µg/l fuel burnt   (1291)
in exhaust condensate of gasoline engine: 0.12–0.33 mg/l gasoline consumed
(1070)
—Ambient air quality:
glc's at Botrange (Belgium): woodland at 20–30 km from industrial area, June–July 1977: 0.40; 0.70 ng/cu m (*n* = 2)   (1223)

glc's in Birkenes (Norway), Jan.–June 1977:
   avg.: 0.28 ng/cu m
   range: n.d.–2.35 ng/cu m (n = 18)
glc's in Rørvik (Sweden), Dec. 1976–April 1977:
   avg.: 0.42 ng/cu m
   range: n.d.–2.88 ng/cu m (n = 21)   (1236)
glc's in The Netherlands

			ng/cu m		
	Delft	The Hague	Vlaardingen	Amsterdam	Rotterdam
summer 1968	n.d.	–	1	–	2
1969	1	2	3	n.d.	3
1970	2	2	4	2	3
1971	2	2	n.d.	1	2
winter 1968	20	12	20	17	23
1969	17	8	15	12	18
1970	7	–	11	–	13
1971	6	2	14	7	6

(1277)

Cleveland (Ohio, USA): max. conc.: 98 ng/cu m (400 values)
   annual geom. mean (all sites): 1.96 ng/cu m
   annual geom. mean of TSP (all sites): 16 ppm (wt)   (556)
glc's in air in Ontario cities (Canada):

	April–June 1975		July–Sept. 1975		Oct.–Dec. 1975		Jan.–March 1976	
location	ng/1000 cu m air	µg/g p.m.	ng/1000 cu m air	µg/g p.m.	ng/1000 cu m air	µg/g p.m.	ng/1000 cu m air	µg/g p.m.
Toronto	5849	65.3	7131	74.8	10528	127.2	4413	56.7
Toronto	5077	67.1	3303	50.5	4693	77.3	9814	71.6
Hamilton	5809	39.9	7183	51.7	7532	108.7	6418	76.7
S. Sarnia	1038	16.8	1049	22.5	2700	51.7	1158	42.6
Sudbury	779	23.9	1104	26.3	2321	104.0	3009	128.7

(999)

glc's in the average American urban atmosphere–1963: 63 µg/g airborne particulates or 8 ng/g cu m air   (1293)
glc's in Budapest 1973: heavy traffic area: winter: 23.4 ng/cu m
   (6 a.m.–8 p.m.)   summer: 19.3 ng/cu m
                low traffic area: winter: 5.6 ng/cu m
                                summer: 9.5 ng/cu m   (1259)

C. WATER POLLUTION FACTORS:
  —Water, sediment and soil quality:
    in Thames river water: at Kew Bridge: 60 ng/l
                           at Albert Bridge: 110 ng/l
                           at Tower Bridge: 160 ng/l   (529)
    in rapid sand filter solids from Lake Constance water: 0.4–1.3 mg/kg
    in river water solids:
      river Rhine: 1.6 mg/kg
      river Aach at Stockach: 1.4–4.8 mg/kg
      river Argen: 0.4 mg/kg

1,2-BENZOPHENANTHRENE

river Schussen:	3.5 mg/kg	(531)
river Gersprenz at Munster (W. Germany):	1.6–12.9 ng/l	
river Danube at Ulm (W. Germany):	9.5 ng/l	
river Main at Seligenstadt (W. Germany):	8.0–84 ng/l	
river Aach at Stockach (W. Germany):	42–105 ng/l	
river Schussen (W. Germany):	46 ng/l	(530)

in groundwater (W. Germany, 1968): 0.7–6.4 μg/cu m ($n = 10$)
in tapwater (W. Germany, 1968): 0.8–3.2 μg/cu m ($n = 6$) (955)
in wells and galleries of an aquifer: Brussels sands (sands covered with a thick loamy layer): <0.1–0.6 ng/l (1066)
in sediment of Wilderness Lake, Colin Scott, Ontario (1976): 28 ppb (dry wt) (932)
in sediments in Severn estuary (U.K.): 0.2–1.5 ppm (dry wt) (1467)
in dried forest soil sample: South of Darmstadt (W. Germany): 10–70 ppb
near Lake Constance: 10–20 ppb (528)

—Manmade sources:
primary and digested raw sewage sludge: 0.1–0.31 ppm
liquors from sewage sludge heat treatment plants: 0.01–0.5 ppm
sludge cake from heat treatment plants: 0.09–0.44 ppm
final effluent of sewage work: 0.03 ppb (1426)
in domestic effluents: 0.040–0.219 ppb
in sewage (high percentage industry): 0.12–0.48 ppb
in sewage during dry weather: 0.004 ppb
in sewage during heavy rain: 3.84 ppb (531)

—Water treatment:

plant location	water source	river water conc. ng/l	drinking water conc. ng/l*	% removal/ transformation
Pittsburgh, Pa	Monongahela river	34.4	0.7	98.0**
Huntington, W.Va.	Ohio river	10.7	2.5	76.6
Philadelphia, Pa.	Delaware river	48.4	4.0	91.7

*treatment provided: lime, ferric salt, A.C., chlorination, fluoridation
**two stages A.C.: powdered and granular carbon (958)

concentration of various stages of water treatment works:

river intake:	0.072 μg/l	
after reservoir:	0.063 μg/l	
after filtration:	0.033 μg/l	
after chlorination:	0.009 μg/l	(434)

	Dec. 1965	May 1966	
raw sewage	8.7 ppb	0.26 ppb	
after mechanical purification	1.2 ppb	0.16 ppb	
after biological purification	0.03 ppb	0.12 ppb	(545)

1,2-benzophenanthrene *see* chrysene

benzo(c)phenanthrene

Manmade sources: in wood preservative sludge: 0.87 g/l of raw sludge (960)
B. AIR POLLUTION FACTORS:
—Manmade sources: in coke oven airborne emissions: 82.7–2,156.1 µg/g of sample
(960)
—Ambient air quality:
  glc's in Birkenes (Norway), Jan.–June 1977:
    avg.: 0.10 ng/cu m
    range: n.d.–0.92 ng/cu m (n = 18)
  glc's in Rørvik (Sweden), Dec. 1976–April 1977:
    avg.: 0.45 ng/cu m
    range: n.d.–5.09 ng/cu m (n = 21) (1236)

benzo(*def*)phenanthrene *see* pyrene

benzophenone-*o*-carboxylic acid *see o*-benzoylbenzoic acid

benzo(*a*)pyrene (3,4-benzopyrene; B(*a*)P)

Manufacturing source: coal tar processing; petroleum refining; shale refining; coal and coke processing kerosene processing; heat and power generation sources.
(347)
Natural sources: quantities synthesized by various bacteria:

species	µg of B(*a*)P produced per kg of dry bacterial biomass	
Mycobacterium smegmatis	60	
Proteus vulgaris	56	
Escherichia coli (strain 1)	50	
Escherichia coli (strain 2)	46	
Pseudomonas fluorescens	30	
Serratia marcescens	20	(931)
Synthesized by algae *Chlorella vulgaris*		(566)

Man caused sources (*air and water*): combustion of tobacco, combustion of fuels; present in run off containing greases, oils, etc.; potential roadbed and asphalt leachate (347)

264 BENZO(a)PYRENE

in gasoline: 0.135 mg/l; 0.143 mg/l; 0.133 mg/l; 8.28 mg/kg
                                                   (380; 385; 1052; 1070; 1220)
        0.09–0.47 mg/kg (n = 13); 0.21–1.00 mg/kg (n = 13)
in fresh motor-oil: 0.02–0.10 mg/kg   (379; 591; 1220)
in used motor oil: 5.8 mg/l   (591)
in used motor-oil after 5,000 km: 83.2–162.0 mg/kg
in used motor-oil after 10,000 km: 110.0–242.4 mg/kg   (1220)
in Kuwait crude oil: 2.8 ppm
in South Louisiana crude oil: 0.75 ppm   (1015)
in crude oils: Brega (Lybia):   1.32 mg/l
              Tia-Juana (Venezuela): 1.66 mg/l
              Safanya (Persian Gulf): 0.40 mg/l   (591)
in diesel oil (gasoil): 0.026 mg/l   (591)
in asphalt up to 0.0027 wt%
in coal tar pitch up to 1.25 wt%   (1380)

A. PROPERTIES: yellowish crystals m.w. 252.3; m.p. 179°C; b.p. 311°C at 10 mm; solub. 0.003 mg/l; in seawater at 22°C: 0.005–0.010 mg/l
B. AIR POLLUTION FACTORS:
   —Manmade sources:
     in cigarette smoke: 2.5 μg/100 cigarettes   (1298)
     emissions from typical European gasoline engine (1608 cu cm)—following European driving cycles—using leaded and unleaded commercial gasolines: 1.9–26.0 μg/l fuel burnt   (1291)
     in exhaust condensate of gasoline engine: 0.05–0.08 mg/l gasoline consumed
                                                                              (1070)
                                     340 μg/g   (1069)
     emission from combustion of fuel oil:
       large furnaces (>1000 hp): 0.13 mg/ton fuel
       small furnaces (<1000 hp): 10.6 mg/ton fuel   (518)
     emissions from space heating installations burning:
       coal (underfeed stoker):   10 mg/10^6 Btu input
       gasoil: 0.9 mg/10^6 Btu input
       gas: 0.2 mg/10^6 Btu input   (954)
     emission from combustion of natural gas:
       industrial boilers: 44 × 10^{-6} lb/10^6 cu ft of gas
       domestic and commercial heating units: 290 × 10^{-6} lbs/10^6 cu ft of gas (519)

source	g/cu m of emitted gas	
refuse burning	11.0	
power station, coal	0.3	
power station, gas	0.1	
automotive diesel	5.0	
coke oven volatiles	35.0	
home furnace, coal	100.0	(491; 1382)

in a typical coke oven emission (benzene solubles): 0.41 wt %   (1381)
emissions from asphalt hot-mixing plant: 3–20 ng/cu m
(in high volume particulate matter)   avg. 11 ng/cu m   (491; 1379)
in outlet of water spray tower of asphalt hot-road-mix process: <100 ng/cu m
in outlet of asphalt air-blowing process: <4,000 ng/cu m   (1292)

Pathways for the pyrosynthesis of benzo(a)pyrene (1287; 1288)

—Ambient air quality:
  comparison of glc's inside and outside houses at sub-urban sites:
    outside: 2.9 ng/cu m or 72.0 µg/g SPM
    inside: 2.1 ng/cu m or 41.0 µg/g SPM (1234)
  in the average American urban atmosphere—1963: 46 µg/g airborne particulates
    or 5.7 ng/g cu m air (1293)
  Cleveland (Ohio, USA): max. conc.: 126 ng/cu m (432 values)
    (1971/1972)   annual geom. mean for all sites: 1.04 ng/cu m
                        annual geom. mean of TSP for all sites: 8 ppm (wt) (556)
  organic fraction of suspended matter: (+ benzo(e)pyrene + perylene)
    Bolivia at 5200 m altitude (Sept.–Dec. 1975): 0.031–0.065 µg/1000 cu m
    Belgium, residential area (Jan.–April 1976): 3.0–23 µg/1000 cu m (428)
  in air of Ontario cities (Canada):

location	April–June 1975		July–Sept. 1975		Oct.–Dec. 1975		Jan.–March 1976	
	ng/1000 cu m air	µg/g p.m.	ng/1000 cu m air	µg/g p.m.	ng/1000 cu m air	µg/g p.m.	ng/1000 cu m air	µg/g p.m.
Toronto	789	8.8	1047	11.0	1674	20.2	720	9.2
Toronto	657	8.7	408	6.2	729	11.7	814	5.9

BENZO(a)PYRENE

Hamilton	1404	9.6	2351	16.9	3498	50.6	1934	23.1
S. Sarnia	338	5.5	114	2.4	596	11.4	190	7.0
Subury	175	5.4	111	2.6	342	15.3	444	19.0

		ng/cu m		ng/cu m
				(999)
Oslo	winter '56	15	summer '56	1
				(1278)
Copenhagen	winter '56	17	summer '56	5.4
				(1279)
London	winter '56	50	summer '56	12
				(1280)
Liege	winteraverage '58/62	113	summeraverage '58/62	14.5
				(1281)
Liverpool	24.10.'56–10.4'57	166	summer '56	33
				(1282)
Hamburg	winteraverage '61/63	183	summeraverage '61/63	17
				(1283)
Milan	winter '58	198	summer '59	1
				(1285)
Paris (rue de Dantzig)	winter '58	300–500 (results of 48 measurements)	summer '58	(1286)
Düsseldorf	winter '63	75	summer '63	3.5
				(1284)
Detroit (Michigan)	winter '59	31	summer '59	3.4
				(1284)
Chatanooga (Tennessee)	winter '59	31	average over the year '60	8.3
				(1284)
Hammond (Indiana)	winter '59	39	average over the year '60	3.9
				(1284)
Charlotte (N. Carolina)	winter '59	39	average over the year '60	5.7
				(1284)
Richwood (Virginia)	winter '59	45		(1284)
St. Louis (Missouri)	winter '59	54	average over the year '60	5.3
				(1284)
Nashville (Tennessee)	winter '59	55	summer '59	1.4
				(1284)
Altoona (Pennsylvania)	winter '59	61		(1284)
Birmingham (Alabama)	winter '59	74	summer '59	1.6
				(1284)

Cleveland (average):	1959:	24 ng/cu m	(557)
Cleveland (NASN site):	1966:	3.2 ng/cu m	(558; 559)
Cleveland (site 4):	1972:	0.6 ng/cu m	(556)
Urban sites (maximum):	1966:	11.2 ng/cu m	(558; 559)
Urban sites (maximum):	1972:	16.2 ng/cu m	(556)
Urban sites (average):	1966:	2.8 ng/cu m	(558; 559)
Urban sites (average):	1971/1972:	0.9 ng/cu m	(556)
Cleveland (16 sites):	1971/1972:	0.87 ng/cu m	(556)
Los Angeles (4 sites):	1971/1972:	1.3 ng/cu m	(560)

Lyon (1 site): 1972: 1.05 ng/cu m (561)
Rome (1 site): 1970/1971: 1.3 ng/cu m (562)
Budapest (1 site): 1971/1972: 2.68 ng/cu m (563)

Period October 1972–May 1973: 14 sites in Belgium

location	geom. mean ng/cu m	95% ile ng/cu m	no. of values
Antwerp—residential area	25.1	104	28
Antwerp—industrial area	12.9	66	29
Gent—residential area	20.7	130	28
Gent—industrial area	12.4	107	31
Brussels	35.4	144	31
Charleroi	63.8	238	30
Liège—residential area	26.7	98	31
Liège—industrial area	38.3	182	30
Ploegsteert—agricultural area	16.5	121	28
Mechelen	16.3	131	26
Houffalize—woodland	5.9	27	27
Zeebrugge	9.4	77	28
Dourbes—woodland	9.2	48	31

(565)

glc's in residential area Belgium—october 1976:
  (BaP + BeP + perylene)   in particulate sample: 20.1 ng/cu m
                           in gas phase sample: 2.69 ng/cu m (1289)
glc's at Botrange (Belgium)—June–July 1977: woodland at 20–30 km from industrial area: 1.01; 1.22 ng/cu m ($n = 2$) (+(e)isomer)
  at Wilrijk (Belgium)—October–December 1976: residential area: 12.2; 62.5 ng/cu m (n = 2) (+(e)isomer) (1233)
glc's in Birkenes (Norway), Jan–June 1977:
    avg.: 0.32 ng/cu m
    range: n.d.–2.35 ng/cu m ($n = 18$)
  in Rørvik (Sweden), Dec. 1976–April 1977:
    avg.: 0.54 ng/cu m
    range: n.d.–4.32 ng/cu m (n = 21) (1236)
  in Budapest 1973: heavy traffic area: winter: 29.7 ng/cu m
  (6–20 hr)   summer: 31.2 ng/cu m
                    low traffic area: winter: 16.9 ng/cu m
                                        summer: 20.1 ng/cu m (1259)

ground level concentrations (average values):
ground level concentrations (average values):

country	measuring point	ng BaP/m^3	year	ref.
E. Germany	East Berlin (street crossings)	4–50	1970	(1247)
Poland	Warsaw	29	1966–67	(1248)
Poland	Zabrze	117	1972	(1248)
W. Germany	Düsseldorf	6–37	1972	(1249)
W. Germany	Cologne	2–188	1970–73	(1250)
Hungary	Budapest	74	1966	(1251)

BENZO(a)PYRENE

country	measuring point					ref.
USSR	Leningrad (heavy traffic)	19–64		1967		(1253)
W. Germany	West Berlin	8		1966–67		(1254)
USA	Washington	9		1966		(1254)
USA	32 measuring points	3		1966–70		(1260)
USA	Cincinnati	6–9		1973		(1254)
Japan	Osaka	50		1967		(1254)

glc's in ng/cu m

country	measuring point	spring	summer	autumn	winter	year	ref.
W. Germany	Hamburg	14–72	1–26	66–296	94–388	1964	(1255)
Great Britain	London						
	(heavy traffic)	20	11	57	68	1965	
	(low traffic)	11	1	38	42	1965	
CSSR	Prague	–	13–36	–	53–145	1963	(1255)
Hungary	Budapest	–	17–32	–	72–141	1963	(1255)
Hungary	Budapest (heavy traffic load)	–	31.2	–	29.7	1972	(1259)
Hungary	Budapest	–	20.1	–	16.9	1972	(1259)
The Netherlands	Rotterdam	–	2	–	23	1971	(1256)
W. Germany	Bonn	–	4.4	–	133	1965	(1257)
Switzerland	Zurich						
	(square with a heavy load of traffic)	–	2.4	–	13	1971	(1258)
	(dwelling area in the centre of the city)	–	2.1	–	12.3	1971	(1258)
	(dwelling area on the outskirts)	–	1.2	–	8.1	1971	(1258)
USA	New York						
	(business area)	0.5–1.8	0.7–3.9	1.5–6	0.5–94		(1255)
	(motorway)	0.1–0.8	0.1–0.7	3.3–3.5	0.7–1.3		(1255)
	(residential area)	0.1–0.6	0.1–0.3	0.6–0.8	0.5–0.7		(1255)
Australia	Sidney	0.6–2.4	0.6–1.8	2.5–7.4	3.8–8.2	1965	(1255)
USA	Detroit						
	(business)	7.2	–	–	0.5–17	1965	(1255)
	(motorway)	–	4–6	3.4–7.3	9.2–13.7		(1255)
	(residential area)	–	2	–	0.9–1.8		(1255)

glc's in East Berlin (ng/cu m):
  points of dense traffic (summer):   11.5 ($n = 26$)
  point of low traffic (summer):   4.5 ($n = 26$)
  points of dense traffic (winter):   24.6 ($n = 24$)

point of low traffic (winter):
   district heating: 7.3 (n = 15)
   individual household heating: 21.1 (n = 10)
   point of dense traffic (all values): 21.4 (n = 59)
   point of low traffic (all values): 8.6 (n = 51) (1246)

Benzo(a)pyrene content of urban air:

city		spring	summer	fall	winter	ref.
			Benzo(a)pyrene, ng/cu m			
New York						
commercial		0.5–8.1	0.7–3.9	1.5–6.0	0.5–9.4	
freeway		0.1–0.8	0.1–0.7	3.3–3.5	0.7–1.3	
residential		0.1–0.6	0.1–0.3	0.6–0.8	0.5–0.7	(1263)
Detroit:						
central sites:	commercial	7.2	–	–	5.0–17.0	
	freeway	–	4.0–6.0	3.4–7.3	9.2–13.7	
residential		–	0.2	...	0.9–1.8	(1264)
Atlanta		2.1–3.6	1.6–4.0	12–15	2.1–9.9	
Birmingham		6.3–18	6.1–10	20–74	23–34	
Cincinnati		2.0–2.1	1.3–3.9	14–18	18–26	
Detroit		3.4–12	4.1–6.0	18–20	16–31	
Los Angeles		0.4–0.8	0.4–1.2	1.2–13	1.1–6.6	
Nashville		2.1–9.0	1.4–6.6	30–55	26	
New Orleans		2.6–5.6	2.0–4.1	3.6–3.9	2.6–6.0	
Philadelphia		2.5–3.4	3.5–19	7.1–12	6.4–8.8	
San Francisco		0.8–0.9	0.2–1.1	3.0–7.5	1.3–2.4	(1265)
Pittsburgh		...	0–23	2.9–37	8.2	(1266)
Hamburg, Germany		14–72	10–26	66–296	94–388	(1267)
London, England		25–48	12–21	44–122	95–147	
Sheffield, England		20–44	21–33	56–63	64–78	
Cannock, England		4–16	6–11	27–31	27–32	(1268)
London (in traffic)		20	11	57	68	
London (background)		11	1	38	42	(1269)
Milan, Italy		12	3	25	150	(1270)
Copenhagen, Denmark		6	5	14	15	(1271)
Prague, Czechloslovakia		–	13–36	–	53–145	(1272)
Budapest, Hungary		–	17–32	–	72–141	(1273)
Pretoria, South Africa		–	10	–	22–28	
Johannesburg, South Africa		–	–	–	22–49	
Durban, South Africa		–	–	–	5–28	(1274)
Osaka, Japan						
commercial		5.7	2.7	9.4	14	
residential		3.3	1.4	3.8	6.7	(1275)
Sidney, Australia		0.6–2.4	0.6–1.8	2.5–7.4	3.8–8.2	(1276)

glc's in The Netherlands (ng/cu m):

	Delft	The Hague	Vlaardingen	Amsterdam	Rotterdam
summer 1968	n.d.	–	3	–	3
1969	3	4	4	2	1

270 BENZO(a)PYRENE

1970	1	4	5	2	3
1971	3	n.d.	9	n.d.	2
winter 1968	20	23	35	22	15
1969	18	12	32	18	23
1970	12	–	13	5	19
1971	6	13	16	8	23

(1277)

C. WATER POLLUTION FACTORS:
  —Biodegradation:

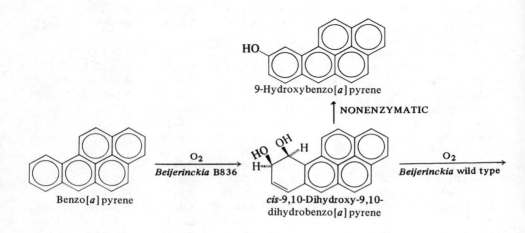

oxidation for benzo(a)pyrene by *Beijerinckia* B836 (425)

Average degradation by soil bacteria after 8 days culture:

	amount of extracted B(a)P μg	amount of B(a)P destroyed (%)
soil not inoculated with bacteria (control)	191	0
soil + N 5 bacterial strain	90	53
soil + N 13 bacterial strain	61	66
soil + N 13 bacterial strain*	33	82

*Before the experiment this strain was cultured in a medium containing B(a)P for 110 days (544)

Microbial degradation to CO_2 in seawater at $12°C$ in the dark after 48 hr incubation at 16 μg/l (n.s.i.): 0 μg/l/day; after addition of water extract of fuel oil 2, after 24 hr incubation: 0.01 μg/l/day—turnover time: 1400 days (477)

Biodegradation to CO_2 in estuarine water:

concentration (μg/l)	month	incubation time (hr)	degradation rate (μg/l/day) $\times 10^3$	turnover time (days)
5	Jan.	24	0	–
5	June	24	0	–
5	May	96	2 ± 1	3500

(381)

degradation in seawater by oil oxidizing micro-organisms (in presence of 0.365 mg/l pyrene and 0.35 mg/l fluorene at 10°C): initial conc. 190 μg/l; after 12 days: 90 μg/l: 53% decrease (1237)

transformation by soil micro-organisms: (934)

		% transformed		
strain	2 days	3 days	4 days	8 days
N - 13	13	43	53	69
N - 2/II	11	23	33	–
N - 5	0	0	0	55
N - 2/I	0	0	0	–
N - 13 cultivated on B(a)P				84

—Manmade sources:

B(a)P levels in industrial effluents:

industry	conc. ppb	
shale oil: after dephenolization	2; 312	(536; 537)
coke by-products: after biochemical treatment	12–16	(538)
gasworks: after filtration through coke bed	20	(539)
after dephenolization	130–290	(540)
oil refineries: coking residue after direct distillation of oil	0.48–5.0	(541)
catalytic cracking of:		
kerosene	0.14	
gasoil fraction at 450°C	0.11–0.19	
	0.05–0.29	
catalytic cracking	0.07–0.11	
thermal cracking 700–800°C	0.09–0.23	
	0.10–3.0	
pyrolysis of ethane–ethylene fractions, 700–800°C	3.6	(541)
after settling ponds of various refineries	up to 0.22	(542)

in wood preservative sludge: 3.59 g/l of raw sludge (960)
in domestic effluent: 0.038–0.074 ppb
in sewage (high percentage industry): 0.1–0.368 ppb
in sewage during dry weather: 0.001 ppb
in sewage during heavy rain: 1.84 ppb
primary and digested raw sewage sludge: 0.27–0.57 ppm
liquors from sewage sludge heat treatment plants: 0.03–0.84 ppm
sludge cake from heat treatment plants: 0.31–0.52 ppm
final effluent of sewage work: 0.03 ppb (1426)

BENZO(a)PYRENE

—Aquatic reactions: evaporation:

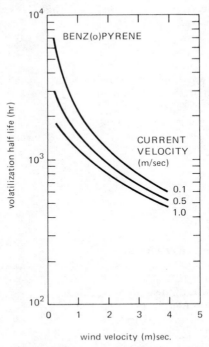

Variation in predicted volatilization rates of benzo(a)pyrene at 25°C under varying conditions of wind and current velocity in a stream 1.0 m in depth (1391)

—Adsorption:
  in estuarine water: at 3 µg/l, 71% adsorbed on particles after 3 hr (381)
  after 3 hr incubation in natural seawater, 75% of 2 µg/l were taken up by suspended aggregates of dead phytoplankton cells and bacteria (957)
—Water, sediment and soil quality:
  in Thames River Water: at Kew Bridge: 130 ng/l
                       at Albert Bridge: 160 ng/l
                       at Tower Bridge: 350 ng/l (529)
  River Gersprenz at Münster (W. Germany): 9.6 ng/l
  River Danube at Ulm (W. Germany): 0.6 ng/l
  River Main at Seligenstadt (W. Germany): 1.1–43 ng/l
  River Aach at Stockach (W. Germany): 4–16 ng/l
  River Schussen (W. Germany): 10 ng/l (530)
  in groundwater (W. Germany—1968): 0.4–3.8 µg/cu m ($n = 10$)
  in tapwater (W. Germany—1968): 0.5–4.0 µg/cu m ($n = 6$) (955)
  in wells and galeries of an aquifer: Brussels sands (sands covered with a thick loamy layer): <0.1–0.4 ng/l (1066)
  In rapid sand filter solids from Lake Constance Water: 0.05–0.2 mg/kg

In river water solids: river Rhine: 0.3 mg/kg
river Aach at Stockach: 0.5–2.0 mg/kg
river Argen: 0.1 mg/kg
river Schussen: 0.4 mg/kg (531)

Typical levels in marine sediments

location	depth m	Conc. ppb dry wt	
Greenland	0.2	5	(532)
Italy (Bay of Naples)	15–45	1000–3000	
Highly industrialized areas	13	7.5	
	2–65	10–530	
Near volcanic pollution	55	260–960	
	120	1.4	
Island affected by pollution	–	100–560	(533)
French Mediterranean coast	14	400	
	16	1500	
	48	75	
	58	traces	
	82	400	
estuary	–	34	
	1	20	
	4	15	
	5	25	
	102	not detected	(534)

in sediment of Wilderness Lake, Colin Scott, Ontario (1976): 13 ppb (dry wt)
(932)

in forest soil: South of Darmstadt (W. Germnay):
1.5–4.0 ppb of dried soil sample
near Lake Constance:
1.5–2.5 ppb of dried soil sample (528)

average content in soil (1966):
at oil refinery: 191,000 γ/kg soil
district of old Moscow: 268–346 γ/kg soil
district of new housing: 104 γ/kg soil
settlement near Moscow: 81 γ/kg soil
field near Moscow: 79 γ/kg soil
Kljasma water storage reservoir: 0 γ/kg soil (934)

"normal" level in plants: 10 μg/kg

in plant-bearing soil approximately 10 μg/kg is found; 1.0 μg/cu m is dissolved in the water (566)

—Water treatment methods:

plant location	water source	river water conc. μg/l	drinking water conc. μg/l*	% removal/ transformation
Pittsburgh, Pa.	Monongahela river	42.1	0.4	99.0**
Huntington, W. Va.	Ohio river	5.6	0.5	91.1
Philadelphia, Pa.	Delaware river	41.1	0.3	99.3

*treatment provided: lime, ferric salt, A.C., chlorination, fluoridation
**two stages A.C.: powdered and granular carbon (958)

chlorination: 6 mg/l chlorine for 6 hr: initial conc.: 53.14 ppb: 98% reduction
(549)
oxidation by ClO_2 produces dichloro-3,4-benzopyrenes (590)
Effect of chlorination on levels of B(a)P in water:

initial chlorine dose (mg/l)	initial B(a)P conc. (ppb)	time (hr)	% reduction	
0.3	5	3	50	(547)
0.5	5	2	50	
		22	20	
0.5	2	2	50	
		13	100	(547)
0.5	1	0.5	81	(548)
		2	94	
0.3	1	0.5	82	
		2	92	(548)

ozonation: after 1 min. contact time with ozone: residual amount: 39% (550)

	Dec. 1965	May 1966	
raw sewage	34.5 ppb	0.43 ppb	
after mechanical purification	1.4 ppb	0.15 ppb	
after biological purification	0.02 ppb	0.07 ppb	(545)

concentration at various stages of water treatment works (+ benzo(e)pyrene):
  after reservoir: 0.051 µg/l
  after filtration: 0.030 µg/l
  after chlorination: 0.009 µg/l (434)
—Sampling and analysis: recovery with open pore polyurethane (OH/NCO = 2.2): 95 ± 6%—quantitative elution with methanol (929)

D. BIOLOGICAL EFFECTS:
  —Residues
Content of foodstuffs:

	benzo(a)pyrene ppb	Ref.
Meat		
Raw		
hamburger	ND(0.1)[a]	(1022)
Cooked		
sausage	0.11, 0.15	(1023; 1024)
hamburger	2.5	(1022)
Smoked		
ham	0.17, 0.7, 0.7, 3.2	(1023; 1025; 1026)
mean of 47 products	0.225	(1023)
fat bacon	0.22	(1023)
chicken	0.7	(1026)
Barbecued		
ribs	6–10	(1027)
beef in sause	3.3	(1026)
pork	5.0	(1026)

Charcoal Broiled		
beefsteak	8, 0.6–0.8	(1027; 1028)
bratwurst	8–12	(1028)
pork loin	4–8	(1028)
Vegetable oils		
margarine	0.4, 0.6, 0.4–3.7	(1038; 1039)
vegetable	ND(0.2)	49 (1038)
sunflower	10.6. 5.2, ND(3)	(1040; 1038)
peanut	4, 0.6, 1.9	(1025; 1040; 1041)
peanut (fried)	3	(1041)
soybean	1.4, 1.7	(1025; 1040)
cottonseed	0.4, 1.4	(1025; 1040)
Miscellaneous		
butter	ND(0.2, 0.5)	(1038)
coffee	0.2–0.3, 0.2–6.0	(1042; 1043)
instant coffee	0.02–0.06	(1044)
coffee substitute	0.1–0.3	(1042)
coffee beans (aqueous infusions)	0.003	(1044)
tea	0.0–16[b]	(1045)
barley, rye, wheat	0.2, 0.2, 0.3, 0.5, 0.3, 1.1, 0.2, 0.3	(1046; 1047)
Fish		
Smoked		
herring	2.6, 0.3, 1.3	(1023)
whitefish	4.3, 6.9	(1026)
mackerel	1.0, 0.53, 0.76	(1029; 1023)
Vegetables		
lettuce	12, 2.9–12.8, 0.2–1.3, 0.7, 1.1 150[c]	(1030; 1031; 1032) (1022; 1033)
leek	12.6–24.5	(1031)
spinach	3.3, 20, 50	(1031; 1034)
cabbage	7.4	(1031)
endive	6.7	(1030)
carrots, peeled	0.07–0.14	(1032)
Fruits		
pruned (dried edible portion)	0–6.2	(1035)
citric acid (7 brands)	ND(0.1)	(1036)
apples, bananas, pineapple, mandarin oranges	mean = 0.03	(1037)

[a] Not detected at the concentration shown in the parentheses.
[b] <3% is extracted by hot water.
[c] Dry weight basis.

marine tissue	source	benzo(a)pyrene ppb	
mussel	Toulon Roads, France	0.2–3.0	(1016)
mussel	Falmouth, Mass.	ND, 0.5	(1017)
mussel	Vancouver, B.C., Canada		

	Remote area	0.0–0.2	
	Outer area	2.0±0.3	
	Around wharf, marina, docks	18	
	Inner harbor	42	(1018)
oyster	Norfolk Harbor, Va.	2–6	(1019)
oyster	Long Island Sound	2	(1017)
oyster	Chincoteague, Va.	0.2	(1017)
oyster	French Coast	0.1–7.0	(1020)
codfish	West Coventry, Greenland	1.5	(1021)
codfish	Atlantic Ocean, off N.J.	ND	(1017)
clam	Chincoteague, Va.	0.3	(1017)
clam	Darien, Conn.: Scott's Cove	ND	(1017)
clam	Fish market, Linden, N.J.	ND	(1017)
crab	Chesapeake Bay	ND	(1017)
crab	Raritan Bay	3	(1017)
menhaden	Raritan Bay	1.5	(1017)
shrimp	Palacios, Texas	ND	(1017)
lake trout	Lake Maskinonge, Ont., Canada	<1	(1017)

typical B(a)P levels in marine fauna and flora:

plankton (Greenland):	5 ppb	(532)
(Italy):	6–21 ppb	(533)
(French Channel Coast):	400 ppb	(535)
algae (sample at 40 m) (Greenland):	60 ppb	(532)
(sample at seabed) (Greenland):	60 ppb	(532)
(Italy):	2 ppb	(533)
Mollusc (Greenland):	60 ppb	
Mollusc: shell (Greenland):	18 ppb	
body (Greenland):	55 ppb	(532)
Mollusc: shell (Italy):	11 ppb	
body (Italy):	130; 540 ppb	(533)
Mollusc (Italy):	2 ppb	
Sardine (Italy):	65 ppb	(533)

mussel (*Mytilus edulis*): conc. in tissue: 0.05 µg/g after chronic pollution, depuration half life: 16 days (580)

in *Mytilus edulis* from 13 sites in Yaguina Bay (Oregon) 1977–1978:

10%: 0.4 µg/kg
50%: 3.2 µg/kg
90%: 15 µg/kg (n = 150) (1671)

—Bioaccumulation:

uptake and depuration by Oysters (*Crassostrea virginica*) from oil treated enclosure:

time of exposure days	depuration time days	concentration oysters µg/g	concentration water µg/l	accumulation factor oysters/water
2	–	0.36	1.9	190
8	–	0.30	0.1	3000
2	7	0.40	–	–

8	7	0.20	—	—
8	23	0.12	—	—

half-life for depuration: 18 days (957)

—Toxicity:

Neanthes arenaceodentata: 96 hr TL_m in seawater at 22°C: >1 ppm (initial conc. in static assay) (995)

atlantic salmon (*Salmo salar*) eggs: 168 hr BCF 70.7 (static test) (1507)

—Carcinogenicity: +

—Mutagenicity in the Salmonella test: +

    121 revertant colonies/nmol

    2398 revertant colonies at 5 µg/plate (1883)

mutagenicity: induced significant mutation to 8-azaguanine resistance in *S. typhimurium* at concentrations as low as 4 µM (1723)

benzo(e)pyrene (1,2-benzpyrene; B(e)P)

Manmade sources:

in gasoline: low octane number:	0.18–0.87 mg/kg (n = 13)	(385)
high octane number:	0.45–1.82 mg/kg (n = 13)	(385)
	6.6 mg/kg	(1220)

in gasoline: 0.307 mg/l; 0.137 mg/l; 0.320 mg/l (380; 1052; 1070)

in bitumen: 1.62–6.56 ppm (506)

in lubricating motor oils: 0.07–0.49 ppm (6 samples) (379; 1220)

in used motor oil after 5,000 km: 92.2–182.0 mg/kg

in used motor oil after 10,000 km: 140.9–278.4 mg/kg (1220)

in asphalts: up to 0.0052 wt %

in coal tar pitches: up to 0.70 wt % (1380)

in Kuwait crude oil: 0.5 ppm

in South Louisiana crude oil: 2.5 ppm (1015)

A. PROPERTIES: m.w. 252; solub. 0.004 mg/l at 25°C

B. AIR POLLUTION FACTORS:

    —Manmade sources:

      in cigarette smoke: 0.3 µg/100 cigarettes (1298)

      in exhaust condensate of gasoline engine: 0.037–0.059 mg/l gasoline consumed (1070)

    emissions from typical European gasoline engine (1608 cu cm)—following European driving cycles—using leaded and unleaded commercial gasolines: 3.6–38.9 µg/l fuel burnt (1291)

    emissions from space heating installations burning:

      coal (underfeed stoker): 7.9 mg/10^6 Btu input

gas: 0.49 mg/10^6 Btu input (954)
in coke oven airborne emissions: 103.86 µg/g of sample (960)
emissions from asphalt hot-mixing plant: 14–40 ng/cu m
  (in high volume particulate matter): avg. 26 ng/cu m (1379)
—Ambient air quality:
Cleveland (16 sites) 1971/72:   1.2 ng/cu m (556)
Los Angeles (4 sites) 1971/72:   2.0 ng/cu m (560)
Lyon (1 site) 1972:   1.8 ng/cu m (561)
Rome (1 site) 1970/71:   0.9 ng/cu m (562)
Budapest (1 site) 1971/72:   1.52 ng/cu m (563)
glc's in The Netherlands (ng/cu m):

		Delft	The Hague	Vlaardingen	Amsterdam	Rotterdam	
summer	1968	n.d.	–	3	–	4	
	1969	2	2	4	–	2	
	1970	2	1	5	2	3	
	1971	5	2	6	1	3	
winter	1968	28	14	31	21	25	
	1969	23	23	31	17	24	
	1970	13	–	22	10	14	
	1971	5	9	20	6	14	(1277)

glc's in Birkenes (Norway), Jan.–June 1977:
  avg.: 0.38 ng/cu m
  range: n.d.–2.81 ng/cu m ($n = 18$)
in Rørvik (Sweden), Dec. '76–April '77:
  avg. 0.51 ng/cu m
  range: n.d.–3.30 ng/cu m ($n = 21$) (1236)
in Budapest 1973: heavy traffic area: winter: 8.4 ng/cu m
  (6–20 hr)   summer: 9.7 ng/cu m
  low traffic area: winter: 19.2 ng/cu m
  summer: 4.7 ng/cu m (1259)
in the average American urban atmosphere, 1963: 42 µg/g airborne particulates or 5 ng/g cu m air (1293)
comparison of glc's inside and outside houses at suburban sites:
  outside: 2.19 ng/cu m or 46.5 µg/g SPM
  inside: 2.28 ng/cu m or 40.8 µg/g SPM (1234)
in air of Ontario cities (Canada):

| | april–june 1975 | | july–sept. 1975 | | oct.–dec. 1975 | | jan.–march 1976 | |
| | ng/1000 | µg/g | ng/1000 | µg/g | ng/1000 | µg/g | ng/1000 | µg/g |
location	cu m air	p.m.	cu m air	p.m.	cu m air	p.m.	cu m air	p.m.
Toronto	440	4.9	519	5.4	1294	15.6	781	10.0
Toronto	478	6.3	375	5.7	400	6.6	791	5.8
Hamilton	606	4.2	1407	10.1	3771	54.4	1607	19.2
S. Sarnia	118	1.9	52	1.1	603	11.5	64	2.4
Sudbury	23	0.7	45	1.1	255	11.4	317	13.6

(999)

Cleveland (Ohio, USA): max. conc.: 78 ng/cu m (464 values)
  (1971/1972)   annual geom. mean (all sites): 1.45 ng/cu m

annual geom. mean of TSP (all sites): 11 ppm (wt)
(556)
C. WATER POLLUTION FACTORS:
—Manmade sources: in wood preservative sludge: 2.48 g/l of raw sludge (960)
—Sediment quality: in sediment of Wilderness Lake - Colin Scott - Ontario (1976): 28 ppb (dry wt) (932)
D. BIOLOGICAL EFFECTS:
—Carcinogenicity: weakly carcinogenic
—Mutagenicity in the *Salmonella* test: +
>60 revertant colonies/nmol
143 revertant colonies at 60 µg/plate (1883)

benzo(b)pyridine *see* quinoline

1,2-benzopyrone *see* coumarin

benzo(b)pyrrole *see* indole

benzoquinhydrone *see* quinhydrone

benzo(f)quinoline (5,6-benzoquinoline)

A. PROPERTIES: m.w. 179.22; m.p. 90–93°C; b.p. 349°C/721 mm
—Manmade sources: in wood preservative sludge: 7.11 g/l of raw sludge (n.s.i.)
(993)
B. AIR POLLUTION FACTORS:
—Manmade sources: in coke oven emissions: 8.3–79.8 µg/g of sample (n.s.i.) (993)
—Ambient air quality: in the average American urban atmosphere—1963: 2 µg/g airborne particulates or 0.2 ng/cu m air (1293)

benzo(h)quinoline (7,8-benzoquinoline)

A. PROPERTIES: m.w. 179.22; m.p. 49–51°C; b.p. 338°C/719 mm
B. AIR POLLUTION FACTORS:
—Ambient air quality: in the average American urban atmosphere—1963: 3 µg/g airborne particulates or 0.3 ng/cu m air (1293)

p-benzoquinone (2,5-cyclohexadien-1,4-dione; 1,4-benzoquinone; quinone)

Use: manufacture of dyes and hydroquinone

A. PROPERTIES: yellow crystals; m.w. 108.09; m.p. 115/124°C decomposes; b.p. sublimes; v.p. 0.09 mm at 20°C; sp.gr. 1.318 at 20°C; log P_{oct} 0.20
B. AIR POLLUTION FACTORS: 1 mg/cu m = 0.226 ppm, 1 ppm = 4.49 mg/cu m
   —Odor: characteristic: pungent
       absolute perception limit: 0.1 ppm
       100% recogn.: 0.15 ppm (54)
       O.I. at 20°C: 790 (316)
C. WATER POLLUTION FACTORS:
   —Reduction of amenities: taste: average: 0.71 mg/l
                                     range: 0.016 to 4.3 mg/l (30)
     tainting of fish flesh: 0.5 mg/l (81)
   —Impact on biodegradation processes:
       at: 0.2 mg/l inhibition of degradation of glucose by *Pseudom. fluorescens*
       at: 55 mg/l inhibition of degradation of glucose by *E. coli* (293)
D. BIOLOGICAL EFFECTS:
   —Bacteria: *E. coli*: toxic: 55 mg/l (30)
   —Algae: blue algae: toxic: <1 mg/l
       Scenedesmus: toxic: 6 mg/l (30)
   —Inhibition of photosynthesis of a fresh water non-axenic uni-algal culture of *Selenastrum capricornutum*:
       at   0.1 mg/l: 37%   carbon-14 fixation (vs. controls)
       at   1 mg/l: 17%   carbon-14 fixation (vs. controls)
       at   10 mg/l: 7–13% carbon-14 fixation (vs. controls)
       at 100 mg/l: 1%   carbon-14 fixation (vs. controls)
       at 1000 mg/l: 2%   carbon-14 fixation (vs. controls) (1690)
   —Arthropoda:
     Daphnia: toxic: 0.4 mg/l
   —Fish: fathead minnows (*Pimephales promelas*): probable toxic conc.: <0.1 mg/l (after 120 hr) (935)
   —Mammalia: rat: oral LD_{50}: 130–296 mg/kg
               rat: parenteral LD_{50}: 25 mg/kg
               rat: inhalation LC_{100}: 230–270 mg/cu m, 2 hr
               mouse: parenteral LD_{50}: 94 mg/kg
               mouse: inhalation: LC_{50}: 250 mg/cu m, 2 hr
               mouse: inhalation: LC_{100}: 320–340 mg/cu m, 2 hr (54)
   —Man: eye irritation from 0.1 ppm (54)

N-(2-benzothiazolyl)-N'-methylurea (benzthiazuron; gatnon)

$$CH_3-NH-\overset{\overset{O}{\|}}{C}-NH-\underset{S}{\overset{N}{\diagdown}}\bigcirc$$

A. PROPERTIES: white powder which decomposes at 287°C; solub. 12 ppm at 20°C; v.p. 1×10^{-5} mm at 90°C

D. BIOLOGICAL EFFECTS:
—Fish: harlequin fish (*Rasbora hetermorpha*):

mg/l	24 hr	48 hr	96 hr	3 m	
$LC_{10}(F)$	850	700	200		
$LC_{50}(F)$	1300	920	400	100	(331)

—Mammals:
acute oral LD_{50} (rats): 1280 mg/kg
dermal applications of 500 mg/kg gave no symptoms
in diet: all rats tested in 60 days feeding tests at 130 mg/kg survived (1855)

benzotrichloride (phenylchloroform; toluenetrichloride; benzenyltrichloride; benzoic trichloride; benzyltrichloride)

$$\bigcirc-CCl_3$$

A. PROPERTIES: colorless to yellowish liquid; m.p. -5°C; b.p. 213–214°C; sp.gr. 1.38
B. AIR POLLUTION FACTORS:
—Odor: penetrating odor (187)
D. BIOLOGICAL EFFECTS:
—Toxicity threshold (cell multiplication inhibition test):
  bacteria (*Pseudomonas putida*): >100 mg/l
  green algae (*Scenedesmus quadricauda*): >100 mg/l
  protozoa (*Entosiphon sulcatum*): 56 mg/l (1900)
  protozoa (*Uronema parduczi Chatton-Lwoff*): >80 mg/l (1901)

benzoylaminoacetic acid see hippuric acid

o-benzoylbenzoic acid (benzophenone-*o*-carboxylic acid)

$$\bigcirc-\overset{C=O}{\underset{\bigcirc-COOH}{}}$$

A. PROPERTIES: m.w. 226.22; m.p. 93°C+H_2O, 127°C anh.
C. WATER POLLUTION FACTORS:
—BOD_5: 0.001 std. dil. sewage (n.s.i.) (258)(165)

D. BIOLOGICAL EFFECTS:
—Mammals: mouse: daily I.P. dose: tolerated 31 mg/kg (211)

benzoylchloride (benzenecarbonylchloride)

C₆H₅—C(=O)Cl

A. PROPERTIES: m.w. 140.57; m.p. -1°C; b.p. 197°C; v.p. 0.4 mm at 20°C, 1 mm at 32°C; v.d. 4.88; sp.gr. 1.22 at 15/15°C; solub. decomposes; sat.conc. 3.1 g/cu m at 20°C, 5.2 g/cu m at 30°C
B. AIR POLLUTION FACTORS: 1 mg/cu m = 0.17 ppm, 1 ppm = 5.84 mg/cu m
—Sampling and analysis: photometry: min.full scale: 61 ppm (53)
D. BIOLOGICAL EFFECTS:
—Fish:

	24 hr LC_{50}	48 hr LC_{50}	96 hr LC_{50}
fathead minnows (*Pimephales promelas*)	43 mg/l	35 mg/l	35 mg/l
grass shrimp (*Palaemonetes pugio*)	—	—	180 mg/l

(1904)

benzoylglycin *see* hippuric acid

benzoylglycocoll *see* hippuric acid

benzthiazuron *see* N-(2-benzothiazolyl)-N'-methylurea

benzylalcohol (phenylcarbinol; α-hydroxytoluene; phenylmethanol)

Uses: perfumes and flavors; solvent; intermediate; inks; surfactant
A. PROPERTIES: colorless liquid; m.w. 108.13; m.p. -15°C; b.p. 206°C; v.p. 1 mm at 58°C; v.d. 3.72; sp.gr. 1.05 at 15/15°C; solub. 40,000 mg/l at 17°C, 35,000 mg/l at 20°C; log P_{oct} 1.10
B. AIR POLLUTION FACTORS: 1 mg/cu m = 0.226 ppm, 1 ppm = 4.42 mg/cu m
—Manmade sources: in gasoline exhaust: <0.1 to 0.7 ppm (1053; 195)
C. WATER POLLUTION FACTORS
—BOD_5: 1.55; 156 (1828; 30)
    1.6 std. dil. (27)
    33% of ThOD (220)
    62% of ThOD (274)
    1.55 std.dil.sew. (282)

—BOD_{20}: 1.95 (30)
—COD: 96% of ThOD (220)
        95% of ThOD (0.05 n $K_2Cr_2O_7$) (274)
        Reflux COD = 98.4% recovery
        Rapid COD = 94.8% recovery (1828)
—$KMnO_4$ value: 0.381 (30)
           acid 11% ThOD; alkaline: 8% ThOD (220)
           4% of ThOD (0.01 n $KMnO_4$) (274)
—TOC: 100% of ThOD (220)
—ThOD: 2.519 (30)
—Impact on biodegradation processes:
  inhibition of degradation of glucose by *Pseudomonas fluorescens* at: 350 mg/l
  inhibition of degradation of glucose by *E. coli* at: >1000 mg/l (293)
—Waste water treatment:
  activated sludge acclimated to the following aromatics (influent: 250 mg/l benzylalcohol, aeration: 30 min)
  benzylalcohol: 29% of ThOD
  mandelic acid: 31% of ThOD (92)
  reverse osmosis 0.6% enrichment of benzylalcohol in product water, negative rejection (221)
  A.S., W, 20°C, $\frac{1}{2}$ day observed, feed: 250 mg/l, 25 day acclimation: 29% of ThOD (93)

D. BIOLOGICAL EFFECTS

—Bacteria: *E. coli*: no effect at 1 g/l
—Algae: *Scenedesmus*: 96 hr threshold effect at 24°C: 640 ppm (356)
—Arthropoda: *Daphnia*: 48 hr threshold toxic effect at 23°C: 360 ppm (356)
—Fish:
  Lepomis macrochirus: static bioassay in fresh water at 23°C, mild aeration applied after 24 hr

material added	% survival after				best fit 96 hr LC_{50}
ppm	24 hr	48 hr	72 hr	96 hr	ppm
56	100	40	0	0	
32	60	20	0	0	10
18	100	80	50	20	
10	100	100	60	21	
5	100	100	100	100	

  Menidia beryllina: static bioassay in synthetic seawater at 23°C: mild aeration applied after 24 hr

material added	% survival after				best fit 96 hr LC_{50}
ppm	24 hr	48 hr	72 hr	96 hr	ppm
32	100	30	20	20	
18	90	30	30	20	15
10	90	80	80	80	

(352)

  fathead minnows: static bioassay in Lake Superior water at 18–22°C: LC_{50}: (1; 24; 48; 72; 96 hr): 770; 770; 770; 480; 460 mg/l (350)
—Mammalia: rat: single oral LD_{50}: 3.10 g/kg
            rat: inhalation: LC_0: 200–300 ppm, 2 hr

rat: inhalation: LC_{33}: 200–300 ppm, 4 hr
rat: inhalation: LC_{100}: 200–300 ppm; 8 hr   (211)

benzylamine (α-aminotoluene)

Uses: chemical intermediate for dyes, pharmaceuticals, polymers
A. PROPERTIES: m.w. 107.15; b.p. 185°C; sp.gr. 0.983 at 19/4°C; log P_{oct} 1.09
C. WATER POLLUTION FACTORS:
 —Impact on biodegradation processes:
  inhibition of degradation of glucose by *Pseudomonas fluorescens* at: 400 mg/l
  inhibition of degradation of glucose by *E. coli* at: >1000 mg/l   (293)
  NH_3 oxidation by *Nitrosomonas* sp.: at 100 mg/l: 26% inhibition
                                         at 50 mg/l: 10% inhibition
                                         at 10 mg/l: 0% inhibition   (390)
D. BIOLOGICAL EFFECTS
 —Bacteria: *E. coli*: no effect at 1 g/l
 —Algae: *Scenedesmus*: toxic: 6 mg/l
 —Arthropoda: *Daphnia*: toxic: 60 mg/l   (30)

benzylchloride (α-chlorotoluene)

Use: dyes, intermediate

Uptake, tissue distribution, and release of benzo(a)pyrene-C^{14} in *Rangia cuneata* exposed to 0.035 ppm benzo(a)pyrene-C^{14} for 24 hours

sampling time	mean total ppm benzo(a)pyrene per animal	% of 24-hour concentration remaining	mean percentage total radioactivity per tissue fraction				
			viscera	mantle	gill	adductor	foot
24-hour uptake (1 day)	5.7 ± 4.0 (8)*	—	64.5 ± 10.0	16.2 ± 11.6	12.0 ± 3.4	4.3 ± 1.7	2.94 ± 0.6
144-hour depuration (6 days)	0.34 ± 0.26 (8)*	6.0	95.3 ± 6.4	2.8 ± 3.4	1.8 ± 3.2	>0.05	>0.05
720-hour depuration (30 days)	0.07 ± 0.08 (8)*	1.2	58.5 ± 32.0	19.3 ± 16.0	12.6 ± 19.0	5.4 ± 6.0	8.7 ± 16.0
1392-hour depuration (58 days)	>0.01 (10)*	>0.2	—	—	—	—	—

*Number of clams sampled   (1048)

A. PROPERTIES: m.w. 126.58; m.p. −41/−43°C; b.p. 179°C; v.p. 1 mm at 22°C, 1.7 mm at 30°C; v.d. 4.36; sp.gr. 1.102 at 18/4°C; sat.conc. 6.2 g/cu m at 20°C, 11 g/cu m at 30°C; P_{oct} 2.30
B. AIR POLLUTION FACTORS: 1 mg/cu m = 0.19 ppm, 1 ppm = 5.262 mg/cu m
 —Odor: characteristic: lacrimator aromatic   (129)

threshold values: 0.25 mg/cu m = 0.047 ppm; 0.04 ppm (37)(279)
         $PIT_{50\%}$: 0.01 ppm
         $PIT_{100\%}$: 0.047 ppm (2)
—Sampling and analysis: photometry: min.full scale: 120 ppm (53)

C. WATER POLLUTION FACTORS:
—Reduction of amenities: faint odor: 0.0016 mg/l (129)

D. BIOLOGICAL EFFECTS:
—Bacteria: *Pseudomonas putida*: inhibition of cell multiplication starts at 4.8 mg/l (329)
—Toxicity threshold (cell multiplication inhibition test):
  bacteria (*Pseudomonas putida*):   4.8 mg/l   (1900)
  algae (*Microcystis aeruginosa*):   30 mg/l   (329)
  green algae (*Scenedesmus quadricauda*):   50 mg/l   (1900)
  protozoa (*Entosiphon sulcatum*):   25 mg/l   (1900)
  protozoa (*Uronema parduczi Chatton-Lwoff*): 50 mg/l   (1901)
—Protozoa: *Vorticela campanula*: toxic: 11 mg/l
         Paramaecium caudatum: toxic: 800 mg/l (30)
—Fish: *Trutta iridea*: paralysis: 10 mg/l
     Cyprinus carpio: paralysis: 17 mg/l (30)

	24 hr LC_{50}	48 hr LC_{50}	96 hr LC_{50}
fathead minnows (*Pimephales promelas*)	11.6	7.3	6
white shrimp (*Penaeus setiferus*)	7.1	4.4	3.9

(1904)

—Carcinogenicity: ?
—Mutagenicity in the *Salmonella* test: weakly mutagenic (without liver homogenate):
  0.02 revertant colonies/nmol
  230 revertant colonies at 2000 µg/plate (1883)

benzylcyanide (α-tolunitrile; phenylacetonitrile)

A. PROPERTIES: m.w. 117.14; m.p. -23/-26°C; b.p. 234°C; v.p. 0.1 mm at 20°C, 0.1 mm at 30°C; v.d. 4.05; sp.gr. 1.015 at 18°C; sat. conc. 0.06 g/cu m at 30°C
B. AIR POLLUTION FACTORS: 1 mg/cu m = 0.21 ppm, 1 ppm = 4.87 mg/cu m
C. WATER POLLUTION FACTORS
—Biodegradation by a mutant microorganism: 500 mg/l at 20°C:
  parent: 84% disruption in 48 hr
  mutant: 100% disruption in 12 hr (152)

(5-benzyl-3-furyl)methyl-α-*trans*-(+)-3-(cyclopentylilidene-methyl)-2,2-dimethylcyclopropanecarboxylate *see* RU 11679

(5-benzyl-3-furyl)methyl-*cis*,*trans*-(+)-2,2-dimethyl-3-(2,2-dimethyl-3-(2-methylpropenyl)cyclopropane-carboxylate *see* SBP-1382

benzylhydroquinone *see* hydroquinonemonobenzylether

***p*-benzyloxyphenol** *see* hydroquinonemonobenzylether

benzyltrichloride *see* benzotrichloride

B(*e*)P *see* benzo(*e*)pyrene

BGE *see* *n*-butylglycidylether

B(*ghi*)F *see* benzo(*ghi*)fluoranthene

BHC *see* hexachlorocyclohexane

BHT *see* 2,6-di-*tert*-butyl-4-methylphenol

***p,p'*-bianiline** *see* benzidine

bianisidine

C. WATER POLLUTION FACTORS:
 —Treatment method: W, A.S. from mixed domestic/industrial treatment plant: 98–100% depletion at 20 mg/l after 6 hr at 25°C (n.s.i.) (419)

bibenzyl (dibenzyl; 1,2-diphenylethane)

Use: organic synthesis
A. PROPERTIES: m.w. 182.27; m.p. 50–53°C; b.p. 284°C; sp.gr. 1.014; log P_{oct} 4.79/4.82
C. WATER POLLUTION FACTORS:
 —Water quality:
  in Eastern Ontario drinking waters (June–Oct. 1978): n.d.–0.8 ng/l ($n = 12$)
  in Eastern Ontario raw waters (June–Oct. 1978): 0.3–1.0 ng/l ($n = 12$) (1698)

bichloromethylether *see* dichloromethylether

bidisin (2-chloro-3(4-chlorophenyl)-methylpropionate; chlorfenpropmethyl; Bay 70533)

Use: a postemergence herbicide
A. PROPERTIES: colorless to light-brown liquid; sp.gr. 1.3 at 20°/4°C; solub. 400 mg/l at 20°C
D. BIOLOGICAL EFFECTS:
  —Fish: harlequin fish (*Rasbora heteromorpha*)

mg/l	24 hr	48 hr	96 hr	3 m (extrap.)	
LC_{10} (F)	1.2	1.3	0.8		
LC_{50} (F)	1.85	1.7	1.1	0.6	(331)

  —Mammals:
  acute oral LD_{50} (rat): 1190–1390 mg/kg
  dermal LD_{50} (rat): >1273 mg/kg   (1854)

biisopropyl *see* 2,3-dimethylbutane

binapacryl (2-(1-methyl-*n*-propyl)-4,6-dinitrophenyl-2-methylcrotonate; 2-*sec*-butyl-4,6-dinitrophenyl-3-methyl-2-butenoate; 2(2-butyl-4,6-dinitrophenyl)-3,3-dimethylacrylate; Acricid; Ambox; dinoseb methacrylate; Endosan; Dapacril; Morrocid; BP 855, 736)

$$O=C-CH=C-(CH_3)_2$$

[Structural formula: 2,4-dinitrophenyl ring with O_2N and NO_2 substituents, an ester linkage $-O-$ to the crotonate group above, and a $-CH(CH_3)-CH_2-CH_3$ (sec-butyl) substituent]

Use: contact miticide, fungicide
A. PROPERTIES: crystalline solid; m.p. 65–69°C; m.w. 322; sp.gr. 1.156
D. BIOLOGICAL EFFECTS:
  —Fish: guppies: LD_{100}: 1 ppm
  —Mammals:
  rat: acute oral LD_{50}: 120–165 mg/kg
  mice: acute oral LD_{50}: 1600 mg/kg
  guinea pigs: acute oral LD_{50}: 300 mg/kg
  dogs: 50 mg/kg causes vomiting and hypernoea
  chickens: single oral doses tolerated up to 800 mg/kg
  mice: acute dermal LD_{50}: no death up to 1000 mg/kg as an acacia suspension
  rabbits: acute dermal LD_{50}: 750 mg/kg in acetone
  chronic toxicity:
    rats: 200 ppm for 2 years without reaction
          500 ppm for 2 years, only small reduction in body wt
    dogs: 40 ppm over 2 years without reaction

bioctyl *see* *n*-hexadecane

biphenyl *see* diphenyl

2,2′-bipyridine (2,2′-dipyridyl)

bis-2-AMINOETHYLAMINE

Uses: iron-chelating agent
A. PROPERTIES: m.w. 156.19; m.p. 70–73°C; b.p. 273°C
C. WATER POLLUTION FACTORS:
—Impact on biodegradation processes:
NH$_3$ oxidation by *Nitrosomonas*: at 100 mg/l: 91% inhibition
at 50 mg/l: 81% inhibition
at 10 mg/l: 23% inhibition   (390)

***bis*-2-aminoethylamine**   *see* diethylenetriamine

***bis*(2-chloroethyl)ether**   *see* β,β'-dichloroethylether

***bis*(β-chloroisopropyl)ether**   *see* dichloroisopropylether

bis-2-chloro-1-methyletheylether
ClCH$_2$CH(CH$_3$)OCH(CH$_3$)CH$_2$Cl
A. PROPERTIES: b.p. 187°C
C. WATER POLLUTION FACTORS:
odor thresholds

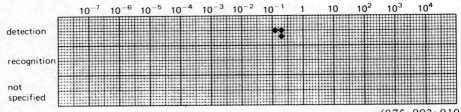

(875; 903; 910)

D. BIOLOGICAL EFFECTS:
—Mammals: rat: inhalation: no effect level: 20 ppm (20 × 6 hr exposures)   (65)

bis(chloromethyl)naphthalene
C$_{12}$H$_{10}$Cl$_2$
A. PROPERTIES: m.w. 225.12; v.d. 7.78
B. AIR POLLUTION FACTORS: 1 mg/cu m = 0.11 ppm, 1 ppm = 9,36 mg/cu m
C. WATER POLLUTION FACTORS:
—Aquatic reactions: evaporation: measured half-life for evaporation from 1 ppm aqueous solution at 25°C, still air and an average depth of 6.5 cm: 45.2 min.
(369)

2,2-*bis*(p-chlorophenyl)-1,1-dichloroethane   *see* DDD

1,1-*bis*(4'-chlorophenyl)-2,2,2-trichloroethanol   *see* kelthane

***bis*(dimethylthiocarbamoyl)disulfide**   *see* thiram

S-(1,2-*bis*(ethoxycarbonyl)ethyl)O,O-dimethylphosphorodithioate *see* malathion

bis-2-ethoxyethylether *see* diethyleneglycoldiethylether

2,2-*bis*-hydroxymethyl-1,3-propanediol *see* pentaerythritol

3,4-*bis*(*p*-hydroxyphenyl)-3-hexene *see* diethylstilbestrol

2,2-*bis*(*p*-metoxyphenyl)-1,1,1-trichloroethane *see* methoxychlor

bis(methylpropyl)amine *see* di-*s*-butylamine

bis(β-methylpropyl)amine *see* di-isobutylamine

bivinyl *see* 1,3-butadiene

B(*j*)F *see* benzo(*j*)fluoranthene

B(*k*)F *see* benzo(*k*)fluoranthene

bladex (2-(4-chloro-6-ethylamino)-S-triazine-2-ylamino-2-propionitrile
  Use: herbicide
D. BIOLOGICAL EFFECTS:
  —Fresh water ectoprocta: no appreciable effect at 2.5 mg/l for 84 hr exposure
                                                                (1902)
  —Fish:

	96 hr LD$_{50}$	96 hr LD$_{100}$	
Sarotherodon mossambicus	24.5 mg/l	64 mg/l	
Cirrhinus mrigala	12.6 mg/l	40 mg/l	(1902)

borane-*tert*-butylaminecomplex *see* *t*-butylamineborane

borneol (bornyl alcohol; 2-camphanol; 2-hydroxycamphane)

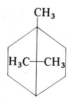

Manufacturing source: organic chemical industry; extraction and recovery from certain woods.   (347)
Users and formulation: flotation agent; textile industry (wet processing of cotton, silk, rayon, wool) soap mfg. (solvent and bactericide); preservative for casein and other proteins in water paints; mfg. of camphor; perfume and incense mfg.; mfg. of chemical esters; mfg. of flavorings and medicinals.   (347)
Man caused sources (*water and air*): general use of soaps, paints, perfumes, and flavors.   (347)

290 BORNYLALCOHOL

A. PROPERTIES: m.w. 154.25; m.p. 206–208°C; f.p. 65°C; d. 1.011; white translucent lumps; camphor-like odor
B. AIR POLLUTION FACTORS:
 —Odor threshold: detection: 0.052 mg/cu m   (840)
C. WATER POLLUTION FACTORS:
 —Biodegradation: adapted A.S.—product as sole carbon source—90.3% COD removal at 8.9 mg COD/g dry inoculum/hr   (327)

bornylalcohol see borneol

BPMC see o-sec-butylphenyl-N-methylcarbamate

bravo see chlorothalonil

bromacil (5-bromo-6-methyl-3-(1-methylpropyl)uracil; 5-bromo-3-sec-butyl-6-methyluracil; Borea; Hyvar X; Hyvar X-L)

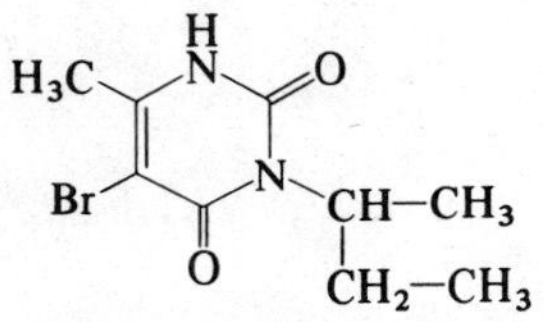

Use: herbicide used in amounts up to 12 kg per acre to control a wide range of grasses and broad-leaf weeds
A. PROPERTIES: m.p. 158–159°C; solub. 815 ppm at 25°C
C. WATER POLLUTION FACTORS:
 —Aqueous reactions: photodecomposition: the action of 4 months of sunlight on dilute (1–10 ppm) aqueous solutions of bromacil resulted in the formation of only one detectable photoproduct, 5-bromo-6-methyluracil in very low yield. The N-dealkylated photoproduct proved to be much less stable toward sunlight wavelengths, forming principally 6-methyluracil

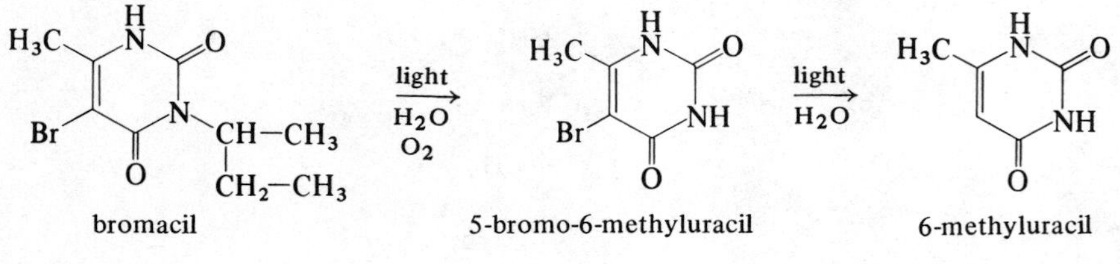

bromacil $\xrightarrow[O_2]{\text{light}, H_2O}$ 5-bromo-6-methyluracil $\xrightarrow{\text{light}, H_2O}$ 6-methyluracil

(1638)

D. BIOLOGICAL EFFECTS:
 —Acute oral LD_{50} (rat): 5200 mg/kg   (1854)

brominecyanide see cyanogenbromide

bromobenzene (phenylbromide)

Users and formulation: solvent (fats, waxes, or resins); intermediates in specialty organic chemicals synthesis; additive to motor oil and fuels. (347)

Man caused sources (water and air): general lab use; use as solvent; discharge of waste motor oils to water; roal surface runoff. (347)

A. PROPERTIES: m.w. 157.02; m.p. -31°C; b.p. 156°C; v.p. 3.3 mm at 20°C; v.d. 5.41; sp.gr. 1.50 at 15/15°C; solub. 500 mg/l at 20°C, 446 mg/l at 30°C, log P_{oct} 2.99 at 20°C

B. AIR POLLUTION FACTORS: 1 mg/cu m = 0.15 ppm, 1 ppm = 6.53 mg/cu m
   —Odor thresholds: 30.5 mg/cu m (748)
   recognition 1.7–2.1 mg/cu m (610)
   —Sampling and analysis: photometry: min. full scale: 120 ppm (53)

D. BIOLOGICAL EFFECTS:
   —Carcinogenicity: ?
   —Mutagenicity in the *Salmonella* test: none
   <0.01 revertant colonies/nmol
   <70 revertant colonies at 750 µg/plate (1883)

5-bromo-3-*sec*-butyl-6-methyluracil *see* bromacil

bromodichloromethane
CHBrCl$_2$

Users and formulations: fire-extinguisher fluid ingredient; solvent (fats, waxes, resins); synthesis intermediate; heavy liquid for mineral and salt separations. (347)

Man caused sources (water and air): results from chlorination of finished water; use of fire extinguishers, lab use. (347)

A. PROPERTIES: colorless liquid; m.w. 163.8, b.p. 90°C; m.p. -55°C; sp.gr. 1.971 at 25/25°C

C. WATER POLLUTION FACTORS:
   —Water quality: in N.W. England tap waters (1974): 1–27 ppb (933)
   —Water treatment: experimental water reclamation plant: sand filter effluent: 63 ng/l
   after chlorination: 82 ng/l (928)

bromoform (tribromomethane)
CHBr$_3$

Users and formulation: pharmaceutical mfgs.; ingredient in fire-resistant chms.; gage fluid; heavy liquid in solid separations based on differences in specific gravity, geological assaying, solvent for waxes, greases and oils. (347)

A. PROPERTIES: colorless liquid; m.w. 252.77; m.p. 6/7°C; b.p. 149°C; v.p. 5.6 mm at 25°C; v.d. 8.7; sp.gr. 2.89 at 20/4°C; solub. 3,190 mg/l at 30°C

B. AIR POLLUTION FACTORS: 1 ppm = 10,34 mg/cu m, 1 mg/cu m = 0.0966
 —Odor: characteristic: chloroformlike, sweetish (211)

odor thresholds — mg/cu m

	10^{-7}	10^{-6}	10^{-5}	10^{-4}	10^{-3}	10^{-2}	10^{-1}	1	10	10^2	10^3	10^4
detection								● ●				
recognition									●	●		
not specified											●	

(610; 671; 777; 795)

C. WATER POLLUTION FACTORS:
 —Odor threshold: detection: 0.3 mg/kg (894)
 —Water quality: in N.W. England tap waters (1974): <0.01–2.5 ppb (933)
 —Water treatment: experimental water reclamation plant:

	sample 1	sample 2	
sand filter effluent	211 ng/l	154 ng/l	
after chlorination	1723 ng/l	3711 ng/l	
final water after A.C.	175 ng/l	135 ng/l	(928)

D. BIOLOGICAL EFFECTS:
 —Larvae of eastern oyster (*Crassostrea virginica*): LD_{50}, 48 hr: 1 mg/l, initial conc., static test (after 48 hr only approx. 30% of original conc. was still present)

(1545)

bromomethane *see* methylbromide

2-bromo-4-nitrophenol

D. BIOLOGICAL EFFECTS
 —Fish: larvae of a sea lamprey: LD_{100} = 5 mg/l (226)
  rainbow trout: LD_{10} = 13 mg/l (226)
  brown trout: LD_{10} = 11 mg/l (226)

3-bromo-4-nitrophenol

D. BIOLOGICAL EFFECTS:
  —Fish: larvae of sea lamprey: LD_{100}: 3 mg/l (226)
  rainbow trout: LD_{10}: 5 mg/l (226)
  brown trout: LD_{10}: 5 mg/l (226)

o-bromophenol

A. PROPERTIES: m.w. 173.02; m.p. 5.6°C; b.p. 195°C; sp.gr. 1.49 at 20/4°C; $\log P_{oct}$ 2.35
C. WATER POLLUTION FACTORS
  —Biodegradation:
  decomposition rate in soil suspensions: 14 days for complete disappearance
  (175)
  —Waste water treatment:
  degradation by *Pseudomonas*: 200 mg/l at 30°C
    ring disruption: parent: 100% in 85 hr
                    mutant: 100% in 14 hr (152)
D. BIOLOGICAL EFFECTS
  —Algae: *Chlorella pyrenoidosa*: 78 mg/l: toxic (41)

m-bromophenol

A. PROPERTIES: m.w. 173.02; m.p. 33°C; b.p. 236°C; $\log P_{oct}$ 2.63
B. AIR POLLUTION FACTORS:
  —Odor threshold: recognition: 0.000007 mg/cu m (712)
C. WATER POLLUTION FACTORS
  —Biodegradation:
  decomposition rate in suspended soils: >72 days for complete disappearance
  (175)
    degradation by *Pseudomonas*: 200 mg/l at 30°C:
    ring disruption: parent: 51% in 96 hr
                    mutant: 100% in 25 hr (152)
D. BIOLOGICAL EFFECTS
  —Algae: *Chlorella pyrenoidosa*: 36 mg/l: toxic (41)

p-bromophenol

BRUCINE

A. PROPERTIES: m.w. 173.02; m.p. 63.5°C; b.p. 238°C; sp.gr. 1.840 at 15°C; solub. 14,200 mg/l at 15°C; log P_{oct} 2.59

C. WATER POLLUTION FACTORS:
  —Biodegradation:
  decomposition rate in soil suspensions: 16 days for complete disappearance
  (175)

  degradation by pseudomonas: 200 mg/l at 30°C:
  ring disruption: parent: 87% in 85 hr
                   mutant: 100% in 22 hr (152)

D. BIOLOGICAL EFFECTS
  —Algae: *Chlorella pyrenoidosa*: toxic: 36 mg/l (41)

brucine (dimethyoxystrychnine)
  Uses: medicine; denaturing alcohol; lubricant additive

A. PROPERTIES: white crystalline alkaloid; very bitter taste; m.w. 394.45; m.p. 178°C

C. WATER POLLUTION FACTORS:
  —Impact on biodegradation processes: at 100 mg/l no inhibition of NH_3 oxidation by *Nitrosomonas* sp. (390)

D. BIOLOGICAL EFFECTS:
  —Fish:
  Lepomis macrochirus: static bioassay in fresh water at 23°C, mild aeration applied after 24 hr

material added ppm	% survival after				best fit 96 hr LC_{50} ppm
	24 hr	48 hr	72 hr	96 hr	
63	0	—	—	—	36
40	40	30	20	20	
32	90	80	60	40	
25	100	100	100	100	(352)

  Menidia beryllina: static bioassay in synthetic seawater at 23°C, mild aeration applied after 24 hr

material added ppm	% survival after				best fit 96 hr LC_{50} ppm
	24 hr	48 hr	72 hr	96 hr	
32	80	60	40	20	
18	100	100	80	70	20
10	100	100	100	90	(352)

busan 25
  Use: microbicide; active ingredients; (2-(thiocyanomethylthio)-benzothiazol(13%) and 2-hydroxypropylmethanethiolsulphonate (11.7%)

D. BIOLOGICAL EFFECTS: harlequin fish (*Rasbora heteromorpha*):

mg/l	24 hr	48 hr	96 hr	3 m (extrap.)	
LC_{10} (F)	0.6	0.43	0.34		
LC_{50} (F)	1.0	0.57	0.42	0.07	(331)

busan 70
Use: microbicide; active ingredient: butanethiol sulphonate
D. BIOLOGICAL EFFECTS: harlequin fish (*Rasbora heteromorpha*):

mg/l	24 hr	48 hr	96 hr	3 m (extrap.)	
LC_{10} (F)	0.48	0.37	0.36		
LC_{50} (F)	0.76	0.47	0.43	0.3	(331)

busan 72
Use: microbicide; active ingredient: 2-(thiocyanomethylthio)benzothiazol (60%)
D. BIOLOGICAL EFFECTS: harlequin fish (*Rasbora heteromorpha*):

mg/l	24 hr	48 hr	96 hr	3 m (extrap.)	
LC_{10} (F)	0.08	0.044	0.031		
LC_{50} (F)	0.13	0.075	0.036	0.006	(331)
active ingredient: (15%)					
LC_{10} (F)	1.4	0.64	0.34		
LC_{50} (F)	1.7	0.88	0.46	0.1	(331)

busan 74
Use: microbicide: active ingredients: 2-(thiocyanomethylthio)benzothiazol (40%) and 2-hydroxypropylmethanethiolsulphonate (35%)
D. BIOLOGICAL EFFECTS: harlequin fish (*Rasbora hetermorpha*):

mg/l	24 hr	48 hr	96 hr	3 m (extrap.)	
LC_{10} (F)	0.12	0.052	0.035		
LC_{50} (F)	0.21	0.084	0.045	0.001	(331)

busan 76
Use: microbicide: active ingredient: β-cyanoethyl-2,3-dibromo-propionate (60%)
D. BIOLOGICAL EFFECTS: harlequin fish (*Rasbora heteromopha*):

mg/l	24 hr	48 hr	96 hr	3 m (extrap.)	
LC_{10} (F)	0.31	0.29	0.29		
LC_{50} (F)	0.47	0.35	0.31	0.26	(331)

busan 77
Use: microbicide: active ingredient: quaternary ammonium compound
D. BIOLOGICAL EFFECTS: harlequin fish (*Rasbora heteromorpha*):

mg/l	24 hr	48 hr	96 hr	3 m (extrap.)	
LC_{10} (F)	0.47	0.32	—		
LC_{50} (F)	0.66	0.39	0.17	0.01	(331)

1,3-butadiene (vinylethylene; divinyl; bivinyl; pyrrolylene; biethylene; erythrene)
$CH_2=CH-CH=CH_2$

1-BUTANAL

Uses: principally in styrene-butadiene rubber; in latex paints; resins; organic intermediate

A. PROPERTIES: colorless gas; m.w. 54.09; m.p. -108.9°C; b.p. -4.41°C; v.p. 2.5 mm at 20°C; 3.3 mm at 30°C; v.d. 1.87; sp.gr. 0.6211 at 20°C liquified; solub. 735 mg/l at 20°C; THC 618 kcal/mole, LHC 587 kcal/mole

B. AIR POLLUTION FACTORS: 1 mg/cu m = 0.45 ppm, 1 ppm = 2.25 mg/cu m
  —Odor: characteristic: quality: undefined
    hedonic tone: unpleasant to neutral

(19; 279; 307; 637; 737; 761; 794)

Odor index (100% recognition): 770,000 (19)

  —Atmospheric reactions:
    R.C.R.: 4.31 (49)
    reactivity: HC consumption: ranking: 1 (63)
    estimated lifetime under photochemical smog conditions in S.E. England: 0.48 hr
    (1699; 1700)

  —Manmade sources:
    glc's downtown Los Angeles 1967: 10% ile: 1 ppb
      average: 2 ppb
      90% ile: 5 ppb (64)
    expected glc's in USA urban air: range: 0.5 to 10 ppb (102)

  —Sampling and analysis: photometry: min.full scale: 67 ppm (53)
    scrubbers: liquid lift type; 15 ml of liquid, 4.5 l/min gas flow, 4 min scrubbing

trap method:	% removal
H_2O, 0°C	0
NH_2OH soln, 0°C	0
H_2SO_4 (conc.), 55°C	29
H_2SO_4, 4% Ag_2SO_4 (conc.), 55°C	87
open tube, -80°C	0
Ethanol, -80°C	37

(311)

C. WATER POLLUTION FACTORS:
  —Biodegradation: the overall catabolic pathway of butadiene as sole carbon and energy source by Nocardia sp. 249 is believed to be the following:

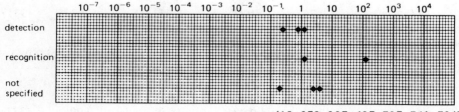

$$CH_3-CHOH-COOH \leftarrow\!\!\!-\!\!\!- CH_2=CH-COOH \underset{CO_2}{\leftarrow\!\!\rightarrow} CH_2=CH-\overset{O}{\underset{\|}{C}}-COOH$$

$$\downarrow$$

$$CH_3-\overset{O}{\underset{\|}{C}}-COOH \underset{CO_2}{-\!\!\!\rightarrow} CH_3COOH \qquad (427)$$

D. BIOLOGICAL EFFECTS
 —Fish: pinperch: TLm (24 hr) 71.5 mg/l (41)
 —Man: EIR: 6.9 (49)
  irritation of the respiratory system: 10,000 ppm, 1 min (186)
  slight irritation of the eyes and upper respiratory tract, no other effects: 8,000 ppm, 8 hr (211)

1-butanal *see* butyraldehyde

butanamide *see* butyramide

n-butane
$CH_3CH_2CH_2CH_3$
A. PROPERTIES: colorless gas; m.w. 58.14; m.p. $-135/-138°C$; b.p. $-1°C$; v.p. 1823 mm at $25°C$, 2.9 atm at $30°C$; v.d. 2.01; sp.gr. 0.60 liquified; solub. 61 mg/l at $20°C$, 30 mg/l at $15°C$, 21 mg/l at $38°C$; LHC 636 kcal/mole
B. AIR POLLUTION FACTORS: 1 mg/cu m = 0.41 ppm, 1 ppm = 2.42 mg/cu m
 —Odor: T.O.C.: 12,100 mg/cu m = 4960 ppm
     not detectable <5,000 ppm (211)
     recognition: 6,160 mg/cu m (761)
     3,000 mg/cu m (737)
     5.5 ppm (279)
 —Atmospheric reactions: R.C.R.: 0.79 (49)
     reactivity: NO ox.: ranking: 0.1 (63)
  estimated lifetime under photochemical smog conditions in S.E. England: 15 hr
     (1699; 1705)
 —Manmade sources:
  glc's at Pt Barrow, Alaska, Sept 1967: 0.03 to 0.19 ppb (101)
  glc's downtown Los Angeles 1967: 0.132 to 0.304 ppm C (52)
     10% ile: 20 ppb
     average: 46 ppb
     90% ile: 80 ppb (64)
  glc's East San Gabriel Valley 1967: 0.098 to 0.175 ppm C (52)
  expected glc's in USA urban air: range: 0.05 to 0.45 ppm (102)
  in flue gas of municipal incinerator: <0.4 ppm (196)
  exhaust gas of diesel engine: 5.3% of emitted HC.s (72)
  combustion gas of household central heating—appr. 50 ppm at 7% CO_2 system on gasoil:

3.3 g/kg gasoil at 6% CO_2 (182)
1.6 g/kg gasoil at 7% CO_2 (312)
in gasoline: 4.31–5.02 vol %
in auto exhaust—gasoline engine:
    62-car survey: 5.3 vol % of total exhaust HC's (391)
    15-fuel study: 4 vol % of total exhaust HC's (392)
    engine variable study: 2.3 vol % of total exhaust HC's (393)
evaporation from gasoline fuel tank: 16.5–48.5 vol % of total evaporated HC's
evaporation from carburetor: 9.1–23.0 vol % of total evaporated HC's
(398; 399; 400; 401; 402)

C. WATER POLLUTION FACTORS:
—incubation with natural flora in the groundwater—in presence of the other components of high-octane gasoline (100 μl/l): biodegradation: 0% after 192 hr at 13°C (initial conc. 0.63 μl/l) (956)

D. BIOLOGICAL EFFECTS:
—Man: EIR: 0 (49)
    drowsiness, no other effects: 10,000 ppm, 10 min (211)

1,4-butanedicarboxylic acid *see* adipic acid

butanedinitrile *see* succinonitrile

butanedioic acid *see* succinic acid

1,4-butanediol (tetramethyleneglycol; 1,4-dihydroxybutane)
$(CH_2CH_2OH)_2$
A. PROPERTIES: m.w. 90.12; m.p. 16°C; b.p. 230°C; sp.gr. 1.020 at 20/4°C
C. WATER POLLUTION FACTORS:
—Biodegradation: adapted A.S.—product as sole carbon source—98.7% COD removal at 40.0 mg COD/g dry inoculum/hr (327)
—Waste water treatment:
    reverse osmosis: 65.9% rejection from 0.01 M solution (221)

butane dioxime *see* dimethylglyoxime

butanenitrile *see* butyronitrile

butanethiol *see* n-butylmercaptan

butanethiolsulfonate *see* busan 70

n-butanoic acid *see* n-butyric acid

butanoic anhydride *see* butyric anhydride

n-butanol (n-butylalcohol; propylcarbinol; 1-butanol)
$CH_3-CH_2-CH_2-CH_2OH$

n-BUTANOL

A. PROPERTIES: colorless liquid; m.w. 74.12; m.p. -89.9°C; b.p. 117.7°C; v.p. 4.4 mm at 20°C, 6.5 mm at 25°C, 10 mm at 30°C; v.d. 2.55; sp.gr. 0.810 at 20/4°C; solub. 77,000 mg/l; THC 638.2 kcal/mole, LHC 597.0 kcal/mole; sat.conc. 20 g/cu m at 20°C, 39 g/cu m at 30°C; log P_{oct} 0.88

B. AIR POLLUTION FACTORS: 1 mg/cu m = 0.33 ppm, 1 ppm = 3.03 mg/cu m
   —Odor: characteristic: quality: rancid, sweet
                          hedonic tone: neutral to pleasant

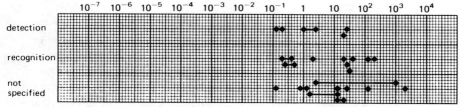

(73; 210; 278; 291; 307; 610; 627; 634; 641; 663; 704; 706; 707; 708; 709; 715; 737; 749; 756; 776; 785; 788; 804)

Odor index: (100% detection): 2600	(19)
threshold for unadapted panellists: 50 ppm	
threshold after adaption with pure odorant: 10,000 ppm	(204)
distinct odor: 48 mg/cu m = 16 ppm	(278)
—Manmade sources:	
glc's at Pt Barrow Alaska Sept 1967: 51 to 126 ppb	(101)
concentrations of 5 to 100 ppm have been reported in workrooms	(211)
—Control methods:	
wet scrubber: water at pH 8.5: outlet: 150 odor units/scf	
$KMnO_4$ at pH 8.5: outlet: 40 odor units/scf	(115)

—Comparison of catalyst performance at space velocity of 80,000/hr:

catalyst	feed conc. ppm	reactor inlet temp. °C	% odor removal ASTM	% conversion GC
0.5% Pt on γ-Al_2O_3	1405	225	96	100
0.1% Pt on γ-Al_2O_3	1405	236	95	100
0.5% Pd on γ-Al_2O_3	1405	376	96	100
10% Cu on γ-Al_2O_3	1405	411	94	100

(1221)

C. WATER POLLUTION FACTORS:

—BOD_5:	1.5	(30)
	1.1–2.04 warb. sewage	(41; 271; 280)
	33% of ThOD	(220)
	1.5 to 2.0 std.dil.sew.	(27)
—BOD_{20}:	1.89	(30)
—COD:	1.9	(27)
	92% of ThOD	(220)

sec-BUTANOL

- $KMnO_4$: 0.228 (30)
  acid 6% ThOD; alkaline: 2% ThOD (220)
- ThOD: 2.594 (30)
- TOC: 97% of ThOD (220)
- Biodegradation: adapted A.S.: product as sole carbon source: 98.8% COD removal at 84.0 mg COD/g dry inoculum/hr (327)
- Impact on biodegradation processes:
  50% inhibition of NH_3 oxidation in *Nitrosomonas* at 8200 mg/l (407)

odor thresholds   mg/kg water

	10^{-7}	10^{-6}	10^{-5}	10^{-4}	10^{-3}	10^{-2}	10^{-1}	1	10	10^2	10^3	10^4
detection								♦♦♦				
recognition												
not specified									♦			

(97; 181; 294; 326; 875; 883; 889; 894; 906; 907)

20% of the population still able to detect odor at 1.5 ppm
10% of the population still able to detect odor at 1.2 ppm
1% of the population still able to detect odor at 0.44 ppm
0.1% of the population still able to detect odor at 0.16 ppm (321)

- Manmade sources: in year old leachate of artificial sanitary landfill: 0.21 g/l (1720)

- Waste water treatment:
  A.C.: adsorbability: 0.107 g/g C, 53.4% reduction, infl.: 1,000 mg/l, effl.: 466 mg/l (32)
  reverse osmosis: 41.3% rejection from a 0.01 M solution (221)
  stabilization pond design: toxicity correction factor: 2.0 at 4,000 mg/l influent (179)
  anaerobic lagoon:
   22 lb COD/day/1,000 cu ft: infl.: 170 mg/l, effl.: 75 mg/l
   48 lb COD/day/1,000 cu ft: infl.: 170 mg/l, effl.: 80 mg/l (37)
  A.S.: after 6 hr: 16.5% of ThOD
     12 hr: 30.8% of ThOD
     24 hr: 36.1% of ThOD (88)

methods	temp °C	days observed	feed mg/l	days acclim.		
A.S., W	20	1	500	24	44% theor. oxidation	
A.S., BOD	20	$\frac{1}{3}$–5	333	30	96% removed	(93)

A.C.: (n.s.i.):	influent ppm	carbon dosage	effluent ppm	% reduction
	1,000	10×	249	75
	500	10×	163	67
	100	10×	52	48

(192)

D. BIOLOGICAL EFFECTS:
- Toxicity threshold (cell multiplication inhibition test):
  bacteria (*Pseudomonas putida*):   650 mg/l (1900)

sec-BUTANOL 301

algae (*Microcystis aeruginosa*):	100 mg/l	(329)
green algae (*Scenedesmus quadricauda*):	875 mg/l	(1900)
protozoa (*Entosiphon sulcatum*):	55 mg/l	(1900)
protozoa (*Uronema parduczi Chatton-Lwoff*):	8.0 mg/l	(1901)

—Algae: *Chlorella pyrenoidosa*: toxic: 8,500 mg/l　　　　　　　　　　　　　(41)
—Fish:
　creek chub: LD_0:　　1,000 mg/l, 24 hr in Detroit river water　　　　　(243)
　　　　　　　LD_{100}: 1,400 mg/l, 24 hr in Detroit river water　　　　(243)
　fathead minnow:
　　static bioassay in Lake Superior Water at 18–22°C: LC_{50} (1; 24; 48; 72; 96 hr):
　　1,950; 1,950; 1,950; 1,950; 1,910 mg/l
　　static bioassay in reconstituted water at 18–22°C: LC_{50} (1; 24; 48; 72; 96 hr):
　　1,940; 1,940; 1,940; 1,940; 1,940 mg/l　　　　　　　　　　　　　　　　　(350)
—Mammalia: rat: oral LD_{50}: 4.36 g/kg　　　　　　　　　　　　　　　　　(211)
　　　　　　rabbit: oral LD_{50}: 4.25 mg/kg　　　　　　　　　　　　　　(211)
　　　　　　mouse: inhalation: no effect: 1,650 ppm, 420 min　　　　　　　(211)
—Man:　mild irritation of nose, throat, and eyes: 25 ppm
　　　　pronounced irritation　　　　　　50 ppm　　　　　　　　　　　　(211)
—Carcinogenicity: none?
—Mutagenicity in the *Salmonella* test: none:
　　<0.0005 revertant colonies/nmol
　　<70 revertant colonies at 10^4 µg/plate　　　　　　　　　　　　　　　(1883)

sec-butanol　(2-butanol; methylethylcarbinol; 2-butylalcohol; SBA)
$CH_3-CHOH-CH_2-CH_3$

A. PROPERTIES: colorless liquid; m.w. 74.12; m.p. -89/-108°C; b.p. 99.5/107.7°C; v.p. 12 mm at 20°C, 24 mm at 30°C; v.d. 2.55; sp.gr. 0.808 at 20/4°C; solub. 125,000 mg/l at 20°C, 201,000 mg/l at 20°C; sat.conc. 52 g/cu m at 20°C, 94 g/cu m at 30°C; log P_{oct} 0.61

B. AIR POLLUTION FACTORS: 1 mg/cu m = 0.330 ppm, 1 ppm = 3.03 mg/cu m
—Odor: characteristic: quality: sweet
　　　　　　　　　　　hedonic tone: pleasant to neutral
odor thresholds　　　　　mg/cu m

	10^{-7}	10^{-6}	10^{-5}	10^{-4}	10^{-3}	10^{-2}	10^{-1}	1	10	10^2	10^3	10^4
detection								• •	• • •			
recognition									•			
not specified							•	• •				

　　　　　　　　　　　　　　　　　　　　　　　　(57; 279; 307; 708; 709; 737)
Odor index (100% recognition): 28,000　　　　　　　　　　　　　　　　　(19)

C. WATER POLLUTION FACTORS:
—BOD_5:　1.87　　　　　　　　　　　　　　　　　　　　　　　　　　　　(277)
　　　　33% of ThOD　　　　　　　　　　　　　　　　　　　　　　　　　　(220)
　　　　0% of ThOD　　　　　　　　　　　　　　　　　　　　　　　　　　(79)
　10 days: 44.2% of ThOD　　　　　　　　　　　　　　　　　　　　　　　　(79)

302 t-BUTANOL

15 days:	69.2% of ThOD	(79)
20 days:	72.3% of ThOD	(79)
30 days:	73.2% of ThOD	(79)
50 days:	77.9% of ThOD	(79)

—COD: 2.47
  91% of ThOD (220)
—$KMnO_4$: acid: 9% ThOD; alkaline: 3% ThOD (220)
—TOC: 96% of ThOD (220)
—Biodegradation: adapted A.S.—product as sole carbon source—98.5% COD removal at 55.0 mg COD/g dry inoculum/hr (327)
—Waste water treatment:
  A.S.: after 6 hr: 4.2% of ThOD
    12 hr: 6.5% of ThOD
    24 hr: 9.3% of ThOD (88)
  reverse osmosis: 58% rejection from a 0.01 M solution (221)

methods	temp °C	days observed	feed mg/l	days acclim.		
A.S., W	20	1	500	24	58% theor. oxidation	
A.S., BOD	20	$\frac{1}{3}$ to 5	333	30	96% removed	(93)

D. BIOLOGICAL EFFECTS:
—Toxicity threshold (cell multiplication inhibition test):
  bacteria (*Pseudomonas putida*): 500 mg/l (1900)
  algae (*Microcystis aeruginosa*): 312 mg/l (329)
  green algae (*Scenedesmus quadricauda*): 95 mg/l (1900)
  protozoa (*Entosiphon sulcatum*): 1280 mg/l (1900)
  protozoa (*Uronema parduczi Chatton-Lwoff*): 1416 mg/l (1901)
—Algae: *Chlorella pyrenoidosa*: toxic: 8,900 mg/l (41)
—Fish: goldfish: LD_{50} (24 hr) 4300 mg/l (277)
—Mammalia: rat: single oral LD_{50}: 6.48 g/kg
    rabbit: single oral MLD: 6.0 ml/kg
    mouse: inhalation: no effect: 1,650 ppm, 420 min (211)

t-butanol (3-butanol; 2-methyl-2-propanol; trimethylcarbinol; *tert*-butylalcohol) $(CH_3)_3COH$

Use: as blending agent up to 7% to increase the octane rating of unleaded gasoline

A. PROPERTIES: m.w. 74.1; m.p. 25°C; v.p. 31 mm at 20°C; 42 mm at 25°C; 56 mm at 30°C; v.d. 2.55; sp.gr. 0.788 at 20/4°C; sat.conc. 121 g/cu m at 20°C, 219 g/cu m at 30°C; log P_{oct} 0.37; b.p. 83°C

B. AIR POLLUTION FACTORS: 1 mg/cu m = 0.330 ppm, 1 ppm = 3.03 mg/cu m
  odor thresholds   mg/cu m

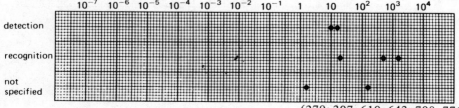

(279; 307; 610; 643; 708; 776)

C. WATER POLLUTION FACTORS:
  -BOD_5: 0   (36)
             0 of ThOD   (220)
  -BOD_5^{20}: nil at 10 mg/l, unadapted sew.   (554)
  -BOD_{20}^{20}: nil at 10 mg/l, unadapted sew.   (554)
  -COD: 2.18   (36)
             80% of ThOD   (220)
  -$KMnO_4$: acid 9% ThOD; alkaline: 0% ThOD   (220)
  -TOC: 2.59   (36)
  -Biodegradation: adapted A.S.–product as sole carbon source: 98.5% COD removal at 30.0 mg COD/g dry inoculum/hr (327)
    slight inhibition of microbial growth after 24 hr exposure at 100 ppm   (523)

  odor thresholds   mg/kg water

  detection: ● ● (around 10 to 10^2)
  recognition:
  not specified: ● (around 10^3)
  scale: 10^{-7} to 10^4
                                                                         (874; 907; 908)

  -Waste water treatments:
    A.C.: adsorbability: 0.059 g/g C; 29.5% reduction, infl.: 1,000 mg/l, effl.: 705 mg/l (32)
    A.S.: after 6 hr: 0.5% of ThOD
              12 hr: 0.7% of ThOD
              24 hr: 0.8% of ThOD   (88)
    reverse osmosis: 90% rejection from a 0.01 M solution   (221)

methods	temp °C	days observed	feed mg/l	days acclim.	% removed	
A.S., W	20	1	500	34	21% theor. oxidation	
A.S., COD	20	$\frac{1}{3}$	333	30	30% removed	(93)

D. BIOLOGICAL EFFECTS:
  -Algae: *Chlorella pyrenoidosa*: toxic: 24,200 mg/l   (41)
  -Fish:
    creek chub: LD_0:   3,000 mg/l, 24 hr in Detroit river water
                 LD_{100}: 6,000 mg/l, 24 hr in Detroit river water   (243)
    guppy (*Poecilia reticulata*): 7 d LC_{50}: 3,550 ppm   (1833)
  -Mammals: rat: single oral LD_{50}: 3.5 g/kg
            rabbit: MLD: 6.0 ml/kg   (211)

2-butanone *see* methylethylketone

2-butanoneoxime (ethylmethylketoxime; methylethylketoxime)
$CH_3 C(NOH) C_2 H_5$
A. PROPERTIES: m.w. 87.12; m.p. -29.5°C; b.p. 152°C; sp.gr. 0.923 at 20/4°C; solub. 100,000 mg/l

D. BIOLOGICAL EFFECTS:
  —Bacteria: *Pseudomonas*: still toxic at 0.63 g/l
  —Algae: *Scenedesmus*: still toxic at 1 g/l
  —Protozoa: *Colpoda*: still toxic at 2.5 g/l (30)

2-butenal *see* crotonaldehyde

1-butene *see* α-butylene

cis-2-butene
$CH_3CH=CHCH_3$
A. PROPERTIES: m.w. 56.10; b.p. 4°C; v.d. 1.94; sp.gr. 0.6 liquified
B. AIR POLLUTION FACTORS: 1 mg/cu m = 0.43 ppm, 1 ppm = 2.33 mg/cu m
  —Odor thresholds: 4.8 mg/cu m (n.s.i.) (710)
  recognition: 28.5 mg/cu m (761)
  —Atmospheric reactions: R.C.R.: 4.83 (49)
  —Estimated lifetime under photochemical smog conditions in S.E. England: 0.6 hr
  (1699; 1700)
  —Manmade sources:
  in gasoline: 0.09–0.35 vol % (312)
  evaporation from gasoline fuel tank: 4.2 vol % of total evaporated HC's
  evaporation from carburetor: 0.2–0.3 vol % of total evaporated HC's
  (398; 399; 400; 401; 402)
  —Sampling and analysis: PMS: det. lim.: 3–7 ppm (n.s.i.) (200)
D. BIOLOGICAL EFFECTS:
  —Man: EIR: 1.6 (49)

trans-2-butene (β-butylene; *sym*-dimethylethylene)
$CH_3CHCHCH_3$
A. PROPERTIES: m.w. 56.10; m.p. −105.4; b.p. 1°C; v.p. 760 mm at 0.9°C; v.d. 1.94; sp.gr. 0.64 liquified
B. AIR POLLUTION FACTORS: 1 mg/cu m = 0.43 ppm, 1 ppm = 2.33 mg/cu m
  —Odor threshold: recognition: 2,700 mg/cu m (761)
  —Atmospheric reactions: R.C.R.: 6.55 (49)
  reactivity HC cons.: ranking: 6–8
  NO ox.: ranking: 2 (63)
  —Manmade sources: in gasoline: 0.07–0.35 vol % (312)
  evaporation from gasoline fuel tank: 4.8 vol % of total evaporated HC's
  evaporation from carburetor: 0.3–0.5 vol % of total evaporated HC's
  (398; 399; 400; 401; 402)
  in exhaust of diesel engine: 0.6% of emitted HC.s (72)
  expected glc's in USA urban air: 5–10 ppb (102)
  —Odor: T.O.C. = 1,3 mg/cu m = 0.6 ppm
  O.I. at 20°C: 3,333,000 (316)

cis-butenedioic acid *see* maleic acid

trans-butenedioic acid *see* fumaric acid

cis-butenedioic anhydride   *see* maleic anhydride

3-butenoic acid   (vinylacetic acid; β-butenoic acid)
CH_2CHCH_2COOH
A. PROPERTIES: m.w. 86.09; m.p. -39°C; b.p. 163°C; sp. gr. 1.013 at 15/15°C
D. BIOLOGICAL EFFECTS:
  —Algae: *Chlorella pyrenoidosa*: toxic: 280 mg/l (n.s.i.)   (41)

2-buten-1-ol
$CH_2OH-CH=CH-CH_3$
B. AIR POLLUTION FACTORS:
  —Manmade sources:
    in gasoline exhaust: 0.1 to 3.6 ppm (+C_5H_8O)
C. WATER POLLUTION FACTORS:
  —Waste water treatment:
    reverse osmosis: 18.3% rejection from a 0.01 M solution   (221)

1-butoxybutane   *see* n-butylether

1-butoxy-2,3-epoxypropane   *see* n-butylglycidylether

3-butoxy-1,2-epoxypropane   *see* n-butylglycidylether

2-butoxyethanol   *see* butylcellosolve

2(β-butoxyethoxy) ethanol   *see* diethyleneglycolmonobutylether

α-(2-(2-butoxyethoxy)ethoxy)-4,5-methylene-2-propyl-toluene   *see* piperonylbutoxide

(butoxymethyl)-oxirane   *see* n-butylglycidylether

butter yellow   *see* 4-dimethylaminoazobenzene

n-butylacetate   (butylethanoate)
$CH_3COO(CH_2)_3CH_3$
  Use: solvents in production of lacquers, perfumes, natural gums and synthetic resins
A. PROPERTIES: m.w. 116.2; m.p. -76.8°C; b.p. 124/127°C; v.p. 10 mm at 20°C, 15 mm at 25°C; v.d. 4.0; sp.gr. 0.882 at 20/4°C; solub. 14,000 mg/l at 20°C; 5,000 mg/l at 25°C
B. AIR POLLUTION FACTORS: 1 mg/cu m = 0.211 ppm, 1 ppm = 4.75 mg/cu m
  —Odor: characteristic: quality: sweet, ester
            hedonic tone: pleasant

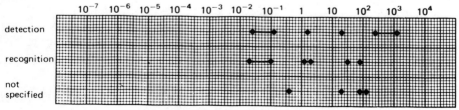

(19; 73; 210; 298; 307; 610; 643; 665; 709; 727; 749; 786; 804)

306 n-BUTYLACETATE

    —Odor index (100% recognition): 284,000   (19)
        USSR: human odor perception: non perception: 0.5 mg/cu m
                                      perception: 0.6 mg/cu m
              human reflex response: no response: 0.1 mg/cu m
                              adverse response: 0.13 mg/cu m
              animal chronic exposure: no effect: 0.1 mg/cu m
                                  adverse effect: 20.0 mg/cu m   (170)
        distinct odor. 55 mg/cu m = 11 ppm   (278)
C. WATER POLLUTION FACTORS:
    —BOD_5:   7% of ThOD   (274)
                 1.020   (30)
    —BOD_{20}: 1.450   (30)
    —BOD_5:   0.15 std.dil.sew.   (282)
                 0.52 std.dil.sew.   (255)
    —COD: 78% of ThOD (0.05 n $K_2Cr_2O_7$)   (274)
    —$KMnO_4$: 0.046   (30)
                 1% of ThOD (0.01 n $KMnO_4$)   (274)
    —ThOD: 2.207   (30)
    —Impact on biodegradation processes:
        0.1 mg/l affects the self purification of surface waters   (181)
    —Odor threshold: detection: 0.066 mg/kg   (889)
                                    0.043 mg/kg   (911)
    —Waste water treatment:
        A.C.: adsorbability: 0.169 g/g C; 84.6% reduction, infl. 1,000 mg/l, effl.: 154 mg/l   (32)
D. BIOLOGICAL EFFECTS:
    —Bacteria: *E. coli*: no toxic effect: 1 g/l
    —Toxicity threshold (cell multiplication inhibition test):
        bacteria (*Pseudomonas putida*):   115 mg/l   (1900)
        algae (*Microcystis aeruginosa*):   280 mg/l   (329)
        green algae (*Scenedesmus quadricauda*):   21 mg/l   (1900)
        protozoa (*Entosiphon sulcatum*):   321 mg/l   (1900)
        protozoa (*Uronema parduczi-Chatton Lwoff*): 574 mg/l   (1901)
    —Algae:
        Scenedesmus: 96 hr TLm at 24°C: 320 ppm   (356)
    —Arthropoda:
        Daphnia: 48 hr TLm at 23°C: 44 ppm   (356)
    —Fish:
        Lepomis macrochirus: static bioassay in fresh water at 23°C, mild aeration applied after 24 hr

material added ppm	% survival after 24 hr	48 hr	72 hr	96 hr	best fit 96 hr LC_{50} ppm
250	0	—	—	—	
180	0	—	—	—	100
125	100	90	50	0	
100	100	100	90	50	
79	100	100	100	100	(352)

Menidia beryllina: static bioassay in synthetic seawater at 23°C: mild aeration applied after 24 hr

material added ppm	% survival after 24 hr	48 hr	72 hr	96 hr	best fit 96 hr LC_{50} ppm
320	0	–	–	–	
240	0	–	–	–	
180	64	60	50	50	185
132	100	80	80	80	
100	100	100	100	100	(352)

—Mammalia: guinea pig: eye irritation: 3,300 ppm, 13 hr
  cat: inhalation: some deaths: 17,500 ppm, 30 min
  cat: inhalation: loss of weight: 4,200 ppm, 6 hr/day, 6 days (211)
—Man: mild eye and nose irritation: 200–300 ppm (211)
  unsatisfactory: >200 ppm
  symptoms of illness: 500 ppm
  severe toxic effects: 2,000 ppm, 60 min (185)

tert-butylacetate
$CH_3COOC(CH_3)_3$
A. PROPERTIES: colorless liquid; b.p. 96°C; sp.gr. 0.896 at 20°C
D. BIOLOGICAL EFFECTS:
  —Toxicity threshold (cell multiplication inhibition test):
    bacteria (*Pseudomonas putida*): 78 mg/l (1900)
    algae (*Microcystis aeruginosa*): 420 mg/l (329)
    green algae (*Scenedesmus quadricauda*): 3700 mg/l (1900)
    protozoa (*Entosiphon sulcatum*): 970 mg/l (1900)
    protozoa (*Uronema parduczi Chatton-Lwoff*): 1850 mg/l (1901)

n-butylacrylate
$CH_2CHCOOC_4H_9$
A. PROPERTIES: m.w. 128.2; m.p. −64°C; b.p. 145°C; v.p. 4 mm at 20°C, 10 mm at 36°C; v.d. 4.42; sp.gr. 0.90 at 20/4°C, solub. 1,600 mg/l at 20°C
B. AIR POLLUTION FACTORS: 1 mg/cu m = 0.19 ppm, 1 ppm = 5.33 mg/cu m
C. WATER POLLUTION FACTORS:
  —Reduction of amenities: organoleptic limit: 0.015 mg/l
  —Waste water treatment:
    A.C.: adsorbability: 0.193 g/g C; 95.9% reduction, infl.: 1,000 mg/l, effl.: 43 mg/l (32)
D. BIOLOGICAL EFFECTS:
  —Threshold conc. of cell multiplication inhibition of the protozoa *Uronema parduczi Chatton-Lwoff*: 21 mg/l (1901)
  —Mammals: rat: acute oral LD_{50}: 3.7 g/kg
    rat: inhalation: 5/6: 1,000 ppm; 4 hr
    rat: inhalation: sat. vap., 30 min. (211)

n-butylalcohol see *n*-butanol

2-butylalcohol *see sec*-butanol

***tert*-butylalcohol** *see t*-butanol

butylaldehyde *see* butyraldehyde

n-butylamine (l-aminobutane)
$CH_3(CH_2)_3NH_2$

Uses: intermediate for emulsifying agents; pharmaceuticals; insecticides; dyes; tanning agents

A. PROPERTIES: m.w. 73.1; m.p. −50°C; b.p. 78°C; v.p. 72 mm at 20°C; sp.gr. 0.74; $\log P_{oct}$ 0.68/0.88

B. AIR POLLUTION FACTORS: 1 mg/cu m = 0.334 ppm, 1 ppm = 2.99 mg/cu m
  −Odor: characteristic: quality: sour, ammoniacal
  hedonic tone: unpleasant to pleasant

T.O.C.: abs. perc. limit:	0.08 ppm	
50% recogn.:	0.24 ppm	
100% recogn.:	0.24 ppm	
O.I.: 100% recogn.:	449,166	(19)
O.I. at 20°C:	395,000	(316)

C. WATER POLLUTION FACTORS:

−BOD$_{5\ days}$:	26.5% of ThOD	
10 days:	48.2% of ThOD	
−BOD$_{15\ days}$:	50.0% of ThOD	
20 days:	48.8% of ThOD	
30 days:	48.8% of ThOD	
50 days:	52.3% of ThOD	(79)

  −BOD$_{10}^{20°C}$: 48.8% ThOD at 2.5 mg/l in mineralized dilution water with settled sewage seed (405)
  −Waste water treatment:
  A.C.: adsorbability: 0.103 g/g C; 52.0% reduction; infl.: 1,000 mg/l, effl.: 480 mg/l (32)
  degradation by *Aerobacter*: 200 mg/l at 30°C:
   parent: 100% in 22 hr
   mutant: 100% in 7 hr (152)
  reverse osmosis: 39.2% rejection from a solution of 0.01 M (221)

D. BIOLOGICAL EFFECTS:
  −Toxicity threshold (cell multiplication inhibition test):

bacteria (*Pseudomonas putida*):	800 mg/l	(1900)
algae (*Microcystis aeruginosa*):	0.14 − 0.19 mg/l	(329)
green algae (*Scenedesmus quadricauda*):	0.53 mg/l	(1900)
protozoa (*Entosiphon sulcatum*):	9 mg/l	(1900)
protozoa (*Uronema parduczi Chatton-Lwoff*):	1752 mg/l	(1901)

  −Arthropoda: brine shrimp: 24 hr TLm: 30–70 ppm (static) (355)
  −Fish: creek chub: critical range: 30–70 mg/l; 24 hr (226)
  Lepomis macrochirus: static bioassay in fresh water at 23°C, mild aeration applied after 24 hr

material added ppm	% survival after				best fit 96 hr LC_{50} ppm
	24 hr	48 hr	72 hr	96 hr	
79	80	20	10	10	
50	80	70	50	20	32
32	80	50	60	50	
10	100	100	100	100	(352)

Menidia beryllina: static bioassay in synthetic seawater at 23°C: mild aeration applied after 24 hr

material added ppm	% survival after				best fit 96 hr LC_{50} ppm
	24 hr	48 hr	72 hr	96 hr	
100	100	100	0	—	
50	95	60	0	—	
32	100	33	33	33	24
18	100	100	75	65	
10	100	100	100	100	(352)

—Mammalia: rat: acute oral LD_{50}: 0.5 g/cu m
           rat: acute oral LD_{50}: 0.2–0.4 g/cu m (10% soln)
           rat: inhalation: deaths: 4,000 ppm, 4 hr
           rat: inhalation: survived: 2,000 ppm, 4 hr
           rat: inhalation: survived: sat. vap., 2–5 min   (211)
—Man: irritating: 5–10 ppm   (211)

sec-butylamine (2-aminobutane)
$CH_3CH_2CH(CH_3)NH_2$

A. PROPERTIES: m.w. 73.1; m.p. −104°C; b.p. 62/63°C; v.d. 2.52; sp.gr. 0.72
B. AIR POLLUTION FACTORS: 1 mg/cu m = 0.334 ppm, 1 ppm = 2.99 mg/cu m
C. WATER POLLUTION FACTORS:
   —Impact on biodegradation processes: at 100 mg/l no inhibition of NH_3 oxidation by *Nitrosomonas* sp.   (390)
D. BIOLOGICAL EFFECTS:
   —Fish: creek chub: critical range: 20–60 mg/l; 24 hr   (226)
   —Mammalia: rat: inhalation: discomfort, lethargy, retarded weight gain; autopsy: organs normal: 233 ppm, 13 × 6.5 hr   (64)

t-butylamine borane (borane-*tert*-butylamine complex)

$$CH_3-\underset{\underset{CH_3}{|}}{\overset{\overset{CH_3}{|}}{C}}-NH_2 \cdot BH_3$$

Use: fogging agent in photoprocessing
A. PROPERTIES: m.w. 86.97; m.p. 98°C (decomposes)
C. WATER POLLUTION FACTORS:

theoretical	analytical
TOD = 2.95	Reflux COD = 38.7% recovery
COD = 2.21	Rapid COD = 24.6% recovery
NOD = 0.74	TKN = 92.3% recovery

$BOD_5 = 0.39$
$BOD_5/COD = 0.176$
BOD_5 (acclimated) = 0.31

—Both COD analytical methods exhibited poor recovery of the theoretical COD. The compound exhibited a moderate response to the BOD_5 analysis. (1828)
—Impact on conventional biological treatment systems:

	chemical conc. mg/l	effect	
unacclimated system	50	inhibitory	
	500	inhibitory	
acclimated system	134	biodegradable	(1828)

D. BIOLOGICAL EFFECTS:
—Algae: *Selenastrum capricornutum*: 0.1 mg/l no effect (1828)
    1.0 mg/l inhibitory
—Arthropoda: *Daphnia magna*: LC_{50} = 0.7 mg/l (1828)
—Fish: *Pimephales promelas*: LC_{50} between 10 and 18 mg/l (1828)

butylated hydroxytoluene   see 2,6-di-*tert*-butyl-4-methylphenol

n-**butylbenzene**   (1-phenylbutane)

$$\text{C}_6\text{H}_5-CH_2-CH_2-CH_2-CH_3$$

Manufacturing source: petroleum refining. (347)
Users and formulation: organic synthesis; pesticide mfg.; solvent for coating compositions; plasticizer; surface active agents; polymer linking agent; asphalt component; naphtha constituent. (347)
A. PROPERTIES: m.w. 134.21; m.p. -81°C; b.p. 183°C; v.p. 1 mm at 23°C; v.d. 4.62; sp.gr. 0.860 at 20°C
B. AIR POLLUTION FACTORS:
—Atmospheric reactions: R.C.R.: 1.03 (49)
—Manmade sources: in gasoline (high octane number): 0.08 wt % (387)
C. WATER POLLUTION FACTORS:
—$BOD_{35}^{25°C}$: 1.96 (n.s.i.) in seawater/inoculum: enrichment cultures of hydrocarbon oxydizing bacteria (521)
—ThOD: 3.22 (521)
—Reduction of amenities: organoleptic limit: 0.1 mg/l (n.s.i.) (181)
—Waste water treatment:

methods	temp °C	days observed	feed mg/l	days acclim.	% theor. oxidation	
A.S., Sd, BOD	20	5		14	14	
A.S., W	20	¼	50–100	14	6	(93)

D. BIOLOGICAL EFFECTS:
—Man: EIR: 6.4 (49)

sec-butylbenzene (2-phenylbutane)

A. PROPERTIES: m.w. 134.21; m.p. -83°C; b.p. 173°C; v.p. 1.1 mm at 20°C; v.d. 4.62; sp.gr. 0.862 at 20°C
B. AIR POLLUTION FACTORS:
  —Atmospheric reactions: R.C.R.: 1.31 (49)
D. BIOLOGICAL EFFECTS:
  —Man: EIR: 1.8 (49)

tert-butylbenzene (2-methyl-2-phenylpropane)

A. PROPERTIES: m.w. 134.21; m.p. -58°C; b.p. 169°C; v.p. 1.5 mm at 20°C; v.d. 4.62; sp.gr. 0.87 at 20°C; log P_{oct} 4.11
B. AIR POLLUTION FACTORS:
  —Atmospheric reactions: R.C.R.: 0.66 (49)
  —Ambient air quality:
   Los Angeles 1966: glc's: avg.: 0.002 ppm ($n = 136$)
                    highest value: 0.006 ppm (1319)
C. WATER POLLUTION FACTORS:
  —Reduction of amenities: T.O.C. = 0.05 mg/l (295)
  —Waste water treatment:

methods	temp °C	days observed	feed mg/l	days acclim.	% theor. oxidation	
A.S., W	20	5	50-200	14	1	
A.S., Sd, BOD	20	5		14	1	(93)

D. BIOLOGICAL EFFECTS:
  —Man: EIR: 0.9 (49)

p-tert-butylbenzoic acid (PTBBA)

A. PROPERTIES: colorless crystalline powder; m.p. 166°C; sp.gr. 1.142 at 20/4°C
C. WATER POLLUTION FACTORS:
  —BOD_5: 0.26 NEN-3235-5.4 (277)
  —COD: 2.37 NEN-3235-5.3 (277)

D. BIOLOGICAL EFFECTS:
- Fish: goldfish: LD_{50} (24 hr): 4 mg/l at pH 5
  (96 hr): 4 mg/l at pH 5
  (24 hr): 33 mg/l at pH 7
  (96 hr): 33 mg/l at pH 7 (277)
- Mammalia: rat: acute oral LD_{50}: 735 mg/kg
  mouse: acute oral LD_{50}: 568 mg/kg (277)

butylbenzylphthalate   (santicizer 160; BBP)

[Chemical structure: phthalate with -C(=O)-O-CH₂-CH₂-CH₂-CH₃ and -C(=O)-O-CH₂-phenyl groups]

Use: to plasticize or flexibilize synthetic resins, chiefly polyvinylchloride

A. PROPERTIES: clear oily liquid; m.w. 312.4; m.p. $<-35°C$; b.p. 370°C; sp.gr. 1.1 at 25/25°C; v.p. 8.6×10^{-6} mm at 20°C, 1.9 mm at 200°C; v.d. 10.8; solub. 2.9 ± 1.2 mg/l (in deionized water); $\log P_{oct}$ 4.78

C. WATER POLLUTION FACTORS:
- Water, soil and sediment quality:
  soil adsorption coefficient: 68 – 350 (1830)
  residue in natural water (U.S. rivers):
  avg.: approx. 0.35 µg/l
  range: n.d.–4.1 µg/l (n = 53)
  residue in sediments of natural waters:
  BBP detected in 7 out of 28 samples: avg. 136 ng/g
  range: n.d.–567 ng/g (1830)
  residue in Delaware river (U.S.A.): conc.range:
  winter: 0.4–1 ppb
  summer: 0.3 ppm (1051)
- Degradation

	% degradation		time	half life	
	primary*	ultimate**	days	days	
biodegradation					
activated sludge	93–99		1		
CO_2 evolution, aerobic		96	28		
gasproduction, anaerobic		<10	28		
river water	100		9	2	
lake water microcosm	>95		7	<4	
lake water microcosm		51–65	28		(1830)
photodegradation	<5		28	>100	(1830)
chemical degradation (hydrolysis)	<5		28	>100	(1830)

*disappearance of BBP as measured by gas chromatography
**mineralization under aerobic conditions to CO_2, under anaerobic conditions to H_2, CH_4 and CO_2

—Water treatment methods:
   domestic activated sludge plant: inlet: 8.0 µg/l
                                            outlet: 1.3 µg/l
                    outlet aerated lagoon: 1.0 µg/l   (1830)
   biodegradation: A.S. 48 hr: 99%   (1840)
                      continuous A.S.: 96%   (1841)
                  in river water, 1 week: 80%   (1841)

D. BIOLOGICAL EFFECTS:
—Acute lethality, measured in static tests:

species	96 hr EC_{50} or LC_{50} mg/l	no effect conc. mg/l	
algae*			
Microcystis	1000	560	
Dunaliella	1.0	0.3	
Navicula	0.6	0.3	
Skeletonema	0.6	0.1	
Selenastrum	0.4	0.1	
invertebrates			
Daphnia magna**	3.7	1.0	
mysid shrimp	0.9	0.4	
fish			
fathead minnow	2.1–5.3	1.0–2.2	
bluegill	1.7	0.38	
rainbow trout	3.3	<0.36	
sheepshead minnow	3.0	1.0	(1830)

*LC_{50} based on cell counts
**48 hr EC 50

—Chronic toxicity:
  fish: fathead minnows: 4 d LC_{50}: 2.32 mg/l (flow through test)
                               14 d LC_{50}: 2.25 mg/l (flow through test)   (1830)
        fathead minnow post-hatch embryo-larval stage: MATC: 0.14–0.36 mg/l
                                                                                (1830)
  invertebrate: *Daphnia magna*: MATC: 0.26–0.76 mg/l   (1830)

—Bio-concentration:
  bluegill: BCF: 663 (based on ^{14}C determinations)
            depuration half life: <2 days   (1830)

n-butylcarbinol   *see* n-pentanol

butylcarbitol   *see* diethyleneglycolmonobutylether

butylcarbitolacetate (butyldigolacetate)
$C_4H_9OCH_2CH_2OCH_2CH_2OOCCH_3$
A. PROPERTIES: m.p. −32°C; b.p. 246°C; v.p. 0.04 mm at 20°C; v.d. 7.02; sp.gr. 0.98 at 20°C; solub. 65,000 mg/l at 20°C
C. WATER POLLUTION FACTORS:
  —BOD_5:   13.3% of ThOD
    10 days: 18.4% of ThOD
    15 days: 24.6% of ThOD
    20 days: 67.6% of ThOD   (79)

p-tert-butylcatechol (4-*tert*-butylpyrocatechol; 4-*tert*-butyl-1,2-dihydroxybenzene)

[Structure: benzene ring with two OH groups (1,2-positions) and a C(CH$_3$)$_3$ group]

Uses: polymerization inhibitor for styrene-butadiene and other olefins

A. PROPERTIES: white crystalline solid; m.w. 166.22; m.p. 56–58°C; b.p. 285°C; d. 1,049 at 60/25°C; v.p. 0.0028 mm at 25°C; sp.gr. 1.05 at 60/25°C; solub. 0.2 wt % at 25°C

C. WATER POLLUTION FACTORS:
—Odor threshold: detection: 1.0 mg/l   (998)

D. BIOLOGICAL EFFECTS:
—Mammals: rat: single oral LD$_{50}$: 2820 mg/kg
          guinea pig: single oral LD$_{50}$: 200–800 mg/kg   (1546)

butylcellosolve (butylglycolether; glycol monobutylether; butylglycol; 2-butoxyethanol; ethyleneglycolmono-*n*-butylether; butyl"Oxitol")
C$_4$H$_9$OCH$_2$CH$_2$OH

A. PROPERTIES: colorless liquid; m.w. 118.17; m.p. <−40°C; b.p. 170°C; v.p. 0.6 mm at 20°C; v.d. 4.07; sp.gr. 0.903 at 20/4°C;

B. AIR POLLUTION FACTORS:
—Odor: characteristic: quality: sweet, ester
                    hedonic tone: pleasant

T.O.C.:		
abs. perc. limit:	0.10 ppm	
50% recogn.:	0.35 ppm	
100% recogn.:	0.48 ppm	
O.I. 100% recogn.:	2,729	(19)
O.I. at 20°C:	1,650	(316)

C. WATER POLLUTION FACTORS:
—BOD: 0.71 -NEN 3235-5.4
        1.68 adapted sew. -NEN 3235-5.4   (277)
—COD: 2.20 NEN 3235-5.3   (277)
—Waste water treatment:
    A.C.: adsorbability: 0.112 g/g carbon, 55.9% reduction, infl.: 1,000 mg/l, effl.: 441 mg/l   (32)

D. BIOLOGICAL EFFECTS:
—Toxicity threshold (cell multiplication inhibition test):

bacteria (*Pseudomonas putida*):	700 mg/l	(1900)
algae (*Microcystis aeruginosa*):	35 mg/l	(329)
green algae (*Scenedesmus quadricauda*):	900 mg/l	(1900)
protozoa (*Entosiphon sulcatum*):	91 mg/l	(1900)
protozoa (*Uronema parduczi Chatton-Lwoff*):	463 mg/l	(1901)

—Arthropoda: brown shrimp (*Crangon crangon*)
            LC$_{50}$ (48 hr): avg. 800 mg/l
                      range: 600–1000 mg/l

LC_{50} (96 hr): avg. 775 mg/l
range: 550–950 mg/l (310)
—Fish: goldfish: LD_{50} (24 hr): 1650 mg/l—modified ASTM D 1345 (277)
guppy (*Poecilia reticulata*): 7 d LC_{50}: 983 ppm (1833)
Lepomis macrochirus: static bioassay in fresh water at 23°C, mild aeration applied after 24 hr

material added ppm	24 hr	% survival after 48 hr	72 hr	96 hr	best fit 96 hr LC_{50} ppm
2,400	40	40	20	0	
1,800	50	50	50	30	
1,000	100	100	90	80	1,490
790	100	100	100	100	
320	100	100	100	100	(352)

Menidia beryllina: static bioassay in synthetic seawater at 23°C, mild aeration applied after 24 hr

material added ppm	24 hr	% survival after 48 hr	72 hr	96 hr	best fit 96 hr LC_{50} ppm
1,800	90	70	30	20	
1,320	100	100	90	30	1,250
1,000	100	100	100	70	(352)

—Mammalia: rat: inhalation: no effects: 20 ppm, 15 × 6 hr (65)
rat: inhalation: 0/5, severely affected kidneys: 80 ppm, 4 hr (211)
rat: single oral dose LD_{50}: 2.5 g/kg
rabbit: single oral dose: no effects: 0.5 ml/kg
rabbit: single oral dose: LD_{50}: 0.32 g/kg
mouse: single oral dose: LD_{50}: 1.2 g/kg
guinea pig: single oral dose: LD_{50}: 1.2 g/kg
rat: repeated oral dose: no effect: 0.125% in diet (211)

butylcellosolveacetate (ethyleneglycolmonobutyletheracetate)
$C_4H_9OCH_2CH_2OOCCH_3$
Use: high boiling solvent for nitrocellulose lacquers, epoxy resins
A. PROPERTIES: colorless liquid; fruity odor; b.p. 192.3°C; sp.gr. 0.94 at 20/20°C
B. AIR POLLUTION FACTORS:
—Odor: characteristic: quality: sweet, ester
hedonic tone: pleasant
T.O.C.: abs. perc. limit: 0.10 ppm
50% recogn.: 0.35 ppm
100% recogn.: 0.48 ppm
O.I.: 2,729 (19)

n-butylchloride (1-chlorobutane)
$CH_3(CH_2)_2CH_2Cl$
A. PROPERTIES: m.w. 92.57; m.p. −123°C; b.p. 78°C; v.p. 80.1 mm at 20°C; v.d. 3.20; sp.gr. 0.884; solub. 660 mg/l at 12°C; $\log P_{oct}$ 2.39
B. AIR POLLUTION FACTORS:
—Odor: characteristic: quality: pungent
hedonic tone: unpleasant

T.O.C.: abs. perc. limit: 8.82 ppm
　　　　 50% recogn.: 13.3 ppm
　　　　100% recogn.: 16.7 ppm
　　　　O.I.: 6,377 (19)
C. WATER POLLUTION FACTORS:
　—Waste water treatment:
　　A.S.: after 6 hr: 0.9% of ThOD
　　　　　　12 hr: 1.3% of ThOD
　　　　　　24 hr: 2.6% of ThOD (88)
D. BIOLOGICAL EFFECTS:
　—Fish: guppy (*Poecilia reticulata*) 7 d LC_{50}: 97 ppm (1833)

butylchloroacetate
$C_4H_9OOCCH_2Cl$
C. WATER POLLUTION FACTORS:
　—Manmade sources: 60 ml/min pure water passed through 25 ft, $\frac{1}{2}$ inch I.D. tube of general chemical grade PVC contained 0.66 ppb butylchloroacetate which constituted 6.14% of total contaminant concentration (430)

n-butylcyanide　see valeronitrile

butyldiglycol　see diethyleneglycolmonobutylether

butyldigol　see diethyleneglycolmonobutylether

butyldigolacetate　see butylcarbitolacetate

4-*tert*-butyl-1,2-dihydroxybenzene　see *p-tert*-butylcatechol

2-*tert*-butyl-4,6-dinitrophenylacetate　(dinoterpacetate)

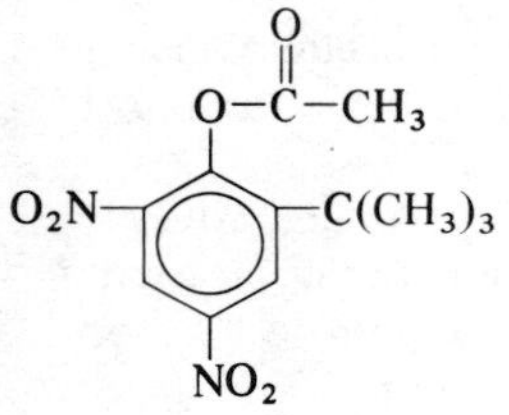

Use: herbicide
A. PROPERTIES: pale yellow crystals; m.p. 133–134.5°C
D. BIOLOGICAL EFFECTS:
　—Harlequin fish (*Rasbora heteromorpha*)

mg/l	24 hr	48 hr	96 hr	3 m (extrapolated)	
LC_{10} (F)	0.045	0.038	0.031		
LC_{50} (F)	0.068	0.051	0.039	0.03	(331)

　—Mammals: acute oral LD_{50} (hens):　>4,000 mg/kg
　　　　　　 acute oral LD_{50} (rabbits):　100 mg/kg

acute oral LD_{50} (rat):   62 mg/kg
acute oral LD_{50} (rat):   >2,000 mg/kg   (1854)

2(2-butyl-4,6-dinitrophenyl)-3,3-dimethylacrylate *see* binapacryl

butyldioxitol *see* diethyleneglycol monobutylether

α-butylene (1-butene; ethylethylene)
$CH_3CH_2CHCH_2$
A. PROPERTIES: m.w. 56.10; m.p. $-130°C$; b.p. $-6°C$; v.p. 400 mm at $-21.7°C$, 760 mm at $-6.3°C$; v.d. 1.94; sp.gr. 0.67 liquefied
B. AIR POLLUTION FACTORS: 1 ppm = 1.23 mg/cu m; 1 mg/cu m = 430 ppm
  —Odor: characteristic: quality: gashouse odor
    T.O.C.: 0.160 mg/cu m = 69 ppb   (307)
    faint odor: 50–59 mg/cu m   (129)
    O.I. at 20°C: 43,480,000   (316)
    T.O.C.: recognition: 39.2 mg/cu m   (761)
      2.1 mg/cu m   (710)
      1.2 mg/cu m   (724)
  —Atmospheric reactions: R.C.R.: 2.26 (n.s.i.)   (49)
  —Manmade sources:
    in diesel engine exhaust gas: 1.8% of emitted HC.s (n.s.i.)   (72)
    expected glc's in USA urban air: range: 1–20 ppb (n.s.i.)   (102)
    in exhaust of gasoline engines:
      62-car survey: 1.8 vol % of total exhaust HC's   (391)
      15-fuel study: 3 vol % (+ isobutylene) of total exhaust HC's   (392)
      engine-variable study: 6.0 vol % (+ isobutylene) total exhaust HC's   (393)
      evaporation from gasoline fuel tank: 4.6 vol % of total evaporated HC's
      evaporation from carburetor: 0–0.3 vol % of total evaporated HC's
                                                         (398; 399; 400; 401; 402)
  —Sampling and analysis:
    PMS: detection limit: 2–9 ppm   (200)
C. WATER POLLUTION FACTORS:
  —Reduction of amenitites: organoleptic limit: 0.2 mg/l   (181)
D. BIOLOGICAL EFFECTS:
  —Plants: tomato: epinasty in petiole: 50,000 ppm, 2 days   (109)
  —Man: EIR: 1.3 (n.i.s.)   (49)

β-butylene *see trans*-2-butene

γ-butylene *see* isobutene

1,2-butyleneoxide

$(CH_3)_2CCH_2O$

A. PROPERTIES: m.w. 72.1; m.p. $-60°C$; b.p. $65°C$; v.d. 2.49; sp.gr. 0.83; solub. 82,400 mg/l at 25°C

318 BUTYL-2,3-EPOXYPROPYLETHER

B. AIR POLLUTION FACTORS: 1 mg/cu m = 0.340 ppm, 1 ppm = 2.94 mg/cu m
  —Odor: characteristic: quality: sweet, alcohol
  hedonic tone: pleasant
  T.O.C.: abs. perc. limit: 0.07 ppm
          50% recogn.: 0.71 ppm
         100% recogn.: 0.71 ppm
         O.I.: 260,563   (19)
D. BIOLOGICAL EFFECTS:
  —Mammalia: rat: single oral LD_{50}: 0.5 g/kg   (221)
  rats, guinea pigs, rabbits: inhalation: repeated 7 hr exposures at 400 ppm can be tolerated for prolonged periods   (211)

butyl-2,3-epoxypropylether   *see n*-butylglycidylether

butylethanoate   *see n*-butylacetate

n-butylether   (di-*n*-butylether; 1-butoxybutane)
$C_4H_9OC_4H_9$
  Use: solvent for hydrocarbons; fatty materials; extracting agent
A. PROPERTIES: m.w. 130.2; m.p. -95°C; b.p. 141°C; v.p. 4.8 mm at 20°C; v.d. 4.5; sp.gr. 0.769 at 20/20°C; solub. 300 mg/l at 20°C;
B. AIR POLLUTION FACTORS: 1 mg/cu m = 0.188 ppm; 1 ppm = 5.33 mg/cu m
  —Odor: characteristic: quality: fruity, sweet
  hedonic tone: pleasant
  T.O.C.: abs. perc. limit: 0.07 ppm
          50% recogn.: 0.24 ppm
         100% recogn.: 0.47 ppm
         O.I.: 13,978   (19)
         detection: 8 mg/cu m   (691)
C. WATER POLLUTION FACTORS:
  —Waste water treatment:
  A.C.: adsorbability: 0.039 g/g C, 100.0% reduction, infl.: 1,000 mg/l, effl.: nil   (32)
D. BIOLOGICAL EFFECTS:
  —Threshold conc. of cell multiplication inhibition of the protozoa *Uronema parduczi Chatton-Lwoff*: >40 mg/l   (1901)
  —Mammals: rat: acute oral LD_{50}: 7.4 g/kg   (211)
         rat: inhalation: 2/6: 4,000 ppm, 4 hr   (211)
  —Man: irritation of eyes and nose: 200 ppm   (211)

N,butyl-N-ethyl-2,6-dinitro-4-trifluoromethylaniline   *see* balan

butylethylene   *see* 1-hexene

butylethylketone   *see* 3-heptanone

N,butyl-N-ethyl-α,α,α-trifluoro-2,6-dinitro-*p*-toluidine   *see* balan

n-butylformate
HCOOC$_4$H$_9$
A. PROPERTIES: m.w. 102.13; m.p. -90°C; b.p. 106.8°C; v.p. 30 mm at 25°C; v.d. 3.5; sp.gr. 0.8885 at 20/4°C
B. AIR POLLUTION FACTORS: 1 mg/cu m = 0.240 ppm, 1 ppm = 4.17 mg/cu m
   —Odor:
   detection: 6.0 mg/kg (894)
   T.O.C.: 17 ppm = 70 mg/cu m (210)
   distinct odor: 60 mg/cu m = 20 ppm (278)
C. WATER POLLUTION FACTORS:
   —Reduction of amenities: T.O.C. = 6.0 mg/l (299)
D. BIOLOGICAL EFFECTS:
   —Mammalia: cat: inhalation: death: 10,000 ppm, 60 min
               dog: inhalation: narcosis: 10,000 ppm, 60 min (211)
   —Man: intolerable irritation: 10,000 ppm, <1 min (211)

n-butylglycidylether (BGE, 1-butoxy-2,3-epoxypropane; 3-butoxy-1,2-epoxypropane; 2,3-expoxypropylbutylether; ERL 0810; butyl-2,3-epoxypropylether; glycidylbutylether; (butoxymethyl)oxirane)

$$C_4H_9OCH_2CH\underset{O}{-}CH_2$$

Use: component of epoxy resin systems. The epoxy group of the glycidylether reacts during the curing process and glycidylethers are therefore generally no longer present in completely cured products
A. PROPERTIES: colorless liquid; m.w. 130.21; b.p. 164/168°C; v.p. 3.2 mm at 25°C; v.d. 3.78/4.50; sp.gr. 0.908 at 25/4°C; solub. 20,000 mg/l at 20°C
B. AIR POLLUTION FACTORS: 1 mg/cu m = 0.188 ppm, 1 ppm = 5.32 mg/cu m
   —Odor: characteristic: hedonic tone: irritating, not unpleasant
D. BIOLOGICAL EFFECTS:
   —Mammalia: mouse: acute intragastric LD$_{50}$: 1.53 g/kg
              rat: acute intragastric LD$_{50}$: 2.26 g/kg
              rat: inhalation: no signs of toxicity: 75 ppm, 50 × 7 hr
              rat: inhalation: retarded growth: 150 ppm, 50 × 7 hr
              rat: inhalation: chronic toxicity: 300 ppm, 50 × 7 hr
              rat: inhalation: LC$_{50}$:   1,030,8 hr   (211)
              rat: inhalation: 7 hr/d, 5 d/w, 50 exposures:
                   75 ppm: slight patchy atrophy of the testes in 1 of
                           10 animals
                   300 ppm: atrophic testes in 5 of 10 animals   (1407)
              rat: i.m. injections for 3 consecutive days: 400 mg/kg: increased
                   leucocyte count   (1405)

butylglycol *see* butylcellosolve

butylglycolether *see* butylcellosolve

n-butylmercaptan (butanethiol)
C$_4$H$_9$SH

BUTYLMETHYLCARBINOL

A. PROPERTIES: m.w. 90.18; m.p. $-116°C$; b.p. $98°C$; v.d. 3.1; sp.gr. 0.84 at $20°C$; solub. 590 mg/l at $22°C$; $\log P_{oct}$ 2.28
B. AIR POLLUTION FACTORS: 1 mg/cu m = 0.27 ppm, 1 ppm = 3.75 mg/cu m
—Odor: characteristic: strong, unpleasant

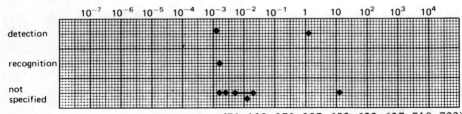

(71; 129; 279; 307; 602; 622; 637; 710; 723)

O.I. at $20°C$: 49,000,000 (316)
—Control methods:
wet scrubber: water: effluent: 200,000 odor units/scf
$KMnO_4$: effluent: 33 odor units/scf (115)
C. WATER POLLUTION FACTORS:
—Reduction of amenities: T.O.C.: average: 0.006 mg/l
range: 0.001 to 0.06 mg/l (294)(97)
D. BIOLOGICAL EFFECTS:
—Fish: *Lepomis macrochirus*: TLm (24 hr): 7.4 mg/l
(48 hr): 5.5 mg/l (30)

butylmethylcarbinol *see* 2-hexanol

butylmethylketone (2-hexanone; methylbutylketone)
$C_4H_9COCH_3$
A. PROPERTIES: colorless liquid; m.w. 100.2; m.p. $-57°C$; b.p. $128°C$; v.d. 3.45; sp.gr. 0.830 at $0/4°C$; solub. 35,000 at $20°C$; v.p. 2 mm at $20°C$; $\log P_{oct}$ 1.38
B. AIR POLLUTION FACTORS:
—Odor threshold: 0.28–0.35 mg/cu m (610)
—Sampling and analysis: photometry: min. full scale: 1,750 ppm (53)
C. WATER POLLUTION FACTORS:
—Waste water treatment:
A.C.: adsorbability: 0.159 g/g C, 80.7% reduction, infl.: 998 mg/l, effl.: 191 mg/l (32)

butyloctylfumarate
$C_4H_9OOC-CH=CHCOOC_8H_{17}$
Use: plasticizer
60 ml/min. pure water passed through 25 ft, $\frac{1}{2}$ in. I.D. tube of general chemical grade PVC contained 1.45 ppb butyloctylfumarate which constituted 13.49% of total contaminant concentration (430)

butyloxitol *see* butylcellosolve

p-tert-**butylphenol** (4-(α,α-dimethylethylphenol))

A. PROPERTIES: white aromatic flake; solub. 700 mg/l; m.w. 150.21; m.p. 99°C; b.p. 236°C; sp.gr. 0.908 at 114/4°C, log P_{oct} 3.31

C. WATER POLLUTION FACTORS:
 —Reduction of amenities: odor threshold: detection: 0.8 mg/l (998)
 approx. conc. causing adverse taste in fish: 0.03 mg/l (41)(998)
 —Impact on biodegradation processes:
 inhibition of degradation of glucose by *Pseudomonas fluorescens* at: 25 mg/l
 inhibition of degradation of glucose by *E. coli* at: >100 mg/l (293)

D. BIOLOGICAL EFFECTS:
 —Bacteria: *E. coli*: LD_0: >100 mg/l
 —Algae: *Scenedesmus*: LD_0: 10 mg/l
 —Arthropoda: *Daphnia*: LD_0: 8 mg/l (30)
 —Mammals: guinea pigs: single oral LD_{50}: 400–2000 mg/kg (1546)

o-sec-**butylphenyl-N-methylcarbamate** (BPMC)
Use: BPMC is one of several carbamate insecticides applied in large quantities in Japan to control planthoppers and leafhoppers on rice plants

C. WATER POLLUTION FACTORS:
 —Aquatic reactions:
 disappearance in Saga soil at 30°C humidity:
 at 0.2 ppm initial concentration 55% BPMC remained in paddy soil after 50 days
 at 1.0 ppm initial concentration 45% BPMC remained in paddy soil after 50 days
 at 10 ppm initial concentration 5% BPMC remained in paddy soil after 50 days
 Since the disappearance rate of BPMC in soils was retarded by addition of sodium azide, it was suggested that soil microorganisms participated in the degradation of BPMC (1324)

butylphosphate (*n*-butylphosphoric acid; acid butylphosphate)
A. PROPERTIES: water white liquid, sp.gr. 1.120–1.125 at 25/4°C; log P_{oct} 0.28
D. BIOLOGICAL EFFECTS:
 —Bacteria: *Pseudomonas putida*: inhibition of cell multiplication starts at >100 mg/l (329)
 —Algae: *Microcystis aeruginosa*: inhibition of cell multiplication starts at 4.1 mg/l (329)

n-**butylphosphoric acid** *see* butylphosphate

butylphthalate *see* di-*n*-butylphthalate

4-*tert*-butylpyrocatechol *see p-tert-*butylcatechol

n-butylsulfide (dibutylsulfide; butylthiobutane)
$(CH_3CH_2CH_2CH_2)_2S$
A. PROPERTIES: m.w. 146.29; m.p. $-79.7°C$; b.p. $182°C$; sp.gr. 0.839 at $16/0°C$; v.p. 1 mm at $21.7°C$, 10 mm at $66.4°C$, 40 mm at $96.0°C$
B. AIR POLLUTION FACTORS:
 —Odor: hedonic tone: unpleasant
   T.O.C.: 0.012 mg/cu m = 2 ppb (307)
   O.I. at $20°C$: 658,000 (316)
C. WATER POLLUTION FACTORS:
 —Reduction of amenities: faint odor: at 0.0011 mg/l (129)

butylsulfonate, sodium
$C_4H_9SO_3Na$
D. BIOLOGICAL EFFECTS:
 —Fish: *Daphnia magna*: TLm, 24 hr: 8,000 mg/l
   TLm, 48 hr: 8,000 mg/l
   TLm, 72 hr: 5,400 mg/l
   TLm, 96 hr: 2,700 mg/l (153)

butylthiobutane *see n-*butylsulfide

***p-tert*-butyltoluene** (1-methyl-4-*tert*-butylbenzene; 8-methylparacymene; PTBT)

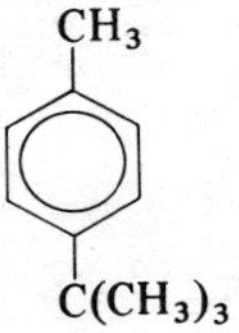

$$\underset{C(CH_3)_3}{\overset{CH_3}{\bigcirc}}$$

A. PROPERTIES: m.w. 148.25; m.p. $-62.53°C$; b.p. $192.8°C$; sp.gr. 0.857 at $20/20°C$
B. AIR POLLUTION FACTORS: 1 mg/cu m = 0.16 ppm, 1 ppm = 6.05 mg/cu m
 —Odor: immediate recognition at 5 ppm (211)
C. WATER POLLUTION FACTORS:
 —BOD_5: 0.06 NEN 3235-5.4
   0.19 adapted sew. NEN 3235-5.4 (277)
 —COD: 250 NEN 3235-5.3 (277)
D. BIOLOGICAL EFFECTS:
 —Fish: goldfish: LD_{50} (24 hr): 3 mg/l—modified ASTM-D1345 (277)
 —Mammalia: rat: inhalation: LC_{50}: 165 ppm, 8 hr
   rat: inhalation: LC_{50}: 248 ppm, 4 hr
   rat: inhalation: LC_{50}: 934 ppm, 1 hr
     no unusual behaviour: 50 ppm 7 hr/day, 25 days
   (211)
   rabbit: percutane LD_{50}: 13.8-27.8 ml/kg
 —Man: moderate eye irritation: 80 ppm, 5 min (211)

sec-butylxanthate (butylxanthic acid; butyldithiocarbonic acid)

$$CH_3-\underset{\underset{CH_3}{|}}{CH}-CH_2-O-CS-SH$$

D. BIOLOGICAL EFFECTS:
 —Fish: rainbow trout (*Salmo gairdneri*) 96 hr, LC_{50}: (1087)
   Cyanamid C 301—sodium salt: 100-166 mg/l (static test) (1087)
   Dow Chemical Z 12—sodium salt: 320 mg/l (static test) (1087)
 —Mammals: ingestion: rats: single oral LD_{50}: >2000 mg/kg (sodium salt) (1546)

butyraldehyde (1-butanal; butylaldehyde; butyric aldehyde)
$CH_3CH_2CH_2CHO$
A. PROPERTIES: m.w. 72.1; m.p. -97/-99°C; b.p. 75/76°C; v.p. 71 mm at 20°C; v.d. 2.48; sp.gr. 0.817 at 20/4°C; solub. 37,000 mg/l, 71,000 mg/l; $\log P_{oct}$ 1.18
B. AIR POLLUTION FACTORS: 1 ppm = 2.9 mg/cu m, 1 mg/cu m = 0.340 ppm
 —Odor: characteristic: quality: sweet, rancid
                       hedonic tone: unpleasant
 —T.O.C.: abs. perc. limit: 0.0046 ppm
          50% recogn.:   0.0092 ppm
          100% recogn.:   0.039 ppm
          O.I.: 2,984,615 (19)
 —T.O.C.: recognition: 0.013-0.014 mg/cu m (610)
          15 mg/cu m (788)
          0.042 mg/cu m (842)
 —Manmade sources:
   in municipal sewer air: 10-100 ppb (212)
   in diesel exhaust: 0.3 ppm (311)
   in exhaust of gasoline engine: 0.4-4.3 vol % of total exhaust aldehydes
     (394; 395; 396; 397)
   in exhaust of a 1970 Ford Maverick gasoline engine operated on a chassis dynamometer following the 7-mode California cycle:
     from API #7 gasoline: 8.3 ppm (incl. unknown compound)
     from API #8 gasoline: 4.1 ppm (incl. unknown compound) (1053)
 —Control methods: catalytic incineration over commercial Co_3O_4 (1.0-1.7 mm) granules, catalyst charge 13 cm^3, space velocity 45000 h^{-1} at 100 ppm in inlet:
   80% decomposition at 200°C
   99% decomposition at 225°C
   80% conversion to CO_2 at 210°C
   95% conversion to CO_2 at 245°C (346)
 —Comparison of catalyst performance at space velocity of 80,000 hr^{-1}:

catalyst	feed conc. ppm	reactor inlet temp. °C	% Odor removal ASTM	% conversion GC
0.5% Pt on γ-Al_2O_3	1537	160	69	64
0.1% Pt on γ-Al_2O_3	1537	193	93	94
0.5% Pd on γ-Al_2O_3	1537	290	99	100
10% Cu on γ-Al_2O_3	1537	381	97	100

(1221)

BUTYRAMIDE

C. WATER POLLUTION FACTORS:
- $BOD_{5\ days}$: 1.06 std.dil.sew. (255)
  - 1.6 std.dil.sew. (27)
  - 28% of ThOD (220)
  - 43.4% ThOD (79)
- $BOD_{10\ days}$: 59.8% ThOD (79)
  - 15 days: 61.5% ThOD (79)
  - 20 days: 66.4% ThOD (79)
- $BOD_{30\ days}$: 64% ThOD (79)
  - 40 days: 72.2% ThOD (79)
  - 50 days: 68.9% ThOD (79)
- COD: 99% of ThOD (220)
- $KMnO_4$: acid: 4, alkaline 8% ThOD (220)
- ThOD: 2.44
- Reduction of amenities: T.O.C.: 0.009 mg/l (305)
  - detection: 0.0373 mg/kg (874)
- Waste water treatment:
  - A.C.: adsorbability: 0.106 g/g carbon; % reduction = 52.8, influent: 1,000 mg/l
    effluent: 472 mg/l (32)

	influent	effluent
anaerobic lagoon: 22 lb COD/day/1,000 cu ft:	190 mg/l;	50 mg/l
48 lb COD/day/1,000 cu ft:	190 mg/l;	35 mg/l (32)

  - A.S.: after 6 hr: 14.2% of ThOD
    - 12 hr: 21.7% of ThOD
    - 24 hr: 22.8% of ThOD (88)
  - aeration by compressed air (stripping effect): 85% removal in 8 hr (30)
  - reverse osmosis: 72.1% rejection from a 0.01 M solution (221)

D. BIOLOGICAL EFFECTS:
- Toxicity threshold (cell multiplication inhibition test):
  - bacteria (*Pseudomonas putida*): 100 mg/l (1900)
  - algae (*Microcystis aeruginosa*): 19 mg/l (329)
  - green algae (*Scenedesmus quadricauda*): 83 mg/l (1900)
  - protozoa (*Entosiphon sulcatum*): 4.2 mg/l (1900)
  - protozoa (*Uronema parduczi Chatton-Lwoff*): 98 mg/l (1901)
- Mammalia: rat: inhalation: no effect 1,000 ppm, 12 × 6 hr (65)
  - rat: inhalation: 1/6   8,000 ppm, 14 hr (104)
  - rat: inhalation: LC_{50} 60,000 ppm, 0.5 hr (104)
  - rat: acute oral LD_{50}: 5.9 g/kg (211)

butyramide (butanamide; butyric amide)
$CH_3CH_2CH_2CONH_2$

A. PROPERTIES: m.w. 87.12; m.p. 116°C; b.p. 216°C; sp.gr. 1.032 at 20/4°C; solub. 162,800 mg/l at 15°C; log P_{oct} −0.21

C. WATER POLLUTION FACTORS:
- Waste water treatment:
  - A.S.: after 6 hr: 1.3% of ThOD
    - 12 hr: 3.8% of ThOD
    - 24 hr: 6.4% of ThOD (89)
  - reverse osmosis: 40.5% rejection from a 0.01 M solution (221)

n-butyric acid (*n*-butanoic acid; ethylacetic acid)
CH₃(CH₂)₂COOH
A. PROPERTIES: colorless liquid; m.w. 88.1; m.p. -5.5/-8°C; b.p. 163.7°C at 757 mm; v.p. 0.43 mm at 20°C, 1.4 mm at 30°C; v.d. 3.04; sp.gr. 0.959 at 20/4°C; solub. 56,200 mg/l at -1.1°C; sat. conc. 2.9 g/cu m at 20°C, 5.6 g/cu m at 30°C; log P_{oct} 0.79
B. AIR POLLUTION FACTORS: 1 mg/cu m = 0.27 ppm; 1 ppm = 3.66 mg/cu m
  —Odor: quality: sour

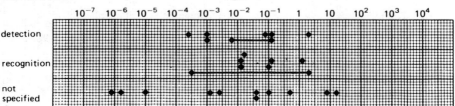

(57; 151; 210; 279; 291; 307; 602; 610; 623; 635; 667; 683; 684; 706; 707; 708; 713; 753; 762; 765; 778; 779; 831; 837)

—Control methods:
  A.C.: retentivity: 35 wt% of adsorbent (83)
  thermal incineration for odor control: min. temp.: 1,425°F (94)
  catalytic incineration over commercial Co_3O_4 (1.0-1.7 mm) granules, catalyst charge 13 cm³, space velocity 45000 h⁻¹ at 140 ppm inlet conc.:
    80% decomposition at 210°C
    90% decomposition at 230°C
    99% decomposition at 275°C
    80% conversion to CO_2 at 270°C
    90% conversion to CO_2 at 360°C (346)
  —Comparison of catalyst performance at space velocity of 80,000/hr

catalyst	feed conc. ppm	reactor inlet temp. °C	% odor removal ASTM	% conversion GC
0.5% Pt on γ-Al₂O₃	1473	225	100	98
0.1% Pt on γ-Al₂O₃	1473	240	75	81
0.5% Pd on γ-Al₂O₃	1473	251	97	100
10% Cu on γ-Al₂O₃	1473	400	96	100

C. WATER POLLUTION FACTORS: (1221)
  —BOD₅: 0.34 (41)
    0.89 (282)
    0.90 std.dil.sew. (284)
    0.34 std.dil./spec.culture (283)
    1.16 std.dil./acclimated (41)
  —BOD₁₀: 0.74 std.dil./spec.culture (283)
  —BOD₂₀: 1.45 (30)
  —BOD$_{35}^{25}$: 1.38 in seawater/inoculum: enrichment cultures of hydrocarbon oxidizing bacteria (521)
  —COD: 1.65 (41)
    1.75 (27)

BUTYRIC ALDEHYDE

  –$KMnO_4$: 0.006 (30)
  –ThOD: 1.818 (30)

odor thresholds   mg/kg water

[odor threshold chart: detection points at ~10^{-2}, ~10^{-1} range; not specified points at ~10^{-2} and ~10^{0}]

(295; 296; 898; 896; 907; 924)

– Manmade sources:
  contents in domestic sewages: 0.4–17 mg/l (85)
  average content in secondary sewage effluents: 30.7 µg/l (86)
  in domestic sewage effluent: 5 µg/l (227)
  in year old leachate of artificial sanitary landfill: 48.8 g/l (1720)
– Waste water treatment:
  A.C.: adsorbability: 0.119 g/g carbon; % reduction: 59.5; infl. = 1,000 mg/l,
    effl.: = 405 mg/l (32)
  A.S.: after 6 hr: 17.6% of ThOD
          12 hr: 25.2% of ThOD
          24 hr: 27.6% of ThOD (88)
  reverse osmosis: 16.4% rejection from a 0.01 M solution (221)
  powdered carbon: at 100 mg/l sodium salt (pH 7.5)–carbon dosage 1000 mg/l:
    4% absorbed (520)

D. BIOLOGICAL EFFECTS:
– Toxicity threshold (cell multiplication inhibition test):
  bacteria (*Pseudomonas putida*): 875 mg/l (1900)
  algae (*Microcystis aeruginosa*): 318 mg/l (329)
  green algae (*Scenedesmus quadricauda*): 2600 mg/l (1900)
  protozoa (*Entosiphon sulcatum*): 26 mg/l (1900)
  protozoa (*Uronema parduczi Chatton-Lwoff*): 129 mg/l (1901)
– Algae: *Chlorella pyrenoidosa*: toxic: 340 mg/l (41)
       Scenedesmus: toxic: 200 mg/l (30)
– Protozoa: *Vorticella campanula*: toxic: 10 mg/l (30)
         Paramaecium caudatum: toxic: 250 mg/l (30)
– Arthropoda: *Daphnia*: toxic: 60 mg/l (30)
           D.manga: TLm: 61 mg/l, 48 hr (153)
– Mollusca: *Limnea ovata*: toxic: 50 mg/l (30)
– Fish: *L.macrochirus*: TLm: 200 mg/l, 24 hr (153)
     L.macrochirus: TLm, 24 hr: 5,000 mg/l (sodium salt) (1294)
     Salmo irideus: TLm: 400 mg/l (30)
     trout: TLm: 20–40 mg/l (30)
– Mammalia: rat: single oral dose LD_{50}: 2.9–3.8 g/kg (211)
          rat: inhalation: no deaths: sat.vap., 8 hr (211)

butyric aldehyde *see* butyraldehyde

butyric amide *see* butyramide

butyrolactam *see* 2-pyrrolidone

butyrone *see* 4-heptanone

butyronitrile (butanenitrile; n-propylcyanide; cyanopropane)
$CH_3CH_2CH_2CN$
A. PROPERTIES: colorless liquid; m.w. 69.10; m.p. $-112.6°C$; b.p. $118°C$; v.p. 10 mm at $15°C$, 40 mm at $38°C$; v.d. 2.4; sp.gr. 0.80 at $20°C$;
C. WATER POLLUTION FACTORS:
 —Waste water treatment:
  A.S.: after 6 hr: 1.2% of ThOD
   12 hr: 1.5% of ThOD
   24 hr: 1.7% of ThOD (89)
 —Biodegradation by a mutant microorganism: 500 mg/l at $20°C$:
 parent: 100% disruption in 13 hr
 mutant: 100% disruption in 4 hr (152)
D. BIOLOGICAL EFFECTS:
 —Mammalia: rat: acute oral LD_{50}: 50–100 mg/kg (211)
  rat: inhalation: no toxic signs, daily urinary thiocyanate: 6 μg (normal: 0.07 μg): autopsy: organs normal: 200 ppm, 20 × 6 hr (65)

C

cabacryl *see* acrylonitrile

cacodylhydride *see* dimethylarsine

cacodylic acid (hydroxydimethylarsine oxide; dimethylarsinic acid; tradenames: Phytar; Ansar)
$(CH_3)_2 AsO_2 H$

Uses: contact herbicide; cotton defoliant; nonselective contact herbicide on noncrop areas; Phytar 138 contains 65.6% cacodylic acid; Fisher purified cacodylic acid contains 95.5% cacodylic acid

A. PROPERTIES: colorless, odorless crystals; m.p. 192–198°C; solub. 2000 g/l at 25°C

C. WATER POLLUTION FACTORS:
—the *degradation* of cacodylic acid in soils proceeds by two mechanisms. Under anaerobic conditions 61% was converted to a volatile organoarsenical within a 24-week period and was lost from soil system.

Under aerobic conditions 35% was converted to a volatile organoarsenical compound and 41% to CO_2 and AsO_4^{3-} within the same 24-week period.

The ultimate environmental fate of the arsenic from cacodylic acid appears to be metabolized to inorganic arsenate which is bound as insoluble compounds in the soil. (1299)

D. BIOLOGICAL EFFECTS:
—Residues in growth media and some natural waters (salt):

diatoms:	1.3–2.4 nmol/l
coccolithophorids:	n.d.–1.9 nmol/l
dinoflagellates:	0.06–0.1 nmol/l
green algae:	1.9 nmol/l
sterile medium:	0.02 nmol/l
surface seawater:	1.6 nmol/l
Salton sea, surface:	67.5 nmol/l

(1933)

—Bioaccumulation in model ecosystem (multi organism experiment)
after 32 days of exposure:
conc. of cacodylic acid in water: 1st day: 10.6 ppb
                                    32rd day: 6.1 ppb
bioaccumulation ratio after 32 days: algae: 1635
                                      snails: 419
                                               110 (after 16 days in clean water)

	daphnia:	1658		
	fish:		21 (after 3 days of exposure)	
bioaccumulation ratio at:	0.1 ppm	1.0 ppm	10.0 ppm cacodylic	
(one organism experiment)			acid in water	
algae: after 2 days exposure	45	17	7	
daphnia after 2 days exposure	39	42	25	
fish: after 2 days exposure	1.4	0.9	1.1	
snails: after 7 days exposure	20	68.5	6.8	(1300)

—Inhalation:
rat: male: exposure to dust atmospheres LC_{50}, 2 hr: >6.9 mg/l*
female: exposure to dust atmospheres LC_{50}, 2 hr: 3.9 mg/l*
mouse: male and female: exposure to dust atmospheres LC_{50}, 2 hr: >6.4 mg/l*
*Phytar® 138 used for exposures: concentrations are expressed as active ingredient.
mouse: respiratory irritant potential: RD_{50} = 3.15 mg/l
—Intraperitoneal administration:
rat: male: LD_{50}: 720 mg/kg
female: LD_{50}: 520 mg/kg
mouse: male: LD_{50}: 520 mg/kg
female: LD_{50}: 600 mg/kg
acute oral LD_{50} (rat): 1350 mg/kg (technical product)   (1855)

caffeine   (theine; methyltheobromine; 1,3,7-trimethylxanthine)
$C_8H_{10}N_4O_2 \cdot H_2O$ (bicyclic compound)
Natural source: coffee beans; tea leaves; kola nuts.
Uses: beverages; medicine
A. PROPERTIES: 212.11; m.p. 236.8; solub. 13,500 mg/l at 16°C, 455,500 mg/l at 65°C; log P_{oct} −0.07
B. AIR POLLUTION FACTORS:
  —Manmade sources:
    concentrations in particulate organic matter in New York city, Jan.–March 1975: 0.7–7.0 µg/1000 cu m (equivalent concentration); ($n = 8$)
    emissions from coffee-roasting plants are implicated as contributors to ambient air levels   (429)
C. WATER POLLUTION FACTORS:
  —Manmade sources: in primary domestic sewage plant effluent: 0.010–0.046 mg/l   (517)
D. BIOLOGICAL EFFECTS:
  —Rats: oral LD_{50}: 200 mg/kg
  —Carcinogenicity: none?
  —Mutagenicity in the *Salmonella* test: none
    <0,002 revertant colonies/nmol
      <70 revertant colonies at 6000 µg/plate   (1883)

2-camphanol   *see* borneol

2-camphanone   *see* camphor

camphor (gum camphor; 2-camphanone)

Manufacturing source: organic chemical industry; woodprocessing industry. (347)
Users and formulation: odorant/flavorant in household; pharmaceutical and industrial products; plasticizer for cellulose esters and ethers; insect repellant and incense mfg.; lacquers and varnishes; explosives; embalming fluid; plastics mfg.; chemical intermediate. (347)
Natural sources (*water and air*): major component of pine oil (leaves, twigs, stems of camphor tree of China, Formosa, Japan); present in forest runoff. (347)

A. PROPERTIES: colorless or white crystals; m.p. 174–179°C; v.p. 1 mm at 41.5°C, 400 mm at 182°C, 700 mm at 209.2°C; sp.gr. 0.99
B. AIR POLLUTION FACTORS:
   —Odor: quality: penetrating aromatic odor

odor thresholds   mg/cu m

	10^{-7}	10^{-6}	10^{-5}	10^{-4}	10^{-3}	10^{-2}	10^{-1}	1	10	10^2	10^3	10^4
detection						● ●	●		●——●—●			
recognition								●	●══●══●			
not specified							●		●	● ●		

                  (210; 279; 307; 610; 672; 690; 753; 755; 771; 774; 775; 837; 871)
   odor index at 20°C: 40

C. WATER POLLUTION FACTORS:
   —Reduction of amenities: T.O.C. in water at room temp.: 1.29 ppm, range 0.25–3.83
      20% of population still able to detect odor at 0.33 ppm
      10% of population still able to detect odor at 0.041 ppm
      1% of population still able to detect odor at 0.0092 ppm
      0.1% of population still able to detect odor at 0.021 ppb   (321)
   —Manmade Sources: in Zurich lake: at surface 12 ppt; at 30 m depth 2 ppt
                      in Zurich area: spring water 2 ppt; tapwater 2 ppt   (513)

D. BIOLOGICAL EFFECTS
   —Fish: fathead minnows: static bioassay in Lake Superior water at 18–22°C: LC_{50} (1; 24; 48; 72; 96 h): 145; 112; 111; 110; 110 mg/l (350)

n-capric acid (decanoic acid; n-decoic acid; n-decylic acid)
$CH_3(CH_2)_8COOH$

Uses: esters for perfumes and fruit flavors, base for wetting agents; intermediate; plasticizer
A. PROPERTIES: mw 172.26; mp 31.5°C; bp 268–270°C; vp 1 mm at 125°C; sp gr 0.886 at 40/4°C; log P_{oct} 4.09
B. AIR POLLUTION FACTORS
   –Odor: threshold odor conc. 0.014 mg/cu m - 1.96 ppb (307)
                                    detection 0.05 mg/cu m (778; 779)
                                    recognition 0.08–0.09 mg/cu m (610)
C. WATER POLLUTION FACTORS:
   –BOD_5: 9% of ThOD
   –COD: 85% of ThOD
   –$KMnO_4$ value: acid: 3% of ThOD
                        alkaline: 0% of ThOD (220)
   –Waste water treatment:
      A.S.: after 6 hr: 10.9% of ThOD
              12 hr: 18.9% of ThOD
               24 hr: 23.4% of ThOD (89)
   –Odor threshold: detection: 10.0 mg/kg (886)
D. BIOLOGICAL EFFECTS:
   –Fish: *Lepomis macrochirus*: chemical is too insoluble in water to be toxic (1294)

caproaldehyde *see n*-hexaldehyde

caproic acid (hexanoic acid; *n*-hexoic acid)
$CH_3(CH_2)_4COOH$
A. PROPERTIES: oily liquid; m.w. 116.2; m.p. -6/-2°C; b.p. 204/208°C; v.p. 0.2 mm at 20°C, 1 mm at 70°C; v.d. 4.01; sp.gr. 0.945 at 0/0°C; solub. 11,000 mg/l, LHC 831.0 kcal/mole; log P_{oct} 1.88/1.92
B. AIR POLLUTION FACTORS: 1 mg/cu m = 0.21 ppm, 1 ppm = 4.83 mg/cu m
   –Odor: characteristic: like limburger cheese (211)

odor thresholds — mg/cu m

	10^{-7}	10^{-6}	10^{-5}	10^{-4}	10^{-3}	10^{-2}	10^{-1}	1	10	10^2	10^3	10^4
detection						◆	◆◆					
recognition							◆	◆◆				
not specified							◆ ◆◆					

                               (307; 610; 647; 683; 684; 778; 779; 840; 872)
   –Control methods: thermal incineration: min. temp. for odor control: 774°C (94)
C. WATER POLLUTION FACTORS:
   –BOD_5:   44% of ThOD (220)
   –$BOD_{2\ days}^{25°C}$: 1.77 (substrate conc.: 3.5 mg/l; inoculum: soil microorganisms)
   –$BOD_{5\ days}$: 2.11
   –$BOD_{10\ days}$: 2.11 (1304)
   –COD: 100% of ThOD
   –$KMnO_4$:   acid 7% of ThOD
                alkaline 3% of ThOD (220)

332 CAPROIC NITRILE

   —Manmade sources:
   average content of secondary effluents: 47.9 μg/l (86)
   —Odor threshold: detection: 3.0 mg/kg (886)
   —Waste water treatment:
   A.S.: after 6 hr: 12.8% of ThOD
   12 hr: 22.0% of ThOD
   24 hr: 39.0% of ThOD (89)
   A.S.: BOD, 20 C, 1–5 d observed, feed: 333 mg/l, 15 days acclimation: 99% removed (93)
   A.C.: adsorbability: 0.194 g/g carbon, 97.0% reduction, infl. 1,000 mg/l, effl.: 30 mg/l (32)
   stabilization pond design:
   toxic correction factor: 1.3 at 200 mg/l in pond influent
   5.0 at 300 mg/l in pond influent (179)
D. BIOLOGICAL EFFECTS:
   —Arthropoda: *Daphnia magna*: TLm (24 hr): 22 mg/l (246)
   —Fish: *L. macrochirus*: TLm (24 hr): 15–200 mg/l (153)
   fathead minnows: static bioassay in Lake Superior water at 18–22°C: LC_{50} (1; 24; 48; 72; 96 hr): 140; 88; 88; 88; 88 mg/l (350)
   —Mammals: rat: oral LD_{50}: 6.44 g/kg (211)
   rat: inhalation: no deaths: sat. vap., 8 hr (211)

caproic nitrile *see* hexanenitrile

α-caprolactam (cyclohexanoneisooxime)

Users and formulation: nylon mfg. and processing; mfg. of plastics, bristles, film, coatings, synthetic leather, plasticizers and paint vehicles, cross linking agent for curing polyurethanes; synthesis of amino acid lysine (347)
A. PROPERTIES: m.w. 113.16; m.p. 69.2°C; b.p. 139°C at 12 mm; v.p. 0.001 mm at 20°C, 0.0035 mm at 30°C; v.d. 3.91; sat.conc. 0.006 g/cu m at 20°C, 0.021 g/cu m at 30°C
B. AIR POLLUTION FACTORS: 1 mg/cu m = 0.21 ppm, 1 ppm = 4.70 mg/cu m
   —Odor: T.O.C. = 0.3 mg/cu m = 63 ppb (730; 57)
C. WATER POLLUTION FACTORS:
   —BOD_{20}: 0.6
   —$KMnO_4$ value: 0.032
   —Biodegradation: adapted A.S.—product as sole carbon source—94.3% COD removal at 16.0 mg COD/g dry inoculum/hr (327)
   —Reduction of amenities: T.O.C. in water at room temp.: 59.7 ppm, range 36.0–100.0 ppm, 8 judges
   20% of population still able to detect odor at 25 ppm

10% of population still able to detect odor at 16 ppm	
1% of population still able to detect odor at 3.8 ppm	
0.1% of population still able to detect odor at 0.92 ppm	(321)
—Water quality: 1.0 mg/l affects the self purification	(181)
nitrification decreases from 100 mg/l onwards	(30)

D. BIOLOGICAL EFFECTS:
—Fish: catfish: no effect: 1 g/l, 30 days
     catfish: toxic: 5 g/l after 18 hr
     catfish: toxic: 10 g/l after 10 hr   (30)

capronitrile *see* hexanenitrile

capryalcohol *see* 2-octanol

caprylaldehyde *see* octanal

caprylic acid (octanoic acid; *n*-octoic acid; *n*-octylic acid)
$CH_3(CH_2)_6COOH$

A. PROPERTIES: colorless liquid or solid; m.w. 144.21; m.p. 16°C; b.p. 237°C; v.p. 1 mm at 92°C; v.d. 5.00; sp.gr. 0.910 at 20/4°C; solub. 2,500 mg/l at 100°C

B. AIR POLLUTION FACTORS: 1 ppm = 5.994 mg/cu m, 1 mg/cu m = 0.167 ppm
—Odor: characteristic: unpleasant, irritating

odor thresholds   mg/cu m

	10^{-7}	10^{-6}	10^{-5}	10^{-4}	10^{-3}	10^{-2}	10^{-1}	1	10	10^2	10^3	10^4
detection						●		●				
recognition				●								
not specified						● ●						

(211; 307; 610; 708; 778; 779)

C. WATER POLLUTION FACTORS:
—$BOD_{2\ days}^{25°}$: 1.07 (substrate conc.: 2.9 mg/l; inoculum: soil microorganisms)
—$BOD_{5\ days}$: 1.28
—$BOD_{10\ days}$: 1.79
—$BOD_{20\ days}$: 2.48   (1304)
—Odor thresholds: T.O.C.. detection: 3.0 mg/kg   (886)
—Waste water treatment: A.S.: after 6 hr: 9.8% of ThOD
                          12 hr: 20.4% of ThOD
                          24 hr: 32.8% of ThOD   (89)
  powdered carbon: at 100 mg/l sodium salt (pH 7.5)—carbon dosage 1000 mg/l: 54% adsorbed   (520)
—Sampling and analysis: extraction efficiency of macro-reticular resins: pH: 5.7; conc. 10 ppm: X-AD-2. 58%
                    X-AD-7: 81%   (370)

D. BIOLOGICAL EFFECT:
  —Fish: *Lepomis macrochirus*: chemical is too insoluble in water to be toxic (1294)
  —Mammals: dog; oral doses: 1–5% in diet causes diarrhea (211)

captan (N-trichloromethylthiotetrahydrophthalimide; *cis*-N-((trichloromethyl)thio)-4-cyclohexene-1,2-dicarboximide; Merpan; Orthocide; Vondcaptan)

Use: protectant-eradicant fungicide
A. PROPERTIES: pure chemical white solid; m.p. 175°C; log P_{oct} 2.35 solub. <0.5 ppm at 20°C
D. BIOLOGICAL EFFECTS:
  —Fish:
    harlequin fish (Rasbora heteromorpha) (89% active ingredient):

mg/l	24 hr	48 hr	96 hr	3 m (extrap.)	
LC_{10} (F)	0.23	0.14	—		
LC_{50} (F)	0.46	0.33	0.3	0.2	(331)

  —Mammals:
    acute oral LD_{50} (rat): 9000 mg/kg (1854)
    in diet: no effect level from 2 years dietary tests on rats: 1,000 mg/kg (1855)
  —Carcinogenicity: +
  —Mutagenicity in the *Salmonella* test: + (without liver homogenate):
    25 revertant colonies/nmol
    820 revertant colonies at 10 µg/plate (1883)

captax *see* 2-mercaptobenzothiazole

carbamide *see* urea

carbanil *see* phenylisocyanate

carbaryl (1-naphthyl-N-methylcarbamate; sevin)

Use: contact insecticide
A. PROPERTIES: white crystalline solid; m.w. 201; m.p. 142°C; sp.gr. 1.232 at 20/20°C; v.p. <0.005 mm at 26°C; solub. 40 ppm at 30°C

C. WATER POLLUTION FACTORS:

— Aquatic reactions:
comparison of calculated hydrolysis and photolysis under given conditions:

pH	hydrolysis* half-life, days	direct photolysis half life, days
5	1500	6.6
7	15	6.6
9	0.15	—

*Calculation based on neutral and alkaline hydrolysis assuming pseudo-first-order kinetics (1070)

biolysis by bacteria: half life: >30,000 days (minimum value assuming a bacterial population of 0.1 mg/l) (1071)

— Mechanism for alkaline hydrolysis:

$$O=C-NH-CH_3 \text{ (naphthyl ester)} \underset{OH^-}{\overset{H_2O}{\rightleftarrows}} C-N^{\ominus}CH_3 \text{ (naphthyl ester)} \xrightarrow{H_2O}$$

$$\text{naphthol-OH} + CH_3N=C=O \xrightarrow{H_2O} CH_3NH-C\begin{smallmatrix}O\\OH\end{smallmatrix} \downarrow CO_2 + CH_3NH_2$$

— Persistence in river water in a sealed glass jar under sunlight and artificial fluorescent light - initial conc. 10 µg/l

after	1 hr	1 wk	2 wk	4 wk	8 wk	
	90	5	0	0	0	(1309)

D. BIOLOGICAL EFFECTS:

— Bacteria: *Pseudomonas putida*: inhibition of cell multiplication starts at >50 mg/l
(329)

— Algae: *Microcystis aeruginosa*: inhibition of cell multiplication starts at 0.03 mg/l
(329)

Dunaliella euchlora: 1000 ppb .65 O.D. expt/O.D. control, 10 day growth test
Phaeodactylum tricornutum: 100 ppb .00 O.D. expt/O.D. control, 10 day growth test
Monochrysis lutheri: 1000 ppb .00 O.D. expt/O.D. control, 10 day growth test
Chlorella sp.: 1000 ppb .80 O.D. expt/O.D. control, 10 day growth test
Chlorella sp.: 10,000 ppb .00 O.D. expt/O.D. control, 10 day growth test
Protococcus sp.: 1000 ppb .74 O.D. expt/O.D. control, 10 day growth test
Protococcus sp.: 10,000 ppb .00 O.D. expt/O.D. control, 10 day growth test
(2347)

CARBARYL

—Crustaceans

Species			Concentration	Duration	Effect	Method	Ref
Gammarus lacustris:			16 µg/l	96 hr	no effect level		(2124)
Gammarus fasciatus:			26 µg/l	96 hr	no effect level		(2126)
Palaemonetes kadiakensis:			5.6 µg/l	96 hr	no effect level		(2126)
Orconectes nais:			8.6 µg/l	96 hr	no effect level		(2126)
Asellus brevicaudus:			240 µg/l	96 hr	no effect level		(2126)
Simocephalus serrulatus:			7.6 µg/l	48 hr	no effect level		(2127)
Daphnia pulex:			6.4 µg/l	48 hr	no effect level		(2127)
Daphnia magna:			5.0 µg/l	63 day	no effect level		(2135)
Gammarus pulex: 98 hr, LC_{50}: 0.029 ppm							(1323)
Cancer magister	Dungeness crab	egg/prezoeal	6 ppb		prevention of hatching and molting	24 hr static lab bioassay	
Cancer magister	Dungeness crab	zoea	10 ppb		prevention of molting and death	96 hr static lab bioassay	
Cancer magister	Dungeness crab	juvenile	280 ppb		death or paralysis	96 hr static lab bioassay	
Cancer magister	Dungeness crab	adult	180 ppb		death or paralysis	96 hr static lab bioassay	(2321)
Palaemon macrodactylus	Korean shrimp		12.0 (8.5–13.5) ppb		TL_{50}	96 hr static lab bioassay	(2353)
Palaemon macrodactylus	Korean shrimp		7.0 (1.5–28) ppb		TL_{50}	96 hr intermittent-flow l	(2352)
Upogebia pugettensis	mud shrimp		40 (30–60) ppb		TLm	48 hr static lab bioassay	(2346)
Callianassa californiensis	ghost shrimp		30 ppb		TLm	48 hr static lab bioassay	(2346)
Callianassa californiensis	ghost shrimp	adult	130 ppb		TLm	24 hr static lab bioassay	(2346)
Cancer magister	Dungeness crab	juvenile (male)	600 (590–610) ppb		EC_{50} (Paralysis or death) loss of equilibrium	24 hr static lab bioassay	(2346)
Hemigrapsis oregonensis	shore crab	adult (female)	270 (60–690) ppb		EC_{50} (Paralysis loss of equilibrium or death)	24 hr static lab bioassay	(2346)

—Molluscs:

Species			Concentration	Effect	Method	Ref
Crassostrea gigas	Pacific oyster	larvae	2200 (1500–2700) ppb	EC_{50} prevention of development to straight linge shell stage.	48 hr static lab bioassay	(2346)
Crassostrea virginica	American oyster	eggs	3,000 ppb	TLm	48 hr static lab bioassay	(2324)
Crassostrea virginica	American oyster	larvae	3,000 ppb	TLm	14-day static lab bioassay	(2324)
Mercanaria mercenaria	hard clam	eggs	3,820 ppb	TLm	14 hr static lab bioassay	(2324)
Mercenaria mercenaria	hard clam	larvae	>2,500 ppb	TLm	14 day static lab bioassay	(2324)
Clinocardium nuttalli	cockle clam	adults	7,300 ppb	TLm	24 hr static lab bioassay	(2346)
Clinocardium nuttalli	cockle clam	juvenile	3,850 ppb	TLm	96 hr static lab bioassay	(2322)
Mytilus edulis	bay mussel	larvae	2,300 (1400–2900) ppb	EC_{50} prevention of development to straight linge shell stage.	96 hr static lab bioassay	(2346)

Gastropod (*Hymnea stagnalis*): 48 hr, LC_{50}: 21 ppm (1323)
growth rate decreases from 1 mg/l onwards (1910)

—Insects:

Pteronarcyc californica: 4.8 µg/l, 96 hr no effect level (2128)
Pteronarcys dorsata: 23.0 µg/l, 30 day LC_{50}; 11.5 µg/l 30 day no effect level (2134)

CARBARYL 337

Pteronarcella badia: 1.7 µg/l, 96 hr no effect level (2128)
Claassenia sabulosa: 5.6 µg/l, 96 hr no effect level (2128)
Acroneuria lycorias: 2.2 µg/l 30 day LC_{50}; 1.3 µg/l 30 day no effect level
(2134)
Hydropsyche bettoni: 2.7 µg/l 30 day LC_{50}; 1.8 µg/l 30 day no effect level
(2134)
insect larvae: *Chaoborus*: 48 hr, LC_{50}: 0.296 ppm (1323)
Cloeon: 48 hr, LC_{50}: 0.48 ppm (1323)
fourth instar larval *Chironomus riparius*: 24 hr, LC_{50}: 104.5 ppb (1853)
rice-field spider (*Oedothorax insecticeps*): LD_{50}: 840 ppm (1814)
bees: 48 hr, LC_{50}: 3.8–4.5 ppm in food (1683)
Fish:
Pimephales promelas 9000 µg/., 96 hr no effect level; 680 µg/l (decline survival
 and reproduction 6 months); 210 µg/l no effect (6 month) (2136)
Lepomis macrochirus 6760 µg/l, 96 hr no effect level (2121)
Lepomis microlophus 11200 µg/l, 96 hr no effect level (2121)
Micropterus salmoides 6400 µg/l, 96 hr no effect level (2121)
Salmo gairdneri 4340 µg/l, 96 hr no effect level (2121)
Salmo trutta 1950 µg/l, 96 hr no effect level (2121)
Oncorhynchus kisutch 764 µg/l, 96 hr no effect level (2121)
Perca flavescens 745 µg/l, 96 hr no effect level (2121)
Ictalurus punctatus 15800 µg/l, 96 hr no effect level (2121)
Ictalurus melas 20000 µg/l, 96 hr no effect level (2121)

	96 hr, LC_{50} mg/l		96 hr, LC_{50} mg/l	
catfish	15.8	bluegill	6.8	
bullhead	20.0	bass	6.4	
goldfish	13.2	rainbow	4.3	
minnow	14.6	brown	2.0	
carp	5.3	coho	0.76	
sunfish	11.2	perch	0.75	(1934)

Channa punctatus: 180 d, LC_{50}: 2.0 mg/l (1508)
Mosquitofish (*Gambusia affinis*): 24 hr, LC_{50}, S: 40.0 mg/l (1104)
Mosquitofish: 48 hr, LC_{50}, S: 35.0 mg/l (1104)
Mosquitofish: 96 hr, LC_{50}, S: 31.8 mg/l (1104)
Carp (*Cyprinus carpio*): 24 hr, LC_{50}, S: 13.51 mg/l (1106)
Carp: 48 hr, LC_{50}, S: 11.74 mg/l (1106)
Carp: 72 hr, LC_{50}, S: 10.36 mg/l (1106)

Parophrys vetulus	English sole	juvenile	4,100 (3200–5000) ppb TLm	24 hr static lab bioassay (2346)
Cymatogaster aggregata	shiner perch	juvenile	3,900 (3800–4000) ppb TLm	24 hr static lab bioassay (2346)
Gasterosteus aculeatus	three-spine stickleback	juvenile	6,700 (5500–7700) ppb TLm	24 hr static lab bioassay (2346)

Gasterosteus aculeatus	three-spine stickleback	3,990 ppb	TLm	96 hr static lab bioassay (2333)
Leiostomus xanthurus	spot	100 ppb	65% survived in experimental and control test	5 months continuous flow chronic lab bioassay (2339)
Onchorynchus keta	chum salmon	juvenile 2,500 (2200–2700) ppb	TLm	96 hr static lab bioassay (2343)

    bluegill: 24 hr, LC_{50}: 3.4 ppm
    rainbow trout: 24 hr, LC_{50}: 3.5 ppm   (1878)
—Bioaccumulation:
    BCF: algae: 4000
          duckweed: 3600
          snails: 300
          catfish: 140
          crayfish: 260   (1891; 1892; 1893)
—Mammals:
    acute oral LD_{50} (female rat):   500 mg/kg   (1854)
    acute oral LD_{50} (male rats):   850 mg/kg
    acute dermal LD_{50} (rats):   >4000 mg/kg
    acute dermal LD_{50} (rabbits):   >2000 mg/kg   (1855)
    dietary feeding: rats fed for 2 years on a diet containing 200 ppm suffered no ill effects   (1855)
—Carcinogenicity: negative
—Mutagenicity in the *Salmonella* test: negative
    <0.008 revertant colonies/nmol
    <70 revertant colonies at 2000 µg/plate   (1883)

carbathione *see* methyldithiocarbamate

carbazole (dibenzopyrrole)

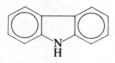

A. PROPERTIES: m.w. 167.21; m.p. 245–246°C; b.p. 355°C; sp.gr. 1.10 at 18/4°C; v.p. 400 mm at 323.0°C; log P_{oct} 3.29
B. AIR POLLUTION FACTORS:
    —Manmade sources: in coal tar pitch fumes: 9.6 wt % (also methylphenanthrene)   (516)
D. BIOLOGICAL EFFECTS:
    —*Daphnia pulex*: bioaccumulation coefficient: 115 (wet wt.)   (1570)

carbetamex (d-N-ethyllactamide carbanilate; carbetamide)
D. BIOLOGICAL EFFECTS: harlequin fish (*Rasbora heteromorpha*):

mg/l	24 hr	48 hr	96 hr	3 m (extrap.)	
LC_{10} (F)	170	150	125		
LC_{50} (F)	220	190	165	100	(331)

carbetamide *see* carbetamex

carbitol *see* diethyleneglycolmonoethylether

carbitolacetate *see* diethyleneglycolmonoethyletheracetate

carbofuran (furadan; NIA 10,242; ENT 27,164; 2,3-dihydro-2,2-dimethyl-7-benzofuranylmethylcarbamate)

Use: systemic insecticide widely used to control corn rootworms (1664; 1665)
A. PROPERTIES: odorless, white crystalline solid; m.w. 221.3; m.p. 150–152°C; sp.gr. 1.18 at 20/20°C; v.p. 2×10^{-5} mm at 33°C; solub. 700 ppm at 25°C
C. WATER POLLUTION FACTORS:
  —Persistence in soils:
    in 4 soils with known insecticide use: the technical carbofuran had a calculated half-life of 11–13 days (pH 6.5) and the granular formulation had a half-life of 60–75 days (pH 6.5) (1663)
    time for 95% disappearance from the soil 3.1–5.6 kg/ha:
      145 to 434 days (1650)
    volatilization at 25°C from soils in the laboratory:
      sandy loam: 0.5% after 60 days
      sand: 36.2% after 60 days (1650)
D. BIOLOGICAL EFFECTS:
  —Blue-green algae: *Nostoc muscorum* at 25 mg/l enhanced survival; growth and nitrogen fixation; gradual inhibition from 50–1000 mg/l; at 1200 mg/l algicidal (1352)
  —Fish:
    Mosquitofish (*Gambusia affinis*): 72-hr, S, LC_{50}: 0.52 mg/l
    Green sunfish (*Lepomis cyanellus*): 72-hr, S, LC_{50}: 0.16 mg/l (1203)
    Cyprinodon variegatus: 96 hr, LC_{50}-FT: 386 µg/l (1331)
  —Mammals
    acute oral LD_{50} (rat): 11 mg/kg
    dermal LD_{50} (rabbit): 10,200 mg/kg (1854)

carbondisulfide (carbonbisulfide)
CS_2

Users and formulation   mfg. rayon, cellophane, carbon tetrachloride, rubber chemicals and flotation chemicals, soil disinfectants, electronic vacuum tubes; solvent (phosphorus, sulfur, bromine, iodine, selenium, fats, resins, rubbers); mfg. grain fumigants, soil conditioners, herbicides; paper mfg.; pharmaceutical mfg.   (347)

A. PROPERTIES: colorless liquid; m.w. 76.14; m.p. $-108.6°C/-116.6°C$; b.p. $46.3°C$; v.p. 260 mm at $20°C$, 430 mm at $30°C$; 200 mm at $10°C$; v.d. 2.64; sp.gr. 1.263 at $20/4°C$; solub. 2,300 mg/l at $22°C$; log P_{oct} 1.84/2.16 (calculated)

B. AIR POLLUTION FACTORS: 1 mg/cu m = 0.315 ppm, 1 ppm = 3.17 mg/cu m
   —Odor: characteristic: quality: vegetable sulfide, aromatic
             hedonic tone: slightly pungent

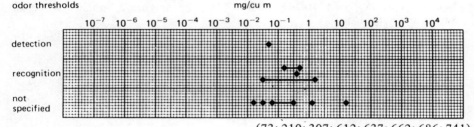

(73; 210; 307; 612; 637; 662; 686; 741)

O.I. 100% recognition: 1,600,000   (2)
USSR: human odor perception: non perception: 0.04 mg/cu m
                             perception: 0.05 mg/cu m = 16 ppb
      human reflex response: no response: 0.03 mg/cu m
                             adverse response: 0.04 mg/cu m   (170)

—Natural sources:
biogenic carbon disulfide emissions from soils in U.S.A.:

sampling sites	soil orders		avg. sulfur flux g S/m²/yr
Wadesville, IN	alfisol	Sept.–oct. 1977	0.002
Philo, OH	inceptisol	Oct. 1977	0.0012
Dismal Swamp, NC	histosol	Oct. 1977	0.0001
Ceder Island, NC	freshly clipped marsh		1.131
Cox's Landing, NC	freshly clipped marsh		0.0215 (1385)

—Control methods:
catalytic combustion: platinized ceramic honeycomb catalyst; ignition temp.: $350°C$, inlet temp. for 90% conversion: $375-400°C$ (91)

—Sampling and analysis:
non dispersive I.R.: min. full scale: 37 ppm   (55)
photometry: min. full scale: 2000 ppm   (53)
detector tubes: UNICO: 10 ppm = det. limit   (59)
              DRAGER: 13 ppm = det. limit
4 f.abs.app., 60 l air/30 min: VLS: det.lim.: 50 µg/cu m/30 min, reagent: solution of diethylamine and coppersulfate   (208)

C. WATER POLLUTION FACTORS:
—Impact on biodegradation processes:
75% reduction of the nitrification process in activated sludge at 35 mg/l (30)
—Reduction of amenities: faint odor: 0.0026 mg/l (129)
organoleptic limit: 1.0 mg/l (181)
D. BIOLOGICAL EFFECTS:
—Fish: mosquito fish: TLm (24-96 hr) 162-135 mg/l (41)
—Man: severe toxic effects: 500 ppm = 1600 mg/cu m, 60 min
symptoms of illness: 150 ppm = 480 mg/cu m
unsatisfactory: >10 ppm = 32 mg/cu m (185)

carbonhexachloride *see* hexachloroethane

carbonoxysulfide *see* carbonylsulfide

carbontetrachloride (tetrachloromethane)
CCl_4
Users and formulation: fire extinguisher mfg; dry cleaning operations; mfg. of refrigerants, aerosols and propellants; mfg. of chlorofluoromethanes; extractant; solvent; veterinary medicine; metal degreasing; fumigant; chlorinating organic compounds
A. PROPERTIES: colorless liquid; m.w. 153.82; mp. -23°C; b.p. 76.7°C; v.p. 90 mm at 20°C, 56 mm at 10°C, 113 mm at 25°C, 137 mm at 30°C; v.d. 5.5; sp.gr. 1.59 at 20°C; solub. 1,160 mg/l at 25°C, 800 mg/l at 20°C; sat.conc. 754 g/cu m at 20°C, 1109 g/cu m at 30°C; log P_{oct} 2.64 at 20°C (calculated)
B. AIR POLLUTION FACTORS: 1 mg/cu m = 0.16 ppm, 1 ppm = 6.39 mg/cu m
odor thresholds   mg/cu m

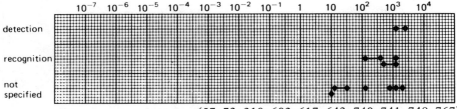

(57; 73; 210; 602; 617; 643; 740; 741; 749; 767)
—Control methods: A.C.: retentivity: 45 wt% of adsorbent (83)
—Sampling and analysis:
I.R.: det. lim.: 0.005 ppm (56)
photometry: min.ful scale: <1,000 ppm (53)
test tubes: UNICO: det. lim.: 5 ppm (59)
DRAGER: det. lim.: 10 ppm (58)
C. WATER POLLUTION FACTORS:
—BOD_5:   0 std.dil.sew. (275)(27)
—$KMnO_4$: 0.001 (30)
—ThOD: 0.21 (27)
—Reduction of amenities: odor threshold: 50 mg/l (84)

CARBONTETRACHLORIDE

chlorination of CS_2:   50% response: 10.0 ppm
                      100% response: 21.4 ppm
chlorination of CH_4:   50% response: 46.8 ppm
                      100% response: 100.0 ppm   (2)
distinct odor: 1600 mg/cu m = 250 ppm   (278)

—Manmade sources:
glc's in Los Angeles county: 0.12–1.63 ppb   (46)
glc's in rural Washington: Dec. 74–Feb. 75: 120 ppt   (315)
waters from upland reservoirs in N.W. England (1974):
     during dry cloudy weather: 1–2 ppb
     during prolonged heavy rain: 13–24 ppb   (933)
in river Maas at Eysden (Netherlands) in 1976:
     median: 1.2 $\mu g/l$; range: 0.1 to 3.7 $\mu g/l$
in river Maas at Keizersveer (Netherlands) in 1976:
     median: 1.3 $\mu g/l$; range: n.d. to 4.2 $\mu g/l$ (1368)

sampling date	sampling location	sampling height m	measured conc. pptv	
Dec. 1971	Northern lat. 0–50°N	sea level	71.2 ± 6.86	(374)
Jan. 1974	Greenland Sea and Arctic Ocean	30–5500	173 ± 39	(376)
June/July 1974	W. Ireland	sea level	110.9 ± 10.7	(375)
October 1973	North Atlantic	sea level	137.6 ± 14.7	(375)
Dec. 1974/Feb. '75	Pullman, Washington State	not reported	120 ± 15	(315)
March 1975	Farnborough Aberporth, U.K.	2000–6100	53 ± 8	(371)
March 1975	Boscombe Down Exeter, U.K.	3000–6100	66 ± 4	(371)
January 1976	off S.W. Wales	900–7300	66 ± 6	(371)
February 1976	off N.E. England	400–5200	64 ± 2	(371)

—Aquatic reactions:
measured half life for evaporation from 1 ppm aqueous solution, still air and an average depth of 6.5 cm: at 25°C: 28.8 min.   (369)

—Waste water treatment:
evaporation from water at 25°C of 1 ppm solution: 50% after 29 min
                                                               90% after 97 min   (313)

experimental water reclamation plant:

	sample 1	sample 2	
sand filter effluent	122 ng/l	19 ng/l	
after clorination	685 ng/l	308 ng/l	
final water after A.C.	152 ng/l	20 ng/l	(928)

—Sampling and analysis:
GC–EC: after n-pentane extraction: det.lim.: 0.01 $\mu g/l$ (84)

D. BIOLOGICAL EFFECTS:
—Toxicity threshold (cell multiplication test):
     bacteria (*Pseudomonas putida*):   30 mg/l   (1900)
     algae (*Microcystis aeruginosa*):   105 mg/l   (1329)

green algae (*Scenedesmus quadricauda*): >600 mg/l (1900)
protozoa (*Entosiphon sulcatum*): 770 mg/l (1900)
protozoa (*Uronema parduczi Chatton-Lwoff*): 616 mg/l (1901)
—Residues: concentrations in various organs of molluscs and fish collected from the relatively clean waters of the Irish Sea in the vicinity of Port Erin, Isle of Man (only organs with highest and lowest concentrations are mentioned - concentrations on dry weight basis):
molluscs: *Baccinum undatum*: muscle: 5 ng/g
digestive gland: 8 ng/g
Modiolus modiolus: digestive tissue: 20 ng/g
mantle: 114 ng/g
Pecten maximus: mantle, testis: 2–3 ng/g
ovary, gill: 14–16 ng/g
fish: *Conger conger* (eel): gill: 3 ng/g
liver: 51 ng/g
Gadus morhua (cod): stomach, skeletal tissue, muscle, liver 4–7 ng/g
brain: 29 ng/g
Pollachius birens (coalfish): muscle: 7 ng/g
gill: alimentary canal: 32–35 ng/g
Scylliorhinus canicula (dogfish): spleen: 3 ng/g
gill: 55 ng/g
Trisopterus luscus (bib): gut, liver: 16–18 ng/g
gill: 209 ng/g (1092)
—Bioconcentration:
fish: *Salmo gairdneri*: $\log P_{oct}$ 2.64
log bioconcentration factor = 1.24 (193)
—Toxicity:
guppy (*Poecilia reticulata*) 14 d LC_{50}: 67 ppm (1833)
Lepomis macrochirus: static bioassay in fresh water at 23°C, mild aeration applied after 24 hr

material added ppm	% survival after 24 hr	48 hr	72 hr	96 hr	best fit 96 hr LC_{50} ppm
320	0	–	–	–	
200	0	–	–	–	
(narcosis) 125	70	60	60	50	125
(narcosis) 100	30	20	20	20	
75	100	100	100	100	

(352)

Menidia beryllina: static bioassay in synthetic seawater at 23°C, mild aeration applied after 24 hr

material added ppm	% survival after 24 hr	48 hr	72 hr	96 hr	best fit 96 hr LC_{50} ppm
320	0	–	–	–	
180	100	80	60	50	
100	100	60	60	60	150
75	100	100	100	90	

(352)

—Mammalia:
rat: oral LD_{50}: 2.92 g/kg (211)
rat: inhalation: no effect: 3,000 ppm, 6 min

rat: inhalation: no effect: 800 ppm, 30 min
rat: inhalation: no effect: 50 ppm, 7 hr
rat: inhalation: no effect: 5 ppm, chronic exposure (211)
guinea pig: inhalation: slight increase in liver weight after a chronic exposure at 5 ppm (211)
—Man: severe toxic effects: 2,000 ppm = 12,800 mg/cu m, 60 min
    symptoms of illness: 500 ppm = 3,200 mg/cu m
    unsatisfactory: >50 ppm = 320 mg/cu m (185)
—Carcinogenicity: +
—Mutagenicity in the *Salmonella* test: negative
  <0.001 revertant colonies/nmol
  <70 revertant colonies at 10 μg/plate (1893)

carbonylchloride *see* phosgene

carbonylsulfide (carbonoxysulfide)
COS
A. PROPERTIES: m.w. 60.07; m.p. -138°C; b.p. -50.2°C; v.d. 2.1; sp.gr. 1.24 at -87°C; solub. 1000 cu cm/l
B. AIR POLLUTION FACTORS:
  —Odor: characteristic: typical sulfide odor except when pure
  —Sampling and analysis:
    I.R. spectrometry: MIRAN (WILKS): 0.002 ppm (56)
    du Pont 400 photometry: min. ful scale: 460 ppm (53)
  —Natural sources:
    biogenic carbonylsulfide emissions from soils in U.S.A.:

sampling sites	soil orders	sampling dates	avg. sulfur flux g S/m^2/yr
Wadesville, IN	Alfisol	9/20–10/3 1977	0.002
Philo, OH	Inceptisol	10/7–10/10 1977	0.0022
Cedar Island, NC	Saline Swamp	10/19–10/28 1977	0.0016
Cox's Landing, NC	Saline Marsh	11/1–11/9 1977	6.36
Clarkedale, AR	Alluvial Clay	11/16–11/20 1977	0.0014
Cedar Island, NC	freshly clipped marsh		0.013
Cox's Landing, NC	freshly clipped marsh		0.0005

(1385)

carbophenothion (S-((*p*-chlorophenylthio)methyl)O,O-diethylphosphorodithioate; S-(4-chlorophenylthiomethyl)diethylphosphorothiolothionate; trithion)

$$(C_2H_5O)_2 \overset{S}{\underset{\parallel}{P}}-S-CH_2-S-\underset{}{\bigcirc}-Cl$$

Use: insecticide; acaricide
A. PROPERTIES: off-white to amber liquid with a mild mercaptan-like odor; sp.gr. 1.275–1.3 at 20/20°C; b.p. 82°C at 0.01 mm; v.p. 3×10^{-7} mm at 20°C; solub. <40 mg/l

C. WATER POLLUTION FACTORS:
  −Degradation: 50% degradation in soil occurs in 100 days or longer, depending on soil type (1855)
  −Persistence in river water in a sealed glass jar under sunlight and artificial fluorescent light−initial conc. = 10 µg/l:

	% of original compound found				
after 1 hr	1 wk	2 wk	4 wk	8 wk	
90	25	10	0	0	(1309)

D. BIOLOGICAL EFFECTS:
  −Crustacean:
  Gammarus lacustris: 96 hr, LC_{50}: 5.2 µg/l (2124)
  Palaemonetes kadiakensis: 96 hr, LC_{50}: 1.2 µg/l (2126)
  Asellus brevicaudus: 96 hr, LC_{50}: 1100 µg/l (2126)
  −Mammals:
  acute oral LD_{50} (male rat): 32.2 mg/kg (1554)
  acute dermal LD_{50} (rabbits): 1270 mg/kg (1555)

o-carboxybenzenesulfonic acid *see* 2-sulfobenzoic acid

p-carboxybenzenesulfonic acid *see* 4-sulfobenzoic acid

carboxymethyltartronate (CMT; trisodium-2-oxa-1,1,3-propanetricarboxylate, trisodiumcarboxymethyltartronate)

$$\begin{array}{c} \text{COONa} \\ | \\ \text{HC}-\text{O}-\text{CH}_2-\text{COONa} \\ | \\ \text{COONa} \end{array}$$

Commercial-grade CMT contains: 78% trisodium carboxymethyltartronate
4% tetrasodium ditartronate
8% disodium diglycolate
10% water

C. WATER POLLUTION FACTORS:
  −Screening tests for CMT biodegradability

screening test	CMT conc. mg/l	acclim. time weeks	% biodegradation	
semi-continuous A.S.	50	4−8	>95	
river water	5	4−6	>95	
CO_2 evolution	20	−	15−40*	
	20	−	65−90**	(1072)

*raw sewage employed as inoculum·
**acclimated activated sludge mixed liquor or soil suspension (0.5%) used as inoculum

6-carboxyuracil *see* orotic acid

catechol (1,2-dihydroxybenzene; 1,2-benzenediol; pyrocatechol; pyrocatechin)

CATECHOL

Uses: Important uses include the preparation of dyes, pharmaceuticals, and in the production of anti-oxidants for rubber and lubricatory oils. It is also used in photography in rubber, in fur dyeing and in specialty inks as an agent for oxygen removal. (367)

A. PROPERTIES: colorless leaflets; m.w. 110.11; m.p. 105°C; b.p. 240°C decomposes; v.d. 3.79; sp.gr. 1.371 at 15°C; solub. 451,000 mg/l at 20°C; log P_{oct}; 0.88/1.01

B. AIR POLLUTION FACTORS: 1 mg/cu m = 0.22 ppm, 1 ppm = 4.50 mg/cu m

C. WATER POLLUTION FACTORS:
   —Biodegradation rates: adapted A.S. at 20°C—product is sole carbon source: 96.0% COD removal at 55.5 mg COD/g dry inoculum/hr (327)
   —Biodegradation: decomposition by a soil microflora: 1 day (176)

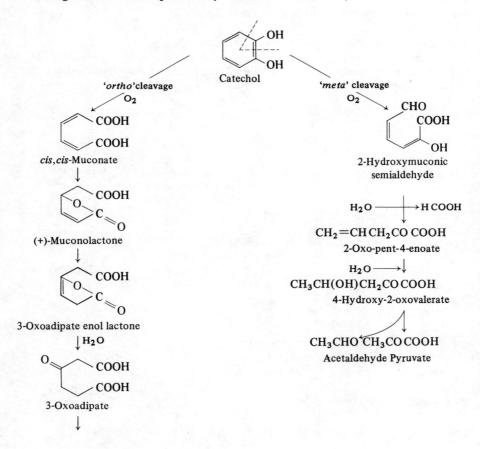

$CoASCOCH_2COCH_2CH_2COOH$
3-Oxoadipyl CoA
↓
$CH_3COOSCoA + HOOCCH_2CH_2COOH$
Acetyl-CoA   Succinate

Routes for degradation of catechol by micro-organisms (1328)
—COD: 1.89 (27)
—ThOD: 1.89 (27)
—Reduction of amenities:
  taste in fish (carp): at 2.5 mg/l (41)
  tainting of fish flesh: 2-5 mg/l (81)
  odor threshold (detection): 8.0 mg/l (998)
—Waste water treatment:
  A.S.: acclimated to the following aromatics:
  phenol: 13% theor. oxidation at 250 mg/l catechol after 30 min.
  o-cresol: 34% theor. oxidation at 250 mg/l catechol after 30 min.
  m-cresol: 22% theor. oxidation at 250 mg/l catechol after 30 min.
  p-cresol: 30% theor. oxidation at 250 mg/l catechol after 30 min.
  benzylalcohol: 22% theor. oxidation at 250 mg/l catechol after 30 min.
  mandelic acid: 26% theor. oxidation at 250 mg/l catechol after 30 min.
  anthranilic acid: 10% theor. oxidation at 250 mg/l catechol after 30 min.
  A.S. W, 20 C, $\frac{1}{2}$ day observed, feed: 500 mg/l, 25 day acclimation: 26% theor.
  oxidation (93)
—ozonation: catechol + O_3 → o-quinone

$$\text{o-quinone} + O_3 + H_2O \longrightarrow \text{(diene)}\begin{matrix}COOH + O_2\\ COOH + 4O_3\end{matrix}$$

$$4O_2 + 2\begin{matrix}COOH\\ |\\ COOH\end{matrix} \xleftarrow{+4O_3} \begin{matrix}COOH\\ |\\ \\ COOH\end{matrix} + \begin{matrix}COOH\\ |\\ COOH\end{matrix} + 4O_2 \quad (251)$$

—Autoxidation at 25°C: $t_{1/2}$: 447 hr at pH 7.0
                            412 hr at pH 9.0 (1908)
—Manmade sources:
  in secondary domestic sewage plant effluent: 0.001 mg/l (517)
—Sampling and analysis
  photometry: min. full scale: 0.67×10^{-6} mole/l (53)

D. BIOLOGICAL EFFECTS:
—Bacteria: *E. coli*: LD_0: 90 mg/l (30)
—Algae: *Scenedesmus*: LD_0: 6 mg/l (30)
—Protozoa: *Vorticella campanula*: LD_0: 1.6 mg/l (30)
           Paramaecium caudatum: LD_0: 35 mg/l (30)
—Arthropoda: *Daphnia*: LD_0: 4 mg/l (30)
—Fish: *Trutta iridea*:   perturbation level: 3 mg/l

Cyprinus carpio: perturbation level: 2.8 mg/l (30)
Goldfish: approx. fatal conc.: 14 mg/l, 48 hr (226)

CDT *see* 1,5,9-cyclododecatriene

cellosolve *see* ethyleneglycolmonoethylether

cellosolveacetate *see* ethyleneglycolmonoethyletheracetate

cerinic acid *see* n-hexacosanoic acid

cerotic acid *see* n-hexacosanoic acid

cetab *see* cetyltrimethylammoniumbromide

cetane *see* n-hexadecane

cetylalcohol *see* 1-hexadecanol

cetylpyridiniumbromide (hexadecylpyridiniumbromide)

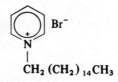

Use: surface active agent; germicide
A. PROPERTIES: cream colored waxy solid
C. WATER POLLUTION FACTORS:
  —Biodegradation: at 18 mg/l (inoculum = sew.) lag period = 6 days; after 17 days complete degradation (488)
D. BIOLOGICAL EFFECTS:
  —*Staphylococcus aureus*: at 20 mg/l bacteriolytic action after 5 hr
   Escherichia coli: at 20 mg/l slight reduction of growth rate
        at 5 mg/l no significant reduction of growth rate (488)

cetyltrimethylammoniumbromide (hexadecyltrimethylammoniumbromide; cetab)
$C_{16}H_{33}(CH_3)_3NBr$
Use: surface active agent; germicide
A. PROPERTIES: white powder; m.w. 364.46; m.p. >230°C decomposes
C. WATER POLLUTION FACTORS:
  —Biodegradation: at 15 mg/l:
   not adapted sew.: lag time: 2 days, complete degradation after 4 days
   adapted sew.: complete degradation after 2 days (488)
D. BIOLOGICAL EFFECTS:
  —*Escherichia coli*: at 20 mg/l no significant reduction of growth rate (488)

cetyltrimethylammoniumchloride (hexadecyltrimethylammoniumchloride)
$C_{16}H_{33}(CH_3)_3NCl$
  Use: surface active agent
C. WATER POLLUTION FACTORS:
  —Degradation: at 15 mg/l:
    not adapted sew.: lag time: 4 days, complete degradation after 7 days
    adapted sew.: at 18 mg/l, complete degradation after 2 days   (488)
D. BIOLOGICAL EFFECTS:
  —*Staphylococcus aureus*: at 20 mg/l no growth   (488)

chandor
  Use: herbicide; active ingredients: trifluralin 24%, linuron 12%
D. BIOLOGICAL EFFECTS:
  —Harlequin fish (*Rasbora heteromorpha*):

mg/l	24 hr	48 hr	96 hr	3 m (extrap.)	
LC_{10} (F)	0.87	0.58	0.58		
LC_{50} (F)	1.1	0.74	0.6	0.3	(331)

chinone *see p*-benzoquinone

chloral (trichloroethanal; trichloroacetaldehyde)
CCl_3CHO
  Use: manufacture of DDT: organic synthesis
A. PROPERTIES: colorless liquid; m.w. 147.40; m.p. $-57.5°C$; b.p. $98°C$; v.p. 35 mm at $20°C$; v.d. 5.1; sp.gr. 1.512 at $20/4°C$
B. AIR POLLUTION FACTORS: 1 mg/cm m = 0.166 ppm, 1 ppm = 6.0 mg/cu m
  —Odor: characteristic: sweet
    T.O.C.: $PIT_{50\%}$: 0.047 ppm
      $PIT_{100\%}$: 0.047 ppm   (2)
      O.I. at $20°C$: 980,000   (316)
      recognition: 0.035-0.050 mg/cu m (hydrate)   (610)
D. BIOLOGICAL EFFECTS:
  —Toxicity threshold (cell multiplication inhibition test) (chloral hydrate):

bacteria (*Pseudomonas putida*):	1.6 mg/l	(1900)
algae (*Microcystis aeruginosa*):	78.0 mg/l	(329)
green algae (*Scenedesmus quadricauda*):	2.8 mg/l	(1900)
protozoa (*Entosiphon sulcatum*):	79 mg/l	(1900)
protozoa (*Uronema parduczi Chatton-Lwoff*):	86 mg/l	(1901)
—Mammals: rat: oral LD_{50}: 0.05-0.4 g/kg		(211)

chloramben *see* 3-amino-2,5-dichlorobenzoic acid

chloramphenicol (chloromycetin; d-(-)-threo-2,2-dichloro-N-[β-hydroxy-α-(hydroxymethyl-p-nitrophenethyl] acetamide)
$Cl_2CHCONHCH(CH_2OH)CH(OH)C_6H_4NO_2$
  Use: an antibiotic derived from *Streptomyces venezuelae* or by organic synthesis; antifungal agent

A. PROPERTIES: m.w. 323.13; m.p. 148-150°C
C. WATER POLLUTION FACTORS:
  —Biodegradation rates: adapted A.S. at 20°C—product is sole carbon source: 86.2% COD removal at 3.3 mg COD/g dry inoculum/hr (327)
D. BIOLOGICAL EFFECTS:
  —Carcinogenicity: ?
  —Mutagenicity in the *Salmonella test*: ?
    <4.5 revertant colonies/nmol
    <70 revertant colonies at 5 μg/plate (1883)

chloraniformethan *see* imugan

chloranil (tetrachloroquinone; tetrachloro-*p*-benzoquinone; spergon)

Uses: agricultural fungicide; dye intermediate; electrodes for pH measurements; vulcanizing agent
A. PROPERTIES: yellow leaflets, m.p. 290°C; sp.gr. 1.97; m.w. 245.9; solub. 250 ppm at room temp.
D. BIOLOGICAL EFFECTS:
  —Fish: fathead minnows (*Pimephales promelas*): 96 hr TLm: 1-0.01 mg/l (static bioassay) (935)
  —Mammals: rats: oral LD_{50}: 4 g/kg (1554)
    in diet: no ill effects were suffered by rats fed diet containing 0.5% (1555)

chlordane (1,2,4,5,6,7,8,8-octachloro-4,7-methano-3a,4,7,7a-tetrahydroindane; Octaklor; 1068; Velsicol 1068; Dowklor)

Uses: insecticide, non-systemic

A. PROPERTIES: colorless to amber, odorless, viscous liquid
  —Technical chlordane consist of 60 to 75% isomers of chlordane and 25 to 40% of related compounds including 2 isomers of heptachlor and one each of enneachloro- and decachlorodicyclopentadiene.
  Two isomers of octachlorodicyclopentadiene have been isolated from chlordane of which α-chlordane is the endo-*cis* and β-chlordane is the endo-*trans* isomer.

The commercial product known as γ-chlordane is substantially the α-isomer.
—Technical chlordane is a mixture of 26 organochlorine compounds whose H_2O
solubility has been reported as 9 μg/l (1393)
—*cis*:*trans* (75:25) chlordane: water solubility: 0.056 ppm (1396)
—Approx. composition of technical chlordane:

fraction	% present	
cis(α)chlordane ($C_{10}H_6Cl_8$)	19 ± 3	
trans(γ)chlordane ($C_{10}H_6Cl_8$)	24 ± 2	
chlordene (4 isomers) ($C_{10}H_6Cl_6$)	21.5 ± 5	
heptachlor ($C_{10}H_5Cl_7$)	10 ± 3	
nonachlor ($C_{10}H_5Cl_9$)	7 ± 3	
($C_{10}H_{7-8}Cl_{6-7}$)	8.5 ± 2	
hexachlorocyclopentadiene (C_5Cl_6)	>1	
octachlorocyclopentadiene (C_5Cl_8)	1 ± 1	
Diels–Alder adduct of cyclopentadiene and pentachlorocyclopentadiene ($C_{10}H_6Cl_5$)	2 ± 1	
others	6 ± 5	(1661)

C. WATER POLLUTION FACTORS:
—Reduction of amenities: T.O.C. in water: 2.5 ppb (326)
  detection: 0.5 ppb (915)
—Aquatic reactions: persistence in river water in a sealed glass jar under sunlight and artificial fluorescent light—initial conc. 10 μg/l:

after	1 hr	1 wk	2 wk	4 wk	8 wk	
	100	90	85	85	85	(1309)

—75–100% disappearance from soils: 3–5 years (1815; 1816)
—Water and sediment quality:
Hawaii: α-chlordane: sediments: 400–5270 ppt
γ-chlordane: sediments: 1330–5120 ppt (1174)

D. BIOLOGICAL EFFECTS:
—Bacteria: oral *Viridans streptococci*: total inhibition at 3.0 ppm (1661)
—Algae:
Scenedesmus quadricauda: 0.1–100 μg/l: stimulation of growth (1395)
Chlamydomonas sp.: 0.1–50 μg/l: stimulation of growth
(soil alga) 100 μg/l: inhibition of cell division (1395)
Oedogonium (filamentous green alga): bioconcentration factor: 98,000 using ^{14}C-*cis*:*trans* (75:25) chlordane (1396)
Scenedesmus quadricauda: bioconcentration factor of 6,000 to 15,000 for *cis*(α) and *trans*(γ)chlordane at treatment levels of 0.1 to 100 μg/l water. Bioconcentration was rapid, occuring within the first 24 hr. (1392)
—Arthropoda:
Pelecypod (*Crassostrea virginica*): 96-hr EC_{50}-FT: 6.2 μg/l
Decapod (*Penaeus duorarum*): 96-hr LC_{50}-FT: 0.4 μg/l
Decapod (*Palaemonetes pugio*): 96-hr LC_{50}-FT: 4.8 μg/l (1132)
—Crustaceans
Gammarus lacustris: 26 μg/l: 96 h LC_{50} (2124)
Gammarus fasciatus: 40 μg/l: 96 h LC_{50} (2126)
Palaemonetes kadiakensis: 4.0 μg/l: 96 h LC_{50}; 2.5 μg/l (120 hour LC_{50}) (2126)

Simocephalus serrulatus: 20 µg/l: 48 h LC_{50} (2127)
Daphnia pulex: 29 µg/l: 48 h LC_{50} (2127)
Lobster (*Homarus americanus*): Canadian East Coast, lipid: 0.078–0.100 µg/g
(74; 1479)
—Insect:
Pteronarcys californica: 96 hr, LC_{50}: 15 µg/l (2128)
—Ni_2B-catalyzed dechlorination of technical grade chlordane yielded a mixture of partially dechlorinated products: the major one was 4,5,6,7,8-pentachloro-2,3,3a,4,7,7a-hexahydro-8-antihydromethanoindene, which contains five chlorine atoms (1413)
—Acute toxicity of chlorinated and dechlorinated chlordane:

species	formulation	LC_{50} (µg/l)[a] chlorinated	dechlorinated	detoxification factor[b]
bluegill	technical	41	582	14
	72% EC[c]	62	800	13
Daphnia	technical	97	813	8
	72% EC	156	1174	8

[a] 96 hr LC_{50} for bluegill, 48 hr EC_{50} for *Daphnia*; calculated on basis of insecticide present in formulation
[b] factor: LC_{50} dechlorinated/LC_{50} chlorinated
[c] EC = emulsifiable concentrate (1413)
—Molluscs: BCF: oyster: 7300 (1924)
—Fish:
fathead minnow (*Pimephales promelas*): 96-hr LC_{50}: 36.9 µg/l (FT)
brook trout: 96-hr LC_{50}:   47.0 µg/l (FT)
            Lowest effect level: 0.32 µg/l (FT)
bluegill: 96-hr LC_{50}: 59.0 µg/l (FT)
         MATC: 0.54 µg/l (FT) (1204)
bluegill: 96 hr, LC_{50}: 0.022 ppm
rainbow trout: 96 hr, LC_{50}: 0.022 ppm (1878)
Lagodon rhomboides: 96 hr, LC_{50}-FT: 6.4 µg/l
Cyprinodon variegatus: 96 hr, LC_{50}-FT: 24.5 µg/l (1132)
Saccobranchus fossilis: 96 hr, LC_{50}-S: 0.42 mg/l (1509)
Cyprinodon variegatus: 96 hr, LC_{50}-FT: 12.5 µg/l (1472)
goby fish (*Acanthogobius flavimanus*) collected at the seashore of Keihinjima along Tokyo Bay—Aug. 1978: residue level: 6 ppb (*cis*) (1721)
goby fish (*Acanthogobius flavimanus*) collected at the seashore of Keihinjima along Tokyo Bay—Aug. 1978: residue level: 9 ppb (*trans*) (1721)
herring (*Clupea harengus*): Canadian East Coast, lipid: 0.039–0.114 µg/g (1479)
Carassius auratus: adsorption of ^{14}C-*cis*-chlordane from water by goldfish is very rapid. Over 99% of the radioactivity recovered from the fish on day 10 and day 15 post-treatment times was unchanged, *cis*-chlordane indicating its inert storage in body tissues (1569)
BCF following exposure to 5 ppb (*cis*-) in a static systems:

	BCF	time for maximum absorption	
frogs (*Xenopus laevis*)	108	96 hr	
bluegills (*Lepomis macrochirus*)	322	24 hr	
goldfish (*Carassius auratus*)	990	16 hr	(1839)

Elimination half-life after maximum absorption in a static system (*cis*-):
goldfish: 4.4 weeks (extrap.)
tropical fish (*Cichlasoma*): 20 weeks (extrap.)
bluegill (*Lepomis macrochirus*): 16 weeks (extrap.)
frog (*Xenopus*): 3.3 weeks (1839)
—Mammals:
acute oral LD_{50} (rats): 457–590 mg/kg (1855)
in diet: rats fed for 104 weeks on a diet containing 15 ppm γ-chlordane suffered no higher mortality than the controls but histopathological changes in the liver were apparent (1855)

chlordene

Use: intermediate in the manufacturing of chlordane

D. BIOLOGICAL EFFECTS:
—Toxicity ratios of chlordene (*C*) to photochlordene (*PC*), calculated from respective 24 hr LC_{50}^* and LT_{50} values:

	C/PC
Daphnia pulex (waterflea)	0.91
Musca domestica (housefly)	0.78 (0.70)*
bluegill:	0.72 (0.63)*

(1681*; 1684; 1685)

—bluegill: 24 hr LC_{50}: 218 ppb
mosquito (late 3rd instar *Aedes aegypti* larvae) 24 hr LC_{50}: 130 ppb
housefly (3 day old female *Musca*) LD_{50}: 158 μg/fly (1681)

chlordimeform *see* chlorphenamidine

chlorfenpropmethyl *see* bidisin

chlorfenvinphos (2-chloro-1-(2,4-dichlorophenyl)vinyldiethylphosphate; O,O-diethyl-O-1-(2',4'-dichlorophenyl)-2-chlorovinylphosphate; Birlane)

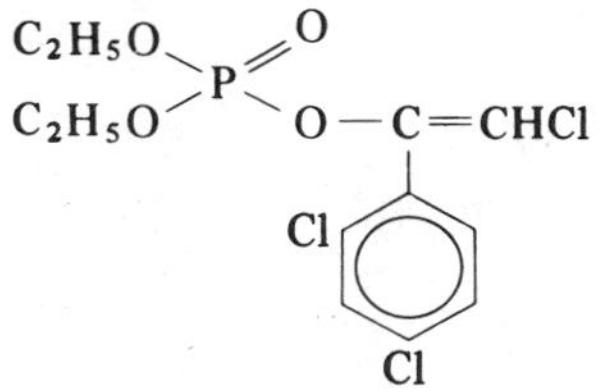

Use: insecticide

354   CHLORINATED CAMPHENE

A. PROPERTIES: amber-colored liquid; m.p. -16/-22°C; b.p. 168-170°C at 0.5 mm; v.p. 1.7×10^{-7} mm at 25°C; sp.gr. 1.36 at 15.5/15.5°C

C. WATER POLLUTION FACTORS:
—Persistence in soil at 10 ppm initial concentration:

	weeks incubation to	
	50% remaining	5% remaining
sterile sandy loam	>24	
sterile organic soil	>24	
non-sterile sandy loam	<1	5
non-sterile organic soil	1	9 (1433)

D. BIOLOGICAL EFFECTS:
acute oral LD_{50} (dogs):   >12,000 mg/kg
           (rabbits):   280-400 mg/kg
           (mice):   117-200 mg/kg   (1855)
           (rat):   10-39 mg/kg
acute dermal LD_{50} (rabbit): 3200-4700 mg/kg   (1854)
           (rats):   31-108 mg/kg   (1855)
in diet: 90 day feeding tests with rats showed that diets containing up to 300 ppm had no adverse effects on growth or food consumption   (1855)

chlorinated camphene   *see* toxaphene

chlordecone   *see* kepone

chlorhexidine   *see* 1,6-di(4-chlorophenyldiguanido)hexane

chlorine cyanide   *see* cyanogenchloride

chlormephos (S-chlormethyl-O,O-diethylphosphorothiolothionate)

$$\begin{array}{c} C_2H_5O \\ \diagdown \\   P \\ \diagup \diagdown \\ C_2H_5O S{-}CH_2Cl \end{array}$$

Use: insecticide

A. PROPERTIES: sp.gr. 1.26
D. BIOLOGICAL EFFECTS:
—Harlequin fish (*Rasbora heteromorpha*):

mg/l	24 hr	48 hr	96 hr	3 m (extrap.)
LC_{10} (F)	3.5	2.8	—	
LC_{50} (F)	4.8	3.5	2.5	1.5 (331)

—Mammals:
acute oral LD_{50} (rat):   7 mg/kg
acute dermal LD_{50} (rat):   27 mg/kg

S-chlormethyl-O,O-diethylphosphorothiolothionate   *see* chlormephos

α-chloroacetophenone   (phenacylchloride)

CH₂Cl
|
C=O
(phenyl ring)

A. PROPERTIES: m.w. 154.59; m.p. 59–60°C; b.p. 244–247°C; v.p. 0.004 mm at 20°C, 0.014 mm at 30°C; v.d. 5.32; sp.gr. 1.324 at 15/4°C; sat.conc. 0.034 g/cu m at 20°C, 0.11 g/cu m at 30°C
B. AIR POLLUTION FACTORS: 1 mg/cu m = 0.16 ppm, 1 ppm = 6.43 mg/cu m
   −Odor: characteristic: quality: apple blossom odor
                          hedonic tone: strong lacrimator
   T.O.C.: 0.1 mg/cu m = 0.016 ppm; 0.1–0.7 mg/cu m   (57)(710)
   O.I. at 20°C = 330   (316)
C. WATER POLLUTION FACTORS:
   −Reduction of amenities: faint odor: 0.0085 mg/l   (129)
D. BIOLOGICAL EFFECTS:
   −Mice: RD_{50} (respiratory rate): 0.96 ppm   (1330)

N-chloroalanine
Use: a rapidly formed chlorination product of alanine.
N-chloroalanine decomposes rapidly in water to form acetaldehyde, ammonia, carbon dioxide, and chloride ion or, depending on pH, pyruvic acid, ammonia, and chloride ion. The half-life of the reaction in the 5–9 pH range is 46 min. at 25°C. The rate constant shows a marked temperature dependence at all pH values, changing by a factor of more than 3 for a 10°C temperature change.   (1906)

chloroallylene *see* allylchloride

o-chloroaniline (2-chlorophenylamine; o-aminochlorobenzene)

NH₂ — (phenyl ring) — Cl

A. PROPERTIES: m.w. 127.57; m.p. α–14°C, β–3.5°C; b.p. 208.8°C; sp.gr. 1.213 at 20/4°C; log P_{oct} 1.90
C. WATER POLLUTION FACTORS:
   −Biodegradation rates:
      degradation by *Aerobacter*: 500 mg/l at 30°C:
            parent: 100% ring disruption in 60 hr
            mutant: 100% ring disruption in 18 hr   (152)
      decomposition by a soil microflora: >64 days   (176)
      adapted A.S. at 20°C−product is sole carbon source: 98.0% COD removal at
         at 16.7 mg COD/g dry inoculum/hr   (327)
      W, A.S. from mixed domestic/industrial treatment plant: 22–41% depletion at
         20 mg/l after 6 hr at 25°C   (419)

***m*-chloroaniline** (3-chlorophenylamine)

NH₂—C₆H₄—Cl

Uses: intermediate for azo dyes and pigments; pharmaceuticals; insecticides; agricultural chemicals

A. PROPERTIES: m.w. 127.57; m.p. -10.4°C; b.p. 229.8°C; v.p. <0.1 mm at 30°C; v.d. 4.41; sp.gr. 1.216 at 20/4°C; log P_{oct} 1.88

B. AIR POLLUTION FACTORS: 1 mg/cu m = 0.19 ppm, 1 ppm = 5.30 mg/cu m

C. WATER POLLUTION FACTORS:
—Biodegradation rates: adapted A.S. at 20°C—product is sole carbon source: 97.2% COD removal at 6.2 mg COD/g dry inoculum/hr (327)
—Waste water treatment:
degradation by *Aerobacter*: 500 mg/l at 30°C:
parent: 100% ring disruption in 68 hr
mutant: 100% ring disruption in 16 hr (152)
decomposition period by a soil microflora: >64 days (176)
W, A.S. from mixed domestic/industrial treatment plant: 14-25% depletion at 20 mg/l after 6 hr at 25°C (419)

***p*-chloroaniline** (4-chlorophenylamine)

Uses: dye intermediate; pharmaceuticals; agricultural chemicals

A. PROPERTIES: rhombic prisms; m.w. 127.57; m.p. 70/72°C; b.p. 231/232°C; v.p. 0.015 mm at 20°C, 0.05 mm at 30°C; v.d. 4.41; sp.gr. 1.427 at 19/4°C; sat.conc. 0.01 g/cu m at 20°C, 0.34 g/cu m at 30°C; log P_{oct} 1.83

B. AIR POLLUTION FACTORS: 1 mg/cu m = 0.19 ppm, 1 ppm = 5.30 mg/cu m
—Odor: hedonic tone: sweet

C. WATER POLLUTION FACTORS:
—Biodegradation rates: adapted A.S. at 20°C—product is sole carbon source: 96.5% COD removal at 5.7 mg COD/g dry inoculum/hr (327)
—Waste water treatment:
degradation by *Aerobacter*: 500 mg/l at 30°C:
parent: 100% ring disruption in 59 hr
mutant: 100% ring disruption in 12 hr (152)
decomposition period by a soil microflora: >64 days (176)
W, A.S. from mixed domestic/industrial treatment plant: 9-34% depletion at 20 mg/l after 6 hr at 25°C (419)

D. BIOLOGICAL EFFECTS:
—Fish:

rainbow trout: 96-h LC_{50} 14 mg/l, S
fathead minnow: 96-h LC_{50} 12 mg/l, S
channel catfish: 96-h LC_{50} 23 mg/l, S
bluegill: 96-h LC_{50} 2 mg/l, S (1510)
—Man: severe toxic effects: 8 ppm = 44 mg/cu m, 1 min
symptoms of illness: 4 ppm = 22 mg/cu m
unsatisfactory: >2 ppm = 11 mg/cu m (185)

o-chlorobenzaldehyde

A. PROPERTIES: m.w. 140.57; m.p. 10–11.5°C; b.p. 209–215°C; sp.gr. 1.248
—Hydrolysis product of o-chlorobenzylidene-malononitrile
D. BIOLOGICAL EFFECTS:
—Fish: rainbow trout: 12 hr, LC_{50}: 5.2 mg/l
24 hr, LC_{50}: 3.6 mg/l
48 hr, LC_{50}: 2.8 mg/l
96 hr, LC_{50}: 2.5 mg/l (1913)

chlorobenzene (phenylchloride)

Users and formulations: solvent recovery plants; intermediate in dyestuffs mfg.; mfg. aniline, insecticide, phenol, chloronitrobenzene. (347)
A. PROPERTIES: colorless liquid; m.w. 112.56; m.p. -45°C; b.p. 132°C; v.p. 8.8 mm at 20°C, 11.8 mm at 25°C, 15 mm at 30°C; v.d. 3.88; sp.gr. 1.1066 at 20/4°C; solub. 500 mg/l at 20°C, 488 mg/l at 30°C; sat.conc. 54 g/cu m at 20°C, 89 g/cu m at 30°C; log P_{oct} 2.84 at 20°C
B. AIR POLLUTION FACTORS: 1 mg/cu m = 0.217 ppm, 1 ppm = 4.678 mg/cu m
—Odor: characteristic: quality: chlorinated mothballs, aromatic

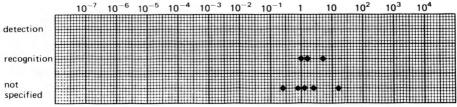

(57; 73; 279; 298; 307; 610; 748; 815; 838)

CHLOROBENZENE

  USSR: human odor perception: 0.4 mg/cu m = 0.09 ppm
     human reflex response: no response: 0.1 mg/cu m
              adverse response: 0.2 mg/cu m
     animal chronic exposure: no effect: 0.1 mg/cu m
               adverse effect: 1.0 mg/cu m   (170)
  O.I. at 20°C: 52,600   (316)
 —Sampling and analysis:
  photometry: min. full scale: 220 ppm   (53)
C. WATER POLLUTION FACTORS:
 —BOD_5: 0.03 std.dil.sew.   (41)(27)(298)
 —COD: 0.41   (27)
 —ThOD: 2.06   (27)
 —Impact on biodegradation processes: at 100 mg/l, no inhibition of NH_3 oxidation by *Nitrosomonas* sp.   (390)
 —Reduction of amenities: T.O.C.: 0.1 mg/l   (296; 903)
 —Water quality:
  in Zürich lake: at surface: 3 ppt; at 30 m depth: 12 ppt
  in Zürich area: in groundwater: 14 ppt; in tapwater 6 ppt   (513)
  in Delaware river (U.S.A.): conc. range: winter: 7.0 ppb
                   summer: n.d.   (1051)
  in river Maas at Eysden (Netherlands) in 1976: median: n.d.: range: n.d.–1.9 µg/l
  in river Maas at Keizersveer (Netherlands) in 1976: median: n.d.: range: n.d.–9.6 µg/l   (1368)
 —Waste water treatment:
  air stripping constant: $K = 0.969$ days^{-1} at 100 mg/l   (82)
  ring disruption by *Pseudomonas* at 200 mg/l:
      parent: 100% in 58 hr
      mutant: 100% in 14 hr   (152)
 —Sampling and analysis:
  photometry: min. full scale: 4.1×10^{-6} mole/l   (53)
D. BIOLOGICAL EFFECTS:
 —Toxicity threshold (cell multiplication inhibition test):
  bacteria (*Pseudomonas putida*):   17 mg/l   (1900)
  algae (*Microcystis aeruginosa*):   120 mg/l   (329)
  green algae (*Scenedesmus quadricauda*): >390 mg/l   (1900)
  protozoa (*Entosiphon sulcatum*):   >390 mg/l   (1900)
  protozoa (*Uronema parduczi Chatton-Lwoff*): >392 mg/l   (1901)
 —Fish:
  fatheads: 24–96 hr, TLm: 29–39 mg/l
  bluegills: 24–96 hr, TLm: 24 mg/l
  goldfish: 24–96 hr, TLm: 51–73 mg/l
  guppies: 24–96 hr, TLm: 45 mg/l   (41)
  rainbow trout: 24 hr, LD_{50}: 1.8 ml/kg   (1528)
  guppy (*Poecilia reticulata*): 14 d LC_{50}: 19 ppm   (1833)
 —Mammalia:
  rabbits: oral dose: no effect, 14.4 mg/kg, 192 days
  rabbits: oral dose: slight dip in growth, 144 mg/kg, 192 days   (121)

guinea pigs: inhalation: no effect, 200 ppm, 7 hr/day, 5 days/w, 44 days (211)
cat: inhalation: death after 7 hr, 3700 ppm
cat: inhalation: tolerated for 1 hr 220–660 ppm (211)
rat: single oral LD_{50}: 2190 mg/kg
rabbit: single oral LD_{50}: 2830 mg/kg (54)
rat: inhalation LC_{50}: 12,000 ppm, 30 min
cat: inhalation LC_{50}: 8,000 ppm, 30 min (54)
—Man: severe toxic effects: 400 ppm: 1872 mg/cu m, 60 min
   symptoms of illness: 200 ppm: 936 mg/cu m, 60 min
   unsatisfactory: >75 ppm: 35.1 mg/cu m, 60 min (185)

p-chlorobenzenesulfonic acid

A. PROPERTIES: m.w. 192.62; m.p. 68°C; b.p. 147/148°C at 25 mm
C. WATER POLLUTION FACTORS:
  —Biodegradation:
    decomposition period by a soil microflora: 16 days (176)
D. BIOLOGICAL EFFECTS:
  —*Daphnia magna*: 24 hr, TLm: 8,600 mg/l (sodium salt)
                48 hr, TLm: 7,659 mg/l (sodium salt)
                72 hr, TLm: 3,964 mg/l (sodium salt)
                96 hr, TLm: 2,150 mg/l (sodium salt) (153)
              100 hr, TLm: 2,394 mg/l (sodium salt) (1295)
  —fish:
    Lepomis macrochirus: 24 hr, TLm: <3,219 mg/l (sodium salt) (1295)
  —snail eggs: *Lymnaea* sp.: 24 hr, TLm: 8,600 mg/l (sodium salt)
                      48 hr, TLm: 7,633 mg/l (sodium salt)
                      72 hr, TLm: 6,343 mg/l (sodium salt)
                      96 hr, TLm: 5,053 mg/l (sodium salt) (153)

o-chlorobenzoic acid

A. PROPERTIES: m.w. 156.57; m.p. 142°C; b.p. sublimes; sp.gr. 1.544 at 20/4°C; solub. 2,100 mg/l at 25°C; log P_{oct} 1.98
C. WATER POLLUTION FACTORS:
  —Biodegradation:
    decomposition period by a soil microflora: >64 days (176)

m-chlorobenzoic acid

A. PROPERTIES: m.w. 156.57; m.p. 158°C; b.p. sublimes; sp.gr. 1.496 at 25/4°C; solub. 400 mg/l at 0°C; log P_{oct} 2.68

C. WATER POLLUTION FACTORS:
—Biodegradation:
decomposition period by a soil microflora: 32 days (176)
lag period for degradation of 16 mg/l by waste water or by soil at pH 7.3 and 30°C: 7–14 days

Bacterial degradation pathway (1219)

D. BIOLOGICAL EFFECTS:
—Mammals: rabbit: acute oral LD_{50}: >500 mg/kg (211)

p-chlorobenzoic acid

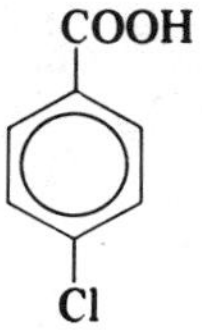

[structure: benzene ring with COOH at top and Cl at bottom (para)]

A. PROPERTIES: m.w. 156.57; m.p. 243°C; b.p. sublimes; sp.gr. 1.541 at 24/4°C; solub. 77 mg/l at 25°C; $\log P_{oct}$ 2.65
C. WATER POLLUTION FACTORS:
—Biodegradation:
decomposition period by a soil microflora: 64 days (176)
lag period for degradation of 16 mg/l by waste water or soil suspension at pH 7.3 and 30°C: more than 25 days (1096)

o-chlorobenzylidene malononitrile (OCBM)

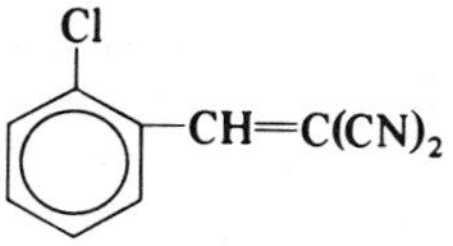

[structure: o-chlorobenzene with –CH=C(CN)$_2$ substituent]

A. PROPERTIES: m.w. 188.62; m.p. 95°C; b.p. 310/315°C; v.d. 6.52;
B. AIR POLLUTION FACTORS: 1 mg/cu m = 0.13 ppm, 1 ppm = 7.84 mg/cu m
C. WATER POLLUTION FACTORS:
—hydrolysis in water:

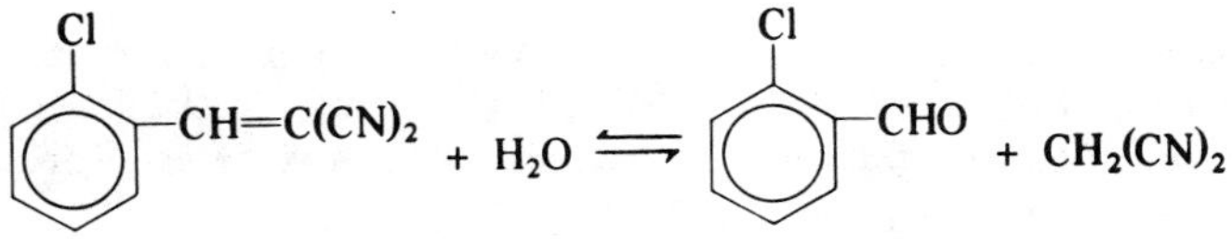

[reaction: o-Cl-C$_6$H$_4$-CH=C(CN)$_2$ + H$_2$O ⇌ o-Cl-C$_6$H$_4$-CHO + CH$_2$(CN)$_2$]

o-chlorobenzaldehyde   malononitrile

D. BIOLOGICAL EFFECTS:
—Fish:
rainbow trout: LC_{50}, 12 hr: 1.28 mg/l
24 hr: 0.45 mg/l
48 hr: 0.42 mg/l
96 hr: 0.22 mg/l (1913)
rainbow trout exposed to equimolar mixture of o-chlorobenzaldehyde and malononitrile (products of hydrolysis):
LC_{50}, 12 hr: 4.7 mg/l
24 hr: 2.0 mg/l
48 hr: 1.2 mg/l
96 hr: 1.1 mg/l (1913)

2-chlorobiphenyl

—Mammals:
   mice: RD_{50} (respiratory rate): 0.52 ppm (1330)

2-chlorobiphenyl

A. PROPERTIES: solub. 5.8 mg/l
C. WATER POLLUTION FACTORS:
   —Biodegradation:
   100% degradation after 1 hour by *Alcaligenes* Y42 (cell number 2×10^9/ml) and *Acinetobacter* P6 (cell number 4.4×10^8/ml) at 9.3 mg/l initial concentration; trimethylsilyl derivative of monochlorobenzoic acid was detected in the metabolite (1086)
D. BIOLOGICAL EFFECTS:
   —Marine yeast *Rhodotorula rubra*: bioconcentration coefficient:
      737 in whole cells
      ±37,000 in their lipid portion (1566)

3-chlorobiphenyl

A. PROPERTIES: solub. 3.3 mg/l
C. WATER POLLUTION FACTORS:
   —Biodegradation:
   100% degradation after 1 hour by *Alcaligenes* Y42 (cell number 2×10^9/ml) and *Acinetobacter* P6 (cell number 4.4×10^8/ml) at 9.3 mg/l initial concentration; trimethylsilyl derivative of monochlorobenzoic acid was detected in the metabolite (1086)
D. BIOLOGICAL EFFECTS:
   —Marine yeast *Rhodotorula rubra*: bioconcentration coefficient:
      1180 in whole cells
      ±59,000 in their lipid portion (1566)

4-chlorobiphenyl

A. PROPERTIES: m.w. 188.66; m.p. 76–78°C; b.p. 282°C; solub. 0.8 mg/l
C. WATER POLLUTION FACTORS:
   —Biodegradation:

$$\text{[biphenyl]}-Cl \rightarrow \text{[phenyl]}-Cl \qquad (1219)$$
bacterial degradation

Eighteen lichens from a variety of habitats were treated with 4-CB, all were shown to partially convert 4-CB to 4-chloro-4'-hydroxybiphenyl—it took between six and 22 hours for the hydroxy derivative to appear.

Only one species (*Pseudocyphellaria crocata*) produced a further metabolite: 4-chloro-4'-methoxybiphenyl. (1651)

100% degradation after 1 hour by *Alcaligenes* Y42 (cell number 2×10^9/ml) and *Acinetobacter* P6 (cell number 4.4×10^8/ml) at 9.3 mg/l initial concentration; trimethylsilyl derivative of monochlorobenzoic acid was detected in the metabolite. (1086)

D. BIOLOGICAL EFFECTS:
—Marine yeast *Rhodotorula rubra*: bioconcentration coefficient:
1,550 in whole cells
±77,500 in their lipid portion (1566)

2-chloro-4,6-*bis*(ethylamino)-*s*-triazine (princep; simazine)

$$C_2H_5NH-\underset{N}{\overset{Cl}{\underset{N}{\bigcirc}}}-NHC_2H_5$$

Use: herbicide
A. PROPERTIES: m.p. 225–227°C; solub. 5 ppm at 20°C
C. WATER POLLUTION FACTORS:
75–100% disappearance from soils: 12 months (1815)
D. BIOLOGICAL EFFECTS:
—Algae:

technical acid	*Chlorococcum* sp.	2.5×10^3 ppb	50% decrease in O_2 evolution	f
technical acid	*Chlorococcum* sp.	2×10^3 ppb	50% decrease in growth	Measured as ABS. (525 mu) after 10 days
technical acid	*Dunaliella tertiolecta*	4×10^3 ppb	50% decrease in O_2 evolution	f
technical acid	*Duneliella tertiolecta*	5×10^3 ppb	50% decrease in growth	Measured as ABS. (525 mu) after 10 days
technical acid	*Isochrysis galbana*	600 ppb	50% decrease in O_2 evolution	f
technical acid	*Isochrysis galbana*	500 ppb	50% decrease in growth	Measured as ABS. (525 mu) after 10 days
technical acid	*Phaeodactylum tricornutum*	600 ppb	50% decrease in O_2 evolution	f

2-CHLORO-1,3-BUTADIENE

technical acid	*Phaeodactylum tricornutum*	500 ppb	50% decrease in growth	Measured as ABS. (525 mu) after 10 days (2348)

—Crustaceans
Gammarus lacustris	13000 µg/l 96 h, LC_{50}	(2124)
Gammarus fasciatus	no effect at 100,000 µg/l 48 hr	(2125)
Daphnia magna	1000 µg/l 48 h, LC_{50}	
Cypridopsis vidua	3200 µg/l 48 h, LC_{50}	
Asellus brevicaudus	no effect at 100,000 µg/l 48 hr	(2125)
Palaemonetes kadiakensis	no effect at 100,000 µg/l 48 hr	(2125)
Orconectes nais	no effect at 100,000 µg/l 48 hr	(2125)

—Fish:
Oncorhynchus kisutch	6600 µg/l 48 h, LC_{50}	(2106)
bluegill: 48 hr LC_{50}: 130 ppm		
rainbow trout: 48 hr LC_{50}: 85 ppm		(1878)

—Mammals:
acute oral LD_{50} (rat): > 5000 mg/kg	(1854)
acute dermal LD_{50} (rabbit): >10,200 mg/kg (80% formulation)	(1855)
in diet (rat): 100 ppm for 2 years: nontoxic	(1855)

2-chloro-1,3-butadiene *see* chloroprene

1-chlorobutane *see* n-butylchloride

4-chloro-o-cresol

Microbial metabolite of MCPA.	(1463)
Impurity in technical grade MCPA (up to 4%).	(1465)

D. BIOLOGICAL EFFECTS:
—Fish:
trout: LD_{50} 24 hr (S): 2.12 mg/l (1.88–2.39 mg/l)	(1463)
exposure: 0.5–1.5 mg/l for 3 weeks: bioaccumulation: 4–5X in the wet weight of the fish	(1463)

—Mammals:
Wistar rats: IP LD_{50} = 794 mg/kg	
OR LD_{50} = 1194 mg/kg	(1390)

p-chloro-m-cresol (6-chloro-3-hydroxytoluene; 4-chloro-3-hydroxytoluene; PCMC; 4-chloro-3-methylphenol; 4-chloro-1-hydroxy-3-methylbenzene)

Uses: external germicide; preservative for glues, gums, inks, textile and leather goods

A. PROPERTIES: odorless crystals (when pure); m.w. 142.6; m.p. 66°C; b.p. 235°C: log P_{oct} 3.10

C. WATER POLLUTION FACTORS:
—Biodegradation:
after 3 weeks of adaptation at 20 mg/l at 22°C: 30% degradation when product is sole carbon source/100% degradation with synthetic sewage—aerobic conditions; under anaerobic conditions no degradation with or without synthetic sewage (512)
—Odor threshold: detection: 0.1 mg/kg (892)

D. BIOLOGICAL EFFECTS:
—Fish:
fathead minnows (*Pimephales promelas*): 96 hr TLm: 0.1–0.01 mg/l (static bioassay) (935)

1-chloro-2-(β-chloroethoxy)ethane *see* β,β'-dichloroethylether

1-chloro-2-(β-chloroisopropoxy)propane *see* dichloroisopropylether

chloro(chloromethoxy)methane *see* dichloromethylether

2-chloro-3(4-chlorophenyl)methylpropionate *see* bidisin

1-chlorodecane (*prim-n*-decylchloride)
$CH_3(CH_2)_9Cl$
A. PROPERTIES: m.w. 176.73; m.p. 34°C; b.p. 223°C; sp.gr. 0.868
B. AIR POLLUTION FACTORS:
—Odor threshold: recognition: 21.7 mg/cu m (761)
C. WATER POLLUTION FACTORS:
—Waste water treatment: A.S.: after 6 hr: 1.6% of ThOD,
after 12 hr: 3.2% of ThOD,
after 24 hr: 5.9% of ThOD

5-chloro-2(2,4-dichlorophenoxy)phenol (triclosan)

C. WATER POLLUTION FACTORS:
—Biodegradation:
after 3 weeks of adaption at 1–5 mg/l at 22°C:
aerobic degradation: product is sole carbon source: 0% degradation
+ synthetic sewage: 50% degradation
anaerobic degradation: product is sole carbon source: 0% degradation
+ synthetic sewage: 50% degradation

2-chloro-1-(2,4-dichlorophenyl)vinyldiethylphosphate *see* chlorfenvinphos

2-chloro-α-(((diethoxyphosphinothioyl)oxy)imino)-benzeneacetonitrile *see* chlorphoxim

4-chloro-2,5-dihydroxydiphenylsulfone

D. BIOLOGICAL EFFECTS:
—Fish: goldfish: approx. fatal conc.: 35 mg/l, 48 hr (226)

1-chlorododecane (*prim-n*-dodecylchloride)
$CH_3(CH_2)_{10}CH_2Cl$
C. WATER POLLUTION FACTORS:
—Waste water treatment:
A.S.: after 6 hr: 3.5% of ThOD
12 hr: 6.2% of ThOD
24 hr: 11.8% of ThOD (88)

1-chloro-2,3-epoxypropane *see* epichlorohydrin

chloroethanoic acid *see* monochloroacetic acid

2-chloroethanol (β-chloroethylalcohol; ethylenechlorohydrin; glycolchlorohydrin)
$ClCH_2CH_2OH$
A. PROPERTIES: colorless liquid; m.w. 80.52; m.p. 67.5/71°C; b.p. 128.8°C; v.p. 4.9 mm at 20°C, 10 mm at 30°C; v.d. 2.78; sp.gr. 1.121 at 20/4°C; sat.cons. 24 g/cu m at 20°C, 42 g/cu m at 30°C
B. AIR POLLUTION FACTORS: 1 mg/cu m = 0.304 ppm, 1 ppm = 3.29 mg/cu m
C. WATER POLLUTION FACTORS:
—BOD_5: 0% of ThOD (79)
0.5 (41)
—BOD_{10}: 16.1% of ThOD (79)
0.5 std.dil.sew. (256)
—BOD_{15}: 74.1% of ThOD (79)
—BOD_{20}: 88.6% of ThOD (79)

—Waste water treatment:

methods	temp °C	days observed	feed mg/l	days acclim.	% removed	
NFG, BOD	20	1-10	200	365 + P	nil	
NFG, BOD	20	1-10	400	365 + P	nil	
NFG, BOD	20	1-10	1,000	365 + P	3	(93)

—Sampling and analysis:
photometry: min. full scale: 625×10^{-6} mole/l (53)

D. BIOLOGICAL EFFECTS:
—Mammalia: rat: single oral LD_{50}: 72-95 mg/kg
rat: inhalation LC_{50}: 0.11 mg/l, 4 hr
rat: inhalation survived: 3.0 mg/l, 0.25 hr
guinea pig: single oral LD_{50}: 110 mg/kg
mouse: inhalation: LC_{17}: 1.2 mg/l, 2 hr
mouse: inhalation: survived: 1.0 mg/l, 2 hr (211)
—Man: readily penetrates the skin
severe toxic effects: 20 ppm = 68 mg/cu m, 60 min
symptoms of illness: 10 ppm = 34 mg/cu m
unsatisfactory: 2 ppm = 7 mg/cu m (85)

chloroethene *see* vinylchloride

β-chloroethylalcohol *see* 2-chloroethanol

2-chloro-4-ethylamino-6-isopropylamino-s-triazine *see* atrazine

2-(4-chloro-6-ethylamino)-s-triazine-2-ylamino-2-propionitrile *see* bladex

chloroethylene *see* vinylchloride

chloroform (trichloromethane)
$CHCl_3$
Manufacturing source: organic chemical industry. (347)
Users and formulation: mfg. fluorocarbon refrigerants and propellants and plastics; mfg. anesthetics and pharmaceuticals, primary source for chlorodifluoromethane; fumigant; solvent; sweetener; fire extinguisher mfg.; electronic circuitry mfg.; analytical chemistry; insecticide. (347)

A. PROPERTIES: colorless liquid; m.w. 119.38; m.p. $-64°C$; b.p. $+62°C$; v.p. 160 mm at 20°C, 245 mm at 30°C; v.d. 4.12; sp.gr. 1.489 at 20°C; solub. 8,000 mg/l at 20°C, 9,300 mg/l at 25°C, 10,000 mg/l at 15°C; sat.conc. 1027 g/cu m at 20°C, 1540 g/cu m at 30°C; log P_{oct} 1.97 at 20°C

B. AIR POLLUTION FACTORS: 1 mg/cu m = 0.20 ppm, 1 ppm = 4.96 mg/cu m
odor thresholds

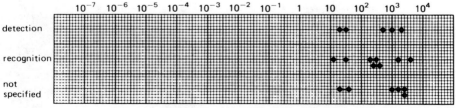

(57; 211; 602; 610; 643; 671; 704; 740; 753; 777; 804; 805; 840)

CHLOROFORM

odor index at 20°C: 70 (316)
—Manmade sources: in rural Washington: Dec 1974–Feb 1975: 20 ppt (315)
—Sampling and analysis:
photometry: min.full scale: <1,000 ppm (53)

C. WATER POLLUTION FACTORS:
—BOD_5: 0–0.0008 std.dil.sew. (41)
       0.02 (275)
—$KMnO_4$: 0.020 (30)
—ThOD: 1.346 (30)
       0.33 (27)
—Reduction of amenities:
odor thresholds: 20 mg/l; 0.1 mg/kg (84; 894)
—Aquatic reactions:
measured half life for evaporation from 1 ppm aqueous solution, still air, and an average depth of 6.5 cm:
at 1–2°C: 34.5 min
at 25°C: 18.5–25.7 min (369)
—Water quality:
in N.W. England tapwaters (1974): 0.7–38 ppb (933)
waters from upland reservoirs in N.W. England (1974):
during dry, cloudy weather: <0.1–0.1 ppb
during prolonged heavy rain: 11–21 ppb (933)
in river Maas at Eysden (Netherlands) in 1976:
median: 1.2 µg/l, range: n.d.–5.6 µg/l
in river Maas at Keizersveer (Netherlands) in 1976:
median: 0.75 µg/l: range: 0.1–2.1 µg/l (1368)
—Waste water treatment:
evaporation from water at 25°C of 1 ppm solution:
50% after 18–25 min
90% after 62–83 min (313)
experimental water reclamation plant:

	sample 1	sample 2	
sand filter effluent	749 ng/l	563 ng/l	
after chlorination	33623 ng/l	40647 ng/l	
final water after A.C.	4068 ng/l	572 ng/l	(928)

—Sampling and analysis:
GC–EC: after n-pentane extraction: det.lim.: 0.1 µg/l (84)

D. BIOLOGICAL EFFECTS:
—Toxicity threshold (cell multiplication inhibition test):
bacteria (*Pseudomonas putida*): 125 mg/l (1900)
algae (*Microcystis aeruginosa*): 185 mg/l (329)
green algae (*Scenedesmus quadricauda*): 1100 mg/l (1900)
protozoa (*Entosiphon sulcatum*): >6560 mg/l (1900)
protozoa (*Uronema parduczi Chatton-Lwoff*): >6560 mg/l (1901)
—Molluscs:
larvae of eastern oyster (*Crassostrea virginica*): 48 hr, LD_{50}: 1 mg/l initial conc., static test (after 48 hr only approx. 15% of original conc. was still present)
(1545)

—Concentrations in various organs of molluscs and fish collected from the relatively clean water of the Irish Sea in the vicinity of Port Erin, Isle of Man (only organs with highest and lowest concentrations are mentioned—concentrations on dry weight basis):

molluscs: *Baccinum undatum*:	digestive gland:	117 ng/g
	muscle:	129 ng/g
Modiolus modiolus:	digestive tissue:	56 ng/g
	mantle:	438 ng/g
Pecten maximus:	mantle:	224 ng/g
	gill:	1040 ng/g
fish: *Conger conger* (eel):	gut:	43 ng/g
	liver:	474 ng/g
Gadus morhua (cod):	stomach:	7 ng/g
	muscle, brain:	167 ng/g
Pollachius birens (coal fish):	alimentary canal:	51 ng/g
	liver:	851 ng/g
Scylliorhinus carnicula (dogfish):	liver, spleen:	76–80 ng/g
	gill:	755 ng/g
Trisopterus luscus (bib):	liver, skeletal tissue:	48–50 ng/g
	gill:	212 ng/g (1092)
Poecilia reticulata (guppy) 14 d LC_{50}: 102 ppm		(1833)

—Man:
severe toxic effects: 2,000 ppm = 9,960 mg/cu m, 60 min
symptoms of illness: 500 ppm = 2,490 mg/cu m
unsatisfactory: >50 ppm =   249 mg/cu m   (185)

chloroformylchloride   *see* phosgene

chlorofos   *see* dimethyl (2,2,2-trichloro-1-hydroxyethyl)phosphonate

1-chloroheptane   (heptylchloride)
$CH_3(CH_2)_6Cl$
A. PROPERTIES: m.w. 134.65; m.p. -69.5°C; b.p. 159.5°C; sp.gr. 0.8725 at 20/0°C
B. AIR POLLUTION FACTORS:
 —Odor: T.O.C.: 0.330 mg/cu m = 59 ppb   (307)
                    0.06 ppm   (279)
        recognition: 157 mg/cu m   (761)
C. WATER POLLUTION FACTORS:
 —Waste water treatment:
  A.S.: after 6 hr: 0.9% of ThOD
         12 hr: 4.3% of ThOD
         24 hr: 4.4% of ThOD   (88)

1-chlorohexane   (*prim-n*-hexylchloride)
$CH_3(CH_2)_4CH_2Cl$
A. PROPERTIES: m.w. 120.62; m.p. -83°C; b.p. 132°C; v.d. 4.2; sp.gr. 0.90 at 20°C
C. WATER POLLUTION FACTORS:
 —Waste water treatment:

A.S.: after 6 hr: 3.9% of ThOD
12 hr: 4.8% of ThOD
24 hr: 6.4% of ThOD   (88)

1-chloro-2-hydroxybenzene   *see* o-chlorophenol

1-chloro-3-hydroxybenzene   *see* m-chlorophenol

1-chloro-4-hydroxybenzene   *see* p-chlorophenol

4-chloro-1-hydroxy-3-methylbenzene   *see* p-chloro-m-cresol

4-chloro-3-hydroxytoluene   *see* p-chloro-m-cresol

6-chloro-3-hydroxytoluene   *see* p-chloro-m-cresol

3-chloro-isobutene   *see* methallylchloride

4-chloro-4'-isopropylbiphenyl

Cl—⟨◯⟩—⟨◯⟩—CH(CH$_3$)$_2$

C. WATER POLLUTION FACTORS:
—Biodegradation:

metabolism by a mixed culture of aerobic bacteria from activated sludge   (1215)

chloromethane   *see* methylchloride

chloromethoxymethane   *see* chloromethylether

N'-(3-chloro-4-methoxyphenyl)-N,N-dimethylurea   *see* dosanex

4-chloro-2-methylaniline   *see* 4-chloro-o-toluidine

3-chloro-4-methylbenzenaminehydrochloride   (DRC-1339; 3-chloro-*p*-toluidine hydrochloride; CPTH; starlicide)
   Use: avian toxicant currently used to control starlings at cattle and poultry feed lots
D. BIOLOGICAL EFFECTS:
   —Crustacean:
      shrimp (19% pink shrimp, *Penaeus duorarum* and 81% white shrimp *Penaeus setiferus*): 96 hr TLm: 10.8 ppm (static test)
      blue crab (*Callinectes sapidus*): 96 hr TLm: 16.0 ppm (static test)   (1400)

2-chloro-1-methylbenzene   *see o*-chlorotoluene

5-chloro-3-methylcatechol

Use: microbial metabolite of MCPA
D. BIOLOGICAL EFFECTS:
   2 month old male Wistar rats: LD_{50}: 1412 mg/kg (1371–1453 mg/kg) subcronic experiment: 100 mg/kg for four weeks: no traces of the compound were found in the tissues due probably to a rapid decomposition of the compound in the tissues   (1460)

3-chloro-4-methyl-7-coumarinyldiethylphosphorothionate   *see* CO-RAL

chloromethylcyanide   *see* chloroacetonitrile

chloromethylether   *see* chloromethylmethylether

5-chloro-2-methyl-4-isothiazoline-3-onecalciumchloride
D. BIOLOGICAL EFFECTS:
   —Fish: bluegill: 49-d BCF 22–27 (whole fish minus viscera) F
                  49-d BCF 157–300 (viscera) F
                  43-d BCF 30 (whole fish minus viscera) F
                  43-d BCF 204 (viscera) F   (1511)

chloromethylmethylether   (chloromethoxymethane; monochlorodimethylether; chloromethylether; CMME)
$ClCH_2OCH_3$
   Users and formulation: mfg. irritant gases (lacrymators); chloromethylating agent; chemical; intermediate.   (347)
   Can be contaminated with 1 to 8% dichloromethylether   (1349; 1350)
A. PROPERTIES: colorless liquid; m.w. 80.52; m.p. −103.5°C; b.p. 59.5°C; sp.gr. 1.0625 at 10/4°C; solub. decomposes;

B. AIR POLLUTION FACTORS: 1 mg/cu m = 0.304 ppm, 1 ppm = 3.29 mg/cu m
C. WATER POLLUTION FACTORS:
   —Aquatic reactions: hydrolysis
   CMME in aqueous solution is hydrolyzed very fast with half-life in the order of
   <1 sec.; extrapolated to pure water (1364)
   the hydrolytic reaction can be depicted as follows:

$$ClCH_2-O-CH_3 \xrightarrow{H_2O} CH_3OH + HCl + CH_2O \quad (1365)$$

D. BIOLOGICAL EFFECTS:
   —Mammalia: rat: single order dose: survival: 0.3 g/kg
                                      death: 1.0 g/kg
                                      LD_{50}: 0.5 g/kg
              rat: inhalation: dangerous: 2,000 ppm, 30 min
                                          100 ppm, >4 hr
   —Man: carcinogenic: W. Germany 1977 (487)
                       USA 1974 (77)
   because the technical grade contains dichloromethylethers which are carcinogenic
   (202)

4-chloro-3-methylphenol see p-chloro-m-cresol

3-chloro-2-methyl-1-propene see methallylchloride

N-chloromethylsulfo-pentachloro-2-aminodiphenylether, sodium salt see eulan

chloromycetin see chloramphenicol

1-chloronaphthalene

A. PROPERTIES: m.w. 162.61; b.p. 250–280°C; sp.gr. 1.194 at 20/4°C
C. WATER POLLUTION FACTORS:
   —Biodegradation:
   metabolic pathway:

d-8-chloro-1,2-dihydroxynaphthalene ⟶ 3-chlorosalicylic acid (177)

D. BIOLOGICAL EFFECTS:
—Mammalia: rat: inhalation: no effect: 37 ppm, 15 × 6 hr   (65)

o-chloronitrobenzene (*o*-nitrochlorobenzene)

A. PROPERTIES: needles; m.w. 157.56; m.p. 32.5°C; b.p. 245.7°C; sp.gr. 1.368 at 22/4°C, log P_{oct} 2.24
C. WATER POLLUTION FACTORS:
—Water quality:
  in river Maas (The Netherlands): avg. in 1973: n.d.–0.03 µg/l (n.s.i.)   (342)
—Biodegradation: decomposition period by a soil microflora: >64 days   (176)
—Odor threshold: detection: 0.015–0.020 mg/kg (n.s.i.)   (903)
D. BIOLOGICAL EFFECTS:
—Fish:
Lepomis macrochirus: static bioassay in fresh water at 23°C, mild aeration applied after 24 hr (technical grade):

material added	% survival after				best fit 96 hr LC_{50}
ppm	24 hr	48 hr	72 hr	96 hr	ppm
5	67	67	0	—	
3.2	90	90	80	20	
2	100	100	0	0	1.2
1.5	90	30	20	0	
1	100	95	90	70	

Menidia beryllina: static bioassay in synthetic seawater at 23°C: mild aeration applied after 24 hr (technical grade):

material added	% survival after				best fit 96 hr LC_{50}
ppm	24 hr	48 hr	72 hr	96 hr	ppm
1.5	90	50	40	10	
1.0	90	30	0	—	0.55
0.5	90	90	80	40	
0.25	100	100	100	100	

(352)

m-chloronitrobenzene (*m*-nitrochlorobenzene)

Users and formulation: dyestuffs mfg.; intermediate in organic chemicals synthesis.
Man caused sources (water and air): formed in small quantities during chlorination, when nitrobenzene is present.   (347)
A. PROPERTIES: pale yellow crystals; m.w. 157.56; m.p. unst. 23.7°C, stab. 44.4°C; b.p. 235–236°C; sp.gr. 1.534 at 20/4°C; log P_{oct} 2.41/2.46

B. AIR POLLUTION FACTORS: Odor threshold: 0.02 mg/cu m (605)
C. WATER POLLUTION FACTORS:
—Biodegradation: decomposition period by a soil microflora: >64 days (176)
D. BIOLOGICAL EFFECTS:
—Fish: bluegill: 96 hr, LC_{50} (S): 1.2 mg/l

p-chloronitrobenzene (p-nitrochlorobenzene)

A. PROPERTIES: m.w. 157.56; m.p. 83.5°C; b.p. 239–242°C; v.d. 5.44; sp.gr. 1.520 at 18/4°C; log P_{oct} 2.39/2.41
B. AIR POLLUTION FACTORS: 1 mg/cu m = 0.15 ppm, 1 ppm = 6.55 mg/cu m
C. WATER POLLUTION FACTORS:
—Biodegradation period by a soil microflora: >64 days (176)
—Inhibition of biodegradation: at 100 mg/l, no inhibition of NH_3 oxidation by *Nitrosomonas* sp. (390)

2-chloro-4-nitrophenol

C. WATER POLLUTION FACTORS:
—Biodegradation rates: adapted A.S. at 20°C-product is sole carbon source: 71.5% COD removal at 5.3 mg COD/g dry inoculum/hr (327)
—Odor threshold: detection: 5 mg/l (998)

3-chloro-6-nitrophenol *see* 5-chloro-2-nitrophenol

5-chloro-2-nitrophenol (3-chloro-6-nitrophenol)

A. PROPERTIES: m.w. 173.56; m.p. 38.9°C; b.p. sublimes
D. BIOLOGICAL EFFECTS:
—Fish: larvae of sea lamprey: LD_{100}: 3 mg/l (30)

O-(2-chloro-4-nitrophenyl)O,O-dimethylphosphorothioate *see* dicapthon

1-chloro-1-nitropropane
$C_2H_5CH(NO_2)Cl$
A. PROPERTIES: m.w 123.5; b.p. 139–1.43°C; v.p. 5.8 mm at 25°C; v.d. 4.26; sp.gr. 1.21 at 20/20°C; solub. 6 mg/l at 20°C
B. AIR POLLUTION FACTORS: 1 mg/cu m = 0.198 ppm, 1 ppm = 5.05 mg/cu m
D. BIOLOGICAL EFFECTS:
—Mammalia: rabbit: oral lethal dose: 0.05–0.10 g/kg
rabbit, guinea pig: inhalation:

0/2	1/2	2,178 ppm, 1 hr	
0/2	0/2	1,069 ppm, 1 hr	
0/2	0/2	693 ppm, 2 hr	
1/1	0/2	393 ppm, 6 hr	(211)

1-chloropentane *see* amylchloride

o-chlorophenol (1-chloro-2-hydroxybenzene)

Use: organic synthesis
A. PROPERTIES: colorless liquid; m.w. 128.56; m.p. 7°C, 0°C, 4.1°C; b.p. 175.6°C; v.p. 40 mm at 82°C, 100 mm at 106°C; sp.gr. 1.241 at 18/15°C; solub. 28,500 mg/l at 20°C; log P_{oct} 2.15/2.19
B. AIR POLLUTION FACTORS:
—Odor: characteristic: quality: medicinal
    threshold concentrations: 0.019 mg/cu m   (710)
    3.6 ppb   (279)
    recognition: 0.0005 mg/cu m   (712)
—Control methods:
wet scrubber: water at pH 8.5: outlet: 200 odor units/scf
    $KMnO_4$ at pH 8.5: outlet: 1 odor units/scf   (115)
C. WATER POLLUTION FACTORS:
—Biodegradation:
decomposition rate in soil suspensions: 14 days for complete disappearance
     (175)
decomposition period by a soil microflora: >64 days   (176)
adapted A.S. at 20°C–product is sole carbon source: 95.6% COD removal at 25.0 mg COD/g dry inoculum/hr   (327)
—Reduction of amenities:
approx. conc. causing adverse taste in fish: 0.015 mg/l   (41)
faint odor: 0.00018 mg/l   (129)
taste threshold: 0,0001 to 0,006 mg/l   (998; 226)
adult bluegills: taste to the flesh at 2.0 mg/l   (226)

o-CHLOROPHENOL

odor thresholds — mg/kg water

[Chart showing odor thresholds from 10^{-7} to 10^4 mg/kg water, with detection and not specified categories marked]

(97; 226; 300; 304; 879; 894; 925; 998)

– Impact on biodegradation processes:
  inhibition of degradation of glucose by *Pseudomonas fluorescens* at: 30 mg/l
  inhibition of degradation of glucose by *E. coli* at: 400 mg/l (293)
– Manmade sources:
  in sewage effluents: 1.7 µg/l (237)
– Waste water treatment:

method	temp °C	days observed	feed mg/l	days acclim.	% removed
SEW, CA	20	25	1	23	4
RW, CA	20	15	1	6+	100

(93)

– degradation by *Pseudomonas*: 200 mg/l at 30°C:
  parent: 100% ring disruption in 52 hr
  mutant: 100% ring disruption in 26 hr (152)
– Lag period for degradation of 16 mg/l by:
  waste water at pH 7.3 and 30°C: 14–25 days
  soil suspension at pH 7.3 and 30°C: >25 days (1096)
– W, unadapted A.S.: at 1 mg/l: 100% removal after 3 hr
  10 mg/l: 97% removal after 6 hr
  100 mg/l: 20% removal after 6 hr (1639)
– Methods of analysis:
  extraction efficiency of macro-reticular resins: pH: 5.7: conc.: 10 ppm; (n.s.i.)
  X-AD 2: 70%; X-AD 7: 85% (370)

D. BIOLOGICAL EFFECTS:
– Algae: *Chlorella pyrenoidosa*: 96 mg/l; toxic (41)
  Scenedesmus: toxic: 60 mg/l
– Bacteria: *Pseudomonas*: toxic: 30 mg/l
– Protozoa: *Colpoda*: toxic: 30 mg/l (30)
– Fish:
  bluegill fingerlings: (96 hr) TLm: 8.4 mg/l (226)
  bluegill sunfish: TLm (24 hr): 8.2 mg/l in std.ref.water (253)
  fatheads: TLm (24–96 hr): 22–11 mg/l
  bluegills: TLm (24–96 hr): 12–8 mg/l
  goldfish: TLm (24–96 hr): 15–12 mg/l
  guppies: TLm (24–96 hr): 23–20 mg/l (158)
  goldfish: 24 hr LC_{50}: 16 ppm
    amount found in dead fish at 20 ppm: 128 µg/g
    BCF (at 20 ppm) = 6.4 (1850)
  guppy (*Poecilia reticulata*): 24 hr LC_{50}: 11 ppm at pH 7.3 (1833)

Idus idus melanotus: LC$_0$, 48 hr: 5 mg/l
                        LC$_{50}$, 48 hr: 8.5 mg/l
                        LC$_{100}$, 48 hr: 10 mg/l   (998)
—Mammals: rat: oral LD$_{50}$: 0.67 g/kg   (211)

m-chlorophenol (1-chloro-3-hydroxybenzene)

A. PROPERTIES: needles; m.w. 128.56; m.p. 32.8°C; b.p. 214°C; v.p. 5 mm at 72°C, 40 mm at 118°C; sp.gr. 1.245 at 45°C; solub. 26,000 mg/l at 20°C; log P_{oct} 2.47/2.50

B. AIR POLLUTION FACTORS:
   —Control methods:
      wet scrubber: water at pH 8.5: 45 odor units/scf
                     KMnO$_4$ at pH 8.5: 25 odor units/scf   (115)

C. WATER POLLUTION FACTORS:
   —Biodegradation:
      biological degradation in 7 days tests; 0% degradation for original, 1st, 2nd and 3rd subculture   (87)
      decomposition rate in soil suspensions: >72 days for complete disappearance
                                                                        (175)
      decomposition period by a soil microflora: >64 days   (176)
      lag period for degradation of 16 mg/l by
         waste water at pH 7.3 and 30°C: 14-25 days
         soil suspension at pH 7.3 and 30°C: >25 days   (1096)
   —Reduction of amenities:
      odor threshold: 0.100 to 0.200 mg/l; detection: 0.05 mg/l   (226)
      taste threshold: 0.900 to 1.0 mg/l; 0.0001 mg/l   (226)(998)
   —Waste water treatment:
      ion exchange: adsorption on Amberlite X AD-4 at 25°C, influent: 350 ppm; solute adsorbed:
         zero leakage: 2.40 lb/cu ft
         100 ppm leakage: 2.53 lb/cu ft   (40)
      degradation by *Pseudomonas*: 200 mg/l at 30°C:
            parent: 100% ring disruption in 72 hr
            mutant: 100% ring disruption in 28 hr   (152)
   —W, unadapted A.S.: at 1 mg/l: 100% removal after 6 hr
                          10 mg/l: 40% removal after 6 hr
                          100 mg/l: 0% removal after 6 hr   (1639)
   —Sampling and analysis:
      photometry: min. full scale: 8.3 × 10^{-6} mole/l   (53)

D. BIOLOGICAL EFFECTS:
   —Algae: *Chlorella pyrenoidosa*: 40 mg/l: toxic   (41)

—Fish: *Idus idus melanotus*: 48 hr, LC_0: 1 mg/l
　　　　　　　　　　　　　48 hr, LC_{50}: 3 mg/l
　　　　　　　　　　　　　48 hr, LC_{100}: 6 mg/l　　　　　　　　　　(998)
　　　guppy (*Poecilia reticulata*): 24 hr, LC_{50}: 6.5 ppm at pH 7.3　　(1833)
—Mammals: rat: oral LD_{50}: 0.57 g/kg　　　　　　　　　　　　　　(211)

p-chlorophenol (1-chloro-4-hydroxybenzene)

A. PROPERTIES: m.w. 128.56; m.p. 43°C; b.p. 217°C; v.p. 0.10 mm at 20°C, 0.25 mm at 30°C, 10 mm at 92°C, 40 mm at 125°C; v.d. 4.4; sp.gr. 1.306; solub. 27,100 mg/l at 20°C; sat.conc. 0.70 g/cu m at 20°C, 1.39 g/cu m at 30°C; log P_{oct} 2.39/2.44

B. AIR POLLUTION FACTORS: 1 mg/cu m = 0.19 ppm, 1 ppm = 5.34 mg/cu m
　—Odor: recognition: 0.001 mg/cu m　　　　　　　　　　　　　　(712)
　　　　　　　　　　1.2 ppm　　　　　　　　　　　　　　　　　　(279)
　—Control methods:
　　wet scrubber: water at pH 8: outlet: 5 odor units/scf　　　　　　(115)
　　　　　　　　$KMnO_4$ at pH 8: outlet: 1 odor unit/scf

C. WATER POLLUTION FACTORS:
　—Biodegradation:
　　decomposition rate in soil suspensions: 9 days for complete disappearance (175)
　　decomposition period by a soil microflora: 16 days　　　　　　　　(176)
　　adapted A.S. at 20°C—product is sole carbon source: 96.0% COD removal at 11.0 mg COD/g dry inoculum/hr　　　　　　　　　　　　　(327)
　—Impact on biodegradation processes:
　　inhibition of degradation of glucose by *Pseudomonas fluorescens* at: 20 mg/l
　　inhibition of degradation of glucose by *E. coli* at: 200 mg/l　　　(293)
　—Reduction of amenities:
　　approx. conc. causing adverse taste in fish: 0.05 mg/l　　　　　　(41)

odor thresholds　　　　　　　　　　mg/kg water

	10^{-7}	10^{-6}	10^{-5}	10^{-4}	10^{-3}	10^{-2}	10^{-1}	1	10	10^2	10^3	10^4
detection						●		● ● ●				
recognition												
not specified							●—●					

　　　　　　　　　　　　　　　　　　　　　　　　(226; 875; 879; 894; 998)
—Taste threshold: average: 1.24 mg/l
　　　　　　　　range: 0.02–20.4 mg/l　　　　　　　　　　　　(294)(30)
　　　　　　　　0.25 mg/l　　　　　　　　　　　　　　　　　　(304)

taste threshold: 1.0 to 1.35 mg/l; 0.0001 mg/l (226; 998)
—Waste water treatment:

methods	temp °C	days observed	feed mg/l	days acclim.	% removed	
SEW, CA	20	25	1	15	33	
RW, CA	20	36	1	23	100	
RW, CA	20	13	1	6+	100	(93)

—degradation by *Pseudomonas*: 200 mg/l at 30°C
    parent: 100% ring disruption in 96 hr
    mutant: 100% ring disruption in 33 hr (152)
adapted culture: 27% removal after 48 hr incubation, feed = 200 mg/l (292)
lag period for degradation of 16 mg/l by
  waste water at pH 7.3 and 30°C: 7–14 days
  soil suspension at pH 7.3 and 30°C: 14–25 days (1096)
—Sampling and analysis:
  photometry: min.full scale: 6.3×10^{-6} mole/l (53)

D. BIOLOGICAL EFFECTS:
—Bacteria: *Pseudomonas*: toxic: 20 mg/l
—Algae: *Scenedesmus*: toxic: 20 mg/l
    Chlorella pyrenoidosa: toxic: 40 mg/l (41)
—Protozoa: *Colpoda*: toxic: 5 mg/l (30)
—Fish:
goldfish: 24 hr LC_{50}: 9.0 ppm
  amount found in dead fish at 10 ppm = 101 µg/g
  BCF (at 10 ppm) = 10 (1850)
Idus idus melanotus: 48 hr LC_0:   2 mg/l
                          48 hr LC_{50}: 3.5 mg/l
                          48 hr LC_{100}: 5.5 mg/l (998)
—Mammals: rat: oral LD_{50}: 0.67 g/kg (211)

chlorophenotane *see* DDT

4-chlorophenoxyacetic acid (4-CPA)

$$\underset{Cl}{\underset{|}{\bigcirc}}\text{—OCH}_2\text{COOH}$$

A. PROPERTIES: m.w. 186.59; m.p. 157–159°C
Technical grade contains 4-chloro-*o*-cresol (~4%) (1606)
C. WATER POLLUTION FACTORS:
—Biodegradation: 11 days for ring cleavage in soil suspension (1827)
—Waste water treatment:
  A.C. type BL (pittsburgh Chem. Co.): % adsorbed by 10 mg A.C. from $10^{-4}M$
  aqueous solution at pH 3.0: 55.0%
                   pH 7.0: 17.7%
                   pH 11.0: 9.8% (1313)

—Aquatic reactions:

[Structure: 4-Cl-C6H4-OCH2COOH + H2O/Light → 4-HO-C6H4-OCH2COOH + C6H5-OCH2COOH + 4-Cl-C6H4-OH]

Photolysis of 4-CPA.

D. BIOLOGICAL EFFECTS:
—Fish:
   sea trout (*Salmo trutta*): 24 hr, LD_{50}: 147 ppm (1606)

4-chlorophenoxy-ω-butyric acid

[Structure: 4-Cl-C6H4-O-CH2CH2CH2COOH]

C. WATER POLLUTION FACTORS:
—Biodegradation: 53 days for ring cleavage in soil suspension (1827)

2(4-chlorophenoxy)-2-methylpropionic acid

[Structure: (CH3)2C(COOH)-O-C6H4-Cl]

C. WATER POLLUTION FACTORS:
—Manmade sources: in sewage effluent: 0.001 mg/l (233)

3-(*p*-(*p*-chlorophenoxy)phenyl)1,1-dimethylurea *see* chloroxuron

4-chlorophenoxy-α-propionic acid

$$\text{CH}_3-\text{CH}-\text{COOH}$$

[Structure: 4-chlorophenoxy group attached via O to CH of CH₃-CH-COOH; para-Cl on ring]

C. WATER POLLUTION FACTORS:
  —Biodegradation: >205 days for ring cleavage in soil suspension (1827)

4-chlorophenoxy-ω-propionic acid

$$\text{O}-\text{CH}_2-\text{CH}_2-\text{COOH}$$

[4-chlorophenyl ring with para-Cl, O linked to CH₂-CH₂-COOH]

C. WATER POLLUTION FACTORS:
  —Biodegradation: 11 days for ring cleavage in soil suspension (1827)

4-chlorophenoxy-α-valeric acid

$$\text{CH}_3-\text{CH}_2-\text{CH}_2-\text{CH}-\text{COOH}$$

[4-chlorophenoxy group via O on CH; para-Cl on ring]

C. WATER POLLUTION FACTORS:
  —Biodegradation: >81 days for ring cleavage in soil suspension (1827)

2-chlorophenylamine *see o-chloroaniline*

3-chlorophenylamine *see m-chloroaniline*

4-chlorophenylamine *see p-chloroaniline*

1-(4-chlorophenyl)-3-(2,6-difluorobenzoyl)urea *see diflubenzuron*

3-(p-chlorophenyl)-1,1-dimethylurea (N'(4-chlorophenyl)-N,N-dimethylurea; CMU; monuron; chlorfenidim; telvar)

3-(p-CHLOROPHENYL)-1,1-DIMETHYLUREA

$$Cl-C_6H_4-NH-\underset{\underset{O}{\|}}{C}-N(CH_3)_2$$

Uses: herbicide, inhibitor of photosynthesis and is absorbed via the roots, sugarcane flowering suppressant

A. PROPERTIES: odorless crystalline solid of m.p. 174–175°C; v.p. 5×10^{-7} mm at 25°C; sp.gr. 1.27 at 20/20°C; solub. 230 ppm at 25°C

C. WATER POLLUTION FACTORS:
—Aquatic reactions: 75–100% disappearance from soils: 10 months (1815) persistence in river water in a sealed glass jar under sunlight and artificial fluorescent light—initial conc. 10 µg/l:

% or original compound found

after	1 hr	1 wk	2 wk	4 wk	8 wk
	80	40	30	20	0	(1309)

—Waste water treatment:
powdered A.C.: Freundlich adsorption parameters: K: 0.066; $1/n = 0.333$
carbon dose to reduce 5 mg/l to a final conc. of 0.1 mg/l: 160 mg/l
carbon dose to reduce 1 mg/l to a final conc. of 0.1 mg/l: 29 mg/l (594)

D. BIOLOGICAL EFFECTS:
—Algae:
Protococcus sp.: 1 ppb: .90 OD expt/OD control: 10 day growth test (2347)
Protococcus sp.: 20 ppb: .00 OD expt/OD control: 10 day growth test (2347)
Chlorella sp.: 1 ppb: .30 OD expt/OD control: 10 day growth test (2347)
Chlorococcum sp.: 100 ppb: 54 percent inhibition of growth: 10 day growth test (2349)
Chlorococcum sp.: 100 ppb: 50 percent decrease O_2 evolution: (2348)
Chlorococcum sp.: 100 ppb: 50 percent decrease in growth: 10 day growth test (2348)
Dunaliella tertiolecta: 90 ppb: 50 percent decrease O_2 evolution: (2348)
Dunaliella tertiolecta: 150 ppb: 50 percent decrease growth: 10 day growth test (2348)
Dunaliella euchlora: 1 ppb: 1.00 OD expt/OD control: 10 day growth test (2347)
Dunaliella euchlora: 20 ppb: .00 OD expt/OD control: 10 day growth test (2348)
Isochrysis galbana: 100 ppb: 50 percent decrease O_2 evolution: (2348)
Isochrysis galbana: 130 ppb: 50 percent decrease in growth: 10 day growth test (2348)
Monochrysis lutheri: 1 ppb: .83 OD expt/OD control: 10 day growth test (2347)
Phaeodactylum tricornutum: 90 ppb: 50 percent decrease O_2 evolution: f (2348)
Phaeodactylum tricornutum: 100 ppb: 50 percent decrease in growth: 10 day growth test (2348)
Phaeodactylum tricornutum: 1 ppb: .65 OD expt/OD control: 10 day growth test (2347)

Phaeodactylum tricornutum: 20 ppb: .00 OD expt/OD control: 10 day growth
test (2347)
—Fish: *Oncorhynchus kisuth*: 48 hr, LC_{50}: 110,000 µg/l (2106)
—Mammals:
acute oral LD_{50} (rat): 3600 mg/kg (1854)
in diet: rats and dogs: no effect level: 250-500 ppm (1855)

S-((chlorophenylthio)methyl)-O,O-diethylphosphorodithioate *see* carbophenothion

4-chlorophenylurea

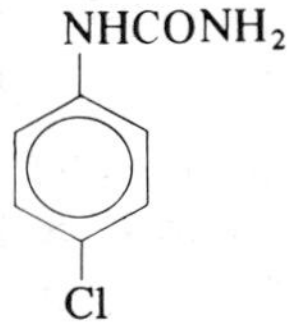

NHCONH$_2$–C$_6$H$_4$–Cl

D. BIOLOGICAL EFFECTS:
—Fish:
rainbow trout:	96-h LC_{50} 72 mg/l	(S)
fathead minnow:	96-h LC_{50} >100 mg/l	(S)
channel catfish:	96-h LC_{50} >100 mg/l	(S)
bluegill:	96-h LC_{50} >100 mg/l	(S)

(1510)

6-chloropicolinic acid

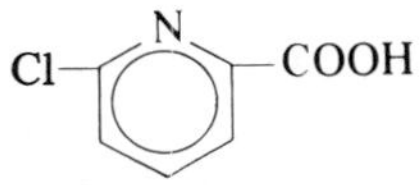

Cl–(pyridine)–COOH

C. WATER POLLUTION FACTORS:
—Manmade source:
sole metabolite other than carbon dioxide from degradation of 2-chloro-6-(trichloromethyl)pyridine
—Degradation: in soil at 1.0 ppm w initial conc. after 35 days incubation:

soil temperature °C	% decomposition
34–35	47.7
18–21	35.8
2–3	17.5

the most important factor influencing the decomposition rate is soil temperature
(1610)

chloropicrin (trichloronitromethane; nitrochloroform)
CCl_3NO_2
Uses: organic synthesis; dye-stuffs; fumigants; fungicides; insecticides; rat exterminator; poison gas
A. PROPERTIES: m.w. 164.39; m.p. -64°C; b.p. 112°C; v.p. 16.9 mm at 20°C, 33 mm

at 30°C; v.d. 5.7; sp.gr. 1.651 at 20/4°C; sat.conc. 170 g/cu m at 20°C; 286 g/cu m at 30°C; solub. 2000 mg/l
B. AIR POLLUTION FACTORS: 1 mg/cu m = 0.149 ppm. 1 ppm = 6.72 mg/cu m
  —Odor: T.O.C.: 1.1 ppm (211)
         7.3 mg/cu m = 1.09 ppm (57)
  O.I. at 20°C: 22,200 (316)
C. WATER POLLUTION FACTORS:
  —Reduction of amenities: faint odor: 0.0073 mg/l (129)
D. BIOLOGICAL EFFECTS:
  —Mammalia: dog: inhalation: LD_{43}: 117–140 ppm, 30 min
                                tolerated: 48 ppm, 15 min
             mouse: inhalation: death: 50–125 ppm, 15 min
                                tolerated: 25 ppm, 15 min
             cat: inhalation: survived 7 days: 38 ppm, 21 min (211)
             mice: RD_{50} (respiratory rate): 7.98 ppm (1330)
             rat: single oral LD_{50}: 250 mg/kg (1546)
  —Carcinogenicity:
  The bioassay of chloropicrin using Osborne-Mendel rats did not permit an evaluation of carcinogenicity because of the short survival time of dosed animals. The bioassay of chloropicrin using B6C3F1 mice did not provide conclusive statistical evidence for the carcinogenicity of this compound. (1732)
  —Man: lowest irritant conc.: 1.3 ppm
        intolerable: 7.5 ppm, 10 min
                     15.0 ppm, 1 min
        lethal: 119 ppm, 30 min
                297 ppm; 10 min (211)

chloroprene (2-chloro-1,3-butadiene)
$CH_2=CCl-CH=CH_2$
A. PROPERTIES: colorless liquid; m.w. 88.5; m.p. −130°C; b.p. 59.4°C; v.p. 118 mm at 10°C, 275 mm at 30°C, 200 mm at 20°C; v.d. 3.06; sp.gr. 0,958 at 20°C; sat.conc. 964 g/cu m at 20°C
B. AIR POLLUTION FACTORS: 1 mg/cu m = 0.27 ppm, 1 ppm = 3.68 mg/cu m
  —Odor: characteristic: quality: slightly etheric
    T.O.C.: recogn.: 0.40 mg/cu m = 0.11 ppm (73; 754)
    USSR: human odor perception: non perception: 0.25 mg/cu m
                                 perception: 0.4 mg/cu m
          human reflex response: adverse response: 0.4 mg/cu m
          animal chronic exposure: no effect: 0.22 mg/cu m
                                   adverse effect: 0.48 mg/cu m (170)
  O.I. at 20°C: 2,390,000 (316)
D. BIOLOGICAL EFFECTS:
  —Mammalia: rat: inhalation: LD_{100}: 829 ppm, 1 hr
                              LD_0:   277 ppm, 1 hr
             rat: oral LD_{100} 0.67 g/kg (211)
  —Man: loss of hair

1-chloropropane (propylchloride)
$CH_3CH_2CH_2Cl$
A. PROPERTIES: colorless liquid; m.w. 78.54; m.p. $-122.8°C$; b.p. $47.2°C$; v.p. 350 mm at $25°C$; v.d. 2.71; sp.gr. 0.890 at $20/4°C$; solub. 2,700 mg/l at $20°C$
B. AIR POLLUTION FACTORS: 1 ppm = 3.21 mg/cu m
C. WATER POLLUTION FACTORS:
  —Waste water treatment:
  A.S.: after 6 hr: 0.7% of ThOD
        12 hr: 0.8% of ThOD
        24 hr: 1.9% of ThOD (88)

2-chloropropane (isopropylchloride)
$CH_3CHClCH_3$
A. PROPERTIES: colorless liquid; m.w. 78.54; m.p. $-117°C$; b.p. $36.5°C$; v.p. 523 mm at $25°C$; v.d. 2.7; sp.gr. 0.859 at $20/4°C$; solub. 3,440 mg/l at $12.5°C$, 3,100 mg/l at $20°C$
B. AIR POLLUTION FACTORS: 1 ppm = 3.21 mg/cu m
D. BIOLOGICAL EFFECTS:
  —Mammalia: rat: inhalation: no effect: 250 ppm, 20 × 6 hr (65)
      guinea pig: single oral dose: survival: 3 g/kg
                                  death: 10 g/kg (211)
  —Man: very little skin irritation (211)

1-chloro-1-propene
$CHCl:CHCH_3$
C. WATER POLLUTION FACTORS:
  —Waste water treatment: evaporation from water at $25°C$ of 1 ppm solution:
      50% after 16 min
      90% after 59 min (313)

2-chloro-1-propene
CH_2CClCH_3
C. WATER POLLUTION FACTORS:
  —Waste water treatment: evaporation from water at $25°C$ of 1 ppm solution:
      50% after 29 min
      90% after 110 min (313)

3-chloro-1-propene *see* allylchloride

2-chloro-1-propylene
$CH=CCl-CH_3$
Measured half-life for evaporation from 1 ppm aqueous solution at $25°C$, still air, and an average depth of 6.5 cm: 33.1 min. (369)

3-chloro-1-propylene *see* allylchloride

chloropropylene-oxide *see* epichlorohydrin

chlorothalonil (bravo; tetrachloroisophthalonitrile)

Use: broad spectrum fungicide
A. PROPERTIES: white crystalline solid; m.p. 250–251°C; solub. 0.6 ppm
D. BIOLOGICAL EFFECTS:
  acute oral LD_{50} (albino rats): >10,000 mg/kg
  acute dermal LD_{50} (albino rats): >10,000 mg/kg (1854)
  It is concluded that under the conditions of this bioassay, technical-grade chlorothalonil was carcinogenic to Osborne-Mendel rats, producing tumors of the kidney. Chlorothalonil was not carcinogenic for B6C3F1 mice. (1770)

chlorothion (chlorthion; O,O-dimethyl-O-(3-chloro-4-nitrophenyl)phosphorothioate)
Use: insecticide
D. BIOLOGICAL EFFECTS:
  —Crustacean
  Daphnia magna: 4.5 μg/l, 48 h, LC_{50}
  —Fish:
  Pimephales promelas: 2800 μg/l, 96 h, LC_{50} (2123)
  Lepomis macrochirus: 700 μg/l, 96 h, LC_{50} (2123)
  —Mammals:
  acute oral LD_{50} (male rat): 880 mg/kg (1854)

α-chlorotoluene see benzylchloride

o-chlorotoluene (2-chloro-1-methylbenzene)

Uses: solvent and intermediate for organic chemicals and dyes
A. PROPERTIES: m.w. 126.58; m.p. -36.5/-34°C; b.p. 159°C; v.p. 2.7 mm at 20°C, 5 mm at 30°C; v.d. 4.37; sp.gr. 1.0817 at 20/4°C; sat. conc. 18.6 g/cu m at 20°C, 33.3 g/cu m at 30°C; log P_{oct} 3.42
B. AIR POLLUTION FACTORS: 1 mg/cu m = 0.19 ppm, 1 ppm = 5.26 mg/cu m
  —Sampling and analysis: photometry: min.full scale: 130 ppm (53)
C. WATER POLLUTION FACTORS:
  —Water quality:
  in river Maas at Eysden (The Netherlands) in 1976: median: n.d.: range: n.d.–0.1 μg/l

in river Maas at Keizersveer (The Netherlands) in 1976: median: n.d.: range:
n.d.-0.1 µg/l (1368)
in Delaware river (U.S.A.): conc. range (n.s.i.):
winter: 3 ppb
summer: n.d. (1051)
D. BIOLOGICAL EFFECTS:
−Toxicity threshold (cell multiplication inhibition test):
bacteria (*Pseudomonas putida*): 15 mg/l
green algae (*Scenedesmus quadricauda*): >100 mg/l
protozoa (*Entosiphon sulcatum*): > 80 mg/l (1900)
protozoa (*Uronema parduczi Chatton-Lwoff*): > 80 mg/l (1901)

m-chlorotoluene

A. PROPERTIES: m.w. 126.59; m.p. −48°C; b.p. 160–162°C; sp.gr. 1.072, log P_{oct} 3.28
D. BIOLOGICAL EFFECTS:
−Fish: guppy (*Poecilia reticulata*): 7 d LD_{50}: 18 ppm (1833)

p-chlorotoluene

Use: solvent and intermediate for organic chemicals and dyes
A. PROPERTIES: m.w. 126.59; b.p. 162–166°C; f.p. 6.5°C; sp.gr. 1.066 (25/15°C); log P_{oct} 3.33
C. WATER POLLUTION FACTORS:
−Water quality:
in river Maas at Eysden (The Netherlands) in 1976: median: 0.1 µg/l: range:
n.d.-0.3 µg/l
in river Maas at Keizersveer (The Netherlands) in 1976: median: 0.1 µg/l; range:
n.d.-0.2 µg/l (1368)
D. BIOLOGICAL EFFECTS:
−Fish: guppy (*Poecilia reticulata*) 14 d LC_{50}: 5.9 ppm (1833)

2-chlorotoluene-4-sulfonate, sodium

Uses: synthesis of dyes; intermediates and drugs

D. BIOLOGICAL EFFECTS:
—Fish: *Lepomis macrochirus*: TLm, 24 hr: <1,374 mg/l (1295)

2-chlorotoluene-5-sulfonate, sodium

D. BIOLOGICAL EFFECTS:
—*Daphnia magna*:

	young	adult
25 hr, TLm:	0.8 mg/l	3.3 mg/l
50 hr, TLm:	0.6 mg/l	1.3 mg/l
100 hr, TLm:	0.4 mg/l	

(1295)

—Snail eggs, *Lymnaea* sp.: 25 hr, TLm: 30 mg/l
—Fish: *Mollienesia latipinna*: 25 hr, TLm: 115.2 mg/l
    50 hr, TLm: 66.1 mg/l (1295)

4-chloro-*o*-toluidine (4-chloro-2-methylaniline)

A. PROPERTIES: m.w. 141.6; m.p. 27°C; b.p. 241°C
—Biodegradation product of chlorphenamidine (1547)

3-chloro-*p*-toluidinehydrochloride *see* 3-chloro-4-methylbenzenaminehydrochloride

N′-(4-chloro-*o*-tolyl)-N,N-dimethylformamidine *see* chlorphenamidine

2-chloro-6-(trichloromethyl)pyridine (nitrapyrin)

Use: a potent inhibitor of nitrification now in use with ammonium fertilizers active ingredient in N-Serve ® nutrient stabilizer

A. PROPERTIES: solub. 40 mg/l; v.p. 0.0028 mm at 20°C; m.p. 62–63°C technical grade contains approximately 10% related chlorinated pyridines

C. WATER POLLUTION FACTORS:
—Degradation:
6-chloropicolinic acid is the sole detectable metabolite, other than carbon dioxide in soil
—Aquatic reactions:
hydrolysis: in buffered, distilled water followed simple first-order kinetics over the concentration range 6.2×10^{-7} to 8.7×10^{-5} M
half-life times ranged from 1.7 to 4.0 days at 35°C depending on concentration.
photolysis: in a natural water followed simple first-order kinetics over the concentration range 7.1×10^{-6} to 7.5×10^{-6} M
half-life under these conditions was 0.5 day
the products of this reaction were 6-chloropicolinic acid, 6-hydroxypicolinic acid and unidentified polar material (1654)
—Inhibition of degradation processes:
~50% inhibition of NH_3 oxidation in *Nitrosomonas* at 11 mg/l (407)

D. BIOLOGICAL EFFECTS:
—rats: single oral LD_{50}: 1230 mg/kg
mice: single oral LD_{50}: 710 mg/kg
rabbits: single oral LD_{50}: 500–1000 mg/kg (1546)

chloroxuron (tenoran; 3-(*p*-(*p*-chlorophenoxy)phenyl)-1,1-dimethylurea)

Use: herbicide absorbed by both roots and leaves

A. PROPERTIES: white crystals, meltingpoint 149–150°C; solub. 3 ppm

D. BIOLOGICAL EFFECTS:
—Fish: *Lepomis macrochirus*: 48 hr, LC_{50}: 25000 µg/l (2116)
—Mammals: acute oral LD_{50} (rat): 3700 mg/kg
acute oral LC_{50} (rabbit): >10,000 mg/kg (1854)

chlorphenamidine (chlordimeform; N'-(4-chloro-*o*-tolyl)-N,N-dimethylformamidine)

Use: broad spectrum insecticide which is effective for all stages of insects and mites including eggs and adults. Chlorphenamidine hydrochloride has been used for the control of the rice stem borer in Japan (1547)

CHLORPHOXIM

A. PROPERTIES: Buff colored crystals; m.p. 32°C; solub. 250 ppm at 20°C

C. WATER POLLUTION FACTORS:
—Degradation: the major metabolites by plants are:
N'-(4-chloro-*o*-tolyl)N-methylformamidine (desmethylchlorphenamidine)
N-formyl-4-chloro-*o*-toluidine
4-chloro-*o*-toluidine (1548; 1549; 1550)

Residues of chlorphenamidine and three degradation products in rice grains, straws and soil:

	chlorphenamidine ppb	desmethyl-chlorphenamidine ppb	N-formyl-4-chloro-*o*-toluidine ppb	4-chloro-*o*-toluidine ppb
rice grains	4–48	0.2–1	10–38	3–61
straws	260–9700	10–180	67–500	80–6900
soil,				
0–5 cm depth	35–2900	5–15	8–380	2–68
5–10 cm depth	33–150	1–4	4–9	1–20

(1547)

D. BIOLOGICAL EFFECTS:
—Mammals:
acute oral LD_{50} (male rat): 127–352 mg/kg (tech. grade)
dermal LD_{50} (rabbit): >3000 mg/kg

chlorphoxim (2-chloro-α-(((diethoxyphosphinothioyl)oxy)imino)benzeneacetonitrile; (*o*-chlorophenyl)glyoxylonitrileoxime-O,O-diethylphosphorothioate)

$$\begin{array}{c} C_2H_5O \\ \\ C_2H_5O \end{array} \overset{S}{\underset{}{P}} -O-N=C \begin{array}{c} CN \\ \\ \end{array} \text{-C}_6\text{H}_4\text{-Cl}$$

Use: pesticide: effective against the larval stages of *Simulium damnosum* (blackfly), the insect vector of human onchocerciasis in Africa (1692); effective against adult mosquitoes (1693) and agricultural insects (1694; 1695)

C. WATER POLLUTION FACTORS:
—Method of analysis:
high concentrations of chlorphoxim form under gas-chromatographic conditions several trialkylphosphates, among which:
o,o-diethyl-O-methylphosphorothioate (DEMTP), and
o,o-diethyl-S-methylphosphorothioate (1695)
ag.c.—method is based on the derivatization of chlorphoxim to DEMTP with a 99% efficiency of conversion (1691)

D. BIOLOGICAL EFFECTS:
—Fish: bluegill: approx. 0.020 ppm/24 hr BCF: approx. 150 (chlorphoxim emulsifiable concentrate) (1691)

chlorpropham *see* isopropyl-N-(3-chlorophenyl)carbamate

chlorpyrifos (lorsban; dursban; O,O-diethyl-O-(3,5,6-trichloro-2-pyridyl) phosphorothioate)

Cl─⟨pyridyl⟩─Cl, OC₂H₅ structure: 3,5,6-trichloro-2-pyridyl-O-P(=S)(OC₂H₅)₂

Use: insecticides

A. PROPERTIES: amber solid cake with amber oil; m.p. 41.5–43.5°C; v.p. 1.87×10^{-5} mm at 20°C; sp.gr. (liq.) 1.398 at 43.5°C; decomposition temp. approx. 160°C; solub. 0.4 ppm at 23°C; log P_{oct}: 5.11 at 20°C

B. AIR POLLUTION FACTORS:
   —Manmade sources:
   in ambient air of storage and office rooms of commercial pest control buildings in a 4 hr period:
   storage room: avg: 220 ng/cu m; range: 83–595 ng/cu m
   office room: avg: 126 ng/cu m; range: 26–357 ng/cu m   (1868)

C. WATER POLLUTION FACTORS:
   —Degradation: persistence in soil at 10 ppm initial concentration:

	weeks incubation to	
	50% remaining	5% remaining
sterile sandy loam	17	–
sterile organic soil	>24	–
non sterile sandy loam	< 1	1
non sterile organic soil	2.5	8 (1433)

   —Hydrolysis: in buffered distilled water at 25°C:
   half-life: 22.8 days at pH 8.1
              35.3 days at pH 6.9
              62.7 days at pH 4.7
   a 16-fold rate enhancement was demonstrated in canal and pond water at 25°C
   qualitatively the products of hydrolysis were:
   3,5,6-trichloro-2-pyridinol,
   O-ethyl, O-hydrogen-O-(3,5,6-trichloro-2-pyridyl)phosphorothioate
   O,O-dihydrogen-O-(3,5,6-trichloro-2-pyridyl)phosphorothioate   (1652)

D. BIOLOGICAL EFFECTS:
   —Crustaceans
   Gammarus lacustris: 96 h, LC_{50}: 0.11 µg/l   (2124)
   Gammarus fasciatus: 96 h, LC_{50}: 0.32 µg/l   (2126)
   shiner perch (*Cymatogaster aggregata*):   3.5 ppb TLm: 96 hr static lab
      bioassay   (2343)
   shiner perch (*Cymatogaster aggregata*):   3.7 ppb TLm: 96 hr flowing water
      lab bio.   (2343)
   Korean shrimp (*Palaemon macrodactylus*): 0.25 (0.10–0.63) ppm TL_{50}: 96 hr
      static lab bioassay   (2353)
   Korean shrimp (*Palaemon macrodactylus*): 0.01 (0.002–0.046) TL_{50}: 96 hr
      intermittent flow lab bioassay   (2353)

—Amphipoda: *Gammarus pulex*: 1 hr, LC_{90-95}: 0.05–0.1 ppm (1653)
—Insects:
   Pteronarcys californica: 96 h, LC_{50}: 10 µg/l (2128)
   Pteronarcella badia: 96 h, LC_{50}: 0.38 µg/l (2128)
   Classenia sabulosa: 96 h, LC_{50}: 0.57 µg/l (2128)
   Ephimeroptera: *Bactis* rhodani: 1 hr, LC_{90-95}: 0.01–0.02 ppm
   Trichoptera: *Brachycentrus subnubilis*: 1 hr, LC_{90-95}: 0.2–0.5 ppm
              Hydropsyche pellucidula: 1 hr, LC_{90-95}: >0.5 ppm
   Odonata: *Agrion*: 1 hr, LC_{90-95}: 0.2 ppm
   Diptera: *Simulium ornatum*: 1 hr, LC_{90-95}: 0.05–0.1 ppm (1653)
—Fish
   Lepomis macrochirus: 96 h, LC_{50}: 2.6 µg/l (2137)
   Salmo gairdneri: 96 h, LC_{50}: 11 µg/l (2137)
   Fundulus heteroclitus: 96 hr, TL_{50}: 0.0047 mg/l (FT)
                       96 hr, TL_{50}: 0.0122 mg/l (1615)
   green sunfish: 72 hr, LC_{50}: 0.04 mg/l (S)
   mosquitofish: 72 hr, LC_{50}: 0.26 mg/l (S) (1203)
   rainbow trout (*Salmo gairdneri*): log bioconcentration factor: 2.67
—Mammals:
   acute oral LD_{50} (male rat): 163 mg/kg
                (female rat): 135 mg/kg (1546)
                (guinea pig): 500 mg/kg
                (rabbit): 1000–2000 mg/kg (1554)

chlorpyrifosmethyl *see* O,O-dimethyl-O-(3,5,6-trichloro-2-pyridyl)phosphorothioate)

5β-cholestan-3β-ol *see* coprostanol

5-cholesten-3β-ol *see* cholesterol

cholesterin *see* cholesterol

cholesterol (cholesterin; 5-cholesten-3β-ol)
$C_{27}H_{16}O$
A. PROPERTIES: white, or faintly yellow, almost odorless pearly granules or crystals; m.p. 148.5°C; b.p. 360°C (decomposes); sp.gr. 1.067 at 20/4°C; m.w. 386.66
   —Natural occurrence: egg yolk, liver, kidneys etc.
C. WATER POLLUTION FACTORS:
   —Biodegradation:
      BOD_{35}^{25}: 0.83 in seawater/inoculum: enrichment cultures of hydrocarbon oxidizing bacteria
      ThOD: 3.12 (521)
   —River water quality:
      in Delaware river (U.S.A.): conc. range winter: 5–10 ppb
                                    summer: 3–8 ppb (1051)

chrysene (1,2-benzophenanthrene)

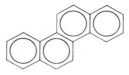

Natural sources (*water and air*): coal tar
Manmade sources:
in gasoline: 0.052 mg/l; 0.650 mg/l (1052; 1070)
    low octane number: 0.23–1.06 mg/kg (n = 13)
    high octane number: 0.58–2.96 mg/kg (n = 13) (385)
                           0.370 mg/l (380)
in high octane gasoline: 6.7 mg/kg (+ cyclopenteno(*cd*)pyrene) (1220)
in fresh motor-oil: 0.56 mg/kg (+ cyclopenteno(*cd*)pyrene)
in used motor-oil after 5,000 km: 86.2–190.4 mg/kg (+ cyclopenteno(*cd*)pyrene)
in used motor-oil after 10,000 km: 128.8–236.6 mg/kg (+ cyclopenteno(*cd*)pyrene)
                                                                                (1220)
in lubricating motor-oils: 0.41–2.61 ppm (6 samples) (379)
in bitumen: 1.64–5.14 ppm (506)
in Kuwait crude oil: 6.9 ppm
in South Louisiana crude oil: 17.5 ppm (1015)

A. PROPERTIES: crystals; m.w. 228.2; m.p. 254°C; b.p. 488°C; sp.gr. 1.274 at 20/4°C;
    solub.   0.0015 mg/l at 15°C
                  0.006 mg/l at 25°C
    in seawater at 22°C: 0.001–0.05 ppm
                  0.017 mg/l at 24°C (practical grade) (1666)

B. AIR POLLUTION FACTORS:
    —Manmade sources: in tail gases of gasoline engine: 27–318 μg/cu m (340)
        emissions from typical European gasoline engine (1608 cu cm)—following European driving cycles—using leaded and unleaded commercial gasolines: 29.8–64.9 μg/l fuel burnt (1291)
    in exhaust condensate of gasoline engine: 0.085–0.123 mg/l gasoline consumed
                                                                                      (1070)
                                                      410 μg/g (1069)
    in cigarette smoke: 6.0 μg/100 cigarettes (1298)
    —Ambient air quality:
        glc's in Budapest 1973: heavy traffic area: winter: 121.2 ng/cu m
        (6 a.m.–8 p.m.)   summer: 78.3 ng/cu m
                                low traffic area: winter: 30.4 ng/cu m
                                                           summer: 56.1 ng/cu m

glc's in The Netherlands (ng/cu m)

	Delft	The Hague	Vlaardingen	Amsterdam	Rotterdam
summer 1970	2	1	5	2	3
1971	4	2	4	2	5
winter 1968	15	21	28	26	21
1969	20	14	38	17	18
1970	10	7	16	4	11
1971	6	4	13	8	18

                                                                                                   (1277)

394 CINNAMALDEHYDE

    glc's in Birkenes (Norway), Jan.–June 1977: avg.: 0.81 ng/cu m
      (+ triphenylene)   range: n.d.–6.18 ng/cu m
                                                                ($n = 18$)
    glc's in Rørvik (Sweden) Dec. '76–April '77: avg.: 1.57 ng/cu m
      (+ triphenylene)   range: 0.21–11.59 ng/cu m
                                                                 ($n = 21$)   (1236)
    comparison of glc's inside and outside houses at suburban sites:
      outside: 4.56 ng/cu m or 91.2 µg/g SPM
      inside: 3.98 ng/cu m or 72.5 µg/g SPM   (1234)
C. WATER POLLUTION FACTORS:
  —Water and sediment quality:
    in sediments in Severn estuary (U.K.): 1.1–5.6 ppm dry wt. (includes perylene)
                                                                                     (1467)
    in sediment of Wilderness Lake, Coline Scott, Ontario (1976): 23 ppb (dry wt.)
                                                                                             (932)
    river Main at Seligenstadt (W. Germany): 11.8–38.4 ng/l   (530)
    in rapid sand filter solids from Lake Constance water: 0.5 mg/kg
    in River Argen water solids: 1.6 mg/kg   (531)
D. BIOLOGICAL EFFECTS:
  —Fish:
    Neanthes arenaceodentata: 96 hr TLm in seawater at 22°C: >1 ppm (initial
      conc. in static assay)   (995)
  —Carcinogenicity: weak
  —Mutagenicity in the *Salmonella* test: +
    38 revertant colonies/nmol
    1670 revertant colonies at 10 µg/plate   (1883)

cinnamaldehyde (3-phenylpropenal; β-phenylacrolein; cinnamic aldehyde; cinnamylaldehyde)

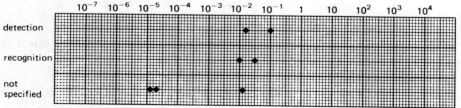

A. PROPERTIES: m.w. 132.15; m.p. -7.5°C; b.p. 251°C; v.p. 1 mm at 76.1°C, 40 mm at 152°C; sp.gr. 1.112 at 15/4°C; log P_{oct} 1.88
B. AIR POLLUTION FACTORS: 1 mg/cu m = 0.182 ppm, 1 ppm = 5.493 mg/cu m
  odor thresholds

                                                          (307; 607; 610; 614; 793; 840)
  O.I. at 20°C: 53,000   (316)

C. WATER POLLUTION FACTORS:
 —Waste water treatment:
  RW, Sd, BOD, 20 C, 10 days observed, 100 days acclimation: 88% removed (93)

cinnamene *see* styrene

cinnamic aldehyde *see* cinnamaldehyde

cinnamylaldehyde *see* cinnamaldehyde

ciodrin (dimethyl-1-methyl-2-(1-phenylethoxycarbonyl)vinylphosphate; crotoxyphos)

D. BIOLOGICAL EFFECTS:
 —Crustaceans
  Gammarus lacustris: 96 hr, LC_{50}: 15 µg/l   (2124)
  Gammarus fasciatus: 96 hr, LC_{50}: 11 µg/l   (2126)
 —Fish
  Lepomis macrochirus: 96 hr, LC_{50}: 250 µg/l   (2137)
  Micropterus salmoides: 96 hr, LC_{50}: 1100 µg/l   (2137)
  Salmo gairdneri: 96 hr, LC_{50}: 55 µg/l   (2137)
  Ictalurus punctatus: 96 hr, LC_{50}: 2500 µg/l   (2137)
 —Mammals:
  acute oral LD_{50} (rat):   approx. 125 mg/kg
  acute dermal LD_{50} (rabbit): approx. 385 mg/kg   (1854)

CIPC *see* isopropyl-N-(3-chlorophenyl)carbamate

citrazinic acid (2,6-dihydroxyisonicotinic acid)

Use: competing coupler in color developer solutions
A. PROPERTIES: m.w. 155.11; m.p. >300°C
C. WATER POLLUTION FACTORS:
 —Oxidation parameters:

theoretical	analytical
TOD = 1.34	Reflux COD = 100.0% recovery
COD = 0.93	TKN = 99.4% recovery
NOD = 0.41	BOD_5 = 0.086

$BOD_5/COD = 0.092$
BOD_5 (acclimated) $= 0.10$ (1828)
—Impact on conventional biological treatment systems:

	chemical conc. mg/l	effect	
unacclimated System	320	no effect	(1828)
	500	no effect	
acclimated System	1,010	no effect	(1828)

The compound did not affect either the unacclimated or acclimated systems; however, there was no evidence to indicate that the compound was biodegradable under either condition. (1828)

D. BIOLOGICAL EFFECTS:
—Algae:
Selenastrum capricornutum: 1 mg/l, no effect
10 mg/l, no effect
100 mg/l, inhibitory (1828)
—Crustacean:
Daphnia magna: $LC_{50} = 32$ mg/l (1828)
—Fish:
Pimephales promelas: $LC_{50} > 100$ mg/l (1828)

citric acid (2-hydroxy-1,2,3-propanetricarboxylic acid; β-hydroxytricarballylic acid)

$$\begin{array}{c} COOH \\ | \\ CH_2 \\ | \\ HO-C-COOH \\ | \\ CH_2 \\ | \\ COOH \end{array}$$

A. PROPERTIES: m.w. 192.12; m.p. 153°C, $-H_2O$: 70/75°C; b.p. decomposes; sp.gr. 1.542 at 18/4°C; solub. 1,330 g/l cold; THC 474.5 kcal/mole; log P_{oct} -1.72
C. WATER POLLUTION FACTORS:
—BOD_5:   0.420 (30)
          0.40 Warburg, sewage (282)(163)
—BOD_{20}: 0.610 (30)
—$KMnO_4$ value: 1.822 (30)
—ThOD: 0.686 (30)
—Manmade sources:
  excreted by man in urine: 3–17 mg/kg body wt/day
               in sweat: 0.2 mg/100 mg/100 ml (203)
—Waste water treatment:
  A.S.: after 6 hr: 1.7% of ThOD
             12 hr: 1.0% of ThOD
             24 hr: 13.2% of ThOD (89)
  A.S., Resp, BOD, 20 C, 1–15 days observed, feed: 720 mg/l, <1 day acclimation: 30% theor. oxidation, 98% removed (93)

—Impact on biodegradation:
  at 100 mg/l, no inhibition of NH_3 oxidation by Nitrosomonas sp.   (390)
—Methods of analysis:
  recovery of citric acid by iron coprecipitation:

	% citric acid recovered by coprecipitation[a]	
pH	Distilled water	Lake Mendota water
8.0	...	47
8.5	...	44
9.0	70	38
9.5	70	33
10.0	75	25

[a]Initial concentrations: $6.5 \times 10^{-7} M$ of citric acid and $0.01 M$ $FeCl_3$.

recovery of citric acid by iron coprecipitation from Lake Mendota water:

	% Recovery of citric acid by coprecipitation[a]		
$FeCl_3$, M	pH 8	pH 9	pH 10
0.005	48	36	20
0.01	47	38	25
0.015	49	38	...
0.02	49	38	20

[a]Initial citric acid concentration: $6.5 \times 10^{-7} M$.   (1310)

Automated fluorometric method based on the Furth-Herman reaction, 10 samples per hour without any separation or preconcentration, detection limit: 10 µg/l   (217)

D. BIOLOGICAL EFFECTS:
—Toxicity threshold (cell multiplication inhibition test):
  bacteria (*Pseudomonas putida*):   >10,000 mg/l   (1900)
  algae (*Microcystis aeruginosa*):   80 mg/l   (329)
  green algae (*Scenedesmus quadricauda*):   640 mg/l   (1900)
  protozoa (*Entosiphon sulcatum*):   485 mg/l   (1900)
  protozoa (*Uronema parduczi Chatton-Lwoff*): 622 mg/l   (1901)
—Arthropoda:
  Daphnia magna: LD_0: 80 mg/l, long time exposure in soft water   (245)
             LD_{100}: 120 mg/l, long time exposure in soft water   (245)
  Daphnia: toxic: 100 mg/l   (30)
—Fish:
  goldfish: period of survival: 4–48 hr: 894 ppm at pH 4.0 days: 625 ppm at pH 4.5   (157)
  goldfish: LD_0: 625 mg/l, long time exposure in hard water   (245)
           LD_{100}: 894 mg/l, long time exposure in hard water   (245)

clopidol see 3,5-dichloro-2,6-dimethyl-4-pyridinol

CMME see chloromethylmethylether

CMT *see* carboxy methyltartronate

COD *see* 1,5-cyclooctadiene

compound 497 *see* dieldrin

co-op brushkiller 112
  Use: herbicide
  active ingredient: iso-octyl esters of 2,4-D and 2,4,5-T
C. WATER POLLUTION FACTORS:
  —Persistence:
    persistence of 2,4-D and 2,4,5-T iso-octyl esters in soil (0–10 cm depth)

	ppm ester					
application rate:	7.8 kg/ha		15.7 kg/ha		31.4 kg/ha	
residue of ester:	2,4-D	2,4,5-T	2,4-D	2,4,5-T	2,4-D	2,4,5-T
days after application						
1	ND	ND	.01	.02	.11	.30
14	ND	ND	ND	.01	.01	.12
28	ND	ND	ND	ND	ND	.03
42	ND	ND	ND	ND	ND	Trace

ND = not detected; sensitivity of method: 0.005 ppm
Trace = detected at level <0.01 ppm

—The major residues in the soil following the application of the 2,4-D/2,4,5-T iso-octyl ester formulation were the respective free phenoxy acids. The persistence of these residues is shown in table below.

2,4-D and 2,4,5-T residues in soil following application of a 2,4-D/2,4,5-T iso-octyl ester formulation

	ppm acid									
application rate:	7.8 kg/ha				15.7 kg/ha				31.4 kg/ha	
residue of ester:	2,4-D		2,4,5-T		2,4-D		2,4,5-T		2,4-D	2,4,5-T
days after application	0–10 cm	10–20 cm	0–10 cm	10–20 cm	0–10 cm	10–20 cm	0–10 cm	10–20 cm	0–10 cm	10–20 cm	0–10 cm	10–20 cm
1	.59	–	.28	–	.82	–	.42	–	3.66	–	1.16	–
14	.87	–	1.05	–	1.47	–	1.32	–	12.4	–	7.1	–
28	.14	–	.92	–	.28	–	3.60	–	.60	–	8.9	–
42	.06	.03	.26	.22	.50	.14	2.39	.76	.39	.03	10.9	.22
56	.01	.06	.84	.18	.05	.09	1.20	.59	.28	.06	7.2	.18
70	.03	.02	.11	.10	.11	.03	.81	.19	.24	.02	7.5	.10
265	.01	ND	.02	.01	.05	Tr	.11	.01	.10	ND	1.2	.01
385	ND	ND	Tr	Tr	ND	ND	Tr	Tr	ND	ND	.02	ND

coprostanol (5β-cholestan-3β-ol)
Coprostanol is produced in the intestine of mammals by the microbial reduction of cholesterol which is the main steroid in the tissue of vertebrates. (1795)
Coprostanol is thus excreted by mammals along with cholesterol and other steroids, although coprostanol is generally the dominant one. (1796)

C. WATER POLLUTION FACTORS:
  −Sediment and water quality:
    in sediments from the Veracruz harbor (Mexico 1979): 0.006–0.44 ppm (1794)
    in sediments near the city of Mazatlan (Mexico 1979): 0.020–0.20 ppm (1794)
    in sediments from the Clyde estuary near the city of Glasgow (1977): 0.19–14 ppm (1797)
    in sediments from the New York Bight (1978): 0.056–5.2 ppm (1798)
    Delaware river (U.S.A.): conc. range: winter: 4–9 ppb
                  (n.s.i.)   summer: 1–2 ppb (1051)
    Old Deerfield river (U.S.A.) conc.: 1.2 ppb
    Chicopee river, Ludlow: 3.6 ppb
    Turners Falls: 0.9 ppb
    Connecticut river, Hadley: non-detectable
    Connecticut river, Sunderland: non-detectable (527)
    Amherst sewage treatment plant effluent: 85–504 μg/l
    Mill river, Amherst (U.S.A.): 130–180 μg/l
    Mill river, Millers Falls: 7 μg/l
    Lake Warner: 0.45 μg/l (527)
  −Water treatment method:
    Old Deerfield, sewage treatment plant influent: 170 μg/l
    Old Deearfield, sewage treatment plant effluent: 4.6–6.8 μg/l (527)
  −Methods of analysis:
    extraction by XAD-2 resin at 3 ml/min, pH 6.5, at 200 μg/l: 97% recovery (527)

co-ral (Muscatox; Resistox; Coumaphos; Bay 21/199; Asuntol; Baymix; Meldane; O,O-diethyl-O-(3-chloro-4-methyl-2-oxo(2H)-1-benzopyran-7-yl)-phosphorothioate; 3-chloro-4-methyl-7-coumarinyl diethyl phosphorothionate)

$$C_2H_5O\text{―}\underset{C_2H_5O}{\overset{S}{\underset{\|}{P}}}\text{―}O\text{―[benzopyran-2-one with Cl and }CH_3\text{]}$$

Use: livestock insecticide

A. PROPERTIES: tan crystalline solid melting at 90–92°C—hydrolyses slowly under alkaline conditions; v.p. 1×10^{-7} mm at 20°C; sp.gr. 1.47 at 20/4°C

D. BIOLOGICAL EFFECTS:
  −Crustaceans
    Gammarus lacustris: 96 h, LC_{50}: 0.07 μg/l (2124)
    Gammarus fasciatus: 96 h, LC_{50}: 0.15 μg/l (2126)
    Daphnia magna: 48 h, LC_{50}: 1.0 μg/l
  −Molluscs:
    eastern oyster (*Crassostrea virginica*):   egg: 110 ppb TLm, 48 hr static lab bioassay (2324)
    eastern oyster (*Crassostrea virginica*):   larvae: >1000 ppb TLm, 14 day static lab bioassay (2324)

hard clam (*Mercenaria mercenaria*): egg: 9120 ppb TLm, 48 hr static lab bioassay (2324)
hard clam (*Mercenaria mercenaria*) larvae: 5210 ppb TLm, 12 day static lab bioassay (2324)
threespine stickleback (*Gasterosteus aculeatus*): 1470 ppb TLm, 96 hr static lab bioassay (2333)

—Insects:
Hydropsyche sp.:	LC_{50}, 24 hr:	5 µg/l	(2110)
Hexagenia sp.:	LC_{50}, 24 hr:	430 µg/l	(2110)

—Fish:
Pimephales promelas:	LC_{50}, 96 hr:	18000 µg/l	(2119)
Lepomis macrochirus:	LC_{50}, 96 hr:	180 µg/l	(2119)
Salmo gairdneri:	LC_{50}, 96 hr:	1500 µg/l	(2119)
Oncorhynchus kisutch:	LC_{50}, 96 hr:	15000 µg/l	(2119)

—Mammals
acute oral LD_{50} (rat): about 56–230 mg/kg (techn. grade) (1854)
(male rats): 41 mg/kg
(female rats): 15.5 mg/kg (1855)
acute dermal LD_{50} (male rats): 860 mg/kg (1855)
in diet: rats tolerated for 2 years a diet containing 100 ppm (1855)

coronene

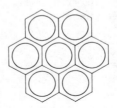

Manmade sources:
in gasoline: 630 µg/kg; 0.165 mg/l (340; 1070)
    low octane number: 0.06–0.85 mg/kg (n = 13) (385)
    high octane number: 0.12–1.11 mg/kg (n = 13) (385)
    0.278 mg/l (380)
    1.9 mg/kg (1220)
in fresh motor-oil: 0.00 mg/kg (1220)
in used motor-oil after 5,000 km: 24.6–36.7 mg/kg
in used motor-oil after 10,000 km: 28.4–63.0 mg/kg (1220)

A. PROPERTIES: m.w. 300.36; m.p. >360°C; b.p. 525°C
B. AIR POLLUTION FACTORS:
—In tailgases of gasoline engine: 14–209 µg/cu m (340)
  emissions from typical European car engine (1608 cu cm)—following European driving cycles—using leaded and unleaded commercial gasolines: 1.4–52.0 µg/l fuel burnt (1291)

—In coke oven emissions: 766.6–864.6 µg/g of sample (960)
—Emission from space heating installation burning:
    coal (underfeed stoker): 0.33 mg/10^6 Btu input
    gasoil: 2.1 mg/10^6 Btu input
    gas: 5.3 mg/10^6 Btu input (954)
—In stack gases of municipal incinerator, after spray tower and electrostatic precipitator: 0.04 mg/1000 cu m; in residues: <20 µg/kg (341)
—Ambient air quality:
  glc's in The Netherlands (ng/cu m)

	Delft	The Hague	Vlaardingen	Amsterdam	Rotterdam
summer 1968	1	–	1	–	n.d.
1969	1	<1	1	n.d.	1
1970	–	<1	1	<1	<1
1971	2	1	n.d.	n.d.	1
winter 1968	n.d.	n.d.	2	2	4
1969	2	1	3	2	3
1970	1	–	2	2	2
1971	n.d.	n.d.	2	1	2

(1277)

glc's in Budapest 1973: heavy traffic area: winter: 0.8 ng/cu m
(6 a.m.–8 p.m.)   summer: 4.7 ng/cu m
               low traffic area: winter: 1.1 ng/cu m
                               summer: 11.1 ng/cu m (1259)
glc's in the average American urban atmosphere—1963:
  15 µg/g airborne particulates or
  2 ng/g cu m air (1293)
comparison of glc's inside and outside houses at suburban sites:
  outside: 0.92 ng/cu m or 21.8 µg/g SPM
  inside: 0.43 ng/cu m or 7.3 µg/g SPM (1234)

C. WATER POLLUTION FACTORS:
—Manmade sources: in effluent spray tower of stack gases of municipal incinerator:
  <0.01 µg/l (341)
—Sediment quality:
  in sediment of Wilderness Lake, Colin Scott, Ontario (1976): 6 ppb (dry wt.)
(932)
  in sediments in Severn estuary (U.K.): traces–1.1 ppm dry wt. (1467)

o-coumaric acid lactone *see* coumarin

coumarin (1,2-benzopyrone; o-coumaric acid lactone; coumarinic lactone; cumarin; tonka bean camphor)

A. PROPERTIES: colorless crystals, flakes or powder; m.w. 146.14; m.p. 67–68°C; b.p. 290–302°C; v.p. 1 mm at 106°C, 40 mm at 189°C; sp.gr. 0.935 at 20/4°C; solub. 100 mg/l at 25°C; log P_{oct} 1.39

COUMARINIC LACTONE

B. AIR POLLUTION FACTORS:
   —Odor: quality: vanilla
          hedonic tone: pleasant

odor thresholds

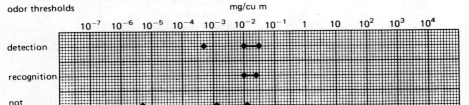

(279; 307; 607; 610; 710; 774; 775; 840)

C. WATER POLLUTION FACTORS:

odor thresholds

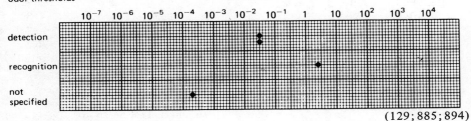

(129; 885; 894)

D. BIOLOGICAL EFFECTS:
   —Mammals:
      rats: single oral LD_{50}: 235–290 mg/kg depending on solvent   (1546)

coumarinic lactone *see* coumarin

coumarone *see* benzofuran

CPTH *see* 3-chloro-4-methylbenzenaminehydrochloride

creosote
   A mixture of phenols and phenol derivatives obtained by the destructive distillation of wood-tar. One of the active constituents is creosol (2-methoxy-4-methylphenol) (from beechwood-tar).
   Also obtained by fractional distillation of coal-tar; contains substantial amounts of naphthalene and anthracene.
   Use: wood preservative
D. BIOLOGICAL EFFECTS:
   —Crustacean:
      adult lobster (*Homarus americanus*): 10°C: 96 hr LD_{50}: 1.76 mg/l
      larval lobster (*Homarus americanus*): 20°C: 96 hr LD_{50}: 0.02 mg/l
      Crangon: 10°C: 96 hr LD_{50}: 0.13 mg/l
      Crangon: 20°C: 96 hr LD_{50}: 0.11 mg/l
   The solutions were aerated gently and renewed at 48 hr interval—the concentra-

tion of creosote in water decreased exponentially with time according to the equation $C = ae^{-bt}$ (C = relative concentration, t = time in hr). The a and b coefficients for creosote at 5 mg/l nominal concentration were 0.511 and −0.022 for the 1 l volume and were 0.841 and −0.067 for 30 l. (1557)

o-cresol (o-cresylic acid; 2-hydroxytoluene; 2-methylphenol)

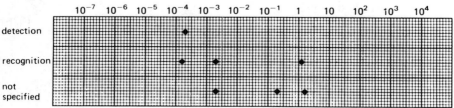

Manufacturing source: coal tar refining; petroleum refining; organic chemical mfg.; wood processing. (347)

Users and formulation: disinfectant; food antioxidant; perfume mfg.; dye mfg.; plastics and resins mfg.; herbicide mfg. (98%-DNOC, UCPA); tricresylphosphate mfg.; ore flotation, textile scouring agent, organic intermediate; mfg. of salicylaldehyde, coumarin; surfactant; cresylic acid constituent. (347)

Natural sources (water and air): coal, petroleum, constituent in wood, constituent in natural runoff. (347)

Man caused sources (water and air): automobile exhaust, roadway runoff, runoff from asphalt, general use of plastics, petroleum distillates, fuels, perfumes, oils, lubricants, metal cleaning and scouring compounds, laboratory chemical; constituent of domestic sewage. (347)

A. PROPERTIES: yellowish liquid; m.w. 108.13; m.p. 31°C; b.p. 191°C; v.p. 0.24 mm at 25°C, 5 mm at 64°C; v.d. 3.7; sp.gr. 1.041; solub. 31,000 mg/l at 40°C, 56,000 mg/l at 100°C; THC 883.4 kcal/mole, 856 kcal/mole; sat. conc. 1.2 g/cu m at 20°C, 2.8 g/cu m at 30°C

B. AIR POLLUTION FACTORS: 1 mg/cu m = 0.22 ppm, 1 ppm = 4.50 mg/cu m

odor thresholds   mg/cu m

	10^{-7}	10^{-6}	10^{-5}	10^{-4}	10^{-3}	10^{-2}	10^{-1}	1	10	10^2	10^3	10^4
detection				◆								
recognition			◆		◆			◆				
not specified						◆		◆	◆			

(57; 73; 279; 291; 610; 712; 829)

—Manmade sources:
   in exhaust of a 1970 Ford Maverick gasoline engine operated on a chassis dynamometer follwing the 7-mode California cycle:
      from API #7 gasoline: 0.5 ppm (including unknown compound)
      from API #8 gasoline: 1.0 ppm (including unknown compound) (1053)

—Control methods:
   wet scrubber: water at pH 8.5: outlet: 20 odor units/scf
                KMnO$_4$ at pH 8.5: outlet: 1 odor unit/scf (115)
   two step catalytic combustion on aluminium oxide (n.s.i.)

influent: 10,000 mg/cu m
after 1st step: 840 mg/cu m
after 2nd step: 23 mg/cu m (190)
—Sampling and analysis:
I.R. spectrometry: det.lim.: 2 mg/cu m (n.s.i.) (190)
C. WATER POLLUTION FACTORS:
—BOD_5: 1.6; 1.76 (30; 36)
1.69–1.74 std.dil.sewage (41)
65% of ThOD (274)
—BOD_{20}: 1.8 (30)
—$BOD_{24hr}^{30°C}$: at 15 mg/l: 53% ThOD (seed water from phenol-degradation plant)
—BOD_{2days}: at 15 mg/l: 61% ThOD (seed water from phenol-degradation plant)
—BOD_{5days}: at 15 mg/l: 77% ThOD (seed water from phenol-degradation plant)
(564)
—BOD_5^{20}: 1.76 at 10 mg/l, unadapted sew.; lag period: 1 day
—BOD_{20}^{20}: 2.16 at 10 mg/l, unadapted sew.; lag period: 1 day (554)
—COD: 2.38; 2.39 (36; 41)
92% of ThOD (0.05 n $K_2Cr_2O_7$) (274)
—$KMnO_4$: 6.166 (30)
68% of ThOD (0.01 n $KMnO_4$) (274)
—ThOD: 2.52 (36)
—Impact on biodegradation processes:
inhibition of nitrification in activated sludge at: 11–16 mg/l (75% reduction)
(n.s.i.) (43)
inhibition of degradation of glucose by *Pseudomonas fluorescens* at: 50 mg/l
(293)
inhibition of degradation of glucose by *E. coli* at: 600 mg/l (293)
—Reduction of amenities:
odor threshold (tentative): average: 0.65 mg/l
range: 0.016–4.1 mg/l (294)(97)
T.O.C. in water: 0.09 ppm
0.65 ppm (326)
T.O.C. in water: 0.26 ppm (325)
—Biodegradation:
% degradation in 7 days tests:
original culture: average: >99.5, range: >99.5
1st subculture: >99.5   98.5->99.5
2nd subculture: >99.5   99.0->99.5
3rd subculture: >99.5   >99.5 (87)
decomposition period by a soil microflora: 1 day (176)
adapted culture: 98% elimination after 48 hr incubation, initial conc. = 500 mg/l
(292)
adapted A.S. at 20°C—product is sole carbon source: 95.0% COD removal at
54.0 mg COD/g dry inoculum/hr (327)
—Waste water treatment:
ion exchange: adsorption on Amberlite X AD-2: 100% retention, infl.: 0.3
ppm, effl.: 0 ppm (40)

W, unadapted A.S.: at 1 mg/l: 100% removal after 3 hr
10 mg/l: 100% removal after 3 hr
100 mg/l: 17% removal after 6 hr (1639)
oxidation by activated sludges acclimated to the following aromatics: 250 mg/l
o-cresol after 30 min

phenol:	34% theor. oxidation
o-cresol:	35% theor. oxidation
m-cresol:	29% theor. oxidation
p-cresol:	6% theor. oxidation
mandelic acid:	9% theor. oxidation

methods	temp °C	days observed	feed mg/l	% removed	
RW, CA	20	2	1	100	
RW; CA	4	7	1	98	(93)

photo-oxidation: an aqueous solution of cresol (n.s.i.) is destroyed by photo-oxidation using visible light as a direct energy source and methyleneblue as a dye-sensitizer (188)
autoxidation at 25°C: $t_{1/2}$: 11090 hr at pH 9.0 (1908)
—Odor threshold: detection: 1.4 mg/l (998)
—Taste threshold conc.: 0.003 mg/l (998)
—Manmade emissions:
60 ml/min pure water passed through 25 ft, $\frac{1}{2}$ inch I.D. tube of general grade PVC contained 4.62 ppb o-cresol which constituted 43.01% of total contaminant concentration (430)
—Sampling and analysis:
photometry: min. full scale: 0.66×10^{-6} mole/l (53)
extraction efficiency of macro-reticular resins: pH: 5.7; conc.: 10 ppm; X-AD-2: 59%; X-AD-7: 67% (370)

D. BIOLOGICAL EFFECTS:
—Bacteria: *E. coli*: LD_0: 60 mg/l
—Algae: *Scenedesmus*: LD_0: 40 mg/l
—Toxicity threshold (cell multiplication inhibition test):

bacteria (*Pseudomonas putida*):	33 mg/l	(1900)
algae (*Microcystis aeruginosa*):	6.8 mg/l	(329)
green algae (*Scenedesmus quadricauda*):	11 mg/l	(1900)
protozoa (*Entosiphon sulcatum*):	17 mg/l	(1900)
protozoa (*Uronema parduczi Chatton-Lwoff*):	31 mg/l	(1901)
ciliate (*Tetrahymena pyriformis*): 24 hr LC_{100}:	3.7 mmole/l	(1662)

—Arthropoda: *Daphnia*: LD_0: 16 mg/l (30)
—Fish: goldfish: TLm (24–96 hr) soft water: 49.1–19 mg/l
bluegills: TLm (24–96 hr) soft water: 22.2–20.8 mg/l
fatheads: TLm (24–96 hr) hard water: 18.0–13.4 mg/l (158)
guppies: TLm (24–96 hr) hard water: 50–18 mg/l (41)
crucian carp: TLm (24 hr): 30 mg/l
roach: TLm (24 hr): 16 mg/l
tench: TLm (24 hr): 15 mg/l
"trout" embryos: TLm (24 hr): 2 mg/l (222)

406 *m*-CRESOL

—Amphibians:
Mexican axolotl (3–4 w after hatching): 48 hr LC_{50}: 40 mg/l
clawed toad (3–4 w after hatching): 48 hr LC_{50}: 38 mg/l (1823)
—Mammals: rat: lethal single oral dose: 1.35 g/kg
rabbit: lethal single oral dose. 0.8 g/kg (211)

***m*-cresol** (*m*-cresylic acid; 3-hydroxytoluene; 3-methylphenol)

A. PROPERTIES: yellowish liquid; m.w. 108.13; m.p. 12°C; b.p. 202°C; v.p. 0.04 mm at 20°C; 0.12 mm at 30°C, 5 mm at 76°C; v.d. 3.72; sp.gr. 1.038 at 20/4°C; solub. 23,500 mg/l at 20°C, 58,000 at 100°C; THC 880 kcal/mole; sat.conc. 0.24 g/cu m at 20°C, 0.68 g/cu m at 30°C; log P_{oct} 1.96/2.01

B. AIR POLLUTION FACTORS: 1 mg/cu m = 0.22 ppm, 1 ppm = 4.50 mg/cu m

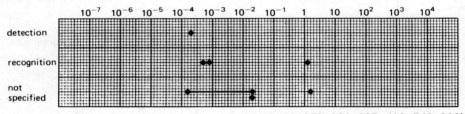

(57; 73; 279; 291; 307; 610; 762; 829)

—Manmade sources:
in exhaust of a 1970 Ford Maverick gasoline engine operated on a chassis dynamometer following the 7-mode California cycle:
from API #7 gasoline: 0.2 ppm
from API #8 gasoline: 0.4 ppm (1053)
—Sampling and analysis:
I.R. spectrometry: det. lim.: 2 mg/cu m (n.s.i.) (190)

C. WATER POLLUTION FACTORS:
—BOD_5: 1.7 std. dil. sew. (27)
1.70–1.88 std. dil. acclimated sewage (41)
68% of ThOD (274)
—$BOD_{24\ hr}^{30°C}$: at 20 mg/l: 46% ThOD (seed water from phenol-degradation plant)
—$BOD_{2\ days}$: at 20 mg/l: 62% ThOD (seed water from phenol-degradation plant)
—$BOD_{5\ days}$: at 20 mg/l: 80% ThOD (seed water from phenol-degradation plant)
(564)
—COD: 2.4 (41, 27)
100% of ThOD (0.05 n $K_2Cr_2O_7$) (274)
—$KMnO_4$: 68% of ThOD (274)
—ThOD: 2.52

—Impact on biodegradation processes:
  75% inhibition of nitrification process in non adapted activated sludge at 11.4
    mg/l   (30)
  inhibition of degradation of glucose by *Pseudomonas fluorescens* at: 40 mg/l
  inhibition of degradation of glucose by *E. coli* at: 600 mg/l   (293)
—Reduction of amenities:
  approx. conc. causing taste in trout and carp: 10 mg/l   (41)
  approx. conc. causing taste in fish: 0.2 mg/l   (41)
  odor threshold (tentative): average: 0.2 mg/l
                              range: 0.016–4.0 mg/l   (97)
  T.O.C. in water: 0.68 ppm; 0.25 ppm   (325; 326)
                 detection: 0.8 mg/l   (998)
  taste threshold conc.: 0.002 mg/l   (998)
—Biodegradation:
  decomposition period by a soil microflora: 1 day   (176)
  adapted A.S. at 20°C—product is sole carbon source: 95.5% COD removal at
    55.0 mg COD/g dry inoculum/hr   (327)
—Waste water treatment:
  activated sludge acclimated to the following aromatics at 250 mg *m*-cresol/l after
    30 min:
    phenol: 37% theor. oxidation
    o-cresol: 41% theor. oxidation
    m-cresol: 38% theor. oxidation
    p-cresol: 2% theor. oxidation
    mandelic acid: 13% theor. oxidation   (92)

methods	temp °C	days observed	feed mg/l	days acclim.	% theor. oxidation	% removed	
RW, CA	20	2	1	nat.		100	
RW, CA	4	7	1	nat.		99	
A.S., W	20	½	250	25+	38		(93)

  W, unadapted A.S.: at 1 mg/l: 100% removal after 3 hr
                       10 mg/l: 100% removal after 3 hr
                      100 mg/l: 4% removal after 6 hr   (1639)
—Sampling and analysis:
  photometry: min. full scale: 0.66×10^{-6} mole/l   (53)
D. BIOLOGICAL EFFECTS:
  —Bacteria: *E. coli*: LD_0: 600 mg/l
  —Toxicity threshold (cell multiplication inhibition test):
    bacteria (*Pseudomonas putida*):   53 mg/l   (1900)
    algae (*Microcystis aeruginosa*):   13 mg/l   (329)
    green algae (*Scenedesmus quadricauda*):   15 mg/l   (1900)
    protozoa (*Entosiphon sulcatum*):   31 mg/l   (1900)
    protozoa (*Uronema parduczi Chatton-Lwoff*): 62 mg/l   (1901)
  —Algae: *Chlorella pyrenoidosa*: toxic: 148–171 mg/l; 800 mg/l (n.s.i.)   (41)
           Scenedesmus: LD_0: 40 mg/l
  —Protozoa: *Vorticella campanula*: perturbation level: 0.5 mg/l
              Paramaecium caudatum: perturbation level: 0.9 mg/l

ciliate (*Tetrahymena pyriformis*): 24 hr LC_{100}: 3.5 mmole/l (1662)
—Arthropoda: *Daphnia*: LD_0: 28 mg/l
        Gammarus pulex: perturbation level: 0.7 mg/l
—Mollusca: *Glossosiphonia complanata*: perturbation level: 1.1 mg/l (30)
—Fish: mosquito fish: TLm (24–96 hr): 24 mg/l
    bluegill: TLm (96 hr): 10–13.6 mg/l (41)
    crucian carp: TLm (24 hr): 25 mg/l
    roach: TLm (24 hr): 23 mg/l
    tench: TLm (24 hr): 21 mg/l
    "trout" embryos: TLm (24 hr): 7 mg/l (222)
—Mammalia: rat: lethal single oral dose: 2.02 g/kg
        rabbit: lethal single oral dose: 1.1 g/kg (211)

p-cresol (*p*-cresylic acid; 4-hydroxytoluene; 4-methylphenol)

A. PROPERTIES: yellowish liquid; m.w. 108.13; m.p. 34.8°C; b.p. 202°C; v.p. 0.04 mm at 20°C, 0.11 mm at 25°C, 1 mm at 53°C; v.d. 3.72; sp.gr. 1.0347 at 20/4°C; solub. 24,000 mg/l at 40°C, 53,000 mg/l at 100°C; THC 880.0 kcal/mole; sat. conc. 0.24 g/cu m at 20°C, 0.74 g/cu m at 30°C; log P_{oct} 1.92/1.94

B. AIR POLLUTION FACTORS: 1 mg/cu m = 0.22 ppm, 1 ppm = 4.50 mg/cu m
—Odor: quality: tarlike
    hedonic tone: pungent

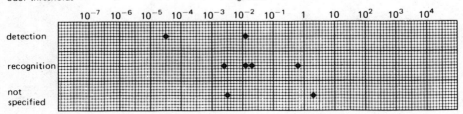

(2; 57; 73; 279; 291; 307; 610; 614; 741; 829)

—Manmade sources:
  in exhaust of a 1970 Ford Maverick gasoline engine operated on a chassis dynamometer following the 7-mode California cycle:
    from API #7 gasoline: 0.4 ppm
    from API #8 gasoline: 0.7 ppm (1053)
—Sampling and analysis:
    I.R. spectrometry: det. lim.: 2 mg/cu m (n.s.i.) (190)

C. WATER POLLUTION FACTORS:
—BOD_5: 1.45 std. dil. sew. at conc. <440 mg/l (27)
    1.4–1.76 std. dil. acclimated sewage (41)

- $BOD_{24\ hr}^{30°C}$: at 15 mg/l: 55% ThOD (seed water from phenol-degradation plant)
- $BOD_{2\ days}$: at 15 mg/l: 61% ThOD (seed water from phenol-degradation plant)
- $BOD_{5\ days}$: at 15 mg/l: 81% ThOD (seed water from phenol-degradation plant)
(564)
- COD: 2.4 (41)
- ThOD: 2.52 (27)
- Impact on biodegradation processes:
 75% reduction of the nitrification process in non adapted activated sludge at: 16.5 mg/l (30)
 inhibition of degradation of glucose by *Pseudomonas fluorescens* at: 30 mg/l
 inhibition of degradation of glucose by *E. coli* at: >1000 mg/l (293)
- Reduction of amenities: T.O.C. in water: 0.055 ppm (325)
- Biodegradation: decomposition period by a soil microflora: 1 day (176)
 adapted A.S. at 20°C—product is sole carbon source: 96.0% COD removal at 55.0 mg COD/g dry inoculum/hr (327)

accumulation of dihydroxybenzoic acid by cells of *Pseudomonas* sp. (1219)

- Manmade sources:
 in sewage effluents: 0.090 mg/l (227)
 in primary domestic sewage plant effluent: 0.020-0.029 mg/l
 in secondary domestic sewage plant effluent: 0.020-0.090 mg/l (517)
- Waste water treatment:
 activated sludges acclimated to the following aromatics: 250 mg/l *p*-cresol, after 30 min:

phenol:	33% theor. oxidation
o-cresol:	34% theor. oxidation
m-cresol:	35% theor. oxidation
p-cresol:	39% theor. oxidation
mandelic acid:	20% theor. oxidation (92)

methods	temp °C	days observed	feed mg/l	days acclim.	% theor. oxidation	% removed
RW, CA	20	6	1	nat.	–	95
RW, CA	4	19	1	nat.	–	95
AS, W	20	½	250	25+	39	– (93)

- W, unadapted A.S.: at 1 mg/l: 100% removal after 3 hr
 10 mg/l: 100% removal after 3 hr
 100 mg/l: 3% removal after 6 hr (1639)
- Taste threshold conc.: 0.002 mg/l (998)
- Odor threshold: detection: 0.2 mg/l (998)
- Sampling and analysis:
 photometry: min. full scale: 0.66×10^{-6} mole/l (53)

410 o-CRESYLIC ACID

D. BIOLOGICAL EFFECTS:
 —Protozoa:
  ciliate (*Tetrahymena pyriformis*): 24 hr, LC_{100}: 3.7 mmole/l   (1662)
 —Algae: *Scenedesmus*: LD_0: 6 mg/l
 —Arthropoda: *Daphnia*: LD_0: 12 mg/l   (30)
 —Fish:
  crucian carp: LC_{50} (24 hr): 21 mg/l
  roach: LC_{50} (24 hr): 17 mg/l
  tench: LC_{50} (24 hr): 16 mg/l
  "trout" embryos: LC_{50} (24 hr): 4 mg/l   (222)
  rainbow trout: toxic: 5 mg/l   (156)
  fathead minnows: static bioassay in Lake Superior water at 18-22°C: LC_{50}
   (1; 24; 48; 72; 96 hr): >30; 26; 21; 21; 19 mg/l   (350)
 —Mammals: rat: lethal single oral dose: 1.8 g/kg
           rabbit: lethal single oral dose: 1.1 g/kg   (211)

o-cresylic acid see o-cresol

m-cresylic acid see m-cresol

p-cresylic acid see p-cresol

crotonaldehyde (β-methylacrolein; 2-butenal; crotonic aldehyde; propylene aldehyde) $CH_3CH=CH-CHO$
A. PROPERTIES: m.w. 70.1; m.p. -75°C; b.p. 99/104°C; v.p. 19 mm at 20°C; v.d. 2.41; sp.gr. 0.85; solub. 155,000 mg/l
B. AIR POLLUTION FACTORS: 1 mg/cu m = 0.349 ppm, 1 ppm = 2.8 mg/cu m
 —Odor: T.O.C.: 0.6 mg/cu m = 0.2 ppm   (57)
        0.1 mg/cu m = 0.035 ppm   (307)
 —T.O.C.: 0.18-0.57 mg/cu m   (710)
          0.420 mg/cu m   (842)
 —Manmade sources:
  in exhaust of a 1970 Ford Maverick gasoline engine operated on a chassis dynamometer following the 7-mode California cycle:
   from API #7 gasoline: 0.9 ppm
   from API #8 gasoline: 0.1 ppm   (1053)
  in exhaust of gasoline engine: 0.4-1.4 vol.% of total exhaust aldehydes
                                                                           (395; 396; 397)
C. WATER POLLUTION FACTORS:
 —BOD_5:   37% of ThOD
 —BOD_{10}: 1.30 std.dil.sew.   (256)
 —COD: 97% of ThOD std.dil.sew.
 —$KMnO_4$: acid: 42% of ThOD; alkaline: 43% of ThOD   (220)
 —Reduction of amenities: faint odor: 0.021 mg/l   (129)
 —Waste water treatment:
  activated carbon: adsorbability: 0.092 g/g C, 45.6% reduction, infl.: 1,000 mg/l,
   effl.: 544 mg/l   (32)

methods	temp °C	days observed	feed mg/l	days acclim.	% theor. oxidation	% removed
NFG, BOD	20	1-10	1,000	365 + P	–	27
NFG, BOD	20	1-10	50-250	365 + P	–	inhibition
NFG, BOD	20	1-10	1,600	365 + P	–	36
RW, Sd, BOD	20	10	?	100	70	–

(93)

—Sampling and analysis:
  photometry: min. full scale: 3.5×10^{-6} mole/l (53)

D. BIOLOGICAL EFFECTS:
—Fish:
  Lepomis macrochirus: static bioassay in fresh water at 23°C, mild aeration applied after 24 hr (85% aqueous):

material added ppm	% survival after				best fit 96 hr LC_{50} ppm
	24 hr	48 hr	72 hr	96 hr	
7.5	0	–	–	–	
5.6	0	–	–	–	
4.2	0	–	–	–	
3.2	90	90	90	70	3.5
1.8	100	100	100	100	

  Menidia beryllina: static bioassy in synthetic seawater at 23°C: mild aeration applied after 24 hr:

material added ppm	% survival after				best fit 96hr LC_{50} ppm
	24 hr	48 hr	72 hr	96 hr	
3.2	0	–	–	–	
1.8	90	90	20	10	1.3
1.0	100	100	100	90	

(352)

—Mammals:
  rat: inhalation: LC_{50}: 1,400 ppm, 30 min
                0/6: sat. vap., 1 min. (104)
      acute oral LD_{50}: 0.16-0.3 g/kg (211)

crotonic aldehyde *see* crotonaldehyde

crotoxyphos *see* ciodrin

crotylmercaptan
$CH_3CH=CH-CH_2SH$
A. PROPERTIES: m.w. 88;
B. AIR POLLUTION FACTORS:
  —Odor: characteristic: skunk odor (51)
  T.O.C.: 0.0075 ppb; 56 ppt; 0.00043-0.0014 mg/cu m (10; 279; 710)
        0.0002 mg/cu m = 0.055 ppb (307)
C. WATER POLLUTION FACTORS:
  —Reduction of amenities: faint odor: at 0.000029 mg/l (129)

cumarin *see* coumarin

cumene *see* isopropylbenzene

cumenehydroperoxide *see* isopropylbenzenehydroxyperoxide

p-cumylphenol (*p*-(2-phenylisopropyl)phenol)

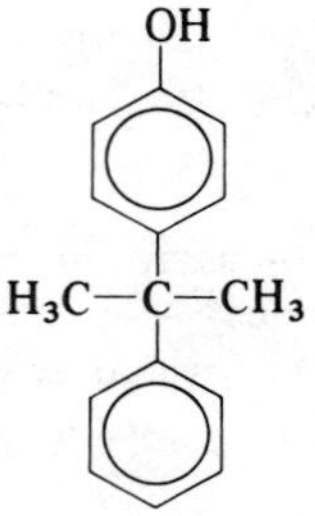

Use: intermediate for resins, insecticides, lubricants
A. PROPERTIES: white to tan crystals; m.w. 212.29; m.p. 70–73°C; b.p. 335°C
B. AIR POLLUTION FACTORS:
   —Odor: characteristic: phenol odor
C. WATER POLLUTION FACTORS:
   —COD: 2.6 (27)
   —ThOD: 2.8 (27)

cupric acetate
$(CH_3-COO)_2Cu$
D. BIOLOGICAL EFFECTS:
   —Fish:

	24 hr LC_{50}	48 hr LC_{50}	96 hr LC_{50}
fathead minnows (*Pimephales promelas*)	0.48 mg/l	0.42 mg/l	0.39 mg/l
grass shrimp (*Palaemonetes pugis*)			37.0 mg/l

(1904)

cutrine
  Use: algicide
  Composition: copper-triethanolamine complexes (331)
D. BIOLOGICAL EFFECTS:
  decapod: *Penaeus californiensis*: 96 hr LC_{50}–FT: 1000 mg/l (1134)
  harlequin fish (*Rasbora heteromorpha*):

mg/l	24 hr	48 hr	96 hr	3 m (extrap.)
LC_{10}(F)	0.7	0.26		
LC_{50}(F)	1.2	0.35	0.24	0.01

(331)

cyanatrine (2-(4-ethylamino-6-methylthio-*s*-triazin-2-ylamino)-2-methylpropionitrile)
  Use: herbicide
  Active ingredient: triazine (50%) (331)

D. BIOLOGICAL EFFECTS:
 —Snail:
  Lymnaea peregra: adult: 20 ppm: 60% mortality after 2 days (1414)
  10 ppm: no mortality after 8 days (1415)
  eggs: 20 ppm: high incidence of embryonic deformities and all failed to hatch (1414)
  0.2 ppm: no effect
 —Amphibian:
  Rana temporaria tadpoles: 96 hr, LC_{50}: 30 ppm (1415)
 —Crustacean:
  Daphnia longispina: 96 hr, LC_{50}: 15.4 ppm (1415)
  Daphnia pulex: 2 ppm: 25% mortality after 3 days
  0.2 ppm: 10% mortality after 3 days (1414)
 —Fish:
  harlequin fish (*Rasbora heteromorpha*):

mg/l	24 hr	48 hr	96 hr	3 m (extrap.)	
$LC_{10}(F)$	15	9	9		
$LC_{50}(F)$	35	18	15	5	(331)

5-cyanoacenaphthene

Use: in soots generated by the combustion of aromatic hydrocarbon fuels doped with pyridine

D. BIOLOGICAL EFFECTS:
 —Mutagenicity: induces significant mutation to 8-azaguanine resistance in *S. typhimurium* at concentrations as low as 1100 µM (1723)

1-cyanoacenaphthylene

Use: in soots generated by the combustion of aromatic hydrocarbon fuels doped with pyridine (1723)

D. BIOLOGICAL EFFECTS:
 —Mutagenicity: induces significant mutation to 8-azaguanine resistance in *S. typhimurium* at concentrations as low as 2800 µM (1723)

5-cyanoacenaphthylene

Use: in soots generated by the combustion of aromatic hydrocarbon fuels doped with pyridine (1723)

D. BIOLOGICAL EFFECTS:
—Mutagenicity: induces significant mutation to 8-azaguanine resistance in *S. typhimurium* at concentrations as low as 1250 μM (1723)

β-cyanoethyl-2,3-dibromopropionate see busan 76

cyanoethylene see acrylonitrile

cyanogenbromide (bromine cyanide)
BrCN

Uses: organic synthesis; parasiticide; fumigating compositions; rat exterminants

A. PROPERTIES: crystals; penetrating odor; slowly decomposed by cold water; m.w. 105.93; m.p. 49–51°C; b.p. 61–62°C; sp.gr. 2.015 at 20°/4°C; v.p. 100 mm at 22.6°C

D. BIOLOGICAL EFFECTS:
—Fish:
Lepomis macrochirus: static bioassay in fresh water at 23°C, mild aeration applied after 24 hr

material added ppm	% survival after 24 hr	48 hr	72 hr	96 hr	best fit 96 hr LC_{50} ppm
0.42	0	—	—	—	
0.32	70	70	70	70	
0.18	100	100	80	70	0.24
0.10	100	100	100	100	

Menidia beryllina: static bioassay in synthetic seawater at 23°C, mild aeration applied after 24 hr

material added ppm	% survival after 24 hr	48 hr	72 hr	96 hr	best fit 96 hr LC_{50} ppm
0.56	0	—	—	—	
0.42	85	85	85	85	
0.32	100	100	100	100	0.47
0.18	100	100	100	100	(352)

cyanogenchloride (chlorinecyanide)
CNCl

A. PROPERTIES: colorless liquid or gas; m.w. 61.48; m.p. -6.5°C; b.p. 12.5/13°C; v.p. 1,000 mm at 20°C; v.d. 2; sp.gr. 1.218 at 4/4°C; solub. 30,000 mg/l; log P_{oct}; 0.64 (calculated)

B. AIR POLLUTION FACTORS: 1 mg/cu m = 0.398 ppm, 1 ppm = 2.51 mg/cu m
 —Odor: characteristic: quality: bitter almonds
                        hedonic tone: pungent
   T.O.C.: recogn.: 2.5 mg/cu m = 1 ppm (73)
   O.I. at 20°C = 1,300,000 (316)
C. WATER POLLUTION FACTORS:
 —Reduction of amenities: faint odor: 0.0025 mg/l (129)
D. BIOLOGICAL EFFECTS:
 —Fish: goldfish: lethal conc.: 1 mg/l, 6–8 hr (154)
 —Mammalia:
   mouse: inhalation: some deaths: 120 ppm, 3–5 min
   rabbit: inhalation: fatal: 1,200 ppm, 2 min
   cat: inhalation: fatal: 120 ppm, 3–5 min
   dog: inhalation: recovered: 20 ppm, 20 min
                    fatal: 48 ppm, 6 hr
                    severe injury: 120 ppm, 8 min.
   goat: inhalation: fatal after 70 hr: 1,000 ppm, 3 min. (211)
 —Man: fatal after 10 min: 159 ppm
              30 min: 48 ppm
   tolerable for 1 min: 20 ppm
              10 min: 2 ppm
   lowest irritant conc. for 10 min: 1 ppm (211)

1-cyanonaphthalene

Use: in soots generated by the combustion of aromatic hydrocarbon fuels doped with pyridine (1723)

D. BIOLOGICAL EFFECTS:
 —Mutagenicity: did not induce significant mutation to 8-azaguanine resistance in S. typhimurium at concentrations up to 1300 μM, the limit of solubility under the assay conditions (1723)

2-cyanonaphthalene

Use: in soots generated by the combustion of aromatic hydrocarbon fuels doped with pyridine (1723)

D. BIOLOGICAL EFFECTS:
 —Mutagenicity: did not induce significant mutation to 8-azaguanine resistance in S. typhimurium at concentrations up to 1300 μM, the limit of solubility under the assay conditions (1723)

(+)-cyano-3-phenoxybenzyl-(+)-α-(4-chlorophenyl)isovalerate *see* fenvalerate

(+)α-cyano-3-phenoxybenzyl-(+)*cis,trans*-2,2-dichlorovinyl-2,2-dimethylcyclopropane-carboxylate *see* cypermethrin

cyanophos *see* cyanox

cyanopropane *see* butyronitrile

cyanox   (O-*p*-cyanophenyl-O,O-dimethylphosphorothioate)

$$\begin{array}{c} CH_3O \\ CH_3O \end{array} \!\! \overset{\overset{\displaystyle S}{\|}}{P}\!\!-\!O\!-\!\!\!\bigcirc\!\!\!-\!CN$$

Use: insecticide
A. PROPERTIES: clear amber liquid, m.p. 14–15°C
D. BIOLOGICAL EFFECTS:
 —Fish:
  harlequin fish (*Rasbora heteromorpha*):

mg/l	24 hr	48 hr (40% active ingredient)	
LC_{10} (F)	20	6.7	
LC_{50} (F)	36	14	(331)

 —Mammals:
  acute oral LD_{50} (rats): 565 mg/kg   (1854)

cyanurotriamide *see* melamine

cycloate *see* ro-neet

1,5,9-cyclododecatriene (CDT)

A. PROPERTIES: sp.gr. 0.8925 at 20/20°C; m.w. 162.28; m.p. −15°C; b.p. 231°C
C. WATER POLLUTION FACTORS:
 —BOD_5: 0.02 NEN 3235-5.4   (277)
 —COD: 3.02 NEN 3235-5.3   (277)
D. BIOLOGICAL EFFECTS:
 —Fish: goldfish: LD_{50} (24 hr): 4 mg/l modified ASTM-D-1345   (277)
 —Mammalia: rat: acute oral LD_{50}: 2–6 ml/kg   (277)

1,3,5-cycloheptatriene (tropilidene)

A. PROPERTIES: dark yellow liquid; m.w. 92.14; m.p. -79.5°C; b.p. 110-130°C; sp.gr. 0.89 at 20/4°C
C. WATER POLLUTION FACTORS:
   —BOD_5: 0.10 = 3% of ThOD (277)
   —COD: 2.30 = 74% of ThOD (277)
D. BIOLOGICAL EFFECTS:
   —Fish: goldfish: LD_{50} (24 hr): 15 mg/l (277)
   —Mammalia: rat: acute oral dose LD_{50}: 57 mg/kg (277)
              rat: acute dermal dose LD_{50}: 442–884 mg/kg (277)

cycloheptene

A. PROPERTIES: m.w. 96.17; b.p. 112-113°C; sp.gr. 0.824
D. BIOLOGICAL EFFECTS:
   —Threshold conc. of cell multiplication inhibition of the protozoan *Uronema parduczi Chatton-Lwoff*: >40 mg/l (1901)

1,3-cyclohexadiene

A. PROPERTIES: m.w. 80.13; b.p. 80°C; sp.gr. 0.841
B. AIR POLLUTION FACTORS:
   —Odor threshold: detection: 0.0025–0.0066 mg/cu m (653)
D. BIOLOGICAL EFFECTS:
   —Fish:
      young Coho salmon: mortality: 10/30 at 50 ppm after 24, 48, 72, 96 hr in artificial sea water at 8°C (317)

2,5-cyclohexadiene-1,4-dione *see p*-benzoquinone

cyclohexane (hexahydrobenzene; hexamethylene)

418 1,2-CYCLOHEXANEDIOL

A. PROPERTIES: colorless; m.w. 84.16; m.p. 6.3°C; b.p. 81°C; v.p. 77 mm at 20°C, 120 mm at 30°C; v.d. 2.90; sp.gr. 0.779 at 20/4°C; solub. 55 mg/l at 20°C, 45 mg/l at 15°C; THC 936 kcal/mole, LHC 881 kcal/mole; sat. conc. 357 g/cu m at 20°C, 532 g/cu m at 30°C

B. AIR POLLUTION FACTORS: 1 mg/cu m = 0.29 ppm, 1 ppm = 3.49 mg/cu m

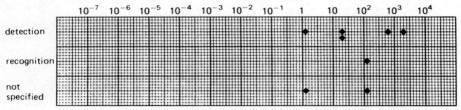

(54; 279; 307; 601; 643; 708; 737; 805; 828)

O.I. at 20°C = 203,000
—Sampling and analysis:

I.R. spectrometry:	det.lim.:	4 ppb	(56)
detector tubes: Unico:	det.lim.:	100 ppm	(59)
Dräger:	det.lim.:	100 ppm	(58)
Auer:	det.lim.:	20 ppm	(57)

C. WATER POLLUTION FACTORS:
—BOD_{35}^{25}: 2.39 in seawater/inoculum: enrichment cultures of hydrocarbon oxidising bacteria (521)
—ThOD: 3.42
—Biodegradation:

Microbial degradation of cyclohexane   (1244; 1245) incubation with natural flora in the groundwater—in presence of the other components of high-octane gasoline (100 µl/l): biodegradation: 45% after 192 hr at 13°C (initial conc. 0.12 µl/l) (956)

Rotating disk contact aerator: infl. 231.4 mg/l, effl. 0.2 mg/l; elimination: >99% or 19077 mg/m^2/24 hr or 5151 g/cu m/hr (406)

Reduction of amenities: T.O.C. = 0.02 mg/l

D. BIOLOGICAL EFFECTS:
—Protozoa:
Threshold conc. of cell multiplication inhibition of the protozoa *Uronema parduczi Chatton-Lwoff*: >50 mg/l (1901)
—Mussels:
mussel larvae (*Mytilus edulis*): 10–20% increase of growth rate at 1 to 100 ppm (475)
—Fish:
guppy (*Poecilia reticulata*): 7 d LC$_{50}$: >84 (1833)
fathead minnows: static bioassay in Lake Superior water at 18–22°C: LC$_{50}$ (1; 24; 48; 72; 96 hr): 95; 93; 93; 93; 93 mg/l
fathead minnows: static bioassay in reconstituted water at 18–22°C: LC$_{50}$ (1; 24; 48; 72; 96 hr): 126; 117; 117; 117; 117 mg/l (350)
mosquito fish: TLm (24 hr): 15,500 mg/l in turbid Oklahoma water (244)
fatheads: TLm (24–96 hr): 43–32 mg/l
bluegills: TLm (24–96 hr): 43–34 mg/l
goldfish: TLm (24–96 hr): 42.3 mg/l
guppies: TLm (24–96 hr): 57.7 mg/l (158)
young Coho salmon: no significant mortalities up to 100 ppm after 96 hrs in artificial sea water at 8°C (317)
—Mammalia:
mouse: inhalation: LC$_{50}$: ±18,000 ppm, 1 hr
rabbit: inhalation: no effect level: 786 ppm, 10 weeks (54)
monkey: inhalation: no effect level: 1243 ppm, 6 hr/day, 50 days (211)
—Man: irritating to the eyes and mucous membranes: 300 ppm (211)
—Carcinogenicity: none?
—Mutagenicity in the *Salmonella* test: none
  <0.006 revertant colonies/nmol
  <70 revertant colonies at 1000 µg/plate (1883)

1,2-cyclohexanediol

A. PROPERTIES: m.w. 116.16; m.p. 72.5–75°C (mixture of *cis* and *trans*); b.p. 118–120°C/10 mm

420 CYCLOHEXANOL

C. WATER POLLUTION FACTORS:
   —Biodegradation:

alternative routes for the conversion of cyclohexan-1,2-diol to adipate (426)

adapted A.S.—product as sole carbon source: 95.0% COD removal at 66.0 mg COD/g dry inoculum/hr (327)

cyclohexanol (hexahydrophenol; cyclohexylalcohol; hexalin; adronol; hydrophenol; hydralin; anol)

A. PROPERTIES: colorless liquid; m.w. 100.16; m.p. 24°C; b.p. 161°C; v.p. 1 mm at 20°C, 3.5 mm at 34°C; v.d. 3.45; sp.gr. 0.95 at 25/4°C; solub. 56,700 mg/l at 15°C, 36,000 mg/l at 20°C; THC 890.3 kcal/mole, LHC 838 kcal/mole; sat.conc. 4.9 g/cu m at 20°C, 10.0 g/cu m at 30°C; log P_{oct} 1.23
B. AIR POLLUTION FACTORS: 1 mg/cu m = 0.244 ppm, 1 ppm = 4.163 mg/cu m
   —Odor: T.O.C.: 0.21 mg/cu m = 0.05 ppm (639; 57; 291)

USSR: human reflex response: no response: 0.04 mg/cu m
       animal chronic exposure: no effect: 0.059 mg/cu m
                                  adverse effect: 0.61 mg/cu m   (170)

O.I. at 20°C = 26,300   (316)

C. WATER POLLUTION FACTORS:
- BOD_5: 0.379   (30)
  - 0.08 warburg, sewage   (41)
  - 0.08 std.dil.sew.   (27)
  - 4% of ThOD   (220)
- BOD_{20}: 1.967   (30)
- COD: 2.15   (27)
  - 95% of ThOD   (220)
- $KMnO_4$: 0.144   (30)
  - acid 9, alkaline 5% of ThOD   (220)
- TOC: 96% of ThOD   (220)
- ThOD: 2.828   (30)
- Biodegradation pathways:

alternative routes for the metabolic conversion of cyclohexanol to adipate   (426)

CYCLOHEXANONE

— Odor thresholds: detection: 3.5 mg/kg (886)
0.4 mg/kg (894)
— Waste water treatment:
reserve osmosis: 68% rejection from 0.01 M solution adapted A.S.—product as sole carbon source—96.0% COD removal at 28.0 mg COD/g dry inoculum/hr (327)

D. BIOLOGICAL EFFECTS:
— Fish:
Lepomis macrochirus: static bioassay in fresh water at 23°C, mild aeration applied after 24 hr:

material added ppm	24 hr	% survival after 48 hr	72 hr	96 hr	best fit 96 hr LC_{50} ppm
1,350	30	30	30	10	
1,000	80	70	70	64	1,100
790	100	100	100	100	

Menidia beryllina: static bioassay in synthetic seawater at 23°C: mild aeration applied after 24 hr:

material added ppm	24 hr	% survival after 48 hr	72 hr	96 hr	best fit 96 hr LC_{50} ppm
1,000	100	90	50	30	
750	75	35	25	25	720
500	100	100	100	100	(352)

fathead minnows: static bioassay in Lake Superior water at 18-22°C: LC_{50} (1; 24; 48; 72; 96 hr): 1,550; 1,033; 1,033; 1,033; 1,033 mg/l (350)
— Mammals
rabbit: single oral LD_{50}: 2.2-2.6 g/kg (211)
rat: single oral LD_{50}: 2.06 ml/kg (54)
rabbit: inhalation: no effect: 145 ppm, 300 hr
dog: inhalation: no effect: sat.vap., 10 min/day, 7 days (211)
— Man: irritation of eyes, nose and throat: 100 ppm, 3-5 min (211)

cyclohexanone (ketohexamethylene; pimelic ketone; sextone, anone)

Use: solvent for various insecticides
A. PROPERTIES: m.w. 98.2; m.p. -26/-38°C; b.p. 157°C; v.p. 4 mm at 20°C, 6.2 mm at 30°C; v.d. 3.38; sp.gr. 0.95 at 20/4°C; solub. 23,000 mg/l at 20°C, 24,000 mg/l at 31°C; sat.conc. 19 g/cu m at 20°C, 32 g/cu m at 30°C; log P_{oct} 0.81
B. AIR POLLUTION FACTORS: 1 mg/cu m = 0.25 ppm, 1 ppm = 4.08 mg/cu m
— Odor: characteristic: quality: sweet, sharp
hedonic tone: pleasant

odor thresholds — mg/cu m

(19; 57; 636; 639; 727; 739; 827)

USSR: human reflex response: no response: 0.06 mg/cu m
animal chronic exposure: no effect: 0.042 mg/cu m
adverse effect: 0.46 mg/cu m (170)
O.I. at 20°C = 21,900 (316)

C. WATER POLLUTION FACTORS:
—BOD_5: 1.232 (30)
32% of ThOD (220)
—BOD_{20}: 2.00 (30)
—$KMnO_4$: 0.480 (30)
acid 18, alkaline 22% of ThOD (220)
—COD: 100% of ThOD (220)
—ThOD: 2.605 (30)
—Biodegradation: adapted A.S.—product as sole carbon source: 96.0% COD removal at 30.0 mg COD/g dry inoculum/hr (327)
—Waste water treatment:
A.C. adsorbability: 0.134 g/g C, 66.8% reduction, infl.: 1,000 mg/l, effl.: 332 mg/l (32)
—Sampling and analysis: photometry: min. full scale: 77×10^{-6} mole/l (53)

D. BIOLOGICAL EFFECTS:
—Bacteria:
Pseudomonas: still toxic at 500 mg/l
—Algae:
Scenedesmus: not toxic at 1 g/l (30)
—Toxicity threshold (cell multiplication inhibition test):
bacteria (*Pseudomonas putida*): 180 mg/l (1900)
algae (*Microcystis aeruginosa*): 52 mg/l (329)
green algae (*Scenedesmus quadricauda*): 370 mg/l (1900)
protozoa (*Entosiphon sulcatum*): 545 mg/l (1900)
protozoa (*Uronema parduczi Chatton-Lwoff*): 280 mg/l (1901)
—Mammals:
rat: single oral LD_{50}: 1.62 ml/kg (54)
mouse: single oral dose, intragastric: LD_0: 1.3–1.5 g/kg
rabbit: single oral dose, intragastric: LD_{100}: 1.6–1.9 g/kg (211)
rabbit: inhalation: no effect (liver alteration?): 190 ppm, 10 w
rat: inhalation: survived: 4,000 ppm, 4 hr
death: 8,000 ppm, 8 hr
monkeys, rabbits: inhalation:

no detectable effect: 190 ppm, 50 × 6 hr
slight eye irritation: 309 ppm, 50 × 6 hr
slightly increased mortality: 3,082 ppm, 50 × 6 hr (211)
—Man: objectionable: 50 ppm
satisfactory: <25 ppm (211)

cyclohexanoneisooxime *see* α-caprolactam

cyclohexanoneoxime
A. PROPERTIES: m.w. 113.16; m.p. 90°C; b.p. 206/210°C; v.d. 3.91
B. AIR POLLUTION FACTORS: 1 mg/cu m = 0.21 ppm, 1 ppm = 4.70 mg/cu m
C. WATER POLLUTION FACTORS:
   —BOD_5: 0.030 (30)
   —BOD_{20}: 0.130 (30)
   —$KMnO_4$ value: 1.896 (30)
D. BIOLOGICAL EFFECTS:
   —Bacteria: *Pseudomonas*: toxic: 30 mg/l
   —Algae: *Scenedesmus*: toxic: 480 mg/l
   —Protozoa: *Colpoda*: toxic: 60 mg/l
   —Arthropoda: *Daphnia*: toxic: 120 mg/l (30)

cyclohexene (1,2,3,4-tetrahydrobenzene)

A. PROPERTIES: m.w. 82.14; m.p. -104°C; b.p. 83°C; v.p. 160 mm at 38°C; v.d. 2.9; sp.gr. 0.81 at 20/4°C; solub. 213 mg/l at 20°C
B. AIR POLLUTION FACTORS: 1 mg/cu m = 0.29 ppm, 1 ppm = 3.41 mg/cu m
   —Odor threshold: detection: 0.6 mg/cu m (637)
C. WATER POLLUTION FACTORS:
   —Biodegradation:

cyclohexene → cyclohexanone

(203)
   —Waste water treatment:
      Rotating disk contact aerator: infl. 123.3 mg/l, effl. 0.4 mg/l; elimination: >99% or 10139 mg/m²/hr or 2737 g/cu m/hr (406)
D. BIOLOGICAL EFFECTS:
   —Bacteria: *Pseudomonas putida*: inhibition of cell multiplication starts at 17 mg/l
(329)

—Algae: *Microcystis aeruginosa*: inhibition of cell multiplication starts at >160 mg/l (329)
—Protozoa:
threshold conc. of cell multiplication inhibition of the protozoa *Uronema parduczi Chatton-Lwoff*: >50 mg/l (1901)
—Fish:
young Coho salmon: no significant mortalities up to 100 ppm after 96 hr in artificial sea water at 8°C (317)

cyclohexylacetate (hexaline acetate)

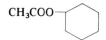

A. PROPERTIES: m.w. 142.19; m.p. 177°C; v.d. 4.9; sp.gr. 1.0
D. BIOLOGICAL EFFECTS:
—Toxocity threshold (cell multiplication inhibition test):
bacteria (*Pseudomonas putida*):	83 mg/l	(1900)
algae (*Microcystis aeruginosa*):	46 mg/l	(329)
green algae (*Scenedesmus quadricauda*):	5.3 mg/l	(1900)
protozoa (*Entosiphon sulcatum*):	120 mg/l	(1900)
protozoa (*Uronema parduczi Chatton-Lwoff*):	>400 mg/l	(1901)

cyclohexylalcohol *see* cyclohexanol

cyclohexylamine (hexahydroaniline; aminocyclohexane)

A. PROPERTIES: m.w. 99.17; m.p. -18°C; b.p. 134°C; v.d. 3.42; sp.gr. 0.8191 at 20/4°C
B. AIR POLLUTION FACTORS: 1 mg/cu m = 0.247 ppm, 1 ppm = 4.06 mg/cu m
D. BIOLOGICAL EFFECTS:
—Toxicity threshold (cell multiplication inhibition test):
bacteria (*Pseudomonas putida*):	420	mg/l	(1900)
algae (*Microcystis aeruginosa*):	0.02	mg/l	(329)
green algae (*Scenedesmus quadricauda*):	0.51	mg/l	(1900)
protozoa (*Entosiphon sulcatum*):	0.6	mg/l	(1900)
protozoa (*Uronema parduczi Chatton-Lwoff*):	>200	mg/l	(1901)

—Mammals:
rat: acute oral LD_{50}: 0.4–0.8 g/kg (5% soln.)
    inhalation: 3/3: 12,000 ppm, 6 hr
                0/3: 1,000 ppm, 6 hr (211)

cyclohexylchloride *see* chlorocyclohexane

cyclohexylmethane *see* methylcyclohexane

1,3-cyclooctadiene

Use: intermediate
C. WATER POLLUTION FACTORS:
  —Waste water treatment methods:
    rotating disk contact aerator: infl. 60.1 mg/l, effl. 0.7 mg/l; elimination: 99% or 4906 mg/m²/24 hr or 1325 g/cu m/24 hr (406)

1,5-cyclooctadiene (COD)

Uses: resin intermediate
A. PROPERTIES: m.w. 108.18; m.p. -69.5°C; b.p. 150.9°C; v.p. 5 mm at 20°C; sp.gr. 0.8803 at 20/20°C
C. WATER POLLUTION FACTORS:
  —BOD_5: 0.19 NEN 3235-5.4 (277)
  —COD: 2.62 NEN 3235-5.4 (277)
D. BIOLOGICAL EFFECTS:
  —Fish: goldfish: LD_{50} (24 hr): 14 mg/l modified ASTM D 1345 (277)
  —Mammalia: rat: acute oral LD_{50}: 2.7 ml/kg (277)

cyclooctane

A. PROPERTIES: m.w. 112.22; m.p. 10–13°C; b.p. 151°/740 mm; sp.gr. 0.834
B. AIR POLLUTION FACTORS:
  —Odor threshold: detection: 3.6 mg/cu m (828)
C. WATER POLLUTION FACTORS:
  —Rotating disk contact aerator: infl. 241.2 mg/l, effl. 0.5 mg/l; elimination: >99% or 19866 mg/m²/hr or 5364 g/cu m/hr (406)

cyclooctene

A. PROPERTIES: m.w. 110.2; m.p. -16°C; b.p. 145-146°C; sp.gr. 0.846
C. WATER POLLUTION FACTORS:
—Waste water treatment methods:
Rotating disk contact aerator: infl. 61.2 mg/l, effl. 0.6 mg/l; elimination: 99% or 5054 mg/m²/hr or 1365 g/cu m/hr (406)

10,H-cyclopenta(*mno*)benzo(*a*)pyrene   see 10,11-methylenebenzo(*a*)pyrene

11,H-cyclopenta(*grs*)benzo(*e*)pyrene   see 8,9-methylenebenzo(*e*)pyrene

cyclopentane (pentamethylene)

Use: solvent for cellulose ethers
A. PROPERTIES: colorless liquid; m.w. 70.14; m.p. -94°C; b.p. 50°C; sp.gr. 0.751; v.p. 400 mm at 31°C
C. WATER POLLUTION FACTORS:
—Biodegradation:
incubation with natural flora in the groundwater—in presence of the other components of high-octane gasoline (100 µl/l): biodegradation: 0% after 192 hr at 13°C (initial conc.: 0.17 µl/l) (956)
D. BIOLOGICAL EFFECTS:
—Fish:
young Coho salmon: no significant mortalities up to 100 ppm after 96 hr in artificial sea water at 8°C (317)

cyclopentanol (cyclopentylalcohol)

Uses: perfume and pharmaceutical solvent; intermediate for dyes, pharmaceuticals and other organics
A. PROPERTIES: m.w. 86.13; b.p. 139-140°C; mp. -19°C; sp.gr. 0.949
B. AIR POLLUTION FACTORS:
—Odor threshold: detection: 4,200-8,700 mg/cu m (727)
C. WATER POLLUTION FACTORS:
—Biodegradation:

adapted A.S.—product as sole carbon source: 97.0% COD removal at 55.0 mg
COD/g dry inoculum/hr (327)
D. BIOLOGICAL EFFECTS:
—Toxicity threshold (cell multiplication inhibition test):
bacteria (*Psuedomonas putida*):	250 mg/l	(1900)
algae (*Microcystis aeruginosa*):	28 mg/l	(329)
green algae (*Scenedesmus quadricauda*):	255 mg/l	(1900)
protozoa (*Entosiphon sulcatum*):	290 mg/l	(1900)
protozoa (*Uronema parduczi Chatton Lwoff*):	>800 mg/l	(1901)

cyclopentanone

Use: intermediate for pharmaceuticals, insecticides and rubber chemicals
A. PROPERTIES: b.p. 125–126°C/630 mm; sp.gr. 0.943
B. AIR POLLUTION FACTORS:
—Odor threshold: detection: 31–1,120 mg/cu m (727)
C. WATER POLLUTION FACTORS:
—Biodegradation:
adapted A.S.—product as sole carbon source: 95.4% COD removal at 57.0
mg/COD/g dry inoculum/hr (327)

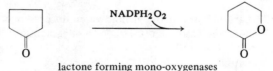

lactone forming mono-oxygenases
metabolic biodegradation (436)

D. BIOLOGICAL EFFECTS:
—Toxicity threshold (cell multiplication inhibition test):
bacteria (*Pseudomonas putida*):	175 mg/l	(1900)
algae (*Microcystis aeruginosa*):	63 mg/l	(329)
green algae (*Scenedesmus quadricauda*):	1900 mg/l	(1900)
protozoa (*Entosiphon sulcatum*):	232 mg/l	(1900)
protozoa (*Uronema parduczi Chatton-Lwoff*):	1210 mg/l	(1901)

cyclopentene

Use: organic synthesis
A. PROPERTIES: m.w. 68.12; m.p. −135°C; b.p. 44°C; sp.gr. 0.77

D. BIOLOGICAL EFFECTS:
—Fish: young Coho salmon: no significant mortalities up to 100 ppm after 96 hr in artificial sea water at 8°C (317)

cyclopenteno(*cd*)pyrene

A. PROPERTIES: m.w. 226
—Manmade sources:
in gasoline: >0.2 g/l
in exhaust condensate of gasoline engine: 0.75–0.99 mg/l gasoline consumed
(1070)
in exhaust condensate of gasoline engine: 2.0 mg/g (1069)

3-*p*-cymenol *see* thymol

cypermethrin ((+)α-cyano-3-phenoxybenzyl-(+)*cis*,*trans*-2,2-dichlorovinyl-2,2-dimethylcyclopropanecarboxylate)

Use: experimental photostable pyrethroid, insecticide (1577)
A. PROPERTIES: solub. 0.041 ppm at room temp.; log P_{oct} 4.47 (1577)
D. BIOLOGICAL EFFECTS:
—Fish:
rainbow trout: technical grade: 24 hr LC_{50}: 55 ppb active ingredient (static test)
formulated product: 24 hr LC_{50}: 11 ppb active ingredient (static test) (1577)

l-cysteine (*l*-2-amino-3-mercaptopropanoic acid; *l*-β-mercaptoalanine)
$HSCH_2CH(NH_2)COOH$
A. PROPERTIES: m.w. 121.15
C. WATER POLLUTION FACTORS:
—A.S.: after 6 hr: 7.5% of ThOD
12 hr: 8.6% of ThOD
24 hr: 11.2% of ThOD (89)

l-cystine (dicysteine; *l*-β,β',-dithiodialanine; *l*-3,3'-dithio-*bis*(2-amino propanoic acid)
$(SCH_2CH(NH_2)COOH)_2$

l-CYSTINE

A. PROPERTIES: m.w. 240.29; m.p. 258/261°C decomposes; solub. 110 mg/l at 25°C, 520 mg/l at 75°C

C. WATER POLLUTION FACTORS:
 —Manmade sources:
  excreted by man: in urine: 1.5-2.4 mg/kg body wt/day (203)
 —Waste water treatment:
  A.S.: after 6 hr: 1.5% of ThOD
        12 hr: 2.4% of ThOD
        24 hr: 4.7% of ThOD (89)
  A.S. Resp, BOD, 20°C, 1-5 days observed, feed: 1,000 mg/l,
   <1 day acclimation: nil % theor. oxidation
                       nil % removed (93)

D

2,4-D *see* 2,4-dichlorophenoxyacetic acid

2,4-D(BEE) *see* 2,4-dichlorophenoxyacetic acid, butoxyethanolester

DBN *see* di-*n*-butylnitrosamine

2,4-D(butylthio) *see* 2,4-dichlorophenoxyacetic acid, butylthio

2,4-D(diethylamine salt) *see* 2,4-dichlorophenoxyacetic acid, diethylamine salt

2,4-D(DMA) *see* 2,4-dichlorophenoxyacetic acid, dimethylamine salt

2,4-D(ethylthio) *see* 2,4-dichlorophenoxyacetic acid, ethylthio

2,4-D(IOE) *see* 2,4-dichlorophenoxy-acetic acid, isooctylester; *see also* co-op brushkiller 112

DAA *see* diacetone alcohol

dacthal (dimethyltetrachloroterephthalate; chlorthaldimethyl; DAC 893; Fatal; chlorthalmethyl; DCPA)

$$\underset{\text{COOCH}_3}{\overset{\text{COOCH}_3}{\text{Cl} \underset{\text{Cl}}{\overset{\text{Cl}}{\bigcirc}} \text{Cl}}}$$

Use: selective preemergence herbicide
A. PROPERTIES: white crystals; m.p. 156°C; v.p. <0.5 mm at 40°C; solub. 0.5 ppm at 25°C
D. BIOLOGICAL EFFECTS:
—Fish: *Lepomis macrochirus*: 48 hr, LC_{50}: 700 mg/l   (2116)
—Mammals: acute oral LD_{50} (rat): >3,000 mg/kg
    acute dermal LD_{50} (albino rabbit): >10,000 mg/kg   (1854)

dactinol
  Use: insecticide; active ingredient: Rotenol (5%)
D. BIOLOGICAL EFFECTS:
  —Fish:
    harlequin fish (*Rasbora heteromorpha*): 24 hr, $LC_{50}(F)$: 9.5 mg/l
    rainbow trout (*Salmo gairdneri*): 24 hr, $LC_{50}(F)$: 7.3 mg/l (hard water)
      (yearling)   0.58 mg/l (soft water)
                         48 hr, $LC_{50}(F)$: 5.8 mg/l (hard water)
                                               0.47 mg/l (soft water)   (331)

dalapon *see* dichloropropionic acid, sodium salt

DAP *see* diallylphthalate

dazomet (3,5-dimethyltetrahydro-2-thio-1,3,5-thiadiazine; tetrahydro-3,5-dimethyl-2H-1,3,5-thiadiazine-2-thione; DMTT; Crag fungicide 974; Mylone)

$$CH_3-N \underset{}{\overset{S\diagup\diagdown S}{\diagup\diagdown}} N-CH_3$$

Use: a soil fumigant and fungicide. Its activity is due to decomposition to methylisothiocyanate   (1439)
A. PROPERTIES: white, almost odorless crystals; m.p. 99.5°C (dec.); solub. 0.12% at 30°C
D. BIOLOGICAL EFFECTS:
  —Mammals:
    acute oral LD_{50} (rats): 500 mg/kg
                         (mice): 400 mg/kg   (1855)

DBCP *see* 1,2-dibromo-3-chloropropane

DBP *see* di-*n*-butylphthalate

DBPC *see* 2,6-di-*tert*-butyl-4-methylphenol

DCPD *see* α-dicyclopentadiene

DDA *see* dioctyladipate

DDD (2,2-*bis*(*p*-chlorophenyl)-1,1-dichloroethane; dichlorodiphenyldichloroethane; TDE; rhothane; tetrachlorodiphenylethane)

$$Cl-C_6H_4-\underset{ClCH}{\overset{H}{C}}-C_6H_4-Cl$$

Uses: nonsystemic contact and stomach insecticide. The technical grade contains related compounds in small amounts, the greatest being the o,p'-isomer

A. PROPERTIES: m.w. 320.1; m.p. 112°C; v.d. 11; solub. 0.160 mg/l at 24°C (99% purity)

C. WATER POLLUTION FACTORS:
—Persistence:
persistence in river water in a sealed glass jar under sunlight and artificial fluorescent light—initial conc. 10 μg/l:

	% of original compound found				
after	1 hr	1 wk	2 wk	4 wk	8 wk
	100	100	100	100	100 (1309)

—Water and bottom sediment quality:
North sea and Norwegian depression: bottom sediments: 0.02–0.45 ppb (varied with depth) (1172)

D. BIOLOGICAL EFFECTS:
—Residues:
residues in Canadian human milk: 1967: avg. 4 ng/g whole milk
                                          1970: avg. 3 ng/g whole milk (1376)
mean residues in eggs of American crocodile in Everglades National Park:
    1972: 0.11 ppm wet wt. (1572)
    1977: 0.04 ppm wet wt. (1571)
in aquatic vascular plants and Lake Päijänne, Finland (1972-1973): 0.5 μg/kg dry weight. (1055)
Seal, Baltic Sea, blubber: 10 mg/g (1181)
residues in animals in different trophic levels from the Weser estuary—1976:
    bivalvia: common edible cockle (*Cerastoderma edule. L.*):
                                       0.8 ± 0.2 ng/g wet tissue
        soft clam (*Mya arenaria L.*): 5.0 ± 1.2 ng/g wet tissue
    polychaeta: lugworm (*Arenicola marina L.*): 3.7 ± 0.6 ng/g wet tissue
    crustacea: brown shrimp (*Crangon crangon*): 0.7 ± 0.2 ng/g wet tissue
    pisces: sommon sole (*Solea solea*): 9.9 ± 2.5 ng/g wet tissue (1440)
in marine animals from the Ligurian Sea (1977-1978):
    shrimp (*Nephros norvegicus*): avg. 0.66 ppb fresh wt.
    mussel (*Mytilus galloprovinciales*): avg. 4.0 ppb fresh wt.
    fish: anchovy (*Engraulis encrasicholus*): avg. 3.2 ppb fresh wt.
        striped mullet (*Mullus barbatus*): avg. 16.0 ppb fresh wt.
        Euthymnus alletteratus: avg. 3.8 ppb fresh wt.
        Sarda sarda: avg. 34.0 ppb fresh wt. (1775)
in marine animals from the Central Mediterranean (1976-1977): fish:
    anchovy (*Engraulis encrasicholus*): avg: 3.3 ppb fresh wt ($n = 12$)
                                                    range: 0.8–13.4 ppb fresh wt.
    stripped mullet (*Mullus barbatus*): avg: 4.5 ppb fresh wt. ($n = 10$)
                                                    range: 0.6–15.1 ppb fresh wt.
    tuna (*Thunnus thynnus thynnus*): avg: 1.6 ppb fresh wt. ($n = 5$)
                                                    range: 0.3–2.9 ppb fresh wt.
    shrimp (*Nephrops norvegicus*): avg: 0.2 ppb fresh wt. ($n = 7$)
                                                   range: 0.1–0.4 ppb fresh wt.

mussel (*Mytilus galloprovincialis*): avg: 5.8 ppb fresh wt. ($n = 4$)
range: 5.3–6.5 ppb fresh wt. (1774)

—Bioaccumulation:
BCF in aquatic model ecosystem:
alga: 6210
snail (*Cipangopaludina japonica* Martens): 4460
carp (*Cyprinus carpio*): 2710 (1834)
pelecypod (*Mytilus edulis*): exposure: 0.05–2.18 µg/l: 50 hr;
bioaccumulation: 9,120 × (1477)

—Distribution and metabolism of radiolabeled DDD, after 33 days in a model ecosystem:

		concentration, ppm	
			mosquito larva	fish
	H_2O	snail (*Physa*)	dry wt.	(*Gambusia*)
DDE–total ^{14}C	0.006	5.6	5.8	39.1
DDD	0.0004	3.3	3.43	33.4
DDD ethylene	–	0.24	–	2.08
unknown I	–	0.14	–	1.54
unknown II	–	0.87	–	–
polar metabolites	0.0056	1.1	–	2.0

(1312)

—Toxicity
—Crustaceans
Gammarus lacustris: 96 hr LC_{50}: 0.64 µg/l (2124)
Gammarus fasciatus: 96 hr LC_{50}: 0.86 µg/l (2126)
Palaemonetes kadiakensis: 96 hr LC_{50}: 0.68 µg/l (2126)
Asellus breviacaudus: 96 hr LC_{50}: 10.0 µg/l (2126)
Simocephalus serrulatus: 48 hr LC_{50}: 4.5 µg/l (2127)
Daphnia pulex: 48 hr LC_{50}: 3.2 µg/l (2127)
Korean shrimp *Palaemon macrodactylus*: TL_{50}, 96 hr, S: 8.3 (4.8–14.4) ppb
Korean shrimp *Palaemon macrodactylus*: TL_{50}, 96 hr, FT: 2.5 (1.6–4.0) ppb
(2352)

—Insects:
Pteronarcys californica: 96 hr LC_{50}: 380 µg/l (2128)
—Mammals:
rats: acute oral LD_{50}: 3.4 g/kg (1854)

p,p'-DDE (dichlorodiphenyldichloroethylene)

Cl–C₆H₄–C(ClCCl)–C₆H₄–Cl

Uses: military product, DDT impurity (347)
Man caused sources (*air and water*): agricultural runoff degradation of DDT, lab use. (347)
A. PROPERTIES: solub. 0.040 mg/l at 20°C; 0.065 mg/l at 24°C; 0.0013 mg/l; log P_{oct} 4.28; 5.69

B. AIR POLLUTION FACTORS:
- In marine atmosphere in the Northwest Gulf of Mexico, March–April 1977:
  avg.: 0.049 ng/cu m
  range: 0.009–0.180 ng/cu m ($n = 10$) (1724)
- Atmospheric levels in agricultural area: n.d.–14.2 ng/cu m (1308)

C. WATER POLLUTION FACTORS:
- Degradation: degrades further to DDA (*bis*(chlorophenyl)acetic acid) by loss of two more molecules HCl (1854)
- Persistence in river water in a sealed glass jar under sunlight and artificial fluorescent light—initial conc. 10 µg/l:

after	1 hr	1 wk	2 wk	4 wk	8 wk	
	100	100	100	100	100	(1309)

% of original compound found

- Sediment quality:
  North Sea and Norwegian depression: bottom sediments: 64–290 ppt (1172)
  Hawaii: sediments: 110–11420 ppt (1174)

D. BIOLOGICAL EFFECTS:
- Residues:
  Euphausid (*Meganyctiphanes norvegica*), Med. Sea: residue: 26 µg/kg (1480)
  in Canadian human milk: 1967: avg.: 103 ng/g whole milk
                                    1970: avg.: 56 ng/g whole milk
                                    1975: avg.: 35 ng/g whole milk (1376)
  in aquatic vascular plants of Lake Päijänne, Finland (1972–1973):
    mean: 2 µg/kg dry weight ($n = 114$)
    S.D.: 3 µg/kg dry weight
    min.: 0 µg/kg dry weight
    max.: 12 µg/kg dry weight (1055)
  mean residues in eggs of American crocodile in Everglades National Park:
    1972: 1.84 ppm wet wt. (1572)
    1977–1978: 1.19 ppm wet wt. (1571)
  goby fish (*Acanthogobius flavimanus*) collected at the seashore of Keihinjima along Tokyo Bay—Aug. 1978: residue level: 29 ppb (1721)
  seal, Baltic Sea, Blubber: 138 mg/g
  Alaskan seal, Alaska, Fat: 0.110 mg/g (1182)
  sea duck, Alaska, Fat: 0.17 mg/g
  in marine animals from the Central Mediterranean (1976–1977): fish:
    anchovy (*Engraulis encrasicholus*): avg.: 11.3 ppb fresh wt. ($n = 12$)
                                           range: 4.9–27.6 ppb fresh wt.
    striped mullet (*Mullus barbatus*): avg.: 16.5 ppb fresh wt. ($n = 10$)
                                           range: 1.9–32.3 ppb fresh wt.
    tuna (*Thunnus thynnus thynnus*): avg.: 11.2 ppb fresh wt. ($n = 5$)
                                           range: 3.3–28.5 ppb fresh wt.
    shrimp (*Nephrops norvegicus*): avg.: 1.5 ppb fresh wt. ($n = 7$)
                                           range: 0.5–2.7 ppb fresh wt.
    mussel (*Mytilus galloprovincialis*): avg.: 9.0 ppb fresh wt. ($n = 4$)
                                           range: 8.2–9.5 ppb fresh wt. (1774)
  in marine animals from the Ligurian Sea (1977–1978):
    shrimp (*Nephrops norvegicus*): avg.: 4.25 ppb fresh wt.
    mussel (*Mytilus galloprovincialis*): avg.: 5.5 ppb fresh wt.

fish: anchovy (*Engraulis encrasicholus*): avg.: 12.0 ppb fresh wt.
 striped mullet (*Mullus barbatus*): avg.: 34.2 ppb fresh wt.
 Euthymnus alletteratus: avg.: 101.0 ppb fresh wt.
 Sarda sarda: avg.: 445.0 ppb fresh wt. (1775)
—Residue in animals in different trophic levels from the Weser estuary (1976):
 bivalvia: common edible cockle (*Cerastoderma edule L.*):
  0 ng/g wet tissue
  soft clam (*Mya arenaria L.*): 0.9 ± 0.2 ng/g wet tissue
 polychaeta: lugworm (*Arenicola marina L.*): 1.0 ± 0.2 ng/g wet tissue
 crustacea: brown shrimp (*Crangon crangon*): 0.9 ± 0.0 ng/g wet tissue
 pisces: common sole (*Solea solea L.*): 3.8 ± 0.8 ng/g wet tissue (1440)
—Bioaccumulation:
 BCF in aquatic model ecosystem:
  alga: 2720
  snail (*Cipangopaludina japonica Martens*): 13700
  carp (*Cyprinus carpio*): 8450 (1834)
—Distribution and metabolism of radiolabeled DDE after 33 days in a model ecosystem:

	H_2O	snail (*Physa*)	mosquito larva dry wt.	fish (*Gambusia*)
DDE–total ^{14}C	0.008	121.6	168.9	149.8
DDE	0.0053	103.5	159.5	145.0
polar metabolites	0.0027	18.1	9.4	4.8

(1312)

—Bioaccumulation factor (BCF) and biodegradability index (B.I.) in a laboratory model ecosystem after 33 days at 26°C:

	BCF	B.I.*
alga (*Oedogonium cardiacum*)	11,251	0.069
snail (*Physa*)	36,342	0.049
mosquito (*Culex pipiens quinquefasciatus*)	59,390	0.033
fish (*Gambusia affinis*)	12,037	0.050

*B.I. = ppm polar degradation-products/ppm nonpolar products (1643)
—Carcinogenicity: +
—Mutagenicity in the *Salmonella* test: negative
 <0.004 revertant colonies/nmol
 <70 revertant colonies at 5000 µg/plate (1883)

DDM see 4,4'-diaminodiphenylmethane

DDVP see dichlorovinyl-dimethylphosphate

***o,p'*-DDT** (*o,p'*-dichloro-diphenyl-trichloroethane)

Use: minor (up to 30%) constitutent of technical DDT, of which *p,p'*-DDT is the predominant component.

D. BIOLOGICAL EFFECTS:
- Residues in Canadian human milk: 1967: avg.: 5 ng/g whole milk
1970: avg.: 5 ng/g whole milk
1975: avg.: 3 ng/g whole milk   (1376)

DDT (dichloro-diphenyl-trichloroethane; chlorophenotene; dicophane; 1,1,1-trichloro-2,2-bis(p-chlorophenyl)ethane)

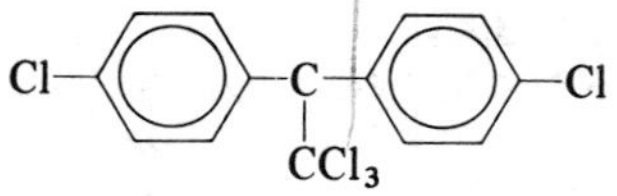

Cl—⟨⟩—C—⟨⟩—Cl
       |
      CCl$_3$

Use: nonsystemic stomach and contact insecticide

A. PROPERTIES: m.w. 354.5; solub. 0.0031–0.0034 mg/l at 25°C; m.p. 108°C; v.p. 1.9×10^{-7} mm at 20°C; log P_{oct} 6.19 at 20°C. DDT is the common name for the technical product of which p,p'-DDT is the predominant component. The product contains up to 30% of the o,p'-isomer.

WARNING: it is not always clearly indicated whether only p,p'-DDT was measured, or the sum of all DDT isomers or also the degradation products!

B. AIR POLLUTION FACTORS:
- Ambient air quality:
  in marine atmosphere in the Northwest Gulf of Mexico, March–April 1977:
    avg.: 0.034 ng/cu m
    range: 0.010–0.078 ng/cu m ($n = 10$)   (1724)
- Atmospheric levels of p,p'-DDT in agricultural areas ranged from ~10 ng/cu m in winter to ~1000 ng/cu m in summer   (1308)

C. WATER POLLUTION FACTORS:
- Degradation:

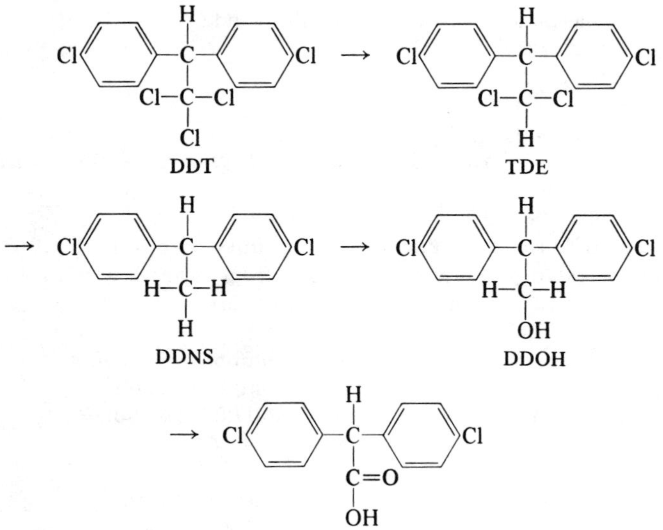

DDT → TDE → DDNS → DDOH →

Metabolic pathway of DDT under oceanic conditions

75-100% disappearance from soils: 4-30 years (1815; 1816)
photooxidation by u.v. light in aqueous medium at 90-95°C: time for the formation of CO_2 (% of theoretical): 25%: 25.9 hr
50%: 66.5 hr
75%: 120.0 hr (1628)
calculated half-life in water at 25°C and 1 m depth, based on evaporation rate of 9.34×10^{-3} m/hr: 73.9 hr (437)
persistence in river water in a sealed glass jar under sunlight and artificial fluorescent light—initial conc. 10 µg/l:

% of original compound found

after	1 hr	1 wk	2 wk	4 wk	8 wk	
	100	100	100	100	100	(1309)

—Water and sediment quality:
England: bottom and suspended sediments: 0.2-20 ng/g (1171)
Northsea and Norwegian depression: bottom sediments: 0.15-0.44 ppb (1172)
Hawaii: sediments: 0.25-6.42 ppb (1174)
Los Angeles Harbor: bottom sediments: 0.115-3.212 mg/g (1175)
San Antonio Bay, Texas: suspended and bottom (1 cm) sediments: 8-60 ppb
(1176)
Southern California: bottom sediments: 10->150 mg/g (1177)
Western Baltic: surface water: 0.17×10^{-9} g/l
in Northern Mississippi water: avg.: 7.03 ng/l;
range: 2.73-12.31 ng/l
coloring colloids in natural surface water can concentrate ^{14}C-DDT from 0.168 ppb by wt. in water to 15,800 times that concentration in coloring colloids
(1306)
—Odor threshold: detection: 0.35 mg/kg water (915)

D. BIOLOGICAL EFFECTS:
—Residues:
in aquatic vascular plants of Lake Päijänne, Finland (1972-1973):
mean: 1 µg/kg dry weight ($n = 114$)
S.D.: 1 µg/kg dry weight
min.: 0 µg/kg dry weight
max.: 7 µg/kg dry weight
mean residues in eggs of American crocodile in Everglades National Park:
1972: 0.33 ppm wet wt. (1572)
1977: 0.09 ppm wet wt. (1571)
In lake trout of accurately known age, residues of p,p'-DDE, p,p'-DDD, and p,p'-DDT increased progressively with age from about 1 ppm at 1 year to concentrations of about 14 ppm or higher at 12 years (1302)

ringed seal	Finland		
pregnant		blubber: 75 mg/kg	
non-pregnant		blubber: 130 mg/kg	
mussel (*Mytilus* sp.)	Mediterranean Coast, Finland, Southern California	40 000-690 000 X	(1183)
Alaskan seal	Alaska	fat: 0.02 mg/g	(1182)
invertebrates	Irish Sea	0.5-0.3 ng/g	(1184)

demersal fish	Irish Sea	0–13 ng/g	(1184)
bottom fish	Irish Sea	0–150 ng/g	
black mullet (*Mugil cephalus*)	South Florida, East and West Coast	flesh: 0.002–0.064 mg/g liver: 0.002–0.156 mg/g milt: 0.001–0.075 mg/g	(1185)
silver mullet (*Mugir curema*)	South Florida, East and West Coast	flesh: 0.014–0.101 mg/g liver: 0.016–0.300 mg/g milt: 0.045–0.246 mg/g	
estuarine carnivores	New Zealand	muscle: 0.59 mg/kg	(1186)
marine carnivores	New Zealand	muscle: 0.32 mg/kg	
dolphin	New Zealand	liver: 1.26 mg/kg	
seal	New Zealand	liver: 5.28 mg/kg	
seal	Baltic Sea	blubber: 78 mg/g	(1181)
seston	Annapolis	0.23 ppt	(1187)
	Baltimore Harbor	0.27 ppt	
	Head of Chesapeake Bay	1.03 ppt	

residues in Canadian human milk: 1967: avg. 33 ng/g whole milk
1970: avg. 15 ng/g whole milk
1975: avg. 6 ng/g whole milk   (1376)

mean concentration in gamebird muscle in upper Tennessee (U.S.A.):

	grouse	quail	woodcock
	mg/kg fresh weight ± st. error		
Johnson county	2.88 ± 0.34 (n = 12)	4.28 ± 0.35 (n = 6)	9.10 ± 0.42 (n = 6)
Carter county	9.19 ± 0.42 (n = 9)	9.12 ± 0.37 (n = 7)	9.95 ± 0.46 (n = 6)
Washington county	9.26 ± 0.35 (n = 10)	9.08 ± 0.43 (n = 11)	9.03 ± 0.41 (n = 6)

quail: acceptable LD_{50}: 500 mg/kg
FAO/WHO residue tolerance limit: 0.5–7.0 mg/kg/day

comparative average concentrations of Σ-DDT residue (wet tissue basis) in brown shrimp (*Crangon crangon*) from various areas of the North Sea (1974–1976):

area	year	No. of analyses	avg. conc.	(ppb) range	
S.E. England	1975	2	13	(13–<14)	(1441)
Netherlands	1974	10	<21	(<16–<26)	(1442)
	1975	12	<30		(1441)
	1976	12	<40		(1441)
Germany	1974	6	33	(13–70)	(1442)
	1975	4	38	(19–56)	(1441)
	1976	6	17	(14–21)	(1441)
	1976	5	2		(1440)

in marine animals from the Ligurian Sea (1977–1978): ΣDDT
   shrimp (*Nephrops norvegicus*):   avg.:   5.1 ppb fresh wt.
   mussel (*Mytilus galloprovincialis*):   avg.:   34.5 ppb fresh wt.
   fish: anchovy (*Engraulis encrasicholus*):   avg.:   27.0 ppb fresh wt.
       striped mullet (*Mullus barbatus*):   avg.:   81.7 ppb fresh wt.
       Euthymnus alletteratus: avg.:   162.0 ppb fresh wt.
       Sarda sarda: avg.:   802.0 ppb fresh wt.   (1775)

in marine animals from the Central Mediterranean (1976–1977): fish: ΣDDT

anchovy (*Engraulis encrasicholus*): avg.: 30.6 ppb fresh wt. ($n = 12$)
range: 11.3–82.7 ppb fresh wt.
striped mullet (*Mullus barbatus*): avg.: 39.5 ppb fresh wt. ($n = 10$)
range: 3.9–85.5 ppb fresh wt.
tuna (*Thunnus thynnus thynnus*): avg.: 22.8 ppb fresh wt. ($n = 5$)
range: 6.6–51.3 ppb fresh wt.
shrimp (*Nephrops norvegicus*): avg.: 2.5 ppb fresh wt. ($n = 7$)
range: 1.2–4.5 ppb fresh wt.
mussel (*Mytilus galloprovincialis*): avg.: 33.0 ppb fresh wt. ($n = 4$)
range: 29.8–35.6 ppb fresh wt. (1774)

residue in great crested grebe (fish-eating bird): Σ-DDT
Great Britain 1963–1966, 1968:
  in liver: avg. 12 ppm, range 0.1–81 ppm (n = 15)
  in liver: avg. 19 ppm, range 10–30 ppm (n = 4) (1778)
Clyde, Scotland, 1971–1975, (bird found dead):
  in pectoral muscle: 4.7 ppm
  in liver: 4.2 ppm (1779)
Lake Päijänne, Finland, 1972–1974 (bird shoot in summer, after breeding season):
  in pectoral muscle: avg. 3.7 ppm, range 0.19–7.6 ppm ($n = 12$)
  in liver: avg. 6.4 ppm, range 0.06–17 ppm ($n = 12$) (1780)

residues in fish eating birds from Gdánsk Bay 1975–1976: ΣDDT

		ppm in fresh tissue		
		pectoral muscle	liver	adipose tissue
great crested grebe	avg.	4.65 ($n = 17$)	9.77 ($n = 18$)	57.5 ($n = 10$)
(*Podiceps cristatus L.*)	range	0.39–20	0.12–37	7.1–120
slavonian grebe		3.1 ($n = 1$)	3.3 ($n = 1$)	
(*Podiceps auritus L.*)				
black guillemot	avg.	0.51 ($n = 2$)	0.87 ($n = 2$)	13 ($n = 2$)
(*Cepphus grylle L.*)	range	0.39–0.64	0.78–0.96	9.0–18.0
goosander		1.5 ($n = 1$)	1.6 ($n = 1$)	47 ($n = 1$)
(*Mergus merganser L.*)				
black-throated diver		4.0 ($n = 1$)	1.1 ($n = 1$)	49 ($n = 1$)
(*Gavia arctica*)				

in marine animals from the Central Mediterranean (1976–1977): fish: pp'DDT
anchovy (*Engraulis encrasicholus*): avg.: 1.4 ppb fresh wt. ($n = 12$)
range: 0.5–2.8 ppb fresh wt.
striped mullet (*Mullus barbatus*): avg.: 0.3 ppb fresh wt. ($n = 10$)
range: 0.1–0.9 ppb fresh wt.
tuna (*Thunnus thynnus thynnus*): avg.: 0.8 ppb fresh wt. ($n = 5$)
range: 0.2–1.3 ppb fresh wt.
shrimp (*Nephrops norvegicus*): avg.: 0.13 ppb fresh wt. ($n = 7$)
range: 0.1–0.3 ppb fresh wt.
mussel (*Mytilus galloprovincialis*): avg.: 3.3 ppb fresh wt. ($n = 4$)
range: 1.7–4.8 ppb fresh wt. (1774)
                                                      residues
sea-skater   Baja California   180–400 ng/g (1483)

(*Halobates sobrinus*)			
mussel (*Mytilus edulis*)	NW Atlantic	0.015 µg/g	(1484)
red crab (*Geryon quinquedues*)	NW Atlantic	0.061 µg/g	(1484)
cod (*Gadus morhua*)	NW Atlantic	0.024 µg/g	(1484)
striped Bass (*Roccus saxatilis*)	NW Atlantic	0.63 µg/g	(1484)
"biota"	Gulf of Mexico	11.4 µg/g	(1485)
periwinkle (*Littorina littorea*)	Medway Estuary, Kent, U.K.	1.0–6.3 µg/g	(1481)
shore crab (*Carcinus maenas*)	Medway Estuary, Kent, U.K.	4.8–25.7 µg/g	(1481)
flounder (*Platichthys flesus*)	Medway Estuary, Kent, U.K.	10.1–650.3 µg/g	(1481)
fish	Baltic Sea	0.06–4.00 mg/kg	(1486)
dover sole (*Microstomus pacificus*)	Southern California Bight	13–36 mg/kg (with eroded fins) 1.8–25 mg/kg (without eroded fins)	(1487)
harbor seal (*Phoca vitulina*)	German North Sea	1.15–4.26 mg/kg (blubber)	(1482)
grey seal (*Halichoerus grypus*)	Nova Scotia	14.0 µg/g	(1488)

    in marine animals from the Ligurian Sea (1977–1978): pp'DDT
        shrimp (*Nephrops norvegicus*) (1 value):   2.7 ppb fresh wt.
        mussel (*Mytilus galloprovincialis*):   avg. 25.9 ppb fresh wt.
        fish: anchovy (*Engraulis encrasicholus*): avg. 15.0 ppb fresh wt.
            striped mullet (*Mullus barbatus*): avg. 31.5 ppb fresh wt.
            Euthymnus alletteratus: avg. 56.7 ppb fresh wt.
            Sarda sarda: avg. 318.0 ppb fresh wt.   (1775)
    in marine animals from the Central Mediterranean (1976–1977): fish: pp'DDT
        anchovy (*Engraulis encrasicholus*): avg.: 15.1 ppb fresh wt. ($n = 12$)
                                      range: 4.6–39.3 ppb fresh wt.
        striped mullet (*Mullus barbatus*): avg.: 17.1 ppb fresh wt. ($n = 10$)
                                        range: 1.3–37.2 ppb fresh wt.
        tuna (*Thunnus thynnus thynnus*): avg.: 8.8 ppb fresh wt. ($n = 5$)
                                        range: 2.6–18.6 ppb fresh wt.
        shrimp (*Nephrops norvegicus*): avg.: 0.7 ppb fresh wt. ($n = 7$)
                                        range: 0.2–1.3 ppb fresh wt.
        mussel (*Mytilus galloprovincialis*): avg.: 15.0 ppb fresh wt. ($n = 4$)
                                        range: 14.3–17.9 ppb fresh wt.   (1774)

—Bioaccumulation:
    BCF: 5 aquatic molluscs: 1200–9000
            eastern oyster: 700–70,000   (1870)
    BCF in aquatic model ecosystem:

alga: 4720
Daphnia pulex: 2560
snail (*Cipangopaludina japonica Martens*): 3660
carp (*Cyprinus carpio*): 2390 (1834)
BCF: fishes: 12,000–40,000 (1924)
      croakers: 20,000 (1897)
      trout: 200 (1898)
      oysters: 15,000 (sum of DDT) (1921)
      pinfish: 12,000 (1897)
BCF: eastern oyster: 700–70,000 (1923)
     5 aquatic molluscs: 1,200–9,000 (1870)
BCF*: brine shrimp: 6,184 (at 0.5 ppb in water)
    mosquito larvae: 16,700–21,500 (at 0.85–1.4 ppb in water)
    fish (silverside): 218 (at 2.1 ppb in water)
*DDT introduced into system in the form of residues on sand (1880)

	exposure	bioaccumulation	
copepod (*Calanus finmarchicus*)	36.3 µg/g: 20 days	11.4 µg/g	
	355 µg/g: 20 days	107.3 µg/g	
	1 337 µg/g: 7 days	104.31 µg/g	(1163)
fish (*Brevoortia tyrannus*)	Diet— 48 days		
	0.58 ppb	1.1 ppb	
	9.0 ppb	11 ppb	
	93 ppb	120 ppb	(1164)

diatom: *Thalassiosira fluviatilis*: bioconcentration coefficient: 25,000 in whole cells

dinoflagellate: *Amphidinium carteri*: bioconcentration coefficient: 80,000 in whole cells (1568)

—Distribution and metabolism of radiolabeled DDT after 33 days in a model ecosystem:

| | | | concentration, ppm | |
| | | | mosquito larva | |
	H_2O	snail (*Physa*)	dry wt.	fish (*Gambusia*)
DDT–total ^{14}C	0.004	22.9	8.9	54.2
DDT	0.00022	7.6	1.8	18.6
DDE	0.00026	12.0	5.2	29.2
DDD	0.00012	1.6	0.4	5.3
polar metabolites	0.0032	0.98	1.5	0.85

(1312)

fresh water mussel: *Anodonta grandis*: bioconc. factor: 2400
                                   half-life: 13.6 days (1680)

—Toxicity:
—algae:

Dunaliella euchlora	1000 ppb	42% reduction in O_2 evolution (2325)
Dunaliella euchlora	100 ppb	32% reduction in O_2 evolution (2325)
Dunaliella euchlora	10 ppb	30% reduction in O_2 evolution (2325)

Phaeodactylum tricornutum		1000 ppb	35% reduction in O_2 evolution (2325)
Skeletonema costatum		1000 ppb	39% reduction in O_2 evolution (2325)
Skeletonema costatum	O_2 production measured by Winkler Light-and-Dark	100 ppb	32% reduction in O_2 evolution (2325)
Skeletonema costatum	Bottle Technique. Length of test 4 hr.	10 ppb	36% reduction in O_2 evolution (2325)
Cyclotella nana		1000 ppb	33% reduction in O_2 evolution (2325)
Cyclotella nana		100 ppb	33% reduction in O_2 evolution (2325) Effect of toxicant on growth of phytoplankton

—Crustaceans
  Gammarus lacustris: 96 hr, LC_{50}: 1.0 µg/l   (2124)
  Gammarus fasciatus: 96 hr, LC_{50}: 0.8 µg/l   (2126)
  Palaemonetes kadiakensis: 96 hr, LC_{50}: 2.3 µg/l   (2126)
  Orconectes nais: 96 hr, LC_{50}: 0.24 µg/l   (2126)
  Asellus brevicaudus: 96 hr, LC_{50}: 4.0 µg/l   (2126)
  Simocephalus serrulatus: 48 hr, LC_{50}: 2.5 µg/l   (2127)
  Daphnia pulex: 48 hr, LC_{50}: 0.36 µg/l   (2127)
  pink shrimp (*Penaeus duorarum*): 28 day flowing lab bioassay: TL_{50} (technical grade): 0.12 ppb   (2355)
  Korean shrimp (*Palaemon macrodactylus*): 96 hr static lab bioassay: TL_{50} (77%): 0.86 (0.47-1.59) ppb   (2353)
  Korean shrimp (*Palaemon macrodactylus*): 96 hr flowing lab bioassay: TL_{50} (77%): 0.17 (0.09-0.32) ppb   (2353)
  sand shrimp (*Crangon septemspinosa*): 96 hr static lab bioassay: LC_{50} (p,p'-isomer): 0.6 ppb   (2327)
  grass shrimp (*Palaemonetes vulgaris*): 96 hr static lab bioassay: LC_{50} (p,p'-isomer): 2. ppb   (2327)
  blue crab (*Callinectes sapidus*): 96 hr static lab bioassay: TLm: 19. (9.–36.) ppb   (2342)
  blue crab (*Callinectes sapidus*): 96 hr static lab bioassay: TLm: 35. (21-57) ppb   (2342)
  hermit crab (*Pagurus longicarpus*): 96 hr static lab bioassay: LC_{50} (p,p'-isomer): 6 ppb   (2327)
  Daphnia magna: 24 hr, LC_{50}: 4.4 µg/l
              48 hr, LC_{50}: 0.36 µg/l   (1002; 1004)

—Insects
  Pteronarcys californica: 96 hr, LC_{50}: 7.0 µg/l   (2128)
  Pteronarcella badia: 96 hr, LC_{50}: 1.9 µg/l   (2128)
  Claassenia sabutosa: 96 hr, LC_{50}: 3.5 µg/l   (2128)
  Pteronarcys: 48 hr, LC_{50}: 7 µg/l   (1003)
  Hydropsyche larvae: significant modification of net construction after 48 hr exposure: 1.0 µg/l   (1006; 1007; 1008)

fourth instar larval *Chironomus riparius*: 24 hr, LC_{50}: 4.7 µg/l (1853)
—Fish:
Notopterus notopterus (9-21 cm): 24 hr, LC_{50}, S: 0.084 mg/l
N. notopterus: 48 hr, LC_{50}, S: 0.062 mg/l
N. notopterus: 96 hr, LC_{50}, S: 0.043 mg/l
Colisa fasciatus (5-8 cm): 24 hr, LC_{50}, S: 0.150 mg/l
C. fasciatus: 48 hr, LC_{50}, S: 0.132 mg/l
C. fasciatus: 96 hr, LC_{50}, S: 0.126 mg/l (1105)
Salmo gairdneri: 96 hr, LC_{50}: 9.6 µg/l (1001)
mummichog (*Fundulus heteroclitus*): 96 hr static lab bioassay: LC_{50} (p,p'-isomer): 3 ppb (2338)
mummichog (*Fundulus heteroclitus*): 96 hr static lab bioassay: LC_{50} (p,p'-isomer): 5 ppb (2339)
striped kilifish (*Fundulus majalis*): 96 hr static lab bioassay: LC_{50} (p,p'-isomer): 1 ppb (2339)
atlantic silverside (*Menidia menidia*): 96 hr static lab bioassay: LC_{50} (p,p'-isomer): .4 ppb (2339)
striped mullet (*Mugil cephalus*): 96 hr static lab bioassay: LC_{50} (p,p'-isomer): .9 ppb (2339)
american eel (*Anguilla rostrata*): 96 hr static lab bioassay: LC_{50} (p,p'-isomer): 4 ppb (2339)
bluehead (*Thalassoma bifasciatum*): 96 hr static lab bioassay: LC_{50} (p,p'-isomer): 7 ppb (2339)
northern puffer (*Sphaeroides maculatus*): 96 hr static lab bioassay: LC_{50} (p,p'-isomer): 89. ppb (2339)
Pimephales promelas: 96 hr, LC_{50}: 19 µg/l (2121)
Lepomis macrochirus: 96 hr, LC_{50}: 8 µg/l (2121)
Lepomis microlophus: 96 hr, LC_{50}: 5 µg/l (2121)
Micropterus salmoides: 96 hr, LC_{50}: 2 µg/l (2121)
Salmo gairdneri: 96 hr, LC_{50}: 7 µg/l (2121)
Salmo gairdneri: 15 day LC_{50}: 0.26 µg/l (2137)
Salmo trutta: 96 hr, LC_{50}: 2 µg/l (2121)
Oncorhynchus kisutch: 96 hr, LC_{50}: 4 µg/l (2121)
Perca flavescens: 96 hr, LC_{50}: 9 µg/l (2121)
Ictalurus punctatus: 96 hr, LC_{50}: 16 µg/l (2121)
Ictalurus melas: 96 hr, LC_{50}: 5 µg/l (2121)
threspine stickleback (*Gasterosteus aculeatus*): 96 hr static lab bioassay: TLm (p,p'-isomer): 11.5 ppb (2333)
shiner perch (*Cymatogaster aggregata*): 96 hr static lab bioassay: TL_{50} (technical grade): 7.6 ppb (2354)
dwarf perch (*Micrometrus minimus*): 96 hr static lab bioassay: TL_{50} (technical grade): 4.6 ppb (2354)
shiner perch (*Cymatogaster aggregata*): 96 hr inter. flow lab bioassay: TL_{50} (p,p'-isomer): .45 (0.21–0.94) ppb (2354)
dwarf perch (*Micrometrus minimus*): 96 hr inter. flow lab bioassay: TL_{50} (p,p'-isomer): 0.26 (0.13–0.52) ppb (2354)
Pagurus longicarpus: 96 hr, LC_{50}: 6 µg/l (1575)

rainbow trout fry (*Salmo gairdneri*):

mg/l	24 hr	48 hr	96 hr	3 m (extrapolated)	
LC_{10} (F)	0.008	0.0019	0.0008		
LC_{50} (F)	0.103	0.0058	0.0024	0.0001	(331)

96 hr, LC_{50}			96 hr, LC_{50}	
catfish	16 µg/l	bluegill	8 µg/l	
bullhead	5 µg/l	bass	2 µg/l	
goldfish	21 µg/l	rainbow	7 µg/l	
minnow	19 µg/l	brown	2 µg/l	
carp	10 µg/l	coho	4 µg/l	
sunfish	5 µg/l	perch	9 µg/l	(1934)

bluegill: 96 hr, LC_{50}: 0.016 ppm
rainbow trout: 96 hr, LC_{50}: 0.018 ppm (1878)
susceptible mosquito fish: 48 hr, LC_{50}: 19 ppb
resistant mosquito fish: 48 hr, LC_{50}: 96 ppb (1851)

—Ni_2B-catalyzed dechlorination of p,p'-DDT yielded a product mixture consisting principally of: 1,1-*bis*(*p*-chlorophenyl)-2-chloroetane; 1,1-*bis*(*p*-chlorophenyl)-ethane; 1-*p*-chlorophenyl-1-phenylethane; and 1,1-diphenylethane.

—Acute toxicity of chlorinated and dechlorinated p,p'-DDT:

		LC_{50} µg/l[a]		detoxification
species	formulation	chlorinated	dechlorinated	factor[b]
bluegill	technical	3.4	3472	1021
	25% emulsifiable concentrate	9.0	1519	169
daphnia	technical	1.1	170	155
	25% emulsifiable concentrate	1.7	101	59

[a] 96 hr LC_{50} for bluegill, 48 hr EC_{50} for daphnia; calculated on basis of insecticide present in formulation
[b] factor = LC_{50} dechlorinated/LC_{50} chlorinated (1413)

—Mammals:
  acute oral LD_{50} (rat): 113 mg/kg (techn. grade) (1854)
  acute dermal LD_{50} (female rat): 2510 mg/kg (1855)
  Though stored in the body fat and excreted in the milk, seventeen human volunteers ate 35 mg/man/day, about 0.5 mg/kg/day for 21 months without suffering ill effects. (1855)

DEA *see* diethanolamine

decachloro-octahydro-1,3,4-metheno-2H-cyclobuta-(c,d)-pentalen-2-one *see* kepone

decahydronaphthalene *see* decalin

dearcide 706
  Use: microbicide; active ingredient propylenediamineacetate (33%)

D. BIOLOGICAL EFFECTS:
—Harlequin fish (*Rasbora heteromorpha*):

mg/l	24 hr	48 hr	96 hr	3 m (extrap.)	
LC_{10} (F)	3.8	2.5	2.5		
LC_{50} (F)	5.6	4.6	3.3	1.4	(331)

decalin (decahydronaphthalene)

Uses: solvent for oils, fats, waxes, resins, rubbers etc.
A. PROPERTIES: m.w. 138.25; m.p. −43°C(*cis*), −31°C(*trans*); b.p. 194°C(*cis*), 186°C-(*trans*); v.p. 1 mm at 23°C; v.d. 4.76; sp.gr. 0.8967 at 20°C; THC 1,502 kcal/mole, LHC 1,417 kcal/mole
B. AIR POLLUTION FACTORS: 1 mg/cu m = 0.17 ppm, 1 ppm = 5.75 mg/cu m
C. WATER POLLUTION FACTORS:

—BOD_5:	0% of ThOD	(274)
—COD:	5% of ThOD (0.05 n Cr_2O_7)	(274)
—$KMnO_4$:	0% of ThOD (−0.01 n $KMnO_4$)	(274)
	0.02	(30)
—ThOD:	3.362	(30)

—Biodegradation: degradation in seawater by oil oxidizing micro-organisms: 13.6% breakdown after 21 days at 22°C in stoppered bottles containing a 1000 ppm mixture of alkanes, cycloalkanes, and aromatics (1237)
—Reduction of amenities: T.O.C.: 0.1 mg/l (301)
D. BIOLOGICAL EFFECTS:
—Mussel larvae (*Mytilus edulis*):
±20% reduction of growth rate at 10 ppm and 50 ppm
±5% reduction of growth rate at 100 ppm (475)
—Mammalia: rat: inhalation: no effect level: 200 ppm, 20 × 6 hr (65)
guinea pig: inhalation: 1/3: 319 ppm, 8 hr/day, 1 day
2/3: 319 ppm, 8 hr/day, 21 days
3/3: 319 ppm, 8 hr/day, 23 days (210)

decamethrin (*see also* pyrethroids)
D. BIOLOGICAL EFFECTS:
—Fish:
rainbow trout: 48-h LC_{50} 0.5 µg/l, S (1514)
desert Pupfish (*Cyprinodon macularius*): 48-h LC_{50} 0.6 µg/l, S
Mosquitofish: 48-h LC_{50} 1.0 µg/l, S
Tilapia mossambica: 48-h LC_{50} 0.8 µg/l, S

n-decane
$CH_3(CH_2)_8CH_3$
Manufacturing source: petroleum refining. (347)
Users and formulation: organic synthesis; solvent; standardized hydrocarbon; jet fuel research; mfg. paraffin products; rubber industry; paper processing industry;

constituent in polyolefin manufacturing wastes. (347)
Natural sources (water and air): constituent in paraffin fraction of petroleum.
(347)
A. PROPERTIES: m.w. 142.28; m.p. -32/-30°C; b.p. 173/174°C; v.p. 2.7 mm at 20°C; v.d. 4.90; sp.gr. 0.730 at 20/4°C; solub. 0.009 mg/l at 20°C in distilled water, 0.087 mg/l at 20°C in salt water
B. AIR POLLUTION FACTORS:
  —Manmade sources:
  glc's in carpark: 140 µg/cu m
   on motorway: 1,060 µg/cu m (48)
  —Odor threshold: 11.3 mg/cu m (737)
  —Evaporation: first order evaporation constants of *n*-decane in 3 mm layer No. 2 fuel oil—darkened room—wind speed: 21 km/hr: at 5°C: 1.19×10^{-3} min^{-1}
  10°C: 1.87×10^{-3} min^{-1}
  20°C: 3.44×10^{-3} min^{-1}
  30°C: 6.98×10^{-3} min^{-1}
  (438)
C. WATER POLLUTION FACTORS:
  —Biodegradation: degradation in seawater by oil-oxidizing micro-organisms: 100% breakdown after 21 days at 22°C in stoppered bottles containing a 1000 ppm mixture of alkanes, cyclo-alkanes, and aromatics (1237)
  —Waste water treatment: A.S.: after 6 hr: 1.3% of ThOD
   12 hr: 2.6% of ThOD
   24 hr: 4.7% of ThOD (88)
  Rotating disk contact aerator: infl. 203.7 mg/l, effl. 0.7 mg/l; elimination: >99% or 16,757 mg/m^2/24 hr or 4,525 g/cu m/24 hr (406)
  —Reduction of amenities: T.O.C. = 10 mg/l (295)
D. BIOLOGICAL EFFECTS:
  —Mussel larvae (*Mytilus edulis*):
   no significant alteration of growth rate at 10 ppm
   ±80% increase of growth rate at 50 to 100 ppm (475)

decanoic acid   *see n-capric acid*

1-decanol   (*n*-decylalcohol; nonylcarbinol)
$CH_3(CH_2)_8CH_2OH$
A. PROPERTIES: m.w. 158.28; m.p. -7°C; b.p. 231°C; v.p. 1 mm at 70°C; v.d. 5.34; sp.gr. 0.83 at 20/4°C
B. AIR POLLUTION FACTORS: 1 mg/cu m = 0.154 ppm, 1 ppm = 6.47 mg/cu m
   odor thresholds   mg/cu m

(307; 610; 761; 787)

O.I. at 20°C: 31,000 (316)
C. WATER POLLUTION FACTORS:
 —A.S.: after 6 hr: 0.9% of ThOD
    12 hr: 9.2% of ThOD
    24 hr: 29.3% of ThOD (88)
D. BIOLOGICAL EFFECTS:
 —Mammalia: mouse: single oral LD_{50}: 6.4–12.8 g/kg (*n*-decyl + *sec*.decyl)
    rat: single oral LD_{50}: 12.8–25.6 g/kg (*n*-decyl + *sec*.decyl)
    rat: inhalation: no deaths: 905 ppm, 6 hr (*n*-decyl + *sec*.decyl)
 (211)

1-decene (decylene)
$CH_2CH(CH_2)_7CH_3$
A. PROPERTIES: colorless liquid; m.p. −66.3°C; b.p. 172°C; sp.gr. 0.7396 at 20/4°C; v.p. 1 mm at 14.7°C, 10 mm at 53.7°C; 40 mm at 83.3°C
B. AIR POLLUTION FACTORS:
 —Odor: T.O.C. = 0.066 mg/cu m = 11.3 ppb (307)
    = 0.12 ppm (279)
  O.I. at 20°C = 23,000,000 (316)
 —Sampling and analysis:
  scrubbers: liquid lift type; 15 ml of liquid, 4.5 l/min gas flow; 4 min. scrubbing

trap method	% removal
H_2O, 0°C	0
NH_2OH soln, 0°C	0
H_2SO_4 (conc.), 55°C	69
H_2SO_4, 4% Ag_2SO_4 (conc.), 55°C	86
open tube, −80°C	89
ethanol, −80°C	90

(311)
  Rotating disk contact aerator: infl. 203.3 mg/l, effl. 0.8 mg/l; elimination: >99% or 16,721 mg/m²/24 hr or 4,515 g/cu m/24 hr (406)

***n*-decoic acid** *see n*-capric acid

***n*-decylalcohol** *see* 1-decanol

***prim-n*-decylchloride** *see* 1-chlorodecane

decylene *see* 1-decene

***n*-decylic acid** *see n*-capric acid

DEF *see* S,S,S-tributylphosphorotrithioate

DEG *see* diethyleneglycol

DEHP *see* dioctylphthalate

dehydroabietic acid
 Uses: basis for thermoplastic resins

D. BIOLOGICAL EFFECTS:
—Fish:
coho salmon:	96 hr, LC_{50}: 1.38–1.76 mg/l (static bioassay)	
rainbow trout:	96 hr, LC_{50}: 1.03–1.74 mg/l (static bioassay)	
sockeye salmon:	96 hr, LC_{50}: 1.18–1.38 mg/l (static bioassay)	(952)
	96 hr, LC_{50}: 0.5 mg/l (flow through bioassay)	(953)

rainbow trout: flow through bioassay:
24 hr TLm: 1.35–1.99 mg/l at 20–25°C
96 hr TLm: 0.65–0.92 mg/l at 20–25°C (442)
rainbow trout: static bioassay: 96 hr TLm: 1.1 mg/l (441)
coho salmon juvenile: 96 hr LC_{50}: 0.75 mg/l (1495)

dehydroabietol
D. BIOLOGICAL EFFECTS:
—Fish: rainbow trout: 96 hr, LC_{50}: 0.8 mg/l, S (441; 1495)

dehydrojuvabione
D. BIOLOGICAL EFFECTS:
—Fish: rainbow trout: renewal bioassay: 96 hr TLm: 1.8 mg/l at 8°C (449)

Δ4′-dehydrojuvabione
D. BIOLOGICAL EFFECTS:
—Fish: rainbow trout: renewal bioassay: 96 hr TLm: 0.8 mg/l at 8°C (449)

delnav *see* dioxathion

demeton (common name for a mixture of demeton-O and Demeton-S; demeton-O = diethyl-2-(ethylthio)ethylphosphorothionate(I); demeton-S = diethyl-2-(ethylthio)-ethylphosphorothiolate(II); systox)

  (I) (II)
  $(C_2H_5O)_2PS \cdot OCH_2CH_2SC_2H_5$ $(C_2H_5O)_2PO \cdot SCH_2CH_2SC_2H_5$

Use: systemic insecticide–acaricide
A. PROPERTIES: demeton-O is a colorless oil of b.p. 123°C at 1 mm; sp.gr. 1.12; v.p. 2.6×10^{-4} mm at 20°C; solub. 2000 ppm at room temp.
D. BIOLOGICAL EFFECTS:
—Crustaceans
Gammarus fasciatus: 96 hr, LC_{50}: 27 µg/l (2126)
—Fish
Pimephales promelas: 96 hr, LC_{50}: 3200 µg/l (2123)
Lepomis macrochirus: 96 hr, LC_{50}: 100 µg/l (2123)
—Mammals:
acute oral LD_{50} (male rats): 30 mg/kg of (I)
1.5 mg/kg of (II) (1855)

DEP *see* diethylphthalate

devrinol (2-(α-naphthoxy)-N,N-diethylpropionamide)

450 DEXON

$$H_3C-CH-C\underset{O}{\overset{O}{\diagdown}}\underset{C_2H_5}{\overset{C_2H_5}{N}}$$

(attached to 1-naphthyloxy)

Use: herbicide
Formulations: 50W-A, a wettable powder containing 50% active ingredient by weight; =2E, an emulsifiable liquid containing 2 lbs active ingredient per gallon; 5G and 10G, granular formulations containing 5 and 10% active ingredient; 42 + 3.5 WP, 40 + 5 WP, and 42 + 7 WP, wettable powders containing 42% + 3.5%, 40% + 5% and 42% + 7% active ingredient of devrinol + simazine, respectively.

A. PROPERTIES: brown solid, m.p. 69.5°C; m.w. 271.36; solub. 73 ppm at 20°C
D. BIOLOGICAL EFFECTS:
 —Fish: goldfish: 96 hr, LC_{50}: >10.0 ppm (495)
 —Mammals:
  male and female rat: acute oral LD_{50}: technical grade: >5000 mg/kg
   50 W: >4640 mg/kg
  rabbit: acute dermal LD_{50}: technical grade: >2000 mg/kg
   50 W: >3640 mg/kg
   acute ocular irritancy: non irritating
   dermal irritancy: non irritating (495)

dexon (4-(dimethylamino)phenyl)diazenesulfonate; sodium fenaminosulf)

$$(CH_3)_2N-\bigcirc-N=N-SO_3Na$$

A. PROPERTIES: yellow brown powder, decomposes above 200°C; solub. 2-3% at 25°C
C. WATER POLLUTION FACTORS:
 —Aquatic reactions:

$$(CH_3)_2N-\bigcirc-N=NSO_3Na \xrightarrow{H_2O} (CH_3)_2N-\bigcirc-OH + (CH_3)_2N-\bigcirc$$
$$+ N_2/H_2$$

photolysis of dexon

D. BIOLOGICAL EFFECTS:
 —Crustacean: *Gammarus lacustris*: 96 hr, LC_{50}: 3.7 mg/l (2124)
 —Insect: *Pteronarcys californica*: 96 hr, LC_{50}: 24.0 mg/l (2128)
 —Mammals: acute oral LD_{50} (rat): 64 mg/kg
   dermal LD_{50} (rat): 100 mg/kg (1854)

dextronic acid *see* d-gluconic acid

diacetic ester *see* ethylacetoacetate

diacetone *see* diacetonealcohol

diacetonealcohol   (4-hydroxy-4-methyl-2-pentanone; diacetonylalcohol; diacetone; dimethylacetonylcarbinol; pyranton A; DAA)
$(CH_3)_2C(OH)CH_2COCH_3$
A. PROPERTIES: colorless liquid, becomes yellow on aging; m.w. 116.16; m.p. $-57/-43°C$; b.p. $166/169°C$; v.p. 1 mm at $20°C$. 1.7 mm at $30°C$; v.d. 4.00; sp.gr. 0.93 at $25/4°C$, sat.conc. 5.7 g/cu m at $20°C$, 10 g/cu m at $30°C$
B. AIR POLLUTION FACTORS: 1 mg/cu m = 0.216 ppm, 1 ppm = 4.75 mg/cu m
   —Odor: characteristic: quality: sweet
                         hedonic tone: unpleasant to pleasant
   T.O.C.: absolute: 0.28 ppm
          50% recogn.: 1.1 ppm
         100% recogn.: 1.7 ppm
         O.I. 100% recogn.: 776   (19)
         O.I. at $20°C$ = 774   (316)
C. WATER POLLUTION FACTORS:
   —BOD_5: 0.07 = 3% of ThOD   (277)
            0.68 = 31% of ThOD after adaptation   (277)
   —COD: 2.11 = 95% of ThOD   (277)
   —Reduction of amenities: T.O.C. in water at room temp.: 44.12 ppm, range: 5.63–269 ppm, 9 judges
     20% of population still able to detect odor at 4.6   ppm
     10% of population still able to detect odor at 1.4   ppm
       1% of population still able to detect odor at 0.032 ppm
     0.1% of population still able to detect odor at 0.76   ppb
D. BIOLOGICAL EFFECTS:
   —Toxicity threshold (cell multiplication inhibition test):
     bacteria (*Pseudomonas putida*):   825 mg/l   (1900)
     algae (*Microcystis aeruginosa*):   530 mg/l   (329)
     green algae (*Scenedesmus quadricauda*): 3000 mg/l   (1900)
     protozoa (*Entosiphon sulcatum*):   1400 mg/l   (1900)
   —Fish:
   Lepomis macrochirus: static bioassay in fresh water at $23°C$, mild aeration applied after 24 hr:

material added		% survival after			best fit 96 hr LC_{50}
ppm	24 hr	48 hr	72 hr	96 hr	ppm
560	90	80	20	10	
420	50	50	50	50	420
320	100	100	100	100	

   Menidia beryllina: static bioassay in synthetic seawater at $23°C$: mild aeration applied after 24 hr:

material added		% survival after			best fit 96 hr LC_{50}
ppm	24 hr	48 hr	72 hr	96 hr	ppm
560	40	20	20	20	
420	80	40	30	30	420
320	100	100	100	100	(352)

     goldfish: LD_{50} (24 hr): >5000 mg/l   (277)

DIACETYLMETHANE

—Mammalia: rat: single oral dose: LD_{50}: 4.0 g/kg
repeated oral dose; no effect: 0.04 g/kg/day, 30 days
inhalation: no deaths: 1,500 ppm; 8 hr (211)
—Man: irritation of eyes, nose and throat at 100 ppm (211)
estimated lethal dose: 30 g (277)

diacetylmethane *see* 2,4-pentanedione

dialkyltetralinindanesulfonate
D. BIOLOGICAL EFFECTS:
—Fish:
fathead minnow: 24-hr LC_{50}, S: 87.0 mg/l   alkyl = C-10
48-hr LC_{50}, S: 86.1 mg/l
fathead minnow: 24-hr LC_{50}, S: 24.8 mg/l   alkyl = C-12
48-hr LC_{50}, S: 21.5 mg/l
fathead minnow: 24-hr LC_{50}, S: 8.1 mg/l   alkyl = C-14
48-hr LC_{50}, S: 5.3 mg/l   (1191)

diallate (S-2,3-dichloroallyl-N,N-di-isopropylthiocarbamate)

$$(CH_3)_2CH \diagdown \overset{O}{\underset{\|}{}} \\ N\overset{\|}{C}-S-CH_2CCl=CCl \\ (CH_3)_2CH \diagup$$

Use: a thiolcarbamate herbicide
C. WATER POLLUTION FACTORS:
—Degradation:
over 50% loss after 4 weeks incubation in microbiologically active soils with rapid formation of CO_2
losses from sterile soils where much slower 30–45% after 20 weeks with only very slow CO_2 formation
representative fungi isolated from soils *Phoma eupyrena*, *Penicillium janthinellum* and *Trichoderma hazzianum* could degrade at least 20% of the applied (2.5 ppm) herbicide after 10 days incubation (1625)
D. BIOLOGICAL EFFECTS:
—Mammals: acute oral LD_{50} (rat): 395 mg/kg (1854)

diallylamine (di-2-propenylamine)
$(CH_2CHCH_2)_2NH$
A. PROPERTIES: m.w. 97.16; m.p. -88.4°C; b.p. 111/112°C; v.d. 3.35; sp.gr. 0.763 at 10/4°C; solub. 86 g/l
B. AIR POLLUTION FACTORS: 1 mg/cu m = 0.252 ppm, 1 ppm = 3.97 mg/cu m
—Odor threshold: 8 mg/cu m (688)
C. WATER POLLUTION FACTORS:
—Biodegradation rates:
degradation by *Aerobacter*: 200 mg/l at 30°C:
parent: 62% degradation in 105 hr
mutant: 100% degradation in 17 hr (152)

D. BIOLOGICAL EFFECTS:
—Mammalia: rat: acute oral LD_{50}: 578 mg/kg
  inhalation: LC_{50}: 2,755 ppm, 4 hr
  inhalation: LC_{50}: 795 ppm, 8 hr
 rat: inhalation: deaths: 200 ppm, 50 × 7 hr
  inhalation: change in liver and kidneys: 20 ppm, 50 × 7 hr
(211)

diallylphthalate (DAP)

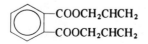

A. PROPERTIES: nearly colorless oily liquid; m.p. -70°C; b.p. 160°C at 4 mm; v.d. 8.3; sp.gr. 1.12 at 20/20°C
D. BIOLOGICAL EFFECTS:
—Toxicity threshold (cell multiplication inhibition test):
bacteria (*Pseudomonas putida*):	>100	mg/l	(1900)
algae (*Microcystis aeruginosa*):	0.65	mg/l	(329)
green algae (*Scenedesmus quadricauda*):	2.9	mg/l	(1900)
protozoa (*Entosiphon sulcatum*):	13	mg/l	(1900)
protozoa (*Uronema parduczi Chatton-Lwoff*):	22	mg/l	(1901)

diamidafos
C. WATER POLLUTION FACTORS:
—Hydrolysis:
  at pH 7.1 and 7.9, no detectable chemical hydrolysis took place within a period of 22 days
  at pH 4.1 the chemical hydrolysis rate was such that at 25°C the half-life was 39.8 days
  acid hydrolysis yielded N-methyl-hydrogen phosphoro-amidate and dihydrogen phenylphosphate
(1624)

1,2-diaminobenzene see *o*-phenylenediamine

1,3-diaminobenzene see *m*-phenylenediamine

1,4-diaminobenzene see *p*-phenylenediamine

4,4'-diaminobiphenyl see benzidine

d-α,ε-diaminocaproic acid see *l*-lysine

2,2'-diaminodiethylamine see diethylenetriamine

diaminoethane see ethylenediamine

di-*p*-aminodi-*m*-methoxydiphenyl see *o*-dianisidine

4,4'-diaminodiphenylmethane (DDM; p,p'-methylenedianiline; MDA)

$$H_2N-\langle\bigcirc\rangle-CH_2-\langle\bigcirc\rangle-NH_2$$

Technical grade contains 4% 2,4'-methyleneaniline

Uses: in preparation of isocyanates and polyisocyanates; an epoxy hardening agent; a raw material in the production of polyurethane elastomers; in the rubber industry; a curative for neoprene; an anti-frosting agent (anti-oxidant) in footwear; a raw material in the preparation of poly(amide-imide)resins (used in magnet wire enamels)

A. PROPERTIES: m.w. 198.26; m.p. 93°C; b.p. 231°C at 11 mm
C. WATER POLLUTION FACTORS:
 —KMnO$_4$ value: 12.64
 —ThOD: 2.869 (30)
D. BIOLOGICAL EFFECTS:
 —Bacteria: *Pseudomonas*: toxic: 15 mg/l
 —Algae: *Scenedesmus*: toxic: 30 mg/l
 —Protozoa: *Colpoda*: toxic: 124 mg/l
 —Arthropoda: *Daphnia*: toxic: 0.25 mg/l (30)
 —Mammals: rats: single oral LD$_{50}$: 120–250 mg/kg (1546)
  16 rats: 4–5 doses of 20 mg DDM by stomach tube over 8 months: after 18 months: 1 rat developed a hepatoma and a heamangioma-like tumor of the kidney
  after 24 months: 1 rat showed an adenocarcinoma of the uterus (433)
  50 rats: 50% developed tumors (4 hepatomas) compared with 26% of a control group
  48 rats: intragastric doses 5 times weekly: all developed liver cirrhosis, four developed hepatomas (2 benign) and others miscellaneous tumors (432)
  Hepatotoxic in humans: no reported human cancers (383)

1,6-diaminohexane see hexamethylenediamine

d-2,6-diaminohexanoic acid see *l*-lysine

2,4-diaminophenol

$$\underset{NH_2}{\underset{|}{\bigcirc}}\overset{OH}{\overset{|}{}}-NH_2$$

C. WATER POLLUTION FACTORS:
 —Biodegradation: adapted A.S. at 20°C—product is sole carbon source: 83.0% COD removal at 12.0 mg COD/g dry inoculum/hr (327)

2,4-diaminophenol hydrochloride (amidol)

[Structure: benzene ring with OH, NH$_2\cdot$HCl, and NH$_2\cdot$HCl substituents]

A. PROPERTIES: grayish white crystals
D. BIOLOGICAL EFFECTS:
 —Fish: goldfish: approx. fatal conc. 80 mg/l, 48 hr (226)

1,3-diamino-2-propanoltetraacetic acid (Dapta)
Use: chelating agent
A. PROPERTIES: m.w. 322.0
C. WATER POLLUTION FACTORS:
 —Oxidation parameters:

theoretical	analytical
TOD = 1.34	Reflux COD = 95.9 percent recovery
COD = 0.94	TKN = 78.9 percent recovery
NOD = 0.40	BOD$_5$ = 0.079
	BOD$_5$/COD = 0.088
	BOD$_5$ (acclimated) = 0.05

The BOD$_5$ analyses, using both an unacclimated and acclimated seed, measured minimal responses by DAPTA. (1828)

 —Impact on conventional biological treatment systems:

	chemical conc. mg/l	effect
unacclimated system	1,000	inhibitory
acclimated system	584	biodegradable (1828)

Only slight inhibition was exhibited by the unacclimated biomass. The compound, at the concentration tested, was found amenable to moderate biodegradation. (1828)

D. BIOLOGICAL EFFECTS:
 —Algae:
 Selenastrum capricornutum: 1 mg/l, no effect
 10 mg/l, inhibitory
 100 mg/l, inhibitory
 —Crustacean:
 Daphnia magna: LC$_{50}$ >100 mg/l
 —Fish:
 Pimephales promelas: LC$_{50}$ >300 mg/l (1828)

2,6-diaminopyridine

[Structure: pyridine ring with H$_2$N– and –NH$_2$ substituents at the 2,6 positions]

A. PROPERTIES: crystals; m.p. 121°C; b.p. 285°C
C. WATER POLLUTION FACTORS:
 —BOD_5: nil, std. dil. sew. (282)
 —$KMnO_4$: 4.677 (n.s.i.) (30)
 —ThOD: 1.470 (30)

diaminotolyl see o-tolidine

2,5-diaminovaleric acid see ornithine

di-n-amylamine (di-n-pentylamine)
$(C_5H_{11})_2NH$
A. PROPERTIES: colorless liquid; m.w. 157.29; b.p. 202°C at 745 mm; v.d. 5.42; sp.gr. 0.78 at 20°C;
D. BIOLOGICAL EFFECTS:
 —Fish: creek chub: critical range: 5–20 mg/l, 24 hr (226)

2,4-diamylphenol (1-hydroxy-2,4-diamylbenzene)

C_5H_{11}—⌬(OH)—C_5H_{11}

Commercial form is a mixture of isomers including both secondary and tertiary amyl groups mainly in 2,4-positions.
Uses: synthetic resins; lubricating oil additives; rust preventives; plasticizers; synthetic detergents; anti-oxidants; rodenticide; fungicide
A. PROPERTIES: light straw-colored liquid with mild phenolic odor; boiling range: (ASTM 5–95%) 280–295°C; sp.gr. 0.930 at 20°C
C. WATER POLLUTION FACTORS:
 —In river water: 0.001–0.005 ppm; in river sediment: 0.3–10 ppm; (2,4-di-t-amylphenol) (555)

o-dianisidine (di-p-amino-di-m-methoxydiphenyl; 3,3′-dimethoxybenzidine)

H_2N—⌬(OCH_3)—⌬(OCH_3)—NH_2

A. PROPERTIES: colorless crystals; m.p. 137°C;
C. AIR POLLUTION FACTORS:
 —Biodegradation rates:
   degradation by Aerobacter: 500 mg/l at 30°C
   % ring disruption: parent: 78% in 120 hr
                      mutant: 100% in 36 hr (152)

diazinon (diethyl-2-isopropyl-6-methyl-4-pyrimidinylphosphorothionate; O,O-diethyl-O-(2-isopropyl-4-methyl-6-pyrimidinylphosphorothioate)

$$\begin{array}{c}\text{CH}_3\\\text{CH}_3\text{>CH}-\overset{\text{N}}{\underset{\text{N}}{\bigcirc}}-\text{O}-\overset{\text{S}}{\underset{\|}{\text{P}}}\text{<}^{\text{OC}_2\text{H}_5}_{\text{OC}_2\text{H}_5}\\\text{CH}_3\end{array}$$

Uses: organophosphorus insecticide used in the rice paddy fields.
Diazinon formulations contain sulfotepp as impurity (1.4–6.9 ppm) (1591)

A. PROPERTIES: decomposes above 120°C; a colorless oil; sp.gr. 1.116; solub. 40 mg/l at room temp.

B. AIR POLLUTION FACTORS:
   —In ambient air of storage and office rooms of commercial pest control buildings in in a 4 hr period:
   storage room: avg.: 284 ng/cu m
                 range: 85–837 ng/cu m
   office room: avg.: 163 ng/cu m
                 range: 31–572 ng/cu m   (1868)

C. WATER POLLUTION FACTORS:
   —Aquatic reactions:
   persistence in soil at 10 ppm initial concentration:

	wk incubation to	
	50% remaining	5% remaining
sterile sandy loam	12.5	—
sterile organic soil	6.5	—
non-sterile sandy loam	<1	1
non-sterile organic soil	2	7 (1433)

   75–100% disappearance from soils: 12 weeks   (1815)
   acid hydrolysis (pH ~ 2) reduced diazinon concentration by >99.9%; hydrolysis products are 6-isopropyl-4-methyl-2-pyrimidinol and thiophosphate
                                                                   (1592; 1593)
   —Water and sediment quality: Mississippi: residues not found in bottom sediments
                                                                       (1170)

D. BIOLOGICAL EFFECTS:
   —Bioconcentration:
   bioconcentration ratios of diazinon by various species of freshwater organisms— exposed to 10 ppb for 7 days:

	bioconcentration ratio	
	individual	average
topmouth gudgeon (*Pseudorasbora parva*)	136, 166, 154	152
silver crucian carp (*Cyprinus auratus*)	32.7, 37.4, 39.4, 36.7	36.6
carp (*Cyprinus carpio*)	80.3, 46.0, 69.1	65.1
guppy (*Lebistes reticulatus*)	four bodys gathered	17.5
crayfish (*Procambarus clarkii*)	5.3, 4.5	4.9
red snail (*Indoplanorbis exustus*)	16.9, 16.4, 17.7	17.0
pond snail (*Cipangopoludina malleata*)	3.1, 6.2, 8.5	5.9

                                                                       (1098)
   topmouth gudgeon: bioconcentration ratio increased with body weight (exposure

at 10 ppb for 7 days) from 50 to 175 for body weights between 2 and 6 g
(1098)
BCF: fish (*Fundulus heteroclitus*): 10 (1925)
—Effect of the diazinon concentration in test water on the bioconcentration ratios by topmouth gudgeon:

diazinon in water, ppb	days after exposure	diazinon in fish, ppb	bioconcentration ratio
13.5	0		
15.0	1	1450	96.7
11.8	3	2008	170.2
11.5	7	1359	118.2
11.5	14	1725	150.0
	days after return to clean water		
(clean water)	1	719	
(clean water)	2	410	
(clean water)	4	80	
(clean water)	8	5	
	days after exposure		
53.8	0		
55.0	1	5967	108.0
51.3	2	7458	146.0
53.8	4	11299	210.0
52.5	7	10793	206.0
	days after return to clean water		
(clean water)	1	4787	
(clean water)	2	1832	
(clean water)	4	179	
(clean water)	7	26	(1098)

—Toxicity:
—Crustacean:
  copepod (*Acartia tonsa*): 96 hr, LC_{50} (S): 2.57 µg/l (1129)
  Gammarus pseudolimneaus: 30 day, LC_{50}: 0.27 µg/l; 30 day no effect level: 0.20 µg/l (2134)
  Gammarus lacustris: 96 hr, LC_{50}: 200 µg/l (2124)
  Simocephalus serrulatus: 48 hr, LC_{50}: 1.4 µg/l (2127)
  Daphnia pulex: 48 hr, LC_{50}: 0.90 µg/l (2127)
  Daphnia magna: 21 day no effect level: 0.26 µg/l (2135)
  daphnids: 48 hr, EC_{50}: 0.0020 mg/l (technical grade) (1591)
  Daphnia magna: 48 hr, LC_{50}: 0.00122–0.00125 mg/l (1927)
—Insects:
  Pteronarcys californica: 96 hr LC_{50}: 25 µg/l (2128)
  Pteronarcys dorsata: 30 day LC_{50}: 4.6 µg/l; 30 day no effect level: 3.29 µg/l
(2134)
  Acroneuria lycorias: 96 hr LC_{50}: 1.7 µg/l; 30 day LC_{50}: 1.25 µg/l; 30 day no effect level: 0.83 µg/l

Ophiogomphus rupinsulensis: 30 day LC_{50}: 2.2 µg/l; 30 day no effect level: 1.29 µg/l
Hydropsyche bettoni: 30 day LC_{50}: 3.54 µg/l; 30 day no effect level: 1.79 µg/l
Ephemerelia subvaria: 30 day LC_{50}: 1.05 µg/l; 30 day no effect level: 0.42 µg/l
rice field spider: *Oedothorax insecticeps*: LD_{50}: 2450 ppm   (1814)
—Highly toxic to bees   (1855)
—Fish:
  bluegills: 24 hr, LC_{50}: 0.052 ppm
  rainbow trout: 24 hr, LC_{50}: 0.380 ppm   (1099)
  flagfish (*Jordanella floridae*): 96-hr, LC_{50}: 1.5–1.8 mg/l (F)
  bluegill: 96-hr, LC_{50}: 0.44–0.48 mg/l (F)
  brook trout: 96-hr LC_{50}: 0.45–1.05 mg/l (F)
    240-d MATC 3.2 µg/l (F)
  fathead minnow: 96-h LC_{50}: 6.8–10.0 mg/l (F)
    274-d MATC 3.2 µg/l (F)   (1515)
  carp: 24 hr LC_{50}: 3.18 mg/l (S)
    48 hr LC_{50}: 3.14 mg/l (S)
    72 hr LC_{50}: 3.11 mg/l (S)   (1103)
  fathead minnow: 96 hr, LC_{50}: 3.7–10.0 mg/l
  bluegill: 96 hr, LC_{50}: 0.17–0.53 mg/l   (1927)
  fathead minnow: 96 hr, LC_{50}: 10.3 mg/l (techn. grade)
  bluegill: 96 hr, LC_{50}: 0.12 mg/l (techn. grade)
  rainbow trout: 96 hr, LC_{50}: 1.35 mg/l (techn. grade)   (1591)
—Mammals:
  acute oral LD_{50} (rat): 300–400 mg/kg (techn. grade)   (1854)
  acute dermal LD_{50} (male rats): 900 mg/kg
    (female rats): 455 mg/kg   (1855)

dibenz(*a,h*)acridine   (1,2–5,6-dibenzacridine)

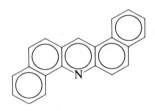

A. PROPERTIES: m.w. 278.35; m.p. 266–267°C; b.p. 524°C
B. AIR POLLUTION FACTORS:
  —Ambient air quality:
    in the average American urban atmosphere—1963:
    0.6 µg/g airborne particulates or
    0.08 ng/g cu m air   (1293)

dibenz(*a,j*)acridine   (1,2–7,8-dibenzacridine)

460 DIBENZ(a,c)ANTHRACENE

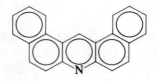

C. AIR POLLUTION FACTORS:
  —Ambient air quality:
    in the average American urban atmosphere—1963:
      0.3 µg/g airborne particulates or
      0.04 ng/g cu m air (1293)
D. BIOLOGICAL EFFECTS:
  —Carcinogenicity: +
  —Mutagenicity in the *Salmonella* test: +
      18 revertant colonies/nmol
      318 revertant colonies at 5 µg/plate (1883)

dibenz(a,c)anthracene

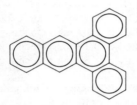

C. WATER POLLUTION FACTORS:
  —Water and sediments quality: in sediments in Severn estuary (U.K.): 0.2–1.7 ppm dry wt. (1467)
D. BIOLOGICAL EFFECTS:
  —Carcinogenicity: +
  —Mutagenicity in the *Salmonella* test: +
      175 revertant colonies/nmol
      6280 revertant colonies at 10 µg/plate (1883)

dibenz(a,h)anthracene (1,2-5,6-dibenzanthracene)

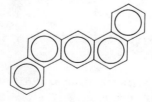

A. PROPERTIES: m.w. 278.35; m.p. 266–267°C; b.p. 524°C
  —Manmade sources:
    in wood preservative sludge (n.s.i.): 0.07 g/l of raw sludge (960)

in gasoline (high octane number): 0.167 mg/l (380)
in coal tar (n.s.i.): 2.13 mg/g of sample (993)
B. AIR POLLUTION FACTORS:
—Manmade sources:
in cigarette smoke: 0.4 µg/100 cigarettes (1298)
in exhaust condensate of gasoline engine: 96 µg/g (1069)
in airborne coal tar emissions (n.s.i.): 0.41 mg/g of sample or
14 µg/cu m of air sampled (993)
in coke oven airborne emissions: 84-124 µg/g of sample (960)
—Ambient air quality:
glc's in Birkenes (Norway)—Jan.–June 1977:
avg.: 0.05 ng/cu m
range: n.d.–0.52 ng/cu m ($n = 18$)
in Rørvik (Sweden)—Dec. 1976–April 1977:
avg.: 0.05 ng/cu m
range: n.d.–0.43 ng/cu m ($n = 21$) (1236)
in Budapest 1973: heavy traffic area: winter: 19.5 ng/cu m
(6 a.m.–8 p.m.) summer: 25.8 ng/cu m
low traffic area: winter: 15.4 ng/cu m
summer: 5.9 ng/cu m (1259)
C. WATER POLLUTION FACTORS:
—Treatment methods: ozonation: after 1 min. contact time with ozone: residual amount: 3.6% (550)
D. BIOLOGICAL EFFECTS:
—*Neanthes arenaceodentata*: 96 hr TLm in seawater at 22°C: >1 ppm (initial conc. in static assay)
—Carcinogenicity: +
—Mutagenicity in the *Salmonella* test: +
11 revertant colonies/nmol
401 revertant colonies at 10 µg/plate (1883)

dibenzocarbazole
D. BIOLOGICAL EFFECTS:
—*Daphnia pulex*: bioaccumulation coefficient: 7126 (wet wt.) (1570)

dibenzo(*b,def*)chrysene (1,2-6,7-dibenzopyrene; dibenzo(*a,h*)pyrene; 3,4-8,9-dibenzopyrene)

462   DIBENZO(def,mno)CHRYSENE

B. AIR POLLUTION FACTORS:
 —Manmade sources:
  in gasoline: 16 µg/l
  in exhaust condensate of gasoline engine: <0.2 µg/l gasoline consumed   (1070)
 —Ambient air quality: in air in Ontario cities (Canada):

location	April–June 1975		July–Oct. 1975		Oct.–Dec. 1975		Jan.–March 1976	
	ng/1000 cu m air	µg/g p.m.	ng/1000 cu m air	µg/g p.m.	ng/1000 cu m air	µg/g p.m.	ng/1000 cu m air	µg/g p.m.
Toronto	313	3.6	183	1.9	446	5.4	213	2.7
Toronto	210	2.8	109	1.7	229	3.8	490	3.6
Hamilton	331	2.3	915	6.6	1132	16.3	704	8.4
S. Sarnia	508	8.2	81	1.7	213	4.1	107	3.9
Sudbury	149	4.6	47	1.1	54	2.4	130	5.6

  p.m. = particulate matter   (999)

dibenzo(def,mno)chrysene   see anthanthrene

dibenzo(def,p)chrysene   (dibenzo(a,l)pyrene; 1,2-9,10-dibenzopyrene)

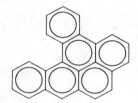

C. AIR POLLUTION FACTORS:
 —Ambient air quality: in air in Ontario cities (Canada):

location	April–June 1975		July–Oct. 1975		Oct.–Dec. 1975		Jan.–March 1976	
	ng/1000 cu m air	µg/g p.m.	ng/1000 cu m air	µg/g p.m.	ng/1000 cu m air	µg/g p.m.	ng/1000 cu m air	µg/g p.m.
Toronto	46	0.5	44	0.5	234	2.8	34	0.4
Toronto	65	0.9	38	0.6	102	1.7	162	1.2
Hamilton	70	0.5	128	0.9	369	5.3	150	1.8
S. Sarnia	23	0.4	8	0.2	44	0.8	7	0.3
Sudbury	8	0.2	9	0.2	37	1.7	32	1.4

                                                                                                   (999)

  glc's in Budapest 1973: heavy traffic area: winter: 108 ng/cu m
  (6 a.m.–8 p.m.)   summer: 114 ng/cu m
                        low traffic area: winter: 3.2 ng/cu m
                                           summer: 3.3 ng/cu m

dibenzo-p-dioxine   see dioxin

dibenzopyran   see xanthene

dibenzo(a,h)pyrene *see* dibenzo(b,def)chrysene

dibenzo(a,l)pyrene *see* dibenzo(def,p)chrysene

dibenzo(cd,jk)pyrene *see* anthanthrene

dibenzo(a,i)pyrene
B. AIR POLLUTION FACTORS:
—Manmade sources:
in coke oven emissions: 493.3–693.2 µg/g of sample (n.s.i.) (960)
D. BIOLOGICAL EFFECTS:
—Carcinogenicity: +
—Mutagenicity in the *Salmonella* test: +
20 revertant colonies/nmol
1290 revertant colonies at 20 µg/plate (1883)

3,4-9,10-dibenzopyrene (dibenzo(cd,l)pyrene)

B. AIR POLLUTION FACTORS:
—Ambient air quality:
glc's in Budapest 1973: heavy traffic area: winter: 14.4 ng/cu m
(6 a.m.–8 p.m.) summer: 11.5 ng/cu m
winter: 0.9 ng/cu m
summer: 0.3 ng/cu m (1259)

dibenzopyrrole *see* carbazole

dibenzothiophene
C. WATER POLLUTION FACTORS:
—Manmade sources:
in Philadelphia stormwater runoff: 44.2–62.3 ng/l (1321)
D. BIOLOGICAL EFFECTS:
—Inhibition of photosynthesis of a freshwater, non axenic unialgal culture of Selenastrum *capricornutum*:
at 1% saturation: 94% carbon-14 fixation (vs. controls)
10% saturation: 101% carbon-14 fixation (vs. controls)
100% saturation: 80% carbon-14 fixation (vs. controls) (1690)

dibenzyl *see* bibenzyl

dibromochloromethane
$CHBr_2Cl$

Users and formulations: mfg. fire extinguishing agents; mfg. aerosol propellants; mfg. refrigerants; mfg. pesticides; organic synthesis. (347)

A. PROPERTIES: clear colorless to pale yellow heavy liquid; m.w. 208.3; m.p. $<-20°C$; b.p. $116-122°C$; sp.gr. 2.38

C. WATER POLLUTION FACTORS:
—In N.W. England tap waters (1974): $<0.01-3$ ppb (933)
—Treatment methods:
experimental water reclamation plant:

	sample 1	sample 2	
sand filter effluent	83 ng/l	84 ng/l	
after chlorination	14,547 ng/l	15.068 ng/l	
final water after A.C.	407 ng/l	165 ng/l	(928)

1,2-dibromo-3-chloropropane (BBCP; DBCP; 3-chloro-1,2-dibromopropane)
$BrCH_2CHBrCH_2Cl$

Uses: soil fumigant; nematocide; intermediate in organic synthesis; LV grade (low in volatiles) commercial preparations of the flame retardant *tris*[(2,3-dibromopropyl)phosphate] contained DBCP in the order of 0.05% (1346; 1347)

A. PROPERTIES: amber to dark brown liquid with a mildly pungent odor; b.p. $196°C$; v.p. 0.8 mm at $21°C$; sp.gr. 2.08 at $20/20°C$; solub. 1000 mg/l at room temp.

C. WATER POLLUTION FACTORS:
—Aquatic reactions:
DBCP is converted by soil water cultures to *n*-propanol. The maximum conversion rate observed ($Br^-/2DBCP_0$) is 63% in the course of 4 weeks—initial conc. $10^{-3} M$ DBCP:

$$CH_2Br-CHBr-CH_2Cl \xrightarrow[\text{pH 7.5}]{\text{soil-}H_2O} CH_3CH_2CH_2OH + 2\ Br^- + Cl^- \quad (1318)$$

—Persistence:
fields treated with DBCP contained still 2 to 5 ppb in the topsoil after 2 to 4 years (1866)
—Water quality:
in well water—California, 1979: n.d. in 64% of the wells
0.1-0.9 ppb in 13% of the wells
1.0-9.9 ppb in 19% of the wells
10-19.9 ppb in 3% of the wells
>20 ppb in 0.8% of the wells

D. BIOLOGICAL EFFECTS:
—Mussels:
hard clam (*Mercenaria mercenaria*): egg: 48 hr, TLm S: 10 mg/l
hard clam (*Mercenaria mercenaria*): larvae: 12 d, TLm S: 0.78 mg/l (2324)
—Mammals:
acute oral LD_{50} (male rats): 173 mg/kg
(male mice): 257 mg/kg
(female mice): 270-620 mg/kg
acute dermal LD_{50} (rabbits): 1,420 mg/kg

in diet: in 90 day feeding tests with rats, the lowest level causing a decrease in growth rate was 150 ppm for females and 450 ppm for males (1855)
—Man: occurrence of primary disruption of spermatogenesis at the testicular level for all users who had extensive exposure to the compound—among them formulators, custom applicators, and farmers (1676)

1,2-dibromoethane *see* ethylenebromide

2,4-dibromophenol

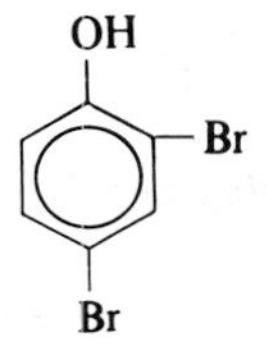

A. PROPERTIES: m.w. 251.92; m.p. 40°C; b.p 177°C at 17 mm; solub. 1,900 mg/l at 15°C; log P_{oct} 2.56 (calculated)
C. WATER POLLUTION FACTORS:
—Waste water treatment:
degradation by *Pseudomonas*: 200 mg/l at 30°C
parent: 75% ring disruption in 72 hr
mutant: 100% ring disruption in 10 hr (152)

2,5-dibromophenol

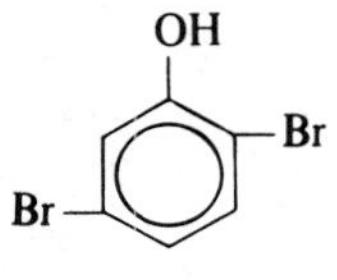

C. WATER POLLUTION FACTORS:
—Waste water treatment:
degradation by *Pseudomonas*: 200 mg/l at 30°C
parent: 58% ring disruption in 120 hr
mutant: 100% ring disruption in 35 hr (152)

2,3-dibromopropylphosphate
D. BIOLOGICAL EFFECTS:
—Goldfish: 5 days LC_{100}: 1.0 mg/l (static bioassay) (950)

2,5-dibromotoluene

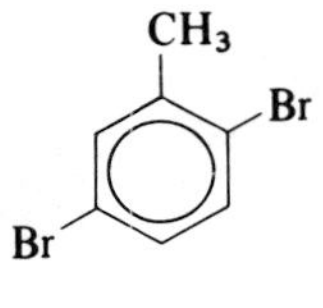

D. BIOLOGICAL EFFECTS:
—Juvenile atlantic salmon: BCF at 14 μg/l: 470
excretion half-life: 90 hr (1852)

2,5-dibromoxylene
D. BIOLOGICAL EFFECTS:
—Juvenile atlantic salmon: BCF at 30 μg/l: 1430
excretion half-life: 86 hr (1852)

di-N-butylamine
$(CH_3CH_2CH_2CH_2)_2NH$
A. PROPERTIES: m.w. 129.3; m.p. $-51°C$; b.p. $160°C$; v.d. 4.46; sp.gr. 0.76; solub. 3,100 mg/l; log P_{oct} 2.68
B. AIR POLLUTION FACTORS: 1 mg/cu m = 0.189 ppm, 1 ppm = 5.29 mg/cu m
—Odor: characteristic: quality: fish, amine
hedonic tone: unpleasant to neutral
T.O.C.: absolute recogn.: 0.08 ppm
50% recogn.: 0.27 ppm
100% recogn.: 0.48 ppm
O.I. 100% recogn.: 5,479 (19)
C. WATER POLLUTION FACTORS:
—Waste water treatment:
A.C.: adsorbability: 0.174 g/g C, 87.0% reduction; infl.: 1,000 mg/l, effl.: 130 mg/l (32)
D. BIOLOGICAL EFFECTS:
—Fish: creek chub: critical range: 20–60 mg/l; 24 hr (226)
—Mammalia: rat: acute oral LD_{50}: 0.5 g/kg
inhalation: 6/6: 500 ppm, 4 hr
0/6: 250 ppm, 4 hr (211)
rat: oral LD_{50}: 0.55 g/kg; 95% conf.: 0.48–0.62 g/kg (226)

di-s-butylamine (bis(2-methylpropyl)amine)
$(CH_3-CH_2-CH)_2NH$
           |
          CH_3
A. PROPERTIES: b.p. $135/134°C$; v.d. 4.46; sp.gr. 0.75
D. BIOLOGICAL EFFECTS:
—Fish: creek chub: critical range: 15–40 mg/l, 24 hr (226)
—Mammalia: rat: inhalation: restlessness, initial tremors, uncoordination, no weight gain, autopsy: organs normal: 150 mg/l, 19 × 6.5 hr (65)

dibutyl-1,1-benzenedicarboxylate see di-n-butylphthalate

2,6-di-t-butylbenzoquinone

Manufacturers: organic chemical industry (347)
Users and formulations: oxidant, polymerization catalyst (347)
A. PROPERTIES: yellow crystals; m.w. 220.3; m.p. 65–67°C; v.d 7.6
C. WATER POLLUTION FACTORS:
 —In river water: 0.001–0.011 ppm; in river sediment: 0.1–40 ppm (555)

2,6-di-*tert*-butyl-*p*-cresol see 2,6-di-*tert*-butyl-4-methylphenol

di-*n*-butylether see *n*-butylether

dibutylketone see 5-nonanone

2,6-di-*tert*-butyl-4-methylphenol (2,6-di-*tert*-butyl-*p*-cresol; butylated hydroxytoluene; BHT; "ionol" CP-antioxidant; DBPC)

$$(CH_3)_3-C-\underset{CH_3}{\overset{OH}{\bigcirc}}-C-(CH_3)_3$$

Use: antioxidant for petroleum products, jet fuels, rubber, plastics and food products; food packaging; animal feeds
A. PROPERTIES: m.w. 220.36; m.p. 69.85°C; b.p. 265°C; sp.gr. 1.048 at 20/4°C; solub. 0.4 mg/l at ±20°C; v.d. 7.6
C. WATER POLLUTION FACTORS:
 —BOD_5: 0.51 NEN 3235-5.4 (277)
 —COD: 2.27 NEN 3235-5.3 (277)
 —Odor threshold: detection: 1.0 mg/l (998)
 —Water and sediment quality: in river water: 0.001–0.002 ppm;
    in river sediment: 1–60 ppm (555)
D. BIOLOGICAL PRODUCTS:
 —Fish: goldfish: not toxic in saturated solution (0.4 mg/l) (277)
 —Mammals: rat: acute oral LD_{50}: ±1800 mg/kg (277)

di-*n*-butylnitrosamine (DBN)
B. AIR POLLUTION FACTORS:
 —Manmade sources:
  emitted during the compounding, forming and curing operations of elastomeric parts by reaction of accelerators/stabilizers used such as zinc dibutyldithiocarbamate, nickel dibutyldithiocarbamate, and tetrabutylthioram polysulfide; emissions ranging from 15 to 120 g DBN/billion g rubber stock have been reported (1800)
D. BIOLOGICAL EFFECTS:
 —Carcinogenicity: +
 —Mutagenicity in the *Salmonella* test: +
  015 revertant colonies/nmol
  384 revertant colonies at 395 μg/plate (1883)

2,4-di-*t*-butylphenol

$$\text{2,4-di-tert-butylphenol structure: phenol with } C(CH_3)_3 \text{ groups at 2 and 4 positions}$$

Uses: intermediate; antioxidant; stabilizer; germicide
A. PROPERTIES: tan crystalline solid; b.p. 152–157°C at 25 mm; sp.gr. 0.907 at 60/4°C
C. WATER POLLUTION FACTORS:
  —Odor threshold: detection: 0.5 mg/l (998)
  —Water and sediment quality: in river water: 0.001–0.005 ppm;
                              in river sediment: 0.1–100 ppm (555)

2,6-di-*t*-butylphenol

$$\text{2,6-di-tert-butylphenol structure: phenol with } C(CH_3)_3 \text{ groups at 2 and 6 positions}$$

Uses: intermediate; antioxidant
A. PROPERTIES: light straw crystalline solid; m.p. 37°C; sp.gr. 0.914 at 20°C; b.p. 253°C; m.w. 206.33
C. WATER POLLUTION FACTORS:
  —Odor threshold: detection: 0.2 mg/l (998)
  —Water and sediment quality: in river water: 0.001–0.006 ppm;
                              in river sediment: 0.1–150 ppm (555)

di-*n*-butylphthalate (butylphthalate; dibutyl-1,1-benzenedicarboxylate; DBP)

$$\text{benzene ring with two } COOC_4H_9 \text{ groups}$$

Manufacturing source: organic chemical industry. (347)
Users and formulation: plasticizer mfg.; plastics mfg. recycling and processing; cosmetics; diluent in polysulfide dental impression materials; industrial stains mfg.; explosive (propellant) component used in fuel matrix of double-base rocket propellant; textile lubricating agent; used in safety glass; insecticides, printing inks, paper coatings, adhesives. (347)
Man caused sources (water and air): general use of plastics and above listed products (leaches from tubings, dishes, paper, containers, etc.) lab use, microcontaminant in lab chemicals, food, detergents, etc. also from lipsticks, applications of paints, coatings, and adhesives; evaporates from perfumes, inks and insecticides.
(347)

DI-n-BUTYLPHTHALATE   469

A. PROPERTIES: colorless oily liquid; m.w. 278.34; m.p. -35°C; b.p. 340°C; v.p. 0.1 mm at 115°C, 2.0 mm at 150°C; sp.gr. 1.0465; solub. 400 mg/l at 25°C, 4,500 mg/l at 25°C, 28 mg/l at 26°C; 50 mg/l in synthetic seawater; 10.6 mg/l

B. AIR POLLUTION FACTORS:
—Ambient air quality:
glc's in residential area (Belgium), Oct. 1976:
in particulate sample: 101 ng/cu m
in gasphase sample: 353 ng/cu m (1289)
organic fraction of suspended matter:
Bolivia at 5200 m altitude (Sept.–Dec. 1975): 19–36 µg/1000 cu m
Belgium, residential area (Jan.–April. 1976):   24–75 µg/1000 cu m (428)
in marine atmosphere in the North West Gulf of Mexico, March–April 1977:
avg.:   1.30 ng/cu m
range: 0.65–3.71 ng/cu m (n = 10) (1724)

C. WATER POLLUTION FACTORS:
—BOD_5: 0.43 std. dil. sew. (285)
—ThOD: 2.24 (27)
—Biodegradation:
aerobic degradation in freshwater hydrosoil:
53% after 24 hr incubation
98% after 5 days incubation (309)
biodegradation in freshwater hydrosoil:

incubation duration (days)	% degradation*	
	aerobic	anaerobic
1	5	0
5	97	31
7	95	41
14	92	61
30	97	98

*based on % recovery of ^{14}C from hydrosoil (1842)
degradation products identified were mono-n-butylphthalate and phthalic acid after aerobic incubation; the latter compound was not present after anaerobic incubation (1842)
—Water quality:
in Tama river (Japan), Jan. 1973: 0.71–3.14 µg/l (218)
              June 1973: 0.38–6.61 µg/l (218)
in Missouri river: 0.09 µg/l (218)
in Delaware river (U.S.A.): conc. range: winter: 0.2–0.6 ppb
                                        summer: 0.1–0.4 ppb (1051)
self purification is affected from 0.2 mg/l onwards (181)
—Ground water:
percolation water at 30–500 m from waste dumping ground contained: 1–12 ppm (183)
in well water (Japan): <0.2 µg/l (218)
—Drinking water:
in tap water (Japan): average: 2.34 µg/l
                       range: 1.43–3.31 µg/l (218)
—Manmade sources: in domestic sewer effluent: 0.2 mg/l (227)

60 ml/min pure water passed through 25 ft, $\frac{1}{2}$ inch, I.D. tube of general chemical grade P.V.C. contained 0.05 ppb dibutylphthalate which constituted 0.46% of total contaminant concentration (430)

—Waste water treatment:

	influent µg/l	carbon conc. mg/l	% removal	
powered activated carbon:	50	50–100	65–72	
	100	50–100	84–85	
		alum. conc.		
aluminum sulfate:	50	25–50	6–79	
	25	25–50	38–77	
	100	25–50	36–44	(218)

—Sampling and analysis: GC-FID: det.lim.: 0.2 µg/l (218)

D. BIOLOGICAL EFFECTS:
—Marine dinoflagellate *Gymnodium breve*: 96 hr TLm: 0.02–0.6 ppm
96 hr EC_{50}: 0.0034–0.2 ppm*

*EC_{50}: median growth limit concentration causing a 50% growth reduction
(1057)

—Crustacean:
larvae of grass shrimp (*Palaemonetes Pugio Holthuis*):
17 days, LC_{50}: 100 ppb–1 ppm
24 hr, LC_{50}: 10–50 ppm
no significiant increase in mortality at 500 ppb after 32 days (551)
grass shrimp (*Palaemonetes kadiakensis*): at 0.08 ppb: 3 day biomagnification factor = 5000 (0.4 ppm) computed on the basis of ^{14}C incorporation (552)

di-sec-butylphthalate

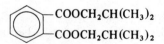

B. AIR POLLUTION FACTORS:
—Organic fraction of suspended matter:
Bolivia at 5200 m altitude (Sept.–Dec. 1975): 0.32–0.40 µg/1000 cu m
Belgium, residential area (Jan.–April 1976): 2.4–7.2 µg/1000 cu m (428)
—glc's in residential area (Belgium), october 1976:
in particulate sample: — ng/cu m
in gas phase sample: 60.1 ng/cu m (428)

dibutylsulfide *see n-butylsulfide*

1,3-dibutylthiourea
$C_4H_9NHCSNHC_4H_9$
A. PROPERTIES: white to light tan powder; m.p. 59–69°C
D. BIOLOGICAL EFFECTS:
—Fish: creek chub: critical range: 30–100 mg/l; 24 hr (226)

dicamba (Banvel; Dianat; Mediben; 3,6-dichloro-*o*-anisic acid)

[Structure: benzene ring with COOH, Cl, OCH₃, Cl substituents]

A. PROPERTIES: white crystalline solids; m.p. 114–116°C; solub. 7900 mg/l
D. BIOLOGICAL EFFECTS:
—Crustacean
Gammarus lacustris: 96 hr, LC_{50}: 3900 µg/l	(2124)
Gammarus fasciatus: 48 hr, no effect level: 100,000 µg/l	(2125)
Daphnia magna: 48 hr, no effect level: 100,000 µg/l	(2125)
Cypridopsis vidua: 48 hr, no effect level: 100,000 µg/l	(2125)
Asellus brevicaudus: 48 hr, no effect level: 100,000 µg/l	(2125)
Palaemonetes kadiakensis: 48 hr, no effect level: 100,000 µg/l	(2125)
Orconectes nais: 48 hr, no effect level: 100,000 µg/l	(2125)

—Fish
Lepomis macrochirus: 48 hr, LC_{50}: 20000 µg/l	(2114)

—Mammals:
acute oral LD_{50} (rat): 2900 ± 800 mg/kg	(1854)
in diet: rats: no effect after 2 years on diets containing 500 ppm	
dogs: no effect after 2 years on diets containing 50 ppm	(1855)

dicapthon (O-(2-chloro-4-nitrophenyl)O,O-dimethylphosphorothioate; di-Captan)

[Structure: (CH₃O)₂P(=S)–O–phenyl with Cl and NO₂ substituents]

Use: nonsystemic insecticide and acaricide
A. PROPERTIES: solub. 6.25 ppm at 20°C; log P_{oct} 3.58 at 20°C
D. BIOLOGICAL EFFECTS:
—Mussels:
hard clam (*Mercenaria mercenaria*): eggs: 48 hr TLm S: 3.34 mg/l	
hard clam (*Mercenaria mercenaria*): larvae: 12 d TLm S: 5.74 mg/l	(2324)

—Mammals:
acute oral LD_{50} (male rat): 400 mg/kg	(1854)
(female rat): 330 mg/kg	(1855)
(albino mice): 475 mg/kg	
in diet: rats maintained for 1 year on diets containing 25, 100 and 250 ppm showed retarded growth only at the higher level	(1855)

1,2-dicarboxy-3,6-endoxo-cyclohexane *see* aquathol k

dichlobenil (Casoron; 2,6-dichlorobenzonitrile)

Use: herbicide
A. PROPERTIES: m.p. 144–145°C; v.p. 5×10^{-4} mm at 25°C; solub. 18 ppm at 20°C
D. BIOLOGICAL EFFECTS:
—Algae:

Chlorococcum sp.	9×10^4 ppb	50% decrease in O_2 evolution	f	technical acid
Chlorococcum sp.	6×10^4 ppb	50% decrease in growth	Measured as ABS. (525 mu) after 10 days	technical acid
Dunaliella tertiolecta	1.25×10^5 ppb	50% decrease in O_2 evolution	f	technical acid
Dunaliella tertiolecta	6×10^4 ppb	50% decrease in growth	Measured as ABS. (525 mu) after 10 days	technical acid
Isochrysis galbana	1×10^5 ppb	50% decrease in O_2 evolution	f	technical acid
Isochrysis galbana	6×10^4 ppb	50% decrease in growth	Measured as ABS. (525 mu) after 10 days	technical acid
Phaeodactylum tricornutum	1.5×10^5 ppb	50% decrease in O_2 evolution	f	technical acid
Phaeodactylum tricornutum	2.5×10^4 ppb	50% decrease in growth	Measured as ABS. (525 mu) after 10 days	technical acid (2348)

—Crustacean
  Gammarus lacustris: 96 hr, LC_{50}: 11000 µg/l (2124)
  Gammarus fasciatus: 96 hr, LC_{50}: 10000 µg/l (2125)
  Hyallella azteca: 96 hr, LC_{50}: 8500 µg/l (2133)
  Simocephalus serrulatus: 48 hr, LC_{50}: 5800 µg/l (2128)
  Daphnia pulex: 48 hr, LC_{50}: 3700 µg/l (2128)
  Daphnia magna: 48 hr, LC_{50}: 10000 µg/l (2125)
  Cypridopsis vidua: 48 hr, LC_{50}: 7800 µg/l (2125)
  Asellus brevicaudus: 48 hr, LC_{50}: 34000 µg/l (2125)
  Palaemonetes kadiakensis: 48 hr, LC_{50}: 9000 µg/l (2125)
  Orconectes nais: 48 hr, LC_{50}: 22000 µg/l (2125)
—Insects
  Pteronarcys californica: 96 hr, LC_{50}: 7000 µg/l (2128)
  Tendipedidae: 96 hr, LC_{50}: 7800 µg/l (2133)
  Callibaetes sp.: 96 hr, LC_{50}: 10300 µg/l (2133)
  Limnephilus: 96 hr, LC_{50}: 13000 µg/l (2133)
  Enallegma: 96 hr, LC_{50}: 20700 µg/l (2133)

—Fish
Lepomis macrochirus: 48 hr, LC_{50}: 20000 µg/l (2133)
bluegill: 48 hr LC_{50}: 20.0 ppm
rainbow trout: 48 hr LC_{50}: 22.0 ppm (1878)
harlequin fish (Rasbora heteromorpha):

	24 hr	48 hr	96 hr	3 m (extrap.)	
mg/l					
LC_{10} (F)	4.3	3.4	3.3		
LC_{50} (F)	6.2	4.7	4.2	4.0	(331)

guppies: 48 hr LD_{50}: 18 ppm
Daphnia magna: IC 50: 9.8 ppm (1855)
acute oral LD_{50} (rat): 3160 mg/kg (1854)
  (male mice): >2460 mg/kg
  (guinea pigs): 501 mg/kg
acute dermal LD_{50} (albino rabbits): 1,350 mg/kg (1855)
in diet: rats fed on diets containing 1000 mg/kg gained weight normally but showed changes in organ weight (1855)

dichlofluanid see euparen

dichlone (2,3-dichloro-1,4-naphthoquinone; phygon)

A. PROPERTIES: yellow crystals; m.p. 193°C; solub. 0.1 ppm at 25°C
D. BIOLOGICAL EFFECTS:
 —Crustacean
  Gammarus lacustris: 96 hr, LC_{50}: 1100 µg/l (2124)
  Gammarus fasciatus: 96 hr, LC_{50}: 100 µg/l (2125)
  Daphnia magna: 48 hr, LC_{50}: 25 µg/l (2125)
  Cypridopsis vidua: 48 hr, LC_{50}: 120 µg/l (2125)
  Asellus brevicaudus: 48 hr, LC_{50}: 200 µg/l (2125)
  Palaemonetes kadiakensis: 48 hr, LC_{50}: 450 µg/l (2125)
  Orconectes nais: 48 hr, LC_{50}: 3200 µg/l (2125)
  Daphnia magna: IC 50: 14 µg/l (1855)
 —Fish:
  Lepomis macrochirus: 48 hr, LC_{50}: 70 µg/l (2106)
  Micropterus salmoides: 48 hr, LC_{50}: 120 µg/l (2114)
 —Mammals:
  acute oral LD_{50} (rats): 1300 mg/kg
  in diet: rats fed for 2 years on a diet containing 1500 ppm suffered no ill effects (1855)

dichloroacetic acid (dichloroethanoic acid)
$CHCl_2COOH$

s-2,3-DICHLOROALLYL-N,N-DIISOPROPYLTHIOLCARBAMATE

A. PROPERTIES: colorless liquid; m.w. 128.95; m.p. 5/6°C; b.p. 194°C; v.p. 1 mm at 44°C; v.d. 4.40; sp.gr. 1.563 at 20/4°C; solub. 86,300 mg/l; log P_{oct} -0.14/1.39 (calculated)
B. AIR POLLUTION FACTORS:
   —Odor threshold: recognition: 0.23 mg/cu m (610)
C. WATER POLLUTION FACTORS:
   —BOD_5: nil std. dil. sew. (161)
D. BIOLOGICAL EFFECTS:
   —Mammals: rat: single oral LD_{50}: 4.5 g/kg
   mouse: single oral LD_{50}: 5.5 g/kg (211)

s-2,3-dichloroallyl-N,N-diisopropylthiolcarbamate *see* diallate

1-(3,4-dichloroaniline)-1-formylamino-2,2,2-trichloroethane *see* imugan

3,6-dichloro-o-anisic acid *see* dicamba

2,4-dichloroaniline

D. BIOLOGICAL EFFECTS:
   —Guppy (*Poecilia reticulata*): 14 d LC_{50}: 11.7 ppm (1833)

3,6-dichloro-o-anisic acid (3,6-dichloro-2-methoxybenzoic acid; dicamba, dianat; Banvel D (= dimethylamine salt))

Use: herbicide
A. PROPERTIES: m.p. 114–116°C; solub. 7.9 g/l at 25°C;
   380 g equivalent/l (sodium salt)
   720 g equivalent (dimethylamine salt)
C. WATER POLLUTION FACTORS:
   —Persistence in soil:
   residues in soil following application of the dimethylamine salt (ppm):

	Dicamba acid (ppm)					
	1.1 kg/ha		2.2 kg/ha		4.5 kg/ha	
days after application	0-10 cm	10-20 cm	0-10 cm	10-20 cm	0-10 cm	10-20 cm
1	.43	–	.75	–	.85	–
14	.23	–	.61	–	3.08	–
28	.05	–	.10	–	.50	–
42	.05	.02	.08	.03	.16	.12
56	.03	.01	.08	.03	.24	.12
70	.02	.01	.09	.02	.13	.04
265	.01	.01	.03	Tr	.07	.03
385	Tr	Tr	Tr	ND	Tr	Tr

(1088)

D. BIOLOGICAL EFFECTS:
—Mammals:
  acute oral LD_{50} (rat): 2900 ± 800 mg/kg
                        1040 mg/kg (dimethylamine salt)   (1854)
  in diet: (rat): 2 years at 500 ppm: no effect   (1855)
          (dog): 2 years at 50 ppm: no effect   (1855)

o-dichlorobenzene (1,2-dichlorobenzene)

Uses: manufacture of 3,4-dichloro-aniline; solvent; dye mfg; fumigant and insecticide; metal polishes; industrial odor control; technical grade contains p-dichlorobenzene (17%) and m-dichlorobenzene (2%)

A. PROPERTIES: colorless liquid; m.w. 147.01; m.p. −16.7/−18°C; b.p. 179°C; v.p. 1 mm at 20°C, 1.5 mm at 25°C, 1.9 mm at 30°C; v.d. 5.07; sp.gr. 1.305 at 20/4°C; solub. 100 mg/l at 20°C, 145 mg/l at 25°C; sat.conc. 8.0 g/cu m at 20°C, 15 g/cu m at 30°C: log P_{oct} 3.38

B. AIR POLLUTION FACTORS: 1 mg/cu m = 0.116 ppm, 1 ppm = 6.01 mg/cu m
  —Odor: recognition: 0.12 mg/cu m   (610)
        T.O.C.: avg. 50 ppm = 305 mg/cu m   (211; 279)
                     300 mg/cu m   (57; 291)
                       O.I. at 20°C = 26   (316)
  —Sampling and analysis:
    photometry: min.full scale: 100 ppm   (53)
    4 f.abs.app.: 30 l air/30 min: UVS: det.lim.: 300 µg/cu m/30 min   (20)

C. WATER POLLUTION FACTORS:
  —Reduction of amenities:
    adverse taste in fish: <0.25 mg/l water   (41)
    organoleptic limit USSR 1970: 0.002 mg/l   (181)
    T.O.C. = 0.01 mg/l   (225)(903)

- Water quality:
  in river Maas (The Netherlands): average in 1973: 0.13 µg/l (o- + p-isomer) (342)
  in Delaware river (U.S.A.): conc. range: winter: 0.4 ppb
      (n.s.i.)   summer: N.D. (1051)
  in Zürich lake: at surface: 16 ppt; at 30 m depth: 26 ppt (n.s.i.)
  in tap water (Zürich): 4 ppt (n.s.i.) (513)
- Waste water treatment:
  degradation by *Pseudomonas*: at 200 mg/l
      parent: 100% ring disruption in 72 hr
      mutant: 100% ring disruption in 26 hr (152)

D. BIOLOGICAL EFFECTS:
- Toxicity threshold (cell multiplication inhibition test):
  bacteria (*Pseudomonas putida*):   15 mg/l (1900)
  algae (*Microcystis aeruginosa*):   53 mg/l (329)
  green algae (*Scenedesmus quadricauda*):   >100 mg/l (1900)
  protozoa (*Entosiphon sulcatum*):   >64 mg/l (1900)
  protozoa (*Uronema parduczi Chatton-Lwoff*):   80 mg/l (1901)
  marine plankton: growth was stopped at 13 ppm (360)
  hard clam eggs: 48 hr, LC_{50}: >100 ppm
  hard clam larvae: 288 hr, LC_{50}: >100 ppm (358)
- Fish:

Lepomis macrochirus: static bioassay in fresh water at 23°C, mild aeration applied after 24 hr:

material added ppm	% survival after 24 hr	48 hr	72 hr	96 hr	best fit 96 hr LC_{50} ppm
50	0	—	—	—	
32	70	50	30	20	
24	90	90	90	90	27
18	90	90	90	90	

Menidia beryllina: static bioassay in synthetic seawater at 23°C, mild aeration applied after 24 hr:

material added ppm	% survival after 24 hr	48 hr	72 hr	96 hr	best fit 96 hr LC_{50} ppm
18	0	—	—	—	
10	20	10	10	10	7.3
5	100	100	100	100	

(352)

guppy (*Poecilia reticulata*): 14 d LC_{50}: 5.9 ppm (1833)

fathead minnows: 72 hr, LC_{100}: 10 ppm
    72 hr, LC_0: 3 ppm (359)

	24 hr LC_{50}	48 hr LC_{50}	96 hr LC_{50}
fathead minnows (*Pimephales promelas*)	~105 mg/l	76 mg/l	57 mg/l
grass shrimp (*Palaemonetes pugio*)	14.3 mg/l	10.3 mg/l	9.4 mg/l

(1904)

—Mammalia:
  rat: inhalation: survived: 977 ppm, 2 hr
                   succumbed: 977 ppm, 7 hr
  rat; guinea pig, rabbit: inhalation: no effect at 93 ppm, 7 hr/day, 5 day/w, 6-7 months
  guinea pig: single oral dose: survived: 0.8 g/kg
                                succumbed: 2.0 g/kg   (211)
  rats: no effect level (138 doses by intubation in 192 days) in the range of 18.8 to 188 mg/kg/day   (1546)
—Man: severe toxic effects: 300 ppm = 1,836 mg/cu m, 60 min
      symptoms of illness: 100 ppm = 612 mg/cu m
      unsatisfactory: 25 ppm = 153 mg/cu m   (185)
      strong and irritating odor: 100 ppm
      no injury at average: 15 ppm
                     range: 1-44 ppm   (211)

m-dichlorobenzene (1,3-dichlorobenzene)

A. PROPERTIES: m.w. 147.01; m.p. −24.8°C; b.p. 172°C; sp.gr. 1.288 at 20/4°C; solub. 123 mg/l at 25°C; 69 mg/l at 22°C; log P_{oct} 3.38
C. WATER POLLUTION FACTORS:
  —Waste water treatment:
    degradation by *Pseudomonas*: at 200 mg/l at 30°C
                       parent: 100% ring disruption in 96 hr
                       mutant: 100% ring disruption in 28 hr   (152)
  —Reduction of amenities: T.O.C. = 0.02 mg/l   (225)(903)
D. BIOLOGICAL EFFECTS:
  —Fish:
    guppy (*Poecilia reticulata*): 14 d, LC_{50}: 7.4 ppm   (1833)

p-dichlorobenzene (1,4-dichlorobenzene)

Manufacturing source: organic chemical industry.   (347)
Users and formulation: mfg. moth repellants, mfg. air deodorizers, mfg. dyes and intermediates; pharmaceuticals mfg; soil fumigant; pesticide.   (347)

2,5-DICHLOROBENZENESULFONATE, SODIUM

A. PROPERTIES: colorless or white crystals; m.w. 147.01; m.p. 53°C; b.p. 173.4°C; v.p. 0.6 mm at 20°C, 1.8 mm at 30°C; v.d. 5.07; sp.gr. 1.458 at 20/4°C; solub. 49 mg/l at 22°C, 79 mg/l at 25°C; sat.conc. 4.8 g/cu m at 20°C, 14 g/cu m at 30°C; log P_{oct} 3.39 at 20°C

B. AIR POLLUTION FACTORS: 1 mg/cu m = 0.166 ppm, 1 ppm = 6.01 mg/cu m
 —Odor: characteristic: quality: very distinctive aromatic odor
    T.O.C.: 15-30 ppm (211)
    O.I. at 20°C = 26 (316)

C. WATER POLLUTION FACTORS:
 —Reduction of amenities:
    T.O.C. = 0.03 mg/l (225; 903)
        = 0.0003 mg/kg (894)
 —Water quality:
    in river Maas at Eysden (Netherlands) in 1976:
      median: 0.1 µg/l; range: n.d.–3.5 µg/l
    in river Maas at Keizersveer (Netherlands) in 1976:
      median: n.d.; range: n.d.–0.4 µg/l (1368)
 —Waste water treatment:
    degradation by *Pseudomonas*: 200 mg/l at 30°C
           parent: 100% ring disruption in 92 hr
           mutant: 100% ring disruption in 25 hr (152)

D. BIOLOGICAL EFFECTS:
 —Bioaccumulation:
    BCF in bacteria (*Sidero capsa treubii*): approx. 20 (1833)
    BCF in guppies (*Poecilia reticulata*): 1800 (on lipid content)
    trout (*Salmo gairdneri*): log. bioconcentration factor = 2.33 (193)
 —Toxicity:
    guppy (*Poecilia reticulata*): 14 d, LC_{50}: 4.0 ppm (1833)

	24 hr LC_{50}	48 hr LC_{50}	96 hr LC_{50}
fathead minnows (*Pimephales promelas*)	35.4 mg/l	35.4 mg/l	33.7 mg/l
grass shrimp (*Palaemonetes pugio*)	–	129 mg/l	69 mg/l

(1904)

 —Mammalia:
    rat, guinea pig, rabbit, mouse: no effect: 96 ppm, 6–7 months
    rat: single oral dose: survived: 1 g/kg
                 succumbed: 4 g/kg
       repeated oral doses: no effect at: 18.8 mg/kg fed for 192 days (211)
    guinea pigs: single oral LD_{50}: 1600–2800 mg/kg (1546)
 —Man: painful to eyes and nose: 50–80 ppm
       severe discomfort: 160 ppm
       no complaints and no injury at: average: 45 ppm
                                     range: 15–85 ppm
       very little irritating effect upon the skin (211)

2,5-dichlorobenzenesulfonate, sodium

D. BIOLOGICAL EFFECTS:
 —*Daphnia magna*: 24 hr, TLm: 4,931 mg/l
                   48 hr, TLm: 4,931 mg/l
                   72 hr, TLm: 2,490 mg/l
                   96 hr, TLm: 938 mg/l   (153)
                  100 hr, TLm: 1,468 mg/l   (1295)
 —Fish: *Lepomis macrochirus*: 24 hr, TLm: <3.750 mg/l   (1295)
 —Snail eggs: *Lymnaea* sp.:   24 hr, TLm: 4,981 mg/l
                               48 hr, TLm: 4,513 mg/l
                               72 hr, TLm: 3,984 mg/l
                               96 hr, TLm: 3,144 mg/l   (153)

3,3'-dichlorobenzidine (3,3'-dichloro-4,4'-diaminobiphenyl; DCB)

Use: intermediate in the manufacture of azo pigments, curing agent for isocyanate terminated resins, for urethane resins

A. PROPERTIES: m.w. 253.13; gray to purple crystalline solid; solub. 3.99 ppm at pH 6.9 and 22°C (DCB.2HCl)

B. AIR POLLUTION FACTORS:
 —Method of analysis:
  air is drawn through a glass-fiber filter followed by a bed of silica gel to collect these substances as either particles or vapors. The compounds are extracted from the samples and analyzed by HPLC with sensitivities in the range of 3 µg/cu m for 48-l air samples   (1686)

C. WATER POLLUTION FACTORS:
 —Aquatic reactions:
  degradation: after 1 month incubation period, at 21°C in the dark: 75% of the original DCB was still present—no metabolites were detected   (1216)
  photolysis of DCB: irradiation with light (from a Hanovia 450-W high-pressure lamp filtered through Pyrex), leads to MCB and benzidine:

min. of irradiation	DCB · 2HCl ppm	MCB · 2HCl ppm	benzidine · 2HCl ppm
0	0.64	0	
1	0.60	0.034	
2	0.56	0.060	
3	0.42	0.068	
4	0.24	0.114	

min. of irradiation	DCB · 2HCl ppm	MCB · 2HCl ppm	benzidine · 2HCl ppm
5	0.078	0.044	0.044
10		0.044	0.072
15			0.086
45			0.072
			(1218)

distribution coefficient (at equilibrium) for a number of aquatic sediments ranged from 2,670 to 12,800 (1216)

D. BIOLOGICAL EFFECTS:
—Bioaccumulation:
bluegill sunfish: BCF: about 500 (based on ^{14}C residues), in water containing 5 ppb or 0.1 ppm, equilibria achieved in 96 to 168 hr depuration: a rapid initial rate of elimination was followed by a low rate with 5–18% of initial ^{14}C remaining after 14 days of exposure to fresh water (1907)
bluegills (*Lepomis macrochirus*):

conc. in water	in whole body	exposure time (hr)	
2.0	265	48	
0.5	277	120	(1216)

—Toxicity:
Man: animal carcinogen: U.S.A. 1974 (77)
West Germany 1977 (487)

2,4-dichlorobenzoic acid

C. WATER POLLUTION FACTORS:
—Treatment methods:
A.C. type BL (Pittsburgh Chem. Co.): % adsorbed by 10 mg A.C. from 10^{-4} M aqueous solution at pH 3.0: 62.2%
pH 7.0: 14.8%
pH 11.0: 6.5% (1313)

2,5-dichlorobenzoic acid

A. PROPERTIES: m.w. 191.01; m.p. 151–154°C; b.p. 301°C

C. WATER POLLUTION FACTORS:
  −Biodegradation:
    lag period for degradation of 16 mg/l by wastewater or soil suspension at pH 7.3 and 30°C: >25 days (1096)

2,6-dichlorobenzonitrile *see* dichlobenil

2,3-dichlorobiphenyl

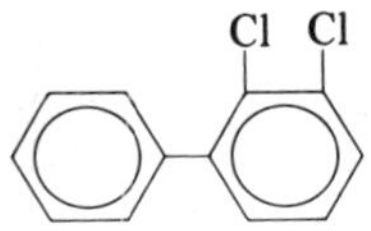

C. WATER POLLUTION FACTORS:
  −Biodegradation:
    100% degradation after 1 hr by *Alcaligenes* Y42 (cell number 2×10^9/ml)
    92.8% degradation after 1 hr by *Acinetobacter* P6 (cell number 4.4×10^8/ml) at 11.1 mg/l initial conc.; dichlorobenzoic acid derivatives were detected in the metabolite. (1086)

2,4-dichlorobiphenyl

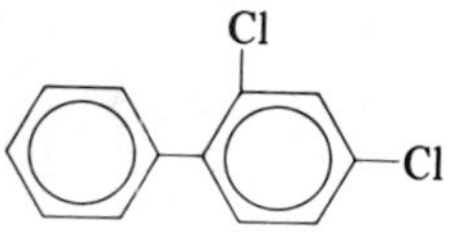

C. WATER POLLUTION FACTORS:
  −Biodegradation:
    100% degradation after 1 hour by *Alcaligenes* Y42 (cell number 2×10^9/ml) and *Acinetobacter* (cell number 4.4×10^8/ml) at 11.1 mg/l initial conc.; dichlorobenzoic acid derivatives were detected in the metabolite (1086)

2,5-dichlorobiphenyl

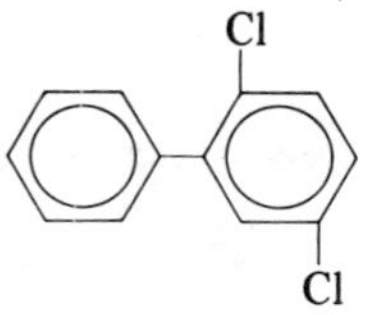

C. WATER POLLUTION FACTORS:
  −Biodegradation:
    100% degradation after 1 hour by *Alcaligenes* Y42 (cell number 2×10^9/ml) and *Acinetobacter* P6 (cell number 4.4×10^8/ml) at 11.1 mg/l initial conc.; dichlorobenzoic acid derivatives were detected in the metabolite. (1086)

2,6-dichlorobiphenyl

C. WATER POLLUTION FACTORS:
 —Biodegradation:
  no degradation after 1 hour by *Alcaligenes* Y42 (cell number 2×10^9/ml), 8.2% degradation after 1 hour by *Acinetobacter* P6 (cell number 4.4×10^8/ml) at 11.1 mg/l initial conc. (1086)

3,4-dichlorobiphenyl

C. WATER POLLUTION FACTORS:
 —Biodegradation:
  100% degradation after 1 hr by *Alcaligenes* Y42 (cell number 2×10^9/ml)
  88.2% degradation after 1 hr by *Acinetobacter* P6 (cell number 4.4×10^8/ml), at 11.1 mg/l initial conc.; dichlorobenzoic acid derivatives were detected in the metabolites. (1086)

3,5-dichlorobiphenyl

C. WATER POLLUTION FACTORS:
 —Biodegradation:
  100% degradation after 1 hr by *Alcaligenes* Y42 (cell number 2×10^9/ml)
  96.4% degradation after 1 hr by *Acinetobacter* P6 (cell number 4.4×10^8/ml), at 11.1 mg/l initial conc.; dichlorobenzoic acid derivatives were detected in the metabolites. (1086)

2,2'-dichlorobiphenyl

C. WATER POLLUTION FACTORS:
  −Biodegradation:
  12.6% degradation after 1 hr by *Alcaligenes* Y42 (cell number 2×10^9/ml)
  28.0% degradation after 1 hr by *Acinetobacter* P6 (cell number 4.4×10^8/ml),
    at 11.1 mg/l initial conc.; trimethylsilyl derivative of monochlorobenzoic acid
    was detected in the metabolite. (1086)

2,4'-dichlorobiphenyl

C. WATER POLLUTION FACTORS:
  −Biodegradation:
  96.4% degradation after 1 hr by *Alcaligenes* Y42 (cell number 2×10^9/ml)
  98.2% degradation after 1 hr by *Acinetobacter* P6 (cell number 4.4×10^8/ml)
    at 11.1 mg/l initial conc.; Trimethylsilyl derivative of monochlorobenzoic acid
    was detected in the metabolite. (1086)
  −Aquatic reactions:
  Freundlich adsorption constants:

	n	K	
illite clay	0.84	20.0	
Woodburn soil	1.18	7.0	
humic acid	0.86	39.8	(1879)

D. BIOLOGICAL EFFECTS:
  −Amphipod (*Gammarus fasciatus*): 4 days LC_{50} (S): 0.12 mg/l (1617)

3,3'-dichlorobiphenyl

C. WATER POLLUTION FACTORS:
  −Biodegradation:
  37% degradation after 1 hour by *Acinetobacter* P6 (cell number 4.4×10^8/ml)
    at 11.1 mg/l initial conc.; trimethylsilyl derivative of monochlorobenzoic acid
    was detected in the metabolite. (1086)

4,4'-dichlorobiphenyl

A. PROPERTIES: m.w. 223.1; m.p. 142–145°C; b.p. 315–319°C; solub. 0.062 mg/l at 20°C, log P_{oct} 5.58

C. WATER POLLUTION FACTORS:
   —Biodegradation:

[structure: 4,4'-dichlorobiphenyl] → → [structure: chlorobenzoyl with COOH, OH, Cl substituted diene] →→ SLOW ?

bacterial degradation (1219)

32.4% degradation after 1 hr by *Alcaligenes* Y42 (cell number 2×10^9/ml)
50.4% degradation after 1 hr by *Acinetobacter* P6 (cell number 4.4×10^8/ml)
at 11.1 mg/l initial conc.; trimethylsilyl derivative of monochlorobenzoic acid
was detected in the metabolite (1086)

D. BIOLOGICAL EFFECTS:
   —Amphibod (*Gammarus fasciatus*): 4 days LC_{50} (S): 0.10 mg/l (1617)

trans-1,4-dichlorobutene-2
$ClH_2C-CH=CH-CH_2Cl$
A. PROPERTIES: colorless liquid; m.p. 3.5°C; b.p. 158°C; sp.gr. 1.1858 at 25/4°C
B. AIR POLLUTION FACTORS:
   —Sampling and analysis: photometry: min. full scale: 11,000 ppm (n.s.i.) (53)
D. BIOLOGICAL EFFECTS:
   —Guppy (*Poecilia reticulata*): 7 d, LC_{50}: $\ll 40$ mg/l (1833)

3,4-dichloro-ω-butyric acid
$CH_2Cl-CHCl-CH_2-COOH$
C. WATER POLLUTION FACTORS:
   —Biodegradation:
   >205 days for ring cleavage in soil suspension (1827)

dichlorodehydroabietic acid
D. BIOLOGICAL EFFECTS:
   —Fish: juvenile rainbow trout: 96 hr, LC_{50} (S): 0.6 mg/l (314)

dichlorodifluoromethane (F 12; R 12)
   Uses: refrigerant and air conditioner; aerosol propellant; plastics; blowing agent;
   low-temperature solvent; leak-detecting agent
A. PROPERTIES: colorless gas; m.w. 120.92; m.p. -111°C, -158°C; b.p. -29.8°C;
   v.p. 4250 mm at 20°C, 7.6 atm at 30°C; v.d. 4.18; sp.gr. 1.329 at 20°C; solub.
   280 mg/l at 25°C,
B. AIR POLLUTION FACTORS: 1 mg/cu m = 0.20 ppm, 1 ppm = 5.026 mg/cu m
   —Atmospheric reactions: no photolysis in vitro after 60 days (201)
   —Ambient air quality:
       Los Angeles: July 1970: average glc's: 0.7 ppb
                    Feb. 1973: average glc's: 0.53 ppb

inside homes and public buildings: 0.8 ppb–0.5 ppm (201)
in rural Washington, Dec. 74–Feb. 75: 0.23 ppb (315)

sampling date	sampling location	sampling height m	measured conc. pptv	
Feb.–Aug. 1973	S. California	6700	90	(378)
23 May 1974	New Mexico	6400	120	(373)
June/July 1974	W. Ireland	sea-level	101.7 ± 27.3	(375)
October 1974	North Atlantic	sea-level	115 ± 33.1	(375)
November 1974	Pullman, Washington State	0–3600	230 ± 8	(372)
August 1975	West of N. Ireland E. of Scotland	7000–8000	231 ± 36	(371)
January 1976	off S.W. Wales	900–7300	236 ± 21	(371)
January 1976	off S.W. Wales	900–4900	258 ± 14	(371)
February 1976	off N.E. England	400–5200	225 ± 5	(371)

D. BIOLOGICAL EFFECTS:
—Mammalia: mouse: inhalation: LD_{50}: 760,000 ppm, 30 min
guinea pig: inhalation: LD_{50}: >800,000 ppm, 30 min
rabbit: inhalation: LD_{50}: >800,000 ppm, 30 min
rat: inhalation: LD_{50}: >800,000 ppm, 30 min (74)
2/5: 795 ppm, 90 days
guinea pig: inhalation: 1/15: 795 ppm, 90 days (54)
rat: single oral LD_{50}: >1 g/kg (54)
—Man: very low toxicity (211)

5,5′-dichloro-2,2′-dihydroxydiphenylmethane (dichlorophen; preventol GD (Bayer); antiphen; 2,2′-methylene bis(4-chlorophenol); di-(5-chloro-2-hydroxyphenyl)methane))

Uses: fungicide; bactericide; herbicide
A. PROPERTIES: white odorless crystals; m.p. 177–178°C; solub. 30 ppm at 25°C
C. WATER POLLUTION FACTORS:
—Biodegradation:
after 3 weeks of adaptation at 10–20 mg/l at 22°C: under aerobic conditions, 25% degradation when product is sole carbon source, 50% degradation with synthetic sewage; under anaerobic conditions, no degradation, even with synthetic sewage. (512)
D. BIOLOGICAL EFFECTS:
—Fish: harlequin fish (*Rasbora heteromorpha*):

mg/l (sodium salt)	24 hr	48 hr	96 hr	3 m (extrap.)	
LC_{10} (F)	4.4	3.8	2.7		
LC_{50} (F)	5.4	4.8	3.6	3.4	(331)

486 β,β'-DICHLORODIISOPROPYLETHER

—Mammals:
acute oral LD_{50} (guinea pigs): 1250 mg/kg (1854)
(dogs): 2000 mg/kg
in diet: rats fed for 90 days on a diet containing 2000 ppm showed no evidence of toxicity (1855)

β,β'-dichlorodiisopropylether *see* dichloroisopropylether

dichlorodimethylether *see* dichloromethylether

dichlorodiphenyldichloroethane *see* DDD

dichlorodiphenyldichloroethylene *see* p,p'-DDE

dichlorodiphenyltrichloroethane *see* DDT

1,1-dichloroethane (ethylidenechloride; ethylidenedichloride)
$CHCl_2CH_3$
  Uses: vinylchloride; chlorinated solvent intermediate; coupling agent in antiknock gasoline; paint, varnish and finish removers; metal degreasing; organic synthesis; ore flotation
A. PROPERTIES: colorless liquid; m.w. 98.96; m.p. $-97.4°C$; b.p. $57.3°C$; v.p. 70 mm at $0°C$, 180 mm at $20°C$, 234 mm at $25°C$, 270 mm at $30°C$; v.d. 3.42; sp.gr. 1.174 at $20/4°C$; solub. 5,500 mg/l at $20°C$; sat.conc. 986 g/cu m at $20°C$, 1,406 g/cu m at $30°C$
B. AIR POLLUTION FACTORS: 1 mg/cu m = 0.24 mm, 1 ppm = 4.11 mg/cu m
  —Odor: characteristic; quality: chloroformlike
    threshold concentration: 120 ppm (n.s.i.)
    200 ppm (n.s.i.) (279)
  —Manmade sources: glc's in rural Washington Dec. '74–Febr. '75: <5 ppt (315)
C. WATER POLLUTION FACTORS:
  —BOD_5: 0.002
  —BOD_{10}: 0.05
  —Waste water treatment:
    evaporation from water at $25°C$ of 1 ppm solution: 50% after 22 min
    90% after 109 min (313)
  —Aquatic reactions: evaporation:
    measured half-life for evaporation from 1 ppm aqueous solution at $25°C$, still air, and an average depth of 6.5 cm: 32.2 min. (369)
D. BIOLOGICAL EFFECTS:
  —Arthropods: brine shrimp: 24 hr TLm: 320 mg/l
  —Fish: pinperch: 24 hr TLm: 160 mg/l (23)
    Lepomis macrochirus: static bioassay in fresh water at $23°C$, mild aeration applied after 24 hr (n.s.i.):

material added		% survival after			best fit 96 hr LC_{50}
ppm	24 hr	48 hr	72 hr	96 hr	ppm
1,000	20 (2 hr)	20	0	—	
560	57	43	43	39	550
420	100	100	100	100	
320	100	100	100	90	

Menidia beryllina: static bioassay in synthetic seawater at 23°C: mild aeration applied after 24 hr (n.s.i.):

material added		% survival after			best fit 96 hr LC_{50}
ppm	24 hr	48 hr	72 hr	96 hr	ppm
560	0	—	—	—	
420	50	50	50	30	480
320	90	90	90	90	
180	100	100	100	100	

                        (352)

guppy (*Poecilia reticulata*): 7 d, LC_{50}: 202 ppm   (1833)

—Mammals:
 mouse: inhalation: LD_{50}: 17,300 ppm, 2 hr
 rat: inhalation: LD_{50}: 16,000 ppm, 8 hr
      no effect: 1,000 ppm, 5 × 6 hr/w, 3 months   (54)

1,2-dichloroethane *see* ethylenedichloride

dichloroethanoic acid *see* dichloroacetic acid

1,1-dichloroethene *see* 1,1-dichloroethylene

1,2-dichloroethene *see* 1,2-dichloroethylene

dichloroethylene *see* ethylenedichloride

1,1-dichloroethylene (vinylidenechloride; 1,1-dichloroethene; uns. dichloroethylene) CH_2CCl_2
 Uses: adhesives; component of synthetic fibers
 Contains: 0.02% of monomethylether of hydroquinone as inhibitor   (243)
A. PROPERTIES: colorless liquid; m.w. 96.95; m.p. −122.5°C; b.p. 31.9°C; v.p. 500 mm at 20°C, 591 mm at 25°C, 720 mm at 30°C; v.d. 3.25; sp.gr. 1.218 at 20/4°C; sat.conc. 2,640 g/cu m at 20°C, 3,675 g/cu m at 30°C
B. AIR POLLUTION FACTORS: 1 mg/cu m = 0.25 ppm, 1 ppm = 3.97 mg/cu m
 —Odor: characteristic; quality: sweet, chloroformlike
  T.O.C.: 500 ppm   (211)
  O.I. at 20°C = 1,300   (316)
  T.O.C.: 5,500 mg/cu m; 2,000–4,000 mg/cu m   (698; 704)
    0.085 ppm (n.s.i.)
 —Manmade sources: glc's in rural Washington: Dec. 74–Feb. '75: <5 ppt (315)
C. WATER POLLUTION FACTORS:
 —Waste water treatment:

evaporation from water at 25°C of 1 ppm solution:
50% after 22 min
90% after 89 min (313)
measured half life for evaporation from 1 ppm aqueous solution at 25°C, still air, and an average depth of 6.5 cm: 27.2 min. (369)

D. BIOLOGICAL EFFECTS:
—Fish:
Lepomis macrochirus: static bioassay in fresh water at 23°C, mild aeration applied after 24 hr:

material added ppm	% survival after 24 hr	48 hr	72 hr	96 hr	best fit 96 hr LC_{50} ppm
750	0 (5 hr)	—	—	—	
560	0 (8 hr)	—	—	—	
320	0 (<24 hr)	—	—	—	220
180	100	80	70	70	
132	100	100	100	100	

Menidia beryllina: static bioassay in synthetic seawater at 23°C, mild aeration applied after 24 hr:

material added ppm	% survival after 24 hr	48 hr	72 hr	96 hr	best fit 96 hr LC_{50} ppm
320	10	10	0	—	
250	100	90	80	70	250
180	100	90	90	80	(352)

—Mammalia:
rat: slight nose irritation; autopsy, organs normal: 200 ppm, 20 × 6 hr (65)
animals: significant injury to liver and kidney: 25 ppm, 8 hr/day, several months (211)
—Man: irritating to the skin after contact of a few minutes (211)

1,2-dichloroethylene (1,2-dichloroethene; acetylenedichloride)
CHClCHCl

Users and formulation: solvent for fats, phenols, camphor, etc; retard fermentation; rubber mfg.; refrigerant; additive to dye and lacquer solutions; low temperature solvent for heat sensitive substances (e.g., caffeine); constituent of perfumes, thermoplastics; used in organic synthesis and medicine. (347)

A. PROPERTIES: colorless liquid; m.w. 96.95; m.p. −81°C (*cis*), −50°C (*trans*); b.p. 60°C (*cis*), 48°C (*trans*); v.p. 200 mm at 25°C (*cis*), 200 mm at 14°C (*trans*); v.d. 3.34; sp.gr. 1.28 (*cis*), 1.26 (*trans*); solub. 800 mg/l at 20°C (*cis*), 600 mg/l at 20°C (*trans*);

B. AIR POLLUTION FACTORS: 1 mg/cu m = 0.25 ppm, 1 ppm = 3.97 mg/cu m
—Odor: characteristic: quality: ethereal
T.O.C.: 1,100 mg/cu m (*trans*-isomer) (740)

C. WATER POLLUTION FACTORS:
—Reduction of amenities: faint odor: 0.0043 mg/l (129)
—Waste water treatment:
evaporation from water at 25°C of 1 ppm solution,
cis: 50% after 18 min
90% after 64 min

trans: 50% after 24 min
       90% after 83 min (313)
measured half-life for evaporation from 1 ppm aqueous solution at 25°C, still air, and an average depth of 6.5 cm:
cis: 19.4 min
trans: 24.0 min (369)

D. BIOLOGICAL EFFECTS:
—Mammalia:
trans isomer is twice as toxic as the cis isomer
rat: inhalation: no deaths at 8,000 ppm, 4 hr
                 all death 16,000 ppm, 4 hr
cats and rabbits: inhalation: pathological changes: 1,600–1,900 ppm (211)

β,β'-dichloroethylether (sym-dichloroethylether; 2,2'-dichloroethylether; 1-chloro-2-(β-chloroethyoxy)ethane; bis(2-chloroethyl)ether)
$ClCH_2CH_2OCH_2CH_2Cl$

Users and formulation: fumigants; processing fats, waxes, greases, cellulose esters; general solvent; insecticide mfg.; textile mfg. (scour textiles) and cleaning; mfg. butadiene, medicinals and pharmaceuticals; selective solvent; constituent in paints, lacquers, varnishes. (347)

Man caused sources (water and air): formed by chlorination of drinking water when ethyl ether is present. (347)

A. PROPERTIES: m.w. 143.02; m.p. -50°C; b.p. 178°C; v.p. 0.71 mm at 20°C, 1.4 mm at 25°C; v.d. 4.93; sp.gr. 1.22 at 20/4°C; solub. 10,200 mg/l;

B. AIR POLLUTION FACTORS: 1 mg/cu m = 0.17 ppm, 1 ppm = 5.85 mg/cu m
—Odor: characteristic: quality: fruity, pungent
  T.O.C.: <15 ppm (211)
  O.I. at 20°C = 60 (316)

C. WATER POLLUTION FACTORS:
—Water quality:
Kanawha river: 13.11.1963: raw: 55 ppb
                           sand filtered: 10 ppb
               21.11.1963: raw: 154 ppb
                           after aeration: 104 ppb
                           carbon filtered: not detectable (159)
—Reduction of amenities:
approx.conc. causing adverse taste in fish: 1.0 mg/l (41)
T.O.C.: 0.36 mg/l (910; 403)
—Treatment methods:
conventional municipal treatment: infl. 0.055 mg/l; effl. 0.010 mg/l; 82% removal
concentional + A.C.:   infl. 0.154 mg/l; effl. n.d.; ±100% removal
(404)

D. BIOLOGICAL EFFECTS:
—Mammalia: rat: single oral LD_{50}: 75–105 mg/kg
           mouse: single oral LD_{50}: 136 mg/kg
           rabbit: single oral LD_{50}: 126 mg/kg
           guinea pig: inhalation: dangerous: 100–260 ppm, 10–15 hr
                                   no serious disturbance: 100 ppm, 1 hr

rat: inhalation: 3/6: 1,000 ppm, 45 min
rat: guinea pig: inhalation: no effect: 69 ppm, 93 X 7 hr (211)
—Man: irritating >200 ppm (211)

N-(dichlorofluoromethylthio)-N,N-dimethylbenzenesulfonamide *see* euparen

D-(−)-threo-2,2-dichloro-N-(β-hydroxy-α-(hydroxymethyl)-*p*-nitro-phenethyl)-acetamide *see* chloramphenicol

dichloroisopropylether (bis(β-chloroisopropyl)ether; β,β′-dichlorodiisopropylether; dichlorodiisopropylether; 1-chloro-2(β-chloroisopropoxy)propane) [$CH_2Cl(CH_3)CH]_2O$

Manufacturing source: organic chemicals industry: constituent in waterborne wastes from propylene glycol manufacture. (347)
Users and formulation: processing fats, waxes, greases; textiles mfg.; cleaning solution mfg.; intermediate in synthesis; extractant; paint and varnish; spotting agents. (347)
A. PROPERTIES: m.w. 171.07; m.p. −102/−97°C; b.p. 189°C; v.p. 0.85 mm at 20°C, 44.5 mm at 100°C; v.d. 6.0; sp.gr. 1.11; solub. 1,700 mg/l
B. AIR POLLUTION FACTORS: 1 mg/cu m = 0.143 ppm, 1 ppm = 7.0 mg/cu m (65)
—Odor threshold: 0.32 ppm (279)
C. WATER POLLUTION FACTORS:
—Reduction of amenities:
odor threshold: average: 0.32 mg/l, range: 0.017–1.1 mg/l (97)
0.20 mg/l (403)
—Waste water treatment:
A.C.: absorbability: 0.20 g/g C; 100,0% reduction, at 1,008 mg/l infl. (32)
conventional municipal treatment: infl. 0.024 mg/l; effl. n.d.
conventional + A.C.: infl. 0.048 mg/l; effl. n.d. (404)
—Water quality:
Kanawha river: 13.11.1963: raw: 24 ppb
sand filtered: not detectable
21.11.1963: raw: 48 ppb
after aeration: 23 ppb
carbon filtered: not detectable (159)
in river Maas (The Netherlands): in 1973: below detection limit (342)
D. BIOLOGICAL EFFECTS:
—Mammalia:
rat: single oral LD_{50}: 0.24 g/kg
repeated oral doses: decrease in growth rate: 0.01 g/kg 31 days
inhalation: 0/10: 350 ppm, 6 hr
2/5: 350 ppm, 8 hr
1/4: 175 ppm, 8 hr (211)

dichloromethane *see* methylenechloride

3,6-dichloro-2-methoxybenzoic acid *see* 3,6-dichloro-*o*-anisic acid

dichloromethylether (dichloro-dimethylether; chloro(chloromethoxy)methane; BCME; bichloromethylether)
CH_2ClOCH_2Cl

A. PROPERTIES: m.w. 114.96; m.p. −41.5°C; b.p. 104°C; v.d. 3.97; sp.gr. 1.315 at 20/4°C
B. AIR POLLUTION FACTORS: 1 mg/cu m = 0.21 ppm, 1 ppm = 4.75 mg/cu m
   —Atmospheric reactions:
   BCME was stable in air at 10 and 100 ppm levels in 70% relative humidity for at least 18 hr (1366)
C. WATER POLLUTION FACTORS:
   —Aqueous reactions:
   although BCME is considered reactive, it is not readily decomposed by water or aqueous base because of its low solubility (1367)
   BCME in aqueous solution is hydrolyzed very fast with half-life in the order of 10–20 sec, extrapolated to pure water. (1304)
   The hydrolytic reaction can be depicted as follows:

$$Cl-CH_2-O-CH_2-Cl \xrightarrow{H_2O} 2\ HCl + 2\ CH_2O \quad (1365)$$

D. BIOLOGICAL EFFECTS:
   —Man: carcinogenic: all isomers USA 1974 (77)
                                  West Germany 1974 (202)

2,3-dichloro-1,4-naphthoquinone

A. PROPERTIES: m.w. 227.05; m.p. 193°C; v.d. 7.84
B. AIR POLLUTION FACTORS: 1 mg/cu m = 0.11 ppm; 1 ppm = 9.44 mg/cu m
D. BIOLOGICAL EFFECTS:
   —Algae: *Microcystis aeruginosa*: toxic: 0.25 mg/l (30)
   —Protozoa: *Plectonema nostrocorum*: toxic: 0.25 mg/l
   —Fish: *Pimephales promelas*: 24 hr, TLm: 0.24 mg/l
                                  96 hr, TLm: 0.15 mg/l (30)

3,4-dichloronitrobenzene

1,1-DICHLORO-1-NITROETHANE

A. PROPERTIES: m.w. 192; m.p. 43°C; b.p. 255°C; v.d. 6.63
B. AIR POLLUTION FACTORS: 1 mg/cu m = 0.13 ppm, 1 ppm = 7.98 mg/cu m (57)
C. WATER POLLUTION FACTORS:
 —Water quality:
  percolation water at 30–500 m from municipal dumping ground: 1–5 ppb (n.s.i.)
  (183)
 —Reduction of amenities: T.O.C.: ±1 mg/l (2,3-dichloronitrobenzene) (225)

1,1-dichloro-1-nitroethane
$CH_3C(NO_2)Cl_2$

A. PROPERTIES: m.w. 143.9; b.p. 124°C; v.p. 16 mm at 25°C; v.d. 4.97; sp.gr. 1.427 at 20/20°C; solub. 3.5 mg/l at 20°C
B. AIR POLLUTION FACTORS: 1 mg/cu m = 0.170 ppm, 1 ppm = 5.89 mg/cu m
D. BIOLOGICAL EFFECTS:
 —Mammalia: rabbit: oral lethal dose: 0.15–0.20 g/kg
       rabbit, guinea pig: inhalation

2/2	2/2	:	100 ppm, 6 hr
0/2	0/2	:	60 ppm, 2 hr
2/2	0/2	:	52 ppm, 18 hr
0/2	0/2	:	34 ppm, 4 hr
0/2	0/2	:	25 ppm, 204 hr

(211)

2,5-dichloro-4-nitrophenol

D. BIOLOGICAL EFFECTS:
 —Fish: larvae of sea lamprey: LD_{100}: 3 mg/l (226)
    rainbow trout: LD_{10}: 13 mg/l (226)
    brown trout: LD_{10}: 7 mg/l (226)

1,5-dichloropentane
$CH_2Cl-CH_2-CH_2-CH_2-CH_2Cl$

D. BIOLOGICAL EFFECTS:
 —Fish: guppy (*Poecilia reticulata*): 7 d, LC_{50}: <11 mg/l (1833)

dichlorophen see 5,5'-dichloro-2,2'-dihydroxydiphenylmethane

2,3-dichlorophenol

A. PROPERTIES: m.w. 163; m.p. 57-58°C; b.p. 206°C
C. WATER POLLUTION FACTORS:
  —Taste threshold conc.: 0.00004 mg/l (998)
  —Odor threshold conc.: detection: 0.03 mg/l (998)
  —River water quality:
    in Delaware river (U.S.A.): conc. range winter: 0.3 ppb
                     (n.s.i.) summer: n.d. (1051)

2,4-dichlorophenol

Use: organic synthesis
Technical grade contains 2,6-dichlorophenol (8.0%)
A. PROPERTIES: m.w. 163.01; m.p. 45°C; b.p. 210°C; sp.gr. 1.383 at 60/25°C; solub. 4,500 mg/l at 25°C, 4,600 mg/l at 20°C
B. AIR POLLUTION FACTORS:
  —Odor threshold: 0.21 ppm (279)
C. WATER POLLUTION FACTORS:
  —Impact on biodegradation processes
    75% inhibition of nitrification process in non acclimated activated sludge (30)
  —Reduction of amenities:
    approx.conc. causing taste in fish: 0.005 mg/l (41)
    odor threshold: average: 0.21 mg/l
                    range: 0.02-1.35 mg/l (294; 97)
    T.O.C. = 0.002 mg/l (304)
    odor threshold: 0.00065 to 0.0065 mg/l
    taste threshold: 0.008 to 0.02 mg/l (226)
  —Odor thresholds: detection: 0.04 mg/l; 0.002 mg/l (879; 998)
                    0.0003 mg/l (998)
  —Biodegradation:
    decomposition rate in soil suspensions: 9 days for complete disappearance (175)
    degradation in 7 days tests:
                      % degradation
                    average   range
    original culture: 8.0   0-40.0
    1st subculture: 47.0   0-87.0
    2nd subculture: 91.5   72.0-99.5
    3rd subculture: 98.5   97.0-99.5 (87)
    adapted culture: 1% removal after 48 hr incubation, feed = 200 mg/l (292)
    adapted A.S. at 20°C—product is sole carbon source: 98.0% COD removal at 10.5 mg COD/g inoculum/hr (327)
  —Waste water treatment:
    ion exchange: adsorption on Amberlite X AD-4 (25°C): influent: 430 ppm:

2,5-DICHLOROPHENOL

solute adsorbed for zero leakage: 5.09 lbs/cu ft
          10 ppm leakage: 5.49 lbs/cu ft (40)
degradation by *Pseudomonas*: 200 mg/l at 30°C:
        parent: 100% ring disruption in 96 hr
        mutant: 100% ring disruption in 34 hr (152)
oxidation with $KMnO_4$ at pH 7.0, 15 min contact time:

$KMnO_4$ ppm	residual phenol ppm	phenol oxidized ppm	phenol removed %	
0.0	0.98	0.0	0.0	
0.1	0.93	0.05	5.1	
0.5	0.55	0.43	43.9	
0.8	0.33	0.65	66.4	
1.0	0.20	0.78	79.6	
1.2	0.044	0.94	96.0	
2.0	0.00	0.98	100.0	(99)

D. BIOLOGICAL EFFECTS:
  —Toxicity threshold (cell multiplication inhibition test):
    bacteria (*Pseudomonas putida*):   6 mg/l   (1900)
    algae (*Microcystis aeruginosa*):   2 mg/l   (327)
    green algae (*Scenedesmus quadricauda*):   3.6 mg/l   (1900)
    protozoa (*Entosiphon sulcatum*):   0.5 mg/l   (1900)
    protozoa (*Uronema parduczi Chatton-Lwoff*):   1.6 mg/l   (1901)
  —Fish:
    guppy (*Poecilia reticulata*): 24 hr LC_{50}: 4.2 ppm at pH 7.3   (1833)
    goldfish: 24 hr LC_{50}: 7.8 ppm
        amount found in dead fish at 8 ppm: 268 µg/g
        BCF (at 8 ppm): 34   (1850)
  —Mammals:
    rat: single oral LD_{50}: 0.58 g/kg   (211)
    guinea pigs: single oral LD_{50}: 500–1000 mg/kg   (1546)
    ingestion: mice: LD_{50}: 1134 mg/kg   (1609)

2,5-dichlorophenol

A. PROPERTIES: m.w. 163.01; m.p. 58°C; b.p. 211°C at 744 min;
C. WATER POLLUTION FACTORS:
  —Biodegradation:
    decomposition rate in soil suspensions: >72 days for complete disappearance (175)
    degradation by *Pseudomonas*: 200 mg/l at 30°C,
        parent: 60% ring disruption in 120 hr
        mutant: 100% ring disruption in 38 hr (152)

—Reduction of amenities: odor threshold: 0.00045 to 0.0033 mg/l (226)
　　　　　　　　　　　　　　　　　　　　　0.0005 mg/l (998)
　　　　　　　　　　　　detection: 0.03 mg/l (998)
D. BIOLOGICAL EFFECTS:
　—Fish:
　　guppy (*Poecilia reticulata*): 24 hr LC_{50}: 2.7 ppm at pH 7.3 (1833)

2,6-dichlorophenol

natural occurrence: sex pheromone of the lone star tick, *Amblyomma americanum*
A. PROPERTIES: m.w. 163; m.p. 65–66°C; b.p. 218–220°C
B. AIR POLLUTION FACTORS:
　—Taste threshold conc.: 0.0002 mg/l (998)
　　　　　　　　　　　　　0.002 mg/l (226)
　—Odor threshold: detection: 0.003 mg/kg (226; 879)
　　　　　　　　　　　　　　0.0075 mg/kg (894)
　　　　　　　　　　　　　　0.2 mg/l (998)
D. BIOLOGICAL EFFECTS:
　—Fish:
　　Idus idus melanotus: 48 hr, LC_0: 2 mg/l
　　　　　　　　　　　　　48 hr, LC_{50}: 4 mg/l
　　　　　　　　　　　　　48 hr, LC_{100}: 5 mg/l (998)

dichlorophenolindophenol
C. WATER POLLUTION FACTORS:
　—~50% inhibition of NH_3 oxidation in *Nitrosomonas* at 250 mg/l (407)

2,4-dichlorophenoxyacetic acid (2,4-D)

Use: systemic herbicide
A. PROPERTIES: white powder; solub. 890 ppm at 25°C; sp.gr. (powder) 1.416 (25°C); m.w. 221.04; m.p. 136–140°C; v.d. 7.63; log P_{oct} 2.81 at 20°C
C. WATER POLLUTION FACTORS:
　—Biodegradation: 26 days for ring cleavage in soil suspension (1827)
　—Metabolic pathways in plants:
　　2,4-D in plants very rapidly undergoes various transformations and its predominant metabolic pathways and rates vary with different species. In bean and

soybean plants major metabolites are 4-O-β-D-glucosides of 4-hydroxy-2,5-dichloro- and 4-hydroxy-2,3-dichlorophenoxy-acetic acids; in addition considerable amounts of N-(2,4-dichlorophenoxy-acetyl)-L-aspartic and N-(2,4-dichlorophenoxy-acetyl)-L-glutamic acids are accumulating in them. Among metabolites in cereals there prevailed 1-O-(2,4-dichlorophenoxy-acetyl)-β-D-glucose while the glycoside of 2,4-dichlorophenol prevailed in strawberry plants.
(1609)

—Odor threshold: detection: 3.13 mg/kg (915)

—Treatment methods:

A.C. type BL (Pittsburgh Chem. Co.): % adsorbed by 10 mg A.C. from $10^{-4} M$ aqueous solution at pH 3.0: 60.1%
pH 7.0: 18.8%
pH 11.0: 14.3% (1313)

D. BIOLOGICAL EFFECTS:

—Residues:

	conc. (ppm)	time of collection after herbicide spraying
residues in lingonberry	<0.05–7.0	2–13 weeks
(N.E. Finland) wild mushroom	<0.05–1.2	2–13 weeks
birch and aspen foliage	0.15–31	13–43 weeks (1598)

—Toxicity:

freshwater ectoprocta: no appreciable effect at 2.5 mg/l for 84 hr exposure
(1902)

—Algae:

Chlorococcum sp.	6×10^4 ppb	50% decrease in O_2 evolution	f	technical acid
Chlorococcum sp.	5×10^4 ppb	50% decrease in growth	measured as ABS. (525 mu) after 10 days	technical acid
Dunaliella tertiolecta	9×10^4 ppb	50% decrease in O_2 evolution	f	technical acid
Dunaliella tertiolecta	7.5×10^4 ppb	50% decrease in growth	measured as ABS. (525 mu) after 10 days	technical acid
Isochrysis galbana	6×10^4 ppb	50% decrease in O_2 evolution	f	technical acid
Isochrysis galbana	5×10^4 ppb	50% decrease in growth	measured as ABS. (525 mu) after 10 days	technical acid (2348)
Phaeodactylum tricornutum	6×10^4 ppb	50% decrease in O_2 evolution	f	technical acid
Phaeodactylum tricornutum	5×10^4 ppb	50% decrease in growth	measured as ABS. (525 mu) after 10 days	technical acid (2348)

blue green alga (*Anabaenopsis Raciborskii*):
at 10 mg/l: stimulated growth and nitrogen fixation
100 mg/l: no significant inhibition of growth
1000 mg/l: complete inhibition of growth (1616)

—Worms: Tubifex tubifex: toxic: 80 mg/l (30)
—Molluscs:
American oyster (*Crassostrea virginica*): Egg: 48 hr static lab bioassay: TLm (Ester): 8×10^3 ppb
American oyster (*Crassostrea virginica*): Larvae: 14 day static lab bioassay: TLm (Ester): 740 ppb
American oyster (*Crassostrea virginica*): Egg: 48 hr static lab bioassay: TLm (Salt): 2.044×10^4 ppb
American oyster (*Crassostrea virginica*): Larvae: 14 day static lab bioassay: TLm (Salt): 6.429×10^4 ppb (2324)
—Fish:
bluegill: 48 hr, LC_{50}: 0.9 ppm
rainbow trout: 48 hr, LC_{50}: 1.1 ppm (1878)
striped bass: 96-hr LC_{50}: 70.1 mg/l (S)
banded killifish: 96-hr LC_{50}: 26.7 mg/l (S)
pumpkinseed: 96-hr LC_{50}: 94.6 mg/l (S)
white perch: 96-hr LC_{50}: 40.0 mg/l (S)
american eel: 96-hr LC_{50}: 300.6 mg/l (S)
carp: 96-hr LC_{50}: 96.5 mg/l (S)
guppy: 96-hr LC_{50}: 70.7 mg/l (S)
bleak (*Alburnus alburnus*): sodium salt

LC_{50} mg/l	12 hr	24 hr	36 hr	48 hr	
embryos	159	129	64	13	
larvae	111	71	62	52	(1903)

—Mammals:
mice: 29% inhibition of testicular DNA synthesis at 200 mg/kg ($p < 0.05$)
(1325)
rats: acute oral LD_{50}: 300–1000 mg/kg
mice: acute oral LD_{50}: 375 mg/kg
dog: acute oral LD_{50}: 100 mg/kg (1546)
mice: ingestion: LD_{50}: 521 mg/kg (1609)

3,4-dichlorophenoxyacetic acid

Cl—⟨benzene ring⟩—O—CH$_2$COOH
   |
   Cl

C. WATER POLLUTION FACTORS:
—Biodegradation: >205 days for ring cleavage in soil suspension (1827)

2,4-dichlorophenoxyacetic acid, amine 80
Use: herbicide
active ingredients: mixed amines of 2,4-D
C. WATER POLLUTION FACTORS:
—2,4-D residues in soil following application of 2,4-D amine formulation (ppm):

	5.6 kg/ha		11.2 kg/ha		22.4 kg/ha	
days after treatment	0–10 cm	10–20 cm	0–10 cm	10–20 cm	0–10 cm	10–20 cm
1	0.87	–	1.61	–	4.41	–
14	1.84	–	4.82	–	6.94	–
28	.31	–	.96	–	2.15	–
42	.10	.02	.12	.21	.42	.56
56	.10	.08	.15	.22	.48	.55
70	.04	.02	.22	.07	.18	.18
265	ND	ND	.04	Tr	.08	.01
385	ND	ND	.01	ND	.05	Tr

(1088)

Tr = trace = <0.01 ppm
ND = <0.005 ppm

2,4-dichlorophenoxyacetic acid, butoxyethanolester (2,4-D(BEE))

$$Cl-C_6H_3(Cl)-O-CH_2-C(=O)-O-CH_2CH_2OCH_2CH_2CH_2CH_3$$

A. PROPERTIES: amber liquid; sp.gr. 1.225 (68/68°F)
D. BIOLOGICAL EFFECTS:
 —Algae:

	conc. ppb	methods of assessment	test procedure
Chlorococcum sp.	1×10^5	50% decrease in O_2 evolution	f
Chlorococcum sp.	7.5×10^4	50% decrease in growth	Measured as ABS. (525 mu) after 10 days
Dunaliella tertiolecta	1×10^5	50% decrease in O_2 evolution	f
Dunaliella tertiolecta	7.5×10^4	50% decrease in growth	Measured as ABS. (525 mu) after 10 days
Isochrysis galbana	1×10^5	50% decrease in O_2 evolution	f
Isochrysis galbana	7.5×10^4	50% decrease in growth	Measured as ABS. (525 mu) after 10 days
Phaeodactylum tricornutum	2×10^5	50% decrease in O_2 evolution	f
Phaeodactylum tricornutum	1.5×10^5	50% decrease in growth	Measured as ABS. (525 mu) after 10 days

 —Crustaceans
  Gammarus lacustris: 96 hr, LC_{50}: 440 µg/l   (2124)
  Gammarus fasciatus: 96 hr, LC_{50}: 5900 µg/l   (2125)
  Daphnia magna: 48 hr, LC_{50}: 5600 µg/l   (2125)

Cypridopsis vidua: 48 hr, LC_{50}: 1800 µg/l (2125)
Asellus brevicaudus: 48 hr, LC_{50}: 3200 µg/l (2125)
Palaemonetes kadiakensis: 48 hr, LC_{50}: 1400 µg/l (2125)
Orconectes nais: 48 hr no effect level: 100,000 µg/l (2125)
—Insects
Pteronarcys californica: 96 hr, LC_{50}: 1600 µg/l (2128)
—Fish
Pimephales promelas: 96 hr, LC_{50}: 5600 µg/l
  1500 µg/l lethal to eggs in 48 hour exposure
  10 mo. no effect level: 300 µg/l (2122)
—Mammals:
  male rats: acute oral LD_{50}: 940 mg/kg
  rabbits: skin absorption: LD_{50}: in the range of 4000 mg/kg (1546)

2,4-dichlorophenoxyacetic acid, butylthio
D. BIOLOGICAL EFFECTS:
—Mosquitofish: 48 hr LC_{50} (S): 0.98 mg/l (susceptible population)
  48 hr LC_{50} (S): 1.70 mg/l (resistant population)

2,4-dichlorophenoxyacetic acid, diethylamine salt
D. BIOLOGICAL EFFECTS:
—Crustaceans
Gammarus fasciatus: 48 hr no effect level: 100,000 µg/l (2125)
Daphnia magna: 48 hr, LC_{50}: 4000 µg/l (2125)
Cypridopsis vidua: 48 hr, LC_{50}: 8000 µg/l (2125)
Asellus brevicaudus: 48 hr no effect level: 100,000 µg/l (2125)
Palaemonetes kadiakensis: 48 hr no effect level: 100,000 µg/l (2125)
Orconectes nais: 48 hr no effect level: 100,000 µg/l (2125)

2,4-dichlorophenoxyacetic acid, dimethylamine salt
A. PROPERTIES: m.p. 85–87°C
C. WATER POLLUTION FACTORS:
—experimental concentrations of 0.5 mg/l can significantly taint the flesh of rainbow trouts to make them unpalatable (1788)
D. BIOLOGICAL EFFECTS:
—Lowest observed avoidance conc.:
  mayfly nymphs (*Ephemerella walkers*): >10 mg/l
  rainbow trout (*Salmo gairdneri*): 1.0 mg/l (1621)

2,4-dichlorophenoxyacetic acid, ethylthio
D. BIOLOGICAL EFFECTS:
—Mosquitofish: 48-hr LC_{50}: 1.34 mg/l (susceptible population) (S)
  48-hr LC_{50}: 1.58 mg/l (resistant population) (S) (1516)

2,4-dichlorophenoxyacetic acid, isooctylester (2,4-D(IOE))
A. PROPERTIES: dark liquid; b.p. 317°C; solub. 10 mg/l; sp.gr. 1.152
D. BIOLOGICAL EFFECTS:

-Crustacean: *Gammarus lacustris*: 96 hr, LC_{50}: 2400 µg/l (2124)
-Mammals:
   rat: single oral LD_{50}: in the range of 500–1000 mg/kg
   rabbit: skin absorption: LD_{50}: >4000 mg/kg

2,4-dichlorophenoxyacetic acid, sodium salt, monohydrate

$$Cl-C_6H_3(Cl)-OCH_2COONa \cdot H_2O$$

A. PROPERTIES: solub. 10% at 30°C
D. BIOLOGICAL EFFECTS:
 —Mammals:
   ingestion: rat: single oral LD_{50}: 650–800 mg/kg
              rabbit: single oral LD_{50}: 800 mg/kg
              mice: single oral LD_{50}: 375 mg/kg
              guinea pig: single oral LD_{50}: 550–1000 mg/kg (1546)

N-(2,4-dichlorophenoxyacetyl)-/-aspartic acids
Metabolite of 2,4-D in plants
D. BIOLOGICAL EFFECTS: mice: LD_{50}: ~600 mg/kg (1609)

N-(2,4-dichlorophenoxyacetyl)-/-glutamic acids
Metabolite of 2,4-D in plants (1609)
D. BIOLOGICAL EFFECTS: mice: LD_{50}: ~600 mg/kg (1609)

2,4-dichlorophenoxy-ω-butyric acid

$$Cl-C_6H_3(Cl)-OCH_2CH_2CH_2COOH$$

C. WATER POLLUTION EFFECTS:
 —Biodegradation: 11 days for ring cleavage in soil suspension (1827)

2,4-dichlorophenoxy-α-propionic acid

$$Cl-C_6H_3(Cl)-O-CH(COOH)-CH_3$$

C. WATER POLLUTION FACTORS:
 —Biodegradation: >205 days for ring cleavage in soil suspension (1827)

D. BIOLOGICAL EFFECTS:
—Mice: 14% inhibition of testicular DNA synthesis at 200 mg/kg (n.s.i.) (1325)

2,4-dichlorophenoxy-ω-propionic acid

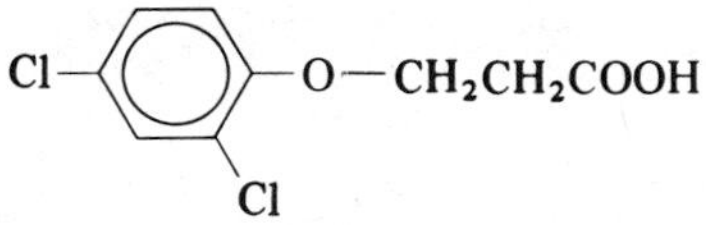

Cl—⟨C₆H₃(Cl)⟩—O—CH₂CH₂COOH

C. WATER POLLUTION FACTORS:
—Biodegradation: 4 days for ring cleavage in soil suspension (1827)

2,4-dichlorophenoxy-α-valeric acid

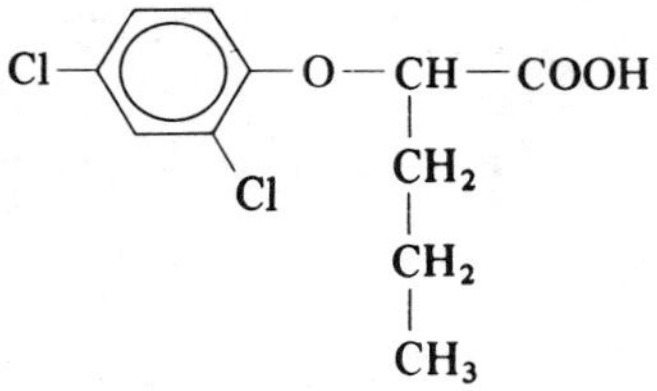

Cl—⟨C₆H₃(Cl)⟩—O—CH(COOH)—CH₂—CH₂—CH₃

C. WATER POLLUTION FACTORS:
—Biodegradation: >81 days for ring cleavage in soil suspension (1827)

2,4-dichlorophenylacetic acid

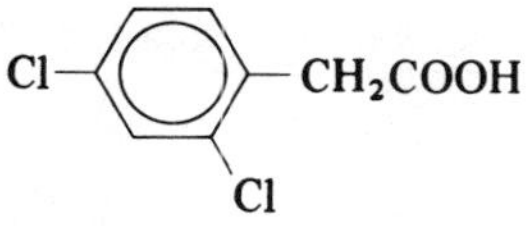

Cl—⟨C₆H₃(Cl)⟩—CH₂COOH

C. WATER POLLUTION FACTORS:
—Treatment methods:
A.C. type BL (Pittsburgh Chem. Co.): % adsorpbed by 10 mg A.C. from $10^{-4}M$ aqueous solution at pH 3.0: 61.3%
pH 7.0: 21.8%
pH 11.0: 11.5% (1313)

1,6-di(4-chlorophenyldiguanido)hexane (chlorhexidine)

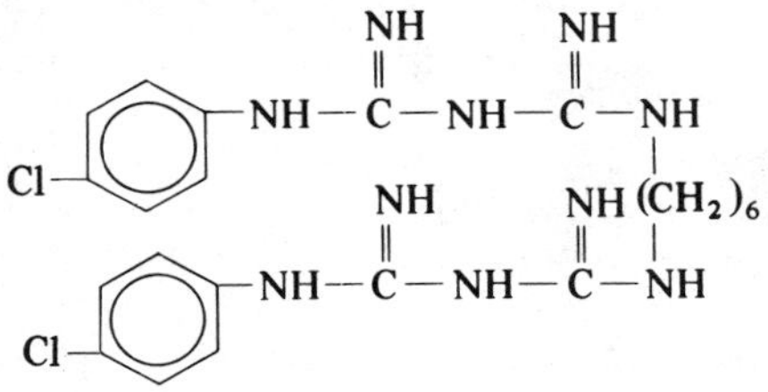

C. WATER POLLUTION FACTORS:
—Degradation: after 3 weeks of adaptation at 12 mg/l at 22°C:
    aerobic conditions: product is sole carbon source: 0% degradation
                                + synthetic sewage: 60–100% degradation
    anaerobic conditions: product is sole carbon source: 0% degradation
                                  + synthetic sewage: 0% degradation   (512)

D. BIOLOGICAL EFFECTS:
—Crustacean:
Daphnia magna: 24 hr, LC_{50}: 740 µg/l
                48 hr, LC_{50}: 250 µg/l   (1002, 1004)
—Insects:
Pteronarcys: 48 hr, LC_{50}: 0.5 µg/l   (1503)
Hydropsyche larvae: significant modification of net construction after 48 hr
    exposure: ±0.5 µg/l   (1006; 1007; 1008)
—Fish:
Salmo gairdneri: 96 hr, LC_{50}: 1.3 µg/l   (1001)

3-(3,4-dichlorophenyl)-1,1-dimethylurea (Diuron; N'-(3,4-dichlorophenyl)-N,N-dimethylurea; DMU; DCMU)

$$Cl-\underset{Cl}{C_6H_3}-NH-\underset{\underset{O}{\|}}{C}-N(CH_3)_2$$

Use: herbicide

A. PROPERTIES: m.p. 158–159°C; solub. 42 mg/l at 25°C in distilled water; v.p. 3.1 × 10^{-6} at 50°C

C. WATER POLLUTION FACTORS:
—Water treatment method:
    powdered A.C.: Freundlich adsorption parameters: K: 0.095; $1/n$ = 0.109
    carbon dose to reduce 5 mg/l to a final conc. of 0.1 mg/l: 66 mg/l
    carbon dose to reduce 1 mg/l to a final conc. of 0.1 mg/l: 12 mg/l   (594)

D. BIOLOGICAL EFFECTS:
—Algae:

Protococcus	0.02 ppb	.52 OPT. DEN. expt/ OPT DEN control	10 day growth test	(2347)
Chlorella sp.	4.00 ppb	.34 OPT. DEN. exp/ OPT DEN control	10 day growth test	(2347)
Chlorococcum sp.	10 ppb	61% inhibition of growth	10 day growth test	(2347)
Chlorococcum sp.	20 ppb	50% reduction O_2 evolution	10 day growth test	(2348)
Chlorococcum sp.	10 ppb	50% reduction in growth	10 day growth test	(2348)
Dunaliella tertiolecta	10 ppb	50% reduction O_2 evolution	10 day growth test	(2348)

Dunaliella tertiolecta	20 ppb	50% reduction in growth	10 day growth test	(2348)
Dunaliella euchlora	0.4 ppb	.44 OPT. DEN. expt/ OPT. DEN control	10 day growth test	(2347)
Isochrysis galbana	10 ppb	50% reduction O_2 evolution	10 day growth test	(2348)
Isochrysis galbana	10 ppb	50% reduction in growth	10 day growth test	(2348)
Monochrysis lutheri	0.02 ppb	.00 optical density expt/ optical density control	10 day growth test	(2347)
Phaeodactylum tricornutum	0.4 ppb	.79 OPT. DEN expt/ OPT DEN control	10 day growth test	(2347)
Phaeodactylum tricornutum	4.0 ppb	.00 OPT. DEN expt/ OPT DEN control	10 day growth test	(2347)
Phaeodactylum tricornutum	10. ppb	50% reduction O_2 evolution	10 day growth test	(2348)
Phaeodactylum tricornutum	10. ppb	50% reduction in growth	10 day growth test	(2348)

    blue green alga: *Agmenellum quadruplicatum* (strain PR 6): lethal at 2×10^{-6} M (n.s.i.)   (1093)

—Crustaceans
  Gammarus lacustris: 96 hr, LC_{50}: 160 µg/l   (2124)
  Gammarus fasciatus: 96 hr, LC_{50}: 700 µg/l   (2125)
  Simocephalus serrulatus: 48 hr, LC_{50}: 2000 µg/l   (2127)
  Daphnia pulex: 48 hr, LC_{50}: 1400 µg/l   (2127)
  Daphnia magna: 24 hr, LC_{50}: 1400 µg/l   (1002, 1004)

—Insects:
  Pteronarcys: 48 hr, LC_{50}: 1.2 µg/l   (1003)
  Pteronarcys californica: 96 hr, LC_{50}: 1200 µg/l   (2124)

—Fish:
  Oncorhynchus kisutch: 48 hr, LC_{50}: 16000 µg/l   (2106)
  harlequin fish (*Rasbora heteromorpha*):

mg/l	24 hr	48 hr	
LC_{10} (F)	110	150	
LC_{50} (F)	200	190	(331)

  bluegill: 48 hr, LC_{50}: 7.4 ppm
  rainbow trout: 48 hr, LC_{50}: 4.3 ppm   (1878)

—Mammals:
  acute oral LD_{50} (rat): 3400 mg/kg   (1854)
  in diet: no effect on rats and dogs fed for 2 and 1 year respectively on diets containing 250 and 500 ppm   (1855)

dichlorophenylisocyanate

C. WATER POLLUTION FACTORS:
  —Biodegradation rate:
    degradation by mutant organisms: 500 mg/l at 20°C:
      parent: 92% ring disruption in 48 hr
      mutant: 48% ring disruption in 8 hr   (152)

3-(3,4-dichlorophenyl)-1-methoxy-1-methylurea (linuron; lorox)

$$\text{Cl-C}_6\text{H}_3(\text{Cl})\text{-NH-C(=O)-N(CH}_3\text{)(OCH}_3\text{)}$$

Use: pesticide
A. PROPERTIES: m.p. 93–94°C; m.w. 249.1; v.p. 1.1×10^{-5} mm at 24°C; solub. 75 mg/l
C. WATER POLLUTION FACTORS:
—Wastewater treatment methods:
powdered A.C.: Freundlich adsorption parameters; K: 0.080; $1/n = 0.157$
carbon dose to reduce 5 mg/l to a final conc. of 0.1 mg/l: 88 mg/l
carbon dose to reduce 1 mg/l to a final conc. of 0.1 mg/l: 16 mg/l   (594)
D. BIOLOGICAL EFFECTS:
—Mammals:
acute oral LD_{50} (rats): 1500–4000 mg/kg
            (dogs):   500 mg/kg   (1855)

3-(3,4-dichlorophenyl)-1-methyl-1-*n*-butylurea (Neburon)

$$\text{Cl}_2\text{C}_6\text{H}_3\text{-NH-C(=O)-N(CH}_3\text{)(C}_4\text{H}_8\text{)}$$

Use: pesticide
A. PROPERTIES: m.p. 102–103°C; solub 6 mg/l
C. WATER POLLUTION FACTORS:
—Water treatment methods:
powdered A.C.: Freundlich adsorption parameters: K: 0.080; $1/n$: 0.301
carbon dose to reduce 5 mg/l to a final conc. of 0.1 mg/l: 122 mg/l
carbon dose to reduce 1 mg/l to a final conc. of 0.1 mg/l: 22 mg/l   (594)
D. BIOLOGICAL EFFECTS:
—Algae

Protococcus sp.	40 ppb	.41 OD expt/OD control	10 day growth test	(2347)
Chlorella sp.	40 ppb	.31 OD expt/OD control	10 day growth test	(2347)
Chlorococcum sp.	30 ppb	68% inhibition in growth	10 day growth test technical acid	(2349)
Chlorococcum sp.	20 ppb	50% decrease O_2 evolution	f technical acid	(2348)

Chlorococcum sp.	30 ppb	50% decrease growth	10 day growth test	technical acid	(2348)
Dunaliella tertiolecta	20 ppb	50% decrease O_2 evolution	f	technical acid	(2348)
Dunaliella tertiolecta	40 ppb	50% decrease growth	10 day growth test	technical acid	(2348)
Dunaliella euchlora	40 ppb	.47 OD expt/OD control	10 day growth test		(2347)
Isochrysis galbana	20 ppb	50% decrease O_2 evolution	f	technical acid	(2348)
Isochrysis galbana	30 ppb	50% decrease growth	10 day growth test	technical acid	(2348)
Monochrysis lutheri	40 ppb	.00 OD expt/OD control	10 day growth test		(2347)
Phaeodactylum tricornutum	40 ppb	.10 OD expt/OD control	10 day growth test		(2347)
Phaeodactylum tricornutum	40 ppb	50% decrease O_2 evolution	f	technical acid	(2348)
Phaeodactylum tricornutum	30 ppb	50% decrease growth	10 day growth test	technical acid	(2348)

—Mammals:
  acute oral LD_{50} (rats): >11,000 mg/kg (1855)

N-(3,4-dichlorophenyl)-2-methylpentanamide (karsil-pentanamide; karsil-oxathün)
Use: pesticide
A. PROPERTIES: solub. 8 mg/l
C. WATER POLLUTION FACTORS:
  —Water treatment methods:
    powdered A.C.: Freundlich adsorption parameters: K: 0.10; $1/n$: 0.238
      carbon dose to reduce 5 mg/l to a final conc. of 0.1 mg/l: 84 mg/l
      carbon dose to reduce 1 mg/l to a final conc. of 0.1 mg/l: 15 mg/l   (594)

2,4-dichlorophenyl-*p*-nitrophenylether

$$Cl\text{-}C_6H_3(Cl)\text{-}O\text{-}C_6H_4\text{-}NO_2$$

Use: herbicide
D. BIOLOGICAL EFFECTS:
  —Fish:
    Rasbora trilineata: 96 hr, LC_{50} (S): 0.756 mg/l
    guppy: 96 hr, LC_{50} (S): 1.59 mg/l   (1513)

N-(3,4-dichlorophenyl)propionamide (3,4-dichloropropionanilide; Stam F-34; propanil)

$$\text{NH} - \overset{\overset{\displaystyle O}{\|}}{C} - C_2H_5$$

(on 3,4-dichlorophenyl ring)

Use: post-emergence contact herbicide

A. PROPERTIES: m.p. 90.6–91.6°C; v.p. 9×10^{-5} mm Hg at 60°C; solub. 50–225 ppm at room temp.

C. WATER POLLUTION FACTORS:
 —Waste water treatment methods:
  powdered A.C.: Freundlich adsorption parameters: K: 0.060; $1/n$: 0.222
   carbon dose to reduce 5 mg/l to a final conc. of 0.1 mg/l: 136 mg/l
   carbon dose to reduce 1 mg/l to a final conc. of 0.1 mg/l: 25 mg/l (594)

D. BIOLOGICAL EFFECTS:
 —Crustaceans:
  Gammarus fasciatus: 96 hr, LC_{50}: 16000 µg/l (2125)
  Daphnia magna: IC_{50}: 4.8 ppm (1855)
 —Fish:
  mosquitofish: 72-hr LC_{50} (S): 7.62 mg/l
  green sunfish: 72-hr LC_{50} (S): 5.85 mg/l (1203)
  Channel catfish: 96-hr LC_{50} (S): 3,796 µg/l (1202)
  mosquitofish: 24 hr LC_{50} (S): 11.3 mg/l
  mosquitofish: 48 hr LC_{50} (S): 11.0 mg/l
  mosquitofish: 96 hr LC_{50} (S): 9.46 mg/l (1104)
 —Mammals:
  acute oral LD_{50} (rat): 1384 mg/kg (technical product) (1854)
  acute dermal LD_{50} (rabbits): 7080 mg/kg (1855)

1,2-dichloropropane (propylenechloride; propylenedichloride)
$CH_2ClCHClCH_3$

Uses: intermediate for perchloro-ethylene and carbontetrachloride; lead scavenger for antiknock fluids; solvent; soil fumigant for nematodes

A. PROPERTIES: colorless liquid; m.w. 112.99; m.p. $-100/-80°C$; b.p. 96.8°C; v.p. 42 mm at 20°C, 50 mm at 25°C, 66 mm at 30°C; v.d. 3.9; sp.gr. 1.16 at 20/20°C; solub. 2,700 mg/l at 20°C; sat.conc. 258 g/cu m at 20°C, 393 g/cu m at 30°C

B. AIR POLLUTION FACTORS: 1 mg/cu m = 0.21 ppm, 1 ppm = 4.62 mg/cu m
 —Odor: characteristic: quality: sweet
     hedonic tone: pleasant
  T.O.C.: 235 mg/cu m = 50 ppm (57)
  O.I.: 100% recognition: 87,096 (19)
  O.I. at 20°C = 1,100 (316)

C. WATER POLLUTION FACTORS:
  −Odor threshold: detection: 0.0014 mg/kg (n.s.i.) (915)
  −river water quality:
    in river Maas at Eysden (Netherlands) in 1976: 0.1 µg/l; range: n.d. to 28.7 µg/l
    in river Maas at Keizersveer (Netherlands) in 1976: 0.1 µg/l; range: n.d. to 0.3 µg/l (n.s.i.) (1368)
  −Waste water treatment:
    activated carbon: adsorbability: 0.183 g/g C, 92.9% reduction, infl.: 1,000 mg/l; effl.: 71 mg/l (32)
D. BIOLOGICAL EFFECTS:
  −Crustacean: shrimp: TLm 48 hr: >100 ppm (362)
  −Fish:
    guppy (*Poecilia reticulata*): 7 d LC_{50}: 116 ppm (1833)
    Lepomis macrochirus: static bioassay in fresh water at 23°C, mild aeration applied after 24 hr:

material added		% survival after			best fit 96 hr LC_{50}
ppm	24 hr	48 hr	72 hr	96 hr	ppm
560	0	−	−	−	
320	60	50	50	50	320
180	100	100	100	100	

  Menidia beryllina: static bioassay in synthetic seawater at 23°C: mild aeration applied after 24 hr:

material added		% survival after			best fit 96 hr LC_{50}
ppm	24 hr	48 hr	72 hr	96 hr	ppm
320	10	10	10	0	
240	100	80	50	50	240
180	70	70	70	70	
100	100	100	100	100	(352)

  −Mammals:
    rats, guinea pigs, dogs: no ill effects and no histological changes, inhalation at 400 ppm, 7 hr/day, 5 day/w, 128−140 days (211)
    guinea pigs: oral lethal dose: 2−4 g/kg (211)
  −Man: mild skin iritation (211)

1,3-dichloropropane
$CH_2Cl-CH_2-CH_2Cl$
D. BIOLOGICAL EFFECTS:
  −Fish:
    guppy (*Poecilia reticulata*): 7 d LC_{50}: 84 ppm (1833)

1,3-dichloro-1-propene (1,3-dichloro-1-propylene)
$CHCl=CHCH_2Cl$
  Use: soil fumigant; nematocide
  1 to 2% epichlorohydrin often added as stabilizer and corrosion inhibitor
A. PROPERTIES: m.w. 110.97; b.p. 104°C(*cis*), 112°C(*trans*); v.d. 3.83; solub. *cis* 2700 mg/l, *trans* 2800 mg/l; v.p. *cis* 43 mm at 25°C, *trans* 34 mm at 25°C; a clear light straw-colored liquid, sharp, sweet, penetrating and irritating odor

2,3-DICHLORO-1-PROPENE

B. AIR POLLUTION FACTORS: 1 mg/cu m = 0.22 ppm, 1 ppm = 4.61 mg/cu m (57)
   —Odor:
     at 3 ppm: odor detected by 7 of 10 volunteers
     at 1 ppm: odor detected by 7 of 10 volunteers, but noticeably fainter than 3 ppm (992)
C. WATER POLLUTION FACTORS:
   —Waste water treatment:
     evaporation from water at 25°C of 1 ppm solution: 50% after 31 min
     (*cis* and *trans*): 90% after 98 min (313)
D. BIOLOGICAL EFFECTS:
   —Algae: *Scenedesmus*: deterioration from 40 mg/l onwards (n.s.i.)
   —Arthropoda: *Daphnia*: deterioration from 40 mg/l onwards (n.s.i.)
   —Protozoa: *Colpoda*: deterioration from 100 mg/l onwards (n.s.i.) (30)
   —Mammals:
     rats: inhalation: 1 ppm, 4 hr/day, 6 months: no effect
     24 guinea pigs, 6 rabbits, 2 dogs: 125–130 exposures in a period of 185 days at 3 ppm: no effect (492)
     rats: male: acute oral LD_{50}: 713 mg/kg
           female: acute oral LD_{50}: 470 mg/kg
           inhalation: 0/6: 1000 ppm, 1 hr
                       4/4: 1000 ppm, 2 hr
                       0/5: 400 ppm, 7 hr
     guinea pigs: 5/5: 400 ppm, 7 hr
     rats, guinea pigs:
       inhalation: 19 repeated 7 hr exposures in a period of 28 days at 50 ppm: marked liver and kidney changes; necrosis
       inhalation: 27 repeated 7 hr exposures in a period of 39 days at 11 ppm: slight to marked kidney and liver changes, evidence of kidney injury in most of the guinea pigs.
     rats, inhalation: 3 ppm, 4 hr/day, 6 months: very slight cloudy swelling of the renal tubular epithelium (492)

2,3-dichloro-1-propene (2,3-dichloro-1-propylene)
$CH_2=CCl-CH_2Cl$

A. PROPERTIES: straw liquid pungent odor; b.p. 201°F; solub. 2,150 mg/l; v.p. 53 mm at 25°C; v.d. 3.8; sp.gr. 1.20
C. WATER POLLUTION FACTORS:
   —Waste water treatment:
     evaporation from water at 25°C of 1 ppm solution:
       50% after 20 min
       90% after 68 min (313)
D. BIOLOGICAL EFFECTS:
   —Fish:
     guppy (*Poecilia reticulata*): 7 d LC_{50}: $\ll 100$ (1833)
   —Mammals:
     rats: single oral LD_{50}: in range of 250 mg/kg when diluted to 20% or less (1546)

2,2-dichloropropionic acid, sodium salt (DPA; Basfapon; Dalapon-Na; Ded-Weed; Dowpon; M. Gramevin; Radapon; Unipon)

$$CH_3-\underset{\underset{Cl}{|}}{\overset{\overset{Cl}{|}}{C}}-COONa$$

Use: selective herbicide
Composition: commercial products contain 85% sodium salt or mixed sodium and magnesium salts

A. PROPERTIES: solub. 500 g/l at 25°C; decomp. at 166°C
C. WATER POLLUTION FACTORS:
  −75–100% disappearance from soils: 8 weeks (1815)
D. BIOLOGICAL EFFECTS:
  −Algae:

Chlorococcum sp.	2.5×10^4 ppb	50% decrease in O_2 evolution	f
Chlorococcum sp.	5×10^4 ppb	50% decrease in growth	Measured as ABS. (525 mu) after 10 days
Dunaliella tertiolecta	2.5×10^4 ppb	50% decrease in O_2 evolution	f
Dunaliella tertiolecta	$1. \times 10^5$ ppb	50% decrease in growth	Measured as ABS. (525 mu) after 10 days
Isochrysis galbana	4×10^4 ppb	50% decrease in O_2 evolution	f
Isochrysis galbana	2×10^4 ppb	50% decrease in growth	Measured as ABS. (525 mu) after 10 days
Phaeodactylum tricornutum	2.5×10^4 ppb	50% decrease in O_2 evolution	f
Phaeodactylum tricornutum	2.5×10^4 ppb	50% decrease in growth	Measured as ABS. (525 mu) after 10 days

  −Crustaceans
  Simocephalus serrulatis: 48 hr, LC_{50}: 16000 μg/l (2127)
  Daphnia pulex: 48 hr, LC_{50}: 11000 μg/l (2127)
  −Insect
  Pteronarcys californica: 96 hr no effect level: 100,000 μg/l (2128)
  −Fish
  Pimephales promelas: 96 hr, LC_{50}: 290000 μg/l (2131)
  Lepomis macrochirus: 96 hr, LC_{50}: 290000 μg/l (2131)
  Oncorhynchus kisutch: 48 hr, LC_{50}: 340000 μg/l (2106)
  Rasbora trilineata: 96 hr LC_{50} (S): 135 mg/l
  guppy: 96 hr LC_{50} (S): 223 mg/l (1513)
  bluegill: 48 hr LC_{50}: 115 mg/l (1878)
  lowest observed avoidance conc.:
    mayfly nymphs (*Ephemerella walkeri*): >10 mg/l (1621)
    rainbow trout (*Salmo gairdneri*): 1.0 mg/l (1621)

510   1,3-DICHLORO-1-PROPYLENE

—Mammals:
acute oral LD_{50} (male rats): 9330 mg/kg
(female rats): 7570 mg/kg
in diet: in rats fed 50 mg/kg/day for 2 years, there was a slight increase in kidney weight but no effects were observed at 15 mg/kg/day   (1855)

1,3-dichloro-1-propylene   see 1,3-dichloro-1-propene

2,3-dichloro-1-propylene   see 2,3-dichloro-1-propene

9,10-dichlorostearic acid
D. BIOLOGICAL EFFECTS:
—Fish: rainbow trout juvenile: 96 hr LC_{50} (S): 2.5 mg/l   (1495)

2,2-dichlorovinyl-O,O-dimethyl phosphate   (DDVP; dichlorvos; Vapona)

$$\begin{array}{c} CH_3O \\ \diagdown O \\ \| \\ P{-}O{-}CH{=}CCl_2 \\ \diagup \\ CH_3O \end{array}$$

Use: a contact and stomach insecticide
A. PROPERTIES: colorless to amber liquid; sp.gr. 1.44; b.p. 35°C at 0.05 mm, 120°C at 14 mm; v.p. 1.2×10^{-2} mm at 20°C; solub. approx. 1% at room temp.
B. AIR POLLUTION FACTORS:
—In ambient air of storage and office rooms of commercial pest control buildings in a 4 hr period:
storage room: avg.: 617 ng/cu m
                range: 147–1501 ng/cu m
office room: avg.: 41 ng/cu m
                range: 19–66 ng/cu m   (1868)
D. BIOLOGICAL EFFECTS:
—Crustaceans
Gammarus lacustris: 96 hr, LC_{50}: 0.50 µg/l   (2124)
Gammarus faciatus: 96 hr, LC_{50}: 0.40 µg/l   (2126)
Simocephalus serrulatus: 48 hr, LC_{50}: 0.26 µg/l   (2127)
Daphnia pulex: 48 hr, LC_{50}: 0.07 µg/l   (2127)
sand shrimp (*Crangon septemspinosa*): 96 hr static lab bioassay: 4 ppb LC_{50}
                (2327)
grass shrimp (*Palaemonetes vulgaris*): 96 hr static lab bioassay: 15 ppb LC_{50}
                (2327)
hermit crab (*Pagurus longicarpus*): 96 hr static lab bioassay: 45 ppb LC_{50}
                (2327)
—Insects:
Pteronarcys californica: 96 hr, LC_{50}: 0.10 µg/l   (2128)
—Fish:
Lepomis macrochirus: 96 hr, LC_{50}: 869 µg/l   (2137)
bluegills: 24 hr, LC_{50}: 1000 ppm   (1855)

mummichog (*Fundulus heteroclitus*): 96 hr static lab bioassay: 3700 ppb LC_{50}
(2328)
mummichog (*Fundulus heteroclitus*): 96 hr static lab bioassay: 2680 ppb LC_{50}
(2329)
striped killifish (*Fundulus majalis*): 96 hr static lab bioassay: 2300 ppb LC_{50}
(2329)
Atlantic silverside (*Menidia menidia*): 96 hr static lab bioassay: 1250 ppb LC_{50}
(2329)
striped mullet (*Mugil cephalus*): 96 hr static lab bioassay: 200 ppb LC_{50} (2329)
American eel (*Anguilla rostrata*): 96 hr static lab bioassay: 1800 ppb LC_{50}
(2329)
bluehead (*Thalassoma bifasciatum*): 96 hr static lab bioassay: 1440 ppb LC_{50}
(2329)
Northern puffer (*Sphaeroides maculatus*): 96 hr static lab bioassay: 2250 ppb LC_{50} (2329)
—Mammals:
acute oral LD_{50} (rat): 56–80 mg/kg
acute dermal LD_{50} (male rat): 107 mg/kg (1854)
in diet: rats fed ninety days on a diet containing 1000 ppm showed no signs of intoxication (1855)
Under the conditions of this study, dichlorvos was not demonstrated to be carcinogenic (1737)

2,4-dichlorotoluene

D. BIOLOGICAL EFFECTS:
—Fish:
guppy (*Poecilia reticulata*): 14 d LC_{50}: 4.6 ppm (1833)

3,4-dichlorotoluene

D. BIOLOGICAL EFFECTS:
—Fish:
guppy (*Poecilia reticulata*): 7 d LC_{50}: 5 ppm (1833)

α,α'-dichloro-m-xylene

<chemical structure: m-xylene with two CH₂Cl groups>

A. PROPERTIES: m.w. 175.06; m.p. 34–37°C; b.p. 250–255°C; sp.gr. 1.202
D. BIOLOGICAL EFFECTS:
—Fish:
guppy (*Poecilia reticulata*): 14 d LC_{50}: 0.12 ppm

dichlorvos see dichlorovinyldimethylphosphate

dicofol see kelthane

dicophane see DDT

1,2-dicyanobenzene see phthalonitrile

1,4-dicyanobutane see adiponitrile

1,5-dicyanopentane see pimelonitrile

dicyclohexyl (bicyclohexyl)

Uses: high-boiling-point solvent and penetrant
A. PROPERTIES: colorless oil; pleasant odor; m.w. 166.3; m.p. 2°C; b.p. 240°C; sp.gr. 0.883 at 25/15.6°C; v.d. 5.73
C. WATER POLLUTION FACTORS:
—Biodegradation:
degradation in seawater by oil oxidizing micro-organisms: 15.4% breakdown after 21 days at 22°C in stoppered bottles containing a 1000 ppm mixture of alkanes, cycloalkanes and aromatics   (1237)

α-dicyclopentadiene (3a,4,7,7a-tetrahydro-4,7-methanoindene;   DCPD;   tricyclo-(5,2,1,0)-3; 8-decadiene)
$C_{10}H_{12}$
A. PROPERTIES: colorless crystals; m.w. 132.20; m.p. 33°C; b.p. 170/172°C; v.p. 10 mm at 47.6°C; v.d. 4.57; sp.gr. 0.976 at 35°C;
B. AIR POLLUTION FACTORS: 1 mg/cu m = 0.18 ppm, 1 ppm = 5.49 mg/cu m
—Odor: characteristic: quality: sweet, sharp
hedonic tone: unpleasant
odor threshold: 0.016 mg/cu m   (720)

abs. perc. lim.: 0.011 ppm
50% recogn.: 0.02 ppm
100% recogn.: 0.02 ppm
O.I. 100% recogn.: 440,500 (19)
D. BIOLOGICAL EFFECTS:
—Mammalia: rat: inhalation: no effect level: 100 ppm, 15 × 6 hr (65)
: 4/6: 2,000 ppm, 4 hr (211)
percutane LD_{50}: 6.72 ml/kg (211)
single oral LD_{50}: 0.41 g/kg (277)
dermal LD_{50}: 4.46 ml/kg (277)

dicysteine *see l*-cystine

didecyldimethylammoniumbromide *see* nonidet GSC/nonidet BX

2,5-dideoxypentonic acid
C. WATER POLLUTION FACTORS:
—In primary domestic sewage plant effluent: 0.006 mg/l (517)

3,4-dideoxypentonic acid
C. WATER POLLUTION FACTORS:
—In primary domestic sewage plant effluent: 0.013 mg/l (517)

dieldrin (1,2,3,4,10,10-hexachloro-6,7-epoxy-1,4,4a,5,6,7,8,8a-octahydro-1,4,5,8-dimethanonaphthalene; Heod; compound 497; Octalox; ENT 16,225)

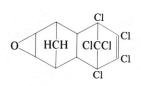
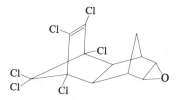

Use: insecticide, stereo-isomer of endrin, obtained by oxidation of aldrin; insecticide; wool processing industry (347)
A. PROPERTIES: m.w. 381; m.p. 176–177°C; v.d. 13.2; sp.gr. 1.75; v.p. 1.8×10^{-7} mm at 25°C; solub. 0.1 mg/l
C. WATER POLLUTION FACTORS:
—Odor threshold: detection: 0.041 mg/kg (915)
—Aquatic reactions:
time for 95% disappearance from the soil (3.1–5.6 kg/ha): 12.8 years (1650)
volatilization at 25°C from soils in the laboratory:
sandy loam: 8.9% after 60 days
sand: 34.2% after 60 days (1650)
75–100% disappearance from soils: 3–25 years (1815; 1816)
photo-oxidation by u.v. light in aqueous medium at 90–95°C: time for the formation of CO_2 (% of theoretical):

25%: 2.9 hr
50%: 4.8 hr
75%: 12.5 hr (1628)
persistence in river water in a sealed glass jar under sunlight and artificial fluorescent light—initial conc. 10 µg/l:

% of original compound found

after	1 hr	1 wk	2 wk	4 wk	8 wk
	100	100	100	100	100 (1309)

calculated half-life in water at 25°C, based on evaporation rate of 5.33×10^{-5} m/hr: 12,940 hr (330; 437)
—Water and sediment quality:
85-90% of dieldrin adsorbed on aquifer sand at 5°C after 3-100 hr equilibrium time
desorption of dieldrin from aquifer sand:

ng dieldrin adsorbed/g of aquifer sand	% leached of total adsorbed lindane			
	wash 1	wash 2	wash 3	total
29.6	4.36	6.95	6.12	17.4
30.0	6.13	5.30	4.90	16.3
29.7	6.74	7.78	4.15	18.8

(1303)
Mississippi: bottom sediments: residues not found (1170)
Irish sea: bottom and suspended sediments: 0.2-140 ng/g (1171)
Hawaii: sediments: 2-39.5 ppb (1174)
Los Angeles Harbor: bottom sediments: 0.6-4.5 ppb (1175)
San Antonio Bay, Texas: suspended and bottom (1 cm) sediments: 0-20 ppb
(1176)
Western Baltic: surface water: 0.17×10^{-9} g/l (1168)

D. BIOLOGICAL EFFECTS:
—Residues:
in aquatic vascular plants of Lake Päijänne, Finland (1972-1973): <0.5 µg/kg dry weight (1055)
mean residues in eggs of American crocodile in Everglades National Park:
1972: 0.01 ppm wet wt. (1572)
1977: 0.02 ppm wet wt. (1571)
mean concentration in gamebird muscle in upper Tennessee (U.S.A.):

mg/kg fresh weight ± standard error

	grouse	quail	woodcock
Johnson county	3.29 ± 0.77 (n = 12)	9.61 ± 1.18 (n = 6)	7.13 ± 0.99 (n = 6)
Carter county	7.78 ± 0.89 (n = 9)	7.13 ± 0.92 (n = 7)	8.48 ± 1.19 (n = 6)
Washington county	13.92 ± 1.90 (n = 10)	9.53 ± 1.12 (n = 11)	8.13 ± 1.33 (n = 6)

quail: acceptable LD_{50}: 35 mg/kg
FAO/WHO residue tolerance limit: 0.03-0.3 mg/kg/day
comparative average concentrations of dieldrin residue (wet tissue basis) in brown shrimp (*Crangon crangon*) from various areas of the North Sea 1974-1976:

area	year	No. of analyses	avg. conc. (ppb)	range	
S.E. England	1975	2	9	(4-13)	(1441)
Netherlands	1974	10	<7	(<5-12)	(1442)
	1975	12	<10		(1441)
	1976	12	<10		(1441)

Germany	1974	6	8	(4-17)	(1442)
	1975	4	4	(2-8)	(1441)
	1976	6	3	(2-3)	(1441)
	1976	5	0.5	(0.4-0.6)	(1440)

residue in animals in different trophic levels from the Weser estuary (1976):
  bivalvia: common edible cockle (*Cerastoderma edule L.*): 1.5 ± 0.6 ng/g wet tissue
    soft clam (*Mya arenaria L.*): 4.5 ± 1.6 ng/g wet tissue
  polychaeta: lugworm (*Arenicola marina L.*): not detected
  crustacea: brown shrimp (*Crangon crangon L.*): 0.5 ± 0.1 ng/g wet tissue
  pisces: common sole (*Solea solea L.*): 4.2 ± 1.0 ng/g wet tissue (1440)

invertebrates:	Irish Sea:	0.5-3 ng/g	(1184)
demersal fish:	Irish Sea:	0-4 ng/g	
bottom fish:	Irish Sea:	0-46 ng/g	
Seston:	Annapolis:	0.81 ppt	(1187)
	Baltimore Harbor:	0.62 ppt	
	Head of Chesapeake Bay:	0.72 ppt	

Periwinkle (*Littorina littorea*): Medway Estuary, Kent, U.K.: 0.1-1.6 µg/g
    (1481)
Shore Crab: (*Carcinus maenas*): Medway Estuary, Kent, U.K.: 0.8-10.7 µg/g
    (1481)
Flounder (*Platichthys flesus*): Medway Estuary, Kent, U.K.: 7.4-210.0 µg/g
    (1481)
Harbour Seal (*Phoca vitulina*): German North Sea: 0.15-0.54 mg/kg
    (blubber)
    (1482)
in marine animals from the Central Mediterranean (1976-1977): fish:
  anchovy (*Engraulis incrasicholus*): avg.: 0.28 ppb fresh wt. ($n = 12$)
    range: 0.1-0.8 ppb fresh wt.
  striped mullet (*Mullus barbatus*): avg.: 0.96 ppb fresh wt. ($n = 10$)
    range: 0.1-2.5 ppb fresh wt.
  tuna (*Thunnus thynnus thynnus*): avg.: 0.2 ppb fresh wt. ($n = 5$)
    range: 0.1-0.4 ppb fresh wt.
  shrimp (*Nephrops norvegicus*): avg.: 0.13 ppb fresh wt. ($n = 7$)
    range: 0.1-0.2 ppb fresh wt.
  mussel (*Mytilus galloprovincialis*): avg.: 0.72 ppb fresh wt. ($n = 4$)
    range: 0.2-1.3 ppb fresh wt. (1774)
residues in Canadian human milk: 1967: avg.: 5 ng/g whole milk
    1970: avg.: 5 ng/g whole milk
    1975: avg.: 2 ng/g whole milk (1376)

—Bioaccumulation:
  pelecypod (*Crassostrea virginica*): exposure: 0.5 µg/l; 168 hr: BCF 2880 (1162)
  pelecypod (*Crassostrea virginica*): exposure: 1-100 µg/l; 43 d bioaccumulation:
    5100-5500 X (1478)
  pelecypod (*Mytilus edulis*): exposure: 0.17-1.97 µg/l; 50 hr bioaccumulation:
    1570 X (1477)
  5 aquatic molluscs: BCF 700-1800 (1870)
  oyster (*Crassostrea virginica*): bioaccumulation factor after 168 hr exposure:
    2070-2880

biological half-life following 168 hr exposure: 67–104 hr (in whole body)
(1629)
trout: BCF: 3300 (1899)
—Toxicity:
toxicity ratios of dieldrin (D) to photodieldrin (PD), calculated from respective 24 hr LC_{50} (*) and LT_{50} values:

	D/PD
crustacea	
Daphnia pulex (water flea)	1.27
Gammarus spp. (amphipod)	12.22
Asellus spp. (isopod)	1.60 (1.7)*
insects	
Aedes aegypti larvae (mosquito)	2.30 (2.0)*
Musca domestica (house fly)	1.18 (1.2)*
fish	
Lebistes reticulatus (guppy)	2.57
Pimephalus promelas (bass)	1.11 (2.4)*
Lepomis macrochirus (blue gill)	5.14 (5.7)*

(1681,* 1684, 1685)

—Insects:
Pteronarcys californica: 96 hr, LC_{50} 0.5 µg/l (2128)
Pteronarcys californica: 96 hr, LC_{50} 39 µg/l, 30 day LC_{50}: 2.0 µg/l (2118)
Acroneuria pacifica: 96 hr, LC_{50} 24 µg/l, 30 day LC_{50}: 0.2 µg/l (2118)
Pteronarcella badia: 96 hr, LC_{50} 0.5 µg/l (2128)
Classenia sabulosa: 96 hr, LC_{50} 0.58 µg/l (2128)
fourth instar larval *Chironomus riparius*: 24 hr LC_{50}: 0.5 µg/l (1853)
mosquito (late 3rd instar *Aedes aegypti* larvae) 24 hr LC_{50}: 6 ppb
housefly (3 day old female *Musca*) LD_{50}: 9.8 µg/fly (1681)

—Crustacean
Gammarus lacustris: 96 hr, LC_{50}: 460 µg/l (2124)
Gammarus fasciatus: 96 hr, LC_{50}: 600 µg/l (2126)
Palaeomonetes kadiakensis: 96 hr, LC_{50}: 20 µg/l (2126)
Orconectes nais: 96 hr, LC_{50}: 740 µg/l (2126)
Asellus brevicaudus: 96 hr, LC_{50}: 5 µg/l (2126)
Simocephalus serrulatus: 48 hr, LC_{50}: 190 µg/l (2127)
Daphnia pulex: 48 hr, LC_{50}: 250 µg/l (2127)
Korean shrimp (*Palaemon macrodactylus*): 96 hr static lab bioassay: 16.9 (10.8–33.4) ppb (2353)
Korean shrimp (*Palaemon macrodactylus*): 96 hr inter. flow lab bioassay: 6.9 (3.7–13.1) ppb (2358)
sand shrimp (*Crangon septemspinosa*): 96 hr static lab bioassay: LC_{50}: 7 ppb (2327)
grass shrimp (*Palaemonetes vulgaris*): 96 hr static lab bioassay: LC_{50}: 50 ppb (2327)
hermit crab (*Pagurus longicarpus*): 96 hr static lab bioassay: LC_{50}: 18 ppb (2327)

American oyster (*Crassostrea virginica*): 48 hr static lab bioassay: TLm: 640 ppb (2324)
mud snail (*Nassa obsoleta*): 96 hr exposure to 1.0 ppm then 133 day post exposure in clean water: No. egg cases deposited significant less than control. (2330)

—Fish

Pimephales promelas: 96 hr, LC_{50}: 16 µg/l (2113)
Lepomis macrochirus: 96 hr, LC_{50}: 8 µg/l (2113)
Salmo gairdneri: 96 hr, LC_{50}: 10 µg/l (2119)
Oncorhynchus kisutch: 96 hr, LC_{50}: 11 µg/l (2119)
Oncorhynchus tschawytscha: 96 hr, LC_{50}: 6 µg/l (2119)
Poecilia latipinna: 3.0 µg/l (19 week LC_{50}) (2120)
Poecilia latipinna: 0.75 µg/l (reduced growth & reproduction—34 week) (2120)
Lepomis gibbosus: 96 hr, LC_{50}: 6.7 µg/l; 1.7 µg/l (affect swimming ability and oxygen consumption—100-day) (2109)
Ictalurus punctatus: 96 hr, LC_{50}: 4.5 µg/l (2137)
mummichog (*Fundulus heteroclitus*): 96 hr static lab bioassay: LC_{50}: 5 ppb (2338)
mummichog (*Fundulus heteroclitus*): 96 hr static lab bioassay: LC_{50}: 5 ppb (100%) (2329)
striped killifish (*Fundulus majalis*): 96 hr static lab bioassay: LC_{50}: 4 ppb (100%) (2329)
Atlantic silverside (*Menidia menidia*): 96 hr static lab bioassay: LC_{50}: 5 ppb (100%) (2329)
striped mullet (*Mugil cephalus*) 96 hr static lab bioassay: LC_{50}: 23 ppb (100%) (2329)
American eel (*Anguilla rostrata*): 96 hr static lab bioassay: LC_{50}: .9 ppb (100%) (2329)
bluehead (*Thalassoma bifasciatum*): 96 hr static lab bioassay: LC_{50}: 6 ppb (100%) (2329)
northern puffer (*Sphaeroides maculatus*): 96 hr static lab bioassay: LC_{50}: 34. ppb (100%) (2329)
threespine stickleback (*Gasterosteus aculeatus*): 96 hr static lab bioassay: TLm: 13.1 ppb (technical) (2333)
shiner perch (*Cymatogaster aggregata*): 96 hr static lab bioassay: TL_{50}: 3.7 ppb (technical) (2354)
dwarf perch (*Micrometrus minimus*): 96 hr static lab bioassay: TL_{50}: 5. ppb (technical (2354)
sailfin mollie (*Poecilia latipinna*): 34 wk flowing water: 7.5 ppb Reduced reproduction control 65 Exp.—young born 37 (0.012% W/V) (2335)
sailfin mollie (*Poecilia latipinna*): 47 hr flowing water test: SGOT activity increase: 6 ppb (0.012% W/V) (2336)
shiner perch (*Cymatogaster aggregata*): 96 hr inter. flow lab bioassay: TL_{50}: 1.5 (0.73–3.20) ppb (technical) (2354)
dwarf perch (*Micrometrus minimus*): 96 inter. flow lab bioassay: TL_{50}: 2.44 (1.16–5.11) ppb (technical) (2354)

518 DIETHANOLAMINE

Pagurus longicarpus: 96 hr LC_{50}: 18 µg/l		(1575)
susceptible mosquito fish: 48 hr LC_{50}: 8 ppb		
resistant mosquito fish: 48 hr LC_{50}: 434 ppb		(1851)
bluegill: 96 hr LC_{50}: 0.008 ppm		
rainbow trout: 96 hr LC_{50}: 0.019 ppm		(1878)
bluegill: 24 hr LC_{50}: 170 ppb		
minnow: 24 hr LC_{50}: 24 ppb		(1681)

—Mammals:
  acute oral LD_{50} (rat): 46–63 mg/kg
  acute dermal LD_{50} (rat): 52–117 mg/kg   (1854)
—Carcinogenicity: +
—Mutagenicity in the *Salmonella* test: negative
  <0.003 revertant colonies/nmol
  <10 revertant colonies at 10^4 µg/plate   (1883)

diethanolamine (diethylolamine; 2,2′-aminodi-ethanol; β,β′-dihydroxydiethylamine; DEA)
$NH(CH_2CH_2OH)_2$
  Uses: liquid detergents for emulsion paints, cutting oils, shampoos, cleavers and polishers; chemical intermediate for resins, plasticizers etc.
A. PROPERTIES: m.w. 105.14; m.p. 28°C; b.p. 269.1°C; v.p. <0.01 mm at 20°C; v.d. 3.6; sp.gr. 1.092 at 30/20°C; solub. 954,000 mg/l; log P_{oct} ⁻1.43
B. AIR POLLUTION FACTORS: 1 mg/cu m = 0.232 ppm, 1 ppm = 4.3 mg/cu m
  —Odor: characteristic: quality: ammoniacal
C. WATER POLLUTION FACTORS:

—BOD_5:	0.10 std.dil.sew.	(41)
	0.9% of ThOD	(79)
—BOD_{10}:	1.4% of ThOD	(79)
—BOD_{15}:	3.5% of ThOD	(79)
—BOD_{20}:	6.8% of ThOD	(79)
—BOD_5:	0.03	(277)
	1.17 adapted sludge	(277)
	0.984 acclimated	(1828)
—COD:	1.52	(277)
—ThOD:	2.13	(27)

—Biodegradation: adapted A.S.—product as sole carbon source—97.0% COD removal at 19.5 mg COD/g dry inoculum/hr   (327)
  biodegradation by natural communities in stream water: % converted to CO_2 in 4 days: initial conc. 21 ng/l: 30%
              0.21 mg/l: 50%
              21 mg/l: 5%   (1875)
  unacclimated system: diethanolamine was examined at concentrations of 1,230, 1,000, and 123 mg/l. Significant anomalies occurred during these analyses, which prevented reaching a valid conclusion. Examination of the limited data which became available indicated the compound to be apparently amenable to biodegradation.   (1828)
  acclimated system: at 123 mg/l: biodegradable

Diethanolamine will be degraded by an acclimated biomass at a significantly high rate. (1828)
—Impact on biodegradation processes:
at 100 mg/l, no inhibition of NH_3 oxidation by *Nitrosomonas* sp. (390)
—Waste water treatment:
activated carbon: adsorbability: 0.057 g/g C, 27.5% reduction, infl.: 996 mg/l, effl.: 722 mg/l (32)

methods	temp °C	days observed	feed mg/l	days acclim.	removed %
NFG, BOD	20	1-10	200-800	365 + P	10
NFG, BOD	20	1-10	1,000-1,200	365 + P	inhibition
RW, BOD	20	1-10	50	10	90
TF, Sd, BOD	20	1-10	?	?	90

(93)

D. BIOLOGICAL EFFECTS:
—Toxicity threshold (cell multiplication inhibition test):
bacteria (*Pseudomonas putida*): >10,000 mg/l (1900)
algae (*Microcystis aeruginosa*): 16 mg/l (329)
green algae (*Scenedesmus quadricauda*): 4,4 mg/l (1900)
protozoa (*Entosiphon sulcatum*): 160 mg/l (1900)
protozoa (*Uronema parduczi Chatton-Lwoff*): 1,720 mg/l (1901)
—Algae:
Selenastrum capricornutum: 1 mg/l: no effect
10 mg/l: no effect
100 mg/l: inhibitory (1828)
—Crustacean:
Daphnia magna: LC_{50}: 1.4 mg/l (1828)
—Fish:
mosquito fish: TLm (24 hr): 1,800 mg/l, in turbid Oklahoma water (244)
bluegill sunfish: TLm (24 hr): 2,100 mg/l in tap water (247)
goldfish: LD_{50} (24 hr): 800 mg/l at pH 9.6
>5000 mg/l at pH 7 (277)
Pimephales promelas: LC_{50} >100 mg/l (1828)
—Mammalia: rat: acute oral LD_{50}: 1.82 g/kg (277)
rabbit: acute dermal LD_{50}: 11.9 ml/kg

diethylamine
$(C_2H_5)_2NH$

A. PROPERTIES: m.w. 73.14; m.p. -48/-50°C; b.p. 57°C; v.p. 200 mm at 20°C, 290 mm at 30°C; v.d. 2.53; sp.gr. 0.711 at 18/4°C; solub. 815,000 mg/l at 14°C; THC 722 kcal/mole, LHC 672 kcal/mole; sat.conc. 757 g/cu m at 20°C; 1,117 g/cu m at 30°C; log P_{oct} 0.43/0.57

Uses: rubber chemicals; textile specialties; selective solvent; dyes; flotation agents; resins; pesticides; polymerization inhibitors; pharmaceuticals; corrosion inhibitors

B. AIR POLLUTION FACTORS: 1 mg/cu m = 0.334 ppm, 1 ppm = 2.91 mg/cu m
—Odor: characteristic: quality: musty, fishy, amine
hedonic tone: unpleasant

520 2-DIETHYLAMINOETHANOL

odor thresholds   mg/cu m

10^{-7} 10^{-6} 10^{-5} 10^{-4} 10^{-3} 10^{-2} 10^{-1} 1 10 10^2 10^3 10^4

detection

recognition

not specified

(19; 151; 279; 291; 635; 664; 675; 726)

O.I. 100% recogn.: 4,250.000 (19)

C. WATER POLLUTION FACTORS:
—Impact on biodegradation processes:
at 100 mg/l, no inhibition of NH_3 oxidation by *Nitrosomonas* sp. (390)

D. BIOLOGICAL EFFECTS:
—Bacteria: *E. coli*: not toxic: 1 g/l
—Algae: *Scenedesmus*: toxic: 4 mg/l
—Arthropoda: *Daphnia*: toxic: 100 mg/l
—Fish: creek chub: LD_0: 70 mg/l, 24 hr in Detroit river water
  LD_{100}: 100 mg/l, 24 hr in Detroit river water (243)
—Mammalia: rat: acute oral LD_{50}: 0.54 g/kc
  inhalation: 3/6: 4,000 ppm, 4 hr
  no deaths: sat. vap.; 5 min (211)

2-diethylaminoethanol (diethylethanolamine; β-diethylaminoethylalcohol; 2-hydroxytriethylamine)

$(C_2H_5)_2NC_2H_4OH$

A. PROPERTIES: colorless liquid; m.w. 117.19; b.p. 163°C; v.p. 1.4 mm at 20°C; v.d. 4.04; sp.gr. 0.88 at 20/4°C; log P_{oct} 0.31/0.46 (calculated)

B. AIR POLLUTION FACTORS: 1 mg/cu m = 0.21 ppm, 1 ppm = 4.87 mg/cu m
—Odor characteristic: quality: amine
  hedonic tone: unpleasant
abs. perc. lim.: 0.011 ppm
50% recogn.: 0.04 ppm
100% recogn.: 0.04 ppm
O.I. 100% recogn.: 33,000 (19)

D. BIOLOGICAL EFFECTS:
—Fish: creek chub: critical range: 80–120 mg/l; 24 hr (226)

diethylaminehydrochloride

D. BIOLOGICAL EFFECTS:
—Fish: creek chub: critical range: 4000–6000 mg/l, 24 hr (226)

N,N-diethylaniline (N-phenyldiethylamine)

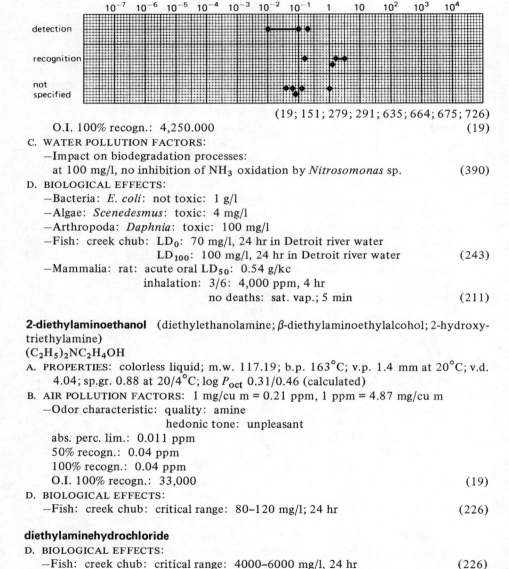

A. PROPERTIES: colorless, yellow or brownish oil; m.w. 149.23; m.p. −38.8°C; b.p. 215.5°C; sp.gr. 0.935 at 20/4°C; solub. 14,400 mg/l at 12°C
C. WATER POLLUTION FACTORS:
 −BOD_5: 0 std. dil. sew. (n.s.i.) (281; 282)
 −COD: 2.591
 −ThOD: 2.730 (30)

o-diethylbenzene (1,2-diethylbenzene)

$$\underset{}{\bigcirc}\begin{array}{l}-CH_2CH_3\\-CH_2CH_3\end{array}$$

Manmade sources: in gasoline (high octane number): 0.10 wt % (387)
A. PROPERTIES: colorless liquid; m.w. 134.21; m.p. <−20°C; b.p. 183.5°C; v.d. 4.62;
B. AIR POLLUTION FACTORS: 1 mg/cu m = 0.18 ppm, 1 ppm = 5.58 mg/cu m
 −Atmospheric reactions: reactivity: NO ox.: ranking: 0.4 (63)
D. BIOLOGICAL EFFECTS:
 −Toxicity threshold (cell multiplication inhibition test):
  bacteria (*Pseudomonas putida*): >20 mg/l
  green algae (*Scenedesmus quadricauda*): >20 mg/l
  protozoa (*Entosiphon sulcatum*): 6.9 mg/l (1900)
  protozoa (*Uronema parduczi Chatton-Lwoff*): >16 mg/l (1901)

m-diethylbenzene

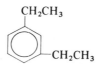

Manmade sources: in gasoline (high octane number): 0.09 wt % (387)

p-diethylbenzene (1,4-diethylbenzene)

$$\underset{CH_2CH_3}{\overset{CH_2CH_3}{\bigcirc}}$$

Manmade sources: in gasoline (high octane number): 0.03 wt % (387)
A. PROPERTIES: colorless liquid; m.w. 134.21; m.p. −35°C; b.p. 183.7°C; v.d. 4.62; sp.gr. 0.862 at 20°C
B. AIR POLLUTION FACTORS: 1 mg/cu m = 0.18 ppm, 1 ppm = 5.58 mg/cu m
 −Atmospheric reactions: reactivity: HC. conc.: ranking: 1.5
   NO. ox.: ranking: 0.4 (63)

diethylcarbitol *see* diethyleneglycoldiethylether

diethylcellosolve *see* ethylenediethylether

O,O-diethyl-O,1-(2',4'-dichlorophenyl)-2-chlorovinylphosphate *see* chlorfenvinphos

diethyldisulfide (ethyldisulfide; ethyldithioethane)
$C_2H_5SSC_2H_5$
A. PROPERTIES: oil; m.w. 122.24; b.p. 153°C; sp.gr. 0.992 at 20/4°C
B. AIR POLLUTION FACTORS:
   odor thresholds   mg/cu m

10^{-7} 10^{-6} 10^{-5} 10^{-4} 10^{-3} 10^{-2} 10^{-1} 1 10 10^2 10^3 10^4

detection
recognition
not specified

(71; 307; 682; 863; 871)

—Control methods:
   wet scrubber: water at pH 8.5:   effluent: 1,200 odor units
                 $KMnO_4$ at pH 8.5: effluent: 4 odor units   (76)
C. WATER POLLUTION FACTORS:
   —Odor threshold: detection: 0.00002 mg/kg   (894)

1,4-diethylenedioxide *see* 1,4-dioxane

diethyleneglycol (2,2'-oxydiethanol; 2,2'-dihydroxyethylether; *bis*(2-hydroxyethyl)-ether; DEG)
$O(CH_2CH_2OH)_2$
   Uses: polyurethane and unsaturated polyester resins; triethylene glycol; textile softener; petroleum solvent extraction; plasticizers and surfactants; cosmetics; solubilizing agent for developer components
A. PROPERTIES: colorless liquid; m.w. 106.12; m.p. -10/-8°C; b.p. 245°C; v.p. <0.01 mm at 20°C; v.d. 3.66; sp.gr. 1.118 at 20°C; log P_{oct} -1.98 (calculated)
B. AIR POLLUTION FACTORS: 1 ppm = 4.35 mg/cu m
C. WATER POLLUTION FACTORS:
   —BOD_5: 0.12   (277)
              0.15   (36)
              1.5% of ThOD   (79)
              0.05 "special grade" DGE   (277)
              0.10 "special grade" DGE adapted sew.   (277)
              0.02 std.dil.sew.   (255)(1828)
              0.06 std.dil.sew.   (290)
   —BOD_5^{20}: 0.15 at 10 mg/l, unadapted sew.: no lag period (554)
   —BOD_{10}: nil std.dil.sew.   (256)
              : 5.6% of ThOD   (79)
   —BOD_{15}: 9.0% of ThOD   (79)
   —BOD_{20}: 18.8% of ThOD   (79)

BOD_{20}^{20}: 0.32 at 10 mg/l, unadapted sew.: no lag period (554)
—COD: 1.06 (36)
   1.49 (277)
   1.51 "special grade" DGE (277)
—ThOD: 1.51 (36)
—Biodegradation: adapted A.S.—product as sole carbon source—95.0% COD removal at 13.7 mg COD/g dry inoculum/hr (327)
   unacclimated system: at 1,320 mg/l: no effect
   acclimated system: at 2,085 mg/l: biodegradable (1828)
—Waste water treatment:
   A.C.: adsorbability: 0.053 g/g C; 26.2% reduction, infl.: 1,000 mg/l, effl.: 738 mg/l (32)
   A.S.: after 6 hr: 2.4% of ThOD
           12 hr: 5.7% of ThOD
           24 hr: 10.0% of ThOD (88)

methods	temp °C	days observed	feed mg/l	days acclim.	removed %
NFG, BOD	20	1–10	200–1,000	365 + P	inhibition
AS, BOD	20	$\frac{1}{3}$–5	333	30+	40
RW, BOD	20	1–10	50	80	20

(93)

D. BIOLOGICAL EFFECTS:
—Toxicity threshold (cell multiplication inhibition test):
   bacteria (*Pseudomonas putida*): 8000 mg/l (1900)
   algae (*Microcystis aeruginosa*): 1700 mg/l (329)
   green algae (*Scenedesmus quadricauda*): 2700 mg/l (1900)
   protozoa (*Entosiphon sulcatum*): 10745 mg/l (1900)
   protozoa (*Uronema parduczi Chatton-Lwoff*): >8000 mg/l (1901)
—Algae:
   Selenastrum capricornutum: 1 mg/l: no effect
                                 10 mg/l: no effect
                                 100 mg/l: no effect (1828)
—Crustacean:
   Daphnia magna: LC_{50}: between 0.3 and 1 mg/l (1828)
—Fish:
   mosquito fish: TLm (24 hr): >32,000 mg/l in turbid Oklahoma water (244)
   goldfish: LD_{50} (24 hr): >5000 mg/l (277)
             LD_{50} (24 hr): >5000 mg/l "special grade" DGE (277)
   Pimephales promelas: LC_{50}: >100 mg/l (1828)
   guppy (*Poecilia reticulata*): 7 d LC_{50}: 61,072 ppm (1833)
—Mammalia: rat: single oral LD_{50}: 14.8 ml/kg; 20.7 g/kg
           guinea pig: single oral LD_{50}: 7.7 ml/kg; 13.2 g/kg
           mouse: single oral LD_{50}: 23.7 ml/kg (211)
—Man: not readily absorbed through the skin
      low acute oral toxicity
      not irritating to eyes and skin
      single lethal oral dose: approx.: 1 ml/kg (211)
      lethal dose: 0.5–2 g/kg (277)

diethyleneglycoldiethylether (bis-2-ethoxyethylether; 1-ethoxy-2-(β-ethoxyethoxy)-ethane; diethyl "carbitol")
$(C_2H_5OCH_2CH_2)_2O$

A. PROPERTIES: colorless liquid; m.w. 162.23; b.p. 188°C; sp.gr. 0.907 at 20/4°C;

C. WATER POLLUTION FACTORS:
—Waste water treatment:
NFG, BOD, 20°C; 1–10 days observed, feed: 200–1,000 mg/l, acclimation: 365
P: 3% removed (93)
—BOD_{10}: 0.10 std.dil.sew. (256)

D. BIOLOGICAL EFFECTS:
—Mammalia: rat: inhalation: restlessness; autopsy: organs normal
: 400 ppm, 17 × 7 hr (65)

diethyleneglycolmonobutylether (butyldigol; butylcarbitol; 2-(β-butoxyethoxy)-ethanol; butyldiglycol; "dowanol DB"; butyl "dioxitol")
$C_4H_9OCH_2CH_2OCH_2CH_2OH$

A. PROPERTIES: colorless liquid; m.w. 162.23; m.p. -68°C; b.p. 231°C; v.p. 0.02 mm at 20°C; v.d. 5.58; sp.gr. 0.96 at 20/4°C; solub completely miscible; log P_{oct} 0.15/0.40 (calcul.)

B. AIR POLLUTION FACTORS: 1 mg/cu m = 0.151 ppm, 1 ppm = 6.64 mg/cu m

C. WATER POLLUTION FACTORS:
—BOD: 0.25 NEN 3235–5.4 (277)
—COD: 2.08 NEN 3235–5.3 (277)
—Waste water treatment:
A.C.: adsorbability: 0.166 g/g C, 82.7% reduction, infl.: 1,000 mg/l, effl.: 173 mg/l (32)

D. BIOLOGICAL EFFECTS:
—Toxicity threshold (cell multiplication inhibition test):

bacteria (*Pseudomonas putida*):	255 mg/l	(1900)
algae (*Microcystis aeruginosa*):	53 mg/l	(329)
green algae (*Scenedesmus quadricauda*):	1,000 mg/l	(1900)
protozoa (*Entosiphon sulcatum*):	73 mg/l	(1900)
protozoa (*Uronema parduczi Chatton-Lwoff*):	420 mg/l	(1901)

—Fish:
goldfish: LD_{50} (24 hr): 2700 mg/l modified ASTM-D-1345 (277)
(1833)

Lepomis macrochirus: static bioassay in fresh water at 23°C, mild aeration applied after 24 hr:

material added ppm	24 hr	% survival after 48 hr	72 hr	96 hr	best fit 96 hr LC_{50} ppm
3,200	33	0	–	–	
2,400	50	50	50	10	
1,800	100	100	80	20	1,300
1,000	100	100	90	70	
100	100	100	100	100	

Menidia beryllina: static bioassay in synthetic seawater at 23°C, mild aeration applied after 24 hr

DIETHYLENEGLYCOLMONOETHYLETHER   525

material added ppm	% survival after				best fit 96 hr LC_{50} ppm
	24 hr	48 hr	72 hr	96 hr	
2,400	100	100	50	0	
1,800	100	100	100	80	2,000
1,000	100	100	100	100	(352)

—Mammalia: rat: single oral LD_{50}: 5.5–6.6 g/kg
  guinea pig: single oral LD_{50}: 2.00 g/kg
  rat: repeated oral doses: no effect: 0.051 g/kg, 30 days
  inhalation: 0/3: sat. vap., 7 hr   (211)

diethyleneglycolmonoethylether (ethyldigol; carbitol; 2-(β-ethoxyethoxy)ethanol; diglycol; "dowanol DE"; "dioxitol")
$C_2H_5(OCH_2CH_2)_2OH$

A. PROPERTIES: colorless liquid; m.w. 134.2; b.p. 202°C; v.p. 0.13 mm at 25°C; v.d. 4.62; sp.gr. 0.990 at 20/4°C; log P_{oct} -0.79/-0.93 (calculated)

B. AIR POLLUTION FACTORS: 1 mg/cu m = 0.188 ppm, 1 ppm = 5.49 mg/cu m
  —Odor: characteristic: quality: sweet, musty
                hedonic tone: neutral
  T.O.C.: abs. perc. lim.: <0.21 ppm
         50% recogn.:   1.10 ppm
         100% recogn.:   1.10 ppm
         O.I. 100% recogn.: 600   (19)
         O.I. at 20°C = 120   (316)

C. WATER POLLUTION FACTORS:
  —BOD_5: 0.20 NEN 3235-5.4
         0.58 adapted sew. NEN 3235-5.4   (277)
  —COD: 1.85 NEN 3235-5.3   (277)
  —Waste water treatment:
    A.C.: adsorbability: 0.087 g/g C, 43.6% reduction, infl.: 1,010 mg/l, effl.: 570 mg/l   (32)

methods	temp °C	days observed	feed mg/l	days acclim.	theor. oxidation	% removed
NFG, BOD	20	1–10	200–1,000	365 + P		nil
RW, BOD	20	1–10	50	59		84
AS., Sd, BOD	20	1–5	?	14+		34 (93)

D. BIOLOGICAL EFFECTS:
  —Fish: goldfish: LD_{50} (24 hr): >5000 mg/l-modified ASTM-D 1345   (277)
  Lepomis macrochirus: static bioassay in fresh water at 23°C, mild aeration applied after 24 hr:

material added ppm	% survival after				best fit 96 hr LC_{50} ppm
	24 hr	48 hr	72 hr	96 hr	
10,000	90	90	90	90	>10,000
3,200	100	100	100	100	

  Menidia beryllina: static bioassay in synthetic seawater at 23°C: mild aeration applied after 24 hr:

material added ppm	% survival after				best fit 96 hr LC_{50} ppm	
	24 hr	48 hr	72 hr	96 hr		
10,000	100	80	80	80	>10,000	(352)

—Mammalia:
rat: single oral dose: LD_{50}: 8.69–9.74 g/kg
guinea pig: single oral dose: LD_{50}: 3.67–4.97 g/kg
cat: single oral dose: lethal: 1 ml/kg
rat: repeated oral doses: no adverse effect: 0.49 g/kg
rabbit, cat, guinea pig, mouse: inhalation: no injury: sat. vap., 12 daily exposures (211)

diethyleneglycolmonoethyletheracetate (carbitolacetate)
$C_2H_5(OCH_2CH_2)_2OOCCH_3$
A. PROPERTIES: m.w. 176.22; m.p. $-11°C$, $-25°C$; b.p. $218°C$; v.p. 0.05 mm at $20°C$; v.d. 6.07; sp.gr. 1.01 at $20°C$
B. AIR POLLUTION FACTORS: 1 mg/cu m = 0.14 ppm, 1 ppm = 7.20 mg/cu
 —Odor: characteristic: quality: sweet
  hedonic tone: pleasant to unpleasant
 threshold values: abs. perc. lim.: 0.026 ppm
  50% recogn.: 0.157 ppm
  100% recogn.: 0.263 ppm
  O.I.: 100% recogn.: 498 (19)
C. WATER POLLUTION FACTORS:
 —BOD_5: 23.1% of ThOD
 10 days: 44.0% of ThOD
 15 days: 82.4% of ThOD
 20 days: 90.1% of ThOD
 30 days: 94.6% of ThOD
 40 days: 100% of ThOD (79)
 —Waste water treatment:
 NFG, BOD, $20°C$, 1–10 days observed, feed: 200 mg/l, acclimation: 365 + P: 42% removed (93)
D. BIOLOGICAL EFFECTS:
 —Mammalia: rat: single oral dose: 11.00 g/kg
  guinea pig: single oral dose: 4.9 g/kg (211)

diethyleneglycolmonomethylether
$CH_3(OCH_2CH_2)_2OH$
A. PROPERTIES: colorless, mild smelling liquid; b.p. $381°F$; v.p. 0.25 mm at $25°C$; v.d. 4.16; sp.gr. 1.021 at $25/25°C$; solub. completely miscible; log P_{oct} $-1.14/-0.93$ (calculated)
D. BIOLOGICAL EFFECTS:
 —Fish:
 Lepomis macrochirus: static bioassay in fresh water at $23°C$, mild aeration applied after 24 hr

material added ppm	% survival after				best fit 96 hr LC_{50} ppm	
	24 hr	48 hr	72 hr	96 hr		
10,000	40	0	—	—	7,500	
5,600	100	100	100	100		(352)

—Mammals:
ingestion: lab animals: single oral LD_{50}: ranging from 4000 to 9000 mg/kg or more (1546)

diethyleneglycolmonomethyletheracetate (methyl "carbitol" acetate) $CH_3COOC_2H_4OC_2H_4OCH_3$
A. PROPERTIES: colorless liquid; b.p. 209.1°C; v.p. 0.12 mm at 20°C; sp.gr. 1.04 at 20/20°C
C. WATER POLLUTION FACTORS:
—BOD_{10}: 1.10 std.dil.sew. (256)

diethyleneimideoxide *see* morpholine

diethyleneoxide *see* tetrahydrofuran

diethylenetriamine (*bis*-2-aminoethylamine; 2,2'-diaminodiethylamine) $(NH_2C_2H_4)_2NH$
A. PROPERTIES: colorless, yellow liquid; m.w. 103.2; m.p. -39°C; b.p. 207°C; v.p. 0.2 mm at 20°C; v.d. 3.48; sp.gr. 0.958 at 20/20°C; sat.conc. 2.1 g/cu m at 20°C
B. AIR POLLUTION FACTORS: 1 mg/cu m = 0.23 ppm, 1 ppm = 4.29 mg/cu m
C. WATER POLLUTION FACTORS:
—Waste water treatment:
A.C.: adsorbability: 0.062 g/g C, 29.4% reduction, infl.: 1,000 mg/l, effl.: 706 mg/l (32)
D. BIOLOGICAL EFFECTS:
—Mammalia:
rat: inhalation: no toxic signs, autopsy: organs normal, 130 ppm, 15 × 6 hr
(65)
ingestion: rats: single oral LD_{50}: 1400 mg/kg (1546)

diethylenetriaminepentaacetic acid (DTPA)
$C_{14}H_{23}O_{10}N_3$
Use: chelating agent
A. PROPERTIES: m.w. 393.85
C. WATER POLLUTION FACTORS:
—BOD_5: 0.01 (acclimated)
0.015 (unacclimated)
—COD: 1.02
—NOD: 0.49
—ThOD: 1.51
—Biodegradation:
unacclimated System: 100 mg/l: no effect
1,000 mg/l: inhibitory
acclimated System: 490 mg/l: biodegradable
DTPA was found to be biodegradable by an acclimated biomass at a concentration above the levels estimated to be in photoprocessing wastes. (1828)
D. BIOLOGICAL EFFECTS:
—Algae:
Selenastrum capricornutum: 1 mg/l: inhibitory

528 DIETHYLETHANOLAMINE

       10 mg/l: inhibitory
       100 mg/l: inhibitory: (1828)
—Crustacean:
 Daphnia magna: LC_{50}: between 10–100 mg/l
—Fish: *Pimephales promelas*: LC_{50}: >300 mg/l
   bluegill (*Lepomis macrochirus*) static 96 hr LC_{50}: 1115 mg/l
   (penta sodium salt): 96 hr: no adverse effect level: 750 mg/l (1869)

diethylethanolamine *see* 2-diethylamino-ethanol

diethylether *see* ethylether

O,O-diethyl-S-(2-(ethylthio)ethyl)phosphorodithioate *see* disulfoton

O,O-diethyl-S-(ethylthio)-methylphosphorodithioate *see* phorate

diethylfumarate
$C_2H_5O_2CCH=CHCO_2C_2H_5$
A. PROPERTIES: m.w. 172.18; m.p. 1–2°C; b.p. 218–219°C; sp.gr. 1.052
D. BIOLOGICAL EFFECTS:
 —Fish:
  eastern mudminnow (*Umbra pygmaeo*):
    static bioassay: 96 hr TLm: 8.5 mg/l at 16°C
    flow through bioassay: 96 hr TLm: 4.2 mg/l at 16°C   (450)

di-2-ethylhexyladipate *see* di-octylaldipate

di-(2-ethyl)hexylphosphoric acid
D. BIOLOGICAL EFFECTS:
 —Fish:
  rainbow trout: 96 hr LC_{50} (S): 48.0–54.0 mg/l   (1500)

di-2-ethylhexylphthalate *see* di-octylphthalate

N,N-diethylhydroxylamine

$$HO-N\begin{matrix}CH_2CH_3\\CH_2CH_3\end{matrix}$$

 Use: anti-oxidant in developing solutions   (1828)
A. PROPERTIES: m.w. 89.14
C. WATER POLLUTION FACTORS:
 —BOD_5: 0.49
 —Reflux COD: 88% recovery
 —rapid COD: 94% recovery
 —ThOD: 2.70
 —NOD: 0.72   (1828)

—Biodegradation: unacclimated system at 700 mg/l: biodegradable (1828)

diethyl-2-isopropyl-6-methyl-4-pyrimidinylphosphorothionate see diazinon

diethylketone (3-pentanone; *sym*-dimethylacetone; propione; ethylketone)
$C_2H_5COC_2H_5$

A. PROPERTIES: colorless liquid; m.w. 86.13; m.p. -42°C; b.p. 102°C; v.p. 13 mm at 20°C; v.d. 2.96; sp.gr. 0.816 at 19/4°C; solub. 47,000 mg/l at 20°C, 38,000 mg/l at 100°C;

B. AIR POLLUTION FACTORS:

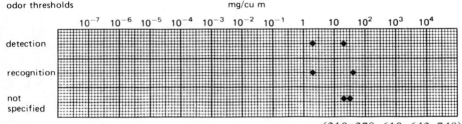

(210; 278; 610; 643; 749)

O.I. at 20°C = 1900 (316)

C. WATER POLLUTION FACTORS:
 —BOD_{10}: 1.00 std. dil. sew. (256)
 —BOD_5: 0% of ThOD
  10 days: 12.3% of ThOD
 —BOD_{10}^{20}: 12.3% of ThOD at 2.5 mg/l in mineralized dilution water with settled sewage seed (405)
  15 days: 50.8% of ThOD
  20 days: 56.9% of ThOD (79)
 —Waste water treatment:
  NFG, BOD, 20°C, 1-10 days observed, feed: 100-1,000 mg/l, acclimation: 365 + P: 28% removed (93)
  activated sludge: after 6 hr: 0.2% of ThOD
           12 hr: 0.3% of ThOD
           24 hr: 0.5% of ThOD (88)

diethylleadchloride
D. BIOLOGICAL EFFECTS:
 —Fish: plaice: 96 hr, LC_{50}: 75 mg/l (Pb) (573)

diethylmethylmethane see 3-methylpentane

diethyl-4-(methylsulphinyl)phenylphosphorothionate see fensulfothion

***O,O*-diethyl-*o,p*-nitrophenylthionophosphate** see parathion

***O,O*-diethyl-*o,p*-nitrophenylphosphorothioate** see parathion

diethylnitrosamine (N-nitrosodiethylamine; nitrous diethylamide)
$(C_2H_5)_2NNO$

Manmade sources: emitted during the compounding, forming and curing operations of elastomeric parts by reaction of accelerators/stabilizers used such as tetra-ethylthiuram disulfide and tetramethylthiuram disulfide. Emissions up to 270 g DEN/ billion g rubber stock have been reported (1800)

A. PROPERTIES: yellow liquid; m.w. 102.14; b.p. 177°C; sp.gr. 0.942 at 20/4°C

D. BIOLOGICAL EFFECTS:
 —Fish: creek chub: critical range: 900–1100 mg/l; 24 hr (226)
 —Carcinogenicity: +
 —Mutagenicity in the *Salmonella* test: weak,
   0.01 revertant colonies/nmol
   380 revertant colonies at 4,080 µg/plate (1883)

diethylolamine *see* diethanolamine

diethyloxalate
$C_2H_5OOC-COOC_2H_5$

Use: solvent; dye intermediate; organic synthesis

A. PROPERTIES: m.w. 146.14; m.p. -41°C; b.p. 185°C; sp.gr. 1.076; gradually decomposed by water

B. AIR POLLUTION FACTORS:
 —Odor threshold: recognition: 0.5 mg/cu m (610)

D. BIOLOGICAL EFFECTS:
 —Bacteria: *Pseudomonas putida*: inhibition of cell multiplication starts at >10,000 mg/l (329)
 —Algae: *Microcystis aeruginosa*: inhibition of cell multiplication starts at 9.0 mg/l (329)

diethyloxide *see* ethylether

O,O-diethyl-S-'4-oxo-3H-1,2,3-benzotriazine-3-yl)methyldithiophosphate *see* azinphosethyl

diethylphosphite
$(C_2H_5O)_2HPO$

A. PROPERTIES: water white liquid; b.p. 138°C; sp.gr. 1.069 at 25°C

C. WATER POLLUTION FACTORS:
 —Photo-oxidation by u.v. in aqueous medium at 90–95°C: time for the formation of CO_2 (% of theoretical): 25%: 10.7 hr
   50%: 28.8 hr
   75%: 64.0 hr (1628)

D. BIOLOGICAL EFFECTS:
 —Algae: *Scenedesmus*: deterioration: 250 mg/l (30)
 —Protozoa: *Colpoda*: deterioration: 100 mg/l (30)
 —Arthropoda: *Daphnia*: TLm: 25 mg/l (48 hrs) (30)

diethylphthalate (ethylphthalate; DEP)
$C_6H_4(CO_2C_2H_5)_2$

Manufacturing source: organic chemical industry.
Users and formulation: plasticizer mfg.; plastics mfg. and processing; explosive (propellant) component; suitable for food packaging application (FDA); dye application agent; diluent in polysulfide dental impression materials solvent; wetting agent; camphor substitute, perfumery, alcohol denaturant, component in insecticidal sprays; mosquito repellant.
Man caused source: general use of plastics and above listed products, evaporates from perfumes, inks, insecticides, repellants, alcohols, dyes.; from combustion of rocket propellants and explosives. (347)

A. PROPERTIES: water-white, stable, odorless liquid; m.w. 222.2; m.p. -40.5°C; b.p. 298°C; v.p. 14 mm at 163°C, 30 mm at 182°C; 734 mm at 295°C v.d. 7.66; sp.gr. 1.120 at 25/25°C; solub. 210 mg/l

B. AIR POLLUTION FACTORS:
  —Air quality:
    organic fraction of suspended matter:
      Bolivia at 5200 m altitude (Sept.–Dec. 1975): 0.51–0.80 μg/1000 cu m
      Belgium, residential area (Jan.–April 1976): 2.1–5.9 μg/1000 cu m (428)

C. WATER POLLUTION FACTORS:
  —Manmade sources:
    60 ml/min pure water passed through 25 ft, $\frac{1}{2}$ inch I.D. tube of general chemical grade P.V.C., contained 0.08 ppb diethylphthalate which constituted 0.74% of total contaminant concentration (430)
  —Toxicity threshold (cell multiplication inhibition test):
    bacteria (*Pseudomonas putida*):   >400 mg/l   (1900)
    algae (*Microcystis aeruginosa*):   15 mg/l   (329)
    green algae (*Scenedesmus quadricauda*):   10 mg/l   (1900)
    protozoa (*Entosiphon sulcatum*):   19 mg/l   (1900)
    protozoa (*Uronema parduczi Chatton-Lwoff*): 48 mg/l   (1901)
    marine dinoflagellate, *Gymnodium breve*: TLm 24 hr: 23.5 ppm
                                              TLm 96 hr: 33.0 ppm
                                              EC_{50}: 3.0–6.1 ppm*

  *EC_{50}: median growth limit concentration causing a 50% growth reduction
  (1057)

O,O-diethyl-o-quinoxalinyl-(2)-thionophosphate see ekalux

diethylstilbestrol (stilbestrol; 3,4-*bis*(*p*-hydroxyphenyl)-3-hexene)

HO—⟨⟩—C(CH₂CH₃)=C(CH₂CH₃)—⟨⟩—OH

Uses: medicine; research; animal feed.
a nonsteroid, synthetic estrogen, always in the *trans*-form.

A. PROPERTIES: white, odorless crystalline powder; m.p. 169–172°C; solub. 12,484–13,184 mg/l; m.w. 268.3

532 DIETHYLSULFIDE

C. WATER POLLUTION FACTORS:
  —Aquatic reactions:
    degradation in drinking water after 10 days: 38.7% remained unchanged (413)
  —Water quality:
    in drinking water in S.W. Germany, avg.: 0.11–0.24 ng/l
    range: 0–0.80 ng/l (413)
D. BIOLOGICAL EFFECTS:
  —Carcinogenicity: +
  —Mutagenicity in the *Salmonella* test: ?
    <0.38 revertant colonies/nmol
    <70 revertant colonies at 50 µg/plate (1883)

diethylsulfide (ethylsulfide; thioether; ethylthioethane)
$(C_2H_5)_2S$

A. PROPERTIES: colorless liquid; m.w. 90.18; m.p. -103°C; b.p. 92°C; v.p. 10 mm at -8°C, 40 mm at 16.1°C, 100 mm at 35.0°C; v.d. 3.1; sp.gr. 0.84 at 20/4°C; solub. 3,130 mg/l at 20°C; log P_{oct} 1.95

B. AIR POLLUTION FACTORS: 1 mg/cu m = 0.267 ppm, 1 ppm = 3.747 mg/cu m
  —Odor: characteristic; quality: garlic like, foul
    hedonic tone: nauseating

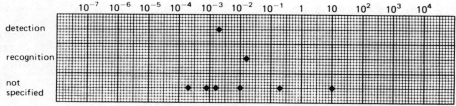

(10; 71; 210; 279; 307; 602; 637; 696; 710; 863)

  O.I. at 20°C = 14,000,000 (316)
  —Control methods:
    wet scrubber: water: effluent: 1,600 odor units (76)
    $KMnO_4$: effluent: 2 odor units

C. WATER POLLUTION FACTORS:
  —Reduction of amenities: faint odor: 0.00025 mg/l (129)
    odor threshold: 0.11 mg/kg (90)

diethylthiophosphate
D. BIOLOGICAL EFFECTS:
  —Fish: bluegill: 96 hr LC_{50}: 100 mg/l
  —Crustaceans: daphnids: 48 hr EC_{50}: 100 mg/l (1591)

1,3-diethylthiourea
$C_2H_5NHCSNHC_2H_5$
A. PROPERTIES: m.p. 68–71°C
D. BIOLOGICAL EFFECTS:
  —Fish: creek chub: critical range: 100–300 mg/l; 24 hr (226)

O,O-diethyl-O-(3,5,6-trichloro-2-pyridyl)phosphorothioate *see* chlorpyrifos

N^4,N^4-diethyl-*a,a,a*-trifluoro-3,5-dinitrotoluene-2,4-diamine *see* dinitramine

difenthos *see* abate

diflubenzuron (N-((4-chlorophenyl)amino)carbonyl)-2,6-difluorobenzamide; dimilin)

$$\text{F-C}_6\text{H}_3\text{F-C(O)-NH-C(O)-NH-C}_6\text{H}_4\text{-Cl}$$

Use: insect growth regulator; inhibits molting of larvae of mosquitoes, houseflies, stable flies and black flies by interfering with chitin synthesis

A. PROPERTIES: m.p. 239°C; solub. 200 ppb

C. WATER POLLUTION FACTORS:
—Diflubenzuron hydrolyzes in water to *p*-chlorophenylurea (1332)

D. BIOLOGICAL EFFECTS:
—cladocerans, clam shrimps and tadpole shrimps were killed at concentration below 0.01 µg/l or less in 24 to 48 hr in laboratory toxicity tests (1555)
—Brine shrimp *Artemia salina*: reproduction declined upon exposure to 2.0 µg/l (1556)
—Estuarine crustacean *Mysidopsis bahia*:
   96 hr LC$_{50}$: 2.1 µg/l
   21 d LC$_{50}$: 1.24 µg/l
   significant decline in reproduction at 0.075 µg/l (1554)
—Bioaccumulation:
   white crappies: exposure time 24 hr: 10 ppb in water, 822 ppb in fish
   bluegill: exposure time 24 hr: 10 ppb in water, 848 ppb in fish
   bluegills exposed to 10 ppb for 24 hr shows residues of 107 ppb, when placed in untreated flow-through rinse tanks for 24, 48, 72 and 96 hr showed no detectable tissue residues after any of these intervals.
   diflubenzuron is metabolized by the fish at a high rate; the excretory products are neither the parent compound, nor *p*-chlorophenylurea
   rainbow trout: 96-hr LC$_{50}$ 250 mg/l (S)
   fathead minnow: 96-hr LC$_{50}$ 430 mg/l (S)
   channel catfish: 96-hr LC$_{50}$ 370 mg/l (S)
   bluegill: 96-hr LC$_{50}$ 660 mg/l (S) (1510)
   coho salmon, juvenile rainbow trout: 96 hr LC$_{50}$: >150 mg/l
                                 1 g/l, 15 min: no mortality (1059)
—Mammals:
   acute oral LD$_{50}$ (rat): >10,000 mg/kg
                    (mouse): >4,640 mg/kg (1854)

2,6-difluorobenzoic acid

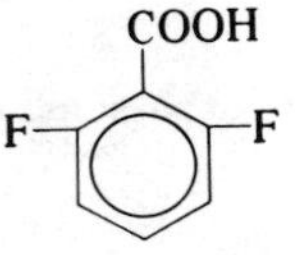

A. PROPERTIES: m.w. 158.11; m.p. 158–160°C
D. BIOLOGICAL EFFECTS:
  —rainbow trout: 96 hr LC_{50} (S): >100 mg/l
  fathead minnow: 96 hr LC_{50} (S): 69 mg/l
  channel catfish: 96 hr LC_{50} (S): >100 mg/l
  bluegill: 96 hr LC_{50} (S): >100 mg/l (1510)

difolitan (N-(1,1,2,2-tetrachloroethylthio)cyclohex-4-ene-1,2-dicarboximide; captafol)

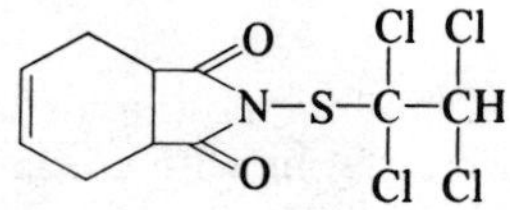

Use: fungicide
A. PROPERTIES: m.p. 159–161°C
D. BIOLOGICAL EFFECTS:
  —Crustacean:
  Gammarus lacustris: 96 hr LC_{50}: 800 µg/l (2124)
  —Insects:
  Pteronarcys californica: 96 hr LC_{50}: 40 µg/l (2128)
  —Mammals:
  acute oral LD_{50} (rats): 4600–6200 mg/kg (1854; 1855)
  in diet: rats fed for nine weeks with diets containing less than 2500 ppm were unaffected (1855)

diglycol *see* diethyleneglycolmonoethylether

diglycoldiacetate
$(CH_3COOC_2H_4)_2O$
C. WATER POLLUTION FACTORS:
  —Waste water treatment:

methods	temp °C	days observed	feed mg/l	days acclim.	% removed	
NFG, BOD	20	1–10	200	365 + P	17	
NFG, BOD	20	1–10	1,000	365 + P	3	(93)

diglyme *see* diethyleneglycoldimethylether

1,8-dihydroacenaphthylene *see* acenaphthene

9,10-dihydroanthracene

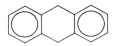

A. PROPERTIES: m.w. 180.25; m.p. 108-110°C; b.p. 312°C, sp.gr. 0.88
—Manmade sources:
 in wood preservative sludge (n.s.i.): 1.18 g/l of raw sludge
 in coke oven emissions (n.s.i.): 22-233 μg/g of sample (993)

dihydroazirine *see* ethyleneimine

dihydrobenzofluorenes
C. AIR POLLUTION FACTORS:
 —Air quality:
 glc's in Birkenes (Norway) Jan.–June 1977: avg.: 0.16 ng/cu m
  (a/b) fluorenes   range: n.d.–0.99 ng/cu m (n = 18)
 glc's in Rørvik (Sweden) Dec. '76–April '77: avg.: 0.29 ng/cu m
  (a/b) fluorenes   range: n.d.–1.89 ng/cu m (n = 21)
 (1236)
 —Manmade sources:
 in coke oven emissions: 24.0–791.4 μg/g of sample (n.s.i.) (960)

dihydrobenzo(c)phenanthrene
—Manmade sources:
 in wood preservative sludge: 0.44 g/l of raw sludge (960)

dihydrobutadienesulfone *see* sulfolane

9,10-dihydro-9,10-diketoanthracene *see* anthraquinone

1,4-dihydro-1,4-diketonaphthalene *see* α-naphthoquinone

2,3-dihydro-2,2-dimethyl-7-benzofuranylmethylcarbamate *see* carbofuran

dihydrofluoranthene
—Manmade sources:
 in wood preservative sludge (n.s.i.): 1.11 g/l of raw sludge
 in code oven airborne emissions (n.s.i.): 31.8–115.1 μg/g of sample (960)

dihydrofluorene
—Manmade sources:
 in wood preservative sludge (n.s.i.): 1.47 g/l of raw sludge
 in coke oven emissions (n.s.i.): 13.6–49.1 μg/g of sample (993)

dihydroheptachlor
Use: insecticide

4,5-DIHYDRO-2-MERCAPTOIMIDAZOLE

D. BIOLOGICAL EFFECTS:
harlequin fish (*Rasbora heteromorpha*):

mg/l	24 hr	48 hr	96 hr	3 m (extrapolated)	
LC_{10} (F)	0.05	0.036	0.031		
LC_{50} (F)	0.071	0.056	0.044	0.04	(331)

4,5-dihydro-2-mercaptoimidazole *see* ethylenethiourea

5,6-dihydro-2-methyl-1,4-oxathün-3-carboxanilide (vitavax; carboxin)

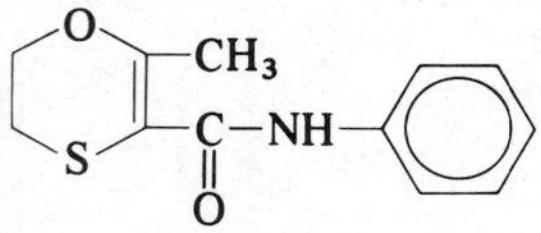

Use: pesticide
A. PROPERTIES: solub. 20 mg/l
C. WATER POLLUTION FACTORS:
—Waste water treatment:
powdered A.C.: Freundlich adsorption parameters: K: 0.023; $1/n$: 0.451
carbon dose to reduce 5 mg/l to a final conc. of 0.1 mg/l: 602 mg/l
carbon dose to reduce 1 mg/l to a final conc. of 0.1 mg/l: 110 mg/l (594)
D. BIOLOGICAL EFFECTS:
—Mammals:
acute oral LD_{50} (rat): 3820 mg/kg
acute dermal LD_{50} (rabbit): >8000 mg/kg (1854)

S-(3,4-dihydro-4-oxobenzo(*d*)-(1,2,3)-triazin-3-ylmethyl)-diethylphosphorothiolothionate *see* azinphosethyl

S-(3,4-dihydro-4-oxobenzo(*d*)-(1,2,3)-triazin-3-ylmethyl)-dimethylphosphorodithioate *see* azinphosmethyl

9,10-dihydrophenanthrene

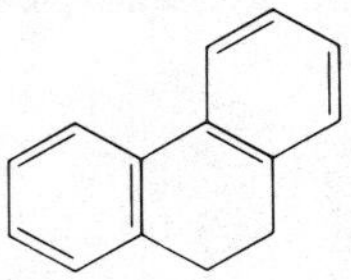

A. PROPERTIES: m.w. 180.25
—Manmade sources:
in wood preservative sludge (n.s.i.): 1.33 g/l of raw sludge
in coke oven emissions (n.s.i.): 79.6–587 µg/g of sample (993)

dihydropyrene
Manmade sources:
in wood preservative sludge (n.s.i.): 1.60 g/l of raw sludge

in coke oven airborne emissions (n.s.i.): 52.3–575.1 µg/g of sample (960)

1,2-dihydroxybenzene *see* catechol

1,3-dihydroxybenzene *see* resorcinol

1,4-dihydroxybenzene *see* hydroquinone

2,5-dihydroxybenzoic acid *see* gentisic acid

1,4-dihydroxybutane *see* 1,4-butanediol

β,β'-dihydroxydiethylamine *see* diethanolamine

β,β'-dihydroxydi-n-propylether *see* dipropyleneglycol

1,2-dihydroxyethane *see* ethyleneglycol

2,2'-dihydroxyethylether *see* diethyleneglycol

N,N'-di(2-hydroxyethyl)glycine, sodium salt (versene Fe-3 specific)
  Use: chelating agent
D. BIOLOGICAL EFFECTS:
  —Bluegill (*Lepomis macrochirus*): static 96 hr LC_{50}: 3092 mg/l
    96 hr no adverse effect level: 1800 mg/l (1869)

2,6-dihydroxyisonicotinic acid *see* citrazinic acid

2,3-dihydroxypropanal *see* glyceraldehyde

α,β-dihydroxypropionaldehyde *see* glyceraldehyde

diisoamylether *see* isoamylether

diisobutylamine (*bis*(β-methylpropyl)amine)
$[(CH_3)_2CHCH_2]_2-NH$
A. PROPERTIES: colorless liquid; m.w. 129.24; m.p. −70°C; b.p. 134°C; sp.gr. 0.745 at 20/4°C; log P_{oct} 2.84/3.04 (calculated)
D. BIOLOGICAL EFFECTS:
  —Fish: creek chub: critical range: 20–40 mg/l; 24 hr (226)

diisobutylcarbinol
$(CH_3)_2CHCH_2CHOHCH_2CH(CH_3)_2$
  Manufacturing source: organic chemical industry. (347)
  Users and formulation: defoamer; reaction medium for production of hydrogen peroxide; surface-active agents, lubricant additives; rubber chemicals; flotation agents. (347)

β-DIISOBUTYLENE

A. PROPERTIES: m.p. $-65°C$; b.p. $194°C$; v.p. 0.3 mm at $20°C$; v.d. 4.98; sp.gr. 0.83; solub. 1,000 mg/l at $20°C$

B. AIR POLLUTION FACTORS:
—Odor: characteristic: quality: sweet, alcohol
hedonic tone: pleasant
T.O.C.: abs. perc. lim.: 0.032 ppm
50% recogn.: 0.048 ppm
100% recogn.: 0.16 ppm
O.I. 100% recogn.: 8,187 (19)

C. WATER POLLUTION FACTORS:
—Reduction of amenities: T.O.C. = 1.3 mg/l (mean value) (297)(403)

D. BIOLOGICAL EFFECTS:
—Mammalia: rat: single oral LD_{50}: 3.56 g/kg
inhalation: no deaths: sat. vap., 8 hr (211)

β-diisobutylene *see* 2,4,4-trimethyl-2-pentene

diisobutylketone (2,6-dimethyl-4-heptanone; isovalerone; valerone; *s*-diisopropylacetone; DIBK)
$(CH_3)_2CHCH_2COCH_2CH(CH_3)_2$

A. PROPERTIES: colorless oil; m.w. 142.24; m.p. $-46/-42°C$; b.p. $165/168°C$; v.p. 1.7 mm at $20°C$ 2.3 mm at $30°C$; v.d. 4.9; sp.gr. 0.806 at $20/4°C$; solub. 500 mg/l; sat.conc. 9.3 g/cu m at $20°C$, 17 g/cu m at $30°C$

B. AIR POLLUTION FACTORS: 1 mg/cu m = 0.172 ppm, 1 ppm = 5.81 mg/cu m
—Odor: characteristic: quality: sweet, ester
hedonic tone: pleasant
T.O.C.: 295 mg/cu m = 50.7 ppm (57)
abs. perc. lim.: <0.11 ppm
50% recogn.: 0.31 ppm
100% recogn.: 0.31 ppm
O.I. 100% recogn.: 4,258 (19)

C. WATER POLLUTION FACTORS:
—Waste water treatment: A.C.: adsorbability: 0.060 g/g C, 100% reduction, infl.: 300 mg/l, effl.: nil (32)

D. BIOLOGICAL EFFECTS:
—Mammalia: rat: single oral dose: LD_{50}: 5.8 g/kg
inhalation: 5/6: 20,000 ppm, 8 hr
: no effect level: 125 ppm, 30 × 7 hr (211)
—Man: objectional odor: 25 ppm
unsatisfactory: 100 ppm (211)

diisobutylphthalate

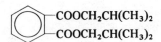

Uses: plasticizer

A. PROPERTIES: m.w. 278.35; m.p. $-64°C$; sp.gr. 1.039–1.043; v.d. 9.59 solub. 6.2 mg/l at 24°C (practical grade) (1666)
B. AIR POLLUTION FACTORS:
  —Air quality:
  organic fraction of suspended matter:
    Bolivia at 5200 m altitude (Sept.–Dec. 1975): 8.9–9.3 ng/cu m
    Belgium, residential area (Jan.–April 1976): 23–50 ng/cu m (428)
    glc's in residential area (Belgium) october 1976:
      in particulate sample: 1.73 ng/cu m
      in gas phase sample: 32.8 ng/cu m (1289)
C. WATER POLLUTION FACTORS:
  —Manmade sources:
  60 ml/min pure water passed through 25 ft., $\frac{1}{2}$ inch I.D. tube of general chemical grade P.V.C. contained 0.08 ppb di-isobutylphthalate which constituted 0.74% total contaminant concentration (430)

diisopropanolamine
$(CH_3)_2COHNHCOH(CH_3)_2$
A. PROPERTIES: m.w. 133.2; m.p. 42°C; b.p. 249°C; v.p. 0.02 mm at 42°C; v.d. 4.59; sp.gr. 0.99; solub. 870,000 mg/l
C. WATER POLLUTION FACTORS:
  —Waste water treatment: A.C.: adsorbability: 0.091 g/g C, 53.3% reduction, infl.: 1,000 mg/l, effl.: 467 mg/l (32)
D. BIOLOGICAL EFFECTS:
  —Ingestion: rat: single oral LD_{50}: in the range of 2000 to 4000 mg/kg (1546)

diisopropoxyde   *see* diisopropylether

s-diisopropylacetone   *see* diisobutylketone

diisopropylamine
$(CH_3)_2CHNHCH(CH_3)_2$
A. PROPERTIES: colorless liquid; m.w. 101.19; m.p. $-96.3°C$; b.p. 83.4°C; v.p. 70 mm at 20°C; v.d. 3.49; sp.gr. 0.722 at 22/0°C; log P_{oct} 1.84 (calculated)
B. AIR POLLUTION FACTORS: 1 mg/cu m = 0.242 ppm, 1 ppm = 4.14 mg/cu m
  —Odor characteristic: quality: fishy, amine, ammoniacal
                        hedonic tone: unpleasant to pleasant
    T.O.C.: abs. perc. lim.: 0.13 ppm
            50% recogn.: 0.38 ppm
            100% recogn.: 0.85 ppm
            O.I. 100% recogn.: 92,823 (19)
D. BIOLOGICAL EFFECTS:
  —Fish: creek chub: critical range: 40–60 mg/l; 24 hr (226)
         creek chub: LD_0: 40 mg/l, 24 hr in Detroit river water
                     LD_{100}: 60 mg/l, 24 hr in Detroit river water (243)
  —Mammalia: rat: inhalation: 2/6: 1,000 ppm, 4 hr (211)
             rat: oral LD_{50}: 0.77 g/kg; 95% conf.: 0.61–0.94 g/kg (226)

diisopropylether (isopropylether; 2-isopropoxypropane; IPE; DIPE; diisopropyloxyde) $(CH_3)_2CHOCH(CH_3)_2$

Uses: solvent for oils, waxes and resins; paint and varnish removers; rubber cements

A. PROPERTIES: colorless liquid; m.w. 102.2; m.p. $-86/-60°C$; b.p. $69°C$; v.p. 130 mm at $20°C$; v.d. 3.52; sp.gr. 0.73 at $20/4°C$; solub. 9,000 mg/l at $20°C$,

B. AIR POLLUTION FACTORS: 1 mg/cu m = 0.177, 1 ppm = 5.65 mg/cu m
 —Odor: characteristic: quality: sweet
 hedonic tone: pleasant
 T.O.C.: abs.perc.lim.: 0.017 ppm
 50% recogn.: 0.053 ppm
 100% recogn.: 0.053 ppm
 O.I. 100% recogn.: 2,924,528 (19)

C. WATER POLLUTION FACTORS:
 —BOD: 0.19 NEN 3235-5.4 (277)
 —COD: NEN 3235-5.3 (277)
 —$KMnO_4$ value: 0.013
 —ThOD: 2.833 (30)
 —Waste water treatment: A.C.: absorbability: 0.162 g/g C; 80% reduction, infl.: 1,023 mg/l, effl.: 203 mg/l (32)

D. BIOLOGICAL EFFECTS:
 —Bacteria: *Pseudomonas*: no effect: 125 mg/l
 —Algae: *Scenedesmus*: no effect: 30 mg/l
 —Protozoa: *Colpoda*: no effect: 125 mg/l (30)
 —Fish: goldfish: LD_{50} (24 hr): 380 mg/l-modified ASTM D1345 (277)
 Lepomis macrochirus: static bioassay in fresh water at $23°C$, mild aeration applied after 24 hr:

material added ppm	% survival after				best fit 96 hr LC_{50} ppm
	24 hr	48 hr	72 hr	96 hr	
10,000	30	30	30	30	
79,000	100	100	100	100	7,000
5,000	80	70	70	70	
1,000	100	100	100	100	

 Menidia beryllina: static bioassay in synthetic seawater at $23°C$: mild aeration applied after 24 hr:

material added ppm	% survival after				best fit 96 hr LC_{50} ppm
	24 hr	48 hr	72 hr	96 hr	
10,000	0	—	—	—	
7,500	0	—	—	—	6,600
5,000	20	20	20	20	
3,200	100	100	100	100	(352)

 —Mammalia:
 monkey: inhalation: LD_{100}: 60,000 ppm
 guinea pig: inhalation: LD_{100}: 60,000 ppm
 rabbit: inhalation: LD_{100}: 60,000 ppm
 rat: inhalation: lethal: 16,000 ppm, 4 hr
 monkey: inhalation: no effect: 3,000 ppm, 2 hr
 guinea pig: inhalation: no effect: 3,000 ppm, 2 hr
 rabbit: inhalation: no effect: 3,000 ppm, 2 hr

monkey, rabbit, guinea pig: inhalation: no effect: 3,000 ppm, 20 × 2 hr/day
1,000 ppm, 20 × 3 hr/day
rabbit: single oral dose: lethal: 5–6.5 g/kg (211)
—Man: irritation of eyes and nose: 800 ppm (211)

diisopropylideneacetone *see* phorone

diisopropylmethane *see* 2,4-dimethylpentane

dimethoate (fosfamid; O,O-dimethyl-S-(N-methylcarbamoylmethyl)phosphorodithioate)

$$\begin{array}{c} CH_3O \\ \diagdown \\ P-S-CH_2-CO-NHCH_3 \\ \diagup \\ CH_3O \end{array}$$
(with S double-bonded to P)

A. PROPERTIES: m.p. 48–52°C; solub. 25 g/l at room temp.; mercaptan odor; v.p. 0.025 mm at 25°C; sp.gr. 1.28 at 65°C; log P_{oct} 2.71

C. WATER POLLUTION FACTORS:
—Persistence in river water in a sealed glass jar under sunlight and artificial fluorescent light—initial conc. 10 µg/l:

after	1 hr	1 wk	2 wk	4 wk	8 wk	
	100	100	85	75	50	(1309)

% of original compound found

D. BIOLOGICAL EFFECTS:
—Fish:
Chingatta: 96 hr LC_{50} (S): 4.48 mg/l (30% EC) (1494)
bluegill: 24 hr LC_{50}: 28.0 ppm
rainbow trout: 24 hr LC_{50}: 20.0 ppm (1878)
—Mammals:
acute oral LD_{50} (male albino rat): 320–380 mg/kg
acute dermal LD_{50} (guinea pig): 650 mg/kg (1854)
—Carcinogenicity:
Tremors and hyperexcitability, both indications of demethoate toxicity, were observed in the treated animals. However, it is considered that the low-dose group of rats and both dose groups of mice survived long enough to permit an evaluation of carcinogenicity. Pathologic evaluation revealed no statistically significant increase in tumors associated with dimethoate treatment in either species of animal, and it is concluded that there was no carcinogenic effect under the conditions of the experiment. (1740)

dimethoxane *see* giv-gard DXN

1,2-dimethoxybenzene (veratrole; pyrocatecholdimethylether)

(benzene ring with two $-OCH_3$ groups in ortho positions)

542   1,3-DIMETOXYBENZENE

Manufacturing source: organic chemical industry. (347)
Users and formulation: mfg. flavors and perfumes; medicine (antiseptic-veratrole).
(347)
A. PROPERTIES: m.w. 138.16; m.p. 22.5°C; b.p. 206/207°C; sp.gr. 1.084 at 25/25°C
C. WATER POLLUTION FACTORS:
—Biodegradation: decomposition by a soil microflora in 8 days (176)

1,3-dimetoxybenzene (resorcinoldimethylether)

A. PROPERTIES: m.w. 138.16; m.p. -52/-55°C; b.p. 217/218°C; sp.gr. 1.0552 at 25/25°C
C. WATER POLLUTION FACTORS:
—Biodegradation: decomposition by a soil microflora in >32 days (176)

1,4-dimethoxybenzene (hydroquinonedimethylether)

A. PROPERTIES: m.w. 138.16; m.p. 56°C; b.p. 212.6°C; sp.gr. 1.036 at 66/4°C; log P_{oct} 3.14/3.2 (calculated)
B. AIR POLLUTION FACTORS: 1 mg/cu m = 0.177 ppm, 1 ppm = 5.65 mg/cu m
C. WATER POLLUTION FACTORS:
—Biodegradation: decomposition by a soil microflora in 8 days (176)
D. BIOLOGICAL EFFECTS:
—Mammalia: rat: single oral LD_{50}: 8.5 g/kg (211)

dimethoxymethane (methylal; formaldehyde dimethylacetal; formal; methylenedimethylether)
$(CH_3O)_2CH_2$
A. PROPERTIES: colorless liquid; m.w. 76.09; m.p. -104.8°C; b.p. -44/-41.2°C; v.p. 330 mm at 20°C, 400 mm at 25°C; v.d. 2.6; sp.gr. 0.856; solub. 330,000 mg/l; sat.conc. 1,367 g/cu m at 20°C; log P_{oct} 0.0
B. AIR POLLUTION FACTORS: 1 mg/cu m = 0.32 ppm, 1 ppm = 3.16 mg/cu m
C. WATER POLLUTION FACTORS:
—Waste water treatment:
A.S., COD, 20°C, $\frac{1}{3}$ day observed, feed: 333 mg/l, 30 days acclimation: 88% removed (93)
D. BIOLOGICAL EFFECTS:
—Mammalia: rat: inhalation: no effect level: 4,000 ppm, 8 × 6 hr (65)

mouse: inhalation: LC_{50}: 18,000 ppm, 7 hr
: 6/5: 11,000 ppm, 15 × 7 hr
guinea pig: subcutane LD_{50}: 5 g/kg (211)

dimethoxystrychnine *see* brucine

dimethrin (2,4-dimethylbenzyl-2,2-dimethyl-3-(2-methylpropenyl)cyclopropanecarboxylate; 2,4-dimethylbenzylchrysanthemamate; a pyrethroid)
Use: insecticide
A. PROPERTIES: amber liquid; sp.gr. 0.986 at 20°C; b.p. 175°C at 3.8 mm
D. BIOLOGICAL EFFECTS:

	96 hr LC_{50} (μg/l)*	
	static test	flow through test
channel catfish	1140	165
bluegill	37.5	22.3
yellow perch	28	—

*toxicity calculated for formulation with 96% of active ingredient (444; 1626)

N,N-dimethylacetamide
$CH_3CON(CH_3)_2$
A. PROPERTIES: m.w. 87.12; m.p. -20/+27°C; b.p. 166°C; v.p. 1.3 mm at 25°C, 9 mm at 60°C; v.d. 3.01; sp.gr. 0.94; log P_{oct} -0.77
B. AIR POLLUTION FACTORS: 1 mg/cu m = 0.28 ppm, 1 ppm = 3.62 mg/cu m
—Odor: characteristic: quality: amine, burnt, oily
T.O.C.: 165 mg/cu m = 46.2 ppm (57)
$PIT_{50\%}$: 21.4 ppm
$PIT_{100\%}$: 46.8 ppm (2)
O.I. at 20°C = 37 (316)
D. BIOLOGICAL EFFECTS:
—Mammalia: rats and dogs: inhalation: liver damage: 100–200 ppm, repeated
(211)

dimethylacetic acid *see* isobutyric acid

***sym*-dimethylacetone** *see* diethylketone

dimethylacetonylcarbinol *see* diacetone alcohol

***sym*-dimethylallene** *see* 2,3-pentadiene

dimethylamine
$(CH_3)_2NH$
A. PROPERTIES: m.w. 45.08; m.p. -92/-96°C; b.p. 7.4°C; v.p. 1,520 mm at 10°C, 1.7 atm at 20°C. 2.5 atm at 30°C; v.d. 1.55; sp.gr. 0.68 at 0/4°C; log P_{oct} -0.38/-0.02 (calcul.)
B. AIR POLLUTION FACTORS: 1 mg/cu m = 0.542 ppm, 1 ppm = 1.84 mg/cu m

544 DIMETHYLAMINE HYDROCHLORIDE

[odor thresholds chart: detection, recognition, not specified across 10^{-7} to 10^4 mg/cu m]

(2; 57; 73; 210; 279; 307; 664; 821; 879)

O.I. at 20°C = 280,000   (316)

—Manmade sources:
  at 40 m downwind from a fishmean plant: glc's: 0.53 ppb   (50)

—Control methods:
  wet scrubber: water   at pH 8.5: outlet: 1,300 odor units/scf
                $KMnO_4$ at pH 8.5: outlet: 20 odor units/scf   (115)

—Sampling and analysis:
  photometry: min. full scale: 7,600 ppm   (53)

C. WATER POLLUTION FACTORS:
  —BOD_5:   1.3
  —BOD_{20}: 2.0
  —$KMnO_4$: 0.024
  —ThOD: 2.006   (30)

—Reduction of amenities:
  taste and odor at 0.6 mg/l   (41)
  odor threshold; average: 23.2 mg/l
               range: 0.001–42.5 mg/l   (30)

D. BIOLOGICAL EFFECTS:
  —Bacteria: *Psuedomonas*: no effect: 1 g/l
  —Algae: *Scenedesmus*: toxic: 250 mg/l
  —Protozoa: *Colpoda*: toxic: 250 mg/l   (30)
  —Fish: creek chub: critical range; 30–50 mg/l, 24 hr   (226)

dimethylamine hydrochloride
$(CH_3)_2NH \cdot HCl$

A. PROPERTIES: m.w. 81.55; m.p. 170–173°C

C. WATER POLLUTION FACTORS:
  —Impact on biodegradation:
    at 100 mg/l, no inhibition of NH_3 oxidation by *Nitrosomonas* sp.   (390)

dimethylaminoaniline   *see p-aminodimethylaniline*

4-dimethylaminoazobenzene   (methyl yellow; butter yellow)

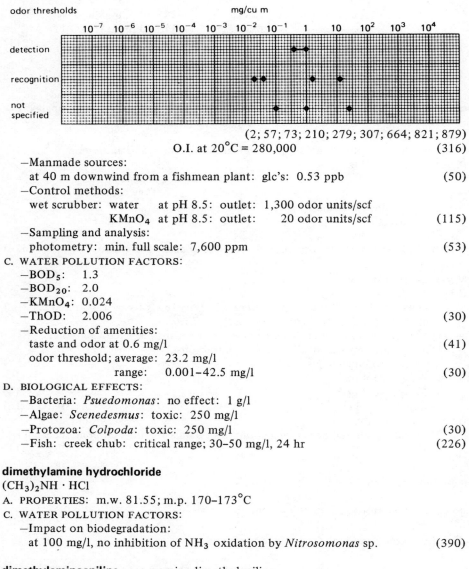

Uses: indicator in volumetric analysis; also in test for peroxidized fats; coloring agent
A. PROPERTIES: m.w. 225.3; m.p. 111°C decomposes; log P_{oct} 4.58; v.d. 7.78
C. WATER POLLUTION FACTORS:
—Impact on biodegradation processes:
at 100 mg/l, no inhibition of NH_3 oxidation by *Nitrosomonas* sp. (390)
D. BIOLOGICAL EFFECTS:
—Man: carcinogenic: USA 1974 (77)
—Carcinogenicity: +
—Mutagenicity in the *Salmonella* test: +
0.12 revertant colonies/nmol
120 revertant colonies at 210 μg/plate (1883)

4-(dimethylamino)-3-methylphenyl-N-methylcarbamate(ester) *see* aminocarb

4-(dimethylamino)phenyl)diazenesulfonate, sodium *see* dexon

4-dimethylamino-*m*-tolylmethylcarbamate *see* aminocarb

4-dimethylamino-3,5-xylenol

D. BIOLOGICAL EFFECTS:
—Fish: bluegill: 96 hr LC_{50} (S): 7.2 mg/l (1500)

N,N-dimethylaniline

Uses: dyes; intermediates; solvent; manufacture of vanillin; stabilizer
A. PROPERTIES: yellow liquid; m.w. 121.18; m.p. 2°C; b.p. 192°C; v.p. 1.1 mm at 30°C, 0.5 mm at 20°C; v.d. 4.17; sp.gr. 0.96 at 20/4°C; sat.conc. 3.3 g/cu m at 20°C, 7.0 g/cu m at 30°C; log P_{oct} 2.31/2.62
B. AIR POLLUTION FACTORS: 1 mg/cu m = 0.20 ppm, 1 ppm = 5.04 mg/cu m

2,3-DIMETHYLANILINE

odor thresholds   mg/cu m

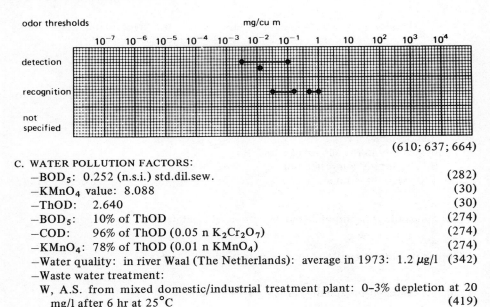

(610; 637; 664)

C. WATER POLLUTION FACTORS:
- BOD_5: 0.252 (n.s.i.) std.dil.sew.   (282)
- $KMnO_4$ value: 8.088   (30)
- ThOD: 2.640   (30)
- BOD_5: 10% of ThOD   (274)
- COD: 96% of ThOD (0.05 n $K_2Cr_2O_7$)   (274)
- $KMnO_4$: 78% of ThOD (0.01 n $KMnO_4$)   (274)
- Water quality: in river Waal (The Netherlands): average in 1973: 1.2 µg/l (342)
- Waste water treatment:
   W, A.S. from mixed domestic/industrial treatment plant: 0–3% depletion at 20 mg/l after 6 hr at 25°C   (419)

2,3-dimethylaniline

A. PROPERTIES: m.w. 121.18; m.p. 2.5°C; b.p. 221–222°C; sp.gr. 0.993
C. WATER POLLUTION FACTORS:
- Biodegradation: adapted A.S. at 20°C—product is sole carbon source: 96.5% COD removal at 12.7 mg COD/g dry inoculum/hr   (327)

2,4-dimethylaniline (2,4-xylidine)

A. PROPERTIES: m.w. 121.18; b.p. 218°C; sp.gr. 0.98, log P_{oct} 1.85 (calculated)
D. BIOLOGICAL EFFECTS:
- Toxicity threshold (cell multiplication inhibition test):
bacteria (*Pseudomonas putida*):	8 mg/l	
green algae (*Scenedesmus quadricauda*):	5 mg/l	
protozoa (*Entosiphon sulcatum*):	9.8 mg/l	(1900)
protozoa (*Uronema parduczi Chatton-Lwoff*):	12 mg/l	(1901)

2,5-dimethylaniline (2,5-xylidine)

A. PROPERTIES: oil; m.w. 121.18; m.p. 15.5°C; b.p. 217°C; v.d. 4.18; sp.gr. 0.98 at 15°C
B. AIR POLLUTION FACTORS: 1 mg/cu m = 0.20 ppm, 1 ppm = 5.04 mg/cu m
—Odor: T.O.C.: 0.024 mg/cu m (n.s.i.) (307)
      recognition: 0.25–0.30 mg/cu m (610)
    O.I. to 20°C = 82,000 (316)
C. WATER POLLUTION FACTORS:
—BOD_5: nil std. dil. sew. (n.s.i.) (282)
—Biodegradation: adapted A.S. at 20°C—product is sole carbon source: 96.5% COD removal at 3.6 mg COD/g dry inoculum/hr (327)
D. BIOLOGICAL EFFECTS:
—Man: severe toxic effects: 40 ppm = 200 mg/cu m, 60 min
    symptoms of illness: 10 ppm = 50 mg/cu m
    unsatisfactory: 5 ppm = 25 mg/cu m (all isomers) (185)

2,6-dimethylaniline (2,6-xylidine)

A. PROPERTIES: m.w. 121.18; m.p. 10–12°C; b.p. 214°/739 mm, sp.gr. 0.984; log P_{oct} 1.96 (calculated)
C. WATER POLLUTION FACTORS:
—Waste water treatment:
Warburg app., activated sludge from mixed domestic/industrial treatment plant: 24–30% depletion at 20 mg/l after 6 hr at 25°C (419)
D. BIOLOGICAL EFFECTS:
—Ciliate (*Tetrahymena pyriformis*): 24 hr LC_{100}: 1.24 mmole/l (1662)

3,4-dimethylaniline (3,4-xylidine)

A. PROPERTIES: m.w. 121.18; m.p. 49–51°C; b.p. 226°C; log P_{oct} 1.70 (calculated)
C. WATER POLLUTION FACTORS:
—Biodegradation: adapted A.S. at 20°C—product is sole carbon source: 76.0% COD removal at 30.0 mg COD/g dry inoculum/hr (327)

3,5-dimethylaniline (3,5-xylidine)

$$\underset{H_3C}{}\text{—}\underset{}{\bigcirc}\text{—}\underset{CH_3}{}\quad NH_2$$

A. PROPERTIES: m.w. 121.18; b.p. 104–105°C at 14 mm, sp.gr. 0.972
C. WATER POLLUTION FACTORS:
 —Waste water treatment:
  Warburg app., activated sludge from mixed domestic/industrial treatment plant: 46–60% depletion of 20 mg/l after 6 hr at 25°C (419)

dimethylarsine (DMA; cacodylhydride)
$(CH_3)_2AsH$
 Use: herbicide
A. PROPERTIES: m.w. 105.99
D. BIOLOGICAL EFFECTS:
 —Bioaccumulation in model ecosystem (multi-organism experiments), after 32 days of exposure: conc. of DMA in water: 1st day: 7.0 ppb
  32nd day: 3.9 ppb
  bioaccumulation ratio after 32 days: algae: 1248–1605
     snails: 299–446
      129–176 (after 16 days in clean water)
    daphnia: 736–2175
    fish: 19–49 (after 3 days of exposure)

dimethylarsinic acid *see* cacodylic acid

1,2-dimethylbenzene *see* *o*-xylene

1,3-dimethylbenzene *see* *m*-xylene

1,4-dimethylbenzene *see* *p*-xylene

dimethyl-1,2-benzenedicarboxylate *see* *o*-dimethylphthalate

dimethyl-1,4-benzenedicarboxylate *see* dimethylterephthalate

3,3-dimethylbenzidine *see* *o*-tolidine

9,10-dimethylbenzo(*a*)anthracene
A. PROPERTIES: solub. 0.055 mg/l at 24°C (99% purity) (1666)
C. WATER POLLUTION FACTORS:
 —Waste water treatment:
  ozonation: after 1 min. contact time with ozon: residual amount: 0% (550)

α,α-dimethylbenzylalcohol *see* 2-phenyl-2-propanol

2,4-dimethylbenzylchrysanthemamate *see* dimethrin

2,4-dimethylbenzyl-2,2-dimethyl-3-(2-methylpropenyl)-cyclopropanecarboxylate *see* dimethrin

2,2-dimethylbutane (neohexane)
$C_2H_5C(CH_3)_3$
 Use: intermediate for agricultural chemicals
A. PROPERTIES: m.w. 86.18; m.p. -100°C; b.p. 50°C; sp.gr. 0.649; v.d. 3.00
B. AIR POLLUTION FACTORS:
 —Manmade sources:
  evaporation from gasoline fuel tank: 0.1 vol % of total evaporated H.C.'s
  evaporation from carburetor: 0.0–0.1 vol. % of total evaporated H.C.'s
  (398; 399; 400; 401; 402)
C. WATER POLLUTION FACTORS:
 —Biodegradation:
  incubation with natural flora in the groundwater—in presence of the other components of high-octane gasoline (100 µl/l): biodegradation: 25% after 192 hr at 13°C (initial conc. 0.28 µl/l) (956)

2,3-dimethylbutane (isopropyldimethylmethane; biisopropyl)
$(CH_3)_2CH-CH(CH_3)_2$
A. PROPERTIES: m.w. 86.17; m.p. -135°C; b.p. 57°C; v.p. 200 mm at 20°C, v.d. 3.00; sp.gr. 0.67 at 17/4°C;
B. AIR POLLUTION FACTORS:
 —Manmade sources: diesel engine exhaust: 0.5% of emitted HC's (72)
  evaporation from gasoline fuel tank:
   1.6–2.6 vol % of total evaporated H.C.'s
  evaporation from carburetor:
   3.0–3.2 vol % of total evaporated H.C.'s (398; 399; 400; 401; 402)
C. WATER POLLUTION FACTORS:
 —Biodegradation:
  incubation with natural flora in the groundwater—in presence of the other components of high-octane gasoline (100 µl/l): biodegradation: 0% after 192 hr at 13°C (initial conc. 0.86 µl/l) (956)

2,3-dimethylbutene-2 (tetramethylethylene)
$(CH_3)_2C=C(CH_3)_2$
A. PROPERTIES: m.w. 84.16; m.p. -74.3°C; b.p. 73.2°C; sp.gr. 0.712 at 20/4°C;
B. AIR POLLUTION FACTORS:
 —Atmospheric reactions:
  reactivity: NO ox.: ranking 10 (63)
         HC cons. ranking 10 (63)
         RCR: 29.3 (49)
  estimated lifetime under photochemical smog conditions in S.E. England: 0.33 hr
  (1699; 1700)

dimethylcarbinol *see* isopropanol

O,O-dimethyl-S-(α-(carboethoxy)benzyl)phosphorodithioate *see* phenthoate

O,O-dimethyl-O-(3-chloro-4-nitrophenyl)phosphorothioate *see* chlorothion

1,2-dimethylcyclohexane

A. PROPERTIES: m.w. 112.22; b.p. 124°C; sp.gr. 0.778
C. WATER POLLUTION FACTORS:
 —Biodegradation:
  incubation with natural flora in the groundwater—in presence of the other components of high-octane gasoline (100 μl/l): biodegradation: 26% after 192 hr at 13°C (initial conc. 0.16 μl/l) (956)

dimethylcyclohexanol

C. WATER POLLUTION FACTORS:
 —Biodegradation: adapted A.S.—product is sole carbon source: 92.3% COD removal at 21.6 mg/COD/g dry inoculum/hr (327)

1,1-dimethylcyclopentane

C. WATER POLLUTION FACTORS:
 —Biodegradation:
  incubation with natural flora in the groundwater—in presence of the other components of high-octant gasoline (100 μl/l: biodegradation: 25% after 192 hr at 13°C (initial conc.: 0.12 μl/l) (956)

dimethyldiethyllead

D. BIOLOGICAL EFFECTS:
 —Residues: taken from various lakes and rivers in Ontario (1979):
  in fish: n.d.–7.1 ng/g wet wt.
  in water, vegetation, weeds and sediments: n.d. (1793)

1,1'-dimethyl-4,4'-dipyridyliumdichloride *see* paraquat

dimethyldioxane

$$H_3C-\underset{O}{\overset{O}{\bigcirc}}-CH_3$$

A. PROPERTIES: m.w. 116.16; b.p. 117°C; v.d. 4.0; sp.gr. 0.90
B. AIR POLLUTION FACTORS: 1 mg/cu m = 0.21 ppm, 1 ppm = 4.83 mg/cu m
C. WATER POLLUTION FACTORS:
  —Impact on biodegradation processes:
    slight inhibition of microbial growth after 24 hr exposure at 500 ppm (523)

dimethyldisulfide Methyldisulfide; methyldithiomethane)
CH_3SSCH_3
A. PROPERTIES: m.w. 94.19; b.p. 112/118°C; sp.gr. 1.057 at 16/4°C; log P_{oct} 1.77
B. AIR POLLUTION FACTORS:
  —Odor: T.O.C.: 0.005 mg/cu m = 0.001 ppm (279)
    50% recogn.: 5.6 ppb (71)
    recognition: 0.029 mg/cu m (863)
    detection: 0.003–0.014 mg/cu m (742)
  —Biogenic dimethyldisulfide emissions from soils in U.S.A.

sampling sites	soil orders	sampling sites	avg. sulfur flux g S/m²/yr
Wadesville, IN	Alfisol	Sept.–Oct. 1977	0.002
Philo, OH	Inceptisol	Oct. 1977	0.0014
Dismal Swamp, NC	Histosol	Oct. 1977	0.0001
Cox's Landing, NC—freshly clipped march			0.0039

(1385)

  —Control methods:
    wet scrubber: water at pH 8.5: effluent: 1,500 odor units/scf
                  $KMnO_4$ at pH 8.5: effluent: 1 odor units/scf (76)
C. WATER POLLUTION FACTORS:
  odor thresholds   mg/kg water

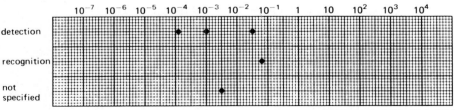

(300; 874; 908; 914; 925)

D. BIOLOGICAL EFFECTS:
  —Mammals: rat: inhalation: no effect level: 100 ppm, 20 X 6 hr (65)

dimethyleneimine *see* ethyleneimine

dimethyleneoxide *see* ethyleneoxide

dimethylethylcarbinol see 2-methyl-2-butanol

sym-dimethylethylene see trans-2-butylene

uns-dimethylethylene see isobutene

4-(α,α-dimethylethylphenol see p-tert-butylphenol

O,O-dimethyl-S-(2-(ethylthio)ethyl)phosphorodithioate see ekatin

N,N-dimethylformamide
$OHCN(CH_3)_2$
A. PROPERTIES: m.w. 73.10; m.p. $-61°C$; b.p. $152/153°C$; v.p. 2.7 mm at $20°C$; v.d. 2.51; sp.gr. 0.95; sat.conc. 12 g/cu m at $20°C$; log P_{oct} $-0.87/-0.59$ (calculated)
B. AIR POLLUTION FACTORS: 1 mg/cu m = 0.33 ppm, 1 ppm = 3.04 mg/cu m
  —Odor: characteristic; quality: fishy
                     hedonic tone: pungent
     T.O.C.: 0.14 mg/cu m = 0.046 ppm   (57; 769)
             $PIT_{50}\%$: 100 ppm   (2)
             $PIT_{100}\%$: 100 ppm   (2)
             O.I. at $20°C$ = 35   (316)
     USSR: human odor perception: non perception: 0.14 mg/cu m
                                    perception: 0.88 mg/cu m
                                                          = 0.29 ppm
            human reflex response: no response: 0.03 mg/cu m
                                    adverse response: 0.055 mg/cu m
            animal chronic exposure: no effect: 0.03 mg/cu m
                                    adverse effect: 0.5   mg/cu m   (170)
  —Control methods:
    catalytic combustion: platinized ceramic honeycomb catalyst; ignition temp: $200°C$, inlet temp for 90% conversion: $350-400°C$ (91)
  —Sampling and analysis: photometry: min. full scale: 470 ppm   (53)
C. WATER POLLUTION FACTORS:
  —$KMnO_4$ value: 0.042
  —ThOD: 1.863   (30)
  —Water quality: the selfpurification is affected from 10.0 mg/l   (181)
     photometry: min. full scale: $<25 \times 10^{-6}$ mole/l   (53)

2,2-dimethylglutaric acid
$HOOC-CH_2-CH_2-C(CH_3)_2COOH$
A. PROPERTIES: m.w. 160.17; m.p. $83-85°C$
C. WATER POLLUTION FACTORS:
  —$BOD_2^{25°C}{}_{days}$: 0.06 (substrate conc. 4.9 mg/l; inoculum; soil microorganisms)
     5 days: 0.06
     10 days: 0.12
     20 days: 0.55   (1304)

3,3-dimethylglutaric acid
$HOOC-CH_2C(CH_3)_2CH_2COOH$

A. PROPERTIES: m.w. 160.17; m.p. 100-102°C
C. WATER POLLUTION FACTORS:
—BOD$_{2\,days}^{25°C}$: 0.041 (substrate conc. 4.9 mg/l; inoculum; soil microorganisms)
 5 days: 0.02
 10 days: 0.86
 20 days: 1.47   (1304)

dimethylglyoxime (butane dioxime)
CH$_3$C(NOH)C(NOH)CH$_3$
Uses: analytical chemistry especially as a reagent for nickel; biochemical research
A. PROPERTIES: white crystals or powder; m.p. 240-241°C; m.w. 116.12; log P_{oct} -2.16/-0.29 (calculated)
C. WATER POLLUTION FACTORS:
—Impact on biodegradation processes:
 NH$_3$ oxidation by *Nitrosomonas* sp.: at 100 mg/l: 30% inhibition
            50 mg/l: 9% inhibition
           140 mg/l: ~50% inhibition   (390)

2,2-dimethylheptane
CH$_3$(CH$_2$)$_4$C(CH$_3$)$_3$
A. PROPERTIES: m.w. 128.26; m.p. -113°C; b.p. 130°C; sp.gr. 0.71
C. WATER POLLUTION FACTORS:
—Reduction of amenities: T.O.C. = 5 mg/l   (295)
—Biodegradation:
 incubation with natural flora in the groundwater—in presence of the other components of high-octane gasoline (100 µl/l): biodegradation: 62% after 192 hr at 13°C (initial conc.: 0.09 µl/l)   (956)

2,6-dimethyl-4-heptanone *see* diisobutylketone

2,6-dimethyl-3-heptene
(CH$_3$)$_2$CHCH$_2$CHCHCH(CH$_3$)$_2$
C. WATER POLLUTION FACTORS:
—Waste water treatment: A.S.: after 6 hr: 0.7% of ThOD
           12 hr: 0.9% of ThOD
          24 hr: 1.5% of ThOD   (88)

2,2-dimethylhexane
(CH$_3$)$_3$C(CH$_2$)$_3$CH$_3$
C. WATER POLLUTION FACTORS:
—Biodegradation:
 incubation with natural flora in the groundwater—in presence of the other components of high-octane gasoline (100 µl/l): biodegradation: 75% after 192 hr at 13°C (initial conc. 0.05 µl/l)   (956)

2,3-dimethylhexane
(CH$_3$)$_2$CH(CH$_3$)CH(CH$_2$)$_2$CH$_3$
A. PROPERTIES: sp.gr. 0.72; solub. 0.13 mg/l at 25°C
C. WATER POLLUTION FACTORS:

554   2,4-DIMETHYLHEXANE

—Biodegradation:
incubation with natural flora in the groundwater—in presence of the other components of high-octane gasoline (100 µl/l): biodegradation: 19% after 192 hr at 13°C (initial conc. 0.54 µl/l) (956)

2,4-dimethylhexane

$$CH_3-\underset{\underset{CH_3}{|}}{CH}-CH_2-\underset{\underset{CH_3}{|}}{CH}-CH_2CH_3$$

C. WATER POLLUTION FACTORS:
—Biodegradation:
incubation with natural flora in the groundwater—in presence of the other components of high-octane gasoline (100 µl/l): biodegradation: 0% after 192 hr at 13°C (initial conc.: 0.46 µl/l) (956)

2,5-dimethylhexane
$(CH_3)_2CH(CH_2)_2CH(CH_3)_2$
A. PROPERTIES: m.w. 114.23; b.p. 108°C; sp.gr. 0.694
C. WATER POLLUTION FACTORS:
—Biodegradation:
incubation with natural flora in the groundwater—in presence of the other components of high-octane gasoline (100 µl/l): biodegradation: 20% after 192 hr at 13°C (initial conc.: 0.53 µl/l) (956)

3,4-dimethylhexane
$CH_3CH_2CH(CH_3)CH(CH_3)CH_2CH_3$
C. WATER POLLUTION FACTORS:
—Biodegradation:
incubation with natural flora in the groundwater—in presence of the other components of high-octane gasoline (100 µl/l): biodegradation: 84% after 192 hr at 13°C (initial conc.: 0.09 µl/l) (956)

1,1-dimethylhydrazine (uns. dimethylhydrazine; UDMH)
$NH_2N(CH_3)_2$
Uses: component of jet and rocket fuels; chemical synthesis; stabilizer for organic peroxide fuel additives; absorbent for acid gases
A. PROPERTIES: m.w. 60.10; m.p. −58°C; b.p. 62.5°C at 717 mm; v.p. 157 mm at 25°C; v.d. 2.07; sp.gr. 0.80; sat.conc. 505 g/cu m at 25°C
B. AIR POLLUTION FACTORS: 1 mg/cu m = 0.40 ppm, 1 ppm = 2.5 mg/cu m
—Odor: characteristic: quality: fishy, aminelike
T.O.C.: <6 ppm
    O.I. at 20°C = 41,300 (316)
T.O.C.: 15–35 mg/cu m (701)
    <0.75 mg/cu m (797)
D. BIOLOGICAL EFFECTS:
—Fish:
guppy: comparison between previously exposed and unexposed guppies preexposure period of 14 days at 1/20th of concentrations listed in column 1:

DIMETHYLMALONIC ACID   555

toxicant concentration	pre-exposed fish		unexposed fish	
	no. of fish	mean survival time (hr)	no. of fish	mean survival time (hr)
192 mg/l in soft water at 22–24.5°C	18	9.53	20	16.03
64 mg/l in soft water at 22–24.5°C	20	7.29	20	41.72
64 mg/l in hard water at 22–24.5°C	19	4.52	20	35.21 (474)

    guppy (*Lebistes reticulatus*): static bioassay:
      24 hr LC_{50}: 78.4 mg/l in hard water at 22–24.5°C
      48 hr LC_{50}: 29.9 mg/l in hard water at 22–24.5°C
      72 hr LC_{50}: 17.2 mg/l in hard water at 22–24.5°C
      96 hr LC_{50}: 10.1 mg/l in hard water at 22–24.5°C
      24 hr LC_{50}: 82.0 mg/l in soft water at 22–24.5°C
      48 hr LC_{50}: 45.5 mg/l in soft water at 22–24.5°C
      72 hr LC_{50}: 32.4 mg/l in soft water at 22–24.5°C
      96 hr LC_{50}: 26.5 mg/l in soft water at 22–24.5°C   (474)
—Mammals: carcinogenic   (54)
    rat: single oral LD_{50}: 122 mg/kg
    mouse: single oral LD_{50}: 265 mg/kg
    rat: inhalation: LC_{50}: 252 ppm, 4 hr
                   0/30: 75 ppm, 7 weeks
                   1/20: 140 ppm, 6 weeks
    mouse: inhalation: 8/30: 75 ppm, 7 weeks
                    29/30: 140 ppm, 6 weeks
    dog: inhalation: mild toxic effect: 0/3: 5 ppm, 36 weeks
                                  1/3: 25 ppm, 13 weeks   (211)
    animal carcinogen   (487)

2,5-dimethyl-4-ketoheptadiene *see* phorone

dimethylketol *see* acetoin

dimethylketone *see* acetone

dimethylleadchloride
D. BIOLOGICAL EFFECTS: plaice: 96 hr LC_{50}: 300 mg/l (Pb)   (573)

dimethylmalonic acid

$$\text{HOOC}-\underset{\underset{\text{CH}_3}{|}}{\overset{\overset{\text{CH}_3}{|}}{\text{C}}}-\text{COOH}$$

A. PROPERTIES: m.w. 132.12; m.p. 192° (dec); log P_{oct} 0.26/0.60 (calculated)
C. WATER POLLUTION FACTORS:
    —$BOD_{2\,days}^{25°C}$: 0.015 (substrate conc.: 6.6 mg/l; inoculum; soil micro-organisms)

5 days: 0.0
10 days: 0.045
20 days: 0.11 (1304)

4,5-dimethyl-2-mercaptothiazole
D. BIOLOGICAL EFFECTS:
—Fish: goldfish: approx. fatal conc.: 56 mg/l, 48 hr (226)

O,O-dimethyl-S-(N-methylcarbamoylmethyl)phosphorodithioate *see* dimethoate

O,O-dimethyl-O-(3-methyl-4-(methylthio)phenyl)phosphorothioate *see* baytex

dimethyl-1-methyl-2-(1-phenylethoxycarbonyl)-vinylphosphate *see* ciodrin

3,5-dimethyl-4-(methylthio)phenolmethylcarbamate *see* mesurol

O,O-dimethyl-O((4-methylthio)m-tolyl)phosphorothioate *see* baytex

1,3-dimethylnaphthalene

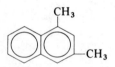

Manufacturing source: petroleum refining; coke processing. (347)
Users and formulation: impurity in naphthalene and its subsequent uses; asphalt constituent; naphtha constituent. (347)
Natural sources (water and air): petroleum, coal tar. (347)
Man caused sources (water and air): general uses of asphalt and naphtha; general laboratory use; use as a solvent; evaporation from moth balls. (347)
A. PROPERTIES: m.w. 156.23; b.p. 263°C
B. AIR POLLUTION FACTORS:
—Manmade emissions:
  in coal tar pitch fumes: 1.6 wt % (n.s.i.) (516)
C. WATER POLLUTION FACTORS:
—in Eastern Ontario drinking waters, June–October 1978:
  0.7–5.2 ng/l ($n = 12$)*
  in Eastern Ontario raw waters, June–October 1978:
  1.3–17.3 ng/l ($n = 2$)*
  *sum of 1,2-, 1,3-, 1,4-, 1,5-, 1,6-, 2,3-, 2,6-dimethylnaphthalenes and 2-ethyl-naphthalene (1698)
D. BIOLOGICAL EFFECTS:
—Uptake and depuration by oysters (Crassostrea virginica) from oil-treated enclosures:

time of exposure days	dupuration time days	concentration oysters μg/g	(n.s.i.) water μg/l	accumulation factor oysters/water
2	–	84	10	8,400
8	–	72	2	36,000
2	7	8	–	–
8	7	4	–	–
8	23	<0.5	–	–

half-life for depuration: 2 days (957)

2,6-dimethylnaphthalene

A. PROPERTIES: m.w. 156.23; m.p. 108–110°C; b.p. 262°C; solub. in seawater at 22°C: 2.4 ppm

C. WATER POLLUTION FACTORS:
—Reduction of amenities: T.O.C.: 0.01 mg/l (302; 878)

D. BIOLOGICAL EFFECTS:
—Fish: *Neanthes arenaceodentata*: 96 hr TLm in seawater at 22°C: 2.6 ppm (initial conc. in static assay) (995)

N,N-dimethyl-1-naphthylamine

C. WATER POLLUTION FACTORS:
—Waste water treatment:
W, A.S. from mixed domestic/industrial treatment plant: 3–8% depletion at 20 mg/l after 6 hr at 25°C (419)

N,N-dimethyl-p-nitroaniline

A. PROPERTIES: yellow fluorescent needles from alcohol; m.w. 116.18; m.p. 163°C

C. WATER POLLUTION FACTORS:
—Impact on biodegradation processes:
75% reduction of nitrification in activated sludge at 19 mg/l (38)

O,O-dimethyl-O-p-nitrophenylphosphorothioate see dimethylparathion

N-dimethylnitrosamine (N-nitrosodimethylamine; nitrous dimethylamide) $(CH_3)_2NNO$
A. PROPERTIES: yellow oily liquid; m.w. 74.08; b.p. 152/153°C; v.d. 2.56; sp.gr. 1.005 at 18/4°C;
B. AIR POLLUTION FACTORS: 1 mg/cu m = 0.33 ppm, 1 ppm = 3.08 mg/cu m
   —Manmade emissions:
      emitted during the compounding, forming and curing operations of elastomeric parts by reaction of accelerators/stabilizers used such as
         tetramethylthiuram disulfide,
         tetramethylthiuram monosulfide, and
         tetraethylthiuram disulfide
      emissions ranging from 15 to 810 g DMN/billion g rubber stock have been reported
      (1800)
C. WATER POLLUTION FACTORS:
   —% biodegradation at 0.01 mg/l initial concentration

	in sea water		in fresh water	
	in daylight	in dark	in daylight	in dark
after 3 days	75	0	75	0
6 days	75	70	92	0
10 days	92	80	98	80
15 days	100	91	100	90

(1085)

D. BIOLOGICAL EFFECTS:
   —Fish: rainbow trout: 10 d LD_{50} (IP): 1770 mg/kg   (1517)
   —Man: carcinogenic: West Germany 1977   (487)
                       USA 1974   (77)
   —Carcinogenicity: +
   —Mutagenicity in the *Salmonella* test: weak
      0.02 revertant colonies/nmol
      1100 revertant colonies at 4440 µg/plate   (1883)

O,O-dimethyl-O-(4-nitro-m-tolyl)phosphorothioate see fenitrothion

2,2-dimethyloctanoic acid (isodecanoic acid)
$(CH_3)_3C-(CH_2)_5-COOH$
C. WATER POLLUTION FACTORS:
   —$BOD_{2\ days}^{25°C}$: 0.09 (substrate conc. 3.5 mg/l; inoculum: soil microorganisms
      5 days: 0.06
      10 days: 0.11
      20 days: 0.26   (1304)

O,O-dimethyl-5,4-oxo-1,2,3-benzotriazin-3(4H)-ylmethyl-phosphorodithioate see guthion

O,O-dimethyl-S-((4-oxo-1,2,3-benzotriazin-3(4H)-yl)methyl)phosphorodithioate see azinphosmethyl

2,6-dimethyl-4-oxo-2,5-heptadiene *see* phorone

dimethylparathion (O,O-dimethyl-O-*p*-nitrophenylphosphorothioate; methylparathion; metaphos)

$$\begin{array}{c} CH_3O \\ CH_3O \end{array} \!\!\! P\!-\!O\!-\!\!\!\bigcirc\!\!\!-\!NO_2 \quad (S)$$

Use: pesticide
A. PROPERTIES: sp.gr. 1.36 at 20/4°C; solub. 55–60 mg/l; log P_{oct} 2.04
B. AIR POLLUTION FACTORS:
 —Ambient air quality:
  atmospheric levels in agricultural area: n.d.–71.0 ng/cu m   (1308)
  concentrations in the air near fields treated with the pesticide applied in a microencapsulated (ENCAP) form to plant foliage, compared with that applied in a conventional emulsifiable concentrate (EC) form:

	formulation	
	EC	ENCAP
day	ng/cu m	ng/cu m
0	7400	3800
1	3300	330
3	580	110
6	36	25
	54	19
9	13	16 (1331)

C. WATER POLLUTION FACTORS:
 —persistence in river water in a sealed glass jar under sunlight and artificial fluorescent light—initial conc. 10 µg/l:

	% of original compound found				
after	1 hr	1 wk	2 wk	4 wk	8 wk
	80	25	10	0	0 (1309)

D. BIOLOGICAL EFFECTS:
 —Copepod: *Acartia tonsa*: 96 hr LC_{50} (S): 890 µg/l   (1129)
 —Crustaceans:
  sand shrimp (*Crangon septemspinosa*): 96 hr static lab bioassay: LC_{50}: 2 ppb
  grass shrimp (*Palaemonetes vulgaris*): 96 hr static lab bioassay: LC_{50}: 3 ppb
  hermit crab (*Pagurus longicarpus*): 96 hr static lab bioassay: LC_{50}: 7 ppb
 —Fish:

Pimephales promelas:	96 hr LC_{50}: 8900 µg/l	(2121)
Lepomis macrochirus:	96 hr LC_{50}: 5720 µg/l	(2121)
Lepomis microlophus:	96 hr LC_{50}: 5170 µg/l	(2121)
Micropterus salmoides:	96 hr LC_{50}: 5220 µg/l	(2121)
Salmo gairdneri:	96 hr LC_{50}: 2750 µg/l	(2121)
Salmo trutta:	96 hr LC_{50}: 4740 µg/l	(2121)
Oncorhynchus kisutch:	96 hr LC_{50}: 5300 µg/l	(2121)
Perca flavescens:	96 hr LC_{50}: 3060 µg/l	(2121)
Italurus punctatus:	96 hr LC_{50}: 5710 µg/l	(2121)
Italurus melas:	96 hr LC_{50}: 6640 µg/l	(2121)

bluegill:	24 hr LC_{50}: 5.7 ppm	
rainbow trout:	24 hr LC_{50}: 2.7 ppm	(1878)
Striped bass:	96-hr LC_{50}: 14.0 mg/l (S)	
Banded killifish:	96-hr LC_{50}: 15.2 mg/l (S)	
Pumpkinseed:	96-hr LC_{50}: 3.6 mg/l (S)	
White perch:	96-hr LC_{50}: 14.0 mg/l (S)	
American eel:	96-hr LC_{50}: 6.3 mg/l (S)	
Carp:	96-hr LC_{50}: 14.8 mg/l (S)	
Guppy:	96-hr LC_{50}: 6.2 mg/l (S)	(1193)

mummichog (*Fundulus heteroclitus*): 96 hr static lab bioassay: LC_{50}: 8,000 ppb
mummichog (*Fundulus heteroclitus*): 96 hr static lab bioassay: LC_{50}: 58,000 ppb
striped killifish (*Fundulus majalis*): 96 hr static lab bioassay: LC_{50}: 13,800 ppb
Atlantic silverside (*Menidia menidia*): 96 hr static lab bioassay: LC_{50}: 5,700 ppb
striped mullet (*Mugil cephalus*): 96 hr static lab bioassay: LC_{50}: 5,200 ppb
American eel (*Anguilla rostrata*): 96 hr static lab bioassay: LC_{50}: 16,900 ppb
bluehead (*Thalassoma bifasciatum*): 96 hr static lab bioassay: LC_{50}: 12,300 ppb
northern puffer (*Sphaeroides maculatus*): 96 hr static lab bioassay: LC_{50}: 75,800 ppb (2327)
—Mammals:
acute oral LD_{50} (rat): approx. 9–25 mg/kg
acute dermal LD_{50} (rabbit): 300–400 mg/kg (1854)

2,2-dimethylpentane
$CH_3CH_2CH_2C(CH_3)_3$
A. PROPERTIES: m.w. 100.21; b.p. 78°C at 743 mm; sp.gr. 0.674
C. WATER POLLUTION FACTORS:
 —Reduction of amenities: T.O.C.: 100 mg/l (295)
 —Biodegradation:
  incubation with natural flora in the groundwater—in presence of the other components of high-octane gasoline (100 µl/l): biodegradation: 9% after 192 hr at 13°C (initial conc.: 0.42 µl/l) (956)

2,3-dimethylpentane (ethylisopropylmethylmethane)
$C_2H_5CH(CH_3)CH(CH_3)_2$
A. PROPERTIES: colorless liquid; m.w. 100.20; b.p. 89.8°C; sp.gr. 0.695 at 20°C
C. WATER POLLUTION FACTORS:
 —Manmade sources: in diesel engine exhaust: 0.9% of emitted H.C.'s (72)
 —Biodegradation:
  incubation with natural flora in the groundwater—in presence of the other components of high-octane gasoline (100 µl/l): biodegradation: 0% after 192 hr at 13°C (initial conc.: 0.48 µl/l) (956)

2,4-dimethylpentane (diisopropylmethane)
$(CH_3)_2CH-CH_2-CH-(CH_3)_2$
A. PROPERTIES: colorless liquid; m.w. 100.20; m.p. -123°C; b.p. 80.5°C; sp.gr. 0.67 at 20°C;
B. AIR POLLUTION FACTORS:
 —T.O.C.: detection: 3,000 mg/cu m (643)

—Manmade sources: in diesel engine exhaust: 0.3% of emitted H.C.'s (72)
C. WATER POLLUTION FACTORS:
   —Biodegradation:
      incubation with natural flora in the ground water—in presence of the other components of high-octane gasoline (100 µl/l): biodegradation: 11% after 192 hr at 13°C (initial conc.: 0.53 µl/l) (956)

3,3-dimethylpentane
$CH_3-CH_2-C(CH_3)_2-CH_2-CH_3$
A. PROPERTIES: m.w. 100.21; m.p. 135°C; b.p. 86°C; sp.gr. 0.693
C. WATER POLLUTION FACTORS:
   —Biodegradation:
      incubation with natural flora in the groundwater—in presence of the other components of high-octane gasoline (100 µl/l): biodegradation: 45% after 192 hr at 13°C (initial conc. 0.04 µl/l) (956)

2,3-dimethylphenol   *see* 2,3-xylenol

2,4-dimethylphenol   *see* 2,4-xylenol

2,5-dimethylphenol   *see* 2,5-xylenol

2,6-dimethylphenol   *see* 2,6-xylenol

3,4-dimethylphenol   *see* 3,4-xylenol

3,5-dimethylphenol   *see* 3,5-xylenol

dimethyl-*p*-phenylenediamine   *see* *p*-aminodimethylaniline

dimethylphosphorodithioic acid
D. BIOLOGICAL EFFECTS:
   —Fish:
      eastern mudminnow: flow through bioassay: 14 d TLm: 14.5 mg/l at 16°C
      static bioassay: 96 hr TLm: 17.0 mg/l at 16°C
(450)

O,O-dimethylphosphorothioate-O,O-diester with 4,4-thiophenol   *see* abate

2,2-dimethylpropionic acid   *see* pivalic acid

dimethylpropylmethane   *see* 2-methylpentane

***o*-dimethylphthalate**   (dimethyl-1,2-benzenedicarboxylate; methylphthalate)

$$\text{C}_6\text{H}_4(\text{COOCH}_3)_2$$

2,6-DIMETHYLPYRIDINE

Manufacturing source: organic chemical industry. (347)
Users and formulation: mfg. of plasticizer, latex, cellulose acetate film; plastics mfg. and processing; used in fuel matrix of double-base rocket propellant; fluidized-bed coating in mfg. of poly(vinylidene fluoride); plasticizer in cellulose acetate and nitrocellulose, resins, rubber; constituent of lacquers, plastics, rubber, coating agents, safety glass, molding powders, insect repellants, perfumes. (347)
Man caused sources (water and air): general use of plastics and above listed products (leaches from tubings, dishes, paper, containers); lab use. (347)

A. PROPERTIES: colorless liquid; m.w. 194.18; m.p. 0.0°C; b.p. 282°C; v.p. 1 mm at 100°C, <0.01 mm at 20°C; v.d. 6.69; sp.gr. 1.19 at 25/25°C; solub. 5,000 mg/l at 20°C; solub. 1744 mg/l

D. BIOLOGICAL EFFECTS:
—Marine dinoflagellate *Gymnodium breve*: 96 hr TLm: 125–185 ppm
EC_{50}: 54–96 ppm*
*EC_{50}: median growth limit concentration causing a 50% growth reduction
(1057)

larvae of grass shrimp (*Palaemonetes pugio Holthius*):
LC_{50} 8 days: 100 ppm
no significant increase in mortality at 1 ppm after 26 days (551)
—Mammalia: mouse: acute oral LD_{50}: 7.2 g/kg
rat: acute oral LD_{50}: 6.9 g/kg
guinea pig: acute oral LD_{50}: 2.4 g/kg
rat: repeated oral dose: slight growth effect: 8% in diet, 2 years
cat: inhalation: lethal: 1,213 ppm, 6.5 hr
nasal irritation: 2 mg/l (211)

2,6-dimethylpyridine (2,6-lutidine)

A. PROPERTIES: m.w. 107.16; m.p. −6°C; b.p. 143–145°C; sp.gr. 0.92; log P_{oct} 1.65 (calculated)
D. BIOLOGICAL EFFECTS:
—Ciliate (*Tetrahymena pyriformis*): 24 hr LC_{100}: 32.7 mmole/l (1662)

2,6-dimethylquinoline

A. PROPERTIES: m.w. 157.22; m.p. 57–59°C
D. BIOLOGICAL EFFECTS:
—Ciliate (*Tetrahymena pyriformis*): 24 hr LC_{100}: 1.27 mmole/l (1662)

—Embryos of South African clawed frog (*Xenopus laevis*):
  96 hr LC_{50}: 6.5 mg/l (1418)

2,2-dimethylsuccinic acid

$$HOOC-\underset{\underset{CH_3}{|}}{\overset{\overset{CH_3}{|}}{C}}-CH_2-COOH$$

C. WATER POLLUTION FACTORS:
 —$BOD_{2\ days}^{25°C}$: 0.0 (substrate conc.: 5.9 mg/l; inoculum: soil microorganisms)
  5 days: 0.0
  10 days: 0.0
  20 days: 0.085 (1304)

dimethylsulfate (DMS; methylsulfate)
$(CH_3O)_2SO_2$

Use: methylating agent for amines and phenols

A. PROPERTIES: colorless liquid; m.w. 126.13; m.p. -32°C; b.p. 188°C decomposes v.p. <1 mm at 20°C; v.d. 4.4; sp.gr. 1.33 at 20/4°C; solub. 28,000 mg/l hydrolyzes;

B. AIR POLLUTION FACTORS: 1 mg/cu m = 0.19 ppm, 1 ppm = 5.24 mg/cu m
 —Odor: characteristic: quality: faint onionlike

D. BIOLOGICAL EFFECTS:
 —Fish:
  Lepomis macrochirus: static bioassay in fresh water at 23°C, mild aeration applied after 24 hr (99%):

material added ppm	24 hr	% survival after 48 hr	72 hr	96 hr	best fit 96 hr LC_{50} ppm
32	0	—	—	—	
18	0	—	—	—	
10	0	—	—	—	7.5
7.5	90	90	90	90	
5	70	70	70	70	

  Menidia beryllina: static bioassay in synthetic seawater at 23°C, mild aeration applied after 24 hr (99%):

material added ppm	24 hr	% survival after 48 hr	72 hr	96 hr	best fit 96 hr LC_{50} ppm
18	6	0	—	—	
15	100	90	90	50	15
10	100	100	100	100	(352)

 —Mammalia:
  rabbit: acute oral toxicity: 50 mg/kg
  rat: acute oral toxicity: 440 mg/kg (211)
  guinea pig: inhalation: LC_{50}: 32 ppm, 1 hr
  rat: inhalation: LC_{50}: 64 ppm, 1 hr
           : 75 ppm, 26 min

564   DIMETHYLSULFIDE

mouse: inhalation: LC_{50}: 75 ppm, 17 min
: 98 ppm, 1 hr
rat: inhalation; no effect level: 0.5 ppm, 2 × 6 hr/week
carcinogenic (54)
—Man: carcinogenic: W. Germany 1973 (70)
USA 1974 (77)

dimethylsulfide (methylsulfide; methylthiomethane)
$(CH_3)_2S$
 Uses: gas odorant; solvent for many inorganic substances; catalyst impregnator
A. PROPERTIES: colorless liquid; m.w. 62.13; m.p. -83.2°C; b.p. 38°C; v.p. 420 mm at 20°C, 620 mm at 30°C; v.d. 2.1; sp.gr. 0.846 at 21/4°C; sat. conc. 1,420 g/cu m at 20°C, 2,030 g/cu m at 30°C; solub. 6300 mg/l
B. AIR POLLUTION FACTORS: 1 mg/cu m = 0.39 ppm, 1 ppm = 2.58 mg/cu m
 —Natural sources:
 biogenic dimethylsulfide emissions from soils in U.S.A. (1977):

sampling sites	soil orders	sampling dates	avg. sulfur flux g S/m²/yr
Wadesville, IN	alfisol	Sept.–Oct.	0.001
Philo, OH	inceptisol	Oct.	0.0002
Dismal Swamp, NC	histosol	Oct.	0.0007
Cedar Island, NC	saline swamp	Oct.	0.0069
Clarkedale, AR	alluvial clay	Nov.	0.0001
			(1385)
Cox's Landing, NC	freshly clipped marsh		0.927
			(1385)

 —Manmade sources:
 glc's: in yard of oil refinery in Dwase (Japan: 3.5 ppb)
 in yard of pulpmill in Niigata: 2.8 ppb
 Fuji health center: 0.23 ppb
 Fujima in Fuji: 0.33 ppb
 —Odor: characteristic: quality: decayed vegetables
 odor thresholds   mg/cu m

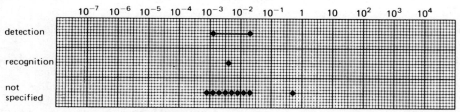

(2; 57; 71; 151; 279; 307; 635; 674; 696; 710; 736; 737; 741; 742; 766; 863)
O.I. 100% recogn.: 2,700,000 (316)
—Control methods:
 wet scrubber: water: effluent: 1,600 odor numbers/scf
 $KMnO_4$: effluent: 1 odor number/scf (76)
C. WATER POLLUTION FACTORS:
 —Reduction of amenities:
 faint odor; 0.0011 mg/l (129)

odor thresholds   mg/kg water

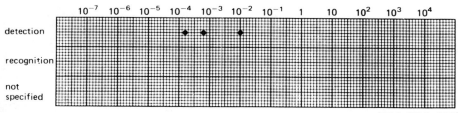

(883; 894; 908)

dimethylsulfoxide (methylsulfoxide; methylsulfinylmethane)
CH_3SOCH_3

    Manufacturing source: by-product of wood pulp mfg. for paper industries; organic chemical industry. (347)
    Users and formulation: solvent for acetylene, SO_2, other gases; pharmaceutical mfg.; antifreeze, hydraulic fluid mfg.; paint and varnish remover; solvent for polymerization and cyanide reactions, analytical reagent, synthetic fibers mfg.; industrial cleaner mfg.; pesticide mfg.; preservation of cells at low temperature. (347)

A. PROPERTIES: crystals; m.w. 78.13; m.p. 8°C; 18.5°C; b.p. 100°C, 189°C decomposes; v.p. 0.42 mm at 20°C; v.d. 2.71; sp.gr. 1.10; log P_{oct} −2.03
B. AIR POLLUTION FACTORS: 1 mg/cu m = 0.31 ppm, 1 ppm = 3.25 mg/cu m
  −Control methods:
    wet scrubber: water: effluent: 400 odor units/scf
                 $KMnO_4$: effluent: 2 odor units/scf   (76)
D. BIOLOGICAL EFFECTS:
  −Carcinogenicity: none
  −Mutagenicity in the *Salmonella* test: none
    <0.00001 revertant colonies/nmol
    70 revertant colonies at 5×10^5 μg/lplate   (1883)

dimethylterephthalate (dimethyl-1,4-benzenedicarboxylate)

Uses: polyester resins for film and fiber production especially polyethyleneterephthalate; intermediate

A. PROPERTIES: m.w. 194.18; m.p. 140°C; b.p. sublimes 300°C; v.p. 16 mm at 100°C, 140 mm at 150°C; solub. 3,300 mg/l (hot)
C. WATER POLLUTION FACTORS:
  −Reduction of amenities: organoleptic limit: 1.5 mg/l   (181)
    in Delaware river (U.S.A.): conc. range: winter 0.06 ppb
                                     summer: n.d.   (1051)
D. BIOLOGICAL EFFECTS:
  −Mammals: rat: acute oral LD_{50}: >3.2 g/kg   (211)

dimethyltetrachloroterephthalate *see* dacthal

3,5-dimethyltetrahydro-2-thio-1,3,5-thiadiazine *see* dazomet

1,3-dimethylthiourea
$CH_3NHCSNHCH_3$
A. PROPERTIES: m.w. 104.18
D. BIOLOGICAL EFFECTS:
   —Fish: creek cub: critical range: 7,000–15,000 mg/l, 24 hr   (226)

N,N-dimethyl-*p*-toluidine

A. PROPERTIES: colorless to brown oil; m.w. 135.21; b.p. 211°C; sp.gr. 0.935
D. BIOLOGICAL EFFECTS:
   —ingestion: rats: single oral LD_{50}: between 1,260–2.520 mg/kg   (1546)
   skin absorption: rabbit: LD_{50}: greater than 1,000 mg/kg   (1546)

dimethyl-(2,2,2-trichloro-1-hydroxyethyl)phosphonate (chlorofos; trichlorphon; dipterex; dylox)

Use: nonsystemic contact and stomach insecticide
A. PROPERTIES: m.p. 81–82°C; solub. 154 g/l at 25°C; v.p. 7.8×10^{-6} mm at 20°C; sp.gr. 1.73 at 20/4°C
D. BIOLOGICAL EFFECTS:
   —Algae:

Dunaliella euchlora	50,000 ppb	.54 (O.D. expt/ O.D. cont.)	organisms grown in test media con-	50% soluble powder
Phaeodactylum tricornutum	50,000 ppb	.85 (O.D. expt/ O.D. cont.)	taining pesticide for 10 days optical	soluble powder
Phaedactylum tricornutum	100,000 ppb	.39 (O.D. expt/ O.D. cont.)	density measured at 530 m_μ	soluble powder
Protococcus sp.	100,000 ppb	.54 (O.D. expt/ O.D. cont.)		soluble powder
Chlorella sp.	50,000 ppb	.70 (O.D. expt/ O.D. cont.)		soluble powder
Chlorella sp.	500,000 ppb	.00 (O.D. expt/ O.D. cont.)		soluble powder

| *Monochrysis luteri* | 50,000 ppb | .55 (O.D. expt/ O.D. cont.) | soluble powder |

(2347)

—Molluscs:
   Crassostrea virginica: american oyster, larvae: 48 hr static lab bioassay, TLm:
   1,000 ppb   (2324)
—Crustaceans:
   Gammarus lacustris: 96 hr LC_{50}: 40 µg/l   (2124)
   Simocephalus serrulatus: 48 hr LC_{50}: 0.32 µg/l   (2126)
   Daphnia pulex: 48 hr LC_{50}: 0.18 µg/l   (2126)
—Insects:
   Pteronarcys californica: 30 day LC_{50}: 9.8 µg/l: 96 hr LC_{50}: 69 µg/l   (2117)
   Pteronarcys californica: 96 hr LC_{50}: 35 µg/l   (2128)
   Acroneuria pacifica: 30 day LC_{50}: 8.7 µg/l: 96 hr LC_{50}: 16.5 µg/l (2117)
   Pteronarcella badia: 96 hr LC_{50}: 11 µg/l   (2128)
   Classenia sabulosa: 96 hr LC_{50}: 22 µg/l   (2128)
—Fish:
   Pimephales promelas: 96 hr LC_{50}: 109000 µg/l   (2122)
   Lepomis macrochirus: 96 hr LC_{50}: 3800 µg/l   (2122)
   rainbow trout: 96 hr LC_{50} (S): 4.85 mg/l   (1101)
—Mammals:
   acute oral LD_{50} (rat): 450–530 mg/kg   (1855)
   acute dermal LD_{50} (rat): >2,000 mg/kg   (1854)
   in diet (rat): 1 year at 500 ppm: no ill effect   (1855)

O,O-dimethyl-O,2,4,5-trichlorophenylphosphorothioate *see* ronnel

O,O-dimethyl-O-(3,5,6-trichloro-2-pyridyl)phosphorothioate (chlorpyrifosmethyl; Reldan)
   Use: insecticidal chemical   (1546)
A. PROPERTIES: amber solid cake; m.p. 45.5–46.5°C; v.p. 4.22 × 10^{-5} mm at 25°C; sp.gr. 1.39 at 50°C; solub. 4 mg/l
   Hazardous decomposition products: SO_2, HCl and various methylsulfides (1546)
D. BIOLOGICAL EFFECTS:
   —Mammals:
      ingestion: male rats: acute oral LD_{50}: 2,140 mg/kg

2,2-dimethylvaleric acid

$$CH_3-CH_2-CH_2-\underset{\underset{CH_3}{|}}{\overset{\overset{CH_3}{|}}{C}}-COOH$$

C. WATER POLLUTION FACTORS:
   —$BOD_{2\ days}^{25°C}$: 0.0 (substrate conc. 3.3 mg/l; inoculum: soil microorganisms)
      5 days: 0.09
      10 days: 0.03   (1304)

4,4-dimethylvaleric acid
$(CH_3)_3C-CH_2CH_2COOH$
C. WATER POLLUTION FACTORS:
—$BOD_2^{25°C}{}_{days}$: 0.06 (substrate conc. 3.3 mg/l; inoculum: soil microorganisms)
    5 days: 0.03
    10 days: 0.09   (1304)

dimilin *see* diflubenzuron

dinitramine (N^4,N^4-diethyl-a,a,a-trifluoro-3,5-dinitrotoluene-2,4-diamine)
Use: herbicide
A. PROPERTIES: crystalline solid, yellow; m.p. 98–99°C; solub. 1 ppm at 25°C; v.p. 3.6×10^{-6} mm at 25°C
C. WATER POLLUTION FACTORS:
—Persistence in soils:
    half-lives in sandy loam soils in Nova Scotia (Canada) were 50–70 days following a spring application (May) and approxim. 150 days following a winter application (November)   (1789)
    half-life under green-house conditions in a sandy loam soil: 33 days   (1792)
D. BIOLOGICAL EFFECTS:
—Mammals:
    acute oral LD_{50} (rat): 3,700 mg/kg
    acute dermal LD_{50} (rabbit): 2,000 mg/kg   (1854)

4,6-dinitro-*o-sec*-amylphenol (DNAP, dinosam)

Use: herbicide
D. BIOLOGICAL EFFECTS:
—Fish:
    Salmo salar, juvenile: lethal threshold, S: 30.0 mg/l   (1130)

o-dinitrobenzene

A. PROPERTIES: colorless to yellowish needles; m.w. 168.11; m.p. 118°C; b.p. 319°C at 773 mm; v.d. 5.79; sp.gr. 1.565 at 17/4°C; solub. 100 mg/l (cold), 3,800 mg/l at 100°C; log P_{oct} 1.58
B. AIR POLLUTION FACTORS: 1 mg/cu m = 0.14 ppm, 1 ppm = 6.99 mg/cu m

C. WATER POLLUTION FACTORS:
- Reduction of amenities: organoleptic limit: 0.5 mg/l (n.s.i.) (181)
- Biodegradation: decomposition by a soil microflora: period: >64 days (176)

m-dinitrobenzene

A. PROPERTIES: colorless yellowish needles: m.w. 168.11; m.p. 89.8°C; b.p. 300-302°C at 770 mm; sp.gr. 1.571 at 0/4°C; solub. 469 mg/l at 15°C, 3,200 mg/l at 100°C; log P_{oct} 1.49

C. WATER POLLUTION FACTORS:
- Biodegradation: decomposition by a soil microflora in >64 days (176) adapted A.S. at 20°C—product is sole carbon source: 0% COD removal after 20 days (327)

D. BIOLOGICAL EFFECTS:
- Toxicity threshold (cell multiplication inhibition test):

bacteria (*Pseudomonas putida*):	14 mg/l	(1900)
algae (*Microcystis aeruginosa*):	0.1 mg/l	(329)
green algae (*Scenedesmus quadricauda*):	0.7 mg/l	(1900)
protozoa (*Entosiphon sulcatum*):	0.76 mg/l	(1900)
protozoa (*Uronema parduczi Chatton-Lwoff*):	0.79 mg/l	(1901)

p-dinitrobenzene   (1,4-dinitrobenzene)

A. PROPERTIES: colorless to yellowish needles; m.w. 168.11; m.p. 173°C; b.p. 299°C at 777 mm (sublimes); sp.gr. 1.625 at 20/4°C; solub. 1,800 mg/l at 100°C; log P_{oct} 1.46/1.49

C. WATER POLLUTION FACTORS:
- Biodegradation: decomposition by a soil microflora in >64 days (176) adapted A.S. at 20°C—product is sole carbon source: 0% COD removal after 20 days (327)

2,4-dinitro-1,3-benzenediol *see* 2,4-dinitroresorcinol

3,5-dinitrobenzoic acid

2,4-DINITRO-6-sec-BUTYLPHENOL

A. PROPERTIES: m.w. 212.12; m.p. 205-207°C; log P_{oct} 1.04/2.22 (calculated)
C. WATER POLLUTION FACTORS:
—Biodegradation: adapted A.S. at 20°C—product is sole carbon source: 50% COD removal (327)

2,4-dinitro-6-sec-butylphenol (Dinoseb; DNBP)

Use: herbicide, corn yield enhancer
A. PROPERTIES: reddish brown liquid or dark brown solid; v.p. 1 mm at 151°C; sp.gr. 1.29 at 30°C; solub. 50 ppm
technical grade contains related butyl nitrophenols (max. 6%)
D. BIOLOGICAL EFFECTS:
—Bacteria: *Pseudomonas putida*: inhibition of cell multiplication starts at >40 mg/l (329)
—Algae: *Microcystis aeruginosa*: inhibition of cell multilplication starts at 5.7 mg/l (329)
—Crustaceans:
Gammarus fasciatus: 96 hr LC_{50}: 1800 µg/l (n.s.i.) (2125)
decapod: *Homarus americanus*, larvae: lethal threshold, S: 7.50 µg/l
decapod: *Homarus americanus*, adult: lethal threshold, S: 300 µg/l (1130)
—Fish:
cutthroat trout: *Salmo clarki*: static bioassay: 96 hr TLm: 0.041-1.35 mg/l at 10°C
lake trout: static bioassay: 96 hr TLm: 0.032-1.4 mg/l at 10°C (452)
Salmo salar, juvenile: lethal threshold, S: 70.0 µg/l (1130)
channel catfish: 96 hr LC_{50}, S: 118 µg/l (1202)
—Mammals:
ingestion: rats: acute oral LD_{50}: 58 mg/kg (1546)
skin absorption: guinea pigs: acute dermal LD_{50}: 100-200 mg/kg (1546)

4,6-dinitro-o-cresol (2-methyl-4,6-dinitrophenol)

Use: dormant ovicidal spray for fruit trees (highly phytotoxic and can not be used successfully on actively growing plants)
A. PROPERTIES: yellow prisms; m.w. 198.13; m.p. 85.8°C; v.d. 6.84; sat.conc. 0.001 g/cu m at 25°C

B. AIR POLLUTION FACTORS: 1 mg/cu m = 0.12 ppm, 1 ppm = 8.24 mg/cu m
  —Odor:
    USSR: animal chronic exposure: no effect: 0.0002 mg/cu m
                                   adverse effect: 0.036 mg/cu m   (170)
    threshold: conc. 0.004–0.021 mg/cu m (n.s.i.)   (734)
C. WATER POLLUTION FACTORS:
  —Biodegradation: adapted culture: 1% removal after 48 hr incubation, feed: 207
    mg/l   (292)
  —Impact on biodegradation processes:
    inhibition of degradation of glucose by *Pseudomonas fluorescens* at: 30 mg/l
    inhibition of degradation of glucose by *E. coli* at: 100 mg/l   (293)
  —Odor threshold: 1.3 mg/l (n.s.i.)   (998)
D. BIOLOGICAL EFFECTS:
  —Toxicity threshold (cell multiplication inhibition test):
    bacteria (*Pseudomonas putida*):   16   mg/l   (1900)
    algae (*Microcystis aeruginosa*):   0.15 mg/l   (329)
    green algae (*Scenedesmus quadricauda*):   13   mg/l   (1900)
    protozoa (*Entosiphon sulcatum*):   5.4   mg/l   (1900)
    protozoa (*Uronema parduczi Chatton-Lwoff*):   0.012 mg/l   (1901)
  —Bacteria: *E. coli*: toxic: 100 mg/l
  —Algae: *Scenedesmus*: toxic: 36 mg/l
  —Arthropods: *Daphnia*: toxic: 8 mg/l   (30)
  —Fish: *Salmo salar*, juvenile: lethal threshold, S: 200 µg/l (n.s.i.) (1130)
  —Man: unsatisfactory: >0.5 mg/cu m (n.s.i.)   (185)

2,4-dinitrophenol

A. PROPERTIES: yellow rhombic crystals or needles; m.w. 184.11; m.p. 111–114°C;
   v.d. 6.36; sp.gr. 1.683 at 24°C; solub. 5,600 mg/l at 18°C, 43,000 mg/l at 100°C;
   log P_{oct} 1.51/1.54
B. AIR POLLUTION FACTORS: 1 mg/cu m = 0.13 ppm, 1 ppm = 7.65 mg/cu m
C. WATER POLLUTION FACTORS:
  —Biodegradation:
    adapted culture: 2% removal after 48 hr incubation, feed: 200 mg/l (292)
    adapted A.S. at 20°C—product is sole carbon source: 85.0% COD removal at 6.0
      mg COD/g dry inoculum/hr   (327)
  —Impact on biodegradation processes:
    inhibition of degradation of glucose by *Psuedomonas fluorescens* at: 3 mg/l
    inhibition of degradation of glucose by *E. coli* at: >100 mg/l   (293)
    ~50% inhibition of NH_3 and NO_2 oxidation at 37 mg/l   (407; 409)
D. BIOLOGICAL EFFECTS:
  —Toxicity threshold (cell multiplication inhibition test):

bacteria (*Pseudomonas putida*): 115 mg/l (1900)
algae (*Microcystis aeruginosa*): 33 mg/l (329)
green algae (*Scenedesmus quadricauda*): 16 mg/l (1900)
protozoa (*Entosiphon sulcatum*): 20 mg/l (1900)
protozoa (*Uronema parduczi Chatton-Lwoff*): 0.22 mg/l (1901)
—Bacteria: *E. coli*: toxic: 100 mg/l
—Algae: *Scenedesmus*: toxic: 40 mg/l
—Arthropods: *Daphnia*: toxic: 6 mg/l (30)
—Fish: *Salmo salar*, juvenile: lethal threshold, S: 700 µg/l (n.s.i.) (1130)
—Man: unsatisfactory: >1 mg/cu m (n.s.i.) (185)

2,5-dinitrophenol

A. PROPERTIES: m.w. 184.11; m.p. 106–109°C; log P_{oct} 1.75
C. WATER POLLUTION FACTORS:
  —Biodegradation: adapted A.S. at 20°C—product is sole carbon source: 0% COD
    removal after 20 days (327)
  —Odor threshold: detection: 2.4 mg/l (998)

2,6-dinitrophenol

A. PROPERTIES: m.w. 184.11; m.p. 58–63°C; log P_{oct} 1.18/1.25
C. WATER POLLUTION FACTORS:
  —Biodegradation: adapted A.S. at 20°C—product is sole carbon source: 0% COD
    removal after 20 days (327)
  —Odor threshold: detection: 10 mg/l (998)

2,4-dinitroresorcinol (2,4-dinitro-1,3-benzenediol)

A. PROPERTIES: light brown powder; m.w. 200.11; m.p. 162–163°C; b.p. decomposes
    v.d. 6.79;
C. WATER POLLUTION FACTORS:
  —Biodegradation:
    adapted culture: 36% removal after 48 hr incubation, feed: 183 mg/l (292)

3,5-dinitrosalicylic acid

$$O_2N-C_6H_2(COOH)(OH)-NO_2$$

C. WATER POLLUTION FACTORS:
 —Biodegradation: adapted A.S. at 20°C—product is sole carbon source: 0% COD removal after 20 days (327)

3,5-dinitro-o-toluamide (zoalene)
$(O_2N)_2C_6H_2(CH_3)CONH_2$
 Use: coccidiostat; permissible food additive
A. PROPERTIES: yellowish solid; m.p. 177°C; solub. 1000 ppm
D. BIOLOGICAL EFFECTS:
 —Fish:
  guppy: static bioassay: 48 hr TLm: >200 mg/l at 24°C
  rainbow trout: static bioassay: 48 hr TLm: >200 mg/l at 15°C (446)
 —Mammals:
  ingestion: male rats: acute oral LD_{50}: 560 mg/kg
       female rats: acute oral LD_{50}: 650 mg/kg (1547)

2,3-dinitrotoluene

$$C_6H_3(CH_3)(NO_2)_2$$

A. PROPERTIES: m.w. 182.14; m.p. 59–61°C
D. BIOLOGICAL EFFECTS:
 —Toxicity threshold (cell multiplication inhibition test):
  bacteria (*Pseudomonas putida*):   9 mg/l (1900)
  algae (*Microcystis aeruginosa*):   0.22 mg/l (329)
  green algae (*Scenedesmus quadricauda*): 0.83 mg/l (1900)
  protozoa (*Entosiphon sulcatum*):   5.9 mg/l (1900)
  protozoa (*Uronema parduczi Chatton-Lwoff*): 1.6 mg/l (1901)

2,4-dinitrotoluene (1-methyl-2,4-dinitrobenzene)

$$C_6H_3(CH_3)(NO_2)_2$$

A. PROPERTIES: yellow needles; m.w. 182.13; m.p. 70°C; b.p. 300°C sl. decomposition; v.d. 6.27; sp.gr. 1.521 at 15°C; solub. 270 mg/l at 22°C

2,6-DINITROTOLUENE

B. AIR POLLUTION FACTORS: 1 mg/cu m = 0.13 ppm, 1 ppm = 7.57 mg/cu m
D. BIOLOGICAL EFFECTS:
—Toxicity threshold (cell multiplication inhibition test):
bacteria (*Pseudomonas putida*):	57 mg/l	(1900)
algae (*Microcystis aeruginosa*):	0.13 mg/l	(329)
green algae (*Scenedesmus quadricauda*):	2.7 mg/l	(1900)
protozoa (*Entosiphon sulcatum*):	0.98 mg/l	(1900)
protozoa (*Uronema parduczi Chatton-Lwoff*):	0.55 mg/l	(1901)

2,6-dinitrotoluene

$$O_2N\text{-}C_6H_3(CH_3)\text{-}NO_2$$

Manufacturing source: explosives mfg.; organic chemical industry. (347)
Users and formulation: mfg. TNT; urethane polymers, flexible and rigid foams and surface coatings, dyes; organic synthesis. (347)

A. PROPERTIES: m.w. 182.14; m.p. 64–66°C
C. WATER POLLUTION FACTORS:
—Odor threshold: 0.1 mg/l (225)
  detection: 0.05–1.0 mg/kg (903)
D. BIOLOGICAL EFFECTS:
—Toxicity threshold (cell multiplication inhibition test):
bacteria (*Pseudomonas putida*):	26 mg/l	
green algae (*Scenedesmus quadricauda*):	12 mg/l	
protozoa (*Entosiphon sulcatum*):	11 mg/l	(1900)
protozoa (*Uronema parduczi Chatton-Lwoff*):	23 mg/l	(1901)

dinocap (karathane; crotothane)

$$O=C-CH=CH-CH_3$$

(I) 2,4-dinitro-6-s-octylphenol ester; (II) 2,6-dinitro-4-s-octylphenol ester

A mixture of three 2,4-dinitro-6-*s*-octylphenols (I) and three 2,6-dinitro-4-*s*-octylphenols (II) in which the side chains are 1-methylheptyl, 1-ethylhexyl and 1-propylpentyl (1855)

D. BIOLOGICAL EFFECTS:
—Fish:
Salmo salar, juvenile: lethal threshold, S: 20.0 μg/l (1130)
—Mammals:

acute oral LD_{50}: male rat: 980 mg/kg (techn. grade)   (1854)
female rat: 1190 mg/kg   (1855)
in diet: dogs fed one year on a diet containing 50 ppm suffered no loss of weight, but diets containing 25 ppm produced cataracts in white Pekin ducks   (1855)

dinonylnaphthalenesulfonate
D. BIOLOGICAL EFFECTS:
—Fish:
rainbow trout (50% solution): 96 hr LC_{50} (S): 10 mg/l   (1500)

dinosam *see* 4,6-dinitro-*o-sec*-amylphenol

dinoseb *see* 2,4-dinitro-6-*sec*-butylphenol

dinoterp acetate *see* 2-*tert*-butyl-4,6-dinitrophenylacetate

dioctyladipate (di-2-ethylhexyladipate; DOA)

$$\begin{array}{c} \overset{O}{\underset{\|}{C}}\overset{C_2H_5}{\underset{|}{}} \\ CH_2-CH_2-C-O-CH_2-CH-C_4H_9 \\ | \\ CH_2-CH_2-C-O-CH_2-CH-C_4H_9 \\ \underset{\|}{}\underset{|}{} \\ O C_2H_5 \end{array}$$

Manufacturing source: organic chemical industry   (347)
Users and formulation: plasticizer mfg; plastics mfg. and processing; plasticizer for cellulose-based liquid lipsticks; commonly blended with DOP and DIOP in processing polyvinyl and other polymers; solvent; aircraft cubes.   (347)
Man caused source: general use of plastics and above listed products (leaches from tubings, dishes, paper, containers, etc.); lab use; general use as a solvent, from aircraft lubrication, lipsticks, application of paints and coatings.   (347)
A. PROPERTIES: light colored oily liquid; m.w. 370; m.p. -79°C; b.p. 214°C at 5 mm v.p. 2.60 mm at 20°C, sp.gr. 0.925 at 20/20°C
C. WATER POLLUTION FACTORS:
—River water quality:
in Delaware river (U.S.A.): conc. range winter: 0.08–0.3 ppb
summer: 0.02–0.3 ppb   (1051)

dioctylphthalate (di-2-ethylhexylphthalate; DOP; DEHP)

$$\text{phthalate diester with two } -O-nC_8H_{17} \text{ groups}$$

Manufacturing source: organic chemical industry.   (347)

576   DIOCTYLPHTHALATE

Users and formulation: plasticizer mfg.; plastics mfg. and recycling, processing; organic pump fluid. (347)
Man caused sources (water and air): general use of plastics and above listed products (leaches from tubings, dishes, paper, containers, etc.); lab use. (347)

A. PROPERTIES: light-colored liquid; m.p. -55°C; b.p. 385°C; v.p. 1.2 mm at 200°C; v.d. 13.45; sp.gr. 0.99 at 20/20°C; solub. 0.285 mg/l at 24°C (technical grade)

B. AIR POLLUTION FACTORS:
—Ambient air quality:
  organic fraction of suspended matter:
    Bolivia at 5200 m altitude (Sept.–Dec. 1975): 17–20 µg/1000 cu m
    Belgium, residential area (Jan.–April 1976): 26–132 µg/1000 cu m (428)
  glc's in residential area (Belgium) Oct. 1976:
    in particulate sample: 54.1 ng/cu m
    in gas phase sample: 127 ng/cu m (1289)
  glc's in marine atmosphere in the North West Gulf of Mexico, March–April 1977:
    avg.: 1.16 ng/cu m
    range: 0.72–1.92 ng/cu m (n = 10) (1724)

C. WATER POLLUTION FACTORS:
—Biodegradation:
  aerobic degradation in fresh water hydrosoil: 50% after 14 days incubation (309)
  biodegradation: A.S., 48 hr: 91% (1840)
                  continuous A.S.: 74% (1841)
                  in river water, 1 week: 10% (1841)
                  in fresh water hydrosoil*:

	aerobic	anaerobic
after 7 days	0%	0%
14 days	47%	0%
30 days	56%	0%

*based on recovery of ^{14}C from hydrosoil (1842)
degradation in the water of a model ecosystem: half-lifetime: 5 days

[Structural formulas: di-n-octyl phthalate → mono-n-octyl phthalate → phthalic acid]

—Water quality:
  the self purification is affected from 1.0 mg/l onwards (181)
  in river water: 0.0001–0.05 ppm; in river sediment: 0.2–56 ppm (555)
  in Delaware river (U.S.A.) conc. range: winter: 3–5 ppb
                                          summer: 0.06–2 ppb (1051)

D. BIOLOGICAL EFFECTS:
—Aquatic plants: *Elodea canadiensis*: concentration of DEHP in elodea in water containing 10 and 0.1 ppm for various intervals:

at 10 ppm in water, after 1 hr, in elodea: 37 ppm
                                 6 hr   293 ppm
                                12 hr   1338 ppm
                                24 hr   1138 ppm
                                48 hr   290 ppm
at 0.1 ppm in water, after 1 hr, in elodea: 1.98 ppm
                                6 hr   7.72 ppm
                                12 hr   15.46 ppm
                                24 hr   27.48 ppm
                                48 hr   23.24 ppm   (339)

—Algae: *Oedogonium*: after 33 days in model ecosystem, the water contained 0.00034 ppm, the alga contained 18.32 ppm, bioconcentration factor: 53.890X
(339)

—Arthropoda: *Daphnia magna*: concentration of DEHP in *Daphnia* in water containing 10 and 0.1 ppm for various intervals:
at 10 ppm in water, after 1 hr, in *Daphnia*: 592 ppm
                               6 hr   532 ppm
                               12 hr   893 ppm
                               24 hr   306 ppm
                               48 hr   1551 ppm
at 0.1 ppm in water, after 1 hr, in *Daphnia*: 42.1 ppm
                               6 hr   19.61 ppm
                               12 hr   15.54 ppm
                               24 hr   17.62 ppm
                               48 hr   18.26 ppm   (339)

waterflea (*Daphnia*): of the DEHP accumulated by the water flea, less than 2% (0.16 ppm) was present as the unmetabolized ester (553)

larvae of grass shrimp (*Palaemonetes Pugio Holthius*): no significant increase in mortality at 1 ppm after 26 days (551)

—Mollusca: *Physa*: after 33 days in ecosystem, the water contained 0.00034 ppm, the snail contained 7.30 ppm, bioconcentration factor: 21,480X (339)

snail *Physa*: concentration of DEHP in snail in water containing 10 and 0.1 ppm for various intervals:
at 10 ppm in water, after 1 hr, in snail: 3586 ppm
                             6 hr   4020 ppm
                             12 hr   2834 ppm
                             24 hr   2350 ppm
                             48 hr   487 ppm
at 0.1 ppm in water, after 1 hr, in snail: 12.06 ppm
                             6 hr   45.08 ppm
                             12 hr   45.45 ppm
                             24 hr   64.34 ppm
                             48 hr   85.75 ppm   (339)

—Insects: *Culex pipiens quinquefasciatus*: concentration of DEHP in *Culex* larvae and pupae in water containing 10 and 0.1 ppm for various intervals:
Culex larvae: at 10 ppm in water, after 1 hr, in larvae: 596 ppm
                                                      6 hr   2634 ppm

1,4-DIOXANE

12 hr	5978 ppm
24 hr	11873 ppm
48 hr	3657 ppm

Culex pupae: at 10 ppm in water, after 1 hr, in pupae: 2272 ppm

6 hr	2578 ppm
12 hr	3144 ppm
24 hr	3962 ppm
48 hr	4346 ppm

Culex larvae: at 0.1 ppm in water, after 1 hr, in larvae: 23.2 ppm

6 hr	91.5 ppm
12 hr	132.02 ppm
24 hr	31.80 ppm
48 hr	16.37 ppm

Culex larvae: at 0.1 ppm in water, after 1 hr, in pupae: 0.73 ppm

6 hr	1.51 ppm
12 hr	0.97 ppm
24 hr	2.03 ppm (339)

Culex pipiens quinquefasciatus larvae: after 33 days in model ecosystem, the water contained 0.00034 ppm, the Culex larvae contained 36.61 ppm, bioconcentration factor: 107,670X (339)

—Fish: Gambusia affinis: after 33 days in model ecosystem, the water contained 0.00034 ppm, the fish contained 0.044 ppm, bioconcentration factor: 130X (339)

Gambusia affinis: concentration of DEHP in fish in water containing 10 and 0.1 ppm for various intervals:

at 10 ppm in water, after 1 hr, in Gambusia: 152 ppm

6 hr	1033 ppm
12 hr	1294 ppm
24 hr	145 ppm
48 hr	469 ppm

at 0.1 ppm in water, after 1 hr, in Gambusia: 0.85 ppm

6 hr	7.23 ppm
12 hr	5.61 ppm
24 hr	8.53 ppm
48 hr	26.53 ppm (339)

—bioaccumulation factors in a laboratory model ecosystem after 3 and 33 days:

	BCF	
	3 d/3.5 ppb	33 d/64 ppt
alga (Oedogonium cardiacum):	660	28500
Daphnia	9426	2600
mosquito fish:	1.16	9400
mosquito larvae (Culex pipiens):	5300	9400
snails (Physa):	438	13600 (1644)

—Mammals: rat: teratogenic (345)

1,4-dioxane (glycolethylenether; 1,4-diethylenedioxide)

1,4-DIOXANE

Uses: solvent for cellulosics and wide range of organic products; lacquers; paints; varnishes; paint and varnish removers; cleaning and detergent preparations; cements; cosmetics; deodorants; fumigants

A. PROPERTIES: colorless liquid; m.w. 88.20; m.p. 10°C; b.p. 101°C; v.p. 30 mm at 20°C, 37 mm at 25°C; 50 mm at 30°C; v.d. 3.03; sp.gr. 1.033 at 20/4°C; THC 561 kcal/mole, LHC 527 kcal/mole; sat.conc. 148 g/cu m at 20°C, 232 g/cu m at 30°C; solub. completely miscible; log P_{oct} -0.42

B. AIR POLLUTION FACTORS: 1 mg/cu m = 0.278 ppm, 1 ppm = 3.6 mg/cu m
 —Odor: characteristic: quality: sweet, alcohol
 hedonic tone: pleasant

T.O.C.: 9.8 mg/cu m = 2.7 ppm	(307)
recognition: 620 mg/cu m = 172 ppm	(73)
170 ppm = 620 mg/cu m	(298)(210)
abs. perc. lim.: 0.80 ppm	
50% recogn.: 1.8 ppm	
100% recogn.: 5.7 ppm	
O.I. 100% recogn.: 6,228	(19)
distinct odor: 1000 mg/cu m = 270 ppm	(278)
O.I. at 20°C = 230	(316)
detection: 45-9,400 mg/cu m	(727)
detection: 270 mg/cu m	(643)

 —Sampling and analysis:

I.R. spectrometry: det. lim.: 0.02 ppm	(56)
photometry: min. full scale: 30,000 ppm	(53)
test tubes: UNICO: det. lim.: 100 ppm	(59)

C. WATER POLLUTION FACTORS:
 —BOD_{10}: nil std. dil. sew. (258)
 —Waste water treatment:
 NFG, BOD, 20°C, 1-10 days observed, feed: 100-900 mg/l, acclimation: 365 +
 P: nil % removed (93)

D. BIOLOGICAL EFFECTS:
 —Bacteria: *Pseudomonas putida*: inhibition of cell multiplication starts at 2700 mg/l (329)
 —Algae: *Microcystis aeruginosa*: inhibition of cell multiplication starts at 575 mg/l (329)
 —Threshold conc. of cell multiplication inhibition of the protozoan *Uronema parduczi Chatton-Lwoff*: 5620 mg/l (1901)
 —Fish:
 Lepomis macrochirus: static bioassay in fresh water at 23°C, mild aeration applied after 24 hr:

material added ppm	24 hr	% survival after 48 hr	72 hr	96 hr	best fit 96 hr LC_{50} ppm
10,000	100	100	100	100	>10,000
7,900	90	90	90	90	

Menidia beryllina: static bioassay in synthetic seawater at 23°C: mild aeration applied after 24 hr:

material added ppm	% survival after				best fit 96 hr LC_{50} ppm
	24 hr	48 hr	72 hr	96 hr	
10,000	70	20	10	10	
7,900	80	30	10	0	6,700
5,000	100	90	90	90	(352)

—Mammalia:
  mouse: single oral LD_{50}: 5.66 g/kg
  rat: single oral LD_{50}: 5.17 g/kg
  guinea pig: single oral LD_{50}: 3.90 g/kg
  cats, rabbits, guinea pigs: inhalation: no effect: <1,350 ppm, 45 × 8 hr   (211)
—Carcinogenicity:
  It is concluded that under the conditions of this bioassay, 1,4-dioxane induced hepatocellular adenomas in female Osborne-Mendel rats. 1,4-Dioxane was carcinogenic in both sexes of rats, producing squamous-cell carcinomas of the nasal turbinates, and in both sexes of B6C3F1 mice, producing hepatocellular carcinomas.   (1757)
—Man: irritation to eyes, nose and throat: 300 ppm
      objectionable: 500 ppm   (211)

2,3-*p*-dioxanedithiol-5,5-*bis*-(O,O-diethylphosphorodithioate)   *see* dioxathion

3,6-dioxaoctane-1,8-diol   *see* triethyleneglycol

dioxathion   (delnav; 2,3-*p*-dioxanedithiol-S,S-*bis*-(O,O-diethylphosphorodithioate); 1,4-dioxane-2,3-ylidene-*bis*(O,O-diethylphosphorothiolothionate))

$$\begin{array}{c} \text{structure: 1,4-dioxane ring with substituents} \\ \text{S-P(=S)-(OC}_2\text{H}_5)_2 \text{ at both 2 and 3 positions} \end{array}$$

Use: insecticide, miticide; mixture of *cis*- and *trans*-isomers
A. PROPERTIES: brown liquid; sp.gr. 1.26 at 26/4°C
D. BIOLOGICAL EFFECTS:
—Crustaceans:
  Gammarus lacustris: 96 hr, LC_{50}: 270 µg/l   (2124)
  Gammarus fasciatus: 96 hr, LC_{50}: 8.6 µg/l   (2126)
  sand shrimp (*Crangon septemspinosa*): 96 hr, LC_{50}, S: 38 ppb
  grass shrimp (*Palaemonetes vulgaris*): 96 hr, LC_{50}, S: 285 ppb
  hermit crab (*Pagurus longicarpus*): 96 hr, LC_{50}, S: 82 ppb   (2327)
—Fish:
  Pimephales promelas: 96 hr, LC_{50}: 9300 µg/l   (2123)
  Lepomis macrochirus: 96 hr, LC_{50}: 34 µg/l   (2123)
  Lepomis cyanellus: 96 hr, LC_{50}: 61 µg/l   (2123)
  Micropterus salmoides: 96 hr, LC_{50}: 36 µg/l   (2123)

—Mammals:
  acute oral LD_{50} (male rat): 43–110 mg/kg (1854)
  (female rat): 23 mg/kg (1855)
  acute dermal LD_{50} (albino male rat): 235 mg/kg
  (albino female rat): 63 mg/kg (1855)
—Carcinogenicity: under the conditions of this bioassay, dietary administration of dioxathion was not carcinogenic in Osborne-Mendel rats or B6C3F1 mice.
(1742)

dioxins (dibenzo-*p*-dioxins; *see also* tetrachlorodibenzo-*p*-dioxin and octachlorodibenzo-*p*-dioxine)

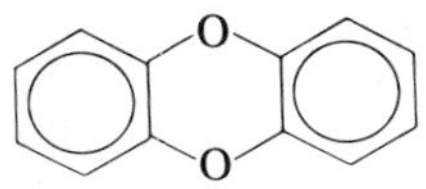

Dioxine can be chlorinated, singly or in multiples at any of eight positions. Thus, theoretically there are 75 possible chlorinated species:
  1 octachloro isomer
  2 monochloro isomers
  2 heptachloro isomers
  10 dichloro isomers
  10 hexachloro isomers
  14 trichloro isomers
  14 pentachloro isomers
  22 tetrachloro isomers (1334)

Manmade sources: chlorinated phenols and especially their alkali salts can condense above 300°C to form polyphenoxyphenols or—in a very specific reaction—to form dibenzo-*p*-dioxins:

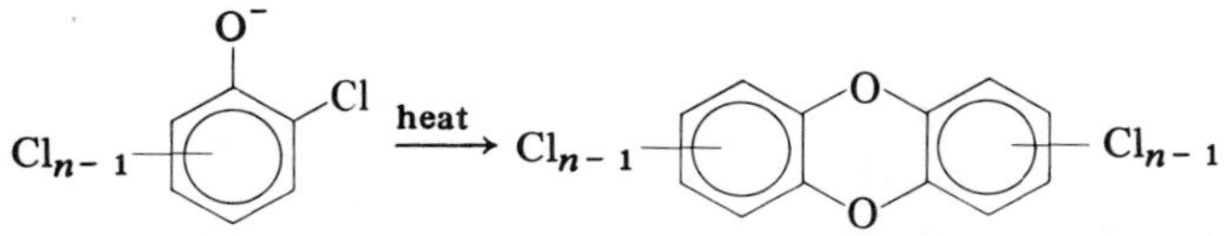

1,3,6,8-tetrachlorodibenzo-p-dioxin and 1,3,7,9-tetrachlorodibenzo-p-dioxin was formed during pyrolysis of 2,4,6-trichlorophenate, sodium (1600)

dioxitol *see* diethyleneglycol mono-ethylether

1,1-dioxo-2,3,4,5-tetrahydrothiophene *see* sulfolane

2,4-dioxy-5-methylpyrimidine *see* thymine

2,4-dioxypyrimidine *see* uracil

dipe *see* di-isopropylether

di-*n*-pentylamine *see* di-*n*-amylamine

diphenamid (N,N-dimethyl-2,2-diphenylacetamide)

A. PROPERTIES: white crystals; m.p. 124–135°C; solub. 260 ppm at 25°C
D. BIOLOGICAL EFFECTS:
—Crustacean:

Gammarus fasciatus:	no effect:	100,000 μg/l, 48 hr	(2125)
Daphnia magna:	48 hr, LC_{50}:	56000 μg/l	(2125)
Cypridopsis vidua:	48 hr, LC_{50}:	50000 μg/l	(2125)
Asellus brevicaudus:	no effect:	100,000 μg/l, 48 hr	(2125)
Palaemonetes kadiakensis:	48 hr, LC_{50}:	58000 μg/l	(2125)
Orconectes nais:	no effect:	100,000 μg/l, 48 hr	(2125)

—Mammals:
acute oral LD_{50} (rat): about 1000 mg/kg (1854)
(mice): about 600 mg/kg
(rabbits): about 1500 mg/kg
(dogs and monkeys): 1000 mg/kg (1855)
in diet: dogs and rats fed for 2 years on a diet containing 2000 ppm suffered no unusual effects on their physiology or fertility (1855)

diphenyl (biphenyl; phenylbenzene)

Use: organic synthesis; fungicide
Manmade sources:
in coal tar: 2.72 mg/g of sample
in wood preservative sludge: 3.64 g/l of rat sludge (993)
A. PROPERTIES: m.w. 154.20; m.p. 70°C; b.p. 254°C; sp.gr. 1.18 at 0/4°C; solub. 7.5 mg/l at 25°C, 8.5 mg/l at 24°C (99% purity); log P_{oct} 3.16/4.09
B. AIR POLLUTION FACTORS:
—Manmade emissions:
in airborne coal tar emissions: 0.29 mg/g of samples or
10 μg/cu m of air sampled (993)
—Odor threshold: 0.06 mg/cu m (817)
C. WATER POLLUTION FACTORS:
—BOD_5: 1.08
—ThOD: 3.01 (275)
—Biodegradation:

Pathways of microbial degradation of biphenyl (1242; 1243)

Biphenyl → 2,3-Dihydro-2,3-dihydroxybiphenyl → 2,3-Dihydroxybiphenyl

Pseudomonad pathway: → 2-Hydroxy-6-oxo-6-Phenylhexa-2,4-dienoate → 2-Oxopenta-4-enoate + Benzoic acid

Gram negative isolate pathway: → 2-Hydroxy-3-phenyl-6-Oxohexa-2,4-dienoate → Phenylpyruvic acid (CH_2COCO_2H)

biodegradation at 1.0 mg/l:

	normal sewage	adapted sewage	
after 24 hr	0%	87%	
after 135 hr	79%	100%	(997)

impact on biodegradation processes:
effect on respiration of glucose by mixed culture derived from activated sludge:

conc. (mg/l)	increase in log period (hrs)	respiration rate (%)	
1	0	100	
10	0	100	
100	110	0	
1,000	>200	0	(997)

—Water quality:
in Eastern Ontario drinking waters (June–Oct. 1978):
0.1–1.7 ng/l ($n = 12$)
in Eastern Ontario raw waters (June–Oct. 1978):
0.3–0.7 ng/l ($n = 2$)   (1698)
in river water: 0.001–0.015 ppm, in river sediment: 1–2 ppm   (555)
—Odor threshold: detection: 0.0005 mg/kg   (894)
—Control methods:
aqueous chlorination of biphenyl at pH values above 6.2 and under conditions of waste water treatment processes, proceeds very slowly, with production of o- and p-chlorobiphenyl. It appears that chlorinemonoxide rather than HOCl may be the active chlorinating species in reactions with such unreactive organic compounds as biphenyl   (1876)
calculated half-life in water at 25°C and 1 m depth, based on evaporation rate of 0.092 m/hr: 7.52 hr   (437)

—Sampling methods:
recovery with open pore polyurethane (OH/NCO = 2,2): 98 ± 5%—quantitative elution with methanol (929)
D. BIOLOGICAL EFFECTS:
—Marine yeast *Rhodotorula rubra*: bioconcentration coefficient:
307 in whole cells
15,000 in their lipid portion (1566)
—Fish: *Salmo gairdneri* (trout): log. bioconcentration factor: 2.64 (193)
—Mammals: ingestion rats: single oral LD_{50}: greater than 4,000 mg/kg (1546)
skin absorption: rabbits: LD_{50}: greater than 4,000 mg/kg (1540)

N-diphenylamine (DPA; N-phenylaniline)

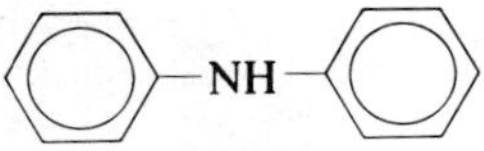

C₆H₅—NH—C₆H₅

Uses: rubber anti-oxidants and accelerators; stabilizers for plastics; solid rocket propellants; pesticides; explosives; dyes; pharmaceuticals

A. PROPERTIES: colorless to grayish crystals; m.w. 169.23; m.p. 52.8°C; b.p. 302°C; sp.gr. 1.159; log P_{oct} 3.22/3.50; solub. 300 mg/l at 25°C
B. AIR POLLUTION FACTORS:
—COD: 90% of ThOD (0.05 n $K_2Cr_2O_7$) (274)
—$KMnO_4$: 88% of ThOD (0.01 n $KMnO_4$) (274)
—ThOD: 2.39 (274)
—Impact on biodegradation processes:
at 100 mg/l, no inhibition of NH_3 oxidation by *Nitrosomonas* sp. (390)

p-diphenylamine-sulfuric acid
C. WATER POLLUTION FACTORS:
—Impact on biodegradation:
at 100 mg/l, no inhibition of NH_3 oxidation by *Nitrosomonas* sp. (390)

sym-diphenylcarbazide see 1,5-diphenylcarbohydrazide

diphenylcarbazone (phenylazoformic acid 2-phenylhydrazide)

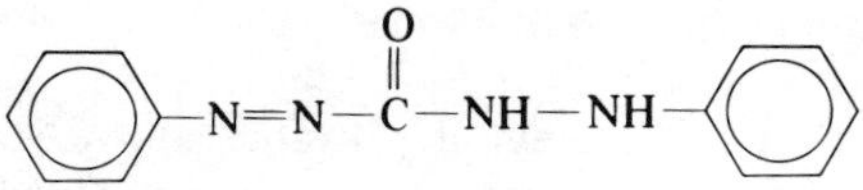

C₆H₅—N=N—C(=O)—NH—NH—C₆H₅

A. PROPERTIES: m.w. 240.27; m.p. 156–159°C
C. WATER POLLUTION FACTORS:
—Impact on biodegradation:
at 100 mg/l, no inhibition of NH_3 oxidation by *Nitrosomonas* sp. (390)

1,5-diphenylcarbohydrazide (*sym*-diphenylcarbazide)

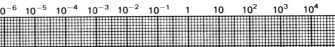

Use: analytical chemistry
A. PROPERTIES: m.w. 242.28; m.p. 166-168°C; white crystals or flakes
C. WATER POLLUTION FACTORS:
  —Impact on biodegradation:
    at 100 mg/l, no inhibition of NH_3 oxidation by *Nitrosomonas* sp. (390)

diphenyldiimide *see* azobenzene

1,2-diphenylethane *see* bibenzyl

diphenylether (diphenyloxide; phenylether; phenoxybenzene)

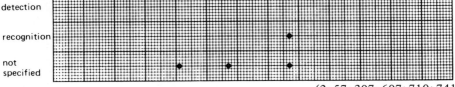

Uses: organic synthesis, perfumery, particularly soaps; heat transfer medium; resins for laminated electrical insulation
A. PROPERTIES: colorless liquid; m.w. 170.20; m.p. 28°C; b.p. 257-259°C; v.p. 0.02 mm at 25°C, 0.12 mm at 30°C; v.d. 5.86; sp.gr. 1.073 at 20°C; sat. conc. 0.56 g/cu m at 20°C, 1.1 g/cu m at 30°C; solub. 21 ppm at 25°C; log P_{oct} 4.20 at 20°C
B. AIR POLLUTION FACTORS: 1 mg/cu m = 0.143 ppm, 1 ppm = 7.0 mg/cu m
  —Odor: characteristic: quality: geranium odor
                         hedonic tone: pleasant

odor thresholds   mg/cu m

10^{-7} 10^{-6} 10^{-5} 10^{-4} 10^{-3} 10^{-2} 10^{-1} 1 10 10^2 10^3 10^4

detection

recognition

not specified

(2; 57; 307; 607; 710; 741)
  O.I. at 20°C = 130   (316)
  —Sampling and analysis: photometry: min. full scale: 200 ppm   (53)
C. WATER POLLUTION FACTORS:
  —Reduction of amenities:
    faint odor: 0.069 µg/l   (129)
    approx. conc. causing taste in fish: 0.05 mg/l   (41)
    taste and odor: 0.013 mg/l   (41)

586   DIPHENYLMETHANE

  T.O.C. = 0.015 mg/l (225)
—Water quality:
  in Zürich lake: at surface: 48 ppt; at 30 m depth: 8 ppt
  in Zürich area: in groundwater: 3 ppt; in tapwater: 3 ppt (513)
  in Bedford basin, Nova Scotia, Canada—Dec. 1976–March 1977:
    range: 15–32 ng/l
         54 ng/l: after rainstorm
         179 ng/l: fire at dump 1.5 km upwind
         103 ng/l: in freshly fallen snow (479)
D. BIOLOGICAL EFFECTS:
  —Fish:
    Salmo gairdneri (trout): log. bioconcentration factor: 2.29 (193)
    in tissue samples of fishes sampled in Nova Scotia area (Canada): 0.3–4 µg/g
      lipid (479)
  —Mammals: rat: single oral LD_{50}: 3.99 g/kg (211)

diphenylmethane

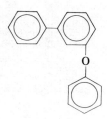

A. PROPERTIES: m.w. 168.24; m.p. 22–24°C; b.p. 264; sp.gr. 1.006; solub. 3.0 mg/l at 24°C (practical grade); log P_{oct} 4.14
D. BIOLOGICAL EFFECTS:
  —Threshold conc. of cell multiplication inhibition of the protozoan *Uronema parduczi Chatton-Lwoff*: 2.2 mg/l (1901)

diphenyloxide   *see* diphenylether

2-diphenylphenylether

A. PROPERTIES: log P_{oct} 5.55
D. BIOLOGICAL EFFECTS:
  —Fish:
    rainbow trout (*Salmo gairdneri*): bioconcentration factor: 552 ± 107
                    log bioconcentration factor: 2.74 (193)

diphenylthiocarbazone   (dithizone)

$$\text{C}_6\text{H}_5-\text{N}=\text{N}-\overset{\overset{\text{S}}{\|}}{\text{C}}-\text{NH}-\text{NH}-\text{C}_6\text{H}_5$$

A. PROPERTIES: m.w. 256.33; m.p. 168°C (dec)
C. WATER POLLUTION FACTORS:
  −Impact on biodegradation:
    ∼50% inhibition of NH_3 oxidation in *Nitrosomonas* at 7.5 mg/l   (407)

di-2-propenylamine *see* di-N-allylamine

di-N-propylamine
$(C_3H_7)_2NH$
A. PROPERTIES: colorless liquid; m.w. 101.2; m.p. −39.6°C; b.p. 109°C to 111°C; v.p. 30 mm at 25°C; v.d. 3.5; sp.gr. 0.738 at 20/4°C; log P_{oct} 1.73
B. AIR POLLUTION FACTORS: 1 mg/cu m = 0.242 ppm, 1 ppm = 4.41 mg/cu m
  −Odor: characteristic: quality: ammoniacal, amine
              hedonic tone: neutral to unpleasant
  T.O.C.: abs. perc. lim.: 0.02 ppm
       50% recogn.: 0.10 ppm
       100% recogn.: 0.10 ppm
       O.I. 100% recogn.: 270,600   (19)
C. WATER POLLUTION FACTORS:
  −Reduction of amenities:
    organoleptic limit 0.5 mg/l   (181)
  −Waste water treatment:
    A.C.: adsorbability: 0.174 g/g C, 80.2% reduction, infl.: 1,000 mg/l, effl.: 198 mg/l   (32)
    degradation by *Aerobacter*: 200 mg/l at 30°C:
              parent: 100% degradation in 26 hr
              mutant: 100% degradation in 12 hr   (152)
D. BIOLOGICAL EFFECTS:
  −Fish: creek chub: critical range: 20–60 mg/l; 24 hr   (226)
  −Mammalia: rat: acute oral LD_{50}: 0.2–0.4 g/kg   (211)

dipropyleneglycol (1,1′-oxydi-2-propanol; β,β'-dihydroxydi-*n*-propylether)
$(CH_3CHOHCH_2)_2O$
A. PROPERTIES: colorless liquid; m.w. 134.2; m.p. supercools; b.p. 229/232°C; solub. completely miscible; v.p. 1 mm at 74°C, <0.01 mm at 20°C; v.d. 4.63; sp.gr. 1.03 at 20/20°C; log P_{oct} −1.17/−1.23 (calculated)
B. AIR POLLUTION FACTORS: 1 mg/cu m = 0.182 ppm, 1 ppm = 5.49 mg/cu m
C. WATER POLLUTION FACTORS:
  −Waste water treatment:
    A.C.: adsorbability: 0.033 g/g C, 16.5% reduction, infl.: 1,000 mg/l, effl.: 835 mg/l   (32)
D. BIOLOGICAL EFFECTS:
  −Mammalia:

588   DIPROPYLDISULFIDE

    rat: single oral LD_{50}: 14.8 g/kg
            repeated oral doses: no effect: 5% in drinking water 77 days   (211)

dipropyldisulfide (propyldisulfide)
$C_3H_7-S-S-C_3H_7$
B. AIR POLLUTION FACTORS:
   —Control methods:
     wet scrubber: water: effluent: 3,000 odor units/scf
                  $KMnO_4$: effluent: 10 odor units/scf   (76)
   —Odor threshold: ~0.053 mg/cu m   (770)
                recognition: 0.11 mg/cu m   (863)
C. WATER POLLUTION FACTORS:
   —Odor threshold: detection: 0.0022–0.004 mg/kg   (877)

dipropylketone *see* 4-heptanone

dipropylphthalate (DPP)

                          C₆H₄(−$COOC_3H_7$)(−$COOC_3H_7$)

Manufacturing sources: organic chemical industry.   (347)
Users and formulation: plasticizer mfg.; plastics mfg. and processing.   (347)
Man caused sources (*water and air*): general use of plastics and above listed products (leaches from tubings, dishes, paper, containers, etc.); lab use.   (347)
A. PROPERTIES: colorless liquid, b.p. 130°C at 1 mm; solub. 150 mg/l; sp.gr. 1.071 at 25°C; solub. 56 ppm
D. BIOLOGICAL EFFECTS:
   —Marine dinoflagellate *Gymnodium breve*:
     96 hr TLm: 1.3–6.5 ppm
     EC_{50}: 0.9–2.4 ppm*
   *EC_{50}: median growth limit concentration causing a 50% growth reduction
                                                            (1057)

di-*n*-propylsulfide (propylsulfide; 1-propylthiopropane)
$C_3H_7SC_3H_7$
A. PROPERTIES: m.w. 118.23; m.p. 102°C; b.p. 142°C; sp.gr. 0.814 at 17°C
B. AIR POLLUTION FACTORS:
   —Odor: characteristic: hedonic tone: foul, nauseating
     T.O.C.: 0.0076 mg/cu m = 1.5 ppb   (307)
           50% recogn.: 19 ppb   (71)
           0.011 ppm   (279)
   —Control methods:
     wet scrubber: water: effluent: 12,000 odor units/scf
                    $KMnO_4$: effluent: 10 odor units/scf   (76)
C. WATER POLLUTION FACTORS:
   —Reduction of amenities: faint odor: 0.81 µg/l   (129)

dipterex *see* dimethyl-(2,2,2-trichloro-1-hydroxyethyl)phosphonate

2,2'-dipyridil *see* 2,2'-bipyridine

diquat (1,1'-ethylene-2,2'-bipyridylium cation; 9,10-dihydro-8a,10a-diazoniaphenanthrene; 6,7-dihydrodipyridol(1,2-a : 2',1'-c)pyrazidinium bromide; reglon; diquat dibromide)
diquat refers to the cation only

A. PROPERTIES: solub. 700 g/l at 20°C
D. BIOLOGICAL EFFECTS:
—Algae:

Chlorococcum sp.	$>.5 \times 10^6$ ppb	50% decrease in O_2 evaluation	*f*
Chlorococcum sp.	$2. \times 10^5$ ppb	50% decrease in growth	Growth measured as ABS. (525 mu) after 10 days
Dunaliella tertiolecta	$>5. \times 10^6$ ppb	50% decrease in O_2 evolution	*f*
Dunaliella tertiolecta	3×10^4 ppb	50% decrease in growth	Growth measured as ABS. (525 mu) after 10 days
Isochrysis galbana	$>5 \times 10^6$ ppb	50% decrease in O_2 evolution	*f*
Isochrysis galbana	1.5×10^4 ppb	50% decrease in growth	Measured as ABS. (525 mu) after 10 days
Phaeodactylum tricornutum	$>5 \times 10^6$ ppb	50% decrease in O_2 evolution	*f*
Phaeodactylum tricornutum	1.5×10^4 ppb	50% decrease in growth	Measured as ABS. (525 mu) after 10 days

(2348)

—Crustacean:
  Hyallella azteca: 96 hr, LC_{50}: 48 µg/l (2133)
—Insects:
  Callibaetes sp.: 96 hr, LC_{50}: 16400 µg/l (2133)
  Limnephilus: 96 hr, LC_{50}: 33000 µg/l (2133)
  Tendipedidae: 96 hr, LC_{50}: >100000 µg/l (2133)
  Enallagma: 96 hr, LC_{50}: >100000 µg/l (2133)
  mayfly nymphs (*Ephemerella walkeri*): 1.0 mg/l (lowest observed avoidance conc.) (1621)
—Fish:
  Pimephales promelas: 96 hr, LC_{50}: 14000 µg/l (2131)
  Lepomis macrochirus: 96 hr, LC_{50}: 35000 µg/l (2112)
  Micropterus salmoides: 96 hr, LC_{50}: 7800 µg/l (2131)

Esox lucius:	48 hr, LC_{50}: 16000 µg/l	(2112)
Stizostedion vitreum vitreum:	96 hr, LC_{50}: 2100 µg/l	(2112)
Salmo gairdneri:	48 hr, LC_{50}: 11200 µg/l	(2112)
Oncorhynchus tshawytscha:	48 hr, LC_{50}: 28500 µg/l	(2106)

Rasbora trilineata: 96 hr LC_{50} (S): 29.9 mg/l
guppy: 96 hr LC_{50} (S): 50.2 mg/l (1513)
brown trout yearlings: 48 hr LC_{50} (S): 570 mg/l (1113)
brown trout: 96 hr LC_{50} (FT): 300 mg/l (1113)
bluegill: 48 hr LC_{50}: 19.0 ppm
rainbow trout: 48 hr LC_{50}: 20.0 ppm (1878)
rainbow trout (*Salmo gairdneri*): >10 mg/l; lowest observed avoidance conc.
(1621)

—Mammals:
acute oral LD_{50} (rat): 400–440 mg diquat/kg
(rabbit): 187 mg/kg
(cows): 37 mg/kg
(dogs): >192 mg/kg
acute dermal LD_{50} (rabbits): >750 mg/kg

disulfoton (O,O-diethyl-S-(2-(ethylthio)ethyl)phosphorodithioate; thiodemeton; disyston)

$$\begin{array}{c} C_2H_5O \\ \diagdown \\   P-S-CH_2-CH_2-S-C_2H_5 \\ \diagup \\ C_2H_5O \end{array}$$

with $\overset{S}{\underset{\parallel}{}}$ on the P

Use: systemic insecticides and acaricide

A. PROPERTIES: pale yellow liquid; sp.gr. 1.14 at 20/4°C; b.p. 62°C at 0.01 mm; solub. 25 ppm, v.p. 1.8 mm at 20°C

D. BIOLOGICAL EFFECTS:
—Crustaceans:

Gammarus lacustris:	96 hr, LC_{50}:	52	µg/l	(2124)
Gammarus fasciatus:	96 hr, LC_{50}:	21	µg/l	(2126)
Palaemonetes kadiakensis:	96 hr, LC_{50}:	38	µg/l	(2126)

—Insects:

Pteronarcys californica:	96 hr, LC_{50}:	5	µg/l	(2128)
Pteronarcys californica:	96 hr, LC_{50}:	24	µg/l	(2117)
	30 day LC_{50}:	1.9 µg/l		
Acroneuria pacifica:	96 hr, LC_{50}:	8.2 µg/l		(2117)
	30 day LC_{50}:	1.4 µg/l		

—Fish:

Pimephales promelas:	96 hr, LC_{50}:	3700	µg/l	(2123)
Lepomis macrochirus:	96 hr, LC_{50}:	63	µg/l	(2123)

—Molluscs:
American oyster (*Crassostrea virginica*): Eggs: TLm: 5860 ppb (48 hr static lab bioassay)
American oyster (*Crassostrea virginica*): Larvea: TLm: 3670 ppb (14 day static lab bioassay)

hard clam (*Mercenaria mercenaria*): Eggs: TLm: 55280 ppb (48 hr static lab bioassay)
hard clam (*Mercenaria mercenaria*): Larvae: TLm: 1390 ppb (12 day static lab bioassay) (2334)
—Mammals:
acute oral LD_{50} (rat): 2.6–12.5 mg/kg
acute dermal LD_{50} (rat): about 20 mg/kg (1854)

disyston *see* disulfoton

***l*-3,3'-dithio-*bis*-(2-aminopropanoic acid)** *see l*-cystine

***l*-β,β'-dithiodi-alanine** *see l*-cystine

diundecylphthalate

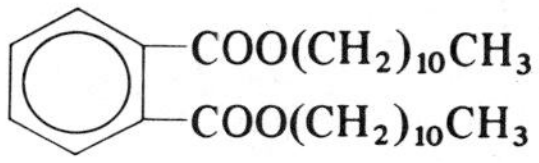

$$\text{C}_6\text{H}_4(\text{COO}(\text{CH}_2)_{10}\text{CH}_3)_2$$

C. WATER POLLUTION FACTORS:
—Biodegradation: in continuous A.S.: 37%
in river water, 1 week: 10% (1841)

diuron *see* 3-(3,4-dichlorophenyl)-1,1-dimethylurea

divinyl *see* 1,3-butadiene

DMDT *see* methoxychlor

DMK *see* acetone

DMS *see* dimethylsulfate

DNBP *see* 2,4-dinitro-6-*sec*-butylphenol

dobane 83 (alkylbenzene; alkylate)

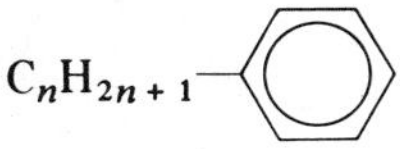

$$C_nH_{2n+1}\text{–}C_6H_5$$

n = 8–13 (mainly linear alkyls)
Use: manufacturing of highly biodegradable detergents

A. PROPERTIES: m.w. 233; sp.gr. 0.867 (15.5/15.5°C) b.p. 271–312°C; solub. <0.001 mg/l

C. WATER POLLUTION FACTORS:
—COD: 1.7 = 50% ThOD (277)

D. BIOLOGICAL EFFECTS:
  —*Carassius auratus*: not toxic in saturated solution (<0.001 mg/l) (277)
  —Rat: acute oral LD_{50}: 3 ml/kg (n.s.i.) (277)

dobane JN (alkylbenzene; alkylate)

$$C_nH_{2n+1}-\bigcirc$$

$n = 10$–13 (mainly linear alkyls)
Use: manufacturing of highly biodegradable detergents
A. PROPERTIES: m.w. 243; sp.gr. 0.869 (15.5/15.5°C); b.p. 283–313°C; solub. <0.01 mg/l
C. WATER POLLUTION FACTORS:
  —COD: 1.6 = 50% ThOD (277)
D. BIOLOGICAL EFFECTS:
  —*Carassius auratus*: not toxic in saturated solution (<0.01 mg/l) (277)

dobane 055 (alkylbenzene; alkylate)

$$C_nH_{2n+1}-\bigcirc$$

$n = 10$–15 (mainly linear alkyls)
Use: manufacturing of highly biodegradable detergents
A. PROPERTIES: m.w. 256; sp.gr. 0.864 (15.5/15.5°C); b.p. 290–331°C; solub. <0.01 mg/l
C. WATER POLLUTION FACTORS:
  —COD: 2.3 = 70% ThOD (277)
D. BIOLOGICAL EFFECTS:
  —*Carassius auratus*: not toxic in saturated solution (<0.01 mg/l) (277)

dobanic acid (alkylbenzene sulfonic acid; sulfonic acid; dodecylbenzene-sulfonic acid)

$$C_nH_{2n+1}-\bigcirc-SO_3H$$

$n = 10$–13
A. PROPERTIES: m.w. 323; sp.gr. 1.05 at (15.5/15.5°C); pourpoint: −12°C
C. WATER POLLUTION FACTORS:
  —COD: 2.41 = 92% ThOD (277)
D. BIOLOGICAL EFFECTS:
  —*Carassius auratus*: 24 hr, LD_{50}: 5 mg/l ⎫
      96 hr, LD_{50}: 5 mg/l ⎬ pH 6
      24 hr, LD_{50}: 7 mg/l ⎫
      96 hr, LD_{50}: 5 mg/l ⎬ pH 7 (277)

dobanol 23 (neodol; linear primary alcohols)
$C_nH_{2n+1}OH$
  $n = 12, 13$
  Use: manufacturing of detergents and shampoos
A. PROPERTIES: colorless liquid at room temperature; m.w. 194 ± 5; p.p.: 17-20°C; b.p. 265-275°C; sp.gr. 0.832 g/ml at 25°C; solub. 2.4 mg/l
C. WATER POLLUTION FACTORS:
  $-BOD_5$: 1.59 = 51% ThOD
  $-COD$:   2.81 = 90% ThOD (277)
D. BIOLOGICAL EFFECTS:
  $-Carassius\ auratus$: not toxic in saturated solution (2-4 mg/l) (277)
  $-Rat$: acute oral LD_{50}: >1 ml/kg (277)

dobanol 25 (neodol; linear primary alcohols)
$C_nH_{2n+1}OH$
  $n = 12 \ldots 15$
  Use: manufacturing of highly biodegradable detergents
A. PROPERTIES: colorless liquid at room temp.; m.w. 270 ± 5; p.p. 20-23°C; b.p. 265-295°C; sp.gr. 0.829 g/ml at 25°C; solub. 0.8 mg/l
C. WATER POLLUTION FACTORS:
  $-BOD_5$: 1.95 = 62% ThOD
         2.41 = 77% ThOD (after adaptation)
  $-COD$:   2.81 = 90% ThOD (277)
D. BIOLOGICAL EFFECTS:
  $-Carassius\ auratus$: not toxic in saturated solution (0.8 mg/l) (277)
  $-Rat$: acute oral LD_{50}: >1 ml/kg (277)

dobanol 45 (neodol; linear primary alcohols)
$C_nH_{2n+1}OH$
  $n = 14, 15$
A. PROPERTIES: m.w. 218 ± 6; p.p. 29-31°C; b.p. 285-295°C; sp.gr. 0.824 g/ml at 25°C; solub. 0.7 mg/l
C. WATER POLLUTION FACTORS:
  $-BOD_5$: 1.84 = 59% ThOD
  $-COD$:   2.66 = 85% ThOD (277)
D. BIOLOGICAL EFFECTS:
  $-Carassius\ auratus$: not toxic in saturated solution (0.7 mg/l) (277)
  $-Rat$: acute oral LD_{50}: >1 mg/kg (277)

dobanol ethoxylates

Dobanol ethoxylates are non-ionic surface active substances, obtained by ethoxylation of linear primary alcohols (dobanols). The linear primary alcohols consist of 80% straight chain and 20% 2-alkyl isomers (mainly 2-methyl isomers).

A. PROPERTIES:
  —dobanol 91-5: sp.gr. 0.979 g/ml at 20°C; m.w. 379
   dobanol 91-6: sp.gr. 0.99 g/ml at 20°C; point of solidification 6°C; m.w. 423
   dobanol 91-8: sp.gr. 1.016 g/ml at 20°C; pourpoint: 15°C; m.w. 529

dobanol 25-9:   pourpoint 25–28°C; m.w. 603
dobanol 45-11: sp.gr. 0.989 g/ml at 20°C; pourpoint 28–31°C; m.w. 705
C. WATER POLLUTION FACTORS:
—Biodegradability (OECD test method):
  dobanol 91-5:   99%
  dobanol 91-6:   99%
  dobanol 91-8:   99%
  dobanol 25-9:   98%
  dobanol 45-11: 99.1%   (277)

dobanol ethoxysulfate 25-3S/27 (sodium salt of sulfonated dobanol 25-3 ethoxylate)
A. PROPERTIES: m.w. 441; sp.gr. 1.04 at 20/20°C
C. WATER POLLUTION FACTORS:
—Biodegradability: 99.2 (OECD test method)   (277)

n-docosane
$C_{22}H_{46}$
Manufacturing source: petroleum refining.   (347)
Users and formulation: organic synthesis; standardized hydrocarbon; mfg; paraffin products; rubber industry; paper processing industry; paraffin industry; calibration, temperature sensing device; constituent in waterborne waste of polyolefin manufacture.   (347)
Natural sources: constituent in paraffin fraction of petroleum.   (347)
Man caused source (*water and air*): component in municipal waste and waste water from general use of paraffins; also lab use; highway runoff, automobile exhaust, motorboat exhaust, and from general use of petroleum oils, tars, etc.   (347)
A. PROPERTIES: solid, m.p. 45.7°C; b.p. 230°C at 15 mm sp.gr. 0.778 at 45/4°C
B. AIR POLLUTION FACTORS:
—Ambient air quality:
  organic fraction of suspended matter:
    Bolivia at 5200 m altitude (Sept.–Dec. 1975): 0.32–0.40 µg/1000 cu m
    Belgium, residential area (Jan.–April 1976):   0.52–2.4 µg/1000 cu m   (428)
  glc's in residential area (Belgium) oct. 1976:
    in particulate phase: 2.33 ng/cu m
    in gasphase: 4.23 ng/cu m   (1289)
  glc's in Botrange (Belgium)—woodland at 20–30 km from industrial area: June–July 1977: 0.22; 0.22 ng/cu m ($n = 2$)
  glc's in Wilrijk (Belgium)—residential area: Oct.–Dec. 1976: 1.00; 12.47 ng/cu m ($n = 2$)   (1233)
  glc's in the average American urban atmosphere—1963:
    480 µg/g airborne particulates, or
    60 ng/g cu m air   (1293)

docosanoic acid   *see* behenic acid

1-docosene
$CH_3(CH_2)_{19}CH{=}CH_2$

n-DODECANE

A. PROPERTIES: m.w. 308.59; m.p. 37.5-38.5°C; sp.gr. 0.794; b.p. 367°C
C. WATER POLLUTION FACTORS:
  —Adsorption on smectite clay particles from simulated seawater at 25°C:
   experimental conditions: 100 µg n-docosene/l water; 50 mg smectite/l
   adsorption: 1.04 µg/mg: 52% adsorbed (1009)

cis-13-docosenoic acid see cis-erucic acid

n-docosoic acid see behenic acid

1,1a,2,2,3a,4,5,5,5a,5b,6-dodecachloro-octahydro-1,3,4-metheno-1H-cyclobuta(cd)-pentalene see mirex

dodecanal (lauraldehyde)
$CH_3(CH_2)_{10}CHO$
A. PROPERTIES: m.w. 184.31; m.p. 44.5°C; b.p. 185°C at 100 mm; sp.gr. 0.835 at 15/4°C
C. WATER POLLUTION FACTORS:
  —Waste water treatment:
   A.S.: after 6 hr: 8.0% of ThOD
              12 hr: 14.6% of ThOD
              24 hr: 19.9% of ThOD (88)

n-dodecane (bihexyl; dihexyl)
$CH_3(CH_2)_{10}CH_3$
  Manufacturing source: refineries running alkylations, petroleum refining industry. (347)
  Users and formulation: organic synthesis; solvent; standardized hydrocarbon; jet fuel research; mfg. paraffin products; rubber industry; paper processing industry. (347)
  Natural sources (water and air): constituent in paraffin fraction of petroleum. (347)
  Man caused sources (water and air): component in municipal waste and waste water from general use of paraffins; lab use; highway runoff, automobile exhaust, motorboat exhaust, and from general use of petroleum oils, tars, etc. (347)
A. PROPERTIES: colorless liquid; m.w. 170.33; m.p. -2.3/-9.6°C; b.p. 216.2°C; v.p. 0.3 mm at 20°C, 1 mm at 48°C; v.d. 5.96; sp.gr. 0.766 at 0/4°C; solub. in salt water 0.005 mg/l at 20°C, in sea water 0.0029 mg/l at 25°C, in distilled water 0.0037 mg/l at 25°C; THC 1,934 kcal/mole, LHC 1,810 kcal/mole (228, 230)
B. AIR POLLUTION FACTORS:
  —Odor: threshold conc. 37 mg/cu m (737)
C. WATER POLLUTION FACTORS:
  —Degradation in seawater by oil oxidizing micro-organisms: 95.1% breakdown after 21 days at 22°C in stoppered bottle containing a 1000 ppm mixture of alkanes, cycloalkanes and aromatics (1237)
  —Waste water treatment:
   A.S.: after 6 hr: 1.6% of ThOD
              12 hr: 4.1% of ThOD
              24 hr: 7.4% of ThOD (88)

A.S., Sd, BOD: 14 days acclimation: 8% of ThOD
A.S., W: 14 days acclimation: 4% of ThOD; feed: 50-200 mg/l (93)
rotating disk contact aerator: infl. 34.8 mg/l, effl. 1.1 mg/l; elimination: 97% or 2783 mg/m^2/24 hr or 752 g/cu m/24 hr (406)

—Aquatic reactions:
first order evaporation constant of n-undecane in 3 mm layer No. 2 fuel oil: in darkened room at windspeed of 21 km/hr at
5°C: 1.57×10^{-4} min^{-1}
10°C: 2.86×10^{-4} min^{-1}
20°C: 5.25×10^{-4} min^{-1}
30°C: 1.28×10^{-3} min^{-1} (438)

—Groundwater: percolation water at 30-500 m from dumping ground: 35 ppb (183)

—Reduction of amenities: T.O.C.: 100 mg/l (295)

dodecanoic acid *see* lauric acid

1-dodecanol (laurylalcohol)
$CH_3(CH_2)_{11}OH$

A. PROPERTIES: crystalline solid; m.w. 186.33; m.p. 22.6/24°C; b.p. 255/259°C; v.p. 1 mm at 91°C, 10 mm at 134.7°C, 100 mm at 192.0°C; v.d. 6.43; sp.gr. 0.831 at 24/4°C; log P_{oct} 5.13

B. AIR POLLUTION FACTORS: 1 mg/cu m = 0.131 ppm, 1 ppm = 7.62 mg/cu m
—Odor: T.O.C.: 0.054 mg/cu m = 7.1 ppb (307)
O.I. at 20°C = 1800 (316)
recognition: 0.02 mg/cu m (610)
25.5 mg/cu m (761)
2.1 mg/cu m (787)

C. WATER POLLUTION FACTORS:
—Waste water treatment:
A.S.: after 6 hr: 4.5% of ThOD
12 hr: 10.1% of ThOD
24 hr: 13.4% of ThOD (88)

methods	temp °C	days observed	feed mg/l	days acclim.	
A.S., BOD	20	1-5		14	30% removed
A.S., W	20	$\frac{1}{4}$	50	14	15% theor. oxidation

(93)

D. BIOLOGICAL EFFECTS:
—Mammalia: rat: single oral dose: LD$_{50}$: >12.8 g/kg
>36.0 ml/kg
rabbit: single oral dose: LD$_{50}$: >36.0 ml/kg (211)

1-dodecene
$CH_3(CH_2)_9CH=CH_2$
A. PROPERTIES: m.w. 168.32; m.p. 35°C; sp.gr. 0.758; b.p. 213°C
C. WATER POLLUTION FACTORS:
—Treatment methods:

rotating disk contact aerator: infl. 36.5 mg/l, effl. 0.9 mg/l; elimination: 98% or 2.939 mg/m²/24 hr or 794 g/cu m/24 hr   (406)

n-dodecylamine (1-aminododecane)
$CH_3(CH_2)_{11}NH_2$
A. PROPERTIES: colorless crystals; m.w. 185.35; m.p. 27°C; b.p. 259°C; v.p. 64 mm at 170°C
C. WATER POLLUTION FACTORS:
  —Biodegradation:
    degradation by *Aerobacter*: 200 mg/l at 30°C:
      parent: 100% degradation in 18 hr
      mutant: 100% degradation in 5 hr   (152)
  —Impact on biodegradation:
    degree of inhibition of NH_3 oxidation by *Nitrosomonas* sp.:
      at 100 mg/l: 96% inhibition
      50 mg/l: 95% inhibition
      <1 mg/l: ~50% inhibition   (390)

prim-n-**dodecylchloride** *see* 1-chlorododecane

N-dodecyldi(aminoethyl)glycine
Use: microbicide
C. WATER POLLUTION FACTORS:
  —Biodegradation:
    after 3 weeks of adaptation at 500 mg/l at 22°C:
      aerobic degradation: product is sole carbon source: 10% degradation
                           + synthetic sewage: 95% degradation
      anaerobic degradation: product is sole carbon source: 50% degradation
                             + synthetic sewage: 80% degradation (512)
D. BIOLOGICAL EFFECTS:
  —Harlequin fish (*Rasbora heteromorpha*)

mg/l	24 hr	48 hr	96 hr	3 m (extrap.)	
LC_{10} (F)	6.8	3.7	1.95		
LC_{50} (F)	8.7	5.9	3.2	1.0	(331)

n-dodecylguanidineacetate (laurylguanidine-acetate; dodine; doguadine; guanidine; melprex-65)

$$C_{12}H_{25}-NH-\overset{NH}{\overset{\|}{C}}-NH_2 \cdot CH_3-COOH$$

D. BIOLOGICAL EFFECTS:
  —Fish: harlequin fish (*Rasbora heteromorpha*):

mg/l	24 hr	48 hr	96 hr	3 m (extrap.)	
LC_{10} (F)	1.3	0.43	0.29		
LC_{50} (F)	1.7	0.82	0.6	0.1	(331)

  —Mammals:
    acute oral LD_{50} (male rats): 1000–2000 mg/kg

acute dermal LD_{50} (rabbits): 2100 mg/kg
in diet: rats: 800 ppm for 1 year: slight retardation of growth (1855)

dodecylsulfate, sodium *see* laurylsulfate, sodium

donax
Use: hydraulic brake fluid
A. PROPERTIES: b.p. 210°C (dry)
C. WATER POLLUTION FACTORS:
—BOD_5: 0.07
—COD: 0.55 (277)
D. BIOLOGICAL EFFECTS:
—*Carassius auratus*: LD_{50}, 24 hr: 1050 mg/l

DOP *see* di-*n*-octylphthalate

dosanex (N'-(3-chloro-4-methoxyphenyl)-N,N-dimethylurea; metoxuron)

$$CH_3O-\underset{}{\underset{}{\bigcirc}}-NH-\underset{\underset{O}{\|}}{C}-N\underset{CH_3}{\overset{CH_3}{\diagup}}$$
(with Cl on the ring)

Use: herbicide
D. BIOLOGICAL EFFECTS:
—Fish: harlequin fish (*Rasbora heteromorpha*):

mg/l	24 hr	48 hr	96 hr	3 m (extrap.)	
LC_{10} (F)	105	37			
LC_{50} (F)	200	54	40	20	(331)

—Mammals:
acute oral LD_{50} (rat): 3200 mg/kg (1854)

n-dotriacontane
$CH_3(CH_2)_{30}CH_3$
A. PROPERTIES: m.w. 450.88; m.p. 68–70°C; b.p. 467°C
B. AIR POLLUTION FACTORS:
—Ambient air quality:
organic fraction of suspended matter:
Bolivia at 5200 m altitude (Sept.–Dec. 1975): 0.16–0.40 µg/1000 cu m
Belgium, residential area (Jan.–April 1976):   1.5–1.9 µg/1000 cu m (428)
glc's in Botrange (Belgium): woodland at 20-30 km from industrial area: June–July 1977: 0.65; 0.42 ng/cu m (*n* = 2)
glc's in Wilrijk (Belgium): residential area: Oct.–Dec. 1976: 1.19; 3.20 ng/cu m (*n* = 2) (1233)

dowanol DB *see* diethyleneglycolmonobutylether

dowanol DE *see* diethyleneglycolmonoethylether

dowanol EE *see* ethyleneglycolmonoethylether

dowanol EM *see* ethyleneglycolmonomethylether

dowicide A
   Composition: 97% O-phenylphenate, sodium (tetrahydrate)
D. BIOLOGICAL EFFECTS:
   —Algae:
      Protococcus sp.: 2.5×10^4 ppb: .75 OD expt/OD control, 10 day growth test
      Chlorella sp.: 5×10^4 ppb: .74 O.D. expt/O.D. control, 10 day growth test
      Dunaliella euchlora: 5×10^4 ppb: .52 O.D. expt/O.D. control, 10 day growth test
      Phaeodactylum tricornutum: 2.5×10^4 ppb: .48 O.D. expt/O.D. control, 10 day growth test
      Monochrysis lutheri: 2.5×10^4 ppb: .22 O.D. expt/O.D. control, 10 day growth test (2347)
   —Molluscs:
      hard clam (*Mercenaria mercenaria*): eggs: 48 hr static lab bioassay: TLm: 1×10^5 ppb
      hard clam (*Mercenaria mercenaria*): larvae: 12 day static lab bioassay: TLm: 750 ppb (2324)

dowicide G
   Composition: 79% pentachlorophenate, sodium (monohydrate) + 11% sodium salts of other chlorophenols
D. BIOLOGICAL EFFECTS:
   —Molluscs:
      hard clam (*Mercenaria mercenaria*): eggs: 48 hr static lab bioassay: TLm: <250 ppb
      hard clam (*Mercenaria mercenaria*): larvae: 12 day static lab bioassay: TLm: <250 ppb
   —Fish:
      Cyprinodon variegatus: 2-wk fry: 96-h LC_{50}-FT: 516 µg/l
      Lagodon rhomboides: 48-h prolarvae: 96-h LC_{50}-S: 66 µg/l (1973)

dowklor *see* chlordane

2,4-DP *see* 2,4-dichlorophenoxypropionic acid

DPA *see* N-diphenylamine

DPP *see* dipropylphthalate

DRC 1339 *see* 3-chloro-4-methylbenzenaminehydrochloride

DTPA *see* diethylenetriaminepenta-acetic acid

durene *see* 1,2,3,4-tetramethylbenzene

dutrex
Dutrex is the trade name for a number of aromatic extracts obtained during the refining of lubricating oils
Uses: process oils and extender oils for natural and synthetic rubber; secondary softeners for dark-colored PVC products; base oils for printing inks; base oils for agro-chemicals and sprays.

A. PROPERTIES:
  —dutrex 217 UK:
   sp.gr. 0.983 at 289/289K; pour point 247 K; solub. 5.8 mg/l
   composition: asphaltenes: <0.1%; polar components: 3.9% aromatics: 83.4%; alifatics: 12.7%
  —dutrex 719 UK:
   sp.gr. 1.025 at 289/289 K; pour point 275 K; solub. 1.4 mg/l
   composition: asphaltenes: <0.1%; polar components: 9.1% aromatics: 87.7%; alifatics: 3.2%

C. WATER POLLUTION FACTORS:

	dutrex 729 HP	dutrex 726 UK	dutrex 719 UK	dutrex 217 UK
—BOD$_5$:	0.36 = 12% ThOD	0.11 = 3% ThOD	0.14 = 5% ThOD	0.10 = 3% ThOD
—COD:	3.01 = 97% ThOD	2.97 = 93% ThOD	2.87 = 93% ThOD	2.68 = 85% ThOD

(277)

D. BIOLOGICAL EFFECTS:
  —*Carassius auratus*:
   dutrex 729 HP: not toxic in saturated solution (1.4 mg/l)   (277)
   dutrex 726 UK: not toxic in saturated solution (4.9 mg/l)
   dutrex 719 UK: not toxic in saturated solution (1.4 mg/l)
   dutrex 217 UK: not toxic in saturated solution (5.8 mg/l)   (277)

dylox *see* dimethyl-(2,2,2-trichloro-1-hydroxyethyl)phosphonate

E

ECH *see* epichlorohydrin

ECHH *see* epichlorohydrin

EDC *see* ethylenedichloride

EDTA *see* ethylene diamine tetra acetic acid

eicosane
$C_{20}H_{42}$

    Manufacturing source: petroleum refining. (347)
    Users and formulation: organic synthesis; solvent; standardized hydrocarbon; jet fuel research; mfg. paraffin products; rubber industry; paper processing industry; cosmetics, lubricants, plasticizers (347)
    Natural sources (water and air): constituent in paraffin fraction of petroleum. (347)
    Man caused sources (water and air): component in municipal waste and waste water from general use of paraffins.; lab use; highway runoff, automobile exhaust, and from general use of petroleum oils, tars, etc. (347)

A. PROPERTIES: white crystalline solid; m.p. 36.7°C; b.p. 205°C at 15 mm sp.gr. 0.788 at melting point

B. AIR POLLUTION FACTORS:
    —Ambient air quality:
      glc's in residential area (Belgium)—Oct. 1976:
        in particulate sample: 0.85 ng/cu m
        in gasphase sample: 7.55 ng/cu m (1289)
      organic fraction of suspended matter:
        Bolivia at 5200 m altitude (Sept.–Dec. 1975): 0.19–0.20 µg/1000 cu m
        Belgium, residential area (Jan.–April 1976):   0.48–0.70 µg/1000 cu m (428)
      glc's in Botrange (Belgium): woodland at 20–30 km from industrial area: June–July 1977: 0.20; 0.29 ng/cu m ($n = 2$) (1233)
      glc's in the average American urban atmosphere—1963:
        180 µg/g airborne particulates or
        23 ng/g cu m air (1293)

C. WATER POLLUTION FACTORS:
    —Biodegradation:
      degradation in seawater by oil-oxidizing micro-organisms: 44.2% breakdown

after 21 days at 22°C in stoppered bottles containing a 1000 ppm mixture of alkanes, cyclo-alkanes and aromatics (1237)
—Aquatic reactions:
adsorption on smectite clay particles from simulated seawater at 25°C; experimental conditions: 100 µg n-eicosane/l water; 50 mg smectite/l adsorption: 0.72 µg/mg = 36% adsorbed (1009)

eicosanoic acid (arachidic acid)
$CH_3(CH_2)_{18}COOH$
A. PROPERTIES: m.w. 312.54; m.p. 74–76°C
B. AIR POLLUTION FACTORS:
—Ambient air quality:
Organic fraction of suspended matter:
Bolivia at 5200 m altitude (Sept.–Dec. 1975): 0.21–0.42 µg/1000 cu m
Belgium, residential area (Jan.–April 1976):   2.5–5.1 µg/1000 cu m
glc's in residential area (Belgium)—Oct. 1976:
in particulate sample: 9.04 ng/cu m
in gasphase sample: 3.00 ng/cu m (1289)
glc's in Botrange (Belgium)—woodland at 20–30 km from industrial area; June–July 1977: 3.76; 3.06 ng/cu m ($n = 2$)
glc's in Wilrijk (Belgium)—residential area: Oct.–Dec. 1976: 7.9; 9.4 ng/cu m ($n = 2$) (1233)
C. WATER POLLUTION FACTORS:
—Reduction of amenities: T.O.C. in water: 20 ppm (326; 886)

1-eicosene
$CH_3(CH_2)_{17}CH=CH_2$
A. PROPERTIES: m.w. 280.54; m.p. 27–29°C; b.p. 151° at 1.5 mm
C. WATER POLLUTION FACTORS:
—Aquatic reactions:
adsorption on smectite clay particles from simulated seawater at 25°C; experimental condctions: 100 µg n-eicosene/l water: 50 mg smectite/l adsorption: 0.83 µg/mg: 41% adsorbed (1009)

ekalux (quinalphos; bayrusil; O,O-diethyl-O-quinoxalinyl-(2)-thionophosphate; O,O-diethyl-O-(2-chinoxalyl)phosphorothioate)

D. BIOLOGICAL EFFECTS:
—Fish:
Saccobranchus fossilis: 96 hr LC_{50} (S): 1.55 mg/l (1509)
—Mammals:
oral LD_{50} (rat):   66 mg/kg
dermal LD_{50} (rat): approx. 340 mg/kg (1854)

ekatin (O,O-dimethyl-S-(2-(ethyl-thio)ethyl)phosphorodithioate; thiometon; dithiometon)

$$\begin{array}{c} CH_3O \\ \diagdown \\ CH_3O\diagup \end{array} \overset{\overset{\displaystyle S}{\|}}{P} - S - CH_2 - CH_2 - S - C_2H_5$$

Uses: insecticide and acaricide

A. PROPERTIES: b.p. 121°C at 1 mm; v.p. 3×10^{-4} mm at 20°C; sp.gr. 1.2 at 20/4°C; solub. 200 mg/l at 25°C

D. BIOLOGICAL EFFECTS:
  —Fish: harlequin fish (*Rasbora heteromorpha*):

mg/l	24 hr	48 hr	96 hr	3 m (extrap.)	
LC_{10} (F)	3.3	2.2	2.3		
LC_{50} (F)	4.7	3.7	3.2	1.2	(331)

  Saccobranchus fossilis: 96 hr LC_{50} (S): 11.0 mg/l (1509)
  —Mammals:
    acute oral LD_{50} (rat): 100–120 mg/kg
    dermal LD_{50} (rat): 680–730 mg/kg (1854)

elon *see* p-methylaminophenol

enanthal *see* heptanal

enanthaldehyde *see* heptanal

enanthic acid (heptanoic acid; enanthylic acid; *n*-heptoic acid; *n*-heptylic acid) $CH_3(CH_2)_5COOH$

Uses: organic synthesis; production of special lubricants for aircraft and brake fluids

A. PROPERTIES: colorless oily liquid; m.w. 130.18; m.p. -10/-7°C; b.p. 223.5°C; v.p. 1 mm at 78°C, 10 mm at 113.2°C, 100 mm at 160°C; sp.gr. 0.915 at 25/4°C; solub. 2,410 mg/l at 15°C; log P_{oct} 2.72 (calculated)

B. AIR POLLUTION FACTORS: 1 mg/cu m = 0.185 ppm, 1 ppm = 5.412 mg/cu m
  —Odor: T.O.C.: 0.082 mg/cu m = 15 ppb (307)
    O.I. at 20°C = 900 (316)
    detection: 0.3 mg/cu m (778; 779)
    recognition: 2.6–3.3 mg/cu m (610)

C. WATER POLLUTION FACTORS:
  —A.S.: after 6 hr: 12.8% of ThOD
         12 hr: 25.4% of ThOD
         24 hr: 42.6% of ThOD (89)
    A.S., BOD, 20°C, 1–5 days observed, feed: 333 mg/l, 15 days acclimation, 99% removed (93)
  —Odor threshold: detection: 3.0 mg/kg (886)

D. BIOLOGICAL EFFECTS:
  —Mammals: mouse: death in 2–4 days: 125 mg/kg/day (211)

endosan *see* binapacryl

endosulfan (6,7,8,9,10,10-hexachloro-1,5,5a,6,9,9a-hexahydro-6,9-methano-2,3,4-benzodioxyanthiepin-3-oxide)

Uses: insecticide for vegetable crops

A. PROPERTIES: technical endosulfan is a mixture of 2 isomers with m.p. of 108 and 206°C; brown crystals; m.p. 70–100°C (pure m.p. 106°C)

Structure of (a) α-endosulfan; (b) β-endosulfan; (c) endosulfan sulfate; (d) endosulfan diol; (e) endosulfan ether; (f) endosulfan α-hydroxy ether; (g) endosulfan lactone.

C. WATER POLLUTION FACTORS:
  —Aquatic reactions:
    irradiation with u.v. light yields endosulfandiol as a major product along with the α-hydroxyether, lactone, and ether of endosulfan. No endosulfan sulfate was produced by irradiation (1560)
  —Photo-oxidation by u.v. light in aqueous medium at 90–95°C: time for the formation of CO_2 (% of theoretical): 25%: 5.0 hr
  50%: 9.5 hr
  75%: 31.0 hr (1628)
  —Persistence in river water in a sealed glass jar under sunlight and artificial fluorescent light—initial conc. 10 μg/l:

% of original compound found

after	1 hr	1 wk	2 wk	4 wk	8 wk	
	100	30	5	0	0	(1309)

—Biodegradation:

^{14}C endosulfan incubated with active soil fungi formed endosulfan sulfate as the major metabolite while the endodiol was the major product formed by active soil bacteria (1561)

in soils and under aerobic conditions, endosulfan sulfate was the major metabolite produced

in soils incubated under N_2/CO_2, the sulfate was also the major product but the conversion rate was smaller than under aerobic conditions

under flooded conditions the diol was produced in greater amounts than the sulfate and the hydroxy-ether was also formed (1562)

stability of endosulfan metabolites in sterile nutrient (control) medium and in medium inoculated with mixed culture of soil micro-organisms (aqueous medium):

	time for 50% degradation	
	control	inoculated
endosulfan lactone	5.5 hr	5.5 hr
α-endosulfan	12.5 wk	1.1 wk
β-endosulfan	5.7 wk	2.2 wk
endosulfan ether	>20 wk	6 wk
endosulfan-α-hydroxy-ether	>20 wk	8 wk
endosulfan sulfate	>20 wk	11 wk
endosulfan diol	>20 wk	14 wk (1559)

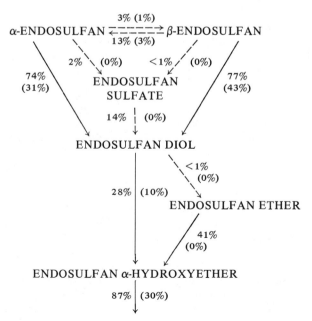

ENDOSULFAN LACTONE

$\Big\downarrow$ 100% disappearence
(100%)

Conversion of α- and β-endosulfan and metabolites in aqueous nutrient medium inoculated with a mixed culture of soil microorganisms and in sterile medium (bracketed numbers). (1559)

D. BIOLOGICAL EFFECTS:
—Residues: residue in animals in different trophic levels from the Weser estuary (1976): wet tissue:
bivalvia:
common edible cockle (*Cerastoderma edule L.*): 0.3 ± 0.1 ng/g
soft clam (*Mya arenaria L.*): 0.7 ± 0.2 ng/g
polychaeta: lugworm (*Arenicola marina L.*): 0.4 ± 0.2 ng/g
crustacea: brown shrimp (*Crangon Crangon L.*): 0.3 ± 0.1 ng/g
pisces: common sole (*Sola solea L.*): 1.3 ± 0.4 ng/g (1440)
—Bioaccumulation:
decapod: *Palaemonetes pugio*: 96 hr, 0.16–1.75 µg/l: 81–245X (1137)
pelecypod: *Mytilus edulis*: 50 hr, 0.14–2.05 µg/l: 600X (1477)
fish: *Lagodon rhomboides*: 96 hr, 0.15–0.26 µg/l: 1046–1299X
Leiostomus xanthus: 96 hr, 0.05–0.31 µg/l: 620–895X
Mugil cephalus: 96 hr, 0.32–0.49 µg/l: 1000–1344X (1137)
—Toxicity:
—Insect: *Pteronarcys californica*: 96 hr LC_{50}: 2.3 µg/l (2128)
Ischnura sp.: 96 hr LC_{50}: 71.8 µg/l (2129)
—Decapod: *Penaeus duorarum*: 96 hr LC_{50} (FT): 0.04 µg/l
Palaemonetes pugio: 96 hr LC_{50} (FT): 1.31 µg/l (1137)
—Crustacean:
Gammarus fasciatus: 96 hr LC_{50}: 5.8 µg/l (2124)
Daphnia magna: 96 hr LC_{50}: 52.9 µg/l (2129)
Korean shrimp (*Palaemon macrodactylus*):
96 hr static lab bioassay; TL_{50}: 17.1 (8.4–39.8) µg/l
Korean shrimp (*Palaemon macrodactylus*):
96 hr flowing water lab bioassay, TL_{50}: 3.4 (1.8–6.5) µg/l (2353)
Daphnia magna: 24 hr, LC_{50}: 240 µg/l
48 hr, LC_{50}: 60 µg/l (1002; 1004)
Pteronarcys: 48 hr, LC_{50}: 2.3 µg/l (1003)
—Fish:
Salmo gairdneri: 96 hr, LC_{50}: 0.3 µg/l
Catastomus commersoni: 96 hr, LC_{50}: 3.0 µg/l (2129)
Lagodon rhomboides: 96 hr, LC_{50} (FT): 0.30 µg/l
Leiostomus xanthurus: 96 hr, LC_{50} (FT): 0.09 µg/l
Mugil cephalis: 96 hr, LC_{50} (FT): 0.38 µg/l (1137)
fathead minnow: flow through bioassay: incipient TLm: 0.86 µg/l at 25°C
flow through bioassay: MATC: 0.20 µg/l (443)
chingatta: 96 hr, LC_{50} (S): 0.011 mg/l (1519)

—Mammals:
It is concluded that under the conditions of this bioassay, technical grade endosulfan was not carcinogenic in female Osborne-Mendel rats or in female B6C3F1 mice. (1764)

acute oral LD_{50} (ringnecked pheasants): 620–1000 mg/kg
                (mallard ducks): 200–750 mg/kg (1855)
                (rat): 30–110 mg/kg
acute dermal LD_{50} (rabbit): 359 mg/kg (1854)
in diet: rats fed 2 years on a diet containing 30 ppm suffered no ill effects.
(1855)

endothal *see* aquathol K

3,6-endoxohexahydrophthalic acid *see* aquathol K

endrin (1,2,3,4,10,10-hexachloro-6,7-epoxy-1,4,4a,5,6,7,8,8a-octahydro-1,4-*endo-endo*-5,8-dimethanonaphthalene)

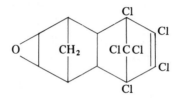

a stereo-isomer of dieldrin, which is the *endo-exo*-isomer (347)
Uses: insecticide, minor constituent in dieldrin

A. PROPERTIES: m.p. approx. 200°C; v.p. 2×10^{-7} mm at 25°C; log P_{oct} 5.6 (calculated)
B. AIR POLLUTION FACTORS:
—Ambient air quality:
atmospheric levels in agricultural area: n.d.–6.5 ng/cu m (1308)
C. WATER POLLUTION FACTORS:
—Aquatic reactions:
persistence in river water in a sealed glass jar under sunlight and artificial fluorescent light—initial conc. 10 µg/l:

% of original compound found

after	1 hr	1 wk	2 wk	4 wk	8 wk	
	100	100	100	100	100	(1309)

photo-oxidation by u.v. light in aqueous medium at 90–95°C time for the formation of CO_2 (% of theoretical): 25%: 15.0 hr
                                 50%: 41.0 hr
                                 75%: 172.0 hr (1628)
—Odor thresholds: 0.041 mg/kg
                     0.018 mg/kg (detection) (326; 915)
D. BIOLOGICAL EFFECTS:
—Residues:

Mean concentration in gamebird muscle in upper Tennessee (U.S.A.):
mg/kg fresh weight ± standard error

	grouse	quail	woodcock
Johnson county	0.35 ± 0.16 ($n = 12$)	1.20 ± 0.15 ($n = 6$)	0.94 ± 0.20 ($n = 6$)
Carter county	1.08 ± 0.14 ($n = 9$)	0.94 ± 0.18 ($n = 7$)	1.28 ± 0.26 ($n = 6$)
Washington county	1.85 ± 0.23 ($n = 10$)	1.13 ± 0.12 ($n = 11$)	1.44 ± 0.17 ($n = 6$)

(1338)

—Bioaccumulation:
—Mussels:
  pelecypod: *Mytilus edulis*: 50 hr, 0.17–1.78 µg/l: 1920X   (1477)
  BCF: 5 aquatic molluscs: 500–1250X   (1870)
  oyster (*Crassostrea virginica*):
    bioaccumulation factor after 168 hr exposure: 1670–2780 (in whole body)
    biological half-life following 168 hr exposure: 67 hr (in whole body)   (1629)
  freshwater mussel: *Lampsilis siliquoidea*: bioconc. factor: 1200
                        half-life: 4.7 days   (1680)
—Fish:
  BCF: mosquito fish: 10 (25 min. exposure)   (1895)
        fathead minnows: 10,000   (1896)
    Cyprinodon variegatus
        Juvenile   .027–.31 µg/l: 4 wk   2 500 X
        Adult   .027–.31 µg/l: 23 wk   6 400 X
        Eggs   (from above adults)   5 700 X   (1136)
  rainbow trout (*Salmo gairdneri*): log bioconcentration factor: 3.17   (193)
  fathead minnows: BCF: from food: 0.8
                      from water: 13,000
        in the food at 0.63 ppm; significant reduction of survival
                                                        (1660)
—Toxicity:
—Insects:
Pteronarcys californica: 96 hr, LC_{50}: 0.25 µg/l   (2128)
Pteronarcys californica: 96 hr, LC_{50}: 2.4 µg/l   (2118)
                          30 day LC_{50}: 1.2 µg/l
Acroneuria pacifica: 96 hr, LC_{50}: 0.32 µg/l   (2128)
                          39 day LC_{50}: 0.03 µg/l
Pteronarcella badia: 96 hr, LC_{50}: 0.54 µg/l   (2128)
Claassenia sabulosa: 96 hr, LC_{50}: 0.76 µg/l   (2128)
—Molluscs:
American oyster (*Crassostrea virginica*): 48 hr static lab bioassay, TLm: 790 ppb
                                                                                  (2324)
—Crustaceans:
Gammarus lacustris: 96 hr, LC_{50}:   3.0 µg/l   (2124)
Gammarus fasciatus: 120 hr, LC_{50}: 0.9 µg/l   (2126)
Palaemonetes kadiakensis: 120 hr, LC_{50}: 0.4 µg/l   (2126)
Orconectes nais: 96 hr, LC_{50}:   3.2 µg/l   (2126)

Asellus brevicaudus: 96 hr, LC_{50}: 1.5 µg/l (2126)
Simocephalus serrulatus: 48 hr, LC_{50}: 26 µg/l (2127)
Daphnia pulex: 48 hr, LC_{50}: 20 µg/l (2127)
Korean shrimp (*Palaemon macrodactylus*): 96 hr static lab bioassay: TL_{50}: 4.7 (2.3-9.4) ppb (2353)
Korean shrimp (*Palaemon macrodactylus*): 96 hr inter. flow lab bioassay: TL_{50}: .12 (0.05-0.25) ppb (2353)
sand shrimp (*Crangon septemspinosa*): 96 hr static lab bioassay: LC_{50}: 1.7 ppb (2327)
grass shrimp (*Palaemonetes vulgaris*): 96 hr static lab bioassay: LC_{50}: 1.8 ppb (2327)
hermit crab (*Pagurus longicarpus*): 96 hr static lab bioassay: LC_{50}: 12 ppb (2327)

—Fish:
Cyprinodon variegatus:
   fry: 96 hr LC_{50} (FT): 0.37 µg/l
   juvenile: 96 hr LC_{50} (FT): 0.34 µg/l
   adult: 96 hr LC_{50} (FT): 0.36 µg/l
largemouth bass (*Micropterus salmoides*): static bioassay: 48 hr TLm: 0.27 µg/l at 19°C (453)
O. punctatus: 96 hr LC_{50}: 0.033 ppm (1576)
Pimephales promelas: 96 hr, LC_{50}: 1.0 µg/l (2113)
Lepomis macrochirus: 96 hr, LC_{50}: 0.6 µg/l (2113)
Salmo gairdneri: 96 hr, LC_{50}: 0.6 µg/l (2119)
Oncorhynchus kisutch: 96 hr, LC_{50}: 0.5 µg/l (2119)
Oncorhynchus tschawytscha: 96 hr, LC_{50}: 1.2 µg/l (2119)
mummichog (*Fundulus heteroclitus*): 96 hr static lab bioassay: LC_{50}: 0.6 ppb (2328; 2329)
striped killifish (*Fundulus majalis*): 96 hr static lab bioassay: LC_{50}: 0.3 ppb (2329)
longnose killifish (*Fundulus similis*): 24 hr flowing lab bioassay: LC_{50}: 0.23 ppb (2328)
menhaden (*Brevoortia patronus*): 24 hr static flowing lab bioassay: LC_{50}: 0.8 ppb (2328)
striped mullet (*Mugil cephalus*): 24 hr flowing lab bioassay: LC_{50}: 2.6 ppb (2328)
striped mullet (*Mugil cephalus*): 96 hr static lab bioassay: LC_{50}: 0.3 ppb (2329)
Atlantic silverside (*Menidia menidia*): 96 hr static lab bioassay: LC_{50}: 0.05 ppb (2329)
bluehead (*Thalassoma bifasciatum*): 96 hr static lab bioassay: LC_{50}: 0.1 ppb (2329)
American eel (*Anguilla rostrata*): 96 hr static lab bioassay: LC_{50}: 0.6 ppb (2329)
Northern puffer (*Sphaeroides maculatus*): 96 hr static lab bioassay: LC_{50}: 3.1 ppb (2329)
threespine stickleback (*Gasterosteus aculeatus*): 96 hr static lab bioassay: TLm: 0.5 ppb (2333)

threespine stickleback (*Gasterosteus aculeatus*): 96 hr static lab bioassay: TLm: 1.5 ppb (2334)
sheepshead minnow (*Cyprinodon vareigatus*): 24 hr flowing lab bioassay: LC_{50}: 0.32 ppb (2338)
spot (*Leiostomus xanthurus*): 24 hr flowing lab bioassay: LC_{50}: 0.45 ppb (2338)
shiner perch (*Cymatogaster aggregata*): 96 hr static lab bioassay: TLm: 0.8 ppb (2354)
dwarf perch (*Micrometrus minimus*): 96 hr static lab bioassay: TLm: 0.6 ppb (2354)
shiner perch (*Cymatogaster aggregata*): 96 hr intermittent flow lab bioassay: TLm: 0.12 (0.06–0.25) ppb (2354)
dwarf perch (*Micrometrus minimus*): 96 hr intermittent flow lab bioassay: TLm: 0.13 (0.06–0.27) ppb (2354)
walking catfish (*Clarius batrachus*): 48 hr LC_{50} (S): 0.005 mg/l (1106; 1340)
bluegill: 96 hr LC_{50}: 0.006 ppm
rainbow trout: 96 hr LC_{50}: 0.007 ppm (1878)
susceptible mosquito fish: 48 hr LC_{50}: 0.6 ppb
resistant mosquito fish: 48 hr LC_{50}: 314 ppb (1851)

flagfish:	96-hr LC_{50}: 0.85 µg/l (F)	
	110-d MATC: 0.22–0.30 µg/l (F)	
	30-d BCF: 9,100 (F)	
	65-d BCF: 15,000 (F)	
	110-d BCF: 7,100 (F)	
	30-d BCF: 12,000 (F)	(1518)
chingatta	96-h LC_{50}: 0.007 mg/l (S)	
	96-h LC_{50}: 0.007 mg/l (S)	
	96-h LC_{50}: 0.007 mg/l (S)	
	96-h LC_{50}: 0.006 mg/l (S)	(1519)
Coregonus lavaretus:		
embryo:	76-h LC_{50}: 3.4 µg/l (S)	
larvae:	72-h LC_{50}: 0.06 µg/l (S)	
Coregonus peled:		
embryo:	72-h LC_{50}: 766 µg/l (S)	
larvae:	72-h LC_{50}: 35 µg/l (S)	
	72-h LC_{50}: 0.6 µg/l (S)	
rainbow trout		
embryo:	72-h LC_{50}: 7.7 µg/l (S)	
	72-h LC_{50}: 7.8 µg/l (S)	
	72-h LC_{50}: 0.6 µg/l (S)	
larvae:	72-h LC_{50}: <0.25 µg/l (S)	
	72-h LC_{50}: 0.07 µg/l (S)	
	72-h LC_{50}: <0.06 µg/l (S)	
	72-h LC_{50}: >0.12–<0.25 µg/l (S)	(1520)
carp		
embryo:	72-h LC_{50}: >256 µg/l (S)	
	72-h LC_{50}: 49 µg/l (S)	

larvae:	72-h LC_{50}: 31 µg/l (S)
juvenille:	72-h LC_{50}: 0.9 µg/l (S)
Carassius carassius	
juvenile:	72-h LC_{50}: 28 µg/l (S)
Tench (*Tinca Tinca*)	
juvenile:	24-h LC_{50}: 2.6 µg/l (S)
	72-h LC_{50}: 16 µg/l (S)
	72-h LC_{50}: 9.1 µg/l (S)
Leucaspius delineatus:	72-h LC_{50}: 21 µg/l (S)
	72-h LC_{50}: 13 µg/l (S)
	72-h LC_{50}: >2-<4 µg/l (S)
Gobio gobio:	24-h LC_{50}: 7.3 µg/l (S)
	72-h LC_{50}: 1.5 µg/l (S)
guppy:	72-h LC_{50}: 0.5 µg/l (S)
	72-h LC_{50}: 0.3 µg/l (S)
	72-h LC_{50}: 0.12 µg/l (S)
	72-h LC_{50}: 0.08 µg/l (S)
	72-h LC_{50}: 0.06 µg/l (S)
Stizostedion lucioperca	
juvenile:	24-h LC_{50}: 7.5 µg/l (S)
	72-h LC_{50}: 4.8 µg/l (S) (1520)

—Mammals:
endrin is carcinogenic for rats, and most likely also for mice and dogs (1425)
acute oral LD_{50} (rats): 7.5–17.5 mg/kg
acute dermal LD_{50} (female rats): 15 mg/kg (1855)

enenthole *see* heptanal

ENI 27,164 *see* carbofuran

ENT 16,225 *see* dieldrin

EO *see* ethylene-oxide

epichlorohydrin (α-epichlorohydrin; 1-chloro-2,3-epoxypropane; γ-chloropropylene-oxide; (chloromethyl)oxirane; ECH)

$$CH_2CHCH_2Cl$$
$$\diagdown\!\diagup$$
$$O$$

Uses: in the manufacturing of epoxy resins, surface active agents, pharmaceuticals, insecticides, agricultural chemicals, textile chemicals, coatings, adhesives, ion-exchange resins, solvents, plasticizers, glycidyl esters, ethynyl-ethylenic alcohol, and fatty acid derivatives (1416)

A. PROPERTIES: colorless liquid; m.w. 92.53; m.p. −57, −25.6°C; b.p. 116/117°C; v.p. 12 mm at 20°C, 22 mm at 30°C; v.d. 3.3; sp.gr. 1.18 at 20/4°C; solub. 60,000 mg/l at 20°C; sat. conc. 60 g/cu m at 20°C, 107 g/cu m at 30°C

612   EPICHLOROHYDRIN

B. AIR POLLUTION FACTORS: 1 mg/cu m = 0.265 ppm, 1 ppm = 3.78 mg/cu m
   —Odor: characteristic: quality: chloroformlike
                          hedonic tone: irritating
   T.O.C.: average: 10 ppm
           100% response: 100 ppm   (211)
           0.3 mg/cu m = 0.08 ppm   (57)(661)
   USSR: human odor perception: non perception: 0.2 mg/cu m
                                perception: 0.3 mg/cu m
         human reflex response: no response: 0.2 mg/cu m
                                adverse response: 0.3 mg/cu m
         animal chronic exposure: no effect: 0.2 mg/cu m
                                  adverse effect: 2.0 mg/cu m
         O.I. at 20°C = 160

C. WATER POLLUTION FACTORS:
   —BOD: 0.03   NEN 3235-5.4   (277)
         0.16 adapted sew. NEN 3235-5.4   (277)
   —COD: 1.16   NEN 3235-5.3   (277)

D. BIOLOGICAL EFFECTS:
   —Toxicity threshold (cell multiplication inhibition test):
     bacteria (*Pseudomonas putida*):   55   mg/l   (1900)
     algae (*Microcystis aeruginosa*):   6.0 mg/l   (329)
     green algae (*Scenedesmus quadricauda*):   5.4 mg/l   (1900)
     protozoa (*Entosiphon sulcatum*):   35   mg/l   (1900)
     protozoa (*Uronema parduczi Chatton-Lwoff*):   57   mg/l   (1901)
   —Fish:
     goldfish: 24 hr LD_{50}: 23 mg/l modified ASTM D 1345   (277)
     Lepomis macrochirus: static bioassay in fresh water at 23°C, mild aeration
       applied after 24 hr:

material added ppm	% survival after				best fit 96 hr LC_{50} ppm
	24 hr	48 hr	72 hr	96 hr	
56	0	—	—	—	
42	50	0	—	—	
37	100	90	80	60	35
32	100	90	80	70	
10	100	100	75	75	

   Menidia beryllina: static bioassay in synthetic seawater at 23°C: mild aeration
     applied after 24 hr:

material added ppm	% survival after				best fit 96 hr LC_{50} ppm
	24 hr	48 hr	72 hr	96 hr	
32	100	30	0	—	
18	100	90	70	50	18
10	90	90	90	90	(352)

     harlequin fish: 48 hr LC_{50}: 36 ppm (static and flow through)   (358)
     sheephead minnow: 96 hr TLm: 11.8 ppm   (359)
   —Mammalia: rat: single oral dose LD_{50}: 0.04 g/kg
              guinea pig: single oral dose LD_{50}: 0.178 g/kg
              mouse: single oral dose LD_{50}: 0.238 g/kg

mouse: inhalation: 0/30: 2,370 ppm
20/30: 8,300 ppm
rat: inhalation: LC_{50}: 250 ppm, 8 hr
500 ppm, 4 hr
guinea pig: inhalation: LC_{50}: 561 ppm, 4 hr
rabbit: inhalation: LC_{50}: 445 ppm, 4 hr (211)
mice: RD_{50} (respiratory rate): 687 ppm (1330)
—Man: nose and eye irritation: 100 ppm (211)
after a comprehensive review of the literature, NIOSH concluded that risks from exposure to epichlorohydrin may include carcinogenesis, mutagenesis, and sterility, as well as damage to the kidneys, liver, respiratory tract and skin
(1410)
transient burning of the eyes and nasal passages at 20 ppm (1416)

epihydric alcohol *see* glycidol

EPN (O-ethyl-O-*p*-nitrophenylphenylphosphonothioate; ethyl-*p*-nitrophenyl-thionobenzenephosphate)

Uses: insecticide; acaricide
A. PROPERTIES: light yellow crystals; m.p. 36°C; sp.gr. 1.5978 at 30°C; v.p. 3×10^{-4} mm at 100°C; m.w. 323.2
D. BIOLOGICAL EFFECTS:
—Crustaceans:
Gammarus lacustris: 96 hr, LC_{50}: 15 µg/l (2124)
Gammarus fasciatus: 96 hr, LC_{50}: 7 µg/l (2126)
Palaemonetes kadiakensis: 96 hr, LC_{50}: 0.56 µg/l (2126)
—Fish:
Pimephales promelas: 96 hr, LC_{50}: 110000 µg/l (2130)
Lepomis macrochirus: 96 hr, LC_{50}: 100 µg/l (2128)
—Mammals:
acute oral LD_{50} (rat): 9–45 mg/kg (1855)
(mouse): 43 mg/kg (1854)
(dogs): 20–45 mg/kg
acute dermal LD_{50} (rats): 25–230 mg/kg (1855)

1,2-epoxy-3-allyloxypropane *see* allylglycidylether

1,2-epoxyethane *see* ethyleneoxide

1,2-epoxyethylbenzene *see* styreneoxide

1,2-epoxy-3-phenoxypropane *see* phenylglycidylether

1,2-epoxypropane *see* propeneoxide

2,3-epoxy-1-propanol *see* glycidol

9,10-epoxystearic acid

$$CH_3(CH_2)_7\underset{\underset{O}{\diagdown\diagup}}{CHCH}(CH_2)_7COOH$$

Contaminant in pulp mill effluents (*cis*-isomer) (146)
D. BIOLOGICAL EFFECTS:
  —Fish:
    rainbow trout: 96 hr LC_{50} (S): 1.5 mg/l (949)
  —Mammals:
    male Sprague-Dawley rats: epoxystearic acid was rapidly absorbed after oral administration and well distributed in all tissues, mainly in the fat and liver of rats; since this compound was metabolized to CO_2 and was removed in the expired air, little or no bioaccumulation could be expected (1461)

eptam (S-ethyldipropylthiocarbamate; EPTC)

$$CH_3-CH_2-S-\overset{\overset{O}{\|}}{C}-N(CH_2CH_2CH_3)_2$$

Use: selective herbicide
  eptam 6E: emulsifiable liquid containing 6 lbs eptam/U.S. gal.
  eptam 2.3G: granular formulation containing 2.3 wt % eptam
  eptam 10G: granular formulation containing 10 wt % eptam
A. PROPERTIES: m.w. 189.32; solub. 370 ppm; sp.gr. 0.966 at 20°C; v.p. 0.035 mm at 25°C
C. WATER POLLUTION FACTORS:
  —Aquatic reactions:
    75–100% disappearance from soils: 4 weeks (1815)
D. BIOLOGICAL EFFECTS:
  —Crustacean:
    Gammarus fasciatus: 96 hr LC_{50}: 23 mg/l (2125)
    brown shrimp (*P. aztecus*): 48 hr LC_{50}: 0.63 ppm (techn. eptam)
    blue crab (*C. sapidus*): 24 hr LC_{50}: >20 ppm (techn. eptam) (494)
  —Molluscs:
    commercial oyster (*C. virginica*): 96 hr EC_{50} (shell growth inhibition): ca. 5 ppm (techn. eptam) (494)
  —Fish:
    killifish (*F. similis*): 48 hr LC_{50}: >10 ppm (techn. eptam)
    bluegill sunfish (*L. macrochirus*): 96 hr LC_{50}: 27 ppm (techn. eptam)
    rainbow trout (*S. gairdneri*): 96 hr LC_{50}: 19 ppm (techn. eptam) (494)
    bobwhite quail: LC_{50} (7 day feed treatment): 20,000 ppm (494)

—Mammals:
    technical eptam: rat: acute oral LD_{50}: 1,642 mg/kg
                       mouse: acute oral LD_{50}: 3,160 mg/kg
    eptam 6E: rat: acute oral LD_{50}: 1,367 mg/kg
    technical eptam: rabbits: acute dermal LD_{50}: ca. 10,000 mg/kg
    eptam 6E: rabbits: acute dermal LD_{50}: >4,640 mg/kg
    technical eptam: rabbits: acute ocular irritancy: moderate irritation, with iritis and corneal dullness, clearing within 4 days
    eptam 6E: rabbits: moderate irritant as technical eptam, clearing after 4 days
    eptam 6E: rats: acute inhalation 1 hr LC_{50}: >31.5 mg/l   (494)

erithane
  Use: fungicide; active ingredient: triphenyltin hydroxide (20%)   (311)
D. BIOLOGICAL EFFECTS:
  —Harlequin fish (*Rasbora heteromorpha*):

mg/l	24 hr	48 hr	96 hr	3 m (extrap.)	
LC_{10} (F)	0.25	0.16	0.035		
LC_{50} (F)	0.36	0.23	0.07	0.004	(331)

ERL 0810 *see n*-butylglycidylether

cis-erucic acid (*cis*-13-docosenoic acid)
$CH_3(CH_2)_7CHCH(CH_2)_{11}COOH$
A. PROPERTIES: colorless needles; m.w. 338.56; m.p. 33.5°C; b.p. 281°C at 33 mm; sp.gr. 0.860 at 55/4°C
C. WATER POLLUTION FACTORS:
  —Waste water treatment:
    A.S.: after 6 hr: 4.2% of ThOD
                 12 hr: 5.8% of ThOD
                 24 hr: 11.0% of ThOD   (89)

erythrene *see* 1,3-butadiene

erythrite *see* erythritol

erythritol (tetrahydroxybutane; erythrite)
$CH_2OH-CHOH-CHOH-CH_2OH$
  Use: mfg. of erythrityl tetranitrate
A. PROPERTIES: white sweet crystals; m.p. 121–122°C; sp.gr. 1.45; b.p. 329–331°C $\log P_{oct}$ -2.53 (calculated)
C. WATER POLLUTION FACTORS:
  —*Manmade sources*: in primary domestic sewage plant effluent: 0.005 mg/l (517)

ethanal *see* acetaldehyde

ethanamide *see* acetamide

ethane (bimethyl; methylmethane; dimethyl)
CH_3CH_3
A. PROPERTIES: colorless gas; m.w. 30.07; m.p. $-172/-183°C$; b.p. $-89°C$; v.p. 38.5 atm at 20°C, 48 atm at 30°C; v.d. 1.04; sp.gr. 0.561 at $-100°C$; solub. 60.4 mg/l at 20°C; THC 372.8 kcal/mole, LHC 341 kcal/mole
B. AIR POLLUTION FACTORS: 1 mg/cu m = 0.80 ppm, 1 ppm = 1.25 mg/cu m
   —Odor: T.O.C.: 1900 mg/cu m = 1520 ppm   (307)
                 150 ppm   (279)
                 recognition: 899,000 mg/cu m   (761)
                 25,000 mg/cu m   (737)
         O.I. at 20°C = 25,300   (316)
   —Atmospheric reactions: reactivity: NO ox.: ranking: 0.03   (63)
      estimated lifetime under photochemical smog conditions in S.E. England: 137 hr
                                                                      (1699; 1703)
   —Manmade sources:
     diesel engine: 1.8% of emitted hydrocarbons   (72)
     reciprocating gasoline engine: 2.0% of emitted hydrocarbons   (78)
     rotary gasoline engine: 1.3% or emitted hydrocarbons   (78)
     expected glc's in USA urban air: range: 0.05–0.50 ppm   (102)
     in flue gas of municipal incinerator: <0.4–0.5 ppm   (156)
     in gasoline engine exhaust: 62-car survey: 1.8 vol.% of total exhaust H.C.'s (391)
                                engine variable study: 2.3 vol.% of total exhaust H.C.'s (393)
   —Sampling and analysis:
     nondispersive I.R.: min. full scale: 50 ppm   (55)
C. WATER POLLUTION FACTORS:
   —BOD_{35}^{25}: 2.45   (62)
   —ThOD: 3.73   (62)
D. BIOLOGICAL EFFECTS:
   —Man: no effect level: <50,000 ppm   (211)

1,2-ethanediamine *see* ethylenediamine

ethanedioic acid *see* oxalic acid

1,2-ethanediol *see* ethyleneglycol

ethanenitrile *see* acetonitrile

ethanethioacetamide *see* thioacetamide

ethanethiol *see* ethylmercaptan

ethanoic acid *see* acetic acid

ethanoic anhydride *see* acetic anhydride

ethanol (ethylalcohol; methylcarbinol)
C_2H_5OH

ETHANOL

Manufacturing source: alcohol, whiskey, and gin mfg.; organic chemical industry; wood products industry (347)

Users and formulation: mfg. acetaldehyde, acetic acid, ethylacetate, ethylchloride, ethylether, butadiene, ethylene dibromide, pharmaceuticals, plastics and plasticizers, lacquers, perfumes, cosmetics, rubber, aerosols, mouthwash products, alcoholic beverages, soaps and cleaning preparations, solvent, dyes, explosives (347)

A. PROPERTIES: colorless liquid; m.w. 46.07; m.p. -114/-117°C; b.p. 78.4°C; v.p. 43.9 mm at 20°C, 50 mm at 25°C, 75 mm at 30°C; v.d. 1.6; sp.gr. 0.789 at 20/4°C; THC 326.5 kcal/mole, LHC 305 kcal/mole: sat. conc. 105 g/cu m at 20°C, 182 g/cu m at 30°C; log P_{oct} -0.32

B. AIR POLLUTION FACTORS: 1 mg/cu m = 0.52 ppm, 1 ppm = 1.92 mg/cu m
 —Odor: characteristic: quality: sweet

odor thresholds — mg/cu m

10^{-7} 10^{-6} 10^{-5} 10^{-4} 10^{-3} 10^{-2} 10^{-1} 1 10 10^2 10^3 10^4

detection

recognition

not specified

(73; 210; 278; 279; 291; 307; 610; 643; 671; 678; 704; 709; 761; 773; 776; 788; 804)

—Manmade sources:
  in gasoline exhaust: 0.1 to 0.6 ppm (195)
  in workrooms: concentrations up to 5,000 ppm have been reported (211)
  from whiskey fermentation vats: average: 182.2 g/cu m grain input (47)

—Control methods:
  catalytic combustion: platinized ceramic honey comb catalyst; ignition temp; 160°C, inlet temp. for 90% conversion: 250-300°C (91)

—Sampling and analysis:
  I.R. spectrometry: det. lim.: 0.04 ppm (56)
  non dispersive I.R.: min. full scale: 250 ppm (55)
  photometry: transparent material (53)
  detector tubes: UNICO: 400 ppm (59)
  AUER: 100 ppm (57)
  DRAGER: 100 ppm (58)

C. WATER POLLUTION FACTORS:

references	(79)	(23)	(23)*	(220)
—BOD_5:	44.2% ThOD	74% ThOD	45% ThOD	37% ThOD
10 days:	65.4% ThOD	74% ThOD	68% ThOD	
15 days:	71.2% ThOD	95% ThOD	72% ThOD	
20 days:	78.9% ThOD	84% ThOD	75% ThOD	
30 days:	78.9% ThOD			
40 days:	78.2% ThOD			
50 days:	77.4% ThOD			

*dilution water is synthetic sea water

ETHANOLAMINE

—BOD_5^{20}: 1.58 at 10 mg/l, unadapted sew., no lag period
—BOD_{20}^{20}: 1.78 at 10 mg/l, unadapted sew. (554)
—BOD_5: 63% of ThOD (274)
    0.93; 1.46; 1.67 Warburg, acclimated sewage (41)
    0.93; to 1.67   (255; 285; 282; 281; 280; 271; 269; 260; 250)
—BOD_{20}: 1.80 (30)
—COD: 90% of ThOD (0.05 n $K_2Cr_2O_7$) (274)
    1.99; 2.0; 2.11 (23, 36, 41)
    97% of ThOD (220)
—$KMnO_4$: 0.065 (30)
    acid: 6% of ThOD; alkaline: 3% of ThOD (220)
—TOD: 2.08 (26)
—TOC: 100% of ThOD (220)
—ThOD: 2.10 (23)
—Manmade sources:
    in year old leachate of artificial sanitary landfill: 0.6 g/l (1720)
—Impact on biodegradation:
    ~50% inhibition of NH_3 oxidation in *Nitrosomonas* at 4100 mg/l (407)

odor thresholds   mg/kg water

	10^{-7}	10^{-6}	10^{-5}	10^{-4}	10^{-3}	10^{-2}	10^{-1}	1	10	10^2	10^3	10^4
detection										●● ●		
recognition												
not specified										●●		

(299; 889; 894; 896; 907; 908)

—Waste water treatment:

                                                      infl. mg/l   effl. mg/l
    anaerobic lagoon: 13 lb COD/day/1,000 cu ft: 80   35
                                      22   270   120
                                      48   270   130   (37)
    air stripping constant: K = 0.302 days^{-1} at 2,220 mg/l (82)
    A.S.: after 6 hr: 12.9% of ThOD
                   12 hr: 25.9% of ThOD
                   24 hr: 37.3% of ThOD (88)

methods	temp °C	days observed	feed mg/l	days acclim.	% theor. oxidation	% removed
AS, Resp, BOD	20	1–50	1,000	<1	24	99
AS, W	20	1	500	24	51	
NFG, BOD	20	1–10	200–1,000	365 + P		60

(93)

    reverse osmosis: 38% rejection from a 0.01 M solution (221)
    anaerobic sludge digestion is inhibited at 1.6 g/l (30)
—Sampling and analysis:
    GC: det. lim.: 1 mg/l (208)

D. BIOLOGICAL EFFECTS:
  −Toxicity threshold (cell multiplication inhibition test):
bacteria (*Pseudomonas putida*):	6500 mg/l	(1900)
algae (*Microcystis aeruginosa*):	1450 mg/l	(329)
green algae (*Scenedesmus quadricauda*):	5000 mg/l	(1900)
protozoa (*Entosiphon sulcatum*):	65 mg/l	(1900)
protozoa (*Uronema parduczi Chatton-Lwoff*):	6120 mg/l	(1901)

  −Fish;
  creek chub: LD_0:   7,000 mg/l, 24 hr in Detroit river water
  $\qquad\qquad$ LD_{100}: 9,000 mg/l, 24 hr in Detroit river water $\qquad$ (243)
  fingerling trout: 24 hr LC_{50}: 11,200 mg/l (flow through system) $\qquad$ (592)
  fathead minnows: static bioassay in Lake Superior water at 18-22°C: LC_{50} (1; 24; 48; 72; 96 hr): >18,000; >18,000; 13,480; 13,480; 13,480 mg/l (350)
  creek chub (*Semotilus atromaculatus*): 24 hr LC_{50}: >7,000 mg/l $\qquad$ (593)
  guppy (*Poecilia reticulata*): 7 d LC_{50}: 11,050 ppm $\qquad$ (1833)

  −Mammalia:
  rat: single oral LD_{50}: 13.7 ml/kg
  rabbit: single oral LD_{50}: 12.5 ml/kg
  guinea pig: inhalation: no signs of intoxication: 6,400 ppm, 8 hr; 3,000 ppm, 64 × 4 hr
  rat: inhalation: no signs of intoxication: 10,750 ppm, 0.5 hr; 3,260 ppm, 6 hr
  $\qquad$ (211)

  −Man:
  severe toxic effects: 8,000 ppm, 60 min
  symptoms of illness: 2,000 ppm
  unsatisfactory: >1,000 ppm $\qquad$ (185)
  headache and slight numbness after 33 min for subjects unaccustomed to alcohol: after 39 min inhalation of 1,380 ppm
  slight headache after 20 min for subjects accustomed to alcohol, after 120 min inhalation of 5,030 ppm $\qquad$ (211)

  −Carcinogenicity: none
  −Mutagenicity in the *Salmonella* test: none
    <0,003 revertant colonies/nmol
    70 revertant colonies at 10^4 µg/plate $\qquad$ (1883)

ethanolamine (2-aminoethol; β-aminoethylalcohol; ethylolamine; β-hydroxyethylamine; monoethanolamine; MEA)
$NH_2CH_2CH_2OH$

  Uses: scrubbing acid gases from gas streams; non-ionic detergent used in dry cleaning, wool treatment, emulsion paints, polishes, agricultural sprays; chemical intermediates; pharmaceuticals; corrosion inhibitor; rubber accelerator

A. PROPERTIES: colorless liquid; m.w. 61.08; m.p. 11°C; b.p. 172°C; v.p. 0.4 mm at 20°C, 6 mm at 60°C; v.d. 2.1; sp.gr. 1.02 at 20/4°C; $\log P_{oct}$ −1.31

B. AIR POLLUTION FACTORS: 1 mg/cu m = 0.39 ppm, 1 ppm = 2.54 mg/cu m
  −Odor: characteristic: quality: mildly ammoniacal
T.O.C.: 50% detection: 3-4 ppm	(211)
$\qquad$ detection: 7.5-10 mg/cu m	(860)
O.I. at 20°C: 130	(316)

620 ETHENE

- Sampling and analysis:
  photometry: min. full scale: 650 ppm (53)
C. WATER POLLUTION FACTORS:
  - BOD_5: 0.93 NEN 3235-5.4 (277)
    0.8–1.1 std. dil. at 2.2 mg/l (27)
    0.78 std. dil. sew. (41)
    0.83 (36)
    0% of ThOD
    10 days: 58.4%
    15 days: 61.2%
    20 days: 64.0%
    30 days: 66.7%
    50 days: 75.6% of ThOD (79)
  - $BOD_{10}^{20°C}$: 58.4% ThOD at 2.5 mg/l in mineralized dilution water with settled sewage seed (405)
  - BOD_5^{20}: 0.83 at 10 mg/l, unadapted sew.; lag period: 2 days
  - BOD_{20}^{20}: 1.00 at 10 mg/l, unadapted sew. (554)
  - COD: 1.27 (36)
    1.28 NEN 3235-5.3 (277)
  - ThOD: 2.49 (36)
  - Inhibition of biodegradation:
    ~50% inhibition of NH_3 oxidation in *Nitrosomonas* at 12,200 mg/l (407)
    NH_3 oxidation by *Nitrosomonas* sp. at 100 mg/l: 16% inhibition (390)
  - Waste water treatment:
    A.C.: adsorbability: 0.015 g/g C, 7.2% reduction, infl.: 1,012 mg/l, effl.: 939 mg/l (32)
    NFG, BOD, 20°C, 1–10 days observed, feed: 200–600 mg/l, acclimated 365+, 63% removed
    NFG, BOD, 20°C, 1–10 days observed, feed: 1,000 mg/l, acclimated 365 + 7% removed (93)
D. BIOLOGICAL EFFECTS:
  - Toxicity threshold (cell multiplication inhibition test):
    bacteria (*Pseudomonas putida*): 6300 mg/l (1900)
    algae (*Microcystis aeruginosa*): 1.6 mg/l (329)
    green algae (*Scenedesmus quadricauda*): 0.75 mg/l (1900)
    protozoa (*Entosiphon sulcatum*): 300 mg/l (1900)
    protozoa (*Uronema parduczi Chatton-Lwoff*): 2945 mg/l (1901)
  - Fish: goldfish: LD_{50} (24 hr): 190 mg/l at pH 10.1
    LD_{50} (96 hr): 170 mg/l at pH 10.1
    LD_{50} (24 hr): >5000 mg/l at pH 7—modified ASTM D 1345 (277)
  - Mammalia:
    rat: acute oral LD_{50}: 2.74 g/kg (211)
    inhalation: 1/61: 104 ppm, 5 × 7 hr
    guinea pig: inhalation: some deaths: 100 ppm, 24–30 days (211)

ethene *see* ethylene

ethenedichloride *see* ethylenedichloride

etheneoxide *see* ethyleneoxide

ethenylbenzene *see* styrene

ethyenylethanoate *see* vinylacetate

ether *see* ethylether

ethinyloestradiol (ethynylestradiol; 19-nor-17α-pregna-1,3,5(10)-trien-20-yne-3,17-diol)
    Use: medicine, an estrogen (female sex hormone)
A. PROPERTIES: m.p. 142–146°C; solub. 5681–5944 mg/l
C. WATER POLLUTION FACTORS:
    —Degradation in drinking water after 10 days: 77.6% remained unchanged   (413)
    —Water quality:
      in drinking water in South West Germany: avg.: 0.69–3.18 ng/l
                                                    range: 0–22.50 ng/l   (413)

ethion (O,O,O,O-tetraethyl-S,S-methylenebisphosphorodithioate; diethion; *bis*(S-diethoxyphosphinothioyl)mercaptomethane)

$$(C_2H_5O)_2-\overset{\overset{S}{\|}}{P}-S-CH_2-S-\overset{\overset{S}{\|}}{P}(OC_2H_5)_2$$

A. PROPERTIES: sp.gr. 1.21–1.23 at 20°C; m.p. −12 to −15°C; v.p. 1.5×10^{-6} mm at 25°C
C. WATER POLLUTION FACTORS:
    —Aquatic reactions:
    persistence in soil at 10 ppm initial concentration:

	weeks incubation to	
	50% remaining	5% remaining
sterile sandy loam	>24	
sterile organic soil	>24	
non-sterile sandy loam	7	>24
non-sterile organic soil	8	>24 (1433)

    persistence in river water in a sealed glass jar under sunlight and artificial fluorescent light—initial conc. 10 µg/l:

		% of original compound found			
after	1 hr	1 wk	2 wk	4 wk	8 wk
	100	90	75	50	50 (1309)

D. BIOLOGICAL EFFECTS:
    —Crustaceans:
    Gammarus lacustris: 96 hr, LC_{50}: 1.8 µg/l   (2124)
    Gammarus fasciatus: 96 hr, LC_{50}: 9.4 µg/l   (2126)
    Palaemonetes kadiakensis: 96 hr, LC_{50}: 5.7 µg/l   (2126)
    —Insects:
    Pteronarcys californica: 96 hr, LC_{50}: 2.8 µg/l   (2128)

—Fish:

Lepomis macrochirus:	96 hr, LC_{50}:	220 µg/l	(2137)
Micropterus salmoides:	96 hr, LC_{50}:	150 µg/l	(2137)
Salmo gairdneri:	96 hr, LC_{50}:	560 µg/l	(2137)
Salmo clarkii:	96 hr, LC_{50}:	720 µg/l	(2137)
Ictalurus punctatus:	96 hr, LC_{50}:	7500 µg/l	(2137)

—Mammals:
acute oral LD_{50} (rat): 208 mg/kg (pure)
96 mg/kg (technical) (1854)
acute dermal LD_{50} (rabbit): 915 mg/kg
in diet: female albino rats survived a 28 day feeding test at 300 ppm without any effect on growth rate though there was evidence of cholinesterase inhibition at 10 ppm (1855)

ethoquad
Trademark for a series of polyethoxylated quaternary ammonium chlorides derived from long-chain amines
Uses: antistatic agents; electroplating bath additives
C. WATER POLLUTION FACTORS:
—Biodegradation:
at 18 mg/l (inoculum = sewage): lag period = 15 days, then biodegradation rate = 1 mg/l/day; residual conc. after 30 days: 3 mg/l: 85% degradation (488)
D. BIOLOGICAL EFFECTS:
—*Staphylococcus aureus*: at 20 mg/l no growth (488)

2(β-ethoxy-ethoxy)ethanol *see* diethyleneglycol mono-ethylether

1-ethoxy-2-(β-ethoxy-ethoxy)ethane *see* diethyleneglycol-di-ethylether

ethoxytriglycol
$C_2H_5O(C_2H_4O)_3H$
A. PROPERTIES: colorless liquid; m.w. 178.2; m.p. −18.7°C; b.p. 255.4°C; v.p. <0.01 mm at 20°C; sp.gr. 1.02 at 20/20°C
C. WATER POLLUTION FACTORS:
—Waste water treatment:
A.C.: adsorbability: 0.139 g/g C; 69.7% reduction, infl.: 1,000 mg/l, effl.: 303 mg/l (32)

ethoxol *see* ethyleneglycolmonoethylether

ethoxyethane *see* ethylether

2-ethoxyethanol *see* ethyleneglycolmonoethylether

2-ethoxyethylacetate *see* ethyleneglycolmonoethyletheracetate

ethylacetate (acetic ether; ethylethanoate)
$CH_3COOC_2H_5$

A. PROPERTIES: colorless liquid; m.w. 88.1; m.p. $-83.6/-82°C$; b.p. $77°C$; v.p. 72.8 mm at $20°C$, 115 mm at $30°C$; v.d. 3.04; sp.gr. 0.901 at $20/4°C$; solub. 79,000 mg/l at $20°C$, 86,000 mg/l at $20°C$, 74,000 mg/l at $35°C$; THC 539 kcal/mole, LHC 505 kcal/mole; sat.conc. 336 g/cu m at $20°C$, 533 g/cu m at $30°C$; $\log P_{oct}$ 0.66/0.73

B. AIR POLLUTION FACTORS: 1 mg/cu m = 0.27 ppm, 1 ppm = 3.66 mg/cu m
—Odor: characteristic: quality: sweet; ester
hedonic tone: pleasant

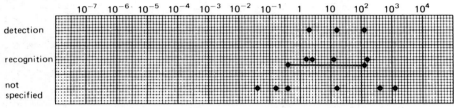

(57; 73; 210; 278; 307; 602; 610; 665; 704; 708; 709; 737)

USSR:
human odor perception: 0.6 mg/cu m, non perception: 0.5 mg/cu m (19)
human reflex response: no response: 0.18 mg/cu m, adverse resp.: 0.3 mg/cu m (170)

—Manmade sources:
from whiskey fermentation vats: average: 0.593 g/cu m of grain input (47)
—Control methods: (83)
catalytic combustion: platinized ceramic honeycomb catalyst: ignition temp.: $275°C$, inlet temp. for 90% conversion: $400-450°C$ (91)

C. WATER POLLUTION FACTORS:
—BOD_4: 0.76-1.14 at 1000 ppm, Sierp, sewage (280)
—BOD_5: 16% of ThOD (274)
0.1-0.6 std. dil. at 1,000 mg/l (27)
0.293 (30)
0.29-0.86 std. dil. sew. (41)
1.24 (41)
36% of ThOD
—BOD_{10}: 0.5 std. dil. sew. (256)
10 days: 50.4% of ThOD
15 days: 51.6% of ThOD
20 days: 53.8% of ThOD (79)
—BOD_5^{20}: 1.24 at 10 mg/l, unadapted sew.: lag period: 1 day
—BOD_{20}^{20}: 1.43 at 10 mg/l, unadapted sew. (554)
—COD: 1.54 (36)
83% of ThOD (0.05 n $K_2Cr_2O_7$) (274)
—$KMnO_4$: 0.077 (36)
1% of ThOD (0.01 n $KMnO_4$) (274)

ETHYLACETIC ACID

—ThOD: 1.82 (36)
—Inhibition of biodegradation:
 ~50% inhibition of NH_3 oxidation in *Nitrosomonas* at 18,000 mg/l (407)

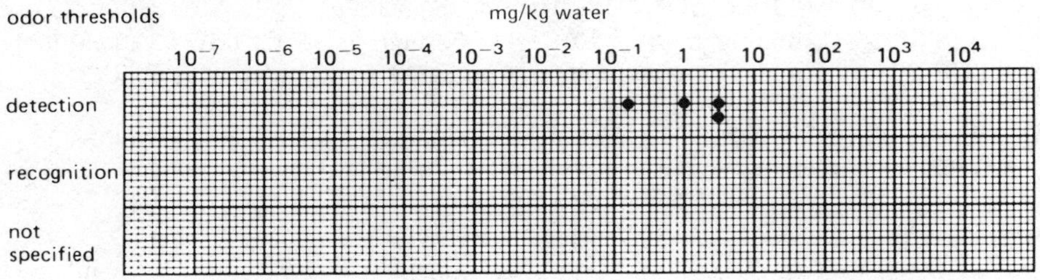

odor thresholds — mg/kg water

[Chart showing detection points between 10^{-1} and 10 on a scale from 10^{-7} to 10^4 mg/kg water; rows labeled detection, recognition, not specified]

(889; 894; 908; 911)

—Waste water treatment:
 A.C.: adsorbability: 0.100 g/g C; 50.5% reduction, infl.: 1,000 mg/l effl. 495 mg/l (32)
 reverse osmosis: 91.1% rejection from a 0.01M solution (221)
—Toxicity threshold (cell multiplication inhibition test):
 bacteria (*Pseudomonas putida*): 650 mg/l (1900)
 algae (*Microcystis aeruginosa*): 550 mg/l (329)
 green algae (*Scenedesmus quadricauda*): 15 mg/l (1900)
 protozoa (*Entosiphon sulcatum*): 202 mg/l (1900)
 protozoa (*Uronema parduczi Chatton-Lwoff*): 1620 mg/l (1901)
—Amphibian:
 mexican axolotl (3-4 w after hatching): 48 hr LC_{50}: 150 mg/l
 clawed toad (3-4 w after hatching): 48 hr LC_{50}: 180 mg/l (1823)
—Man: severe toxic effects: 2,000 ppm, 60 min
 symptoms of illness: 800 ppm
 unsatisfactory: >400 ppm (185)

ethylacetic acid *see n*-butyric acid

ethylacetoacetate (diacetic ester; acetoacetic ester)
$CH_3COCH_2COOC_2H_5$ keto form
$CH_3C(OH)CHCOOC_2H_5$ enol form
 At room temperature a mixture of about 93% keto form and 7% enol form
 Uses: organic synthesis; antipyrene; lacquers; dopes; plastics; manufacture of dyes; pharmaceuticals; flavoring
A. PROPERTIES: colorless liquid, fruity odor; sp.gr. 1.02 at 20/4°C; m.p. (enol) -80°C, (keto) -39°C; b.p. 180-181°C, v.p. 0.8 mm at 20°C
D. BIOLOGICAL EFFECTS:
—Toxicity threshold (cell multiplication inhibition test):
 bacteria (*Pseudomonas putida*): 33 mg/l
 green algae (*Scenedesmus quadricauda*): 7.6 mg/l
 protozoa (*Entosiphon sulcatum*): 391 mg/l (1900)

ethylacetone *see* 2-pentanone

ethylacrylate (ethylpropenoate)
$CH_2CHCOOC_2H_5$
A. PROPERTIES: m.w. 100.11; m.p. $<-75°C$; b.p. $100°C$; v.p. 29 mm at $20°C$, 49 mm at $30°C$; v.d. 3.5; sp.gr. 0.924 at $20/4°C$; solub. 20,000 mg/l; sat.conc. 158 g/cu m at $20°C$, 258 g/cu m at $30°C$
B. AIR POLLUTION FACTORS: 1 mg/cu m = 0.24 ppm, 1 ppm = 4.46 mg/cu m
   —Odor: characteristic: quality: sour, pungent, hot plastic
   hedonic tone: unpleasant

   T.O.C.: $PIT_{50\%}$: 0.1 ppb
   $PIT_{100\%}$: 0.47 ppb (2)
   T.O.C. = 0.0004 mg/cu m = 0.96 ppb (307)
   asb. perc. lim.: 0.2 ppb
   50% recogn.: 0.3 ppb
   100% recogn.: 0.36 ppb
   O.I. 100% recogn.: 113,000,000 (19)
   —Sampling and analysis:
   wet scrubber + adsorption column: 8–10% NaOH + 50 ppm methylether: retention time: 0.5 min
   removal efficiency: $<90\%$ when influent <400 ppm ethylacrylate
   97% when influent = 2,000 ppm ethylacrylate
   $>99\%$ when influent $>5,000$ ppm ethylacrylate (207)

C. WATER POLLUTION FACTORS:

	(1)*	(2)*	(3)*	
—BOD_5:	28% of ThOD	66% of ThOD	11% of ThOD	
10 days:	32% of ThOD	74% of ThOD	53% of ThOD	
15 days:	32% of ThOD	76% of ThOD	53% of ThOD	
20 days:	35% of ThOD	79% of ThOD	53% of ThOD	(23)

(1)*: non acclimated, fresh dilute water
(2)*: acclimated, fresh dilution water
(3)*: non acclimated, salt dilution water
—COD: 1.71 (23)
—ThOD: 1.92 (23)
—Reduction of amenities:
0.6 mg/l can cause tainting of the flesh of fish and other aquatic organisms (34)
odor threshold: average: 0.0067 mg/l
range: 0.0018–0.014 mg/l (294, 97)
organoleptic limit: 0.005 mg/l (181)
—Waste water treatment:
A.C.: adsorbability: 0.157 g/g C; 77.7% reduction, infl.: 1,015 mg/l, effl.: 226 mg/l (32)

D. BIOLOGICAL EFFECTS:
—Bacteria: *Pseudomonas putida*: inhibition of cell multiplication starts at 270 mg/l (329)
—Algae: *Microcystis aeruginosa*: inhibition of cell multiplication starts at 14 mg/l (329)
—Mammalia: rat: acute oral LD_{50}: 1.0 g/kg
rabbit: acute oral LD_{50}: 0.4 g/kg

626 ETHYLALDEHYDE

rat: inhalation:	LC_{50}: <1,000 ppm, 4 hr	
	5/6: 2,000 ppm, 4 hr	
	0/6: 1,000 ppm, 4 hr	
	12/18: 540 ppm, 19 days	
	2/29: 70 ppm, 30 hr	(211)

ethylaldehyde *see* acetaldehyde

ethylamine (monoethylamine; aminoethane)
$C_2H_5NH_2$
   Manufacturing source: organic chemical industry. (347)
   Users and formulation: resin mfg.; stabilizer for rubber latex; intermediate for dye mfg.; pharmaceutical mfg.; solvent in petroleum and vegetable oil refining; raw material for mfg. amides; plasticizer; detergents mfg.; organic synthesis. (347)
A. PROPERTIES: colorless liquid; m.w. 45.08; m.p. $-81/-83°C$; b.p. $17°C$; v.p. 400 mm at $2°C$, 1.2 atm at $20°C$, 1.7 atm at $30°C$; v.d. 1.56; sp.gr. 0.71 at $0/4°C$; log P_{oct} $-0.27/-0.08$ (calculated)
B. AIR POLLUTION FACTORS: 1 mg/cu m = 0.53 ppm, 1 ppm = 1.87 mg/cu m
   —Odor: characteristic: quality: sharp, ammoniacal
                   hedonic tone: unpleasant

T.O.C.: 0.05 mg/cu m = 26 ppb	(291)(846)
abs. perc. lim.: 0.27 ppm	
50% recogn.: 0.83 ppm	
100% recogn.: 0.83 ppm	
O.I. 100% recogn.: 622,891	(19)

C. WATER POLLUTION FACTORS:

—Reduction of amenities: T.O.C. = 10 mg/l (detection)	(894)
organoleptic limit: 0.5 mg/l	(181)

D. BIOLOGICAL EFFECTS:
   —Toxicity threshold (cell multiplication inhibition test):

bacteria (*Pseudomonas putida*):	29 mg/l	(1900)
algae (*Microcystis aeruginosa*):	1.3 mg/l	(329)
green algae (*Scenedesmus quadricauda*):	2.3 mg/l	(1900)
protozoa (*Entosiphon sulcatum*):	45 mg/l	(1900)
protozoa (*Uronema parduczi Chatton-Lwoff*):	3500 mg/l	(1901)

   —Algae:
     Scenedesmus: no effect: 10 mg/l

—Arthropoda: *Daphnia*: no effect: 40 mg/l	(30)
—Fish: creek chub: LD_0: 30 mg/l; 24 hr in Detroit river water	
LD_{100}: 50 mg/l, 24 hr in Detroit river water	(243)

   —Mammalia:

rat: oral LD_{50}: 0.40 g/kg; 95% conf.: 0.29–0.56 g/kg	(226)
rat: acute oral LD_{50}: 0.4–0.8 g/kg (70% soln)	
inhalation: 2/6: 8,000 ppm, 4 hr	(211)

ethyl-*p*-aminobenzoate *see* benzocain

2-ethylamino-ethanol *see* N-ethylethanolamine

6-ethylamino-4-isopropylamino-2-methylthio-1,3,5-triazine *see* ametryn

2-(4-ethylamino-6-methylthio-s-triazin-2-ylamino)-2-methylpropionitrile *see* cyanatrine

ethyl-*sec*-amylketone (5-methyl-3-heptanone; ethyl-isoamylketone; octanone-3 $CH_3CH_2COCH_2CH(CH_3)CH_2CH_3$
  Natural sources: produced by *Streptomyces cinnamoneus*-like organisms and contributes to characteristic odors of actinomycete cultures (1315)
A. PROPERTIES: colorless liquid; m.w. 128.22; b.p. 160/162°C; v.p. 2.0 mm at 25°C; sp.gr. 0.85 at 0/4°C;
B. AIR POLLUTION FACTORS: 1 mg/cu m = 0.19 ppm, 1 ppm = 5.33 mg/cu m
  —Odor: T.O.C.: <5 ppm; 31 mg/cu m; 0.0013 mg/cu m   (211; 850; 607)
         0.043 mg/cu m = 8.2 ppb   (307)
         O.I. at 20°C = 660   (316)
C. WATER POLLUTION FACTORS:
  —COD: 1.94 = 67% of ThOD—ASTM procedure/NEN 3235-5.3   (277)
        2.61 = 91% of ThOD—modified "Shell" procedure   (272)
  —BOD: 2.20 NEN 3235-5.4   (277)
  —Waste water treatment:
    A.S.: after 6 hr: 1.1% of ThOD
                12 hr: 2.0% of ThOD
                24 hr: 3.4% of ThOD   (88)
D. BIOLOGICAL EFFECTS:
  —Toxicity threshold (cell multiplication inhibition test):
    bacteria (*Pseudomonas putida*):   25 mg/l   (1900)
    algae (*Microcystis aeruginosa*):   40 mg/l   (329)
    green algae (*Scenedesmus quadricauda*):   53 mg/l   (1900)
    protozoa (*Entosiphon sulcatum*):   256 mg/l   (1900)
    protozoa (*Uronema parduczi Chatton-Lwoff*): 65 mg/l   (1901)
  —Mammalia: guinea pig: single oral LD_{50}: 2.5 g/kg
             mouse: single oral LD_{50}: 3.8 g/kg
             rat: single oral LD_{50}: 3.5 g/kg
             mouse: inhalation: 3/6: 3,000 ppm, 4 hr
             rat: inhalation: 0/6: irritation: 3,000 ppm, 4 hr (211)
  —Man: mild irritation of the nose: 25 ppm
        headache: 100 ppm   (211)

o-ethylaniline (monoethylaniline)

[structure: benzene ring with NH$_2$ and CH$_2$CH$_3$ substituents in ortho position]

A. PROPERTIES: brown liquid; m.p. −44°C; sp.gr. 0.982 at 20°C

ethylanthracene

C. WATER POLLUTION FACTORS:
—BOD_5: 0.048 (n.s.i.) std. dil. sew. (282)

ethylanthracene
B. AIR POLLUTION FACTORS:
—*Manmade sources*: in coke oven emissions: 49-1474 μg/g of sample (n.s.i.) (960)

ethylbenzaldehyde
B. AIR POLLUTION FACTORS:
—Manmade sources: in gasoline engine exhaust: <0.1-0.2 ppm (195)

ethylbenzene (phenylethane)

Manufacturing source: petroleum refining: organic chemical industry (347)
Users and formulations: styrene mfg.; acetophenone mfg.; solvent; asphalt constituent; naphtha constituent (347)
Manmade sources: in gasoline (high octane number): 4.60 wt % (387)
A. PROPERTIES: colorless liquid; m.w. 106.17; m.p. -94.97°C; b.p. 136.2°C; v.p. 7 mm at 20°C, 12 mm at 30°C; v.d. 3.66; sp.gr. 0.867 at 20/4°C; solub. 140 mg/l at 15°C, 152 mg/l at 20°C, 206 mg/l at 15°C; sat. conc. 40 g/cu m at 20°C, 67 g/cu m at 30°C; log P_{oct} 3.15
B. AIR POLLUTION FACTORS: 1 mg/cu m = 0.23 ppm, 1 ppm = 4.35 mg/cu m
—Odor threshold: 2-2.6 mg/cu m (700)
               detection: 0.4 mg/cu m (727)
—Atmospheric reactions: R.C.R.: 1.24 (49)
               reactivity: HC conc.: 1 (63)
   estimated lifetime under photochemical smog conditions in S.E. England: 4.6 hr (1699; 1700)
—Ambient air quality:
   in urban air: 10-50 ppb (29)
   expected glc's in USA urban air: 2-20 ppb (102)
   downtown Los Angeles 1967: 10%ile: 1 ppb, average: 5 ppb, 90%ile: 9 ppb (64)
   in carpark: 500 μg/cu m
   at motorway: 400 μg/cu m
—glc's in The Netherlands:

tunnel Amsterdam 1973:	avg.:	3 ppb (n = 3)
tunnel Rotterdam 1974.10.2:	avg.:	6 ppb (n = 12)
	max.:	8 ppb
The Hague 1974.10.11:	avg.:	5 ppb (n = 12)
	max.:	14 ppb
Roelofarendsveen 1974.9.11:	avg.:	1 ppb (n = 12)
	max.:	1 ppb (1231)

ETHYLBENZENE 629

—glc's in Los Angeles 1966:   avg.:   0.006 ppm ($n = 136$)
                                highest value: 0.022 ppm   (1319)
—Manmade sources:
    in exhaust gas of reciprocating gasoline engine: 0.51% of HC's
    in exhaust gas of rotary gasoline engine: 1.67% of HC's   (78)
    in exhaust gas of diesel engine: 0.7% of HC's   (72)
—Sampling and analysis:
    photometry: min. full scale: 400 ppm   (53)

C. WATER POLLUTION FACTORS:
—$BOD_{35}^{25°C}$: 1.73 in seawater/inoculum: enrichment cultures of hydrocarbon
         oxidizing bacteria   (521)
—ThOD: 3.17   (521)
—Biodegradation:
    incubation with natural flora in the groundwater—in presence of the other components of high-octane gasoline (100 μl/l): biodegradation: 100% after 192 hr at 13°C (initial conc. 1.36 μl/l)   (956)
—Reduction of amenities:
    0.25 mg/l can cause tainting of the flesh of fish and other aquatic organisms (34)
    approx. conc. causing adverse taste in fish: 0.25 mg/l   (41)
—Odor thresholds: detection: 0.2 mg/kg   (886)
                                0.140 mg/l   (403)
—Water quality:
    Kanawha river water: 13.11.1963: raw; 11 ppb
                                                  sand filtered: 11 ppb
                        21.11.1963: not detectable   (159)
—Waste water treatment:
    A.S.: 27% theor. oxidation of 500 ppm ethylbenzene by phenol acclimated activated sludge after 12 hr aeration   (26)
    A.C.: adsorbability: 0.018 g/g C; 84.3% reduction, infl.: 115 mg/l, effl.: 18 mg/l   (32)

methods	temp °C	feed mg/l	days observed	days acclim.	theor. oxidation %
A.S., Sd, BOD	20		5	14+	3
A.S., W	20	50–100	$\frac{1}{4}$	14+	8 (93)

    Conventional municipal treatment: influent: 0.011 mg/l; effluent: 0.011 mg/l; 0% removal   (404)
—Sampling and analysis:
    photometry; min. full scale: 6×10^{-6} mole/l   (53)
    extraction efficiency of macro-reticular resins; sample flow: 20 ml/min; pH: 5.7; conc. 10 ppm: X AD-2: 60%
                                       X AD-7: 59%   (370)

D. BIOLOGICAL EFFECTS:
—Toxicity threshold (cell multiplication inhibition test):

bacteria (*Pseudomonas putida*):	12 mg/l	(1900)
algae (*Microcystis aeruginosa*):	33 mg/l	(329)
green algae (*Scenedesmus quadricauda*):	>160 mg/l	(1900)
protozoa (*Entosiphon sulcatum*):	140 mg/l	(1900)
protozoa (*Uronema parduczi Chatton-Lwoff*):	>110 mg/l	(1901)

630 ETHYLBUTANOATE

—Fish:
fatheads: soft water: TLm (25-96 hr): 48.5 mg/l
fatheads: hard water: TLm (25-96 hr): 42.3 mg/l
bluegills: soft water: TLm (25-96 hr): 35.1-32.0 mg/l
goldfish: soft water: TLm (25-96 hr): 94.4 mg/l
guppies: soft water: TLm (25-96 hr): 97.1 mg/l (158)
young Coho salmon: in artificial sea water at 8°C; mortality:
30/30 at 50 ppm after 24 hrs
2/30 (not significant) at 10 ppm after 24 up to 96 hrs (317)
—Mammals:
ingestion: rat: single oral LD_{50}: in the range of 3500 mg/kg (1546)

ethylbutanoate *see* ethylbutyrate

2-ethyl-1-butanol (3-methylolpentane; pseudohexylalcohol)
$CH_3CH_2(C_2H_5)CHCH_2OH$
A. PROPERTIES: colorless liquid; m.w. 102.2; m.p. -50°C; b.p. 150°C; v.p. 1.8 mm at 20°C; v.d. 3.54; sp.gr. 0.8328 at 20/20°C; solub. 4,300 mg/l at 20°C, 6,300 mg/l at 24°C;.
B. AIR POLLUTION FACTORS: 1 mg/cu m = 0.24 ppm, 1 ppm = 4.17 mg/cu m
—Odor: characteristic: quality: musty, sweet
hedonic tone: neutral
T.O.C.: abs. perc. lim.: 0.07 ppm
50% recogn.: 0.77 ppm
100% recogn.: 0.77 ppm
O.I. 100% recogn.: 1,701 (19)
C. WATER POLLUTION FACTORS:
—Waste water treatment:
A.C.: adsorbability: 0.17 g/g C; 85.5% reduction, infl.: 1,000 mg/l, effl.: 145 mg/l (32)
—Reduction of amenities: T.O.C. = 0.2 mg/l (299)(894)
D. BIOLOGICAL EFFECTS:
—Mammalia: rat: single oral LD_{50}: 1.85 g/kg
inhalation: no deaths: sat. vap.: 8 hr (211)

ethylbutyrate (ethylbutanoate)
$CH_3CH_2CH_2COOC_2H_5$
A. PROPERTIES: colorless liquid; m.w. 116.16; m.p. -93°C; b.p. 121°C; v.p. 11.3 mm at 20°C; v.d. 4.00; sp.gr. 0.88 at 20/4°C; solub. 6,800 mg/l at 25°C
B. AIR POLLUTION FACTORS: 1 mg/cu m = 0.193 ppm, 1 ppm = 5.184 mg/cu m
—Odor: T.O.C.: 0.039 mg/cu m = 7.5 ppb (307)
detection: 0.28 mg/cu m (840)
recognition: 0.13-0.23 mg/cu m (610)
O.I. at 20°C = 1,982,000 (316)
C. WATER POLLUTION FACTORS:
—Odor threshold: detection: 0.001 mg/kg (889)
—Waste water treatment:
ion exchange: adsorption on Amberlite X AD-2: 100% retention, infl.: 100 ppm, effl.: nil

D. BIOLOGICAL EFFECTS:
—Toxicity threshold (cell multiplication inhibition test):

bacteria (*Pseudomonas putida*):	140 mg/l	(1900)
algae (*Microcystis aeruginosa*):	700 mg/l	(329)
green algae (*Scenedesmus quadricauda*):	47 mg/l	(1900)
protozoa (*Entosiphon sulcatum*):	236 mg/l	(1900)
protozoa (*Uronema parduczi Chatton-Lwoff*):	916 mg/l	(1901)

ethylchloride (monochloroethane; hydrochloric ether; muriatic ether)
C_2H_5Cl
Uses: manufacture of TEL and ethylcellulose; anesthetic; organic synthesis; alkylating agent; refrigeration; analytical reagent; solvent

A. PROPERTIES: colorless liquid or gas; m.w. 64.5; m.p. -138.3°C; b.p. 12.4°C; v.p. 457 mm at 0°C, 700 mm at 10°C, 1,000 mm at 20°C, 1.9 atm at 30°C; v.d. 2.23; sp.gr. 0.92 at 0/4°C; solub. 3,330 mg/l at 0°C, 5,740 mg/l at 20°C; log P_{oct} 1.54 (calculated)

B. AIR POLLUTION FACTORS: 1 mg/cu m = 0.37 ppm, 1 ppm = 2.69 mg/cu m
—Odor: characteristic; quality: etheric
threshold: recognition: 10–12 mg/cu m (610)
—Manmade sources:
glc's in rural Washington Dec. 74–Feb. 75: <5 ppt (315)

C. WATER POLLUTION FACTORS:
—Waste water treatment:
evaporation from water at 25°C of 1 ppm solution (still air, avg. depth 6.5 cm)
50% after 21 min
90% after 79 min (369)

D. BIOLOGICAL EFFECTS:
—Mammalia:
guinea pig: inhalation: histopathological changes in liver and kidney at 2,000 ppm, 540 min
normal: 1,000 ppm, 810 min (211)
—Man: after 30 sec quickly increased toxic effect: 33,600 ppm
weak analgesia after 12 min: 19,000 ppm
slight symptoms of poisoning: 13,000 ppm (211)

ethylcyclohexane

⬡—CH₂CH₃

Uses: organic synthesis

A. PROPERTIES: colorless liquid; m.w. 112.22; m.p. -111°C; b.p. 131.8°C; sp.gr. 0.787

C. WATER POLLUTION FACTORS:
—Biodegradation:
incubation with natural flora in the groundwater—in presence of the other components of high-octane gasoline (100 µl/l): biodegradation: 95% after 192 hr at 13°C (initial conc.: 0.06 µl/l) (956)

co-oxidation to cyclohexane acid by *Pseudomonas* using hexane as the growth substrate

D. BIOLOGICAL EFFECTS:
 —Fish: young Coho salmon: no significant mortalities up to 100 ppm after 96 hr in artificial seawater at 8°C (317)

ethylcyclopentane

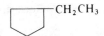

A. PROPERTIES: colorless liquid; b.p. 103.5°C; sp.gr. 0.766
C. WATER POLLUTION FACTORS:
 —Biodegradation:
  incubation with natural flora in the groundwater—in presence of the other components of high-octane gasoline (100 µl/l): biodegradation: 31% after 192 hr at 13°C (initial conc.: 0.11 µl/l) (956)
D. BIOLOGICAL EFFECTS:
 —Fish: young Coho salmon: no significant mortalities up to 100 ppm after 96 hr in artificial seawater at 8°C (317)

ethyldiethanolamine
$C_2H_5N(CH_2CH_2OH)$
A. PROPERTIES: water white liquid; b.p. 246–252°C; sp.gr. 1.02 at 20/20°C;
D. BIOLOGICAL EFFECTS:
 —Fish: creek chub: critical range: 160–200 mg/l, 24 hr (226)

ethyldigol *see* diethyleneglycolmonoethylether

ethyldimethylmethane *see* isopentane

S-ethyldipropylthiocarbamate *see* eptam

ethyldisulfide *see* diethyldisulfide

ethyldithioethane *see* diethyldisulfide

ethylene (ethene; etherin; elayl)
C_2H_4
A. PROPERTIES: colorless gas; m.w. 28.05; m.p. −169°C; b.p. −104°C; v.p. >40 atm at 20°C; v.d. 0.978; sp.gr. 0.566 at −102/4°C; solub. 256 cucm/l at 0°C, 131 mg/l at 20°C; THC 373 kcal/mole, LHC 352 kcal/mole
B. AIR POLLUTION FACTORS: 1 mg/cu = 0.86 ppm, 1 ppm = 1.17 mg/cu m

ETHYLENE

—Odor: characteristic: quality: olefinic
　　　　　　　hedonic tone: unpleasant to neutral

odor thresholds　　　　　mg/cu m

	10^{-7}	10^{-6}	10^{-5}	10^{-4}	10^{-3}	10^{-2}	10^{-1}	1	10	10^2	10^3	10^4
detection										◆ ◆		
recognition												
not specified									◆		◆	

　　　　　　　　　　　　　　　　　　　　　　　　　(19; 637; 729; 737)

O.I. 100% recognition: 1,428　　　　　　　　　　　　　　　　(19)
O.I. at 20°C = 57,100　　　　　　　　　　　　　　　　　　　(316)

—Atmospheric reactions:
R.C.R.: 1.0　　　　　　　　　　　　　　　　　　　　　　　(49)
reactivity: HC conc.: ranking: 0.1–0.3
　　　　　　NO ox.: ranking: 0.3–0.4　　　　　　　　　　　(63)
estimated lifetime under photochemical smog conditions in S.E. England: 7.2 hr
　　　　　　　　　　　　　　　　　　　　　　　　　　(1699; 1706)

—Natural sources:
maximum rates of ethylene production by various fruits

fruit	variety	temp °C	$\mu l\ C_2H_4$/kg/hr
apple	McIntosh	20	112
pear	Bartlett	20	122
pear	Bosc	20	29
peach	Hale	20	36

　　　　　　　　　　　　　　　　　　　　　　　　　　　　　(29)

—Manmade sources:
glc's in urban air: 12 to 250 ppb　　　　　　　　　　　　　　(29)
glc's downtown Los Angeles: 20–102 ppb
glc's East San Gabriel Valley: 15–37 ppb
in exhaust gases of:
reciprocating gasoline engine: 14.1% of emitted hydrocarbons
rotary gasoline engine: 8.1% of emitted hydrocarbons　　　　　(72)
diesel engine: 14.5% of emitted hydrocarbons　　　　　　　　(78)
emissions from burning agricultural wastes:
　average: 1.7 lb/ton of material
　range:　10% ile: 0.5 lb/ton
　　　　　90% ile: 2.5 lb/ton　　　　　　　　　　　　　　(114)
in flue gas of municipal incinerator: 1.1–1.5 ppm　　　　　　(116)
in auto exhaust-gasoline engine:
　62-car survey:　　　　14.5 vol. % of total exhaust H.C.'s　　(391)
　15-fuel study:　　　　17 vol. % of total exhaust H.C.'s　　　(392)
　engine-variable study: 19.0 vol. % of total exhaust H.C.'s　　(393)
—emission by burning of wood chips and green bush: 1–2.7 kg/ton　(1214)
　　　　　　　burning of wheat straw: 189–364 g/ton dry material　(1213)

—Sampling and analysis:
non dispersive I.R.: min. full scale: 500 ppm　　　　　　　　(55)

634 ETHYLENE

    detector tubes: UNICO: 0.5 ppm (59)
                        AUER: 10 ppm (57)
                      DRAGER: 5,000 ppm (58)
    effectiveness of scrubbers at 325–375°F: for inlet conc.: 50 ppm
    (1): $HgSO_4/H_2SO_4$: 60% adsorption
    (2): $PdSO_4/H_2SO_4$: 96% adsorption
    (3): (1) + (2): 96% adsorption (198)
    PMS: det. lim.: at 80.3 nm: 6 ppm
                  at 95.0 nm: 7 ppm
                  at 104.8 nm: 10 ppm (200)

C. WATER POLLUTION FACTORS:
    —Organoleptic limit: 0.5 mg/l (181)

D. BIOLOGICAL EFFECTS:

Seed-plants	Concentration ppm	Exposure time, hr	Effects or comments	
Vanda orchid	1	24	Fading of flowers	(117)
Antirrhinum majus	0.5	1	Abscission of flowers	(118)
Cattleya orchid, buds	0.01	24	Sepal tissue collapse	(118)
	0.05	6		
	0.3	1		
Cattleya orchid	0.002	24	Sepal tissue collapse	(119)
	0.1	8		
Chenopodium album	0.05		Epinasty	(120)
Dianthus caryophyllus	0.10	6	Inhibited flower opening	(118)
Fagopyrum sagittatum (*F. esculentum*)	0.05		Epinasty	(120)
Gossypium hirsutum	0.6	720	Reduction in growth and yield	(121)
Itelianthus annuus	0.05		Epinasty	(120)
Lathyrus odoratus	0.2	72	Epinasty	(120)
Lilium regale	4.0		Growth retardation and epinasty	(122)
Lycopersicon esculentum	0.1	48	Growth retardation	(120)
Narcissus species	2.0	72	Growth retardation	(122)
	4.0	72	Growth retardation and leaf curl	

Rosa species	0.33	120	Epinasty and leaf abscission at room temperature	(123)
	10.0	24	Petal fall at 70°F, none at 32°F, 40°F and 50°F	
	40.0	24	Epinasty at 70°F	
	40.0	48	Epinasty and leaf abscission at 70°F	
	40.0	168	No abscission at 41°F	
Solamon tuberosum	0.05	16	Epinasty	(124)
Tagetes patula	0.05		Epinasty	(120)
Tulipa gesneriana	4.0		Leaf roll	(122)
Tomato	0.1	48	Leaf epinasty	(125)
	0.04	3–4	Leaf epinasty of mature leaves	
African marigold	0.001		Leaf epinasty	(125)
Lemon	0.025–0.05		Epinasty	(126)
Datura stramonium	0.1		Close to limit for response	(127)
Lycopersicum esculentum	0.2		Epinasty of leaves	(127)
Begonia luminosa	8		Slight epinasty	(127)
Sweet pea	0.1		Inhibited elongation of the epicotyl	(128)
	0.4		Production of triple response; horizontal mutation and swelling	(128)

ethylene-*bis*-iminodiacetic acid *see* ethylenediaminetetracetic acid

ethylenebromide (1,2-dibromoethane; ethylenedibromide; glycoldibromide; EDB) CH_2BrCH_2Br

Uses: scavenger for lead in gasoline, grain and fruit fumigant, general solvent, waterproofing preparations, organic synthesis, insecticide, medicin.

A. PROPERTIES: colorless liquid; m.w. 187.88; m.p. 9.97°C; b.p. 131.6°C; v.p. 11 mm at 20°C, 17 mm at 30°C; v.d. 6.5; sp.gr. 2.701 at 25/4°C; solub. 4.310 mg/l at 30°C; sat. conc. 113 g/cu m at 20°C, 168 g/cu m at 30°C

B. AIR POLLUTION FACTORS: 1 mg/cu m = 0.13 ppm, 1 ppm = 7.81 mg/cu m

—Odor: characteristic; quality: chloroformlike

 T.O.C.: <200 mg/cu m = 26 ppm (291, 57)
 O.I. at 20°C = 550 (316)

—Atmospheric reactions:

EDB is resistant to atmospheric oxidation by peroxides and ozone, typically the half-life for these reactions is in excess of 100 days (1221)

—Manmade sources:

in tailpipe exhaust of gasoline engine using leaded gasoline (0.51 g Pb/l)

636   ETHYLENECARBOXYLIC ACID

ECE driving test cycle conditions: cold start: 11.6-19.8 ppb
90-154 µg/cu m
hot start: 10.2-10.9 ppb
80-85 µg/cu m

constant speed test conditions
vehicle speed

mph	ppb	µg/cu m	
idle	9	70	
10	10	78	
30	8	66	
40	5	61	
50	0.3	2	(1221)

evaporation from the fuel tank and carburetor of cars operated on leaded fuel ranges from 2 to 25 mg/day for 1972 through 1974 model-year cars in the U.S.A. (1343)

—glc's in ambient air or in the workplace:
London, exhibition road—5 m from roadside at 1 m above ground level (1976):
  range: 0.001-0.17 µg/cu m—hourly averages (1221)
  garage: 5 m downwind of petrol pumps at height of 1 m: 1.2-1.8 µg/cu m
    hourly averages
open air car park: 0.02-0.05 µg/cu m, hourly averages. (1221)
glc's: in vicinity of gasoline stations along traffic arteries in 3 major U.S. cities:
  0.07-0.11 µg/cu m
  at an oil refinery: 0.2-1.7 µg/cu m
  at EDB manufacturing sites in the USA: 90-115 µg/cu m (1343)
  at 1 mile from the U.S. Dept. of Agriculture's fumigation center: 96 µg/cu m (1344)
citrus fumigation centers: avg. glc's inside: 370-3100 µg/cu m
    avg. glc's outside: 0.1-29 µg/cu m
    (to which site personnel were exposed) (1345)
glc's (U.S.A.):

	EDB µg/cu m
rural areas	0.05-0.10
metropolitan areas	0.10-0.40
near "clusters" of gas stations (2 sites)	0.18-0.50
$\frac{1}{8}$-1 mile from above	0.08-0.20
near gasoline bulk loading terminals (4 sites)	0.13-0.20
near pipeline pumping stations	same as background
near lead mix storage plants	same as background
near lead mix blending plants	same as background (1719)

—Sampling and analysis: photometry: min. full scale: 1,750 ppm (53)

C. WATER POLLUTION FACTORS:
—Hydrolysis
In water EDB hydrolyzes to ethyleneglycol and bromoethanol; under neutral conditions and ambient temperature the half-life of this reaction is 5-10 days (1221)

—Degradation in soil:
in about 2 months, EDB is converted almost completely and quantitatively to ethylene in a soil culture at initial conc. of $10^{-3} M$ EDB (1318)

D. BIOLOGICAL EFFECTS:
—Fish:
Bluegill sunfish (*Lepomis macrochirus*): LC_{50}, 48 h: 18 mg/l
(freshwater) (577)
—Mammals
animal carcinogen: West Germany 1977 (487)
female mice: single oral LD_{50}: 0.420 g/kg
male rats: single oral LD_{50}: 0.148 g/kg
female rats: single oral LD_{50}: 0.117 g/kg
guinea pigs: single oral LD_{50}: 0.110 g/kg
chicken: single oral LD_{50}: 0.079 g/kg
female rabbits: single oral LD_{50}: 0.055 g/kg
rat: max. survival period: 3,000 ppm for 6 min
400 ppm for 30 min
200 ppm for 2 hr
max. exposure without adverse effect: 800 ppm for 6 min.
100 ppm for 2.5 hr.
50 ppm for 7 hr.
rat: repeated inhalation: LD_{50}: 50 ppm, 7 h/day, 5 days/w, 6 months (211)
mortality
rats: inhalation of 20 ppm, 6 hr/d, 5 d/week: male: 15/48
female: 9/48
inhalation of 20 ppm, 6 hr/d, 5 d/week,
+ diet containing 0.05 wt % disulfiram: male: 45/48
female: 47/48 (1719)
man: NIOSH recommends that no worker be exposed to both ethylenedibromide
and disulfiram (1710)
—Carcinogenesis bio-assay:
female rats: exposed during 61 wks to a TWA of 37–39 mg/kg
male rats: exposed during 49 wks to a TWA of 38–41 mg/kg
male and female mice: exposed during 78 wks to a TWA of 107 mg/kg
male and female mice: exposed during 90 wks to a TWA of 62 mg/kg
conclusions: male and female rats: squamous cell carcinomas of the forestomach.
male rats: hemangiosarcomas
femals rats: liver cancer
male and female mice: squamous cell carcinomas of the forestomach; respiratory
tract cancers (1314)
—Man: severe skin irritant
serious toxic interaction between inhaled EDB and ingested disulfiram (Antabuse, Ro-Fulfiran, or Tetraethylthiuramdisulfide) (1314)
—Carcinogenicity: +
—Mutagenicity in the *Salmonella* test: weakly mutagenic (without liver homogenate):
0.06 revertant colonies/nmol
128 revertant colonies at 4344 µg/plate (1883)

ethylenecarboxylic acid *see* acrylic acid

ethylenechloride *see* ethylenedichloride

ethylenechlorohydrin see 2-chloroethanol

ethylenecyanide see succinonitrile

ethylenediacetate see glycoldiacetate

ethylenediamine (diaminoethane; 1,2-ethanediamine)
$NH_2CH_2CH_2NH_2(H_2O)$

Uses: medicine; neutralizing oils; corrosion inhibitor in anti-freeze solutions; textile lubricants; dyes; rubber accelerators; organic synthesis; adhesives

A. PROPERTIES: colorless liquid; m.w. 78.12 hydrate, 60.1 anhydrous; m.p. 10°C hydr., 8.5°C anhydr.; b.p. 118°C hydr.; v.p. 116 mm at 20°C anhydr.; 9 mm at 20°C, 16 mm at 30°C; v.d. 2.1 anhydr.; sp.gr. 0.963 at 21/4°C hydr., 0.8994 at 20/4°C anhydr.; sat. conc. 29 g/cu m at 20°C, 51 g/cu m at 30°C

B. AIR POLLUTION FACTORS: 1 mg/cu m = 0.40 ppm, 1 ppm = 2.50 mg/cu m
 —Odor: characteristic: quality: ammoniacal, musty
  hedonic tone: unpleasant
 T.O.C.: abs. perc. lim.: 1.0 ppm
  50% recogn.: 3.4 ppm
  100% recogn.: 11.2 ppm
  O.I. 100% recogn.: 1,178   (19)
 —Sampling and analysis:
  gas washing bottle: medium: 200 ml water, sampling rate: 0.12 cu ft/min
   test conc.: 103 ppm: absorption efficiency: 99+%   (103)

C. WATER POLLUTION FACTORS:

theoretical	analytical
ThOD = 3.45	reflux COD = 78.9% recovery
COD = 1.33	rapid COD = 11.9% recovery
NOD = 2.12	TKN = 76.8% recovery
	BOD_5 = 0.01
	BOD_5/COD = 0.008
	BOD_5 (acclimated) > 1.0

The rapid COD method was found inadequate because of volatilization of the compound or compound oxidation products in the open digestion flask at the high test temperatures. The results of the BOD_5 analyses suggest that ethylenediamine is degradable only under acclimated conditions.

—Impact on conventional biological treatment systems:

	chemical conc. mg/l	effect
unacclimated system	108.5	no effect
	1,085	inhibitory
acclimated system	225	biodegradable

The Warburg studies verified the BOD_5 analyses. The compound, which was readily degraded by an acclimated system, did not affect an unacclimated biomass. (1828)

NH_3 oxidation by *Nitrosomonas*: at 100 mg/l: 73% inhibition
  10 mg/l: 41% inhibition   (390)

—Biodegradation: adapted A.S.—product as sole carbon source—97.5% COD removal at 9.8 mg COD/g dry inoculum/hr   (327)

−Waste water treatment:
A.C.: adsorbability: 0.021 g/g C; 10.7% reduction, infl.: 1,000 mg/l, effl.: 893 mg/l (32)
RW, BOD, 20°C, 1–10 days observed, feed: 50 mg/l, 28 days acclimation: 81% removed (93)
−Sampling and analysis:
EDA is adsorbed on activated silicagel, desorbed with 0.5% aqueous cupric chloride, and analyzed by GC using 2% KOH on a Chromosorb 103 column. The method can detect 1.0 ppmv EDA in samples collected for 4.5 hours at a 300 cc per minute flow. Other amines do not interfere with the determination of EDA (1420)

D. BIOLOGICAL EFFECTS:
−Toxicity threshold (cell multiplication inhibition test):
bacteria (*Pseudomonas putida*): 0.85 mg/l (1900)
algae (*Microcystis aeruginosa*): 0.08 mg/l (329)
green algae (*Scenedesmus quadricauda*): 0.85 mg/l (1900)
protozoa (*Entosiphon sulcatum*): 1.8 mg/l (1900)
protozoa (*Uronema parduczi Chatton-Lwoff*): 52 mg/l (1901)
−Bacteria: *E. coli*: LD_0: 200 mg/l
−Algae: *Selenastrum capricornutum*: 1 mg/l: no effect
10 mg/l: no effect
100 mg/l: no effect (1828)
−Arthropods: *Daphnia*: LD_0: 8 mg/l (30)
Daphnia magna: LC_{50}: 0.88 mg/l (1828)
−Fish: creek cub: critical range: 30–60 mg/l, 24 hr (226)
rainbow trout yearlings: 48 hr LC_{50}: 230 mg/l (static bioassay) (939)
Pimephales promelas: LC_{50}: >1000 mg/l (1828)
−Mammalia: rat: acute oral LD_{50}: 1.16 g/kg
inhalation: 6/6: 4,000 ppm, 8 hr
0/6: 2,000 ppm, 8 hr
no effect: 59 ppm, 30 × 7 hr (211)

ethylenediaminetetraacetate, copper chelate (Versene AG: 7.5% Cu, 45.5% active ingredient)
D. BIOLOGICAL EFFECTS:
−Bluegill (*Lepomis macrochirus*): static 96 hr LC_{50}: 555 mg/l
96 hr no adverse effect level: 320 mg/l (1869)

ethylenediaminetetraacetate, diammonium salt (Versene $(NH_4)_2$EDTA)
Use: chelating agent
D. BIOLOGICAL EFFECTS:
−Bluegill (*Lepomis macrochirus*): static 96 hr LC_{50}: 2340 mg/l
96 hr no adverse effect level: 1350 mg/l (1869)

ethylenediaminetetraacetate, magnesium chelate (Versene AG: 2.5% Mg, 40.0% active ingredient)
D. BIOLOGICAL EFFECTS:
−Bluegill (*Lepomis macrochirus*): static 96 hr LC_{50}: 2520 mg/l
96 hr no adverse effect level: 1350 mg/l (1869)

ethylenediaminetetraacetate, tetraammonium salt (Versene $(NH_4)_4$EDTA)
Use: chelating agent
D. BIOLOGICAL EFFECTS:
—Bluegill (*Lepomis macrochirus*): static 96 hr LC_{50}: 705 mg/l
  96 hr no adverse effect level: 240 mg/l (1869)

ethylenediaminetetraacetate, tetrasodium salt

$$\begin{array}{c} NaOOC-CH_2 \\ NaOOC-CH_2 \end{array} \!\!\!\! N-CH_2CH_2-N \!\!\!\! \begin{array}{c} CH_2-COONa \\ CH_2-COONa \end{array}$$

Use: chelating agent
A. PROPERTIES: white solid; solub. 90 wt % at 25°C
D. BIOLOGICAL EFFECTS:
—Fish:
  bluegill (*Lepomis macrochirus*): static 96 hr LC_{50}: 486 mg/l (Versene powder)
  1030 mg/l (Versene 100)
  96 hr, no adverse effect level: 456 mg/l (Versene powder)
  870 mg/l (Versene 100)
  (1869)
—Mammals:
  ingestion: rats: single oral LD_{50}: in the range of 2000–4000 mg/kg (1546)

ethylenediaminetetraacetate trihydrate, trisodium

$$\begin{array}{c} HOOCCH_2 \\ NaOOCCH_2 \end{array} \!\!\!\! NCH_2CH_2N \!\!\!\! \begin{array}{c} CH_2COONa \\ CH_2COONa \end{array} \cdot (H_2O)_3$$

A. PROPERTIES: m.w. 412.24
C. WATER POLLUTION FACTORS:

theoretical	analytical
ThOD = 0.97 gm/gm	reflux COD = 91.9% recovery
COD = 0.66 gm/gm	rapid COD = 74.1% recovery
NOD = 0.31 gm/gm	TKN = 89.7% recovery
	BOD_5 = 0.004 gm/gm
	BOD_5/COD = 0.007
	BOD_5 (acclimated) = 0.006 gm/gm

The response to either of the COD analytical methods was erratic for several replicates. The same erratic results were exhibited by the TKN analysis. The results of the BOD_5 analyses, which used both unacclimated and acclimated seeds, suggested that trisodium EDTA · $3H_2O$ is resistant to biodegradation
  (1828)
—Impact on biodegradation:
  Unacclimated system at: 500 mg/l: inhibitory
  1,000 mg/l: inhibitory
  5,000 mg/l: inhibitory
  Acclimated system at: 800 mg/l: no effect (1828)

There was no conclusive evidence however that the compound itself is amenable to biodegradation at 800 mg/l (1828)
D. BIOLOGICAL EFFECTS:
 —Algae:
 Selenastrum capricornutum: 1 mg/l: no effect
                              10 mg/l: inhibitory
                              100 mg/l: inhibitory (1828)
 —Crustacean:
 Daphnia magna: LC_{50}: >100 mg/l (1828)
 —Fish:
 Pimephales promelas: LC_{50}: >300 mg/l (1828)
 —Mammals:
 No compound-related signs of clinical toxicity were noted. Although a variety of tumors occurred among test and control animals of both species, no tumors were related to treatment. Since survival was satisfactory and showed no consistent variation among test and control groups, the absence of treatment-related tumors could not be attributed to early mortality. (1771)

ethylenediaminetetraacetate, zinc chelate (Versene AG: 1 lb Zn, 53% active ingredient)
D. BIOLOGICAL EFFECTS:
 —Fish:
   bluegill (*Lepomis macrochirus*): static 96 hr LC_{50}: 685 mg/l
                                    96 hr no adverse effect level: 320 mg/l (1869)

ethylenediaminetetraacetic acid (EDTA; ethylene*bis*iminodiacetic acid; ethylenedinitrilotetraacetic acid; Versene acid)

$$\begin{array}{c} HOOC-CH_2 \diagdown \diagup CH_2COOH \\ NCH_2CH_2N \\ HOOC-CH_2 \diagup \diagdown CH_2COOH \end{array}$$

Use: acid chelating agent
A. PROPERTIES: colorless crystals.; m.p. decomposing at 240°C; solub. <1000 mg/l at 25°C
C. WATER POLLUTION FACTORS:
 —BOD_5: 0.01 std. dil. sew. (169; 289)
 —Manmade sources: in sewage effluents: 0.10–0.550 mg/l (227)
D. BIOLOGICAL EFFECTS:
 —Toxicity threshold (cell multiplication inhibition test):
   bacteria (*Pseudomonas putida*):   105 mg/l   (1900)
   algae (*Microcystis aeruginosa*):   76 mg/l   (326)
   green algae (*Scenedesmus quadricauda*):   11 mg/l   (1900)
   protozoa (*Entosiphon sulcatum*):   36 mg/l   (1900)
   protozoa (*Uronema parduczi Chatton-Lwoff*):   17 mg/l   (1901)
 —Fish:
   bluegill (*Lepomis macrochirus*): static 96 hr LC_{50}: 159 mg/l
                                    96 hr no adverse effect level: 100 mg/l (1869)

- Mammals:
    ingestion: rats: single oral LD_{50}: 4000 mg/kg (1546)
- Carcinogenicity: none ?
- Mutagenicity in the *Salmonella* test: none
    <0.002 revertant colonies/nmol
    <70 revertant colonies at 10^4 μg/plate (1883)

ethylenediaminetetraacetic acid, ammonium ferric
D. WATER POLLUTION FACTORS:

theoretical
$ThOD = 1.28$
$COD = 0.75$
$NOD = 0.53$

analytical
reflux COD = 73.2% recovery
rapid COD = 70.0% recovery
TKN = 94.4% recovery
$BOD_5 = 0.015$
$BOD_5/COD = 0.027$
BOD_5 (acclimated) = 0.06

The response to either COD method was both erratic and incomplete. The results of the BOD_5 analysis, which used both acclimated and unacclimated seeds, indicated the compound to be resistant to biodegradation.

- Impact on biodegradation:
    unacclimated system at: 468 mg/l: no effect
                             563 mg/l: no effect
                             4,680 mg/l: no effect
    acclimated system at: 518 mg/l: no effect

The compound proved to have no effect on an unacclimated biomass at the tested concentration. The compound was found to be resistant to biodegradation, even under acclimated conditions. (1828)

D. BIOLOGICAL EFFECTS:
- Algae:
    Selenastrum capricornutum: 1 mg/l: no effect
                                  10 mg/l: no effect
                                  100 mg/l: no effect (1828)
- Crustacean:
    Daphnia magna: LC_{50}: 2.8 mg/l (1828)
- Fish:
    Pimephales promelas: LC_{50}: 190 mg/l (1828)

ethylenediaminetetraacetic acid dihydrate, calcium chelate of the disodium salt
Use: chelating agent
A. PROPERTIES: solub. 100 g/100 g at 25°C
D. BIOLOGICAL EFFECTS:
- Fish:
    bluegill (*Lepomis macrochirus*): static 96 hr LC_{50}: 2340 mg/l
                                      96 hr no adverse effect level: 1000 mg/l (1869)
- Mammals:
    ingestion: rats: single oral LD_{50}: 10,000 mg/kg
               rabbits: single oral LD_{50}: 7,000 mg/kg
               dogs: single oral LD_{50}: 12,000 mg/kg (1546)

ethylenedibromide *see* ethylenebromide

***trans*-1,2-ethylenedicarboxylic acid** *see* fumaric acid

(*cis*)-1,2-ethylenedicarboxylic acid *see* maleic acid

ethylenedichloride (1,2-dichloroethane; dichloroethylene; EDC; ethylenechloride; glycoldichloride; *sym*-dichloroethane; ethenedichloride)
  Manufacturing source: organic chemical industry. (347)
  Users and formulation: mfg. of vinyl chloride; mfg of tetraethyllead; intermediate insecticidal fumigant (peachtree borer, japanese beetle, root-knot nematodes); tobacco flavoring; constituent in paint, varnish and finish removers, metal degreaser, constituent in soaps and scouring compounds, wetting and penetrating agents; used in chemical synthesis and ore flotation. (347)
  Uses: Ingredient in cosmetics (nail lacquers) and as a food additive as a result of its use in extracting spices such as annatto, paprika, and turmeric
A. PROPERTIES: colorless liquid; m.w. 99; m.p. $-35.4°C$; b.p. $83.5°C$; v.p. 40 mm at $10°C$, 61 mm at $20°C$, 105 mm at $30°C$; sp.gr. 1.25 at $20/4°C$; solub. 8,690 mg/l at $20°C$, 9,200 mg/l at $0°C$; sat. conc. 350 g/cu m at $20°C$, 537 g/cu m at $30°C$
B. AIR POLLUTION FACTORS: 1 mg/cu m = 0.24 ppm, 1 ppm = 4.43 mg/cu m
  —Odor: characteristic: quality: sweet, chloroformlike, aromatic
    hedonic tone: unpleasant to neutral

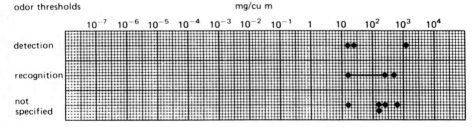

(29; 57; 210; 278; 307; 624; 642; 698; 708; 804)
  O.I. 100% recogn.: 2,037
  USSR: human odor perception: non perception: 12.2 mg/cu m
                                   perception: 23.2 mg/cu m
        human reflex respone: adverse response: 6.0 mg/cu m (170)
  —Manmade sources:
    glc's in rural Washington Dec. '74–Febr. '75: <5 ppt (315)
  —Sampling and analysis: PMS: det. lim.: 7–46 ppm (220)
C. WATER POLLUTION FACTORS:

	1*	2*	
—BOD_5: 0.002 std. dil. sew.	0%	7% of ThOD	(41, 23)
10 days: 0.05	18%	7% of ThOD	(26)
15 days:		7% of ThOD	
20 days:		20% of ThOD	(23)

  1* fresh dilution water; 2* salt dilution water

- COD: 1.025 (23)
- ThOD: 0.97 (23)
- Reduction of amenities:
  faint odor: 0.025 mg/l (129)
  odor threshold: 20 mg/l; 29 mg/kg (84)(873)
- Aquatic reactions:
  measured half-life for evaporation from 1 ppm aqueous solution at 25°C, still air, and an average depth of 6.5 cm: 28.0 min. (369)
- Water quality:
  in river Maas at Eysden (Netherlands) in 1976: median: 8.2 µg/l; range: n.d. to 59.5 µg/l
  in river Maas at Keizersveer (Netherlands) in 1976: median: 6.9 µg/l; range: 0.1 to 47.4 µg/l (1368)
  U.S.A.: in 11 raw water locations: <0.2–31 µg/l
  in 26 finished water locations (32.9% of total): 0.2–6 µg/l (1342)
- Waste water treatment:
  A.C.: adsorbability: 0.163 g/g C; 81.1% reduction, infl.: 1,000 mg/l; effl.: 189 mg/l (32)

methods	temp °C	days observed	feed mg/l	days acclim.	% removed
NFG, BOD	20	1–10	200	365 + P	45
NFG, BOD	20	1–10	400	365 + P	30
NFG, BOD	20	1–10	1,000	365 + P	9

(93)

  evaporation from water at 25°C of 1 ppm solution:
  50% after 29 min
  90% after 96 min (313)
- Sampling and analysis:
  GC–EC after n-pentane extraction; det. lim.: 10 µg/l (84)

D. BIOLOGICAL EFFECTS:
- Toxicity threshold (cell multiplication inhibition test):
  bacteria (*Pseudomonas putida*): 135 mg/l (1900)
  algae (*Microcystis aeruginosa*): 105 mg/l (329)
  green algae (*Scenedesmus quadricauda*): 710 mg/l (1900)
  protozoa (*Entosiphon sulcatum*): 1127 mg/l (1900)
  protozoa (*Uronema parduczi Chatton-Lwoff*): 1050 mg/l (1901)
- Arthropoda:
  Crangon crangon in sea water at 15°C:

exposure time	LC_{50} mg/l
3 min	±2000
9 min	±630
27 min	400
1 hr	345
8 hr	125
24 hr	75
48 hr	65
96 hr	65

(328)

  Crangon crangon in sea water at 15°C:

	EC_{50} after recovery period in unpolluted sea water	
exposure time	0 min	1 to 3 hr
1 min	±2,000 ppm	±4,000 ppm
3 min	±2,000 ppm	±2,000 ppm
9 min	±650 ppm	±1,400 ppm (328)

prawn (*Palaemon serratus*): no effect at 25 mg/l after 96 hr exposure (576)

—Fish: *Gobius minutus* in sea water at 15°C:

	EC_{50} after recovery period in unpolluted sea water	
exposure time	0 min	1 to 3 hr
1 min	±1,400 ppm	±1,700 ppm
3 min	±1,400 ppm	±1,400 ppm
9 min	±400 ppm	±400 ppm (328)

Gobius minutus: LC_{50} in sea water at 15°C:

exposure period	mg/l
3 min	±1400
9 min	±400
27 min	225
60 min	185
3 hr	185
up to 96 hr	185 (328)

fathead minnows: LC_{50}: 500 ppm

rainbow trout, bluegill: no effect level: 5 ppm, 24 hr (361)

dab (*Limanda limanda*): no effect at 60 mg/l after 96 hr exposure (576)

guppy (*Poecilia reticulata*): 7 d LC_{50}: 106 ppm (1833)

—Mammalia:

rat: single oral LD_{50}: 0.68 g/kg (211)
    inhalation: LC_{50}: 12,000 ppm, 31.8 min
        3,000 ppm, 165 min
        1,000 ppm, 432 min (54)

rat and guinea pig: inhalation: survival: 100 ppm, 7 hr/day, 5 day/w, many months (211)

rats and mice: inhalation 7 hr daily, 5 times weekly at 5, 10, 50, and 150 ppm after 104 weeks of an 18 month study produced no "specific" type of tumor (1348)

—Carcinogenic to Osborne-Mendel rats and B6C3F1 mice, utilized forced-feeding (directly into the stomach) for the route of exposure

—Weak mutagen in bacteria and mutagenic in fruit flies (1010)

—Man: severe toxic effects: 500 ppm = 2,050 mg/cu m, 60 min
    symptoms of illness: 100 ppm = 410 mg/cu m
    unsatisfactory: >50 ppm = 205 (185)
    no effect level: 200 ppm, 7 hr
        1,000 ppm, 60 min
        3,000 ppm, 6 min (54)

ethylenedinitrate *see* ethyleneglycoldinitrate

ethylenedinitrilotetraacetic acid   *see* ethylenediaminetetraacetic acid

2,2(ethylenedioxy)diethanol   *see* triethyleneglycol

ethyleneglycol   (1,2-ethanediol; 1,2-dihydroxyethane; MEG) $(CH_2OH)_2$
  Use: coolant and anti-freeze; asphalt emulsion paints; heat transfer agent; brake fluids; glycol di-acetate; polyester fibers and films; solvent; ingredient of deicing fluid for airport runways

A. PROPERTIES: colorless liquid; m.w. 62.1; m.p. $-17/-12.6°C$; b.p. $198°C$; v.p. 0.05 mm at $20°C$, 0.2 mm at $30°C$; v.d. 2.14; sp.gr. 1.113 at $20/4°C$; sat. conc. <0.34 g/cu m at $20°C$, 0.65 g/cu m at $30°C$; $\log P_{oct} -1.93$

B. AIR POLLUTION FACTORS: 1 mg/cu m = 0.39 ppm, 1 ppm = 2.58 mg/cu m
  —Odor: T.O.C.: 25 ppm = 90 mg/cu m   (210)
            0.23 mg/cu m = 0.08 ppm   (279)
            O.I. at $20°C$ = 3   (316)
  —Sampling and analysis:
    4f. abs. app.: 40 l air/30 min: VLS: det. lim.: 50 μg/cu m/30 min   (208)

C. WATER POLLUTION FACTORS:
  —BOD_5: 0.47 NEN 3235-5.4   (277)
        0.81 adapted sew.   (27)
        0.6 std. dil. sew. at 10 mg/l   (27)
        0.36; 0.40   (36; 30)
        0.16–0.68 Warburg, sewage   (41)
        12% of ThOD; 10 days: 51.8%; 15 days: 71.0%   (79)
    20 days: 1.08   (30)
        78% of ThOD   (79)
  —BOD_5: 0.15 "special grade" NEN 3235-5.4
        0.67 "special grade" adapted sew.   (277)
  —BOD_5^{20}: 0.36 at 10 mg/l, unadapted sew.; lag period: 4 days
  —BOD_{20}^{20}: 0.91 at 10 mg/l, unadapted sew.   (554)
  —COD: 1.29 "special grade" NEN 3235-5.3   (277)
        1.24   NEN 3235-5.3   (277)
        1.21   (36)
  —$KMnO_4$: 0.404   (30)
  —ThOD: 1.26   (36)
  —BOD_5: 38% of ThOD   (274)
  —COD: 94% of ThOD (0.05 n $K_2Cr_2O_7$)   (274)
  —$KMnO_4$: 5% of ThOD (0.01 n $KMnO_4$)   (274)
  —Biodegradation: adapted A.S.—product as sole carbon source—96.8% COD removal at 41.7 mg COD/g dry inoculum/hr   (327)
  —Manmade sources:
    in primary domestic sewage plant effluent: 0.003 mg/l   (517)
  —Waste water treatment:
    A.C.: adsorbability: 0.0136 g/g C; 6.8% reduction, infl.: 1,000 mg/l, effl.: 932 mg/l   (32)
    anaerobic lagoon:

lb COD/day/1,000 cu ft	infl. mg/l	effl. mg/l	
13	135	30	
22	755	155	
48	755	190	(37)

—A.S.: after 6 hr: 10.3% of ThOD
12 hr: 21.8% of ThOD
24 hr: 40.2% of ThOD (88)

methods	temp °C	days observed	feed mg/l	days acclim.	
A.S., BOD	20	$\frac{1}{3}$–5	333	30+	84% removed
A.S., Resp, BOD	20	1–5	484	1	8% theor. oxidation
NFG; BOD	20	1–10	200–1,000	365 + P	35% removed
					(93)

—Sampling and analysis:
  VLS: det. lim.: 0.2 mg/l (208)

D. BIOLOGICAL EFFECTS:
  —Toxicity threshold (cell multiplication inhibition test):
    bacteria (*Pseudomonas putida*):   >10,000 mg/l   (1900)
    algae (*Microcystis aeruginosa*):   2,000 mg/l   (329)
    green algae (*Scenedesmus quadricauda*):   >10,000 mg/l   (1900)
    protozoa (*Entosiphon sulcatum*):   >10,000 mg/l   (1900)
    protozoa (*Uronema parduczi Chatton-Lwoff*): >10,000 mg/l   (1901)
  —Bacteria: *Pseudomonas*: toxic: 250 mg/l
  —Algae: *Chlorella pyrenoidosa*: toxic: 180,000 mg/l   (41)
        Scenedesmus: no effect: 1 g/l   (30)
  —Fish:
    goldfish: 24 hr LD_{50}: >5,000 mg/l—modified ASTM D 1345   (277)
    guppy (*Poecilia reticulata*): 7 d LC_{50}: 49,300 ppm   (1833)
  —Mammalia:
    rat: single oral LD_{50}: 5.50 ml/kg; 8.54 g/kg
    guinea pig: single oral LD_{50}: 7.35 ml/kg; 6.61 g/kg
    mouse: single oral LD_{50}: 13.1 ml/kg; 13.7 g/kg
    rat: inhalation: no injury: 140–160 ppm, 8 hr/day, 16 weeks   (211)
  —Man: single lethal oral dose: approx. 1.4 ml/kg
       no significant skin irritation   (211)
  —Carcinogenicity: none
  —Mutagenicity in the *Salmonella* test: none
    <0.0004 revertant colonies/nmol
    <70 revertant colonies at 10^4 µg/plate   (1883)

ethyleneglycolacetate
D. BIOLOGICAL EFFECTS:
  —Toxicity threshold (cell multiplication inhibition test):
    bacteria (*Pseudomonas putida*):   875 mg/l
    green algae (*Scenedesmus quadricauda*):   9 mg/l
    protozoa (*Entosiphon sulcatum*):   34 mg/l   (1900)
    protozoa (*Uronema parduczi Chatton-Lwoff*): 5910 mg/l   (1901)

ethyleneglycoldiacetate *see* glycoldiacetate

ethyleneglycoldiethylether (diethylcellosolve) $(CH_2OC_2H_5)_2$
A. PROPERTIES: m.w. 118.2; b.p. 170°C; v.p. 0.88 mm at 25°C; v.d. 4.0; sp.gr. 0.900
B. AIR POLLUTION FACTORS: 1 mg/cu m = 0.207 ppm, 1 ppm = 4.84 mg/cu m
C. WATER POLLUTION FACTORS:
—Waste water treatment:
NFG, BOD, 20°C, 1–10 days observed, feed: 200–1,000 mg/l, acclimation: 365 + P: 2% removed (93)
—BOD_{10}: 0.10 std. dil. sew. (250)
D. BIOLOGICAL EFFECTS:
—Mammalia:
guinea pig: single oral LD_{50}: 2.44 g/kg
rat: single oral LD_{50}: 4.39 g/kg
guinea pig: inhalation: no evident injury: 500 ppm, 12 × 8 hr
rabbit: inhalation: 1/2: 500 ppm, 12 × 8 hr
cats: inhalation: 2/2: 500 ppm, 12 × 8 hr
mouse: inhalation: no evident injury: 500 ppm, 12 × 8 hr (211)

ethyleneglycoldinitrate (ethylenedinitrate; glycoldinitrate; ethylenenitrate) (CH_2ONO_2)
A. PROPERTIES: yellow liquid; m.w. 152.07; m.p. −20°C; b.p. 114/116°C explodes; v.p. 0.05 mm at 20°C; v.d. 5.25; sp.gr. 1.483 at 8°C;
B. AIR POLLUTION FACTORS: 1 mg/cu m = 0.161 ppm, 1 ppm = 6.24 mg/cu m
D. BIOLOGICAL EFFECTS:
—Fish: perch: lower toxic limit: 5 mg/l, 3–4 days (30)
—Mammalia:
cat: inhalation: marked blood changes only: 21 ppm, 1,000 days
temporary blood changes: 2 ppm, 1,000 days (211)
—Man:
severe toxic effects: 20 ppm = 128 mg/cu m, 60 min
symptoms of illness: 1 ppm = 6.4 mg/cu m,
unsatisfactory: >0.5 ppm = 3.2 mg/cu m (185)
minimal dose for headache: 1.8–3.5 cc of 1% solution of skin (211)

ethyleneglycolmono-*n*-butylether *see* butylcellosolve

ethyleneglycolmonobutyletheracetate *see* butylcellosolve acetate

ethyleneglycolmonoethylether (ethoxol; ethylglycol; cellosolve; glycolmonoethylether; 2-ethoxyethanol; Dowanol EE; Oxitol)
$C_2H_5OCH_2CH_2OH$
Uses: solvent for nitrocellulose; natural and synthetic resins; mutual solvent for formulation of soluble oils; lacquers and lacquer thinners; dyeing and printing textiles; varnish removers; cleaning solutions; leather; anti-icing additive for aviation fuels.

ETHYLENEGLYCOLMONOETHYLETHER

A. PROPERTIES: colorless liquid; m.w. 90.1; b.p. 135°C; v.p. 3.8 mm at 20°C, 7 mm at 30°C; v.d. 3.10; sp.gr. 0.93 at 20°C; sat. conc. 18 g/cu m at 20°C, 33 g/cu m at 30°C; log P_{oct} −0.54

B. AIR POLLUTION FACTORS: 1 mg/cu m = 0.27 ppm, 1 ppm = 3.75 mg/cu m
  −Odor: characteristic: quality: sweet, musty
  hedonic tone: unpleasant to pleasant
  T.O.C.: 90 mg/cu m = 25 ppm (279)(291)
  abs. perc. lim.: 0.30 ppm
  50% recogn.: 0.55 ppm
  100% recogn.: 1.33 ppm
  O.I. 100% recogn.: 3.909 (19)
  distinct odor: 180 mg/cu m = 50 ppm (278)
  −Sampling and analysis:
  6f. abs. app., 60 l air/30 min: GC: det. lim.: 2 mg/cu m/30 min (208)

C. WATER POLLUTION FACTORS:
  −BOD_5: 1.03 NEN 3235-5.4
  1.27 adapted sludge NEN 3235-5.4 (277)
  1.58 at 5-10 ppm, std. dil. sew. (286)
  −BOD_{10}: 1.1 std. dil. sew. (256)
  −COD: 1.92 (277)
  −Water quality:
  1.0 mg/l affects the self purification (181)
  −Waste water treatment:
  A.C.: adsorbability: 0.063 g/g C; 31.0% reduction, infl.: 1,022 mg/l, effl.: 705 mg/l (32)

methods	temp °C	feed mg/l	days observed	days acclim.	
A.S., Sd, BOD	20	?	5	14+	54% theor. oxidation
A.S., W	20	50	$\frac{1}{4}$	14+	52% theor. oxidation
NFG, BOD	20	900	1-10	365+	48% removed

(93)

D. BIOLOGICAL EFFECTS:
  −Fish: goldfish: LD_{50}: (24 hr): >5000 mg/l-modified ASTM D 1345 (277)
  Lepomis macrochirus: static bioassay in fresh water at 23°C, mild aeration applied after 24 hr:

material added ppm	% survival after				best fit 96 hr LC_{50} ppm
	24 hr	48 hr	72 hr	96 hr	
10,000	100	100	100	100	>10,000
1,000	100	100	100	100	

Menidia beryllina: static bioassay in synthetic seawater at 23°C: mild aeration applied after 24 hr:

material added ppm	% survival after				best fit 96 hr LC_{50} ppm
	24 hr	48 hr	72 hr	96 hr	
10,000	100	100	100	100	>10,000

(352)

guppy (*Poecilia reticulata*): 7 d LC_{50}: 16,400 ppm (1833)
—Mammalia: guinea pig: single oral dose: lethal: 1.4-2.8 g/kg
rat: single oral dose: lethal: 3-5.5 g/kg
rabbit: single oral dose: lethal: 3.1 g/kg
mouse: single oral dose: lethal: 4.3 g/kg
inhalation: LC_{50}: 1820 ppm, 7 hr (211)

ethyleneglycolmonoethyletheracetate (cellosolveacetate; ethylglycolacetate; 2-ethoxyethylacetate)
$CH_3COOCH_2CH_2OC_2H_5$

A. PROPERTIES: colorless liquid; m.w. 132.16; m.p. -62/-58°C; b.p. 156°C; v.p. 1.2 mm at 20°C, 3.8 mm at 30°C; v.d. 4.72; sp.gr. 0.97 at 20/4°C; solub. 230,000 mg/l at 20°C; sat. conc. 14 g/cu m at 20°C, 26 g/cu m at 30°C

B. AIR POLLUTION FACTORS: 1 mg/cu m = 0.18 ppm, 1 ppm = 5.49 mg/cu m
—Odor: characteristic: quality: sweet, musty
hedonic tone: pleasant
T.O.C.: abs. perc. lim.: 0.056 ppm
50% recogn.: 0.138 ppm
100% recogn.: 0.250 ppm
O.I. 100% recogn.: 10,520 (19)

C. WATER POLLUTION FACTORS:
—Waste water treatment:
A.C.: adsorbability: 0.132 g/g C; % reduction: 65.8; infl.: 1,000 mg/l, effl.: 342 mg/l (32)

D. BIOLOGICAL EFFECTS:
—Mammalia:
guinea pig: single oral LD_{50}: 1.91 g/kg
rat: single oral LD_{50}: 5.10 g/kg
rabbit: single oral LD_{50}: 1.95 g/kg
guinea pig, mouse: inhalation: no deaths: 450 ppm, 12 × 8 hr
dog: inhalation: no effect: 600 ppm, 120 × 7 hr
rabbit: inhalation: 1/2: 450 ppm, 12 × 8 hr
cat: inhalation: 2/2: 450 ppm, 12 × 8 hr (211)

ethyleneglycolmonohexylether (*n*-hexylcellosolve)
$C_6H_{13}OCH_2CH_2OH$

A. PROPERTIES: water-white liquid; m.w. 146.2; m.p. -50.1°C; b.p. 208.1°C; v.p. 0.05 mm at 20°C; sp.gr. 0.888 at 20/20°C; solub. 9,900 mg/l

C. WATER POLLUTION FACTORS:
—Waste water treatment:
A.C.: adsorbability: 0.170 g/g C; 87.1% reduction, infl.: 975 mg/l, effl.: 126 mg/l (32)

ethyleneglycolmonoisopropylether *see* 2-isopropoxyethanol

ethyleneglycolmonomethylether (methylglycol; methylcellosolve; 2-methoxyethanol; glycolmonomethylether; Dowanol EM; methyl Oxitol)
$CH_3OCH_2CH_2OH$

ETHYLENEGLYCOLMONOMETHYLETHER

Uses: solvent for nitrocellulose, cellulose acetate, alcohol-soluble dyes, natural and synthetic resins; solvent mixtures; lacquers; enamels; varnishes; leather; perfume fixative; jet fuel deicing additive

A. PROPERTIES: colorless liquid; m.w. 76.1; m.p. $-85°C$; b.p. $124°C$; v.p. 6.2 mm at $20°C$, 14 mm at $30°C$; v.d. 2.62; sp.gr. 0.97 at $20/4°C$; sat. conc. 33 g/cu m at $20°C$, 56 g/cu m at $30°C$

B. AIR POLLUTION FACTORS: 1 mg/cu m = 0.32 ppm, 1 ppm = 3.16 mg/cu m
  —Odor: characteristic: quality: sweet, alcohol
                         hedonic tone: pleasant
      T.O.C.: 190 mg/cu m = 60 ppm   (279)(278)
               distinct odor: 280 mg/cu m = 90 ppm   (278)
               abs. perc. lim.: 0.09 ppm
                 50% recogn.: 0.22 ppm
               100% recogn.: 0.40 ppm
               O.I. 100% recogn.: 19,725   (19)
               O.I. at $20°C$ = 140   (316)

C. WATER POLLUTION FACTORS:
  —BOD_5: 0.12   NEN 3235-5.4
             0.50 adapted sludge NEN 3235-5.4   (277)
  —BOD_{10}: 1.10 std. dil. sew.   (256)
  —COD: 1.69   NEN 3235-5.3   (277)
  —Waste water treatment:
      A.C.: adsorbability: 0.028 g/g C; 13.5% reduction, infl.: 1,024 mg/l, effl.: 886 mg/l   (32)
      NFG, BOD, $20°C$, 1–10 days observed, feed: 200–1,000 mg/l, acclimation: 365 + P: 15% removed   (93)

D. BIOLOGICAL EFFECTS:
  —Toxicity threshold (cell multiplication inhibition test):
      bacteria (*Pseudomonas putida*):   >10,000 mg/l   (1900)
      algae (*Microcystis aeruginosa*):   100 mg/l   (329)
      green algae (*Scenedesmus quadricauda*):   >10,000 mg/l   (1900)
      protozoa (*Entosiphon sulcatum*):   1,715 mg/l   (1900)
      protozoa (*Uronema parduczi Chatton-Lwoff*): >10,000 mg/l   (1901)
  —Fish:
      goldfish: 24 hr LD_{50}: >5000 mg/l–modified ASTM D 1345   (277)
      Lepomis macrochirus: static bioassay in fresh water at $23°C$, mild aeration applied after 24 hr:

material added ppm	24 hr	% survival after 48 hr	72 hr	96 hr	best fit 96 hr LC_{50} ppm
10,000	100	100	100	100	>10,000
3,200	100	100	100	100	

    Menidia beryllina: static bioassay in synthetic seawater at $23°C$: mild aeration applied after 24 hr:

material added ppm	24 hr	% survival after 48 hr	72 hr	96 hr	best fit 96 hr LC_{50} ppm
10,000	100	95	90	60	>10,000
5,000	100	100	100	90	

                                                                                                                  (352)

guppy (*Poecilia reticulata*): 7 d LC_{50}: 17,400 ppm (1833)
rainbow trout fingerlings: 96 hr LC_{50} at 12°C: 15,520 ppm
    96 hr LC_0 at 12°C: 12,610 ppm (363)
—Mammalia: guinea pig: single oral dose: lethal: 0.95 g/kg
    rat: single oral dose: lethal: 3.4 g/kg
    rabbit: single oral dose: lethal: 0.89 g/kg
    mouse: inhalation: LC_{50}: 1,480 ppm, 7 hr (211)

ethyleneglycolmonomethyletheracetate (methylcellosolveacetate; methylglycolacetate; glycolmonomethyletheracetate; 2-methoxyethylacetate)
$CH_3OC_2H_4OCOCH_3$

Uses: solvent for nitrocellulose, cellulose acetate, various gums, resins, waxes, oils; textile printing; photographic film; lacquers; dopes

A. PROPERTIES: m.w. 118.13; m.p. −65°C; b.p. 144°C; v.p. 7 mm at 20°C; v.d. 4.07; sp.gr. 1.005 at 20/20°C; sat.conc. 47 g/cu m at 20°C

B. AIR POLLUTION FACTORS: 1 mg/cu m = 0.20 ppm, 1 ppm = 4.91 mg/cu m
—Odor: characteristic: quality: sweet, ester
    hedonic tone: pleasant
T.O.C.: abs.perc.lim.: 0.34 ppm
    50% recogn.: 0.64 ppm
    100% recogn.: 0.64 ppm
    O.I. 100% recogn.: 4,109 (19)

C. WATER POLLUTION FACTORS:
—Waste water treatment:

methods	temp °C	days observed	feed mg/l	days acclim.	% removed
NFG, BOD	20	1–10	1,000	365 + P	22
NFG, BOD	20	1–10	200–1,000	365 + P	26

(93)

Lepomis macrochirus: static bioassay in fresh water at 23°C, mild aeration applied after 24 hr:

material added ppm	24 hr	% survival after 48 hr	72 hr	96 hr	best fit 96 hr LC_{50} ppm
100	0	—	—	—	
75	40	40	30	20	
50	60	40	40	40	45
25	100	100	90	90	
10	100	100	100	100	

Menidia beryllina: static bioassay in synthetic seawater at 23°C, mild aeration applied after 24 hr:

material added ppm	24 hr	% survival after 48 hr	72 hr	96 hr	best fit 96 hr LC_{50} ppm
100	0	—	—	—	
75	20	10	0	—	
50	60	30	30	30	40
25	100	100	100	100	

(352)

—Mammalia: guinea pig: single oral LD_{50}: 1.25 g/kg
    rat: single oral LD_{50}: 3.93 g/kg

rabbit: repeated oral doses: all died: 0.5-1.0 ml/kg/day, 3 days
rat: inhalation: 2/6: 7,000 ppm, 4 hr
cat: inhalation: decrease in blood pigments: 200 ppm, 4-6 hr (211)

ethyleneimine (dimethyleneimine; dihydroazirine)

A. PROPERTIES: oil; m.w. 43.07; m.p. -71°C; b.p. 55/56°C; v.p. 160 mm at 20°C, 250 mm at 30°C; v.d. 1.5; sp.gr. 0.832 at 20/4°C; sat.conc. 375 g/cu m at 20°C, 567 g/cu m at 30°C; solub. completely miscible
B. AIR POLLUTION FACTORS: 1 mg/cu m = 0.56 ppm, 1 ppm = 1.79 mg/cu m
  —Odor: characteristic: quality: ammoniacal
    T.O.C.: 3.5 mg/cu m = 1.96 ppm (291; 629)
    abs.perc.lim.: 2 ppm (54)
    O.I. at 20°C = 105,300 (316)
D. BIOLOGICAL EFFECTS:
  —Toxicity threshold (cell multiplication inhibition test):
    bacteria (*Pseudomonas putida*): 5.5 mg/l (1900)
    algae (*Microcystis aeruginosa*): 0.12 mg/l (329)
    green algae (*Scenedesmus quadricauda*): 0.37 mg/l (1900)
    protozoa (*Entosiphon sulcatum*): 4.3 mg/l (1900)
    protozoa (*Uronema parduczi Chatton-Lwoff*): 27 mg/l (1901)
  —Mammalia: guinea pig: inhalation: LC_{50}: 1,500 ppm, 30 min
    mouse: inhalation: LC_{50}: 1,000 ppm, 30 min
    rat: inhalation: LC_{50}: 250 ppm, 30 min
    rabbit: inhalation: LC_{50}: 50 ppm, 30 min
    cat: inhalation: LC_{50}: 250 ppm, 30 min
    rabbit: single oral LD_{50}: 5 mm^3/kg
    rat: single oral LD_{50}: 6-15 mm^3/kg
    cat: single oral LD_{50}: 10 mm^3/kg (54)
  —Man: irritation of eyes and nose: 100 ppm
    carcinogenic: USA 1974 (77)
    West Germany 1977 (487)
  —Carcinogenicity: +
  —Mutagenicity in the *Salmonella* test: +
    (without liver homogenate): 2.0 revertant colonies/nmol
      469 revertant colonies at 10 μg/plate (1883)

ethylenenaphthalene see acenaphthene

ethylenenitrate see ethyleneglycoldinitrate

ethyleneoxide (1,2-epoxyethane; oxirane; dimethyleneoxide; etheneoxide; EO)

ETHYLENEOXIDE

A. PROPERTIES: colorless liquid or gas; m.w. 44.05; m.p. $-111°C$; b.p. $11°C$; v.p. 1,095 mm at $20°C$, 2.1 atm at $30°C$; v.d. 1.52; sp.gr. 0.887 at $7/4°C$; THC 302 kcal/mole, LHC 289 kcal/mole; solub. completely miscible

B. AIR POLLUTION FACTORS: 1 mg/cu m = 0.55 ppm, 1 ppm = 1.83 mg/cu m
 —Odor: characteristic: quality: sweet, olefinic
 hedonic tone: neutral
 T.O.C.: recognition: 1.5 mg/cu m = 0.8 ppm (73; 868)
 mean det.conc.: 700 ppm (211)
 abs.perc.lim.: 260 ppm
 50% recogn.: 500 ppm
 100% recogn.: 500 ppm
 O.I. 100% recogn.: 2,000 (19)
 USSR: human odor perception: 1.5 mg/cu m
 human reflex response: adverse response: 0.65 mg/cu m (170)
 T.O.C.: 1283 mg/cu m = 700 ppm (279)
 1.5 mg/cu m = 0.82 ppm (291)
 —Sampling and analysis:
 4 f.abs. app., 1 l air/30 min: VLS: det.lim.: 2 mg/cu m/30 min
 GC: det.lim.: 10 mg/cu m/30 min (208)

C. WATER POLLUTION FACTORS:
 —BOD_5: 0.06 NEN 3235-5.4 (277)
 —COD: 1.74 NEN 3235-5.3 (277)
 —Sampling and analysis:
 VLS: det.lim.: 0.2 mg/l
 GC: det.lim.: 1 mg/l (208)

D. BIOLOGICAL EFFECTS:
 —Fish: goldfish: LD_{50} (24 hr): 90 mg/l–modified ASTM D 1345 (277)
 —Mammalia: guinea pig: single intragastrical LD_{50}: 0.27 g/kg
 rat: single intragastrical LD_{50}: 0.33 g/kg
 guinea pig: inhalation: slight respiratory changes: 250–280 ppm/8 hr (211)
 mouse: inhalation: LC_{50}: 835 ppm, 4 hr
 rat: inhalation: LC_{50}: 1,460 ppm, 4 hr
 dog: inhalation: LC_{50}: 960 ppm, 4 hr (54)
 0/3: 710 ppm, 4 hr (211)
 repeated exposure to inhalation:

Species	Pathological findings	Mortality ratio	ppm	Period of exposure
rat	growth depression	0/40	113	122–157 × 4 hr
guinea pig	no findings	0/16	113	122–157 × 4 hr
rabbit	no findings	0/4	113	122–157 × 4 hr
monkey	no findings	0/2	113	122–157 × 4 hr
rat	no findings	3/20	100	130 × 6 hr
mouse	no findings	8/30	100	130 × 6 hr
dog	anemia	0/30	100	130 × 6 hr

(211)
 —Man: no effect level: 5–10 ppm, during 10 years (54)
 severe toxic effects: 250 ppm = 450 mg/cu m, 60 min

symptoms of illness: 100 ppm = 180 mg/cu m,
unsatisfactory: >10 ppm = 18 mg/cu m (185)

ethylenetetrachloride see tetrachloroethylene

ethylenethiourea (4,5-dihydroimidazole-2(3H)-thione; 4,5-dihydro-2-mercaptoimidazole; ETU; imidazolidinethione; imidazoline-2-thiol; mercaptoimidazoline; NA 22; NA.22-D; Pennac CRA; Rhodanin S 62; Sodium-22 Neoprene accelerator; tetrahydro-2H-imidazole-2-thione; 2-thioldihydroglyoxaline; Warecure C)

Uses: extensively used as an accelerator in the curing of polychloroprene (Neoprene) and other elastomers; may be present as a contaminant in the ethylene-*bis*dithiocarbamate fungicides and can also be formed when food containing the fungicides is cooked. (1631)
Manmade source: photolytic degradation product of ethylene*bis*dithiocarbamate fungicides (1799)
A. PROPERTIES: white to pale green crystals; m.p. 199-204°C; log P_{oct} -0.66
B. AIR POLLUTION FACTORS:
  —Sampling and analysis:
    ETU is collected from air using midget impingers containing water, or PVC or cellulose ester membrane filters which are then extracted with water. Pentacyanoamineferrate reagent is added to the filter extract or to the impinger contents to form a colored coordination complex. The absorbance of the solution is measured spectrophotometrically at 590 nm. The detection limit is 0.75 μg/sample (1799)
D. BIOLOGICAL EFFECTS:
  —Fish:
    creek chub: critical range: 6000-8000 mg/l, 24 hr (226)
  —Laboratory animal studies: ETU has been shown to be carcinogenic and teratogenic (causing malformations in offspring) in laboratory animals.
(1631 to 1637)

ethylenetrichloride see trichloroethylene

ethylethanoate see ethylacetate

N-ethylethanolamine (2-ethylaminoethanol; β-hydroxydiethylamine; monoethylethanolamine)
$C_2H_5HNCH_2CH_2OH$
A. PROPERTIES: colorless liquid; m.w. 89.24; b.p. 167-169°C at 751 mm; sp.gr. 0.914 at 20/4°C
D. BIOLOGICAL EFFECTS:
  —Fish: creek chub: critical range: 40-70 mg/l (226)

ethylether (diethylether; ethoxyethane; ether; ethyloxide; diethyloxide)
$C_2H_5OC_2H_5$

Uses: manufacture of ethylene and other chemical synthesis; industrial solvent; analytical chemistry; perfumery; extractant; alcohol denaturant

A. PROPERTIES: colorless liquid; m.w. 74.12; m.p. $-116/-123°C$; b.p. $35°C$; v.p. 442 mm at $20°C$, 921 mm at $40°C$; v.d. 2.56; sp.gr. 0.7135 at $20/4°C$; solub. 69,000 mg/l at $20°C$, 84,300 mg/l at $15°C$, 60,500 mg/l at $25°C$; THC 652 kcal/mole, LHC 606 kcal/mole; log P_{oct} 0.77/0.83

B. AIR POLLUTION FACTORS: 1 mg/cu m = 0.330 ppm, 1 ppm = 3.03 mg/cu m
—Odor: characteristic: quality: sweetish pungent ether odor

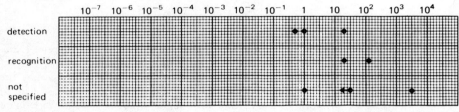

(307; 602; 671; 678; 709; 774; 775; 804)

O.I. at $20°C$ = 1,940,000 (316)

C. WATER POLLUTION:
—BOD_5: 0.03 std.dil.sew. (269)
—BOD_{10}: nil std.dil.sew. (256)
—$KMnO_4$ value: 0.026 (30)
—ThOD: 2.60 (30)
—Reduction of amenities:
organoleptic limit: 0.3 mg/l (181)
—Waste water treatment:
NFG, BOD, $20°C$, 1-10 days observed, feed: 200-1,000 mg/l, acclimation: 365 + P: nil% removed (93)

D. BIOLOGICAL EFFECTS:
—Fish:
guppy (Poecilia reticulata): 14 d LC_{50}: 2138 ppm (1833)
Lepomis macrochirus: static bioassay in fresh water at $23°C$, mild aeration applied after 24 hr:

| material added | % survival after | | | | best fit 96 hr LC_{50} |
ppm	24 hr	48 hr	72 hr	96 hr	ppm
10,000	100	100	100	100	>10,000
7,900	100	100	100	100	

Menidia beryllina: static bioassay in synthetic seawater at $23°C$, mild aeration applied after 24 hr:

| material added | % survival after | | | | best fit 96 hr LC_{50} |
ppm	24 hr	48 hr	72 hr	96 hr	ppm
10,000	90	90	90	90	>10,000

(352)

—Mammalia: mouse: inhalation: LC_{50}: 42,000 ppm, 3 hr

2-ETHYL-1-HEXANOL

	rat:	inhalation:	lethal:	64,000 ppm
	dog:	inhalation:	lethal:	106,000 ppm
	monkey:	inhalation:	lethal:	71,600–192,500 ppm	(211)
—Man:	severe toxic effects:	8,000 ppm = 24,624 mg/cu m, 60 min		
	symptoms of illness:	2,000 ppm = 6,156 mg/cu m		
	unsatisfactory:	>500 ppm = 1,539 mg/cu m		(185)
	nose irritation:	200 ppm		(211)

ethylethylene *see* α-butylene

S-ethylethylcyclohexylthiocarbamate *see* Ro-Neet

ethylformate (ethylmethanoate)
CH_3CH_2OCHO
A. PROPERTIES: colorless liquid; m.w. 74.08; m.p. −79/−80°C; b.p. 54°C; v.p. 192 mm at 20°C, 300 mm at 30°C; v.d. 2.55; sp.gr. 0.924 at 25/4°C; solub. 105,000 mg/l at 20°C, 118,000 mg/l at 25°C; THC 392 kcal/mole, LHC 369 kcal/mole; sat.conc. 774 g/cu m at 20°C, 1,170 g/cu m at 30°C
B. AIR POLLUTION FACTORS: 1 mg/cu m = 0.33 ppm, 1 ppm = 308 mg/cu m
 —Odor threshold: recognition: 54–61 mg/cu m (610)
C. WATER POLLUTION FACTORS:
 —Waste water treatment:
  NFG,BOD, 20°C, 1–10 days observed, feed: 200–1,000 mg/l, acclimation: 365 +
  P: 30% removed (93)
 —Impact on biodegradation processes:
  BOD_{10} test is not affected at 1 g/l (30)
 —Odor threshold: detection: 17 mg/kg (908)
D. BIOLOGICAL EFFECTS:
 —Mammalia: rat: inhalation: no deaths: conc.vap., 5 min
           5/6: 8,000 ppm, 4 hr
       cat: inhalation: no deaths: 4,000 ppm, 4 hr
       rat: single oral LD_{50}: 4.3 g/kg (211)
 —Man: eye and nose irritation: 330 ppm (211)

ethylglycol *see* ethyleneglycolmonoethylether

ethylglycolacetate *see* ethyleneglycolmonoethyletheracetate

ethylguthion *see* azinphosethyl

S-ethylhexahydro-1H-azepine-1-carbothioate *see* molinate

2-ethyl-1-hexanol
$CH_3(CH_2)_3CH(C_2H_5)CH_2OH$
 Uses: plasticizers for PVC resins; defoaming agent; wetting agent; organic synthesis; solvent mixtures for nitrocellulose, paints lacquers, baking finishes; textile finishing compounds; plasticizers; inks; rubber; paper; lubricants; photography; dry cleaning

2-ETHYL-1-HEXANOL

A. PROPERTIES: colorless liquid; m.w. 130.23; m.p. $-76°C$; b.p. $183.5°C$; v.p. 0.05 mm at $20°C$; v.d. 4.49; sp.gr. 0.834 at $20/20°C$; solub. 1,000 mg/l at $20°C$;

B. AIR POLLUTION: 1 mg/cu m = 0.18 ppm, 1 ppm = 5.41 mg/cu m
   —Odor: characteristic: quality: musty
                         hedonic tone: unpleasant to pleasant
   T.O.C.: abs.perc.lim.: 0.075 ppm
          50% recogn.: 0.138 ppm
         100% recogn.: 0.138 ppm
        O.I. 100% recogn.: 949   (19)

C. WATER POLLUTION FACTORS:
   —Impact on biodegradation processes:
   effect on respiration of glucose by mixed culture derived from activated sludge:

concentration mg/l	increase in lag period hours	respiration rate %
1	0	100
10	13	68
100	110	0
1,000	>200	0 (997)

   —Water quality:
   Kanawha river (USA): 13.11.1963: raw: 110 ppb
                                     sand filtered: 8 ppb
                      21.11.1963: raw: not detectable
                                     after aeration: 7 ppb
                                     carbon filtered: not detectable   (59)
   Delaware river (U.S.A.): conc. range: winter: 3–5 ppb
                                          summer: n.d.   (1051)

   —Waste water treatment:
   A.C.: adsorbability: 0.138 g/g; 98.5% reduction, infl.: 700 mg/l, effl. 10 mg/l
                                                                                                  (32)
      conventional municipal treatment: infl. 0.110 mg/l; effl. 0.008 mg/l; 86% removal
                                                                                                         (404)
      Biodegradation at 0.1 mg/l:

	normal sewage	adapted sewage	
after 24 hr	0%	100%	
after 135 hr	100%	100%	(997)

   —Reduction of amenities: T.O.C. = 0.27 mg/l   (297)(403)
   T.O.C. in water at room temp.: 1.3 ppm, range: 0.58–2.08 ppm, 13 judges
     20% of population still able to detect odor at 0.61   ppm
     10% of population still able to detect odor at 0.42   ppm
      1% of population still able to detect odor at 0.12   ppm
    0.1% of population still able to detect odor at 0.035 ppm   (321)

D. BIOLOGICAL EFFECTS:
   —Fish: rainbow trout: 96 hr LC_{50} (S): 32–37 mg/l   (1500)
   —Mammals:
      rat: single oral LD_{50}: 3.2–6.4 g/kg
      mouse: single oral LD_{50}: 3.2–6.4 g/kg
      rat: inhalation: no deaths: 235 ppm, 6 hr   (211)

2-ethylhexylamine

$$CH_3-(CH_2)_3-\underset{\underset{CH_2CH_3}{|}}{CH}-CH_2-NH_2$$

D. BIOLOGICAL EFFECTS:
—Toxicity threshold (cell multiplication inhibition test):
bacteria (*Pseudomonas putida*):	82 mg/l	(1900)
algae (*Mycrocystis aeruginosa*):	0.02 mg/l	(329)
green algae (*Scenedesmus quadricauda*):	0.36 mg/l	(1900)
protozoa (*Entosiphon sulcatum*):	12 mg/l	(1900)
protozoa (*Uronema parduczi Chatton-Lwoff*):	8.0 mg/l	(1901)

2-ethylhexylglycidylether (1-glycidyloxy-2-ethylhexane)

$$\underset{O}{CH_2\!\!-\!\!\!CH}-CH_2-O-CH_2-\underset{\underset{C_2H_5}{|}}{CH}-(CH_2)_3-CH_3$$

A. PROPERTIES: b.p. 118–120°C at 20 mm; sp.gr. 0.893 at 20/4°C; colorless liquid
C. WATER POLLUTION FACTORS:
—BOD_5: 0.14 NEN 3235-5.4 (277)
—COD: 2.46 NEN 3235-5.3 (277)
D. BIOLOGICAL EFFECTS:
Fish: goldfish: LD_{50} (24 hr): 14 mg/l—modified ASTM D 1345 (277)

ethylhydrosulfide see ethylmercaptan

ethylidenechloride see 1,1-dichloroethane

ethylidenecyanohydrin see lactonitrile

ethylidenedichloride see 1,1-dichloroethane

ethylisoamylketone see ethyl-*sec*-amylketone

ethylisobutylmethane see 2-methylhexane

ethylketone see diethylketone

D-N-ethyllactamidecarbanilate see carbetamex

ethylmercaptan (ethanethiol; ethylhydrosulfide; ethylthioalcohol)
C_2H_5SH
  Uses: LPG odorant; adhesive stabilizer; chemical intermediate
A. PROPERTIES: m.w. 62.13; m.p. -121/-144°C; b.p. 36°C; v.p. 440 mm at 20°C, 640 mm at 30°C; v.d. 2.14; solub. 15,000 mg/l; sat.conc. 1,488 g/cu m at 20°C, 2,093 g/cu m at 30°C
B. AIR POLLUTION FACTORS: 1 mg/cu m = 0.388 ppm, 1 ppm = 2.582 mg/cu m

—Odor: characteristic: quality: decayed cabbage, earthy, sulfidy

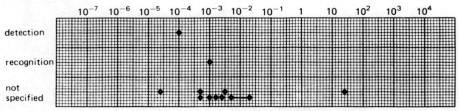

(2; 71; 210; 307; 602; 622; 652; 710; 802; 829; 843; 863)

    threshold: unadapted: 2 ppb; after adaption with pure odorant: 28 ppb   (204)
    O.I. at 20°C = 289,500,000   (316)
—Control methods:
    odor removal: median odor threshold level increases from <7.8 ppm to >125
      ppm in 50% aqueous ethyleneglycol after reaction with chloramine-T   (1821)
    activated carbon: retentivity: 25 wt% of adsorbent   (83)

C. WATER POLLUTION FACTORS:
    —Reduction of amenities:
    taste and odor at 0.00019 mg/l   (51)
    faint odor: 0.00026 mg/l   (98; 129)

ethylmethanoate *see* ethylformate

1-ethyl-4-methylbenzene *see* p-ethyltoluene

N-ethylmorpholine

A. PROPERTIES: m.w. 115.2; m.p. -65/-63°C; b.p. 138/139°C; v.p. 6.1 mm at 20°C; v.d. 4.00; sp.gr. 0.987 at 20/20°C;

B. AIR POLLUTION FACTORS: 1 mg/cu m = 0.21 ppm, 1 ppm = 4.79 mg/cu m
    —Odor: characteristic; quality: ammoniacal
                        hedonic tone: unpleasant to pleasant
    T.O.C.: abs.perc.lim.:   0.08 ppm
             50% recogn.:   0.25 ppm
            100% recogn.:   0.25 ppm
        O.I. 100% recogn.: 26,280   (19)

C. WATER POLLUTION FACTORS:
    —Waste water treatment:
      A.C.: adsorbability: 0.107 g/g C; 52.3% reduction, infl.: 1,000 mg/l, effl.: 467
        mg/l   (32)

D. BIOLOGICAL EFFECTS:
—Mammalia: rat: single oral LD_{50}: 1.78 g/kg
inhalation: no deaths: sat.vap., max 2 hr (211)

ethyl-p-nitrophenylphenylphosphonothioate *see* EPN

ethyl-p-nitrophenylthionobenzenephosphate *see* EPN

ethylolamine *see* ethanolamine

ethyloxide *see* ethylether

ethylphenanthrene
B. AIR POLLUTION FACTORS:
—Manmade sources:
in coke oven emissions: 58.0–1,578,6 µg/g of sample (n.s.i.) (960)

o-ethylphenol

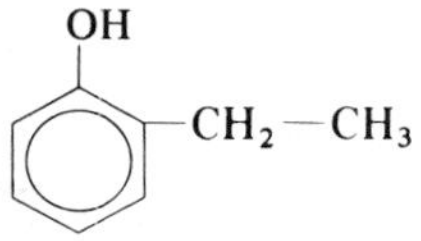

A. PROPERTIES: m.w. 122.17; m.p. -18°C; b.p. 195-197°C; sp.gr. 1.037
B. AIR POLLUTION FACTORS:
—Manmade sources:
in exhaust of a 1970 Ford Maverick gasoline engine operated on a chassis dynamometer following the 7-mode California cycle:
from API #7 gasoline: 0.2 ppm
from API #8 gasoline: 0.4 ppm (1053)
C. WATER POLLUTION FACTORS:
—Taste threshold conc.: 0.03 mg/l (998)
—Odor threshold: detection: 0.3 mg/l (998)

p-ethylphenol

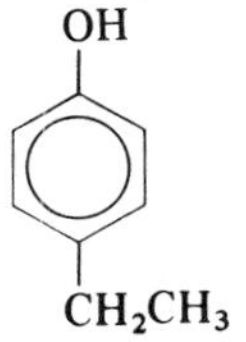

A. PROPERTIES: colorless needles; m.w. 122.17; m.p. 42-45°C; b.p. 218-219°C; $\log P_{oct}$ 2.66/2.81 (calculated)
C. WATER POLLUTION FACTORS:
—Taste threshold conc.: 0.01 mg/l (998)
—Odor threshold: detection: 0.6 mg/l (998)

—Manmade sources:
60 ml/min pure water passed through 25 ft, $\frac{1}{2}$ inch I.D. tube of general chemical grade PVC contained 0.24 ppb p-ethylphenol which constituted 2.23% of total contaminant concentration (430)

D. BIOLOGICAL EFFECTS:
—Ciliate (*Tetrahymena pyriformis*): 24 hr LC_{100}: 2.07 mmole/l (1662)

O-ethyl-S-phenylethylphosphonodithioate (dyfonate; fonofos)

$$C_2H_5O\underset{C_2H_5O}{\overset{}{\diagdown}}\overset{S}{\underset{S-\bigcirc}{\overset{\diagup\diagup}{P}}}$$

Use: soil insecticide

C. WATER POLLUTION FACTORS:
—persistence in soil at 10 ppm initial concentration:

	weeks incubation to	
	50% remaining	5% remaining
sterile sandy loam	>24	–
sterile organic soil	>24	–
nonsterile sandy loam	3	20
nonsterile organic soil	4	20

D. BIOLOGICAL EFFECTS:
—Fish:
bluegill: 24 hr LC_{50}: 0.045 ppm
rainbow trout: 24 hr LC_{50}: 0.110 ppm (1878)
—Mammals:
acute oral LD_{50} (male rat): 8–17.5 mg/kg (techn. grade)
acute dermal LD_{50} (rabbit): 25 mg/kg (1854)

ethylphthalate *see* diethylphthalate

ethylpropenoate *see* ethylacrylate

ethylpropionate
$CH_3CH_2COOC_2H_5$
A. PROPERTIES: $\log P_{oct}$ 1.21
B. AIR POLLUTION FACTORS:
—Odor threshold: recognition: 0.3–0.5 mg/cu m (610)
D. BIOLOGICAL EFFECTS:
—Toxicity threshold (cell multiplication inhibition test):
bacteria (*Pseudomonas putida*):   270 mg/l
green algae (*Scenedesmus quadricauda*):   14 mg/l
protozoa (*Entosiphon sulcatum*):   560 mg/l (1900)
protozoa (*Uronema parduczi Chatton-Lwoff*): 665 mg/l (1901)
—Amphibian:
mexican axolotl (3–4 w after hatching): 48 hr LC_{50}: 54 mg/l
clawed toad (3–4 w after hatching):   48 hr LC_{50}: 56 mg/l (1823)

ethylpropylcarbinol *see* 3-hexanol

ethylpyrophosphate *see* TEPP

ethylsulfide *see* diethylsulfide

ethylthioalcohol *see* ethylmercaptan

N-(2-ethylthio)benzenesulfonamide-S,O,O-diisopropylphosphorodithioate *see* bensulide

ethylthioethane *see* diethylsulfide

o-ethyltoluene

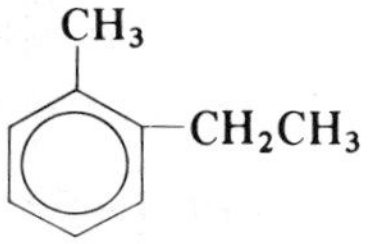

Manufacturing source: organic chemical industry; petroleum refining industry.
Users and formulation: mfg. dyes, medicinals, flavors, perfumes, sweeteners, germicides; asphalt and naphtha constituent.
B. AIR POLLUTION FACTORS:
  —Atmospheric reactions:
    estimated lifetime under photochemical smog conditions in S.E. England: 27 hr
(1699; 1700)

m-ethyltoluene

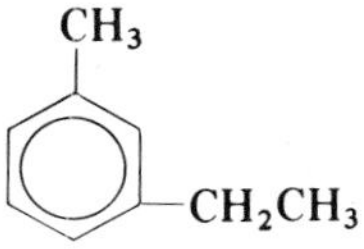

B. AIR POLLUTION FACTORS:
  —Atmospheric reactions:
    estimated lifetime under photochemical smog conditions in S.E. England: 1.9 hr
(1699; 1700)

p-ethyltoluene (1-ethyl-4-methylbenzene)

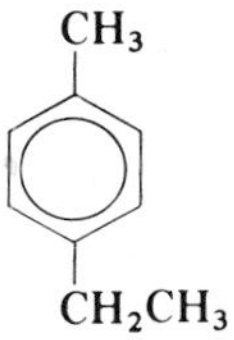

O-ETHYL-O-(2,4,5-TRICHLOROPHENYL)ETHYLPHOSPHONOTHIOATE

A. PROPERTIES: colorless liquid; m.w. 120.19; m.p. $-20°C$; b.p. $162°C$; v.d. 4.15; sp.gr. 0.862

B. AIR POLLUTION FACTORS: 1 mg/cu m = 0.20 ppm, 1 ppm = 5.00 mg/cu m
 —Atmospheric reactions: reactivity: HC. conc.: ranking: 3 (63)
 estimated lifetime under photochemical smog conditions in S.E. England: 2.9 hr
 (1699; 1700)

 —Manmade sources:
 glc's: carpark: 250 μg/cu m
 motorway: 920 μg/cu m (48)
 in diesel engine exhaust: 0.7% of emitted hydrocarbons (72)
 glc's in the Netherlands:
 tunnel Rotterdam 1974.10.12: avg.: 11 ppb ($n = 12$)
 max.: 17 ppb
 The Hague 1974.10.11: avg.: 10 ppb ($n = 12$)
 max.: 30 ppb
 Roelofarendsveen 1974.9.11: avg.: 0.5 ppb ($n = 12$)
 max.: 1 ppb (1231)

O-ethyl-O-(2,4,5-trichlorophenyl)ethylphosphonothioate *see* trichloronat

ethyltriglycol *see* triethyleneglycolmonoethylether

ethylxanthate
$C_2H_5O(S)SH$
A. PROPERTIES: pale yellow powder; b.p. decomposes at $180°C$; solub. $>20\%$
B. AIR POLLUTION FACTORS:
 —Odor threshold: detection: 0.03 mg/cu m (637)
C. WATER POLLUTION FACTORS:
 —Impact on biodegradation processes:
 ~50% inhibition of NH_3 oxidation in *Nitrosomonas* at 12 mg/l (407)
D. BIOLOGICAL EFFECTS:
 —Fish:
 rainbow trout (*Salmo gairdneri*): 96 hr LC_{50} (static test):
 sodium salt: Cyanamid C 325: 29–37 mg/l
 Dow Chemical Z 4: 10–50 mg/l
 potassium salt: Cyanamid C 303: 52 mg/l
 Dow Chemical Z 3: 10–100 mg/l
 LC_{100}:
 sodium salt: 56 mg/l after 4 days exposure (static test)
 1.0 mg/l after 8 days exposure (flow through test)
 potassium salt: 100 mg/l after 2 days exposure (static test)
 0.5 mg/l after 2 days exposure (flow through test) (1087)
 —Mammals:
 ingestion: rats: single oral LD_{50}: between 500 and 1000 mg/kg (1546)

ethynylcarbinol *see* propargylalcohol

ETU *see* ethylenethiourea

eulan
   Use: treatment of textiles
   Active ingredient: N-chloromethylsulfopentachloro-2-aminodiphenylether, sodium salt
C. WATER POLLUTION FACTORS:

	eulan U33	eulan WA neu
—BOD_5	540 mg/g	790 mg/g
—COD	1200 mg/g	1440 mg/g

   —Biological degradation:
   70% elimination at 50 mg/l, after 5 hr aeration (OECD-confirmatory test)
   active ingredient is not degraded but adsorbs on the sludge   (1718)
D. BIOLOGICAL EFFECTS:

		eulan U33	eulan WA neu
—Bacteria:			
Pseudomonas fluoresc.:	no effect at	10 g/l	10 g/l
E. coli:	no effect at	5 g/l	5 g/l
—Crustacean:			
Daphnia pulex:	no effect at	5 g/l	5 g/l
—Fish:			
goldfish: 96 hr LC_0:		100 ppb	1000 ppb
96 hr LC_{50}:		500 ppb	
bluegill: 96 hr LC_{50}:		100 ppb	
rainbow trout: 96 hr LC_{50}:		127 ppb	
—Mammals:			
rats: acute oral toxicity:		600 mg/kg	1000 mg/kg

euparen (dichlorofluanid; N-(dichlorofluoromethylthio)-N,N-dimethylbenzene-sulfonamide)

$$\begin{array}{c} CH_3 \\ \\ CH_3 \end{array} N-\underset{\underset{O}{\|}}{\overset{\overset{O}{\|}}{S}}-N-S-\underset{F}{\overset{Cl}{\underset{|}{\overset{|}{C}}}}-Cl$$

Use: fungicide
A. PROPERTIES: m.p. 105–105.6°C; v.p. 1×10^{-6} mm at 20°C
D. BIOLOGICAL EFFECTS:
   —Residues:
   major product of euparen degradation on strawberry leaves:
   N,N-dimethylbenzensulfonamide
   residues of dichlofluanid in canned products of strawberry:
   0.1–0.59 ppm
   the content of dichlofluanid residues on strawberry treated twice with fungicide
   varies from 0.5 to 1.0 ppm, for strawberry sprayed three times from 1.0 to 2.0
   ppm, respectively   (1351)
   —Toxicity: mammals:
   acute oral LD_{50} (rats): 325–1000 mg/kg (aqueous emulsion)
   acute dermal LD_{50} (rats): 1000 mg/kg
   in diet: rats fed 1000 mg/kg/day for 4 months suffered no ill effects   (1855)

F

F 12 *see* dichlorodifluoromethane

F 72 (ammonium salt of saturated carboxylic acid)
Use: dispersant—oil clean up—improves biodegradation of spilled oils.
D. BIOLOGICAL EFFECTS:
—Algae:
Phaeodactylum tricornutum: at 300 ppm growth decreased by 50%
at 500 ppm growth decreased by 80% (1089)
Dunaliella tertiolecta: 50% reduction of growth rate at 1000 ppm (1090)
—Crustacean:
Daphnia magna: LC_{50}, 24 hr: 280 ppm (1089)
Artemia salina: LD_{50}, 48 hr: $>$1000 ppm (493)
—Molluscs:
Littorina littorea: LD_{50}, 48 hr: $>$1000 ppm
Crassostrea gigas: LD_{50}, 48 hr: 1000 ppm
Mytilus edulis: LD_{50}, 48 hr: 500 ppm (493)
—Fish:
Phoxinus phoxinus: LC_{50}, 24 hr: 300 ppm (1089)
Anguilla anguilla: LD_{50}, 48 hr: 300 mg/l (493)
—Mammals:
Mus musculus: ingestion of emulsions containing up to 5% of F 72 did not produce any effect on the weight of the animals (1089)

fenac *see* 2,3,6-trichlorophenylaceticacid

fenaminosulf *see* dexon

fenchlorphos *see* ronnel

fenethcarb
D. BIOLOGICAL EFFECTS:
—Crustacean:
Daphnia magna: LC_{50}, 24 hr: 1.0 mg/l (1002, 1004)
—Insects:
Hydropsyche larvae: significant deterioration of net construction after 48 hr: 1.0 µg/l (1006, 1007, 1008)

—Fish:
   Salmo gairdneri: LC_{50}, 96 hr: 35,000 µg/l (1001)

fenitrothion (O,O-dimethyl-O-(4-nitro-*m*-tolyl)phosphorothioate; accothion; sumithion)

$(CH_3O)_2PSO$—⟨C₆H₃(CH₃)⟩—NO_2

Use: organophosphorus insecticide. Used to control the spruce budworm (*Choristoneura fumiferana*) in North American forests. (1667)
Formulation contains 5-methyl fenitrothion (2.47%). (1595)
A. PROPERTIES: sp.gr. 1.32, b.p. 140–145°C at 0.1 mm decomp., solub. 30 ppm, v.p. 5.4×10^{-5} mm at 20°C, log P_{oct} 3.38 at 20°C
C. WATER POLLUTION FACTORS:
   —Biodegradation:

[Scheme: fenitrothion → aminofenitrothion → demethyl aminofenitrothion]

possible degradation route (1595)

cellulose degrading microorganisms in forest soil: 1 and 5 ppm is not inhibitory as measured by evolution of $^{14}CO_2$ (1667)
degradation in aquatic model system containing only water and soil showed a large number of metabolites including fenitrooxon, aminofenitrothion, and its N-formyl- and N-acetyl derivatives as well as demethylated products. Several phenolic products are found in water, together with a sulfate conjugate of 3-methyl-4-nitrophenol. 3-Methyl-4-nitrophenol is among the degradation products present in the highest concentrations both in water and in soil. (1834)
   —Aquatic reactions:
   It is hydrolyzed by alkali. The half-lifetime in 0.01N NaOH at 30°C is 272 minutes.
D. BIOLOGICAL EFFECTS:
   —Residues:
      duckweed (*Lemna minor*): conc. in residue   time (hr) post spray

1.44 ppm	1
4.00 ppm	10
0.10 ppm	23
0.03 ppm	192 (1595)

—Bioaccumulation:

	conc. in water mg/l	uptake rate K_1	excretion constants K_2	accumulation coefficient K_1/K_2
clam (*Mya arenaria*)	0.013	10	0.46	22
	0.00043	19	0.55	35
	0.00018	12	0.62	19
mussel (*Mytilus edulis*)	0.013	47	0.60	78
	0.00018	52	0.40	130
freshwater clam (*Anodonta cataractae*)	0.00083	3	0.35	9

(1558)

BCF in underyearling trout: 250
    yearling trout: 230
    minnow: 200 (1834)

BCF in aquatic model ecosystem after 21 days:
  alga: 181
  Daphnia pulex: 69
  snail (*Cipango paludina japonica Martens*): 33
  carp (*Cyprinus carpio*): 98 (1834)

—Metabolism:
  fish metabolism: fenitrothion is metabolized in rainbow trout through oxidation to phosphate, cleavage at the P-O-aryl linkage, O-demethylation, and conjugation with glucuronic acid (1834)

—Toxicity:
—Insect: ricefield spider (*Oedothorax insecticeps*): LD_{50}: 3200 ppm (1814)
—Crustacean: *Daphnia*: 3 hr LC_{50}: 0.0092 ppm (1835, 1836)
—Fish: *Channa punctatus*: 180 hr, LC_{50} (R): 1.5 mg/l (1508)
    carp: 24 hr, LC_{50} (S): 3.31 mg/l
        48 hr, LC_{50} (S): 2.55 mg/l
        72 hr, LC_{50} (S): 2.30 mg/l (1103)
    Saccobranchus fossilis: 96 hr, LC_{50} (S): 12.5 mg/l (1509)
    chingatta: 96 hr, LC_{50} (S): 12.2 mg/l (1494)
    rainbow trout: 48 hr, LC_{50}: 1.28 ppm
    bluegill: 48 hr, LC_{50}: 2.72 ppm
    carp: 48 hr, LC_{50}: 4.4 ppm
    carp: no effect dosage after 4 week exposure: 0.02 ppm (1835; 1836)

—Mammals:
  acute oral LD_{50} (rats): 250 mg/kg (techn. grade)

(mice): 870 mg/kg
acute dermal LD_{50} (mice): >3000 mg/kg (1855)

fenoprop *see* 2(2,4,5-trichlorophenoxy)propionic acid

fenpropanate ((+)α-cyano-3-phenoxybenzyl-2,2,3,3-tetramethylcyclopropane-carboxylate; a pyrethroid; *see also* pyrethroids)

Use: experimental photostable pyrethroid, insecticide (1577)
A. PROPERTIES: solub. 0.026 ppm at room temp., log P_{oct} 3.03 (1577)
D. BIOLOGICAL EFFECTS:
—Fish:
rainbow trout: technical grade: 24 hr, LC_{50}: 76.7 ppb active ingredient
(static test)
formulated product: 24 hr, LC_{50}: 8.6 ppb active ingredient
(static test) (1577)

fensulfothion (dasanit; Terracur P; diethyl-4-(methylsulfinyl)phenylphosphorothionate)

Use: insecticide and nematocide
A. PROPERTIES: oily liquid, yellowish color, b.p. 138–141°C at 0.01 mm, sp.gr. 1.2, solub. about 1600 ppm
C. WATER POLLUTION FACTORS:
—Persistence in soil at 10 ppm initial concentration

	weeks incubation to	
	50% remaining	5% remaining
sterile sandy loam	>24	—
sterile organic soil	>24	—
nonsterile sandy loam	<1	4
nonsterile organic soil	1	6	(1433)

D. BIOLOGICAL EFFECTS:
—Mammalia:
acute oral LD_{50} (male rats): 4.6–10.5 mg/kg
acute dermal LD_{50}–in xylene (rats): 3.5–30 mg/kg
in diet: rats fed 16 weeks on a diet containing 20 ppm showed normal growth rate and food consumption

fenthion *see* Baytex

fentinhydroxide *see* triphenyltinhydroxide

fenuron *see* 3(phenyl)-1,1-dimethylurea

fenvalerate ((+)α-cyano-3-phenoxybenzyl-(+)-α-(4-chlorophenyl)isovalerate; a pyrethroid; *see also* pyrethroids)

Use: experimental photostable pyrethroid, insecticide	(1577)
A. PROPERTIES: solub. 0.085 ppm at room temp., log P_{oct} 4.42	(1577)

D. BIOLOGICAL EFFECTS:
—Fish:
   rainbow trout: technical grade: 24 hr LC_{50}: 76.0 ppb active ingredient
                                                                      (static test)
            formulated product: 24 hr LC_{50}: 21.0 ppb active ingredient
                                                                      (static test)   (1577)

FD Red no. 2 *see* amaranth

fire-trol
  Use: forest fire retardant
D. BIOLOGICAL EFFECTS:

	coho salmon	96-hr LC_{50}	90 to >1 500 mg/l	S
		96-hr LC_{50}	280 mg/l	FT
Fire-Trol 100		20-da LC_{50}	73 mg/l	FT
	rainbow trout	96-hr LC_{50}	150 to >1 000 mg/l	S
		96-hr LC_{50}	>100 mg/l	FT
		20-da LC_{50}	43 mg/l	FT
	bluegill	96-hr LC_{50}	>1 500 mg/l	S
	fathead minnow	96-hr LC_{50}	>1 500 mg/l	S
	largemouth bass	96-hr LC_{50}	>1 500 mg/l	S
Fire-Trol 931	coho salmon	96-hr LC_{50}	580–1 000 mg/l	S
	rainbow trout	96-hr LC_{50}	700–1 000 mg/l	S
	bluegill	96-hr LC_{50}	>1 500 mg/l	S
	fathead minnow	96-hr LC_{50}	>1 500 mg/l	S
	largemouth bass	96-hr LC_{50}	>1 500 mg/l	S

(1197)

fluoranthene (idryl)

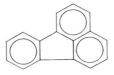

Sources:
in Kuwait crude oil:	2.9 ppm	
in South Louisiana crude oil:	5.0 ppm	(1015)
in commercial coal tar:	20,000 ppm	(1874)
in coal tar:	54.28 mg/g of sample	(993)
in wood preservative sludge:	26.47 g/l of raw sludge	(993)
in gasoline:	1.84 mg/l	(1070)
high octane number:	1.65 mg/l	(380)
	7.56 mg/kg	(1226)
	2.12–10.10 mg/kg (n = 13)	(385)
low octane number:	0.70–3.52 mg/kg (n = 13)	(385)
in lubricating motoroils:	0.06–0.28 ppm (6 samples)	(379)
in motoroil:	0.11 mg/kg	(1226)

in used motoroil: after 18 european driving cycles: 5.4–6.0 mg/kg
  after 5,000 km: 109.0–173.0 mg/kg
  after 10,000 km: 129.0–270.0 mg/kg
  after 10,000 km + 18 european
   driving cycles: 143–302.1 mg/kg (1226)

A. PROPERTIES: m.w. 202; m.p. 107°C; b.p. 250°C; solub: 0.265 ppm at 25°C, in seawater at 22°C: 0.1 ± 0.06 ppm, 0.120 mg/l at 24°C (99% purity)

B. AIR POLLUTION FACTORS:
  —Manmade sources:
   in tail gases of gasoline engine: 39–248 µg/cu m (340)
   emissions from typical European gasoline engine (1608 cu cm)—following European driving cycles—using leaded and unleaded commercial gasolines: 17.8–235.5 µg/l fuel burnt (1291)
   in exhaust condensate of gasoline engine: 1.06–1.66 mg/l gasoline consumed
    (1070)
   in stack gases of municipal incinerator, after spray tower and electrostatic precipitator: 0.58 mg/1000 cu m; in residues: 58 µg/kg (341)
   in coal tar pitch fumes: 11.8 wt % (519)
   in airborne coal tar emissions: 144.8 mg/g of sample or
    5,090 µg/cu m of air sampled
   in coke oven emissions: 269.7–5,979 µg/g of sample (993)
   emissions from space heating installation:
    burning coal (underfeed stoker): 38 mg/10^6 Btu input
     gasoil: 1.9 mg/10^6 Btu input
     gas: 2.9 mg/10^6 Btu input (954)
  —Ambient air quality:

glc's in the Netherlands (ng/cu m):

		Delft	The Hague	Vlaardingen	Amsterdam	Rotterdam
summer	1968	n.d.	—	3	—	1
	1969	5	3	4	3	1
	1970	4	1	3	2	2
	1971	7	2	1	2	2
winter	1968	13	11	27	25	8
	1969	17	6	25	18	8
	1970	6	4	10	9	3
	1971	1	3	4	1	4

(1277)

glc's at Botrange (Belgium): woodland at 20–30 km from industrial area, June–July 1977: 0.41; 0.46 ng/cu m ($n = 2$)

glc's at Wilrijk (Belgium): residential area: Oct.–Dec. 1976: 1.17; 14.87 ng/cu m
($n = 2$) (1233)

glc's in the average American urban atmosphere—1963:
~30 µg/g airborne particulates or
~4 ng/g cu m air (1293)

glc's in Birkenes (Norway), Jan–June 1977: avg.: 1.03 ng/cu m
range: n.d.–6.28 ng/cu m ($n = 18$)

glc's in Rørvik (Sweden), Dec. 76–April '77: avg.: 2.04 ng/cu m
range: 0.19–13.14 ng/cu m ($n = 21$)
(1236)

glc's in Budapest 1973: heavy traffic area: winter: 7.8 ng/cu m
(6 a.m.–8 p.m.) summer: 8.5 ng/cu m
low traffic area: winter: 2.0 ng/cu m
summer: 2.1 ng/cu m (1259)

glc's in residential area (Belgium)—Oct. 1976:
in particulate sample: 2.22 ng/cu m
in gasphase sample: 8.52 ng/cu m (1289)

organic fraction of suspended matter:
Bolivia at 5200 m altitude (Sept.–Dec. 1975): 0.032–0.041 µg/1000 cu m
Belgium, residential area (Jan.–April 1976): 0.72–1.4 µg/1000 cu m (428)

C. WATER POLLUTION FACTORS:
—Manmade sources:
in domestic sewage effluent: 0.00001 mg/l (227)
effluent spray tower of municipal incinerator: 0.54 µg/l (341)
in outlet of waterspray tower of asphalt hot-road-mix process: 2800 ng/cu m
(1292)

in leachate from test panels freshly coated with coal tar:
influent: 0.003 µg/l
effluent: 0.081 µg/l (1874)

primary and digested raw sewage sludge: 0.525–1.2 ppm
liquors from sewage sludge heat treatment plants: 0.04–1.5 ppm
sludge cake from heat treatment plants: 0.47–1.6 ppm
final effluent of sewage works: 0.35 ppb (1426)
—Water and sediment quality:

W. Germany:
    river Gersprenz at Munster: 38.5–71.3 ng/l
    river Danube at Ulm: 61–94 ng/l
    river Main at Seligenstadt: 21.3–694 ng/l
    river Aach at Stockach: 379–761 ng/l
    river Schussen: 358 ng/l   (530)
in rapid sand filter solids from Lake Constance water: 1.0–10.0 mg/kg
in river water solids:
    river Rhine: 2.0 mg/kg
    river Aach at Stockach: 2.6–20.0 mg/kg
    river Argen: 2.0 mg/kg
    river Schussen: 11.0 mg/kg   (531)
in wells and galeries of an aquifer: Brussels sands (sands covered with a thick
    loamy layer): 0.5–8.6 ng/l   (1066)
in domestic effluents: 0.273–2.416 ppb
sewage (high percentage industry): 2.66–3.42 ppb
sewage during dry weather: 0.352 ppb
sewage during heavy rain: 16.35 ppb   (531)
in groundwater (W. Germany, 1968):   6.5–41.5 µg/cu m ($n = 10$)
in tapwater (W. Germany, 1968):   23.0–118.3 µg/cu m ($n = 6$)   (955)
in Eastern Ontario drinking waters (June–Oct. 1978): 0.2–3.9 ng/l ($n = 12$)
in Eastern Ontario raw waters (June–Oct. 1978): 0.7–1.4 ng/l ($n = 2$)   (1698)
in Thames river water: at Kew Bridge: 140 ng/l
                      at Albert Bridge: 200 ng/l
                      at Tower Bridge: 360 ng/l   (529)
in sediments in Severn esturary (U.K.): 0.5–5.2 ppm dry wt.   (1467)
—Waste water treatment method:

	Dec. 1965	May 1966	
in raw sewage	45.3 ppb	3.23 ppb	
after mechanical purification	14.6 ppb	1.22 ppb	
after biological purification	0.26 ppb	0.28 ppb	(545)

concentration at various stages of water treatment works:
    river intake: 0.150 µg/l
    after reservoir: 0.140 µg/l
    after filtration: 0.081 µg/l
    after chlorination: 0.045 µg/l   (434)

plant location	water source	conc. ng/l	drinking water* conc. ng/l	% removal/ transformation
Pittsburgh, Pa.	Monongahela river	408	n.d.	100**
Huntington, W.Va.	Ohio river	38	2.4	89.9
Philadelphia, Pa.	Delaware river	125	9.0	92.2

*treatment provided: lime, ferric sulfate or chloride; activated carbon, chlorination
and fluoridation
**two stages A.C.: powdered carbon and granular carbon   (958)
    —Sampling and analysis:
        recovery with open pore polyurethane (OH/NCO = 2.2): 77–97% depending
        upon load-quantitative elution with methanol   (929)

D. BIOLOGICAL EFFECTS:
—Uptake and depuration by Oysters (*Crassostrea virginica*) from oil treated enclosure.

time of exposure days	depuration time days	concentration oysters μg/g	water μg/l	accumulation factor oysters/water
2	–	5.0	7.2	695
8	–	4.0	0.4	10,000
2	7	1.7	–	–
8	7	1.4	–	–
8	23	0.4	–	–

half-life for depuration: 5 days   (957)
Neanthes arenaceodentata: 96 hr TLm in seawater at 22°C: 0.5 ppm (initial conc. in static assay)   (955)

fluorene

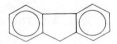

Sources: in commercial coal tar: 1200 ppm   (1874)
            in coal tar: 27.39 mg/g of sample   (993)
            in wood preservative sludge: 6.61 g/l of raw sludge   (993)
A. PROPERTIES: solub. 1.9 mg/l at 25°C, in seawater at 22°C: 0.8 ± 0.2 ppm
B. AIR POLLUTION FACTORS:
—Manmade sources:
  in airborne coal tar emissions: 2.41 mg/g of sample or 85 μg/cu m of air sampled
  in coke oven emissions: 15.4–271.5 μg/g of sample   (993)
  in coal tar pitch fumes: 9.1 wt %   (516)
C. WATER POLLUTION FACTORS:
—Manmade sources:
  in leachate from test panels freshly coated with coal tar:
    influent: 0.001 μg/l
    effluent: 0.021 μg/l   (1874)
—Biodegradation:
  biodegradation to CO_2:

sampling site	concentration	month	incubation time (hr)	degradation rate (μg/l/day) × 10^3	turnover time (days)
control station	50	–	–	0	∞
near oil storage tanks	50	–	–	38 ± 6	1400
	50	–	–	0	∞
Skidaway river	30	Jan.	72	0	∞
	30	Aug.	72	12 ± 9	7500
	30	Aug.	24	0	∞

(381)

degradation in seawater by oil-oxidizing micro-organisms (in presence of 0.19 mg/l 3,4-benzpyrene and 0.365 mg/l pyrene): initial conc. 0.35 mg/l, after 12 days: 0.27 mg/l; 23% decrease (1237)
microbial degradation to CO_2—in seawater at 12°C—in the dark after 48 hr incubation at 30 µg/l: nil, even after addition of water extract of no. 2 fuel oil.
(477)
—Aquatic reactions:
photo-oxidation by U.V. in aqueous medium at 90–95°C:
time for the formation of CO_2 (% of theoretical): 25%: 75.3 hr
50%: 160.6 hr
75%: 297.4 hr (1628)
in estuarine waters: at 15 µg/l, 12% adsorbed on particles after 3 hr (381)
—Water and sediment quality:
in sediment of Wilderness Lake, Colin Scott, Ontario (1976): 38 ppb (dry wt)
(932)
in Eastern Ontario drinking waters (June–Oct. 1978): 0.04–1.8 ng/l ($n = 12$)
in Eastern Ontario raw waters (June–Oct. 1978): 0.4–0.9 ng/l ($n = 2$) (1698)
D. BIOLOGICAL EFFECTS:
—*Neanthes arenaceodentata*: 96 hr TLm in seawater at 22°C: 1.0 ppm (initial conc. in static assay) (995)
—Carcinogenicity: negative
—Mutagenicity in the *Salmonella* test: negative
<0.01 revertant colonies/nmol
<70 revertant colonies at 1000 µg/plate (1883)

fluorenecarbonitrile
Sources: in wood preservative sludge: 7.64 g/l of raw sludge
in coke oven emissions: 11.5–180.3 µg/g of sample (993)

9-fluorenone
C. WATER POLLUTION FACTORS:
—Water quality:
in Eastern Ontario drinking waters (June–Oct. 1978): n.d.–1.9 ng/l ($n = 12$)
in Eastern Ontario raw waters (June–Oct. 1978): 1.2–10.4 ng/l ($n = 2$) (1698)

fluorescent whitening agents (FWA)
D. BIOLOGICAL EFFECTS:
—Fish:
FWA 2A: bluegill: 96 hr LC_{50}, S: 474 mg/l
FWA 3A: bluegill: 96 hr LC_{50}, S: 32 mg/l
FWA 4A: bluegill: 96 hr LC_{50}, S: 26 mg/l
FWA 1A: bluegill: 96 hr LC_{50}, S: 1,000 mg/l (1121)

fluorotrichloromethane (trichlorofluoromethane; F11)
CCl_3F
Manufacturing source: organic chemical industry. (347)
Users and formulation: mfg. aerosol sprays; mfg. commercial refrigeration equip-

676 FLUOROTRICHLOROMETHANE

ment; blowing agent for polyurethane foams; cleaning compounds mfg.; solvent; fire extinguisher (347)
Man caused sources (water and air): general use of spray cans, polyurethane foams, cleaning compounds/solvents, fire extinguishers; lab use. (347)

A. PROPERTIES: colorless liquid; m.w. 137.4; m.p. $-111°C$; b.p. $23.8°C$; v.p. 0.904 atm at $20°C$, 1.29 atm at $30°C$; sp.gr. 1.494 at $17°C$; solub. 1,100 mg/l at $25°C$; sat.conc. 5,111 g/cu m at $20°C$

B. AIR POLLUTION FACTORS: 1 mg/cu m = 0.18 ppm, 1 ppm = 5.7 mg/cu m
 —Odor: characteristic: quality: sweet
   hedonic tone: pleasant to unpleasant
  T.O.C.: abs. perc. lim.: 5.0 ppm
   50% recogn.: 135 ppm
   100% recogn.: 209 ppm
   O.I. 100% recogn.: 84 (19)
 —Atmospheric reactions:
  no photolysis in vitro after 60 days (201)
 —Ambient air quality:

sampling date	sampling location	sampling height m	measured concentration pptv	
Dec. 1971	Northern lat. 0–50°N	sea level	40–80	(374)
Feb.–Aug. 1973	S. California	6700	60	(378)
May 1974	New Mexico	6400	78–82	(373)
Jan. 1974	Greenland Sea and Arctic Ocean	30–5500	118 ± 20	
June/July '74	Western Ireland	sea level	79.8 ± 4.9	(375)
Oct. 1973	North Atlantic	sea level	88.6 ± 4.05	(375)
Sept. 1975	46–80°N	sea level	80–95	(377)
Nov. 1974	Pullman, WA	0–3600	128 ± 5	(372)
Dec. '74/Feb. '75	Pullman, WA	not reported	125 ± 8	(315)
March 1975	Farnborough-Aberporth U.K.	2000–6100	156 ± 21	
March 1975	Boscombe Down-Exeter U.K.	3000–6100	148 ± 8	
June 1975	off S.W. England	150–6100	198 ± 24	
Aug. 1975	West of N. Ireland E. of Scotland	7000–8000	172 ± 33	
Jan. 1976	off S.W. Wales	900–7300	167 ± 3	
Jan. 1976	off S.W. Wales	900–4900	159 ± 9	
Feb. 1976	off N.E. England	400–5200	168 ± 8	

tropospheric flight group data:

flight group	time period	no. of samples	mean conc.	std.dev. of the mean
Alaska	May 1975	7	127 ppt	2.3 ppt
Pacific Northwest	March 1976	14	133 ppt	2.1 ppt
North America	October 1976	82	136 ppt	0.5 ppt
California	April 1977	20	149 ppt	1.1 ppt

(1223)

glc's in Washington, rural area: Dec. 1974–Feb. 1975: 0.125 ppb (315)
glc's Los Angeles County: 0.14–2.2 ppb (46)
   inside homes and public buildings: 0.1 ppb–0.5 ppm (201)
   near Los Angeles: July 1970: avg.: 0.56 ppb
                       Feb. 1973: avg.: 0.40 ppb (201)

D. BIOLOGICAL EFFECTS:

—Concentrations in various organs of molluscs and fish collected from the relatively clean water of the Irish Sea in the vicinity of Port Erin, Isle of Man (only highest and lowest conc. are given):

molluscs: *Baccinum undatum*: 0.3 ng/g (on a dry wt. basis) digestive gland
         Modiolus modiolus: 0.2 ng/g: digestive tissue (dry wt.)
                        4.4 ng/g: mantle (dry wt.)
         Pecten maximus: 0.4 ng/g: muscle and testis (dry wt.)
                        1.4 ng/g: mantle (dry wt.)
fish: *Conger conger* (eel): muscle: 0.1 ng/g (on dry wt. basis)
                         brain: 5.0 ng/g (on dry wt. basis)
     Gadus morhua (cod): liver: 0.5 ng/g (on dry wt. basis)
                        brain: 3.3 ng/g (on dry wt. basis)
     Pollachius birens (coalfish):
                       muscle: 0.9 ng/g (on dry wt. basis)
                       liver: 3.5 ng/g (on dry wt. basis)
     Scyllorhinus canicula (dogfish):
                       spleen: 0.4 ng/g (on dry wt. basis)
                       gill: 4.5 ng/g (on dry wt. basis)
     Trisopterus luscus (bib):
                       gut: 0.1 ng/g (on dry wt. basis)
                       gill: 3.2 ng/g (on dry wt. basis) (1092)

—Mammalia:
  guinea pig: inhalation: LD_{50}: 250,000 ppm, 30 min
  rat: inhalation: LD_{50}: 100,000 ppm, 30 min
  rabbit: inhalation: LD_{50}: 250,000 ppm, 30 min (74)
  rat, mouse, rabbit, guinea pig: inhalation: LD_{50}: no effect: 4,000 ppm, 6 hr/day, 43 days (54)

—Man: 3,000–10,000 ppm inhaled is exhaled entirely after 3–10 min (54)

—Carcinogenicity:

The results of the bioassay of trichlorofluoromethane in Osborne-Mendel rats for possible carcinogenicity are not conclusive because inadequate numbers of rats survived long enough to be at risk from late-developing tumors. Under the conditions of this bioassay, trichlorofluoromethane was not carcinogenic to male or female B6C3F1 mice. (1748)

folpet (N-(trichloromethylthio)phthalimide; phaltan; thiophal)

FONOFOS

Use: fungicide-bactericide for vinyls, paints and enamels
A. PROPERTIES: m.p. 177°C; v.p. $<10^{-5}$ at 20°C
D. BIOLOGICAL EFFECTS:
  —Carcinogenicity: negative
  —Mutagenicity in the *Salmonella* test: positive (without liver homogenate)
      64 revertant colonies/nmol
      2170 revertant colonies at 10 µg/plate   (1883)
  —Birds:
      chick embryo: teratogenic effects at 12 mg/kg in egg, injected after incubation of 4 days   (1058)
  —Mammals:
      acute oral LD_{50} (rats): >10,000 mg/kg   (1854)
      acute dermal LD_{50} (albino rabbits): >22,600 mg/kg   (1855)
      in diet: no-effect level in 17 months feeding tests on rats was 3200 ppm, on dogs 1,500 ppm   (1855)

fonofos *see* O-ethyl-S-phenylethylphosphonodithioate

formal *see* dimethoxymethane

formaldehyde (formalin; oxomethane; methanal)
HCHO
A. PROPERTIES: colorless gas; m.w. 30.0; m.p. -118/-92°C; b.p. -21/-19°C; v.p. -88°C at 10 mm; v.d. 1.03; sp.gr. 0.815 at -20/4°C; THC 134 kcal/mole, LHC 124 kcal/mole; log P_{oct} 0.00 (calculated)
B. AIR POLLUTION FACTORS: 1 mg/cu m = 0.815 ppm, 1 ppm = 1.248 mg/cu m
  —Odor: characteristic: quality: hay, strawlike
                  hedonic tone: pungent

odor thresholds   mg/cu m

	10^{-7}	10^{-6}	10^{-5}	10^{-4}	10^{-3}	10^{-2}	10^{-1}	1	10	10^2	10^3	10^4
detection												
recognition						♦	♦					
not specified					♦♦♦	♦						♦

                        (73; 279; 307; 610; 625; 657; 741; 751; 788; 809; 836)
    USSR: human odor perception: non perception: 0.05 mg/cu m
                             perception: 0.07 mg/cu m
            human reflex response: no response: 0.07 mg/cu m
                             adverse response: 0.084 mg/cu m   (170)
      O.I. at 20°C = 5,000,000   (316)
—Manmade sources:
      glc's downtown Tokyo: Jan–May 1969: 4.2–50.4 µg/cu m/2 hr   (205)
      in diesel exhaust: 18.3 ppm   (311)
      in exhaust of a 1970 Ford Maverick gasoline engine operated on a chassis dynamometer following the 7-mode California cycle:

FORMALDEHYDE 679

    from **API** #7 gasoline: 14.9 ppm
    from **API** #8 gasoline: 11.3 ppm   (1053)
    in exhaust of gasoline engine: 59.9–72.9 vol.% of total exhaust aldehydes
                                                         (394; 395; 396; 397)
—Emission control:
    A.C.: retentivity: 3 wt% of adsorbent   (83)
—Sampling and analysis:
    second derivative spectroscopy: det. lim.: 200 ppb   (42)
    photometry: min. full scale: 9,500 ppm   (53)
    PMS: det. lim.: 1–11 ppm   (200)
    4 f. abs. app.; 60 l air /30 min: VLS: det. lim.:
        det. lim.:   15 μg/cu m/30 min; reagent: pararosaniline
                 150 μg/cu m/30 min; reagent: chromotropic acid
                                   (several interferences)   (208)
C. **WATER POLLUTION FACTORS**:
  —BOD_5:   60% of ThOD   (274)
                0.6–1.07 std. dil. at <260 mg/l   (27)
                0.728   (30)
              0.33–1.06 std. dil. sewage   (41)
              1.06 std. dil. sew.   (260)
              0.64 std. dil. sew.   (282)
              0.33 std. dil. sew. at 2.5–10 ppm   (269)
              0.45 std. dil. sew. at 1.7–20 ppm   (263)
              1.10 manom 50% sew; at 260 ppm   (270)
              0.57 manom   5% sew; at 260 ppm   (270)
              0   Sierp, 10% sew; at 440 ppm   (280)
              1.00 Warburg, 50% sew; at 130 ppm   (281)
              1.10 Warburg, 25–50% sew; 250 ppm   (281)
  —BOD_{20}: 1.228   (30)
  —COD: 100% of ThOD (0.05 n $K_2Cr_2O_7$)   (274)
             1.06   (41, 27)
  —$KMnO_4$: 2.84   (30)
                 94% of ThOD (0.01 n $KMnO_4$)   (274)
  —ThOD: 1.068   (30)
  —Biodegradation:

$$\text{HCHO} \begin{array}{c} \xrightarrow{+\frac{1}{2}O_2} HCOOH \rightarrow CO_2 + H_2O \\ \xrightarrow{+H_2} CH_3OH \rightarrow CO_2 + H_2O \end{array}$$
                                                                                 (171)

—Reduction of amenities:
    taste and odor caused at 50.0 mg/l   (41, 27)
    odor threshold: average: 49.9 mg/l
                            range: 0.8–102 mg/l   (97)
—Impact on biodegradation processes:
    inhibition of anaerobic sludge digestion: at 100 mg/l
                         aerobic degradation: 135–175 mg/l
    methane fermentation can be acclimated up to 15% formaldehyde   (30)

inhibition of degradation of glucose by *Pseudomonas fluorescens*: at 2 mg/l
inhibition of degradation of glucose by *E. coli*: at 1 mg/l (293)
—Waste water treatment:
A.C.: adsorbability: 0.018 g/g C; 9.2% reduction, infl.: 1,000 mg/l, effl.: 908 mg/l (32)

trickling filter: feed ppm	removal of formaldehyde lb/yd^3	% removal efficiency	
110	1.12	23	
184	1.25	16	
266	1.75	15	
300	3.10	23	
360	3.45	28	(96)

methods	temp °C	days observed	feed mg/l	days acclim.		
A.S., Resp, BOD	20	1–5	720	<1	nil % theor. oxidation	
A.S., W	20	$\frac{1}{3}$	1,500	P	52% theor. oxidation	
A.S., BOD	20	$\frac{1}{8}$–5	500	30+	47% removed	
A.S., BOD	20	$\frac{1}{3}$–5	333	30+	94% removed	(93)

—Sampling and analysis:
VLS; det. lim.: 0.1 mg/l; reagent: pararosaniline
1 mg/l; reagent: chromotropic acid
(several interferences) (208)

D. BIOLOGICAL EFFECTS:
—Toxicity threshold (cell multiplication inhibition test) (35% w/w)
    bacteria (*Pseudomonas putida*): 14 mg/l (1900)
    algae (*Microcystis aeruginosa*): 0.39 mg/l (329)
    green algae (*Scenedesmus quadricauda*): 2.5 mg/l (1900)
    protozoa (*Entosiphon sulcatum*): 22 mg/l (1900)
    protozoa (*Uronema parduczi Chatton-Lwoff*): 6.5 mg/l (1901)
—Bacteria:
E. coli: toxic: 1 mg/l
—Algae:
Scenedesmus: toxic: 0.3–0.5 mg/l
—Arthropoda: *Daphnia*: toxic: 2 mg/l (39)
—Fish: sensitive aquatic organisms: TLm (? hr): >10 mg/l (27)
    guppies: TLm (? hr): 50–200 mg/l (30)

species	exposure time	exposure type	effect endpoint	concentration (37% w/w)
rainbow trout (green egg)	96 hr	S	LC$_{50}$	700 mg/l
rainbow trout (green egg)	96 hr	S	LC$_{50}$	565 mg/l
rainbow trout (green egg)	96 hr	S	LC$_{50}$	631 mg/l
rainbow trout (eyed egg)	96 hr	S	LC$_{50}$	198 mg/l
rainbow trout (eyed egg)	96 hr	S	LC$_{50}$	338 mg/l
rainbow trout (eyed egg)	96 hr	S	LC$_{50}$	435 mg/l
rainbow trout (eyed egg)	96 hr	S	LC$_{50}$	289 mg/l
rainbow trout (sac larvae)	96 hr	S	LC$_{50}$	96 mg/l

rainbow trout (sac larvae)	96 hr	S	LC_{50}	89.5 mg/l
rainbow trout (sac larvae)	96 hr	S	LC_{50}	89.5 mg/l
rainbow trout (sac larvae)	96 hr	S	LC_{50}	112 mg/l
rainbow trout fingerlings	96 hr	S	LC_{50}	94.5 mg/l
rainbow trout fingerlings	96 hr	S	LC_{50}	73.5 mg/l
rainbow trout fingerlings	96 hr	S	LC_{50}	106 mg/l
rainbow trout fingerlings	96 hr	S	LC_{50}	61.9 mg/l
rainbow trout	96 hr	S	LC_{50}	440 mg/l
rainbow trout	96 hr	S	LC_{50}	610 mg/l
rainbow trout	96 hr	S	LC_{50}	618 mg/l
rainbow trout	24 hr	S	LC_{50}	7,200 mg/l
rainbow trout	24 hr	S	LC_{50}	249 mg/l
rainbow trout	24 hr	S	LC_{50}	214 mg/l
rainbow trout fingerlings	96 hr	S	LC_{50}	134 mg/l
rainbow trout fingerlings	96 hr	S	LC_{50}	145 mg/l
rainbow trout fingerlings	96 hr	S	LC_{50}	123 mg/l
rainbow trout (green egg)	96 hr	S	LC_{50}	1,020 mg/l

rainbow trout: 96-hr LC_{50}: 118 µl/l (FT)
atlantic salmon: 96-hr LC_{50}: 173 µl/l (FT)
lake trout: 96-hr LC_{50}: 100 µl/l (FT)
black bullhead: 96-hr LC_{50}: 62.1 µl/l (FT)
channel catfish: 96-hr LC_{50}: 65.8 µl/l (FT)
green sunfish: 96-hr LC_{50}: 173 µl/l (FT)
bluegill: 96-hr LC_{50}: 100 µl/l (FT)
smallmouth bass (*Micropterus dolomieui*): 96-hr LC_{50}: 136 µl/l (FT)
largemouth bass (*Micropterus salmoides*): 96-hr LC_{50}: 143 µl/l (FT)   (1192)
—Mammals:
guinea pig: single oral LD_{50}: 0.26 g/kg
rat: single oral LD_{50}: 0.1–0.8 g/kg   (211)
—Inhalation:
  rat: 4 hr LC_{50}:   250 ppm
       30 min LC_{50}: 830 ppm
  cat: 8 hr LC_{50}:   650 ppm
        3.5 hr at 200 ppm: all survived   (104)

	conc. ppm	exposure time	time effects	
guinea pig	15.5	up to 10 hr.	8/20 died	(961)
rabbit	15.5	up to 10 hr.	3/5 died	(961)
mouse	15.5	up to 10 hr.	17/50 died	(961)
C3H mouse	735	2 hr.	death	(962)
cat	200	3.5 hr.	irritation, recovery in a few hours	(963)
	650	4	irritation, recovery in 6 days	
	650–1600	8 & 4	hyperpnea; edema; emphysema; purulent processes of lungs usually fatal	
	4900	3	same as above, and corneal erosion	
C3H mouse	163	3 one-hr/wk × up	severe intoxication and death,	(962)

		to 35 wk.	trachea and major bronchi showed metaplasia to atypical squamous cell epithelium	
	82	3 one-hr/wk × up to 35 wk.	normal weight gain	
	41	3 one-hr/wk × up to 35 wk.	normal weight gain	
dog, rat, monkey, rabbit, guinea pig	3.8	90 day continuous	interstitial inflammation in lungs	(960)
rat	490–1388	30 min.	listlessness; lacrimation; increased nasal secretions; respiratory difficulty; lung hyperemia and edema	(964)
rat	.30–.64 g/kg	subcutaneous injection	same symptoms as above, but less severe	
mouse	.15–.46 g/kg			
rabbit trachea	284–1512	within 1 hr	ciliostatsis	(965)
	30–60	10 min.	ciliostasis	(966)
	60–100	5 min.	ciliostasis	(970)
guinea pig	0.58–49	1 hr.	altered pulmonary dynamics including increased resistance and work, and decreased respiratory rate and compliance	(971)
guinea pig	50	1 hr.	increased pulmonary resistance and tidal	(972)
	1000	1 hr.	volume; decreased respiratory rate and minute volume	
rat	0.5–2.5	2 min.	increased firing of nasopalantine nerve	(973)

—Man: severe toxic effects: 100 ppm = 120 mg/cu m, 1 min
symptoms of illness: 30 ppm = 36 mg/cu m
unsatisfactory: 10 ppm = 12 mg/cu m (185)

conc. ppm	exposure time	effects	
2–3		tingling of eyes, nose and throat	(961)
4–5	10–30 min.	irritation; discomfort; lacrimation; some tolerance develops, tolerable for some, not all	(211)
20	short	discomfort and lacrimation	(974)
12		severe irritation	(975)
13.8	30 min.	nasal and eye irritation and lacrimation	(976)
4	5 min.	severe eye irritation	(977)
1	5 min.	8% of test panel reported eye irritation	(978)
2	12	24% of test panel reported eye irritation	

4	5	100% of test panel reported eye irritation	
2-4	5	33% of test panel reported eye irritation	
5	5	67% of test panel reported eye irritation	
	5-20 min.	asthmatic attacks	(979)
0.9-1.6	occup. exposure	intense irritation; itching of eyes; dry and sore throat; increased thirst, disturbed sleep	(980)
.13-.45	occup. exposure	burning, stinging eyes; headaches; intolerable irritation of eyes, nose and throat, one illness	(981)
.25-1.39	occup. exposure	upper respiratory tract irritations; burning of eyes and nose; sneezing ; coughing; headaches	(982)
0.4-0.8	occup. exposure	chronic airway obstruction; lowered FEV_{10} FVC ratio; acute exposures caused eye, nose throat irritation, and lower respiratory tract symptoms	(983)
	eye irritation threshold: 3-10 ppm		(1297)

formaldehydedimethylacetal *see* dimethoxymethane

formalin *see* formaldehyde

formamide (methanamide)
$HCONH_2$
A. PROPERTIES: colorless liquid; m.w. 45.04; m.p. 2.55°C; b.p. 210°C decomposes; v.p. 0.1 mm at 30°C; v.d. 1.56; sp.gr. 1.134 at 20/4°C; sat. conc. <0.24 g/cu m at 30°C
B. AIR POLLUTION FACTORS: 1 mg/cu m = 0.53 ppm, 1 ppm = 1.87 mg/cu m
C. WATER POLLUTION FACTORS:
  —Waste water treatment:
   A.S. after 6 hr: 1.6% of ThOD
       12 hr: 4.7% of ThOD
       24 hr: 11.8% of ThOD   (89)
  —BOD_5: nil std. dil. sew.   (285)

formic acid (methanoic acid)
HCOOH
A. PROPERTIES: colorless liquid; m.w. 46.03; m.p. 8.4°C; b.p. 101°C; v.p. 35 mm at 20°C, 54 mm at 30°C; v.d. 1.6; sp.gr. 1.220 at 20/4°C; THC 62.9 kcal/mole, LHC 59.4 kcal/mole; sat. conc. 80 g/cu m at 20°C, 131 g/cu m at 30°C; log P_{oct} -1.55/-0.22 (calculated)
B. AIR POLLUTION FACTORS: 1 mg/cu m = 0.52 ppm, 1 ppm = 1.91 mg/cu m
  —Odor: hedonic tone: pungent, penetrating

684 FORMIC ACID

odor thresholds   mg/cu m

[Chart showing odor thresholds from 10^{-7} to 10^4 mg/cu m with detection, recognition, and not specified markers]

                                                                      (307; 610; 777; 778; 805; 871)

    O.I. at $20°C$ = 2200   (316)

    —Sampling and analysis:
      photometry: min. full scale: 40,000 ppm   (53)

C. WATER POLLUTION FACTORS:
    —BOD_5:   0.15–0.19 std. dil.   (282, 285)
                 0.086   (30)
                 0.02–0.27 std. dil. sew.   (287, 259)
    —BOD_{20}: 0.250   (30)
    —$KMnO_4$: 0.126
    —ThOD: 0.35
    —Reduction of amenities: T.O.C.: 800 mg/l   (295)
                                     detection: 450 mg/kg   (919)
                                                        1,500 mg/kg   (873)
    —Manmade sources:
      contents of domestic sewage: 0–1.3 mg/l   (85)
      average content of secondary sewage effluent: 91 μg/l   (65)
      excreted by man: in urine: 0.4–2 mg/kg body wt/day   (203)
    —Waste water treatment:
      A.C.: adsorbability: 0.047 g/g C; 23.5% reduction, infl.: 1,000 mg/l, effl.: 765
        mg/l   (32)
      A.S. after 6 hr: 28.3% of ThOD
                 12 hr: 45.4% of ThOD
                 24 hr: 70.0% of ThOD

methods	temp °C	days observed	feed mg/l	days acclim.	
A.S., BOD	20	1–5	333	23	95% removed
A.S., Resp, BOD	20	1–5	720	1	40% theor. oxidation
A.S., W	20	1	500	24+	36% theor. oxidation

                                                                                                              (93)

      stabilization pond design: toxicity correction factor (C_{tox})
                             = 2.0 for influent = 180 mg/l
                             = 16.0 for influent = 360 mg/l   (179)
      powdered carbon: at 100 mg/l sodium salt (pH 7.5)—carbon dosage 1000 mg/l:
        2% adsorbed   (520)

D. BIOLOGICAL EFFECTS:
    —Slight inhibition of microbial growth after 24 hr exposure at 5000 ppm   (523)

—Bacteria: *E. coli*: no effect: 1 g/l
—Algae: *Scenedesmus*: toxic: 100 mg/l
—Arthropoda: *Daphnia*: toxic: 120 mg/l (30)
            Gammarus pulex: toxic: 2,500 mg/l
—Protozoa: *Paramecium caudatum*: toxic: 6,000 mg/l
          Vorticella campanula: toxic: 500 mg/l
—Fish: *Trutta iridea*: perturbation level: 1,000 mg/l (30)
      Lepomis macrochirus: 24 hr TLm: 5,000 mg/l (sodium salt) (1294)

N-formyl-4-chloro-o-toluidine
Biodegradation product of chlorphenamidine (1547)

fosfamid *see* dimethoate

frescon (*see also* N-tritylmorpholine)
D. BIOLOGICAL EFFECTS:
—Frescon Paste:
  active ingredient: N-tritylmorpholine (50%)

	mg/l	24 hr	48 hr	96 hr	
harlequin fish	$LC_{10}(F)$	0.078	0.094		
(*Rasbora heteromorpha*)	$LC_{50}(F)$	0.135	0.115	0.08	(331)
rainbow trout yearling	$LC_{50}(S)$	8.2	2.5	1.5	(331)
(*Salmo gairdneri*)					

—Frescon E/C:
  active ingredient: N-tritylmorpholine (16.5%):

	mg/l	24 hr	48 hr	96 hr	
harlequin fish	$LC_{10}(F)$	0.2	0.17		
(*Rasbora heteromorpha*)	$LC_{50}(F)$	0.33	0.28	0.1	
rainbow trout yearling	$LC_{50}(S)$	0.74	0.62	0.52	
(*Salmo gairdneri*)					(331)

fumaric acid (*trans*-butenedioic acid; *trans*-1,2-ethylenedicarboxylic acid)
HOOCCHCHCOOH
A. PROPERTIES: colorless crystals; m.w. 116.07; m.p. 276/287°C; b.p. 290°C sublimes; sp.gr. 1.635 at 20/4°C; solub. 7,000 mg/l at 25°C, 98,000 mg/l at 100°C; log P_{oct} 0.07/0.56 (calculated)
C. WATER POLLUTION FACTORS:
—BOD_5: 0.57–0.70 Warburg, sewage (258, 259, 164, 165)
       0.65 std. dil. at 4–8 mg/l (27)
       0.175 (30)
       34% of ThOD (220)
—$KMnO_4$ value: acid: 87% of ThOD; alkaline: 68% of ThOD (220)
—ThOD: 0.827 (30)
—Waste water treatment:
  A.S. after 6 hr: 1.0% of ThOD
            12 hr: 0.1% of ThOD
            24 hr: 1.7% of ThOD (89)

D. BIOLOGICAL EFFECTS:
—Man: oral ingestion: 500 mg/day for a year is tolerated (211)

fumigrain *see* acrylonitrile

furadan *see* carbofuran

fural *see* furfural

2-furaldehyde *see* furfural

furan (furfuran)

Use: organic synthesis, especially for pyrrole, tetrahydrofuran, thiophene; as a solvent for resins and in the formation of lacquers.
Furan is often stablized with butylated hydroxytoluene (BHT) to inhibit the formation of peroxides (502)
A. PROPERTIES: colorless liquid; m.w. 68.07; m.p. -86/-85°C; b.p. 31.3°C at 758 mm; v.d. 2.35; sp.gr. 0.937 at 20/4°C; solub. 10,000 mg/l at 25°C
B. AIR POLLUTION FACTORS: 1 mg/cu m = 0.35 ppm, 1 ppm = 2.83 mg/cu m
—Manmade sources:
in vapor phase of cigarette smoke: 8.4 µg/40 ml puff (1384)
—Odor: characteristic: quality: ethereal
—Sampling and analysis: photometry: min. full scale: <100,000 ppm (53)
C. WATER POLLUTION FACTORS:
—Odor threshold: detection: 6 mg/kg (908)
—Sampling and analysis: photometry: min. full scale: 300×10^{-6} mole/l (53)
D. BIOLOGICAL EFFECTS:
—Mammals:
chronic inhalation: dogs: 10 ppm produced noticeable circulatory disturbance (502)
rats: IP LD_{50}: 5.2 mg/kg
mice: IP LD_{50}: 7.0 mg/kg (1383)
mice: inhalation LC_{50}: 0.12 µg/ml (1383)

furanace
D. BIOLOGICAL EFFECTS:
—Fish:
Atlantic salmon: 96 hr LC_{50} (S): 1.41 mg/l
rainbow trout: 96 hr LC_{50} (S): 1.00 mg/l
fathead minnow: 96 hr LC_{50} (S): 0.820 mg/l
Channel catfish: 96 hr LC_{50} (S): 1.07 mg/l
green sunfish: 96 hr LC_{50} (S): 2.48 mg/l
bluegill: 96 hr LC_{50} (S): 3.00 mg/l (1205)

2-furancarbinol *see* furfuryl alcohol

2-furancarbonal *see* furfural

2-furancarboxylic acid *see* 2-furoic acid

2,5-furandione *see* maleic anhydride

2-furanmethanol *see* furfuryl alcohol

furazolidine
D. BIOLOGICAL EFFECTS:
 —Fish:
  guppy: static bioassay: 96 hr TLm: 25 mg/l at 24°C
  rainbow trout: static bioassay: 96 hr TLm: >30 mg/l at 15°C   (446)

furfural (furfurole; 2-furancarbonal; 2-furaldehyde; fural; furfuraldehyde; furole)

Uses: solvent refining of lubricating oils; solvent for nitrocellulose, cellulose acetate, shoe dyes; intermediate for tetrahydrofuran and furfuryl alcohol; phenolic and furan polymers; weed killer; fungicide; adipic acid and adiponitrile; flavoring.

A. PROPERTIES: colorless to yellow liquid; m.w. 96.08; m.p. -38°C; b.p. 162°C; v.p. 1 mm at 20°C, 3 mm at 30°C, 10 mm at 50°C; v.d. 3.31; sp.gr. 1.16 at 20/4°C; solub. 83,000 mg/l at 20°C, 199,000 mg/l at 90°C; LHC 560.3 kcal/mole; sat. conc. 5.8 g/cu m at 20°C, 12 g/cu m at 30°C

B. AIR POLLUTION FACTORS: 1 mg/cu m = 0.251 ppm, 1 ppm = 3.991 mg/cu m
 —Odor: characteristic: quality: almond
  T.O.C.: recogn.: 1.0 mg/cu m = 0.25 ppm   (856)
             0.008 mg/cu m   (607)
  USSR: human odor perception: 1.0 mg/cu m
     human reflex response: adverse response: 0.084 mg/cu m
     animal chronic exposure: adverse effect: 0.33 mg/cu m   (170)
  O.I. at 20°C = 5260   (316)
 —Emission control:
  photometry: min. full scale: 1.6 ppm   (53)

C. WATER POLLUTION FACTORS:
 —BOD_5: 0.77 std. dil. sewage at 2–20 mg/l   (27, 41)
      0.28 at 440 ppm, Sierp, sew.   (280)
 —ThOD: 1.66   (27)
 —Biodegradation: adapted A.S. at 20°C—product is sole carbon source: 96.3%
   COD removal at 37.0 mg COD/g dry inoculum/hr   (327)
 —Reduction of amenities:
  taste and odor caused at 4.0 mg/l   (41)
  organoleptic limit: 1.0 mg/l   (181)

T.O.C. in water: 0.6 ppm; 3.0 mg/kg (326)(882)
—Waste water treatment:

methods	temp °C	days observed	feed mg/l	days acclim.	% removed	
RW, CA	20	2	1	nat.	100	
RW, CA	20	2	25	9	100	(93)

—Sampling and analysis:
  photometry: min. full scale: 0.065×10^{-6} mole/l (53)

D. BIOLOGICAL EFFECTS:
  —Toxicity threshold (cell multiplication inhibition test):
    bacteria (*Pseudomonas putida*): 16 mg/l (1900)
    algae (*Microcystis aeruginosa*): 2.7 mg/l (329)
    green algae (*Scenedesmus quadricauda*): 31 mg/l (1900)
    protozoa (*Entosiphon sulcatum*): 0.6 mg/l (1900)
    protozoa (*Uronema parduczi Chatton-Lwoff*): 11 mg/l (1901)
  —Protozoa: *Vorticella campanula*: perturbation level: 200 mg/l
              Paramaecium caudatum: perturbation level: 1,200 mg/l (30)
  —Arthropoda: *Gammarus pulex*: perturbation level: 800 mg/l
               Epeorus assimilis: perturbation level: 300 mg/l
  —Fish:
    Gambusia affinis: no effect: 10 mg/l (30)
                       24 mg/l TLm (96 hr) (30)
    fathead minnows: static bioassay in Lake Superior Water at 18-22°C:
         LC_{50} (1; 24; 48; 72; 96 hr): >50; 48; 37; 32; 32; mg/l
         static bioassay in reconstituted water at 18-22°C:
         LC_{50} (1; 24; 48; 72; 96 hr): >50; >50; 37; 33; 32 mg/l
         (350)
    mosquito fish: TLm (24-96 hr): 44-24 mg/l in Turbid Oklahoma (244)
    bluegill: TLm (24-96 hr): 32-1.2 mg/l, in tapwater (247, 41)
  —Mammalia: rat: acute oral LD_{50}: 0.05-0.1 g/kg
             mouse: acute oral LD_{50}: 0.5 g/kg
             dog: acute oral LD_{50}: 0.65 g/kg (211)

furfuralalcohol *see* furfuryl alcohol

furfuraldehyde *see* furfural

furfuran *see* furan

furfurole *see* furfural

furfuryl alcohol (2-furanmethanol; 2-furancarbinol; furfuralcohol; 2-furylcarbinol)

Uses: a resin raw material; solvent in textile printing, viscosity reducer for viscous epoxy resins

A. PROPERTIES: colorless to yellow liquid, turns amber during storage, turns black in the presence of air; m.w. 98.1; m.p. −14.6 stable crystalline form, −29°C metastable cryst. form; b.p. 171°C; v.p. 0.4 mm at 20°C, 1 mm at 31.8°C, 6.3 mm at 60°C, 53.5 mm at 100°C; v.d. 3.37; sp.gr. 1.13 at 20/4°C; sat. conc. 2.1 g/cu m at 20°C, 4.1 g/cu m at 30°C
B. AIR POLLUTION FACTORS: 1 mg/cu m = 0.249 ppm, 1 ppm = 4.01 mg/cu m
  −Odor: T.O.C.: 50% response: 8 ppm (211)
  32 mg/cu m (703)
  O.I. at 20°C = 5260 (316)
C. WATER POLLUTION FACTORS:
  −BOD_5: 0.532
  −BOD_{20}: 1.314 (30)
  −$KMnO_4$: 2.970 (30)
D. BIOLOGICAL EFFECTS:
  −Toxicity threshold (cell multiplication inhibition test):
  bacteria (*Pseudomonas putida*): 180 mg/l (1900)
  algae (*Microcystis aeruginosa*): 5.2 mg/l (329)
  green algae (*Scenedesmus quadricauda*): 25 mg/l (1900)
  protozoa (*Entosiphon sulcatum*): 227 mg/l (1900)
  protozoa (*Uronema parduczi Chatton-Lwoff*): 384 mg/l (1901)
  −Bacteria:
  Pseudomonas: toxic: 50 mg/l
  −Algae:
  Scenedesmus: toxic: 100 mg/l
  −Protozoa: *Colpoda*: toxic: 1,250 mg/l (30)
  −Arthropoda: *Daphnia*: toxic: 1,500 mg/l
  −Mammalia: rat: single oral LD_{50}: 0.275 g/kg
  inhalation: LC_0: 19 ppm, 30 × 6 hr
  LC_{50}: 233 ppm, 4 hr
  mouse: inhalation: LC_0: 19 ppm, 15 × 6 hr
  700 ppm, 10 min (211)

2-furoic acid (2-furancarboxylic acid; pyromucic acid)

A. PROPERTIES: white crystals; m.w. 112.08; m.p. 133°C; b.p. 230°C sublimes; solub. 35,700 mg/l at 15°C; log P_{oct} 0.60/0.73 (calculated)
C. WATER POLLUTION FACTORS:
  −COD: 1.25 (27)
  −ThOD: 1.29 (27)

furole *see* furfural

2-furylcarbinol *see* furfuryl alcohol

FWA *see* fluorescent whitening agents

G

galacitol
C. WATER POLLUTION FACTORS:
—Manmade sources:
in primary domestic sewage plant effluent: 0.002 mg/l (517)

gallic acid (3,4,5-trihydroxybenzoic acid)

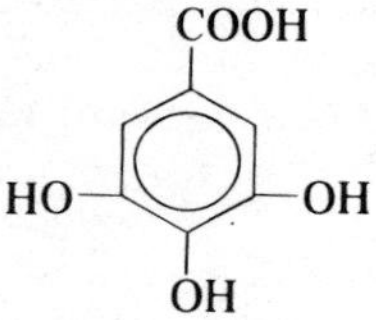

A. PROPERTIES: colorless needles; m.w. 170.12; m.p. 220°C decomposes; sp.gr. 1.694 at 4/4°C; solub. 11,600 mg/l at 25°C, 330,000 mg/l at 100°C; log P_{oct} -0.25/-0.15
C. WATER POLLUTION FACTORS:
—BOD_5: 0.080 std. dil. sew. (282; 163)
        6% of ThOD (274)
—$KMnO_4$ value: 2.690
        : 65% of ThOD (274)
—COD: 69% of ThOD (274)
—ThOD: 1.320 (30)
—Biodegradation: adapted A.S. at 20°C—product is sole carbon source: 90.5% COD removal at 20.0 mg COD/g dry inoculum/hr (327)
D. BIOLOGICAL EFFECTS:
—Mammalia: mouse: I.P.: LD_{50}: >500 mg/kg (211)

gallotannic acid *see* tannic acid

gamlen oil spill remover
D. BIOLOGICAL EFFECTS:
—Fish:

	mg/l	24 hr	48 hr	96 hr	3 m (extrap.)
harlequin fish	LC_{10}(F)	6.0	4.0	2.2	
(*Rasbora heteromorpha*)	LC_{50}(F)	7.5	6.4	4.0	1.0

(331)

garlon *see* 2(2,4,5-trichlorophenoxy)propionic acid

gatnon *see* N-(2-benzothiazolyl)-N'-methylurea

GC compound 1189 *see* kepone

gendriv
D. BIOLOGICAL EFFECTS:
—Fish:
rainbow trout: static bioassay: 96 hr TLm: 218 mg/l at 14–26°C (454)

gensol
D. BIOLOGICAL EFFECTS:
—Fish:
Gensol No. 1: brown trout yearlings: 48 hr LC_{50} (S): 50 mg/l
Gensol No. 12: borwn trout yearlings: 48 hr LC_{50} (S): 80 mg/l (1113)

gentisic acid (2,5-dihydroxybenzoic acid)

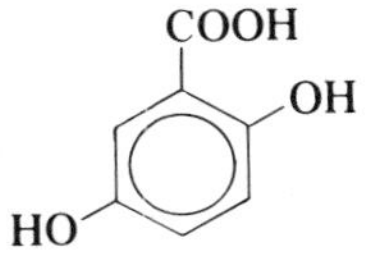

A. PROPERTIES: m.p. 199–200°C
C. WATER POLLUTION FACTORS:
—Biodegradation: adapted A.S. at 20°C—product is sole carbon source: 97.6% COD removal at 80.0 mg COD/g dry inoculum/hr (327)

giv-gard DXN
Use: microbicide; active ingredient dimethoxane (92%)
D. BIOLOGICAL EFFECTS:
—Fish:

	mg/l	24 hr	48 hr	96 hr	3 m (extrap.)
harlequin fish	LC_{10}(F)	74	36	34	
(*Rasbora heteromorpha*)	LC_{50}(F)	92	54	44	30

(331)

glacial acetic acid *see* acetic acid

d-gluconic acid (dextronic acid; *d*-glyconic acid; maltonic acid; glycogenic acid)
$COOH(CHOH)_4CH_2OH$
A. PROPERTIES: light brown syrupy liquid; m.w. 196.16; m.p. 125/126°C; log P_{oct} −2.57 (calculated)
C. WATER POLLUTION FACTORS:
—BOD_5: 0.350
—$KMnO_4$: 2.244
—ThOD: 0.900 (30)

d-glucono-δ-lactone

$$\text{HOH}_2\text{C} \overset{\text{HO}}{\underset{\text{O}}{\bigcirc}} \overset{\text{OH}}{\underset{\text{O}}{\bigcirc}} \overset{\text{OH}}{=}\text{O}$$

A. PROPERTIES: m.w. 117.16; m.p. 155°C; readily soluble in water
C. WATER POLLUTION FACTORS:

theoretical	analytical
ThOD = 0.99	reflux COD = 100.0% recovery
COD = 0.99	BOD_5 = 0.613
	BOD_5/COD = 0.62 (1828)

—Impact on biological treatment systems:
  unacclimated system, conc. 500 mg/l: biodegradable   (1828)

glucose

Uses: ingredient in confectionery, jelly etc.; reducing agent; in alcoholic fermentations; pharmaceuticals; medicine; synthesis of amino acids
A. PROPERTIES: log P_{oct} −3.29 (calculated)
C. WATER POLLUTION FACTORS:
  —$BOD_{35}^{25°C}$: 0.78 in seawater/inoculum: enrichment cultures of hydrocarbon oxidizing bacteria
  —ThOD: 1.07
  —Biodegradation:
   adapted A.S.—product is sole carbon source—98.5% COD removal at 180.0 mg COD/g dry inoculum/hr   (327)
   biodegradation by natural communities in stream water: % converted to CO_2 in 3 days:
    initial conc. 1.8 ng/l: 1
         18 ng/l: 11
         1.8 μg/l: 22
         18 μg/l: 33   (1875)
  —Waste water treatment method:
   powdered carbon: at 100 mg/l, carbon dosage 1000 mg/l: 8% adsorbed   (520)

dl-glutamic acid (2-aminopentanedioic acid; dl-glutaminic acid; dl-α-aminoglutaric acid)
COOH(CH$_2$)$_2$CH(NH$_2$)COOH

A. PROPERTIES: m.w. 147.13; m.p. 225/227°C; sp.gr. 1.4601 at 20/4°C; solub. 26,400 mg/l at 25°C, 81,600 mg/l at 50°C
C. WATER POLLUTION FACTORS:
  —BOD_5: 0.64   std. dil. sew. (261)
           0.42–0.45 std. dil. sew. (263)
  —Manmade sources:
    excreted by man: in urine: 1.8–11.5 mg/kg body wt/day (203)
  —Waste water treatment:
    A.S.: after 6 hr: 14.0% of ThOD
               12 hr: 25.1% of ThOD
               24 hr: 35.8% of ThOD

methods	temp °C	days observed	feed mg/l	days acclim.	
A.S., BOD	20	1–5	333	15	95% removed
A.S., Resp, BOD	20	1–5	500	<1	98% removed
					31% theor. oxidation

(93)

l-glutamine (2-aminopentanedioicamide; d-α-aminoglutaramic acid)
$C_3H_5(NH_2)(CONH_2)COOH$
A. PROPERTIES: m.w. 146.15; m.p. 185°C; solub. 42,500 mg/l at 25°C
C. WATER POLLUTION FACTORS:
  —Waste water treatment:
    A.S.: after 6 hr: 9.2% of ThOD
               12 hr: 19.2% of ThOD
    (l-glut.):   24 hr: 30.7% of ThOD (89)

dl-glutaminic acid see dl-glutamic acid

γ-glutamylcysteinylglycine see glutathione

glutaric acid (pentanedioic acid)
$HOOC(CH_2)_3COOH$
A. PROPERTIES: colorless crystals; m.w. 132.11; m.p. 98°C; b.p. 303°C decomposes; sp.gr. 1.429 at 15/4°C; solub. 640,000 mg/l at 20°C; log P_{oct} −0.47/−0.08 (calculated)
C. WATER POLLUTION FACTORS:
  —BOD_5: 0.433 (30)
           0.72 std. dil. acclimated (41)
  —BOD_{20}: 1.241 (30)
  —$BOD_{2\ days}^{25°C}$: 0.045 (substrate conc.: 6.6 mg/l; inoculum: soil microorganisms)
     5 days: 0.70
   10 days: 0.91
   20 days: 1.08 (1304)
  —COD: 1.21 (41)
  —$KMnO_4$: 0.006 mg/l (30)
  —ThOD: 1.212 (30)

—Waste water treatment:
   A.S.: after 6 hr: 0.9% of ThOD (89)
D. BIOLOGICAL EFFECTS:
   —Bacteria: *Pseudomonas*: no effect: 125 mg/l
   —Algae: *Scenedesmus*: no effect: 1 g/l
   —Protozoa: *Colpoda*: toxic: 16 mg/l (30)
   —Fish: *L. macrochirus*: TLm (24 hr): 330 mg/l (153)

glutaronitrile (pentanedinitrile; trimethylenedicyanide; trimethylenecyanide)
$CN(CH_2)_3CN$
A. PROPERTIES: colorless liquid; m.w. 94.11; m.p. $-29°C$; b.p. $287.4°C$; sp.gr. 0.955 at $15/4°C$;
C. WATER POLLUTION FACTORS:
   Waste water treatment:
      A.S.: after 6 hr: 1.1% of ThOD
              12 hr: 2.0% of ThOD
              24 hr: 2.9% of ThOD (89)

glutathione (γ-glutamylcysteinylglycine)
$C_{10}H_{17}O_6N_3S$
A. PROPERTIES: white crystalline powder; m.p. $190-192°C$;
C. WATER POLLUTION FACTORS:
   —Waste water treatment:
      A.S.: after 6 hr: 5.1% of ThOD
              12 hr: 8.2% of ThOD
              24 hr: 22.0% of ThOD (89)

glyceraldehyde (2,3-dihydroxypropanal; α,β-dihydroxypropionaldehyde)
$CH_2OHCHOHCHO$
A. PROPERTIES: m.w. 90.08; m.p. $138°C$; sp.gr. 1.453 at $18/4°C$
C. WATER POLLUTION FACTORS:
   —Waste water treatment:
      A.S.: after 6 hr: 4.9% of ThOD
              12 hr: 9.4% of ThOD
              24 hr: 20.1% of ThOD (88)

glyceric acid
$HOCH_2CHOHCOOH$
A. PROPERTIES: m.w. 106.08; $\log P_{oct}$ -1.68 (calculated)
C. WATER POLLUTION FACTORS:
   —In primary domestic sewage plant effluent: 0.005 mg/l (517)

glycerine *see* glycerol

glycerol (glycerine; 1,2,3-propanetriol; glycylalcohol; trihydroxypropane)
$HOCH_2CHOHCH_2OH$
   Uses: alkyd resins; cellophane; explosives; ester gums; pharmaceuticals; perfumery;

plasticizer for regenerated cellulose; cosmetics; foodstuffs; special soaps; lubricant and softener; bacteriostat; penetrant; hydraulic fluid; humectant
A. PROPERTIES: colorless liquid; m.w. 92.09; m.p. 18°C; b.p. 291°C; v.p. 0.0025 mm at 50°C; v.d. 3.17; sp.gr. 1.260 at 20/4°C; sat. conc. <0.5 g/cu m at 30°C; log P_{oct} −2.66/−2.47 (calculated)
B. AIR POLLUTION FACTORS: 1 mg/cu m = 0.26 ppm, 1 ppm = 3.83 mg/cu m
C. WATER POLLUTION FACTORS:
  −BOD_5: 0.87 NEN 3235-5.4 (277)
   51% of ThOD (274)
   0.81 at 2 ppm, std. dil./spec. cult. (283)
   0.7 std. dil. at 2-500 mg/l (27)
   0.617 to 0.80 std. dil. (290, 282, 281, 280)
   31% of ThOD (220)
  −BOD_{10}: 0.98 at 2 ppm std. dil./sp. cult. (283)
  −BOD_{20}: 0.940 (30)
  −BOD_{35}^{25}: 0.86 in seawater/inoculum: enrichment cultures of hydrocarbon oxidizing bacteria (521)
  −COD: 1.16 NEN 3235-5.3 (277)
   82% of ThOD (0.05 n $K_2Cr_2O_7$) (274)
   95% of ThOD (22)
  −$KMnO_4$ value: 1.013 (30)
   acid: 25% of ThOD; alkaline: 3% of ThOD (220)
   20% of ThOD (0.01 n $KMnO_4$) (274)
  −T.O.C.: 93% of ThOD (220)
  −ThOD: 1.217 (30)
   1.56 (27)
  −Biodegradation: adapted A.S.—product as sole carbon source—98.7% COD removal at 85.0 mg COD/g dry inoculum/hr (327)
  −Waste water treatment:
   A.S. Resp, 20°C, 1-5 days observed, feed: 720 mg/l, 1 d accl.: 28% theor.ox. (93)
   in primary domestic sewage plant effluent: 0.015-0.019 mg/l
   in secondary domestic sewage plant effluent: 0.004 mg/l (517)
D. BIOLOGICAL EFFECTS:
  −Toxicity threshold (cell multiplication inhibition test):
   bacteria (*Pseudomonas putida*): >10,000 mg/l (1900)
   algae (*Microcystis aeruginosa*): 2,900 mg/l (329)
   green algae (*Scenedesmus quadricauda*): >10,000 mg/l (1900)
   protozoa (*Entosiphon sulcatum*): 3,200 mg/l (1900)
   protozoa (*Uronema parduczi Chatton-Lwoff*): >10,000 mg/l (1901)
  −Fish:
   goldfish: 24 hr LC_{50}: >5,000 mg/l—modified ASTM D 1345 (277)

glycerol trinitrate see nitroglycerine

glyceryl nitrate see nitroglycerin

glyceryl tripalmitate see tripalmitin

glycide *see* glycidol

glycidol (2,3-epoxy-1-propanol; epihydric alcohol; glycide)
O—CH$_2$CHCH$_2$OH
|_____|

A. PROPERTIES: colorless liquid; m.w. 74.08; b.p. 162°C decomposes; v.p. 0.9 mm at 25°C; v.d. 2.15; sp.gr. 1.165 at 0/4°C;
B. AIR POLLUTION FACTORS: 1 mg/cu m = 0.330 ppm, 1 ppm = 3.03 mg/cu m
D. BIOLOGICAL EFFECTS:
—Mammalia: rat: single oral LD$_{50}$: 0.45 g/kg
mouse: single oral LD$_{50}$: 0.85 g/kg (211)
—Carcinogenicity: ?
—Mutagenicity in the *Salmonella* test: + (without liver homogenate)
0.58 revertant colonies/nmol
1730 revertant colonies at 223 µg/plate (1883)

glycidylallylether *see* allylglycidylether

glycidylbutylether *see* n-butylglycidylether

glycidylchloride *see* epichlorohydrin

1-glycidyloxy-2-ethylhexane *see* 2-ethylhexylglycidylether

glycine (photographic) *see* p-hydroxyphenylglycine

glycogenic acid *see* d-gluconic acid

glycol-*bis*(hydroxyethyl)ether *see* triethyleneglycol

glycolchlorohydrin *see* 2-chloroethanol

glycoldiacetate (ethyleneglycoldiacetate; ethylenediacetate)
(CH$_2$OOCCH$_3$)$_2$
Uses: solvent for cellulose esters and ethers; resins; lacquers; printing inks; perfume fixative; nondiscoloring plasticizer for ethyl and benzyl cellulose.
A. PROPERTIES: colorless liquid; m.w. 146.14; m.p. -31°C; b.p. 186/191°C; v.p. 0.4 mm at 20°C; v.d. 5.04; sp.gr. 1.104 at 20/4°C; solub. 143,000 mg/l; log P_{oct} 0.10/0.38 (calculated)
B. AIR POLLUTION FACTORS:
—Odor: characteristic: quality: fruity, acid
hedonic tone: pleasant
T.O.C.: abs.perc.lim.: 0.093 ppm
50% recogn.: 0.312 ppm
100% recogn.: 0.312 ppm
O.I. 100% recogn.: 4,198 (19)
D. BIOLOGICAL EFFECTS:
—Fish:

Lepomis macrochirus: static bioassay in fresh water at 23°C, mild aeration applied after 24 hr:

material added ppm	% survival after				best fit 96 hr LC_{50} ppm
	24 hr	48 hr	72 hr	96 hr	
125	0	—	—	—	
100	31	30	30	30	90
79	95	95	95	85	
50	100	100	100	100	

Menidia beryllina: static bioassay in synthetic seawater at 23°C: mild aeration applied after 24 hr

material added ppm	% survival after				best fit 96 hr LC_{50} ppm
	24 hr	48 hr	72 hr	96 hr	
100	100	40	30	10	
75	100	90	60	60	78
56	100	100	80	80	

(352)

glycoldibromide *see* ethylenebromide

glycoldichloride *see* ethylenedichloride

glycoldinitrate *see* ethyleneglycoldinitrate

glycolethylene-ether *see* 1,4-dioxane

***dl*-glycoleucine** *see dl*-norleucine

glycolic acid (hydroxyethanoic acid; hydroxyacetic acid)
$HOCH_2COOH$
A. PROPERTIES: crystalline solid; m.w. 76.05; m.p. α: 63°C, β: 79°C; b.p. decomposes; sp.gr. 1.27, $\log P_{oct}$ −1.11
C. WATER POLLUTION FACTORS:
 —BOD_5: 0.175 std.dil.sewage (282, 163)
D. BIOLOGICAL EFFECTS:
 —Mammalia: rat: single oral LD_{50}: 1.6–3.2 g/kg
   inhalation: no effect: sat.vap. 6 hr (211)

glycol monobutylether *see* butylcellosolve

glycolmonoethylether *see* ethyleneglycolmonoethylether

glycol monoethyletheracetate *see* ethyleneglycolmonomethyletheracetate

glycol monomethylether *see* ethyleneglycolmonomethylether

***d*-glyconic acid** *see d*-gluconic acid

glycylalcohol *see* glycerol

griseofulvin (7-chloro-4,6-dimethoxycoumaran-3-one-2-*spiro*-1′-(2′-methoxy-6′-methylcyclohex-2′-on-4′-one); grisetin)

Use: fungicide (antibiotic), isolated from cultures of *Penicillium griseofulvum*
A. PROPERTIES: m.p. 222°C, sublimes without decomposition at 210°C
D. BIOLOGICAL EFFECTS:
  —Mussels:
    hard clam (*Mercenaria mercenaria*): egg: 48 hr TLm (S): <250 ppb
    hard clam (*Mercenaria mercenaria*): larvae: 14 d TLm (S): <1. × 10^3 ppb
(2324)

guaiacol *see* 2-methoxyphenol

guanidine *see* n-dodecylguanidineacetate

guanine riboside *see* guanosine

guanosine (guanine riboside)

Nucleoside containing guanine and *d*-ribose
Use: biochemical research
A. PROPERTIES: white crystalline, odorless powder with mild, saline taste m.p. 237–240°C decomposes, m.w. 283.24; log P_{oct} −1.79 (calculated) (1457)
C. WATER POLLUTION FACTORS:
  —In primary domestic sewage plant effluent: 0.004–0.050 mg/l (517)

dl-N-δ-guanylornithine *see* dl-arginine

gum camphor *see* camphor

gusathion
  Use: insecticide
  Active ingredients: azinphosmethyl (25%)
  demeton-S-methylsulfone (7.5%) (331)
D. BIOLOGICAL EFFECTS:
  —Fish:

	mg/l	24 hr	48 hr	96 hr	3 m (extrap.)
harlequin fish	$LC_{10}(F)$	0.27	0.22		
(*Rasbora heteromorpha*)	$LC_{50}(F)$	0.45	0.34	0.17	0.1

(331)

gusathion A *see* azinphosethyl

gusathion M *see* azinphosmethyl

guthion *see* azinphosmethyl

H

halowax

D. BIOLOGICAL EFFECTS:

			exposure	bioaccumulation
halowax 1000	decapod	(*Palaemonetes pugio*)	40 µg/l: 3d	63 X
halowax 1013	decapod	(*Palaemonetes pugio*)	40 µg/l: 3d	187 X
halowax 1099	decapod	(*Palaemonetes pugio*)	40 µg/l: 3d	257 X
				(1474)
				effective conc. µg/l
halowax 1000	decapod	(*Palaemonetes pugio*) postlarvae	96 hr $LC_{50}(S)$	440
halowax 1013	decapod	(*Palaemonetes pugio*) postlarvae	96 hr $LC_{50}(S)$	74
halowax 1099	decapod	(*Palaemonetes pugio*) postlarvae	96 hr $LC_{50}(S)$	69
				(1474)

HCB *see* hexachlorobenzene

HCCH *see* hexachlorocyclohexane

HCE *see* hexachloroethane

HCH *see* hexachlorocyclohexane

HCP *see* 2,2'-methylene-*bis*(3,4,6-trichlorophenol)

hemellitene *see* 1,2,3-trimethylbenzene

hemimellitene *see* 1,2,3,-trimethylbenzene

hemiterpene *see* isoprene

hendecane *see* n-undecane

hendecanoic acid *see* n-undecylic acid

1-hendecanol *see* 1-undecanol

2-hendecanone *see* 2-undecanone

10-hendecenoic acid *see* undecylenic acid

hendecylmethylketone *see* 2-tridecanone

n-heneicosane
$CH_3-(CH_2)_{19}CH_3$
B. AIR POLLUTION FACTORS:
 —Ambient air quality:
  organic fraction of suspended matter:
   Bolivia at 5200 m altitude (Sept–Dec. 1975): 0.19–0.20 µg/1000 cu m
   Belgium, residential area (Jan.–April 1976): 0.38–1.2 µg/1000 cu m   (428)
   glc's in Botrange (Belgium)—woodland at 20–30 km from industrial area: June–July 1977: 0.20; 0.27 ng/cu m ($n = 2$)
   glc's in Wilrijk (Belgium)—residential area: Oct.–Dec. 1976: 0.64; 6.43 ng/cu m
   ($n = 2$)   (1233)
   glc's in the average American urban atmosphere—1963:
    320 µg/g airborne particulates or
    40 ng/cu m air   (1293)
   glc's in residential area (Belgium)—oct. 1976:
    in particulate sample: 1.08 ng/cu m
    in gas phase sample: 4.12 ng/cu m   (1289)

heneicosanoic acid
$CH_3(CH_2)_{19}COOH$
A. PROPERTIES: m.p. 74.3°C
B. AIR POLLUTION FACTORS:
 —Ambient air quality:
  organic fraction of suspended matter:
   Bolivia at 5200 m altitude (Sept–Dec. 1975): 0.05–0.15 µg/1000 cu m;
   residential area, 10 km south of Antwerp, Belgium (Jan.–April 1976): 0.8–1.2 µg/1000 cu m
   glc's in Botrange (Belgium)—woodland at 20–30 km from industrial area:
   June–July 1977: 0.52; 0.59 ng/cu m ($n = 2$)
   glc's in Wilrijk (Belgium)—residential area:
   Oct.–Dec. 1976: 1.2; 2.5 ng/cu m ($n = 2$)   (1233)
   glc's in residential area (Belgium)—Oct. 1976:
    in particulate sample: 2.56 ng/cu m
    in gasphase sample: 1.65 ng/cu m   (1289)

n-hentriacontane
$CH_3(CH_2)_{29}CH_3$
A. PROPERTIES: sp.gr. 0.781 at 68°C; b.p. 302°C at 15 mm; m.p. 68°C
B. AIR POLLUTION FACTORS:

—Ambient air quality:
  organic fraction of suspended matter:
    Bolivia at 5200 m altitude (Sept.–Dec. 1975): 0.48–0.93 µg/1000 cu m;
    Belgium, residential area (Jan.–April 1976): 3.5–5.1 µg/1000 cu m (420)
  glc's in Botrange (Belgium)—woodland at 20–30 km from industrial area:
    June–July 1977: 2.76; 394 ng/cu m ($n = 2$)
  glc's in Wilrijk (Belgium): residential area:
    Oct.–Dec. 1976: 7.52; 9.26 ng/cu m ($n = 2$) (1233)
  glc's in residential area (Belgium)—Oct. 1976:
    in particulate sample: 11.2 ng/cu m
    in gas phase sample: 3.99 ng/cu m (1289)
C. WATER POLLUTION FACTORS:
  —$BOD_{35}^{25°C}$: 2.56 in seawater/inoculum: enrichment cultures of hydrocarbon oxidizing bacteria
  —ThOD: 3.44 (521)

HEOD *see* dieldrin

2,2′,3,4,4′,5,5′-heptabromobiphenyl (HBB$_7$)

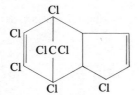

Major component (27% by weight) of flame retardant "Firemaster", a mixture of PBB's averaging 6 bromines per molecule (Michigan Chemical Corp.)
Major component (25% by weight) of 2,2′,3,3′,5,5′,6,6′-octabromobiphenyl (Aldrich Chemical Company).
A. PROPERTIES: m.p. 165–166°C

heptachlor (1,4,5,6,7,8,8-heptachloro-3a,4,7,7a-tetrahydro-4,7-methano-indene; heptachlorodicyclopentadiene)

Technical product contains about 72% heptachlor and 28% of related compounds.
A. PROPERTIES: m.p. 95–96°C; sp.gr. 1.57–1.59; v.p. 3×10^{-4} mm at 25°C
C. WATER POLLUTION FACTORS:
  —Aquatic reactions:
    75–100% disappearance from soils: 2–5 years (1815; 1816)

persistance in river water in a sealed glass jar under sunlight and artificial fluorescent light—initial conc. 10 µg/l:

after	% of original compound found				
	1 hr	1 wk	2 wk	4 wk	8 wk
	100	25	0	0	0	(1309)

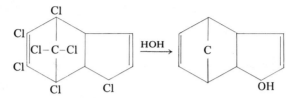

conversion of heptachlor to 1-hydroxychlordene was complete in 2 weeks in river water kept in a sealed jar under sunlight and artificial fluorescent light—initial conc. 10 µg/l. If kept in distilled water, after 4 weeks 60% of the converted heptachlor remained as 1-hydroxychlordene and 40% was further converted to the epoxide (1309)
—Odor threshold: detection: 0.02 mg/kg (915)
—Water quality:
  in Northern Mississippi water: avg.: 0.08 ng/l; range: 0.03–0.18 ng/l (1082)

D. BIOLOGICAL EFFECTS:
—Residues:
  mean concentration in gamebird muscle in upper Tennessee (USA):

	mg/kg fresh weight ± standard error		
	grouse	quail	woodcock
Johnson county	0.79 ± 0.15 ($n=12$)	1.15 ± 0.19 ($n=6$)	0.86 ± 0.17 ($n=6$)
Carter county	1.01 ± 0.20 ($n=9$)	0.86 ± 0.15 ($n=7$)	1.29 ± 0.14 ($n=6$)
Washington county	1.92 ± 0.30 (n-10)	0.98 ± 0.16 ($n=11$)	1.00 ± 0.21 ($n=6$)

quail: acceptable LD_{50}: 125 mg/kg
FAO/WHO residue tolerance limit: 0.03–0.3 mg/kg/day (1338)
ringed seal, Greenland, fat: 0.001 mg/g
bearded seal, Greenland, fat: 0.039 mg/g (1180)
—Bioaccumulation:
—Molluscs: 5 aquatic molluscs: 250 - 2500 X (1870)
         oysters: 17,600 X (1921)
—Crustaceans:
  pelecypod (*Crassostrea virginica*): 0.083–14.0 µg/l, 96 hr; 3900–8500 X
  decapod (*Penaeus duorarum*): 0.04–0.2 µg/l, 96 hr; 200–300 X
  decapod (*Palaemonetes vulgaris*): 0.13–5.0 µg/l, 96 hr; 500–700 X (1138)
—Fish:
  Cyprinodon variegatus: 2.7–8.8 µg/l, 96 hr; 7400–21300 X
  Lagodon rhomboides: 0.20–4.4 µg/l, 96 hr; 2800–7700 X
  Leiostomus xanthurus: 0.50–1.4 µg/l, 96 hr; 3000–13800 X (1165)
  Cyprinodon variegatus: 1.22–4.3 µg/l, 28 d; 3700–4600 X (1476)
  trout: 300 X (1898)
—Toxicity:

toxicity ratios of heptachlor (H) to photoheptachlor (PH), calculated from respective 24 hr LC_{50}* and LT_{50} values:

	H/PH
Crustacea	
Daphnia pulex (water flea)	1.24
Gammarus spp. (amphipod)	1.45
Insects	
Aedes aegypti larvae (mosquito)	2.80 (2.5)*
Musca domestica (house fly)	1.90 (1.96)*
Fish	
Pimephalus promelas (bass)	1.23 (1.63)*
	(1681; 1684; 1685)

—Insects:
Pteronarcys californica: 96 hr, LC_{50}: 1.1 µg/l
Pteronarcella badia: 96 hr, LC_{50}: 0.9 µg/l
Claassenia sabulosa: 96 hr, LC_{50}: 2.8 µg/l (2128)
mosquito (late 3rd instar *Aedes aegypti* larvae): 24 hr LC_{50}: 5 ppb
house-fly (3 day old female *Musca*): LD_{50}: 11 µg/fly (1681)
—Crustaceans:
Gammarus lacustris: 96 hr, LC_{50}: 29 µg/l   (2124)
Gammarus fasciatus: 96 hr, LC_{50}: 40 µg/l   (2126)
Palaemonetes kadiakensis: 96 hr, LC_{50}: 1.8 µg/l   (2126)
Orconectes nais: 96 hr, LC_{50}: 7.8 µg/l   (2126)
Simocephalus serrulatus: 48 hr, LC_{50}: 47 µg/l   (2127)
Daphnia pulex: 48 hr, LC_{50}: 42 µg/l   (2127)
Korean shrimp (*Palaemon macrodactylus*): 96 hr static lab bioassay: TL_{50}: 14.5 (8.2–25.9) ppb (2353)
sand shrimp (*Crangon septemspinosa*):   96 hr static lab bioassay: LC_{50}: 8 ppb (2327)
grass shrimp (*Palaemonetes vulgaris*):   96 hr static lab bioassay: LC_{50}: 440 ppb (2327)
hermit crab (*Pagurus longicarpus*):   96 hr static lab bioassay: LC_{50}: 55 ppb (2327)
isopod (*Asellus*): 24 hr LC_{50}: 100 ppb (1681)
pelecypod (*Crassostrea virginica*): 96 hr EC_{50}-FT: 1.5 g/l
decapod (*Penaeus duorarum*):   96 hr EC_{50}-FT: 0.11 g/l
decapod (*Palaemonetes vulgaris*): 96 hr EC_{50}-FT: 1.06 g/l (1138)
—Fish:
Pimephales promelas: 96 hr, LC_{50}: 56 µg/l   (2113)
Lepomis macrochirus: 96 hr, LC_{50}: 19 µg/l   (2113)
Lepomis microlophus: 96 hr, LC_{50}: 17 µg/l   (2107)
Salmo gairdneri: 96 hr, LC_{50}: 19 µg/l   (2119)
Oncorhynchus kisutch: 96 hr, LC_{50}: 59 µg/l   (2119)
Oncorhynchus tschawytscha: 96 hr, LC_{50}: 17 µg/l   (2189)
mummichog (*Fundulus heteroclitus*): 96 hr static lab bioassay: LC_{50}: 50 ppb (2328)
mummichog (*Fundulus heteroclitus*): 96 hr static lab bioassay: LC_{50}: 50.0 ppb (2329)

striped killifish (*Fundulus majalis*): 96 hr static lab bioassay: LC_{50}: 32 ppb
(2329)
Atlantic silverside (*Menidia menidia*): 96 hr static lab bioassay: LC_{50}: 3 ppb
(2329)
striped mullet (*Mugil cephalus*): 96 hr static lab bioassay: LC_{50}: 194 ppb
(2329)
American eel (*Anguilla rostrata*): 96 hr static lab bioassay: LC_{50}: 10 ppb
(2329)
bluehead (*Thalassoma bifasciatum*): 96 hr static lab bioassay: LC_{50}: .8 ppb
(2329)
bluegill: 96 hr LC_{50}: 0.190 ppm
rainbow trout: 96 hr LC_{50}: 0.150 ppm (1878)
northern puffer (*Sphaeroides maculatus*): 96 hr LC_{50}(S): 188 ppb (2329)
threespine stickleback (*Gasterosteus aculeatus*): 96 hr Tlm (S): 111.9 ppb
(2333)
Cyprinodon variegatus: 96 hr LC_{50} (FT): 3.68 µg/l
Lagodon rhomboides: 96 hr LC_{50} (FT): 3.77 µg/l
Leiostomus xanthurus: 96 hr LC_{50} (FT): 0.85 µg/l (1138)
minnow: 24 hr LC_{50}: 13 ppb (1681)
fathead minnow: flow through bioassay: incipient TLm: 7.02 µg/l at 25°C
MATC: 0.86 µg/l (443)
Pagurus longicarpus: 96 hr LC_{50}: 55 µg/l (1575)
—Mammals:
acute oral LD_{50} (rats): 100–162 mg/kg
acute dermal LD_{50} (rats): 195–250 mg/kg (1855)

heptachlorepoxide (Velsicol 53-CS-17; ENT 25,584; 1,4,5,6,7,8,8-heptachloro-2,3-epoxy-3a,4,7,7a-tetrahydro-4,7-methanoindene)
A degradation product of heptachlor
C. WATER POLLUTION FACTORS:
—Aquatic reactions:
persistence in river water in a sealed glass jar under sunlight and artificial fluorescent light - initial conc. 10 µg/l:

after	1 hr	1 wk	% of original compound found 2 wk	4 wk	8 wk	
	100	100	100	100	100	(1309)

D. BIOLOGICAL EFFECTS:
—Residues:
goby fish (*Acanthogobius flavimanus*) collected at the seashore of Keihinjima along Tokyo Bay—Aug. 1978: residue level: 1 ppb (1721)
residues in Canadian human milk: 1967: avg.: 3 ng/g whole milk
1970: avg.: 4 ng/g whole milk
1975: avg.: 1 ng/g whole milk (1376)
ringed seal, Greenland, fat: 0.026 mg/g (1180)
—Bioaccumulation:
pelecypod (*Mytilus edulis*): exposure: 0.22–1.95 µg/l, 50 hr:
bioaccumulation factor: 1700 X (1477)
—Toxicity:

1,4,5,6,7,8,8-HEPTACHLOR-2,3-EPOXY-3a,4,7,7a-TETRAHYDRO-4,7-METHANOINDENE

decapod (*Penaeus duorarum*): 96 hr LC_{50}(FT); effective conc. 0.04 µg/l;

(1138)

1,4,5,6,7,8,8-heptachlor-2,3-epoxy-3a,4,7,7a-tetrahydro-4,7-methanoindene *see* heptachlorepoxide

heptachloronorbornene
D. BIOLOGICAL EFFECTS:
 —Fish:
  fathead minnow: 96 hr LC_{50} (FT): 85.6 µg/l (1206)

heptachlorostyrene
D. BIOLOGICAL EFFECTS:
 —Residues:
  in fat phase of the following fish from the Frierfjord (Norway), April 1975–Sept. 1976:
   cod, whiting, plaice, eel, sprot: avg.: 12.5 ppm
                             range: 1.4–92 ppm (1065)

1,4,5,6,7,8,8-heptachloro-3a,4,7,7a-tetrahydro-4,7-methano-indene *see* heptachlor

n-heptacosane
$CH_3(CH_2)_{25}CH_3$
B. AIR POLLUTION FACTORS:
 —Ambient air quality:
  organic fraction of suspended matter:
   Bolivia at 5200 m altitude (Sept.–Dec. 1975): 0.37–0.73 µg/1000cu m;
   Belgium, residential area (Jan.–April 1976): 4.5–6.5 µg/1000cu m (428)
  glc's in the average American urban atmosphere—1963:
   260 µg/g airborne particulates or
   32 ng/g cu m air (1293)
  glc's in residential area (Belgium) oct. 1976:
   in particulate sample: 11.1 ng/cu m
   in gas phase sample: 7.99 ng/cu m (1289)
  glc's at Botrange (Belgium)—woodland at 20–30 km from industrial area:
   June–July 1977: 4.86; 4.05 ng/cu m (*n* = 2)
  glc's in Wilrijk (Belgium)—residential area:
   Oct.–Dec. 1976: 7.55; 14.25 ng/cu m (*n* = 2) (1233)

n-heptacosanoic acid
$CH_3(CH_2)_{25}COOH$
B. AIR POLLUTION FACTORS:
 —Ambient air quality:
  glc's at Botrange (Belgium)—woodland at 20–30 km from industrial area:
   June–July 1977: 0.55; 0.63 ng/cu m (*n* = 2) (1233)

n-heptadecane
$CH_3(CH_2)_{15}CH_3$

A. PROPERTIES: m.p. 22.5°C; sp.gr. 0.778; b.p. 303°C
B. AIR POLLUTION FACTORS:
  −Ambient air quality:
    glc's in residential area (Belgium) Oct. 1976:
      in particulate sample: − ng/cu m
      in gas phase sample: 65.6 ng/cu m (1289)
    glc's in the average American urban atmosphere - 1963:
      20 µg/g airborne particulates or
      2.5 ng/cu m air (1293)
    organic fraction of suspended matter:
      Bolivia at 5200 m altitude (Sept.−Dec. 1975): 0.19−0.22 µg/1000 cu m
      residential area, 10 km south of Antwerp, Belgium (Jan.−April 1976): 0.65−1.8 µg/1000 cu m (428)
  −Atmospheric reactions:
    first order evaporation constant of n-heptadecane in 3 mm No. 2 fuel oil layer in dark room at windspeed of 21 km/hr at:
      5°C: 4.00×10^{-7} min^{-1}
      10°C: 5.70×10^{-5} min^{-1}
      20°C: 4.08×10^{-5} min^{-1}
      30°C: 1.11×10^{-4} min^{-1} (438)
C. WATER POLLUTION FACTORS:
  −Biodegradation:
    biodegradation to CO_2:

sampling site	concentration (µg/l)	month	incubation time (hr)	degradation rate (µg/1/day)$\times 10^3$	turnover time (days)
control station	20	−	−	101 ± 5	220
near oil storage tanks	20	−	−	390 ± 77	60
near oil storage tanks	20	−	−	115 ± 16	160
Skidaway river	3	June	24	50 ± 11	59
	6	June	24	90 ± 22	67
	12	June	24	160 ± 30	78
	24	June	24	240 ± 37	100
	10	Aug.	24	171 ± 20	59
estuarine	20	−	−	140 ± 12	140
coastal	20	−	−	34 ± 8	590
Gulf stream	20	−	−	3 ± 4	6700

(381)

  Microbial degradation to CO_2 −in seawater at 12°C−in the dark after 24 hr incubation at 30 µg/l−0.07 µg/l/day−turnover time: 400 days; after addition of water extract of no. 2 fuel oil−after 16 hr incubation: 0.50 µg/l/day−turnover time: 60 days. (477)
  −Water quality:
    in Zürich Lake: at surface: 20 ppt; at 30 m depth; 4 ppt
    in Zürich area: in springwater: 1ppt; in tapwater: 3 ppt (513)

708 HEPTADECANOIC ACID

heptadecanoic acid (margaric acid)
$CH_3(CH_2)_{15}COOH$
A. PROPERTIES: a saturated fatty acid not normally found in natural fats or waxes; m.p. 61°C; sp.gr. 0.8355 (90.6/4°C); b.p. 363.8°C
B. AIR POLLUTION FACTORS:
  —Ambient air quality:
  organic fraction of suspended matter:
   Bolivia at 5200 m altitude (Sept.–Dec. 1975): 0.20–0.26 µg/1000 cu m
   Belgium, residential area (Jan.–April 1976): 1.7–2.5 µg/1000 cu m (428)
   glc's at Botrange (Belgium)—woodland at 20–30 km from industrial area:
   June–July 1977: 0.67; 0.41 ng/cu m ($n = 2$)
   glc's in Wilrijk (Belgium)—residential area:
   Oct.–Dec. 1976: 1.50; 3.32 ng/cu m ($n = 2$) (1233)
   glc's in residential area (Belgium) oct. 1976:
   in particulate sample: 2.85 ng/cu m
   in gas phase sample: 5.71 ng/cu m (1289)
C. WATER POLLUTION FACTORS:
  —Odor threshold: detection: 20.0 mg/kg (886)

heptaldehyde *see* heptanal

heptanal (enanthaldehyde; enanthal; 1-heptylaldehyde; enenthole; heptaldehyde)
$CH_3(CH_2)_5CHO$
A. PROPERTIES: colorless liquid; m.w. 114.18; m.p. -45°C; b.p. 155°C; v.p. 3 mm at 25°C; v.d. 3.9; sp.gr. 0.850 at 20/4°C;
B. AIR POLLUTION FACTORS: 1 mg/cu m = 0.215 ppm, 1 ppm = 4.6 mg/cu m

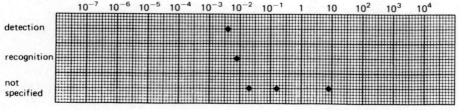

(279; 610; 637; 788; 842)

C. WATER POLLUTION FACTORS:
  —BOD_5:   28% of ThOD
  —$KMnO_4$ value: acid: 9% of ThOD; alkaline; 6% of ThOD
  —COD: 73% of ThOD (220)
  —Reduction of amenities: T.O.C. = 0.003 mg/l; 0.0195 mg/kg (305)(873)
  —Waste water treatment:
   A.S.: after 6 hr: 5.0% of ThOD
           12 hr: 7.0% of ThOD
           24 hr: 14.7% of ThOD (88)
D. BIOLOGICAL EFFECTS:
  —Mammalia: mouse: acute oral LD_{50}: 25 g/kg (221)

n-heptane
$CH_3(CH_2)_5CH_3$

A. PROPERTIES: colorless liquid; m.w. 100.20; m.p. -91°C; b.p. 98°C; v.p. 35 mm at 20°C; 58 mm at 30°C; v.d. 3.45; sp.gr. 0.68 at 20°C; solub. in water 3 mg/l at 20°C, in distilled water 2.4 mg/l at 20°C, in salt water 10.5 mg/l at 20°C; sat.-conc. 196 g/cu m at 20°C, 306 g/cu m at 30°C

B. AIR POLLUTION FACTORS: 1 mg/cu m = 0.24 ppm, 1 ppm ≏ 4.17 mg/cu m

odor thresholds

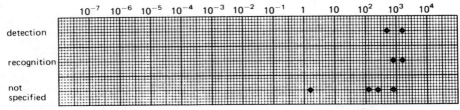

(73; 278; 291; 307; 708; 737; 761; 781)

    O.I. at 20°C = 200   (316)
    —Atmospheric reactions: reactivity: NO ox.: ranking: 0.2   (63)

C. WATER POLLUTION FACTORS:
    —BOD_5: 1.92 NEN 3235-5.4   (277)
    —COD: 0.06 NEN 3235-5.3   (277)
    —Biodegradation:
      incubation with natural flora in the groundwater—in presence of the other components of high-octane gasoline (100 µl/l): biodegradation: 49% after 192 hr at 13°C (initial conc.: 0.37 µl/l)   (956)
    —Reduction of amenities: T.O.C. = 50 mg/l   (295)

D. BIOLOGICAL EFFECTS:
    —Fish:
      Gambusia affinis: no effect: 5,600 mg/l
                       toxic: 10,000 mg/l, 24 hr   (30)
      (mosquito fish): TLm (24-96 hr): 4,924 mg/l   (244)
      goldfish: LD_{50} (24 hr): 4 mg/l — modified ASTM D 1345   (277)
      white roach: lethal in 1 to 4 hrs: 30 mg/l   (226)
      young Coho salmon: no significant mortalities when applied in amounts up to 100 ppm, after 96 hrs in artificial sea water at 8°C   (317)
    —Man: slight vertigo: 1,000 ppm, 6 min
                          2,000 ppm, 4 min   (211)

heptanedioic acid   *see* pimelic acid

heptanenitrile
$CH_3(CH_2)_5CN$

C. WATER POLLUTION FACTORS:
    —Waste water treatment:
      A.S.: after   6 hr: 0.9% of ThOD
                  12 hr: 1.5% of ThOD
                  24 hr: 1.7% of ThOD   (89)

710 1-HEPTANETHIOL

1-heptanethiol   *see* *n*-heptylmercaptan

***n*-heptanoic acid**   *see* enanthic acid

1-heptanol   (*n*-heptylalcohol)
$CH_3(CH_2)_5CH_2OH$

A. PROPERTIES: colorless liquid; m.w. 116.20; m.p. $-34°C$; b.p. $176°C$; v.p. 1 mm at $42°C$; v.d. 4.03; sp.gr. 0.82 at $20/4°C$; solub. 900 mg/l at $18°C$, 2,000 mg/l at $20°C$, 2,800 mg/l at $100°C$; $\log P_{oct}$ 2.41 (calculated)

B. AIR POLLUTION FACTORS: 1 mg/cu m = 0.21 ppm, 1 ppm = 4.83 mg/cu m

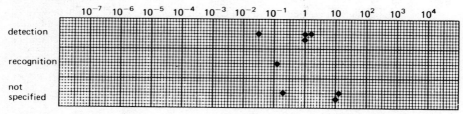

(307; 610; 634; 739; 780; 787; 824; 825; 827)

O.I. at $20°C$ = 23,100 (316)

C. WATER POLLUTION FACTORS:
- $-BOD_5$:   30% of ThOD
- $-COD$ 92% of ThOD
- $-KMnO_4$ value: acid: 7% of ThOD, alkaline: 2% of ThOD (220)
- Waste water treatment:
  A.S.: after   6 hr: 7.9% of ThOD
            12 hr: 20.4% of ThOD
            24 hr: 28.9% of ThOD (88)
  reverse osmosis: 52% rejection from a 0.01 M solution (221)

D. BIOLOGICAL EFFECTS:
- Toxicity threshold (cell multiplication inhibition test):

bacteria (*Pseudomonas putida*):	67 mg/l	(1900)
algae (*Microcystis aeruginosa*):	3.5 mg/l	(329)
green algae (*Scenedesmus quadricauda*):	17 mg/l	(1900)
protozoa (*Entosiphon sulcatum*):	31 mg/l	(1900)
protozoa (*Uronema parduczi Chatton-Lwoff*):	17 mg/l	(1901)
mexican axolotl (3–4 w after hatching):	48 hr LC_{50}: 52 mg/l	
clawed toad (3–4 w after hatching):	48 hr LC_{50}: 44 mg/l	(1823)

2-heptanone   (*n*-amylmethylketone; methyl-*n*-amylketone)
$CH_3COC_5H_{11}$

A. PROPERTIES: colorless liquid; m.w. 114.18; m.p. $-35/-27°C$; b.p. $150°C$; v.p. 2.6 mm at $20°C$; v.d. 3.94; sp.gr. 0.82 at $15/4°C$; solub. 4,300 mg/l; sat. conc. 6.8 g/cu m at $20°C$, 13 g/cu m at $30°C$

B. AIR POLLUTION FACTORS: 1 mg/cu m = 0.214 ppm, 1 ppm = 4.66 mg/cu m

odor thresholds   mg/cu m

	10^{-7}	10^{-6}	10^{-5}	10^{-4}	10^{-3}	10^{-2}	10^{-1}	1	10	10^2	10^3	10^4
detection								♦♦				
recognition												
not specified							♦ ♦♦					

(307; 772; 822; 842)
O.I. at 20°C = 171,000   (316)

C. WATER POLLUTION FACTORS:
—Waste water treatment:
   NFG, BOD, 20°C, 1-10 days observed, feed: 100-1,000 mg/l, acclimation:
   365 + P: 17% removed   (93)
—BOD_{10}: 0.50 std. dil. sew.   (256)
—Odor threshold: detection: 0.14 mg/kg   (882)
                 recognition: 3.0 mg/kg   (914)

D. BIOLOGICAL EFFECTS:
—Mammals:
   guinea pig: inhalation: irritation of the mucous membrane: 1,500 ppm
               death: 4,800 ppm after a few exposures   (211)

3-heptanone (butylethylketone)
$C_2H_5CO(CH_2)_3CH_3$
A. PROPERTIES: colorless liquid; m.w. 114.18; m.p. -37/-39°C; b.p. 148.5°C; v.p.
   1.4 mm at 25°C; v.d. 3.93; sp.gr. 0.818 at 20/4°C; solub. 14,300 mg/l at 20°C
B. AIR POLLUTION FACTORS: 1 mg/cu m = 0.214 ppm, 1 ppm = 4.66 mg/cu m
C. WATER POLLUTION FACTORS:
—Waste water treatment:
   A.S.: after 6 hr: 1.2% of ThOD
               12 hr: 2.1% of ThOD
               24 hr: 3.7% of ThOD   (88)
—Odor threshold: detection: 0.0075 mg/kg   (899)
D. BIOLOGICAL EFFECTS:
—Mammals:
   rat: single oral LD_{50}: 2.79 g/kg
        inhalation: no deaths: 2,000 ppm, 4 hr
                    deaths: 4,000 ppm, 4 hr   (211)

4-heptanone (dipropylketone; butyrone)
$(C_3H_7)_2CO$
A. PROPERTIES: colorless liquid; m.w. 114.18; m.p. -32.6°C; b.p. 143°C; v.p. 1.2
   mm at 25°C; v.d. 3.9; sp.gr. 0.817 at 20/4°C
B. AIR POLLUTION FACTORS: 1 mg/cu m = 0.214 ppm, 1 ppm = 4.66 mg/cu m
C. WATER POLLUTION FACTORS:
—Waste water treatment:

1-HEPTENE

A.S.: after 6 hr: 1.4% of ThOD
12 hr: 2.2% of ThOD
24 hr: 3.8% of ThOD (88)

1-heptene
$CH_2=CH(CH_2)_4 CH_3$
D. BIOLOGICAL EFFECTS:
—Threshold conc. of cell multiplication inhibition of the protozoan *Uronema parduczi Chatton-Lwoff*: 1.8 mg/l (1901)

***n*-heptylalcohol** *see* 1-heptanol

1-heptylaldehyde *see* heptanal

heptylcarbinol *see* 1-octanol

heptylchloride *see* 1-chloroheptane

***n*-heptylmercaptan** (1-heptanethiol)
$C_7H_{15}SH$
A. PROPERTIES: m.w. 132.36; m.p. -43.4°C; b.p. 176.2°C; sp.gr. 0.839 at 25/4°C
B. AIR POLLUTION FACTORS:
—Emission control:
wet scrubber: water at pH 8.5: effluent: 3,200 odor units/scf
$KMnO_4$ at pH 8.5: effluent: 20 odor units/scf (115)

hexachlorobenzene (perchlorobenzene)

Manufacturing source: organic chemical industry; byproduct of tetrachloroethylene (347)

Users and formulation: manufacture of pentachlorophenol, wood preservative; fungicide, seed treatment (control of wheat bunt); used in production of aromatic fluorocarbons; organic synthesis, impregnation of paper; in technical pentachlorophenol up to 13%, in herbicide DCPA up to 10-14%, in pesticide PCNB between 1 and 6%. An impurity of pentachloronitrobenzene, 0.46% has been found in Terraclor® (1805)

Manmade sources: industries identified as possible HCB sources and potential origin of HCB wastes:

industry/type	potential origin of HCB wastes
Basic HCB production/distribution	HCB production operation
Chlorinated solvents production	Reaction side-product in the production of chlorinated solvents, mainly, carbon

industry/type	potential origin of HCB wastes
	tetrachloride, perchloroethylene, trichloroethylene, and dichloroethylene
Pesticide production	Reaction side-product in the production of Dacthal, simazine, mirex, atrazine, propazine, and pentachloronitrobenzene (PCNB)
Pesticide formulation/distribution	Formulation, packaging and distribution of HCB-containing pesticides
Electrolytic chlorine production	Chlorine attack on the graphite anode or its hydrocarbon coating
Ordnance and pyrotechnics production	Use of HCB in the manufacture of pyrotechnics, and trace bullets and other ordnance items
Sodium chlorate production	Similar to electrolytic chlorine production where graphite anodes are used
Aluminium manufacture	Use of HCB as a fluxing agent in aluminum smelting
Seed treatment industry	Use of HCB in seed protectant formulations
Pentachlorophenol (PCP) production	Reaction by-product of PCB production by chlorination of phenol
Wood preservatives industry	Use of HCB as a wood preserving agent
Electrode manufacture	Use of HCB as a porosity control in the manufacture of graphite anodes
Vinyl chloride monomer production	By-product in the manufacture of vinyl chloride monomer
Synthetic rubber production	Use of HCB as a peptizing agent in the production of nitroso and styrene rubbers for tires

(1375)

A. PROPERTIES: m.w. 284.80; m.p. 227/229°C; b.p. 326/322°C; v.d. 9.84; sp.gr. 2.044 at 23°C; white crystalline solid; v.p. 1.089×10^{-5} mm Hg; solub. 0.006 mg/l; 0.11 mg/l at 24°C (99% purity); 0.004 mg/l; log P_{oct} 6.18

B. AIR POLLUTION FACTORS: 1 mg/cu m = 0.08 ppm, 1 ppm = 11.8 mg/cu m
   —Sampling and analysis:
   concentration on chromosorb 101 (60/80 mesh), sampling time: 1 hr at 2 lt per min; extraction with 5% acetone in methanol: trapping efficiency: 87–100%; GC-EC: det.lim.: 28 ng/cu m (150)

C. WATER POLLUTION FACTORS:
   —Reduction of amenities: T.O.C. = 3 mg/l (225)(903)
   Water quality: in water of river Tiber (Italy): from 1969 to 1973: average level: 2.5 ppt based on 120 samples: 34.2% contained less than 1 ppt, 63.8% contained between 1 and 10 ppt and 1.8% over 10 ppt (332)
   —Waste water treatment:
   degradation by *Pseudomonas*: 200 mg/l at 30°C:
       parent: 0% ring disruption in 120 hr
       mutant: 0% ring disruption in 120 hr (152)

714 HEXACHLOROBENZENE

D. BIOLOGICAL EFFECTS:
  —Residues:
   in peanut butter samples: avg.: 7.4 ppb
                          range: 0.97–38 ppb ($n = 11$)   (1805)
   residue in Canadian human milk–1975: avg. 2 ng/g whole milk   (1376)
  —HCB and fat levels in Lake trout (*Salvelinus namaycush*), rainbow trout (*Salmo gairdneri*) and coho salmon (*Oncorhynchus kisutch*) from lake Ontario:

	lake trout	rainbow trout	coho salmon
% fat in tissue mean (range)	16.2 (8.9–22.5)	8.9 (5.3–14.7)	7.5 (3.7–9.7)
HCB in whole fish homogenates ng/g mean (range)	80 (40–120)	62 (30–125)	36 (16–50)
HCB in the fat ng/g mean (range)	500 (200–700)	700 (300–1700)	500 (200–1100) (1563)

   average levels of HCB in fish collected from Lake Ontario:
     alewives (*Alosa pseudoharengus*) and smelt (*Osmerus mordax*): 24 ng/g
     coho salmon muscle: 97 ng/g   (1564)
   mosquito fish from the industrialized region along the lower Mississippi river contained 72 to 380 ng/g   (1565)
   in fat phase of the following fish from the Frierfjord (Norway) April 1975– Sept. 1976: cod, whiting, plaice, eel, sprot: avg.: 30.8 ppm
                           range: 7–141 ppm   (1065)
  —residues in fish-eating birds from Gdánsk Bay 1975–76:   (1776)

(ppm in fresh tissue)

	pectoral muscle	liver	adipose tissue
great crested grebe (*Podiceps cristatus L.*)			
avg.:	0.167 ($n = 17$)	0.34 ($n = 18$)	2.58 ($n = 10$)
range:	0.04–0.59	0.04–1.0	0.72–7.2
slavonian grebe (*Podiceps auritus L.*)	0.06 ($n = 1$)	0.08 ($n = 1$)	
black guillemot (*Cepphus grylle L.*)			
avg.:	0.04 ($n = 2$)	0.08 ($n = 2$)	1.2 ($n = 2$)
range:	0.03–0.06	0.06–0.10	0.63–1.8
goosander (*Mergus merganser L.*)	0.09 ($n = 1$)	0.14 ($n = 1$)	2.7 ($n = 1$)
black throated diver (*Gavia arctica L.*)	0.015 ($n = 1$)	0.05 ($n = 1$)	1.6 ($n = 1$)

   in marine animals from the central Mediterranean (1976–1977):
     fish: anchovy (*Engraulis encrasicholus*):
           avg.: 0.77 ppb fresh wt. (n = 12)
           range: 0.1–2.2 ppb fresh wt.
       striped mullet (*Mullus barbatus*):
           avg.: 1.5 ppb fresh wt. ($n = 10$)
           range: 0.1 - 8.1 ppb fresh wt.

tuna (*Thunnus thynnus thynnus*):
    avg.: 0.12 ppb fresh wt. ($n = 5$)
    range: 0.1–0.2 ppb fresh wt.
shrimp (*Nephrops norvegicus*):
    avg.: 0.23 ppb fresh wt. ($n = 7$)
    range: 0.1–0.6 ppb fresh wt.
mussel (*Mytilus galloprovincialis*):
    avg.: 1.5 ppb fresh wt. ($n = 4$)
    range: 1.19–1.83 ppb fresh wt.   (1774)
—Bioaccumulation:
BCF in bacteria (*Siderocapsa treubii*): approx. 50,000   (1833)
BCF: algae: 320–1570
     snails: 1360–3320
     daphnids: 770–1030
     fish: 1160–3740   (1891; 1892; 1893)
BCF: 5 aquatic molluscs: 10–250 (γ - isomer)   (1870)
BCF in guppies (*Poecilia reticulata*): 290,000 (on lipid content)   (1833)
BCF*: brine shrimp: 95 (at 5.2 ppb in water)
     mosquito larvae: 220 (at 6.6–13.1 ppb in water)
     fish (silverside): 1613 (at 1.8 ppb in water)
*γ-BHC introduced into system in the form of residues on sand   (1880)
—Fish: trout: log. bioconcentration factor: 3.89   (193)
—Toxicity:
—Algae:
Chlorella pyrenoidosa—strain 211-8b:
  inhibition of photosynthetic oxygen evolution of 33.3% at 0.1 ppm
                                         42% at 1.0 ppm
                                         51% at 5.0 ppm
  respiration, however, was inhibited only slightly, if at all at levels of 0.1–5.0 ppm   (1212)
—Insect:
ricefield-spider (*Oedothorax insecticeps*): LD_{50}: 21 ppm   (1814)
—Fish:
guppy (*Poecilia reticulata*): 14 d LC_{50}: >0.32 ppm   (1833)
—Mammalia:

species	minimum lethal single oral dose	average lethal single oral dose	absolute lethal single oral dose
mouse	2000 mg/kg body wt.	±4000 mg/kg body wt.	±7500 mg/kg body wt.
rat	2000 mg/kg body wt.	3500 mg/kg body wt.	6000 mg/kg body wt.
guinea pig	3000 mg/kg body wt.		
rabbit		2600 mg/kg body wt.	
cat		±1700 mg/kg body wt.	

male rats: oral doses during 13 days:
at 2 mg/kg/day: no toxic effects
     6 mg/kg/day: very slight skin twitching and nervousness, significant increase in liver weight.
   20 mg/kg/day: neurotoxic symptoms, increase in liver weight

60 mg/kg/day: neurotoxic symptoms, increase in liver and kidney weight
200 mg/kg/day: neurotoxic symptoms, increase in liver and kidney weight
(332)
rats: daily dose in feed during 51 days: 100 mg/kg/day: 13 deaths out of 33 in 1 month. Neurotoxic symptoms. Increased liver weight. Porphyria. (333)
rats: daily oral dose:
300 mg/kg body wt./day: mortality: 3 out of 10 after 10 days
150 mg/kg body wt./day: mortality: 6 out of 10 after 30 days
50 mg/kg body wt./day: mortality: 3 out of 10 after 30 days (334)
rats: daily oral dose in water: 0.025 mg/kg/day during 4–8 months: No toxic symptoms. Possible effect on conditioned reflexes.
quail: oral daily dose in feed, during 90 days: 0.1 mg/kg/day: no toxic effects
0.5 mg/kg/day: slight increase in liver weight. Minimal liver pathology and porphyria
1.8 mg/kg/day: increased liver weight. Liver and kidney pathology. Decreased egg production. Porphyria.
7.2 mg/kg/day: mortality 5/15, 18 to 62 day period. Neurotoxic symptoms. Decreased egg production and hatchability. Porphyria. Increased liver weight. Liver and kidney path. (336)
white rats: male: oral LD_{50}: 1,250 mg/kg (1802)
HCB has been found to effect reproduction of the rat, with 20 ppm being the no-effect level. Suckling rats were particularly sensitive. Weanling rats from dams fed HCB contained HCB residues, had enlarged livers with increased hepatic aniline hydroxylase activity. No gross abnormalities were observed in the pups.
(1612)

hexachlorobiphenyls (*see also* polychlorobiphenyls)

2,2',4,4',5,5'-hexachlorobiphenyl
D. BIOLOGICAL EFFECTS:
—Algae:
Chlorella pyrenoidosa: bioconcentration coefficient: 10,000 in whole cells
(1567)

2,4,5,2',4',5'-hexachlorobiphenyl
A. PROPERTIES: solub. 0.00095 mg/l at 24°C; log P_{oct} 6.72

2,4,6,2',4',6'-hexachlorobiphenyl
D. BIOLOGICAL EFFECTS:
—Crustacean:
amphipod (*Gammarus fasciatus*): 4 days (S) LC_{50}: 0.15 mg/l (1617)

2,4,5,2',4',5'-hexachlorobiphenyl
C. WATER POLLUTION FACTORS:

—Aquatic reactions:

Freundlich adsorption constants:	n	k	
Illite clay	1.26	200	
Woodburn soil	1.25	320	
humic acid	3.78	33.9	(1879)

hexachlorobutadiene
$Cl_2CCClCCClCl_2$

Manufacturing source: organic chemical industry. (347)

Users and formulation: solvent for natural rubber, synthetic rubber and other polymers; heat transfer liquid, transformer liquid, and hydraulic fluid; washing liquor for removing hydrocarbons. (347)

Man caused sources (*water and air*): roadway runoff (hydraulic fluids and rubber), lab use. (347)

A. PROPERTIES: m.p. $-19°C$ to $-22°C$; b.p. $210°C$ to $220°C$; v.p. 22 mm at $100°C$, 500 mm at $200°C$; sp.gr. 1.675 at $15.5/15.5°C$, clear colorless liquid; solub. 2 ppm

B. AIR POLLUTION FACTORS:
  —Sampling and analysis:
   concentration on chromosorb 101 (60/80 mesh), sampling time: 1 hr at 2 l/min; extraction with 5% acetone in methanol: trapping efficiency: 98.0–103%: GC-EC: det.lim.: 28 ng/cu m (150)

C. WATER POLLUTION FACTORS:
  —Reduction of amenities:
   T.O.C.: 0.006 mg/l (225)(903)
   T.O.C. in water: 6 ppb (316)
  —Water quality:
   percolation water at 30–500 m from dumping grounds: 40–55 ppb (183)
   in river IJssel (Netherlands): 0.13 ppb
   in lake Ketelmeer (Netherlands): 0.13–0.20 ppb
   in lake IJsselmeer (Netherlands): 0.05–0.07 ppb (526)

D. BIOLOGICAL EFFECTS:
  —Bioaccumulation:
   No bioaccumulation in the following incomplete food chain: water of the lake Ketelmeer: 0.13 ppb

detritus	0.20–0.22 ppm
↓	
Oligochaeta*	0.09–0.40 ppm
↓	
Sphaerium	2.41 ppm
↓	
roach	0.23–1.40 ppm
↓	
perch	0.13–0.40 ppm
↓	
pike perch	0.11–1.15 ppm

*mainly *Limnodrilus* sp. (526)

—Toxicity:

—Fish:
  goldfish: renewal bioassay: 96 hr TLm: 0.09 mg/l at 17.5°C　　　(456)
  guppy (*Poecilia reticulata*): 14 d LC_{50}: 0.4 ppm　　　(1833)
—Bird:
  adult japanese quail: in diet for a period of 90 days in concentrations up to 30 ppm (about 5 mg/kg/day): no deleterious effect on reproduction　　　(600)
—Mammals:
  inhalation:
    rats: inhalation: no deaths at 161 ppm for 0.88 hr exposure
                                     34 ppm for 3.5 hr exposure　　　(596; 599)
    rat: inhalation: no effect level: 5 ppm, 15 × 6 hr　　　(65)
    guinea pigs, cats: most died at 161 ppm for 0.88 hr exposure
                                     34 ppm for 7.5 hr exposure　　　(596; 599)
  ingestion:
    rats: in diet at 3 mg/kg/day for 30 days: no effect
                    10 mg/kg/day for 30 days: marginal changes
                    30 mg/kg/day for 30 days: increase in the kidney–body weight ratio as well as renal tubular degeneration, necrosis and regeneration, decreased food consumption and body weight gain　　　(595)
    adult male rat: single oral LD_{50}:　　504–667 mg/kg
    adult female rat: single oral LD_{50}:　　200–400 mg/kg
    21 day old male rat: single oral LD_{50}:　　46–91 mg/kg
    21 day old female rat: single oral LD_{50}:　　26–81 mg/kg　　　(595)
    rats: ingestion in diet for up to 2 years of: 20 mg/kg/day caused decreased body weight gain and survival, increased urinary excretion of coproporphyrin, increased weights of kidneys, increased renal tubular epithelial hyperplasia/proliferation and renal tubular adenomas and adenocarcinomas, some of which metastasized to the lungs.
    2 mg/kg day; lesser degree of toxicity, including an increase in urinary coproporphyrin excretion and an increase in renal tubular epithelial hyperplasia/proliferation, no neoplasms　　　(595)
    guinea pig: single oral LD_{50}: 90 mg/kg　　　(597)
    mouse:　　single oral LD_{50}: 87–116 mg/kg　　　(596; 598)
    rat:　　　single oral LD_{50}: 200–350 mg/kg　　　(596; 598)
  dermal application:
    rabbits: 24 hr: 4/4:　126 mg/kg
             7 hr: 1/2:　126 mg/kg
             4 hr: LD_0:　126 mg/kg
             24 hr: LD_0:　63 mg/kg　　　(596)

α-hexachlorocyclohexane

$$\begin{array}{c} Cl \\ Cl\diagup\!\!\!\diagdown Cl \\ Cl\diagdown\!\!\!\diagup Cl \\ Cl \end{array}$$

A. PROPERTIES: solub. 1.4 mg/l (in salt water)
C. WATER POLLUTION FACTORS:
  −Odor threshold conc. 0.088 mg/kg (915)
D. BIOLOGICAL EFFECTS:
  −Residues:
    residue in animals in different trophic levels from the Weser estuary (1976): wet tissue basis:
      bivalvia: common edible cockle (*Cerastoderma edule L.*): 0.9 ± 0.2 ng/g
                soft clam (*Mya arenaria L.*): 3.1 ± 0.5
      polychaeta: lugworm (*Arenicola marina L.*): 1.5 ± 0.2 ng/g
      crustacea: brown shrimp (*Crangon crangon L.*): 2.3 ± 0.3 ng/g
      pisces: common sole (*Solea solea L.*): 4.1 ± 1.0 ng/g (1440)
    comparative average concentrations of α-HCH residue (wet tissue basis) in brown shrimp (*Crangon crangon*) from various areas of the North Sea (1974–1976):

area	year	no. of analyses	avg. conc. (ppb)	range	
S.E. England	1975	2	2	1–3	(1441)
Netherlands	1974	10	3	<2–6	(1442)
	1975	12	<5	<5–7	(1441)
	1976	12	<5		(1441)
Germany	1976	5	2	2–3	(1440)

  −Bioaccumulation:
    algae:
      in *Chlamydomonas*: 2,700 × on dry weight basis
                           12,000 × on lipid basis
      in *Dunaliella*:   1,500 × on dry weight basis
                           13,000 × on lipid basis
    no difference was found in accumulation behavior between living or dead algae cells. Both accumulation and elimination takes place in less than 30 min.
    (442)
    mussels:
      pelecypod (*Mytilus edulis*), 50 hr at 0.89–2.20 µg/l: BCF 106× (1477)
    crustacea:
      Artemia salina: bioconcentration factor: 60–90 (8000 to 11000 on lipid basis)
        −at conc. in the water 10–250 µg/l
        −equilibrium reached after 24 hr
        −elimination: the initial concentrations were halved within 48–72 hr after transfer into noncontaminated water (1073)
    fish:
      Lebistes: bioaccumulation factor: 500 (17,000 on lipid basis) for concentrations of 10–1400 µg/l:
        −equilibrium reached after 24 hr
        −elimination: the initial concentrations were halved within 10 hr after transfer into noncontaminated water (1073)
  −Toxicity:
    algae:
      Chlamydomonas: no influences on growth for exposures to concentrations up to the solubility limit for 2 days (442)
      Dunaliella: no influence on growth after 2 and 4 days exposure to the max.

solubility (1.4 mg/l) of α-HCH in salt water (1073)
crustacean:
　Artemia salina (3 days old): mortality percentage too low to calculate;
　　　　　　　　　　　LC$_{50}$ 48 hr, within solubility range of α-HCH
　　　　(3 weeks old): LC$_{50}$ 48 hr: >1.4 mg/l
　　　　　　　　　　　LC$_{50}$ 96 hr:　0.5 mg/l (1073)
fish:
　Lebistes Reticulatus: LC$_{50}$ 96 hr: >1.4 mg/l
　　　　　　　　　　　LC$_{10}$ 35 days: 0.5 mg/l
　　　　　　　　　　　*EC$_{50}$ 48 hr: 1.38 mg/l
　　　　　　　　　　　EC$_{50}$ 96 hr: 1.31 mg/l
　*immobilization and mortality. (1073)

β-hexachlorocyclohexane (β-HCH)
C. WATER POLLUTION FACTORS:
　—Odor threshold conc.: 0.00032 mg/kg (915)
D. BIOLOGICAL EFFECTS:
　—Residues:
　　residue in Canadian human milk: 1975: avg. 2 ng/g whole milk (1576)

γ-hexachlorocyclohexane (lindane: benzenehexachloride; BHC; HCCH; HCH; TBH; chloran)

$$\begin{array}{c} Cl \\ Cl \diagup\diagdown Cl \\ Cl \diagdown\diagup Cl \\ Cl \end{array}$$

Use and formulation: medicinal mfg. (scabicide); insecticide mfg. (347)
The main insecticidal component is the γ-isomer
CAVEAT: in literature it is not always clear whether the measured concentrations refer only to the γ-isomer or to the sum of the isomers. Unless mentioned otherwise all concentrations mentioned hereunder refer to the γ-isomer.
A. PROPERTIES: m.w. 290.85; m.p. 112°C; sp.gr. 1.87 at 20/4°C; solub. 17.0 mg/l at 24°C (99% purity)
C. WATER POLLUTION FACTORS:
　—Biodegradation:
　　anaerobe bacteria: up to 90% degraded in 4 days, transformed to chlorine-free
　　　　　　　　　　metabolites (1077)
　　A mixed culture of bacteria with *Pseudomonas aeruginosa* as a main component showed isomerisation of γ-HCH to α-HCH and metabolism to γ-2,3,4,5,6-pentachlorocyclohexane (PCCH) and tetrachlorobenzene (TeCB) and unknown nonpolar metabolites (1468)
　—Odor threshold: detection: 12.0 mg/kg (915)
　—Aquatic reactions:
　　calculated half-life in water at 25°C and 1 m depth, based on evaporation rate of 1.5 × 10^{-4} m/hr: 4590 hr (437)
　　Photo-oxidation by U.V. light in aqueous medium at 90–95°C:

time for the formation of CO_2 (% of theoretical):

	α	γ
25%:	4.2 hr	3.0 hr
50%:	24.2 hr	17.4 hr
75%:	40.0 hr	45.8 hr

(1628)

30–40% of lindane adsorbed on aquifer sand at 5°C after 3–100 hr equilibrium time

desorption of lindane from aquifer sand:

ng lindane adsorbed/g of aquifer sand	% leached of total adsorbed lindane			
	wash 1	wash 2	wash 3	total
22.8	52.0	12.2	3.6	67.8
25.3	29.5	6.7	2.4	38.6
20.9	31.0	29.4	11.9	72.3
26.5	47.3	16.9	9.0	73.2

(1303)

75–100% disappearance from soils: 3–10 years (1815; 1816)

persistence in river water in a sealed glass jar under sunlight and artificial fluorescent light—initial conc. 10 µg/l:

% of original compound found

after	1 hr	1 wk	2 wk	4 wk	8 wk	
	100	100	100	100	100	(1309)

—Water and sediment quality:

in Northern Mississippi water: avg.: 0.077 ng/l; range: 0.02–0.16 ng/l (1082)
Hawaii: sediments: 90–5320 ppt (1174)

D. BIOLOGICAL EFFECTS:

—Residues:

in marine animals from the Ligurian Sea (1977–1978): (α + γ-BHC)
  shrimp (*Nephrops norvegicus*): avg. 0.93 ppb fresh wt.
  mussel (*Mytilus galloprovincialis*): avg. 1.1 ppb fresh wt.
  fish: anchovy (*Engraulis encrasicholus*): avg. 0.81 ppb fresh wt.
     striped mullet (*Mullus barbatus*): avg. 1.04 ppb fresh wt.
     Euthymnus alletteratus: avg. 0.41 ppb fresh wt.
     Sarda sarda: avg. 2.60 ppb fresh wt. (1775)
  hooded seal, Greenland, fat: 0.003 mg/g
  bearded seal, Greenland, fat: 0.053 mg/g (1180)

in marine animals from the Central Mediterranean (1976–1977): ΣBHC
fish:
  anchovy (*Engraulis encrasicholus*): avg.: 1.5 ppb fresh wt. ($n = 12$)
     range: 0.2–3.4 ppb fresh wt.
  striped mullet (*Mullus barbatus*): avg.: 1.5 ppb fresh wt. ($n = 10$)
     range: 0.1–5.0 ppb fresh wt.
  tuna (*Thunnus thynnus thynnus*): avg.: 0.4 ppb fresh wt. ($n = 5$)
     range: 0.1–0.6 ppb fresh wt.
  shrimp (*Nephrops norvegicus*): avg.: 0.47 ppb fresh wt. ($n = 7$)
     range: 0.1–1.0 ppb fresh wt.
  mussel (*Mytilus galloprovincialis*): avg.: 1.5 ppb fresh wt. ($n = 4$)
     range: 0.7–3.0 ppb fresh wt. (1774)

comparative average concentrations of γ-HCH residue (wet tissue basis) in brown shrimp (*Crangon crangon*) from various areas of the North Sea (1974-1976):

area	year	no. of analyses	avg. conc. (ppb)	range	
S.E. England	1975	2	<2	<1-2	(1441)
Netherlands	1974	10	<2	<2-3	(1442)
	1975	12	<5		(1441)
	1976	12	<5		(1441)
Germany	1974	6	39	10-110	(1442)
	1975	4	10	6-15	(1441)
	1976	6	3	2-6	(1441)
	1976	5	2	1-3	(1440)

residue in animals in different trophic levels from the Weser estuary (1976): wet tissue basis:

bivalvia: common edible cockle (*Cerastoderma edule L.*):
2.4 ± 0.4 ng/g

soft clam (*Mya arenaria L.*): 2.8 ± 1.2 ng/g

polychaeta: lugworm (*Arenicola marina L.*): 1.7 ± 0.5 ng/g

crustacea: brown shrimp (*Crangon crangon L.*): 1.8 ± 0.6 ng/g

pisces: common sole (*Solea solea L.*): 3.1 ± 0.9 ng/g wet tissue   (1440)

in aquatic vascular plants of lake Päijänne (1972-1973): <0.5 μg/kg dry wt.
(1055)

in Canadian human milk: 1967: avg.: 3 ng/g whole milk
    1970: avg.: 2 ng/g whole milk   (1376)

in harbour seal (*Phoca vitulana*), German North Sea: 0.29-0.36 mg/kg (blubber)
(1482)

mean concentration in gamebird muscle in upper Tennessee (USA):

mg/kg fresh weight ± standard error

	grouse	quail	woodcock
Johnson county	0.99 ± 0.14 ($n = 12$)	1.69 ± 0.20 ($n = 6$)	1.63 ± 0.20 ($n = 6$)
Carter county	1.81 ± 0.25 ($n = 9$)	1.63 ± 0.16 ($n = 7$)	1.76 ± 0.22 ($n = 6$)
Washington county	2.17 ± 0.15 ($n = 10$)	1.76 ± 0.19 ($n = 11$)	1.83 ± 0.18 ($n = 6$)

quail: acceptable LD_{50}: 250 mg/kg

FAO/WHO residue tolerance limit: 0.1-0.5 mg/kg/day   (1338)

—Bioaccumulation:

decapod (*Penaeus duorarum*):   96 hr, at 0.19-0.68 μg/l: 51-142X   (1471)

decapod (*Penaeus duorarum*):   96 hr, at 0.13-0.62 μg/l: 32-143X   (1471)

decapod (*Palaemonetes pugio*): 96 hr, at 1.0-5.5 μg/l: 25-80X   (1471)

BCF: oysters: 60X   (1921)

pelecypod (*Mytilus edulis*): 50 hr, at 1.24-2.29 μg/l: 100X   (1477)

fish: *Lagodon rhomboides*: 96 hr, at 32-91.3 μg/l: 308-554X   (1471)

Lagodon rhomboides: 96 hr, at 18.4-31.3 μg/l: 167-287X   (1471)

Cyprinodon variegatus: 96 hr, at 41.9-108.7 μg/l: 337-727X   (1471)

—Toxicity:

Ni_2B catalyzed dechlorination of technical lindane yielded a mixture of benzene, cyclohexene and cyclohexane: acute toxicity of chlorinated and dechlorinated lindane:

species	formulation	LC$_{50}$ (μg/l)^a chlorinated	dechlorinated	detoxification factor^b
bluegill	technical	57	82,065	1440
	1% dust	138	69,000	500
Daphnia	technical	516	19,342	37
	1% dust	6442	13,054	2

^{a}96 hr LC$_{50}$ for bluegill, 48 hr EC$_{50}$ for Daphnia; calculated on basis of insecticide present in formulation.
^{b}factor: LC$_{50}$ dechlorinated/LC$_{50}$ chlorinated (1413)
 —Bacteria:
  Pseudomonas putida: inhibition of cell multiplication starts at >5 mg/l (329)
 —Algae:
  Microcystis aeruginosa: inhibition of cell multiplication starts at 0.3 mg/l (329)

Protocuccus sp.	5,000 ppb	0.75 O.D. expt/O.D. control	Test organisms grown in test media containing pesticides for ten days. Absorbance measured at 530 m$_\mu$
Chlorella sp.	5,000 ppb	0.57 O.D. exp/O.D. control	
Dunaliella euchlora	9,000 ppb	0.60 O.D. exp/O.D. control	
Phaeodactylum tricornutum	5,000 ppb	0.30 O.D. expt/O.D. control	
Monochrysis lutheri	5,000 ppb	1.00 O.D. expt/O.D. control	(2347)

 —Crustaceans:
  Gammarus lacustris: 96 hr, LC$_{50}$: 48 μg/l (2124)
  Gammarus fasciatus: 96 hr, LC$_{50}$: 10 μg/l (2126)
  Asellus brevicaudus: 96 hr, LC$_{50}$: 10 μg/l (2126)
  Simocephalus serrulatus: 48 hr, LC$_{50}$: 520 μg/l (2127)
  Daphnia pulex: 48 hr, LC$_{50}$: 460 μg/l (2127)
  Korean shrimp (Palaemon macrodactylus): 96 hr static lab bioassay: TL$_{50}$: 12.5 (4.7-32.7) ppb
  Korean shrimp (Palaemon macrodactylus): 96 hr flowing lab bioassay: TL$_{50}$: 9.2 (5.8-15.0) ppb (2353)
  sand shrimp (Crangon septemspinosa): 96 hr static lab bioassay: LC$_{50}$: 5 ppb (2327)
  grass shrimp (Palaemonetes vulgaris): 96 hr static lab bioassay: LC$_{50}$: 10. ppb (2327)
  hermit crab (Pagurus longicarpus): 96 hr static lab bioassay: LC$_{50}$: 5. ppb (2327)
  Daphnia magna: 24 hr LC$_{50}$: 1250 μg/l
   48 hr LC$_{50}$: 460 μg/l (1002; 1004)
  Gammarus pulex: 48 hr LC$_{50}$: 0.03 ppm
  brown shrimp (Penaeus aztecus): 96 hr TLm (static test): 400 ppb (2323)
 —Molluscs:
  gastropoda (Lymnea stagnalis): 48 hr LC$_{50}$: 7.3 ppm

at 2 ppm: decrease in growth rate from third month; 60% decrease of fecundity
Lymnea stagnalis L.: rearing of larvae is impossible at 2 mg/l (1910)
decapod: *Penaeus duorarum*: 96 hr LC_{50} (FT): 0.34 µg/l (1471)
Penaeus duorarum: 96 hr LC_{50} (FT): 0.17 µg/l
Palaemonetes pugio: 96 hr LC_{50} (FT): 4.44 µg/l (1471)
mysidacea: *Mysidopsis bahia*: 96 hr LC_{50} (FT): 6.28 µg/l (1471)
eastern oyster: *Crassostrea virginica*: egg: 48 hr TLm (static lab bioassay): 9100 ppb
hard clam: *Mercenaria mercenaria*: egg: 48 hr TLm (static lab bioassay): >10000 ppb
hard clam: *Mercenaria mercenaria*: larvae: 12 day TLm (static lab bioassay): >10000 ppb (2324)
mud snail: *Nassa obsoleta*: 96 hr (static lab bioassay). Acute toxicity experiment followed by 133-day post exposure in clean water. Reduced deposition of egg cases from 1473 by control to 749 by expt. at 10000 ppb (2330)

—Insects:
Pteronarcys californica: 96 hr LC_{50}: 4.5 µg/l (2128)
fourth instar larval *Chironomus riparius*: 24 hr LC_{50}: 3.6 µg/l (1853)
insect larvae (*Chaoborus*): 48 hr LC_{50}: 0.008 ppm
(*Cloeon*): 48 hr LC_{50}: 0.092 ppm

—Fish:

	96 hr LC_{50} µg/l		96 hr LC_{50} µg/l
catfish	44	bluegill	68
bullhead	64	bass	32
goldfish	131	rainbow	27
minnow	87	brown	2
carp	90	coho	41
sunfish	83	perch	68

(1934)

bluegill: 96 hr LC_{50}: 0.062 ppm
rainbow trout: 96 hr LC_{50}: 0.060 ppm (1878)
rainbow trout: flow through bioassay: 24 hr TLm: 0.051 mg/l at 20–25°C
96 hr TLm: 0.032 mg/l at 20–25°C

(442)

bluegill: flow through bioassay: 2-y TLm: 0.03 mg/l at 27°C
MATC: 0.0091 mg/l at 27°C
fathead minnow: flow through bioassay: 1-y TLm: 0.069 mg/l at 25°C
MATC: 0.0091 mg/l at 25°C
brook trout: flow through bioassay: 1.5-y TLm: 0.026 mg/l at 9–16°C
MATC: 0.0088 mg/l at 9–16°C
Rainbow trout fry (*Salmo gairdneri Richardson*)
at 20°C, pH = 8.1, hardness = 20 mg $CaCO_3$/l:

	24 hr	48 hr	96 hr	
LC_{10}	0.026 mg/l	0.02 mg/l	0.018 mg/l	
LC_{50}	0.037 mg/l	0.023 mg/l	0.022 mg/l	(331)

Extrapolated LC_{50}, 3 months: 0.01 mg/l

mummichog (*Fundulus heteroclitus*): 96 hr static lab bioassay: LC_{50}: 20 ppb (2328)
mummichog (*Fundulus heteroclitus*): 96 hr static lab bioassay: LC_{50}: 60 ppb (2329)
striped killifish (*Fundulus majalis*): 96 hr static lab bioassay: LC_{50}: 28 ppb (2329)
Atlantic silverside (*Menidia menidia*): 96 hr static lab bioassay: LC_{50}: 9 ppb (2329)
striped mullet (*Mugil cephalus*): 96 hr static lab bioassay: LC_{50}: 66 ppb (2329)
American eel (*Anguilla rostrata*): 96 hr static lab bioassay: LC_{50}: 56 ppb (2329)
bluehead (*Thalossoma bifasciatum*): 96 hr static lab bioassay: LC_{50}: 14 ppb (2329)
northern puffer (*Sphaeroides maculatis*): 96 hr static lab bioassay: LC_{50}: 35 ppb (2329)
threespine stickleback (*Gasterosteus aculeatus*): 96 hr static lab bioassay: TLm: 50 ppb (2333)
Colisa fasciatus: static bioassay: 96 hr TLm: 0.64 mg/l at $18°C$
96 hr TLm: 0.41 mg/l at $33°C$
12 hr TLm: 0.87 mg/l at $18°C$
12 hr TLm: 0.60 mg/l at $33°C$ (458)
Lagodon rhomboides: 96 hr LC_{50} (FT): 86.4 µg/l (1471)
Pimephales promelas: 96 hr, LC_{50}: 87 µg/l (2121)
Lepomis macrochirus: 96 hr, LC_{50}: 68 µg/l (2121)
Lepomis microlophus: 96 hr, LC_{50}: 83 µg/l (2121)
Micropterus salmoides: 96 hr, LC_{50}: 32 µg/l (2121)
Salmo gairdneri: 96 hr, LC_{50}: 27 µg/l (2121)
Salmo trutta: 96 hr, LC_{50}: 2 µg/l (2121)
Oncorhynchus kisutch: 96 hr, LC_{50}: 41 µg/l (2121)
Perca flavescens: 96 hr, LC_{50}: 68 µg/l (2121)
Ictalurus punctatus: 96 hr, LC_{50}: 44 µg/l (2121)
Ictalurus melas: 96 hr, LC_{50}: 64 µg/l (2121)
Pteronarcys: 48 hr LC_{50}: 4.5 µg/l (1003)
Salmo gairdneri: 96 hr LC_{50}: 13 µg/l (1001)
guppy: static bioassay: 48 hr TLm: 0.8 mg/l at $24°C$
rainbow trout: static bioassay: 48 hr TLm: 1.05 mg/l at $12°C$ (457)

	mg/l	24 hr	48 hr	96 hr	3 m (extrap.)
rainbow trout fry	LC_{10} (F)	0.026	0.02	0.018	
(*Salmo gairdneri*)	LC_{50} (F)	0.037	0.023	0.022	0.02

(331)

Cyprinodon variegatus: 96 hr LC_{50} (FT): 103.9 µg/l
Lagodon rhomboides: 96 hr LC_{50} (FT): 30.6 µg/l (1471)
—Mammals:
rats and mice: highly carcinogenic (1601)
acute oral LD_{50} (rats): 88–91 mg/kg
acute dermal LD_{50} (rats): 900–1,000 mg/kg

in diet: rats fed on a diet containing 10 ppm for 12 months suffered no ill effects
(1855)

hexachlorocyclopentadiene (perchlorocyclopentadiene)

Use: key intermediate in the synthesis of stable chlorinated cyclodiene insecticides including aldrin, dieldrin, endrin, endosulfan, heptachlor, chlordane, isodrin, and mirex. Some other products derived from hexachlorocyclopentadiene are non-flammable resins and shock proof plastics, acids, esters, ketones and fluorocarbons.
(1399)
A. PROPERTIES: m.w. 273; m.p. 9/10°C; b.p. 234°C; v.p. 0.08 mm at 25°C; v.d. 9.4
B. AIR POLLUTION FACTORS: 1 mg/cu m = 0.089 ppm, 1 ppm = 11.17 mg/cu m
—Odor: T.O.C.: 0.15 ppm (211; 230)
O.I. at 20°C = 700 (316)
C. WATER POLLUTION FACTORS:
—Reduction of amenities:
odor threshold: 0.0016–0.0014 mg/l (30)
D. BIOLOGICAL EFFECTS:
—Fathead minnows: larvae: 96 hr LC_{50}: 7.0 µg/l (flow through test)
larval and early juvenile stages: 30 d LC_{50}: 6.7 µg/l (flow through test)
bioconcentration factor: <11 (30 day exposure) (1398)
—Mammalia:
guinea pigs: inhalation: max time-conc. for survival:
20.2 ppm, 0.25 hr
7.2 ppm, 1 hr
3.1 ppm, 3.5 hr
1.5 ppm, 7 hr
guinea pigs, rabbits, rats,: inhalation: mild liver and kidney injury: 0.15 ppm, 7 hr/day, 150 days
rats, rabbits: single oral dose: lethal: 0.42–0.62 g/kg (211)
—Man: pronounced pungent odor at 0.33 ppm (211)

1,2,3,4,10,10-hexachloro-6,7-epoxy-1,4,4a,5,6,7,8,8a-octahydro-1,4-*endo-endo*-5,8-dimethanonaphthalene see endrin

hexachloroethane (perchloroethane; carbonhexachloride; phenohep; HCE)
CCl_3CCl_3
Manufacturing source: organic chemical industry. (347)
Users and formulation: mfg. smoke candles and grenades; by-product of industrial chlorination processes; plasticizer for cellulose esters; minor use in rubber and insecticidal formulations; medicinal mfg.; moth repellant; retardant in fermentation processes; fire extinguishing fluids mfg.; camphor substitute in nitro cellulose solvent. (347)

Man caused sources (water and air): general use in veterinary medicine, fire extinguishers, moth repellants, insecticides, and as laboratory chemical. (347)
A. PROPERTIES: solid rhombic crystals; m.w. 236.76; m.p. 187°C sublimes; v.p. 0.4 mm at 20°C, 0.8 mm at 30°C; v.d. 8.16; sp.gr. 2.09 at 20/4°C; solub. 50 mg/l at 22°C; sat. conc. 5.2 g/cu m at 20°C, 10 g/cu m at 30°C
B. AIR POLLUTION FACTORS: 1 mg/cu m = 0.10 ppm, 1 ppm = 9.68 mg/cu m
   —Sampling and analysis:
   photometry: min. full scale: 128 ppm (53)
C. WATER POLLUTION FACTORS:
   —Aquatic reactions:
   measured half life for evaporation from 1 ppm aqueous solution at 25°C, still air, and an average depth of 6.5 cm: 40.7 min (369)
   photo-oxidation by U.V. in aqueous medium at 90–95°C:
   time for the formation of CO_2 (% of theoretical): 25%: 25.2 hr
                                                          50%: 93.7 hr
                                                          75%: 172.0 hr   (1628)
   —Odor threshold: detection: 0.010 mg/kg
                    camphoraceous odor (886)
D. BIOLOGICAL EFFECTS:
   —Mammalis:
   No evidence was provided for the carcinogenicity of the compound in Osborne-Mendel rats. It is concluded that under the conditions of this bioassay, hexachloroethane was carcinogenic in B6C3F1 mice, inducing hepatocellular carcinomas in both sexes. (1760)
   Lethal dosages for animals following single administration:

		diluent	mg/kg
rabbit, male	oral ALD	methylcellulose	>1000
rat, male	I.P. ALD	corn oil	2900
rat, female	oral LD_{50}	corn oil	4460
		methylcellulose	7080
rat, male	oral LD_{50}	corn oil	5160
		methylcellulose	7690
guinea pig male	oral LD_{50}	corn oil	4970
rabbit, male	dermal LC_{50}	water paste	≥32000

(1329)

1,2,3,4,10,10-hexachloro-1,4,4a,5,8,8a-hexahydro-1,4-endohexo-5,8-dimethano-naphthalene *see* aldrin

6,7,8,9,10,10-hexachloro-1,5,5a,6,9,9a-hexahydro-6,9-methano-2,3,4-benzodi-oxathiepin-3-oxide *see* endosulfan

hexachloronorbornadiene
   Use: intermediate in the synthesis of isodrin and endrin
D. BIOLOGICAL EFFECTS:
   —Fathead minnows: larvae: 96 hr LC_{50}: 188 µg/l (flow through test)
     larval and early juvenile fatheads: 30 day LC_{50}: 123 µg/l (flow through test)
     30 day old fatheads: bioconcentration factor: approximately 6400 (1398)

hexachloronorbornene
Use: intermediate in the synthesis of isodrin and endrin (1399)
D. BIOLOGICAL EFFECTS:
  —Fathead minnows: larvae: 96 hr LC_{50}: 85.6 µg/l (flow through test)
    larval and early juvenile fatheads: 30 day LC_{50}: 60.1 µg/l (flow thr. test)
    30 day old fatheads: bioconcentration factor: approximately 11,200 (1398)

hexachlorophene *see* 2,2'-methylene-bis (3,4,6-trichlorophenol)

***n*-hexacosane**
$CH_3(CH_2)_{24}CH_3$
B. AIR POLLUTION FACTORS:
  —Ambient air quality:
    organic fraction of suspended matter:
      Bolivia at 5200 m altitude (Sept.–Dec. 1975): 0.64–1.1 µg/1000 cu m
      Belgium, residential area (Jan.–April 1976): 3.2–4.6 µg/1000 cu m (428)
    glc's in Botrange (Belgium)—woodland at 20–30 km from industrial area; June–July 1977: 1.02; 1.78 ng/cu m ($n = 2$)
    glc's in Wilrijk (Belgium)—residential area; Oct.–Dec. 1976: 5.47; 16.14 ng/cu m ($n = 2$) (1233)
    glc's in residential area (Belgium)—Oct. 1976:
      in particulate sample: 9.73 ng/cu m
      in gas phase sample: 8.70 ng/cu m (1289)
    glc's in the average American urban atmosphere; 1963:
      85 µg/g airborne particulate or
      11 ng/g cu m air

***n*-hexacosanoic acid** (cerotic acid; cerinic acid)
$CH_3(CH_2)_{24}COOH$
A. PROPERTIES: a fatty acid obtained from beeswax, carnauba wax, or Chinese wax; sp.gr. 0.8198 (100/4°C); m.p. 87.7°C
C. AIR POLLUTION FACTORS:
  —Ambient air quality:
    glc's at Botrange (Belgium)—woodland at 20–30 km from industrial area:
      June–July 1977: 3.64; 3.71 ng/cu m ($n = 2$) (1233)
    glc's in residential area (Belgium)—Oct. 1976:
      in particulate sample: 9.12 ng/cu m (1289)

***n*-hexadecane** (cetane; bioctyl)
$CH_3(CH_2)_{14}CH_3$
  Manufacturing source: petroleum refining. (347)
  Users and formulation: organic synthesis; solvent; standardized hydrocarbon; jet fuel research; mfg. paraffin products; rubber industry; paper processing industry; reference for diesel fuels; solvent; organic intermediate, constituent in waterborne waste of polyolefin manufacture. (347)
  Natural sources (*water and air*): constituent in paraffin fraction of petroleum (347)
  Man caused sources (*water and air*): component in municipal waste and waste

water from general use of paraffins; lab use; highway runoff; automobile exhaust; general use of petroleum oils. (347)
A. PROPERTIES: m.w. 226.44; m.p. 18/20°C; b.p. 287°C; sp.gr. 0.7749 at 20/4°C; solub. sea water: 0.0004 mg/l at 25°C, distilled water: 0.0009 mg/l at 25°C
B. AIR POLLUTION FACTORS:
   —Ambient air quality:
   glc's in residential area (Belgium)—Oct. 1976:
   in gas phase sample: 66.5 ng/cu m (1289)
C. WATER POLLUTION FACTORS:
   —$BOD_{35}^{25°C}$: 2.39 in seawater/inoculum: enrichment cultures of hydrocarbon oxidizing bacteria
   —ThOD: 3.46
   —Biodegradation:
   biodegradation to CO_2 in estuarine area:

concentration (μg/l)	month	incubation time (hr)	degradation rate (μg/l/day) $\times 10^3$	turnover time (days)
3	Jan.	24	29 ± 8	103
3	July	24	38 ± 12	79
25	April	24	120 ± 30	210
25	July	24	130 ± 20	190

(381)
   degradation in seawater by oil oxidizing microorganisms: 59.2% breakdown after 21 days at 22°C in stoppered bottles containing a 1000 ppm mixture of alkanes, cycloalkanes and aromatics (1237)
   —Aquatic reactions:
   in estuarine area: at   3 μg/l,   8% adsorbed on particles after 3 hr
                         10 μg/l, 19% adsorbed on particles after 3 hr (381)
   first order evaporation constant of n-hexadecane in 3 mm No. 2 fuel oil layer in dark room at windspeed of 21 km/hr:
   at 5°C: 1.08×10^{-6} mm^{-1}
       10°C: 2.58×10^{-5} mm^{-1}
       20°C: 3.99×10^{-5} mm^{-1}
       30°C: 6.14×10^{-5} mm^{-1} (438)
D. BIOLOGICAL EFFECTS:
   Atlantic salmon egg: 168 hr BCF: 5.6 X (S) (1507)
   mussel larvae (*Mytilus edulis*): slight reduction of growth rate at 10 ppm and 50 ppm
                                     slight increase of growth rate at 100 ppm   (475)

hexadecanoic acid *see* palmitic acid

1-hexadecanol (cetylalcohol; n-hexadecylalcohol; ethol; ethal; palmitylalcohol) $CH_3(CH_2)_{15}OH$
A. PROPERTIES: white crystalline solid; m.w. 242.44; m.p. 49.3°C; b.p. 190°C at 15 mm, 344°C; v.p. 1 mm at 123°C; v.d. 8.36; sp.gr. 0.817 at 50/4°C
B. AIR POLLUTION FACTORS: 1 mg/cu m = 0.101 ppm, 1 ppm = 9.91 mg/cu m
C. WATER POLLUTION FACTORS:
   —Waste water treatment:

A.S.: after 6 hr: 0.3% of ThOD
12 hr: 0.4% of ThOD
24 hr: toxic (88)
D. BIOLOGICAL EFFECTS:
—Mammalia:
mouse: single oral LD_{50}: 3.2–6.4 g/kg
rat: single oral LD_{50}: 6.4–12.8 g/kg
inhalation: all died within two days: 2.22 mg/l, 6 hr; all survived: 0.41 mg/l, 6 hr (211)

n-hexadecylic acid see palmitic acid

hexadecylpyridiniumbromide see cetylpyrimidiniumbromide

hexadecyltrimethylammoniumbromide see cetyltrimethylammoniumbromide

hexadecyltrimethylammoniumchloride see cetyltrimethylammoniumchloride

2,4-hexadienal (sorbic aldehyde)
$CH_3CHCHCHCHCHO$
A. PROPERTIES: m.w. 96.13; sp.gr. 0.871; b.p. 170°C
B. AIR POLLUTION FACTORS:
—Odor threshold: 0.0018 mg/cu m (842)
C. WATER POLLUTION FACTORS:
—Waste water treatment:
RW, Sd, BOD, 20°C, 10 days observed, 100 days acclimation: 87% theor. oxidation (93)
D. BIOLOGICAL EFFECTS:
—Mammals: rat: inhalation: 1/6: 2,000 ppm, 4 hr (104)
acute oral LD_{50}: 0.7 g/kg (211)

hexahydroaniline see cyclohexylamine

hexahydrobenzene see cyclohexane

hexahydro-p-cresol see 4-methylcyclohexanol

hexahydromethylphenol see 4-methylcyclohexanol

hexahydrophenol see cyclohexanol

hexahydrothymol see menthol

hexahydrotoluene see methylcyclohexane

n-hexaldehyde (n-hexanal; caproaldehyde; n-caproic aldehyde; n-hexoic aldehyde)
$CH_3(CH_2)_4CHO$
Use: organic synthesis of plasticizers, rubber chemicals, dyes, synthetic resins; insecticides

A. PROPERTIES: colorless liquid; m.w. 100.16; m.p. -56°C; b.p. 128/131°C; v.p. 10 mm at 20°C; v.d. 3.6; sp.gr. 0.833 at 20/4°C
B. AIR POLLUTION FACTORS:
 —Manmade sources: diesel exhaust: 0.2 ppm (311)
 —Odor threshold: 0.039 mg/cu m (842)
C. WATER POLLUTION FACTORS:
 odor thresholds   mg/kg water

	10^{-7}	10^{-6}	10^{-5}	10^{-4}	10^{-3}	10^{-2}	10^{-1}	1	10	10^2	10^3	10^4
detection					●	●						
recognition								●				
not specified					●							

(305; 873; 882; 888; 889; 914)

D. BIOLOGICAL EFFECTS:
 —Mammals: rat: inhalation: 1/6: 2,000 ppm, 4 hr. (104)

hexalin *see* cyclohexanol

hexaline acetate *see* cyclohexylacetate

hexamethylene *see* cyclohexane

hexamethylenediamine (1,6-hexanediamine; 1,6-diaminohexane)
$NH_2(CH_2)_6NH_2$
 Use: formation of high polymers, e.g., nylon 66
A. PROPERTIES: colorless leaflets; m.w. 116.21; m.p. 39/40°C; b.p. 196/205°C sublimes; v.d. 4.01
B. AIR POLLUTION FACTORS: 1 mg/cu m = 0.210 ppm, 1 ppm = 4.75 mg/cu m
 —Odor: USSR: animal chronic exposure: no effect: 0.001 mg/cu m
                                    adverse effect: 0.04 mg/cu m (170)
     threshold conc.: 0.0032 mg/cu m (732)
C. WATER POLLUTION FACTORS:
 —Inhibition of biodegradation:
  NH_3 oxidation by Nitrosomonas sp.: at 100 mg/l: 52% inhibition
                                          50 mg/l: 45% inhibition
                                          10 mg/l: 27% inhibition
D. BIOLOGICAL EFFECTS:
 —Mammals: rat: inhalation: no effect level: 1 mg/l, 15 × 6 hr (65)

hexamethylenediamineadipate
C. WATER POLLUTION FACTORS:
 —Water quality: self purification is affected at 1.0 mg/l (181)
D. BIOLOGICAL EFFECTS:
 —Arthropod: *Daphnia*: sublethal concentration: 0.1 mg/l (30)

hexamethyleneimine

A. PROPERTIES: clear colorless liquid; m.p. -37°C; b.p. 138°C; sp.gr. 0.88 at 20/4°C
C. WATER POLLUTION FACTORS:
 —BOD_5: 1.31 std. dil. sew.
 —COD: 2.17 (41)

hexamethylenetetramine (methenamine; HMTA; aminoform; hexamine)

Use: catalyst in phenolformaldehyde and resorcinolformaldehyde resins; protein modifier; organic synthesis; pharmaceuticals; ingredient of high explosive cyclonite; fuel tablets
A. PROPERTIES: white crystalline powder, or colorless lustrous crystals; m.w. 140.19; m.p. 280°C sublimes; log P_{oct} -2.13/-2.34 (calculated)
C. WATER POLLUTION FACTORS:
 —BOD_5: 0.015; 0.026 std. dil. sew. (258)
 —Impact on biodegradation: at 100 mg/l, no inhibition of NH_3 oxidation by *Nitrosomonas* sp. (390)

2,6,10,15,19,23-hexamethyl-2,6,10,14,18,22-tetracosahexane *see* squalene

hexamine *see* hexamethylenetetramine

n-hexane
$CH_3(CH_2)_4CH_3$
A. PROPERTIES: colorless liquid; m.w. 86.17; m.p. -94.3°C; b.p. 68.7°C; v.p. 120 mm at 20°C, 190 mm at 30°C; v.d. 2.79; sp.gr. 0.6603 at 20/4°C; solub. in distilled water: 13 mg/l at 20°C, in salt water: 75.5 mg/l at 20°C, in water: 9.5 mg/l at 20°C; sat. conc. 564 g/cu m at 20°C, 862 g/cu m at 30°C
B. AIR POLLUTION FACTORS: 1 mg/cu m = 0.28 ppm, 1 ppm = 3.58 mg/cu m
 —Odor threshold conc.: 875 mg/cu m (781)
  230 mg/cu m (737)
 —Atmospheric reactions: R.C.R.: 1.12 (49)
  reactivity: HC cons.: <0.1 (63)
  estimated lifetime under photochemical smog conditions in S.E. England: 5.9 hr
   (1699; 1700)
 —Manmade sources:
  in diesel engine exhaust gases: 1.2% of emitted hydrocarbons (72)

in gasoline engine exhaust: 62-car survey: 1.2 vol.% of total exhaust H.C.'s
(391)
evaporation from gasoline fuel gank: 0.7–1.8 vol.% of total evaporated H.C.'s
evaporation from carburetor: 0.2–8.7 vol.% of total evaporated H.C.'s
(398; 399; 400; 401; 402)
—Sampling and analysis:
effectiveness of scrubbers at 325–375°F: inlet conc.: 15 ppm;
  (1): $HgSO_4/H_2SO_4$: 1% absorbed
  (2): $PdSO_4/H_2SO_4$:
  (3): (1) + (2):   6% absorbed   (198)
4 f. abs. app.: 60 l air/30 min: GC: det. lim.: 1 mg/cu m/30 min   (208)

C. WATER POLLUTION FACTORS:
  —$BOD_{35}^{25°C}$: 2.49
  —BOD_5:   2.21 NEN 3235-5.4   (277)
  —COD: 0.04 NEN 3235-5.3   (277)
  —ThOD: 3.52   (62)
  —Biodegradation:
    rotating disk contact aerator: infl. 214.8 mg/l, effl. 0.3 mg/l; elimination:
          >99%, or 17,708 mg/m^2/24 hr or 4782
          g/m^3/24 hr   (406)
    incubation with natural flora in the groundwater—in presence of the other components of high-octane gasoline (100 μl/l): biodegradation: 46% after 192 hr
    at 13°C (initial conc. 1.36 μl/l)   (956)
  —Aquatic reactions:
    photo-oxidation by U.V. light in aqueous medium at 50°C:
    50.51% degradation to CO_2 after 24 hr   (1628)
  —Sampling methods:
    extraction efficiency of macro-reticular resins: sample flow: 20 ml/min;
    pH: 5.7; conc. 10 ppm: X AD-2: 82%
                           X AD-3: 83%   (370)

D. BIOLOGICAL EFFECTS:
  —Man: no effect level: 2,000 ppm, 10 min   (49)
        dizziness: 5,000 ppm   (211)
  —Algae: giant kelp (*Macrocystis pyrifera*): little or no effect on the photosynthetic
    activity: 10 mg/l   (226)
  —Fish: goldfish: LD_{50} (24 hr) = 4 mg/l-modified ASTM D 1345 (277)
        : young Coho calmon: no mortalities when applied in amounts up
    to 100 ppm, after 96 hrs in artificial sea water at 8°C   (317)

1,6-hexanediamine *see* hexamethylenediamine

hexanedinitrile *see* adiponitrile

hexanedioic acid *see* adipic acid

hexanedioic acid dinitrile *see* adiponitrile

hexanenitrile (capronitrile; caproic nitrile; n-amylcyanide)

734   1-HEXANETHIOL

$CH_3(CH_2)_4CN$

A. PROPERTIES: colorless liquid; m.w. 97.16; m.p. $-79.4°C$; b.p. $163°C$; sp.gr. 0.809 at $20/4°C$

C. WATER POLLUTION FACTORS:
—Waste water treatment:
A.S.: after   6 hr: 1.5% of ThOD
              12 hr: 2.9% of ThOD
              24 hr: 5.1% of ThOD   (89)

1-hexanethiol   *see prim-n-hexylmercaptan*

hexanoic acid   *see caproic acid*

n-hexanol   (*n*-hexylalcohol; amylcarbinol)
$CH_3(CH_2)_4CH_2OH$

Use: pharmaceuticals; solvent; plasticizer; intermediate for textile and leather finishing agent.

A. PROPERTIES: colorless liquid; m.w. 102.2; m.p. $-51/-44°C$; b.p. $158°C$; v.p. 0.98 mm at $20°C$; v.d. 3.52; sp.gr. 0.82 at $20/4°C$; solub. 5,900 mg/l at $20°C$; log P_{oct} 2.03

B. AIR POLLUTION FACTORS: 1 mg/cu m = 0.24 ppm, 1 ppm = 4.25 mg/cu m
—Odor: characteristic: quality: sweet alcohol
                          hedonic tone: pleasant

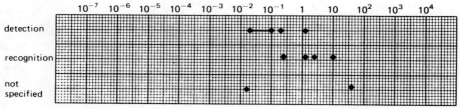

(307; 610; 627; 643; 676; 761; 787; 828)

C. WATER POLLUTION FACTORS:
—BOD_5:   28% of ThOD
—COD: 94% of ThOD
—$KMnO_4$: acid: 8% of ThOD; alkaline; 2% of ThOD
—TOC: 100% of ThOD   (220)

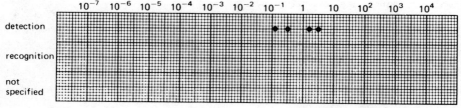

(326; 883; 888; 889; 894; 900)

—Waste water treatment:
A.C.: adsorbability: 0.191 g/g C; 95.5% reduction, infl.: 1,000 mg/l, effl.: 45 mg/l (32)
reverse osmosis: 47% rejection from a 0.01 M solution (221)
anaerobic lagoon: lb COD/day/1,000 cu ft   infl. mg/l   effl. mg/l

	infl. mg/l	effl. mg/l	
22	140	20	
48	140	30	(37)

ion exchange: adsorption on amberlite X AD-2: infl.: 200 ppm
retention efficiency: 85% effl.: 30 ppm (40)
NFG, BOD, 20°C, 1-10 days observed, feed: 200-1,000 mg/l
acclimation: 365 + P: 57% removed (93)

D. BIOLOGICAL EFFECTS:
—Toxicity threshold (cell multiplication inhibition test):

bacteria (*Pseudomonas putida*):	62 mg/l	(1900)
algae (*Microcystis aeruginosa*):	12 mg/l	(329)
green algae (*Scenedesmus quadricauda*):	30 mg/l	(1900)
protozoa (*Entosiphon sulcatum*):	75 mg/l	(1900)
protozoa (*Uronema parduczi Chatton-Lwoff*):	93 mg/l	(1901)

2-hexanol (butylmethylcarbinol)
$CH_3CHOH(CH_2)_3CH_3$
A. PROPERTIES: colorless liquid; m.w. 162.17; m.p. 136/140°C; sp.gr. 0.809 at 20/4°C
B. AIR POLLUTION FACTORS:
—Odor threshold conc.: detection: 50 mg/cu m (643)
C. WATER POLLUTION FACTORS:
—A.S. after   6 hr: 3.4% of ThOD
              12 hr: 2.9% of ThOD
              24 hr: 3.7% of ThOD (88)
D. BIOLOGICAL EFFECTS:
—Toxicity threshold (cell multiplication inhibition test):

bacteria (*Pseudomonas putida*):	63 mg/l	(1900)
algae (*Microcystis aeruginosa*):	32 mg/l	(329)
green algae (*Scenedesmus quadricauda*):	72 mg/l	(1900)
protozoa (*Entosiphon sulcatum*):	116 mg/l	(1900)
protozoa (*Uronema parduczi Chatton-Lwoff*):	335 mg/l	(1901)

3-hexanol (ethylpropylcarbinol)
$CH_3CH_2CHOHCH_2CH_2CH_3$
A. PROPERTIES: colorless liquid; m.w. 102.17; b.p. 135°C; sp.gr. 0.8188 at 20/4°C
C. WATER POLLUTION FACTORS:
—Waste water treatment:
A.S.: after   6 hr: 2.2% of ThOD
              12 hr: 2.4% of ThOD
              24 hr: 2.7% of ThOD (88)
D. BIOLOGICAL EFFECTS:
—Toxicity threshold (cell multiplication inhibition test):
bacteria (*Pseudomonas putida*):   105 mg/l   (1900)

algae (*Microcystis aeruginosa*): 32 mg/l (329)
green algae (*Scenedesmus quadricauda*): 63 mg/l (1900)
protozoa (*Entosiphon sulcatum*): 182 mg/l (1900)
protozoa (*Uronema parduczi-Chatton Lwoff*): 246 mg/l (1901)

hexanone see methylisobutylketone

2-hexanone see butylmethylketone

1-hexene (butylethylene)
$CH_2CH(CH_2)_3CH_2$
Use: synthesis of flavors, perfumes, dyes, resins
A. PROPERTIES: colorless liquid; m.w. 84.16; m.p. $-98.5°C$; b.p. $63°C$; v.d. 3.0; sp.gr. 0.6732 at $20/4°C$; solub. 50 mg/l at $20°C$
B. AIR POLLUTION FACTORS:
—Manmade sources:
emitted by home central heating on gasoil: 20 ppm at $CO_2 = 7\%$
6.2 g/kg gasoil at $CO_2 = 6\%$
1.2 g/kg gasoil at $CO_2 = 7\%$ (182)
—Atmospheric reactions: R.C.R.: 1.66 (49)
C. WATER POLLUTION FACTORS:
—Waste water treatment:
A.S.: after 6 hr: 0.5% of ThOD
12 hr: 0.5% of ThOD
24 hr: 0.7% of ThOD (88)
rotating disk contact aerator: infl. 189.3 mg/l, effl. 0.4 mg/l
elimination: >99%, or 15,585 mg/m^2/24 hr,
or 4208 g/m^3/24 hr (406)
D. BIOLOGICAL EFFECTS:
—Man: EIR: 3.5

2-hexene
$CH_3CHCHCH_2CH_2CH_3$
A. PROPERTIES: colorless liquid; m.p. $-146°C$; b.p. $68°C$; sp.gr. 0.686 at $15/15°C$
B. AIR POLLUTION FACTORS:
—Sampling and analysis:
effectiveness of scrubbers at $325-375°F$: inlet conc.; 25 ppm
(1): $HgSO_4/H_2SO_4$: 100% absorbed
(2): $PdSO_4/H_2SO_4$: 99+% absorbed
(3): (1) + (2): 100% adsorbed (198)

hexobarbital (N-methyl-5-cyclohexenyl-5-methylbarbituric acid)
Use: medicine (sedative)
A. PROPERTIES: white crystals; m.p. $145-147°C$
D. BIOLOGICAL EFFECTS:
—BCF: fish (*C. auratus*): 1.74
frog (*R. pipiens*): 1.50 (1871)

n-hexoic acid *see* caproic acid

hexone *see* methylisobutylketone

n-hexylamine
$CH_3(CH_2)_5NH_2$
A. PROPERTIES: colorless liquid; m.w. 101.19; m.p. -19°C; b.p. 132°C; v.p. 6.5 mm at 20°C; v.d. 3.5; sp.gr. 0.80; solub. 12,000 mg/l; log P_{oct} 1.52/2.34 (calculated)
B. AIR POLLUTION FACTORS: 1 mg/cu m = 0.242 ppm, 1 ppm = 4.14 mg/cu m
C. WATER POLLUTION FACTORS:
—Waste water treatment:
degradation by *Aerobacter*: 200 mg/l at 30°C:
parent: 100% of 20 hr
mutant: 100% of 10 hr (152)
D. BIOLOGICAL EFFECTS:
—Mammalia: rat: acute oral LD_{50}: 0.39 g/kg
inhalation: 2/6: 500 ppm, 4 hr
no deaths: sat. vap., 1 hr (211)
rat: oral LD_{50}: 0.67 g/kg; 95% conf.: 0.62–0.74 g/kg (226)

prim-n-hexylchloride *see* 1-chlorohexane

n-hexylcellosolve *see* ethyleneglycolmonohexylether

hexyleneglycol (2-methyl-2,4-pentanediol; HG)

$$\begin{array}{c} CH_3 \\ | \\ CHOH \\ | \\ CH_2 \\ | \\ CH_3-C-OH \\ | \\ CH_3 \end{array}$$

Use: hydraulic brake fluids; printing inks; fuel and lubricant additive; emulsifying agent; inhibitor of ice formation in carburetor
A. PROPERTIES: m.w. 118.2; m.p. -40°C; b.p. 198°C; v.p. 0.06 mm at 20°C; v.d. 4.0; sp.gr. 0.92; log P_{oct} -0.14 (calculated)
B. AIR POLLUTION FACTORS:
—Odor: T.O.C. = <50 ppm (211)
O.I. at 20°C: <2 (316)
C. WATER POLLUTION FACTORS:
—BOD_5: 0.02 NEN 3235-5.4 (277)
—COD: 2.20 NEN 3235-5.3 (277)

Theoretical	Analytical
TOD = 2.3	reflux COD = 77.4% recovery
COD = 2.3	rapid COD = 69.1% recovery
	BOD_5: <0.004 (1828)

—Waste water treatment:
A.C.: adsorbability: 0.122 g/g C; 61.4% reduction; infl.: 1,000 mg/l, effl.: 386 mg/l (32)
hexyleneglycol at 1.040 mg/l was found amenable to low-rate biodegradation by an unacclimated biomass

D. BIOLOGICAL EFFECTS:
—Fish: goldfish: LD_{50} (24 hr): >5000 mg/l (modified ASTM D 1345)   (277)
Lepomis macrochirus: static bioassay in fresh water at 23°C, mild aeration applied after 24 hr:

material added ppm	% survival after				best fit 96 hr LC_{50} ppm
	24 hr	48 hr	72 hr	96 hr	
10,000	100	100	100	100	>10,000
3,200	100	100	100	100	

Menidia beryllina: static bioassay in synthetic seawater at 23°C, mild aeration applied after 24 hr

material added ppm	% survival after				best fit 96 hr LC_{50} ppm
	24 hr	48 hr	72 hr	96 hr	
10,000	100	70	60	50	
7,900	100	60	60	60	10,000
5,000	100	100	100	100	(352)

—Mammals:
mouse: single oral LD_{50}: 3.8 ml/kg
rat: single oral LD_{50}: 4.8 ml/kg
inhalation: no deaths: sat. vap.: 8 hr   (211)
—Man: nasal irritation: 100 ppm
respiratory discomfort: 1,000 ppm   (211)

prim-n-hexylmercaptan (1-hexanethiol)
$C_6H_{13}SH$

A. PROPERTIES: colorless liquid; m.w. 118.23; m.p. -81.03°C; b.p. 149/150°C; sp.gr. 0.849 at 20/4°C

B. AIR POLLUTION FACTORS:
—Odor threshold conc.: 11.8 mg/cu m   (711)
—Control methods:
wet scrubber: water at pH 8.5: effl. 85,000 odor units/scf
$KMnO_4$ at pH 8.5: effl. 10.5 odor units/scf   (115)

hexylmethylketone *see* 2-octanone

hexylthiocarbam *see* ro-neet

hexylxanthate, potassium salt
D. BIOLOGICAL EFFECTS:
—Fish:
rainbow trout (*Salmo gairdneri*): 96 hr LC_{50}: 10–100 mg/l (static test)   (1087)

HG *see* hexyleneglycol

hippuric acid (benzaminoacetic acid; benzoylaminoacetic acid; benzoylglycocol; benzoylglycin)

$$\text{C}_6\text{H}_5-\overset{\overset{\text{O}}{\|}}{\text{C}}-\text{NH}-\text{CH}_2\text{COOH}$$

A. PROPERTIES: m.p. 188°C; decomposes upon further heating; log P_{oct} -0.07/-0.25 (calculated)

C. WATER POLLUTION FACTORS:
 —Manmade sources: in sewage effluent: 0.002 mg/l (227)
 —Odor threshold conc.: detection: 2.5 mg/l (998)

dl-histidine (α-amino-β-imidazolepropionic acid)

$$\underset{\underset{\text{H}}{|}}{\text{N}}\diagdown\text{N}\diagup\text{CH}_2-\overset{\overset{\text{NH}_2}{|}}{\text{CH}}-\text{COOH}$$

A. PROPERTIES: m.w. 155.16; m.p. 285°C decomposes; b.p. decomposes; log P_{oct} -2.86

C. WATER POLLUTION FACTORS:
 —Waste water treatment:
  A.S.: after 6 hr: 2.8% of ThOD
       12 hr: 6.3% of ThOD
       24 hr: 16.5% of ThOD (89)
 —Manmade sources:
  excreted by man: in urine: 0.98–6.6 mg/kg body wt/day
                   in feces: 1.4–2.1 mg/kg body wt/day
                   in sweat: 6.0–10 mg/100 ml (203)
 —Impact on biodegradation:
  ~50% inhibition of NH_3 oxidation by *Nitrosomonas* at 0.5 mg/l (*l*-isomer) (408)

HMTA *see* hexamethylenetetramine

homoserine (2-amino-4-hydroxybutanoic acid)

$$\text{HOCH}_2-\text{CH}_2-\overset{\overset{\text{NH}_2}{|}}{\text{CH}}-\text{COOH}$$

C. WATER POLLUTION FACTORS:
 —Waste water treatment:
  A.S.: after   6 hr: 1.1% of ThOD
  (*dl*-homoserine) 12 hr: 2.3% of ThOD
                   24 hr: 4.4% of ThOD (89)

hyamine 1622 *see* benzenthionium chloride

hydralin *see* cyclohexanol

hydrazine
H_2NNH_2

Hydrazine compounds have become more frequently and diversely used in industry, due mainly to their strong reducing or antioxidant property. In addition, they have been used extensively in recent years as propellants in aerospace operations.

A. PROPERTIES: liquid; m.w. 32.05; m.p. 2°C; b.p. 113°C; v.p. 16 mm at 20°C, 25 mm at 30°C; v.d. 1.1; sp.gr. 1.01; sat. conc. 28 g/cu m at 20°C, 42 g/cu m at 30°C; log P_{oct} -1.37/-0.60 (calculated)

—Odor: characteristic: quality: ammoniacal
    T.O.C. = 3-4 ppm (701; 703)
    O.I. at 20°C = 5,300 (316)

C. WATER POLLUTION FACTORS:
—Impact on biodegradation processes:
75% inhibition of nitrification process of non acclimated activated sludge: at 48 mg/l (30)
—Sampling and analysis:
photometry: min. full scale: 260×10^{-6} mole/l (53)

D. BIOLOGICAL EFFECTS:
—Fish: trout: perturbated: 0.7 mg/l, 24 hr (30)
       rainbow trout: fatal in 22 to 35 min: 146 mg/l (226)
guppy: survival time compared between previously exposed and unexposed guppies: pre-exposure period of 15 days at one twentieth (1/20) of the concentration listed in column 1

toxicant concentration	preexposed fish no. of fish	preexposed fish mean survival time (hr)	unexposed fish no. of fish	unexposed fish mean survival time (hr)
6.6 mg/l in soft water at 22-24.5°C	2	3.75	10	7.53
2.0 mg/l in soft water at 22-24.5°C	5	36.15	10	28.20
200.0 mg/l in hard water at 22-24.5°C	5	0.95	10	0.95
66 mg/l in hard water at 22-24.5°C	9	1.47	10	1.48
20 mg/l in hard water at 22-24.5°C	10	4.50	10	3.93

(474)

  guppy (*Lebistes reticulatus*):
    static bioassay: 24 hr LC_{50}: 4.6 mg/l in hard water at 22-24.5°C
                    48 hr LC_{50}: 3.98 mg/l in hard water at 22-24.5°C
                    72 hr LC_{50}: 3.85 mg/l in hard water at 22-24.5°C
                    96 hr LC_{50}: 3.85 mg/l in hard water at 22-24.5°C
    static bioassay: 24 hr LC_{50}: 3.32 mg/l in soft water at 22-24.5°C

HYDROGEN CYANIDE 741

  48 hr LC_{50}: 1.58 mg/l in soft water at 22-24.5°C
  72 hr LC_{50}: 0.82 mg/l in soft water at 22-24.5°C
  96 hr LC_{50}: 0.61 mg/l in soft water at 22-24.5°C   (474)
—Mammalia:
 mouse: single oral LD_{50}: 59 mg/kg
 rat: single oral LD_{50}: 60 mg/kg
 mouse: inhalation: LC_{50}: 252 ppm, 4 hr
 rat: inhalation: LC_{50}: 570 ppm, 4 hr   (54)
 mouse: inhalation: 4/11: 900 ppm, 8 × 1 hr/day, 8 days
 rat: inhalation: 16/10: 225 ppm, 5 × 6 hr
 mouse: inhalation: 8/10: 225 ppm, 5 × 6 hr
 mouse: inhalation: 0/20: 100 ppm, 1 hr/day, 6 days/w, 4 weeks
 rat: inhalation: 14/16: 54 ppm, 6 hr/day, 5 days/w, 13 days
 mouse: inhalation: 7/10: 54 ppm, 6 hr/day, 5 days/w, 13 days
 ape: inhalation: 2/10: 1 ppm, 90 days continuously
 rat: inhalation: 48/50: 1 ppm, 90 days continuously
 mouse: inhalation: 98/100: 1 ppm, 90 days continuously
 rat: inhalation: 0/23: 0.4 ppm, 7 months continuously
 rabbit: inhalation: 0/23: 0.7 ppm, 7 months continuously   (54)
—Man: carcinogenic

hydrazinium hydroxide
N_2H_5OH
D. BIOLOGICAL EFFECTS:
 —Threshold concentration of cell multiplication inhibition of the protozoan *Uronema parduczi Chatton-Lwoff*: 0.24 mg/l   (1901)

hydrochloric ether *see* ethylchloride

hydrocyanic acid *see* hydrogen cyanide

hydrogen cyanide (hydrocyanic acid; prussic acid)
HCN
A. PROPERTIES: colorless gas or liquid; m.w. 27.03; m.p. -13.3°C; b.p. 25.6°C; v.p. 620 mm at 20°C, 807 mm at 30°C; v.d. 0.94; sp.gr. 0.6976 at 20/4°C; LHC 154 kcal/mole; log P_{oct} 0.35/1.07 (calculated)
B. AIR POLLUTION FACTORS: 1 mg/cu m = 0.9 ppm, 1 ppm = 1.13 mg/cu m
 —Odor: characteristic: quality: bitter almonds
  TOC.: 0.2-5 ppm   (54; 60)
  recognition: 2-5 ppm   (211)
    1.00 mg/cu m = 0.9 ppm   (73)
    1 ppm; 0.07 ppm   (279)
    6 mg/cu m; <1.1 mg/cu m   (811; 816)
 O.I. at 20°C = 163,000   (316)
—Manmade sources: in cigarette smoke: 1,600 ppm   (66)
 HCN emissions from Pratt & Whitney aircraft jet turbine 9D engine operating at several power settings ranged from 8.4 to 42 ppb as measured in the tailpipe
  (1097)

HYDROGEN SULFIDE

—Sampling and analysis:
  non dispersive I.R.: min. full scale: 125 ppm (55)
  detector tubes: UNICO: 10 ppm (59)
  DRAGER: 10 ppm (58)

C. WATER POLLUTION FACTORS:
  —Waste water treatment;
  sludge digestion: at 25 mg/l: no adverse effect in 24 days
  at 30 mg/l: initial retarding effect for 6 days
  at 50 mg/l: 10% reduction in gas production (29)
  —Reduction of amenities:
  taste and odor at 0.001 mg/l (41)
  faint odor: 0.001 mg/l (129)
  —Biodegradation:
  biological degradation pathways:

$$HCN + \tfrac{1}{2} O_2 \xrightarrow{enzyme} HCNO \xrightarrow{+ H_2O + enzyme} CO_2 + NH_3$$

$$HCN + S \xrightarrow{enzyme} HCNS \xrightarrow{+ H_2O\ soil\ microbes} HSCONH_2$$
$$\rightarrow COS + NH_3 \rightarrow 2\ COS + 3\ O_2 \rightarrow 2\ CO_2 + 2\ SO_2$$

$$HCN + 2\ H_2O \xrightarrow{enzyme} NH_4COOH \xrightarrow{+ 1/2\ O_2} NH_3 + CO_2 + H_2O$$

$$HCN + H_2O \xrightarrow{soil\ microbes} SHCONH_2 \rightarrow COS + NH_3$$
$$\rightarrow 2\ CO_2 + 3\ O_2 \rightarrow 2\ CO_2 + 2\ SO_2 \quad (171)$$

D. BIOLOGICAL EFFECTS:
  —Arthropods:
  Daphnia: 50% immobilization after 48 hr at 1.8 mg/l (29)
  Asellus communis: 96 hr LC_{50}: 2.29 mg/l
       LTC (10–12 d): 1.90 mg/l
  Gammarus pseudolimnaeous: 96 hr LC_{50}: 0.17 mg/l
       LTC (10–12 d): 0.07 mg/l (1387)
  —Fish:
  pinperch: TLm (24 hr): 0,069 mg/l (41)
  sunfish: TLm (24 hr): 0.18 mg/l (29)
  pinperch: TLm (24 hr): 0.05 mg/l in sea water (248)
  trout: toxic: 0.10 to 0.15 mg/l (226)

bluegill, eggs	96-h LTC: 535–693 µg/l	F
swim up fry	96-h LC_{50}: 232–365 µg/l	F
juvenile	96-h LC_{50}: 75–125 µg/l	F
yellow perch (*Perca flavescens*), eggs	96-h LC_{50}: >276->389 µg/l	F
swim up fry	96-h LC_{50}: 295->395 µg/l	F
juvenile	96-h LC_{50}: 76–108 µg/l	F
brook trout, eggs	96-h LC_{50}: >212->242 µg/l	F
sac fry	96-h LC_{50}: 108–518 µg/l	F
swim up fry	96-h LC_{50}: 56–106 µg/l	F
juvenile	96-h LC_{50}: 53–143 µg/l	F
rainbow trout	96-h LC_{50}: 57 µg/l	F

HYDROGEN SULFIDE   743

fathead minnow, eggs: 96-h LC_{50}: 121-352 µg/l   F
swim up fry: 96-h LC_{50}: 82-122 µg/l   F
juvenile: 96-h LC_{50}: 82-137 µg/l   F
juvenile (wild stock): 96-h LC_{50}: 157-191 µg/l F   (1512)
—Mammalia: guinea pig: inhalation: no symptoms: 200 ppm, $1\frac{1}{2}$ hr
                                    fatal: 315 ppm
            rabbit: inhalation: no marked symptoms: 120 ppm
            mouse: inhalation: death: 1,300 ppm, 1-2 min
                                       45 ppm, 2.5-4 hr
            cat: inhalation: death: 180 ppm
                             toxic: 125 ppm, 6-7 min
            dog: inhalation: may be fatal: 0.04-0.07 ppm
                             may be tolerated: 0.035 ppm   (211)
            rat: inhalation: LC_{50}: 142 ppm, 30 min
            mouse: single oral LD_{50}: 8.5 mg/kg   (54)
—Man: deadly doses by ingestion: 0.7-3.5 mg/kg body weight (as-CN) (54)
      severe toxic effects: 40 ppm = 44 mg/cu m, 1 min
      symptoms of illness: 20 ppm = 22 mg/cu m
      unsatisfactory: 10 ppm = 11 mg/cu m   (185)
      immediately fatal: 270 ppm
      fatal after 10 min: 181 ppm
      fatal after 30 min: 135 ppm
      dangerous to life: 110-135 ppm
      tolerated for 1-1.5 hr: 45-54 ppm
      slight symptoms: 18-36 ppm   (211)

hydrogen sulfide
H_2S
A. PROPERTIES: colorless gas: m.w. 34.08; m.p. -83.8/-85.5°C; b.p. -60.2°C; v.p. 10 atm at -0.4°C, 20 atm at 25.5°C; v.d. 1.189 at 0°C
B. AIR POLLUTION FACTORS: 1 ppm = 1.42 mg/cu m, 1 mg/cu m = 0.704 ppm
   —Odor: characteristic: quality: rotten egg odor

odor thresholds   mg/cu m

	10^{-7}	10^{-6}	10^{-5}	10^{-4}	10^{-3}	10^{-2}	10^{-1}	1	10	10^2	10^3	10^4
detection				●—●		●						
recognition					●————————→ ●							
not specified				←● ●●●		●	●					

(71; 73; 170; 210; 279; 612; 616; 630; 644; 652; 677; 678; 710; 733; 741; 742; 744;
                793; 801; 803; 821; 843; 857; 858; 863; 867)
   O.I. at 20°C = 17,000,000   (316)
   —Natural sources:
     estimated emission from natural sources in the USA: 0.07 ton/day/1,000 sq. mile
                                                              (110)

biogenic H₂S emissions from soils in U.S.A. (1977):

sampling sites	soil orders	sampling dates	avg.sulfur flux g S/m²/yr
Wadesville, IN	alfisol	9/20–10/3	0.010
Philo, OH	inceptisol	10/7–10/10	0.0028
Dismal Swamp, NC	histosol	10/14–10/17	0.0178
Cedar Island, NC	saline swamp	10/19–10/28	0.0194
Cox's Landing, NC	saline marsh	11/1–11/9	139.5
Clarkedale, AR	alluvial clay	11/16–11/2'	0.0004
Cedar Island, NC	freshly clipped marsh		0.0556

(1385)

—Manmade sources:
 glc's in yard of oil refinery at Owase: 0.66 ppb (50)
 in cigarette smoke: 40 ppm (66)
 from combustion: source   emission factor
          coal   0.0045 lb/lb coal
          fuel oil   1 lb/1,000 lb oil
          natural gas 0.13   lb/1,000 lb gas
 in municipal sewer air: 0.2–10 ppm; 200–300 ppm (Melbourne) (212)
—Control methods:
 catalytic combustion: platinized ceramic honeycomb catalyst: ignition temp.:
  400°C, inlet temp. for 90% conversion: 400–425°C (91)
—Sampling and analysis:
 photometry: min. full scale: 100 ppm (53)
 6 f.abs.app., 60 l air/30 min: VLS: det.lim.: 200 g/cu m/30 min
          reagent: molybdate solution (208)
 In Smith-Greenburg impinger: 90% of 100–800 ppm H₂S is absorbed by a 0.091
  M solution of zinc acetate at 21°C and 10l/min sampling rate. At 400 ppm
  H₂S:

absorbent	absorption efficiency	
0.091 M Zn acetate	98%	
0.091 M Cd hydroxide	93%	
0.091 M Cd sulfate	62%	
0.0168 M Zn acetate	52%	
0.0168 M Cd hydroxide	40%	
0.0168 M Cd sulfate	30%	(320)

C. WATER POLLUTION FACTORS:
 —Impact on biodegradation processes:
  sludge digestion is inhibited at 70–200 mg/l (30)
 —Reduction of amenities:
  taste and odor: 0.001 mg/l (41)
  faint odor: 0.0011 mg/l (51)
   0.13 mg/l (98)
D. BIOLOGICAL EFFECTS:
 —Arthropoda: *Asellus*: TLm (96 hr): 0.111 mg/l
       Crangonyx: TLm (96 hr): 1.07 mg/l
       Gammarus: TLm (96 hr): 0.84 mg/l

—Insects: *Ephemera*: TLm (96 hr): 0.316 mg/l
(no effect levels are 8–12% of the TLm (96 hr) values for the above arthropoda
and insects) (219)
flies: inhalation: LC_{50}: 380 mg/cu m, >960 min
flies: inhalation: LC_{50}: 1,500 mg/cu m, 7 min
—Fish:
goldfish: toxic: 100 mg/l (155)
        lethal: 10 mg/l, 96 hr (154)
        LD_0: 1 mg/l, long time exposure in hard water (245)
trout: toxic: 0.86 mg/l, 24 hr
minnows, shiners: toxic: 1.0 mg/l, ? hr
king salmon: toxic: 1.0 mg/l, ? hr
cutthroat trout: toxic: 1.0 mg/l, ? hr
fatheads minnows: TLm, 48 hr: 1.38 mg/l
carp: toxic: 3.3 mg/l, 24 hr
goldfish: toxic: 4.3 mg/l, 24 hr
sunfish: toxic: 4.9 to 5.3 mg/l, 1 hr
goldfish: toxic: 5 mg/l, 200 hr
minnows: toxic: 5–6 mg/l, 24 hr
carp: toxic: 6–25 mg/l, 24 hr
trout: toxic: 10 mg/l, 15 min
goldfish: toxic: 25 mg/l, 24 hr
tench: toxic: 100 mg/l, 3 hr (226)
bluegill, eggs: flow through bioassay: 72 hr TLm: 0.0190 mg/l at 21–22°C
  35 d old fry: flow through bioassay: 96 hr TLm: 0.0131 mg/l at 21–22°C
  juveniles: flow through bioassay: 96 hr TLm: 0.0478 mg/l at 21–22°C
  adults: flow through bioassay: 96 hr TLm: 0.0448 mg/l at 21–22°C
(469)

fathead minnow: flow through bioassay:
  96 hr TLm: 0.0071–0.55 mg/l at 6–24°C
  MATC: 0.0037 mg/l at 6–24°C
bluegill: flow through bioassay:
  96 hr TLm: 0.0090–0.0140 mg/l at 20–22°C
  MATC: 0.0004 mg/l at 20–22°C
brook trout: flow through bioassay:
  96 hr TLm: 0.0216–0.0308 mg/l at 8–12.5°C
  MATC: 0.0055 mg/l at 8–12.5°C (468)
—Mammalia: mice: inhalation: LC_{50}: 1,500 mg/cu m, 18 min
                                  380 mg/cu m; 410 min
                                  96 mg/cu m, 804 min
                                  24 mg/cu m; >960 min
        rats: inhalation: LC_{50}: 1,500 mg/cu m, 14 min
                                380 mg/cu m, >960 min (105)
—Man: severe toxic effects: 200 ppm = 280 mg/cu m, 1 min
      symptoms of illness: 50 ppm = 70 mg/cu m
      unsatisfactory: 20 ppm = 28 mg/cu m (185)
      lethal: 600 ppm, 30 min
            800 ppm, immediately (186)

hydrophenol *see* cyclohexanol

hydroquinol *see* hydroquinone

hydroquinone (1,4-dihydroxybenzene; 1,4-benzenediol; quinol; hydroquinol; *p*-hydroxyphenol)

Use: organic reducing agent which is the basis for its use as a photographic developer and oxidation inhibitor; dye intermediate; medicine; inhibitor of polymerization

A. PROPERTIES: hexagonal prisms; m.w. 110.1; m.p. 172°C; b.p. 218.2°C; v.p. 4.0 mm at 150°C, 40 mm at 192°C, 200 mm at 238°C; v.d. 3.81; sp.gr. 1.358 at 20/4°C; solub. 59,000 mg/l at 15°C, 70,000 mg/l at 25°C, 94,000 mg/l at 28°C; sat. conc. 0.00006 g/cu m at 25°C; log P_{oct} 0.50/0.59

B. AIR POLLUTION FACTORS: 1 mg/cu m = 0.222 ppm, 1 ppm = 4.50 mg/cu m

C. WATER POLLUTION FACTORS:
- BOD_5: 0.478 (30)
-        1.00 (36)
-        37% of ThOD (220)
-        25% of ThOD (274)
- BOD_5^{20}: 1.00 at 10 mg/l, unadapted sew.: lag period: 1 day
- BOD_{20}^{20}: 1.15 at 10 mg/l, unadapted sew.: (554)
- BOD_5 = 0.62 (average)
- BOD_5/COD = 0.32

  Variable results were obtained by the BOD_5 analysis. The test is apparently dependent upon the degree of acclimation of the inoculum to the compound
  (1828)
- COD: 90% of ThOD (0.05N Cr_2O_7) (274)
-     1.83 (36)
  reflux COD = 99.5% recovery
  rapid COD = 104.7% recovery (1828)
- $KMnO_4$: 88% of ThOD (0.01N $KMnO_4$) (274)
-        acid: 98% of ThOD; alkaline: 43% of ThOD (220)
-        6.575 (30)
- T.O.C.: 96% of ThOD (220)
- ThOD: 1.89 (36)
- Biodegradation: adapted A.S. at 20°C—product is sole carbon source: 90.0% COD removal at 54.2 mg COD/g dry inoculum/hr (327)
- Impact on biodegradation processes:
  at 0.3 mg/l inhibition of degradation of glucose by *Pseudomonas fluorescens*
  at 50 mg/l inhibition of degradation of glucose by *E. coli* (293)

HYDROQUINONE 747

—Impact on conventional biological treatment systems:

	chemical conc. mg/l	effect	
unacclimated system	10	no effect	
	100	inhibitory	
	530	inhibitory	
acclimated system	100	biodegradable	(1828)

—Natural sources:
  in leaves of blueberry, red whortleberry, cranberry and bearberry up to 1% (211)
—Waste water treatment:
  A.C.: absorbability: 0.167 g/g C, 93.3% reduction, infl.: 1,000 mg/l, effl.: 167 mg/l (32)
  ozonation 95% reduction of COD at 1 mg/l after 24 hr (250)

$$\underset{\text{OH}}{\underset{|}{\bigcirc}}\text{–OH} + O_3 \longrightarrow \underset{\text{O}}{\underset{\|}{\bigcirc}}\text{=O} + H_2O + O_2 \longrightarrow \text{dibasic acids} \quad (251)$$

hydroquinone → p-quinone

—Odor threshold conc.: detection: 5.0 mg/l (998)
—Aquatic reactions:
  photooxidation by U.V. in aqueous medium at 90–95°C:
    time for the formation of CO_2 (% of theoretical) 25%: 10.3 hr
                                                        50%: 22.9 hr
                                                        75%: 43.7 hr (1628)
  autoxidation at 25°C: $t^{1/2}$: 111 hr at pH 7.0
                                    41 hr at pH 8.0
                                    0.8 hr at pH 9.0 (1908)

D. BIOLOGICAL EFFECTS:
  —Toxicity threshold (cell multiplication inhibition test):
    bacteria (*Pseudomonas putida*):   58 mg/l
    green algae (*Scenedesmus quadricauda*): 0.93 mg/l
    protozoa (*Entosiphon sulcatum*):   11 mg/l   (1900)
    protozoa (*Uronema parduczi Chatton-Lwoff*): 21 mg/l (1901)
  —Bacteria: *E. coli*: LD_0: 50 mg/l (30)
  —Algae: *Chlorella pyrenoidosa*: 178 mg/l (41)
    Scenedesmus: LC_0: 4 mg/l
    Selenastrum capricornutum: 0.1 mg/l: no effect
                                  0.4 mg/l: no effect
                                  1.0 mg/l: inhibitory
                                  4.0 mg/l: inhibitory
                                 10.0 mg/l: inhibitory
                                 40.0 mg/l: inhibitory (1828)
  —Ciliate (*Tetrahymena pyriformis*): 24 hr LC_{100}: 15.4 mmole/l (1662)
  —Arthropod: *Daphnia*: LD_0: 0.6 mg/l (30)

Daphnia magna: LC_{50}: 0.05 mg/l (1828)
—Fish: goldfish: approx. fatal conc.: 48 hr: 0.287 mg/l (226)
Pimephales promelas: LC_{50}: between 0.1 and 0.18 mg/l (1828)
—Mammals:
rabbit: single lethal oral dose: 0.2 g/kg (211)
cat: single lethal oral dose: 0.08 g/kg

hydroquinonebenzylether *see* hydroquinone monobenzylether

hydroquinonedimethylether *see* 1,4-dimethoxybenzene

hydroquinonemonobenzylether (*p*-benzyloxyphenol; hydroquinonebenzylether; benzylhydroquinone; algerite alba)

A. PROPERTIES: light tan powder; m.p. 121–122°C; sp.gr. 1.26
D. BIOLOGICAL EFFECTS:
—Fish: goldfish: approx. fatal conc.: 2.5 mg/l, 48 hr (226)

hydroquinonemonomethylether *see* 4-methoxyphenol

hydroquinone monosulfonate, sodium
$NaHSO_3C_6H_3(OH)_2$
Use: A substitution product of hydroquinone, hydroquinone monosulfonate is formed in the developing bath by reaction of hydroquinone with sulfite. The compound functions as a reducing agent. (1828)
A. PROPERTIES: m.w. 212.12
C. WATER POLLUTION FACTORS:
theoretical   analytical:
TOD = 0.98   Reflux COD = 85.1 % recovery
COD = 0.98   Rapid COD = 87.0 % recovery
              BOD_5 = 0.67
              BOD_5/COD = 0.802 (1828)
the BOD_5 analysis indicated that the hydroquinone monosulfonate exerted a significant oxygen demand
—Impact on conventional biological treatment systems:

	chemical conc. mg/l	effect
unacclimated system	100	no effect

	1,000	no effect
acclimated system	118	biodegradable

hydroquinone monosulfonate did not affect an unacclimated system at concentrations as high as 1,000 mg/l (1828)

D. BIOLOGICAL EFFECTS:
 −Algae:
 Selenastrum capricornutum: 1.0 mg/l: no effect
 10.0 mg/l: no effect
 100.0 mg/l: inhibitory (1828)
 −Crustacean:
 Daphnia magna: LC_{50}: 0.47 mg/l (1828)
 −Fish:
 Pimephales promelas: LC_{50}: 10.0 mg/l (1828)

hydroxyacetic acid *see* glycolic acid

o-hydroxyaniline *see* o-aminophenol

m-hydroxyaniline *see* m-aminophenol

p-hydroxyaniline *see* p-aminophenol

o-hydroxy-anisole *see* 2-methoxyphenol

m-hydroxy-anisole *see* 3-methoxyphenol

p-hydroxy-anisole *see* 4-methoxyphenol

o-hydroxybenzaldehyde *see* salicylaldehyde

4-hydroxybenzene sulfonic acid (*p*-phenolsulfonic acid; 1-phenol-4-sulfonic acid)

A. PROPERTIES: orange red crystals; m.w. 174.17
C. WATER POLLUTION FACTORS:
 −Biodegradation: decomposition by a soil microflora in 32 days (176)

m-hydroxybenzoic acid

p-HYDROXYBENZOIC ACID

Use: intermediate for plasticizers; resins; light stabilizers; petroleum additives; pharmaceuticals

A. PROPERTIES: m.w. 138.12; m.p. 201.3/204°C; sp.gr. 1.473 at 4/4°C; solub. 9,200 mg/l at 18°C; log P_{oct} 1.50

C. WATER POLLUTION FACTORS:
—Biodegradation:
decomposition by a soil microflora in 2 days (176)
—Manmade sources:
in liquid manure: average: 35.8 mg/l
             max: 96.0 mg/l
in dungheap effluent: average: 4.8 mg/l
             max: 52.0 mg/l (216)
in primary domestic sewage plant effluent: 0.007–0.040 mg/l (517)
—Reduction of amenities:
lowest conc. producing taste in chlorinated water: 1.0 mg/l (226)

D. BIOLOGICAL EFFECTS:
—Mammals: guinea pig: I.P.: LD_{50}: 2.8 g/kg (211)

p-hydroxybenzoic acid

Use: intermediates; synthetic drugs

A. PROPERTIES: m.w. 138.12; m.p. 213°C; b.p. 76°C sublimes; sp.gr. 1.443 at 20/4°C; solub. 7,900 mg/l at 15°C, 26,000 mg/l at 75°C; log P_{oct} 1.58

C. WATER POLLUTION FACTORS:
—Biodegradation:
decomposition by a soil microflora: 1 day (176)
adapted A.S. at 20°C—product is sole carbon source: 98.7% COD removal at 100.0 mg COD/g dry inoculum/hr
—Manmade sources:
in primary domestic sewage plant effluent: 0.001 mg/l (517)
—Waste water treatment:

methods	temp °C	days observed	feed mol/warburg flask	acclim. days	% th. oxidation	
sel.strain,Sd,W	30	$\frac{1}{12}$	2	no	57	
sel.strain,Sd,W	30	$\frac{1}{24}$	3	2	58	
sel.strain,Sd,W	30	$\frac{1}{12}$	2	2	69	(93)

—Reduction of amenities:
lowest conc. producing taste in chlorinated water: 0.01 mg/l (226)
odor threshold conc.: detection: 5 mg/l (998)
taste threshold conc.: 1.6 mg/l

D. BIOLOGICAL EFFECTS:
—Mammals: rat: acute oral LD_{50}: 400–3,200 mg/kg
guinea pig: I.P. LD_{50}: 3 g/kg (211)

o-hydroxybiphenyl see o-phenylphenol

3-hydroxybutanal (aldol; acetaldol, β-hydroxybutyraldehyde)
$CH_3CHOHCH_2CHO$
A. PROPERTIES: colorless syrupy liquid; m.w. 88.10; m.p. 0°C; b.p. 83°C at 20 mm decomposes; v.p. 21 mm at 20°C; v.d. 3.0; sp.gr. 1.103 at 20/4°C
B. AIR POLLUTION FACTORS: 1 mg/cu m = 0.278 ppm; 1 ppm = 3.6 mg/cu m
C. WATER POLLUTION FACTORS:
—BOD_{10}: 0.90 std.dil.sew. (256)
—Waste water treatment:

methods	temp °C	days observed	feed mg/l	days acclim.	% removed	
NFG,BOD	20	1–10	200–600	365 + P	36	
NFG,BOD	20	1–10	1,000	365 + P	inhibition	(93)

D. BIOLOGICAL EFFECTS:
—Mammalia: rat: inhalation: no deaths: sat. vap., 0.5 hr
2/6: 4,000 ppm, 4 hr (104)
acute oral LD_{50}: 2.2 g/kg (211)

1-hydroxybutanedioic acid see 1-malic acid

3-hydroxy-2-butanone see acetoin

β-hydroxybutyraldehyde see 3-hydroxybutanal

2-hydroxycamphane see borneol

1-hydroxy-2,4-diamylbenzene see 2,4-diamylphenol

2-(2'-hydroxy-3',5'-di-t-amylphenyl)-2H-benzotriazole)

C. WATER POLLUTION FACTORS:
—Water and sediment quality:
in river water: 0.007–0.085 ppm; in river sediment: 1–100 ppm (555)

β-hydroxydiethylamine see N-ethylethanolamine

hydroxydimethylarsineoxide *see* cacodylic acid

hydroxyethanoic acid *see* glycolic acid

N(2-hydroxyethyl)acetate *see* N-acetylethanolamine

β-hydroxyethylamine *see* ethanolamine

N-hydroxyethylenediaminetriacetate, iron chelate (Versenol AG)
 Composition: 5% Fe; 31.3% active ingredient
B. BIOLOGICAL EFFECTS:
 —Fish:
  bluegill (*Lepomis macrochirus*): static 96 hr LC_{50}: 8100 mg/l
   96 hr no adverse effect level: 5600 mg/l (1869)

N-(hydroxyethyl)ethylenediaminetetraacetic acid, trisodium salt (Versenol 120)
 Use: chelating agent
D. BIOLOGICAL EFFECTS:
 —Fish:
  bluegill (*Lepomis macrochirus*): static 96 hr LC_{50}: 808 mg/l
   96 hr no adverse effect level: 560 mg/l (1869)

3-hydroxyindole (3-indolol)

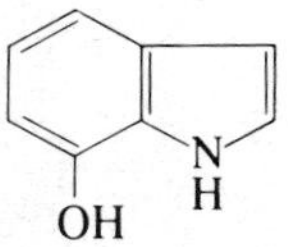

A. PROPERTIES: m.w. 133.15
C. WATER POLLUTION FACTORS:
 —River and sediment quality:
  in primary domestic sewage plant effluent: 0.002 mg/l
  in secondary domestic sewage plant effluent: 0.002 mg/l (517)

2-hydroxyisobutyric acid (2-methyl-lactic acid)
$(CH_3)_2C(OH)COOH$
A. PROPERTIES: m.w. 104.11; m.p. 77–80°C; b.p. 84°C at 1.5 mm; $\log P_{oct}$ −0.36
C. WATER POLLUTION FACTORS:
 —Manmade sources:
  in primary domestic sewage plant effluent: 0.004 mg/l (517)

α-hydroxyisobutyronitrile *see* acetone cyanohydrin

hydroxylaminesulfate
$(NH_2OH)_2 \cdot H_2SO_4$
 Use: antioxidant in developing solutions (1828)
A. PROPERTIES: m.w. 164.14

C. WATER POLLUTION FACTORS:

theoretical:
TOD = 0.585
NOD = 0.585

analytical:
reflux COD = 0.520
rapid COD = 0.524
TKN = 14.7% recovery
BOD_5 = 0.135 m
BOD_5/COD = 0.231
BOD_5 (acclimated) = 0.12 m   (1828)

Note that the oxygen demand of hydroxylamine sulfate is due to the nitrogenous component. According to the definition of COD used in this study, the theoretical COD is equal to zero. Analytically, however, the compound as a strong reducing agent will react with the dichromate and a COD will be measured. The BOD_5 analysis may have also measured, in part, the oxygen demand due to chemical oxidation.   (1828)

—Impact on conventional biological treatment systems:

Although not apparent before these experiments were undertaken, the investigative approach taken in this study was perhaps not applicable to the analysis of hydroxylamine sulfate. Several attempts were made to analyze the impact of the compound on an unacclimated biomass, however no data correlations became evident. What appeared to be taking place was volatilization or partial oxidation of the compound. At the lowest level tested (100 mg/l) these interferences were not as severe. At this concentration the compound appeared to inhibit the activity of the biomass.

The acclimated Warburg analysis did measure significant oxygen consumption and COD reduction, but the previous discrepancies in data did not allow conclusive acceptance of these results. It was evident, however, that the apparent inhibitory effects exhibited by the compound were significantly reduced upon acclimation of the biomass.   (1828)

D. BIOLOGICAL EFFECTS:

—Algae:

Selenastrum capricornutum: 0.1 mg/l: no effect
1.0 mg/l: inhibitory
10.0 mg/l: inhibitory
100.0 mg/l: inhibitory   (1828)

—Crustacean:

Daphnia magna: LC_{50}: 1.2 mg/l   (1828)

—Fish:

Pimephales promelas: LC_{50}: 7.2 mg/l   (1828)

1-hydroxy-2-methoxybenzene *see* 2-methoxyphenol

2-*exo*-hydroxy-2-methylbornane

4-HYDROXYMETHYL-2,6-di-tert-BUTYLPHENOL

Natural source: this camphor smelling compound has been isolated from the culture broth of the organisms *Streptomyces antibioticus* No 5324, *Streptomyces praecox* ATCC 3374 *Streptomyces griseus* ATCC 10137, and has been shown to be the major odorous component produced by these organisms (1316)

4-hydroxymethyl-2,6-*tert*-butylphenol (Ionox 100 antioxidant)

$(CH_3)_3C$ — [phenol ring with OH at top, $C(CH_3)_3$ on right, CH_2OH at bottom]

A. PROPERTIES: white colorless powder; m.w. 236.4; m.p. 141°C; v.p. 0.03 mm at 100°C
C. WATER POLLUTION FACTORS:
 —BOD_5 : 0.35 NEN 3235-5.4 (277)
 —COD : 1.88 NEN 3235-5.4 (277)
D. BIOLOGICAL EFFECTS:
 —Fish: goldfish: LD_{50} (24, 96 hr): 9 mg/l-modified ASTM D 1345 (277)
 —Mammals: rat: acute oral LD_{50} : > 700 mg/kg (277)

4-hydroxy-4-methyl-2-pentanone   *see* diacetone alcohol

2-(2′-hydroxy-5′-methylphenyl)-2H-benzotriazole

[structure: benzotriazole fused to benzene ring linked via N to a phenyl bearing CH (presumably OH) and CH_3]

Use: ultraviolet light absorber in plastic, possesses antioxidant and thermal stabilization properties
A. PROPERTIES: m.p. 129-130°C; b.p. 225°C at 10 mm
C. WATER POLLUTION FACTORS:
 —River and sediment quality:
  in river water: 0.006-0.10 ppm; in river sediment: 2-670 ppm (555)

2-hydroxy-2-methylpropanenitrile   *see* acetone cyanohydrin

1-hydroxynaphthalene *see* α-naphthol

2-hydroxynaphthalene *see* β-naphthol

2-hydroxy-3-naphtoic acid (3-naphthol-2-carboxylic acid; β-oxynaphthoic acid)

A. PROPERTIES: yellow rhombic leaflets; m.p. 217.5/219°C
C. WATER POLLUTION FACTORS:
—Waste water treatment:
ion exchange: adsorption on Amberlite XAD-2: 39% retention, infl.: 0.6 ppm, effl.: 0.37 ppm (40)

12-hydroxy-9-octadecenoic acid   *see* ricinoleic acid

m-**hydroxyphenol**   *see* resorcinol

p-**hydroxyphenol**   *see* hydroquinone

4-hydroxyphenyl acetic acid

A. PROPERTIES: m.w. 152.15; m.p. 149-151°C
C. WATER POLLUTION FACTORS:
—Manmade sources:
in primary domestic sewage plant effluent: 0.016-0.190 mg/l (517)

dl-β-*p*-**hydroxyphenylalanine**   *see dl*-tyrosine

p-**hydroxyphenylglycine**   (glycine(photographic); photoglycin)

NHCH₂COOH

(structure: benzene ring with NHCH₂COOH and OH)

A. PROPERTIES: white to buff crystals or powder; m.p. 240°C, with decomposition
D. BIOLOGICAL EFFECTS:
—Fish: goldfish: approx. fatal conc.: 20 mg/l, 48 hr (226)

3-hydroxyphenylhydracrylic acid
C. WATER POLLUTION FACTORS:
—Manmade sources:
in primary domestic sewage plant effluent: 0.010-0.022 mg/l (517)

3-(p-hydroxyphenyl)propionic acid

[structure: para-hydroxyphenyl ring with CH₂CH₂COOH substituent]

A. PROPERTIES: m.w. 166.18; m.p. 129–131°C
C. WATER POLLUTION FACTORS:
  —Manmade sources:
   in primary domestic sewage plant effluent: 0.006–0.020 mg/l   (517)

2-hydroxyproline (1,2-hydroxy-pyrrolylmethanoic acid)
$C_4H_7N(OH)COOH$
A. PROPERTIES: m.w. 131.13; m.p. 274°C, b.p. 238/241 decomposes; solub. 250,000 mg/l at 0°C
C. WATER POLLUTION FACTORS:
  —Waste water treatment:
   A.S.: after   6 hr: 1.0% of ThOD
         12 hr: 2.9% of ThOD
         24 hr: 18.2% of ThOD (1-hydroxyproline)   (89)
  —Manmade sources:
   excreted by man: in urine: 0.02 mg/kg body wt/day   (203)

2-hydroxypropanenitrile see lactonitrile

2-hydroxy-1,2,3-propanetricarboxylic acid see citric acid

2-hydroxypropylamine see isopropanolamine

2-hydroxypropylmethanethiosulfonate see busan 25, busan 74

1,2-hydroxypyrrolylmethanoic acid see 2-hydroxyproline

8-hydroxyquinoline (8-quinolinol; oxyquinoline; oxine)

A. PROPERTIES: white crystals or powder, darkens when exposed to light; m.p. 73.75°C; b.p. 267°C
C. WATER POLLUTION FACTORS:
  —COD: 80% of ThOD (0.05 n Cr_2O_7) (n.s.i.)
  —$KMnO_4$: 44% of ThOD (0.01 n $KMnO_4$) (n.s.i.)
  —ThOD: 2.02 (n.s.i.)   (274)

—Impact on biodegradation processes:
75% inhibition if nitrification process in non acclimated activated sludge at 73 mg/l   (30)

1-hydroxysuccinic acid *see* l-malic acid

α-hydroxytoluene *see* benzylalcohol

2-hydroxytoluene *see* o-cresol

3-hydroxytoluene *see* m-cresol

4-hydroxytoluene *see* p-cresol

dl-α-hydroxy-α-toluic acid *see* dl-mandelic acid

β-hydroxytricarballylic acid *see* citric acid

2-hydroxytriethylamine *see* 2-diethylamino-ethanol

hypnone *see* acetophenone

hypoxanthine (6(1)-purinone; 6-oxypurine; sarcine)

Use: biochemical research; biological media; an intermediate in the metabolism of animal purines; also widely distributed in the vegetable kingdom.
A. PROPERTIES: m.w. 136.11; m.p. 300°C decomposes at 150°C; solub. 700 mg/l at 19°C; 14,000 mg/l at 100°C; log P_{oct} −1.11
C. WATER POLLUTION FACTORS:
—Manmade sources:
in primary domestic sewage plant effluent: 0.012–0.042 mg/l   (517)

hypoxanthine riboside *see* inosine

I

IA *see* indole-3-acetic acid

IAA *see* indole-3-acetic acid

idryl *see* fluoranthene

igepals (polyoxyethylated alkylphenols)
  Use: nonionic emulsifiers and wetting agents
D. BIOLOGICAL EFFECTS:
  —Fish:
   bluegill: Igepal CO-630; 96 hr LC_{50} (FT): 6.3 mg/l
   Igepal CO-880; 96 hr LC_{50} (S):   >1,000 mg/l   (1125)
   Igepal CO-630; 96 hr LC_{50} (S):   7.9 mg/l
   Igepal CO-520; 96 hr LC_{50} (S):   >2.4, <2.8 mg/l   (1125)

imidan (N-(mercaptomethyl)phthalimide-S-(O,O-dimethylphosphorodithioate); phosmet; PMP)

Use: insecticide
A. PROPERTIES: solub. 25 mg/l at 25°C; log P_{oct} 2.83
D. BIOLOGICAL EFFECTS:
  —Fish:

Chinook salmon (*Oncorhynchus tshawytscha*):	96-hr LC_{50}: 150 μg/l	(S)
bluegill	96-hr LC_{50}: 70 μg/l	(S)
smallmouth bass	96-hr LC_{50}: 150 μg/l	(S)
rainbow trout:		
fingerling:	96-hr LC_{50}: 560 μg/l	(S)
yolk sac fry:	96-hr LC_{50}: >10 000 μg/l	(S)
eyed eggs:	96-hr LC_{50}: >10 000 μg/l	(S)
fathead minnow:	96-hr LC_{50}: 7,300 μg/l	(S)

Channel catfish: 96-hr LC_{50}: 11,000 µg/l (S)
(1207)
—Mammals: acute oral LC_{50} (rat): 300 mg/kg (1854)

2-imidazolidinethione *see* ethylenethiourea

imsol A
Composition: 90% isopropanol
D. BIOLOGICAL EFFECTS:
—Fish:

	mg/l	24 hr	48 hr	96 hr	3 m (extrap.)
harlequin fish	LC_{10} (F)	6,000	3,700	1,500	
(*Rasbora heteromorpha*)	LC_{50} (F)	7,100	4,900	4,200	2,000

(331)

imugan (1-(3,4-dichloro-aniline)-1-formylamino-2,2,2-trichloroethane; chloraniformethane)

Cl—⟨ ⟩—NH—CH—CCl₃
                  |
Cl   NH—CHO

Use: fungicide
A. PROPERTIES: solub. 160 mg/l at 20°C
D. BIOLOGICAL EFFECTS:
—Fish:

	mg/l	24 hr	48 hr	96 hr	3 m (extrap.)
harlequin fish	LC_{10} (F)	5.0	3.0	2.4	
(*Rasbora heteromorpha*)	LC_{50} (F)	7.2	5.2	3.7	1.0

(331)

—Mammals:
acute oral LD_{50} (male rat): >2,500 mg/kg
dermal LD_{50} (rat): >1,000 mg/kg
(1854)

1,2,3-indantrione monohydrate *see* ninhydrin

indene

Manufacturing source: petroleum refining; coke processing. (347)
Users and formulation: paint and coating mfg.; tile mfg.: preparation of coumarine-indene resins; chemical synthesis intermediate; asphalt and naphtha constituent.
(347)
Natural sources (water and air): coal, lignite, crude petroleum. (347)

Man caused sources (water and air): general usage of asphalt and naphtha; general lab use; solvent. (347)
A. PROPERTIES: colorless liquid; m.p. $-35°C$; b.p. $182°C$; sp.gr. 1.006 at $20/4°C$; $\log P_{oct}$ 2.92
B. AIR POLLUTION FACTORS:
 —Odor threshold: detection: 0.02 mg/cu m (637)
C. WATER POLLUTION FACTORS:
 —Reduction of amenities:
  T.O.C. = 0.001 mg/l (295)
D. BIOLOGICAL EFFECTS:
 —Fish:
  fathead minnows: static bioassay in Lake Superior Water at $18-22°C$: LC_{50} (1; 24; 48; 72; 96 hr): 39; 14; 14; 14; 14 mg/l (350)

indeno-(1,2,3-*cd*)fluoranthene
Manmade sources:
in gasoline: 16 µg/l
in exhaust condensate of gasoline engine: 12–32 µg/l gasoline consumed (1070)

indeno(1,2,3-*cd*)pyrene (*o*-phenylenepyrene)

Manmade sources:
in gasoline: 59 µg/l (1070)
in gasoline: low octane number: 0.04–0.18 mg/kg ($n = 13$)
       high octane number: 0.07–0.38 mg/kg ($n = 13$) (385)
in high-octane gasoline: 2.88 mg/kg (1220)
in fresh motor-oil: 0.03 mg/kg (1220)
in used motor-oil after 5,000 km: 34.0–59.4 mg/kg
         10,000 km: 46.7–83.2 mg/kg (1220)
A. PROPERTIES: m.w. 276.34; m.p. $160-163°C$; b.p. $536°C$
B. AIR POLLUTION FACTORS:
 —Manmade sources:
  in tail gases of gasoline engine: 11–87 µg/cu m (340)
  in stack gases of municipal incinerator, after spray tower and electrostatic precipitator: 0.18 mg/100 cu m; in residues: <10 µg/kg (n.s.i.) (341)
  in exhaust condensate of gasoline engine: 268 µg/g (1069)
  in exhaust condensate of gasoline engine: 32–86 µg/l gasoline consumed (1070)
  emissions from typical European car engine (1608 cu cm)—following European

driving cycles—using leaded and unleaded gasolines: 0.7–25.5 µg/l fuel burnt
(1291)
in coke oven emissions: 101.5 µg/g of sample (960)
in cigarette smoke: 0.4 µg/100 cigarettes (1298)
—Ambient air quality:
glc's in Birkenes (Norway) Jan.–June 1977, avg.: 0.36 ng/cu m
range: n.d.–2.67 ng/cu m ($n = 18$)
glc's in Rørvik (Sweden) Dec. '76–April '77: avg.: 0.46 ng/cu m
range: n.d.–2.94 ng/cu m ($n = 21$)
(1236)
glc's at Botrange (Belgium)—woodland at 20–30 km from industrial area: June–July 1977: 0.13; 0.16 ng/cu m ($n = 2$) (1233)

C. WATER POLLUTION FACTORS:
—Manmade sources:
in effluent spray tower of stack gases of municipal incinerator: <0.01 µg/l
(n.s.i.) (341)
primary and digested raw sewage sludge: 0.21–0.27 ppm
liquors from sewage sludge heat treatment plants: 0.03–0.55 ppm
sludge cake from heat treatment plants: 0.22–0.57 ppm
final effluent of sewage work: 0.03 ppb (1426)
—Water and sediment quality:
in groundwater (W. Germany, 1968): 0.2–1.8 µg/cu m (n.s.i.) ($n = 10$)
in tapwater (W. Germany, 1968): 0.9–3.0 µg/cu m (n.s.i.) ($n = 6$) (955)
in wells and galeries of an aquifer: Brussels sands (sands covered with a thick loamy layer): <0.1–0.7 ng/l (1066)
West Germany:
river Gersprenz at Munster: 5.4–12 ng/l
river Danube at Ulm: 9.5–16.4 ng/l
river Main at Seligenstadt: 12.5–217 ng/l
river Aach at Stockach: 116–188 ng/l
river Schussen: 45 ng/l (530)
in Thames river water: at Kew Bridge: 50 ng/l
at Albert Bridge: 110 ng/l
at Tower Bridge: 210 ng/l (529)
in rapid sand filter solids from Lake Constance Water: 0.5 mg/kg
in river water solids: river Rhine: 0.6 mg/kg
river Aach at Stockach: 1.2–4.3 mg/kg
river Argen: 0.6 mg/kg
river Schussen: 3.6 mg/kg (531)
—Water treatment methods:

	Dec. 1965	May 1966	
raw sewage	15.0 ppb	0.60 ppb	
after mechanical purification	3.0 ppb	0.14 ppb	
after biological purification	0.12 ppb	0.07 ppb	(545)

in domestic effluent: 0.022–0.238 ppb
in sewage (high percentage industry): 0.476–0.93 ppb
in sewage during dry weather: 0.017 ppb
in sewage during heavy rain: 4.98 ppb (531)

concentration at various stages of water treatment works:
river intake: 0.069 µg/l
after reservoir: 0.066 µg/l
after filtration: 0.027 µg/l
after chlorination: 0.009 µg/l (434)

plant location	water source	river water conc. ng/l	drinking water* conc. ng/l	% removal/ transformation
Pittsburgh, Pa.	Monongahela river	60.4	1.2	98.0**
Huntington, W.Va.	Ohio river	9.5	1.2	87.4
Philadelphia, Pa.	Delaware river	72.4	1.7	97.7

*treatment provided: lime, ferric salt, A.C., chlorination, fluoridation
**two stages A.C.: powdered and granular carbon

11H-indeno(1,2-b)quinoline
C. AIR POLLUTION FACTORS:
 —Ambient air quality:
  in the average American urban atmosphere—1963:
   1 µg/g airborne particulates, or
   0.1 ng/cu m air (1293)

indicane
C. WATER POLLUTION FACTORS:
 —Manmade sources:
  in primary domestic sewage plant effluent: 0.001–0.002 mg/l (517)

indole (benzo(b)pyrrole)

Use: chemical reagent; perfumery; medicine; flavoring agent
A. PROPERTIES: white to yellowish scales, turning red on exposure to light and air; m.w. 117.14; m.p. 52.5°C; b.p. 254°C; log P_{oct} 2.00/2.25
B. AIR POLLUTION FACTORS:
 —Odor threshold: detection: 0.0006 mg/cu m (840)
 —Contol methods:
  wet scrubber: water at pH 8.5: outlet: 5 odor units/scf
   $KMnO_4$ at pH 8.5: outlet: 1 odor units/scf (115)
C. WATER POLLUTION FACTORS:
 —BOD_5: 2.07 std.dil.sew. (256)
 —COD: 2.460 (223)
 —$KMnO_4$ value: 8.792 (30)
 —ThOD: 2.46 (30)
 —BOD_5: 84% of ThOD (274)
 —COD: 97% of ThOD (0.05 n Cr_2O_7) (274)

—$KMnO_4$: 91% of ThOD (0.01 n $KMnO_4$) (274)
—Manmade sources:
  contents of domestic sewages: 0.25 µg/l (85)
—Odor threshold conc.: detection: 0.3 mg/l (998)
—Taste threshold conc.: 0.5 mg/l (998)

indole-3-acetic acid (IA; IAA; β-indolylacetic acid)

Use: plant growth hormone; agriculture and horticulture
A. PROPERTIES: m.w. 175.19; m.p. 165–169°C
C. WATER POLLUTION FACTORS:
—Manmade sources:
  in secondary domestic sewage plant effluent: 0.013 mg/l (517)

3-indolol see 3-hydroxyindole

β-indolylacetic acid see indole-3-acetic acid

inosine (hypoxanthine riboside)

Use: biochemical research
an important intermediate in animal purine metabolism
A. PROPERTIES: m.w. 268.23; m.p. 212–213°C decomposes; log P_{oct} −2.08
C. WATER POLLUTION FACTORS:
—Manmade sources:
  in primary domestic sewage plant effluent: 0.011–0.050 mg/l
  in secondary domestic sewage plant effluent: 0.020 mg/l (517)

iodomethane see methyliodide

ionol CP antioxidant see 2,6-di-*tert*-butyl-4-methylphenol

ionox 100 antioxidant see 4-hydroxymethyl-2,6-di-*tert*-butylphenol

IPA *see* isopropanol

IPC *see* isopropyl-N-phenylcarbamate

IPE *see* diisopropylether

prim-**isoamylalcohol** (3-methyl-1-butanol; *prim*-isobutylcarbinol; isopentanol) $(CH_3)_2CHCH_2CH_2OH$
A. PROPERTIES: m.w. 88.15; m.p. −117.2°C; b.p. 131/132°C; v.p. 2.3 mm at 20°C, 4.8 mm at 30°C; v.d. 3.04; sp.gr. 0.812; solub. 30,000 mg/l at 20°C, 26,720 mg/l at 22°C; sat. conc. 11 g/cu m at 20°C, 22 g/cu m at 30°C; log P_{oct} 1.16
B. AIR POLLUTION FACTORS: 1 mg/cu m = 0.27 ppm, 1 ppm = 3.66 mg/cu m
  —Odor: characteristic: quality: sweet
          hedonic tone: pleasant
  T.O.C.: 0.027 mg/cu m = 7.3 ppb   (307)
          recogn.: 38 mg/cu m = 10.2 ppm
          10 ppm = 35 mg/cu m (n.s.i.)   (210)
          abs.perc.lim.: 0.12 ppm
          50% recogn.: 1.0 ppm
          100% recogn.: 1.0 ppm
          O.I. 100% recogn.: 13,150
  —Manmade sources:
    from whiskey fermentation vats: 0.166 g/cu m of grain input
  —Sampling and analysis:
    photometry: min. full scale: 5,000 ppm   (53)
C. WATER POLLUTION FACTORS:
  —BOD$_5$: 0.150 std.dil.sew.   (260)
           0.162 std.dil.sew.   (27, 30, 282)
           59% of ThOD   (274)
  —COD: 77% of ThOD (0.05 n $K_2Cr_2O_7$)   (274)
  —KMnO$_4$: 0.164   (30)
           2% of ThOD (0.01 n KMnO$_4$)   (274)
  —ThOD: 2.740   (30)
  —Waste water treatment:
    A.S. BOD, 20°C, 1−5 days observed, feed: 333 mg/l, 30 days acclimation, 79%
      removed   (93)
    A.S.: after 6 hr: 10.2% of ThOD
               12 hr: 20.4% of ThOD
               24 hr: 29.8% of ThOD   (88)

isoamylaldehyde *see* 3-methylbutanal

α-isoamylene *see* 3-methyl-1-butene

β-isoamylene *see* 2-methyl-2-butene

isoamylether (3-methyl-1-(γ-methylbutoxy)butane; diisoamylether) $(CH_3)_2CH(CH_2)_2O(CH_2)_2CH(CH_3)_2$

A. PROPERTIES: m.w. 158.28; b.p. 172°C; sp.gr. 0.78 at 15/15°C
C. WATER POLLUTION FACTORS:
 —BOD$_5$: nil std.dil.sew. (282)
 —ThOD: 3.01 (30)

isobenzan *see* telodrin

isobutanal *see* isobutyraldehyde

isobutane (2-methylpropane; trimethylmethane) (CH$_3$)$_3$CH
A. PROPERTIES: m.w. 58.12; m.p. -145°C; b.p. -12°C; v.p. 1 atm at -11.7°C, 2 atm at +7.5°C, 5 atm at 39°C; v.d. 2.01; sp.gr. 0.60, liquified; solub. in water: 49 mg/l at 20°C; LHC 635 kcal/mole
B. AIR POLLUTION FACTORS: 1 mg/cu m = 0.41 ppm, 1 ppm = 2.42 mg/cu m
 —Manmade sources:
  glc's downtown Los Angeles: 1967: 10% ile: 5 ppb
  average: 12 ppb
  90% ile: 20 ppb (64)
  expected glc's in USA urban air: range: 0.05 to 0.30 ppm (102)
  in exhaust gas of diesel engine: 0.8% of emitted HC.s (72)
  in flue gas of municipal incinerator: <0.4 ppm (196)
  in gasoline: 0.69–1.04 vol% (312)
  evaporation from gasoline fuel tank:
  2.7–6.5 vol.% of total evaporated H.C.'s
  evaporation from carburetor:
  0.1–0.6 vol.% of total evaporated H.C.'s (398; 399; 400; 401; 402)
 —Atmospheric reactions:
  estimated lifetime under photochemical smog conditions in S.E. England: 17 hr
  (1699; 1704)
 —Odor: T.O.C. = 2.9 mg/cu m = 1.2 ppm (307)
  recognition: 1,370 mg/cu m (761)
C. WATER POLLUTION FACTORS:
 —Biodegradation:
  incubation with natural flora in the groundwater—in presence of the other components of high-octane gasoline (100 µl/l): biodegradation: 0% after 192 hr at 13°C (initial conc.: 0.11 µl/l) (956)

isobutanol (isopropylcarbinol; 2-methylpropanol-1) (CH$_3$)$_2$CHCH$_2$OH
A. PROPERTIES: colorless liquid; m.w. 74.1; m.p. -108°C; b.p. 107.9°C; v.p. 10.0 mm at 25°C; v.d. 2.55; sp.gr. 0.798 at 25/4°C; solub. 95,000 mg/l at 18°C; log P_{oct} 0.65/0.83
B. AIR POLLUTION FACTORS: 1 mg/cu m = 0.330 ppm, 1 ppm = 3.03 mg/cu m
 —Odor: characteristic: quality: sweet, musty
  hedonic tone: unpleasant to pleasant

ISOBUTENE

odor thresholds   mg/cu m

10^{-7} 10^{-6} 10^{-5} 10^{-4} 10^{-3} 10^{-2} 10^{-1} 1   10 10^2 10^3 10^4

detection / recognition / not specified

(210; 278; 279; 291; 298; 307; 610; 675; 676; 708; 737; 749; 776; 871)
- O.I. 100% recogn.: 5,131   (19)
- Manmade sources:
  from whisky fermentation vats: average emission: 0.051 g/cu m grain input   (47)

C. WATER POLLUTION FACTORS:
- BOD_5:   64% of ThOD   (274)
           0.07 std. dil. sew.   (285)
           1.66 std. dil. sew.   (282)
- COD: 100% of ThOD (0.05 n $K_2Cr_2O_7$)   (274)
- $KMnO_4$:   2% of ThOD (0.10 n $KMnO_4$)   (274)
- ThOD: 2.60   (274)
- Manmade sources:
  in year old leachate of artificial sanitary landfill: 0.3 g/l   (1720)
- Water quality:
  the self purification of surface water is affected at 1.0 mg/l   (181)
- Waste water treatment:
  A.C.: adsorbability: 0.084 g/g C: 41.9% reduction, infl.: 1,000 mg/l, effl.: 581 mg/l   (32)

  anaerobic lagoon:

	lb COD/day/1,000 cu ft	infl. mg/l	effl. mg/l	
	22	250	80	
	48	250	85	(37)

  A.S.:   after 6 hr: 8.3% of ThOD
             12 hr: 18.1% of ThOD
             24 hr: 32.5% of ThOD   (88)

methods	temp °C	feed mg/l	days observed	days acclim.	
A.S., W	20	500	1	24	44% theor. oxidation
A.S., BOD	20	333	$\frac{1}{3}$–5	30	98% removed (93)

D. BIOLOGICAL EFFECTS:
- Toxicity threshold (cell multiplication inhibition test):
  - bacteria (*Pseudomonas putida*):   280 mg/l   (1900)
  - algae (*Microcystis aeruginosa*):   290 mg/l   (329)
  - green algae (*Scenedesmus quadricauda*):   350 mg/l   (1900)
  - protozoa (*Entosiphon sulcatum*):   295 mg/l   (1900)
  - protozoa (*Uronema parduczi Chatton-Lwoff*): 169 mg/l   (1901)
- Mammals:
  rat: single oral dose: LD_{50}: 2.46 g/kg
  mouse: inhalation: no effect: 2,125 ppm, 223 × 9.2 hr   (211)
- Man: no eye irritation at 100 ppm, 8 hr   (211)

isobutene (isobutylene; 2-methylpropene; uns. dimethylethylene; γ-butylene)
$CH_2C(CH_3)CH_3$
A. PROPERTIES: colorless gas; m.w. 56.10; m.p. $-146.8°C$; b.p. $-6°C$; v.p. 2.6 atm. at $20°C$, 3.6 atm. at $30°C$; v.d. 1.94; solub. 263 mg/l at $20°C$
B. AIR POLLUTION FACTORS: 1 mg/cu m = 0.43 ppm, 1 ppm = 2.33 mg/cu m
  —Atmospheric reactions:
   reactivity: HC cons.: 1.5–2 ranking
     NO ox.: 1 ranking (63)
  —Manmade sources:
   expected glc's in USA urban air: range: 1–10 ppb (102)
  —Odor: T.O.C.: 1.3 mg/cu m = 0.56 ppm (307)
    3.0 mg/cu m (710)
    recognition: 4,880 mg/cu m (761)
    O.I. at $20°C$ = 4,640,000 (316)

isobutoxyethene *see* isobutylvinylether

isobutylacetate (β-methylpropylethanoate)
$CH_3COOCH_2CH(CH_3)_2$
A. PROPERTIES: m.w. 116.2; m.p. $-98.9°C$; b.p. $116/118°C$; v.p. 10 mm at $16°C$, 20 mm at $25°C$; v.d. 4.0; sp.gr. 0.8712; solub. 6,300 mg/l at $25°C$
B. AIR POLLUTION FACTORS: 1 mg/cu m = 0.211 ppm, 1 ppm = 4.75 mg/cu m
  —Odor: characteristic: quality: sweet, ester
           hedonic tone: pleasant
   T.O.C.: abs. perc. lim.: 0.35 ppm
       50% recogn.: 0.50 ppm
       100% recogn.: 0.50 ppm
       O.I. 100% recogn.: 34,200 (19)
       threshold: 4 ppm = 17 mg/cu m (210)
       distinct odor: 34 mg/cu m = 7 ppm (278)
       recognition: 1.9–2.1 mg/cu m (610)
C. WATER POLLUTION FACTORS:
  —Odor threshold: detection: 0.073 mg/kg (911)
  —Waste water treatment:
   A.C.: adsorbability: 0.164 g/g C; 82.0% reduction, infl.: 1,000 mg/l, effl.: 180 mg/l (32)
D. BIOLOGICAL EFFECTS:
  —Toxicity threshold (cell multiplication inhibition test):
   bacteria (*Pseudomonas putida*):   200 mg/l   (1900)
   algae (*Microcystis aeruginosa*):   205 mg/l   (329)
   green algae (*Scenedesmus quadricauda*):   80 mg/l   (1900)
   protozoa (*Entosiphon sulcatum*):   411 mg/l   (1900)
   protozoa (*Uronema parduczi Chatton-Lwoff*): 727 mg/l (1901)
  —Mammals: rat: inhalation: 6/6: 21,000 ppm, 150 min
             no symptoms: 3,000 ppm, 6 hr (211)

isobutylamine (1-amino-2-methylpropane)
$(CH_3)_2CHCH_2NH_2$

768 ISOBUTYLBENZENE

A. PROPERTIES: m.w. 73.14; m.p. -85.5°C; b.p. 68°C; v.p. 100 mm at 19°C; v.d. 2.52; sp.gr. 0.736; log P_{oct} 0.70 (calculated)
B. AIR POLLUTION FACTORS: 1 mg/cu m = 0.334 ppm, 1 ppm = 2.99 mg/cu m
D. BIOLOGICAL EFFECTS:
  —Fish: creek chub: LD_0:   20 mg/l, 24 hr in Detroit river water
  LD_{100}: 60 mg/l, 24 hr in Detroit river water (243)

isobutylbenzene

$$\text{C}_6\text{H}_5\text{-CH}_2\text{CH(CH}_3)_2$$

A. PROPERTIES: m.p. -51.6°C; b.p. 171.1°C; sp.gr. 0.8532 at 20/4°C
B. AIR POLLUTION FACTORS:
  —Atmospheric reactions: R.C.R.: 0.86 (49)
C. WATER POLLUTION FACTORS:
  —Odor threshold conc.: detection: 0.08 mg/kg (894)
D. BIOLOGICAL EFFECTS:
  —Man: EIR: 5/7 (49)

prim-**isobutylcarbinol** see *prim*-isoamylalcohol

isobutylene see isobutene

isobutylethylene see 4-methyl-1-pentene

isobutyltrimethylmethane see isooctane

isobutylxanthate, sodium
A. PROPERTIES: solid; b.p. decomposes
D. BIOLOGICAL EFFECTS:
  —Fish:
    rainbow trout (*Salmo gairdneri*) 96 hr LC_{50}:
      Cyanamid C 317:   56->100 mg/l (static test)
      Dow Chemical Z 14: 10-100 mg/l (static test) (1087)
  —Mammals:
    ingestion: rats: single oral LD_{50}: approx. 1000 mg/kg (1546)

isobutyraldehyde (isobutanal; 2-methylpropanal; isobutylaldehyde) $(CH_3)_2CHCHO$
  Manufacturing source: organic chemical industry (317)
  Users and formulation: solvent (artificial leather mfg. coated paper, textile mfg. plastics, oil, drug and perfume mfg. industries); mfg. of brake fluid, butyl esters, plasticizers; mfg. resins and rubber chemicals; mfg. organic chemicals (317)
A. PROPERTIES: m.w. 72.10; m.p. -65.9°C; b.p. 61.5°C; v.p. 170 mm at 20°C; v.d. 2.48; sp.gr. 0.7938; solub. 110,000 mg/l

B. AIR POLLUTION FACTORS: 1 mg/cu m = 0.340 ppm, 1 ppm = 2.9 mg/cu m
  —Odor: characteristic: quality: sweet, ester
                         hedonic tone: pleasant to unpleasant
    T.O.C.: abs. perc. lim.: 0.047 ppm
            50% recogn.: 0.141 ppm
            100% recogn.: 0.236 ppm
            O.I. 100% recogn.: 766,949 (19)
  —Sampling and analysis:
    photometry: min. full scale: 1900 ppm (53)
C. WATER POLLUTION FACTORS:
  —BOD_5: 1.6 std. dil. (27)
  —ThOD: 2.44
  —Waste water treatment:
    anaerobic lagoon: influent   effluent
      22 lb COD/day/1,000 cu ft: 210 mg/l   50 mg/l
      48 lb COD/day/1,000 cu ft: 210 mg/l   50 mg/l   (37)
    A.S.: after 6 hr: 8.7% of ThOD
                12 hr: 15.4% of ThOD
                24 hr: 24.3% of ThOD   (88)
  —T.O.C.: detection: 0.01   mg/kg   (908)
                       0.0023 mg/kg   (874)
D. BIOLOGICAL EFFECTS:
  —Mammals:
    rat: inhalation: 1/6: 8,000 ppm, 4 hr   (104)
         inhalation: no deaths: 1,000 ppm, 12 X 6 hr
         slight nose irritation; autopsy, organs normal   (65)

isobutyric acid (2-methylpanioc acid; dimethylacetic acid)
$(CH_3)_2 CHCOOH$
A. PROPERTIES: colorless liquid: m.w. 88.10; m.p. -47.0°C; b.p. 154.4°C; v.p. 1 mm at 14.7°C; v.d. 3.04; sp.gr. 0.949 at 20/4°C; solub. 200,000 mg/l at 20°C; log P_{oct} 0.50/1.13 (calculated)
C. WATER POLLUTION FACTORS:
  —Reduction of amenities:
    T.O.C. = 30 mg/l   (295)
           = 0.05 mg/l   (296)
  —Manmade sources:
    average content of secondary sewage effluents: 26.5 g/l   (86)
    in year old leachate of artificial sanitary landfill: 16.5 g/l (1720)
D. BIOLOGICAL EFFECTS:
  —Mammals: rat: inhalation: single oral LD_{50}: 400->800 mg/kg (211)
  —Algae: Chlorella pyrenoidosa: toxic: 345 mg/l   (41)

isodecyldiphenylphosphate
D. BIOLOGICAL EFFECTS:
  —Fish:
    Lepomis macrochirus: static bioassay in fresh water at 23°C, mild aeration applied after 24 hr:

material added ppm	24 hr	% survival after 48 hr	72 hr	96 hr	best fit 96 hr LC_{50} ppm
10,000	100	100	90	10	
5,000	90	80	80	80	6,700
1,000	100	100	100	100	

Menidia beryllina: static bioassay in synthetic seawater at 23°C: mild aeration applied after 24 hr:

material added ppm	24 hr	% survival after 48 hr	72 hr	96 hr	best fit 96 hr LC_{50} ppm
5,000	30	20	20	10	
3,200	60	20	10	10	1,400
2,000	100	40	30	20	
1,000	100	100	100	70	

(354)

isodrin (an isomer of aldrin)

D. BIOLOGICAL EFFECTS:
—Insects:
  mosquito (late 3rd instar *Aedes aegypti* larvae): 24 hr LC_{50}: 19 ppb
  housefly (3 day old female *Musca*): LD_{50}: 54 µg/fly   (1681)
—Fish:
  bluegill: 24 hr LC_{50}: 12 ppb
  minnow: 24 hr LC_{50}: 6 ppb
—Toxicity ratio of isodrin (I) to photo-isodrin (PI), calculated from respective 24 hr LC^*_{50} and LT_{50} value

	I/PI
Crustacea	
Daphnia pulex (water flea)	0.45
Gammarus spp. (amphipod)	0.43
Asellus spp. (isopod)	0.64
Insects	
Aedes aegypti larvae (mosquito)	0.30 (.33)*
Musca domestica (house fly)	0.41 (.48)*
Fish	
Lebistes reticulatus (guppy)	0.45
Pimephalus promelas (bass)	0.3 (.60)*
Lepomis macrochirus (blue gill)	0.75 (.48)*

(1681*; 1684; 1685)

l-isoleucine (d-2-amino-3-methylpentanoic acid; d-α-amino-β-methylvaleric acid)

$$\begin{array}{c} \text{COOH} \\ | \\ \text{CH}-\text{NH}_2 \\ | \\ \text{CH}-\text{CH}_3 \\ | \\ \text{CH}_2 \\ | \\ \text{CH}_3 \end{array}$$

A. PROPERTIES: m.w. 131.17; m.p. 283°C decomposes; solub. 41,200 mg/l at 25°C, 60,800 mg/l at 75°C

C. WATER POLLUTION FACTORS:
—Manmade sources:
excreted by man: in urine: 0.11–0.6 mg/kg body wt/day
in feces: 3.3–5.5 mg/kg body wt/day
in sweat: 1.0–3.6 mg/100 ml (203)
—Waste water treatment:
A.S.: after 6 hr: 2.4% of ThOD
12 hr: 5.3% of ThOD
24 hr: 14.8% of ThOD (89)

isooctane (2,2,4-trimethylpentane; isobutyltrimethylmethane)

$$\begin{array}{c} \text{CH}_3 \\ | \\ \text{CH}_3-\overset{\displaystyle |}{\text{C}}-\text{CH}_2-\overset{\displaystyle |}{\text{CH}}-\text{CH}_3 \\ | \quad\quad\quad | \\ \text{CH}_3 \quad\; \text{CH}_3 \end{array}$$

Use: organic synthesis; solvent; motor fuel; used with n-heptane to prepare standard mixtures to determine antiknock property of gasoline

A. PROPERTIES: colorless liquid; m.w. 114.23; m.p. −107.4°C; b.p. 99.3°C; v.d. 3.9; sp.gr. 0.6918 at 20/4°C; solub. 0.56 mg/l at 25°C

B. AIR POLLUTION FACTORS:
—Atmospheric reactions:
reactivity: NO ox.: ranking: 0.15 (63)
R.C.R.: 0.81 (49)
—Manmade sources:
diesel engine: 1.0% of emitted HC (72)
in gasoline engine exhaust: 62-car survey: 1.0 vol.% of total exhaust H.C.'s (391)

C. WATER POLLUTION FACTORS:
—Biodegradation:
incubation with natural flora in the groundwater—in presence of the other components of high-octane gasoline (100 μl/l): biodegradation: 13% after 192 hr at 13°C (initial conc. 3.47 μl/l) (956)
—Evaporation:
calculated half-life time in water at 25°C and 1 m depth based on evaporation: 5.55 hr; evaporation rate: 0.124 m/hr (437)

D. BIOLOGICAL EFFECTS:
—Man: EIR: 0.9 (49)

isooctanol (isooctylalcohol)
$C_7H_{15}CH_2OH$
A. PROPERTIES: b.p. 182/195°C; sp.gr. 0.832 at 20/20°C
D. BIOLOGICAL EFFECTS:
—Toxicity threshold (cell multiplication inhibition test):

bacteria (*Pseudomonas putida*):	63 mg/l	(1900)
algae (*Microcystis aeruginosa*):	7.3 mg/l	(329)
green algae (*Scenedesmus quadricauda*):	8.5 mg/l	(1900)
protozoa (*Entosiphon sulcatum*):	30 mg/l	(1900)
protozoa (*Uronema parduczi Chatton-Lwoff*):	55 mg/l	(1901)

isooctaphenone *see* isophorone

isopentane (2-methylbutane; ethyldimethylmethane)
$(CH_3)_2CHCH_2CH_3$
Use: solvent; manufacture of chlorinated derivatives; blowing agent for polystyrene
A. PROPERTIES: colorless liquid; m.w. 72.15; m.p. -160.5°C; b.p. 28°C; sp.gr. 0.62 at 19°C; solub. 48 mg/l at 20°C
B. AIR POLLUTION FACTORS:
—Manmade sources:
glc's downtown Los Angeles 1967: 10%ile: 12 ppb
average: 35 ppb
90%ile: 58 ppb   (64)
expected glc's in USA urban air: range: 50–350 ppb   (102)
in exhaust of reciprocating gasoline engine: 1.7% of emitted HC.s
in exhaust of rotary gasoline engine: 8.6% of emitted HC.s   (78)
in exhaust of diesel engine: 3.7% of emitted HC.s   (72)
in flue gas of municipal incinerator: <0.6–<0.7 ppm   (196)
in gasoline: 9.13–10.99 vol %   (312)
evaporation from gasoline fuel tank:
20.3–26.4 vol.% of total evaporated H.C.'s
evaporation from carburetor:
17.8–45.3 vol.% of total evaporated H.C.'s   (398; 399; 400; 401; 402)
in exhaust of gasoline engines:
62-car survey: 3.7 vol.% of total exhaust H.C.'s   (391)
15-fuel study: 4 vol.% of total exhaust H.C.'s   (392)
engine-variable study: 2.4 vol.% of total exhaust H.C.'s   (393)
—Atmospheric reactions:
estimated lifetime under photochemical smog conditions in S.E. England: 11 hr
(1699; 1700)
C. WATER POLLUTION FACTORS:
—Biodegradation:
incubation with natural flora in the groundwater—in presence of the other components of high-octane gasoline (100 µl/l): biodegradation: 0% after 192 hr at 13°C (initial conc.: 3.29 µl/l)   (956)

isopentanol *see prim*-isoamylalcohol

isophorone (isooctaphenone; 3,5,5-trimethyl-2-cyclohexene-1-one)

Manufacturing source: organic chemical industry. (347)
Users and formulation: solvent; intermediate for alcohols, raw material for 3,5-dimethylaniline; solvent for polyvinyl and nitrocellulose resins; lacquers, finishes mfg.; pesticide mfg. (347)
A. PROPERTIES: m.w. 138.2; m.p. $-8°C$; b.p. $215°C$; v.p. 0.38 mm at $20°C$; v.d. 4.77; sp.gr. 0.92; solub. 12,000 mg/l
B. AIR POLLUTION FACTORS: 1 mg/cu m = 0.18 ppm, 1 ppm = 5.65 mg/cu m
   —Odor: characteristic: quality: sharp, peppermint-like
   T.O.C.: abs. perc. lim.: 0.20 ppm
         50% recogn.: 0.54 ppm
         100% recogn.: 0.54 ppm
         O.I. 100% recogn.: 2,444 (19)
C. WATER POLLUTION FACTORS:
   —Water quality:
   Kanawha river: 13.11.1963: raw: 90 ppb
                              sand filtered: not detectable
                  21.11.1963: raw: 25 ppb
                              after aeration: 16 ppb
                              carbon filtered: not detectable (159)
   —Waste water treatment:
   A.C.: adsorbability: 0.193 g/g C; 96.6% reduction, infl.: 1,000 mg/l, effl.: 34 mg/l (32)
   conventional municipal treatment: infl. 0.090 mg/l, effl. n.d.
   conventional + A.C.:   infl. 0.025 mg/l, effl. n.d. (404)
D. BIOLOGICAL EFFECTS:
   —Man: irritating to eyes, nose and throat: 25 ppm (211)

isophthalic acid (1,3-benzenedicarboxylic acid; *m*-phthalic acid)

A. PROPERTIES: colorless needles; m.w. 166.13; m.p. $330/312°C$; b.p. sublimes; solub. 130 mg/l at $25°C$, 2,200 mg/l if hot; log P_{oct} 1.66
C. WATER POLLUTION FACTORS:
   —Biodegradation:
   decomposition by a soil microflora: 8 days (176)
   adapted A.S. at $20°C$—product is sole carbon source: 95.0% COD removal at 76.0 mg COD/g dry inoculum/hr (327)

D. BIOLOGICAL EFFECTS:
—Mammals: mouse: I.P. LD_{50}: >500 mg/kg (211)

isopimaric acid
D. BIOLOGICAL EFFECTS:
—Fish:
Coho salmon juvenile: 96 hr LC_{50}: 0.22 mg/l (1495)
rainbow trout: static bioassay: 96 hr TLm: 0.4 mg/l at pH 7.0 (441)

isopimarol
D. BIOLOGICAL EFFECTS:
—Fish:
rainbow trout: static bioassay: 96 hr TLm: 0.3 mg/l at pH 7.0 (411)

isoprene (2-methyl-1,3-butadiene; methylbivinyl; hemiterpene)
$CH_2 C(CH_3)CHCH_2$
Use: monomer for manufacture of poly-isoprene
A. PROPERTIES: colorless liquid; m.w. 68.11; m.p. -146°C; b.p. 34.08°C; v.p. 493 mm at 20°C, 700 mm at 30°C; v.d. 2.35; sp.gr. 0.681 at 20/4°C; sat. conc. 178 g/cu m at 20°C, 2,510 g/cu m at 30°C
B. AIR POLLUTION FACTORS: 1 mg/cu m = 0.35 ppm, 1 ppm = 2.83 mg/cu m
—Natural occurence:
emission rate of foliar isoprene: 0.02 to 2.4 ppb/min/in^2 depending on light intensity ranging from 50 to 200 ft cd (197)
—Measured emission rate from *Quercus virginianus*: 33 µg/g dry wt/hr (1729)
C. WATER POLLUTION FACTORS:
—Impact on biodegradation processes:
slight inhibition of microbial growth after 24 hr exposure at saturation concentration (523)
—Odor threshold: detection: 0.005 mg/kg (886)
—Waste water treatment:

A.C.:	infl. ppm	carbon dosage	effl. ppm	% reduction	
	1,000	10X	110	89	
	500	10X	110	78	(192)

D. BIOLOGICAL EFFECTS:
—Fish: fatheads: TLm (24-96 hr): 87-74 mg/l
bluegill: TLm (24-96 hr): 42.5 mg/l
goldfish: TLm (24-96 hr): 180.0 mg/l
guppies: TLm (24-96 hr): 240.0 mg/l
—Mammalia: rat: inhalation: no effect level: 1,670 ppm, 15 X 6 hr (65)

isopropanol (2-propanol; sec-propylalcohol; dimethylcarbinol; perspirit; petrohol; avantine; IPA)
$(CH_3)_2 CHOH$
Use: manufacture of acetone, glycerol and isopropylacetate; solvent; deicing agent for liquid fuels, pharmaceuticals; preservative
A. PROPERTIES: colorless liquid; m.w. 60.10; m.p. -86/-89°C; b.p; 82.4°C; v.p. 32

ISOPROPANOL 775

mm at 20°C, 57 mm at 30°C; v.d. 2.07; sp.gr. 0.785 at 20/4°C; log P_{oct} -0.16/ 0.28 (calculated)

B. AIR POLLUTION FACTORS: 1 mg/cu m = 0.408 ppm, 1 ppm = 2.5 mg/cu m
 —Odor: characteristic: quality: sharp, musty
  hedonic tone: unpleasant, pleasant
  threshold for unadapted panelists: 700 ppm
  threshold after adaption with pure odorant: 20,000 ppm (204)

odor thresholds   mg/cu m

(73; 210; 278; 279; 307; 610; 632; 643; 668; 676; 709; 727; 749; 776; 804)
O.I. 100% recogn.: 1,539 (19)

C. WATER POLLUTION FACTORS:
 —BOD_5: 0.075 at 1,000 ppm, Warburg/sewage (281)
  1.29 to 2; 1.59 (271, 281, 282)
  28% of ThOD* 13% of ThOD*
  10 days: 77% of ThOD 42% of ThOD
  15 days: 80% of ThOD 60% of ThOD
  20 days: 78% of ThOD 72% of ThOD (23)
  *fresh dilution water
  **salt dilution water
 —BOD_5^{20}: 0.16 at 10 mg/l, unadapted sew; lag period: 5 days
 —BOD_{20}^{20}: 1.68 at 10 mg/l, unadapted sew; (554)
 —BOD_5: 60% of ThOD (274)
  1.19, NEN 3235-5.4
  1.72 adapted sew, NEN 3235-5.4 (277)
 —COD: 2.30; 2.21; 1.61 (23; 27; 36)
 —$KMnO_4$: 0.202 (30)
 —TOD: 2.39 (26)
 —ThOD: 2.40 (23; 36)
 —COD: 2.23, NEN 3235-5.3 (277)
  97% of ThOD (0.05 n $K_2Cr_2O_7$) (274)
 —$KMnO_4$: 2% of ThOD (0.01 n $KMnO_4$) (274)
 —Biodegradation: adapted A.S.—product as sole carbon source—99.0% COD removal at 52.0 mg COD/g dry inoculum/hr (327)
 —Manmade sources:
  in year old leachate of artificial sanitary landfill: 0.3 g/l (1720)
 —Waste water treatment:
  A.C.: adsorbability: 0.025 g/g C; 12.6% reduction, infl.: 1,000 mg/l, effl.: 874 mg/l (32)
  anaerobic lagoon:

	lb COD/day/1,000 cu ft	infl. mg/l	effl. mg/l	
	13	60	30	
	22	175	45	
	48	175	55	(37)

A.S.: after 6 hr: 4.8% of ThOD
12 hr: 8.3% of ThOD
24 hr: 10.4% of ThOD (88)

methods	temp °C	days observed	feed mg/l	days acclim.		
A.S., W	20	1	500	24	50% theor. oxidation	
A.S., BOD	20	$\frac{1}{3}$–5	333	30	57% removed	(93)

D. BIOLOGICAL EFFECTS:
  −Toxicity threshold (cell multiplication inhibition test):
   bacteria (*Pseudomonas putida*): 1050 mg/l (1900)
   algae (*Microcystis aeruginosa*): 1000 mg/l (329)
   green algae (*Scenedesmus quadricauda*): 1800 mg/l (1900)
   protozoa (*Entosiphon sulcatum*): 4930 mg/l (1900)
   protozoa (*Uronema parduczi Chatton-Lwoff*): 3425 mg/l (1901)
  −Algae: *Chlorella pyrenoidosa*: toxic: 17,400 mg/l (41)
  −Arthropod: brown shrimp (*Crangon crangon*):
    48 hr, LC_{50}: avg.: 1400 mg/l
       range: 900–1950 mg/l
    96 hr, LC_{50}: avg.: 1150 mg/l
       range: 750–1650 mg/l (310)
−Fish:
  creek chub: LD_0: 900 mg/l, 24 hr in Detroit river water
       LD_{100}: 1,100 mg/l, 24 hr in Detroit river water (243)
  goldfish: 24 hr LD_{50}: >5000 mg/l−modified ASTM D 1345 (277)
  fathead minnows: static bioassay in Lake Superior Water at 18-22°C: LC_{50}
    (1; 24; 48; 72; 96 hr): 11,830; 11,160; 11,130; 11,130; 11,130 mg/l (350)
  guppy (*Poecilia reticulata*): 7 d LC_{50}: 7060 ppm (1833)
−Mammals:
  rabbit: single oral LD_{50}: 10.0 ml/kg
  rat: single oral LD_{50}: 5.84 g/kg
  mouse: inhalation: no reaction: 2,050 ppm, 480 min (211)

isopropanolamine (MIPA; 2-hydroxypropylamine; 1-amino-2-propanol)
$CH_3 CHOHCH_2 NH_2$
  Use: emulsifying agent; dry cleaning soaps; soluble textile oils; wax removers; metal cutting oils; cosmetics; emulsion paints; plasticizers; insecticides
A. PROPERTIES: liquid; slight ammonia odor; m.w. 75.11; sp.gr. 0.9619; m.p. 1.4°C; b.p. 160°C; v.p. <1 mm at 20°C; v.d. 2.6, solub. complete; log P_{oct} −0.96
C. WATER POLLUTION FACTORS:
  −BOD_5: 5.1% of ThOD
   10 days: 34.0% of ThOD
   15 days: 43.4% of ThOD
   20 days: 46.0% of ThOD
   $BOD_{10}^{20°C}$: 34.0% ThOD at 2.5 mg/l in mineralized dilution water with settled
       sewage seed (405)

—Waste water treatment:
 A.C.: adsorbability: 0.040 g/g C; 20.0% reduction, infl.: 1,000 mg/l; effl.: 800
   mg/l   (32)
D. BIOLOGICAL EFFECTS:
 —Mammals: rat: acute oral LD_{50}: 4.26 g/kg   (211)

p-isopropoxydiphenylamine

$$\text{C}_6\text{H}_5-\text{NH}-\text{C}_6\text{H}_4-\text{O}-\text{CH}(\text{CH}_3)_2$$

A. PROPERTIES: dark gray flakes; sp.gr. 1.10
D. BIOLOGICAL EFFECTS:
 —Fish: goldfish: approx. fatal conc.: 5.7 mg/l, 48 hr   (226)

2-isopropoxyethanol (isopropylglycolether; isopropylglycol; ethyleneglycolmonoisopropylether; "isopropyloxitol")
$(CH_3)_2CHOCH_2CH_2OH$
A. PROPERTIES: m.w. 104.15; b.p. 140/144°C; sp.gr. 0.9091 at 20/20°C
C. WATER POLLUTION FACTORS:
 —BOD_5: 0.18 NEN 3235-5.4   (277)
 —COD: 2.08 NEN 3235-5.3   (277)
D. BIOLOGICAL EFFECTS:
 —Fish:
   goldfish: LD_{50} (24 hr): >5000 mg/l-modified ASTM D 1345   (277)
 —Mammals: rat: inhalation: no effect level: 100 ppm, 15 × 6 hr   (65)

2-isopropoxypropane see diisopropylether

isopropylacetate
$CH_3COOCH(CH_3)_2$
 Use: solvent; paints; lacquers and printing inks; organic synthesis
A. PROPERTIES: colorless liquid; m.w. 102.1; m.p. -73°C; b.p. 90°C; v.p. 47.5 mm
   at 20°C, 73 mm at 25°C; v.d. 3.52; sp.gr. 0.877 at 16/4°C; solub. 18,000 mg/l
   at 20°C, 30,900 mg/l at 20°C
B. AIR POLLUTION FACTORS: 1 mg/cu m = 0.240 ppm, 1 ppm = 4.17 mg/cu m
 —Odor: characteristic: quality: sweet, ester
                       hedonic tone: pleasant to unpleasant

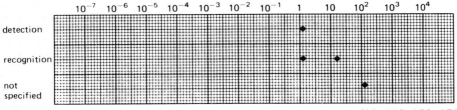

(19; 210; 278; 279)

ISOPROPYLACETIC ACID

    O.I. 100% recogn.: 56,907 (19)
    O.I. at 20°C = 2,100 (316)
C. WATER POLLUTION FACTORS:
   —BOD_5:   0.26 std. dil. sew. (255)
            12.7% of ThOD
    10 days: 40.0% of ThOD
    15 days: 40.0% of ThOD
    20 days: 40.0% of ThOD
    30 days: 42.7% of ThOD
    40 days: 49.1% of ThOD
   —Waste water treatment:
    A.C.: adsorbability: 0.137 g/g C; 68.1% reduction, infl.: 1,000 mg/l, effl.: 319 mg/l (32)
D. BIOLOGICAL EFFECTS:
   —Toxicity threshold (cell multiplication inhibition test):
    bacteria (*Pseudomonas putida*):   190 mg/l   (1900)
    algae (*Microcystis aeruginosa*):   1400 mg/l   (329)
    green algae (*Scenedesmus quadricauda*):   165 mg/l   (1900)
    protozoa (*Entosiphon sulcatum*):   460 mg/l   (1900)
    protozoa (*Uronema parduczi Chatton-Lwoff*): 1602 mg/l   (1901)
   —Mammals: rat: inhalation: 5/6: 32,000 ppm, 4 hr (211)
   —Man: eye irritation: 200 ppm (211)

isopropylacetic acid *see* isovaleric acid

isopropylacetone *see* 4-methyl-2-pentanone

isopropylamine
$(CH_3)_2 CHNH_2$
A. PROPERTIES: colorless liquid; m.w. 59.11; m.p. −101°C; b.p. 32/34°C; v.p. 460 mm at 20°C; v.d. 2.03; sp.gr. 0.690 at 20/4°C; log P_{oct} −0.03
B. AIR POLLUTION FACTORS: 1 mg/cu m = 0.414 ppm, 1 ppm = 2.42 mg/cu m
   —Odor: characteristic: quality: ammoniacal, amine
                    hedonic tone: unpleasant to pleasant
    T.O.C.: abs. perc. lim.: 0.21 ppm
           50% recogn.: 0.72 ppm
          100% recogn.: 0.95 ppm
          O.I. 100% recogn.: 661,052 (19)
C. WATER POLLUTION FACTORS:
   —Reduction of amenities: T.O.C.: detection 5 mg/l (299)(894)
D. BIOLOGICAL EFFECTS:
   —Fish: creek chub: critical range: 40–80 mg/l; 24 hr (226)
   —Mammalia: rat: acute oral LD_{50}:   0.82 g/kg
                             : <0.1 g/kg (40% soln)
              inhalation: 6/6: 8,000 ppm, 4 hr
                     0/6: 4,000 ppm, 4 hr
                     no deaths: sat. vap., 2 min (211)

isopropylbenzene (2-phenylpropane; cumene)

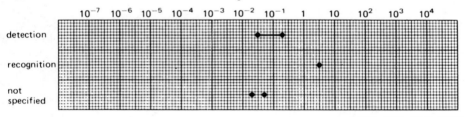

Manufacturing source: petroleum refining; coal tar distillation; organic chemical industry. (347)

Users and formulation: mfg. of acetone, alpha-methylstryene, phenol, polymerization catalysts, diisopropylbenzene; component motor fuel; catalyst for acrylic and polyester-type resins; solvent; asphalt and naphtha constituent; in gasoline (high octane number): 0.16 wt % (347; 387)

Natural sources (water and air): petroleum crudes, coal tar.

A. PROPERTIES: colorless liquid; m.p. -96°C; b.p. 152.7°C; v.p. 3.2 mm at 20°C; v.d. 4.13; sp.gr. 0.8620; solub. 50 mg/l at 20°C; log P_{oct} 3.66

B. AIR POLLUTION FACTORS: 1 mg/cu m = 0.200 ppm, 1 ppm = 4,900 mg/cu m
 —Odor: characteristic: quality: sharp
 hedonic tone: unpleasant
 USSR: human odor perception: 0.06 mg/cu m = 0.012 ppm
 human reflex response: adverse response: 0.028 mg/cu m
 animal chronic exposure: no effect: 0.014 mg/cu m (170)

odor thresholds   mg/cu m

	10^{-7}	10^{-6}	10^{-5}	10^{-4}	10^{-3}	10^{-2}	10^{-1}	1	10	10^2	10^3	10^4
detection							◆━━◆					
recognition									◆			
not specified						◆◆						

(651; 727; 818; 854)

  O.I. 100% recogn.: 83,000 (19)
 —Ambient air quality:
 Los Angeles 1966: glc's: avg.: 0.003 ppm (n = 136)
                 highest value: 0.012 ppm (1319)
 —Atmospheric reactions: R.C.R.: 1.14 (49)
 reactivity: HC cons.: ranking: 1 (63)
 estimated lifetime under photochemical smog conditions in S.E. England: 6.0 hr (1699; 1700)
 —Sampling and analysis:
 photometry: min. full scale: 190 ppm (53)

C. WATER POLLUTION FACTORS:
 —Reduction of amenities:
 T.O.C. = 0.1 mg/l (295)
 approx. conc. causing adverse taste in fish: 0.25 mg/l (41)
 —Control methods: calculated half life time based on evaporative loss for a water depth of 1 m at 25°C: 5.79 hr (330; 437)

—Sampling methods:
  extraction efficiency of macroreticular resins, sample flow: 20 ml/min; pH: 5.7; conc. 10 ppm: X AD-2: 67%
  X AD-7: 67% (370)

D. BIOLOGICAL EFFECTS:
  —Mussels:
    mussel larvae (*Mytilus edulis*): no significant alteration of growth rate at concentrations of 1 to 50 ppm (475)
  —Mammals:
    mouse: inhalation: LD_{50}: 2,000 ppm, 7 hr (211)
    rat: ingestion: single oral LD_{50}: 1400 mg/kg (1546)
    rabbit: skin absorption: LD_{50}: >10,000 mg/kg (1546)

isopropylbenzenehydroxyperoxide (cumene hydroperoxide; α, α-dimethylbenzylhydroperoxide)

$$CH_3-\underset{\underset{\text{Ph}}{|}}{\overset{\overset{OOH}{|}}{C}}-CH_3$$

Use: production of acetone and phenol; polymerization catalyst

D. BIOLOGICAL EFFECTS:
  —Threshold concentration of cell multiplication inhibition of the protozoan *Uronema parduczi Chatton-Lwoff*: 0.35 mg/l (1901)

isopropylcarbinol *see* isobutanol

isopropylchloride *see* 2-chloropropane

isopropyl-N-(3-chlorophenyl)carbamate (CIPC; chlorpropham; chloro-IPC)

$$NH-\overset{\overset{O}{\|}}{C}-O-CH\underset{CH_3}{\overset{CH_3}{<}}$$
(with 3-chlorophenyl group)

Use: pesticide

A. PROPERTIES: m.p. 38–39°C; solub. 88 mg/l

C. WATER POLLUTION FACTORS:
  —Biodegradation:
    biodegradation by fungi (*Aspergillus fumigatus*): half-life: 120 days at pH 7[*]
    biodegradation by bacteria (*Pseudomonas striata*): half-life: 2.9 days at pH 7[**]
    [*]assuming 1 mg/l fungi
    [**]assuming a bacterial population of 0.1 mg/l, calculated from the data of Moe (1075)

—Aquatic reactions: 75-100% disappearance from soils: 8 weeks (1815)
hydrolysis at pH 5-9: half-life: >10,000 days*
direct photolysis at pH 5-9: half life: 121 days**
*calculation based on neutral and alkaline hydrolysis assuming pseudo-first-order kinetics
**minimum direct photolysis half lives calculated by the procedures of Zepp and Cline assuming a quantum efficiency of 1 and for a midsummer day at latitude 40° (1074)
—Waste water treatment:
powdered A.C.: Freundlich adsorption parameters: $K = 0.050$; $1/n = 0.187$
carbon dose to reduce 5 mg/l to a final conc. of 0.1 mg/l: 151 mg/l
carbon dose to reduce 1 mg/l to a final conc. of 0.1 mg/l: 27 mg/l (594)

D. BIOLOGICAL EFFECTS:
—Fish: *Lepomis macrochirus*: 48 hr LC_{50}: 8,000 µg/l (2116)
—Mammals:
acute oral LD_{50} (rats): 3,800 mg/kg (1854)
in diet: no toxic effects were observed when it was fed at dietary levels of 2,000 ppm for 2 years to rats (1855)

isopropyl-*m*-cresol see thymol

isopropylcyanohydrin see acetonecyanohydrin

isopropylether see diisopropylether

isopropylethylene see 3-methyl-1-butene

isopropylethylthionocarbamate
Use: flotation agent
A. PROPERTIES: oil liquid, pale yellow; b.p. decomposes; sp.gr. 1.0; solub. <10%
D. BIOLOGICAL EFFECTS:
—Fish:
rainbow trout: 96 hr LC_{50}: 45-48 mg/l (static bioassay) (947)
—Mammals:
rats: ingestion: single oral LD_{50}: range of 500-1000 mg/kg (1546)

isopropyldimethylmethane see 2,3-dimethylbutane

isopropylglycol see 2-isopropoxyethanol

isopropylglycolether see 2-isopropoxyethanol

isopropylideneacetone see mesityloxide

1-isopropyl-2-methylethylene see 4-methyl-2-pentene

isopropylmethylketone see 3-methyl-2-butanone

6-isopropyl-4-methyl-2-pyrimidinol
D. BIOLOGICAL EFFECTS:
 −Fish:
  bluegill: 96 hr LC_{50}: 1200 mg/l
  daphnids: 48 hr EC_{50}: 1050 mg/l (1591)

isopropyloxitol see 2-isopropoxyethanol

isopropyl-N-phenylcarbamate (IPC; propham; isopropylcarbanilate)

$$\text{C}_6\text{H}_5\text{-NH-C(=O)-O-CH(CH}_3\text{)}_2$$

Use: pesticide
A. PROPERTIES: m.p. 86–88°C; solub. 250 mg/l
C. WATER POLLUTION FACTORS:
 −Biodegradation:
  biodegradation by fungi (*Aspergillus fumigatus*): half-life: 190 days at pH 7*
  biodegradation by bacteria (*Pseudomonas striata*): half-life: 3.2 days at pH 7**
  *assuming 1 mg/l fungi
  **assuming a bacterial population of 0.1 mg/l, calculated from the data of Moe
  (1076)
 −Aquatic reactions:
  75–100% disappearance from soils: 4 weeks (1815)
  comparison of calculated hydrolysis and photolysis under given conditions at pH 5–9: hydrolysis half life: >10,000 days*
      direct photolysis half-life: 254 days**
  *calculation based on neutral and alkaline hydrolysis assuming pseudo-first-order kinetics
  **minimum direct photolysis half lives calculated by the procedures of Zepp and Cline (1977) assuming a quantum efficiency of 1 and for a midsummer day at latitude 40° (1074)
 −Water treatment methods:
  powdered A.C.: freundlich adsorption parameters: K: 0.045
                              $1/n$: 0.467
  carbon dose to reduce 5 mg/l to a final conc. of 0.1 mg/l: 312 mg/l
  carbon dose to reduce 1 mg/l to a final conc. of 0.1 mg/l: 57 mg/l (594)
D. BIOLOGICAL EFFECTS:
 −Crustaceans:

Gammarus lacustris:	96 hr LC_{50}: 10,000 µg/l	(2124)
Gammarus fasciatus:	96 hr LC_{50}: 19,000 µg/l	(2125)
Simocephales serrulatus:	48 hr LC_{50}: 10,000 µg/l	(2127)
Daphnia pulex:	48 hr LC_{50}: 10,000 µg/l	(2127)

 −Mammals:
  acute oral LD_{50} (rat): 5000 mg/kg (1854)

isopropyltrimethylmethane see 2,2,3-trimethylbutane

isopropylxanthate, sodium
Use: exchange flotation agent
A. PROPERTIES: white to yellow powder
D. BIOLOGICAL EFFECTS:
—Fish:
rainbow trout (*Salmo gairdneri*): 96 hr LC_{50}:
Cyanamid C 343: 217 mg/l (static test)
Dow Chemical Z 11: 100–180 mg/l (static test)
(potassium salt): Dow Chemical Z 9: 32–320 mg/l (static test)
LC_{100}: 180 mg/l after 4 days exposure (static test)
0.3 mg/l after 3 days exposure (flow through test)   (1087)
—Mammals:
ingestion: rats: single oral LD_{50}: approx. 1000 mg/kg   (1546)

isoquinoline

Use: manufacture of pharmaceuticals, dyes, insecticides, rubber accelerators and in organic synthesis
A. PROPERTIES: m.w. 129; m.p. 26–28°C; b.p. 242°C; sp.gr. 1.099; log P_{oct} 2.08
D. BIOLOGICAL EFFECTS:
—Crustacean:
Daphnia pulex: bioaccumulation factor: 2.41 (initial conc. in water: 6.2 ppm)
24 hr LC_{50}: 39.9 mg/l
immobilization concentration IC_{50}: 33.6 mg/l   (1050)
—Protozoa:
ciliate (*Tetrahymena pyriformis*): 24 hr LC_{100}: 6.19 mmole/l   (1662)

isovaleraldehyde see 3-methylbutanal

isovaleric acid (3-methylbutanoic acid; isopropylacetic acid)
$(CH_3)_2 CHCH_2 COOH$
A PROPERTIES: colorless liquid; m.w. 102.13; m.p. −37.6°C; b.p. 176.7°C; v.p. 1 mm at 34°C; sp.gr. 0.937 at 15/4°C; solub. 42,000 mg/l at 20°C; P_{oct} 0.93/1.51 (calculated)
B. AIR POLLUTION FACTORS: 1 mg/cu m = 0.236 ppm, 1 ppm = 4.245 mg/cu m
—Odor: hedonic tone: unpleasant

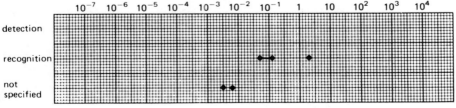

(307; 610; 683; 713; 807)

784 ISOVALERONE

    O.I. at $20°C = 365,500$ (316)
   —Natural occurence: tobacco, valeriana, hop oil   (211)
C. WATER POLLUTION FACTORS:
   —Reduction of amenities: T.O.C. = 5 mg/l   (295)
   —Manmade sources:
     average content in secondary sewage effluent: 73.4 µg/l   (86)
D. BIOLOGICAL EFFECTS:
   —Mammals: rat: single oral dose: LD_{50}: <3.2 g/kg   (211)

isovalerone *see* diisobutylketone

J

juvabiol
D. BIOLOGICAL EFFECTS:
—Fish: rainbow trout: renewal bioassay: 96 hr TLm: 2.0 mg/l at 8°C (449)

juvabione
A hormonelike compound in the wood of balsam fir which prevents insects from developing to the adult stage (1854)
D. BIOLOGICAL EFFECTS:
—Fish: rainbow trout: renewal bioassay: 96 hr TLm: 1.5 mg/l at 8°C (449)

K

karmex *see* 3-(3,4-dichlorophenyl)-1,1-dimethylurea

kathon LP
D. BIOLOGICAL EFFECTS:
 —Bioaccumulation:
  Fish: bluegill: 67 d BCF 165 (whole fish minus viscera) (F)
   67 d BCF 1280 (viscera) (F) (1511)

kelevan
C. WATER POLLUTION FACTORS:
 —Photo-oxidation by U.V. light in aqueous medium at 90–95°C: time for the formation of CO_2 (% of theoretical):
 25%: 1.2 hr
 50%: 9.6 hr
 75%: 19.0 hr (1628)

kelthane (dicofol; 1,1-*bis*(4′-chlorophenyl)2,2,2-trichloroethanol)

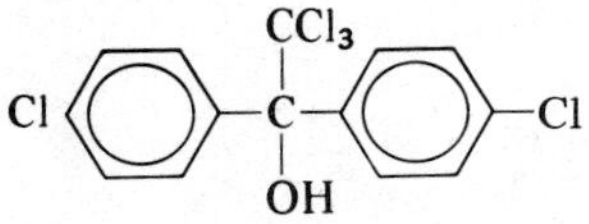

Cl–C₆H₄–C(CCl₃)(OH)–C₆H₄–Cl

Use: miticidal pesticide
A. PROPERTIES: solub. 1.2 mg/l at 24°C (99% purity), 0.8 mg/l at 20°C in distilled water (66)
C. WATER POLLUTION FACTORS:
 —Aquatic reactions:
  degradation in anaerobic sewage sludge to DBP (4,4′-dichlorobenzophenone) (1432)

  hydrolysis to DBP at:
   pH 8.2: half-life: 60 min; initial conc.: 0.4 mg/l
   pH 10.2: half-life: 3 min; initial conc.: 0.4 mg/l (1431)
  conversion of soluble kelthane to DBP in river water
   pH 7.5; duration of experiment 24 hr

river water	% conversion to DBP	% kelthane recovery
filtered	94	60

filtered	88	28	
unfiltered	58	36	
unfiltered	47	43	(1431)

D. BIOLOGICAL EFFECTS:
  —Crustacean:
  decapod (*Crangon franciscorum*): adult; 48 hr: 89–843 ppb; lethal concentration decreased with increasing temperature (1153)
  grass shrimp (*Crangon franciscorum*):
   24 hr LC_{50}: 1.29 ppm; 95% conf. lim.: 0.78–2.14 ppm
   48 hr LC_{50}: 0.59 ppm; 95% conf. lim.: 0.44–0.83 ppm
   100 hr LC_{50}: 0.1 ppm (581)
  —Mammals:
  acute oral LD_{50} (rat): 684–809 mg/kg (techn. product) (1854)
  acute dermal LD_{50} (rabbit): 1870 mg/kg
  in diet: dogs were fed 1 year on a diet containing 300 ppm with no evidence of toxicity (1855)
  Under the conditions of this bioassay, technical-grade dicofol was carcinogenic in male B6C3F1 mice, causing hepatocellular carcinomas. No evidence for carcinogenicity was obtained for this compound in Osborne-Mendel rats of either sex or in female B6C3F1 mice. (1753)

kepone (chlordecone; decachlorooctahydro-1,3,4-metheno-2H-cyclobuta-(*c*,*d*)-pentalen-2-one; GC compound 1189)

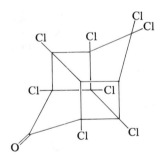

Use: pesticide for control of the banana root borer, tobacco wireworm; bait for control of ants and cockroaches is converted to kelevan.

Highly stable odorless and colorless solid, which sublimes without melting and with some decomposition or rapid sublimation at about 350°C. At atmospheric pressure v.p. $<3 \times 10^{-7}$ mm; solub. at pH 4–9 = 2–4 ppm. At ordinary temperature and humidity it readily forms hydrates and is normally used as a mono- to trihydrate.
Technical grade kepone contains unreacted hexachlorocyclopentadiene, generally in concentrations of less than 0.5%

A. PROPERTIES: solub. 7.6 mg/l at 24°C (99% purity) (1666)
C. WATER POLLUTION FACTORS:
  —Persistence:
  disappearance of kepone after 5 months in soil is minimal (1363)

D. BIOLOGICAL EFFECTS:
 —Bioaccumulation:
  giant cordgrass (*Spartina cynosuroides*) detritus sorbs kepone from contaminated brackish waters during decomposition. Concentrations of kepone on detritus increased from 0.5 μg/g dry wt after 14 days of decomposition to 4.5 μg/g dry wt after 119 days (1820)
  —Algae: *Chlorococcum* sp.:   100 μg/l, 24 hr: 800 ×
         Dunaliella tertiolecta: 100 μg/l, 24 hr: 230 ×
         Nitzschia sp.:   100 μg/l, 24 hr: 410 ×
         Thalassiosira pseudonana: 100 μg/l, 24 hr: 520 ×   (1139)
  —Decapods: *Palaemonetes pugio*: 96 hr at 12–121 μg/l: 698 ×   (1140)
            Callinectes sapidus: 96 hr at 110–210 μg/l: 8.1 ×
  —Fish: *Cyprinodon variegatus*: 96 hr at 7.1–78.5 μg/l: 1 548 ×
        Leiostomus xanthurus: 96 hr at 1.5–15.9 μg/l: 1 211 ×   (1160)
 —Food chains:

algae to pelecypod	food chain: *Chlorococcum* sp. to *Crassostrea virginica*	oyster fed algae containing 34 μg/g for 14 days.	0.21 μg/g
mysid to fish	food chain: *Mysidopsis baha* to *Leiostomus xanthurus*	spot fed mysids containing 1.03 μg/g for 30 days.	1.1 μg/g (1166)

 —Toxicity:
  natural phytoplankton: at 1.0 ppm: ^{14}C uptake reduced by 95% within 4 hr
 —Algae:
  Chlorococcum sp.:   7 day EC_{50} (S): 350 μg/l
  Dunaliella tertiolecta: 7 day EC_{50} (S): 580 μg/l
  Nitzshia sp.:   7 day EC_{50} (S): 600 μg/l
  Thalassiosira pseudomana: 7 day EC_{50} (S): 600 μg/l   (1139)
 —Crustacean:
  decapods: *Palaemonetes pugio*: 96 hr LC_{50} (FT): 120.9 μg/l
           Callinectes sapidus: 96 hr LC_{50} (FT): <210 μg/l   (1140)
  fiddler crab: 24 hr LC_{50}: >1.6 mg/l
                96 hr LC_{50}: 1.47 mg/l
                no effect level: 0.32 mg/l
 —Fish:
  Cyprinodon variegatus: 96 hr LC_{50} (FT): 69.5 μg/l
  Leiostomus xanthurus: 96 hr LC_{50} (FT): 6.6 μg/l
  Cyprinodon variegatus: 28 day, 22%; mortality: 0.80 μg/l
                           28 day, 80%; mortality: 1.9 μg/l   (1141)
  Cyprinodon variegatus adult: 4 μg/l, 10 days: scoliosis, black-tail, loss of equilibrium, sporadic hyperkinesis, and tetanic convulsions. (1159)
  Cyprinodon variegatus fry (from exposed embryo): 6.6 and 33 μg/l, 36 days: diminished activity; loss of equilibrium; cessation of feeding; emaciation.
  (1141)

LC_{50} at specified hours (ppm)

	24 hr	48 hr	96 hr	no effect
sunfish	0.62	0.27	0.14	
trout	0.066	0.038	0.02	

bluegill 0.257 0.051 0.024
rainbow 0.156 0.036 0.04
—Birds:
quail: chronic oral LD_{50}: 500 mg/kg
pheasants: chronic oral LD_{50}: 1000 mg/kg
normal survival: <200 mg/kg (1358)
female chicken: acute oral LD_{50}: 480 mg/kg (1359)
young bobwhite quail: dietary LC_{50}: 600 ppm
adult bobwhite quail: dietary LC_{50}: 530 ppm
young ring-necked pheasants: dietary LC_{50}: 606 ppm
adult ring-necked pheasants: dietary LC_{50}: 115 ppm
young mallards: dietary LC_{50}: 400 ppm
quail: feeding at 50 ppm inhibited reproduction (1360)
—Mammals:
rats: acute toxicity:

dosage mg/kg	no. rats killed	survival times
500	2/2	12 hr/36 hr
250	2/2	28 hr/5 days
100	0/2	all animals still alive at
75	0/2	the end of 10 days
50	0/2	(1361)

rabbit: single oral dosage of 50 mg/kg: mortality 0/10 after 72 hr (1361)
results of acute toxicity tests (based on Allied data):

			LD_{50} in mg/kg (no. of animals)		range of days
species	route	solvent	males	females	to death
rats	oral	oil	132 (40)	126 (40)	2–7
rats	oral	water	96 (40)		1–9
rabbits	oral	oil	71 (40)		2–6
rabbits	oral	water	65 (40)		1–10
rabbits	percutaneous	oil	410 (40)		3–13
rabbits	percutaneous	water	435 (40)		1–30
dogs	oral	oil	ca. 250 (16)		2–8

(1353)
male and female B6C3F1 mice: carcinogenic under test conditions (1362)

α-ketoglutaric acid
C. WATER POLLUTION FACTORS:
—Waste water treatment:
A.S.: after 6 hr: 2.1% of ThOD
12 hr: 3.1% of ThOD
24 hr: 5.9% of ThOD (89)

ketohexamethylene *see* cyclohexanone

konsin
—Composition: diethyleneglycol (45%); triethyleneglycol (55%)

KURON

D. BIOLOGICAL EFFECTS:
 —Fish:

	mg/l	24 hr	48 hr	96 hr	3 m (extrap.)
harlequin fish	LC_{10} (F)	470	370	300	
(*Rasbora heteromorpha*)	LC_{50} (F)	1000	820	600	100
					(331)

kuron *see* 2(2,4,5-trichlorophenoxy)propionic acid

L

dl-lactic acid (2-hydroxypropanoic acid)
CH$_3$CHOHCOOH
A. PROPERTIES: colorless syrup; m.w. 90.08; m.p. 16/18°C; b.p. 122°C at 15 mm; sp.gr. 1.249 at 15°C; log P_{oct} -0.62
B. AIR POLLUTION FACTORS:
   —Odor threshold conc.: 9 mg/cu m   (652)
C. WATER POLLUTION FACTORS:
   —BOD$_5$: 0.63 Warburg, sewage   (282, 163)
            0.64 std. dil. sp. cult.   (283, 162)
            22% of ThOD   (220)
   —BOD$_{10}$: 0.88 std. dil. sp. cult.   (283, 162)
   —COD: 100% of ThOD
   —KMnO$_4$: acid: 25% of ThOD; alkaline: 10% of ThOD   (220)
   —Manmade sources:
     excreted by man: in urine: 40 mg/kg body wt/day
                       in sweat: 45-452 mg/100 ml   (203)
   —Waste water treatment:
     A.S.: after 6 hr: 27.5% of ThOD
               12 hr: 29.4% of ThOD
               24 hr: 33.3% of ThOD   (89)
     A.S., Resp, BOD, 20°C, 1-5 days observed, feed: 720 mg/l, acclimation: <1 day
     69% theor. oxidation   (93)
D. BIOLOGICAL EFFECTS:
   —Protozoa: *Vorticella campanula*: perturbation: 200 mg/l
   —Arthropoda: *Daphnia*: LD$_0$: 170 mg/l, 26-72 hr   (30)
   —Molluscs: *Limnea ovata*: perturbation: 50 mg/l
   —Insects: *Sialis flavilatera*: perturbation: 1,500 mg/l
   —Fish: *Squalis leuciscus*: TLm (? hr): 1,000 mg/l
         Salmo irideus: TLm (? hr): 400 mg/l
         trout: TLm 18 hr: 100 mg/l
         goldfish: period of survival: 6-43 hr: 654 ppm; pH = 4.0
                            days: 430 ppm; pH = 4.6   (157)
   —Mammalia:
     guinea pig: single oral LD$_{50}$: 1.81 g/kg
     rat: single oral LD$_{50}$: 3.73 g/kg
         repeated oral doses: weight loss, anemia: 1.5 g/kg/day   (211)

lactonitrile (2-hydroxypropanenitrile; acetaldehydecyanohydrin; ethylidenecyanohydrin)
$CH_3CHOHCN$
A. PROPERTIES: colorless or strawcolored liquid; m.w. 71.08; m.p. $-40°C$; b.p. $183°C$ slightly decomposes; v.p. 10 mm at $74°C$; v.d. 2.45; sp.gr. 0.992
C. WATER POLLUTION FACTORS:
—Waste water treatment:

methods	temp °C	days observed	feed mg/l	days acclim.	% theor. oxidation	% removed
RW, CO_2, CAN	20	8	10	17	60	100
RW, CO_2, CAN	20	13	50	30+	60	100
RW, CO_2, CAN	5	35	50	30+ at $20°C$	60	100
ASC, CAN	22–25	28	88	28	75+	95+ (93)

D. BIOLOGICAL EFFECTS:
—Fish: pinperch: TLm (24 hr): 0.215 mg/l, in sea water (248)
fathead minnow: TLm (96 hr): 0.9 mg/l (41)
bluegill: TLm (24–96 hr): 4.0–0.90 mg/l (41)
guppies: TLm (96 hr): 1.37 mg/l (41)
—Mammalia: rat: acute oral LD_{50}: 21 mg/kg (211)

landrin
D. BIOLOGICAL EFFECTS:
—Insects:
fourth instar larval *Chironomus riparius*: 24 hr LC_{50}: 51.4 ppb (1853)

LAS *see* linear alkyl sulfonates

lauraldehyde *see* dodecanal

lauric acid (dodecanoic acid)
$CH_3(CH_2)_{10}COOH$
A fatty acid occurring in many vegetable fats as the glyceride, especially in coconut oil and laurel oil
Use: alkyd resins; wetting agents; soaps; detergents; cosmetics; insecticides; food additives
A. PROPERTIES: colorless solid; m.w. 200.31; m.p. $44°C$; b.p. $225°C$ at 100 mm; v.p. 1 mm at $211°C$; sp.gr. 0.871 at $50°C$; log P_{oct} 4.20
B. AIR POLLUTION FACTORS:
—Odor: characteristic: quality: like oil of bay
T.O.C. = 0.028 mg/cu m = 3.4 ppb (307)
detection: 0.1 mg/cu m (778; 779)
recognition: 0.004–0.005 mg/cu m (610)

—Ambient air quality:
  glc's: Detroit freeway interchange Nov. 1963: 13.7 µg/1,000 cu m
    New York high-traffic city location: 39.4 µg/1,000 cu m (100)
  organic fraction of suspended matter:
    Bolivia at 5200 m altitude (Sept.–Dec. 1975): 1.4–1.7 µg/1,000 cu m
    residential area, 10 km south of Antwerp, Belgium (Jan.–April 1976): 4.5–8.4
    µg/1,000 cu m (428)
  glc's in residential area (Belgium) oct. 1976:
    in particulate phase: 0.01 ng/cu m
    in gas phase: 30.3 ng/cu m (1289)
C. WATER POLLUTION FACTORS:
  —Waste water treatment:
    A.S. after 6 hr: 4.1% of ThOD
           12 hr: 4.3% of ThOD
           24 hr: 6.1% of ThOD (89)
D. BIOLOGICAL EFFECTS:
  —Mammals: rat: oral ingestion: no effect: 35% in diet for 2 years (211)

laurylalcohol *see* 1-dodecanol

laurylguanidineacetate *see* n-dodecylguanidineacetate

laurylsulfate, sodium (dodecylsulfate, sodium)
$CH_3(CH_2)_{11}OSO_3Na$
  Use: wetting agent in textiles; detergent in toothpaste; food additive and surfactant
A. PROPERTIES: small white or yellow crystals, m.w. 288.38; m.p. 204–207°C; log P_{oct} 1.60
C. WATER POLLUTION FACTORS:
  —Biodegradation:
    at 17 mg/l (inoculum: sew.): no degradation after 30 days, (no salt mentioned)
    (488)
D. BIOLOGICAL EFFECTS:
  —Toxicity threshold (cell multiplication inhibition test):
    bacteria (*Pseudomonas putida*):   290 mg/l   (1900)
    green algae (*Scenedesmus quadricauda*): 0.02 mg/l   (1900)
    protozoa (*Entosiphon sulcatum*):   40 mg/l   (1900)
    protozoa (*Uronema parduczi Chatton-Lwoff*): 0.75 mg/l (1901)
  —Fish:

	48 hr	LC_{50} (FT) 96 hr	10 d
trout	5.95 mg/l	4.62 mg/l	2.85 mg/l at 15°C
zebrafish (*Brachydario rerio*)	8.81 mg/l	7.97 mg/l	7.97 mg/l at 25°C
flagfish (*Gordanella floridae*)	10.0 mg/l	8.10 mg/l	6.90 mg/l at 25°C

(479)

794   LEADACETATE

Brown trout: median survival time (MST)   (1208)
                0.07-hr 1 000 mg/l
                0.07-hr 560 mg/l
                0.08-hr 320 mg/l
                0.15-hr 180 mg/l
                0.26-hr 150 mg/l
                0.86-hr 120 mg/l
                2.15-hr 100 mg/l
                6.5-hr   56 mg/l
                32-hr   32 mg/l
                45-hr   18 mg/l   (1208)

leadacetate   (sugar of lead)
$(CH_3COO)_2Pb \cdot 3H_2O$

Use and formulation: medicine; dyeing of textiles; waterproofing; varnishes; lead driers; chrome pigments; insecticides; antifouling paints

A. PROPERTIES: white crystals; commercial grades are frequently brown or gray lumps; m.w. 379.35; m.p. 75°C; sp.gr. 2.55

D. BIOLOGICAL EFFECTS:
— Threshold conc. of cell multiplication inhibition of the protozoan *Uronema parduczi Chatton-Lwoff*: 0.07 mg/l   (1901)
— Bacteria:
Pseudomonas putida: inhibition of cell multiplication starts at 1.8 mg/l   (329)
— Algae:
Microcystis aeruginosa: inhibition of cell multiplication starts at 0.45 mg/l (329)

leptophos (phosvel; MBCP; O-(4-bromo-2,5-dichlorophenyl)O-methylphenylphosphonothioate

A. PROPERTIES: m.p. 70°C; sp.gr. 1.53 at 25°C; solub. 0.03 ppm; solub. 0.0047 ppm at 20°C; log P_{oct} 6.31 at 20°C

C. WATER POLLUTION FACTORS:
— Photodegradation of thin dry film:

	by U.V. after 24 hr	by sunlight after 65 days
leptophos	47%	37% of initial leptophos conc.
leptophos phenol	25%	12%
leptophos oxon	9.2%	12%
unknown compounds	5.3%	3.6% (1669)

amount of leptophos remaining in different types of water in closed glass containers after 16 weeks at 25°C (initial conc. 1.18 mg/l):
    in distilled water: 39.0% of initial conc.
    in tap water: 2.8%

	in artesian water:	12.4%	
	in irrigating water:	62.4%	
	in draining water:	105.3%	
	in river Nile water:	43.3%	(1668)

D. BIOLOGICAL EFFECTS:
 —Fish: Chingatta: 96 hr LC_{50} (S): 3.08–31.2 mg/l (1519)
 —Mammals:
  acute oral LD_{50} (male rat): 52.8 mg/kg
  acute dermal LD_{50} (male rat): >10,000 mg/kg (1854)

l-leucin (l-2-amino-4-methylpentanoic acid; l-α-aminoisocaproic acid)

$$\begin{array}{c} COOH \\ | \\ CH-NH_2 \\ | \\ CH_2 \\ | \\ CH \\ / \quad \backslash \\ CH_3 \quad CH_3 \end{array}$$

A. PROPERTIES: m.w. 131.17; m.p. 295°C decomposes; b.p. sublimes; sp.gr. 1.293 at 18/4°C; solub. 24,300 mg/l at 25°C, 38,200 mg/l at 75°C, log P_{oct} −1.71

C. WATER POLLUTION FACTORS:
 —Waste water treatment:
  A.S.: after 6 hr: 1.4% of ThOD
        12 hr: 3.6% of ThOD
        24 hr: 9.9% of ThOD
  A.S., BOD, 20°C, 1–5 days observed, feed: 333 mg/l, acclimation: 15 days: 99% removed (93)
 —Manmade sources:
  excreted by man: in urine: 0.2–0.52 mg/kg body wt/day
                   in feces: 4.3–6.9 mg/kg body wt/day
                   in sweat: 1.2–4.2 mg/100 ml (203)

levopimaric acid *see* pimaric acid

light water brand AFFF 6% concentrate
 Use: synthetic, foam forming liquid designed for use with seawater, brackish water, or fresh water—may be used for flammable fuel fire prevention and extinguishment (505)

A. PROPERTIES: sp.gr. 1.02; f.p. 23°F; solub. miscible with fresh water and seawater in all proportions

C. WATER POLLUTION FACTORS:
 —BOD_5 of 6% concentrate: 210,000 mg/l
 —BOD_{ult} of 6% concentrate: 420,000 mg/l
 —COD of 6% concentrate: 420,000 mg/l (505)

D. BIOLOGICAL EFFECTS:
 —No microbial inhibition of 6% concentrate concentrations <1,000 mg/l (505)

—Crustaceans:
  water-flea (*Daphnia magna*): 48 hr LC_{50}: 5850 mg 6% concentrate/l
  scud (*Gammarus fasciatus*): 48 hr LC_{50}: 5170 mg 6% concentrate/l
  grass shrimp (*Palaemonetes vulgaris*): 96 hr LC_{50}: 280 mg 6% concentrate/l; static test
  fiddler crab (*Uca pugilator*): 96 hr LC_{50}: 3260 mg 6% concentrate/l; static test
  (505)
—Mussels:
  atlantic oyster larvae (*Crassostrea virginica*): 48 hr LC_{50}: >100, <240 mg 6% concentrate/l
  (505)
—Fish:
  fathead minnows (*Pimephales promelas*): 96 hr LC_{50}: 3000 mg 6% concentrate/l; continuous flow test
  rainbow trout (*Salmo gairdneri*): 96 hr LC_{50}: 1800 mg 6% concentrate/l; static test
  (505)

lindane see hexachlorocyclohexane

linear alkyl sulfonates (LAS)
A straight-chain alkylbenzene sulfonate
D. BIOLOGICAL EFFECTS:

	species	results		exposure type	temp. (°C)	test conditions	
0% bio-degraded	bluegill	96-hr LC_{50}	0.72 mg/l	S	20.8	ALK DO HD pH	40.0 5.8 54.0 7.2
36.7% bio-degraded	bluegill	96-hr LC_{50}	0.89 mg/l	S	20.5	ALK DO HD pH	39.5 6.2 62.0 7.4
53.3% bio-degraded	bluegill	96-hr LC_{50}	1.16 mg/l	S	20.6	ALK DO HD pH	40.0 6.2 49.6 7.3
76.0% bio-degraded	bluegill	96-hr LC_{50}	1.64 mg/l	S	21.0	ALK DO HD pH	58.0 4.9 61.0 7.4
							(1190)
commercial	fathead minnow	24-hr LC_{50} 48-hr LC_{50}	1.9 mg/l 1.7 mg/l	S S	— —	HD	100
homolog C_{10}	fathead minnow	24-hr LC_{50} 48-hr LC_{50}	48.0 mg/l 43.0 mg/l	S S	— —	HD	100
homolog C_{11}	fathead minnow	24-hr LC_{50} 48-hr LC_{50}	17.0 mg/l 16.0 mg/l	S S	— —	HD	100
homolog C_{12}	fathead minnow	24-hr LC_{50} 48-hr LC_{50}	4.7 mg/l 4.7 mg/l	S S	— —	HD	100
homolog C_{13}	fathead minnow	24-hr LC_{50} 48-hr LC_{50}	1.7 mg/l 0.4 mg/l	S S	— —	HD	100

	species	results		exposure type	temp. (°C)	test conditions	
homolog C_{14}	fathead	24-hr LC_{50}	0.6 mg/l	S	–	HD	100
	minnow	48-hr LC_{50}	0.4 mg/l	S	–		
model intermediate C_4	fathead	24-hr LC_{50}	10 000 mg/l	S	–	HD	100
	minnow	48-hr LC_{50}	10 000 mg/l	S	–		
model intermediate C_5	fathead	24-hr LC_{50}	10 000 mg/l	S	–	HD	100
	minnow	48-hr LC_{50}	10 000 mg/l	S	–		
model intermediate C_{11}	fathead	24-hr LC_{50}	85.9 mg/l	S	–	HD	100
	minnow	48-hr LC_{50}	76.6 mg/l	S	–		(1191)

linevol (neoflex; linear plasticizer alcohols)
  Use: manufacturing of plasticizers for PVC
A. PROPERTIES:
  —Linevol 79: composition: C_7: 47%; C_8: 36%; C_9: 17%
    sp.gr. 0.827–0.833; boiling range 180–214°C; solub. 800 mg/l at 20°C
  —Linevol 911: composition: C_9: 19%; C_{10}: 48%; C_{11}: 33%
    sp.gr. 0.833–0.839; boiling range 225–248°C; solub. 300 mg/l at 20°C
C. WATER POLLUTION FACTORS:

	linevol 79	linevol 911
—BOD_5	2.28 = 78% of ThOD	2.13 = 70% of ThOD
—COD	2.80 = 96% of ThOD	2.69 = 89% of ThOD

D. BIOLOGICAL EFFECTS:
  —Fish:
    Carassius auratus: linevol 79: 24 hr LD_{50}: 9 mg/l
              linevol 911: 24 hr LD_{50}: 3 mg/l
  —Mammals:
    rat: oral LD_{50}: >20 ml/kg = >16,000 mg/kg

lithic acid *see* uric acid

2,6-lutidine *see* 2,6-dimethylpyridine

l-lysine (d-2,6-diaminohexanoic acid; d-α,ϵ-diaminocaproic acid)

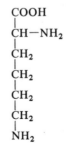

Use: biochemical and nutritional research; pharmaceuticals; culture media
A. PROPERTIES: m.w. 146.19; m.p. 224°C decomposes; log P_{oct} −2.82
C. WATER POLLUTION FACTORS:
  —Waste water treatment:

A.S.: after 6 hr: 1.8% of ThOD
        12 hr: 4.5% of ThOD
        24 hr: 14.1% of ThOD (89)
—Manmade sources:
excreted by man: in urine: 0.48–2 mg/kg body wt/day
                 in feces: 4.5–6.9 mg/kg body wt/day
                 in sweat: 1.4–3.2 mg/100 ml (203)
—Impact on biodegradation:
∼50% inhibition of NH_3 oxidation in *Nitrosomonas* at 4 mg/l (*l*-isomer)   (408)

M

mac *see* methallylchloride

magniflox
D. BIOLOGICAL EFFECTS:
—Fish:
magniflox 512 C:
rainbow trout: static bioassay: 96 hr TLm: 8.70 mg/l at 14–16°C
flow through: 14 d TLm: 1.10 mg/l at 15–17°C (454)
magniflox 905 V:
rainbow trout: static bioassay: 96 hr TLm: >8.0 mg/l at 14–16°C (454)

malaoxon
D. BIOLOGICAL EFFECTS:
—Insects:
fourth instar larval *Chironomus riparius*: 24 hr LC_{50}: 5.4 ppb (1853)

malathion (mercaptothion; carbofos; S-(1,2-*bis*(ethoxycarbonyl)ethyl 0,0-dimethylphosphorodithioate)

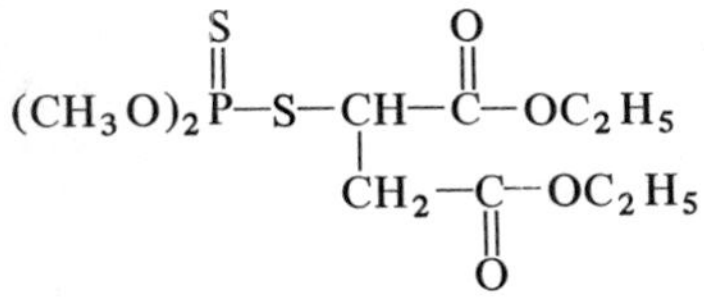

$$(CH_3O)_2\overset{\underset{\|}{S}}{P}-S-\underset{\underset{\underset{O}{\|}}{CH_2-C-OC_2H_5}}{CH}-\overset{\underset{\|}{O}}{C}-OC_2H_5$$

Use: organophosphate insecticide

A. PROPERTIES: solub. 145 ppm at 20°C; log P_{oct} 2.89 at 20°C; m.w. 330.36; b.p. 156–157°C at 0.7 mm; m.p. 2.85°C; sp.gr. 1.2315 at 25°C; v.p. 4.10^{-5} mm at 20°C

C. WATER POLLUTION FACTORS:
—Biodegradation:
biodegradation pathways: malathion is rapidly degraded in vitro by salt-marsh bacteria to malathionmonocarboxylic acid, malathiondicarboxylic acid, and various phosphothionates as a result of carboxyesterase cleavage. In addition, some expected phosphatase activity produces desmethylmalathion, phosphomono- and dithionates, 4-carbon dicarboxylic acids, and the corresponding ethylesters (1453)

—Aquatic reactions:
  hydrolysis:
    between pH 5 and pH 7: no hydrolysis found after 12 days (1447)
    at pH 8: half-life: 36 hr (1447)
    at pH 9: half-life: 12 days (1452)
    at pH 12: instantaneous hydrolysis (1447)
  persistence in river water in a sealed glass jar under sunlight and artificial fluorescent light - initial conc. 10 µg/l:

after	1 hr	1 wk	2 wk	4 wk	8 wk	
			% of original compound found			
	100	25	10	0	0	(1309)

  75–100% disappearance from soils: 1 week (1815)
—Water and sediment quality:
  Mississippi river: residues not found in bottom sediments (1170)
—Odor threshold: detection: 1.0 mg/kg (915)

D. BIOLOGICAL EFFECTS:
—Bioaccumulation:
  Lagodon rhomboides (pinfish): no bioaccumulation when exposed for 24 hr to 0.075 mg/l (1454)
—Toxicity:
—Insects:

	µg/l	LC_{50}	µg/l	no effect level µg/l	
Pteronarcys californica	10	96 hr			(2128)
Pteronarcys dorsata			11.1 (30 day LC_{50})	9.4–30 day	(2134)
Acroneuria lycorias	1.0		0.3 (30 day LC_{50})	0.17–30 day	(2134)
Pteronarcella badia	1.1	96 hr			(2128)
Classenia sabulosa	2.8	96 hr			(2128)
Boyeria vinosa			2.3 (30 day LC_{50})	1.65–30 day	(2134)
Ophiogomphus rupinsulensis			0.52 (30 day LC_{50})	0.28–30 day	(2134)
Hydropsyche bettoni			0.34 (30 day LC_{50})	0.24–30 day	(2134)

  fourth instar larval *Chironomus riparius*: 24 hr LC_{50}: 1.9 ppb (1853)
—Crustaceans:

	µg/l	LC_{50}	µg/l	no effect level µg/l	
Gammarus pseudolimneus			0.023 (30 day LC_{50})	0.008–30 day	(2134)
Gammarus lacustris	1.0	96 hr			(2126)
Gammarus fasciatus	0.76	96 hr	0.5 (120 hour LC_{50})		(2126)
Palaemonetes kadiakensis	12	96 hr	9.0 (120 hour LC_{50})		(2126)
Orconectes nais	180	96 hr			(2126)
Asellus brevicaudus	3000	96 hr			(2126)
Simocephalus serrulatus	3.5	48 hr			(2127)
Daphnia pulex	1.8	48 hr			(2127)
Daphnia magna				0.6–21 day	(2135)

	ppb	measure	method	
Tetrahymena pyriformis	10,000	8.8 % decrease in a population measured as absorbance at 540 mµ	96 hr growth test in *Tetrahymena* broth	(2350)

Palaemon macrodactylus (Korean shrimp)	81.5 (19.6–26.1)	TL$_{50}$	96 hr static lab bioassay (2353)
Palaemon macrodactylus (Korean shrimp)	33.7 (21.3–53.1)	TL$_{50}$	96 hr intermittent flow lab bioassay (2353)
Crangon septemspinosa (sand shrimp)	33	LC$_{50}$	96 hr static lab bioassay (2327)
Palaemonetes vulgaris (grass shrimp)	82	LC$_{50}$	96 hr static lab bioassay
Pagurus longicarpus (hermit crab)	83	LC$_{50}$	96 hr static lab bioassay (2327)

Daphnia magna: 24 hr LC$_{50}$: 0.9 µg/l
 48 hr LC$_{50}$: 1.8 µg/l (1002; 1004)
 1 w LC$_{50}$: 0.003 mg/l (1646)

—Mussels
 fresh water mussel larva (*Glochidium*): 0.0001 mg/l - no effect level on shell closing activity (1645)
Crassostrea virginica (American oyster): egg: static lab bioassay:
 48 hr TLm: 9070 ppb
Crassostrea virginica (American oyster): larvae: static lab bioassay:
 14 d TLm: 2660 ppb
Anodonta cygnea: 24 hr/10 mg l: no significant reduction of periodical activity
 24 hr/1 mg l: no significant change in biological activity

—Fish:

		ppb	
100%	*Fundulus heteroclitus* (mummichog)	70	96 hr LC$_{50}$ (S) (2328)
100%	*Fundulus heteroclitus* (mummichog)	80	96 hr LC$_{50}$ (S) (2329)
100%	*Fundulus majalis* (striped killifish)	250	96 hr LC$_{50}$ (S)
100%	*Menidia menidia* (Atlantic silverside)	125	96 hr LC$_{50}$ (S)
100%	*Mugil cephalus* (striped mullet)	550	96 hr LC$_{50}$ (S)
100%	*Thalassoma bifasciatum* (bluehead)	27	96 hr LC$_{50}$ (S)
100%	*Anguilla rostrata* (American eel)	82	96 hr LC$_{50}$ (S)
100%	*Sphaeroides maculatus* (northern puffer)	3250	96 hr LC$_{50}$ (S) (2329)
57%	*Gasterosteus aculeatus* (threespine stickleback)	76.9	96 hr TLm (S) (2333)

eastern mudminnow: static bioassay: 96 hr TLm: 0.24 mg/l at 16°C
 flow through bioassay: 14 d TLm: 0.14 mg/l at 16°C (450)
guppies (*Lebistes reticulatus*): 1 w LC$_{50}$: 0.819 mg/l (1645)

	µg/l	LC$_{50}$			
Pimephales promelas	9000	96 hr	580 (spinal deformity 10 month)	2–10 month exposure	(2122)
Lepomis macrochirus	110	96 hr	7.4 (spinal deformity several months)	3.6–11 months	(2111)
Lepomis cyanellus	120	96 hr			(2123)

Lepomis microlophus	170	96 hr	(2121)
Micropterus salmoides	285	96 hr	
Salmo gairdneri	170	96 hr	
Salmo trutta	200	96 hr	
Oncorhynchus kisutch	101	96 hr	
Perca flavescens	263	96 hr	
Ictalurus punctatus	8970	96 hr	
Ictalurus melas	12900	96 hr	

Salmo gairdneri: 96 hr LC_{50}: 170 μg/l (1001)
bluegill: 24 hr LC_{50}: 0.120 ppm
rainbow trout: 24 hr LC_{50}: 0.100 ppm (1878)

	96 hr LC_{50}, mg/l		96 hr LC_{50}, mg/l
catfish	9.0	bluegill	0.10
bullhead	12.9	bass	0.29
goldfish	10.7	rainbow	0.17
minnow	8.7	brown	0.20
carp	6.6	coho	0.10
sunfish	0.17	perch	0.26

(1934)

striped bass: 96-hr LC_{50}: 0.039 mg/l (S)
banded killifish: 96-hr LC_{50}: 0.24 mg/l (S)
pumpkinseed: 96-hr LC_{50}: 0.48 mg/l (S)
white perch: 96-hr LC_{50}: 1.1 mg/l (S)
American eel: 96-hr LC_{50}: 0.50 mg/l (S)
carp: 96-hr LC_{50}: 1.9 mg/l (S)
guppy: 96-hr LC_{50}: 1.2 mg/l (S)
Cyprinodon variegatus: 96 hr LC_{50} (FT): 51 μg/l (1131)
flagfish: 96 hr LC_{50} (F): 349 μg/l
        110 d MATC (F): 8.6–11 μg/l (1518)
chingatta: 96 hr LC_{50} (S): 7.60 mg/l
        96 hr LC_{50} (S): 7.35 mg/l
        96 hr LC_{50} (S): 7.05 mg/l
        96 hr LC_{50} (S): 6.95 mg/l (1519)
Cyprinodon variegatus, embryo: 3 and 10 ppm—skeletal malformations associated with behavioral changes (1160)

—Mammals:
buffalo calves (6–9 months) (*Bubalus bubalis*): no effect level: 0.5 mg/kg/day; the administration of 1.0 to 1.5 mg/kg/day produced, however, significant changes in all biochemical parameters studied (1336)
rat: no effect level: 5 mg/kg/day (1337)
man: no effect level: 0.2 mg/kg/day (1337)
rats: LD_{50}: between 480 and 5800 mg/kg
        avg. 1500 mg/kg (1647; 1648; 1649)
acute oral LD_{50} (rat): 1375 mg/kg
acute dermal LD_{50} (rabbit): 4100 mg/kg
in diet: rats fed for 104 weeks at levels as high as 5000 ppm with no gross effects (1854)

It is concluded that under the conditions of this bioassay, there was no clear

evidence of the association of the tumor incidence with the administration of malathion to Osborne-Mendel rats or B6C3F1 mice. (1772)
—Carcinogenicity: negative
—Mutagenicity in the *Salmonella* test: negative
  <0.05 revertant colonies/nmol
  <70 revertant colonies at 500 µg/plate (1883)

maleic acid ((*cis*)-1,2-ethylenedicarboxylic acid; (*cis*)-butenedioic acid)

$$\begin{array}{c} CH-COOH \\ \| \\ CH-COOH \end{array}$$

Use: organic synthesis; dyeing and finishing of cotton, wool, and silk; preservative for oils and fats

A. PROPERTIES: m.w. 116.07; m.p. 133/138°C; b.p. 135°C decomposes; solub. 788 g/l at 25°C, 3,926 g/l at 97.5°C; sp.gr. 1.590 at 20/4°C; log P_{oct} -0.79/-0.32 (calculated)

C. WATER POLLUTION FACTORS:
—BOD_5: 0.38 std. dil. (27)
    0.38 Warburg, sewage (164; 259)
    0.63 std. dil acclim. (41)
    0.64 (36)
    21% ThOD (220)
—BOD_5^{20}: 0.64 at 10 mg/l, unadapted sew.; lag period: 1 day
—BOD_{20}^{20}: 0.76 at 10 mg/l, unadapted sew.; lag period: 1 day (554)
—COD: 0.80; 0.83; 0.93 (36)(27)(41)
    100% of ThOD (220)
—$KMnO_4$: acid 90% ThOD
    alkal 89% ThOD (220)
—ThOD: 0.83 (36)
—Reduction of amenities: organoleptic limit USSR 1970: 1.0 mg/l (181)
—Waste water treatment: A.S. after 6 hr: 2.7% of ThOD
            12 hr: 4.5% of ThOD
            24 hr: 2.7% of ThOD (89)

maleic anhydride ((*cis*)-butenedioic anhydride; 2,5-furandione; toxilic anhydride)

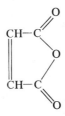

A. PROPERTIES: white needles or flakes; m.w. 98.06; m.p. 53°C; b.p. 199.7/202°C sublimes; v.p. 0.00005 mm at 20°C, 0.0002 mm at 30°C; v.d. 3.38; sat. conc. 0.0003 g/cu m at 20°C, 0.001 g/cu m at 30°C; sp.gr. 0.934 at 20/4°C

B. AIR POLLUTION FACTORS: 1 mg/cu m = 0.25 ppm, 1 ppm = 4.08 mg/cu m
   —Odor: T.O.C.: 1.3–1.7 mg/cu m = 0.325–0.425 ppm   (57; 669)
      USSR: odor perception: 1.0 mg/cu m = 0.25 ppm
      O.I. at 20°C = 0.2   (316)
C. WATER POLLUTION FACTORS:
   —BOD_5: 0.4–0.6 Warburg, sewage   (164; 165; 258; 259)
D. BIOLOGICAL EFFECTS:
   —Fish: mosquito fish: 24–96 hr TLm: 240–230 mg/l   (41)
          bluegill sunfish: 24 hr TLm: 150 mg/l; tap water   (247)
   —Mammals: rat: single oral dose: LD_{50}: 400–800 mg/kg   (211)
   —Man: can produce severe eye and skin burns   (211)

maleic hydrazide (regulox 36)
   Use: herbicide
D. BIOLOGICAL EFFECTS:
   —Fish:

	mg/l	24 hr	48 hr	96 hr	3m (extrap.)
harlequin fish					
(*Rasbora heteromorpha*)	LC_{10}(F)	300			
	LC_{50}(F)	530	150	125	100

                                                                          (331)
   —Carcinogenicity: none
   —Mutagenicity in the *Salmonella* test: none
      <0.0008 revertant colonies/nmol
      <70 revertant colonies at 10^4 µg/plate   (1883)

1-malic acid (1-hydroxybutanedioic acid; 1-hydroxysuccinic acid)

$$\begin{array}{c} COOH \\ | \\ CH_2 \\ | \\ CHOH \\ | \\ COOH \end{array}$$

A. PROPERTIES: colorless needles; m.w. 134.09; m.p. 140°C; b.p. 140°C decomposes; sp.gr. 1.595; $\log P_{oct}$ −1.26
C. WATER POLLUTION FACTORS:
   —Natural occurence: in many fruits
   —BOD_5: 0.468   (30)
              0.34–0.57, Warburg, sewage   (259)(164)
   —COD: 0.68; 0.70   (27)(30)
      $KMnO_4$: 0.602; 2.180   (30)
      ThOD: 0.718   (27)(30)

   —Waste water treatment: A.S. after

	l	dl	
6 hr:	9.6%	6.0%	ThOD
12 hr:	22.9%	4.6%	ThOD
24 hr:	44.8%	20.8%	ThOD

                                                                           (89)
D. BIOLOGICAL EFFECTS:
   —Mammalia: mice, rats: single oral LD_{50}: 1.6–3.2 g/kg (all isomers)   (211)

malonic acid (propanedioic acid; methanedicarboxylic acid)
HOOCCH$_2$COOH
A. PROPERTIES: m.w. 104.06; m.p. 135.6°C; b.p. decomposes; solub. 611 g/l at 0°C, 735 g/l at 20°C, 926 g/l at 50°C; log P_{oct} -0.91/-0.18 (calculated)
C. WATER POLLUTION FACTORS:
  —BOD$_5$: 38% of ThOD
  —BOD$_2^{25°C}$ days: 0.31 (substrate conc.: 13.5 mg/l; inoculum: soil microorganisms
    5 days: 0.36
   10 days: 0.52
   20 days: 0.53   (1304)
  —COD: 100%
  —KMnO$_4$: acid: 60% of ThOD
            alkaline: 5% of ThOD   (220)
  —Waste water treatment: A.S. after 6 hr: 1.2% of ThOD
                                    12 hr: toxic
                                    24 hr: 0.9% of ThOD   (89)

malononitrile
N≡C—CH$_2$—C≡N
Manmade sources: hydrolysis product of O-chlorobenzylidenemalononitrile
A. PROPERTIES: m.w. 66.06; m.p. 32-34°C; b.p. 220°C; sp.gr. 1.049
D. BIOLOGICAL EFFECTS:
  —Fish: rainbow trout: LC$_{50}$, 12 hr: 19.4 mg/l
                               24 hr: 6.2 mg/l
                               48 hr: 4.2 mg/l
                               96 hr: 1.6 mg/l   (1913)

maltonic acid *see d-gluconic acid*

dl-mandelic acid (*dl*-phenylglycolic acid; *dl*-α-hydroxy-α-toluic acid)

A. PROPERTIES: m.w. 152.14; m.p. 118.1°C; b.p. decomposes; solub. 160 g/l at 20°C; sp.gr. 1.361 at 4°C
C. WATER POLLUTION FACTORS:
  —BOD$_5$: 0.31 std. dil./sewage   (282)
  —Waste water treatment: A.S. 42% theor. oxidation, 250 mg/l infl., 30 min aeration

methods sel. strain.	temp °C	days observed	feed mg/l	days acclim.	
Sd. W	30	$\frac{1}{24}$	2 mol/warb. flask	2	58% theor. oxidation

A.S., W   20 $\frac{1}{2}$ 250   25+   42% theor.
                                                          oxidation (93)

manganeseethylene-bis-dithiocarbamate   (dithane; trimangol 80; maneb)

$$\left[\begin{array}{c} CH_2-NH-\overset{S}{\underset{\|}{C}}-S \\ | \\ CH_2-NH-\underset{\|}{\overset{}{C}}-S \\ S \end{array} \right]_x Mn$$

Use: agricultural fungicide

D. BIOLOGICAL EFFECTS:
  —Amphibians:
    Male and female adult newts, *Triturus cristatus carniflex*:
      percutaneous exposure to 5 ppm: teratogenic effects in the regenerating forelimb as well as in amphibian embryos and homeothermic vertebrate embryos
      (1430)
    adult newt *Triturus cristatus carniflex*: percutaneous exposure to maneb 80:

maneb 80 (ppm)	LT_{50} (hr) (95% conf. lim) male	female
125	8.4 (7.6–9.2)	28.5 (16.4–49.6)
125*	8.8 (8.3–9.3)	16.0 (7.6–33.8)
100	28.0 (16.0–49.0)	19.5 (9.5–40.2)
100*	11.0 (8.9–13.5)	16.0 (9.2–27.7)
75	19.0 (7.0–57.3)	25.5 (11.6–55.8)
50	76.0 (57.6–100.3)	168.0 (85.7–329.3)
25	255.0 (147.4–441.1)	**

*second experiment
**some female newts were still alive after 5 months   (1063)
  —Fish:

	mg/l	24 hr	48 hr	96 hr	3m (extrap.)
harlequin fish (*Rasbora heteromorpha*) $LC_{10}(F)$	0.7	0.6	0.31		
$LC_{50}(F)$	0.9	0.77	0.53	0.4	

(331)
  —Mammals:
    acute oral LD_{50} (rat): 6750 mg/kg   (1854)
    in diet: rats fed 2 years on a diet containing 250 ppm suffered no ill effects
    (1855)

manna sugar *see* mannitol

mannite *see* minnitol

mannitol (manna sugar; mannite)

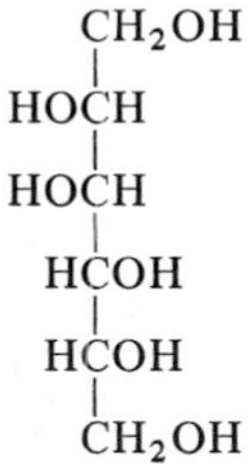

$$\begin{array}{c} CH_2OH \\ | \\ HOCH \\ | \\ HOCH \\ | \\ HCOH \\ | \\ HCOH \\ | \\ CH_2OH \end{array}$$

A. PROPERTIES: white crystalline powder or granules; m.p. 156–167°C; b.p. 290–295°C at 3–3.5 mm; sp.gr. 1.52; $\log P_{oct}$ −3.10
C. WATER POLLUTION FACTORS:
 —BOD_5: 59% ThOD (274)
 —COD: 87% ThOD (0.05 n $K_2Cr_2O_7$) (274)
 —ThOD: 1.15

margaric acid *see* heptadecanoic acid

MBC *see* methyl-2-benzimidazolecarbamate

MBT *see* methylene-*bis*-thiocyanate

MBT *see* 2-mercaptobenzothiazole

MCPA *see* 2-methyl-4-chlorophenoxyacetic acid

MDA *see* 4,4′-diaminodiphenylmethane

MEA *see* ethanolamine

mercarbam (5-(N-ethoxycarbonyl)-N-methylcarbamoylmethyl)-O,O-diethylphosphorodithioate)

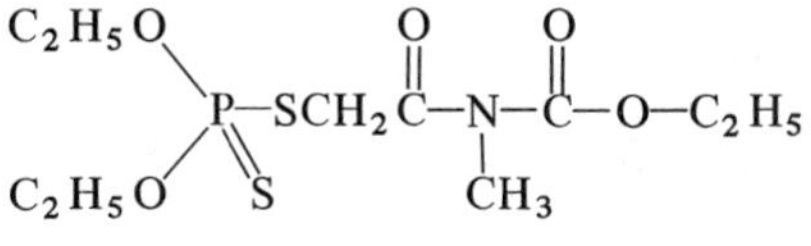

$$\begin{array}{c} C_2H_5O \\ \diagdown \\ P-SCH_2\overset{O}{\overset{\|}{C}}-\underset{\underset{CH_3}{|}}{N}-\overset{O}{\overset{\|}{C}}-O-C_2H_5 \\ \diagup\| \\ C_2H_5OS \end{array}$$

Use: insecticide
A. PROPERTIES: colorless oily liquid; sp.gr. 1.22 at 20°C; b.p. 144°C at 0.02 mm; solub. <1000 ppm at room temp.
D. BIOLOGICAL EFFECTS:
 —Fish:

	mg/l	24 hr	48 hr	96 hr	3m (extrap.)
harlequin fish (*Rasbora heteromorpha*) $LC_{10}(F)$	0.056	0.005			
$LC_{50}(F)$	0.008	0.007	0.004	0.002	

(331)

—Mammals:
  acute oral LD_{50} (mice):   106 mg/kg
  acute oral LD_{50} (rat):   36 mg/kg   (1855)
  acute dermal LD_{50} (rat): 1220 mg/kg   (1854)
  in diet: rats: no ill effect at 1.6 mg/kg/day for 16 weeks, slight depression of the growth rate at 4.56 mg/kg/day for 16 weeks   (1855)

MEG *see* ethyleneglycol

Me-6K *see* 4-methoxy-4-methyl-2-pentanone

MEK *see* methylethylketone

melamine (2,4,6-triamino-*s*-triazine; cyanurotriamide)

Use: melamine resins; organic synthesis; leather tanning
A. PROPERTIES: m.w. 126.13; m.p. >300°C sublimes; b.p. sublimes; v.p. 315°C 50 mm; v.d. 4.34; sp.gr. 1.573 at 250°C
C. WATER POLLUTION FACTORS:
  —BOD_5: nil   (27)
          0.006 std. dil. sew.   (258)
  —ThOD: 3.04   (27)
  —Impact on biodegradation:
    at 100 mg/l, no inhibition of NH_3 oxidation by *Nitrosomonas* sp.   (390)
D. BIOLOGICAL EFFECTS:
  —Mammals:
    mouse: acute oral dose: death: 1.6 g/kg
    rat: acute oral dose: no death: 3.2 g/kg   (211)

melaniline *see* diphenylguanidine

melprex *see* *n*-dodecylguanidineacetate

***p*-menthan-3-ol** *see* menthol

menthol (hexahydrothylmol; methylhydroxyisopropylcyclohexane; *p*-menthan-3-ol; peppermint camphor)

A. PROPERTIES: white crystals; m.p. 41–43°C; v.p. 1 mm at 56°C, 10 mm at 96°C, 100 mm at 149.4°C; log P_{oct} 3.25/3.27 (calculated)
B. AIR POLLUTION FACTORS:
odor thresholds   mg/cu m

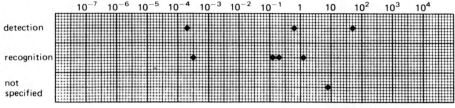

(307; 610; 614; 640; 840)

d-isomer: detection: 0.9 mg/cu m
recognition: 2.1 mg/cu m
l-isomer: detection: 0.9 mg/cu m
recognition: 2.1 mg/cu m (640)
O.I. at 20°C = 100 (316)

C. WATER POLLUTION FACTORS:
—Oxidation parameters: COD: 87% ThOD (0.05 n $K_2Cr_2O_7$) (274)
KMnO$_4$: 1% ThOD (0.01 n KMnO$_4$) (274)
ThOD: 2.97
—Biodegradation: adapted A.S. at 20°C–product as sole carbon source: 95.1% COD removal at 17.7 mg COD/g dry inoculum/hr (327)

mercaptoacetic acid (2-mercaptoethanoic acid; thioglycolic acid; thioranic acid) $HSCH_2COOH$
A. PROPERTIES: colorless liquid; m.w. 92.11; m.p. −16.5°C; b.p. 104/106°C at 11 mm; v.p. 10 mm at 18°C; sp.gr. 1.325 at 20/4°C
B. AIR POLLUTION FACTORS:
—Control methods: wet scrubber:
water at pH 8.5: outlet: 30 odor units/scf
KMnO$_4$: water at pH 8.5: outlet: 1 odor unit/scf (115)
C. WATER POLLUTION FACTORS:
—Waste water treatment: A.S., Resp, BOD, 20°C, 1–5 days observed, feed 662 mg/l acclimation: 1 day: nil % removed or oxidized (93)
D. BIOLOGICAL EFFECTS:
—Mammalia: rat: single oral LD$_{50}$: <50 mg/kg (211)

l-β-**mercaptoalanine**   *see l*-cysteine

2-mercaptobenzothiazole (captax; MBT)

A. PROPERTIES: yellowish powder; m.p. 164–175°C; sp.gr. 152

C. WATER POLLUTION FACTORS:
- Reduction of amenities: T.O.C. in water at room temp.: 1.76 ppm, range 0.40–10.9 ppm, 7 judges
  20% of population still able to detect odor at 0.079 ppm
  10% of population still able to detect odor at 0.016 ppm
  1% of population still able to detect odor at 0.088 ppb
  0.1% of population still able to detect odor at 0.51 ppt (321)
D. BIOLOGICAL EFFECTS:
- Fish: goldfish: approx. fatal conc. 2 mg/l, 48 hr (226)

2-mercaptodiethylsuccinate
D. BIOLOGICAL EFFECTS:
- Fish:
  eastern mudminnow: static bioassay: 96 hr TLm: 47.0 mg/l at 16°C (450)

2-mercaptoethanoic acid see mercaptoacetic acid

2-mercaptoethanol (monothioglycol)
$H_2SCH_2CH_2OH$
A. PROPERTIES: b.p. 157°C; v.p. 1 mm at 20°C; v.d. 2.70; sp.gr. 1.12
B. AIR POLLUTION FACTORS:
- Control methods: wet scrubber:
  water: at pH 8.5: outlet: 30 odor units/scf
  $KMnO_4$: at pH 8.5: outlet: 1 odor unit/scf (115)
C. WATER POLLUTION FACTORS:
- Reduction of amenities:
  odor threshold: average: 0.64 mg/l
              range: 0.07–1.1 mg/l (294)(97)

mercaptophos see baytex

mercuric acetate
$(CH_3CO_2)_2Hg$
A. PROPERTIES: m.w. 318.68; m.p. 179–182°C
D. BIOLOGICAL EFFECTS:
- Fish:

	LC_{50} (mg/l)		
	24 hr	48 hr	96 hr
fathead minnows (*Pimephales promelas*)	0.53	0.42	0.19
grass shrimp (*Palaemonetes pugio*)	–	–	0.06

(1904)

mercuric thiocyanate (mercury(II) thiocyanate)
$Hg(SCN)_2$
A. PROPERTIES: m.w. 316.75; m.p. 165°C (decomposes)
D. BIOLOGICAL EFFECTS:
- Fish:

	LC_{50} (mg/l)		
	24 hr	48 hr	96 hr
fathead minnows (*Pimephales promelas*)	0.39	0.39	0.15
grass shrimp (*Palaemonetes pugio*)	–	–	0.09

merphos (folex; tributylphosphorotrithioite)

$$\begin{array}{c} C_4H_g \\ C_4H_g-P \\ C_4H_g \end{array}$$

Use: defoliant

C. WATER POLLUTION FACTORS:
 —Aquatic reactions:
  was converted to DEF within 1 hr in river water

$$\begin{array}{c} C_4H_g \\ C_4H_g-P \\ C_4H_g \end{array} \xrightarrow{(O)} \begin{array}{c} C_4H_g \\ C_4H_g-P=O \\ C_4H_g \end{array}$$
$$\text{merphos} \qquad \text{DEF}$$

 —Persistence in river water in a sealed glass jar under sunlight and artificial fluorescent light—initial conc. 10 μg/l:

	% of original compound found					
after	1 hr	1 wk	2 wk	4 wk	8 wk	
	0	0	0	0	0	
as DEF	100	50	30	10	<5	(1309)

D. BIOLOGICAL EFFECTS:
 —Mammals:
  acute oral LD_{50}: 1272 mg/kg   (1854)

mesidine (1-amino-1,3,5-trimethylbenzene; 2,4,6-trimethylaniline)

$$\underset{H_3C}{\overset{CH_3}{\underset{CH_3}{\bigcirc}}}NH_2$$

A. PROPERTIES: m.w. 135.20; m.p. −15°C; b.p. 233; sp.gr. 0.963
B. AIR POLLUTION FACTORS:
 —Odor threshold conc. 0.013 mg/cu m   (841)
C. WATER POLLUTION FACTORS:
 —Waste water treatment methods:
  W, A.S. from mixed domestic/industrial treatment plant: 0–9% depletion at 20 mg/l after 6 hr at 25°C   (419)

mesitylene (1,3,5-trimethylbenzene)

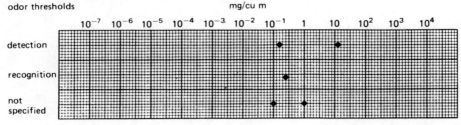

Use: intermediate, including anthraquinone dyes; ultraviolet oxidation stabilizer for plastics
Manmade sources: in gasoline (high octane number): 1.32 wt % (387)
A. PROPERTIES: m.w. 120.19; m.p. -52.7; b.p. 164.7; sp.gr. 0.865 at 20°C
B. AIR POLLUTION FACTORS:

odor thresholds

(279; 610; 637; 643; 724)

—Atmospheric reactions: reactivity: HC cons.: ranking: 8
NO ox.: ranking: 1.2 (63)
estimated lifetime under photochemical smog conditions in S.E. England: 0.7 hr (1699; 1700)
—Manmade sources: in exhaust of diesel engine: 0.4% of emitted hydrocarbons (72)

—Ambient air quality:
glc's in carpark: 0.200 µg/cu m
on motorway: 0.840 µg/cu m (48)
Los Angeles 1966: glc's avg.: 0.003 ppm ($n = 136$)
highest value: 0.011 ppm (1319)
glc's in the Netherlands:
tunnel Rotterdam: 1974.10.12: avg.: 3 ppb ($n = 12$)
max.: 5 ppb
The Hague: 1974.10.11: avg.: 2 ppb ($n = 12$)
max.: 4 ppb
Roelofarendsveen: 1974. 9.11: avg.: 0.5 ppb ($n = 12$)
max.: 0.5 ppb (1231)

C. WATER POLLUTION FACTORS:
—BOD_5: 3% of ThOD
COD: 10% of ThOD
$KMnO_4$: acid: 2% ThOD
alkal.: 0% ThOD (220)
—Reduction of amenities:
odor threshold: average: 0.027 mg/l
range: 0.00024–0.062 mg/l (97)

T.O.C. = 0.5 mg/l (295)
D. BIOLOGICAL EFFECTS:
 —Fish: goldfish: flow through bioassay: 96 hr TLm: 13 mg/l at 17–19°C (471)
 —Man: EIR: 3.1 (49)

mesityloxide (4-methyl-3-penten-2-one; isopropylideneacetone; methylisobutenyl-ketone; MO)
$(CH_3)_2 CCHCOCH_3$
A. PROPERTIES: colorless oily liquid; m.w. 98.14; m.p. −59°C; b.p. 130°C; v.p. 8.7 mm at 20°C; v.d. 3.40; solub. 28 g/l at 20°C
B. AIR POLLUTION FACTORS: 1 mg/cu m = 0.249 ppm, 1 ppm = 4.02 mg/cu m
 —Odor: characteristic: quality: sweet, peppermint-like
         hedonic tone: pleasant
     odor threshold values: −50% detection: 12 ppm
                              −100% detection: 25 ppm (211)
     absolute perception limit: 0.017 ppm
      50% recogn.:   0.051 ppm
     100% recogn.:   0.051 ppm
     odor index (O.I.) 100% recogn.: 210,000 (19)
     O.I. at 20°C = 460 (316)
 —Manmade sources: in exhaust of gasoline engine: <0.1–1.5 ppm (195)
 —Control methods: catalytic combustion: platinized ceramic honeycomb-catalyst:
     ignition temp.: 180°C; inlet temp. for 90% conversion: 250–300°C (91)
C. WATER POLLUTION FACTORS:
 —Oxidation parameters: BOD_5 = 1.91 NEN 3235-5.4 (277)
                        COD = 2.40 NEN 3235-5.3 (277)
D. BIOLOGICAL EFFECTS:
 —Mammalia:
  guinea pig: inhalation: death in a few min: 13,000 ppm
                        : no serious disturbance: 1,000 ppm, 60 min
                        : no serious disturbance: 100 ppm, several hours
  rat: inhalation: no effect: 50 ppm, 30 × 8 hr (211)
       acute oral LD_{50}: 655 mg/kg
  goldfish: LD_{50} (24 hr): 540 mg/l-modified ASTM D 1345 (277)
 —Man: eye irritation: 25 ppm
       nasal irritation, objectional odor, bad taste 50 ppm (211)

mesurol (3,5-dimethyl-4-(methylthio)phenolmethylcarbamate; mercaptodimethur; methiocarb; metmercapturon)

$$CH_3S-\underset{CH_3}{\overset{CH_3}{\bigcirc}}-O-\overset{O}{\underset{\|}{C}}-NH-CH_3$$

A. PROPERTIES: m.p. 121°C
C. WATER POLLUTION FACTORS:

mesurol → 4-methylthio-3,5-dimethylphenol + methylcarbamic acid (via H_2O)

(1309)

—Persistence in river water in a sealed glass jar under sunlight and artificial fluorescent light—initial conc. 10 μg/l:

after	1 hr	1 wk	2 wk	4 wk	8 wk
% of original compound found	90	0	0	0	0

D. BIOLOGICAL EFFECTS:
  —Mammals:
   acute oral LD_{50} (rats): 87–130 mg/kg   (1855)
   acute dermal LD_{50}: 350–500 mg/kg
   inhalation: approx. 535 mg/kg   (1855)

mesylchloride see methanesulfonylchloride

metacil see aminocarb

metaphos see dimethylparathion

methacrolein (methacrylaldehyde; α-methylacrolein; 2-methylpropenal)
$CH_2C(CH_3)CHO$
  Use: copolymers, resins
A. PROPERTIES: colorless liquid; m.w. 70.09; b.p. 68/73.5; v.d. 2.4; sp.gr. 0.830 at 20/4°C; solub. 64,000 mg/l
B. AIR POLLUTION FACTORS: 1 mg/cu m = 0.349 ppm, 1 ppm = 2.8 mg/cu m
  —Manmade sources:
   in exhaust of a 1970 Ford Maverick gasoline engine operated on a chassis dynamometer following the 7-mode California cycle:
     from API #7 gasoline: 1.6 ppm
     from API #8 gasoline: 0.9 ppm   (1053)
C. WATER POLLUTION FACTORS:
  —Waste water treatment:
   RW, Sd, BOD, 20°C, 10 days observed, feed: <0.3 mg/l, 100 days acclimation: 65% theor. oxidation   (91)
D. BIOLOGICAL EFFECTS:
  —Mammals:
   rat: inhalation: 5/6: 250 ppm, 4 hr   (104)
   acute oral LD_{50}: 0.14 g/kg   (211)

methacrylaldehyde see methacrolein

methacrylic acid (2-methylpropenoic acid; α-methylacrylic acid)

$CH_2C(CH_3)COOH$

A. PROPERTIES: colorless crystals; m.w. 86.09; m.p. 16°C; b.p. 163°C; v.p. 0.65 mm at 20°C, 1 mm at 25°C, 1.4 mm at 30°C; v.d. 2.97; sp.gr. 1.015 at 20/4° sat. conc. 3.0 g/cu m at 20°C, 6.3 g/cu m at 30°C

B. AIR POLLUTION: 1 mg/cu m = 0.28 ppm, 1 ppm = 3.58 mg/cu m
 —Odor: characteristic: quality: acrid
 hedonic tone: repulsive

C. WATER POLLUTION FACTORS:
 —Oxydation parameters: BOD_5: 0.89 std. dil. sewage (285)(168)
 ThOD: 1.67 (27)
 —Waste water treatment:
 Sew, CO_2, 20°C, 10 days observed, feed: 11 mg/l
 19 days acclimation: 60% theor. oxidation (93)

D. BIOLOGICAL EFFECTS:
 —Mammalia: rat: inhalation: no toxic signs, organs normal, slight renal congestion: 300 ppm, 20 × 6 hr (65)

methacrylonitrile
$CH_2C(CH_3)CN$

A. PROPERTIES: colorless liquid; m.w. 67.09; m.p. -35.8°C, b.p. 90.3°C; v.p. 40 mm at 13°C, 65 mm at 25°C, 100 mm at 33°C; v.d. 2.31; sp.gr. 0.800 at 20/4°C; solub. 25 g/l; sat. conc. 208 g/cu m at 20°C, 318 g/cu m at 30°C

B. AIR POLLUTION: 1 mg/cu m = 0.365 ppm, 1 ppm = 2.74 mg/cu m
 —Odor: T.O.C. <5 mg/cu m = <1.8 ppm (57)
 O.I. at 20°C = 72,000 (316)

C. WATER POLLUTION FACTORS:
 —biodegradation: by mutant microorganisms: 500 mg/l at 20°C
 parent: 93% disruption in 28 hr
 mutant: 100% disruption in 3.5 hr (152)

D. BIOLOGICAL EFFECTS:
 —Mammalia:
 mouse: inhalation: LC_{50}: 630 ppm, 1 hr
 LC_{50}: 400 ppm, 4 hr
 : 0/6: 75 ppm, 8 hr
 acute oral LD_{50}: 20–25 mg/kg
 rat: acute oral LD_{50}: 25–50 mg/kg
 inhalation LD_{50}: 9,880 ppm, 2 hr (211)

methallylchloride (3-chloro-2-methyl-1-propene; 3-chloroisobutene; methylallylchloride; MAC)

$$CH_2=\underset{\underset{CH_3}{|}}{C}-CH_2-Cl$$

A. PROPERTIES: colorless to pale yellow liquid; b.p. 71.5–75.5°C; sp.gr. 0.924 at 20/4°C

B. AIR POLLUTION:
 —Odor: characteristic: quality: sharp, penetrating odor (277)

C. WATER POLLUTION FACTORS:
 —Oxidation parameters: BOD$_5$: 0.81 NEN 3235-5.4 (272)
 COD: 1.18 NEN 3235-5.4 (272)
D. BIOLOGICAL EFFECTS:
 —Fish: goldfish: LD$_{50}$ (24 hr): 14 mg/l modified ASTM—D 1345 (277)

methamsodium (sodium-N-methyldithiocarbamate)

$$CH_3NHC-SNa$$
$$\|$$
$$S$$

Use: a soil fungicide, nematicide and herbicide with a fumigant action. Its activity is due to decomposition to methylisothiocyanate (MITC):

$$CH_3-NH-\underset{\underset{S}{\|}}{C}-SNa \rightarrow CH_3-N=C=S \qquad (1439)$$

methanal *see* formaldehyde

methanamide *see* formamide

methane (marsh gas, methylhydride)
CH_4
 Manufacturing source: natural gas mfg.; coal processing (347)
 Users and formulation: mfg. carbon black, acetylene, hydrogen, halogenated methanes and ethylene, methanol, hydrogen cyanide, carbon tetrachloride, chloroform, fuel. (347)
 Natural sources (water and air): marsh gas; natural gas; coal. (347)
 Man caused sources (water and air): pipeline and domestic leakage; tobacco smoke; incomplete combustion of natural gas, coal, wood, petroleum fuels; lab use. (347)
A. PROPERTIES: colorless gas; m.w. 16.04°C; m.p. -182/-184°C; b.p. -162°C; v.d. 0.416/0.55 sp.gr. 0.42 at -164°C; solub. 24 mg/l (242)
 THC 212.8 Kcal/mole, LHC 191 Kcal/mole
B. AIR POLLUTION FACTORS: 1 mg/cu m = 1.50 ppm, 1 ppm = 0.67 mg/cu m
 —Odor: odor threshold values: odorless (57)
 —Atmospheric reactions: reactivity NO ox.: <0.01 (63)
 estimated lifetime under photochemical smog conditions in S.E. England: 5290 hr
 (1699; 1701)
 —Natural sources: glc's Pt Barrow Alaska Sept. 1967: 1.35-1.65 ppm (101)
 —Manmade sources: glc's Los Angeles downtown 1967: 10% ile: 1.7 ppm
 avg.: 2.4 ppm
 90% ile: 3.5 ppm (64)
 in exhaust of diesel engine: 16.7% of emitted HC's (72)
 in exhaust of reciprocating gasoline engine: 24.3% of emitted HC's
 in exhaust of rotary gasoline engine: 4.9% of emitted HC's (78)
 expected glc's in USA urban air: range: 1.6-10 ppm (102)
 in combustion gas of household central heating on gasoil: ±20 ppm (182)
 in flue gas of municipal incinerator: <0.4-13.0 ppm (196)

in auto exhaust of gasoline engine:
  62-car survey: 16.7 vol.% total exhaust HC's (391)
  15-fuel survey: 18 vol.% total exhaust HC's (392)
  engine-variable study: 13.8 vol.% total exhaust HC's (393)
—Control methods: catalytic combustion: platinized honeycomb catalyst: ignition temp.: 300°C, inlet temp. for 90% conversion: 400–450°C (91)
—Sampling and analysis: non dispersive I.R.: min. full scale: 50 ppm (55)
effectiveness of scrubbers at 325–375°F: inlet 20 ppm methane:
  (1): $HgSO_4/H_2SO_4$: 0% absorbed
  (2): $PdSO_4/H_2SO_4$: 0% absorbed
  (3): (1) + (2): 0% absorbed (198)
PMS: at 80.3 nm: det. lim.: 4 ppm
     95 nm: det. lim.: 71 ppm (200)

C. WATER POLLUTION FACTORS:
—Oxidation parameters: $BOD_{35}^{25°C}$ 3.04 (62)
                       ThOD: 3.99 (62)

D. BIOLOGICAL EFFECTS:
—Man: tolerable limit: 1,000 ppm = 1,800 mg/cu m (186)

methanecarboxylic acid *see* acetic acid

methanedicarboxylic acid *see* malonic acid

ethanesulfonylchloride (mesylchloride)
CH_3SO_2Cl
Use: intermediate; flame-resistant products; stabilizer for liquid sulfur trioxide; biological chemicals
A. PROPERTIES: pale yellow liquid; sp.gr. 1.485 (20/20°C); b.p. 164°C
D. BIOLOGICAL EFFECTS:
—Fish:
  Lepomis macrochirus: static bioassay in fresh water at 23°C, mild aeration applied after 24 hr:

material added	% survival after				best fit 96 hr LC_{50}
ppm	24 hr	48 hr	72 hr	96 hr	ppm
18	0	—	—	—	
13	0	—	—	—	11
10	100	100	100	100	
7.6	100	100	100	100	

  Menidia beryllina: static bioassay in synthetic seawater at 23°C, mild aeration applied after 24 hr:

material added	% survival after				best fit 96 hr LC_{50}
ppm	24 hr	48 hr	72 hr	96 hr	ppm
24	20	20	10	0	
18	50	50	40	30	15
10	100	100	100	100	(352)

methanethiol *see* methylmercaptan

methanoic acid *see* formic acid

methanol (methylalcohol; carbinol; wood alcohol; wood spirit)
CH_3OH
  Manufacturing source: organic chemical industry; wood processing industry.
  Users and formulation: mfg. formaldehyde, methacrylates, methylamines, dimethylterephthalate, methyl halides, ethyleneglycol, polyformaldehydes, plastics; solvent; denaturant for ethanol; dehydrator for natural gas. (347)
  Natural sources (*water and air*): wood. (347)
A. PROPERTIES: colorless liquid; m.w. 32.04; m.p. $-98°C$; b.p. $65°C$; v.p. 92 mm at $20°C$, 160 mm at $30°C$; v.d. 1.1; sp.gr. 0.796 at $15°C$; THC 173.6 kcal/mole; LHC 162 kcal/mole; sat. conc 166 g/cu m at $20°C$, 270 g/cu m at $30°C$; P_{oct} $-0.82/-0.66$
B. AIR POLLUTION: 1 mg/cu m = 0.764 ppm, 1 ppm = 1.33 mg/cu m
  —Odor: characteristic: quality: sour, sharp, sweet
                hedonic tone: neutral

odor thresholds — mg/cu m (10^{-7} to 10^5)
detection; recognition; not specified

(2; 29; 278; 279; 307; 610; 631; 663; 671; 676; 704; 709; 741; 749; 761; 770; 788; 856; 871)

threshold for unadapted panelists: 2,000 ppm	
after adaptation with pure odorant: 20,000 ppm	(204)
distinct odor = 11,700 mg/cu m = 8,800 ppm	(278)
O.I.: 100% recogn.: 2,393	(19)

  USSR: human odor perception: 4.3 mg/cu m
        human reflex: no response: 1.8 mg/cu m
                      adverse response: 3.3 mg/cu m
        animal chronic exposure: no effect: 0.57 mg/cu m
                                 adverse: 5.3 mg/cu m   (170)

  —Manmade sources:
  in cigarette smoke: 700 ppm (66)
  in exhaust gas of gasoline engine: 0.1–0.6 ppm (195)
  in workrooms conc. of 50–6,000 ppm have been reported (211)
  —Sampling and analysis: PMS: det. lim.: 4–11 ppm (200)

C. WATER POLLUTION FACTORS:
  —Oxidation parameters:

	Ref. 220	Ref. 79	Ref. 23	Ref. 23*
BOD_5:	48% ThOD	53.4% ThOD	76% bio. ox.	69% bio. ox.
10 days:		62.7% ThOD	88% bio. ox.	84% bio. ox.
15 days:		69.4% ThOD	91% bio. ox.	85% bio. ox.
20 days:		67.0% ThOD	95% bio. ox.	97% bio. ox.

METHANOL

30 days:	69.4% ThOD	
40 days:	93.4% ThOD	
50 days:	97.7% ThOD	

*salt dilution water

BOD_5: 0.13 sierp/10% sewage: 500 at 1,500 mg/l (280)
 0.6–1.1 std. dil. at 1–1,000 mg/l (27)
 0.76 to 1.12 (255)(259)(260)(269)(280)(281)(271)
 (282)(284)(26)(30)(285
 0.76; 1.12; 1.24, Warburg, sewage (41)

BOD_{20}: 1.26 (30)
$BOD_{10}^{20°C}$: 62.7% ThOD at 2.5 mg/l in mineralized dilution water with settled sewage seed (405)
BOD_5^{20}: 1.12 at 10 mg/l, unadapted sew.: lag period: 1 day
BOD_{20}^{20}: 1.18 at 10 mg/l, unadapted sew.: lab period: 1 day (554)

COD: 1.05; 1.42; 1.50 (23)(41)(36)
 99% of ThOD (220)
 95% of ThOD (0.05 n $k_2Cr_2O_7$) (274)

$KMnO_4$: 0.058 (30)
 acid: 10% ThOD
 alkaline: 2% ThOD (220)
 1% ThOD (274)

TOD 1.50 (30)
TOC: 95% ThOD (220)
ThOD: 1.50 (23)

—Odor threshold: 1,600 mg/kg (907)
 detection: 10.0 mg/kg (894)

—Impact on biodegradation processes: sludge digestion is inhibited at 800 mg/l (30)
~50% inhibition of NH_3 oxidation in *Nitrosomonas* at 160 mg/l (407)

—Waste water treatment: A.C.: adsorbability: 0.007 g/g C; 3.6% reduction, infl.: 1,000 mg/l, effl.: 964 mg/l (32)

anaerobic lagoon: lb COD/Day/1,000 cu ft; infl. mg/l; effl. mg/l

13	80	35
22	380	135
48	380	145

(37)

air stripping constant: k = 0.263 days^{-1} at 1,360 mg/l (82)
reverse osmosis: 22% rejection from a 0.01 M solution (221)

A.S.: after 6 hr: 5.3% ThOD
 12 hr: 10.5% ThOD
 24 hr: 21.0% ThOD

methods	temp °C	days observed	feed mg/l	days acclim.	% theor. oxidation
A.S., W	20	$\frac{1}{3}$	2,000	P	8
A.S., W	20	$\frac{1}{3}$	1,000	P	15
A.S., W	20	$\frac{1}{3}$	500	P	30
A.S., Resp, BOD	20	1–5	1,000	<1	3
A.S., W	20	1	500	24	55

A.S., BOD	20	$\frac{1}{3}$-5	333	30	90
					(93)

A.C.: infl. ppm; carbon dosage; effl. ppm; % removal

infl. ppm	carbon dosage	effl. ppm	% removal	
1,000	10X	830	17	
200	10X	132	33	
15	10X	10	33	(192)

D. BIOLOGICAL EFFECTS:
  —Toxicity threshold (cell multiplication inhibition test):

bacteria (*Pseudomonas putida*):	6600 mg/l	(1900)
algae (*Microcystis aeruginosa*):	530 mg/l	(329)
green algae (*Scenedesmus quadricauda*):	8000 mg/l	(1900)
protozoa (*Entosiphon sulcatum*):	>10,000 mg/l	(1900)
protozoa (*Uronema parduczi Chatton Lwoff*):	>10,000 mg/l	(1901)

  —Bacteria:
  Pseudomonas: LD_0: 0.6 g/l
  —Algae:
  Chlorella pyrenoidosa: toxic: 31,100 mg/l (41)
  Scenedesmus: LD_0: 10 g/l
  —Protozoa: *Colpoda*: LD_0: 1.25 g/l
  —Arthropoda: *Daphnia*: no effect at 10 g/l, 48 hr (30)
             brine shrimp: TLm (24 hr): 10,000 mg/l
  —Fish:
  trout: TLm (48 hr): 8.000 mg/l (23)
  creek chub: LD_0:   8,000 mg/l, 24 hr in Detroit river water
             LD_{100}: 17,000 mg/l, 24 hr in Detroit river water (243)
  —Mammalia: cat: inhalation: LD_{50}: 33,600 ppm, 6 hr
                          65,700 ppm, 4.5 hr
            mouse: inhalation: LD_{50}: 61,100 ppm, 134 min
  —Man: severe toxic effects: 2,000 ppm = 2,560 mg/cu m, 60 min
        symptoms of illness: 500 ppm = 640 mg/cu m
        unsatisfactory: 200 ppm = 256 mg/cu m

metheneamine see hexamethylenetetramine

dl-methionine (*dl*-α-amino-γ-methylmercaptobutyric acid; *dl*-2-amino-4-methylthiobutanoic acid)

$$\begin{array}{c} COOH \\ | \\ CHNH_2 \\ | \\ CH_2 \\ | \\ CH_2 \\ | \\ S \\ | \\ CH_3 \end{array}$$

Use: pharmaceuticals; cosmetics; nutrition and biochemical studies; food and feed supplement

A. PROPERTIES: m.w. 149.21; m.p. 281; solub. 33,800 mg/l at 25°C, 105,200 mg/l at 75°C
C. WATER POLLUTION FACTORS:
 —Waste water treatment:
  A.S. after 6 hr: 1.5% of ThOD
         12 hr: 2.4
         24 hr: 2.6
  A.S., BOD, 20°C, 1–5 days observed, feed: 333 mg/l,
         15 days acclimation: 80% removed
 —Manmade sources: excreted by man in urine: 0.12–0.17 mg/kg body wt/day
 (203)
 —Impact on biodegradation processes:
  no effect on sludge digestion at 1,000 mg/l
  stimulation of sludge digestion at 100 mg/l (30)
  ~50% inhibition of NH_3 oxidation in *Nitrosomonas* at 9 mg/l (*l*-isomer) (408)

2-methoxyaniline   *see o*-anisidine

3-methoxyaniline   *see m*-anisidine

4-methoxyaniline   *see p*-anisidine

methoxybenzene   *see* methylphenylether

2-methoxybenzoic acid   *see o*-anisic acid

3-methoxybenzoic acid   *see m*-anisic acid

4-methoxybenzoic acid   *see p*-anisic acid

methoxychlor   (methoxy DDT; DMDT; 2,2-*bis*(*p*-methoxyphenyl)-1,1,1-trichloroethane)

$$CH_3O-\text{C}_6H_4-\underset{\underset{CCl_3}{|}}{\overset{\overset{H}{|}}{C}}-C_6H_4-OCH_3$$

Use: insecticide effective against mosquito larvae and house-flies.
The technical product contains 88% of the *p,p'*-isomer, the bulk of the remainder being the *o,p*-isomer
A. PROPERTIES: sp.gr. 1.41 at 25°C; m.p. 98°C; solub. 0.040 mg/l at 24°C (99% purity) (1666)
C. WATER POLLUTION FACTORS:
 —Threshold odor onc. in water: 4.7 mg/kg
D. BIOLOGICAL EFFECTS:
 —Residues:
  mean concentration in gamebird muscle in upper Tennessee (USA):

	mg/kg fresh weight ± st. error		
	grouse	quail	woodcock
Johnson county	0.53 ± 0.10 (n = 12)	0.66 ± 0.12 (n = 6)	0.89 ± 0.17 (n = 6)
Carter county	0.94 ± 0.19 (n = 9)	0.49 ± 0.09 (n = 7)	1.02 ± 0.19 (n = 6)
Washington county	1.02 ± 0.22 (n = 10)	1.43 ± 0.30 (n = 11)	0.94 ± 0.19 (n = 6)

quail: acceptable LD_{50}: 22,000 mg/kg

—Bioaccumulation:
  distribution and metabolism of radiolabeled methoxychlor after 33 days in a model ecosystem:

			concentration ppm	
	H_2O	snail (*Physa*)	mosquito larva dry wt.	fish (*Gambusia*)
methoxychlor–total ^3H	0.0016	15.7	0.48	0.33
methoxychlor	0.00011	13.2	–	0.17
methoxychlorethylene	–	0.7	–	–
mono-OH-ethane	0.00013	1.0	–	–
di-OH-ethane	0.00003	–	–	–
di-OH-ethylene	0.00003	–	–	–
unknowns	0.00009	–	–	–
polar metabolites	0.00125	0.8	–	0.16
				(1312)

—Molluscs:
  BCF: 5 aquatic molluscs: 800–1500   (1870)
  oyster: 5780   (1924)

—Crustaceans:

		exposure	bioaccumulation
decapod	*Cancer magister*		
	adult	1.8 μg/l: 15 days	0.11 μg/g
		7.5 μg/l: 15 days	0.48 μg/g
	juvenile	0.04 μg/l: 12 days	0.11 μg/g
		2.0 μg/l: 12 days	0.88 μg/g
			(1143)
decapod	*Rhithropanopeus harrisii*		
	5 day larvae	1.0 ppb	0.1 μg/g
	1st crab	1.0 ppb	0.1 μg/g
	5 day larvae	2.5 ppb	0.56 μg/g
	1st crab	2.5 ppb	0.23 μg/g
	10 day larvae	4.0 ppb	0.59 μg/g
	1st crab	4.0 ppb	0.76 μg/g
	5 day larvae	5.5 ppb	1.28 μg/g
	1st crab	5.5 ppb	0.55 μg/g
			(1149)

—Amphibian:
  Bioaccumulation by the American toad (*Bufo americanus*): geometric mean

residues of methyoxychlor accumulated from food and water:

dosage group	days of exposure	n	(95% confidence limits) ppm wet weight
0.024 ppm in food	1	4	0.013 (0.003–0.043)
	6	4	0.008 (0.002–0.024)
0.325 ppm in food	1	5	0.033 (0.012–0.088)
0.069 ± 0.036 ppm in water	1	4	0.145 (0.065–0.323)
	6	4	0.244 (0.124–0.482)
	36	4	0.124 (0.048–0.327)
controls	–	12	not detected

(1709)

—Toxicity:
—Insects:
Pteronarcys californica: 96 hr LC_{50}: 1.4 µg/l (2128)
Taeniopteryx nivalis: 96 hr LC_{50}: 0.98 µg/l (2138)
Stenonema spp.: 96 hr LC_{50}: 0.63 µg/l (2138)
—Crustaceans:
Gammarus lacustris: 96 hr, LC_{50}: 0.8 µg/l (2124)
Gammarus fasciatus: 96 hr, LC_{50}: 1.9 µg/l (2126)
Palaemonetes kadiakensis: 96 hr, LC_{50}: 1.0 µg/l (2126)
Orconectes nais: 96 hr, LC_{50}: 0.5 µg/l (2126)
Asellus brevicaudus: 96 hr, LC_{50}: 3.2 µg/l (2126)
Simocephalus serrulatus: 48 hr, LC_{50}: 5 µg/l (2127)
Daphnia pulex: 48 hr, LC_{50}: 0.78 µg/l (2127)
89.5% Palaemon macrodactylus (Korean shrimp): TL_{50}, 96 hr static lab bioassay: 0.44 (0.21–0.93) ppb (2353)
89.5% Palaemon macrodactylus (Korean shrimp): TL_{50}, 96 hr intermittent-flow lab bioassay: 6.7 (4.37–10.7) ppb (2353)
100% Crangon septemspinosa (sand shrimp): LC_{50}, 96 hr static lab bioassay: 4. ppb (2327)
100% Palaemonetes vulgaris (grass shrimp): LC_{50}, 96 hr static lab bioassay: 12. ppb (2327)
100% Pagurus longicarpus (hermit crab): LC_{50}, 96 hr static lab bioassay: 7. ppb (2327)

—Decapod:
Cancer magister: zoeae: 96 hr LC_{50} (S): 0.42 µg/l
juvenile: 96 hr LC_{50} (S): 5.10 µg/l
adults: 96 hr LC_{50} (S): 1.30 µg/l (1143)

	conc.	exposure	effect
Callinectes sapidus: larvae:	0.7–1.9 ppb	continuous	prolonged duration of zoeal megalopa development; increased time from hatch to first crab (1149; 1150)
Rhithropanopeus harrissii: larvae:	1.0–5.0 ppb	continuous	duration of zoeal development prolonged; time from hatch to first crab prolonged (1149; 1150)

Cancer magister: 4.0 µg/l continuous delay in molting
  juveniles: (1143)
  —Fish:
    Cyprinodon variegatus: 96 hr LC_{50} (FT): 49 µg/l (1131)
    Pimephales promelas: 96 hr LC_{50}: 7.5 µg/l (2138)
    Lepomis macrochirus: 96 hr LC_{50}: 62.0 µg/l (2113)
    Salmo gairdneri: 96 hr LC_{50}: 62.6 µg/l (219)
    Oncorhynchus kisutch: 96 hr LC_{50}: 66.2 µg/l (219)
    Oncorhynchus tschawtscha: 96 hr LC_{50}: 27.9 µg/l (219)
    Perca flavescens: 96 hr LC_{50}: 20.0 µg/l (2138)
    atlantic salmon: 24 hr LC_{50} (S): 75–92 µg/l (1189)

			96 hr LC_{50} (S)	
100%	*Fundulus heteroclitus*	mummichog	35 ppb	(2328)
100%	*Fundulus heteroclitus*	mummichog	35 ppb	(2329)
100%	*Fundulus majalis*	striped killifish	30 ppb	
100%	*Menidia menidia*	atlantic silverside	33 ppb	
100%	*Mugil cephalus*	striped mullet	63 ppb	
100%	*Anguilla rostrata*	American eel	12. ppb	
100%	*Thalassoma bifasciatum*	bluehead	13. ppb	
100%	*Sphaeroides maculatus*	northern puffer	150. ppb	(2329)
89.5%	*Gasterosteus aculeatus*	threespine stickleback	69.1 ppb	(2333)

    bluegill: 96 hr LC_{50}: 0.062 ppm
    rainbow trout: 96 hr LC_{50}: 0.020 ppm (1878)
  —Mammals:
    acute oral LD_{50} (rats): 6000 mg/kg
    in diet (rats) no ill effects: 0.02% for 2 years a reduction in growth 0.16% for 2 years (1955)
    Under the conditions of this study, methoxychlor was not found to be carcinogenic in Osborne-Mendel rats or B6C3F1 mice of either sex (1752)

methoxy DDT *see* methoxychlor

2-methoxy-3,6-dichlorobenzoic acid

C. WATER POLLUTION FACTORS:
  —Waste water treatment methods:
    A.C. type BL (Pittsburgh Chem. Co.): % adsorbed by 10 mg A.C. from 10^{-4} M aqueous solution, at pH 3.0: 37.5%
                  7.0: 11.2%
                  11.0: 6.0% (1313)

2-methoxyethanol *see* ethyleneglycolmonomethylether

2-methoxyethylacetate *see* ethyleneglycolmonomethyletheracetate

methoxyhexanone *see* 4-methoxy-4-methyl-2-pentanone

3-methoxy-4-hydroxybenzaldehyde *see* vanillin

4-methoxy-4-methyl-2-pentanone (methoxyhexanone "pentoxane"; "ME-6K")

$$CH_3-\underset{\underset{O-CH_3}{|}}{\overset{\overset{CH_3}{|}}{C}}-CH_3-\underset{\underset{O}{\parallel}}{C}-CH_3$$

A. PROPERTIES: m.w. 130.02; b.p. 159.1°C; sp.gr. 0.899 at 25/25°C
B. AIR POLLUTION FACTORS:
 —Odor: odor threshold values: 25 ppm (277)
C. WATER POLLUTION FACTORS:
 —Oxidation parameters: BOD_5: 0.11 NEN 3235-5.4 (277)
  COD: 2.24 NEN 3235-5.3 (277)
D. BIOLOGICAL EFFECTS:
 —Fish: goldfish: LD_{50} (24 hr): 3,800 mg/l—modified ASTM D 1345 (277)
 —Mammalia: rat: acute oral dose: 3.72 mg/kg (277)

1-methoxy-2-nitrobenzene (*o*-nitroanisole)

Manufacturing source: organic chemical industry. (347)
Users and formulation: dye mfg.; synthesis of guaiacol; organic synthesis; mfg. of pharmaceutical intermediates. (347)
A. PROPERTIES: m.w. 153.13; m.p. 10°C; b.p. 273/277°C; sp.gr. 1.254 at 20/4°C solub. 1.690 at 30°C
C. WATER POLLUTION FACTORS:
 —Biodegradation: decomposition by a soil microflora in >64 days (176)

1-methoxy-3-nitrobenzene (*m*-nitroanisole)

A. PROPERTIES: m.w. 153.13; m.p. 38°C; b.p. 258°C; sp.gr. 1.373 at 18°C; log P_{oct} 2.10
C. WATER POLLUTION FACTORS:
 —Biodegradation: decomposition by a soil microflora: >64 days (176)

1-methoxy-4-nitrobenzene (*p*-nitroanisole)

OCH₃—⟨benzene ring⟩—NO₂

A. PROPERTIES: m.w. 153.13; m.p. 52°C; b.p. 280/260°C; v.d. 5.29 sp.gr. 1.2192 at 60/4°C; solub. 70 mg/l at 15°C, 589 mg/l at 30°C; log P_{oct} 2.03
B. AIR POLLUTION FACTORS: 1 mg/cu m = 0.16 ppm, 1 ppm = 6.36 mg/cu m
C. WATER POLLUTION FACTORS:
 —Biodegradation: decomposition by a soil microflora: >64 days (176)

2-methoxyphenol (pyrocatecholmonomethylether; guaiacol; 1-hydroxy-2-methoxybenzene; *o*-methoxyphenol; *o*-hydroxyanisole; methylcatechol)

OH, OCH₃ on benzene ring

Manufacturing source: organic chemical industry. (347)
Users and formulation: mfg. perfumes and flavors (vanillin); raw material for mfg. of papaverine (medicinal); mfg. catechol and guaiacol compounds. (347)
Natural sources (*water and air*): wood, coal tar, major constituent of beechwood and creosote. (347)
Man caused sources (*water and air*): disposal of wood or coal tars and their decomposition would leach guaiacol; use of perfumes, flavors (vanillin), medicinals. (347)

A. PROPERTIES: colorless to pale yellow liquid; m.w. 124.14; m.p. 28.2/32°C b.p. 205°C; v.p. 0.103 mm at 25°C, 1 mm at 52°C, 100 mm at 144°C; v.d. 4.27; sp.gr. 1.1287 at 21/4°C; solub. 16,000 mg/l at 15°C; log P_{oct} 1.31/2.53 (calculated)
B. AIR POLLUTION FACTORS: 1 mg/cu m = 0.197 ppm, 1 ppm = 5.07 mg/cu m

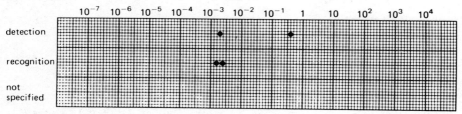

(610; 682; 871; 872)

C. WATER POLLUTION FACTORS:
 —Biodegradation: decomposition by a soil microflora: 4 days (176)

—Oxidation parameters: BOD$_5$:   68% ThOD (274)
                       COD: 100% ThOD (0.05 n k$_2$Cr$_2$O$_7$) (274)
                       KMnO$_4$: 48% ThOD (0.01 n KMnO$_4$) (274)
                       ThOD: 2.06 (274)
—Odor threshold: detection: 0.021 mg/kg (326;923)
                             0.003 mg/kg (883)
                             0.01 mg/l   (998)
—Taste threshold conc.: 0.05 mg/l (998)
D. BIOLOGICAL EFFECTS:
—Mammals: rat: oral dose by intubation: LD$_{50}$: 1.0–2.0 g/kg (211)

3-methoxyphenol (resorcinolmonomethylether; *m*-methoxyphenol; *m*-hydroxyanisole)

A. PROPERTIES: m.w. 124.14; m.p. <17.5°C; b.p. 244.3°C; v.d. 4.29; sp.gr. 1; log P_{oct} 1.58
B. AIR POLLUTION FACTORS: 1 mg/cu m = 0.19 ppm, 1 ppm = 5.16 mg/cu m
C. WATER POLLUTION FACTORS:
—Biodegradation: decomposition by a soil microflora: 16 days (176)

4-methoxyphenol (hydroquinonemonomethylether; *p*-hydroxyanisole)

A. PROPERTIES: m.w. 124.14; m.p. 53°C; b.p. 243°C; solub. 40,000 at 25°C; log P_{oct} 1.34
B. AIR POLLUTION FACTORS: 1 mg/cu m = 0.197 ppm, 1 ppm = 5.08 mg/cu m
C. WATER POLLUTION FACTORS:
—Biodegradation: decomposition by a soil microflora: 8 days (176)
D. BIOLOGICAL EFFECTS:
—Fish: goldfish: aprox. fatal conc.: 200 mg/l; 48 hr (226)
—Mammalia: rat: single oral LD$_{50}$: 1.6 g/kg (211)

methylacetaldehyde *see* propionaldehyde

methylacetate
CH$_3$COOCH$_3$
A. PROPERTIES: colorless liquid; m.w. 74.1; m.p. -99°C; b.p. 58°C v.p. 170 mm at

828 METHYLACETIC ACID

20°C, 235 mm at 25°C, 255 mm at 30°C; v.d. 2.56; sp.gr. 0.93 at 25/4°C; solub. 319,000 mg/l at 20°C, 240,000 mg/l at 20°C THC 381 kcal/mole; LHC 358 kcal/mole; sat. conc. 665 g/cu m at 20°C, 994 g/cu m at 30°C; log P_{oct} 0.18

B. AIR POLLUTION FACTORS: 1 mg/cu m = 0.330 ppm, 1 ppm = 3.03 mg/cu m

odor thresholds

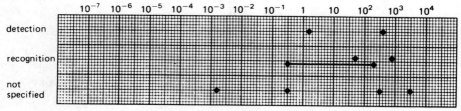

(57; 73; 210; 278; 298; 610; 665; 704; 749; 871)

USSR: human odor perception: nonperception: 0.4 mg/cu m
                                                 perception: 0.5 mg/cu m = 0.165 ppm
      human reflex response: no response: 0.066 mg/cu m
                                      adverse response: 0.08 mg/cu m   (170)
O.I.: at 20°C = 1,100   (316)

C. WATER POLLUTION FACTORS:
—Waste water treatment: A.C.: adsorbability: 0.054 g/g C, 26.2% reduction, infl. 1,030 mg/l, effl. 760 mg/l   (32)

D. BIOLOGICAL EFFECTS:
—Mammalia: cat: inhalation: lethal: 22,000 ppm, 2-3 hr
                                  eye irritation, salivation: 5,000 ppm, 20 min
                                  weight loss: 6,600 ppm, 6 hr/day, 8 days
           mouse: lethal: 11,000 ppm, 3 hr
                  no effect: 5,000 ppm, 20 min
—Man:
severe toxic effects: 500 ppm = 1,540 mg/cu m, 60 min
symptoms of illness: 200 ppm = 616 mg/cu m
unsatisfactory: 100 ppm = 308 mg/cu m   (185)
irritation persisting after exposure: 10,000 ppm for a short time   (211)

methylacetic acid *see* propionic acid

methylacetylene (propyne, allylene, propine)
CH₃CCH

A. PROPERTIES: m.w. 40.06; m.p. 104.7°C; b.p. -23/-27.5°C; v.p. 5.2 atm at 20°C, 6.9 atm at 30°C; v.d. 1.4; sp.gr. 0.678 at -27/4°C; solub. 3640 mg/l at 20°C   (242)

B. AIR POLLUTION FACTORS: 1 mg/cu m = 0.60 ppm, 1 ppm = 1.67 mg/cu m
—Manmade sources: in exhaust of diesel engine: 0.9% of emitted hydrocarbons   (72)

α-**methylacrolein** *see* methacrolein

β-**methylacrolein** *see* crotonaldehyde

methylacrylate
CH$_2$CHCOOCH$_3$
A. PROPERTIES: colorless liquid; m.w. 86.09; m.p. $<-75°$C; b.p. 80°C; v.p. 70 mm at 20°C, 110 mm at 30°C; v.d. 3.0; sp.gr. 0.958 at 20/4°C; solub. 52,000 mg/l; sat. conc. 319 g/cu m at 20°C, 499 g/cu m at 30°C
B. AIR POLLUTION FACTORS: 1 mg/cu m = 0.28 ppm, 1 ppm = 3.58 mg/cu m
 —Odor threshold conc. 0.017 mg/cu m (621)
C. WATER POLLUTION FACTORS:
 —Reduction of amenities: organoleptic limit: USSR 1970: 0.02 mg/l (181)
D. BIOLOGICAL EFFECTS:
 —Threshold conc. of cell multiplication inhibition of the protozoan *Uronema parduczi Chatton-Lwoff*: 64 mg/l (1901)
 —Mammals: rabbit: acute oral LD$_{50}$: 0.2 g/kg
 rat: inhalation: 3/6: 1,000 ppm, 4 hr (211)

α-methylacrylic acid *see* methacrylic acid

methylal *see* dimethoxymethane

methylallylchloride *see* methallylchloride

methylamine (monomethylamine; aminomethane)
CH$_3$NH$_2$
Use: intermediate for accelerators, dyes, pharmaceuticals, insecticides, fungicides, surface active agents; tanning; dyeing of acetate textiles; fuel additive; polymerization inhibitor; component of paint removers; solvent; photographic developer; rocket propellant
A. PROPERTIES: colorless gas; m.w. 31.06; m.p. $-92.5°$C; b.p. $-6.5°$C; v.p. 3.1 atm at 20°C, 4.3 atm at 30°C; v.d. 1.07; sp.gr. 0.769 at $-70/4°$C; solub. 11,539 at 12.5°C; THC 256 kcal/mole; LHC 236 kcal/mole; P_{oct} 0.57
B. AIR POLLUTION FACTORS: 1 mg/cu m = 0.778 ppm, 1 ppm = 1.27 mg/cu m
 —Odor: characteristic: quality: fishy, pungent
 T.O.C.: PIT$_{50\%}$: 0.021 ppm
 PIT$_{100\%}$: 0.021 ppm (2)
 -0.03 mg/cu m = 0.02 ppm (291)(57)
 T.O.C. = 4.3 mg/cu m = 3.3 ppm (279)
 O.I.: at 20°C = 940,000 (316)
C. WATER POLLUTION FACTORS:
 —Impact on biodegradation:
 $\sim 50\%$ inhibition of NH$_3$ oxidation in *Nitrosomonas* at 310 mg/l (407)
 —Reduction of amenities:
 odor threshold: avg.: 3.35 mg/l
 range: 0.65–5.23 mg/l (97; 886)
D. BIOLOGICAL EFFECTS:
 —Bacteria: *E. coli*: LD$_0$: 1 g/l
 —Algae: *Scenedesmus*: LD$_0$: 4 mg/l
 —Arthropoda: *Daphnia*: 480 mg/l (30)

—Fish: creek chub: critical range: 10–30 mg/l; 24 hr in Detroit river water (226)(243)
—Mammalia: rat: acute oral LD_{50}: 0.1–0.2 g/kg (10% soln) (211)

6-methyl-2-aminobenzothiazolehydrochloride
D. BIOLOGICAL EFFECTS:
—Fish:

	mg/l	24 hr	48 hr	
harlequin fish	LC_{10} (F)	2.6	1.2	
(*Rasbora heteromorpha*)	LC_{50} (F)	5.5	2.2	(331)

p-methylaminophenol (metol; elon)

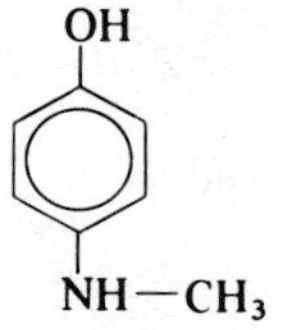

D. BIOLOGICAL EFFECTS:
—Fish: goldfish: approx. fatal conc.: 0.5 mg/l; 48 hr (226)

4-methylaminophenolsulfate

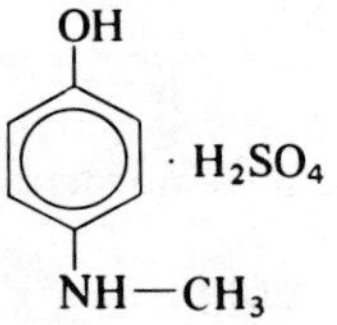

Use: reducing agent used primarily in the development of silver sensitized films and papers (1828)

A. PROPERTIES: m.w. 172.1; m.p. 260° (dec.)

C. WATER POLLUTION FACTORS:

theoretical:
TOD = 1.86
COD = 1.49
NOD = 0.37

analytical:
reflux COD = 79.9% recovery
rapid COD = 84.6% recovery
TKN = 93.9%
BOD_5 = 0.199
BOD_5/COD = 0.167
BOD_5 (acclimated) = 0.554

—The BOD_5 analysis measured a large apparent BOD due to chemical oxidation. Partial oxidation of this type is found to typically occur with substituted aromatic compounds. The Warburg analysis estimated this demand to be approximately 0.09.

	chemical conc. mg/l	effect
unacclimated System	50	inhibitory
	335	inhibitory

acclimated System 117 biodegradable
The compound proved to be strongly inhibitory to an unacclimated system at concentrations as low as 50 mg/l. Acclimation of the biomass to the compound eliminated any inhibitory effects. The data indicated biodegradation of 4-methylaminophenol sulfate proceeded at a moderate rate. (1828)

D. BIOLOGICAL EFFECTS:
 —Algae: *Selenastrum capricornutum*: 0.1 mg/l: no effect
　　　　　　　　　　　　　　　　　1.0 mg/l: no effect
　　　　　　　　　　　　　　　　　10.0 mg/l: no effect (1828)
 —Crustacean: *Daphnia magna*: LC_{50}: 0.019 mg/l (1828)
 —Fish: *Pimephales promelas*: LC_{50}: 0.25 mg/l (1828)

methylamylalcohol (methylisobutylcarbinol; 4-methyl-2-pentanol; MIBC) $(CH_3)_2CHCH_2CHOHCH_3$

Use: solvent for dyestuffs, oils, gums, resins, waxes, nitrocellulose and ethylcellulose; organic synthesis; froth flotation; brake fluids

A. PROPERTIES: colorless liquid; m.w. 102.2; m.p. -90°C; b.p. 132°C; v.p. 5 mm at 20°C, 13 mm at 33°C, 27 mm at 45°C; v.d. 3.52; sp.gr. 0.81 at 20/20°C; solub. 17 g/l at 20°C; sat.conc. 25 g/cu m at 20°C, 45 g/cu m at 30°C

B. AIR POLLUTION FACTORS: 1 mg/cu m = 0.239 ppm, 1 ppm = 4.17 mg/cu m
 —Odor: characteristic: quality: sweet, alcohol
　　　　　　　　　　　hedonic tone: unpleasant to pleasant
　　　　　　　　　　　absolute perception limit: 0.33 ppm
　　　　　　　　　　　50% recogn.: 0.52 ppm
　　　　　　　　　　　100% recogn.: 0.52 ppm
 O.I.: 12,634 (19)

C. WATER POLLUTION FACTORS:
 —Oxidation parameters: BOD_5: 2.12 NEN 3235 5.4
　　　　　　　　　　　　COD: 2.60 NEN 3235 5.3 (277)

D. BIOLOGICAL EFFECTS:
 —Fish: goldfish: LD_{50} (24 hr): 360 mg/l modified ASTM D 1345 (277)
 —Mammalia: rat: single oral LD_{50}: 2.59 g/kg
　　　　　　　　inhalation: 5/6: 2,000 ppm, 8 hr (211)

methyl-*n*-amylketone *see* 2-heptanone

N-methylaniline

Use: organic synthesis; solvent; acid acceptor

A. PROPERTIES: yellow liquid; m.w. 107.15; m.p. -57°C; b.p. 195.7°C v.p. 0.3 mm at 20°C, 0.65 mm at 30°C; v.d. 3.70; sp.gr. 0.986 at 20/4°C; sat. conc. 1.8 g/cu m at 20°C, 3.7 g/cu m at 30°C; log P_{oct} 1.66/1.82

B. AIR POLLUTION FACTORS: 1 mg/cu m = 0.23 ppm, 1 ppm = 4.45 mg/cu m
 —Odor threshold conc.: recognition: 6.9 - 8.6 mg/cu m (610)

C. WATER POLLUTION FACTORS:
  —Impact on biodegradation processes:
    NH_3 oxidation by *Nitrosomonas*: at 100 mg/l: 90% inhibition
                                                at 50 mg/l: 83% inhibition
                                                at 10 mg/l: 71% inhibition
                                                at <1 mg/l: ~50% inhibition

o-methylaniline *see o*-toluidine

m-methylaniline *see m*-toluidine

p-methylaniline *see p*-toluidine

2-methylanthracene

Manmade sources:
in wood preservative sludge (n.s.i.): 7.77 g/l of raw sludge   (960)
in gasoline: 0.74 mg/l   (1070)

B. AIR POLLUTION FACTORS:
  —Manmade emissions:
    in coke oven airborne emissions (n.s.i.): 85–1692 µg/g of sample   (960)
    in exhaust condensate of gasoline engine: 0.09–0.10 mg/l gasoline consumed
                                                                                                        (1070)
  —Ambient air quality:
    glc's in Birkenes (Norway): Jan.–June 1977: avg.: 0.007 ng/cu m
                                                                                            range: n.d.–0.05 ng/cu m
                                                                                               ($n = 18$)
    glc's in Rørvik (Sweden): Dec. '76–April '77: avg.: 0.007 ng/cu m
                                                                                          range: n.d.–0.09 ng/cu m
                                                                                          ($n = 21$)   (1236)
    Organic fraction of suspended matter (n.s.i.):
      Bolivia at 5200 m altitude (Sept.–Dec. 1975): 0.031–0.038 µg/1000 cu m
      (+ methylphenanthrene)
      residential area, 10 km south of Antwerp, Belgium (Jan.–April 1976): 0.37–1,3
      µg/1000 cu m (+ methylphenanthrene)   (960)

methylbenz(*a*)anthracene   (methylB(a)A)

A. PROPERTIES:
   9-methyl B(a)A: solub. 0.037 mg/l at 24°C (99% purity)
   10-methyl B(a)A: solub. 0.011 mg/l at 24°C (99% purity)
B. AIR POLLUTION FACTORS:
   —Manmade sources: in coke oven emissions (n.s.i.): 22.4–1669.7 µg of sample
   (960)
C. WATER POLLUTION FACTORS:
   —Sediment quality: in sediment of Wilderness Lake, Colin Scott, Ontario; 1976:
     1 ppb (dry wt) (n.s.i.)   (932)
D. BIOLOGICAL EFFECTS: (7-methyl B(a)A)
   —Carcinogenicity: positive
   —Mutagenicity in the *Salmonella* test: positive
     22 revertant colonies/nmol
     897 revertant colonies at 10 µg/plate   (1883)

methylbenzene   *see* toluene

4-methylbenzenesulfonic acid   *see* p-toluenesulfonic acid

4-methylbenzenethiol

Manmade sources: in wood preservative sludge: 0.52 g/l of raw sludge

methyl-2-benzimidazolecarbamate   (MBC)
Manmade source: product of hydrolysis of Benomyl in water and soil
   (1373; 1374)

C. WATER POLLUTION FACTORS:
   —Impact on biodegradation:
   effect of MBC on nitrification:
   by *Nitrosomonas* sp.

	µg nitrite recovered / ml medium incubation		
conc. ppm	6 days	15 days	
0	3.8	108.0	
10	4.5	146.4	
100	3.0	104.5	(1372)
by *Nitrobacter agilis*			
0	289.5	0	
10	296.1	0	
100	473.8	0	(1372)

methylbenzoate (niobe oil)

Manufacturing source: organic chemical industry; natural products industry.
(347)

Users and formulation: dye carrier in dyeing of polyester fibers; additives for disinfectants, soy sauce, and pesticides, perfume mfg.; solvent for cellulose esters and ethers, resins and rubber; flavoring. (347)

Natural sources (water and air): occurs in oils of clove, ylang ylang, and turberose.
(347)

A. PROPERTIES: colorless liquid; m.w. 136.14; m.p. $-12°C$; b.p. $199°C$, v.p. 1 mm at $39°C$; v.d. 4.69; sp.gr. 1.088 at $20/4°C$, solub. 157 mg/l at $30°C$; $\log P_{oct}$ 2.12

B. AIR POLLUTION FACTORS:
 —Odor threshold conc.: 0.0025 mg/cu m (607)

D. BIOLOGICAL EFFECTS:
 —Mammals: rat: acute oral LD_{50}: 3.4 g/kg; inhalation: no deaths: sat. vap. 8 hr
(211)

o-methylbenzoic acid see o-toluic acid

m-methylbenzoic acid see m-toluic acid

p-methylbenzoic acid see p-toluic acid

methylbenzo (ghi) perylene
 Manmade sources: in coke oven emissions: 36.8 µg/g of sample (n.s.i.) (960)

methylbenzo(a)pyrene
B. AIR POLLUTION FACTORS:
 —Manmade sources:
  in coke oven emissions (n.s.i.): 5.9–344.1 µg/g of sample (960)
C. WATER POLLUTION FACTORS:
 —Sediment quality:
  in sediment of Wilderness Lake, Colin Scott, Ontario (1976): 10 ppb (dry wt) (n.s.i.) (932)

methylbenzo(e)pyrene
C. WATER POLLUTION FACTORS:
 —Sediment quality:
  in sediment of Wilderness Lake, Colin Scott, Ontario (1976): 6 ppb (dry wt) (n.s.i.) (932)

methylbenzothiazole
C. WATER POLLUTION FACTORS:

—Biodegradation: at 1.0 mg/l:

	normal sewage	adapted sewage	
after 24 hr	0%	0%	
after 135 hr	0%	25%	(997)

—Odor threshold conc. detection: 0.0075 mg/kg (2-methylbenzothiazole) (894)

α-methylbenzylalcohol (methylphenylcarbinol; 1-phenylethanol)

Manufacturing source: organic chemical industry (347)
Users and formulation: perfume and flavoring manufacture; dye manufacture; laboratory reagent (347)
A. PROPERTIES: colorless liquid; m.w. 122.16; b.p. 203/205°C; sp.gr. 1.013 at 20/4°C
C. WATER POLLUTION FACTORS:
—Oxidation parameters:
BOD_{10}: 1.0 std. dil. sew. (256)
2.1 (redistilled α-methylbenzylalcohol)
std. dil. sew. (250)
—Reduction of amenities: T.O.C.: 1,500 mg/l (297)
avg. 1.45 mg/l (403)
—Water quality:
in Kanawha river:
13.11.1963: raw: not detectable
21.11.1963: raw: 26 ppb
after aeration: 12 ppb
carbon filtered: no detectable (159)
—Water treatment method:
conventional municipal treatment + A.C.: infl. 0.026 mg/l, effl. n.d. (404)

methylbiphenyl
Manmade sources:
in airborne coal tar emissions: 0.29 mg/g of samples or 10 µg/cu m of air sampled (n.s.i.)
in coal tar: 3.61 mg/g of sample (n.s.i.)
in wood preservative sludge: 9.79 g/l of raw sludge (n.s.i.) (993)

β-methylbivinyl *see* isoprene

methylbromide (bromomethane)
CH_3Br
Use: soil and space fumigant; organic synthesis
A. PROPERTIES: colorless liquid or gas; m.w. 94.95; m.p. −93°C; b.p. 4.6°C; v.d. 3.27; sp.gr. 1.73 at 0/0°C; solub. 900 mg/l at 20°C

B. AIR POLLUTION FACTORS: 1 mg/cu m = 0.25 ppm, 1 ppm = 3.95 mg/cu
  —Manmade sources: glc's in rural Washington Dec. '74–Feb '75: <5 ppt
D. BIOLOGICAL EFFECTS:
  —Fish:
    Lepomis macrochirus: static bioassay in fresh water at 23°C, mild aeration applied after 24 hr:

material added ppm	24 hr	% survival after 48 hr	72 hr	96 hr	best fit 96 hr LC_{50} ppm
14 (~3.3 ml/l)	30	0	—	—	
11 (~2.5 ml/l)	60	50	50	50	11
7 (~1.7 ml/l)	100	90	90	90	
1.4 (~0.33 ml/l)	100	100	100	100	

  Menidia beryllina: static bioassay in synthetic seawater at 23°C: mild aeration applied after 24 hr:

material added ppm	24 hr	% survival after 48 hr	72 hr	96 hr	best fit 96 hr LC_{50} ppm
14 (~3.3 ml/l)	0	—	—	—	
11 (~2.5 ml/l)	70	60	60	60	12
7 (~1.7 ml/l)	100	100	80	80	(352)

  —Mammals: rat: inhalation: 8 hr survival dose: 260 ppm   (211)

2-methyl-1,3-butadiene   *see* isoprene

3-methylbutanal   (isovaleraldehyde; β-methylbutyraldehyde; isoamylaldehyde) $(CH_3)_2 CHCH_2 CHO$
  Manufacturing source: organic chemical industry; petroleum refining industry.
                                                                                          (347)
  Users and formulation: flavor/perfume mfg.; pharmaceuticals; synthetic resins; rubber accelerators.   (347)
  Natural sources (*water and air*): oils in lemon, orange, peppermint, eucalyptus.
                                                                                          (347)
A. PROPERTIES: colorless liquid; m.w. 86.13; m.p. -51°C; b.p. 92.5°C; sp.gr. 0.803 at 17/4°C
C. WATER POLLUTION FACTORS:
  —Waste water treatment:
    A.S. after 6 hr: 9.2% of ThOD
           12 hr: 14.2% of ThOD
           24 hr: 16.1% of ThOD   (88)
  —Odor threshold conc.: detection: 0.007 mg/kg   (908)
                                   0.00202 mg/kg   (874)

2-methylbutane   *see* isopentane

3-methylbutanoic acid   *see* isovaleric acid

1-methyl-1-butanol   *see sec-act*-amylalcohol

2-methyl-2-butanol   (*tert*-amylalcohol; dimethylethylcarbinol; amylenehydrate; *tert*-pentanol

$CH_3CH_2C(CH_3)(OH)CH_3$

A. PROPERTIES: colorless liquid; m.w. 88.15; m.p. $-11.9°C$; b.p. $101.8°C$; sp.gr. 0.809; solub. 140,000 mg/l at $30°C$; log P_{oct} 0.89

B. AIR POLLUTION FACTORS: 1 mg/cu m = 0.273 ppm, 1 ppm = 3.662 mg/cu m
  - Odor: characteristic: quality: sour, sharp
    hedonic tone: unpleasant to neutral
    threshold values: 8.2 mg/cu m; 2.3 ppm (307; 279)
    detection: 20–40 mg/cu m (776)
    recognition: 2.0–3.0 mg/cu m (610)
    absolute perception limit: 0.04 ppm
    50% recogn.: 0.23 ppm
    100% recogn.: 0.23 ppm
  O.I.: 17,130 (19)

C. WATER POLLUTION FACTORS:
  - Reduction of amenities: T.O.C. = 10 mg/l (2-methylbutanol) (299)
  - Waste water treatment: A.S. after 6 hr: 1.3% of ThOD
                                     12 hr: 1.7% of ThOD
                                     24 hr: 3.7% of ThOD (88)
    Reverse osmosis: 88% rejection from a 0.01 M solution
    (221)

D. BIOLOGICAL EFFECTS:
  - Toxicity threshold (cell multiplication inhibition test):
    bacteria (*Pseudomonas putida*): 410 mg/l (1900)
    algae (*Microcystis aeruginosa*): 105 mg/l (329)
    green algae (*Scenedesmus quadricauda*): 1250 mg/l (1900)
    protozoa (*Entosiphon sulcatum*): 680 mg/l (1900)
    protozoa (*Uronema parduczi Chatton-Lwoff*): 859 mg/l (1901)
  - Fish:
    creek chub: LD_0: 1,300 mg/l, 24 hr in Detroit river water
                LD_{100}: 2,000 mg/l, 24 hr in Detroit river water (243)
  - Mammals:
    ingestion: rats: single oral LD_{50}: 1,000–2,000 mg/kg (1546)

3-methyl-1-butanol *see prim*-isoamylalcohol

3-methyl-2-butanone (isopropylmethylketone)
$CH_3COCH(CH_3)_2$

A. PROPERTIES: colorless liquid; m.w. 86.13; m.p. $-92°C$; b.p. $93°C$; sp.gr. 0.815 at $15/4°C$

B. AIR POLLUTION FACTORS:
  - Odor threshold conc.: recognition: 15–17 mg/cu m (610)

C. WATER POLLUTION FACTORS:
  - Waste water treatment:
    A.S. after 6 hr: 1.2% of ThOD
               12 hr: 1.3% of ThOD
               24 hr: 2.2% of ThOD (88)
  - Odor threshold conc.: detection: 1670 mg/kg (874)

2-methyl-1-butene
$H_2CC(CH_3)CH_2CH_3$
A. PROPERTIES: colorless, volatile liquid, disagreeable odor; b.p. 31.4°C; sp.gr. 0.650 at 20/20°C
B. AIR POLLUTION FACTORS:
  —Manmade sources: in gasoline exhaust: 0.34–0.52 vol% (312)
    evaporation from gasoline fuel tank: 4.3 vol.% of total evaporated H.C.'s
    evaporation from carburetor: 0.6–1.7 vol.% of total evaporated H.C.'s
    (398; 399; 400; 401; 402)

2-methyl-2-butene (trimethylethylene; β-isoamylene)
$(CH_3)_2CCHCH_3$
A. PROPERTIES: colorless liquid; m.w. 70.13; m.p. -124°C; b.p. 38.4°C; sp.gr. 0.668 at 13/4°C
B. AIR POLLUTION FACTORS:
  —Odor threshold conc. 0.25 ppm (279)
  —Manmade sources:
    evaporation from gasoline tank: 2.7–7.3 vol.% of total evaporated H.C.'s
    evaporation from carburetor: 1.2–3.9 vol.% of total evaporated H.C.'s
    (398; 399; 400; 401; 402)
  —Atmospheric reactions:
    estimated lifetime under photochemical smog conditions in S.E. England: 0.46 hr
    (1699; 1700)
C. WATER POLLUTION FACTORS:
  —Waste water treatment:
    A.S. after 6 hr: 0.1% of ThOD
           12 hr: 1.1% of ThOD
           24 hr: 0.9% of ThOD (88)

3-methyl-1-butene (isopropylethylene; α-isoamylene)
$(CH_3)_2CHCHCH_2$
A. PROPERTIES: colorless liquid; m.w. 70.13; m.p. -135°C; b.p. 21/25°C v.d. 2.4; sp.gr. 0.648 at 20/4°C; solub. 130 mg/l at 20°C (242)
B. AIR POLLUTION FACTORS:
  —Atmospheric reactions: R.C.R.: 8.62 (all isomers) (49)
              reactivity: NO ox.: ranking: 3 (all isomers) (63)
  —Manmade sources: in combustion gas of home central heating system on gasoil:
    15 ppm at CO_2 = 7%
    3.5 g/kg gas oil at CO_2 = 6%
    1.9 g/kg gas oil at CO_2 = 7% (182)
    in gasoline: 0.06–0.08 vol% (312)
    evaporation from gasoline fuel tank: 0.8 vol.% of total evaporated H.C.'s
    evaporation from carburetor: 0.1–0.2 vol.% of total evaporated H.C.'s
    (398; 399; 400; 401; 402)
C. WATER POLLUTION FACTORS:
  —Waste water treatment:
    A.S. after 6 hr: toxic
           12 hr: 0.8% of ThOD
           24 hr: 0.6% of ThOD (88)

D. BIOLOGICAL EFFECTS:
—Man: EIR: 1.9 (all isomers) (49)

2-methyl-1-butene-3-one *see* methylisopropenylketone

1-methyl-4-*tert*-butylbenzene *see* p-*tert*-butyltoluene

methyl-1-(butylcarbamoyl)-2-benzimidazole carbamate *see* benomyl

β-methylbutyraldehyde *see* 3-methylbutanal

methylcarbitolacetate *see* diethyleneglycolmonomethyletheracetate

methylcatechol *see* 2-methoxyphenol

methylcellosolve *see* ethyleneglycolmonomethylether

methylcellosolveacetate *see* ethyleneglycolmonomethyletheracetate

methylchloride (chloromethane)
CH_3Cl
  Manufacturing source: organic chemical industry. (347)
  Users and formulation: mfg. silicones, tetraethyllead, synthetic rubber and methyl cellulose; refrigerant mfg.; mfg. of organic chemicals (methylene chloride, chloroform, carbontetrachloride, etc.); mfg. fumigants; low temperature solvent; catalyst carrier in polymerization; medicine; fluid for thermometric or thermostatic equipment; methylating agent; extractant; propellant; herbicide. (347)
A. PROPERTIES: colorless gas; m.w. 51; m.p. $-97.7°C$; b.p. $-24°C$; v.p. 5.0 atm at $20°C$, 6.7 atm at $30°C$; v.d. 1.8; sp.gr. 0.991 at $-25°C$; solub. 4,000 cu cm/l
B. AIR POLLUTION FACTORS: 1 mg/cu m = 0.48 ppm, 1 ppm = 2.09 mg/cu m
  —Odor: odor threshold values: 21 mg/cu m (307)
               recogn.: >22.5 mg/cu m (73)
               $PIT_{50\%}$: >10 ppm
               $PIT_{100\%}$: >10 ppm (2)
  O.I. at $20°C$ = 200,000 (316)
  —Manmade sources: in cigarette smoke: 1,200 ppm (66)
               in rural Washington: Dec '74–Feb '75: 0.530 ppb (316)
  —Control methods: activated carbon: retentivity: 5% wt of adsorbent (83)
C. WATER POLLUTION FACTORS:
  —Waste water treatment: evaporation from water at $25°C$ of a 1 ppm solution: 50% after 27 min, 90% after 91 min (313)
D. BIOLOGICAL EFFECTS:
  —Toxicity threshold (cell multiplication inhibition test):
   bacteria (*Pseudomonas putida*): 500 mg/l (1900)
   algae (*Microcystis aeruginosa*): 550 mg/l (329)
   green algae (*Scenedesmus quadricauda*): 1450 mg/l (1900)
   protozoa (*Entosiphon sulcatum*): 8000 mg/l (1900)

840   METHYLCHLOROFORM

—Fish:
  Lepomis macrochirus: static bioassay in fresh water at 23°C, mild aeration applied after 24 hr:

material added ppm	24 hr	% survival after 48 hr	72 hr	96 hr	best fit 96 hr LC_{50} ppm
1800 (~800 ml/l)	0	died in 5-20 min			
900 (~400 ml/l)	10	10	10	10	550
450 (~200 ml/l)	100	90	90	90	
300 (~133 ml/l)	100	100	100	100	

  Menidia beryllina: static bioassay in synthetic seawater at 23°C: mild aeration applied after 24 hr:

material added ppm	24 hr	% survival after 48 hr	72 hr	96 hr	best fit 96 hr LC_{50} ppm
900 (~400 ml/l)	0-1 hr	—	—	—	
450 (~200 ml/l)	0-1 hr	—	—	—	
300 (~133 ml/l)	40	40	40	40	270
150 (~67 ml/l)	100	100	100	100	

—Mammalia:
  guinea pigs, mice, dogs, rabbits: inhalation: no effect: 500 ppm, 6 hr/day, 6 day/w, 175 days
  most animals are killed in a short time: 150,000–300,000 ppm
  dangerous in 30–60 min: 20,000–40,000 ppm
  max for 60 min without serious effect: 7,000 ppm
  max for 8 hr: 500–1,000 ppm   (211)

methylchloroform   *see* 1,1,1-trichloroethane

2-methyl-4-chlorophenoxyacetic acid   (metaxon; 4-chloro-*o*-toloxyacetic acid; 4-chloro-*o*-cresoxyacetic acid; MCP)

$$Cl-\underset{}{\underset{}{C_6H_3(CH_3)}}-OCH_2COOH$$

(It should be stated which salt or ester is present.)
Use: hormone type herbicide, commercial grade contained 4% of 4-chloro-*o*-cresol
                                                                                            (1390)
A. PROPERTIES: solub. 730 mg/l (acid)
C. WATER POLLUTION FACTORS:
  —75–100% disappearance from soils: 3 months
  —Biodegradation:
    metabolized by bacteria to: 5-chloro-3-methylcatechol   (1460)
                                  4-chloro-*o*-cresol   (1463)
D. BIOLOGICAL EFFECTS:
  —Mussels:
    Crassostrea virginica (American oyster), egg:
      static lab bioassay: 48 hr TLm, 1.562×10^4 ppb

Crassostrea virginica (American oyster), larvae:
  static lab bioassay: 14 day TLm, 3.13×10^4 ppb (2324)
—Fish:
Lepomis macrochirus:	48 hr LC_{50}: 1500 µg/l	(2116)
blue gill:	48 hr LC_{50}: 100 ppm	(1878)
brown trout:	24 hr $LD_{50}(S)$: 147 mg/l	(1522)

—Mammals:
  mice: 54% inhibition of testicular DNA synthesis at 200 mg/kg ($p<0.01$) (1325)
  rat: LD_{50}: 500 mg/kg (1390)

3-methylcholanthrene

B. AIR POLLUTION FACTORS:
  —Ambient air quality:
    glc's in Budapest 1973: heavy traffic area: winter: 18.6 ng/cu m
    (6–20 hr)   summer: 14.9 ng/cu m (1259)
D. BIOLOGICAL EFFECTS:
  —Carcinogenicity: positive
  —Mutagenicity in the *Salmonella* test: positive
    58 revertant colonies/nmol
    2160 revertant colonies at 10 µg/plate (1883)

2-methylchrysene

Manmade sources:
in gasoline: 8 µg/l
in exhaust condensate of gasoline engine: 5 µg/l gasoline consumed (1070)
in wood preservative sludge (n.s.i.): 0.43 g/l of raw sludge
in coke oven airborne emissions: 107.6–1151.6 µg/g of sample (960)

4-methylchrysene

Manmade sources:
in gasoline: 8 µg/l
in exhaust condensate of gasoline engine: 5 µg/l gasoline consumed (1070)

methylcyanide *see* acetonitrile

methyl-2-cyanoacrylate
CH$_2$C(CN)COOCH$_3$
A. PROPERTIES: colorless liquid; v.p. 48–49°C at 2.5 mm; sp.gr. 1.1044 at 27/4°C
B. AIR POLLUTION FACTORS:
　—Control methods:
　　activated carbon: inlet 10 ppm: removal efficiency: 95% up to 0.2 lb/lb carbon—
　　　American Carbon Inc. type G 107　　　　　　　　　　　　　　　　　　(229)
　—Odor threshold conc.: 4.5–13.5 mg/cu m　　　　　　　　　　　　　　　　(750)

methylcyclohexane　(hexahydrotoluene; cyclohexylmethane)

Use: solvent for cellulose ethers; organic synthesis
A. PROPERTIES: colorless liquid; m.w. 98.18; m.p. −126°C; b.p. 101°C; v.p. 144 mm
　at 20°C; v.d. 3.38; sp.gr. 0.77 at 20/4°C; solub. 14.0 mg/l at 20°C; sat. conc.
　192 g/cu m at 20°C, 295 g/cu m at 30°C　　　　　　　　　　　　　　　　(242)
B. AIR POLLUTION FACTORS: 1 mg/cu m = 0.25 ppm, 1 ppm = 4.08 mg/cu m
C. WATER POLLUTION FACTORS:
　—Biodegradation:
　　incubation with natural flora in the groundwater - in presence of the other
　　components of high-octane gasoline (100 µl/l):
　　biodegradation: 75% after 192 hr at 13°C (initial conc.: 0.05 µg/l)　　　(956)
　　methylcyclohexane → 4-methylcyclohexane　　　　　　　　　　　　　　(203)

D. BIOLOGICAL EFFECTS:
　—Fish:
　　golden shiner:　96-h LC$_{50}$: 72.0 mg/l (emulsion) (R)
　　　　　　　　　　96-h LC$_{50}$: 240.0 µl/l (unemulsified) (R)
　　rainbow trout: 42-d BCF: 150 (F)　　　　　　　　　　　　　　　　　　(1253)
　—Mammalia: rabbits: inhalation: 100/100: 10,000 ppm, 6 hr/day, 5 day/w, 2w
　　　　　　　　　　　　LD$_{25}$: 7,300 ppm, 6 hr/day, 5 day/w, 2w
　　　　　　　　　　　　LD$_0$: 5,600 ppm, 6 hr/day, 5 day/w, 4w
　　monkey: no sign of illness: 373 ppm, 6 hr/day, 5 day/w, 10w　　　　　　(211)

4-methylcyclohexanol　(hexahydro-*p*-cresol; methylhexalen; hexahydromethylphenol; methyladronal; methylanol; sextol)

A. PROPERTIES: aromatic colorless liquid; m.w. 114.18; f.p. -50°C; b.p. 174°C; v.p. 1.5 mm at 30°C; v.d. 3.94; sp.gr. 0.912; solub. ±35 g/l at 20°C
B. AIR POLLUTION FACTORS: 1 mg/cu m = 0.214 ppm; 1 ppm = 4.67 mg/cu m
 —Odor: odor threshold values: <500 ppm (211)
C. WATER POLLUTION FACTORS:
 —Biodegradation: adapted A.S. at 20°C—product as sole carbon source: 94.0% COD removal at 40.0 mg COD/g dry inoculum/hr (327)
D. BIOLOGICAL EFFECTS:
 —Mammalia: rabbit: single oral LD_{50}: 1.75–2.0 g/kg
  inhalation: LC_0: 503 ppm, 50 × 6 hr
  : no effect: 232 ppm, 50 × 6 hr (211)

2-methylcyclohexanone (methylanon; sexton B)

A. PROPERTIES: colorless liquid; m.w. 112.17; m.p. -19°C; b.p. 163°C; v.d. 3.86; sp.gr. 0.925 at 18°C
B. AIR POLLUTION FACTORS: 1 mg/cu m = 0.21 ppm, 1 ppm = 4.66 mg/cu m
 —Odor: characteristic: quality: peppermintlike
D. BIOLOGICAL EFFECTS:
 —Toxicity threshold (cell multiplication inhibition test):

bacteria (*Pseudomonas putida*):	60 mg/l	(1900)
algae (*Microcystis aeruginosa*):	26 mg/l	(329)
green algae (*Scenedesmus quadricauda*):	88 mg/l	(1900)
protozoa (*Entosiphon sulcatum*):	160 mg/l	(1900)
protozoa (*Uronema parduczi Chatton-Lwoff*):	349 mg/l	(1901)

 —Mammals:
  rabbit: single oral LD_{100}: 1.0–1.2 g/kg (*p*- and *m*-isomer) (211)
   inhalation: no effect level: 182 ppm, 50 × 6 hr (211)

4-methylcyclohexanone

C. WATER POLLUTION FACTORS:
—Biodegradation: adapted A.S. at 20°C—product as sole carbon source: 96.7% COD removal at 62.5 mg COD/g dry inoculum/hr (327)

N-methyl-5-cyclohexenyl-5-methylbarbituric acid *see* hexobarbital

methylcyclopentadienylmanganesetricarbonyl (MMT)
Use: antiknock agent, used as an octane improver in unleaded gasoline.
A. PROPERTIES: commercial grade MMT is a dark orange liquid, completely soluble in hydrocarbons; solub. in water 70 ppm at 25°C; sp.gr. 1.38; b.p. 232.8°C; v.p. 0.047 mm at 20°C; half-life of a few seconds in air; decomposes rapidly in sunlight
D. BIOLOGICAL EFFECTS:
—Mammals:
rabbits: dermal toxicity: MMT caused a significant toxic effect on numerous body organs—LD_{50}: 140–795 mg/kg
rats: inhalation: one hour: LC_{50}: 247 mg/cu m
four hours: LC_{50}: 76 mg/cu m
rats: single intragastric administration: LD_{50}: 58 mg/kg
mice: single intragastric administration: LD_{50}: 230 mg/kg (1296)

methylcyclopentane

Use: organic synthesis; extractive solvent; azeotropic distillation agent
A. PROPERTIES: colorless liquid; sp.gr. 0.750 (20/4°C); m.p. −142.5°C; b.p. 72°C at 742 mm
C. WATER POLLUTION FACTORS:
—Biodegradation:
incubation with natural flora in the groundwater—in presence of the other components of high-octane gasoline (100 μg/l): biodegradation: 10% after 192 hr at 13°C (initial conc.: 0.41 μl/l) (956)

methyldibenzanthracene
Manmade sources: in coke oven emissions: 89.0 μg/g of sample (n.s.i.). (960)

methyl-3-(3′,5′-di-*t*-butyl-4′-hydroxyphenyl)propionate

C. WATER POLLUTION FACTORS:
—Water and sediment quality:
in river water: 0.025–0.20 ppm; in river sediment: 1.5–170 ppm (555)

1-methyl-2,4-dinitrobenzene *see* 2,4-dinitrotoluene

2-methyl-4,6-dinitrophenol *see* 4,6-dinitro-*o*-cresol

methyldisulfide *see* dimethyldisulfide

methyldithiocarbamate (carbathione)
C. WATER POLLUTION FACTORS:
—Reduction of amenities: organoleptic limit—USSR 1970: 0.02 mg/l (181)
—Waste water treatment: 75% reduction of nitrific. of non acclimated activated sludge: at 0.9 mg/l (30)

methyldithiomethane *see* dimethyldisulfide

1,12-methylenebenzo(*ghi*)perylene (benzo(*ghi*)cyclopenta(*pqr*)perylene)

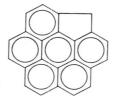

Manmade sources:
in gasoline: <0.2 µg/l
in exhaust condensate of gasoline engine: 19–41 µg/l gasoline consumed (1070)
A. PROPERTIES: m.w. 288

8,9-methylenebenzo(*e*)pyrene (11H-cyclopenta(*qrs*)benzo(*e*)pyrene)

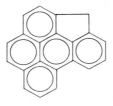

Manmade sources:
in gasoline: 0.013 mg/l
in exhaust condensate of gasoline engine: 0.017–0.043 mg/l gasoline consumed
(1070)

10,11-methylene-benzo(*a*)pyrene (10,H-cyclopenta(*mno*)benzo(*a*)pyrene

846 4,4-METHYLENE-bis-(2-CHLOROANILINE)

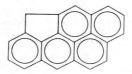

Manmade sources:
in gasoline: 0.005 mg/l
in exhaust condensate of gasoline engine: 0.008–0.018 mg/l gasoline consumed
(1070)

4,4-methylene-bis-(2-chloroaniline) (3,3′-dichloro-4,4′-diaminophenylmethane; p,p′-methylene-bis(o-chloroaniline)

$$NH_2-\underset{Cl}{\bigcirc}-CH_2-\underset{Cl}{\bigcirc}-NH_2$$

Use: prinicpal use is in curing urethane and epoxy resins, and crosslinking urethane foam

Composition: commercial preparations frequently contain also 2-chloroaniline and a compound containing three 2-chloroaniline moieties joined by methylene groups
(1858)

A. PROPERTIES: tan colored pellets; m.p. 99–107°C; sp.gr. 1.44
C. WATER POLLUTION FACTORS:
—Water and sediment quality:
survey of 4,4′-methylene bis (2-chloro-aniline) concentrations in water, sediment and sewage sludge, leading away from a manufacturing plant in Adrian, Michigan (USA):

industrial lagoon sediment:	minimum 1600 ppm (dry weight)
industrial lagoon effluent water:	250 ppb
industrial site deep well water:	1.5 ppb
surface runoff water from industrial site:	1 ppb
sewage treatment plant: influent:	<0.5 ppb
sewage treatment plant: effluent:	<0.5 ppb
sewage treatment plant: activated sludge:	estimated 18 ppm (dry weight)
Raisin river water:	not detected (≤0.1 ppb)

D. BIOLOGICAL EFFECTS:
—Fish:
In two white suckers collected near the Adrian sewage treatment plant outfall contained 13–49 ppb 2-chloroaniline, suggesting that 2-chloroaniline residues are chemically bound to macromolecules of the fish tissue (1858)
—Animal carcinogen (1859)
—Carcinogenicity: positive
—Mutagenicity in the *Salmonella* test: positive
  2.7 revertant colonies/nmol
  1050 revertant colonies at 100 μg/plate (1883)

methylene*bis*thiocyanate (MBT)
C. WATER POLLUTION FACTORS:
  —Impact on biodegradation processes:
    MBT added as a shock treatment to a continuous-flow model waste water treatment system:

ppm MBT in feed	% metabolized in 24 hours
0.3	27
1.0	26
3.0	18

(1079)

  —Waste water treatment methods:
    degradation at 1.0 ppm level in Warburg vessels, in the presence of unadapted proliferating bacteria, and other more conventional nutrients:

time (hr)	% metabolized
3	24
14	54
24	67

(1079)

2,2'methylene-*bis*(3,4,6-trichlorophenol) (hexachlorophene; HCP)

Cl—[ring: Cl, Cl, OH]—CH$_2$—[ring: Cl, Cl, OH]—Cl

Use: HCP is employed primarily in consumer products including soaps, cosmetics and germicidal preparations

C. WATER POLLUTION FACTORS:
  —Water treatment methods: sewage treatment plant (Sept. 1969):

	ppb	% removal
after grit chamber (influent)	22.2	—
after primary clarifier	21.6	2.7
after trickling filter	6.4	71.4
after chlorination	6.8	69.4
out to river (effluent)	7.3	67.1

(1301)

  levels of HCP in 24 hr composite influent and effluent samples and calculated total output from Corvallis, Eugene, and Salem (Oregon, USA) sewage treatment plants:

	influent ppb	effluent ppb	% removal	output g/24 hr	output g/10,000 pop./24 hr
Corvallis	30.8	12.2	60.4	245	66.5
Eugene	20.0	5.6	72.1	244	44.0
Salem	30.0	12.1	59.7	973	122.0

(1301)

  Taylor water treatment plant for city of Corvallis (Oregon, USA):

stage sampled in plant (May 1970)	conc. ppb	% removal
Raw Willamette river water (influent)	0.03	—
after flocculation step	0.02	33

methyleneblue

after sedimentation	0.02	33	
after filtration (final product)	0.01	67	(1301)
—Odor threshold conc. (detection): 10 mg/l			(998)

$$\left[(CH_3)_2N\text{—}\underset{S}{\overset{+}{\bigcirc\bigcirc\bigcirc}}\text{—}N(CH_3)_2\right] Cl^- \cdot H_2O$$
(with N at top center of ring system)

A. PROPERTIES: m.w. 373.9; m.p. 190°C decomposes
D. BIOLOGICAL EFFECTS:
  —Decapod: *Penaeus californiensis*: 96 hr LC$_{50}$ (FT): 100 mg/l   (1134)

methylenechloride (dichloromethane)
CH_2Cl_2

Users and formulations: paint stripping and solvent degreasing; mfg. aerosols, photographic film, synthetic fibers; extraction of naturally-occurring heat sensitive substances; refrigerant in low-pressure refrig. and air-conditioners; fumigant; solvent; textile and leather coatings; pharmaceutical; used in plastics processing; spotting agent; dewaxing; organic synthesis; blowing agent in foams.   (347)

A. PROPERTIES: colorless liquid; m.w. 84.93; m.p. -97°C; b.p. 40 to 42°C v.p. 349 mm at 20°C, 500 mm at 30°C; v.d. 2.93; solub. 20,000 mg/l at 20°C 16,700 mg/l at 25°C, sat.conc. 1,549 g/cu m at 20°C, 2,235 g/cu m at 20°C   (31)

B. AIR POLLUTION FACTORS: 1 mg/cu m = 0.28 ppm, 1 ppm = 3.53 mg/cu m
  —Odor: characteristic: quality: sweetish
         hedonic tone: not unpleasant

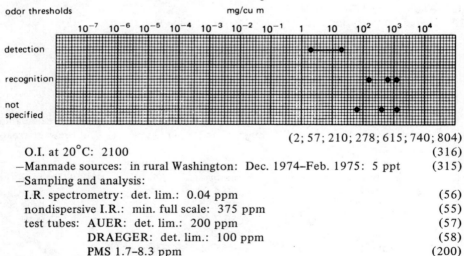

(2; 57; 210; 278; 615; 740; 804)

  O.I. at 20°C: 2100   (316)
  —Manmade sources: in rural Washington: Dec. 1974–Feb. 1975: 5 ppt   (315)
  —Sampling and analysis:
    I.R. spectrometry: det. lim.: 0.04 ppm   (56)
    nondispersive I.R.: min. full scale: 375 ppm   (55)
    test tubes: AUER: det. lim.: 200 ppm   (57)
         DRAEGER: det. lim.: 100 ppm   (58)
         PMS 1.7–8.3 ppm   (200)

C. WATER POLLUTION FACTORS:
  —Reduction of amenities: odor threshold: 100 mg/l   (84)
                organoleptic limit USSR 1970: 7.5 mg/l   (181)

—Water quality:
  in river Maas at Eysden (Netherlands) in 1976:
    median: 3.6 µg/l;
    range: n.d. to 11.3 µg/l
  in river Maas at Keizersveer (Netherlands) in 1976:
    median: 2.6 µg/l;
    range: n.d. to 10.8 µg/l (1368)
—Sampling and analysis: GC-EC: after n-pentane extraction det. lim.: 10 µg/l
 (84)
—Waste water treatment: evaporation from water of 1 ppm solution: 50% after 19–24 min, 90% after 60–80 min (313)
  measured half-life for evaporation from 1 ppm aqueous solution, still air, and an average depth of 6.5 cm: at 1–2°C: 34.9 min
    at 25°C: 18.4–25.2 min (369)

D. BIOLOGICAL EFFECTS:
—Bacteria: *Pseudomonas*: LD_0: 1 g/l
—Algae: *Scenedesmus*: 125 mg/l
—Arthropod: *Daphnia*: 1.25 g/l
—Protozoa: *Colpoda*: 1 g/l (30)
  Threshold conc. of cell multiplication inhibition of the protozoan *Uronema parduczi Chatton-Lwoff*: >16,000 mg/l (1901)
—Fish:
  Pimephales promelas Rafinesque (flow through test):

	effective concentration (EC) value[a] mg/l			lethal concentration (LC) value mg/l		
hrs	EC_{10}	EC_{50}	EC_{90}	LC_{10}	LC_{50}	LC_{90}
24	68.5	112.8	220.1	122.0	268.0	589.0
	(44.2–86.7)[b]	(99.8–150.8)	(175.1–335.4)	(72.7–160.8)	(213.0–346.6)	(432.6–1077.4)
48	66.3	99.0	147.6	94.0	265.0	746.3
	(42.6–79.7)	(83.2–121.5)	(120.5–249.7)	(50.7–130.4)	(202.5–369.7)	(494.7–1712.1)
72	66.3	99.0	147.6	67.3	232.4	802.0
	(42.6–79.7)	(83.2–121.5)	(120.5–249.7)	(32.3–98.9)	(172.4–337.6)	(497.4–2132.6)
96	66.3	99.0	147.6	51.2	193.0	722.1
	(42.6–79.7)	(83.2–121.5)	(120.5–249.7)	(22.5–78.2)	(140.8–277.8)	(447.4–1947.1)

[a]Effective concentration is the concentration producing an adverse effect. In this case the effect noted was loss of equilibrium (1054)
[b]95% confidence limits

  Pimephales promelas Rafinesque: 96 hr, LC_{50}:
    flow through test: 193 mg/l; 95% conf.lim.: 140.8–227.8 mg/l
    static test: 310 mg/l; 95% conf.lim.: 262–391 mg/l (1054)
  guppy (*Poecilia reticulata*): 14 d LC_{50}: 294 ppm (1833)
  fathead minnow: 96 hr LC_{50} (F): 193 mg/l
    96 hr LC_{50} (S): 310 mg/l (1524)
—Mammals:
  mouse: inhalation: LD_0: 11,000 ppm, 8 hr
    LD_{50}: 15,000 ppm, 8 hr
  rabbit, guinea pig, rat, dog: inhalation: no pathology: 5,000 ppm, 7 hr/day, 5 day/w, 6 w (211)
  ingestion: rats: single oral LD_{50}: 1,600 mg/kg (1546)

p,p'-methylenedianiline see 4,4-diaminodiphenylmethane

methylenedimethylether see dimethoxymethane

4,5-methylenephenanthrene
  Manmade sources: in exhaust condensate of gasoline engine: 0.47–0.76 mg/l gasoline consumed. (1070)
  A. PROPERTIES: solub. 1.1 mg/l at 24°C

2-(1-methylethoxy)phenolmethylcarbamate see baygon

methylethylcarbinol see sec-butanol

methylethylene see propylene

sym-**methylethylethylene** see 2-pentene

methylethylketone (MEK; 2-butanone)
$CH_3COCH_2CH_3$
  Manufacturing source: organic chemical industry. (347)
  Users and formulation: solvent or swelling agent of resins; intermediate in mfg. of ketones and amines; flush-off paint stripper; extraction and production of wax from lube oil fractions of petroleum; solvent in nitrocellulose coatings and vinyl films; cements and adhesives; smokeless powder mfg.; cleaning fluids; printing catalyst and carrier. (347)
  Man caused sources (water and air): general use as a solvent, evaporation from applied paints and coatings, cements, adhesives, cleaning fluids; lab use. (347)
  A. PROPERTIES: colorless liquid; m.w. 72.1; m.p. −86.4°C; b.p. 79.6°C; v.p. 77.5 mm at 20°C; v.d. 2.41; sp.gr. 0.805 at 20/4°C; solub. 353 g/l; at 10°C; 190 g/l at 90°C; THC 583 kcal/mole; LHC 550 kcal/mole; log P_{oct} 0.26
  B. AIR POLLUTION FACTORS: 1 mg/cu m = 0.340 ppm, 1 ppm = 2.94 mg/cu m
    —Odor: characteristic: quality: sweet, sharp
              hedonic tone: neutral to unpleasant

T.O.C.: recogn.: 32–80 mg/cu m	(73)
25 ppm = 80 mg/cu m	(278)(210)
30 mg/cu m	(291)
$PIT_{50\%}$: 4.68 ppm	
$PIT_{100\%}$: 10 ppm	(2)
abs. perc. limit: 2.0 ppm	
50% recogn.: 5.5 ppm	
100% recogn.: 6.0 ppm	
O.I. 100% recogn.: 15,350	(19)
T.O.C.: 25 mg/cu m = 8.5 ppm	(279)
14 mg/cu m = 4.7 ppm	(307)
distinct odor: 163 mg/cu m = 50 ppm	(278)

  —Manmade sources:
    in cigarette smoke: 500 ppm (66)
    in gasoline exhaust: <0.1–1.0 ppm (195)

in exhaust of a 1970 Ford Maverick gasoline engine operated on a chassis dynamometer following the 7-mode California cycle:
from API #7 gasoline: 2.6 ppm
from API #8 gasoline: 1.4 ppm (1053)
—Atmospheric reactions:
estimated lifetime under photochemical smog conditions in S.E. England: 1½ hr
(1699; 1700)
—Control methods: catalytic combustion: platinized ceramic honeycomb catalyst:
ignition temp.: 175°C; inlet temp. for 90% conversion: 300–350°C (91)
—Sampling and analysis: photometry: min. full scale: 1,360 ppm (53)

C. WATER POLLUTION FACTORS:
—Oxidation parameters:
BOD$_5$:   1.92 NEN 3235-5.3 (277)
           1.515; 1.81 (30)(36)
           2.24 std. dil. sewage (41)(27)
           46% ThOD (220)
BOD$_5^{20}$: 1.81 at 10 mg/l, unadapted sew.: lag period: 1 day
BOD$_{20}^{20}$: 1.92 at 10 mg/l, unadapted sew. (554)
COD: 2.31 NEN 3235-5.3 (277)
           2.20 (36)
           100% ThOD (220)
KMnO$_4$: 0.029 (30)
           acid: 7% ThOD
           alkaline: 7% ThOD (220)
TOC: 100% ThOD (220)
ThOD: 2.44 (36)
—Odor threshold conc. (detection): 1.0 mg/l (998)
—Manmade sources:
PVC pipe cement used for joint connection caused 4.5 ppm, 6 months after PVC pipe installation and 2.2 ppm after 8 months (at equilibrium condition—reached in about 48 hr) (1670)
—Waste water treatment: A.C.: adsorbability: 0.094 g/g C; 46.8% reduction, infl. 1,000 mg/l, effl.: 532 mg/l; (32)
reverse osmosis: 72.9% rejection from a 0.01 M solution (221)
anaerobic lagoon: lb COD/day/1,000 cu ft   infl. mg/l   effl. mg/l
                           13   10   5   (37)
A.S.: after 6 hr: 0.4% of ThOD
            12 hr: 0.8
            24 hr: 3.1 (88)

D. BIOLOGICAL EFFECTS:
—Toxicity threshold (cell multiplication inhibition test):
bacteria (*Pseudomonas putida*):   1150 mg/l   (1900)
algae (*Microcystis aeruginosa*):   110 mg/l   (329)
green algae (*Scenedesmus quadricauda*):   4300 mg/l   (1900)
protozoa (*Entosiphon sulcatum*):   190 mg/l   (1900)
protozoa (*Uronema parduczi Chatton-Lwoff*): 2830 mg/l   (1901)
—Bacteria:
Pseudomonas: LD$_0$: 2.5 g/l

—Algae:
   Scenedesmus: LD_0: 12.5 g/l
   Protozoa: *Colpoda*: LD_0: 5 g/l (30)
—Fish: mosquito fish: TLM (24–96 hr): 5,600 mg/l;
   bluegill: TLM (24–96 hr): 5,640–1,690 mg/l in tapwater (41)(247)
   goldfish: LD_{50} (24 hr): 5,000 mg/l, modified ASTM D 1345 (277)
—Mammalia:
   rat: single oral lethal dose: 3.3 g/kg
   guinea pig: inhalation: no serious disturbance: 10,000 ppm, 1 hr
   rat: inhalation: no death: 2,000 ppm, 2 hr
      4/6 deaths: 4,000 ppm, 2 hr (211)
—Man: no permanent ill effects: 700 ppm; complaints: >300 ppm (211)

methylethylketoxime *see* 2-butanoneoxime

2-methyl-5-ethylpyridine (aldehydine)

C_2H_5— (pyridine ring) —CH_3

Manmade sources:
in coke oven emissions: 24.4–73.5 µg/g of sample
in coal tar (n.s.i.): 3.12 mg/g of sample (n.s.i.)
in wood preservative sludge: 0.37 g/l of raw sludge (993)

2-methylfluorene
Manmade sources:
in wood preservative sludge: 3.82 g/l of raw sludge
in coke oven emissions (n.s.i.): 8.9–98.7 µg/g of sample (993)

9-methylfluorene
Manmade sources:
in wood preservative sludge: 1.44 g/l of raw sludge
in coke oven emissions: 6.7–102.8 µg/g of sample (993)

methylformate (methylmethanoate)
$HCOO-CH_3$
   Use: organic synthesis; cellulose acetate solvent; military poison gases; fumigant; larvicides
A. PROPERTIES: colorless liquid; m.w. 60.05; m.p. $-99.8°C$; b.p. $32°C$; v.p. 480 mm at $20°C$; 700 mm at $30°C$; v.d. 2.07; sp.gr. 0.975 at $20/4°C$ solub. 304 g/l at $20°C$; satur. conc. 1,569 g/cu m at $20°C$, 2,213 g/cu m at $30°C$ (57)
B. AIR POLLUTION FACTORS: 1 mg/cu m = 0.40 ppm, 1 ppm = 2.50 mg/cu m
   —Odor: odor threshold values: recognition: 5,000 mg/cu m = 2,000 ppm
      165–180 mg/cu m (73)(210)(610)
      distinct odor: 6,900 mg/cu m = 2,750 ppm (278)

O.I. 100% recogn. at 20°C = 300 (316)
—Manmade sources: gasoline engine exhaust: <0.1–0.7 ppm (195)
—Sampling and analysis: photometry: min. full scale: 27,000 ppm (53)
D. BIOLOGICAL EFFECTS:
—Mammalia: guinea pig: inhalation: lethal: 50,000 ppm, 30 min
                                      lethal: 25,000 ppm, 60 min
                                      lethal: 10,000 ppm, 3–4 hr
                              no deaths: 3,500 ppm, 8 hr (211)
—Man: no symptoms: 1,500 ppm, 1 min (211)

2-methylfuran (silvan; sylvan)

Formulation: shipments of methylfuran contain 0.01% hydroquinone as an oxidation inhibitor
Use: chemical intermediate (501)
A. PROPERTIES: colorless liquid; m.w. 82.10; m.p. −88.7; b.p. 63/65.6°C; v.p. 59 mm at 0°C, 95 mm at 10°C, 142 mm at 20°C, 225 mm at 30°C; sp.gr. 0.913 at 20°C, solub. 3,000 mg/l at 20°C
C. WATER POLLUTION FACTORS:
—Odor threshold conc. (detection): 4 mg/kg (908)
D. BIOLOGICAL EFFECTS:
—Toxicity threshold (cell multiplication inhibition test):
  bacteria (*Pseudomonas putida*):   90 mg/l   (1900)
  algae (*Microcystis aeruginosa*):   40 mg/l   (329)
  green algae (*Scenedesmus quadricauda*):   40 mg/l   (1900)
  protozoa (*Entosiphon sulcatum*):   107 mg/l (1900)
  protozoa (*Uronema parduczi Chatton-Lwoff*): 26 mg/l (1901)

methylglycol see ethyleneglycolmonomethylether

methylglycolacetate see ethyleneglycolmonomethyletheracetate

2-methylheneicosane
C. WATER POLLUTION FACTORS:
  —Aquatic reactions:
    adsorption on smectite clay particles from simulated seawater at 25°C—experimental conditions: 100 µg 2-methylheneicosane/l; 50 mg smectite/l; adsorption: 1.87 µg/mg = 92% adsorbed (1009)

2-methylheptane
$(CH_3)_2CH(CH_2)_4CH_3$
A. PROPERTIES: m.w. 114.23; b.p. 116°/761 mm; sp.gr. 0.698
C. WATER POLLUTION FACTORS:
  —Reduction of amenities: T.O.C. 50 mg/l (295)

—Biodegradation:
    incubation with natural flora in the groundwater—in presence of the other components of high-octane gasoline (100 μl/l): biodegradation: 38% after 192 hr at 13°C (initial conc.: 0.35 μl/l) (956)

3-methylheptane
$CH_3-CH_2-CH(CH_3)-(CH_2)_3-CH_3$
C. WATER POLLUTION FACTORS:
—Biodegradation:
    incubation with natural flora in the groundwater—in presence of the other components of high-octane gasoline (100 μl/l): biodegradation: 45% after 192 hr at 13°C (initial conc.: 0.46 μl/l) (956)

4-methylheptane
$CH_3(CH_2)_2CH(CH_3)(CH_2)_2CH_3$
C. WATER POLLUTION FACTORS:
—Biodegradation:
    incubation with natural flora in the groundwater—in presence of the other components of high-octane gasoline (100 μl/l): biodegradation: 48% after 192 hr at 13°C (initial conc. 0.15 μl/l) (956)

5-methyl-3-heptanone see ethyl-sec-amylketone

2-methylhexadecane
C. WATER POLLUTION FACTORS:
—Aquatic reactions:
    adsorption on smectite clay particles from simulated seawater at 25°C—experimental conditions: 100 μg 2-methylhexadecane/l; 50 mg smectite/l; adsorption: 0.95 μg/mg = 48% adsorbed

2-methylhexane (ethylisobutylmethane)
$CH_3(CH_2)_3CH(CH_3)_2$
Use: organic synthesis
A. PROPERTIES: m.w. 100.21; m.p. -118°C; b.p. 90°C; sp.gr. 0.679; colorless liquid
C. WATER POLLUTION FACTORS:
—Biodegradation:
    incubation with natural flora in the groundwater—in presence of the other components of high-octane gasoline (100 μl/l): biodegradation: 23% after 192 hr at 13°C (initial conc.: 0.74 μl/l) (956)

3-methylhexane
$CH_3CH_2CH(CH_3)CH_2CH_2CH_3$
A. PROPERTIES: colorless liquid; b.p. 92°C; sp.gr. 0.692
C. WATER POLLUTION FACTORS:
—Reduction of amenities: T.O.C. = 500 mg/l (295)
—Biodegradation:
    incubation with natural flora in the groundwater—in presence of the other components of high-octane gasoline (100 μl/l): biodegradation: 0% after 192 hr at 13°C (initial conc. 0.66 μl/l) (956)

5-methyl-2-hexanone *see* methylisoamylketone

methylhexylcarbinol *see* 2-octanol

methylhydrazine *see* monomethylhydrazine

methyl-2-hydroxybenzoate *see* methylsalicylate

methylhydroxyisopropylcyclohexane *see* menthol

3-methylindole *see* skatole

1-methylinosine
C. WATER POLLUTION FACTORS:
　—Manmade sources: in sewage effluent: 0.080 mg/l　　　　　　　　　　(277; 517)

methyliodide (iodomethane)
CH_3I
　Use: medicine; organic synthesis; microscopy; testing for pyridine
A. PROPERTIES: colorless to brownish liquid; m.w. 141.95; m.p. -66.1°C; b.p. 42.5°C; v.p. 400 mm at 25°C; v.d. 4.9; sp.gr. 2.279; solub. 14 g/l at 20°C; log P_{oct} 1.69
B. AIR POLLUTION FACTORS: 1 ppm = 5.8 mg/cu m
　—Manmade sources: glc's in rural Washington Dec. '74–Feb-75: <5 ppt　　(315)
D. BIOLOGICAL EFFECTS:
　—Concentrations in various organs of molluscs and fish collected from the relatively clean water of the Irish Sea in the vicinity of Port Erin, Isle of Man (only organs with highest and lowest concentrations are mentioned, concentrations on dry weight basis):
Molluscs:
　Baccinum undatum: digestive gland: 14 ng/g
　Modiolus modiolus: digestive tissue: 10 ng/g
　　　　　　　　　　　mantle: 188 ng/g
　Pecten maximus: ovary, mantle: 3 ng/g
　　　　　　　　　gill: 24 ng/g　　　　　　　　　　　　　　　　　　　　(1092)
Fish:
　Conger conger (eel): gill: 3 ng/g
　　　　　　　　　　　liver: 42 ng/g
　Gadus morhua (cod): skeletal tissue: 0.4 ng/g
　　　　　　　　　　　brain: 54 ng/g
　Pollachius birens (coalfish): muscle: 4 ng/g
　　　　　　　　　　　　　　　brain: 166 ng/g
　Scylliorhinus canicula (dogfish): spleen: 13 ng/g
　　　　　　　　　　　　　　　　　brain: 103 ng/g
　Trisopterus luscus (bib): skeletal tissue: 2 ng/g
　　　　　　　　　　　　　　brain: 137 ng/g　　　　　　　　　　　　　　(1092)
　—Mammalia:
　　mouse: inhalation, rapid narcosis: death after 10 min: 78,693 ppm
　　　　　　　　　　　　　　　　　　death after 30 min exposure:　18,109 ppm

death of all animals within 24 hr: 73–734 ppm
no marked toxic symptoms: 54 ppm
rat: subcutane LD_{50}: 0.15–0.22 mg/kg body wt
oral LD_{50}: 0.15–0.22 mg/kg body wt (211)
—Carcinogenicity: ?
—Mutagenicity in the *Salmonella* test (without liver homogenate): weakly mutagenic (1883)

methylisoamylketone (5-methyl-2-hexanone)
$CH_3COCH_2CH_2CH(CH_3)_2$
A. PROPERTIES: colorless liquid; m.w. 114.2; b.p. 144°C; sp.gr. 0.818 at 17/4°C; solub. 5,400 mg/l
B. AIR POLLUTION FACTORS:
—Odor: characteristic: quality: sweet, sharp
hedonic tone: pleasant
absolute perception limit: 0.012 ppm
50% recogn.: 0.049 ppm
100% recogn.: 0.070 ppm
O.I. 75,142
C. WATER POLLUTION FACTORS:
—Waste water treatment: A.C.: adsorbability: 0.169 g/g C; 85.2% reduction, infl.: 986 mg/l, effl.: 146 mg/l (32)
D. BIOLOGICAL EFFECTS:
—Toxicity threshold (cell multiplication inhibition test):

bacteria (*Pseudomonas putida*):	115 mg/l	(1900)
algae (*Microcystis aeruginosa*):	90 mg/l	(329)
green algae (*Scenedesmus quadricauda*):	125 mg/l	(1900)
protozoa (*Entosiphon sulcatum*):	980 mg/l	(1900)
protozoa (*Uronema parduczi Chatton-Lwoff*):	980 mg/l	(1901)

methylisobutenylketone *see* mesityloxide

methylisobutylcarbinol *see* methylamylalcohol

methylisobutylketone (hexone; 4-methyl-2-pentanone; MIBK; hexanone)
$(CH_3)_2CHCH_2COCH_3$
Use: solvent for paints, varnishes, nitrocellulose lacquers; manufacture of methylamylalcohol; extraction processes; organic synthesis; denaturant for alcohol
A. PROPERTIES: colorless liquid; m.w. 100.2; m.p. -85/-80°C; b.p. 116/119°C; v.p. 6 mm at 20°C; 10 mm at 30°C; v.d. 3.45; sp.gr. 0.8017 at 20/4°C solub. 19,000 mg/l, 17,000 mg/l at 20°C; sat. conc. 27 g/cu m at 20°C, 53 g/cu m at 30°C
B. AIR POLLUTION FACTORS: 1 mg/cu m = 0.244 ppm, 1 ppm = 4.10 mg/cu m
—Odor: characteristic: quality: sweet, sharp
hedonic tone: pleasant to unpleasant

METHYLISOBUTYLKETONE

odor thresholds — mg/cu m (10^{-7} to 10^4)

- detection
- recognition
- not specified

(2; 19; 73; 210; 278; 279; 291; 298; 610; 676; 741; 749; 820; 827; 828)

O.I. (100% recognition): 70,357 (19)

—Atmospheric reactions:
  estimated lifetime under photochemical smog conditions in S.E. England: 2.4 hr
  (1699; 1700)

—Control methods: catalytic combustion: platinized ceramic honeycomb catalyst:
  ignition temp.: 175°C; inlet temp. for 10% conversion: 300–350°C (91)

—Sampling and analysis: photometry: min. full scale: 1,360 ppm (53)

C. WATER POLLUTION FACTORS:
- BOD_5: 2.06 NEN 3235-5.4 (277)
  - 0.60 (30)
  - 4.4% ThOD (79)
  - 0.12 std. dil. sew (255)
  - 2.14 std. dil. sew (260)
- BOD_{10}: 0.6 std. dil. sew (256)
  - 49.3% of ThOD
- $BOD_{10}^{20°C}$: 49.3% ThOD at 2.5 mg/l in mineralized dilution water with settled sewage seed (405)
- BOD_{15}: 55.9% of ThOD
- BOD_{20}: 56.6% of ThOD
- BOD_{50}: 64.8% of ThOD (79)
- COD: 2.16 = 79% ThOD—ASTM procedure
  - 2.46 = 90% ThOD—modified "Shell" procedure (272)
- $KMnO_4$: 0.010 (30)
- ThOD: 2.72 (30)
- Waste water treatment: A.C.: adsorbability: 0.169 g/g C; 84.8% reduction, infl.: 1,000 mg/l, effl: 152 mg/l (32)
  ion exchange: adsorption on Amberlite XAD-2: infl.: 100 ppm, effl.: 0 ppm: retention efficiency: 100% (40)
  NFG, BOD, 20°C, 1–10 days observed; feed: 100–1,000 mg/l, acclimation: 365 + P: 22% removed (93)
  A.S.: after 6 hr: 1.5% of ThOD
  12 hr: 1.8% of ThOD
  24 hr: 3.0% of ThOD (88)

D. BIOLOGICAL EFFECTS:
—Toxicity threshold (cell multiplication inhibition test):
  bacteria (*Pseudomonas putida*): 275 mg/l (1900)
  algae (*Microcystis aeruginosa*): 136 mg/l (329)
  green algae (*Scenedesmus quadricauda*): 725 mg/l (1900)

protozoa (*Entosiphon sulcatum*): 447 mg/l (1900)
protozoa (*Uronema parduczi Chatton-Lwoff*): 941 mg/l (1901)
—Fish: goldfish: LD_{50} (24 hr): 460 mg/l (277)
—Mammalia: rat: single oral LD_{50}: 2.08 g/kg
  inhalation: survived: 2,000 ppm, 4 hr
  death: 4,000 ppm, 4 hr (211)
—Man: complaints: >100 ppm; eye irritation: 200 ppm; nasal irritation: 400 ppm (211)

methylisopropenylketone (2-methyl-1-butene-3-one)
$CH_3COC(CH_3)CH_2$
A. PROPERTIES: m.w. 84.06; m.p. -53.7°C; b.p. 97.7°C; v.p. 42 mm at 25°C sp.gr. 0.855; solub. 47,000 mg/l
B. AIR POLLUTION FACTORS: 1 mg/cu m = 0.291 ppm, 1 ppm = 3.44 mg/cu m
  —Odor: characteristic: quality: very pungent
    odor threshold values: 0.291 ppm = 0.001 mg/l (211)(796)
  O.I. at 20°C = 184,000 (316)
C. WATER POLLUTION FACTORS:
  —Waste water treatment: RW, Sd, BOD, 20°C, 10 days observed, 70 days acclimation: 45% theor. oxidation (93)
D. BIOLOGICAL EFFECTS:
  —Mammalia:
    guinea pig: single oral dose: lethal: 0.06–0.25 g/kg
    rat: single oral dose: LD_{50}: 0.18 g/kg
      inhalation: 5/6: 125 ppm, 4 hr
      inhalation: 0/6: marked irritation: 524 ppm, 90 min
      inhalation: dangerous: 1,455 ppm, 30 min
      inhalation: dangerous: 2,910 ppm, a few min
    rat, guinea pig, rabbit: inhalation: slight increase in mortality, slight kidney injury: 15 ppm, 100 × 7 hr (211)
  —Man: no irritation: 0.3 ppm; eye irritation: 1.45 ppm; strong eye irritation: 14.5 ppm (211)

2-methyl-4-isothiazolin-3-one calcium chloride
D. BIOLOGICAL EFFECTS:
  —Fish:
    bluegill: 60 d BCF (F): 4.8 (whole fish minus viscera)
    60 d BCF (F): 40.1 (viscera) (1511)

methylisothiocyanate (methylmustard oil)
CH_3NCS
Manmade sources: decomposition product of dazomet (1439)
A. PROPERTIES: colorless crystals; m.w. 733.11; m.p. 35°C; b.p. 119°C; v.p. 19 mm at 20°C, 30 mm at 30°C; v.d. 2.53; sp.gr. 1.069 at 37/4°C; sat. conc. 75.6 g/cu m at 20°C, 115 g/cu m at 30°C
B. AIR POLLUTION FACTORS: 1 mg/cu m = 0.33 ppm, 1 ppm = 3.04 mg/cu m
  —Odor threshold conc.: 15 mg/cu m (602)

C. WATER POLLUTION FACTORS:
   —Impact on biodegradation processes: 75% inhibition of nitrification process in nonacclimated activated sludge at 0.8 mg/l (30)
D. BIOLOGICAL EFFECTS:
   —Mammalia: rat: inhalation: no effect level: 2.5 ppm, 15 X 6 hr (65)

2-methyllactic acid see 2-hydroxyisobutyric acid

methylmercaptan (methanethiol, methylthioalcohol)
CH_3SH
A. PROPERTIES: colorless gas; m.w. 48.11; m.p. -123.1°C; b.p. 6-7.6°C; b.p. 400 mm at 7.9°C, 1 atm at 6.8°C, 2 atm at 26.1°C, 5 atm at 55.9°C v.d. 1.7; sp.gr. 0.868 at 20/4°C;
B. AIR POLLUTION FACTORS: 1 mg/cu m = 4.88 ppm, 1 ppm = 2.05 mg/cu m
   —Odor: characteristic; quality: sulfidy, pungent, decayed cabbage
      odor threshold values: 1.1 ppb (210)

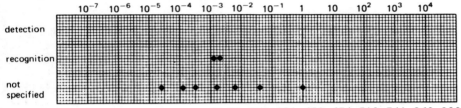

(2; 71; 279; 291; 652; 674; 710; 741; 863; 930)
   O.I. at 20°C = 53,300,000 (316)
   —Biogenic methylmercaptan emissions from soils in USA: Cox's landing, N.C.— saline marsh (Nov. 1977): 6.56 g S/m²/yr (1385)
   —Manmade sources:
      glc's: in oil refinery at Owasa (Japan): 5.6 ppb
      in primary school at Yokkaichi: 2.8 ppb
      in pulpmill at Niigata: 57.0 ppb
      Fujima at Fuji: 0.36 ppb
      lock gate at Tsusen river at Niigata: 550–1,060 ppb (50)
      in municipal sewer air: 10–50 ppb (212)
C. WATER POLLUTION FACTORS:
   —Reduction of amenities: taste and odor caused at 0.0011 mg/l (41)
      odor threshold: 0.0011 mg/l (51)(129)
         0.041 mg/l (98)
D. BIOLOGICAL EFFECTS:
   —Fish: *Salmonides*: TLm: 0.55–0.9 mg/l (30)

methylmercuric chloride
CH_3HgCl
C. WATER POLLUTION FACTORS:
   —Biodegradation:
      Of the 40 microorganisms tested, only 27 were able to tolerate methylmercuric

chloride concentrations of 0.37 mg/l to 2.5 mg/l. Sixteen aerobes that showed growth in the cultures during acclimation to methylmercuric chloride were also positive for demethylation. Demethylation of over 60% of the initial methylmercury concentration occurred within cultures of *Enterobacter aerogenes*, *Serratia marcescens*, *Proteus mirabilis*, *Enterobacter cloacae*, *Providencia* sp., *Citrobacter freundii*, and *Pseudomonas fluorescens*. Demethylation under aerobic conditions ranged from 20 to 84%. *Desulfovibrio desulfuricans* showed 32% demethylation during screening tests under anaerobic conditions and was the only anaerobe with significant demethylating capacity.

methylmercury
C. WATER POLLUTION FACTORS:
— Waste water treatment method:
  removal from river water at conc. of 1–10 ppm:
   by adsorption on ground rubber: 70%
   by adsorption on saw dust: 74%   (1424)
D. BIOLOGICAL EFFECTS:
 — Crustaceans:
  Brine shrimp (*Artemia salina*): Reduced life span: 0.005 mg/l
                                  Inhibition of nauplii production: 0.002 mg/l
                                                                             (1492)
  Amphipod (*Gammarus duebani*): 96-hr LC_{50}:
   100% sea water: 0.15 mg/l
   2% sea water: 0.23 mg/l   (1493)
  Japanese medaka: 96 hr LC_{50} (S): 88 µg/l   (1525)
 — Fish:

Blue gourami (*Trichogaster trichopterus*)					
	24 hr	27°C	LC_{50}	0.123 mg/l	pH = 7.4
	48 hr	27°C	LC_{50}	0.094 mg/l	
	96 hr	27°C	LC_{50}	0.089 mg/l	(1339)

methylmethacrylate
$CH_2C(CH_3)COOCH_3$
A. PROPERTIES: colorless liquid; m.w. 100.11; m.p. −50°C; b.p. 101°C v.p. 28 mm at 20°C, 40 mm at 26°C, 49 mm at 30°C; v.d. 3.45 sp.gr. 0.936 at 20/4°C; sat. conc. 164 g/cu m at 20°C, 258 g/cu m at 30°C
B. AIR POLLUTION FACTORS: 1 mg/cu m = 0.24 ppm, 1 ppm = 4.16 mg/cu m
 — Odor: characteristic: quality: sulfidy, sweet, sharp
          hedonic tone: unpleasant

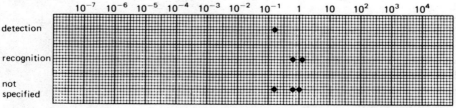

(57; 279; 291; 307; 659; 675; 676; 741)

O.I. = 119,705   (19)

C. WATER POLLUTION FACTORS:
  —Waste water treatment: Sew, CO_2, 26°C, 10 days observed, feed: 10 mg/l, 19
    days acclimation: 47% theor. oxidation (93)
  —Oxidation parameters: BOD_5: 0.14 std. dil. (168)
D. BIOLOGICAL EFFECTS:
  —Toxicity threshold (cell multiplication inhibition test):
    bacteria (*Pseudomonas putida*): 100 mg/l (1900)
    algae (*Microcystis aeruginosa*): 120 mg/l (329)
    green algae (*Scenedesmus quadricauda*): 37 mg/l (1900)
    protozoa (*Entosiphon sulcatum*): 450 mg/l (1900)
    protozoa (*Uronema parduczi Chatton-Lwoff*): 556 mg/l (1901)
  —Fish: fatheads: TLm (24–96 hr): 499–159 mg/l
        bluegill: TLm (24–96 hr): 368–232 mg/l
        goldfish: TLm (24–96 hr): 423–277 mg/l
        guppies: TLm (24–96 hr): 368 mg/l (158)
  —Mammalia: rat: acute oral LD_{50}: 8.4 g/kg
            rabbit: acute oral LD_{50}: 6–7 g/kg
            rat: inhalation: LC_{50}: 3,750 ppm, 8 hr (211)

methylmethanoate *see* methylformate

3-methyl-1-(γ-methylbutoxy)butane *see* isoamylether

2-methyl-2(methylthio)propionaldehyde-O-(methylcarbamoyl)oxime *see* aldicarb

N-methylmorpholine (4-methylmorpholine)

A. PROPERTIES: m.w. 101.2; m.p. −65°C, b.p. 111/115°C, 16.6 mm at 20°C; v.d. 3.5;
   sp.gr. 0.9213 at 20/20°C; log P_{oct} −0.33
B. WATER POLLUTION FACTORS:
  —Waste water treatment: A.C.: adsorbability: 0.085 g/g C; 42.5% reduct. infl.:
    1,000 mg/l, effl.: 575 mg/l (32)
D. BIOLOGICAL EFFECTS:
  —Mammalia: rat: single oral LD_{50}: 2.72 g/kg
            : inhalation: no deaths: sat. vap., max 1 hr (211)

methyl mustard oil *see* methylisothiocyanate

1-methylnaphthalene (α-methylnaphthalene)

2-METHYLNAPHTHALENE

Manufacturing source: petroleum refining; coal processing. (347)
Users and formulation: insecticide mfg.; mfg. phthalic anhydride; solvent organic synthesis; asphalt and naptha constituent. (347)
Natural sources (water and air): coal; petroleum. (347)
Man caused sources (water and air): general uses of asphalt and naphthas; general lab use; use as a solvent; use of certain insecticides. (347)

A. PROPERTIES: colorless liquid; m.w. 142.19; m.p. $-22°C$; b.p. $240/243°C$; v.d. 4.91; spec.gr. 1.025; solub. 26–28 mg/l at $25°C$ (distilled water)
B. AIR POLLUTION FACTORS: 1 mg/cu m = 0.17 ppm, 1 ppm = 5.91 mg/cu m
 —Manmade sources: in coal tar fumes: 0.7 wt % (516)
C. WATER POLLUTION FACTORS:
 —Reduction of amenities:
  T.O.C. in water at room temp.: 0.023 ppm, range: 0.00252–0.17 ppm
  20% of population still able to detect odor at 0.0021 ppm
  10% of population still able to detect odor at 0.75 ppb
  1% of population still able to detect odor at 0.018 ppb
  0.1% of population still able to detect odor at 0.45 ppt (321)
  odor threshold conc. (detection): 0.0075 mg/kg (430)
 —Biodegradation:
  biodegradation at 0.1 mg/l (n.s.i.):

	normal sewage	adapted sewage
after 24 hr	0%	84%
after 135 hr	0%	95%	(997)

  after 3 hr incubation in natural seawater, 2% of 25 µg/l (n.s.i.) were taken up by suspended aggregates of dead phytoplankton cells and bacteria (957)
 —Impact on biodegradation processes:
  inhibition of photosynthesis of a freshwater, non axenic unialgal culture of Selenastrum capricornutum:
   at 1% saturation: 96% carbon-14 fixation (vs. controls)
     10% saturation: 71% carbon-14 fixation (vs. controls)
     100% saturation: 7% carbon-14 fixation (vs. controls) (1690)
  effect on respiration of glucose by mixed culture derived from activated sludge:

concentration (mg/l)	increase in lag period (hr)	respiration rate (%)
1 (n.s.i.)	10	94
10 (n.s.i.)	167	0
1,000 (n.s.i.)	>200	0	(997)

 —Water quality:
  in Eastern Ontario drinking waters (June–Oct. 1978): 0.2–6.4 ng/l ($n = 12$)*
  in Eastern Ontario raw waters (June–Oct. 1978): 1.2–4.1 ng/l ($n = 2$)* (1698)
  *sum of 1- and 2-methylnaphthalenes
 —Waste water treatment method:
  compounds identified from the oxidation with chlorine dioxide: 4-chloro-1-methylnaphthalene; 2-chloro-1-methylnaphthalene; 2,4-dichloro-1-methyl-naphthalene; 1-hydroxymethylnaphthalene; 1-naphthaldehyde; 1-naphthoic acid (1696)

D. BIOLOGICAL EFFECTS:
 —Bioaccumulation:

oysters: uptake and release:

time of exposure days	conc. oyster µg/g	water µg/l	depuration time days	conc. oyster µg/g	water µg/l
2	15.0 (8.8)*	29	14	0.1	0
4	5.7 (1.6)	16	14	0.1	0
9	7.0 (14.6)	6	14	0.2	0
15	7.2	5			

*figures between brackets refer to compartment surrounded by a 60 µm Nitex filter to filter out large particles (579)

uptake and depuration by oysters (*Crassostrea virginica*) from oil-treated enclosure:

time of exposure days	depuration time, days	concentration (n.s.i.) oysters µg/g	water µg/l	accumulation factor oysters/water
2	–	56	8	7,000
8	–	36	3	12,000
2	7	1	–	–
8	7	2	–	–
8	23	<0.5	–	–

half-life for depuration: 2 days (957)

—BCF* in muscle of:

    Coho salmon   and   Starry flounder
    30 after 2 wk exposure   2000 after 1 2k exposure
    70 after 3 wk exposure   330 after 2 wk exposure
    130 after 5 wk exposure   113 after 1 wk depuration
    50 after 6 wk exposure   270 after 2 wk depuration
    n.d. after 1 wk depuration   (after 2 wk exposure)
    (after 6 wk exposure)   (1659)

*exposure to approx. 1 ppm of the water soluble fraction of Prudhoe Bay crude oil

Coho salmon (*O. kisuth*) exposed to 0.02 ppm at 10°C for 5 weeks: 0.40 ppm in muscle tissue (dry weight basis): bioaccumulation factor: 20 (384)

—Toxicity: fish:
fathead minnows; static bioassay in Lake Superior water at 18–22°C: LC_{50} (1; 24; 48; 72; 96 hr): 39; 9; 9; 9; 9 mg/l (350)
brown trout yearlings: 48 hr LC_{50}: 8.4 mg/l (n.s.i.) (static bioassay) (939)

2-methylnaphthalene (β-methylnaphthalene)

A. PROPERTIES: solid; m.p. 34°C; b.p. 241–242°C; sp.gr. 0.994
B. AIR POLLUTION FACTORS:
  —Manmade sources: in coal tar pitch fumes: 1.0 wt % (516)

C. WATER POLLUTION FACTORS:
 —Reduction of amenities:
   T.O.C. in water at room temp.: 0.012 ppm, range: 0.003–0.04 ppm, 10 judges
   20% of population still able to detect odor at 2.0 ppb
   10% of population still able to detect odor at 0.79 ppb
   1% of population still able to detect odor at 0.039 ppb
   0.1% of population still able to detect odor at 0.0016 ppb (321)
 —Manmade sources: in sewage effluent: 0.0014 mg/l (295)
 —Biodegradation:
   biodegradation to CO_2:

sampling site	conc. $\mu g/l$	month	incubation time (hr)	degradation rate $(\mu g/l/day) \times 10^3$	turnover time (days)
control station	30	—	—	120 ± 40	310
near oil storage tanks	30	—	—	1100 ± 300	30
near oil storage tanks	30	—	—	240 ± 30	120
Skidaway river	10	Feb.	24	18 ± 11	560
Skidaway river	30	Feb.	24	45 ± 7	670
Skidaway river	30	April	24	250 ± 60	120
Skidaway river	15	Oct.	24	350 ± 200	43
Skidaway river	30	Oct.	24	460 ± 220	65
estuarine	30	—	—	250 ± 60	120
coastal	30	—	—	0	∞
gulf stream	30	—	—	0	∞

(381)

   microbial degradation to CO_2 —in seawater at 12°C—in the dark after 24 hr incubation at 50 $\mu g/l$: 0.10 $\mu g/l/day$—turnover time: 500 days;
   after addition of aqueous extract of fuel oil 2: degradation rate: 0.26 $\mu g/l/day$—turnover time: 200 days (477)
   degradation in seawater by oil oxidizing micro-organisms: 17.1% breakdown after 21 days at 22°C in stoppered bottles containing a 100 ppm mixture of alkanes, cyclo-alkanes, and aromatics (1237)
 —Aquatic reactions:
   in estuarine waters: at 30 $\mu g/l$, 6% adsorbed on particles after 3 hr (381)
 —Waste water treatment methods:
   compounds identified from the oxidation with chlorinedioxide: 1-chloro-2-methylnaphthalene; 3-chloro-2-methylnaphthalene; 1,3-dichloro-2-methylnaphthalene; 2-hydroxymethylnaphthalene; 2-naphthaldehyde; 2-naphthoic acid; 2-methyl-1,4-naphthoquinone (1696)

D. BIOLOGICAL EFFECTS:
 —Bioaccumulation:
   rainbow trout: 4 wk BCF (bile) (F): 23,500
                  4 wk BCF (other tissues): 40–300 (1526)
 —BCF* in muscle of:

Starry flounder	and	Coho salmon
2800 after 1 wk exposure		30 after 2 wk exposure
470 after 2 wk exposure		100 after 3 wk exposure
110 after 1 wk depuration		190 after 5 wk exposure
200 after 2 wk depuration		70 after 6 wk exposure

(after 2 wk exposure)   n.d. after 1 wk depuration
(after 6 wk exposure)
n.d. = not detectable
*exposure to approx. 1 ppm of the water soluble fraction of Prudhoe Bay crude oil   (1659)
Coho salmon (*O. kisuth*) exposed to 0.02 ppm at 10°C for 5 weeks: 0.56 ppm in muscle tissue (dry weight basis): bioaccumulation factor: 28   (384)

o-methylnitrobenzene (*o*-nitrotoluene)

A. PROPERTIES: yellow liquid; m.w. 137.13; m.p. -10.6°C/-4.1°C b.p. 225°C; v.p. 0.1 mm at 20°C, 0.25 mm at 30°C, 1.6 mm at 60°C; spec. gr. 1,163 at 20/4°C; solub. 652 mg/l at 30°C; sat. conc. 0.75 g/cu m at 20°C, 1.8 g/cu m at 30°C; log P_{oct} 2.30
B. AIR POLLUTION FACTORS: 1 mg/cu m = 0.18 ppm, 1 ppm = 5.70 mg/cu m
C. WATER POLLUTION FACTORS:
  —Biodegradation: decomposition by a soil microflora in >64 days   (176)
    adapted A.S. at 20°C—product is sole carbon source: 98.0% COD removal at 32.5 mg COD/g dry inoculum/hr   (327)
  —Water quality: in river Maas in The Netherlands: average in 1973: 0.025–0.07 μg/l (*o*-+ *p*-isomer)   (342)
  —Odor threshold conc. (detection): 0.13 mg/kg   (894)
D. BIOLOGICAL EFFECTS:
  —Toxicity threshold (cell multiplication inhibition test):

bacteria (*Pseudomonas putida*):	18 mg/l	(1900)
algae (*Microcystis aeruginosa*):	3.1 mg/l	(329)
green algae (*Scenedesmus quadricauda*):	28 mg/l	(1900)
protozoa (*Entosiphon sulcatum*):	46 mg/l	(1900)
protozoa (*Uronema parduczi Chatton-Lwoff*):	24 mg/l	(1901)

  —Fish: fish TLm: 10-2 mg/l;
       Vairon (F): TLm (6 hr): distilled water: 18–20 mg/l
                                 hard water: 35–40 mg/l   (30)
  —Man: severe toxic effect: 200 ppm, 60 min
       symptoms of illness: 40 ppm
       unsatisfactory: >1 ppm   (185)

m-methylnitrobenzene (*m*-nitrotoluene)

Use: organic synthesis

A. PROPERTIES: m.w. 137.13; m.p. 15.5°C; b.p. 231°C; v.p. 0.1 mm at 20°C, 0.25 mm at 30°C, 1.0 mm at 60°C; v.d. 4.73; sp.gr. 1.157 at 20/4°C; solub. 498 mg/l at 30°C; sat. conc. 0.75 g/cu m at 20°C, 1.8 g/cu m at 30°C; log P_{oct} 2.40/2.45
B. AIR POLLUTION FACTORS: 1 mg/cu m = 0.18 ppm, 1 ppm = 5.70 mg/cu m
C. WATER POLLUTION FACTORS:
 —Biodegradation: decomposition by a soil microflora in >64 days (176)
  adapted A.S. at 20°C—product is sole carbon source: 98.5% COD removal at 21.0 mg COD/g dry inoculum/hr (327)
D. BIOLOGICAL EFFECTS:
 —Toxicity threshold (cell multiplication inhibition test):
  bacteria (*Pseudomonas putida*): 10 mg/l (1900)
  algae (*Microcystis aeruginosa*): 1 mg/l (329)
  green algae (*Scenedesmus quadricauda*): 4.4 mg/l (1900)
  protozoa (*Entosiphon sulcatum*): 12 mg/l (1900)
  protozoa (*Uronema parduczi Chatton-Lwoff*): 24 mg/l (1901)
 —Fish:
  fathead minnows: static bioassay in Lake Superior water at 18-22°C: LC_{50} (1; 24; 48; 72; 96 hr): 43; 30; 30; 30; 30 mg/l (350)

p-methylnitrobenzene (*p*-nitrotoluene)

A. PROPERTIES: colorless needles; m.w. 137.13; m.p. 51.3°C; b.p. 238°C; v.p. 0.1 mm at 20°C, 0.22 mm at 30°C, 1.3 mm at 65°C v.d. 4.72; sp.gr. 1.286 at 20°C; solub. 442 mg/l at 30°C; sat. conc. 0.75 g/cu m at 20°C, 1.60 g/cu m 30°C; log P_{oct} 2.37/2.42
B. AIR POLLUTION FACTOR: 1 mg/cu m = 0.18 ppm, 1 ppm = 5.70 mg/cu m
C. WATER POLLUTION FACTORS:
 —Biodegradation: decomposition by a soil microflora in >64 days (176)
  adapted A.S. at 20°C—product is sole carbon source: 98.0% COD removal at 32.5 mg COD/g dry inoculum/hr (327)
 —Odor threshold conc. (detection): 0.003 mg/kg (894)
D. BIOLOGICAL EFFECTS:
 —Toxicity threshold (cell multiplication inhibition test):
  bacteria (*Pseudomonas putida*): 26 mg/l (1900)
  algae (*Microcystis aeruginosa*): 3.3 mg/l (329)
  green algae (*Scenedesmus quadricauda*): 15 mg/l (1900)
  protozoa (*Entosiphon sulcatum*): 8.6 mg/l (1900)
  protozoa (*Uronema parduczi Chatton-Lwoff*): 46 mg/l (1901)
 —Fish:
  Vairon (F): TLm (6 hr): distilled water: 20-22 mg/l
       hard water: 45-50 mg/l

3-methyl-4-nitrophenol see 4-nitro-*m*-cresol

3-methyl-6-nitrophenol see 6-nitro-*m*-cresol

4-methyl-2-nitrophenol see 2-nitro-*p*-cresol

methylnonylketone see 2-indecanone

3-methylolpentane see 2-ethyl-1-butanol

methyloxirane see propeneoxide

methyloxitol see ethyleneglycolmonomethylether

8-methylparacymene see *p-tert*-butyltoluene

methylparathion see dimethylparathion

2-methylpentane (dimethylpropylmethane)
$(CH_3)_2 CHCH_2 CH_2 CH_3$
  Use: chemical synthesis; solvent; gasoline component.
A. PROPERTIES: colorless liquid; m.w. 86.17; m.p. −154°C; b.p. 60°C v.p. 400 mm at 42°C; v.d. 2.97; sp.gr. 0.654
B. AIR POLLUTION FACTORS:
  —Manmade sources:
  diesel engine: 1.5% of emitted HC's (72)
  reciprocating gasoline engine: 0.55% of emitted HC's (78)
  rotary gasoline engine: 1.36% of emitted HC's (78)
  evaporation from gasoline fuel tank: 1.9 vol.% of total evaporated HC's
  evaporation from carburetor: 2.3–2.4 vol.% of total evaporated HC's
  (398; 399; 400; 401; 402)
  in gasoline engine exhaust: 62-car survey: 1.5 vol.% of total exhaust HC's
  (391)
  —Atmospheric reactions:
  estimated lifetime under photochemical smog conditions in S.E. England: 6.9 hr (1699; 1700)
C. WATER POLLUTION FACTORS:
  —Biodegradation:
  incubation with natural flora in the groundwater—in presence of the other components of high-octane gasoline (100 µl/l): biodegradation: 6% after 192 hr at 13°C (initial conc. 1.72 µl/l) (956)

3-methylpentane (diethylmethylmethane)
$CH_3 CH_2 CH(CH_3)CH_2 CH_3$
  Use: organic synthesis; solvent
A. PROPERTIES: colorless liquid; b.p. 64.0°C; sp.gr. 0.6645; m.w. 86.18
B. AIR POLLUTION FACTORS:

—Manmade sources:
 evaporation from gasoline fuel tank: 1.1 vol.% of total evaporated HC's
 evaporation from carburetor: 1.2–1.3 vol.% of total evaporated HC's
 (398; 399; 400; 401; 402)
—Atmospheric reactions:
 estimated lifetime under photochemical smog conditions in S.E. England: 5.2 hr
 (1699; 1700)

C. WATER POLLUTION FACTORS:
 —Biodegradation:
  incubation with natural flora in the groundwater—in presence of the other components of high-octane gasoline (100 µl/l): biodegradation: 7% after 192 hr at 13°C (initial conc.: 1.30 µl/l) (956)

2-methyl-2,4-pentanediol see hexyleneglycol

4-methyl-2-pentanol see methylamylalcohol

4-methyl-2-pentanone see methylisobutylketone

2-methyl-1-pentene (1-methyl-1-propylethene)
$CH_2C(CH_3)CH_2CH_2CH_3$
A. PROPERTIES: m.w. 84.16; m.p. −136°C; b.p. 61.5°C; sp.gr. 0.6817 solub. 78 mg/l at 20°C
C. WATER POLLUTION FACTORS:
 —Waste water treatment: A.S.: after 6 hr: 1.0% of ThOD
  12 hr: 1.1% of ThOD
  24 hr: 1.7% of ThOD (88)

4-methyl-1-pentene (isobutylethylene)
$CH_2CHCH_2CH(CH_3)_2$
A. PROPERTIES: m.w. 84.16; m.p. −153.6°C; b.p. 53.6°C; sp.gr. 0.6646; solub. 48 mg/l at 20°C
C. WATER POLLUTION FACTORS:
 —Waste water treatment: A.S. after 6 hr: 0.9% of ThOD
  12 hr: 0.8% of ThOD
  24 hr: 1.4% of ThOD (88)

4-methyl-2-pentene (1-isopropyl-2-methylethylene)
$CH_3CHCHCH(CH_3)_2$
A. PROPERTIES: m.w. 84.16; m.p. −134.43°C; b.p. 54/58°C; sp.gr. 0.67
C. WATER POLLUTION FACTORS:
 —Waste water treatment: A.S. after 6 hr: 0.6% of ThOD
  12 hr: 0.6% of ThOD
  24 hr: 1.3% of ThOD (88)

4-methyl-3-penten-2-one see mesityloxide

1-methylphenanthrene

Manmade sources:
in wood preservative sludge (n.s.i.): 8.21 g/l of raw sludge (993)
in South Louisiana crude oil: 111 ppm (1015)
in exhaust condensate of gasoline engine: 0.26–0.40 mg/l gasoline consumed
(1070)

A. PROPERTIES: solub. in distilled water at 25°C; 0.073 ± 0.005 mg/l
 solub. in seawater at 22°C: 0.3 ± 0.1 ppm
B. AIR POLLUTION FACTORS:
 —Manmade sources:
 in gasoline: 3.18 mg/l (1070)
 in coke oven emissions (n.s.i.): 44.7–1023.4 µg/g of sample (993)
 in coal tar pitch fumes: 6.0 wt % (n.s.i.) (516)
 —Ambient air quality:
 glc's in Birkenes (Norway): Jan–June 1977:
  avg.: 0.07 ng/cu m
  range: n.d.–0.49 ng/cu m ($n = 18$)
 glc's in Rørvik (Sweden): Dec. '76–April '77:
  avg.: 0.14 ng/cu m
  range: n.d. –1.05 ng/cu m ($n = 21$) (1236)
 glc's in residential area (Belgium): Oct. 1976:
  (+ methylanthracene) in particulate sample: 0.90 ng/cu m
   in gasphase sample: 10.2 ng/cu m (1289)
C. WATER POLLUTION FACTORS:
 —Water quality:
 in Eastern Ontario drinking waters (June–Oct. 1978): n.d.–5.1 ng/l ($n = 12$)
 in Eastern Ontario raw waters (June–Oct. 1978): n.d.–34.4 ng/l ($n = 2$) (1698)
D. BIOLOGICAL EFFECTS:
 —*Neanthes arenaceodentata*: 96 hr TLm in seawater at 22°C: 0.3 ppm (initial conc.
  in static assay) (995)

2-methylphenanthrene

Manmade sources:
in Kuwait crude oil: 89 ppm (1015)
in South Louisiana crude oil: 144 ppm (1015)

in gasoline: 7.73 mg/l (1070)
in exhaust condensate of gasoline engine: 0.27–0.58 mg/l gasoline consumed
(1070)

3-methylphenanthrene

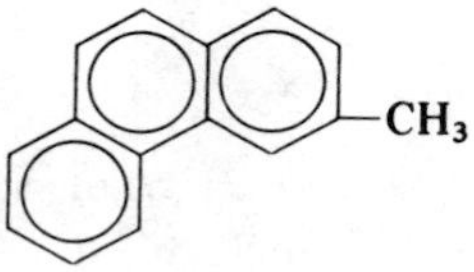

Manmade sources:
in gasoline: 6.87 mg/l
in exhaust condensate of gasoline engine: 0.26–0.51 mg/l gasoline consumed
(1070)

4-methylphenanthrene

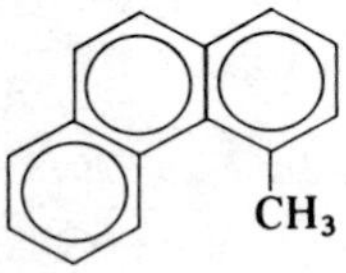

Manmade sources:
in gasoline: 1.24 mg/l
in exhaust condensate of gasoline engine: 0.19–0.33 mg/l gasoline consumed
(1070)

2-methylphenol *see o*-cresol

3-methylphenol *see m*-cresol

4-methylphenol *see p*-cresol

methylphenylcarbinol *see* α-methylbenzylalcohol

methylphenylether (anisole; methoxybenzene)

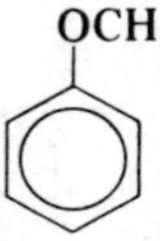

A. PROPERTIES: colorless liquid; m.w. 108.13; m.p. -37°C; b.p. 153.8°C v.p. 3.1 mm at 25°C, 10 mm at 42°C, 40 mm at 70°C; v.d. 3.72 sp.gr. 0.9954 at 20/4°C; log P_{oct} 2.04/2.11
B. AIR POLLUTION FACTORS: 1 mg/cu m = 0.226 ppm, 1 ppm = 4.46 mg/cu m
—Manmade sources: in gasoline exhaust: <0.1 ppm (195)
—Sampling and analysis: -photometry: min. full scale: 167 ppm (53)

C. WATER POLLUTION FACTORS:
  −Biodegradation: decomposition by a soil microflora in 8 days (176)
  −Oxidation parameters: KMnO$_4$ value: 4.044 (30)
  ThOD: 2.522 (30)
  −Odor threshold conc. (detection): 0.05 mg/kg; 0.2 mg/l (891; 893; 998)
  −Waste water treatment:

  acclimation
  NFG, BOD, 20°C, 1-10 days observed, feed: 400 mg/l. 365 + P: 23% removed
  NFG, BOD, 20°C, 1-10 days observed, feed: 1,000 mg/l. 365 + P: 15% (93)

methylphenylketone *see* acetophenone

methylphthalate *see* o-dimethylphthalate

methylpicrylnitramine *see* tetryl

2-methylpropanal *see* isobutyraldehyde

2-methylpropane *see* isobutane

2-methylpropanoic acid *see* isobutyric acid

2-methyl-2-propanol *see* t-butanol

2-methylpropene *see* isobutene

2-methylpropenal *see* methacrolein

2-methylpropenoic acid *see* methacrylic acid

methylpropionate
CH$_3$CH$_2$COOCH$_3$
A. PROPERTIES: colorless liquid; m.w. 88.15; m.p. −87.5°C; v.p. 40 mm at 11°C; v.d. 3.03; sp.gr. 0.937 at 4°C
B. AIR POLLUTION FACTORS:
  −Odor threshold conc. (recognition): 10-12.5 mg/cu m (610)
D. BIOLOGICAL EFFECTS:
  −Toxicity threshold (cell multiplication inhibition test):
  bacteria (*Pseudomonas putida*):   330 mg/l   (1900)
  algae (*Microcystis aeruginosa*):   13 mg/l   (329)
  green algae (*Scenedesmus quadricauda*): 11 mg/l   (1900)
  protozoa (*Entosiphon sulcatum*):   311 mg/l   (1900)
  protozoa (*Uronema parduczi Chatton-Lwoff*): 1573 mg/l (1901)

β-methylpropylethanoate *see* isobutylacetate

methylpropylcarbinol *see* sec-act-amylalcohol

1-methyl-1-propylethene *see* 2-methyl-1-pentene

methyl-n-propylketone *see* 2-pentanone

1-methylpyrene

Manmade sources:
in coal tar (n.s.i.): 2.06 mg/g of sample (993)
in wood preservative sludge (n.s.i.): 3.22 g/l of raw sludge (993)
B. AIR POLLUTION FACTORS:
—Manmade sources:
in airborne coal tar emissions (n.s.i.): 3.88 mg/g of sample or 136 µg/cu m of air sampled (993)
in coke oven emissions (n.s.i.): 137–1872 µg/g of sample (993)
—Ambient air quality:
glc's at Botrange (Belgium): woodland at 20-30 km from industrial area—June-July 1977: 0.05; 0.08 ng/cu m ($n = 2$) (n.s.i.)
glc's at Wilrijk (Belgium): residential area: Oct.–Dec. 1976: 0.76; 2.71 ng/cu m ($n = 2$) (n.s.i.) (1233)
glc's in residential area (Belgium): Oct. 1976 (n.s.i.): in particulate sample: 0.93 ng/cu m (1289)
glc's in Birkenes (Norway): Jan–June 1977:
avg.: 0.04 ng/cu m
range: n.d.–0.35 ng/cu m ($n = 18$)
glc's in Rørvik (Sweden): Dec. '76–April '77:
avg.: 0.04 ng/cu m
range: n.d.–0.48 ng/cu m ($n = 21$) (1236)
C. WATER POLLUTION FACTORS:
—Sediment quality:
in sediment of Wilderness Lake, Colin Scott, Ontario (1976): <2 ppb (dry wt) (n.s.i.) (932)

2-methylpyridine *see* α-picoline

3-methylpyridine *see* β-picoline

N-methyl-2-pyridone-5-carboxamide
C. WATER POLLUTION FACTORS:
—Manmade sources: in sewage effluent: 0.010 mg/l (234)
in primary domestic sewage plant effluent: 0.020–0.025 mg/l. (517)

N'-methyl-4-pyridone-3-carboxamide
—*Manmade sources*: in primary domestic sewage plant effluent: 0.010–0.014 mg/l (517)

N-methyl-2-pyrrolidone

A. PROPERTIES: m.w. 99.13; b.p. 202°C; v.d. 3.4; sp.gr. <1.0
B. AIR POLLUTION FACTORS: 1 mg/cu m = 0.24 ppm, 1 ppm = 4.12 mg/cu m
C. WATER POLLUTION FACTORS:
   —Impact on biodegradation processes: self purification of surface water is affected
      from 0.5 mg/l upwards (n.s.i.) (181)
D. BIOLOGICAL EFFECTS:
   —Bacteria: *Pseudomonas*: LD_0: 5 g/l
   —Algae: *Scenedesmus*: 5 g/l
   —Protozoa: *Colpoda*: 5 g/l (n.s.i.) (30)

2-methylquinoline

A. PROPERTIES: log $P_{oct.}$ 2.23 (calculated; n.s.i.)
D. BIOLOGICAL EFFECTS:
   —Ciliate (*Tetrahymena pyriformis*): 24 hr LC_{100}: 2.79 mmole/l (n.s.i.) (1662)
   —Embryos of South African clawed frog, *Xenopus laevis*: 96 hr LC_{50}: 26.4 mg/l
      (1918)

6-methylquinoline

A. PROPERTIES: log P_{oct} 2.57
C. WATER POLLUTION FACTORS:
   —Manmade sources:
      in coal liquefaction waste water (349)
D. BIOLOGICAL EFFECTS:
   —*Daphnia magna*: 48 hr LC_{50}: 11 mg/l
      in mixture with resorcinol: intra-additive interaction or antag-
         onism depending upon concentration (349)
   —Ciliate (*Tetrahymena pyriformis*): 24 hr LC_{100}: 1.57 mmole/l (1662)

8-methylquinoline

A. PROPERTIES: log P_{oct} 2.60
D. BIOLOGICAL EFFECTS:
 −Ciliate (*Tetrahymena pyriformis*): 24 hr LC_{100}: 1.57 mmole/l (1662)

methylsalicylate (methyl-2-hydroxybenzoate; artificial oil of wintergreen)

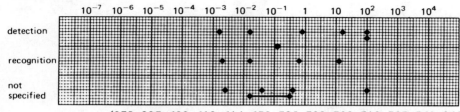

A. PROPERTIES: colorless liquid; m.w. 152.14; m.p. −8.6°C; b.p. 222°C; v.p. 1 mm at 54°C; v.d. 5.24; sp.gr. 1.18 at 20/4°C; solub. 740 mg/l at 30°C, 5,000 mg/l at 30°C
B. AIR POLLUTION FACTORS: 1 mg/cu m = 0.158 ppm, 1 ppm = 6.324 mg/cu m
 −Odor: wintergreen

odor thresholds

(279; 307; 602; 610; 614; 672; 705; 708; 709; 713; 727; 762; 840)
O.I. at 20°C = 113,400 (316)
C. WATER POLLUTION FACTORS:
 −Odor threshold conc. (detection): 0.04 mg/kg (882)
D. BIOLOGICAL EFFECTS:
 −Mammalia: rat: inhalation: no effect level: 120 ppm, 20 × 7 hr (65)
  guinea pig: acute oral LD_{50}: 0.7 g/kg
  rabbit: acute oral LD_{50}: 2.8 g/kg
  dog: acute oral LD_{50}: 2.1 g/kg (211)
  rat: ingestion: single oral LD_{50}: 700 mg/kg (1546)
 −Man: acute oral LD_{50}: 0.5 g/kg (adult) (211)

methylsulfate *see* dimethylsulfate

methylsulfide *see* dimethylsulfide

methylsulfanilylcarbamate *see* asulam

methylsulfinylmethane *see* dimethylsulfoxide

methylsulfonic acid
B. AIR POLLUTION FACTORS:
 −Ambient air quality:
  glc's in the vicinity of Karlsruhe (W. Germany), Oct.−Nov. 1978: in atmospheric aerosol.

0.02–0.43 µg/cu m (2 hr averages)
avg. 0.16 µg/cu m (n = 10) (1730)

4-(methylsulfonyl)-2,6-dinitro-N,N-dipropylaniline (planavin)
D. BIOLOGICAL EFFECTS:
—Freshwater ectoprocta: no appreciable effect at 2.5 mg/l for 84 hr exposure
(1902)

methylsulfoxide *see* dimethylsulfoxide

N-methyl-N-2,4,6-tetranitroaniline *see* tetryl

methyltheobromine *see* caffeine

methylthioalcohol *see* methylmercaptan

methylthiomethane *see* dimethylsulfide

methylthioureasulfate
C. WATER POLLUTION FACTORS:
—Impact on biodegradation processes: 75% inhibition of nitrification process in nonacclimated activated sludge at: 6.5 mg/l (30)

methyltriethyllead
$CH_3Pb(C_2H_5)_3$
D. BIOLOGICAL EFFECTS:
—Residues: taken from various lakes and rivers in Ontario (1979):
in fish: n.d.–4.4 ng/g wet wt.
in water, vegetation, weeds and sediments: n.d. (1793)

methyltriphenylene
Manmade sources:
in wood preservative sludge: 0.38 g/l of raw sludge
in coke oven airborne emissions: 40.8–464.0 µg/g of sample (960)

5-methyluracil *see* thymine

6-methyluracil
Manmade sources: formed by photodecomposition of bromacil. (1638)

2-methylvaleric acid

$$CH_3-CH_2-CH_2-\underset{\underset{CH_3}{|}}{CH}-COOH$$

C. WATER POLLUTION FACTORS:
—$BOD_{2\ days}^{25°C}$: 1.03 (substrate conc. 3.5 mg/l; inoculum: (soil microorganisms)
5 days: 2.09
10 days: 2.14 (1304)

3-methylvaleric acid

$$CH_3-CH_2-CH(CH_3)-CH_2-COOH$$

C. WATER POLLUTION FACTORS:
 —$BOD_{2\ days}^{25°C}$: 0.00 (substrate conc.: 3.5 mg/l;
   inoculum: soil micro-organisms)
 5 days: 1.11
 10 days: 1.89 (1304)

4-methylvaleric acid

$$CH_3-CH(CH_3)-CH_2-CH_2-COOH$$

C. WATER POLLUTION FACTORS:
 —$BOD_{2\ days}^{25°C}$: 0.97 (substrate conc.: 3.5 mg/l;
   inoculum: soil micro-organisms)
 5 days: 1.54
 10 days: 2.09 (1304)

methylvinylketone
$CH_2CHCOCH_3$

A. PROPERTIES: m.w. 70.09; b.p. 81°C; v.d. 2.41; sp.gr. 0.84
B. AIR POLLUTION FACTORS: 1 mg/cu m = 0.34 ppm, 1 ppm = 2.91 mg/cu m
 —Odor threshold conc.: 0.5 mg/cu m (747)
C. WATER POLLUTION FACTORS:
 —Oxidation parameters: BOD_5: 10% ThOD
   COD: 100% ThOD
   $KMnO_4$ value acid: 34% ThOD
        alkaline: 49% ThOD (220)
 —Waste water treatment:

method	temp °C	days observed	feed mg/l	days acclim.	% theor. oxidation
TF, Sd, BOD	20	10	—	—	nil
RW, Sd, BOD	20	10	1.5	70	nil (93)

 —Photooxidation by U.V. light in aqueous medium at 50°C: 16.8% degradation
   to CO_2 after 24 hr (1628)

1-methylxanthine
C. WATER POLLUTION FACTORS:
 —Manmade sources: in sewage effluent: 0.0006–0.017 mg/l (234)
   in primary domestic sewage plant effluent: 0.070 mg/l
   in secondary domestic sewage plant effluent: 0.006 mg/l (517)

7-methylxanthine
Manmade sources:
 in primary domestic sewage plant effluent: 0.002–0.090 mg/l
 in secondary domestic sewage plant effluent: 0.005 mg/l (517)

methyl yellow *see* 4-dimethylaminoazobenzene

metoxuron *see* dosanex

metribuzin *see* 4-amino-6-*t*-butyl-3-methylthio-1,2,4-triazin-5(4H)-one

mevinphos (phosdrin)

$$\begin{array}{c} CH_3O \\ \diagdown \\ CH_3O \end{array} \!\!\!\! \begin{array}{c} O \\ \| \\ P \end{array} \!\!-\!\! O \!-\!\! \begin{array}{c} CH_3 \\ | \\ C \end{array} \!\!=\!\! CH \!-\!\! \begin{array}{c} O \\ \| \\ C \end{array} \!\!-\! OCH_3$$

Trademark for a mixture which contains no less than 60% of the *cis*-isomer of 2-carbomethoxy-1-methylvinyldimethylphosphate.
Use: contact and systemic insecticide and acaricide

A. PROPERTIES: m.w. 224.1; b.p. 99–103°C at 0.03 mm; sp.gr. 1.25 at 20/4°C
D. BIOLOGICAL EFFECTS:

	96 hr LC_{50} (S) $\mu g/l$	
—Crustaceans:		
Gammarus lacustris	130	(2124)
Gammarus fasciatus	2.8	(2126)
Palaemonetes kadiakensis	12	(2126)
Asellus brevicaudus	56	(2126)
Simocephalus serrulatus	0.43	(2127)
Daphnia pulex	0.16	(2127)
Crangon septemspinosa (sand shrimp)	11	(2327)
Palaemonetes vulgaris (grass shrimp)	69	(2327)
Pagurus longicarpus (hermit crab)	28	(2327)
—Insects:		
Pteronarcys californica	5	(2128)
—Fish:		
Lepomis macrochirus	70	(2137)
Micropterus salmoides	110	(2137)
Fundulus heteroclitus (mummichog)	65	(2329)
Fundulus heteroclitus (mummichog)	300	(2329)
Fundulus majalis (striped killifish)	75	(2329)
Menidia menidia (Atlantic silverside)	320	(2329)
Mugil cephalus (striped mullet)	300	(2329)
Anguilla rostrata (American eel)	65	(2329)
Thalassoma bifasciatum (bluehead)	74	(2329)
Sphaeroides maculatus (northern puffer)	800	(2329)
bluegill: 24 hr LC_{50}: 0.041 ppm		
rainbow trout: 24 hr LC_{50}: 0.034 ppm		(1878)

—Mammals:
  acute oral LD_{50} (mice): 7.1–18.0 mg/kg
              (rat): approx. 3.7–12 mg/kg
  acute dermal LD_{50} (rabbit): approx. 16–34 mg/kg

inhalation (rats and mice): no deaths, sat. vap. at 25°C for 4 hr (1854)
in diet: rats and dogs: no gross toxic effects, at 25 ppm in diet (1855)

mexacarbate (zectran; 4-dimethylamino-3,5-xylyl-N-methylcarbamate)

$$(CH_3)_2N-\underset{CH_3}{\overset{CH_3}{C_6H_2}}-O-\underset{\|}{\overset{O}{C}}-NH-CH_3$$

D. BIOLOGICAL EFFECTS:
  —Bioaccumulation:
  BCF*: brine shrimp: 18   (at 5.0 ppb in water)
         mosquito larvae: 0–8   (at 5.5–10.8 ppb in water)
         fish (silverside): 45 (at 4.7 ppb in water) (1880)
  *mexacarbate introduced into system in the form of residues on sand
  —Toxicity:
  insects: fourth instar larval *Chironomus riparius*: 24 hr LC_{50}, 12,2 ppb (1853)
  fish:
  Atlantic salmon: 96-hr LC_{50} (S): 22.3 mg/l
  brown trout: 96-hr LC_{50} (S): 20.0 mg/l
  fathead minnow: 96-hr LC_{50} (S): 23.7 mg/l
  yellow perch: 96-hr LC_{50} (S): 16.2 mg/l (1502)
  coho salmon: 96-hr LC_{50} (S): 4–23.0 mg/l
  bluegill: 96-hr LC_{50} (S): 0.6–22.9 mg/l (1502)
  —Mammals:
  acute oral LD_{50} (male rat): 19 mg/kg
                      (dog): 15–30 mg/kg (1855)

MIBC *see* methylamylalcohol

MIBK *see* methylisobutylketone

MIPA *see* isopropanolamine

mirex (1,1a,2,2,3a,4,5,5,5a,5b,6-dodecachlorooctahydro-1,3,4-metheno-1H-cyclobuta(*cd*)pentalene)

Use: mirex, a chlorinated insecticide, is the active ingredient in bait used to control

the fire ant, harvester ant and Texas leaf-cutting ant. Mirex is also marketed under the tradename Dechlorane for use in flame-retardant coatings for various materials.
A. PROPERTIES: mirex is a snow-white, odorless, free-flowing crystalline solid; m.w. 546; solub. 0.20 mg/l at 24°C (practical grade)
C. WATER POLLUTION FACTORS:
 −River and sediment quality:
  in Tombigbee river (Mississippi): ~0.001 µg/l increasing to 0.03 µg/l during spring flood, throughout the year after treatment of the upper watershed with mirex. (1370)
  Mississippi: bottom sediments: residues not found (1170)
 −Aquatic reactions:
  photooxidation by U.V. light in aqueous medium at 90-95°C: time for the formation of CO_2 (% of theoretical): 25%: 21.3 hr
  50%: 48.4 hr
  75%: 105.1 hr (1628)
  Exposure to sunlight and U.V. light have indicated slow degradation; resulting compounds included chlordecone hydrate, undecachloropentacyclodecane, and nonachloropentacyclodecan-5-one hydrate (483)
D. BIOLOGICAL EFFECTS:
 −Bioaccumulation:
  BCF: algae: 12,200 fish: 2,580
   snails: 4900 crayfish: 16,860–71,400
   daphnids: 14,650 (1891; 1892; 1893)
  uptake of mirex by selected freshwater invertebrates after exposure to different concentrations of mirex for various periods of time: (1922)

species	mirex conc. in water, ppm	period of exposure, hr	mirex residue in animal[a] ppm
Group A			
Erpobdella punctata (leech)	2.0	48	1.92
Placobdella rugosa (leech)	2.0	48	4.91
Glossiphonia sp. (leech)	2.0	48	5.07
Palaemonetes kadiakensis (shrimp)	2.0	48	9.95
Orconectes mississippiensis (crayfish)	2.0	48	11.01
Group B			
Orconectes mississippiensis (crayfish)	2.0	120	21.20
Orconectes mississippiensis (crayfish)	0.5	120	7.12
Placobdella rugosa (leech)	0.5	120	1.95
Group C			
Macromia sp. naiad (dragonfly)	0.5	24	10.37
Macromia sp. naiad (dragonfly)	1.0	24	13.68
Macromia sp. naiad (dragonfly)	1.0	168	92.90
Group D			
Hyallela azteca (amphipod)	0.001	672	2.53
Orconectes mississippiensis (crayfish)	0.001	672	1.06

[a]Values represent mean of 2–3 analyses of whole-body residue pooled from 200 individuals for *Hyallella*, and from 5–10 animals for all others.

bioaccumulation factors after 70 days exposure to 0.038 µg/l:
grass shrimp: 13,100–17,400
sheephead minnows: 28,900–50,000
mud crabs: 15,000–18,700
hermit crabs: 44,800–71,100
ribbed mussels (soft tissue): 42,000–52,600
american oysters (soft tissue): 34,200–73,700
conc. in sand substrate: <0.010–0.012 µg/g after 21 days
0.019–0.029 µg/g after 42 days
0.031–0.042 µg/g after 63 days
0.031–0.050 µg/g after 70 days

accumulation factors: $\dfrac{\text{conc. sand substrate}}{\text{conc. water}}$:

after 70 days: 930–1500 (1630)

blue crab (*Callinectes sapidus*): conc. in water: 0.22 µg/l; in crab hepatopancreas after 6 hr: 31 µg/kg (1371)

blue crabs: conc. in water: 0.03–0.16 µg/l; whole body residue: 20–590 µg/kg after 28 days exposure

grass shrimp: conc. in water: 0.03–1.16 µg/l; whole body residue: 90–2,400 µg/kg after 28 days exposure (1372)

—Biomagnification:
concentrations of mirex in components of two experimental systems (µg/kg):

exposure days	system[a]	shrimp whole-body residue[b]	*Thalassia* detritus	*Thalassia* leaves	water[b]
1	C	100	17	8	0.12
	S	40 (50)			0.88 (0.12)
4	C	160	28	18	0.057
	S	100 (60)			0.10 (0.057)
7	C	260	27	17	0.031
	S	230 (100)			0.073 (0.031)
10	C	140	31	13	0.030
	S	220 (110)			0.058 (0.030)
13	C	130	29	15	0.018
	S	200 (120)			0.031 (0.018)

[a] C = complete system (shrimp, *Thalassia*, sand)
S = shrimp-only system
[b] values in parentheses are mirex concentrations from hypothetically equal water concentrations of mirex in both systems. (1370)

accumulation in the food chain after 6 applications of mirex (1.25 lb of bait per acre) over a 4-hear period:
aquatic plants: contained negligible amounts
snails, crawfish, fish: 0.01–0.75 ppm
softshell turtle fat: 24.82 ppm
birds: 1.2 to 1.91 ppm
fat of vertebrates: up to 73.94 ppm (484)

marine unicellular algae species: no observable effect on population growth following a 7-day exposure to 10–50 ppt

bioconcentration factors ranged from 3200–7300 at 10–50 ppt resulting in 0.032–0.365 ppm (485)
american oysters (*Crassostrea virginica* Gmelin): bioconcentration factor after 43 days of exposure: 5100–5500 (based on oven dry wt.) (1620)

decapod	*Callinectes sapidus*	exposure	bioaccumulation
	5 day larvae	1.0 ppb	301 ppb
	8 day larvae	1.0 ppb	406 ppb
	5 day larvae	10.0 ppb	1 620 ppb
	8 day larvae	10.0 ppb	1 370 ppb
decapod	*Uca pugilator*	0–30 ppb: 42 days	up to 3.6 μg/g (1149)
decapod	*Callinectes sapidus*	0–30 ppb: 42 days	up to 380 ppb (1167)
fish	*Fundulus grandis*	0–30 ppb: 42 days	up to 90 ppb
fish	*Fundulus similis*	0–30 ppb: 42 days	up to 260 ppb (1167)

bioconcentration factors after 4 seasonal experiments, each lasting 28 days, at $\simeq$0.5 ppb: minnows: 40,800 ($\simeq$20.4 ppm)
pink and grass shrimp: 10,000 ($\simeq$5 ppm)
blue crabs: 2300 ($\simeq$1.5 ppm) (486)

—Toxicity:
Photosynthesis of plankton is inhibited by 16, 10, 33 and 19% after exposure to 1 ppb after 5, 10, 15 and 20 days respectively (1682)

freshwater cniderian: *Hydra* sp.: 1 day LC_{50}: >100,000 ppm
2 days:	682 ppm
3 days:	23 ppm
4 days:	4 ppm
5 days:	1 ppm
6 days:	0.5 ppm (1605)

—Algae:
Tetrahymena pyriformis: 0.9 ppb; 16.03% decrease in population size; 96 hr growth test (2351)

—Crustacean:

bait (.3% mirex)	*Penaeus aztecus*	brown shrimp	one particle of mirex bait/shrimp	48% paralysis or death in 4 days	static bioassay (2340)
bait (.3% mirex)	*Palaemonetes pugio*	grass shrimp	one particle of mirex bait/shrimp	63% paralysis/or death in 4 days	static bioassay
technical	*Penaeus duorarum*	pink shrimp	1.0 ppb	100% paralysis/or death in 11 days	flowing water bioassay
technical	*Penaeus duorarum*	pink shrimp	0.1 ppb	36% paralysis/or death in 35 days	flowing water bioassay
bait (0.3% mirex)	*Uca pugilator*	fiddler crab	one particle of mirex bait per crab	73% paralysis/or death in 14 days	flowing water bioassay
bait (0.3%) mirex	*Callinectes sapidus*	blue crab	1 particle of bait/crab	84% paralysis/death in 20 days	96 hr flowing water bioassay

NAPHTHALENE

Manufacturing source: petroleum refining; coal tar distillation (347)
in commercial coal tar: 150 ppm. (1874)
Users and formulation: moth ball mfg.; mfg. alpha and beta naphthols and pesticides, fungicides, dyes, detergents and wetting agents, phthalic anhydride, synthetic resins, celluloids, lampblack, smokeless powder, solvent; lubricants; motor fuel mfg.; cutting fluid, synthetic tanning, preservative; emulsion breaker; asphalt and naphtha constituent, petroleum, coal tar. (347)

A. PROPERTIES: white flakes or powder; m.w. 128.16; m.p. 80.2°C b.p. 217.9°C sublimes; v.p. 1 mm at 53°C; v.d. 4.42; sp.gr. 1.152 solub. 30 mg/l; THC 1,231 kcal/mole; LHC 1,208 kcal/mole; log P_{oct} 3.01/3.45; solub. at 22°C in seawater: 20 ± 2 mg/l; 31–34 mg/l in distilled water at 25°C

B. AIR POLLUTION FACTORS: 1 mg/cu m = 0.191 ppm, 1 ppm = 5.24 mg/cu m
 —Manmade sources: in coal tar pitch fumes: 0.9 wt % (516)

odor thresholds	mg/cu m
	10^{-7} 10^{-6} 10^{-5} 10^{-4} 10^{-3} 10^{-2} 10^{-1} 1 10 10^2 10^3 10^4
detection	
recognition	♦ ♦ ♦
not specified	♦ ♦

(57; 307; 610; 684; 753; 755)
O.I. at 20°C = 2,400 (316)
 —Sampling and analysis: photometry: min. full scale: 7.4 ppm (53)

C. WATER POLLUTION FACTORS:
 —Oxidation parameters:
  BOD_5: 0% of ThOD (274)
   nil std. dil. sewage (41)(27)(275)
   nil% ThOD (220)
  BOD_{35}^{25}: 1.92 in seawater/inoculum: enrichment cultures of hydrocarbon oxidizing bacteria (521)
  COD: 80% ThOD (0.05 n Cr_2O_7) (274)
   22% ThOD (220)
  $KMnO_4$: acid: 3% ThOD
   alkaline: 0% ThOD (220)
   0% ThOD (0.01 n $KMnO_4$) (274)
  ThOD: 2.99 (274)
 —Reduction of amenities: approx. conc. causing adverse taste in fish
   (Rudd): 1.0 mg/l (41)
   tainting of fish flesh: 1.0 mg/l (81)
  T.O.C. = 0.5 mg/l (296)
   = 0.005 mg/l (295)
   = 0.0068 mg/l (297)(403)
   = 0.001 mg/l (301)
  T.O.C. in water: 0.068 ppm (326)

 —Biodegradation: biodegradation to CO_2:

Nocardia sp. 119 and *Micrococcus* sp. 22 r isolated from garden soils and rice field drains degraded molinate completely into various hydroxy and oxidized products (1848; 1849)
—Aquatic reactions:
molinate added to tapwater decreased over a 14 day holding period to 40% based on recovery of ^{14}C. The loss of ^{14}C was attributed primarily to volatilization. The five major organosoluble metabolites in the tapwater were molinate sulfoxide, 3- and 4-hydroxymolinate, 4-ketomolinate, and ketohexamethylene-imine

D. BIOLOGICAL EFFECTS:
—Crustaceans:

	96 hr LC_{50} $\mu g/l$	
Gammarus lacustris	4500	
Gammarus fasciatus	300	(2124)
Daphnia magna	600	(2125)
Asellus brevicaudus	400	
Palaemonetes kadiakensis	1000	
Orconectes nais	5600	

brown shrimp (*P. aztecus*): 48 hr LC_{50}: >1 ppm (498)
—Mussels:
commercial oyster (C. virginica): EC_{50} (96 hr exp.; shell growth inhibition): >1 ppm (498)
—Fish:
BCF in Japanese carp exposed to 0.2 ppm (ring-^{14}C) molinate = 1.13 during 1 to 14 days exposure (based on ^{14}C in whole tissue) (1844)
goldfish: 96 hr LC_{50}: 30 ppm (technical grade) (1845)
mosquito fish (*Gambusia affinis*): 96 hr TLm: 16.4 ppm (1846)
japanese carp (*Cyprinus carpio var. Yamato Kai*): 20 day lethal conc.: 0.18 ppm (1847)
american carp: no toxic effect after 2 weeks at 10 ppm (1844)
stickleback (*G. aculeatus*): 72 hr, LC_{50}: ca. 14 ppm
rainbow trout (*S. Gairdneri*): 96 hr LC_{50}: 0.20 ppm
bluegill sunfish (*L. macrochirus*): 96 hr LC_{50}: 29 ppm
goldfish (*C. auratus*): 96 hr LC_{50}: 30 ppm (498)
carp: 21 d LC_{50} (R): 0.18 mg/l (1527)
mosquitofish: 24 hr LC_{50} (S): 30.70 mg/l
48 hr LC_{50} (S): 21.4 mg/l
96 hr LC_{50} (S): 16.4 mg/l (1104)
mosquitofish: 72-hr LC_{50} (S): 17.0 mg/l
green sunfish: 72-hr LC_{50} (S): 34.5 mg/l (1203)
—Birds:
mallard duckling: LC_{50} (5-day feed treatment): >9,300 ppm
coturnix quail: LC_{50} (9-week feed treatment): >1,000 ppm (998)
—Mammals: acute oral toxicity:
male albino rats and mice
technical ordram: rats: LD_{50}: 720 mg/kg
mice: LD_{50}: 795 mg/kg

884 MONOCHLOROACETIC ACID

    ordram 6E: rats: LD_{50}:   584 mg/kg
                          mice: LD_{50}:   1260 mg/kg
acute dermal toxicity: rabbits
                      technical ordram: LD_{50}: $>$ 2,000 mg/kg
                      ordram 6E: LD_{50}: $>$10,000 mg/kg
acute ocular irritancy: rabbits:
    technical ordram: moderate irritation with iritis and corneal dullness. All signs
                      clearing within 2 to 5 days
    ordram 6E: produced comparable effects clearing within 4 to 6 days
acute inhalation toxicity: rats:
    ordram 6E: LC_{50} (1 hr): $>$2.1 mg/l

monochloroacetic acid   *see* chloroethanoic acid
$CH_2ClCOOH$
A. PROPERTIES: colorless crystals; m.w. 94.50; m.p. α: 63; β: 55–56; γ: 50°C; v.p.
    1 mm at 43°C; v.d. 3.25; sp.gr. 1.58 at 20/20°C
B. AIR POLLUTION FACTORS:
    —Odor: T.O.C.: 0.045 ppm   (279)
            O.I. at 20°C = 1,460   (316)
D. BIOLOGICAL EFFECTS:
    —Worms: *Tubifex tubifex*: perturbation conc.: 150 mg/l
    —Protozoa: *Vorticella campanula*: perturbation conc.: 9 mg/l
               Paramaecium caudatum: toxic: 150 mg/l   (30)
    —Arthropoda: *Gammarus pulex*: perturbation conc.: 30 mg/l
    —Insects: *Chironomus plumosus*: perturbation conc.: 140 mg/l
    —Fish: *Trutta iridea*: perturbation conc.: 20 mg/l
            Cyprinus carpio: perturbation conc.: 14 mg/l   (30)
    —Mammalia: guinea pig: oral LD_{50}: 80 mg/kg
                  mouse: oral LD_{50}: 255 mg/kg
                  rat: oral LD_{50}: 76 mg/kg   (211)

monochlorodehydroabietic acid
D. BIOLOGICAL EFFECTS:
    —Fish:
       juvenile rainbow trout: 96 hr LC_{50} (S): 0.6 mg/l   (314)

monochlorodimethylether   *see* chloromethylmethylether

monochloroethane   *see* ethylchloride

monochlorohydroquinone
$C_6H_3(OH)_2Cl$
    Use: a substitution product of hydroquinone, monochlorohydroquinone serves as
    reducing agent in developing baths
A. PROPERTIES: m.w. 144.56
C. WATER POLLUTION FACTORS:
    —Oxidation parameters:

theoretical	analytical
TOD = 1.33	reflux COD = 97.0% recovery
COD = 1.33	rapid COD = 92.2% recovery
	BOD_5 = 0.031
	BOD_5/COD = 0.024

—Biodegradation:

	chemical conc. mg/l	effect
unacclimated system	10	no effect
	100	inhibitory
	1,000	inhibitory
acclimated system	180	no effect

Monochlorohydroquinone affected an unacclimated system the same way that hydroquinone did. The compound did not affect the system at 10 mg/l, but became inhibitory at 100 mg/l. The inhibition to the higher concentrations was removed by acclimation of the biomass, although the compound was not found to be amenable to biodegradation.

D. BIOLOGICAL EFFECTS:
 —Algae: *Selenastrum capricornutum*: 0.1 mg/l no effect
 0.4 mg/l no effect
 1.0 mg/l no effect
 4.0 mg/l inhibitory
 10.0 mg/l inhibitory
 40.0 mg/l inhibitory
 —Crustacean: *Daphnia magna*: LC_{50}: 0.017 mg/l
 —Fish: *Pimephales promelas*: LC_{50}: 0.19 mg/l (1828)

monocrotophos *see* azodrin

monofluoroacetatic acid
$CH_2F-COOH$
D. BIOLOGICAL EFFECTS:
 —Toxicity threshold (cell multiplication inhibition test):

bacteria (*Pseudomonas putida*):	65 mg/l	(1900)
algae (*Microcystis aeruginosa*):	0.0004 mg/l	(329)
green algae (*Scenedesmus quadricauda*):	0.055 mg/l	(1900)
protozoa (*Entosiphon sulcatum*):	31 mg/l	(1900)
protozoa (*Uronema parduczi Chatton-Lwoff*):	5300 mg/l	(1901)

monolinuron
D. BIOLOGICAL EFFECTS:
 —Bacteria: *Pseudomonas putida*: inhibition of cell multiplication starts at 11 mg/l
 (329)
 —Algae: *Microcystis aeruginosa*: inhibition of cell multiplication starts at 0.14 mg/l
 (329)

monomethylamine *see* methylamine

monomethylhydrazine (methylhydrazine; MMH)
CH_3NHNH_2
  Use: missile propellant; intermediate; solvent
A. PROPERTIES: sp.gr. 0.874 at 25°C; m.p. −52.4°C; b.p. 87.5°C, soluble in water
D. BIOLOGICAL EFFECTS:
  −Guppy: (*Lebistes reticulatus*) static bioassay:
    24 hr LC_{50}: 5.26 mg/l in hard water at 22–24.5°C
    48 hr   4.02 mg/l in hard water at 22–24.5°C
    72 hr   3.67 mg/l in hard water at 22–24.5°C
    96 hr   3.26 mg/l in hard water at 22–24.5°C
    24 hr LC_{50}: 6.69 mg/l in soft water at 22–24.5°C
    48 hr   3.66 mg/l in soft water at 22–24.5°C
    72 hr   2.74 mg/l in soft water at 22–24.5°C
    96 hr   2.58 mg/l in soft water at 22–24.5°C   (474)

monosodiummethanearsenate (MSMA)

$$CH_3-\underset{\underset{ONa}{|}}{\overset{\overset{O}{\|}}{As}}-OH$$

  Use: selective postemergent herbicide in cotton and noncrop areas
D. BIOLOGICAL EFFECTS:
  −Bacteria:
    mixed culture of *Citrobacter frundii*, *Aeromonas* sp. and *Klebsiella* sp. isolated from soil:
    24 hr LC_{10}: 13 mg/l
    24 hr LC_{50}: 27 mg/l; 48 hr LC_{50}: 60 mg/l; 96 hr LC_{50}: 220 mg/l
    24 hr LC_{90}: 65 mg/l   (1722)
  −Mammals:
    white mice: oral LD_{50}: 1,800 mg/kg
    rats: oral LD_{50}:   700 mg/kg
    cattle: a total dose of 100 mg/kg administered at a rate of 10 mg/kg/day is lethal
                                                                       (1422)
    snowshoe hares: LD_{50}: 173 mg/kg   (1423)
    Peromyscus leucopus (mouse): oral LD_{50}: 300 mg/kg   (1421)

monothioglycol see 2-mercaptoethanol

monuron see 3-(*p*-chlorophenyl)-1,1-dimethylurea

morpholine (tetrahydro-1,4-oxazine; diethyleneimide oxide)

Use: rubber accelerator; solvent; organic synthesis; additive to boiler water; waxes and polishes; corrosion inhibitor; optical brightener for detergents.
Impurities: N-ethylmorpholine; ethylene diamine. (R 515)

A. PROPERTIES: colorless hygroscopic liquid; m.w. 87.12; m.p. -3/-5°C; b.p. 127/129°C; v.p. 4.3 mm at 10°C, 8.0 mm at 20°C, 13.4 mm at 30°C; v.d. 3.00; sp.gr. 1.00 at 20/4°C; log P_{oct} -1.08

B. AIR POLLUTION FACTORS: 1 mg/cu m = 0.28 ppm, 1 ppm = 3.62 mg/cu m
—Odor: characteristic: quality: fishy, amine
  hedonic tone: unpleasant
  T.O.C.: abs. perc. limit: 0.01 ppm
    50% recogn.: 0.07 ppm
    100% recogn.: 0.14 ppm
  O.I.: 65,857 (19)

C. WATER POLLUTION FACTORS:
—Oxidation parameters:
  BOD_5: 0.02 std. dil. sew. (255)
    nil (27)
    0.9% ThOD (79)
  BOD_{10}: 0.9% ThOD
    nil std. dil. sew. (256)
  BOD_{15}: 4.0% ThOD (79)
  BOD_{20}: 5.1% ThOD
  ThOD: 2.6 (27)

—Waste water treatment:

method	temp. °C	days observed	feed mg/l	days acclim.	% removed
NFG, BOD	20	1-10	250	365 + P	8
NFG, BOD	20	1-10	500	365 + P	4
NFG, BOD	20	1-10	900	365 + P	nil

(93)

D. BIOLOGICAL EFFECTS:
—Toxicity threshold (cell multiplication inhibition test):
  bacteria (*Pseudomonas putida*): 310 mg/l (1900)
  algae (*Microcystis aeruginosa*): 1.7 mg/l (329)
  green algae (*Scenedesmus quadricauda*): 4.1 mg/l (1900)
  protozoa (*Entosiphon sulcatum*): 12 mg/l (1900)
  protozoa (*Uronema parduczi Chatton-Lwoff*): 815 mg/l (1901)

—Fish:
Lepomis macrochirus: static bioassay in fresh water at 23°C, mild aeration applied after 24 hr:

material added ppm	% survival after				best fit 96 hr LC_{50} ppm
	24 hr	48 hr	72 hr	96 hr	
560	40	0	—	—	
420	90	80	40	10	
370	100	80	50	40	350
320	100	100	100	80	
10	100	100	100	100	

Menidia beryllina: static bioassay in synthetic seawater at 23°C, mild aeration applied after 24 hr:

material added ppm	% survival after				best fit 96 hr LC_{50} ppm
	24 hr	48 hr	72 hr	96 hr	
560	70	30	0	—	
420	100	100	50	20	400
320	100	100	100	100	(352)

—Mammalia: guinea pig: acute oral LD_{50}: 0.9 g/kg
           rat: acute oral LD_{50}: 1.05–1.6 g/kg
             : repeated oral doses: 19/20: 0.8 g/kg, daily × 30
             : repeated oral doses: 8/20: 0.16 g/kg, daily × 30
           guinea pig: repeated oral doses: 16/20: 0.45 g/kg, daily × 30
                           : 3/20: 0.04 g/kg, daily × 30
           rat, guinea pig: inhalation: some deaths: 18,000 ppm, 6 × 8 hr
                         : no deaths: 12,000 ppm, 8 hr   (211)

MSMA *see* monosodiummethanearsonate

mucic acid (2,3,4,5-tetrahydroxyhexanedioic acid)
$COOH(CHOH)_4COOH$
A. PROPERTIES: colorless crystals or white powder; m.w. 210.14; m.p. 206°C decomposes; b.p. 255°C; solub. 3,300 mg/l at 14°C
C. WATER POLLUTION FACTORS:
  —Oxidation parameters: BOD_5: 0.245 Warburg/sewage   (282)(163)
  —Waste water treatment: A.S. after 6 hr: 10.3% of ThOD
                                 12 hr: 20.4% of ThOD
                                 24 hr: 24.0% of ThOD   (89)

muramic acid

$$\begin{array}{c} CH_2OH \\ | \\ \text{[pyranose ring: O, HO, OH, NH}_2\text{]} \\ | \\ CH_3\text{—}CH\text{—}COOH \end{array}$$

A. PROPERTIES: m.w. 251.24; m.p. 148°C decomp.
C. WATER POLLUTION FACTORS:
  —Manmade sources: in domestic sewage: 0.5 mg/l   (85)

muriatic ether *see* ethylchloride

mustard oil *see* allylisothiocyanate

MVC *see* vinylchloride

myristic acid (tetradecanoic acid)
$CH_3(CH_2)_{12}COOH$
  Use: soaps; cosmetics; synthesis of esters for flavors and perfumes; component of food-grade additives
A. PROPERTIES: colorless leaflets; m.w. 228.37; m.p. 54/58°C b.p. 250.5°C at 100 mm; sp.gr. 0.858 at 60°C
B. AIR POLLUTION FACTORS:
  —Manmade sources: glc's: Detroit freeway interchange/Nov. 1963:
  17.5 µg/1,000 cu m
  New York high traffic city location/Feb. '64
  48.6 µg/1,000 cu m (100)
  organic fraction of suspended matter:
  Bolivia at 5200 m altitude (Sept.–Dec. 1975): 1.3–1.9 µg/1000 cu m
  Belgium, residential area (Jan.–April 1976): 6.4–7.8 µg/1000 cu m (428)
  glc's in residential area (Belgium) Oct. 1976:
  in particulate sample: 1.39 ng/cu m
  in gas phase sample: 7.58 ng/cu m (1289)
  glc's in Botrange (Belgium)—woodland at 20–30 km from industrial area: June–July 1977: 0.25; 0.83 ng/cu m ($n = 2$)
  glc's in Wilrijk (Belgium): residential area: Oct.–Dec. 1976: 1.53; 4.99 ng/cu m
  ($n = 2$) (1233)
C. WATER POLLUTION FACTORS:
  —Oxidation parameters: BOD_5: 2% of ThOD
  COD: 30% of ThOD
  $KMnO_4$ value: acid: 0% of ThOD (220)
  alkaline: 1% of ThOD
  —Reduction of amenities: T.O.C. in water: 10 ppb; 10 ppm (326)(886)
  —Waste water treatment: A.S. after 6 hr: 0.6% of ThOD
  12 hr: 1.7% of ThOD
  24 hr: 3.7% of ThOD (89)
D. BIOLOGICAL EFFECTS:
  —Fish:
  goldfish: lethal dose: 8 mg/l (sodium salt) (522)

N

nabam (disodium ethylene-1,2-bisdithiocarbamate; dithane)

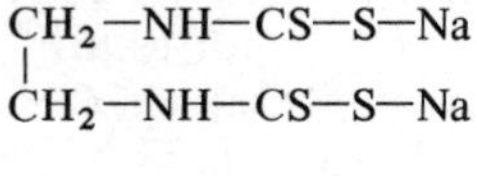

$$CH_2-NH-CS-S-Na$$
$$CH_2-NH-CS-S-Na$$

Use: fungicide
A. PROPERTIES: solub. 20%
D. BIOLOGICAL EFFECTS:
 —Algae:

	ppb		
Protococcus sp.	1×10^3	.53 O.D. expt/O.D. control	10 day growth test
Chlorella sp.	1×10^3	.63 O.D. expt/O.D. control	10 day growth test
Dunaliella euchlora	100	.27 O.D. expt/O.D. control	10 day growth test
Phaeodactylum tricornutum	1×10^3	.00 O.D. expt/O.D. control	10 day growth test
Monochrysis lutheri	100	.48 O.D. expt/O.D. control	10 day growth test

(2347)

 —Mussels:

	ppb		
Mercenaria mercenaria (hard clam), egg:	<500	TLm	48 hr static lab bioassay
Mercenaria mercenaria (hard clam), larvae:	1.75×10^3	TLm	12 day static lab bioassay
Crassostrea virginica (American oyster), egg:	<500	TLm	48 hr static lab bioassay

(2324)

 —Mammals:
 acute oral LD_{50} (rat): 395 mg/kg (1854)

naled (1,2-dibromo-2,2-dichloro-ethyldimethylphosphate)

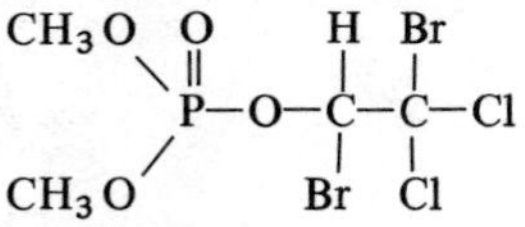

$$\begin{array}{c} CH_3O \\ \diagdown \\ CH_3O \end{array} \overset{O}{\underset{}{||}} P-O-\underset{Br}{\overset{H}{\underset{|}{C}}}-\underset{Cl}{\overset{Br}{\underset{|}{C}}}-Cl$$

Use: insecticide, acaricide
A. PROPERTIES: b.p. 110°C at 0.5 mm, v.p. 2×10^{-3} mm at 20°C

D. BIOLOGICAL EFFECTS:

	96 hr LC_{50} $\mu g/l$	
—Crustaceans:		
Gammarus lacustris	110	(2124)
Gammarus fasciatus	14	(2126)
Palaemonetes kadiakensis	90	(2126)
Orconectes nais	1800	(2126)
Asellus brevicaudus	230	(2126)
Simocephalus serrulatus	1.1	(2127)
Daphnia pulex	0.35	(2127)
—Insects:		
Pteronarcys californica	8.0	(2128)
—Fish:		
Lepomis macrochirus	180	(2137)
Salmo gairdneri	132	(2137)
—Mammals:		
acute oral LD_{50} (rat): 430 mg/kg		
acute dermal LD_{50} (rabbit): 1100 mg/kg		(1854)
in diet: (rats) 30 ppm for 27 days: without toxic effects		(1855)

nalfloc-N206
Use: microbicide; active ingredient methylene-bisthiocyanate

D. BIOLOGICAL EFFECTS:
—Fish:

		mg/l	24 hr	48 hr	96 hr	3m (extrap.)
harlequin fish						
(*Rasbora heteromorpha*)	$LC_{10}(F)$	1.35	1.3	1.0		
	$LC_{50}(F)$	2.5	1.9	1.4	0.5	
						(331)

Na-PCP *see* pentachlorophenate, sodium

naphtha, heavy aromatic
D. BIOLOGICAL EFFECTS:
—fathead minnow: 96 hr LC_{50}: 4.2–20.8 mg/l
—bluegill: 96 hr LC_{50}: 2.1–4.2 mg/l
—*Daphnia magna*: 48 hr LC_{50}: 0.42–2.3 mg/l   (1927)

naphthalene (tar camphor)

882   MMH

	Callinectes sapidus	blue crab	5.6×10^4 ppb $(4.0-7.8) \times 10^4$ ppb	TLm		96 hr static lab bioassay (2342)

striped hermit crab, Clibanarius vittatus: mortalities in 10 to 70 days using an average conc. of 0.038 μg/l (flow through) (1574)

polychaete	*Arenicola cristata*	adult	<0.003-0.062 μg/l	30 days	decreased feeding activity. (1156)
decapod fish	*Palaemonetes vulgaris/ Lagodon rhomboides Prey/ Predator*	adult	0.025-0.046 μg/l	13 days	decreased survival rate of prey. (1154)
decapod	*Callinectes sapidus*	larvae	.01-10.0 ppb	continuous	no mortality until after five days; differential survival in relation to concentration. (1149)

—Mammals:
  laboratory mice: 10 ppm in feed: 100% mortality in 60 days (1377)
  old field mice (*Peromyscus polionotus*):
    at 17.8 ppm in feed: 50% mortality in 105 days
                      92% mortality in 450 days (1378)
  acute oral LD_{50} (albino rats): 306-600 mg/kg (1855)
  acute dermal LD_{50} (rabbits): 800 mg/kg (1855)

MMH see monomethylhydrazine

MMT see methylcyclopentadienylmanganesetricarbonyl

MO see mesityloxide

molinate (S-ethylhexahydro-1H-azepine-1-carbothioate; ordram)

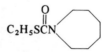

Use: broadleaf and grass weed control in rice culture
Formulations:
ordram 6E: an emulsifiable liq. containing 6 lbs ordram/gal.
ordram 5G & 5GS: granular formulations containing 5 wt% ordram
ordram 10G: a granular formulation containing 10 wt% ordram
A. PROPERTIES: m.w. 187.30; sp.gr. 1.06 at 20°C; b.p. 117°C at 10 mm; solub. 880 ppm; 900 ppm at 21°C
C. WATER POLLUTION FACTORS:
  —Biodegradation:

sampling site	conc. (μg/l)	month	incubation time (hr)	degradation rate (μg/l/day) × 10^3	turnover time (days)
control station	30	–	–	730 ± 70	41
near oil storage tanks	30	–	–	2800 ± 300	11
near oil storage tanks	60	–	–	4700 ± 100	13
near oil storage tanks	30	–	–	840 ± 90	36
Skidaway river	30	Jan.	24	70 ± 8	430
Skidaway river	30	May	24	820 ± 110	37
Skidaway river	15	Aug.	24	420 ± 80	36
Skidaway river	30	Aug.	24	680 ± 90	45
Skidaway river	60	Aug.	24	1200 ± 70	60
estuarine	30	–	–	870 ± 20	34
coastal	30	–	–	330 ± 40	230
gulf stream	30	–	–	12 ± 7	2500

(381)

microbial degradation to CO_2 –in seawater at 12°C–in the dark after 24 hr incubation at 50 μg/l: degradation rate: 0.10 μg/l/day–turnover time 500 days;
after addition of aqueous extract of fuel oil 2: degradation rate: 1.0–5.0 μg/l/day–turnover time: 10–22 days (4077)

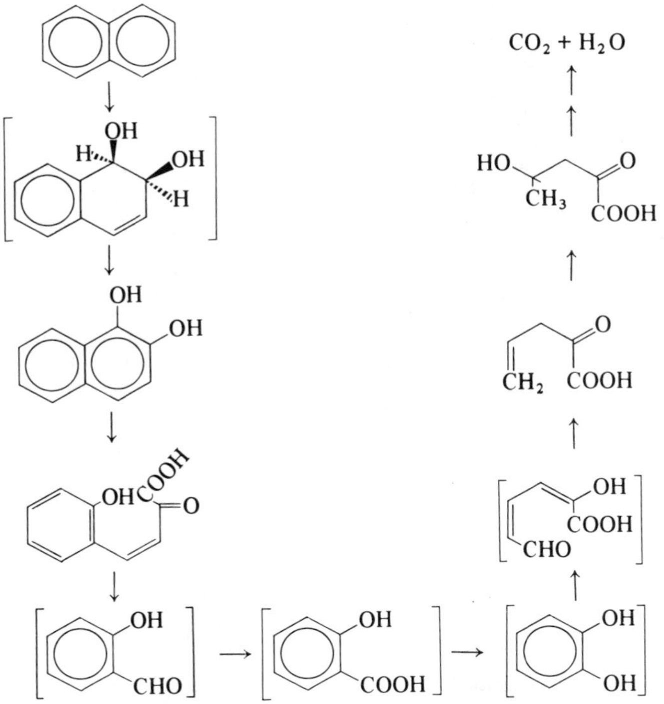

metabolic pathway for the degradation of naphthalene by certain *Pseudomonas* species

(1049)

proposed reaction sequence for the metabolism of naphthalene by *Cunninghamella elegans* (1219)

—Manmade sources:
  in leachate from test panels freshly coated with coal tar:
    influent: 0.004 µg/l
    effluent: 0.025 µg/l (1874)
  60 ml/min pure water passed through 25 ft, $\frac{1}{2}$ In. I.D. tube of general chemical grade P.V.C. contained 1.72 ppb naphthalene which constituted 16.02% of total contaminant concentration (430)
—Aquatic reactions: evaporation:

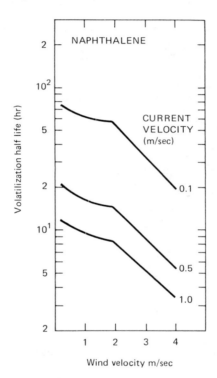

variation in predicted volatilization rates of naphthalene at 25°C under varying conditions of wind and current velocity in a stream 1.0 m in depth (1391)
calculated half-life in water at 25°C and 1 m depth, based on evaporation rate of 0.096 m/hr: 7.15 hr (437)
—Adsorption:
  in estuarine waters: at 30 µg/l, 0.7% is adsorbed on particles after 3 hr (381)
—Water quality:
  in Delaware river (U.S.A.): conc. range: winter: 0.7–0.9 ppb (1051)
  in Zürich lake; at surface: 8 ppt; at 30 m depth: 52 ppt
  in Zürich tap water: 8 ppt (513)

in Eastern Ontario drinking waters (June–Oct. 1978):
0.6–7.5 ng/l ($n = 12$)
in Eastern Ontario raw waters (June–Oct. 1978):
2.4–3.3 ng/l ($n = 2$) (1698)

—Waste water treatment:
ion exchange: adsorption on Amberlite XAD-2: 100% retention eff., infl.: 0.05 ppm, effl,: 0 ppm (40)
calculated half life time based on evaporative loss for a water depth of 1 m at 25°C: 7.15 hr (330)
compounds identified from the oxidation with chlorine dioxide: monochloronaphthalene; 1.4-dichloronaphthalene; phthalic acid (1696)

—Sampling and analysis: photometry: min. full scale: 0.65×10^{-6} mole/l (53)
recovery with open-pore polyurethane (OH/NCO = 2.2): 56–98%, depending upon load-quantitative elution with methanol (929)
extraction efficiency of macro-reticular resins: sample flow: 20 ml/min; pH: 5.7; conc. 10 ppm: X AD-2: 90%
X AD-7: 93% (370)

D. BIOLOGICAL EFFECTS:
—Bioaccumulation:
uptake and depuration by oysters (*Crassostrea virginica*) from oil-treated enclosure:

time of exposure days	depuration time days	concentration oyster µg/g	concentration water µg/g	accumulation factor oyster/water
2	–	30	5	6,000
8	–	12	3	4,000
2	7	1	–	–
8	7	2	–	–
8	23	<0.5	–	–

half-life for depuration: 2 days (957)
oysters: uptake and release:

time of exposure days	conc. in oyster* µg/g	conc. in water µg/l	dep. time days	conc. oyster µg/g	conc. water µg/l
2	8.6 (4.6)	27	14	0.1	0
4	2.7 (0.9)	17	14	0.1	0
9	2.9 (1.8)	5	14	0.2	0
15	2.7	4			

*figures in brackets refer to compartment surrounded by a 60 µm Nitex filter
(579)

—Fish:
BCF* in muscle of:

Starry flounder	and	Coho salmon
700 after 1 wk exposure		20 after 2 wk exposure
240 after 2 wk		50 after 3 wk exposure
100 after 1 wk depuration*		80 after 5 wk exposure
270 after 2 wk depuration		40 after 6 wk exposure

(after 2 wk exposure)   n.d. after 1 wk depuration
                                                         (after 6 wk exposure)
*exposure to approx. 1 ppm of the water soluble fraction of Prudhoe Bay crude
 oil   (1659)
coho salmon (*O. kisuth*) exposed to 0.02 ppm at 10°C for 5 weeks:
   0.24 ppm in muscle tissue (dry weight basis): bioaccumulation factor: 12
                                                                           (384)
Atlantic salmon eggs   168-h BCF 82.5   S   (1507)
rainbow trout   4-wk BCF (bile) 13 000   F   (1526)
                                4-wk BCF 40–300 (other   —
                                tissues)

—Toxicity:
 Algae: *Chlorella vulgaris*: at 33 ppm: 50% reduction of cell numbers vs. controls,
  after 1 day incubation at 20°C   (343)
  inhibition of photosynthesis of a freshwater, non axenic unialgal culture of
  Selenastrum capricornutum at:
     1% saturation: 110% carbon -14 fixation (vs. controls)
    10% saturation: 89% carbon -14 fixation (vs. controls)
   100% saturation: 15% carbon -14 fixation (vs. controls)   (1690)
—Anthropoda: *Calanus helgolandicus Claus* immersed for 24 hr in sea water solutions of [1 - ^{14}C] naphthalene accumulated 3.6 pg/*Calanus* from concentrations as low as 0.1 µg/l and up to 5000 pg/*Calanus* from 1000 µg/l solutions.
 Release of radioactivity by adult female *Calanus* that had accumulated 170 pg
 naphthalene/*Calanus* from solution amounted to 70% after 1 day, 80% after
 2 days, 88% after 4 days and 97% after 10 days. Release of radioactivity by
 female adult *Calanus* accumulated from a diet of *Biddulphia* cells up to 250
 pg naphthalene/*Calanus*: 40% after 1 day, 53% after 2 days, 67% after 4 days
 and 72% after 6 days. *Calanus* fed on labelled diets for 24 hr showed that at
 the end of this period over 90% of the radioactivity in the animals was present
 as unchanged naphthalene. However, more than two-thirds of that released by
 the animals was in some form other than the hydrocarbon, a finding consistent
 with the view that *Calanus* is able to metabolize it   (343)
—Amphipod: *Parhyale hawaiensis* (Dana):
 LC 24 hr static test   open bowl (ppm)   closed bottle
    LC_0 4   1
    LC_{50} 15   6.5
    LC_{80} 7.5
    LC_{95} 20   (1062)
—Fish:
 mosquito fish: 24/96 hr TLm: 220/150 mg/l in turbid Oklahoma   (41, 244)
 static test: 96 hr TLm (ppm) at: 4°C   8°C   12°C
    pink salmon (*Oncorhynchus gorbuscha*): 1.37   1.84   1.24
    shrimp (*Pandalus goniurus*):   2.16   1.02   0.971
 naphthalene concentrations declined through evaporation to nondetectable levels
  after 48 hr at 12°C, 72 hr at 8°C and 96 hr at 4°C   (1397)
 coho salmon (*Oncorhynchus kisuth*): after intraperitoneal injection, the highest
  percentage of metabolites occurred in the gall bladder.

The following metabolites were identified: 1-naphthol; 1,2-dihydro-1,2-dihydroxynaphthalene; 1-naphthylglycoside; 1-naphthylsulfate; N-acetyl-S-(1-naphthyl)-cysteine (mercapturic acid); 1-naphthyl-β-glucuronic acid (1618)
Neanthes arenaceodentata: 96 hr TLm in seawater at 22°C: 3.8 ppm (initial conc. in static assay) (995)
—Carcinogenicity: negative
—Mutagenicity in the *Salmonella* test: negative,
 <0.09 revertant colonies/nmol
 <70 revertant colonies at 100 μg/plate (1883)

naphthalene carboxylic acid
C. WATER POLLUTION FACTORS:
—Glc's in the residential area (Belgium) Oct. 1976:
 in gas phase sample: 4.37 ng/cu m (1289)

1-naphthalenesulfonic acid

A. PROPERTIES: log P_{oct} -1.63 (calculated)
C. WATER POLLUTION FACTORS:
—Biodegradation: adapted A.S. at 20°C—product is sole carbon source: 90.5% COD removal at 18.0 mg COD/g dry inoculum (327)

naphtho(1,2,3,4-*def*)chrysene
B. AIR POLLUTION FACTORS:
—In air in Ontario cities (Canada):

location	April-June 1975		July-Sept. 1975		Oct.-Dec. 1975		Jan.-March 1976	
	ng/1000 cu m air	μg/g ppm	ng/1000 cu m air	μg/g ppm	ng/1000 cu m air	μg/g ppm	ng/1000 cu m air	μg/g p.m.
Toronto	270	3.0	362	3.8	538	6.5	228	2.9
Toronto	410	5.4	201	3.1	300	4.9	2762	20.2
Hamilton	184	1.3	1017	7.3	2027	29.2	903	10.8
S. Sarnia	823	13.3	61	1.3	434	8.3	129	4.7
Sudbury	510	15.6	73	1.7	99	4.4	230	9.8

(999)

naphthoic acid
C. WATER POLLUTION FACTORS:
—Biodegradation: adapted A.S. at 20°C—product is sole carbon source: 90.2% COD removal at 15.5 mg COD/g dry inoculum (327)

α-naphthol (1-hydroxynaphthalene)

Use: dyes; organic synthesis; synthetic perfumes
A. PROPERTIES: yellow crystals; m.w. 144.16; m.p. 96.1°C; b.p. 278/288°C; v.p. 1 mm at 94°C, 10 mm at 142°C, 100 mm at 206°C sp.gr. 1.224 at 4°C; THC 1,185 kcal/mole; LHC 1,162 kcal/mole; log P_{oct} 2.98
B. AIR POLLUTION FACTORS:
 —Odor threshold conc. (recognition): 0.0030–0.0052 mg/cu m (610)
C. WATER POLLUTION FACTORS:
 —Oxidation parameters:
 BOD_5: 1.7 std. dil. sewage (27)
 : 16% of ThOD (n.s.i.) (220)
 COD: 91% of ThOD (220)
 $KMnO_4$ value: acid: 4% ThOD
 alkaline: 3% ThOD (220)
 ThOD: 2.55 (27)
 BOD_5: 60% ThOD (274)
 COD: 99% ThOD (0.05 n Cr_2O_7) (274)
 $KMnO_4$: 43% ThOD (0.01 n $KMnO_4$)
 —Biodegradation: adapted A.S. at 20°C—product is sole carbon source: 92.1% COD removal at 38.4 mg COD/g dry inoculum/hr (327)
 biodegradation to CO_2 in estuarine water (n.s.i.):

concentration $\mu g/l$	month	incubation time (hr)	degradation rate ($\mu g/l/day$) × 10^3	turnover time (days)	
10	June	24	240 ± 40	41	(381)

 —Reduction of amenities:
 tainting of fish flesh: 0.5 mg/l (81)
 taste at fish flesh: 0.5 mg/l (30)
 odor threshold conc. (detection): 4.0 mg/l water (998)
 taste threshold conc.: 0.5 mg/l water (998)
D. BIOLOGICAL EFFECTS:
 —Carcinogenicity: ?
 —Mutagenicity in the *Salmonella* test: negative,
 <0.01 revertant colonies/nmol
 <70 revertant colonies at 1000 μg/plate (1883)

β-naphthol (2-hydroxynaphthalene)

Use: dyes; pigments; anti-oxidants for rubber, fats, oils; insecticide; synthesis of fungicides, pharmaceuticals, perfumes

3-NAPHTHOL-2-CARBOXYLIC ACID

A. PROPERTIES: colorless leaflets; m.w. 144.16; m.p. 122; b.p. 295°C; v.p. 5 mm at 145°C, 100 mm at 234°C; sp.gr. 1.217 at 4°C; solub. 740 mg/l at 25°C, 12,500 mg/l warm; log P_{oct} 2.84

B. AIR POLLUTION FACTORS:
  —Odor threshold conc. (recognition): 0.23–0.30 mg/cu m (610)

C. WATER POLLUTION FACTORS:
  —Oxidation parameters:
  BOD_5: 1.79 at 0.75–3 ppm std. dil. sew. (275)
  ThOD: 2.55 (275)
  BOD_5: 68% ThOD (274)
  COD: 93% ThOD (0.05 n Cr_2O_7) (274)
  $KMnO_4$: 58% ThOD (0.01 n $KMnO_4$) (274)
  —Biodegradation: adapted A.S. at 20°C—product is sole carbon source: 89.0% COD removal at 39.2 mg COD/g dry inoculum/hr (327)
  —Reduction of amenities:
  tainting of fish flesh: 1.0 mg/l (81)
  odor threshold: average: 1.29 ppm
  range: 0.01–11.4 ppm (97)
  detection: 3.0 mg/l (998)

D. BIOLOGICAL EFFECTS:
  —Inhibition of photosynthesis of a freshwater, non axenic unialgal culture of *Selenastrum capricornutum* at:
  1% saturation: 64% carbon-14 fixation (vs. controls)
  10% saturation: 12% carbon-14 fixation (vs. controls)
  100% saturation: 3% carbon-14 fixation (vs. controls) (1690)

3-naphthol-2-carboxylic acid *see* 2-hydroxy-3-naphthoic acid

1-naphthol-2-sulfonic acid

C. WATER POLLUTION FACTORS:
  —Biodegradation: adapted A.S. at 20°C—product is sole carbon source: 91.0% COD removal at 18.0 mg COD/g dry inoculum/hr (327)

1,4-naphthoquinone *see* α-naphthoquinone

α-naphthoquinone (1,4-naphthoquinone; 1,4-dihydro-1,4-diketonaphthalene)

A. PROPERTIES: yellow crystals; m.w. 158.15; m.p. 125°C; b.p. 100°C sublimes; v.d. 5.46; sp.gr. 1.422; log P_{oct} 1.71/1.78
B. AIR POLLUTION FACTORS: 1 mg/cu m = 0.15 ppm, 1 ppm = 6.57
C. WATER POLLUTION FACTORS:
 —Oxidation parameters: BOD_5: 0.81 std. dil. sew. (259)
 ThOD: 2.1 (27)
D. BIOLOGICAL EFFECTS:
 —Algae: blue algae: toxic: 0.3–0.6 mg/l (30)
 inhibition of photosynthesis of a freshwater, non axenic unialgal culture of *Selenastrum capricornutum* at:
 0.01% saturation: 83% carbon-14 fixation (vs. controls)
 0.1% saturation: 59% carbon-14 fixation (vs. controls)
 1% saturation: 9–14% carbon-14 fixation (vs. controls)
 10% saturation: 3% carbon-14 fixation (vs. controls)
 100% saturation: 2% carbon-14 fixation (vs. controls) (1690)
 —Fish: TLm (24–48 hr) 0.3–0.6 mg/l (30)

2-(α-naphthoxy)-N,N-diethylpropionamide *see* devrinol

α-naphthylamine

Use: dyes and dye intermediates
A. PROPERTIES: yellow rhombic needles; m.w. 143.18; m.p. 50°C; b.p. 301°C; v.d. 4.93; sp.gr. 1.131; solub. 1,700 mg/l; log P_{oct} 2.22 (calculated)
B. AIR POLLUTION FACTORS:
 —Odor threshold conc. (recognition): 0.14–0.29 mg/cu m (610)
C. WATER POLLUTION FACTORS:
 —Oxidation parameters: BOD_5: 0.89 (n.s.i.) std. dil. sew. (282)
 —Biodegradation: adapted A.S. at 20°C—product is sole carbon source: 0% COD removal after 20 days (327)
 W, A.S. from mixed domestic/industrial treatment plant: 80–84% depletion at 20 mg/l after 6 hr at 25°C (419)
 —Impact on biodegradation processes:
 NH_3 oxidation by *Nitrosomonas*: at 100 mg/l: 81% inhibition
 50 mg/l: 81% inhibition
 10 mg/l: 45% inhibition (390)
 —Sampling and analysis: photometry: min. full scale: 0.192×10^{-6} mole/l (53)
 —Reduction of amenities: approx. conc. causing adverse taste in fish (Rudd): 3.0 mg/l (41)
D. BIOLOGICAL EFFECTS:
 —Inhibition of photosynthesis of a freshwater, non-axenic unialgal culture of *Selenastrum capricornutum* at:
 0.01% saturation: 119% carbon-14 fixation (vs. controls)

0.1% saturation:	98% carbon-14 fixation (vs. controls)	
1.0% saturation:	63% carbon-14 fixation (vs. controls)	
10% saturation:	16% carbon-14 fixation (vs. controls)	
100% saturation:	3% carbon-14 fixation (vs. controls)	(1690)

—Fish: Gardon (F): TLm: 6 mg/l (30)
—Man: carcinogenic: USA 1974 (77)
    unsatisfactory: >0.01 mg/cu m (185)
—Carcinogenicity: negative
—Mutagenicity in the *Salmonella* test: positive
  0.42 revertant colonies/nmol
  290 revertant colonies at 100 μg/plate (1883)

β-naphthylamine

Use: dyes

A. PROPERTIES: leaflets; m.w. 143.18; m.p. 110.2°C; b.p. 306.1°C; v.d. 4.95; sp.gr. 1.061 at 98/4°C; log P_{oct} 2.25 (calculated)

B. AIR POLLUTION FACTORS: 1 mg/cu m = 0.17 ppm, 1 ppm = 5.95 mg/cu m
—Odor threshold conc. (recognition): 1.4–1.9 mg/cu m (610)

C. WATER POLLUTION FACTORS:
—Oxidation parameters: BOD_5: 57% ThOD (274)
                    COD: 92% ThOD (0.05 n $K_2Cr_2O_7$) (274)
                    $KMnO_4$: 42% ThOD (0.01 n $KMnO_4$) (274)
                    ThOD: 2.57 (274)
—Sampling and analysis: photometry: min. full scale: 0.195 × 10^{-6} mole/l (53)

D. BIOLOGICAL EFFECTS:
—Man: carcinogenic (70; 77; 487)
—Carcinogenicity: positive
—Mutagenicity in the Salmonella test: positive
  8.5 revertant colonies/nmol
  590 revertant colonies at 10 μg/plate (1883)

N-α-naphthylethylenediamine dihydrochloride

$NH-CH_2CH_2NH_2 \cdot 2HCl$

A. PROPERTIES: colorless crystals
C. WATER POLLUTION FACTORS:
—Impact on biodegradation processes:

NH$_3$ oxidation by *Nitrosomonas*: at 100 mg/l: 93% inhibition (n.s.i.)
                                  50 mg/l: 79% inhibition (n.s.i.)
                                  10 mg/l: 29% inhibition (n.s.i.)   (390)

1-naphthyl-N-methylcarbamate *see* carbaryl

NDPA *see* N-nitroso-di-N-propylamine

neodol *see* dobanol

neoflex *see* lineval

neohexane *see* 2,2-dimethylbutane

neopentanoic acid *see* pivalic acid

neste A
Use: an emulsifier–surfactant agent
Composition: talestol + mixture of aromatic hydrocarbons
D. BIOLOGICAL EFFECTS:
  —Fish:
    northern pike yolk sac larvae: 48 hr LC$_{50}$ (S): 28 mg/l
    northern pike yolk sac larvae: 48 hr LC$_{50}$ (S): 32 mg/l
    northern pike–free-swimming: 48 hr LC$_{50}$ (S): 5.2 mg/l
    northern pike–1 mo. old: 48 hr LC$_{50}$ (S): 10.0 mg/l
    neste A + oil:
      northern pike yolk sac larvae: 48 hr LC$_{50}$ (S): 66 mg/l
      northern pike free-swimming: 48 hr LC$_{50}$ (S): 4.4 mg/l   (1116)

new blitane
Use: fungicide;
active ingredients: copper oxychloride (64.7%)
                      maneb (14.4%)
D. BIOLOGICAL EFFECTS:
  —Fish:

	mg/l	24 hr	48 hr	96 hr
harlequin fish				
(*Rasbora heteromorpha*)	LC$_{10}$ (F)	8.2		
	LC$_{50}$ (F)	16.5	19	9.6	(331)

NIA 10,242 *see* carbofuran

nimitox *see* abate

ninhydrin (triketohydrindene hydrate; 1,2,3-indantrione monohydrate) [structure: benzene fused ring with three C=O groups] · H$_2$O

Use: chemical intermediate
A. PROPERTIES: white crystals or powder; m.p. 240–245°C decomposes; m.w. 178.14
C. WATER POLLUTION FACTORS:
 —Impact on biodegradation processes:
  NH$_3$ oxidation by *Nitrosomonas* sp. at 100 mg/l: 30% inhibition
   50 mg/l: 26% inhibition
   10 mg/l: 31% inhibition (390)

niobe oil *see* methylbenzoate

nitrapyrin *see* 2-chloro-6-(trichloromethyl)pyridine

nitrilotriacetic acid (NTA, triglycine, TGA, triglycollamic acid) N(CH$_2$COOH)$_3$
 Use: synthesis; chelating agent; eluting agent in purification of rare earth elements.
A. PROPERTIES: white crystalline powder; m.p. 240°C, with decomposition; m.w. 191.14
C. WATER POLLUTION FACTORS:
 —Manmade sources: in sewage effluent: 0.17–1.1 mg/l (236)
  BOD$_5^{20°C}$: nil—nonacclimated seed organisms
  BOD$_{20}^{20°C}$: 0.65–0.72 at conc. of 2–10 mg/l - acclimated organisms (1317)

theoretical	analytical
TOD = 1.08	Reflux COD = 90.7% recovery
COD = 0.75	TKN = 97.3% recovery
NOD = 0.33	BOD$_5$ = 0.014
	BOD$_5$/COD = 0.021 (1828)

 —Biodegradation:

H$^+$N—(CH$_2$COO$^-$)$_3$
nitrilotriacetic acid
↓   ⤷NAD$^+$ / NADH+H$^+$
H$_2$N(CH$_2$COO$^-$)$_2$ + OHC—COO$^-$
iminodiacetic acid
↓   ⤷NAD$^+$ / NADH+H$^+$
H$_2$N—CH$_2$—COO$^-$ + OHC—COO$^-$ ⟶ glycerate
glycine
↓
NH$_3$ + OHC—COO$^-$
    glyoxylate (588)

—Conventional A.S.:

	mg/l	result
unacclimated system	1,000	no effect
acclimated system	1,222	no effect

The data indicated that NTA is resistant to biodegradation. There was, however, no evidence of detrimental effects to the efficiency of an activated sludge biomass at these concentration (1828)

D. BIOLOGICAL EFFECTS:
—Toxicity threshold (cell multiplication inhibition test):

bacteria (*Pseudomonas putida*):	>10,000	mg/l	(1900)
algae (*Microcystis aeruginosa*):	510	mg/l	(329)
green algae (*Scenedesmus quadricauda*):	8.3	mg/l	(1900)
protozoa (*Entosiphon sulcatum*):	800	mg/l	(1900)
protozoa (*Uronema parduczi Chatton-Lwoff*):	>800	mg/l	(1901)

—Algae:
Selenastrum capricornutum: 1 mg/l: no effect
    10 mg/l: no effect
    100 mg/l: marginal inhibition (1828)
—Crustacean:
Daphnia magna: LC_{50}: >100 mg/l (1828)
—Fish:
Pimephales promelas: LC_{50}: >100 mg/l (1828)
—Mammals:
rats: oral LD_{50}: 1.1–2.4 g/kg body wt (for Na_3NTA, Na_2NTA, K_3NTA)
dogs: oral LD_{50}: >5 g/kg body wt (for Na_3NTA, Na_2NTA, K_3NTA)
apes: oral LD_{50}: ± 0.75 g/kg body wt (for Na_3NTA, Na_2NTA, K_3NTA)
rats: no effect dose: 30 mg/day/kg body wt for 2 years (589)

nitrilotriacetic acid, monohydrated sodium salt
D. BIOLOGICAL EFFECTS:

NITRILOTRIACETIC ACID, MONOHYDRATED SODIUM SALT

organism tested	common name	life stage or size (mm)	conc. (ppb act. ingred.) in water
Cyclotella nana			5×10^3
Tisbe furcata			2.7×10^5
Acartia clausi			1.35×10^6
Trigriopus japonicus			3.2×10^6
Pseudodiaptimus coronatus			7×10^5
Eurytemora affinis			1.25×10^6
	crab zoea		1.65×10^6
Nereis virens	sand worm	adult	5.5×10^6
Nereis virens	sand worm	adult	5.5×10^6
Palaemonetes vulgaris	grass shrimp	adult	4.1×10^6
Palaemonetes vulgaris	grass shrimp	adult	1.8×10^6
Palaemonetes vulgaris	grass shrimp		1.0×10^6
Penaeus setiferus	white shrimp	subadult	1×10^6
Penaeus setiferus	white shrimp	subadult	5×10^6
Homarus americanus	American lobster	subadult (292 grams)	3.8×10^6
Homarus americanus	American lobster	subadult (292 grams)	3.15×10^6
Homarus americanus	American lobster	first larval stage	1×10^5
Uca pugilator	fiddler crab	adult	1×10^7
Uca pugilator	fiddler crab	adult	1×10^6
Pagurus longicarpus	hermit crab	adult	5.5×10^6
Pagurus longicarpus	hermit crab	adult	1.8×10^6
	oyster	larvae	3.5×10^5
Nassa obsoleta	mud snail	adult	5.5×10^6
Nassa obsoleta	mud snail	adult	5.1×10^6
Mytilus edulis	bay mussel	adult	6.1×10^6
Mytilus edulis	bay mussel	adult	3.4×10^6
Mercenaria mercenaria	hard clam	adult	$>1 \times 10^7$
Mercenaria mercenaria	hard clam	adult	$>1 \times 10^7$
Asterias forbesi	starfish	subadult	3×10^6
Asterias forbesi	starfish	subadult	3×10^6
Fundulus heteroclitus	mummichog	adult	5.5×10^6
Fundulus heteroclitus	mummichog	adult	5.5×10^6
Fundulus heteroclitus	mummichog	adult	1×10^3
Stenotomus chrysops	scup	subadult	3.15×10^6
Stenotomus chrysops	scup	subadult	3.15×10^6
Roccus saxatilis	striped bass	juvenile (65 mm)	5.5×10^6
Roccus saxatilis	striped bass	juvenile (65 mm)	5.5×10^6
Roccus saxatilis	striped bass	juvenile (65 mm)	3×10^6
Roccus saxatilis	striped bass	juvenile (65 mm)	10×10^6

methods of assessment	test procedure	
38% growth as compared to controls	72 hr static lab bioassay	(2331)
TL_{50}	72 hr static lab bioassay	(2344)
TL_{50}	72 hr static lab bioassay	
TL_{50}	72 hr static lab bioassay	
TL_{50}	72 hr static lab bioassay	
TL_{50}	72 hr static lab bioassay	
TL_{50}	72 hr static lab bioassay	
TL_{50}	96 hr static lab bioassay	
TL_{50}	168 hr static lab bioassay	
TL_{50}	96 hr static lab bioassay	
TL_{50}	168 hr static lab bioassay	
subjected to histopathologic examination	168 hr static lab bioassay	
78% mortality	22 day chronic flowing lab bioassay	
90% mortality	96 hr static lab bioassay	
TL_{50}	96 hr static lab bioassay	
TL_{50}	168 hr static lab bioassay	
100% mortality	7 day static lab bioassay	
25% mortality	96 hr static lab bioassay	
46% mortality	45 day chronic flowing lab bioassay	
TL_{50}	96 hr static lab bioassay	
TL_{50}	168 hr static lab bioassay	
46% mortality	24 hr static lab bioassay	
TL_{50}	96 hr static lab bioassay	
TL_{50}	168 hr static lab bioassay	
TL_{50}	96 hr static lab bioassay	
TL_{50}	168 hr static lab bioassay	
TL_{50}	96 hr static lab bioassay	
TL_{50}	168 hr static lab bioassay	
TL_{50}	96 hr static lab bioassay	
TL_{50}	168 hr static lab bioassay	
TL_{50}	96 hr static lab bioassay	
TL_{50}	168 hr static lab bioassay	
examined for histopathology	168 hr static lab bioassay	
TL_{50}	96 hr static lab bioassay	
TL_{50}	168 hr static lab bioassay	
TL_{50}	96 hr static lab bioassay	
TL_{50}	168 hr static lab bioassay	
TL_{100}, histopathology	168 hr static lab bioassay	
TL_0	168 hr static lab bioassay	
		(2344)

p-nitro-acetophenone

$$\text{COCH}_3\text{-C}_6\text{H}_4\text{-NO}_2$$

C. WATER POLLUTION FACTORS:
 —Biodegradation: adapted A.S. at 20°C—product is sole carbon source: 98.8% COD removal at 5.2 mg COD/g dry inoculum/hr (327)

o-nitro-aniline (1-amino-2-nitrobenzene)

$$\text{NH}_2\text{-C}_6\text{H}_4\text{-NO}_2$$

A. PROPERTIES: orange rhombic needles; m.w. 138.12; m.p. 71.5°C; b.p. 284.11°C, 270°C decomposes; v.p. <0.1 mm at 30°C; v.d. 4.77; sp.gr. 1.442 at 20/4°C; sat.conc. <0.7 g/cu m at 30°C; solub. 1260 mg/l at 25°C; log P_{oct} 1.44/1.83
B. AIR POLLUTION FACTORS: 1 mg/cu m = 0.17 ppm, 1 ppm = 5.74 mg/cu m
C. WATER POLLUTION FACTORS:
 —Biodegradation: decomposition by a soil microflora in >64 days (176)
 adapted A.S. at 20°C—product is sole carbon source: 0% removal after 20 days (determined photometrically) (327)

m-nitroaniline (1-amino-3-nitrobenzene)

$$\text{NH}_2\text{-C}_6\text{H}_4\text{-NO}_2$$

A. PROPERTIES: yellow needles; m.w. 138.12; m.p. 111.8°C; b.p. 270/306°C decomposes; sp.gr. 1,430 at 20/4°C; solub 890 mg/l at 25°C; log P_{oct} 1.37
B. AIR POLLUTION FACTORS:
 —Odor: characteristic: quality: burning sweet
C. WATER POLLUTION FACTORS:
 —Biodegradation: decomposition by a soil microflora in >64 days (176)
 adapted A.S. at 20°C—product is sole carbon source: 0% removal after 20 days (determined photometrically) (327)

p-nitroaniline (1-amino-4-nitrobenzene)

Use: intermediate for dyes and antioxidants; gasoline gum inhibitors; medicinals for poultry; corrosion inhibitor

A. PROPERTIES: yellow monoclinic needles; m.w. 138.12; m.p. 147.5°C; b.p. 260/332°C decomposes; v.p. 0.0015 mm at 20°C, 0.007 mm at 30°C; v.d. 4.77; spec.grav. 1.424; solub. 800 mg/l at 19°C, 22,000 mg/l at 100°C; sat. conc. 0.011 g/cu m at 20°C, 0.051 g/cu m at 30°C
B. AIR POLLUTION FACTORS: 1 mg/cu m = 0.17 ppm, 1 ppm = 5.74 mg/cu m
C. WATER POLLUTION FACTORS:
—Biodegradation: decomposition by a soil microflora in >64 days (176)
adapted A.S. at 20°C—product is sole carbon source: 0% removal after 20 days (photometrical determination) (327)
—Impact on biodegradation processes: inhibition of degradation of glucose by *Pseudomonas fluorescens*: 20 mg/l
inhibition of degradation of glucose by *E. coli*: >100 mg/l (293)
NH_3 oxidation by *Nitrosomonas*:
at 100 mg/l: 64% inhibition
50 mg/l: 52% inhibition
10 mg/l: 46% inhibition
—Sampling and analysis: photometry: min. full scale: 0.22×10^{-6} mole/l (53)
D. BIOLOGICAL EFFECTS:
—Bacteria: E. coli.: no toxic effect: >100 mg/l
—Algae: Scenedesmus: LD_0: 20 mg/l
—Arthropods: daphnia: LD_0: 24 mg/l (30)
—Toxicity threshold (cell multiplication inhibition test):

bacteria (*Pseudomonas putida*):	4 mg/l	(1900)
algae (*Microcystis aeruginosa*):	0.35 mg/l	(329)
green algae (*Scenedesmus quadricauda*):	11 mg/l	(1900)
protozoa (*Entosiphon sulcatum*):	6.9 mg/l	(1900)
protozoa (*Uronema parduczi Chatton-Lwoff*):	3.1 mg/l	(1901)

o-**nitroanisole**   *see* 1-methoxy-2-nitrobenzene

m-**nitroanisole**   *see* 1-methoxy-3-nitrobenzene

p-**nitroanisole**   *see* 1-methoxy-4-nitrobenzene

o-**nitrobenzaldehyde**

o-nitrobenzaldehyde

A. PROPERTIES: m.w. 151.12; m.p. 43–46°C; b.p. 153°C at 23 mm
B. AIR POLLUTION FACTORS:
  —Odor threshold conc. (recognition): 4.5 mg/cu m (610)
C. WATER POLLUTION FACTORS:
  —Biodegradation: adapted A.S. at 20°C—product is sole carbon source: 97.0% COD removal at 13.8 mg COD/g dry inoculum/hr (327)

m-nitrobenzaldehyde

A. PROPERTIES: m.w. 151.12; m.p. 57–59°C
B. AIR POLLUTION FACTORS:
  —Odor threshold conc. (recognition): 3.0 mg/cu m (610)
C. WATER POLLUTION FACTORS:
  —Biodegradation: adapted A.S. at 20°C—product is sole carbon source: 94.0% COD removal at 10.0 mg COD/g dry inoculum/hr (327)

p-nitrobenzaldehyde

A. PROPERTIES: m.w. 151.12; m.p. 105–108°C
C. WATER POLLUTION FACTORS:
  —Biodegradation: adapted A.S. at 20°C—product is sole carbon source: 97.0% COD removal at 13.8 mg COD/g dry inoculum/hr (327)
  —Impact on biodegradation processes:
  NH_3 oxidation by *Nitrosomonas* sp. at 100 mg/l: 76% inhibition
  50 mg/l: 32% inhibition
  10 mg/l: 29% inhibition (390)

nitrobenzene (oil of mirbane)

NITROBENZENE

Manufacturing source: organic chemical industry.
Users and formulation: mfg. aniline and dyestuffs; solvent recovery plants; mfg. rubber chemicals, drugs, photographic chemicals; refining lubricants oils; solvent in TNT production; solvent for cellulose ethers; cellulose acetate mfg.; constituent in metal polish and shoe polish formulations.

A. PROPERTIES: yellow liquid; m.w. 123.1; m.p. 6°C; b.p. 211°C v.p. 0.15 mm at 20°C; 0.35 mm at 30°C v.d. 4.25; sp.gr. 1.20 at 25/4°C; solub. 1,900 mg/l at 20°C, 8,000 mg/l at 80°C; THC 739 kcal/mole; LHC 727 kcal/mole; sat. conc.: 1.0 g/cu m at 20°C, 2.3 g/cu m at 30°C; log P_{oct} 1.85/1.88
B. AIR POLLUTION FACTORS: 1 mg/cu m = 0.20 ppm, 1 ppm = 5.12 mg/cu m
  —Odor: characteristic: quality: shoe polish, bitter almonds

  (21; 57; 279; 291; 306; 602; 604; 606; 610; 678; 682; 704; 710; 741; 793; 871)
  USSR: human odor perception: 0.0182 mg/cu m = 3.6 ppb
        human reflex response: no response: 0.008 mg/cu m
                               adverse response: 0.0129 mg/cu m
        animal chronic exposure: no effect: 0.008 mg/cu m
                                 adverse effect: 0.008 mg/cu m   (170)
  O.I.: at 20°C = 200   (316)
  —Sampling and analysis: photometry: min. full scale: 3 ppm   (53)
C. WATER POLLUTION FACTORS:
  —BOD_5: nil std. dil. sewage at <440 mg/l   (41)(27)
  ThOD: 1.95   (27)
  —Impact on biodegradation processes:
    inhibition of degradation of glucose by *Pseudomonas fluorescens* at 30 mg/l
    inhibition of degradation of glucose by *E. coli* at 600 mg/l   (293)
  —Reduction amenities: faint odor at 0.03 mg/l   (129)
    odor threshold conc. (detection): 0.2 mg/kg   (894)
    in river Maas (The Netherlands): average in 1973: n.d. to 0.07 µg/l   (342)
  —Biodegradation: decomposition by a soil microflora in >64 days   (176)
    adapted A.S. at 20°C—product is sole carbon source: 98.0% COD removal at
    14.0 mg COD/g dry inoculum/hr   (327)
  —Waste water treatment: A.C.: adsorbability: 0.196 g/g carbon;
                          95.6% reduction, infl.: 1,023 mg/l,
                          effl.: 44 mg/l   (32)
                          Air stripping constant: k = 0.843 days^{-1}
                            at C_0 = 250 mg/l   (82)
  —Sampling and analysis: photometry: min. full scale: 0.115 × 10^{-6} mole/l (53)
D. BIOLOGICAL EFFECTS:
  —Toxicity threshold (cell multiplication inhibition test):
    bacteria (*Pseudomonas putida*):   7 mg/l   (1900)
    algae (*Microcystis aeruginosa*):   1.9 mg/l   (329)

green algae (*Scenedesmus quadricauda*):	33 mg/l	(1900)
protozoa (*Entosiphon sulcatum*):	1.9 mg/l	(1900)
protozoa (*Uronema parduczi Chatton-Lwoff*):	15 mg/l	(1901)

—Bacteria: *E. coli*: LD_0: 600 mg/l
—Algae: *Scenedesmus*: LD_0: 40 mg/l
—Arthropoda: *Daphnia*: LD_0: 28 mg/l (30)
—Fish: *Vairon* (F) distilled water: TLm (6 hr): 20-24 mg/l
      hard water: TLm (6 hr): 90-100 mg/l (30)
—Man: severe toxic effects: 200 ppm = 1,020 mg/cu m, 60 min
    symptoms of illness: 40 ppm = 204 mg/cu m
    unsatisfactory: >1 ppm = 5.1 mg/cu m (185)

m-nitrobenzenesulfonate, sodium

[Structure: benzene ring with NO_2 and SO_3Na in meta positions]

D. BIOLOGICAL EFFECTS:
—Crustacean: *Daphnia magna*: TLm, 24 hr: 8,665 mg/l
        48 hr: 8,665 mg/l
        72 hr: 6,017 mg/l
        96 hr: 5,067 mg/l (153)
        100 hr: 2,235 mg/l (1295)
—Fish: *Lepomis macrochirus*: TLm, 24 hr: <1,350 mg/l (1295)

o-nitrobenzenesulfonic acid

[Structure: benzene ring with NO_2 and SO_3H in ortho positions]

A. PROPERTIES: leaflets; m.w. 203.17; m.p. 70°C; b.p. decomposes
C. WATER POLLUTION FACTORS:
    —Biodegradation: decomposition by a soil microflora in >64 days (176)

m-nitrobenzenesulfonic acid

[Structure: benzene ring with NO_2 and SO_3H in meta positions]

C. WATER POLLUTION FACTORS:
    —Biodegradation: decomposition by a soil microflora in >64 days (176)

p-nitrobenzenesulfonic acid

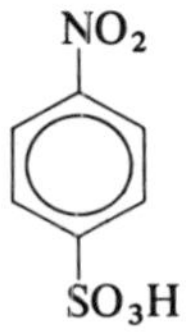

[structure: benzene ring with NO$_2$ and SO$_3$H para]

C. WATER POLLUTION FACTORS:
  —Biodegradation: decomposition by a soil microflora in >64 days (176)

o-nitrobenzoic acid

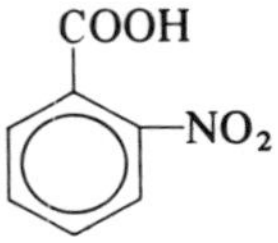

[structure: benzene ring with COOH and NO$_2$ ortho]

A. PROPERTIES: needles; m.w. 167.12; m.p. 147.5°C; sp.gr. 1.575 at 4/4°C solub. 6,800 mg/l at 20°C; log P_{oct} 1.04/1.52 (calculated)
C. WATER POLLUTION FACTORS:
  —Biodegradation: decomposition by a soil microflora in 8 days (176)
    adapted A.S. at 20°C—product is sole carbon source: 93.4% COD removal at 20.0 mg COD/g dry inoculum/hr (327)
    lag period for degradation of 16 mg/l by:
    waste water at pH 7.3: 30°C: 3-5 days
    soil suspension at pH 7.3: 30°C: 14-25 days (1096)
D. BIOLOGICAL EFFECTS:
  —Mammals: rabbit: acute oral LD_{50}: >200 mg/kg (211)

m-nitrobenzoic acid

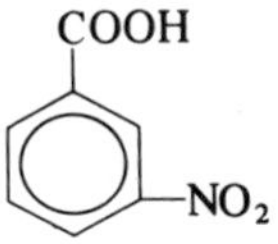

[structure: benzene ring with COOH and NO$_2$ meta]

A. PROPERTIES: leaflets; m.w. 167.12; m.p. 141.4; sp.gr. 1.494 at 4/4°C, solub. 3,100 mg/l at 20°C; log P_{oct} 1.83
C. WATER POLLUTION FACTORS:
  —Biodegradation: decomposition by a soil microflora in >64 days (176)
    adapted A.S. at 20°C—product is sole carbon source: 93.4% COD removal at 7.0 mg COD/g dry inoculum/hr (327)
    lag period for degradation of 16 mg/l by:
    waste water at pH 7.3 and 30°C: 7-14 days
    soil suspension at pH 7.3 and 30°C: >25 days (1096)
D. BIOLOGICAL EFFECTS:
  —Mammalia: rabbit: acute oral LD_{50}: >200 mg/kg (211)

p-nitrobenzoic acid

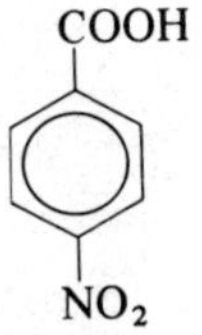

p-nitrobenzoic acid

[Structure: benzene ring with COOH at top and NO₂ at bottom]

A. PROPERTIES: leaflets; m.w. 167.12; m.p. 242.4°C, b.p. sublimes; sp.gr. 1.550 at 32/4°C solub. 240 mg/l at 25°C; log P_{oct} 1.89

C. WATER POLLUTION FACTORS:
—Biodegradation: decomposition by a soil microflora in 4 days (176)
adapted A.S. at 20°C—product is sole carbon source: 92.0% COD removal at 19.7 mg COD/g dry inoculum/hr (327)
lag period for degradation of 16 mg/l by waste water or by soil at pH 7.3 and 30°C: 3–5 days (1096)

D. BIOLOGICAL EFFECTS:
—Mammals: rabbit: acute oral LD_{50}: >200 mg/kg (211)

o-nitrochlorobenzene *see o*-chloronitrobenzene

m-nitrochlorobenzene *see m*-chloronitrobenzene

p-nitrochlorobenzene *see p*-chloronitrobenzene

nitrochloroform *see* chloropicrin

4-nitrochlorobenzene-2-sulfonate, sodium

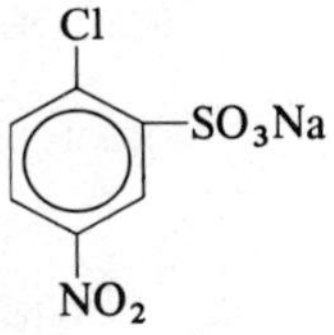

[Structure: benzene ring with Cl at top, SO₃Na at right, NO₂ at bottom]

D. BIOLOGICAL EFFECTS:
—Crustacean:

Daphnia magna: TLm,	24 hr:	4,698 mg/l		
	48 hr:	3,483 mg/l		
	72 hr:	948 mg/l		
	96 hr:	948 mg/l	(153)	
	100 hr:	1,474 mg/l	(1295)	
—Fish: *Lepomis macrochirus*: TLm, 25–100 hr:		6,375 mg/l	(1295)	
—Snail eggs, *Lymnaea sp.*:	TLm, 24 hr:	3,532 mg/l		
	48 hr:	4,439 mg/l		
	72 hr:	3,736 mg/l		
	96 hr:	3,208 mg/l	(153)	

2-nitro-p-cresol (4-methyl-2-nitrophenol)

A. PROPERTIES: yellow needles; m.w. 153.13; m.p. 36.5°C; b.p. 125°C at 25 mm; sp.gr. 1.23 g at 38/4°C
B. AIR POLLUTION FACTORS:
 —Odor threshold conc. (recognition): 0.13–0.14 mg/cu m (610)
C. WATER POLLUTION FACTORS:
 —Odor threshold conc. (detection): 0.8 mg/l (998)
D. BIOLOGICAL EFFECTS:
 —Toxicity threshold (cell multiplication inhibition test):

bacteria (*Pseudomonas putida*):	4 mg/l	(1900)
algae (*Microcystis aeruginosa*):	32 mg/l	(329)
green algae (*Scenedesmus quadricauda*):	3.8 mg/l	(1900)
protozoa (*Entosiphon sulcatum*):	0.42 mg/l	(1900)
protozoa (*Uronema parduczi Chatton-Lwoff*):	5.8 mg/l	(1901)

4-nitro-m-cresol

C. WATER POLLUTION FACTORS:
 —Odor threshold conc. (detection): 5 mg/l (998)
D. BIOLOGICAL EFFECTS:
 —Toxicity threshold (cell multiplication inhibition test):

bacteria (*Pseudomonas putida*):	6 mg/l	
green algae (*Scenedesmus quadricauda*):	6.8 mg/l	
protozoa *(Entosiphon sulcatum)*:	5.8 mg/l	(1900)
protozoa (*Uronema parduczi Chatton-Lwoff*):	0.26 mg/l	(1901)

6-nitro-m-cresol

C. WATER POLLUTION FACTORS:

3-NITRO-2,5-DICHLOROBENZOIC ACID

—Odor threshold conc. (detection): 0.2 mg/l (998)

D. BIOLOGICAL EFFECTS:
—Toxicity threshold (cell multiplication inhibition test):
bacteria (*Pseudomonas putida*): 7 mg/l
green algae (*Scenedesmus quadricauda*): 7 mg/l
protozoa (*Entosiphon sulcatum*): 1.3 mg/l (1900)
protozoa (*Uronema parduczi Chatton-Lwoff*): 5.3 mg/l (1901)

3-nitro-2,5-dichlorobenzoic acid

C. WATER POLLUTION FACTORS:
—Waste water treatment methods:
A.C. type BL (Pittsburgh Chem. Co.): % adsorbed by 10 mg A.C. from 10^{-4} M aqueous solution, at pH 3.0: 49.0%
7.0: 12.0%
11.0: 7.0% (1313)

nitroglycerine (glyceroltrinitrate, trinitroglycerine, glycerylnitrate, trinitrin, glonoin)
$C_3H_5(ONO_2)_3$

A. PROPERTIES: colorless to yellow liquid; m.w. 227.09; m.p. 3/13°C; b.p. 260°C explodes; v.p. 0.00025 mm at 20°C; v.d. 7.8; sp.gr. 1.60; solub. 1,800 mg/l at 20°C; THC 367 kcal/mole, LHC 355 kcal/mole; sat. conc. conc. 0.0031 kcal/mole
B. AIR POLLUTION FACTORS: 1 mg/cu m = 0.108 ppm, 1 ppm = 9.29 mg/cu m
D. BIOLOGICAL EFFECTS:
—Bacteria: *Pseudomonas*: LD_0: 2 mg/l
—Algae: *Scenedesmus*: LD_0: 6.5 mg/l
—Arthropoda: *Daphnia*: LD_0: 26 mg/l (30)

nitroglycol
$C_2H_4(ONO_2)_2$

A. PROPERTIES: m.w. 152.06; m.p. −22.3°C; b.p. 105.5°C at 19 mm Hg; v.p. 0.04 mm at 20°C, 0.11 mm at 30°C; v.d. 5.25
B. AIR POLLUTION FACTORS: 1 mg/cu m = 0.16 ppm, 1 ppm = 6.32 mg/cu m
D. BIOLOGICAL EFFECTS:
—Bacteria: *Pseudomonas*: LD_0: 190 mg/l
—Algae: *Scenedesmus*: LD_0: 100 mg/l
—Arthropoda: *Daphnia*: LD_0: 190 mg/l (30)

nitromethane
CH_3NO_2

A. PROPERTIES: colorless liquid; m.w. 61.04; b.p. 101°C; v.p. 27.8 mm at 20°C,

46 mm at 30°C; v.d. 2.11; sp.gr. 1.13 at 20/4°C; solub. 90–100 mg/l; sat. conc. 90 g/cu m at 20°C, 148 g/cu m at 30°C; log P_{oct} 0.17

B. AIR POLLUTION FACTORS: 1 mg/cu m = 0.40 ppm, 1 ppm = 2.50 mg/cu m
 —Odor: odor threshold values: <250 mg/cu m = <100 ppm (57)
   O.I.: at 20°C = >460 (316)
 —Manmade sources: in exhaust of gasoline engine: <0.8–5.0 ppm (195)

D. BIOLOGICAL EFFECTS:
 —Mammalia: rabbits: single oral dose: lethal: 0.75–1.0 g/kg
  dog: single oral dose: 0/12: 0.125 g/kg
   : single oral dose: 12/12: 0.25–1.5 g/kg
  mouse: single oral dose: 1/5: 1.2 g/kg
   : single oral dose: 6/10: 1.5 g/kg
  rabbit: 2/2: 30,000 ppm, 2 hr
   1/2: 5,000 ppm, 6 hr
   0/2: 500 ppm, 140 hr
  guinea pig: 4/4: 10,000 ppm, 3–6 hr
   0/2: 10,000 ppm, 1 hr
   0/2: 500 ppm, 140 hr
  monkey: 1/1: 1,000 ppm, 48 hr
   0/1: 500 ppm, 140 hr (211)
 —Man: severe toxic effects: 800 ppm = 2,028 mg/cu m, 60 min
  symptoms of illness: 500 ppm = 1,268 mg/cu m
  unsatisfactory: 200 ppm = 507 mg/cu m (185)

2-nitronaphthalene

Manmade sources: by-product of the commercial preparation of α-naphthylamine. (348)

D. BIOLOGICAL EFFECTS:
 —Carcinogenicity: positive
 —Mutagenicity in the *Salmonella* test: positive
  8.7 revertant colonies/nmol
  5250 revertant colonies at 100 μg/plate (1883)
 —Metabolic precursor of BNA in dogs and monkeys (348)

o-nitrophenol

Use: intermediate in organic synthesis; indicator

918 *m*-NITROPHENOL

A. PROPERTIES: light yellow needles or prisms; m.w. 139.11; m.p. 45°C; b.p. 214/217°C; v.p. 20 mm at 105°C, 100 mm at 146°C; sp.gr. 1.657 at 20°C; solub. 2,100 mg/l at 20°C, 10,800 mg/l at 100°C

B. AIR POLLUTION FACTORS:
—T.O.C.: recognition: 0.11–0.13 mg/cu m (610)
  detection: 0.0012 mg/cu m (829)

C. WATER POLLUTION FACTORS:
—Biodegradation: decomposition by a soil microflora in >64 days (176)
 degradation in 7 day tests:
 original culture: average: 81.0%; range: 14.0–98.5%
 1st subculture: average: 84.5%; range: 22.0–99.0%
 2nd subculture: average: 93.5%; range: 70.0–99.0%
 3rd subculture: average: 98.5%; range: 97.5–99.5% (87)
 adapted A.S. at 20°C—product is sole carbon source: 97.0% COD removal at 14.0 mg COD/g dry inoculum/hr (327)
 lag period for degradation of 16 mg/l by:
  waste water at pH 7.3; 30°C: 3–5 days
  soil suspension at pH 7.3; 30°C: 7–14 days (1096)
—Impact on biodegradation processes: inhibition of degradation of glucose by *Pseudomonas fluorescens* at: 20 mg/l
 inhibition of degradation by *E. coli*: >1000 mg/l (293)
—Reduction of amenities: T.O.C. = 10 mg/l (n.s.i.) (296)
       detection: 1.7 mg/l (998)
       taste threshold conc.: 0.001 mg/l (998)

D. BIOLOGICAL EFFECTS:
—Toxicity threshold (cell multiplication inhibition test):
 bacteria (*Pseudomonas putida*):   0.9 mg/l   (1900)
 algae (*Microcystis aeruginosa*):   27.0 mg/l (329)
 green algae (*Scenedesmus quadricauda*): 4.3 mg/l   (1900)
 protozoa (*Entosiphon sulcatum*):   0.4 mg/l   (1900)
 protozoa (*Uronema parduczi Chatton-Lwoff*): 2.9 mg/l (1901)
—Bacteria: *E. coli*: LD_0: 1,000 mg/l
—Algae: *Scenedesmus*: LD_0: 36 mg/l
—Arthropod: *Daphnia*: LD_0: 60 mg/l (30)
—Fish: bluegill: TLm (24–48 hr): 67–46.3 mg/l (41; 253)
 Vairon (F): TLm (6 hr): distilled water: 14–18 mg/l
       hard water: 125–130 mg/l (30)

m-nitrophenol

$$\underset{}{\text{OH}}\text{NO}_2$$

A. PROPERTIES: colorless to yellow crystals; m.w. 139.11; m.p. 96°C; b.p. 194°C at 70 mm; sp.gr. 1.2797 at 10/4°C; solub. 13,500 mg/l at 25°C, 133,000 mg/l at 90°C

B. AIR POLLUTION FACTORS:
  −Odor threshold conc.: 3.0 mg/cu m (829)
C. WATER POLLUTION FACTORS:
  −Biodegradation: decomposition by a soil microflora in 4 days (176)
    adapted A.S. at 20°C−product is sole carbon source: 95.0% COD removal at 17.5 mg COD/g dry inoculum/hr (327)
    lag period for degradation of 16 mg/l by waste water or by soil at pH 7.3 and 30°C: 3−5 days (1096)
  −Impact on biodegradation processes:
    Inhibition of degradation of glucose by *Pseudomonas fluorescens* at: 20 mg/l
    Inhibition of degradation of glucose by *E. coli*: 300 mg/l (293)
  −Odor threshold conc. (detection): 0.6 mg/l (998)
D. BIOLOGICAL EFFECTS:
  −Toxicity threshold (cell multiplication inhibition test):
    bacteria (*Pseudomonas putida*):   7 mg/l   (1900)
    algae (*Microcystis aeruginosa*):   17.0 mg/l   (329)
    green algae (*Scenedesmus quadricauda*):   7.6 mg/l   (1900)
    protozoa (*Entosiphon sulcatum*):   0.97 mg/l   (1900)
    protozoa (*Uronema parduczi Chatton-Lwoff*): 3.4 mg/l (1901)
  −Bacteria: *E. coli*: LD_0: 300 mg/l
  −Algae: *Scenedesmus*: LD_0: 28 mg/l
  −Arthropod: *Daphnia*: LD_0: 24 mg/l
  −Fish: *Vairon* (F): TLm (6 hr): distilled water: 9−10 mg/l
                                   hard water: 20−22 mg/l (30)

p-nitrophenol

Use: intermediate in organic synthesis; production of parathion; fungicide for leather
A. PROPERTIES: colorless to yellowish crystals; m.w. 139.11; mp. 114°C; b.p. 279°C decomposes; v.p. 2.2 mm at 146°C, 18.7 mm at 186°C; sp.gr. 1.479 at 20°C; solub. 16 g/l at 25°C, 269 g/l at 90°C;
B. AIR POLLUTION FACTORS:
  −Odor threshold conc. (detection): 2.3 mg/cu m (829)
C. WATER POLLUTION FACTORS:
  −Biodegradation:
    adapted culture: 2% removal after 48 hr incubation, feed: 200 mg/l (292)
    decomposition by a soil microflora in 16 days (176)
    adapted A.S. at 20°C−product is sole carbon source: 95.0% COD removal at 17.5 mg COD/g dry inoculum/hr (327)
    lag period for degradation of 16 mg/l by:

waste water at pH 7.3; 30°C: 3-5 days
soil suspension at pH 7.3; 30°C: 7-14 days (1096)
—Impact on biodegradation process:
inhibition of degradation of glucose by *Pseudomonas fluorescens* at 20 mg/l
inhibition of degradation of glucose by *E. coli*: 100 mg/l (293)
—Waste water treatment: ion exchange: adsorption on Amberlite XAD-2:
100% retention eff.; infl.: 0.2 ppm, effl.: nil (40)
reaction of excess hypochlorous acid with dilute aqueous solutions at initial pH's of 6.0 and 3.5:

	(1)	(2)	(3)	(4)	(5)
yields at pH 6.0	20%	1%	0.3%	39%	
yields at pH 3.5	26%	—	—	52%	

(1) = 2,6-dichlorobenzoquinone
(2) = 2,6-dichloro-4-nitrophenol
(3) = 2,3,4,6-tetrachlorophenol
—Odor threshold conc. (detection): 2.5 mg/l (998)

D. BIOLOGICAL EFFECTS:
—Toxicity threshold (cell multiplication inhibition test):
bacteria (*Pseudomonas putida*): 4 mg/l (1900)
algae (*Microcystis aeruginosa*): 56.0 mg/l (329)
green algae (*Scenedesmus quadricauda*): 7.4 mg/l (1900)
protozoa (*Entosiphon sulcatum*): 0.83 mg/l (1900)
protozoa (*Uronema parduczi Chatton-Lwoff*): 0.89 mg/l (1901)
—Bacteria: *E. coli*: LD_0: 100 mg/l
—Algae: *Chlorella pyrenoidosa*: toxic: 9-14 mg/l (n.s.i.) (41)
   Scenedesmus: LD_0: 42 mg/l
—Arthropod: *Daphnia*: LD_0: 14 mg/l (30)
—Fish: *Vairon* (F): TLm 6 hr: distilled water: 4-6 mg/l
             hard water: 30-33 mg/l (30)
—Carcinogenicity: ?
—Mutagenicity in the Salmonella test: negative
   <0.02 revertant colonies/nmol
   70 revertant colonies at 500 µg/plate (1883)

N-nitrosodiethanolamine
Manmade sources: in cutting fluids: (Canada): 0-5.53 mg/g (1060)
                        (U.S.A.): 0.2-29.9 mg/g (1061)

N-nitrosodiethylamine *see* diethylnitrosamine

N-nitrosodimethylamine *see* N-dimethylnitrosamine

N-nitrosodi-N-propylamine (NDPA)
Contaminant of herbicide Treflan (Trifluralin) in concentrations up to 150 ppm
(1579, 1580)
C. WATER POLLUTION FACTORS:
  —Dissipation of ^{14}C NDPA from soil: 571 µg NDPA/M^2 was applied to soil, after 2 hours only 52.6% of the amount originally applied was still present. Approximately 92% of the radioactivity was extractable with methanol. Analysis confirmed that all of the extractable radioactivity was present as NDPA. The level of soil radioactivity continued to decline and after 8 and 49 days had decreased to 16.9 and 11% respectively, of the amount initially applied. At eight days only 16% of the soil radioactivity was extractable with methanol.
  —Soybeans harvested from plants grown in such treated soil are free of residues of NDPA and NDPA transformation products (1578)

o-**nitrotoluene** *see o*-methylnitrobenzene

m-**nitrotoluene** *see m*-methylnitrobenzene

p-**nitrotoluene** *see p*-methylnitrobenzene

4-nitrotoluene-2-sulfonate, sodium

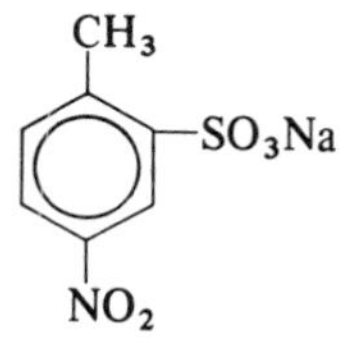

D. BIOLOGICAL EFFECTS:
  —Fish: *Lepomis macrochirus*: 24 hr TLm: <1,440 mg/l (1295)

nitrourea
$NH_2CONHNO_2$
  Use: explosives
A. PROPERTIES: white crystalline powder; m.p. 158–159°C
C. WATER POLLUTION FACTORS:
  —Impact on biodegradation processes:
    ~50% inhibition of NO_2^- oxidation at 1 mg/l (409)

nitrous diethylamide *see* diethylnitrosamine

nitrous dimethylamide *see* N-dimethylnitrosamine

3-nitro-*o*-xylene

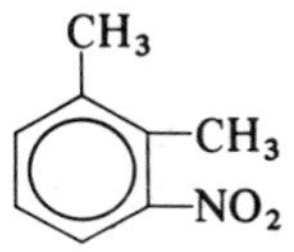

A. PROPERTIES: needles; m.w. 151.16; m.p. 7-9°C; b.p. 245-250°C; v.d. 5.22; sp.gr. 1.147 at 15°C
B. AIR POLLUTION FACTORS: 1 mg/cu m = 0.16 ppm, 1 ppm = 6.28 mg/cu m (57)
C. WATER POLLUTION FACTORS:
—Water quality:
in Delaware river (U.S.A.): conc. range: winter: 0.3 ppb
(n.s.i.) summer: n.d. (1051)

cis-nonachlor (1,2,3,4,5,6,7,8,8-nonachloro-3a,4,7,7a-tetrahydro-4,7-methanoindan)
Use: insecticide
D. BIOLOGICAL EFFECTS:
—Fish:
goby fish (Acanthogobius flavimanus) collected at the seashore of Keihinjima along Tokyo Bay—Aug. 1978: residue level: 8 ppb (1721)

trans-nonachlor
Major constituent of technical chlordane (1673) and technical heptachlor (1674)
D. BIOLOGICAL EFFECTS:
—Residues:
goby fish (*Acanthogobius flavimanus*) collected at the seashore (Aug. 1978) along Tokyo Bay contained 18 ppb (1672)
residue in Canadian human milk: 1975: avg.: 1 ng/g whole milk (1376)

n-nonacosane
$CH_3(CH_2)_{27}CH_3$
B. AIR POLLUTION FACTORS:
—Ambient air quality:
organic fraction of suspended matter:
Bolivia at 5200 m altitude (Sept.-Dec. 1975): 0.53-0.98 µg/1000 cu m
Belgium, residential area (Jan.-April 1976): 4.5-6.6 µg/1000 cu m (428)
glc's in residential area (Belgium) oct. 1976:
in particulate sample: 15.8 ng/cu m
in gas phase sample: 6.56 ng/cu m (1289)
glc's in Botrange (Belgium)—woodland at 20-30 km from industrial area—June-July 1977: 3.63; 5.00 ng/cu m (n = 2)
glc's in Wilrijk (Belgium)—residential area: Oct.-Dec. 1976: 12.68; 12.98 ng/cu m (n = 2) (1233)

n-nonadecane
$CH_3(CH_2)_{17}CH_3$
Use: organic synthesis
A. PROPERTIES: m.p. 32°C; b.p. 330°C; sp.gr. 0.777; m.w. 268.53
B. AIR POLLUTION FACTORS:
—Ambient air quality:
organic fraction of suspended matter:
Bolivia at 5200 m altitude (Sept.-Dec. 1975): 0.24-0.34 µg/1000 cu m
Belgium, residential area (Jan.-April. 1976): 0.86-1.6 µg/1000 cu m (428)

in the average American urban atmosphere, 1963:
160 µg/g airborne particulates or
20 ng/g cu m air (1293)
glc's in residential area (Belgium) oct. 1976:
in particulate sample: 0.80 ng/cu m
in gas phase sample: 15.1 ng/cu m (1289)
glc's in Botrange (Belgium)—woodland at 20-30 km from industrial area, June–July 1977: 0.27–0.34 ng/cu m ($n = 2$) (1233)

nonadecanoic acid
$CH_3(CH_2)_{17}COOH$
A. PROPERTIES: a saturated fatty acid normally not found in natural vegetable fats or waxes; m.p. 68.7°C; b.p. 297°C at 100 mm
B. AIR POLLUTION FACTORS:
  —Ambient air quality:
  organic fraction of suspended matter:
    Bolivia at 5200 m altitude (Sept.–Dec. 1975): 0.07–0.14 µg/1000 cu m
    Belgium, residential area (Jan.–April 1976): 0.77–1.5 µg/1000 cu m (428)
  glc's in residential area (Belgium), Oct. 1976:
    in particulate sample: 1.91 ng/cu m
    in gas phase sample: 1.02 ng/cu m (1289)
  glc's Botrange (Belgium)—woodland at 20-30 km from industrial area, June–July 1977: 0.36; 0.40 ng/cu m ($n = 2$)
  glc's in Wilrijk (Belgium), residential area, Oct.–Dec. 1976: 1.1; 1.7 ng/cu m ($n = 2$) (1233)
C. WATER POLLUTION FACTORS:
  —Odor threshold conc. (detection): 20.0 mg/kg (886)

n-nonanal (pelargonaldehyde; pelargonic aldehyde; n-nonaldehyde)
$C_8H_{17}CHO$
A. PROPERTIES: colorless liquid; sp.gr. 0.822
B. AIR POLLUTION FACTORS:

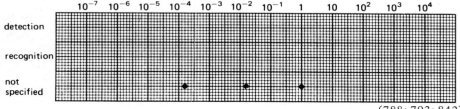

(788; 793; 842)

C. WATER POLLUTION FACTORS:
  —Reduction of amenities: T.O.C. = 0.001 mg/l (305)
  —Waste water treatment: A.S. after 6 hr: 8.4% of ThOD
                                    12 hr: 13.5% of ThOD
                                    24 hr: 21.1% of ThOD (88)

n-nonane
$CH_3(CH_2)_7CH_3$

Manufacturing source: petroleum refinery. (347)
Users and formulation: organic synthesis; solvent; standardized hydrocarbon; jet fuel research; mfg. paraffin products; rubber industry; paper processing industry; biodegradable detergents; distillation chaser. (347)
Natural sources (water and air): constituent in paraffin fraction of petroleum. (347)
Man caused sources (water and air): component in municipal waste and waste water from general use of paraffins; lab use. (347)

A. PROPERTIES: colorless liquid; m.w. 128.25; m.p. -54°C; b.p. 151°C; v.p. 3.22 mm at 20°C; v.d. 4.41; sp.gr.: 0.72 at 20°C; solub. (distilled): 0.07 mg/l at 20°C, (salt water) 0.43 mg/l at 20°C

B. AIR POLLUTION FACTORS: 1 mg/cu m = 1.88 ppm, 1 ppm = 5.33 mg/cu m
— Odor: odor threshold values: 2.3 mg/cu m = 0.43 ppm (307)
　　　　　　　　　　　　　　recognition: 108 mg/cu m (761)
　　　　　　　　　　　　　　60 mg/cu m (737)
　O.I.: at 20°C = 9,800 (316)
— Atmospheric reactions: reactivity: NO ox.: ranking: 0.15 (63)
　first order evaporation constant of n-nonane in 3 mm layer of No. 2 fuel oil in darkened room, wind speed of 21 km/hr; at 5°C: 3.49×10^{-3} min^{-1} (438)

C. WATER POLLUTION FACTORS:
— Biodegradation:
　degradation in seawater by oil oxidizing microorganisms:
　100% breakdown after 21 d at 22°C in 1000 ppm mixture of alkanes, cycloalkanes, and aromatics contained in stoppered bottles (1237)
— Reduction of amenities: T.O.C. = 10 mg/l (295)
— Waste water treatment: A.S. after　6 hr: 0.2% of ThOD
　　　　　　　　　　　　　　　　　12 hr: 0.4% of ThOD
　　　　　　　　　　　　　　　　　24 hr: 1.1% of ThOD (88)

nonanoic acid　*see* pelargonic acid

1-nonanol　(*n*-nonylalcohol)
$C_9H_{19}OH$

A. PROPERTIES: colorless liquid; m.w. 144.25; m.p. -5°C; b.p. 194/213°C; v.p. 0.3 mm at 20°C; v.d. 4.98; sp.gr. 0.827 at 20/4°C; solub. 1,000 mg/l at 20°C
B. AIR POLLUTION FACTORS: 1 mg/cu m = 0.17 mm, 1 ppm = 6.00 mg/cu m
— Odor threshold conc.: 41 mg/cu m (787)
C. WATER POLLUTION FACTORS:
— Waste water treatment: A.S.: after　6 hr: toxic
　　　　　　　　　　　　　　　　12 hr: 0.9% of ThOD
　　　　　　　　　　　　　　　　24 hr: 9.9% of ThOD (88)
D. BIOLOGICAL EFFECTS:
— Mammals:
　mouse: single oral dose: LD_{50}: 6.4–12.8 g/kg
　rat:　 single oral dose: LC_{50}: 3.2–6.4 g/kg
　　　　inhalation: no deaths:　730 ppm, 6 hr (211)

2-nonanone (heptylmethylketone)
$CH_3CO(CH_2)_6CH_3$
A. PROPERTIES: m.w. 142.24; m.p. $-19/-8.2°C$; b.p. $194/196°C$, sp.gr. 0.8317
B. AIR POLLUTION FACTORS:
  —Odor threshold conc.: 0.075 mg/cu m (842)
  —Sampling and analysis: scrubbers: liquid lift type: 15 ml of liquid, 4.5 l/min gas flow, 4 min scrubbing

trap method	% removal	
H_2O, 0°C	0	
NH_2OH soln, 0°C	0	
H_2SO_4 (concd), 55°C	0	
H_2SO_4, 4% Ag_2SO_4 (conc.), 55°C	0	
open tube, $-80°C$	88	
ethanol, $-80°C$	86	(311)

5-nonanone (dibutylketone)
$(CH_3CH_2CH_2CH_2)_2CO$
A. PROPERTIES: colorless liquid; m.w. 142.24; m.p. $-5.9°C$; b.p. $186°C$; sp.gr. 0.827 at $13/4°C$
C. WATER POLLUTION FACTORS:
  —Waste water treatment: A.S. after 6 hr: 0.9% of ThOD
                                     12 hr: 2.6% of ThOD
                                     24 hr: 7.9% of ThOD (88)

nonene *see* propylene trimer

nonidet G2C/nonidet BX
  Composition: contains didecyldimethylammoniumbromide
  Use: bactericide
C. WATER POLLUTION FACTORS:
  —COD: 0.54 = 96% ThOD (277)
D. BIOLOGICAL EFFECTS:
  Carassius auratus: 24 hr LD_{50}: 15 mg/l
                       96 hr LD_{50}: 14 mg/l (277)

nonidet NP 50 (alkylphenolethyleneoxide condensate)

$$C_nH_{2n}H + \text{—}\bigcirc\text{—}O\text{—}(CH_2CH_2O)_n \cdot H$$

A. PROPERTIES: sp.gr. 1.06 at $15.5/15.5°C$;
C. WATER POLLUTION FACTORS:
  —Biodegradability: 20% (OECD method) (277)
   COD: 2.08 = 94% ThOD (277)
D. BIOLOGICAL EFFECTS:
  —*Carassius auratus*: 24 hr LD_{50}: 8 mg/l
                        96 hr LD_{50}: 7 mg/l (277)

nonylcarbinol *see* 1-decanol

***n*-nonylic acid** *see* pelargonic acid

nonylphenol

$$\underset{\text{OH}}{\text{C}_6\text{H}_4}-\text{C}_9\text{H}_{19}$$

Use: nonionic surfactant; lube oil additives; stabilizers; petroleum-demulsifiers; fungicides; bactericides; dyes; drugs; adhesives; rubber chemicals; phenolic resins and plasticizers.

A mixture of isomeric monoalkylphenols.

A. PROPERTIES: sp.gr. 0.950; b.p. 315°C; f.p. −10°C (sets to glass below this temperature)

C. WATER POLLUTION FACTORS:
—Biodegradation at 1.0 mg/l:

	normal sewage	adapted sewage	
after 24 hr	0%	0%	
after 135 hr	0%	45%	(997)

D. BIOLOGICAL EFFECTS:
—BCF in mussel (*Mytilus edulis*): 10 (on wet weight) (1864)

nonylphenolethoxylate
(Contains 9–10 EO)

D. BIOLOGICAL EFFECTS:
—Fish:

	48 hr LC_{50}	96 hr LC_{50}	
Rasbora heteromorpha	11.3 mg/l	8.6 mg/l	
Salmo trutta	2.7 mg/l	1.0 mg/l	
Idus idus	7.4–11.3 mg/l	7.0–11.2 mg/l	
Carassius auratus	4.9 mg/l	—	(1905)

***dl*-norleucine** (*dl*-2-aminohexanoic acid; *dl*-α-aminocaproic acid; *dl*-glycoleucine)
$CH_3(CH_2)_3CH(NH_2)COOH$

A. PROPERTIES: shiny leaflets; m.w. 131.17; m.p. 327°C decomposes; solub. 11,800 mg/l at 25°C, 28,800 mg/l at 75°C; log P_{oct} −1.22 (calculated)

C. WATER POLLUTION FACTORS:
—Waste water treatment: A.S. after 6 hr: 2.3% of ThOD
                                   12 hr: 5.2% of ThOD
                                   24 hr: 12.9% of ThOD (89)

norphytane *see* pristane

NRDC 107 (*see also* pyrethroids and pyrethrins)
Use: insecticide; synthetic pyrethroid

D. BIOLOGICAL EFFECTS:
  —Fish:

	mg/l	24 hr	48 hr	96 hr	3m (extrap.)
harlequin fish *(Rasbora heteromorpha)* $LC_{10}(F)$		0.018	0.015		
$LC_{50}(F)$		0.025	0.025	0.014	0.01

(331)

NTA *see* nitrilotriacetic acid

O

OCBM see *o*-chlorobenzylidenemalonitrile

OCDD see octachlorodibenzo-*p*-dioxin

2,2′,3,3′,4,4′,5,5′-octachlorobiphenyl
D. BIOLOGICAL EFFECTS:
  —Algae:
    Chlorella pyrenoidosa: bioconcentration coefficient: 11000 in whole cells
(1567)

octachlorodibenzo-*p*-dioxine (OCDD)
Technical PCP contains typically 500–1500 ppm OCDD as impurity (1782)
Manmade sources: photolytical condensation of pure PCP to form OCDD has been reported in conc. up to 70 ppm in Southern yellow pine treated with pure PCP and exposed to both natural and artificial sunlight (1781)

Generation of dioxins from *o*-chlorophenates such as the sodium salt of PCP

1,2,4,5,6,7,8,8-octachloro-4,7-methano-3a,4,7,7a-tetrahydro-indane see chlordane

octachlorostyrene
D. BIOLOGICAL EFFECTS:
  —In fat phase of the following fish from the Frierfjord (Norway), April 1975–Sept. 1976: cod, whiting, plaice, eel, sprot:
      average: 39 ppm
      range: 5–361 ppm (1065)

n-octacosane
$CH_3(CH_2)_{26}CH_3$
A. PROPERTIES: m.w. 394.77; m.p. 61–63°C; b.p. 278°C/15 mm

B. AIR POLLUTION FACTORS:
—Ambient air quality:
organic fraction of suspended matter:
Bolivia at 5200 m altitude (Sept.–Dec. 1975): 0.32–0.71 µg/1000 cu m
Belgium, residential area (Jan.–April 1976):   2.4–4.2 µg/1000 cu m (428)
in the average American urban atmosphere, 1963:
340 µg/g airborne particulates or
43 ng/g cu m air (1293)
glc's in residential area (Belgium) oct. 1976:
in particulate sample: 8.10 ng/cu m
in gas phase sample: 7.80 ng/cu m (1289)
glc's in Botrange (Belgium)—woodland at 20–30 km from industrial area: June–July 1977: 1.33; 1.01 ng/cu m ($n = 2$)
glc's in Wilrijk (Belgium): residential area: Oct.–Dec. 1976: 4.38; 11.22 ng/cu m ($n = 2$) (1233)

C. WATER POLLUTION FACTORS:
—Aquatic reactions:
adsorption on smectite clay particles from simulated seawater at 25°C, experimental conditions: 100 µg n-octacosane/l water; 50 mg smectite/l; adsorption: 1.41 µg/mg = 70% adsorbed (1009)

octacosanoic acid
$CH_3(CH_2)_{26}COOH$

B. AIR POLLUTION FACTORS:
—Ambient air quality:
glc's at Botrange (Belgium)—woodland at 20–30 km from industrial area: June–July 1977: 2.86; 3.34 ng/cu m ($n = 2$) (1233)

octadecanamide *see* stearamide

n-octadecane
$CH_3(CH_2)_{16}CH_3$

Manufacturing source: petroleum refining. (347)
Users and formulation: organic synthesis; solvent; standardized hydrocarbon; jet fuel research; mfg. paraffin products; rubber industry; paper processing industry; calibration. (347)
Natural sources (water and air): constituent in paraffin fraction of petroleum. (347)
Man caused sources (water and air): component in municipal waste and waste water from general use of paraffins; lab use, highway runoff, automobile exhaust; general use of petroleum oils. (347)

A. PROPERTIES: m.w. 254.49; m.p. 28°C; b.p. 317°C; sp.gr. 0.777 at 28/4°C; solub.
in sea water 0.0008 mg/l at 25°C (230)
in distilled water: 0.0021 mg/l at 25°C (230)
in water: 0.007 mg/l at 25°C (254)

B. AIR POLLUTION FACTORS:
—Ambient air quality:
in the average American urban atmosphere, 1963:

930 OCTADECANOIC ACID

    110 µg/g airborne particulates or
    14 ng/g cu m air   (1293)
  organic fraction of suspended matter:
    Bolivia at 5200 m altitude (Sept.–Dec. 1975): 0.24–0.34 µg/1000 cu m
    Belgium, residential area (Jan.–April 1976):   0.76–1.3 µg/1000 cu m   (428)
    glc's in residential area (Belgium), Oct. 1976: in gasphase sample: 42.8 ng/cu m
        (1289)
    glc's in Botrange (Belgium)—woodland at 20–30 km from industrial area, June–July 1977: 0.57; 0.69 ng/cu m ($n = 2$)
C. WATER POLLUTION FACTORS:
  —Biodegradation:
    degradation in seawater by oil-oxidizing microorganisms: 46.3% breakdown after 21 days at 22°C in stoppered bottles containing a 1000 ppm mixture of alkanes, cycloalkanes, and aromatics   (1237)
    microbial degradation to CO_2—in seawater at 12°C—in the dark after 24 hr incubation at 30 µg/l: degradation rate: 0.16 µg/l/day—turnover time: 200 days
        (477)
  —Aquatic reactions: adsorption on smectite clay particles from simulated seawater at 25°C: experimental conditions: 100 µg n-octadecane/l water; 50 mg smectite/l: adsorption: 0.74 µg/mg = 38% adsorbed   (384)
    in estuarine water: at 5 µg/l, 22% adsorbed on particles after 3 hr   (381)
    first-order evaporation constant of n-octadecane in 3 mm No. 2 fuel oil layer in dark room at wind speed of 21 km/hr:
      at 10°C: 2.20×10^{-5}
          20°C: 4.00×10^{-5} (438)
  —Waste water treatment: A.S. after 6 hr: 0.5% of ThOD
                               12 hr: 1.1% of ThOD
                               24 hr: 3.2% of ThOD   (88)

octadecanoic acid *see* stearic acid

1-octadecanol (n-octadecylalcohol)
$CH_3(CH_2)_{16}CH_2OH$
  Use: perfumery; cosmetics-intermediate; surface active agents; lubricants; resins
A. PROPERTIES: leaflets; m.w. 270.49; m.p. 59°C; b.p. 210°C at 15 mm; sp.gr. 0.812 at 59/4°C
C. WATER POLLUTION FACTORS:
  —Waste water treatment: A.S. after 6 hr: 0.3% of ThOD
                               12 hr: 0.5% of ThOD
                               24 hr: 0.3% of ThOD   (88)
  in river water: 0.007–0.16 ppm; in river sediment: 0.5 ppm   (55)

octadecyl-3-(3'-5'-di-t-butyl-4'-hydroxyphenyl)propionate

$$HO-\underset{(CH_3)_3C}{\overset{(CH_3)_3C}{\bigcirc}}-CH_2CH_2\overset{O}{\overset{\|}{C}}OC_{18}H_{37}$$

C. WATER POLLUTION FACTORS:
—in river water: 0.008–0.2 ppm; in river sediment: 8–220 ppm (555)

n-octadecylic acid *see* stearic acid

1,2,3,4,5,6,7,8-octahydroanthracene

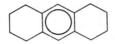

B. AIR POLLUTION FACTORS:
—Manmade sources:
in coke oven emissions: 8.8–70.3 µg/g of sample (n.s.i.) (960)
C. WATER POLLUTION FACTORS:
—Odor threshold conc.: 0.001 mg/kg (904)

octahydrofluoranthene
Manmade sources:
in airborne coal tar emissions: 0.15 mg/g of sample or 5 µg/cu m of air sampled
in coal tar: 0.23 mg/g of sample (993)

octahydrophenanthrene
Manmade sources: in coke oven emissions: 31.85 µg/g of sample (n.s.i.) (960)

octahydropyrene
Manmade sources:
in airborne coal tar emissions: 0.15 mg/g of sample or 5 µg/cu m of air sampled
in coal tar: 0.21 mg/g of sample (993)

octaklor *see* chlordane

octalox *see* dieldrin

octanal (caprylaldehyde; caprylic aldehyde; *n*-octylaldehyde
$CH_3(CH_2)_6CHO$
  Use: perfumery; flavors
A. PROPERTIES: colorless liquid; m.w. 128.21; b.p. 163/168°C; v.d. 4.4; sp.gr. 0.82 at 20/4°C
B. AIR POLLUTION FACTORS:
—T.O.C.: −5 mg/cu m (788)
            −0.015 mg/cu m (842)
C. WATER POLLUTION FACTORS:
—Reduction of amenities: T.O.C. = 0.0007 mg/l (305)(880)
—Waste water treatment: A.S. after   6 hr: 2.7% of ThOD
                                     12 hr: 2.9% of ThOD
                                     24 hr: 4.9% of ThOD

***n*-octane**
$CH_3(CH_2)_6CH_3$
  Manufacturing source: petroleum refining. (347)
  Users and formulation: organic synthesis; solvent; rubber industry; paper processing ind. (347)
  Natural sources (water and air): constituent in paraffin fraction of petroleum. (347)
A. PROPERTIES: m.w. 114.23; m.p. -56.5°C; b.p. 125.7°C; v.p. 11 mm at 20°C; 18 mm at 30°C; v.d. 3.93; sp.gr. 0.70 at 25/4°C; solub. 0.66 mg/l at 20°C (242)
  in distilled water: 0.7 mg/l at 20°C (228)
  in salt water: 2.5 mg/l at 20°C;
  sat. conc. 62 g/cu m at 20°C, 108 g/cu m at 30°C
B. AIR POLLUTION FACTORS: 1 mg/cu m = 0.21 ppm, 1 ppm = 4.75 mg/cu m
  odor thresholds

(73; 210; 278; 307; 708; 737; 749)
  O.I. at 20°C = 100 (316)
C. WATER POLLUTION FACTORS:
  —$BOD_{35\,d}^{25°C}$: 2.33 in seawater/inoculum: enrichment cultures of hydrocarbon oxidizing bacteria (521)
  —ThOD: 3.50
  —Biodegradation:
    incubation with natural flora in the groundwater—in presence of the other components of high-octane gasoline (100 µl/l): biodegradation: 54% after 192 hr at 13°C (initial conc. 0.34 µl/l) (956)
  —Reduction of amenities: T.O.C. = 10 mg/l (295)
  —Control methods: calculated half life time based on evaporative loss for a water depth of 1 m at 25°C: 5.55 hr; 0.124 m/hr (330)
    rotating disk contact aerator: infl. 214.6 mg/l, effl. 0.3 mg/l; elimination: >99% or 17,687 mg/m²/24 hr or 4,776 g/m³/24 hr (406)
D. BIOLOGICAL EFFECTS:
  —Fish: young Coho salmon: no significant mortalities up to 100 ppm, in artificial sea water after 96 hrs at 8°C (317)
  —Mammalia: mice: inhalation:
    narcosis: 6,600–13,700 ppm, 30–90 min
    no deaths or convulsions: 13,700 ppm (211)

octane dioic acid *see* suberic acid

octanethiol *see* octylmercaptan

octanoic acid *see* caprylic acid

1-octanol (heptylcarbinol)
$CH_3(CH_2)_7OH$
A. PROPERTIES: colorless liquid; m.w. 130.23; m.p. $-17°C$; b.p. $195°C$; v.p. 1 mm at $54°C$; v.d. 4.48; sp.gr. 0.824 at $20/4°C$; solub. 300 mg/l at $20°C$; $\log P_{oct}$ 3.15
B. AIR POLLUTION FACTORS: 1 mg/cu m = 0.19 ppm, 1 ppm = 5.41 mg/cu m

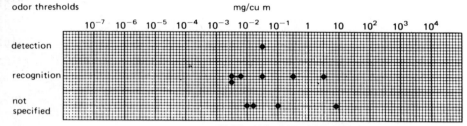

(307; 610; 627; 663; 761; 786; 787; 795; 827)

O.I.: at $20°C$ = 33,000

C. WATER POLLUTION FACTORS:
  —Oxidation parameters:
    BOD_5:   38% ThOD   (274)
          1.088   (30)
          1.1 std. dil.   (27)
          33% ThOD   (220)
    BOD_{35}^{25}: 1.45 in seawater/inoculum: enrichment cultures of hydrocarbon oxidizing bacteria   (521)
    COD: 98% ThOD   (220)
          82% ThOD (0.05 n $K_2Cr_2O_7$)   (274)
    $KMnO_4$ value: 0.110   (30)
          acid: 6% ThOD
          alkaline: 1% ThOD
    ThOD: 2.95   (220)
                (30)
  —Reduction of amenities: T.O.C. = 0.13 mg/l   (294)
    taste and odor threshold: (tentative): 0.13 mg/l   (27)
                    range: 0.0087–0.56 mg/l   (30)
  T.O.C. in water: 0.13 ppm   (326)
  —Waste water treatment: reverse osmosis: 68% rejection from a 0.01 M solution   (221)

D. BIOLOGICAL EFFECTS:
  —Toxicity threshold (cell multiplication inhibition test):
    bacteria (*Pseudomonas putida*):   >50 mg/l   (1900)
    algae (*Microcystis aeruginosa*):   1.9 mg/l   (329)
    green algae (*Scenedesmus quadricauda*):   6.3 mg/l   (1900)
    protozoa (*Entosiphon sulcatum*):   44 mg/l   (1900)
    protozoa (*Uronema parduczi Chatton-Lwoff*):   23 mg/l   (1901)
  —Fish: sea lamprey: no effect at 5 mg/l, 24 hr   (30)

2-octanol (methylhexylcarbinol, caprylalcohol)
$CH_3CHOH(CH_2)_5CH_3$

934 4-OCTANOL

A. PROPERTIES: colorless oily liquid; m.w. 130.33; m.p. -38.6°C; b.p. 178.5°C; v.p. 1 mm at 32.8°C; 40 mm at 98°C; v.d. 4.49; sp.gr. 0.819; solub. <1,000 mg/l
B. AIR POLLUTION FACTORS: 1 mg/cu m = 0.188 ppm, 1 ppm = 5.32 mg/cu m
—Odor: odor threshold values: 0.0014 mg/cu m = 0.26 ppb   (307)
   2.6 ppb   (279)
   detection: 0.05 mg/cu m   (776)
   recognition: 0.013–0.016 mg/cu m   (610)
   (d)-isomer: detection: 0.0011 mg/cu m
   (l)-isomer: detection: 0.0003 mg/cu m   (829)
 O.I.: at 20°C = 506,000   (316)
C. WATER POLLUTION FACTORS:
—Reduction of amenities:
 T.O.C.: average: 0.13 ppm
   range: 0.0087–0.56 ppm   (97)
—Waste water treatment: A.S.: after 6 hr: 1.1% of ThOD
   12 hr: 2.9% of ThOD
   24 hr: 2.7% of ThOD   (88)
 Stabilization pond design: toxicity correction factor:
   = 2.0 at 150 mg/l pond influent
   = 4.0 at 200 mg/l pond influent (n.s.i.)   (179)
D. BIOLOGICAL EFFECTS:
—Mammalia: rat: single oral LD_{50}: >3.2 g/kg   (211)

4-octanol
$CH_3(CH_2)_2CHOH(CH_2)_3CH_3$
C. WATER POLLUTION FACTORS:
—Waste water treatment: after 6 hr: 1.0% of ThOD
   12 hr: 1.5% of ThOD
   24 hr: 1.4% of ThOD   (88)

2-octanone (hexylmethylketone)
$CH_3COC_6H_{13}$
A. PROPERTIES: colorless liquid; m.w. 128.21; m.p. -16/-21°C; b.p. 173.5°C; v.p. 0.75 mm at 20°C, 1.2 mm at 25°C, 1.5 mm at 30°C; v.d. 4.43; sp.gr. 0.836 at 25/25°C; solub. 900 mg/l; sat. conc. 5.2 g/cu m at 20°C, 10 g/cu m at 30°C
B. AIR POLLUTION FACTORS: 1 mg/cu m = 0.19 ppm, 1 ppm = 5.33 mg/cu m
—Odor: odor threshold values: 1.304 mg/cu m = 248 ppm   (307)
   recognition: 0.06–0.1 mg/cu m   (610)
   detection: 0.78 mg/cu m   (824; 825)
 O.I.: at 20°C = 4   (316)
C. WATER POLLUTION FACTORS:
—Waste water treatment: A.S. after 6 hr: 1.4% of ThOD
   12 hr: 2.3% of ThOD
   24 hr: 4.6% of ThOD   (88)

octanone-3 see ethyl-*sec*-amylketone

***cis*-9-octedecenoic acid** see oleic acid

1-octene
$C_6H_{13}CHCH_2$
Use: organic synthesis; plasticizer; surfactants
A. PROPERTIES: b.p. 121°C; v.d. 3.9; sp.gr. 0.70; solub. 2.7 mg/l at 20°C (242)
B. AIR POLLUTION FACTORS:
—Odor threshold conc. (detection): 5 mg/cu m (643)
C. WATER POLLUTION FACTORS:
—Waste water treatment: A.S. after 6 hr: 0.7% of ThOD
12 hr: 0.6% of ThOD
(octene-1, and -2 mixture) 24 hr: 0.9% of ThOD (88)
rotating disk contact aerator: infl. 61.7 mg/l, effl. 0.5 mg/l, elimination: >99%
or 5,056 mg/m^2/24 hr, or 1,365 g/m^3/24 hr (406)
—Odor threshold conc. (detection): 0.0005 mg/kg (814)

2-octene
$CH_3(CH_2)_4CHCHCH_3$
Use: organic synthesis; lubricants
B. AIR POLLUTION FACTORS:
—Odor threshold conc. (detection): 5 mg/cu m (643)
C. WATER POLLUTION FACTORS:
—Waste water treatment methods:
rotating disk contact aerator: infl. 65.6 mg/l, effl. 0.7 mg/l, elimination: 99%, or
5,363 mg/m^2/24 hr or 1,448 g/m^3/24 hr (406)

n-octoic acid *see* caprylic acid

p-octyldiphenylamine
D. BIOLOGICAL EFFECTS:
—Fish: goldfish: approx. fatal conc.: >40 mg/l (48 hr) (226)

n-octylic acid *see* caprylic acid

octylmercaptan (octanethiol)
$CH_3(CH_2)_7SH$
A. PROPERTIES: water white liquid; b.p. 199°C; sp.gr. 0.8395 at 25/4°C
B. AIR POLLUTION FACTORS:
—Emission control:
wet scrubber: water at pH 8.5: effluent: 3,500 odor units/scf
$KMnO_4$ at pH 8.5: effluent: 6.5 odor units/scf (115)

oestradiol-17β
A. PROPERTIES: solub. 1559–1764 mg/l
C. WATER POLLUTION FACTORS:
—Degradation in drinking water after 10 days: 61.1% remained unchanged (413)
—In drinking water in S.W. Germany: avg.: 0.12–0.42 ng/l
range: 0–0.94 ng/l (413)

oil of bitter almonds *see* benzaldehyde

oil of wintergreen, artificial see methylsalicylate

olefins see amylene dimer

oleic acid (9-octadecenoic acid (*cis* form))
$CH_3(CH_2)_7CHCH(CH_2)_7COOH$
A. PROPERTIES: colorless needles; m.w. 282.46; m.p. 14°C; b.p. 360°C; v.p. 1 mm at 177°C; v.d. 9.74; sp.gr. 0.89 at 18/4°C
B. AIR POLLUTION FACTORS:
—Manmade sources: glcs:
Detroit freeway interchange Oct–Nov 1963: 32.2 µg/1,000 cu m
New York High Traffic city location: 122.8 µg/1,000 cu m (100)
glc's in residential area (Belgium), Oct. 1976: in particulate sample: 2.06 ng/cu m
(1289)
C. WATER POLLUTION FACTORS:
—Oxidation parameters:
BOD_5: 6% ThOD (220)
COD: 2.25 (27)
100% ThOD (220)
$KMnO_4$ value: 0.677 (30)
    acid: 5% ThOD
    alkaline: 11% ThOD (220)
ThOD: 2.89 (27)
—Manmade sources: in sewage effluents: 1.3 µg/l (233)
D. BIOLOGICAL EFFECTS:
—Fish:
goldfish: lethal dose: 8 mg/l (sodium salt) (522)
fathead minnows: static bioassay in Lake Superior water at 18-22°C: LC_{50} (1; 24; 48; 72; 96 hr): >1000; 285; 252; 205; 205 mg/l (350)

omacide-24
Use: fungicide; active ingredient: sodium salt of pyrithione
D. BIOLOGICAL EFFECTS:
—Fish:

	mg/l	24 hr	48 hr	96 hr	3 m (extrap.)
harlequin fish (*Rasbora heteromorpha*)	LC_{10} (F)	0.09	0.068	—	
	LC_{50} (F)	0.13	0.082	0.054	0.008

(331)

opogard
Use: herbicide; active ingredients terbutryne (35%) and terbutrylazine (15%)
D. BIOLOGICAL EFFECTS:
—Fish:

	mg/l	24 hr	48 hr	96 hr	3 m (extrap.)
harlequin fish (*Rasbora heteromorpha*)	LC_{10} (F)	5.6	3.2	3.0	
	LC_{50} (F)	9.5	6.0	5.0	0.3

(331)

ordram *see* molinate

ornithine (2,5-diaminovaleric acid)

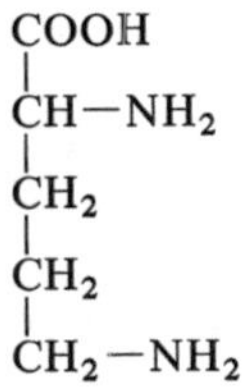

$$\begin{array}{c} COOH \\ | \\ CH-NH_2 \\ | \\ CH_2 \\ | \\ CH_2 \\ | \\ CH_2-NH_2 \end{array}$$

A. PROPERTIES: $\log P_{oct}$ −2.89 (calculated)
C. WATER POLLUTION FACTORS:
  —Manmade sources: excreted by man: 0.15 mg/kg body wt/day

orotic acid (uracil-6-carboxylic acid; 6-carboxy-uracil)

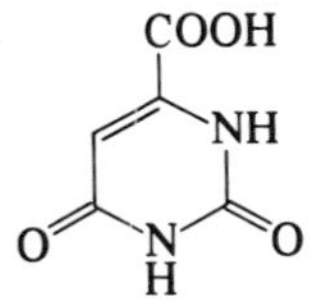

A. PROPERTIES: crystals; m.p. 345–346°C
C. WATER POLLUTION FACTORS:
  —In primary domestic sewage plant effluent: 0.002–0.005 mg/l

7-oxabicyclo-(2,2,1)-heptane-2,3-dicarboxylic acid *see* aquathol K

oxalacetic acid

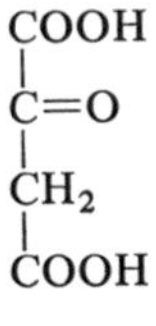

$$\begin{array}{c} COOH \\ | \\ C=O \\ | \\ CH_2 \\ | \\ COOH \end{array}$$

A. PROPERTIES: m.w. 132.07; m.p. 161°C (dec.)
C. WATER POLLUTION FACTORS:
  —Waste water treatment: A.S. after   6 hr: 32.7% of ThOD
                                        12 hr: 40.0% of ThOD
                                        24 hr: 40.7% of ThOD   (89)

oxalate, sodium

$$\begin{array}{c} COONa \\ | \\ COONa \end{array}$$

Use: reagent; textile finishing; pyrotechnics; leather finishing; blueprinting.
A. PROPERTIES: white crystalline powder; sp.gr. 2.34; m.p. 250–270°C (dec.)

D. BIOLOGICAL EFFECTS:
  —Fish: *Lepomis macrochirus*: 24 hr TLm: 4,000 mg/l   (1294)

oxalic acid (ethanedioic acid)
$(COOH)_2 \; 2H_2O$

A. PROPERTIES: colorless crystals; m.w. 126.07; m.p. 101°C; 189°C anh. b.p. 150°C, sublimes; v.p. 0.0003 mm at 30°C; sp.gr. 1.653; solub. 95,000 mg/l at 15°C, 1,290,000 mg/l at 90°C; sat. conc. 0.0015 g/cu m at 30°C log P_{oct} -0.81/-0.43
B. AIR POLLUTION FACTORS: 1 mg/cu m = 0.27 ppm, 1 ppm = 3.74 mg/cu m
C. WATER POLLUTION FACTORS:
  —Oxidation parameters:

BOD_5:	0.100; 0.12; 0.14 std. dil. sewage	(163)(27)(290)
	0.086; 0.16	(39)(36)
BOD_{20}:	0.115	(30)
BOD_5^{20}:	0.16 at 10 mg/l, unadapted sew; no lag period	(554)
BOD_{20}^{20}:	0.16 at 10 mg/l, unadapted sew; no lag period	(554)
COD:	0.18; 0.126; 100% ThOD	(36)(39)(220)
$KMnO_4$:	0.49	
	acid: 98% ThOD; alkaline: 100% ThOD	(220)
TOD:	0.125	(39)
ThOD:	0.18	(36)

  —Manmade sources: excreted by man: in urine: 0.2–0.5 mg/kg body wt/day
                                                                          (203)
    in primary domestic sewage plant effluent: 0.002 mg/l   (517)
  —Waste water treatment: A.S.: after   6 hr: 4.5% of ThOD
                                       12 hr: 0.4% of ThOD
                                       24 hr: toxic   (89)
    coagulation: 86% reduction of BOD with 3 lb alum/1,000 gal   (95)

methods	temp °C	days observed	feed mg/l	days acclim.	% theor. oxidation	% removed
AS, Resp, BOD	20	1–5	720	<1	12	40
AS, CA, BOD	–	84/85	333	<30	–	99
						(93)

    Sampling and analysis: photometry: min. full scale: 23×10^{-6} mole/l   (53)
D. BIOLOGICAL EFFECTS:
  —Toxicity threshold (cell multiplication inhibition test):

bacteria (*Pseudomonas putida*):	1550 mg/l	(1900)
algae (*Microcystis aeruginosa*):	80 mg/l	(329)
green algae (*Scenedesmus quadricauda*):	790 mg/l	(1900)
protozoa (*Entosiphon sulcatum*):	222 mg/l	(1900)

  —Arthropod: *Gammarus pulex*: perturbation level: 25 mg/l
  —Protozoa: *Vorticella campanula*: perturbation level: 50 mg/l
              Paramaecium caudatum: perturbation level: 50 mg/l
  —Worm: *Tubifex tubifex*: perturbation level: 80 mg/l
  —Mollusca: *Limnea ovate*: perturbation level: 60 mg/l
  —Insect: *Sialis flavilatera*: perturbation level: 1,000 mg/l   (30)
  —Fish:

goldfish: period of survival: 0.4–0.5 hr, 1,000 ppm, pH: 2.6
: 4 days, 200 ppm, pH: 5.3 (157)
fishes: LD_0: approx. 20 mg/l (30)

oxirane *see* ethyleneoxide

oxitol ethyleneglycolmonoethylether

oxine 8-hydroxquinoline

oxomethane *see* formaldehyde

4-oxopentanoic acid *see* β-acetylpropionic acid

oxychlordane
D. BIOLOGICAL EFFECTS:
 —Residues:
  residue in Canadian human milk, 1975, average: 1 ng/g whole milk (1376)
  gobyfish (*Acanthogobius flavimanus*) collected at the seashore of Keihinjima along Tokyo Bay—Aug. 1978: residue level: 3 ppb (1721)

oxydemetonmethyl (S-(2-(ethylsulfinyl)ethyl)O,O-dimethylphosphorothioate; O,O-dimethyl-S-2-(ethylsulfinyl)ethylphospohorothioate; metasystox-R; metilmercaptofosoksid; demeton-O-methylsulfoxyd)

$$\begin{array}{c} CH_3 \\ \diagdown \\ \diagup \\ CH_3 \end{array} \overset{O}{\underset{\|}{P}} - S - CH_2 - CH_2 - \overset{O}{\underset{\|}{S}} - CH_2 - CH_3$$

Use: systemic and contact insecticide
A. PROPERTIES: sp.gr. 1.28 at 20/4°C; solub. in all proportions; m.p. <-10°C; b.p. 106°C at 0.01 mm
D. BIOLOGICAL EFFECTS:
 —Crustaceans:
Gammarus lacustris:	96 hr, LC_{50}:	190 µg/l	(2124)
Gammarus fasciatus:	96 hr, LC_{50}:	1000 µg/l	(2126)

 —Insects:
Pteronarcys californica:	96 hr, LC_{50}:	35 µg/l	(2128)

 —Fish:
Lepomis macrochirus:	96 hr, LC_{50}:	14000 µg/l	(2137)
Salmo gairdneri:	96 hr, LC_{50}:	4000 µg/l	(2132)

 —Mammals:
  acute i.p. LD_{50} (rats): 20 mg/kg
  acute oral LD_{50} (rat): 56–65 mg/kg
  dermal LD_{50} (rat): 250 mg/kg (1854)
  in diet: (rats): 20 ppm, 12 weeks, a slight cholinesterase depression (1855)

2,2′-oxydiethanol see diethyleneglycol

1,1-oxydi-2-propanol see dipropyleneglycol

oxydipropionitrile
$CNC_2H_4OC_2H_4CN$

C. WATER POLLUTION FACTORS
—Waste water treatment:

methods	temp °C	days observed	feed mg/l	days acclim.	% theor. oxidation	% removed
RW, COD	20	10	10	95		100
RW, CO_2, CAN	20	17	10	125	60	100
RW, CO_2, CAN	20	18	43	150+	60	100
RW, CO_2, CAN	5	75	43	158+ at 20°C	60	100
ASC, CAN	22–25	28	113	28	75+	99+ (93)

D. BIOLOGICAL EFFECTS:
—Fish: fathead minnows: TLm (96 hr): 3,600–3,920 mg/l   (41)
bluegills: TLm (96 hr): 4,200 mg/l
guppies: TLm (96 hr): 4,450 mg/l   (41)

β-oxynaphthoic acid see 2-hydroxy-3-naphthoic acid

6-oxypurine see hypoxanthine

oxyquinoline see 8-hydroxyquinoline

P

palmitic acid (hexadecanoic acid; *n*-hexadecylic acid)
$CH_3(CH_2)_{14}COOH$
   One of the more common fatty acids. It occurs in natural fats and oils and in tall oil, and in most commercial grade stearic acid.
A. PROPERTIES: colorless needles; m.w. 256.42; m.p. 64°C; b.p. 339/356°C; sp.gr. 0.853 at 62°C
B. AIR POLLUTION FACTORS:
   —Manmade sources: glcs:
   Detroit freeway interchange Oct–Nov 1963: 146.8 µg/1,000 cu m
   New York high traffic location Feb. 1964: 220.0 µg/1,000 cu m (100)
   organic fraction of suspended matter:
     Bolivia at 5200 m altitude (Sept.–Dec. 1975): 3.0–4.1 µg/1000 cu m
     Belgium, residential area (Jan.–April 1976): 27–39 µg/1000 cu m (428)
   glc's in residential area (Belgium) oct. 1976:
     in particulate sample: 29.0 ng/cu m
     in gas phase sample: 4.77 ng/cu m (1289)
   glc's Botrange (Belgium)—woodland at 20–30 km from industrial area: June–July 1977: 10.36; 5.73 ng/cu m ($n = 2$)
   glc's in Wilrijk (Belgium), residential area: Oct.–Dec. 1976: 15.37; 43.69 ng/cu m ($n = 2$) (1233)
C. WATER POLLUTION FACTORS:
   —Oxidation parameters:
   BOD_5:   1.07 st. dil. sewage (163; 282)
           2% ThOD
   BOD_{35}^{25}: 2.14 in seawater/inoculum: enrichment cultures of hydrocarbon oxidizing bacteria (521)
   COD: 28% ThOD
   $KMnO_4$ value: acid: 1% ThOD
               alkaline: 1% ThOD (220)
   ThOD: 2.87 (521)
   —Reduction of amenities: T.O.C. in water: 10 ppm (326)
   —Manmade sources: in sewage effluents: 0.10 mg/l (227)
     in primary domestic sewage plant effluent: 0.006–0.012 mg/l (517)
   —Waste water treatment: A.S. after   6 hr: 0.3% ThOD
                                   12 hr: 1.0% ThOD
                                   24 hr: 2.5% ThOD (89)

D. BIOLOGICAL EFFECTS:
—Fish:
goldfish: lethal dose: 11 mg/l (sodium salt) (522)

palmitin see tripalmitin

palmitylalcohol see 1-hexadecanol

palustric resin acid
D. BIOLOGICAL EFFECTS:
—Fish: Rainbow trout juvenile: 96 hr $LC_{50}(S)$: 0.32 mg/l (1495)

PAN see peroxyacetylnitrate

para-acetaldehyde see paraldehyde

paraldehyde (2,4,6-trimethyl-1,3,5-trioxane; paraacetaldehyde)

$$\begin{array}{c} H_3C \quad O \quad CH_3 \\ O \quad O \\ CH_3 \end{array}$$

A. PROPERTIES: colorless liquid; m.w. 132.3; m.p. 13°C; b.p. 125°C; v.p. 25.3 mm at 20°C; v.d. 4.55; sp.gr. 0.994 at 20/4°C; solub. 120,000 mg/l at 13°C, 58,800 mg/l at 100°C; log P_{oct} 0.59/0.95 (calculated)
B. AIR POLLUTION FACTORS:
—Odor threshold conc. (recognition): 0.02–0.025 mg/cu m (610)
C. WATER POLLUTION FACTORS:
—Waste water treatment: A.C.: adsorbability: 0.148 g/g C; 73.9% reduction, infl.: 1,000 mg/l effl.: 260 mg/l (32)
D. BIOLOGICAL EFFECTS:
—Mammalia: rabbit: acute oral LD_{50}: 5.0 g/kg
rat: acute oral LD_{50}: 1.65 g/kg
dog: acute oral LD_{50}: 3.0–4.0 g/kg (211)

paraquat (1,1'-dimethyl-4,4'-dipyridylium dichloride)

$$CH_3-N\bigcirc\!\!-\!\!\bigcirc N-CH_3^{+2} \;\; 2Cl^-$$

Use: nonselective aquatic and terrestial herbicide, usually present as the dichloride salt or the di(methylsulfate) salt
A. PROPERTIES: m.w. 408
C. WATER POLLUTION FACTORS:
—Biodegradation:
no significant degradation in sterile or nonsterile soil incubated aerobically or anaerobically for 90 days at 25°C in the dark—at 15–25 μg/g of soil (994)

D. BIOLOGICAL EFFECTS:
  —Algae:

	ppb		
Chlorococcum sp.:	$>5 \times 10^6$	50% decrease in O_2 evolution	f
Chlorococcum sp.:	5×10^4	50% decrease in growth	Measured as ABS. (525 mu) after 10 days
Dunaliella tertiolecta:	2.5×10^6	50% decrease in O_2 evolution	f
Dunaliella tertiolecta:	2×10^4	50% decrease in growth	Measured as ABS. (525 mu) after 10 days
Isochrysis galbana:	5×10^6	50% decrease in O_2 evolution	f
Isochrysis galbana:	5×10^3	50% decrease in growth	Measured as ABS. (525 mu) after 10 days
Phaeodactylum tricornutum:	3.5×10^6	50% decrease in O_2 evolution	f
Phaeodactylum tricornutum:	$1. \times 10^4$	50% decrease in growth	Measured as ABS. (525 mu) after 10 days (2348)

  —Crustaceans:
  Gammarus lacustris: 96 hr, LC_{50}: 11000 µg/l   (2124)
  Simocephalus serrulatus: 48 hr, LC_{50}: 4000 µg/l   (2127)
  Daphnia pulex: 48 hr, LC_{50}: 3700 µg/l   (2127)
  —Insect
  Pteronarcys californica: no effect at 100,000 µg/l 96 hr.   (2128)
  —Fish:
  rainbow trout: 48 hr LC_{50}: 62.0 ppm   (1878)
  guppy (*Poecilia mexicana*): 24 hr LC_{50} (S): 12.53 mg/l
  brown trout: 48 hr LC_{50} (S): 82   mg/l
  Rasbora trilineata: 96 hr LC_{50} (S): 6.99 mg/l
  guppy: 96 hr LC_{50} (S): 11.5 mg/l   (1513)
  —Mammals:
  acute oral LD_{50} (rats): 155–203 mg/kg (dichloride)
                   (cats):   35 mg/kg (dichloride)
                   (hens):   362 mg/kg (dichloride)
  acute dermal LD_{50} (rabbits): >663 mg/kg (dichloride)
  acute oral LD_{50} (rats): 320 mg/kg (di(methylsulfate))
  acute dermal LD_{50} (rabbits): > 1050 mg/kg (di(methylsulfate))
  in diet (rats): no effect level after 27 mo. feeding test:
               >234 ppm (dichloride)
               >372 ppm (di(methylsulfate))
      (dogs): no effect level after 2 year feeding test:
               >47 ppm (dichloride)
               >75 ppm (di(methylsulfate)   (1855)

parathion (*o,o*-diethyl-*o-p*-nitrophenylphosphorothioate; ethylparathion; *o,o*-diethyl-*p*-nitrophenylthiophosphate; AATP; "Alkron"; compound 3422; DNTP; DDP; E-605; genithion; niran; paradust; paraflow; paraspray; parawet; penphos; phoskil; thiphos; vapophor)

PARATHION

$$(C_2H_5O)_2P(=S)-O-C_6H_4-NO_2$$

Use: insecticide and acaricide

A. PROPERTIES: m.w. 291.3; m.p. 375°C; sp.gr. 1.26 at 25/4°C; b.p. 157–162°C at 0.6 mm; v.p. 0.003 mm at 24°C; solub. 24 ppm, log P_{oct} 3.81 at 20°C

C. WATER POLLUTION FACTORS:
  —Biodegradation:

$$(C_2H_5O)_2P(=S)-O-C_6H_4-NO_2 \rightarrow (C_2H_5O)_2P(=S)-OH + HO-C_6H_4-NO_2$$

                        diethyl-O-thiophosphoric acid   *p*-nitrophenol

  —Aquatic reactions:
  persistence in river water in a sealed glass jar under sunlight and artificial fluorescent light—initial conc. 10 µg/l.

  % of original compound found

after	1 hr	1 wk	2 wk	4 wk	8 wk	
	100	50	30	<5	0	(1309)

  persistence in soil at 10 ppm initial concentration:

	weeks incubation to	
	50% remaining	5% remaining
sterile sandy loam	>24	
sterile organic soil	>24	
nonsterile sandy loam	<1	3
nonsterile organic soil	1.5	10	(1433)

  75–100% disappearance from soils: 1 week   (1815)
  photooxidation by U.V. in aqueous medium at 90–95°C:
  time for the formation of CO_2 (% of theoretical): 25%: 15.1 hr
                                                          50%: 57.6 hr
                                                          75%: 148.8 hr   (1628)

  —Sediment quality:
  Mississippi: residues not found in bottom sediments   (1170)
  —Odor threshold conc. (detection): 0.04 mg/kg   (915)

D. BIOLOGICAL EFFECTS:
  —Bioaccumulation:
  BCF: fish (*Fundulus heteroclitus*): 80
        mussels: 50   (1925)
  —Residues:
  clam (*Mya arenaria*), Massachusetts: 0.02–0.03 mg/g   (1188)
  —Toxicity:
  Pteronarcys: 48 hr LC_{50}: 5.4 µg/l   (1003)
  —Crustaceans:
  Gammarus lacustris   96 hr LC_{50}: 3.5 µg/l   (2124)

Gammarus fasciatus	96 hr LC_{50}: 2.1 µg/l; 1.6 µg/l (120 hour LC_{50})		
			(2126)
Palaemonetes kadiakensis	96 hr LC_{50}: 1.5 µg/l		(2126)
Simocephalus serrulatus	48 hr LC_{50}: 0.37 µg/l		(2127)
Daphnia pulex	48 hr LC_{50}: 0.60 µg/l		(2127)
Orconectes nais	96 hr LC_{50}: 0.04 µg/l		(2126)
Asellus brevicaudus	96 hr LC_{50}: 600 µg/l		(2126)
Daphnia magna:	24 hr LC_{50}: 0.8 µg/l		
	48 hr LC_{50}: 0.37 µg/l		(1002, 1004)

—Insects:

	96 hr LC_{50} µg/l	other µg/l	
Pteronarcys californica	36	2.2 (30 day LC_{50})	(2117)
Pteronarcys dorsata	3.0	0.90 (30 day LC_{50})	(2134)
Pteronarcella badia	4.2		(2128)
Claassenia sabulosa	1.5		(2128)
Acroneuria pacifica	3.0	0.44 (30 day LC_{50})	(2117)
Acroneuria lycorias		0.013 (30 day LC_{50})	(2134)
Ephemerella subvaria	0.16	0.056 (30 day LC_{50})	(2134)
Ophigomphus rupinsulensis	3.25	0.22 (30 day LC_{50})	(2134)
Hydropsyche bettoni		0.45 (30 day LC_{50})	(2134)

fourth instar larval *Chironomus riparius*: 24 hr LC_{50}: 2.5 ppb   (1853)

—Fish:

Pimephales promelas:	96 hr LC_{50}: 1410 µg/l	(2130)
Lepomis macrochirus:	96 hr LC_{50}: 65 µg/l	(2123)
Lepomis cyanellus:	96 hr LC_{50}: 425 µg/l	(2123)
Micropterus salmoides:	96 hr LC_{50}: 190 µg/l	(2123)

mosquito fish: 72 hr LC_{50} (S): 0.20 mg/l
green sunfish: 72 hr LC_{50} (S): 0.02 mg/l   (1203)

—Mammals:
acute oral LD_{50} (rats): 3.6–13 mg/kg
acute dermal LD_{50} (rats): 6.8–21 mg/kg   (1855)

PBNA   *see* N-phenyl-β-naphthylamine

PBN   *see* peroxybutyrylnitrate

PBzN   *see* peroxybenzoylnitrate

PCB's   *see* polychlorobiphenyls

PCMC   *see* p-chloro-m-cresol

PCN's   *see* polychloronaphthalenes

PCNB   *see* pentachloronitrobenzene

pebulate   (S-propyl-N-butyl-N-ethylthiolcarbamate; Tillam)

$CH_3CH_2CH_2CH_2$
CH_3CH_2 $>N-\overset{O}{\underset{\|}{C}}-S-CH_2CH_2CH_3$

A. PROPERTIES: b.p. 142°C at 20 mm; sp.gr. 0.95 at 30/4°C; solub. 92 ppm at 21°C
D. BIOLOGICAL EFFECTS:
 —Crustacean:
  Gammarus fasciatus: 96 hr LC_{50}: 10 mg/l (2125)
 —Mammals:
  acute oral LD_{50} (rats): 1120 mg/kg (techn. product)
  acute dermal LD_{50} (rabbits): >2936 mg/kg (techn. product) (1855)

PEG's *see* polyethyleneglycols

pelargonaldehyde *see n*-nonanal

pelargonic acid (nonanoic acid, *n*-nonylic acid)
$CH_3(CH_2)_7COOH$
A. PROPERTIES: colorless oily liquid; m.w. 158.24; m.p. 13°C; b.p. 254°C; v.p. 1 mm at 108°C, 10 mm at 137°C, 100 mm at 184°C; sp.gr. 0.907 at 20/4°C
B. AIR POLLUTION FACTORS:
 —Odor: odor threshold values: 0.0055 mg/cu m = 0.84 ppb (307)
  detection: 0.02 mg/cu m (778; 779)
  recognition: 0.0016–0.0032 mg/cu m (610)
 O.I. at 20°C = 164,000 (316)
C. WATER POLLUTION FACTORS:
 —Oxidation parameters:
  BOD_5: 0.59 std. dil. sewage (282)(163)
  $KMnO_4$ value: 0.018 (30)
  ThOD: 2.53 (30)
 —Odor threshold conc. (detection): 3.0 mg/kg (886)
D. BIOLOGICAL EFFECTS:
 —Mammals: rat: oral dose: no deaths up to 3.2 g/kg
  inhalation: no symptoms could be produced (211)

pentabromotoluene

D. BIOLOGICAL EFFECTS:
 —Bioconcentration:
  atlantic salmon: 96 hr BCF (water) (S): 23.7
   42 d BCF (food) (F): 0.035 (1532)

pentachloroaniline

Use: a microbial metabolite of pentachloronitrobenzene. (1807)
A. PROPERTIES: m.w. 265.36; m.p. 232–235°C
 —Residues: in peanut butter samples: avg.: 42 ppb
  range: 3.4–140 ppb (1805)

pentachloroanisole

—Residues: in peanut butter samples: avg.: 9.7 ppb
  range: 1.5–33 ppb ($n = 11$) (1805)

pentachlorobenzene

An impurity of pentachloronitrobenzene; 0.17% has been found in Terrachlor.
(1805)
A. PROPERTIES: m.w. 250.34; m.p. 82–85°C; b.p. 275–277°C; sp.gr. 1.609; solub. 0.24 ppm at 22°C
C. WATER POLLUTION FACTORS:
 —Reduction of amenities: T.O.C. = 0.06 mg/l (225)
D. BIOLOGICAL EFFECTS:
 —Residues:
  in peanut butter samples: avg.: 16 ppb
   range: 1.8–62 ppb ($n = 11$) (1805)
  in fat phase of the following fish from the Frierfjord (Norway), April 1975–Sept. 1976: cod, whiting, plaice, eel, sprot:
   average: 3.87 ppm
   range: 0.4–10 ppm (1065)
 —Bioaccumulation:
  BCF in bacteria (*Siderocapsa treubii*): approx. 16,000 (1833)

BCF in guppies (*Poecilia reticulata*): 260,000 (on lipid content) (1833)
—Toxicity:
guppy (*Poecilia reticulata*): 14 d LC_{50}: 0.178 ppm (1833)

pentachlorobenzylalcohol (PCBalc)

Tradename "Blastin"
Use: pesticide, against rice blast

C. WATER POLLUTION FACTORS:
—Biodegradation:
possible metabolic pathways of pentachlorobenzylalcohol

microbial metabolites identified:
   2,3,4,5,6-pentachlorobenzoic acid
   2,3,4,6-tetrachlorobenzoic acid
   2,3,5,6-tetrachlorobenzoic acid
   2,4,6-trichlorobenzoic acid (1817)
—The metabolites showed high phytotoxicity to cucurbitaceae leguminosae and solanaceae (1817)
—Aquatic reactions:
   PCB alc. is easily and completely photodecomposed (1817)
D. BIOLOGICAL EFFECTS:
—No effect on the growth of bacteria and actinomycetes in soil was observed at concentrations of 100 and 1000 ppm in soil (1819)
—Mammals:
   subacute (90 days) toxicity test to albino rats: at 2000 mg/kg/day, no effect on organs and blood function
   accumulation: rats: PCB alc. is rapidly excreted without accumulation in organs (1823)

2,4,5,2′,5′-pentachlorobiphenyl (see also PCB's)

Manufacturing source: organic chemical industry. (347)
Users and formulation: mfg. electrical insulation; fire resistant heat transfer and hydraulic fluids; high temperature lubricants; elastomers; adhesives; paints, lacquers, varnishes, pigments, and waxes; heat sensitive paper. (347)
Man caused sources (*water and air*): general use of PCB containing materials, e.g., paints, coatings, adhesives, paper, electrical insulation, etc.; lab use. (347)
A. PROPERTIES: solub. 0.010 mg/l at 24°C; log P_{oct} 6.11
C. WATER POLLUTION FACTORS:
—Biodegradation:
   1.2% degradation after 2 hours by *Alcaligenes* Y42 (cell number 2×10^9/ml)
   no degradation after 2 hours by *Acinetobacter* P6 (cell number 4.4×10^8/ml) at 16.3 mg/l initial concentration (1086)
D. BIOLOGICAL EFFECTS:
—Bioaccumulation factor (BCF) and biodegradability index (B.I.) in a laboratory model ecosystem after 33 days at 26°C:

	BCF	B.I.*
alga (*Oedogonium cardiacum*)	5,464	0.029
snail (*Physa*)	59,629	0.027
mosquito (*Culex pipiens quinque fasciatus*)	17,345	0.0134
fish (*Gambusia affinis*)	12,152	0.019

$$^*B.I. = \frac{\text{ppm polar degradation product}}{\text{ppm non polar product}}$$

(1643)

—Toxicity:
amphipod (*Gammarus fasciatus*): 4 days (S) LC_{50}: 0.21 mg/l (1617)

2,2'-4,4',5 pentachlorodiphenylether

D. BIOLOGICAL EFFECTS:
—Fish:
Atlantic salmon: 96 hr BCF (water) (S): 1414
42 d BCF (food) (F): 0.36 (1532)

pentachloroethane (pentalin)
$CHCl_2 CCl_3$
Use: as solvent for oil and grease in metal cleaning
A. PROPERTIES: m.w. 202.3; m.p. -29°C; b.p. 162°C; v.p. 3.4 mm at 20°C, 6 mm at 30°C v.d. 7.2; sp.gr. 1.67 at 25/4; sat. conc. 37 g/cu m at 20°C, 64 g/cu m at 30°C
B. AIR POLLUTION FACTORS: 1 mg/cu m = 0.12 ppm, 1 ppm = 8.41 mg/cu m
—Odor: characteristic: quality: sweetish (211)
C. WATER POLLUTION FACTORS:
—Waste water treatment: evaporation from water at 25°C of 1 ppm solution: 50% after 48 min, 90% after >140 min (313)(369)
D. BIOLOGICAL EFFECTS:
—Fish:
guppy (*Poecilia reticulata*): 7 d LC_{50}: 15 ppm (1833)
—Mammals:
cats: inhalation: changes in liver, lungs and kidneys: 121 ppm, 8-9 hr/day, 23 days
dogs: oral lethal dose: 1.75 g/kg (211)

pentachlorofluorobenzene

C. WATER POLLUTION FACTORS:
—Water quality: in percolation water at 30-500 m from dumping ground: 1-5 ppb (183)

2,3,4,5,6-pentachloromandelonitrile

Use: rice blast control agent

C. WATER POLLUTION FACTORS:
  −Aquatic reactions:
  PCMN was instantly decomposed to pentachlorobenzaldehyde by liberating hydrogen cyanide quantitively, followed by further transformation to the corresponding alcohol and acid

(1817)

pentachloronitrobenzene (PCNB; Quintozene, Terrachlor)

Use: a soil fungicide and seed disinfectant on peanuts (1805)
Residues: residues of PCNB have been found in several peanut butter samples: avg. 4.9 ppb, range: 0.8–10 ppb (n = 11) (1805)
Composition: of Terrachlor: pentachloronitrobenzene: 99.2%
  hexachlorobenzene: 0.46%
  pentachlorobenzene: 0.17%
  tetrachloronitrobenzene: 0.06% (1805)

A. PROPERTIES: m.w. 295.34; m.p. 146°C; v.d. 10.2
B. AIR POLLUTION FACTORS: 1 mg/cu m = 0.08 ppm, 1 ppm = 12.3 mg/cu m
C. WATER POLLUTION FACTORS:
  −Degradation:
  PCNB is fairly rapidly converted to pentachloroaniline by various soil microorganisms and to pentachlorothioanisole by a few filamentous fungi
  (1807, 1808)

D. BIOLOGICAL EFFECTS:
acute oral LD_{50} (rat): >12,000 mg/kg (as aqueous suspension) (1854)
It is concluded that under the conditions of this bioassay PCNB was not carcinogenic in either Osborne-Mendel rats or B6C3FA mice. (1765)

pentachlorophenate, sodium

D. BIOLOGICAL EFFECTS:
—Bioaccumulation:
decapod:	*Penaeus aztecus*:	14–195 µg/l, 96 h:	0.11–0.45 X
decapod:	*Palaemonetes pugio*:	32–515 µg/l, 96 h:	0.5–3.0 X
fish:	*Fundulus similis*:	36–306 µg/l, 96 h:	12–41 X
fish:	*Mugil cephalus*:	26–308 µg/l, 96 h:	6.3–79 X (1675)

—Toxicity:
—Crustaceans:
  decapods:
  Palaemonetes pugio: 96 hr LC_{50} (FT): >515.0 µg/l (1475)
  Palaemonetes pugio
  (24 hr larvae): 96 hr LC_{50} (S): 649 µg/l (1474)
  Penaeus aztecus: 96 hr LC_{50} (FT): >195.0 µg/l (1475)
  grass shrimp (*Palaemonetes pugio Holthuis*): toxicity at different molt stages:

molt stage at beginning of exposure*

	C	D_0	D_3–D_4
LC_{50}, 24 hr	4.2 ppm	5.9 ppm	0.5 ppm
95% conf. lim.	2.8–7.7 ppm	3.6–21.0 ppm	0.4–0.6 ppm
LC_{50}, 48 hr	3.5 ppm	3.6 ppm	0.4 ppm
95% conf. lim.	2.2–6.3 ppm	2.1–7.1 ppm	0.4–0.5 ppm
LC_{50}, 72 hr	3.3 ppm	3.1 ppm	0.4 ppm
95% conf. lim.	2.0–6.0 ppm	1.8–5.1 ppm	0.4–0.5 ppm
LC_{50}, 96 hr	2.6 ppm	2.7 ppm	0.4 ppm
95% conf. lim.	1.4–4.4 ppm	1.4–4.8 ppm	0.4–0.5 ppm (1064)

*: C: intermolt stage; D_0: early premolt stage; D_3–D_4: late premolt stages

—Molluscs:

			effective conc. µg/l	
pelecypod	*Crassostrea virginica*, embryo	48 hr EC_{50}-S	40	(1474)
pelecypod	*Crassostrea virginica*	96 hr EC_{50}-FT	76.5	(1475)
—Fish:	*Lagodon rhomboides* 48 hr prolarval	96 hr LC_{50}-S	38	(1474)
	Lagodon rhomboides	96 hr LC_{50}-FT	53.2	(1475)

Mugil cephalus	96 hr LC_{50}-FT	112.1	(1475)
Fundulus similis	96 hr LC_{50}-FT	>306.0	(1475)

fathead minnow: flow through bioassay:
48 hr TLm: 0.21 mg/l at 15°C (463)
0.37 mg/l at 25°C
96 hr TLm: 0.21 mg/l at 15°C
0.34 mg/l at 25°C
0.285 mg/l at 25°C (464)
0.21 mg/l at 25°C (465)
brook trout: flow through bioassay:
96 hr TLm: 0.135 mg/l at 15°C (464)
219 hr TLm: 0.118 mg/l at 15°C
336 hr TLm: 0.118 mg/l at 15°C
juvenile Chinook salmon: incipient 96 hr $LC_{50}(F)$ = 0.078 mg/l (1677)
rainbow trout: 96 hr LC_{50}: 0.048–0.100 mg/l (sodium salt) (static bioassay)
coho salmon: 96 hr LC_{50}: 0.037–0.096 mg/l (sodium salt) (static bioassay)
sockeye salmon: 96 hr LC_{50}: 0.050–0.130 mg/l (sodium salt) (static bioassay)
(952)
fathead minnow: TLm 24: 0.33 mg/l (249)

pentachlorophenol (PCP; penta)

Manufacturing source: organic chemical industry; pesticide mfg. industry. (347)
Users and formulation: mfg. insecticides, algicides, herbicides, and fungicides; preservation of wood and wood products; mfg. of sodium pentachlorophenate.
(347)
Technical PCP has been reported to contain chlorodiphenylethers, chlorodibenzo-*p*-dioxins, chlorodibenzofurans, and hydroxychlorodiphenylethers; the octachloro-dibenzo-*p*-dioxin content is typically 500–1500 ppm. (1782)
Commercial pentachlorophenol (PCP) contains significant quantities of tetrachlorophenol (TCP). Negative chemical ionization mass spectrometry has been used to examine a commercial PCP formulation and a series of environmental and human samples.
The ratio of PCP to TCP in Dowicide G-ST, a commercial PCP formulation was 2.5 ± 0.1. The ratio of m/z 267 to m/z 229 in a jellyfish, *Mnemiopsis macrydi*, from the Gulf of Mexico was 2.7 ± 0.1; in human semen it was 4.1 ± 0.1; and in human adipose tissue it was 15.5 ± 0.1. PCP in the semen samples was concentrated in the sperm cells by a factor of 9. (1863)
Man caused sources (*water and air*): agricultural runoff; general use of treated wood; lab use. (347)
A. PROPERTIES: white monoclinic, crystalline solid, technical grade dark grey to brown; m.w. 266.35; m.p. 188/191°C; b.p. 310°C decomposes; v.p. 0.00011 mm

at 20°C; v.d. 9.20; sp.gr. 1.978; solub: 5 mg/l at 0°C, 14 mg/l at 20°C, 35 mg/l at 50°C, 85 mg/l at 70°C; log P_{oct} 5.01

B. AIR POLLUTION FACTORS: 1 mg/cu m = 0.09 mm, 1 ppm = 11.1 mg/cu m
 —Ambient air quality:
  organic fraction of suspended matter:
   Bolivia at 5200 m altitude (Sept.–Dec. 1975): 0.25–0.93 µg/1000 cu m
   Belgium, residential area (Jan.–April 1976): 5.7–7.8 µg/1000 cu m (428)
   glc's in residential area (Belgium) Oct. 1976: in particulate phase: 2.43 ng/cu m
   (1289)

C. WATER POLLUTION FACTORS:
 —Biodegradation: decomposition rate in soil suspensions: >72 days for complete disappearance (175)
 —Aquatic reactions:
  photooxidation by U.V. in aqueous medium at 90–95°C:
   time for the formation of CO_2 (% of theoretical): 25%: 31.7 hr
                                                        50%: 66.0 hr
                                                        75%: 180.7 hr (1628)
 —Reduction of amenities:
  threshold odor: 0.857 to 12.0 mg/l (226)
           detection: 1.6 mg/l (998)
  taste threshold conc.: 0.03 mg/l (998)
 —Manmade sources: percolation water at 30–500 m from dumping grounds:
  <1–2 ppb (183)
 —Waste water treatment: degradation by *Pseudomonas*: 200 mg/l at 30°C
           parent: 7% ring disruption in 120 hr
           mutant: 26% ring disruption in 120 hr (152)
  sewage treatment plant (Sept. 1969, Oregon, USA):

		% removal
after grit chamber (influent)	0.91 ppm	—
after primary clarifier	1.10 ppm	0.0
after trickling filter	0.51 ppm	44.0
after chlorination	0.49 ppm	46.2
out of river (effluent)	0.33 ppm	63.7 (1301)

 —levels of PCP in 24 hr composite influent and effluent sample and calculated total output from Corvallis, Eugene and Salem (Oregon, USA) sewage treatment plants:

cities	influent ppb	effluent ppb	% removal	output g/24 hr	output g/10,000 pop./24 hr
Corvallis	1.4	1.0	28.6	20	5.5
Eugene	4.1	3.3	19.5	144	26.0
Salem	4.6	4.4	4.4	267	33.6

(1301)

 —Taylor water treatment plant for city of Corvallis (Oregon, USA, May 1970)

stage sampled in plant	conc. ppb	% removal
Raw Willamette river water (influent)	0.17	—
at flocculation step	0.06	65
after sedimentation	0.04	76
after filtration (final product)	0.06	65 (1301)

- Reaction of excess hypochlorous acid with dilute aqueous solutions at initial pH's of 6.0. and 3.5: (1091)

	(1)	(2)	(3)	(4)	(5)
yields at pH 6.0	30%	3%	20%	27%	
yields at pH 3.5	28%	6%	15%	18%	

at pH 6.0 + (octachloro-1,4-cyclohexadiene) 2.7%

(2) = chloranil
(3) = hexachloro, 5-cyclohexadienone

at pH 3.5 + (octachlorodibenzodioxine) 0.24%

D. BIOLOGICAL EFFECTS:

- Residues:
  in peanut butter samples: average: 28 ppb
  range: 12–64 ppb ($n = 11$) (1805)
- Bioaccumulation:
  fish: goldfish: BCF (at 0.2 ppm) = 475 (1850)
- Toxicity:
  algae: *Chlorella pyrenoidosa*: toxic: 0.001 mg/l (41)
  amphibian:
  Mexican axolotl (3–4 w after hatching): 48 hr LC_{50}: 0.30 mg/l
  clawed toad (3–4 w after hatching): 48 hr LC_{50}: 0.26 mg/l (1823)

LC_{50}: mg/l:	24 hr		96 hr		Threshold	
	mean	SD	mean	SD	mean	SD
goldfish:						
small size	0.267	0.042	0.247	0.025	0.240	0.026
large size	0.250	0.053	0.190	0.020	0.190	0.020

PENTACHLOROTHIOANISOLE

LC_{50}: mg/l:	24 hr		96 hr		Threshold		
	mean	SD	mean	SD	mean	SD	
fathead minnows:							
small	0.240	0.036	0.227	0.029	0.227	0.029	
large	0.213	0.025	0.203	0.012	0.203	0.012	
fathead minnows:							
age: 4 weeks	0.222	0.021	0.198	0.017	0.198	0.017	
7 weeks	0.245	0.039	0.230	0.036	0.230	0.036	
11 weeks	0.232	0.052	0.222	0.039	0.222	0.039	
14 weeks	0.200	0.016	0.190	0.012	0.190	0.012	(338)

Idus idus melanotus: 48 hr LC_0: 0.2 mg/l
48 hr LC_{50}: 0.3 mg/l
48 hr LC_{100}: 0.5 mg/l (998)
Cyprinodon variegatus: 1-day fry: 96 hr LC_{50}-S: 329 µg/l (1474)
6-wk fry: 96 hr LC_{50}-S: 223 µg/l (1474)
Cyprinodon variegatus: 96 hr LC_{50}:-FT: 442 µg/l (1472)
guppy (*Poecilia reticulata*): 24 hr LC_{50}: 0.38 ppm at pH 7.3 (1833)
goldfish: flow through bioassay: 96 hr TLm: 0.22 mg/l at 25°C (465)
120 hr TLm: 0.253 mg/l at 25°C (464)
336 hr TLm: 0.189 mg/l at 25°C (464)
bluegill: flow through bioassay: 30 hr TLm: 0.303 mg/l at 25°C (464)
243 hr TLm: 0.251 mg/l at 25°C (464)
406 hr TLm: 0.188 mg/l at 25°C (464)
fathead minnows: static bioassay in Lake Superior Water at 18–22°C: LC_{50} (1; 24; 48; 72; 96 hr): 8; 0.6; 0.6; 0.6; 0.6 mg/l (350)

	48 hr	96 hr	10 d
trout: flow through bioassay LC_{50}	0.25 mg/l	0.23 mg/l	0.23 mg/l at 15°C
zebrafish flow through bioassay LC_{50}	1.24 mg/l	1.13 mg/l	1.08 mg/l at 25°C

(*Brachydanio rerio*)
flagfish flow through bioassay LC_{50} 1.82 mg/l 1.74 mg/l 1.74 mg/l at 25°C
(*Jordanella floridae*) (479)
freshwater fish (*Channa gachua*): static test: LC_{50}, 24 hr: 0.79 mg/l
(test solutions changed after every 24 hr) 48 hr: 0.56 mg/l
72 hr: 0.43 mg/l
96 hr: 0.39 mg/l (1697)
goldfish: 24 hr LC_{50}: 0.27 ppm
amount found in dead fish at 0.2 ppm: 95 µg/g (1850)
—Mammals:
rabbit: oral LD_{50}: 70–100 mg/kg
rat: oral LD_{50}: 27–78 mg/kg (211)
acute oral LD_{50} (rats): 210 mg/kg
in diet (dogs and rats) 28 weeks at 10 mg/kg/day: no fatalities (1855)
—Man: painful irritation to the eyes and upper respiratory tract at 1.0 mg/cu m (211)

pentachlorothioanisole

A metabolite by filimentous fungi of pentachloronitrobenzene (1808)
Residues: in peanut butter samples: avg.: 23 ppb
range: 2.6–61 ppb ($n = 11$) (1805)

n-pentacosane
$CH_3(CH_2)_{23}CH_3$
A. PROPERTIES: m.p. 53°C; b.p. 402°C; sp.gr. 0.80
B. AIR POLLUTION FACTORS:
—Ambient air quality:
organic fraction of suspended matter:
Bolivia at 5200 m altitude (Sept.–Dec. 1975): 0.48–1.1 µg/1000 cu m
residential area, 10 km south of Antwerp, Belgium (Jan.–April 1976): 2.9–5.3 µg/1000 cu m
glc's in residential area (Belgium) oct. 1976:
in particulate sample: 9.50 ng/cu m
in gas phase sample: 5.74 ng/cu m (1289)
glc's in the average American urban atmosphere—1963:
480 µg/g airborne particulates or
60 ng/g cu m air (1293)
glc's in Botrange (Belgium)—woodland at 20–30 km from industrial area: June–July 1977: 1.75; 2.73 ng/cu m ($n = 2$)
glc's in Wilrijk (Belgium), residential area: Oct.–Dec. 1976: 4.48; 19.95 ng/cu m ($n = 2$) (1233)

n-pentacosanoic acid
$CH_3(CH_2)_{23}COOH$
B. AIR POLLUTION FACTORS:
—Ambient air quality:
glc's in residential area (Belgium), Oct. 1976: in particulate sample: 2.55 ng/cu m (1289)
glc's at Botrange (Belgium)—woodland at 20–30 km from industrial area: June–July 1977: 0.76; 0.91 ng/cu m ($n = 2$) (1233)

n-pentadecane
$CH_3(CH_2)_{13}CH_3$
Use: organic synthesis
A. PROPERTIES: m.w. 212.42; m.p. 9.9°C; b.p. 270°; sp.gr. 0.769
B. AIR POLLUTION FACTORS:
—Ambient air quality:
glc's in residential area (Belgium), Oct. 1976: in gas phase sample: 61.0 ng/cu m (1289)

958 PENTADECANOIC ACID

C. WATER POLLUTION FACTORS:
—Biodegradation:
degradation in seawater by oil-oxidizing microorganisms: 48.8% breakdown after 21 days at 22°C in stoppered bottles containing a 1000 ppm mixture of alkanes, cycloalkanes, and aromatics (1237)
—Water quality:
in Zürich lake: at surface: 16 ppt; at 30 m depth: 4 ppt
in Zürich area: springwater: 2 ppt; groundwater: 2ppt; tapwater: 4 ppt (513)

pentadecanoic acid (pentadecylic acid)
$CH_3(CH_2)_{13}COOH$

A saturated fatty acid normally not found in vegetable fats.
Use: organic synthesis.

B. AIR POLLUTION FACTORS:
—Ambient air quality:
organic fraction of suspended matter:
Bolivia at 5200 m altitude (Sept.–Dec. 1975): 0.51–0.71 μg/1000 cu m
Belgium, residential area (Jan.–April 1976): 1.8–3.0 μg/1000 cu m (428)
glc's in residential area (Belgium), oct. 1976:
in particulate sample: 3.60 ng/cu m
in gas phase sample: 5.35 ng/cu m (1289)
glc's in Botrange (Belgium)—woodland at 20–30 km from industrial area: June–July 1977: 0.41; 0.55 ng/cu m ($n = 2$)
glc's in Wilrijk (Belgium); residential area: Oct.–Dec. 1976: 1.14; 2.34 ng/cu m ($n = 2$) (1233)
—Atmospheric reactions:
first order evaporation constant of n-pentadecane in 3 mm No. 2 fuel oil layer in dark room at windspeed of 21 km/hr at
5°C: 5.61×10^{-6} min^{-1}
10°C: 4.28×10^{-5} min^{-1}
20°C: 5.24×10^{-5} min^{-1}
30°C: 1.14×10^{-4} min^{-1} (438)

C. WATER POLLUTION FACTORS:
—Odor threshold conc. (detection): 10.0 mg/kg (886)

pentadecylic acid *see* pentadecanoic acid

2,3-pentadiene (*sym*-dimethylallene)
$CH_3CHCCHCH_3$
A. PROPERTIES: m.w. 68.11; m.p. −125.6°C; b.p. 48.3°C; sp.gr. 0.695 at 20°C
B. AIR POLLUTION FACTORS:
—Manmade sources:
combustion gases of home heating on gasoil: 10 ppm at $CO_2 = 7\%$
5 g/kg gas oil at $CO_2 = 6\%$
2.8 g/kg gas oil at $CO_2 = 7\%$ (182)

pentaerythritol (pentaerythrite, 2,2-bishydroxymethyl-1,3-propanediol)
$C(CH_2OH)_4$

A. PROPERTIES: m.w.; 136.15; m.p. 260/269°C; b.p. sublimes; v.d. 4.70; solub. 55,600 mg/l at 15°C; log P_{oct} -1.70 (calculated)
B. AIR POLLUTION FACTORS: 1 mg/cu m = 0.18 ppm, 1 ppm = 5.66 mg/cu m
C. WATER POLLUTION FACTORS:
—Oxidation parameters: COD: 1.23 (27)(41)
ThOD: 1.3 (27)
—Waste water treatment:
trickling filter followed by activated sludge: 95-96% BOD removal (41)

methods	temp °C	days observed	feed mg/l	acclimation	% theor. oxidation
A.S., W	20	$\frac{1}{3}$	200	P	4
A.S., W	20	$\frac{1}{3}$	1,200	P	nil

(93)

pentamethylene see cyclopentane

pentamethyleneglycol see 1,5-pentanediol

pentanal see n-valeraldehyde

pentanamide see valeramide

n-pentane
$CH_3(CH_2)_3CH_3$

Manufacturing source: petroleum refining; natural gas recovery. (347)
Users and formulation: specialty chemical mfg.; solvent recovery and extraction; natural gas processing plants; blowing agent for plastic foams; production of olefin, hydrogen, ammonia; fuel; production, artificial ice mfg.; low temperature thermometers, pesticide. (347)
Natural sources (water and air): constitutent in paraffin fraction of petroleum. (347)
Man caused sources (water and air): leakage of natural gas-carrying pipelines; leakage due to domestic use. (347)

A. PROPERTIES: colorless liquid; m.w. 72.15; m.p. -130°C; b.p. 36°C; v.p. 430 mm at 20°C, 620 mm at 30°C; v.d. 2.48; sp.gr. 0.626 at 20/4°C; sat. conc. 1,689 g/cu m, at 20°C, 2,355 g/cu m at 30°C
B. AIR POLLUTION FACTORS: 1 mg/cu m = 0.33 ppm, 1 ppm = 3.00 mg/cu m

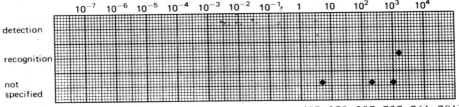

(57; 279; 307; 737; 761; 781)

960 PENTANEDINITRILE

O.I. at $20°C = 570$ (316)
—Atmospheric reactions: reactivity: HC cons.: ranking: <0.01
: NO ox.: ranking: 0.2 (63)
—Manmade sources:
glcs: Pt Barrow, Alaska. Sept 1967: 0.08–3 ppb (101)
glcs downtown Los Angeles 1967: 10% ile: 8 ppb
average: 21 ppb
90% ile: 35 ppb (64)
expected glcs in USA urban air: range: 0.05–0.35 ppm (102)
in exhaust of diesel engine: 2.5% of emitted HC.s (72)
in flue gas of municipal incinerator: <0.6–<0.7 ppm (196)
in gasoline: 2.56–2.71 vol %
evaporation from gasoline fuel tank: 6.9–8.6 vol.% of total evaporated HC's
evaporation from carburetor: 1.5–9.9 vol.% of total evaporated HC's
(398; 399; 400; 401; 402)
in exhaust of gasoline engines: 62-car survey: 2.5 vol.% of total exhaust HC's
(391)

C. WATER POLLUTION FACTORS:
—Biodegradation:
incubation with natural flora in the groundwater—in presence of the other components of high-octane gasoline (100 μl/l); biodegradation: 70% after 192 hr at $13°C$ (initial conc. 0.55 μl/l) (956)
photooxidation by U.V. light in aqueous medium at $50°C$: 31.4% degradation to CO_2 after 24 hr (1628)
—Waste water treatment: A.S. after 6 hr: 0.7% of ThOD
12 hr: 0.5% of ThOD
24 hr: 0.7% of ThOD (88)

D. BIOLOGICAL EFFECTS:
—Fish:
young Coho salmon: no mortalities up to 100 ppm after 96 hrs in artificial sea water at $8°C$ (317)
—Man: no effect level: 5,000 ppm, 10 min (211)

pentanedinitrile *see* glutaronitrile

pentanedioic acid *see* glutaric acid

1,5-pentanediol (pentamethyleneglycol)
$CH_2OH(CH_2)_3CH_2OH$
A. PROPERTIES: thick liquid; m.w. 104.15; b.p. $239.4°C$; spec.gr. 0.994 at $20/20°C$;
log P_{oct} $-0.99/-0.78$ (calculated)
C. WATER POLLUTION FACTORS:
—Oxidation parameters: BOD_{10}: 2.0 std. dil. sew (256)
—Waste water treatment:
NFG, BOD, $20°C$, 1–10 days observed, feed: 200–1,000 mg/l, acclimation: 365 + P: 27% removed (93)

reverse osmosis: 62.0% rejection from a 0.01 M solution (211)

2,4-pentanedione (acetylacetone; acetylmethane)
$CH_3COCH_2COCH_3$
A. PROPERTIES: colorless liquid; m.w. 100.11; m.p. $-23.2°C$; b.p. $139°C$ at 746 mm; sp.gr. 0.976; solub. 125,000 mg/l at $20°C$, 515,000 mg/l at $80°C$; THC 616 kcal/mole; log P_{oct} 1.90/2.25 (calculated)
B. AIR POLLUTION FACTORS:
  —Odor: characteristic: quality: sour rancid
  hedonic tone: unpleasant
  absolute perception limit: 0.010 ppm
  50% recogn.: 0.020 ppm
  100% recogn.: 0.024 ppm
  O.I.: 384.166
C. WATER POLLUTION FACTORS:
  —Oxidation parameters:
  BOD_5: 5.6% of ThOD
  BOD_{10}: 40.0% of ThOD at 2.5 mg/l (405)
  BOD_{15}: 62.8% of ThOD
  BOD_{20}: 69.6
D. BIOLOGICAL EFFECTS:
  —Toxicity threshold (cell multiplication inhibition test):

bacteria (*Pseudomonas putida*):	67 mg/l	(1900)
algae (*Microcystis aeruginosa*):	8.5 mg/l	(329)
green algae (*Scenedesmus quadricauda*):	2.7 mg/l	(1900)
protozoa (*Entosiphon sulcatum*):	11 mg/l	(1900)
protozoa (*Uronema parduczi Chatton-Lwoff*):	5.9 mg/l	(1901)

pentanenitrile *see* valeronitrile

pentanethiol *see* pentylmercaptan

pentanoic acid *see* n-valeric acid

n-pentanol (*n*-butylcarbinol, *prim-n*-amylalcohol)
$C_5H_{11}OH$
  Users and formulation: solvent mfg. of petroleum additives; urea-formaldehyde plastics processing; organic chemicals mfg.; raw material for pharmaceutical preparations. (347)
A. PROPERTIES: colorless liquid; m.w. 88.2; m.p. $-78.9°C$; b.p. $138°C$; v.p. 2.8 mm at $20°C$; v.d. 3.04; sp.gr. 0.824 at 20/20°; solub. 27,000 mg/l at $22°C$; THC 795.7 kcal/mole; LHC 745 kcal/mole; log P_{oct} 1.40
B. AIR POLLUTION FACTORS: 1 mg/cu m = 0.278 ppm, 1 ppm = 3.60 mg/cu m
  —Odor: characteristic: quality: sweet
  hedonic tone: pleasant

962 n-PENTANOL

odor thresholds   mg/cu m

(19; 278; 279; 307; 602; 610; 613; 676; 704; 709; 749; 788; 828)

O.I. at $20°C = 368$

C. WATER POLLUTION FACTORS:
—Oxidation parameters:

BOD_5:	46% ThOD	(274)
	: 0.1265	(30)
	: 1.50; 1.61	(258)(260)
	: 1.59 at 1,000 ppm, Warburg, sew.	(267)
	: 1.10 at 10–20 ppm, std. dil. sew	(269)
BOD_{20}:	0.1732	(30)
COD:	81% ThOD (0.05 n $K_2Cr_2O_7$)	(274)
$KMnO_4$:	0.070	
	0% ThOD (0.01 n $KMnO_4$)	(274)
ThOD:	2.727	

—Waste water treatment: A.C.: adsorbability: 0.155 g/g C; 71.8% reduction, infl.: 1,000 mg/l, effl.: 282 mg/l   (32)

air flotation after chemical addition: 89% removal   (173)

anaerobic lagoon: lb COD/day/1,000 cu ft: infl. mg/l; effl. mg/l

22	315	70
48	315	100 (37)

A.S.: after   6 hr: 16.4% of ThOD
           12 hr: 25.4% of ThOD
           24 hr: 28.0% of ThOD   (88)

reverse osmosis: 40% rejection from 0.01 M solution   (221)

A.S., BOD, $20°C$, 1–5 days observed, feed: 333 mg/l, 30 days acclimation: 84% removed   (93)

D. BIOLOGICAL EFFECTS:
—Toxicity threshold (cell multiplication inhibition test):

bacteria (*Pseudomonas putida*):	220 mg/l	(1900)
algae (*Microcystis aeruginosa*):	17 mg/l	(329)
green algae (*Scenedesmus quadricauda*):	260 mg/l	(1900)
protozoa (*Entosiphon sulcatum*):	17 mg/l	(1900)
protozoa (*Uronema parduczi Chatton-Lwoff*):	144 mg/l	(1901)

—Bacteria: *E. coli*: no effect at 1 g/l
—Algae: *Scenedesmus*: toxic: 280 mg/l
—Arthropod: *Daphnia*: toxic: 440 mg/l   (m30)
—Fish: creek chub: LD_0:   350 mg/l, 24 hr in Detroit river   (243)
                LD_{100}: 500 mg/l, 24 hr in Detroit river   (243)
      rainbow trout: static bioassay: 96 hr TLm: 370–490 mg/l at $10°C$ (451)

2-pentanol *see sec-act-amylalcohol*

3-pentanol
$CH_3-CH_2-CHOH-CH_2-CH_3$
A. PROPERTIES: m.w. 88.15; b.p. 114–115°/749 mm; sp.gr. 0.815; log P_{oct} 1.37 (calculated)
D. BIOLOGICAL EFFECTS:
 —Fish:
  guppy (*Poecilia reticulata*): 7 d LC_{50}: 989 ppm (1833)

***tert*-pentanol** *see* 2-methyl-2-butanol

1-pentanolacetate *see prim*-amylacetate

2-pentanone (methyl-*n*-propylketone, ethylaceton)
$CH_3COC_3H_7$
A. PROPERTIES: colorless liquid; m.w. 86.1; m.p. −78/−83°C; b.p. 102°C; v.p. 12 mm at 20°C, 16 mm at 25°C, 21 mm at 30°C; v.d. 2.96; sp.gr. 0.812 at 15/15°C; solub. 43,000 mg/l; sat. conc. 52 g/cu m at 20°C; 95 g/cu m at 30°C
B. AIR POLLUTION FACTORS: 1 mg/cu m = 0.284 ppm, 1 ppm = 3.58 mg/cu m

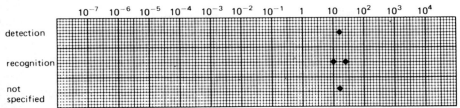

(210; 278; 279; 610; 749)
 O.I.: at 20°C = 2,000 (316)
 —Manmade sources: in exhaust of gasoline engine: <0.1–0.8 ppm (or isopropylketone) (195)
C. WATER POLLUTION FACTORS:
 —Waste water treatment:
  A.S.: after 6 hr: 0.4% of ThOD
        12 hr: 0.8% of ThOD
        24 hr: 1.8% of ThOD (88)
  A.C.: adsorbability: 0.139 g/g C, 69.5% reduction, infl.: 1,000 mg/l, effl.: 305 mg/l (32)
D. BIOLOGICAL EFFECTS:
 —Mammalia: guinea pig: inhalation: no serious disturbance:
            max 5,000 ppm, 1 hr
            max 2,000 ppm, 8 hr (211)

3-pentanone *see* diethylketone

4-pentenal
$CH_2CH(CH_2)_2CHO$

A. PROPERTIES: THC 804 kcal/mole; LHC 757 kcal/mole
C. WATER POLLUTION FACTORS:
  —Waste water treatment:
    RW, Sd, 20°C, 10 days observed, 30 days acclimation: 60% theor. oxidation (93)

1-pentene (α-n-amylene; propylethylene)
$CH_3CH_2CH_2CHCH_2$
  Use: organic synthesis
A. PROPERTIES: colorless liquid; m.w. 70.13; m.p. −138°C; b.p. 30.0°C; v.p. 100 mm at −18°C, 400 mm at 12.8°C, 760 mm at 30.1°C, v.d. 2.4; sp.gr. 0.64 at 20°C
B. AIR POLLUTION FACTORS: 1 mg/cu m = 0.344 ppm, 1 ppm = 2.915 mg/cu m
  —Odor: characteristic: hedonic tone: nauseating at high conc.
          odor threshold values: 0.0062 mg/cu m = 2.1 ppb   (307)
      O.I. = 376,000,000.   (316)
  —Manmade sources: in gasoline: 0.16–0.32 vol %   (312)
      evaporation from gasoline fuel tank: 5.6 vol.% of total evaporated HC's
      evaporation from carburetor: 0.3–0.7 vol.% of total evaporated HC's
                                                              (398; 399; 400; 401; 402)
C. WATER POLLUTION FACTORS:
  —Reduction of amenities: faint odor: 0.0066 mg/l   (129)
  —Waste water treatment: A.S. after 6 hr: toxic
                                     12 hr: 0.8% of ThOD
                                     24 hr: 0.5% of ThOD   (88)

2-pentene (sym-methylethylethylene; β-n-amylene)
$CH_3CH_2CHCHCH_3$
  Use: polymerization inhibitor; organic synthesis
A. PROPERTIES: colorless liquid; m.w. 70.13; m.p. −139°C; b.p. 36.4°C; sp.gr. 0.651; solub 203 mg/l at 20°C   (242)
B. AIR POLLUTION FACTORS:
  —Manmade sources: in gasoline: trans-: 0.41–0.62 vol %
                                 cis-:   0.22–0.39 vol %   (312)
      evaporation from gasoline fuel tank: 1.5–2.9 vol.% (cis + trans) of total evaporated HC's
      evaporation from carburetor: 1.0–2.7 vol.% (cis + trans) of total evaporated HC's
                                                              (398; 399; 400; 401; 402)
C. WATER POLLUTION FACTORS:
  —Waste water treatment: A.S. after 6 hr: 0.3% of ThOD
                                     12 hr: 0.6% of ThOD
                                     24 hr: 0.6% of ThOD   (88)

pentoxane   see 4-methoxy-4-methyl-2-pentanone

pentylamine   see n-amylamine

pentylchloride   see amylchloride

pentylmercaptan   (pentanethiol; amylmercaptan)
$CH_3(CH_2)_4SH$

A. PROPERTIES: m.w. 104.21; m.p. −75.7°C; b.p. 126°C; sp.gr. 0.857°C
B. AIR POLLUTION FACTORS:
  —Odor threshold conc.: 0.0005 mg/cu m   (802)
                              0.0004 mg/cu m   (723)
  —Control methods:
    wet scrubber: water at pH 8.5:   effluent: 100,000 odor units/scf
                : $KMnO_4$ at pH 8.5: effluent: 16 odor units/scf   (115)

pepperminth camphor *see* menthol

perchlorobenzene *see* hexachlorobenzene

perchlorocyclopentadiene *see* hexachlorocyclopentadiene

perchloroethane *see* hexachloroethane

perchloroethylene *see* tetrachloroethylene

perinaphthenone *see* phenalen-1-one

permethrin (3-phenoxybenzyl(+)-*cis*,*trans*-3-(2,2-dichlorovinyl)-2,2-dimethylcyclopropanecarboxylate)

Use: experimental photostable pyrethroid, insecticide   (1577)
A. PROPERTIES: solub. 0.040 ppm at room temp., log P_{oct} 3.48
D. BIOLOGICAL EFFECTS:
  —Fish:
    rainbow trout: 48-h LC_{50}: 6.0 µg/l (S)
    desert pupfish: 48-h LC_{50}: 5.0 µg/l (S)
    mosquitofish: 48-h LC_{50}: 97 µg/l (S)
    Tilapia mossambica: 48-h LC_{50}: 44 µg/l (S)   (1514)
    Salmo salar, exposure: 0.022 mg/l: 96 hr: bioaccumulation: 1.21 µg/g (1144)
    Salmo salar, juvenile lethal threshold-S: 9.00 µg/l   (1144)
    Atlantic salmon, 48 hr LTC-S: 8.8 µg/l   (1196)
    rainbow trout: 24 hr LC_{50}: 135 ppb active ingredient: technical grade
                                 61 ppb active ingredient: formulated product
                                   (static test)   (1577)

peroxyacetylnitrate (PAN)

$$CH_3 \underset{\underset{O}{\|}}{C} ONO_2$$

Photochemical oxidant—indicator of photochemical activity in the atmosphere—produced under influence of sunlight in presence of hydrocarbons and nitrogen oxides.

B. AIR POLLUTION FACTORS:
—Ambient air quality: average values of PAN at Harwell (U.K.)

	monthly avg. ppb	diurnal avg. max (ppb)	min.
Nov. 1974	0.13	0.15	0.10
Dec. 1974	0.06	0.07	0.05
Jan. 1975	0.065	0.07	0.06
Feb. 1975	0.20	0.29	0.13
March 1975	0.13	0.16	0.10
April 1975	0.30	0.44	0.15
May 1975	0.40	0.64	0.23
June 1975	0.44	0.58	0.28
July 1975	0.34	0.72	0.12
Aug. 1975	0.77	1.18	0.43
Sept. 1975	0.19	0.27	0.12
Oct. 1975	0.20	0.34	0.20 (1224)

Individual max. glc's: at Harwell: Aug. '74–Aug. '75: 2.0–5.7 ppb
in London: Aug. '74–Aug. '75: 4.3–16.1 ppb (1224)

Average and maximum concentration of peroxyacetyl nitrate (PAN) recorded at Riverside, California during last 5 months of 1967 and first 4 months of 1968:

month	PAN (ppb) avg. (24 hr)	avg. (a) (10 hr)	max. (month)
August	5.9	7.9	28
September	5.1	6.7	34
October	7.0	7.4	43
November	6.9	8.1	58
December	0.9	1.2	12
January	0.8	1.0	8
February	1.4	1.3	25
March	3.4	3.5	38
April	3.1	4.0	21

(a) Average for the 10-hr period from 8:00 a.m. to 6:00 p.m. is presented because it is the period when maximum concentrations of total oxidants occur.
maximum glc's up to 20 ppb in Delft (The Netherlands) (1232)

—Atmospheric reactions:
deposition velocity: over grass: 0.21 cm s^{-1}
over soil: 0.26 cm s^{-1} (1224)

D. BIOLOGICAL EFFECTS:
—Terrestrial plants: proposed limiting values for foliar injury to vegetation:
0.20 ppm for 0.5 hr exposure
0.10   1.0
0.5   2.0
0.2   80 (1230)

—Relative phytotoxicity on two species of plants:

exposure time (hr)	conc. (ppb)	% injury
Bean (var. Pinto)		
1.0	140	55
4.0	40	90
8.0	20	44

Petunia (var. Rosy morn)
                1.0   140   33   (1262)
—Mammals:
  mice: 2 hr LC_{50}: 106 ppm   (1228)
  rats: 4 hr LC_{50}: 96 ppm   (1227)
  male A-strain mice: exposure to 15 ppm for a period of 6 months resulted in mortality, weight loss, increased lung-heart weights, and severe inflammatory and hyper- and metaplastic epithelial changes in the tracheobronchial tree and alveolar tissue (1224)
  rats: 4 week exposure: no toxic effect level: 0.9 ppm   (1227)
         13 week exposure: 1 ppm: slight irritation of the mucus membranes of the posterior part of the nasal cavity
                           0.2 ppm: no treatment-related changes detected
                                                                 (1227)

peroxybenzoylnitrate (PBzN)

$$\text{C}_6\text{H}_5-\underset{\underset{\text{O}}{\|}}{\text{C}}-\text{ONO}_2$$

Photochemical oxidant—results from the photooxidation of aromatic olefins and benzylic aromatics in the presence of nitrogen oxides.
B. AIR POLLUTION FACTORS:
  —Ambient air quality:
    glc's, sampled over periods of 100 min., varied between 0.2 and 4.6 ppb location Delft (Netherlands)—date 24 June 1976 (1226)
D. BIOLOGICAL EFFECTS:
  —Man:
    slight eye-irritation expected in the range from 5–10 ppb   (1225)

peroxybutyrylnitrate (PBN)
D. BIOLOGICAL EFFECTS:
—Relative phytotoxicity on two species of plants:

	exposure time (hr)	conc. (ppb)	% injury
Bean (var. Pinto)	0.5	100	80
	1.0	30	59
Petunia (var. Rosy morn)	0.5	100	90
	1.0	12	45

                                                                 (1262)

peroxyisobutyrylnitrate (Piso BN)
D. BIOLOGICAL EFFECTS:
  —Relative phytotoxicity on two species of plants:

	exposure time (hr)	conc. (ppb)	% injury
Bean (var. Pinto)	0.5	100	80
Petunia (var. Rosy morn)	0.5	25	18

                                                                 (1262)

peroxypropionylnitrate (PPN)
D. BIOLOGICAL EFFECTS:
—relative phytotoxicity on two species of plants:

	exposure time (hr)	conc. (ppb)	% injury
Bean (var. Pinto)	0.5	100	90
	1.0	24	7
	4.0	10	86
	8.0	5	100
Petunia (var. Rosy morn)	0.5	50	
	1.0	24	(1262)

perylene

A. PROPERTIES: m.w. 252.32; m.p. 277–279°C; b.p. 503°C; solub. 0.0004 ± 0.00002 mg/l at 25°C in distilled water
—Manmade sources:
  in Kuwait crude oil: <0.1 ppm
  in South Louisiana crude oil: 34.8 ppm (1015)
  in high-octane gasoline: 2.07 mg/kg (1220)
  in gasoline: low octane number: 0.01–0.13 mg/kg ($n = 13$)
           high octane number: 0.01–0.16 mg/kg ($n = 13$) (385)
  in gasoline: 0.018 mg/l
  in exhaust condensate of gasoline engine: 0.007–0.014 mg/l gasoline consumed
    (1070)
  in lubricating motoroils: 0.01–0.09 ppm (379)
  in fresh motoroil: 0.03 mg/kg (1220)
  in used motoroil after 5,000 km: 14.3–35.6 mg/kg
                10,000 km: 25.1–57.4 mg/kg (1220)
  in coal tar: 3.51 mg/g of sample (993)
  in bitumen: 0.11–2.29 ppm (506)

B. AIR POLLUTION FACTORS:
—Manmade sources:
  in stack gases of municipal incinerator, after spray tower and electrostatic precipitator: 0.18 mg/1000 cu m; in residues: 82 μg/kg (341)
  in effluent spray tower of stack gases of municipal incinerator: 0.11 μg/l
  in tail gases of gasoline engine: 2–14 μg/cu m (340)
  emission from space-heating installation burning coal (underfeed stoken): 1.6 mg/10^6 Btu input (954)
  in coke oven emissions: 22.1–702.1 μg/g of sample (993)
  emissions from asphalt hot-mixing plant: 5–16 ng/cu m
    (in high volume particulate matter)   avg. 12 ng/cu m (1379)

—Ambient air concentrations:
  in the average American urban atmosphere, 1963:
    5.5 µg/g airborne particulate or
    0.7 ng/g cu m air   (1293)
  glc's in Birkenes (Norway), Jan.–June 1977:
    avg.: 0.06 ng/cu m
    range: n.d.–0.49 ng/cu m ($n = 18$)
  glc's in Rørvik (Sweden), Dec. '76–April '77:
    avg.: 0.06 ng/cu m
    range: n.d.–0.27 ng/cu m ($n = 21$)   (1236)
  glc's in Budapest 1973:
    heavy traffic area: winter: 44.8 ng/cu m
                          summer: 47.8 ng/cu m
    low traffic area: winter: 12.5 ng/cu m
                          summer: 3.9 ng/cu m   (1259)
  glc's at Botrange (Belgium)—woodland at 20–30 km from industrial area: June–July 1977: 0.10; 0.16 ng/cu m ($n = 2$)   (1233)
  Cleveland (Ohio, USA) (1971/1972):
    max. conc.: 9.7 ng/cu m (432 values)
    annual geom. mean (all sites): 0.32 ng/cu m
    annual geom. mean of TSP (all sites): 3 ppm   (556)
  in air in Ontario cities (Canada):

	April–June '75		July–Sept. '75		Oct.–Dec. '75		Jan.–March 1976	
location	ng/1000 cu m air	µg/g p.m.	ng/1000 cu m air	µg/g p.m.	ng/1000 cu m air	µg/g p.m.	ng/1000 cu m air	µg/g p.m.
Toronto	102	1.1	123	1.3	193	2.3	108	1.4
Toronto	99	1.3	57	0.9	136	2.2	51	0.4
Hamilton	141	1.0	283	2.0	403	5.8	347	4.2
S. Sarnia	27	0.4	13	0.3	87	1.7	19	0.7
Sudbury	17	0.5	17	0.4	41	1.8	50	2.1

                                                                                    (999)

C. WATER POLLUTION FACTORS:
  —Manmade sources: in sewage effluent: 0.001 mg/l   (227)
                       in raw sewage: 0.12–3.0 ppb   (545)
  —Water and sediment quality:
    in Thames river water: at Kew Bridge: 40 ng/l
                            at Albert Bridge: 70 ng/l
                            at Tower Bridge: 130 ng/l   (529)
  in sediment of Wilderness Lake, Colin Scott, Ontario (1976): 20 ppb (dry wt)   (932)
—Waste water treatment methods:
  concentration at various stages of water treatment works:
    after reservoir: 0.039 µg/l
    after filtration: 0.024 µg/l
    after chlorination: 0.006 µg/l   (434)

PGE   *see* phenylglycidylether

phaltan *see* folpet

phenacylchloride *see* α-chloroacetophenone

phenalen-1-one (perinaphthenone)

Manmade sources:
in water soluble fraction of a No. 2 fuel oil
in organic extracts of airborne particulates (368)
C. WATER POLLUTION FACTORS:
  —in Eastern Ontario drinking waters (June–Oct. 1978): n.d.–1.1 ng/l ($n = 12$)
  in Eastern Ontario raw waters (June–Oct. 1978): 0.3–2.8 ng/l ($n = 2$) (1698)
D. BIOLOGICAL EFFECTS:
  —Blue green algae: *Agmenellum quadruplicatum* and *Coccochloris elabens*: abrupt toxicity at 5 ppm (independant of wavelength of light)
  —green algae: *Dunaliella tertiolecta* and *Chlorella autotrophica*: no growth at 250 ppb in white light (daylight fluorescent)
  growth affected at 10 ppm (wavelengths restricted to 530 nm and beyond)
  —estuarine diatom: Amphora sp.: no growth at 5 ppm growth rate affected from 0.5 ppm onwards in white light (368)

phenanthrene

Use: dyestuffs; explosives; synthesis of drugs; biochemical research
Manmade sources:
in Kuwait crude oil: 26 ppm
in South Louisiana crude oil: 70 ppm (1015)
in gasoline (high octane number): 20.5 mg/l (380)
in gasoline: 15.7 mg/l
in exhaust condensate of gasoline engine: 2.3–2.9 mg/l gasoline consumed (1070)
in commercial coal tar: 25,000 ppm (phen. + anthracene) (1874)
A. PROPERTIES: colorless leaflets; m.w. 178.22; m.p. 100°C; b.p. 340°C; sp.gr. 1.025; solub. 1.6 mg/l at 15°C, solub. at 8.5°C 0.423 ppm, at 21°C 0.816 ppm, at 30°C 1.277 ppm, in seawater at 22°C 0.6 ± 0.1 ppm; log P_{oct} 4.46
B. AIR POLLUTION FACTORS:
  —Manmade emissions:
    emissions from space-heating installation burning:
      coal (underfeed stoken): 10 mg/10^6 Btu input
      gasoil 3.5 mg/10^6 Btu input (954)

in coke oven emissions: 163.5–2,828.5 µg/g of sample (960)
in coal tar pitch fumes: 36.4 wt % (phenanthrene and/or anthracene) (516)
—Ambient air quality:
  glc's in residential area (Belgium), Oct. 1976:
    in particulate sample: 1.21 ng/cu m
    in gasphase sample: 44.7 ng/cu m (1239)
  glc's in Budapest, 1973 (6–20 hr):
    heavy traffic area: summer: 5.3 ng/cu m
    low traffic area: winter: 1.5 ng/cu m (1259)
  glc's in Birkenes (Norway), Jan–June 1977:
    avg.: 0.74 ng/cu m
    range: n.d.–5.09 ng/cu m ($n = 18$)
  glc's in Rørvik (Sweden), Dec. 1976–April 1977:
    avg.: 0.96 ng/cu m
    range: n.d.–6.00 ng/cu m ($n = 18$) (1236)
—Odor threshold conc.: 0.055–0.06 mg/cu m (610)
C. WATER POLLUTION FACTORS:
  —Manmade sources:
    in leachate from test panels freshly coated with coal tar:
      influent: 0.019 µg/l
      effluent: 0.210 µg/l (1874)
  —Biodegradation:
    degradation of phenanthrene by a pseudomonad:

Phenanthrene → Phenanthrene cis-3,4-dihydrodiol → 3,4-Dihydroxy-phenanthrene → cis-4-(1-Hydroxynaphth-2-yl)-2-oxobut-3-enoic acid ⇌

Further metabolism ← 1,2-Dihydroxynaphthalene ← 1-Hydroxy-2-naphthoic acid ← 1-Hydroxy-2-naphthaldehyde ←[pyruvate] 4-(1-Hydroxynaphth-2-yl)-2-oxo-4-hydroxybutyrate

proposed pathway of phenanthrene metabolism by soil pseudomonas: (543)

(I) → → ↓

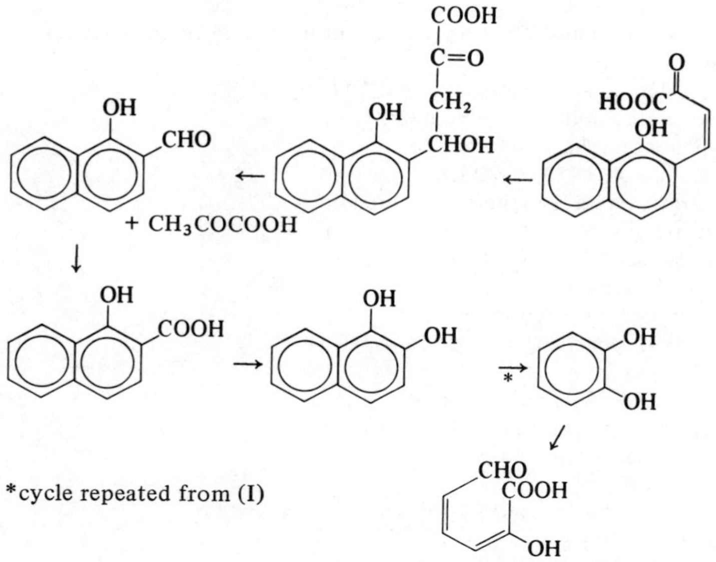

α-hydroxymuconi semialdehyde

*cycle repeated from (I)

—Water quality:
  in Eastern Ontario drinking waters (June–Oct. 1978):
    0.1–4.8 ng/l ($n = 12$) (phen. + anthracene)
  in Eastern Ontario raw waters (June–Oct. 1978):
    5.3–9.8 ng/l ($n = 2$) (phen. + anthracene) (1698)
—Odor threshold conc.: 1 mg/l (301)
—Methods of analysis:
  recovery with open pore polyurethane (OH/NCO = 2.2): 58–92% depending upon load—quantitative elution with methanol (929)

D. BIOLOGICAL EFFECTS:
—Algae:
  inhibition of photosynthesis of a freshwater, non-axenic unialgal culture of *Selenastrum capricornutum* at:
    1% saturation: 96% carbon-14 fixation (vs. controls)
    10%   104%
    100%   76%
—Protozoa: *Vorticella campanula*: perturbation level: 4,000 mg/l
           Paramaecium caudatum: perturbation level: 12,000 mg/l (30)
—Arthropod: *Gammarus pulex*: perturbation level: 8,000 mg/l (30)
—Fish: *Trutta iridea*: perturbation level: 4,500 mg/l (30)
       Neanthes arenaceodentata: 96 hr TLm in seawater at 22°C: 0.6 ppm (initial conc. in static assay) (995)
—Carcinogenicity: negative
—Mutagenicity in the Salmonella test: negative
  <0.25 revertant colonies/nmol
  <70 revertant colonies at 50 µg/plate (1883)

***o*-phenanthroline** (1,10-phenanthroline)

Use: metal chelating agent
A. PROPERTIES: m.w. 180.21; m.p. 114–117°C
C. WATER POLLUTION FACTORS:
   – Impact on biodegradation:
   ~50% inhibition of NH_3 oxidation in *Nitrosomonas* at 9 mg/l   (407)

phenochlor DP-5 (*see also* PCB's)
A polychlorinated biphenyl.
D. BIOLOGICAL EFFECTS:
   – Sediment dwelling polychaete *Nereis diversicolor* accumulated DP-5 from sediments up to a concentration factor of 3.5 (worm/wet sediment)–equilibrium was reached after 40–60 days. A biological half-life of about 27 days was computed   (1388)

phenohep *see* hexachloroethane

phenol (carbolic acid, hydroxybenzene, phenic acid, phenylic acid)
C_6H_5OH
A. PROPERTIES: colorless till brown-black; m.w. 94.11; m.p. 41°C; b.p. 182.0°C; v.p. 0.2 mm at 20°C, 1mm at 40°C; v.d. 3.24; sp.gr. 1.07; solub. 82 g/l at 15°C; sat.conc. 0.77 g/cu m at 20°C, 2.0 g/cu m at 30°C; log P_{oct} 1.46
B. AIR POLLUTION FACTORS: 1 mg/cu m = 0.26 ppm, 1 ppm = 3.92 mg/cu m
   – Odor: characteristic: medicinal, sickening sweet and acrid with a sharp and burning taste

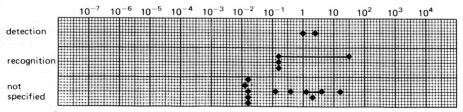

(2; 3; 4; 5; 9; 57; 73; 279; 307; 602; 613; 670; 676; 704; 709; 749; 788; 828)
USSR: human odor perception: non perception: 0.022 mg/cu m
                              perception: 0.184 mg/cu m
      human reflex response: adverse response: 0.015 mg/cu m
      animal chronic exposure: no effect: 0.01 mg/cu m
                              adverse effect: 0.1 mg/cu m   (170)

974 PHENOL

 O.I. at 20°C = 16 (316)
—Manmade sources:
 in exhaust of a 1970 Ford Maverick gasoline engine operated on a chassis dynamometer following the 7-mode California cycle:
  from API #7 gasoline: 1.4 ppm
  from API #8 gasoline: 2.0 ppm (1053)
—Control methods:
 activated carbon: retentivity: 30 wt% of adsorbent (83)
 wet scrubber: water at pH 8.5 = 11 odor units/scf in effluent
     : $KMnO_4$ at pH 8.5 = 1 odor units/scf in effluent (115)
—Sampling and analysis: photometry: min. full scale: 105 ppm (53)

C. WATER POLLUTION FACTORS:
—Oxidation parameters:
 BOD_5: 1.4; 1.66; 1.8 std.dil.sewage, adapted (41)
    0.05 manom, sewage, at 1,000 ppm
    2.0 manom, adapted sewage at 1,000 ppm
    1.88 sierp, adapted sewage at 1,000 ppm
 4 days: 0.54 sierp, sewage at 500 mg/l
    1.84 sierp, adapted sewage at 500 mg/l
 10 days: 2.1 (26)
 BOD_5: 33% ThOD; 90% bio. ox*; 55% bio. ox**
 10 days: 89% bio. ox*; 74% bio. ox**
 15 days: 87% bio. ox. (*); 78%
 20 days: 96% bio. ox. (*); 86% (23)
   *nonacclimated, fresh dilution water
   **nonacclimated, salt dilution water
   bio. ox. = biological oxidation
 BOD_5: 1.68 NEN 3235-5.4 (277)
 COD: 2.33 NEN 3235-5.3 (277)
 COD: 2.28; 3.2; 2.38; 2.30; 2.37 (41)(16)(39)
   100% of ThOD (220)
 $KMnO_4$ value: acid: 84% of ThOD
     alkaline: 80% of ThOD (220)
     9.66 (30)
 TOD: 2.35 (39)
 TOC: 100% of ThOD (220)
 ThOD: 2.26; 2.40 (220)
—Biodegradation:
pathway of breakdown:

$$\text{phenol} \rightarrow \text{catechol} \rightarrow \text{o-quinone} \rightarrow \text{cis,cis-muconic acid} \rightarrow$$

$$\begin{array}{c} CH_2-COOH \\ | \\ CH_2-COOH \end{array} \rightarrow$$

$$CH_3-COOH \rightarrow CO_2 + H_2O \quad (117)$$

—Biodegradation to CO_2 in estuarine water:

PHENOL 975

conc. (μg/l)	month	incubation time (hr)	degradation rate (μg/l/day) $\times 10^3$	turnover time (days)	
5	Jan.	24	270 ± 20	18	
10	Jan.	24	550 ± 10	18	
5	March	24	100 ± 10	25	
10	June	24	580 ± 40	17	(381)

conc. mg/l	BOD at 30°C, in % ThOD*			
	24 hr	2 days	5 days	
25	63	72	76	
220	18	39	57	
2000	0	0	0	

*seed water from a phenol-degradation plant (564)

biological degradation in 7 days tests:
  original culture: average: >99.5%: range: >99.5% degradation
  1st subculture: average: >99.5%: range: 99.5->99.5% degradation
  2nd subculture: average: >99.5%: range: 99.0->99.5% degradation
  3rd subculture: average: >99.5%: range: 99.0->99.5% degradation (87)
decomposition rate in soil suspensions: 2 days for complete disappearance
  (175)
decomposition period by soil microflora in 1 day (176)
adapted culture: 100% after 48 hr incubation feed: 500 mg/l (292)
adapted A.S. at 20°C–product is sole carbon source: 98.5% COD removal at 80.0 mg COD/g dry inoculum/hr (327)

—Impact on biodegradation processes:
inhibition of degradation of glucose by *Pseudomonas fluorescent*: 70 mg/l
inhibition of degradation of glucose by *E. coli*: >1000 mg/l (293)
inhibition of the nitrification process in non adapted activated sludge from 5.6 mg/l upwards (43)

effect on BOD test: BOD days	original sample mg/l	100 ppm phenol added mg/l	
5	5	0	
10	13	2	
15	14	3	
20	15	4	
25	16	7	
30	16	9	(172)

inhibition of cellulose degradation by natural soil populations	at 17 hr	at 200 hr	
500 ppm phenol	72 %	44.0 %	
1000 ppm phenol	97.6 %	56.5 %	
1500 ppm phenol	98.4 %	60.3 %	
2000 ppm phenol	98.6 %	89.1 %	
5000 ppm phenol	98.7 %	99.5 %	(1867)

inhibition of starch degradation by natural soil populations	at 20 hr	at 140 hr	
500 ppm phenol	41 %	40.3 %	
1000 ppm phenol	96 %	52.7 %	
1500 ppm phenol	97.4 %	85.1 %	
5000 ppm phenol	98.4 %	98.4 %	(1867)

inhibition of photosynthesis of a freshwater, non-axenic unialgal culture of
Selenastrum capricornutum:
at 10 mg/l: 105% carbon-14 fixation (vs. controls)
    100 mg/l: 92%
    1000 mg/l: 19% (1690)
—Reduction of amenities:
taste and odor of fish is affected at: 15–25 mg/l (26)
tainting of the flesh of fish and other aquatic organisms:
  phenol: 1–10 mg/l
  phenols in polluted river: 0.02–0.15 mg/l (34)
taste in trout and carp: 25 mg/l; 1.0 mg/l (30)(41)
odor threshold: average: 5.9 ppm
                range: 0.016–16.7 ppm
taste and odor threshold: (tentative): 0.15 mg/l (97)
organoleptic limit: USSR 1970: 0.001 mg/l (181)
T.O.C. in water: 5.9 ppm
                 7.5 ppm (326)
T.O.C. in water: 4.2 ppm (326)
—Manmade sources: excreted by man: in urine: 0.2–6.6 mg/kg body wt/day
                                   in feces: 0–3 mg/kg body wt/day
                                   in sweat: 2–8 mg/100 ml (203)
in primary domestic sewage plant effluent: 0.006–0.012 mg/l (517)
—Waste water treatment:
A.S.: 33% theoretical oxidation of 500 ppm phenol by acclimated activated
  sludge after 12 hr aeration (26)
oxidation by activated sludges acclimated to the following aromatics: 250 mg/l
  influent, 30 min aeration:
  phenol: 39% theor. oxidation
  o-cresol: 34% theor. oxidation
  m-cresol: 37% theor. oxidation
  p-cresol: 20% theor. oxidation
  mandelic acid: 15% theor. oxidation
  anthranilic acid: 20% theor. oxidation (92)

method	temp °C	days observed	feed mg/l	days acclim.	% theor. oxidation	% removed	
RW, CA	20	2	1	nat.	–	100	
RW, CA	4	4	1	–	–	100	
AS, W	20	$\frac{1}{2}$	250	25 +	39	–	
AS, W	20	$\frac{1}{2}$	500	25 +	32	–	
NFG, BOD	20	1–10	40	365 + P	–	42	
NFG, BOD	20	1–10	200	365 + P	–	8	
NFG, BOD	20	1–10	200–1,000	365 + P	–	inhibition	(93)

phenol removal by trickling filtration

influent ppm	effluent ppm	removal ppm	removal %	lb/1,000 cu ft medium
5	0.0	5.0	100	0.31
20	0.0	20.0	100	1.25
25	2.5	22.5	90	1.42

30	0.5	29.5	98.5	1.85
40	3.0	37.0	92.5	2.33
70	8.0	62.0	88.5	3.88
80	8.0	72.0	90.0	4.53
100	30.0	70.0	70.0	4.40
115	40.0	75.0	65.0	4.72
120	60.0	60.0	50.0	3.77
140	80.0	60.0	42.8	3.77
190	132.0	58.0	30.0	3.65
270	224.0	54.0	20.0	3.14

phenol removal by forced draft cooling towers

sample location	flow Mgal/day	conc. mg/l	% removal	
make up	0.840	17.0	—	(26)
blowdown	0.144	0.202	98.8	
make up	1.590	8.6	—	
blowdown	0.875	0.167	98.1	
make up	1.648	12.7	—	
blowdown	0.875	0.261	99.6	
make up	0.865	41.0	—	
blowdown	0.144	0.238	99.4	
make up	1.440	47.8	—	
blowdown	0.720	0.231	99.5	(26)

degradation by *Pseudomonas*: 500 mg/l at 30°C
  parent: 100% ring disruption in 25 hr
  mutant: 100% ring disruption in 8 hr   (152)

A.C.: adsorbability: 0.161 g/g C. 80.6% reduction; incl.: 1,000 mg/l, effl.: 194 mg/l   (32)

A.C.:	Influent	carbon dosage	effluent	% removal	
	1,000 ppm	10X	3 ppm	99+	
	500 ppm	10X	2 ppm	99+	
	100 ppm	10X	1 ppm	99	(192)

ion exchange: adsorption on Amberlite X AD-4 at 25°C
  infl.: 250 ppm: solute adsorbed for zero leakage: 0.78 lb/cu ft
                        for 10 ppm leakage: 0.83 lb/cu ft
    adsorption on amberlite X AD-2: infl.: 0.4 ppm, effl.: 0.22 ppm; 45% retention efficiency
    adsorption on amberlite X AD-7: retention efficiency: 86% infl.: 0.4 ppm, effl.: 0.06 ppm   (40)

solvent extraction

solvent	influent mg/l	effluent mg/l	% phenol removal	remarks
aromatics 75% paraffins 25%	200	0.2	99.9	solvent regenerated with caustic; lab study
aliphatic esters	4,000	60	98.5	regeneration by distill.
benzene	—	—	90–95	regeneration by distill.

benzene	750	34	95.5	caustic regeneration
light cycle oil	>300	±30	90	electrostatic extraction also removes oil from water at the same time
light oil	±3,000	±35	98–99.5	centrifugal extraction caustic regeneration
Tricresylphosphate	3,000	300–150	90–95	steam or caustic regen. very high extraction coefficient (41)

solvent extraction: distribution coefficients for a 2% phenol solution in water

solvent	distribution coefficient	temp. °C	
benzene	2.2	20	
monochlorobenzene	2.0	20	
tricresylphosphate	28.0	20	
diethylether	17.0	20	
diisopropylether	17.0	20	
butanol	19.0	20	
ethylacetate	36.0	22	
isopropylether	45.0	22	
phenosolvan	49.0	22	
phenosolvan	38.0	60	
diphenylxylenylphosphate	26.6	20	(30)

coagulation: NIL% BOD reduction with 3 lb alum/1,000 gal (95)
air flotation after chemical addition: 40% removal (173)
reverse osmosis: 0% rejection from a 0.01 M solution (221)
percent recovery of phenol by iron coprecipitation:

	% phenol recovered by coprecipitation[a]		
$FeCl_3$, M	distilled water	Lake Mendota water	Lake Mary water
0.005	21	60	–
0.01	29	62	80
0.015	32	66	–
0.02	36	68	–
0.025	40	70	–

[a] Initial phenol concentration 1.06×10^{-7} M with a pH of 8.5.
effect of pH on recovery of phenol by iron coprecipitation:

	% phenol recovered by coprecipitation[a]	
pH	Lake Mendota water	Lake Mary water
7.5	57	87
8.0	56	86
8.5	62	80
9.0	63	82
9.5	–	86
10.0	–	87

[a] Initial concentrations: 1.06×10^{-7} M phenol and 1×10^{-2} M $FeCl_3$. (1310)

—W, unadapted A.S.: at 1 mg/l: 100% removal after 3 hr
                            10 mg/l: 100% removal after 3 hr
                            100 mg/l: 20% removal after 6 hr (1639)

—Aquatic reactions:
  photooxidation by U.V. light in aqueous medium at 50°C: 10.96% degradation to CO_2 after 24 hr (1628)
  autoxidation at 25°C: $t_{1/2}$: 286 hr at pH 9.0
  629 hr at pH 7.0 (1908)
—Water quality:
  in Delaware river (U.S.A.): conc. range: winter: 2–4 ppb
  summer: n.d. (1051)
—Sampling and analysis:
  U.V. photometry: min. full scale: 0.7×10^{-6} mole/l (53)
  differential U.V. photometry: min. full scale: 100 ppb (75)
  extraction efficiency of macroreticular resins: sample flow: 20 ml/min; pH 5.7; conc. 10 ppm: X AD-2: 27%
  X AD-7: 45% (370)
  breakthrough capacities with <1 ppm leakage using 3000–3500 ppm phenol:

			adsorption capacity mg/dry g	
			+0.1 mol/l NaCl	+0.1 mol/l Na_2SO_4
4-vinylpyridine-divinylbenzene	(1)	275	263	266
2-vinylpyridine-divinylbenzene	(2)	240	236	249
Amberlite IRA–400	(3)	405	63	42
Amberlite IR–45	(4)	272	0	0
Amberlite XAD–4	(5)	150	155	128
Amberlite XAD–2	(5)	89	58	56
Styrene-divinylbenzene	(6)	0		

pH	2.3	2.7	6.0	7.3	8.5	9.0
4-vinylpyridine-divinylbenzene (1)	266	266	275	255	243	244

(1) pulverized copolymer of 4-vinylpyridine with divinylbenzene, containing 72 mol % 4-vinylpyridine
(2) pulverized copolymer of 2-vinylpyridine with divinylbenzene, containing 71 mol % 2-vinylpyridine
(3) strong base anion exchange resin in the hydroxide form
(4) weak base anion exchange resin in the free base form
(5) porous styrene-divinylbenzene resin with no ion exchange functional group
(6) pulverized copolymer of styrene with divinylbenzene, containing 57 mol % styrene (1877)

D. BIOLOGICAL EFFECTS:
—Toxicity threshold (cell multiplication inhibition test):
  bacteria (*Pseudomonas putida*): 64 mg/l (1900)
  algae (*Microcystis aeruginosa*): 4.6 mg/l (329)
  green algae (*Scenedesmus quadricauda*): 7.5 mg/l (1900)
  protozoa (*Entosiphon sulcatum*): 33 mg/l (1900)
  protozoa (*Uronema parduczi Chatton-Lwoff*): 144 mg/l (1901)
—Bacteria: *E. coli*: >1,000 mg/l
—Algae: *Chlorella pyrenoidosa*: toxic: 233 mg/l; 1.060 mg/l (41)
  Scenedesmus: LD_0: 40 mg/l

	ppb		
Protococcus sp.	3×10^5	.59 O.D. expt/ O.D. control	10 day growth test

Chlorella sp.			3×10^5	.63 O.D. expt/ O.D. control	10 day growth test
Dunaliella euchlora			1×10^5	.51 O.D. expt/ O.D. control	10 day growth test
Phaeodactylum tricornutum			1×10^5	.00 O.D. expt/ O.D. control	10 day growth test
Monochrysis lutheri			1×10^5	.00 O.D. expt/ O.D. control	10 day growth test
Crassostrea virginica	American oyster	Egg	5.825×10^4	TLm	48 hr static lab bioassay
Mercenaria mercenaria	Hard clam	Egg	5.263×10^4	TLm	48 hr static lab bioassay
Mercenaria mercenaria	Hard clam	Larvae	5.5×10^4	TLm	12 day static lab bioassay (2347; 2324)

—Protozoa:
Paramaecium caudatum: perturbation level: 10 mg/l
Vorticella campanula: perturbation level: 3 mg/l (30)
—Ciliate: *Tetrahymena pyriformis*: 24 hr LC_{100}: 6.37 mmole/l (1662)
—Arthropoda: *Daphnia*: LD_0: 16 mg/l
         D. magna: TLm 25–50 hr: 100/100 mg/l
         D. magna young: TLm 25–50 hr: 17/7 mg/l
         D. magna adult: TLm 25–50 hr; 61/21 mg/l
         brine shrimp' TLm 24–48 hr: 157/56 mg/l (23)

Crangon crangon in sea water at 15°C
exposure time; EC_{50} after recovery period in unpolluted sea water

	0 min	1 to 3 hr
1 min	>18,000 ppm	±2,400 ppm
3 min	±56,000 ppm	±1,200 ppm
9 min	±1,000 ppm	±850 ppm (328)

Crangon crangon in sea water at 15°C: exposure time; LC_{50}, mg/l

3 min	±5600
9 min	±1000
27 min	400
1 hr	120
3 hr	80
6 hr	40
24 hr	40
48 hr	30
72 hr	30
96 hr	25 (328)

—Fish: *L. macrochirus*: TLm 25–50 hr: >10/>15 mg/l
     M. latipinna: TLm 25–50 hr: 63/22 mg/l (26)
     arctopsyche grandis: TLm 24–48–96 hr: 61/56/0.001 mg/l
     mosquito fish: TLm 24–48–96 hr: 22.7/22.2/56 mg/l (41)
     bluegill: TLm 24–48–96 hr: 19/19/5.7 mg/l (247)
     fatheads: soft water: TLm 24–48–96 hr: 40.6/40.6/34.3 mg/l
     fatheads: hard water: 24–48–96 hr: 38.6/38.6/32.0

bluegills: soft water: 24-48-96 hr: 25.8/23.9/23.9 mg/l
goldfish: soft water: 24-48-96 hr: 49.9/49.1/44.5 mg/l
guppies: soft water: 24-48-96 hr: 49.9/49.9/39.2 mg/l (158)
rainbow trout: lethal conc.: 5 mg/l, 3 hr (154)
perch: lethal conc.: 9 mg/l, 1 hr
crucian carp: TLm 24 hr: 25 mg/l
roach: TLm 24 hr: 15 mg/l
tench: TLm 24 hr: 17 mg/l
"trout" embryos: TLm 24 hr: 5 mg/l (222)
Carassius auratus (goldfish): TLm 48 hr: 44.5 mg/l
rainbow trout: TLm 18 weeks: 4.0 mg/l (222)
creek chub: LD_0: 10 mg/l in Detroit river water
LD_{100}: 20 mg/l in Detroit river water (243)
blue gill sunfish: TLm 24 hr: 22.7 mg/l in std. ref. water (253)
goldfish: LD_{50}, 24 hr: 46 mg/l - modified ASTM-D 1345 (277)
goldfish: approx. fatal conc.: 28.9 mg/l; 48 hr (226)
Gobius minutus in sea water at 15°C
exposure time; EC_{50} after recovery period in unpolluted sea water

	0 min	1 to 3 hr	24 hr
1 min	>18,000 ppm	1,000 ppm	400 ppm
3 min	>18,000 ppm	400 ppm	400 ppm
9 min	±18,000 ppm		

(328)

Gobius minutus in sea water at 15°C: exposure time; LC_{50}, mg/l

9 min	±18,000
27 min	320
1 hr	85
3 hr	20
6 hr	15
12 hr	13
24 hr	10
48 hr	10
72 hr	9
96 hr	9

(328)

	LC_{50}			
	48 hr mg/l	96 hr mg/l	10 d mg/l	
trout: flow through bioassay:	11.6	11.6	11.6	at 15°C
zebrafish: flow through bioassay: (*Brachydanio rerio*)	30.9	29.0	29.0	at 25°C
flagfish: flow through bioassay: (*Jordanella floridae*)	36.3	36.3	36.3	at 25°C

(479)

fathead minnow:
  flow through bioassay: 24 hr TLm: 8.21 mg/l at 20-25°C
                            96 hr TLm: 5.02 mg/l at 20-25°C (442)
fathead minnow: static bioassay in Lake Superior Water at 18-22°C: LC_{50}
(1; 24; 48; 72; 96 hr): >50; >50; >50; 33; 32 mg/l (350)
Ophicephalus punctatus: 48 hr LC_{50}: 46.0 mg/l (static bioassay) (948)
guppy (*Poecilia reticulata*): 24 hr LC_{50}: 30 ppm at pH 7.3 (1833)

fathead minnow: flow through bioassay:
    48 hr TLm: 41 mg/l at 15°C
                  28 mg/l at 25°C
    96 hr TLm: 36 mg/l at 15°C
                  24 mg/l at 25°C   (463)
goldfish: 24-h LC_{50}: 60–200 mg/l (S)
golden shiner: 24-h LC_{50}: 35–129 mg/l (S)
bluegill: 24-h LC_{50}: 19–160 mg/l (S)
rainbow trout: 24-h LC_{50}: 5.6–11.3 mg/l (S)   (1533)
—Mammalia: rat: single oral LD_{50}: 0.53 g/kg
          rabbit: single oral LD_{50}: 0.4–0.6 g/kg
          cat: single oral LD_{50}: 0.1 g/kg
          dog: single oral LD_{50}: 0.5 g/kg
—Man: oral, ingestion: 1g dose may be lethal   (211)

p-phenolsulfonate, sodium

D. BIOLOGICAL EFFECTS:
  —Crustacean:
  Daphnia magna: 24 hr, TLm: 13,510 mg/l
                  48 hr, TLm: 13,510 mg/l
                  72 hr, TLm: 3,494 mg/l
                  96 hr, TLm: 1,471 mg/l   (153)
  —Molluscs:
  snail eggs, *Lymnaea sp.*: 24 hr, TLm: 10,700 mg/l
                          48 hr, TLm: 9,122 mg/l
                          72 hr, TLm: 8,828 mg/l
                          96 hr, TLm: 8,828 mg/l   (153)
  —Fish:
  Lepomis macrochirus: 100 hr, TLm: 19,616 mg/l   (1295)

1-phenol-4-sulfonic acid   *see* 4-hydroxybenzene sulfonic acid

phenoxyacetic acid

A. PROPERTIES: solub. 12,000 ppm at 10°C; log P_{oct} 1.26 at 20°C

C. WATER POLLUTION FACTORS:

—Waste water treatment methods:

A.C. type BL (Pittsburgh Chem. Co.): % adsorbed by 10 mg A.C. from $10^{-4} M$ aqueous solution, at pH 3.0: 44.0%
7.0: 8.0%
11.0: 4.0% (1313)

phenoxybenzene *see* diphenylether

3-phenyoxybenzyl(+)*cis*,*trans*-3-(2,2-dichlorovinyl)-2,2-dimethylcyclopropanecarboxylate *see* permethrin

phenthoate (O,O-dimethyl-S-(α-(carboethoxy)benzyl)phosphorodithioate; dimethenthoate)

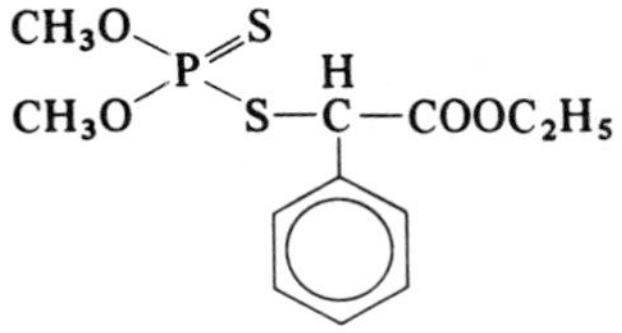

Use: effective broad spectrum organo phosphorus insecticide

A. PROPERTIES: yellowish oil; sp.gr. 1.22; m.p. 17.5°C

C. WATER POLLUTION FACTORS:

—Metabolism in citrus and photodegradation: The products recovered from citrus and glass plates exposed to sunlight were unchanged phenthoate, phenthoate oxon, demethyl phenthoate, mandelic acid, bis(α-(carbo-ethoxy)benzyl)disulfide, O,O-dimethylphosphorothioic and phosphorodithioic acid. Similar products generally were found in citrus leaf and fruit extracts. (1607)

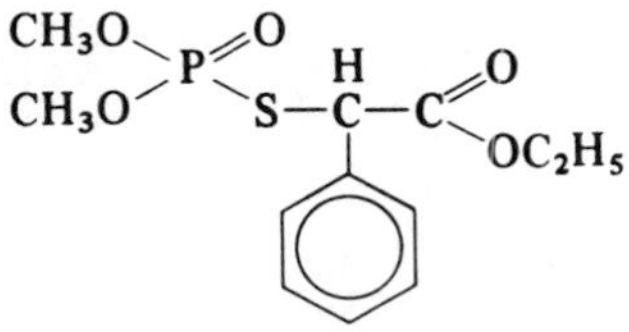

phenthoate oxon

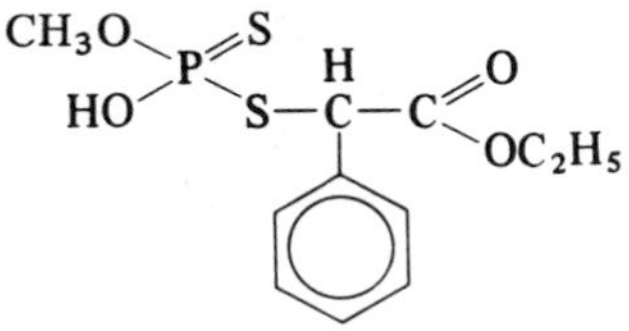

demethylphenthoate

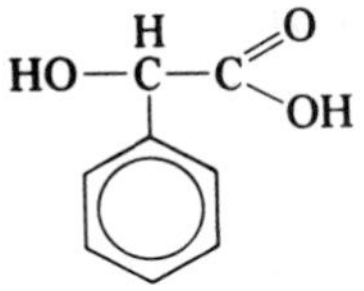

mandelic acid

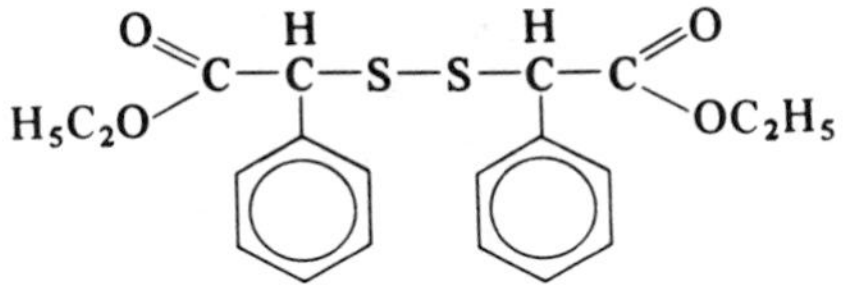

bis[α-(carbo-ethoxy)benzyl] disulfide

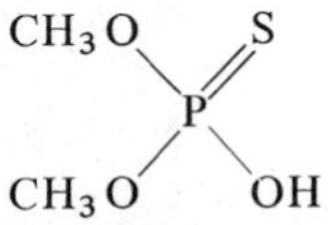

0,0'-dimethylphosphorodithioic acid
metabolites of phenthoate

—Hydrolysis: Phenthoate was fairly stable in phosphate buffered water with a half-life of approximately 12 days at pH 8.0. The major hydrolysis products were phenthoate acid, demethylphenthoate, and demethylphenthoate oxone.

(1607)

D. BIOLOGICAL EFFECTS:
—Insects:
housefly, *Musca domestica L.*: LD_{50}: 5 mg/kg (1607)
—Mammals:
rat: oral LD_{50}: 4728 mg/kg (1608)
acute oral LD_{50} (rat): 300–400 mg/kg (1854)

N-phenylacetamide   *see* acetanilide

phenylacetate

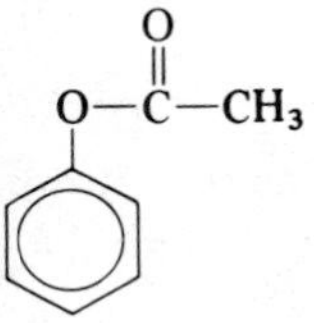

A. PROPERTIES: b.p. 196°C; v.d. 4.7; sp.gr. 1.10; log P_{oct} 1.49
B. AIR POLLUTION FACTORS:
—Odor threshold conc.: detection: 0.2 mg/cu m
recognition: 0.2 mg/cu m (709)
C. WATER POLLUTION FACTORS:
—Waste water treatment:
sel. strain, Sd, W 30°C, 1/2 day observed, 2 μmole/warburg flask, no acclimation:
52% theor. oxidation (93)
D. BIOLOGICAL EFFECTS:
—Toxicity threshold (cell multiplication inhibition test):
bacteria (*Pseudomonas putida*): 115 mg/l (1900)
algae (*Microcystis aeruginosa*): 7.5 mg/l (329)
green algae (*Scenedesmus quadricauda*): 3 mg/l (1900)
protozoa (*Entosiphon sulcatum*): 10 mg/l (1900)
protozoa (*Uronema parduczi Chatton-Lwoff*): 17 mg/l (1901)

phenylacetic acid   (α-toluic acid)

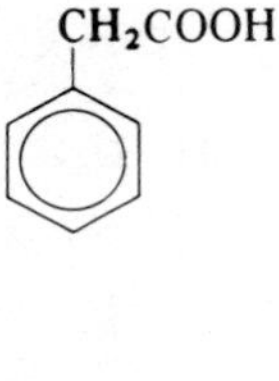

Use: perfume; medicine; mfg of penicillin; fungicide; plant hormones; flavoring; laboratory reagent
A. PROPERTIES: m.w. 136.15; m.p. 77–78.5°C; b.p. 265°C; sp.gr. 1.081; solub. 16,600 ppm at 20°C; log P_{oct} 1.41
C. WATER POLLUTION FACTORS:
  —Manmade sources:
   in primary domestic sewage plant effluent: 0.010 mg/l   (517)
  —Waste water treatment methods:
   A.C. type BL (Pittsburgh Chem. Co.): % adsorbed by 10 mg A.C. from $10^{-4} M$ aqueous solution at pH   3.0: 40.1%
                                    7.0: 12.6%
                                   11.0:   8.5%   (1313)

phenylacetonitrile *see* benzylcyanide

β-phenylacrolein *see* cinnamaldehyde

phenylalanine (2-amino-3-phenylpropionic acid)

$$CH_2-\underset{\underset{NH_2}{|}}{CH}-COOH$$

(phenyl ring attached to CH_2)

Use: biochemical research; dietary supplement; laboratory reagent.
C. WATER POLLUTION FACTORS:
  —Natural sources: in clay soil: 0–500 µg/kg soil   (174)
  —Manmade sources:
   excreted by man: in urine: 0.21–0.6 mg/kg body wt/day
                    in sweat: 1.0–3.5 mg/100 ml   (203)
   in primary domestic sewage plant effluent: 0.050–0.090 mg/l   (517)
  —Waste water treatment:
   A.S.: after   6 hr: 2.3% of ThOD
   (*dl*-phenylal.) 12 hr: 5.6% of ThOD
                    24 hr: 16.4% of ThOD   (89)
   AS, BOD, 20°C, 1–5 days observed, feed: 333 mg/l, 15 days acclimation: 99% removed   (93)
  —Odor threshold conc. (detection): >10 mg/l   (998)

phenylamine *see* aniline

N-phenylaniline *see* N-diphenylamine

N-phenylanthranilic acid

A. PROPERTIES: m.w. 213.24; m.p. 185–188°C
C. WATER POLLUTION FACTORS:
  —Biodegradation: adapted A.S. at 20°C—product is sole carbon source: 28.0% COD removal after 120 hr (327)

p-phenylazoaniline (p-aminoazobenzene; aniline yellow)

Use: dyes; insecticide
A. PROPERTIES: m.w. 197.24; m.p. 123–126°C; b.p. >360°C; log P_{oct} 2.98 (calculated)
C. WATER POLLUTION FACTORS:
  —Impact on biodegradation processes:
    NH_3 oxidation by *Nitrosomonas* sp.: at 100 mg/l: 54% inhibition
                                              50 mg/l: 47% inhibition
                                              10 mg/l: 0% inhibition (390)

phenylazoformic acid 2-phenylhydrazide see diphenylcarbazone

phenylbenzene see diphenyl

phenylbenzoate

Manufacturing source: organic chemical industry. (347)
Users and formulation: plasticizer mfg.; plastics mfg. and processing; mfg. perfume, insecticide. (347)
Man caused sources (*water and air*): general use of plastics and above listed products (leaches from tubings, dishes, paper, etc); lab use, general use of perfumes; insecticides, antiseptics. (347)
A. PROPERTIES: colorless crystals, geranium odor; m.w. 198.2; m.p. 70°C; b.p. 314°C; v.p. 1 mm at 106.8°C; sp.gr. 1.235

phenylbromide see bromobenzene

1-phenylbutane see *n*-butylbenzene

2-phenylbutane see *sec*-butylbenzene

phenylcarbinol *see* benzylalcohol

phenylcarbonimide *see* phenylisocyanate

phenylchloride *see* chlorobenzene

phenylchloroform *see* benzotrichloride

phenylcyanide *see* benzonitrile

N-phenyldiethylamine *see* N,N-diethylaniline

3(phenyl)-1,1-dimethylurea (fenuron; fenidin; PDU)

$$C_6H_5-NH-\overset{O}{\underset{\|}{C}}-N(CH_3)_2$$

Use : weed and brush killer

A. PROPERTIES: m.p. 127–129°C; v.p. 1.6 × 10^{-4} mm at 60°C; solub. 3850 ppm at 25°C

C. WATER POLLUTION FACTORS:
— Aquatic reactions:
persistence in river water in a sealed glass jar under sunlight and artificial fluorescent light - initial conc. 10 µg/l:

after	1 hr	1 wk	2 wk	4 wk	8 wk	
	80	60	20	0	0	(1309)

% of original compound found

— Waste water treatment methods:
powdered A.C.: Freundlich adsorption parameters $K = 0.029$; $1/n = 0.456$
carbon dose to reduce 5 mg/l to a final conc. of 0.1 mg/l: 500 mg/l
carbon dose to reduce 1 mg/l to a final conc. of 0.1 mg/l: 92 mg/l   (594)

D. BIOLOGICAL EFFECTS:
— Algae:

	ppb	method of assessment		
Protococcus sp.	2,900	.33 Opt. Den. Expt/Opt. Den Control	10 day growth test	(2347)
Chlorella sp.	290	.82 Opt. Den. Expt/Opt. Den. Control	10 day growth test	(2347)
Chlorella sp	2,900	.00 Opt. Den. Expt/Opt. Den Control	10 day growth test	(2347)
Chlorococcum sp.	1,000	68% inhibition of growth	10 day growth test	(2349)
Chlorococcum sp.	750	50% decrease in growth	10 day growth test	(2348)
Chlorococcum sp.	2,000	50% decrease in O_2 evolution	f	(2348)
Dunaliella tertiolecta	1,250	50% decrease in O_2 evolution	f	(2348)

Dunaliella tertiolecta	1,500	50% decrease in growth	10 day growth test	(2348)
Dunaliella euchlora	290	.46 Opt. Den. Expt/Opt. Den Control	10 day growth test	(2347)
Isochrysis galbana	1,250	50% decrease O_2 evolution	f	(2348)
Isochrysis galbana	750	50% decrease growth	10 day growth test	(2348)
Monochrysis lutheri	290	.67 Opt. Den. Expt/Opt. Den. Control	10 day growth test	(2347)
Monochrysis lutheri	2,900	.00 Opt. Den Expt/Opt. Den. Control	10 day growth test	(2347)
Phaeodactylum tricornutum	290	.82 Opt. Den. Expt/Opt. Den. Control	10 day growth test	(2347)
Phaeodactylum tricornutum	1,250	50% decrease O_2 evolution	f	(2348)
Phaeodactylum tricornutum	750	50% decrease growth	10 day growth test	(2348)

—Mammals:
acute oral LD_{50} (rat): 6400 mg/kg (1854)
in diet (rats): 90 days at 500 ppm: no apparent effect (1855)

o-phenylenediamine (1,2-diaminobenzene; 1,2-benzenediamine)

A. PROPERTIES: brownish yellow crystals or tablets; m.w. 108.14; m.p.: 102/103°C; b.p. 256/258°C; solub. 41,500 mg/l at 35°C, 7,330,000 mg/l at 81°C; log P_{oct} 0.15

C. WATER POLLUTION FACTORS:
 —Biodegradation: adapted A.S. at 20°C—product is sole carbon source: 33.0% COD removal after 120 hr (photometrical determination) (327)
 —Waste water treatment: decomposition by a soil microflora: in >64 days (176)

m-phenylenediamine (1,3-diaminobenzene; 1,3-benzenediamine)

A. PROPERTIES: colorless needles; m.w. 108.14; m.p. 63/64°C; b.p. 282/284°C; sp.gr. 1.07 at 58/4°C; solub. 351,000 mg/l at 25°C; log P_{oct} 0.00/0.03 (calculated)

C. WATER POLLUTION FACTORS:
 —Biodegradation: decomposition by a soil microflora in >64 days (176)

adapted A.S. at 20°C—product is sole carbon source: 60% removal after 120 hr
(photometrical determination) (327)

p-phenylenediamine (1,4-diaminobenzene; 1,4-benzenediamine)

A. PROPERTIES: colorless crystals; m.w. 108.14; m.p. 140°C; b.p. 267°C sublimes; v.d. 3.73; solub. 38,000 mg/l at 24°C, 6,690,000 mg/l at 107°C
B. AIR POLLUTION FACTORS: 1 mg/cu m = 0.22 ppm, 1 ppm = 4.50 mg/cu m
C. WATER POLLUTION FACTORS:
 —Biodegradation: adapted A.S. at 20°C—product is sole carbon source: 80.0% removal after 120 hr (photometrical determination) (327)
 —Waste water treatment methods:
  ion exchange: adsorption on amberlite XAD-2: infl.: 0.9 ppm effl.: 0.02 ppm; 98% retention efficiency (40)
D. BIOLOGICAL EFFECTS:
 —Fish: goldfish: approx. fatal conc. 5.74 mg/l, 48 hr (226)

o-phenylenepyrene see indeno(1,2,3,cd)pyrene

1-phenyl-1,2-epoxyethane see styrene oxide

phenylethane see ethylbenzene

1-phenylethanol see α-methylbenzylalcohol

phenylether see diphenylether

β-phenylethylamine sulfate
$C_8H_{11}ON \cdot \frac{1}{2}H_2SO_4$
 Use: silver ion complexing agent (1828)
A. PROPERTIES: m.w. 188.22
C. WATER POLLUTION FACTORS:
 —Theoretical:
  TOD = 1.96
  COD = 1.62
  NOD = 0.34
 —Analytical:
  reflux. COD = 113% recovery
  TKN = 104.7% recovery
  BOD_5 = 1.16
  BOD_5/COD = 0.634
  BOD_5 (acclimated) = 1.21 (1828)

unacclimated system at 1,000 mg/l: no effect
acclimated system at 650 mg/l: biodegradable (1828)
The compound proved to have no effect on the unacclimated system. The BOD_5 analysis, however, measured a very significant response. The discrepancy lies in the differences of the seed. The BOD_5 seed source serviced a high industrial input, while the Warburg seed source serviced a domestic waste only. (1828)

D. BIOLOGICAL EFFECTS:
—Algae: *Selenastrum carpicornutum*: 1.0 mg/l: no effect
　　　　　　　　　　　　　　　　　10.0 mg/l: no effect
　　　　　　　　　　　　　　　　　100.0 mg/l: inhibitory (1828)
—Crustacean: *Daphnia magna*: LC_{50} = 2.6 mg/l
—Fish: *Pimephales promelas*: LC_{50} > 100 mg/l (1828)

phenylethylene *see* styrene

phenylglycidylether (2,3-epoxypropoxybenzene; 1,2-epoxy-3-phenoxypropane; 2,3-epoxypropylphenylether; glycidolphenylether; glycidylphenylether; PGE; phenolglycidylether; 1-phenoxy-2,3-epoxypropane; 3-phenoxy-1,2-epoxypropane; 3-phenoxy-1,2-propylene oxide; (phenoxymethyl)oxirane; phenoxypropeneoxide; γ-phenoxypropyleneoxide; phenyl-2,3-epoxypropylether; 3-phenyloxy-1,2-epoxypropane)

$$\text{C}_6\text{H}_5\text{-OCH}_2\text{CHCH}_2\text{O (epoxide ring)}$$

Use: component of epoxy resin systems. The epoxygroup of the glycidylethers reacts during the curing process and glycidylethers are therefore generally no longer present in completely cured products.

A. PROPERTIES: colorless liquid; m.w. 150.17; m.p. 3.5°C; b.p. 245°C v.p. 0.01 mm at 20°C; v.d. 4.37; 5.19; sp.gr. 1.11 at 20/4°C; solub. 2,400 mg/l

C. WATER POLLUTION FACTORS:
—Oxidation parameters: BOD_5: 0.14 NEN 3235-5.4 (277)
　　　　　　　　　　　　COD: 2.18 NEN 3235-5.3 (277)

D. BIOLOGICAL EFFECTS:
—Fish: goldfish: LD_{50} (24, 96 hr): 69/43 mg/l modified ASTM-D-1345 (277)
—Mammalia: mouse: intragastric LD_{50}: 1.40 g/kg
　　　　　　rat: intragastric LD_{50}: 3.85 g/kg
　　　　　　mouse: rat: inhalation, no deaths, sat. vap, 4–8 hr
　　　　　　rat: inhalation: no evidence of toxicity: supersaturated vapor (±15 ppm), 50 × 7 hr (211)
　　rat: inhalation: 6 hr/d for 19 consecutive days: focal degenerative changes involving the seminiferous tubulus in both gonads in: 1 of 8 at 1.75 ppm
　　　　　　　　　　　　　　　　　　　　　　　　　　　　　　　　　　　　　1 of 8 at 5.84 ppm
　　　　　　　　　　　　　　　　　　　　　　　　　　　　　　　　　　　　　3 of 8 at 11.20 ppm
(1409)

N-phenylglycine

COOH
|
CH₂
|
NH
⌬

C. WATER POLLUTION FACTORS:
—Waste water treatment methods:
powdered carbon: at 100 mg/l, carbon dosage 1000 mg/l: 48% adsorbed   (520)

dl-phenylglycolic acid see dl-mandelic acid

phenylisocyanate (phenylcarbonimide, carbanil)

A. PROPERTIES: m.w. 119.12; b.p. 165.6°C; sp.gr. 1.095 at 20/4°C; solub. decomposes
C. WATER POLLUTION FACTORS:
—Biodegradation:
biodegradation by mutant microorganisms: 500 mg/l at 20°C:
    parent: 90% ring disruption in 48 hr
    mutant: 100% ring disruption in 8 hr   (152)

p-(2-phenylisopropyl)phenol see p-cumylphenol

phenylmercaptan see thiophenol

phenylmercury acetate (PMA)

$$\text{C}_6\text{H}_5-\text{HgO}-\overset{\text{O}}{\overset{\|}{\text{C}}}-\text{CH}_3$$

Use: slimicide used by the paper and pulp industry, fungicide.
A. PROPERTIES: m.p. 149–153°C; v.p. 9×10^{-6} mm at 35°C; solub. 4.37 g/l
C. WATER POLLUTION FACTORS:
—Degradation:
soil and aquatic microorganisms: phenylmercuric acetate is quickly degraded, with diphenylmercury as one of the major metabolic products   (1679)
D. BIOLOGICAL EFFECTS:
—guppies: biological half-life for PMA: 7–11 days
                              ^{203}Hg-PMA: 43–58 days
elodea (*Elodea canadensis*):

biological half-life for ^{203}Hg-PMA: 43–58 days
PMA: 7–11 days
snails (*Helisoma campanulata*):
biological half-life for PMA: 10.8 days
coontail (*Ceratophyllum demersum*):
biological half-life for PMA: 43–58 days (1678)
—Mammals:
acute oral LD$_{50}$ of 20% solution: 100 mg/kg (1854)
in diet: as low as 0.5 ppm Hg produced renal lesions (1855)

phenylmethane   *see* toluene

phenylmethanol   *see* benzylalcohol

phenylmethylketone   *see* acetophenone

2-phenylnaphthalene

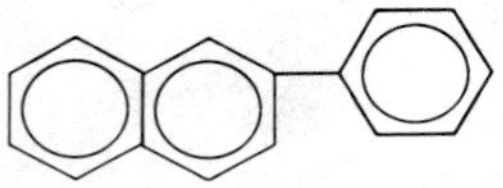

A. PROPERTIES: m.w. 204.27
—Manmade sources:
in gasoline: 0.54 mg/l
in exhaust condensate of gasoline engine: 0.10–0.19 mg/l gasoline consumed
(1070)

N-phenyl-α-naphthylamine   (algerite powder)

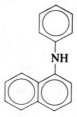

A. PROPERTIES: crystallizes in prisms, white to slightly yellowish
D. BIOLOGICAL EFFECTS:
—Fish: goldfish: approx. fatal conc. 4.4 mg/l (48 hr) (226)

N-phenyl-β-naphthylamine   (PBNA)

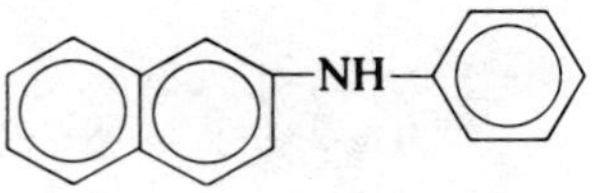

Use: rubber antioxidant; antioxidant for greases and oils; stabilizer during the manufacture of synthetic rubber; intermediate in the synthesis of dyes as well as other antioxidants.
Up to 1% in finished rubber.
Commercial PBNA is contaminated with 20-30 ppm BNA. (348)
A. PROPERTIES: light gray powder, m.p. 107°C; b.p. 395°C; sp.gr. 1.24 m.w. 219.29
D. BIOLOGICAL EFFECTS:
—Metabolic precursor of BNA: PBNA is metabolized to BNA in humans and in dogs (348)

phenylnitrile
C. WATER POLLUTION FACTORS:
—Biodegradation:
biodegradation by mutant microorganisms: 500 mg/l at 20°C
parent: 95% ring disruption in 48 hr
mutant: 100% ring disruption in 10 hr (152)

o-phenylphenate, sodium (tetrahydrate) *see* dowicide

o-phenylphenol (o-hydroxybiphenyl)

Use: intermediate for dyes; germicide; fungicide; rubber chemicals; food packaging.
A. PROPERTIES: needles; m.w. 170.20; m.p. 56°C; b.p. 275°C; v.p. 20 mm at 163°C, 100 mm at 206°C; solub. 0.7 g/l at 25°C; sp.gr. 1.2 at 25/25°C
C. WATER POLLUTION FACTORS:
—Biodegradation:
after 3 weeks of adaption of 10-40 mg/l at 22°C, 100% degradation under aerobic and anaerobic conditions, as sole carbon source or with synthetic sewage (512)
—Water quality:
in Delaware river (U.S.A.): conc. range: winter: 0.3 ppb
(n.s.i.) summer: n.d. (1051)
—Reduction of amenities:
approx. conc. causing adverse taste in fish: 1.0 mg/l (41)
odor threshold. conc. (detection): 0.4 mg/l (998)
—Waste water treatment: coagulation: 0% BOD reduction with 3 lb alum/1,000 gal (95)
D. BIOLOGICAL EFFECTS:
—Mammals:
rat: oral LD_{50}: 2.7 g/kg
dog: repeated oral dose: no effect: 0.5 g/kg for 1 year (211)
ingestion: guinea pigs: single oral LD_{50}: 3500 mg/kg (1546)
white rats: single oral LD_{50}: 2480 mg/kg (1855)
in diet (rats) 2 years at 0.2%: no adverse effects (1855)

1-phenylpropane *see* n-propylbenzene

2-phenylpropane *see* isopropylbenzene

2-phenyl-2-propanol (α,α-dimethylbenzylalcohol)

A. PROPERTIES: m.w. 136.19; m.p. 32–34°C; b.p. 202°C. sp.gr. 0.973
C. WATER POLLUTION FACTORS:
—Water quality:
in Delaware river (U.S.A.): conc. range: winter: 2–3 ppb
summer: n.d. (1051)

3-phenylpropenal *see* cinnamaldehyde

3-phenylpropene *see* allylbenzene

1-phenyl-3-pyrazolidone

Use: reducing agent used primarily in the development of silver sensitized films and papers. (1828)
A. PROPERTIES: m.w. 162.19; m.p. 122–123°C
C. WATER POLLUTION FACTORS:

theoretical	analytical
TOD = 2.29	reflux COD = 105.2% recovery
COD = 1.92	rapid COD = 113.0% recovery
NOD = 0.37	TKN = 129.0 percent recovery
	BOD_5 = 0.14
	BOD_5/COD = 0.069 (1828)

—The high recoveries of COD and TKN in excess of 100% suggest possible error in the assay of the chemical as received, or the presence of reactive impurities. The Warburg analysis measured 0.18 gm/gm oxygen uptake due to initial chemical oxidation. Thus the oxygen demand measured by the BOD_5 test was probably due primarily to chemical action and not biological activity. (1828)
—Biodegradation:
unacclimated system at   100 mg/l: inhibitory
                        1,000 mg/l: inhibitory
acclimated system at   215 mg/l: no effect (1828)

D. BIOLOGICAL EFFECTS:
—Algae: *Selenastrum capricornutum*: 0.1 mg/l: no effect
1.0 mg/l: no effect
10.0 mg/l: no effect (1828)
—Crustacean: *Daphnia magna*: LC_{50} = 10 mg/l
—Fish: *Pimephales promelas*: LC_{50} between 1 and 10 mg/l (1828)

p-phenylsulfonic acid *see* 4-hydroxybenzene sulfonic acid

phloroglucinol (1,3,5-trihydroxybenzene)

A. PROPERTIES: m.w. 126.11; m.p. 222°C anh: 219°C; b.p. sublimes decomposes; solub. 11,350 mg/l at 25°C; log P_{oct} -0.19
C. WATER POLLUTION FACTORS:
—Oxidation parameters:
BOD_5:   0.468
$KMnO_4$ value: 5,260
ThOD: 1.523 (30)
BOD_5:   31% ThOD (274)
COD: 100% ThOD (0.05 n Cr_2O_7) (274)
$KMnO_4$:   87% ThOD (0.01 n $KMnO_4$) (274)
—Biodegradation: adapted A.S. at 20°C—product is sole carbon source: 92.5% COD removal at 22.1 mg COD/g dry inoculum/hr (327)
—Impact on biodegradation processes:
inhibition of degradation of glucose by *Pseudomonas fluorescens* at: 100 mg/l
inhibition of degradation of glucose by *E. coli*: >1,000 mg/l (293)
—Reduction of amenities: tainting of fish: 10.0 mg/l (81)
odor threshold conc. (detection) >10.0 mg/l (998)
D. BIOLOGICAL EFFECTS:
—Bacteria: *E. coli*: LD_0: >1,000 mg/l
—Algae: *Scenedesmus*: LD_0: 200 mg/l
—Arthropoda: *Daphnia*: LD_0: 0.6 mg/l (30)
—Fish: goldfish: approx. fatal conc.: 630 mg/l, 48 hr (226)

phorate (thimet; O,O-diethyl-S-(ethylthiomethyl)phosphorodithioate)

A. PROPERTIES: solub. 50 ppm; b.p. 118–120°C at 0.8 mm; v.p. 8.4 × 10^{-4} mm at 20°C
C. AIR POLLUTION FACTORS:

PHORONE

—In formulating plant: in the worker's breathing zone, concentrations ranged from
0.07 to 14.6 mg/cu m (20 min. avg.) (1689)
D. BIOLOGICAL EFFECTS:
—Crustacean:
Gammarus lacustris: 96 hr LC_{50}: 9 µg/l (2124)
Gammarus fasciatus: 96 hr LC_{50}: 0.60 µg/l (2126)
Orconectes nais: 96 hr LC_{50}: 50 µg/l (2126)
—Mammals:
acute dermal LD_{50} (rats): 2.5–6.2 mg/kg (1855)
acute oral LD_{50} (rat): 1.6–4 mg/kg
dermal toxicity (guinea pig) approx. 630 mg/kg (Thimet 10 G) (1854)
in diet (rats) 90 days: 6 ppm, no effect (92% techn. product) (1855)

phorone (2,6-dimethyl-4-oxo-2,5-heptadiene; 2,5-dimethyl-4-keto-heptadiene; diisopropylideneacetone)
$(CH_3)_2C=CHCOCH=C(CH_3)_2$
A. PROPERTIES: m.p. 27–28°C; b.p. 196°C some decomposition; sp.gr. 0.8714 at 35/38°C
C. WATER POLLUTION FACTORS:
—Oxidation parameters:
BOD_5: 0.19 NEN 3235-5.4 (277)
COD: 2.68 NEN 3235-5.3 (277)
D. BIOLOGICAL EFFECTS:
—Fish: goldfish: LD_{50} (24 hr): 60 mg/l modified ASTM D 1345 (277)

phosgene (chloroformylchloride, carbonylchloride)
$COCl_2$
A. PROPERTIES: colorless gas; m.w. 98.92; m.p. -10/-127°C; b.p. 8.1°C; v.p. 1.6 atm at 20°C, 2.2 atm at 30°C; v.d. 3.42; sp.gr. 1.392 at 19/4°C; solub. decomposes
B. AIR POLLUTION FACTORS: 1 mg/cu m = 0.24 ppm; 1 ppm = 4.11 mg/cu m
—Odor: characteristic: quality: haylike

odor thresholds   mg/kg water

	10^{-7}	10^{-6}	10^{-5}	10^{-4}	10^{-3}	10^{-2}	10^{-1}	1	10	10^2	10^3	10^4
detection							◆◆					
recognition								◆—◆				
not specified								◆ ◆				

(2; 54; 658; 784; 805)
O.I. at 20°C = 1,600,000 (316)
—Man made sources: in flue gas of municipal incinerator: <0.5 ppm (196)
—Sampling and analysis: photometry: min. full. scale: 1,440 ppm (53)
container: GC: det. lim.: 0.04 µg/cu m (208)
C. WATER POLLUTION FACTORS:
—Reduction of amenities: faint odor: 0.0044 mg/l (129)

D. BIOLOGICAL EFFECTS:
 —Mammalia: rabbit: inhalation: mortality: 72%: 100–137 ppm, 30 min
             rabbit: inhalation: mortality: 80–100%: 67 ppm, 30 min
             rat: inhalation: mortality: 100%: 37 ppm, 20 min
             rat: inhalation: mortality: 50%: 25 ppm, 20 min
             rat: inhalation: mortality: 100%: 20 ppm, 30 min
             rat: inhalation: mortality: 75%: 10–15 ppm, 30 min
             rat: inhalation: mortality: 0%: 7.5 ppm, 20 min   (54)
 —Man: severe toxic effects: 5 ppm = 21 mg/cu m, 1 min
       symptoms of illness: 1 ppm = 4.2 mg/cu m,
       unsatisfactory: >0.5 ppm = 2.1 mg/cu m   (185)
       strong smell, irritating: 2 ppm,
       immediate irritation of respiratory tract: 10 ppm   (54)

phosmet see imidan

phosphamidon (2-chloro-2-diethylcarbamoyl-1-methylvinyldimethylphosphate; 2-chloro-N,N-diethyl-3-(dimethyoxyphosphinyloxy)crotonamide; Dimecron)

$$\begin{array}{c} CH_3O \\ \diagdown \\ CH_3O \end{array}\!\!\!\!\overset{O}{\underset{}{\overset{\|}{P}}}\!-\!O\!-\!\underset{Cl}{\overset{CH_3}{\underset{|}{C}}}\!\!=\!\!C\!-\!\overset{O}{\underset{}{\overset{\|}{C}}}\!-\!N\!\!\begin{array}{c} \diagup C_2H_5 \\ \diagdown C_2H_5 \end{array}$$

A. PROPERTIES: b.p. 162°C at 1.5 mm; v.p. 2.5×10^{-5} mm at 20°C; sp.gr. 1.2 at 25/4°C
C. WATER POLLUTION FACTORS:
 —It is hydrolyzed by alkali with $t_{1/2}$ of 13.8 days at pH 7 and 23°C
                                     2.2 days at pH 10   (1855)
D. BIOLOGICAL EFFECTS:

	96 hr LC_{50}	
—Crustaceans:	µg/l	
Gammarus lacustris	2.8	(2124)
Gammarus fasciatus	16	(2126)
Orconectes nais	7500	(2126)
Simocephalus serrulatus	6.6 (48 hr)	(2127)
Daphnia pulex	8.8 (48 hr)	(2127)
—Insects:		
Pteronarcys californica	150	(2128)
—Fish:		
Pimephales promelas	100000	(2137)
Lepomis macrochirus	4500	(2137)
Ictalurus punctatus	70000	(2137)
carp: 24 hr, LC_{50} (S): 177.7 mg/l		
48 hr, LC_{50} (S): 169.3 mg/l		
72 hr, LC_{50} (S): 163.4 mg/l		(1103)
rainbow trout: 24 hr LC_{50}: 5.0 ppm		(1878)

—Mammals:
acute oral LD_{50} (rat): 20–22.4 mg/kg (1854)
acute dermal LD_{50} (rat): 125–530 mg/kg
(rabbits): 267 mg/kg (1855)
in diet (rats) 90 days at 2.5 mg/kg, no effect (1855)

phosphate esters
Commercial phosphate esters are complex mixtures of aryl and alkylarylphosphates that compose oily, viscous liquids with good thermal stability and resistance to oxidation.
They are used as lubricants, oil additives, plasticizers and hydraulic fluids.
Examples of hydraulic fluids are: Houghtosafe and Pydraul; both are tri-arylphosphate hydraulic fluids.
D. BIOLOGICAL EFFECTS:
—Hydralic fluid and species:

	LC_{50}		
	static test		flow through test
	24 hr	96 hr	96 hr
	mg/l	mg/l	mg/l
Houghtosafe 1120			
scud	3.1	0.7	—
rainbow trout	4.2	1.7	0.65
fathead minnow	>90	35	17
channel catfish	130	43	>15
blue gill	32	12	11
Pydraul 50 E			
scud	1.5	0.56	>1.5
rainbow trout	1.3	0.72	0.67
fathead minnow	2.5	1.3	2.1
channel catfish	7.2	3.0	2.4
bluegill	4.4	2.2	2.8
Pydraul 115 E			
rainbow trout	>100	45	
channel catfish	>100	>100	
bluegill	>100	>100	(1603)

phosvel *see* leptophos

photo-aldrin

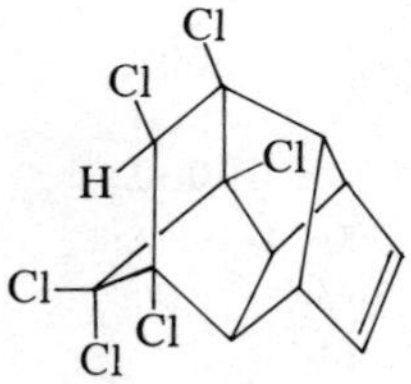

Photodegradation product of aldrin.

D. BIOLOGICAL EFFECTS:
—bluegill: 24 hr LC_{50}: 90 ppb
isopod (*Asellus*): 24 hr LC_{50}: 40 ppb
mosquito (late 3rd instar *Aedes aegypti* larvae): 24 hr LC_{50}: 0.5 ppb
housefly (3 day old female *Musca*) LD_{50}: 6.9 µg/fly   (1681)

photo-*cis*-chlordane
Photodegradation product of chlordane.
D. BIOLOGICAL EFFECTS:
—BCF following exposure to 5 ppb in a static system (time for maximum absorption):
bluegills (*Lepomis macrochirus*): 1180 (24 hr)
goldfish (*Carassius auratus*):   1143 (16 hr)   (1839)
—Elimination half-life after maximum absorption in a static system:
goldfish: 1.1 wk
bluegill: 9.1 wk (extrap.)   (1839)

photochlordene

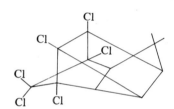

Photodegradation product of chlordene.
D. BIOLOGICAL EFFECTS:
—bluegill: 24 hr LC_{50}: 346 ppb
mosquito (late 3rd instar *Aedes aegypti* larvae): 24 hr LC_{50}: 150 ppb
housefly (3 day old female *Musca*): LD_{50}: 179 µg/fly   (1681)

photodieldrin

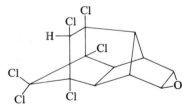

Photodieldrin

A sunlight conversion product of insecticide, dieldrin.
D. BIOLOGICAL EFFECTS:
—Bioconcentration:
aquatic invertebrates with exoskeleton:
clam (*Simpsoniconcha ambigua*): BCF, 6 ppb/96 hr: 4.24

crayfish (*Cambarus*): BCF, 30 ppb/12 hr: 8.13
shrimp (*Palaemonetes*): BCF, 10 ppb/48 hr: 29.7
aquatic invertebrates at lower trophic levels
  Daphnia: BCF, 6 ppb/96 hr: 63,300
  Simocephalus vetulus: BCF, 6 ppb/96 hr: 1,020
  Gammarus: BCF, 10 ppb/96 hr: 1,172
  larvae *Aedes Aegypti*: BCF, 6 ppb/12 hr: 666 (dead after 24 hr)
fresh water fish
  bluegill: BCF, 20 ppb/4d: 78
  minnow: BCF, 20 ppb/6d: 60
  goldfish: BCF, 20 ppb/6d: 325
  guppy: BCF, 20 ppb/6d: 820 (1641)
elimination half-life after maximum absorption in a static system:
  bluegill: 3.2 weeks (1839)
  bluegill: 24 hr LC_{50}: 30 ppb
  minnow: 24 hr LC_{50}: 10 ppb
  mosquito (late 3rd instar *Aedes aegypti* larvae): 24 hr LC_{50}: 3 ppb
  housefly (3 day old female *Musca*): LC_{50}: 8.3 µg/fly (1681)

photoglycine *see p*-hydroxyphenylglycine

photoheptachlor

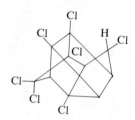

A sunlight degradation product of heptachlor
D. BIOLOGICAL EFFECTS:
  —minnow: 24 hr LC_{50}: 8 ppb
  isopod (*Asellus*): 24 hr LC_{50}: 60 ppb
  mosquito (late 3rd instar *Aedes aegypti* larvae): 24 hr LC_{50}: 2 ppb
  housefly (3 day old female *Musca*): LD_{50}: 5.6 µg/fly (1681)

photoisodrin

A sunlight degradation product of isodrin.

D. BIOLOGICAL EFFECTS:
—bluegill: 24 hr LC_{50}: 25 ppb
minnow: 24 hr LC_{50}: 10 ppb
mosquito (late 3rd instar *Aedes aegypti* larvae):   24 hr LC_{50}: 19 ppb
housefly (3 day old female *Musca*): LD_{50}: 113 µg/fly   (1681)

photomirex   (8-monohydromirex)
Synthetic photomirex contained 4% 2,8-dihydromirex   (1809)
D. BIOLOGICAL EFFECTS:
—Mutagenicity: nonmutagenic in an Ames bacterial test   (1810)
—Toxicity: quails and female rats: 1000 mg/kg as single oral doses to quails or 100 ppm diet to quails and rats caused moderate hepatic changes consisting of an increase in liver weight without indications of severe damage to the livers.
(1809)

phthalandione   *see* o-phthalic anhydride

phthalates
C. WATER POLLUTION FACTORS:
—Biodegradation:
phthalate esters are metabolized by aquatic micro-organisms, several aquatic invertebrates and several species of fish
short chain phthalate esters such as dibutylphthalate are more rapidly metabolized than long chain phthalates such as DEHP, both in aquatic micro-organisms and fish
hydrolysis of phthalate diesters to the respective monoesters appears to be the first and the major biotransformation reaction in all of these species, but subsequent oxidative metabolism also may occur   (1843)

o-phthalic acid   (1,2-benzenedicarboxylic acid)

Use: dyes; medicine; phenolphthalein; phthalimide; anthranilic acid; synthetic perfumes
A. PROPERTIES: m.w. 166.13; m.p. 206/208°C decomposes; b.p. >191°C decomposes; sp.gr. 1.593; solub. 5,400 mg/l at 14°C; 180,000 mg/l at 99°C; THC: 770.2 kcal/mole, LHC 751 kcal/mole; log P_{oct} 0.10/0.41 (calculated)
C. WATER POLLUTION FACTORS:
—Biodegradation: decomposition by a soil microflora in 2 days   (176)
adapted A.S. at 20°C—product is sole carbon source: 96.8% COD removal at 78.4 mg COD/g dry inoculum/hr   (327)
—Oxidation parameters:
BOD_5: 0.85–1.44 std. dil. sewage   (259)(282)(27)(163)(164)
: 1.0   (30)
: 1.4   std. dil. acclimated   (41)

1002 m-PHTHALIC ACID

    COD: 1.37; 1.45   (30)(27)
    $KMnO_4$ value: 0.047   (30)
    ThOD: 1.44   (30)
    —Water quality: self purification affected at 0.5 mg/l   (181)
    —Manmade sources:
        in primary domestic sewage plant effluent: 0.200 mg/l   (517)
    —Sampling and analysis: photometry: min. full scale: $<0.3 \times 10^{-6}$ mole/l   (53)
D. BIOLOGICAL EFFECTS:
    —Mammals: rat: acute oral LD_{50}: 3.2 g/kg (mono K salt)   (211)

m-phthalic acid   *see* isophthalic acid

p-phthalic acid   *see* terephthalic acid

phthalic acid, diethylester   *see* diethylphthalate

o-phthalic anhydride   (phthalandione)

Manufacturing source: plants oxidizing xylenes, naphthalene.   (347)
Users and formulation: plasticizer mfg.; specialty chemical mfg.; mfg. synthetic fibers; mfg. or dyes, pigments, pharmaceuticals, insecticides, chlorinated products.
    (347)
A. PROPERTIES: white flakes or needles; m.w. 148.11; m.p. 130.8°C; b.p. 284.5°C sublimes; v.p. 0.0002 mm at 20°C, 0.001 mm at 30°C; v.d. 5.10; sp.gr. 1.527 at 4°C; sat. conc. 0.0016 g/cu m at 20°C, 0.0078 g/cu m at 30°C
B. AIR POLLUTION FACTORS: 1 mg/cu m = 0.16 ppm, 1 ppm = 6.16 mg/cu m
    —Odor: characteristic: quality: choking
        threshold conc. 0.32 mg/cu m
    —Sampling and analysis: photometry: min. full scale: 29 ppm   (53)
C. WATER POLLUTION FACTORS:
    —Oxidation parameters:
        BOD_5: 0.72; 1.26 std. dil. sewage   (285)(259)(41)
        : 1.2; 1.26 std. dil. sewage
D. BIOLOGICAL EFFECTS:
    —Fish: fatheads: TLm (24–96 hr): not found   (158)
    —Mammalia: rat: single oral LD_{50}: 800–1,600 mg/kg   (211)
    —Man: irritation of the mucous membrane: 25 mg/cu m
        definite conjunctival irritation: 30 mg/cu m   (211)

phthalimide

A. PROPERTIES: m.w. 147.13; m.p. 233-235°C; log P_{oct} 1.15
C. WATER POLLUTION FACTORS:
 —Biodegradation: adapted A.S. at 20°C—product is sole carbon source: 96.2%
   COD removal at 20.8 mg COD/g dry inoculum/hr (327)

phthalonitrile (1,2-dicyanobenzene)

A. PROPERTIES: m.w. 128.13; m.p. 139-141°C
C. WATER POLLUTION FACTORS:
 —Biodegradation:
   biodegradation by mutant microorganisms: 500 mg/l at 20°C
     parent: 68% disruption in 48 hr
     mutant: 100% disruption in 15.5 (152)

phygon (2,3-dichloro-1,4-naphthoquinone; dichlone)

Use: fungicide
A. PROPERTIES: solub. 0.1 ppm at 25°C
D. BIOLOGICAL EFFECTS:
 —Bivalves:

			ppb		method of assessment
Mercenaria mercenaria	hard clam	egg	14	TLm	48 hr static lab bioassay
Mercenaria mercenaria	hard clam	larvae	1.75×10^3	TLm	12 day static lab bioassay
Crassostrea virginica	American oyster	egg	14	TLm	48 hr static lab bioassay
Crassostrea virginica	American oyster	larvae	41	TLm	14 day static lab bioassay (2324)

 —Crustacean: *Daphnia magna*: IC_{50}: 0.014 ppm (1855)
 —Mammals:
   acute oral LD_{50} (rat): 1300 mg/kg (1854)
   in diet (rats) 2 years at 1500 ppm: no ill effects (1855)

picene see benzo(a)chrysene

picloram (4-amino-3,5,6-trichloropicolinic acid; Tordon)

Use: a herbicide used as defoliant in forest warfare
A. PROPERTIES: v.p. 6.16×10^{-7} mm at $35°C$; solub. 430 ppm at $25°C$
C. WATER POLLUTION FACTORS:
 —Persistence:
   75-100% disappearance from soils: 18 months (1815)
D. BIOLOGICAL EFFECTS:
 —Crustacean: *Gammarus lacustris*: 96 hr LC_{50}: 27,000 µg/l (2124)
 —Insect: *Pteronarcys californica*: 96 hr LC_{50}: 48,000 µg/l (2128)
 —Fish:
   cutthroat trout: static bioassay: 96 hr TLm: 3.45-8.6 mg/l at $10°C$
   lake trout: static bioassay: 96 hr TLm: 1.55-4.95 mg/l at $10°C$ (452)
 —Mammals:
   acute oral LD_{50} (rats): 8200 mg/kg
                    (mice): 2000-4000 mg/kg
                    (rabbits): 2000 mg/kg
                    (guinea pigs): about 3000 mg/kg
   acute dermal LD_{50} (rabbits): >4000 mg/kg (1855)

α-picoline (2-picoline; 2-methylpyridine)

A. PROPERTIES: colorless liquid; m.w. 93.12; m.p. $-70°C$; b.p. $129°C$; v.p. 8 mm at $20°C$; v.d. 3.21; sp.gr. 0.95 at $15/4°C$; log P_{oct} 1.06 (calculated)
B. AIR POLLUTION FACTORS:
 —Odor: characteristic: quality: sweet
                       hedonic tone: pleasant
   absolute perception limit: 0.014 ppm
   50% recogn.: 0.023 ppm
   100% recogn.: 0.046 ppm
 O.I.: 114,347
C. WATER POLLUTION FACTORS:
 —Oxidation parameters: $KMnO_4$ value: 0.224
                        ThOD: 2.75 (30)
D. BIOLOGICAL EFFECTS:
 —Protozoa:
   ciliate (*Tetrahymena pyriformis*): 24 hr LC_{100}: 64.4 mmole/l (1662)

β-picoline (3-methylpyridine; 3-picoline)

A. PROPERTIES: odorless liquid; m.w. 93.12; b.p. 143.5°C; sp.gr. 0.9613 at 15/4°C; log P_{oct} 1.20
C. WATER POLLUTION FACTORS:
—Oxidation parameters:
COD: 4% ThOD (0.05 n $K_2Cr_2O_7$) (274)
$KMnO_4$: 2% ThOD (0.01 n $KMnO_4$) (274)
ThOD: 2.75 (274)
—Waste water treatment: RW, CA, 20°C, 2 days observed, feed: 1 mg/l, 16 days acclimation: 100% removed (93)

picramic acid (2-amino-4,6-dinitrophenol)

$$O_2N - \underset{NO_2}{\underset{|}{C_6H_2}} - \underset{OH}{\overset{|}{}} - NH_2$$

Manufacturing source: organic chemical industry (347)
Users and formulation: mfg. red hair preparations; pesticides mfg.; mfg. azo dyes, indicators; reagent for albumin. (347)
A. PROPERTIES: red crystals; m.w. 199.12; m.p. 168°C; solub. 1,400 mg/l at 22°C
C. WATER POLLUTION FACTORS:
—Biodegradation: adapted culture: 2% removal after 48% incubation, feed: 200 mg/l (292)
—Waste water treatment: adsorption on Amberlite X AD-2: infl. 0.4 ppm, effl. 0.22 ppm; retention efficiency: 43% (40)

picric acid (2,4,6-trinitrophenol)

$$O_2N - \underset{NO_2}{\underset{|}{C_6H_2}} - \underset{OH}{\overset{|}{}} - NO_2$$

A. PROPERTIES: yellow leaflets; m.w. 229.11, m.p. 121.8°C; b.p. explodes >300°C; v.d. 7.91; sp.gr. 1.763; solub. 14,000 mg/l at 20°C, 68,000 mg/l at 100°C; log P_{oct} 2.03
B. AIR POLLUTION FACTORS: 1 mg/cu m = 0.11 ppm, 1 ppm = 9.52 mg/cu m
C. WATER POLLUTION FACTORS:
—Biodegradation: adapted culture: 2% removal after 48 hr incubation, feed = 220 mg/l (292)
adapted A.S. at 20°C—product is sole carbon source: 0% COD removal after 20 days (327)
—Oxidation parameters: COD: 0.92
ThOD: 0.98 (27)
—Odor threshold conc. (detection): 4 mg/l (998)

—Impact on biodegradation processes:
biological treatment decreases from 20–30 mg/l onwards (30)
D. BIOLOGICAL EFFECTS:
—Bacteria:
  E. coli: no toxic effect at 1 g/l
  Pseudomonas putida: inhibition of cell multiplication starts at 1020 mg/l for a 50% solution (329)
—Protozoa: Uronema parduczi Chatton-Lwoff: inhibition of cell multiplication starts at 26 mg/l (1901)
—Algae:
  Scenedesmus: LD_0: 240 mg/l
  Microcystis aeruginosa: inhibition of cell multiplication starts at 72 mg/l for a 50% solution (329)
—Arthropoda: Daphnia: LD_0: 88 mg/l
—Fish: LD_0: 30 mg/l
  trout: perturbation level: 0.4 mg/l, 50 min (30)

pimaric acid (levopimaric acid)
  Use: resins
D. BIOLOGICAL EFFECTS:
—Fish:
  Coho salmon juvenile: 96 hr LC_{50}: 0.32 mg/l (1495)
  rainbow trout: static bioassay: 96 hr TLm: 0.8 mg/l at pH: 7.0 (441)

pimelic acid (heptanedioic acid)
$COOH(CH_2)_5COOH$
A. PROPERTIES: m.w. 160.17; m.p. 103°C; b.p. 272°C at 100 mm, sublimes; sp.gr. 1.329 at 15°C; solub. 25.2 g/l at 13°C, 50.0 g/l at 20°C; log P_{oct} 0.14/0.70 (calculated)
C. WATER POLLUTION FACTORS:
—Waste water treatment: A.S. after 6 hr: 1.1% of ThOD
                                    12 hr: 1.0% of ThOD
                                    24 hr: 2.8% of ThOD (89)

pimelic ketone see cyclohexanone

pimelonitrile (heptanedinitrile; dicyanopentane)
$CN(CH_2)_5CN$
A. PROPERTIES: m.w. 122.17; b.p. 175–176°C/14 mm; sp.gr. 0.951
C. WATER POLLUTION FACTORS:
—Waste water treatment: A.S. after 6 hr: 0.6% of ThOD
                                    12 hr: 1.0% of ThOD
                                    24 hr: 1.3% of ThOD (89)

dl-α-pinene (dl-2,6,6-trimethylbicyclo(3,1,1)hept-2-ene)

A. PROPERTIES: colorless liquid; m.w. 136.23; m.p. -55°C; b.p. 155°C; v.p. 5 mm at 25°C; v.d. 4.72; sp.gr. 0.86 at 20/4°C
B. AIR POLLUTION FACTORS:
—Odor: Odor threshold values: 0.064 mg/cu m = 11.4 ppb (307)
0.016 mg/cu m (607)
O.I.: at 20°C = 469,000 (316)
—Natural sources:
rate of foliar α-pinene accumulation in a closed atmosphere from pine trees:
0.2–0.5 ppb/min/g at 17°C
1.2–3.5 ppb/min/g at ?°C (197)
measured emission rate from *Pinus taeda*: 2760 µg/m²/h (1728)
—Control methods: platinized ceramic honeycomb catalyst: ignition temp.: 200°C, inlet temp. for 90% conversion: 250–300°C (91)
C. WATER POLLUTION FACTORS:
—Odor threshold conc. (detection): 0.0025 mg/kg (894)

N,N'-(1,4-piperazinediyl-*bis*(2,2,2-trichloroethylidene))-*bis*(formamide) *see* triforine

piperonylbutoxide (α-(2-(2-butoxy-ethoxy)ethoxy)-4,5-(methylene-dioxy)-2-propyltoluene; 3,4-methylenedioxy-6-propylbenzyl(heptyl)diethyleneglycol; butylcarbitol-6-propylpiperonylether)

Use: as a synergist of pyrethrin insecticide in arable fields and in storehouses of agricultural products
A. PROPERTIES: sp.gr. 1.05 at 20/20°C
D. BIOLOGICAL EFFECTS:
—Residues:
10 species of agricultural products were investigated: piperonylbutoxide was detected from 3 barleys and 3 wheats harvested in U.S.A. and Australia. The residual range was from 0.2 to 1.4 ppm (1632)
—Mammals:
acute oral LD_{50} (rat): >7500 mg/kg (1854)

pirimiphosmethyl *see* actellic

PisoBN *see* peroxyisobutyrylnitrate

pivalic acid (trimethylacetic acid; neopentanoic acid; 2,2-dimethylpropionic acid; "Versatic 5")

$(CH_3)_3CCOOH$

A. PROPERTIES: m.w. 102.13; m.p. 32–35°C; b.p. 163.8°C; sp.gr. 0.902 at 40/20°C

C. WATER POLLUTION FACTORS:
  —Oxidation parameters:
   BOD_5:   0.21 NEN 3235-5.4 (277)
   $BOD_2^{25°C}$: 0.024 (substrate conc.: 4.1 mg/l-inoculum: soil microorganisms)
    5 days: 0.0
    10 days: 0.29
    20 days: 1.76 (1304)
   COD: 1.94 NEN 2235-5.4 (277)
  —Reduction of amenities: T.O.C. = 50 mg/l (295)

D. BIOLOGICAL EFFECTS:
  —Fish: goldfish: LD_{50} (24–96 hr): 400–375 mg/l at pH5
         goldfish: LD_{50} (24 hr): 4500 mg/l at pH7 (277)
  —Mammalia: rat: acute oral LD_{50}: between 500 and 5000 mg/kg (277)

PMA *see* phenylmercuric acetate

polychlorobiphenyls (PCB's)

nCl — ⌬—⌬ — nCl

See also: Aroclor   dichlorobiphenyl
           Phenoclor trichlorobiphenyl
                      tetrachlorobiphenyl
                      pentachlorobiphenyl
                      hexachlorobiphenyl
                      heptachlorobiphenyl
                      octachlorobiphenyl

PCB's are mixtures of chlorinated byphenyls. The degree of chlorination is usually indicated by the commercial nomenclature of these compounds. In the Aroclor series, a four digit code is used, of which the last two digits represent the percentage by weight of chlorine in the mixture; thus Aroclor 1254 is a PCB mixture containing 54% chlorine. Phenoclor DP 6 and Clophen A 60 both indicate biphenyls with an approximate mean content of 6 chlorine atoms per molecule. (1444)

Impurities in commercial PCB's include chlorinated dibenzofurans and chlorinated naphthalenes. (1444)

Use: PCB's are relatively nonflammable, have useful heat exchange and dielectric properties and are used primarily in the electrical industry in capacitors and transformers.

PCB's are also used in the formulation of lubricating and cutting oils, in pesticides, adhesives, plastics, inks, paints, and sealants. (1444)

A. PROPERTIES:
   solubility: the solubility of PCB decreases with increasing chlorination (1446)
               0.04–0.2 ppm (1447)

B. AIR POLLUTION FACTORS:
  —Ambient air quality:
    in marine atmosphere in the Northwest Gulf of Mexico, March–April 1977:
      range: 0.17–0.79 ng/m^3 ($n = 10$)
      average: 0.35 ng/m^3 (1724)
    in Bermuda atmosphere: avg. 0.20 ng/cu m; range: 0.08–0.48 ng/cu m (1725)

C. WATER POLLUTION FACTORS:
  —Water quality: PCB in seawaters:

California coastal (1971):	range:	0.011–0.050 ppb ($n = 2$)	
Mexican coastal (1971):	range:	0.012–0.090 ppb ($n = 4$)	
California current (1972):	range:	0.008 ppb ($n = 1$)	
North Central Pacific Gyre (1972):	range:	0.005–0.006 ppb ($n = 2$)	(1783)
N.W. Atlantic ocean (1972):	avg.:	0.017 ppb ($n = 11$)	(1784)
N.W. Atlantic ocean (1972):	avg.:	0.035 ppb	
	range:	0.004–0.150 ppb ($n = 36$)	(1784)
N.W. Atlantic ocean (1973):	avg.:	0.009 ppb ($n = 6$)	(1785)
	range:	0.450–4.2 ppb ($n = 2$)	(1786)
(1974–75):	avg.:	0.001 ppb ($n = 4$)	(1785)
N.W. Mediterranean coastal (1975):	avg.:	0.013 ppb ($n = 11$)	(1787)
Central Mediterranean coastal (1976):	avg.:	0.297 ppb	
	range:	0.009–1 ppb ($n = 66$)	(1777)
Central Mediterranean coastal (1977):	avg.:	0.135 ppb	
	range:	n.d.–0.380 ppb ($n = 20$)	(1777)

  —Manmade sources:
    in soot deposits on the ice of a lake near Stockholm: 2.8 ppm (dry wt.) (1445)
    soap powder: 1 of 4 brands contained PCB's up to 0.48 µg/g
    toilet paper: 9 of 11 brands contained PCB's ranging from <0.01 to 21 µg/g
                                                                              (1327)
  —Aquatic reactions:
    partition coefficient between sediment and water: ~10^5 (1599)
    Freundlich constants and concentration factors of adsorbing agents:

agent	Freudlich constant, n	conc. factor	
total plankton	1.0	10^4	(1929)
activated sludge	0.6	2×10^4–7×10^4	(1928)
zooplankton (Osaka bay)	–	0.5×10^5	(1928)
phytoplankton (Osaka bay)	1.6	2×10^2–2×10^3	(1928)
mud of Osaka bay	1.4	4×10^2–10^3	(1928)
Iwafune sand	–	10^2	(1928)
granular activated carbon	–	2×10^2	(1928)
woodburn soil	1.1	8–10	(1931)
Illite clay	0.81	1–2	(1931)
Montmorillonite clay	1.0	0.5	(1931)
Kaolinite clay	1.0	0.2	(1931)
Goodrich 1115 foam	–	0.2×10^6	(1930)
Amberlite XAD-4	–	0.2×10^5	(1930)

  —Waste water treatment methods:
    Catalytic dechlorination to biphenyl was achieved with 5% platinum or palladium
    on 60/80 mesh glass beads (1552)

A.S.: 15-10% degradation of 1 mg in 48 hr—increased chlorine in molecule decreases degradation (1078)
—Method of analysis: XAD-2 macroreticular resin can be used to analyze large volumes of water for low ng/l levels of PCB's. Detection limits for individual PCB's are typically in the 1-10 ng/l range for potable water from a river source and ca. 0.04 ng/l for underground water sources low in organic content (1822)
residues: in Ottawa tapwater: nd
             in groundwater: 0.1-0.2 ng/l (as Aroclor 1016) (1822)

D. BIOLOGICAL EFFECTS:
—Residues: mean values and ranges of PCB concentrations in plankton from: *various marine regions*:

region	PCB's ppm in lipid wt.	PCB's ppm wet wt.	
Baltic sea, Stockholm Archipelago	18(3-35)		(1581)
Baltic sea, Turku Archipelago, 1973	25(4-77)	0.19(0.04-0.75)	(1582)
1974	38(26-340)	0.37(0.03-3.3)	(1583)
1976	28(4-66)	0.23(0.04-0.72)	(1583)
South Atlantic Shelf	48(7-120)	0.20(0.02-0.64)	(1584)
N.E. Atlantic	2.3	—	(1585)
N.E. Atlantic	(0.1-5.5)	(0.01-0.12)	(1586)
Firth of Clyde	(0.1-17)	(0.08-2.2)	(1586)
N.E. Pacific Estuarine zones	2.2(1-16)	—	(1587)

fresh water area:
Lake Päijänne   —   0.03(0.002-0.14)   (1588)

   in marine animals from the Ligurian Sea (1977-1978):
      shrimp (*Nephrops norvegicus*):   avg.   73 ppb fresh wt.
      mussel (*Mytilus galloprovincialis*):   avg.   172 ppb fresh wt.
      fish: anchovy (*Engraulis encrasicholus*): avg.   169 ppb fresh wt.
         striped mullet (*Mullus barbatus*): avg.   808 ppb fresh wt.
         (*Euthymnus alletteratus*):   avg.   673 ppb fresh wt.
         (*Sarda sarda*):   avg. 5,606 ppb fresh wt.   (1775)

   mean residues in eggs of american crocodile in Everglades National Park:
      1972: not detected (1572)
      1977: 0.52 ppm wet wt. (1571)

   in aquatic vascular plants of Lake Päijänne, Finland (1972-1973):
      mean: 20 µg/kg dry weight (*n* = 112)
      S.D.:   47 µg/kg dry weight
      min.:   0 µg/kg dry weight
      max.:   331 µg/kg dry weight (1055)

   in tissue of bivalves from Raritan Bay—Lower N.Y. Bay complex:
      Crassostrea virginica: 81 ± 32 ng/g tissue
      Mya arenaria: 149 ± 67 ng/g tissue
      Mercenaria mercenaria: 131 ± 27 ng/g tissue
   in sediments (same locations): 3.4-2035 ng/g dry sediment (1675)

residues in animals in different trophic levels from the Weser Estuary (1976):
   bivalvia: common edible cockle (*Cerastoderma edule L.*):
                                   11 ± 4   ng/g wet tissue
      soft clam (*Mya arenaria L.*): 44 ± 14 ng/g wet tissue

polychaeta: lugworm (*Arenicola marina L.*): 64 ± 16 ng/g wet tissue
crustacea: brown shrimp (*Crangon crangon L.*): 60 ± 4 ng/g wet tissue
pisces: common sole (*Solea solea L.*): 206 ± 49 ng/g wet tissue   (1440)
comparative average concentrations of PCB residue (wet tissue basis) in brown shrimp (*Crangon crangon*) from various areas of the North Sea, 1974–1976:

area	year	no. of analyses	avg. conc. (ppb)	range	
S.E. England	1975	2	90	80–100	(1441)
Netherlands	1974	10	140	50–290	(1442)
	1975	12	110	50–200	(1441)
	1976	12	120	40–180	(1441)
Germany	1974	6	48	36–60	(1442)
	1975	4	180	140–220	(1441)
	1976	6	100	80–140	(1441)
	1976	5	60	56–64	(1440)

goby fish (*Acanthogobius flavimanus*) collected at the sea shore of Keihinjima along Tokyo Bay, Aug. 1978: residue level: 670 ppb   (1721)
in fat phase of the following fish from the Frierfjord (Norway), April 1975–Sept. 1976: cod, whiting, plaice, eel, sprot:
  avg.: 13 ppm
  range: 1.4–70 ppm   (1065)
in marine animals from the Central Mediterranean (1976–1977):
  fish: anchovy (*Engraulis encrasicholus*):
    avg.: 47.0 ppb fresh wt (n = 12)
    range: 9.1–176.7 ppb frest wt.
  striped mullet (*Mullus barbatus*):
    avg.: 107.0 ppb fresh wt (n = 10)
    range: 17.0–373.3 ppb fresh wt.
  tuna (*Thunnus thynnus thynnus*):
    avg.: 22.0 ppb fresh wt (n = 5)
    range: 8.9–44.4 ppb fresh wt.
  shrimp (*Nephrops norvegicus*):
    avg.: 13.0 ppb fresh wt. (n = 7)
    range: 4.4–55.0 ppb fresh wt.
  mussel (*Mytilus galloprovincialis*):
    avg.: 78.0 ppb fresh wt. (n = 4)
    range: 60.5–100.1 ppb fresh wt.   (1774)
residue in great crested grebe (fish eating bird):
Great Britain, 1963–1966: in liver: avg. 36 ppm; range: 28–40 ppm (n = 4)
  (1778)
Clyde, Scotland, 1971–1975–bird found dead:
  in pectoral muscle: 17 ppm
  in liver: 19 ppm   (1779)
Lake Päijänne, Finland, 1972–1974 (bird shoot in summer, after breeding season):
  in pectoral muscle: avg. 3.1 ppm; range: 0.54–11 ppm (n = 12)
  in liver: avg. 5.2 ppm; range: 0.41–10 ppm (n = 12)   (1780)
residues in fish eating birds from Gdansk Bay, 1975–1976 (ppm in fresh tissue):
  (1776)

		pectoral muscle	liver	adipose tissue
great crested grebe,	avg.	3.96 ($n = 17$)	7.91 ($n = 18$)	38.3
(*Podiceps cristatus L.*)	range	0.56–17	0.15–28	9.8–67
slovanion greb		1.8 ($n = 1$)	1.8 ($n = 1$)	–
(*Podiceps auritus L.*)				
black guillemot		1.2 ($n = 2$)	1.9 ($n = 2$)	53 ($n = 2$)
(*Cepphus grylle L.*)		1.1–1.4	1.8–2.0	30–77
goosander		1.9 ($n = 1$)	2.8 ($n = 1$)	62 ($n = 1$)
(*Mergus merganser L.*)				
black throated diver		2.2 ($n = 1$)	0.69 ($n = 1$)	36 ($n = 1$)
(*Gravia arctia L.*)				

concentration of PCB in marine samples (ppm, whole base):

samples	source	PCB
oyster (*Crassostrea*)	Nagasaki	0.012
	Nagasaki	0.016
	Hiroshima	0.009
	Hiroshima	0.015
	Hiroshima	0.038
	Osaka	0.016
	Osaka	0.037
short-necked clam (*Tapes*)	–	0.003
	–	0.012
	Ise.Mie	0.006
	Matsusaka.Mie	0.002
clam (*Meretrix*)	–	0.003
	(Korea)	0.002
corbicula (*Corbicula*)	Shimane	0.015
	Shimane	0.169
turban shell (*Turbo*): muscle	–	nd
intestine	–	0.004
muscle	Mie	0.005
intestine	Mie	0.005
mud-snail (*Viviparus*)	–	0.004
mussel (*Mytilus californianus*)	(CA, USA**)	0.004
*Cellana stearnsii**	Osaka	0.009
*Liolophura japonica**	Osaka	
muscle		0.019
intestine		0.011
Sea cucumber (*Holothuroidae*)	–	0.011
	Mie	0.021
Sea squirt (*Pyuridae*)	Miyagi	0.003

tr < 0.001 ppm, –: indistinct,
*collected directly in the estuaries of Osaka Bay, 1977.
**collected at Pacific Beach in Scripps Institution, 1976. (1067)

    residues in Canadian human milk: 1970: avg. 6 ng/g whole milk
      (PCB expressed as Aroclor 1260) 1975: avg. 12 ng/g whole milk (1376)
    levels in blood of nonoccupationally exposed mothers and their children in Osaka

prefecture (Japan) in 1976:
  mothers: avg.: 2.8 ± 0.8 ppb
           range: 1.7-4.2 ppb
  children: avg.: 3.8 ± 3.6 ppb
           range: n.d.-12.8 ppb   (1389)
—Bioaccumulation:
  PCB's in pelagic organisms—a food chain interrelationship study:

organism	PCB conc. factor (wet wt.)
microplankton	170,000
macroplanktonic enphausiid, *Meganyctiphanes norvegica*	50,000
carnivorous decapod shrimp, *Sergestes arcticus*	47,000
Pasiphaea sivado	20,000
myctophid fish, Myctophus glaciale	6,000

  (surface water conc. 2.5 ng/l)   (1602)
  conclusion: no biomagnification in this foodchain, if whole organisms are considered

polychlorinated naphthalenes (PCN's)

Use: PCN's are marketed in the U.S.A. as Halowaxes, primarily used to impregnate capacitor tissue paper and as engine oil additives

B. AIR POLLUTION FACTORS:
—Ambient air quality:
  PCN's found in air sampled near a PCN's manufacturing site:

location	total PCN conc. (ng/cu m)	
	day 1	day 2
500 m south, in valley	450	1900
500 m east, on hill	49	830
800 m north, in valley	91	2900
800 m west, on hill	25	110
average	150	1400

  PCN's found in air near an electronics components manufacturer (a potential PCN user):

location	total PCN conc. (ng/cu m)	
	day 1	day 2
400 m Northwest	25	9.8
400 m Southwest	31	11
600 m Southeast	9.8	15
700 m Northeast	11	33
average	19	17

                                                                                                                (1000)

poly(dimethylsiloxane)

D. BIOLOGICAL EFFECTS:
—Bluegill: static bioassay: 96 hr TLm: >10,000 mg/l at 24°C
  rainbow trout: static bioassay: 96 hr TLm: >10,000 mg/l at 13°C   (460)

polyethyleneglycols (PEG's)
A. PROPERTIES: PEG 200: m.w. 190–210; p.p. −20°C
                PEG 400: m.w. 380–420; p.p.   7°C
                PEG 800: m.w. 760–840; p.p. 30°C
C. WATER POLLUTION FACTORS:

	PEG 200	PEG 400	PEG 800
BOD_5:	0.02 = 1% ThOD	0.01 = 1% ThOD	0 = 0% ThOD
COD:	1.62 = 98% ThOD	1.71 = 98% ThOD	1.74 = 98% ThOD

                                                                                 (277)

D. BIOLOGICAL EFFECTS:
    —*Carassius auratus*: LD_{50}, 24 hr: >5000 mg/l (PEG 200, 400, 800)   (277)
    —rat: oral LD_{50}: 30–50 g/kg in function of mol. wt.   (277)

polyoxyethylene(dimethylimino)ethylene(dimethylimino)ethylenedichloride)
(WSCP-Busan 77 (Buckman Lab.)

$$\left[O-CH-CH_2-\underset{\underset{CH_3}{|}}{\overset{\overset{CH_3-Cl}{|}}{N}}-CH_2CH_2-\underset{\underset{CH_3Cl}{|}}{\overset{\overset{CH_3}{|}}{N}}-CH_2-CH_2 \right]_n$$

C. WATER POLLUTION FACTORS:
    —Biodegradation:
        after 3 weeks of adaptation at 5–20 mg/l at 22°C:
            aerobic degradation: product is sole carbon source: 0% degradation
                                + synthetic sewage: 10–50% degradation
            anaerobic degradation: product is sole carbon source: 0% degradation
                                  + synthetic sewage: 0% degradation   (512)

polypropyleneglycol
D. BIOLOGICAL EFFECTS:
    —Fish:
    Lepomis macrochirus: static bioassay in fresh water at 23°C, mild aeration applied after 24 hr:

material added	% survival after				best fit 96 hr LC_{50}
ppm	24 hr	48 hr	72 hr	96 hr	ppm
2,400	90	30	10	0	
1,800	100	100	90	70	1,700
1,000	100	80	80	77	

    Menidia beryllina: static bioassay in synthetic seawater at 23°C: mild aeration applied after 24 hr:

material added	% survival after				best fit 96 hr LC_{50}
ppm	24 hr	48 hr	72 hr	96 hr	ppm
1,320	100	20	0	—	
1,000	85	0	—	—	650
750	100	50	50	25	
560	100	100	100	100	(352)

    rainbow trout: static bioassay: 96 hr TLm: >10,000 mg/l at 10°C   (451)

polyram
= coprecipitation of zinc ammonia ethylene bis-dithiocarbamate + poly-ethylene-bis-thiocarbamoyl disulfide (zineb + poly-ethylene thiuram disulphide)

$$\left[\left(\begin{array}{c} CH_2-NH-\overset{S}{\underset{\parallel}{C}}-S \\ | \\ CH_2-NH-\underset{\parallel}{C}-S \\ S \end{array}\right)_n \left(\begin{array}{c} CH_2-NH-\overset{S}{\underset{\parallel}{C}} \\ | \\ CH_2-NH-\underset{\parallel}{C} \\ S \end{array}\right)_m \right]_x$$

where $n:m = 1:3$; x is unknown
Use: fungicide: active ingredient: metiram (80%)

D. BIOLOGICAL EFFECTS:
  —Fish:
    harlequin fish (Rasbora heteromorpha):
    24 hr LC_{50} (F): 1000 mg/l   (331)
    24 hr LC_{50}:   32 ppm
    48 hr LC_{50}:   17 ppm   (1855)
  —Mammals:
    acute oral LD_{50} (rat): >10,000 mg/kg   (1854)
              (male mice): >5,400 mg/kg
              (female guinea pigs): 2,400–4,800 mg/kg   (1855)
    highest non-toxic oral dose, 5x/week for 5 weeks (rabbits): 200 mg/kg
    in diet (dogs): 45 mg/kg/day for 90 days: no ill effects
                   7.5 mg/kg/day for 23 months: no ill effects   (1855)

PP 511 *see* actellic

PPN *see* peroxypropionylnitrate

19-nor-17-pregna-1,3,5(10-trien-20-yne-3,17-diol *see* ethinyloestradiol

presco 5
  Composition: 21.3% Mecoprop, 10% MCPA, 2.8% 2,3,6-trichlorobenzoic acid
  Use: herbicide
D. BIOLOGICAL EFFECTS:
  —Fish:

	mg/l	24 hr	48 hr	96 hr	3 m (extrapolated)
harlequin fish	LC_{10}(F)	600	550	460	
(Rasbora heteromorpha)	LC_{50}(F)	700	630	560	500

(331)

preventol GD *see* 5,5'-dichloro-2,2'-dihydroxydiphenylmethane

princep *see* 2-chloro-4,6-bis(ethylamino)-s-triazine

pristane (2,6,10,14-tetramethylpentadecane; norphytane)

$$(CH_3)_2CH-(CH_2)_3CH(CH_3)(CH_2)_3CH(CH_3)(CH_2)_3CH(CH_3)_2$$

Use: precision lubricant; chromatographic oil; anti-corrosion agent.
C. WATER POLLUTION FACTORS:
—Biodegradation:
degradation in seawater by oil-oxidizing microorganisms: 25.4% breakdown after 21 days at 22°C in stoppered bottles containing a 1000 ppm mixture of alkanes, cycloalkanes and aromatics (1237)

dl-proline (1-pyrrolylmethanoic acid; dl-2-pyrrolidinecarboxylic acid)

[structure: pyrrolidine ring with N–H and COOH]

A. PROPERTIES: m.w. 115.13; m.p. 205°C decomposes
C. WATER POLLUTION FACTORS:
—Waste water treatment: A.S. after 6 hr: 4.8% of ThOD
                                    12 hr: 13.1% of ThOD
   (*l*-prol.)   24 hr: 25.2% of ThOD (89)
—Manmade sources: excreted by man: in urine: 0.3–0.9 mg/kg body wt/day (203)

prometone (2,4-*bis*(isopropylamino)-6-methoxy-*s*-triazine; 2-methoxy-4,6-*bis*(isopropylamino)-*s*-traizine; pramitol; prometone)

[structure: s-triazine ring with OCH$_3$ at 2-position and NHCH(CH$_3$)$_2$ groups at 4 and 6 positions]

A. PROPERTIES: m.p. 91–92°C; solub. 750 ppm at 20°C; v.p. 2.3×10^{-6} mm at 20°C
D. BIOLOGICAL EFFECTS:
—Algae:

	ppb		method of assessment
Phaeodactylum tricornutum	100	50% decrease in O$_2$ evolution	f
Phaeodactylum tricornutum	250	50% decrease in growth	measured as ABS. (525 mu) after 10 days
Chlorococcum sp.	400	50% decrease in O$_2$ evolution	f
Chlorococcum sp.	500	50% decrease in growth	measured as ABS. (525 mu) after 10 days

Dunaliella tertiolecta	2×10^3	50% decrease in O_2 evolution	f
Dunaliella tertiolecta	1.5×10^3	50% decrease in growth	measured as ABS. (525 mu) after 10 days
Isochrysis galbana	1×10^3	50% decrease in O_2 evolution	f
Isochrysis galbana	1×10^3	50% decrease in growth	measured as ABS. (525 mu) after 10 days (2348)

—Mammals:
acute oral LD_{50} (rat): 1750–2980 mg/kg (1854)
acute dermal LD_{50} (rabbit): 2200 mg/kg (25% formulation) (1855)

propanal *see* propionaldehyde

propanamide *see* propionamide

propane
$CH_3CH_2CH_3$
A. PROPERTIES: colorless gas; m.w. 44.09; m.p. $-189.9°C$; b.p. $-42°C$ v.d. 1.52; sp.gr. 0.58 at $-44/4°C$; THC 530.5 kcal/mole, LHC 488 kcal/mole; v.p. 8.5 atm at $20°C$, 11.0 atm at $30°C$
B. AIR POLLUTION FACTORS: 1 mg/cu m = 0.55 ppm, 1 ppm = 1.83
  —Odor threshold values: 36,000 mg/cu m (781)
  22,000 mg/cu m (737)
  O.I. at $20°C$ = 425 (316)
  —Manmade sources: in gasoline: 0.07–0.08 vol. % (312)
  expected glc's in USA urban air: 0.05–0.40 ppm (102)
  in flue gas of municipal incinerator: <0.4–0.5 ppm (196)
  —Atmospheric reactions:
  estimated lifetime under photochemical smog conditions in S.E. England: 31 hr
  (1699; 1704)
  —Sampling and analysis: non dispersive I.R.: min. full scale: 50 ppm (55)
  detector tubes: Auer: 200 ppm: det. limit (47)
  Dräger: 2,000 ppm: det. limit (58)
D. BIOLOGICAL EFFECTS:
  —Man: no effect level: 10,000 ppm for brief exposures
  slight dizziness in a few minutes: 100,000 ppm (211)

propanedioic acid *see* malonic acid

1,2-propanediol *see* propyleneglycol

1-propanethiol *see* n-propylmercaptan

1,2,3-propanetriol *see* glycerol

propanil *see* N-(3,4-dichlorophenyl)propionamide

1018 2-PROPANOIC ACID

2-propanoic acid *see dl*-lactic acid

propanoic acid *see* propionic acid

n-propanol
$CH_3CH_2CH_2OH$

Users and formulation: solvent in mfg. printing inks, nail polishes, polymerization and spinning of acrylnotrile, dyeing of wool, cellulose acetate film, PVC adhesives, metal degreaser; mfg. floor wax, cleaning preparations, solvent for resins, cellulose esters, waxes, vegetable oils; brake fluid mfg.; antiseptic mfg. (347)

Natural sources (water and air): fermentation and spoilage products of many vegetable substances. (347)

A. PROPERTIES: colorless liquid; m.w. 60.09; m.p. $-127°C$; b.p. $97.8°C$; v.p. 14.5 mm at $20°C$, 20.8 mm at $25°C$, 27 mm at $30°C$; v.d. 2.07; sp.gr. 0.804 at $20/4°C$; sat. conc. 46 g/cu m at $20°C$, 85 g/cu m at $30°C$; log P_{oct} 0.34

B. AIR POLLUTION FACTORS: 1 mg/cu m = 0.40 ppm, 1 ppm = 2.50 mg/cu m.
—Odor: characteristic: quality: sweet, alcohol
hedonic tone: pleasant

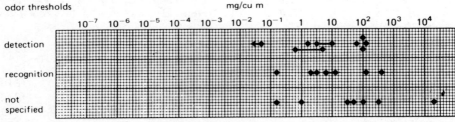

(210; 278; 298; 307; 610; 627; 634; 643; 676; 704; 708; 709; 714; 739; 749; 776; 788; 828)

O.I. at $20°C$ = 470 (316)

C. WATER POLLUTION FACTORS:
—Biodegradation: adapted A.S.—product as sole carbon source—98.8% COD removal at 71.0 mg COD/g dry inoculum/hr (327)
—Oxidation parameters:
BOD: 0.47–1.05 std. dil/sewage (260)(285)
ThOD: 2.4 (27)
—Manmade sources:
in year old leachate of artificial sanitary landfill: 1.0 g/l (1720)

(889; 896; 907; 908)

—Waste water treatment: A.C.: adsorbability: 0.038 g/g C; 18.9% reductn. infl.:
 1,000 mg/l, effl.: 811 mg/l (32)
 reverse osmosis: 42% rejection from a 0.01 M solution (221)
 anaerobic lagoon: lb COD/day/1,000 cu ft infl. mg/l effl. mg/l
                                 22   170   35
                                 48   170   40
 A.S.: 6 hr: 13.9% of ThOD
       12 hr: 26.8% of ThOD
       24 hr: 36.9% of ThOD (88)

methods	temp °C	days observed	feed mg/l	days acclim.	
A.S., W	20	1	500	24	55% theor. oxidation
A.S., BOD	20	$\frac{1}{3}$-5	333	30	94% removed (93)

D. BIOLOGICAL EFFECTS:
 —Toxicity threshold (cell multiplication inhibition test):
   bacteria (*Pseudomonas putida*): 2700 mg/l (1900)
   algae (*Microcystis aeruginosa*): 255 mg/l (329)
   green algae (*Scenedesmus quadricauda*): 3100 mg/l (1900)
   protozoa (*Entosiphon sulcatum*): 38 mg/l (1900)
   protozoa (*Uronema parduczi Chatton-Lwoff*): 568 mg/l (1901)
 —Algae:
   Chlorella pyrenoidosa: toxic: 11,200 mg/l (41)
 —Amphibians:
   mexican axolotl (3-4 wk after hatching): 48 hr LC_{50}: 4000 mg/l
   clawed toad (3-4 wk after hatching): 48 hr LC_{50}: 4000 mg/l (1823)
 —Fish: gudgeon: TLm: 200-500 mg/l (30)
       creek chub: LD_0: 200 mg/l, 24 hr in Detroit river water
                   LD_{100}: 500 mg/l, 24 hr in Detroit river water (243)
 —Mammalia: rat: inhalation: 2/6: 4,000 ppm, 4 hr
               mouse: inhalation: no reaction: 2,050 ppm, 480 min
   rat: single oral LD_{50}: 1.9 g/kg
   rabbit: single oral LD_{50}: 3.5 ml/kg
   mouse: single oral LD_{50}: 4.5 g/kg

2-propanol *see* isopropanol

2-propanone *see* acetone

2-propanoneoxime *see* acetoxime

propargyl alcohol (2-propyn-1-ol, ethynylcarbinol, acetylenylcarbinol, propiolic alcohol)
CHCCH₂OH
A. PROPERTIES: colorless liquid; m.w. 56.06; m.p. -17°C; b.p. 115°C; v.d. 1.93; sp.gr. 0.97 at 20/4°C
B. AIR POLLUTION FACTORS: 1 mg/cu m = 0.43 ppm, 1 ppm = 2.33 mg/cu m
 —Odor threshold conc.: 0.35 mg/cu m (710)

1020 PROPENAL

C. WATER POLLUTION FACTORS:
—Oxidation parameters:
  BOD_5:   2% ThOD
  COD: 97% ThOD
  $KMnO_4$ value: acid: 78% ThOD
                   alkaline: 67% ThOD
  TOC: 99% ThOD

D. BIOLOGICAL EFFECTS:
—Toxicity threshold (cell multiplication inhibition test):
  bacteria (*Pseudomonas putida*):   150 mg/l
  green algae (*Scenedesmus quadricauda*):   18 mg/l
  protozoa (*Entosiphon sulcatum*):   17 mg/l   (1900)
  protozoa (*Uronema parduczi Chatton-Lwoff*):   >800 mg/l   (1901)

propenal *see* acrolein

propene *see* propylene

propeneoxide   (1,2-epoxypropane; propyleneoxide; methyloxirane) CH_3CHCH_2O

A. PROPERTIES: colorless liquid; m.w. 58.08; m.p. $-104.4°C$; b.p. $34/35°C$; v.p. 400 mm at $18°C$; v.d. 2.00; sp.gr. 0.859 at $0/4°C$; solub. 650,000 mg/l at $30°C$, 405,000 mg/l at $20°C$

B. AIR POLLUTION FACTORS: 1 mg/cu m = 0.421 ppm, 1 ppm = 2.37 mg/cu m
—Odor: characteristic: quality: sweet
                        hedonic tone: neutral to pleasant
  T.O.C. = abs. perc. lim.:   9.9 ppm
           50% recogn.:   35.0 ppm
           100% recogn.:   35.0 ppm
         = 473 mg/cu m   (702)
           O.I. 100% recogn.: 16,600

C. WATER POLLUTION FACTORS:
—Waste water treatment:
  A.S., BOD, 1/3–5 days observed, feed: 333 mg/l, acclimation 30+ days, 75% removed   (93)
  A.C.: adsorbability: 0.052 g/g C; 26.1% reduction, infl.: 1,000 mg/l, effl.: 739 mg/l

D. BIOLOGICAL EFFECTS:
—Fish:
  mosquito fish (*Gambusia affinis*): static bioassay: 96 hr TLm: 141 mg/l at $24°C$
  bluegill: static bioassay: 96 hr TLm: 215 mg/l at $24°C$ (466)
—Mammalia:
  guinea pig: intragastrical LD_{50}: 0.69 g/kg
  rat: intragastrical LD_{50}: 1.14 g/kg
  rat: by intubation: no effect: 0.2 g/kg, 5 X/week, 3 weeks
  guinea pig: inhalation: 0/10: 1,800 ppm, 7 hr
                          0/5:   3,600 ppm, 2 hr
                          2/5:   3,600 ppm, 7 hr

rat: inhalation: 0/5 no detect. injury: 900 ppm, 7 hr
               0/10:   1,330 ppm, 4 hr
               0/5 no detect; injury: 1,800 ppm, 2 hr
               0/5:   3,600 ppm, 1 hr
               10/10:   3,600 ppm, 7 hr
               5/10:   7,200 ppm, 1 hr
               10/10:   14,400 ppm, 0.5 hr
mouse: inhalation: 1/10: 1,330 ppm, 4 hr
rat: inhalation: 5/10 eye and respir. irritation: 557 ppm, 79 × 7 hr
               7/40:   195 ppm, 138 × 7 hr
               9/40:   102 ppm, 138 × 7 hr
guinea pig: inhalation: 0/8 slight liver injury: 457 ppm, 110 × 7 hr
               0/16:   195 ppm, 128 × 7 hr
               0/16 no effects: 102 ppm, 128 × 7 hr   (211)

propenamide see acrylamide

2-propenenitrile see acrylonitrile

propene trimer see propylene trimer

propenoic acid see acrylic acid

propenol-3 see allylalcohol

propenylalcohol see allylalcohol

2-propenylamine see allylamine

2-propenylethanoate see allylacetate

2-propenylisothiocyanate see allylisothiocyanate

propham see isopropyl-N-phenylcarbamate

propine see methylacetylene

propiolic alcohol see propargylalcohol

propionaldehyde (propanal; methylacetaldehyde)
CH_3CH_2CHO
   Use: manufacture of polyvinyl acetals and other plastics; disinfectant
A. PROPERTIES: colorless liquid; m.w. 58.1; m.p. -81°C; b.p. 49°C; v.p. 235 mm at
       20°C, 687 mm at 45°C; v.d. 2.0; sp.gr. 0.807 at 20/4°C; solub. 200,000 mg/l
       at 20°C; log P_{oct} 0.83 (calculated)
B. AIR POLLUTION FACTORS: 1 mg/cu m = 0.41 ppm, 1 ppm = 2.41 mg/cu m
   —Odor: characteristic: quality: sweet, ester, acrid, irritating
                         hedonic tone: pleasant

PROPIONALDEHYDE

```
odor thresholds                mg/cu m
          10⁻⁷ 10⁻⁶ 10⁻⁵ 10⁻⁴ 10⁻³ 10⁻² 10⁻¹  1   10  10²  10³  10⁴
detection
recognition
not
specified
```
(610; 676; 724; 788; 842)

    O.I. 100% recogn.: 4,346,000   (19)
  —Sampling and analysis: PMS: det. lim.: 2–7 ppm   (200)
  —Manmade sources:
    in exhaust of a 1970 Ford Maverick gasoline engine operated on a chassis dynamometer following the 7-mode California cycle:
      from API #7 gasoline: 3.2 ppm
      from API #8 gasoline: 1.4 ppm   (1053)
C. WATER POLLUTION FACTORS:
  —BOD_5:   38% of ThOD
  —COD: 97% of ThOD
  —$KMnO_4$: acid: 7% of ThOD; alkaline: 8% of ThOD   (220)
  —Reduction of amenities:
    faint odor: 0.002 mg/l   (129)
    T.O.D. = 0.0095 mg/l   (305)
            detection: 0.145 mg/kg   (874)
  —Waste water treatment:
    A.C.: adsorbability: 0.057 g/g C; 27.6% reduction, infl.: 1,000 mg/l, effl.: 723 mg/l   (32)
    A.S., BOD, 20°C, 1/3–5 days observed, feed: 333 mg/l, acclimation: 30 days; 95% removed   (93)
    A.S.: after   6 hr: 14.4% of ThOD
                  12 hr: 24.9% of ThOD
                  24 hr: 28.8% of ThOD   (88)
    aeration by compressed air (stripping effect): 85% removal in 8 hr   (30)
  —Sampling and analysis:
    scrubbers: liquid lift type; 15 ml of liquid, 4.5 l/min gas flow, 4 min scrubbing

trap method	% removal
H_2O, 0°C:	16
NH_2OH soln, 0°C:	98
H_2SO_4 (concd), 55°C:	97
open tube, −80°C:	3
ethanol, −80°C:	93

(311)

D. BIOLOGICAL EFFECTS:
  —Algae: *Clorella pyrenoidosa*: 3,450 mg/l   (41)
    Lepomis macrochirus: static bioassay in fresh water at 23°C, mil aeration applied after 24 hr:

material added ppm	24 hr	% survival after 48 hr	72 hr	96 hr	best fit 96 hr LC_{50} ppm
180	70	0	—	—	
132	100	80	60	60	130
100	100	71	71	71	
79	100	100	100	90	

Menidia beryllina: static bioassay in synthetic seawater at 23°C: mild aeration applied after 24 hr:

material added ppm	24 hr	% survival after 48 hr	72 hr	96 hr	best fit 96 hr LC_{50} ppm
132	67	30	30	23	
100	95	85	70	55	100
75	100	100	80	80	

(352)

—Mammalia: rat: acute oral LD_{50}: 0.8–1.6 g/kg (211)

    inhalation: no effect level: 90 ppm, 20 × 6 hr (65)

        LC_{50}: 26,000, 30 min

        5/6: 8,000 ppm, 4 hr

        3/3: 60,000 ppm, 20 min (104)

propionamide (propanamide)
$CH_3CH_2CONH_2$

A. PROPERTIES: colorless leaflets; m.w. 73.09; m.p. 79°C; b.p. 213°C; sp.gr. 1.042; log P_{oct} −0.99/−0.70 (calculated)

C. WATER POLLUTION FACTORS:
 —Waste water treatment:
  A.S.: after 6 hr: 2.4% of ThOD
          12 hr: 6.1% of ThOD
          24 hr: 18.1% of ThOD (89)

propionate, sodium
CH_3CH_2COONa

Use: fungicide; mold preventive; medicine; widely used in foods.

D. BIOLOGICAL EFFECTS:
 —*Culex* sp. larvae: 48 hr TLm: 2,320 mg/l (1294)
 —Fish: *Lepomis macrochirus*: 24 hr TLm: 5,000 mg/l (1294)

propione see diethylketone

propionic acid (propanoic acid; methylacetic acid)
CH_3CH_2COOH

A. PROPERTIES: colorless liquid; m.w. 74.1; m.p. −22°C; b.p. 141°C; v.p. 2.9 mm at 20°C; v.d. 2.56; sp.gr. 0.992; log P_{oct} 0.25/0.33

B. AIR POLLUTION FACTORS: 1 ppm = 3.800 mg/cu m, 1 mg/cu m = 0.325 ppm
 —Odor: characteristic: quality: sour
                hedonic tone: unpleasant

1024 PROPIONIC ACID

odor thresholds	mg/cu m

(detection, recognition, not specified — chart with data points across 10^{-7} to 10^4 mg/cu m)

(279; 307; 610; 667; 671; 676; 683; 753; 755; 778; 779; 823; 825; 826)
O.I. 91,500 (19)
—Control methods: thermal incineration for odor control: min. temp.: 1,350°F
(94)
—Sampling and analysis: photometry: min. full scale: 820 ppm (53)
C. WATER POLLUTION FACTORS:
  —BOD_5: 1.3 (30)
         0.36 std. dil. sp. culture (283)
         0.36–0.84 std. dil. sew. (41)
         0.96 std. dil. acclim. (41)
         37% of ThOD (220)
  10 days: 0.69 std. dil. sp. culture (283)
  20 days: 1.4 (30)
  —COD: 1.4 (41)
         99% of ThOD (220)
  —$KMnO_4$ value: 0.014 (30)
         acid: 0% of ThOD; alkaline: 0% of ThOD (220)
  —TOC: 100% of ThOD (220)
  —ThOD: 1.513 (30)
  —Reduction of amenities: T.O.C. = 200 mg/l (295)
  —Manmade sources:
  contents of domestic sewages: 1.2–8 mg/l (85)
  average content of secondary sewage effluents: 13.7 µg/l (86)
  in sewage effluents: 0.190 mg/l (227)
  powdered carbon: at 100 mg/l sodium salt (pH 7.5)—carbon dosage 1000 mg/l:
    4% adsorbed (520)
  —Waste water treatment:
  A.C.: adsorbability: 0.065 g/g C; 32.6% reduction, infl.: 1,000 mg/l, effl.: 674
    mg/l (32)
  A.S.: after 6 hr: 18.6% of ThOD
           12 hr: 30.8% of ThOD
           24 hr: 40.4% of ThOD (88)
  stabilization ponds design: toxicity correction factor: 2.65 for 180 mg/l in
    pond influent (179)
  in year old leachate of artificial sanitary landfill: 4.5 g/l (1720)
D. BIOLOGICAL EFFECTS:
  —Algae: *Chlorella pyrenoidosa*: toxic: 250 mg/l (41)
  —Protozoa: *Vorticella campanula*: perturbation level: 4,000 mg/l
          Paramaecium caudatum: perturbation level: 8,000 mg/l (30)

—Arthropoda: *Gamarus pulex*: perturbation level: 6,000 mg/l (30)
          Daphnia magna: TLm (24 hr): 130 mg/l (246)
          D. magna: TLm (48 hr): 50 mg/l (153)
—Insects: *Culex* sp. larvae: TLm (24, 48 hr): >1,000 mg/l (153)
—Fish: *L. macrochirus*: TLm (24 hr): 188 mg/l (153)
—Mammalia: rat: single oral LD_{50}: >400 mg/kg (211)

propoxur *see* Baygon

n-propylacetate
$CH_3COOCH_2CH_2CH_3$
Users and formulation: flavoring agents; perfumery; solvent for nitrocellulose and other cellulose derivatives; natural and synthetic resins; lacquers; plastics; organic synthesis; lab. reagent.

A. PROPERTIES: colorless liquid; m.w. 102.13; m.p. $-95/-92°C$; b.p. $102°C$; v.p. 25 mm at $20°C$, 35 mm at $25°C$, 42 mm at $30°C$; v.d. 3.52; sp.gr. 0.89 at $20/4°C$, solub. 18,900 mg/l at $20°C$, 26,000 mg/l at $20°C$; sat. conc. 139 g/cu m at $20°C$, 226 g/cu m at $30°C$; log P_{oct} 1.39/1.60 (calculated)

B. AIR POLLUTION FACTORS: 1 mg/cu m = 0.246 ppm, 1 ppm = 4.25 mg/cu m
  —Odor: characteristic: quality: sweet, ester
                  hedonic tone: pleasant

odor thresholds (mg/cu m):
- detection: ~10^{-2} to 10
- recognition: ~10^{-2} to 10^{2}

(610; 676; 709; 749)
O.I. 100% recogn.: 218,666 (19)
O.I. at $20°C$ = 1,600 (316)

C. WATER POLLUTION FACTORS:
  —Waste water treatment:
    A.C.: adsorbability: 0.149 g/g C, 75.2% reduction, infl.: 1,000 ppm, effl.: 248 mg/l (32)

D. BIOLOGICAL EFFECTS:
  —Toxicity threshold (cell multiplication inhibition test):
    bacteria (*Pseudomonas putida*): 170 mg/l (1900)
    algae (*Microcystis aeruginosa*): 530 mg/l (329)
    green algae (*Scenedesmus quadricauda*): 26 mg/l (1900)
    protozoa (*Entosiphon sulcatum*): 97 mg/l (1900)
    protozoa (*Uronema parduczi Chatton-Lwoff*): 843 mg/l (1901)
  —Mammals:
    cat: inhalation: death: 24,500 ppm, 30 min
        eye irritation, salivation: 5,300 ppm, 6 hr/day, 5 days (211)

sec-propylalcohol see isopropanol

n-propylamine
$CH_3CH_2CH_2NH_2$
  Manufacturing source: organic chemical industry. (347)
  Users and formulation: mfg. rubber chemicals, dyestuffs, pharmaceuticals, agricultural chemicals, corrosion inhibitors, textile and leather finishing resins. (347)
A. PROPERTIES: colorless liquid; m.w. 59.11; m.p. -116/-83°C; b.p. 49°C; v.p. 245 mm at 20°C, 400 mm at 31°C; v.d. 2.03; sp.gr. 0.718 at 20/20°C; sat. conc. 788 g/cu m at 20°C, 1,180 g/cu m at 30°C; log P_{oct} 0.15/0.37 (calculated)
B. AIR POLLUTION FACTORS: 1 mg/cu m = 0.41 ppm, 1 ppm = 2.46 mg/cu m
  —Odor threshold conc. 0.022 mg/cu m (821)
  —Control methods: catalytic incineration over commercial Co_3O_4 (1.0–1.7 mm) granules, catalyst charge 13 cm^3, space velocity 45000 h^{-1} at 100 ppm inlet conc.: 80% decomposition at 225°C
    99% decomposition at 240°C
    80% conversion to CO_2 at 240°C
    95% conversion to CO_2 at 275°C (346)
C. WATER POLLUTION FACTORS:
  —Waste water treatment:
    degradation by *Aerobacter*: 200 mg/l at 30°C
      parent: 100% degradation in 31 hr
      mutant: 100% degradation in 9 hr
D. BIOLOGICAL EFFECTS:
  —Fish: creek chub: critical range: 40–60 mg/l; 24 hr (226)
  —Mammalia: rat: acute oral LD_{50}: 0.2–0.4 g/kg (211)

n-propylbenzene (1-phenylpropane)

Manufacturing source: petroleum refining; by-product of cumene mfg. (347)
Users and formulation: mfg. methylstyrene; textile dyeing, printing solvent for cellulose acetate, asphalt and naphtha constituent. (347)
Natural sources (water and air): petroleum (347)
Manmade sources: in gasoline (high octane-number): 0.61 wt %. (387)
A. PROPERTIES: colorless liquid; m.w. 120.19; m.p. -101/-99°C; b.p. 159°C; v.p. 2.5 mm at 20°C; v.d. 4.14; sp.gr. 0.862 at 20°C; solub. 60 mg/l at 15°C; log P_{oct} 3.57/3.68
B. AIR POLLUTION FACTORS:
  —Ambient air quality:
    Los Angeles 1966: glc's avg.: 0.002 ppm (*n* = 136)
      highest value: 0.006 ppm (1319)
  —Atmospheric reactions:

R.C.R.: 0.98
estimated lifetime under photochemical smog conditions in S.E. England: 6.0 hr
(1699; 1700)
reactivity: NO ox.: ranking: 0.2–0.3 (63)
C. WATER POLLUTION FACTORS:
—Reduction of amenities:
organoleptic limit USSR 1970: 0.2 mg/l (181)
—Waste water treatment:
A.S., Sd, BOD, 20°C, 5 days observed, 14 days acclim.: 3% theor. oxidation
A.S., W, 20°C, 1/4 day observed, feed: 50–100 mg/l, 14 days acclim., 1% theor. oxidation (93)
D. BIOLOGICAL EFFECTS:
—Man: EIR: 5.4 (49)

propylcarbinol see *n*-butanol

propylchloride see 1-chloropropane

***n*-propylcyanide** see butyronitrile

S-propyldipropylthiocarbamate (Vernam; Vernolate)

$$CH_3-CH_2-CH_2-S-\underset{\underset{O}{\|}}{C}-N\begin{array}{c}CH_2CH_2CH_3\\ \\ CH_2CH_2CH_3\end{array}$$

Use: selective herbicide
A. PROPERTIES: b.p. 140°C at 20 mm; sp.gr. 0.95 at 20°C; solub. 107 ppm at 21°C
D. BIOLOGICAL EFFECTS:
—Crustaceans:

Gammarus lacustris:	96 hr LC_{50}: 1800 µg/l	(2124)
Gammarus fasciatus:	96 hr LC_{50}: 13000 µg/l	(2125)
Daphnia magna:	48 hr LC_{50}: 1100 µg/l	(2125)
Cypridopsis vidua:	48 hr LC_{50}: 240 µg/l	(2125)
Asellus bravicaudus:	48 hr LC_{50}: 5600 µg/l	(2125)
Palaemonetes kadiakensis:	48 hr LC_{50}: 1900 µg/l	(2125)
Orconectes nais:	48 hr LC_{50}: 24000 µg/l	(2125)

—Mammals:
acute oral LD_{50} (male rat): 1780 mg/kg (technical grade) (1854)
acute oral LD_{50} (rabbit): >2955 mg/kg (1855)

propylene (propene; methylethylene)
CH_2CHCH_3
A. PROPERTIES: colorless gas; m.w. 42.08; m.p. −185.2°C; b.p. −47.8°C; sp.gr. 0.609 at −47/4°C; solub. 200 mg/l at 20°C; LHC 460 kcal/mole
B. AIR POLLUTION FACTORS:
—Odor: characteristic: quality: aromatic (211)
hedonic tone: neutral to unpleasant

1028 PROPYLENE ALDEHYDE

odor thresholds   mg/cu m

[Chart showing odor thresholds on log scale from 10^{-7} to 10^4 mg/cu m:
- detection: ~10²
- recognition: ~10²
- not specified: ~10¹ and ~10²]

(19; 676; 729; 737)

    O.I. 100% recogn.: 14,792   (19)
   —Atmospheric reactions:
    R.C.R.: 2.09
    reactivity: HC conc.: ranking: 1–2
                  NO ox.:   ranking: 0.5–1   (63)
    estimated lifetime under photochemical smog conditions in S.E. England: 1.1 hr
                                                                                             (1699; 1700)
   —Manmade sources:
    glc's: downtown Los Angeles: 0.013–0.064 ppm (as C)
          East San Gabriel Valley: 0.010–0.024 ppm (as C)   (52)
    expected glc's in USA urban air: range: 5–50 ppb   (102)
    glc's downtown Los Angeles: 10%ile: 3 ppb
                                  average: 10.5 ppb
                                  90%ile: 21 ppb   (64)
    in flue gas of municipal incinerator: $<0.4–<1.5$ ppm   (196)
    in exhaust of diesel engine: 6.3% of emitted HC.s   (72)
    in exhaust of reciprocating engine: 7.3% of emitted HC.s
    in exhaust of rotary gasoline engine: 5.3% of emitted HC.s   (78)
    in gasoline: 0.01 vol%   (312)
   —Sampling and analysis:
    non dispersive I.R.: min. full scale: 500 ppm   (55)
    detector tubes: DRÄGER: det. lim.: 1,000 mg/cu m   (58)
    scrubbers: liquid lift type; 15 ml of liquid, 4.5 l/min gal flow, 4 min scrubbing
    trap method   % removal
    H_2O, 0°C: 0
    NH_2OH soln, 0°C: 0
    H_2SO_4 (conc.), 55°C: 24
    H_2SO_4, 4% Ag_2SO_4 (conc.) 55°C: 85
    open tube, −80°C: 0
    ethanol, −80°C: 1   (311)
C. WATER POLLUTION FACTORS:
   —Reduction of amenities: organoleptic limit: 0.5 mg/l   (181)
D. BIOLOGICAL EFFECTS:
   —Seed plants: sweet pea: declination in seedlings: 1,000 ppm, 3 days
                   tomato: epinasty in petiole: 50 ppm, 2 days   (109)
   —Man: EIR: 3.9   (49)

propylene aldehyde   *see* crotonaldehyde

propylenechloride *see* 1,2-dichloropropane

propylenediamineacetate *see* dearcide 706

propylenedichloride *see* 1,2-dichloropropane

propyleneglycol (1,2-propanediol)
$CH_3CHOHCH_2OH$
Use: organic synthesis; especially polypropylene glycol; antifreeze solution; solvent; soft-drink syrups anti-oxidants; coolant in refrigeration systems; plasticizers; suntan lotions; brake fluids; deicing fluids for airport runways
A. PROPERTIES: colorless liquid; m.w. 76.1; m.p. supercools; b.p. 188.2°C; v.p. 0.2 mm at 20°C; v.d. 2.52; sp.gr. 1.0381; log P_{oct} −1.41/−0.30 (calculated)
B. AIR POLLUTION FACTORS: 1 mg/cu m = 0.322 ppm, 1 ppm = 3.11 mg/cu m
C. WATER POLLUTION FACTORS:
 −BOD_5:   0.955   (30)
      2.2% of ThOD
      10 days: 56.7% of ThOD; 15 days: 72.1%; 20 days: 77.8%; 30 days: 79.0%
      40 days: 77.8% of ThOD; 50 days: 80.9% of ThOD   (79)
 −BOD_{20}: 1.225   (30)
 −$KMnO_4$: 0.727   (30)
 −ThOD: 1.685   (30)
 −Waste water treatment:
   A.C.: adsorbability: 0.024 g/g C; 11.6% reduction, infl.: 1,000 mg/l, effl.: 835
   mg/l   (32)
D. BIOLOGICAL EFFECTS:
 −Algae: *Chlorella pyrenoidosa*: toxic: 92,000 mg/l   (41)
 −Fish:
   fingerling trout: at 50,000 mg/l at 10°C: no mortality or apparent signs of stress
    were produced during a 24 hr exposure period (static bioassay)   (592)
   guppy (*Lebistes reticulatus*): 48 hr LC_{50}: >10,000 mg/l   (593)
 −Mammalia:
   rats, monkeys: inhalation: no ill effect: sat. air, 12−18 months
   rat: single oral LD_{50}: 32.5 ml/kg
   rabbit: single oral LD_{50}: 18.5 ml/kg
   dog: single oral LD_{50}: 9.6 ml/kg
   rat: repeated oral doses: no effect: 13.2 g/kg/day, 69 days
                30 ml/kg/day, 6 months   (211)
 −Man: inhalation: no effect, sat. air, prolonged periods no eye irritation   (211)

propyleneoxide *see* propeneoxide

propylene trimer (propene trimer; nonene; PT 3; C_9 olefin mixture)
  Composition: 99.5% olefins (mixture of branched C_9 olefins)
  Use: in manufacturing of isodecanol, nonylbenzene, and nonylphenol
C. WATER POLLUTION FACTORS:
  −BOD_5 = 0.69 = 20% ThOD
  −COD = 1.64 = 48% ThOD   (277)

1030 S-PROPYLETHYLBUTYLTHIOCARBAMATE

D. BIOLOGICAL EFFECTS:
—*Carassius auratus*: 24 hr LD_{50}: 3 mg/l (277)

S-propylethylbutylthiocarbamate *see* tillam

propylethylene *see* 1-pentene

***n*-propylmercaptan** (1-propanethiol)
$CH_3CH_2CH_2SH$
A. PROPERTIES: m.w. 76.15; m.p. $-111.5°C$; b.p. $68°C$; v.p. 40 mm at $-3.2°C$, 100 mm at $15.3°C$, 400 mm at $49.2°C$; sp.gr. 0.836 at $25/4°C$
B. AIR POLLUTION FACTORS: 1 mg/cu m = 0.316 ppm, 1 ppm = 3.165 mg/cu m
—Odor: hedonic tone: unpleasant

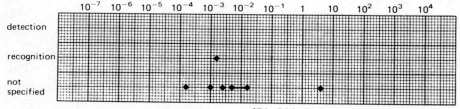

(71; 210; 307; 602; 622; 710; 723; 863)

O.I. at $20°C$ = 263,000,000 (316)
—Control methods:
wet scrubber: water at pH 8.5: effluent: 3,300,000 odor units/scf
$KMnO_4$ at pH 8.5: effluent: 160 odor units/scf (76)

C. WATER POLLUTION FACTORS:
—Reduction of amenities:
faint odor: 0.000075 mg/l (129)

propylsulfide *see* di-*n*-propylsulfide

1-propylthiopropane *see* di-*n*-propylsulfide

propyne *see* methylacetylene

2-propyn-1-ol *see* propargylalcohol

prussic acid *see* hydrogen cyanide

pseudocumene *see* 1,2,4-trimethylbenzene

PTBBA *see* p-*tert*-butylbenzoic acid

PTBT *see* p-*tert*-butyltoluene

6(1)purine *see* hypoxanthine

2,6,8(1,3,9)purinetrione *see* uric acid

pydraul *see* phosphate esters

pyranthon A *see* diacetonealcohol

pyrene (benzo(*def*)phenanthrene)

Manmade sources:
   in gasoline: 1.55 mg/l; 4.7 mg/l; 17.1 mg/kg   (1052; 1070; 1220)
   in bitumen: 0.17–0.80 ppm   (506)
   in coal tar: 46.5 mg/g of sample; 19,000 ppm   (993; 1874)
   in wood preservative sludge: 24.2 g/l of raw sludge   (993)
   in fresh motor-oil: 0.29 mg/kg   (1220)
   in used motor-oil after 18 European driving cycles: 10.7–15.4 mg/kg
   in used motor-oil after 5,000 km: 286–450 mg/kg
   in used motor-oil after 10,000 km: 330–700 mg/kg
   in used motor-oil after 10,000 km and 18 European
           driving cycles: 389–743 mg/kg   (1220)

Natural sources:
   in Kuwait crude oil: 4.5 ppm
   in South Louisiana crude oil: 3.5 ppm   (1015)

A. PROPERTIES: m.w. 202.26; m.p. 149–151°C; solub. 0.16 mg/l at 26°C, 0.032 mg/l at 24°C (practical grade)

B. AIR POLLUTION FACTORS:
  —Manmade sources:
    in tail gases of gasoline engine: 63–484 µg/cu m   (340)
    in stack gases of municipal incinerator, after spray tower and electrostatic precipitator: 1.58 mg/1000 cu m; in residues: 49 µg/kg   (341)
    in exhaust condensate of gasoline engine: 2.15–2.88 mg/l gasoline consumed   (1070)
    emissions from typical European gasoline engine (1608 cu cm)—following European driving cycles—using leaded and unleaded commercial gasolines: 28.2–570.5 µg/l fuel burnt   (1291)
    emissions from space heating installation burning:
      coal (underfeed stoker): 16 mg/10^6 Btu input
      gasoil (underfeed stoker): 6.1 mg/10^6 Btu input
      gas (underfeed stoker): 18 mg/10^6 But input   (954)
    in airborne coal tar emissions: 105.5 mg/g of sample or 3705 µg/cu m of air sampled
    in coke oven emissions: 206–4627 µg/g of sample   (993)
    emission from asphalt hot-mix plant: avg.: 107 µg/1000 cu m
                               range: 44–240 µg/1000 cu m   (491; 1379)

in coal tar pitch fumes: 8.5 wt % (516)
—Ambient air quality:
glc's in the Netherlands (ng/cu m)

		Delft	The Hague	Vlaardingen	Amsterdam	Rotterdam
summer	1968	1	—	4	—	2
	1969	3	3	3	4	1
	1970	2	2	2	1	3
	1971	5	2	2	2	2
winter	1968	15	14	22	22	10
	1969	16	6	24	19	10
	1970	5	2	11	8	3
	1971	1	4	8	4	5

(1277)

glc's in Budapest, 1973: heavy traffic area: winter: 53.6 ng/cu m
   (6–20 hr)   summer: 27.9 ng/cu m
                                 low traffic area: winter: 4.9 ng/cu m
                                                  summer: 11.6 ng/cu m   (1259)

glc's in residential area (Belgium), oct. 1976:
  in particulate sample: 1.64 ng/cu m
  in gas phase sample: 3.36 ng/cu m   (1289)

glc's in Birkenes (Norway), Jan–June 1977:
    avg.: 0.90 ng/cu m
  range: n.d.–4.89 ng/cu m ($n = 18$)

glc's in Rørvik (Sweden), Dec. '76–April '77:
    avg.: 1.33 ng/cu m
    range: 0.13–9.55 ng/cu m ($n = 21$)   (1236)

organic fraction of suspended matter:
  Bolivia at 5200 m altitude (Sept.–Dec. 1975): 0.033–0.034 µg/1000 cu m
  Belgium, residential area (Jan–April 1976): 0.72–1.2 µg/1000 cu m   (428)

glc's in Botrange (Belgium)—woodland at 20–30 km from industrial area: June–July 1977: 0.30; 0.30 ng/cu m ($n = 2$)

glc's in Wilrijk (Belgium)—residential area: Oct–Dec 1976: 0.88; 10.78 ng/cu m ($n = 2$)   (1233)

Cleveland, Ohio, U.S.A. (1971/1972):
  max. conc. 75 ng/cu m (400 values)
  annual geom. mean (all sites): 0.25 ng/cu m
  annual geom. mean of TSP (all sites): 2 ppm (wt)   (556)

Cleveland (16 sites) 1971/1972:   0.20 ng/cu m   (556)
Los Angeles (4 sites) 1971/1972: 1.80 ng/cu m   (560)
Lyon (1 site) 1972:   2.25 ng/cµ m   (561)
Rome (1 site) 1970/1971:   5.0 ng/cu m   (562)
Budapest (1 site) 1971/1972:   2.06 ng/cu m   (563)

in the average American urban atmosphere—1963:
  42 µg/g airborne particulates or
  5 ng/g cu m air   (1293)

comparison of glc's inside and outside houses at suburban sites:
  outside: 2.62 ng/cu m or 54.3 µg/g SPM
  inside: 1.32 ng/cu m or 24.9 µg/g SPM   (1234)

C. WATER POLLUTION FACTORS:
- Biodegradation:
  degradation in seawater by oil-oxidizing microorganisms (in presence of 0.19 mg/l 3,4-benzpyrene and 0.35 mg/l fluorene): initial conc. 0.365 mg/l, after 12 days: 0.055 mg/l; 85% decrease (1237)
- Manmade sources:
  in domestic effluent: 1.763 ppb
  in sewage (high percentage industry): 2.56–3.12 ppb
  in sewage during dry weather: 0.254 ppb
  in sewage during heavy rain: 16.05 ppb (531)
  in outlet of waterspray tower of asphalt hot-road-mix process: 3300 ng/cu m
  in outlet of asphalt air-blowing process: 3,100,000 ng/cu m (1292)
  in leachate from test panels freshly coated with coal tar:
    influent: 0.002 μg/l
    effluent: 0.071 μg/l (1874)
  in effluent spray tower of stack gases of municipal incinerator: 0.46 μg/l (341)
- Water and sediment quality:
  in rapid sand filter solids from Lake Constance water: 0.2 mg/kg
  in river water solids: river Rhine: 1.2 mg/kg
  river Aach at Stockach: 1.4–8.0 mg/kg
  river Argen: 1.1 mg/kg
  river Schussen: 7.9 mg/kg (531)
  in river water: <0.001 ppm, in river sediment: 0.5–75 ppm (555)
  in Eastern Ontario drinking waters (June–Oct. 1978):
    0.04–2.0 ng/l ($n = 12$)
  in Eastern Ontario raw waters (June–Oct. 1978):
    0.2–1.7 ng/l ($n = 2$) (1698)
  in sediments in Severn estuary (U.K.): 0.3–3.5 ppm dry wt. (1467)
  in sediment of Wilderness Lake, Colin Scott, Ontario (1976): 23 ppb (dry wt.)
  (932)
- Aquatic reactions:
  pyrene adsorbed on garden soil for 240 hr and exposed to U.V. radiation undergoes chemical change with the formation of 1,1'-bipyrene, 1,6- and 1,8-pyrenediones and 1,6- and 1,8-pyrenediols, in addition to 3 unknown compounds:

pyrene $\xrightarrow{h\nu}_{\text{Soil, 32°C}}$ 1,1'-bipyrene + 1,6-pyrenedione + O 1,8-pyrenedione adsorption on smectite clay particles from simulated seawater at 25°C—experimental conditions: 100 μg pyrene/l; 50 mg smectite/l—adsorption: 0.39 μg/mg = 19% adsorbed (1009)

—Waste water treatment methods:
ozonation: after 1 min. contact time with ozone: residual amount: 15% (550)
chlorination: 6 mg/l chlorine for 6 hr: initial conc. 27.07 ppb: 24% reduction (549)
double percolating filter: influent: 1.0 ppb, 64% removal
overloaded single percolating filter: infl. 1.55 ppb, 3% removal
infl. 0.43 ppb, 46–65% removal (546)

	Dec. 1965	May 1966	
in raw sewage	11.8 ppb	1.24 ppb	
after mechanical purification	2.3 ppb	0.35 ppb	
after biological purification	0.04 ppb	–	(545)

concentration at various stages of water treatment works:
   river intake: 0.100 μg/l
   after reservoir: 0.075 μg/l
   after filtration: 0.045 μg/l
   after chlorination: 0.018 μg/l (434)

—Sampling methods:
recovery with open pore polyurethane (OH/NCO = 2.2): 89–100% depending upon load-quantitative elution with methanol (929)

D. BIOLOGICAL EFFECTS:
—Fish:
mosquito fish: static bioassay: 96 hr TLm: 0.0026 mg/l at 24–27°C (467)
—Carcinogenicity: negative
—Mutagenicity in the *Salmonella* test: negative
<0.02 revertant colonies/nmol
<100 revertant colonies at 1000 μg/plate (1883)
—Man: carcinogenic

Pyrethrins (pyrethrum; *see also*: RU-11679; SBP-1382; *d-trans*-allethrin; dimethrin; decamethrin; fenpropanate; fenvalerate; permethrin)

Trivial name applied to the insecticidal principles in pyrethrum flowers Chrysanthemum cinerariaefolium. The pyrethrins include pyrethrins I and II, cinerins I and II, and jasmolins I and II.

Synthetic pyrethinlike (pyretroids) compounds produced in attempts to duplicate the activity of natural pyrethrins include allethrin, cyclethrin, dimethrin, resmethrin, etc.

	R'	R
pyrethrin I	CH_3	$CH_2CH=CHCH=CH_2$
cinerin I	CH_3	$CH_2CH=CHCH_3$
pyrethrin II	$\overset{O}{\overset{\|}{C}}-O-CH_3$	$CH_2CH=CHCH=CH_2$
cinerin II	$\overset{O}{\overset{\|}{C}}-O-CH_3$	$CH_2CH=CHCH_3$

D. BIOLOGICAL EFFECTS:
 —Crustaceans:
  Gammarus lacustris: 96 hr LC_{50}: 12 µg/l (2126)
  Gammarus fasciatus: 96 hr LC_{50}: 11 µg/l (2126)
  Simocephalus serrulatus: 48 hr LC_{50}: 42 µg/l (2127)
  Daphnia pulex: 48 hr LC_{50}: 25 µg/l (2127)
 —Insects:
  Pteronarcys californica: 96 hr LC_{50}: 1.0 µg/l (2128)
 —Fish:
  rainbow trout: 24 hr LC_{50}: 56 ppb (1596)
  rainbow trout: 96 hr LC_{50} (S): 52.2 mg/l (1101)

	96 hr LC_{50} (µg/l)*	
	static test	flow through test
coho salmon	39.0	23.0
steelhead trout	24.6	22.5
channel catfish	114	132
bluegill	49	104
yellow perch	<50.1	44.5

*toxicity calculated for formulation with 20% of active ingredient
(1626)
Atlantic salmon: 48 hr LTC (S): 32.0 µg/l (1196)

pyrethroids see RU-16579; SBP-1382; *d-trans*-allethrin; dimethrin; decamethrin; fenpropanate; fenvalerate; permethrin; pyrehtrins

pyrethrum see pyrethrins

pyridine

Use: synthesis of vitamins and drugs; solvent; waterproofing; rubber chemicals.
A. PROPERTIES: colorless liquid; m.w. 79.1; m.p. -42°C; b.p. 115°C; v.p. 14 mm at 20°C, 20 mm at 25°C, 26 mm at 30°C; v.d. 2.73; sp.gr. 0.982; THC 660 kcal/mole, LHC 644 kcal/mole; sat. conc. 65 g/cu m at 20°C, 108 g/cu m at 30°C; log P_{oct} 0.64/1.04
B. AIR POLLUTION FACTORS: 1 mg/cu m = 0.30 ppm, 1 ppm = 3.29 mg/cu m
 —Odor: characteristic: quality: burnt, pungent, diamine

PYRIDINE

odor thresholds   mg/cu m

[odor threshold chart showing detection, recognition, and not specified ranges from 10^{-7} to 10^{4} mg/cu m]

(2; 57; 210; 306; 307; 602; 606; 610; 643; 664; 682; 704; 708; 710; 731; 737; 741; 802; 871)

    undapted: 0.5 ppm
    after adaption with pure odorant: 700 ppm   (204)
  USSR: human odor perception: non perception: 0.084 mg/cu m
                    perception: 0.21 mg/cu m = 0.063 ppm
      human reflex response: no response: 0.079 mg/cu m
                    adverse response: 0.098 mg/cu m
      animal chronic exposure: no effect: 0.1 mg/cu m
                    adverse effect: 1.0 mg/cu m   (170)
    O.I. at 20°C = 2,390   (316)
—Control methods:
  catalytic combustion: platinized ceramic honeycomb catalyst: ignition temp.: 407°C, inlet temp for 90% conversion: 400–450°C   (91)
—Sampling and analysis:
  photometry: min. full scale: 9.0 ppm   (53)

C. WATER POLLUTION FACTORS:
  —BOD_5:   1.15; 0.06   (30, 36)
           0–1.47 std. dil. sew.   (41)
           52% of ThOD   (274)
           0 at 440 ppm, Sierp, sew.   (280)
  —BOD_5^{20}: 0.06 at 10 mg/l, unadapted sew.
  —BOD_{20}^{20}: 2.02 at 10 mg/l, unadapted sew.; lag period: 9 days   (554)
  —COD: 2% ThOD (0.05 n $K_2Cr_2O_7$)   (174)
          0.02; 0.05; 0.037   (27, 36, 223)
  —$KMnO_4$: 0.024   (30)
           0% of ThOD (0.01 n $KMnO_4$)   (274)
  —ThOD: 2.23; 3.03; 3.13   (30, 27, 36)
  —Reduction of amenities:
    approx. conc. causing adverse taste in fish (carp, rudd): 5.0 mg/l   (41)
    faint odor: 0.0037 mg/l   (129)
    taste and odor threshold: 0.8 mg/l (tentative)   (27)
  —Odor thresholds: See top of next page.
    tainting of fish flesh: 5.0 mg/l   (81)
  —Aquatic reactions:
    photooxidation by U.V. light in aqueous medium at 50°C: 23.06% degradation to CO_2 after 24 hr   (1628)
  —Waste water treatment:

odor thresholds   mg/kg water

[Chart showing odor threshold ranges on log scale from 10^{-7} to 10^4 mg/kg water, with rows for "detection", "recognition", and "not specified"]

(97; 294; 296; 876; 901; 907)

A.C.: adsorbability: 0.095 g/g C; 47.3% reduction, infl.: 1,000 mg/l, effl.: 527 mg/l   (32)

methods	temp °C	days observed	feed mg/l	days acclim.	% removed	
RW, CA	20	3	1	6	100	
Sew D, CA	20	7	1	none	100	(93)

A.C.: influent ppm	carbon dosage	effluent ppm	% reduction	
1,000	10X	145	86	
500	10X	71	86	(192)

—Sampling and analysis:
photometry: min. full scale: 0.6×10^{-6} mole/l   (53)

D. BIOLOGICAL EFFECTS:
—Toxicity threshold (cell multiplication inhibition test):
bacteria (*Pseudomonas putida*):   340 mg/l   (1900)
algae (*Microcystis aeruginosa*):   28 mg/l   (329)
green algae (*Scenedesmus quadricauda*):   120 mg/l   (1900)
protozoa (*Entosiphon sulcatum*):   3.5 mg/l   (1900)
protozoa (*Uronema parduczi Chatton-Lwoff*): 183 mg/l   (1901)
—Algae:
inhibition of photosynthesis of a freshwater, nonaxenic unialgal culture of *Selenastrum capricornutum* at:
10 mg/l: 67% carbon-14 fixation (vs. controls)
100 mg/l: 86% carbon-14 fixation (vs. controls)
1000 mg/l: 80% carbon-14 fixation (vs. controls)   (1690)
—Protozoa:
ciliate (*Tetrahymena pyriformis*): 24 hr LC_{100}: 113.8 mmole/l   (1662)
—Amphibian:
mexican axolotl (3–4 wk after hatching): 48 hr LC_{50}:   950 mg/l
clawed toad (3–4 wk after hatching):   48 hr LC_{50}: 1400 mg/l   (1823)
—Fish:
Alburnus: LD_0: 100 mg/l   (30)
mosquito fish: TLm (24 hr): 1,350 mg/l in turbid Oklahoma   (244)
—Mammalia: rat: acute oral LD_{50}: 0.8–1.6 g/kg
           mouse: acute oral LD_{50}: 0.8–1.6 g/kg
              rat: inhalation: 6/6: 23,000 ppm, 1.5 hr
                     5/6:   4,000 ppm, 6 hr

2/3: 3,600 ppm, 6 hr (211)
—Man: transient symptoms: 125 ppm, 4 hr/day, 1–2 weeks (211)

α-pyridylamine *see* 2-aminopyridine

pyrocatechol *see* catechol

pyrocatechin *see* catechol

pyrocatecholdimethylether *see* 1,2-dimethoxybenzene

pyrocatechol monomethylether *see* 2-methoxyphenol

pyrogallic acid *see* pyrogallol

pyrogallol (1,2,3-benzenetriol; γ-trihydroxybenzene; pyrogallic acid)

A. PROPERTIES: needles of leaflets; m.w. 126.11; m.p. 133/134°C decomposes; v.p. 20 mm at 185°C, 400 mm at 281°C; sp.gr. 1.453 at 4/4°C; solub. 625,000 mg/l at 25°C, 400,000 mg/l at 13°C
B. AIR POLLUTION FACTORS: 1 mg/cu m = 0.194 ppm, 1 ppm = 5.15 mg/cu m
  —Odor: hedonic tone: practically odorless
C. WATER POLLUTION FACTORS:
  —BOD_5: 1% of ThOD (274)
  —COD: 95% of ThOD (0.05 n $K_2Cr_2O_7$) (274)
  —$KMnO_4$: 82% of ThOD (0.01 n $KMnO_4$) (274)
  —ThOD: 1.52 (274)
  —Biodegradation: adapted A.S. at 20°C—product is sole carbon source: 40% COD removal after 120 hr (327)
  —Odor threshold conc. (detection): 2.0 mg/l (998)
    tainting of fish flesh: 20.0–30.0 mg/l (81)
  —Impact on biodegradation processes:
    inhibition of degradation of glucose by *Pseudomonas fluorescens*: at 3 mg/l
    inhibition of degradation of glucose by *E. coli*: at 30 mg/l (293)
D. BIOLOGICAL EFFECTS:
  —Algae: *Scenedesmus*: LD_0: 8 mg/l
  —Arthropoda: *Daphnia*: LD_0: 18 mg/l (30)
  —Mammalia: rabbit: approx. lethal oral dose: 1.1 g/kg (211)
  —Fish: goldfish: approx. fatal conc.: 18 mg/l; 48 hr (226)

pyromucic acid *see* 2-furoic acid

dl-2-pyrrolidinecarboxylic acid *see* dl-proline

2-pyrrolidinone *see* 2-pyrrolidone

2-pyrrolidone (2-pyrrolidinone; butyrolactam)

Use: plasticizer for acrylic latexes; solvent for polymers, insecticides, special inks etc.

A. PROPERTIES: m.w. 85.11; b.p. 245°C; soluble in water; sp.gr. 1.1
C. WATER POLLUTION FACTORS:

theoretical	analytical
TOD = 2.44	reflux COD = 95.3% recovery
COD = 1.69	rapid COD = 101.8% recovery
NOD = 0.75	TKN = 90.0% recovery
	BOD_5 = 1.16
	BOD_5/COD = 0.720
	BOD_5 (acclimated) = 1.39

—The results of the BOD_5 analyses suggest 2-pyrrolidone to be readily biodegradable. (1828)
—Biodegradation:

	chemical conc. mg/l	effect
unacclimated system	·126	no effect
	1,260	no effect
acclimated system	137	biodegradable

In contrast to the BOD_5 analysis, the Warburg analysis, using an unacclimated biomass, did not measure any evidence of biodegradation of the compound. This is probably due to the different characteristics of the seeds used in the two analyses. The BOD_5 seed, from a source servicing a high industrial input, readily assimilated the compound. The Warburg seed, from a domestic waste source, was completely unadapted to the compound. The compound was found, however, to be amenable to high rate biodegradation by the acclimated biomass.
(1828)

D. BIOLOGICAL EFFECTS:
—Algae: *Selenastrum capricornutum*: 1 mg/l: no effect
10 mg/l: no effect
100 mg/l: no effect   (1828)
—Crustacean: *Daphnia magna*: LC_{50}: 3.4 mg/l
—Fish: *Pimephales promelas*: LC_{50} >100 mg/l   (1828)

pyrrolylene *see* 1,3-butadiene

1-pyrrolylmethanoic acid *see dl*-proline

Q

quinalphos *see* ekalaux

quinhydrone (benzoquinhydrone)
$C_6H_4O_2C_6H_4(OH)_2$
A. PROPERTIES: black, dark green crystals; m.w. 218.20; m.p. 171°C; b.p. sublimes; sp.gr. 1.401 at 20/4°C
C. WATER POLLUTION FACTORS:
 —Impact on biodegradation processes:
  at 0.2 mg/l inhibition of degradation of glucose by *Pseudomonas fluorescens*
  at 5 mg/l inhibition of degradation of glucose by *E. coli* (293)
D. BIOLOGICAL EFFECTS:
 —Algae: *Scenedesmus*: LD_0: 4 mg/l
 —Bacteria: *E. coli*: LD_0: 5 mg/l
 —Arthropoda: *Daphnia*: LD_0: 0.4 mg/l (30)

quinic acid (1,3,4,5-tetrahydroxycyclohexanecarboxylic acid)

A. PROPERTIES: colorless crystals; m.w. 192.17; m.p. 163°C; b.p. decomposes; sp.gr. 1.637; solub. 400,000 mg/l
C. WATER POLLUTION FACTORS:
 —Waste water treatment:
  sel. strain, Sd, W, 30°C, 1 hr observed, 2 μmole/Warburg flask, 2 days acclim.: 59% theor. oxidation (93)

quinone *see* p-benzoquinone

quinol *see* hydroquinone

quinoline (benzo(b)pyridine; 1-benzazine)

A. PROPERTIES: colorless liquid; m.w. 129.15; m.p. $-19.5°C$; b.p. $237.7°C$; v.p. 1 mm at $59.7°C$; v.d. 4.45; sp.gr. 1.095 at $20/4°C$; solub. 60,000 mg/l; log P_{oct} 2.03/2.06

B. AIR POLLUTION FACTORS:
   odor thresholds

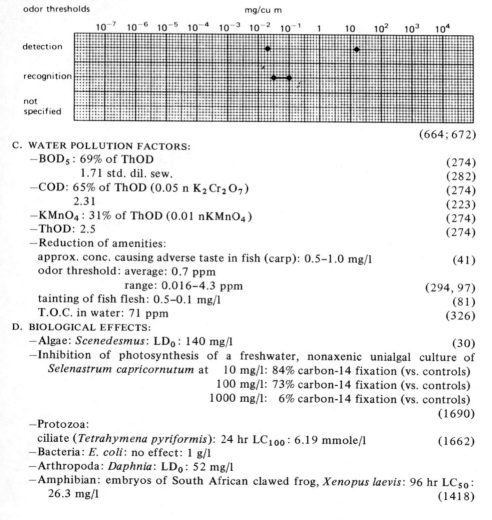

                                                                                   (664; 672)

C. WATER POLLUTION FACTORS:
   —BOD_5: 69% of ThOD   (274)
         1.71 std. dil. sew.   (282)
   —COD: 65% of ThOD (0.05 n $K_2Cr_2O_7$)   (274)
         2.31   (223)
   —$KMnO_4$: 31% of ThOD (0.01 $nKMnO_4$)   (274)
   —ThOD: 2.5   (274)
   —Reduction of amenities:
      approx. conc. causing adverse taste in fish (carp): 0.5–1.0 mg/l   (41)
      odor threshold: average: 0.7 ppm
                       range: 0.016–4.3 ppm   (294, 97)
      tainting of fish flesh: 0.5–0.1 mg/l   (81)
      T.O.C. in water: 71 ppm   (326)

D. BIOLOGICAL EFFECTS:
   —Algae: *Scenedesmus*: LD_0: 140 mg/l   (30)
   —Inhibition of photosynthesis of a freshwater, nonaxenic unialgal culture of *Selenastrum capricornutum* at   10 mg/l: 84% carbon-14 fixation (vs. controls)
                                                        100 mg/l: 73% carbon-14 fixation (vs. controls)
                                                       1000 mg/l: 6% carbon-14 fixation (vs. controls)
                                                                               (1690)
   —Protozoa:
      ciliate (*Tetrahymena pyriformis*): 24 hr LC_{100}: 6.19 mmole/l   (1662)
   —Bacteria: *E. coli*: no effect: 1 g/l
   —Arthropoda: *Daphnia*: LD_0: 52 mg/l
   —Amphibian: embryos of South African clawed frog, *Xenopus laevis*: 96 hr LC_{50}: 26.3 mg/l   (1418)

—Mammals:
  rat: single oral LD_{50}: 0.46 g/kg
       inhalation: 0/3: 17 ppm, 6 hr
                   3/3: 4,000 ppm, 5.5 hr   (211)

8-quinolinol see 8-hydroxyquinoline

quintozene see pentachloronitrobenzene

R

R 12 *see* dichlorofluoromethane

racemic acid *see dl*-tartaric acid

resmethrin *see* SBP-1382

resorcin *see* resorcinol

resorcinol (*m*-hydroxyphenol; 1,3-dihydroxybenzene; 1,3-benzenediol; resorcin)

A. PROPERTIES: white needle shaped crystals or rhombic tablets and pyramids which turn pink upon exposure to light and air; m.w. 110.11; m.p. 276/280°C; v.p. 5 mm at 138°C, 100 mm at 209°C; v.d. 3.79; sp.gr. 1.285 at 15°C; solub. 840 g/l at 0°C, 2,290 g/l at 30°C; log P_{oct} 0.77/0.80

C. WATER POLLUTION FACTORS:
 —Biodegradation: adapted culture: 89% removal after 48 hr incubation, feed
 446 mg/l (292)
 decompositon by a soil microflora in 8 days (176)
 adapted A.S. at 20°C—product is sole carbon source: 90.0% COD removal at
 57.5 mg COD/g dry inoculum/hr (327)
 —Oxidation parameters:
 BOD_5: 1.15 std. dil. (27)
  : 61% ThOD (274)
 COD: 100% ThOD (0.05 n K_2CrO_7) (274)
 $KMnO_4$: 90% ThOD (0.01 n $KMnO_4$) (274)
 ThOD: 1.89 (27)
 —Impact on biodegradation processes:
 at 200 mg/l, inhibition of degradation of glucose by *Pseudomonas fluorescens*
 at >1,000 mg/l, inhibition of degradation of glucose by *E. coli* (293)
 —Aquatic reactions:
 auto-oxidation at 25°C: $t_{1/2}$: 1612 hours at pH 9.0 (1908)

-Reduction of amenities:
   approx. conc. causing adverse taste in fish (carp): 30 mg/l   (41)
   tainting of fish flesh: 30 mg/l   (81)
-Taste threshold conc.: 2 mg/l   (998)
-Odor threshold conc. (detection): 6.0 mg/l   (998)
-Sampling and analysis: photometry: min. full. scale: 0.67×10^{-6} mole/l   (53)

D. BIOLOGICAL EFFECTS:
-Bacteria: *E. coli* LD_0: <1,000 mg/l   (30)
-Algae: *Scenedesmus*: LD_0: 60 mg/l   (30)
-Arthropoda: *Daphnia*: LD_0: 0.8 mg/l   (30)
-Fish: goldfish: approx. fatal conc.: 57.4 mg/l, 48 hr   (226)

	24 hr LC_{50}	48 hr LC_{50}	96 hr LC_{50}
fathead minnows (*Pimephales promelas*)	88.6 mg/l	72.6 mg/l	53.4 mg/l
grass shrimp (*Palaemonetes pugio*)	170 mg/l	78 mg/l	42 mg/l

(1904)

-Mammals: rats, guinea pigs: approx. lethal oral dose: 0.37 g/kg   (211)
   rabbit: approx. lethal oral dose: 0.75 g/kg   (211)
-Carcinogenicity: none ?
-Mutagenicity in the *Salmonella test*: none
   <0.008 revertant colonies/nmol
   <70 revertant colonies at 1000 μg/plate   (1883)

resorcinoldimethylether   *see* 1,3-dimethoxybenzene

resorcinolmonomethylether   *see* 3-methoxyphenol

rhodanin S 62   *see* ethylenethiourea

rhotane   *see* DDD

ricinoleic acid   (12-hydroxy-9-octadecenoic acid, ricinolic acid)
$CH_3(CH_2)_5 CHOHCH_2 CHCH(CH_2)_7 COOH$
A. PROPERTIES: colorless liquid: m.w. 298.46; m.p. α: 7.7°C, β: 16.0°C, γ: 5.0°C; b.p. 250°C at 15 mm; sp.gr. 0.945 at 15°C
C. WATER POLLUTION FACTORS:
-Waste water treatment: A.S. after   6 hr: 5.5% ThOD
   12 hr: 17.6 ThOD
   24 hr: 29.7 ThOD   (89)

roccal   (benzalkonium chloride; alkyldimethyl benzylammonium chloride)
Use: microbicide.
D. BIOLOGICAL EFFECTS:
-Mussels:
   Mercenaria mercenaria (hard clam), egg:
      static lab bioassay: 48 hr TLm: 190 ppb
   Mercenaria mercenaria (hard clam), larvae:
      static lab bioassay: 12 day TLm: 140 ppb   (2324)

—Fish:

	mg/l	24 hr	48 hr	96 hr	3m (extrap.)
harlequin fish					
(*Rasbora heteromorpha*) $LC_{10}(F)$	1.85	0.59	—	—	
$LC_{50}(F)$	2.45	1.1	0.62	0.04	

(331)

rodinal *see p*-aminophenol

ro-neet (S-ethylethylcyclohexylthiocarbamate; cycloate; hexylthiocarbam)

Use: herbicide
A. PROPERTIES: m.w. 215.36; sp.gr. 1.0156 at 20°C; v.p. 0.002 mm at 25°C; b.p. 145–146°C at 10 mm Hg; solub. 100 ppm
  —Formulations:
  Ro-neet 6-E: an emulsifiable concentrate formulation containing 6 lbs ro-neet/gal.
  Ro-neet 10-G: a granular formulation containing 10% Ro-neet by weight
D. BIOLOGICAL EFFECTS:
  —Fish:
  Technical ro-neet: rainbow trout (*S. gairdneri*) 96 hr LC_{50}: 4.5–5.6 ppm (497)
  —Birds:
  bobwhite quail: LC_{50}: >56,000 ppm (7-day feeding) (497)
  —Mammals:
    acute oral toxicity: male rats: technical grade: LD_{50}: 3190 mg/kg
                                   ro-neet 6E: LD_{50}: 3160 mg/kg
    acute dermal toxicity: rabbits: technical grade: LD_{50}: >2000 mg/kg
                                    ro-neet 6E: LD_{50}: >4000 mg/kg
    eye irritation: rabbits: technical grade: single 0.1 ml doses produced mild transient conjunctivitis
                             ro-neet 6E: single 0.1 ml doses produced moderate reversible conjunctivitis
    acute inhalation: female rats: LC_{50}: 90 mg/l for 1 hr exposure (497)

ronnel (O,O-dimethyl-O-(2,4,5-trichlorophenyl)phosphorothioate; fenchlorphos)

1046 ROTENONE

Use: insecticide
A. PROPERTIES: m.p. 40–42°C; v.p. 8×10^{-4} mm at 25°C; solub. 40 ppm at room temp., 44 mg/l, 1.08 mg/l at 20°C; b.p. decomposes; sp.gr. 1.48 (25/4°C); log P_{oct} 4.88
D. BIOLOGICAL EFFECTS:
  —Fish:
    Pimephales promelas: 96 hr LC_{50}: 305 µg/l   (2130)
  —Mammals:
    acute oral: LD_{50}: (rats): 1740 mg/kg
    acute dermal LD_{50}: (rats): 2000 mg/kg   (1855)
    ingestion: rat: single oral LD_{50}:   2500 mg/kg
               guinea pig: single oral LD_{50}: 2800 mg/kg
               rabbit: single oral LD_{50}:   420 mg/kg   (1546)
    skin adsorption: rabbit: LD_{50} in the range of 1000 to 2000 mg/kg   (1546)

rotenone (tubatoxin; noxfish; *see also* dactinol)

[Chemical structure of rotenone]

Rotenone is the trivial name given to the main insecticidal compound in the roots of certain *Derris* and *Lonchocarpus* spp.
Use: a selective nonsystemic insecticide
A. PROPERTIES: m.w. 394.4; m.p. 178–181°C; b.p. 210–220°/0.5 mm; sp.gr. 1.27 at 20°C; m.p. 163°C; solub. 15 ppm at 100°C
C. WATER POLLUTION FACTORS:
  —Odor threshold conc. 0.36 mg/kg   (915)
D. BIOLOGICAL EFFECTS:
  —Crustaceans:
    Gammarus lacustris: 96 hr LC_{50}: 2600 µg/l   (2124)
    Simocephalus serrulatus: 48 hr LC_{50}: 190 µg/l   (2127)
    Daphnia pulex: 48 hr LC_{50}: 100 µg/l   (2127)
  —Insects:
    Pteronarcys californica: 96 hr LC_{50}: 380 µg/l   (2128)
  —Fish: exposed to Noxfish (5% rotenone)
    bowfin: (*Amia calva*) 96-hr LC_{50}: 30.0 µg/l (S)
    Coho salmon: 96-hr LC_{50}: 62.0 µg/l (S)
    Chinook salmon: 96-hr LC_{50}: 36.9 µg/l (S)
    rainbow trout: 96-hr LC_{50}: 46.0 µg/l (S)
    Atlantic salmon: 96-hr LC_{50}: 21.5 µg/l (S)

brook trout:	96-hr LC_{50}: 44.3 µg/l (S)
lake trout:	96-hr LC_{50}: 26.9 µg/l (S)
northern pike:	96-hr LC_{50}: 33.0 µg/l (S)
goldfish:	96-hr LC_{50}: 497 µg/l (S)
carp:	96-hr LC_{50}: 50.0 µg/l (S)
fathead minnow:	96-hr LC_{50}: 142 µg/l (S)
longnose sucker: (*Catostomus catostomus*):	96-hr LC_{50}: 57.0 µg/l (S)
white sucker:	96-hr LC_{50}: 68.0 µg/l (S)
black bullhead:	96-hr LC_{50}: 389 µg/l (S)
channel catfish:	96-hr LC_{50}: 164 µg/l (S)
green sunfish:	96-hr LC_{50}: 141 µg/l (S)
bluegill:	96-hr LC_{50}: 141 µg/l (S)
smallmouth bass:	96-hr LC_{50}: 79.0 µg/l (S)
largemouth bass:	96-hr LC_{50}: 142 µg/l (S)
yellow perch:	96-hr LC_{50}: 70.0 µg/l (S)
walleye:	24-hr LC_{50}: 16.5 µg/l (S)

—Mammals:
acute oral LD_{50} (white rat):   132–1500 mg/kg
(white mice):   350 mg/kg   (1855)
respiratory inhibitor, induces mammary adenomas

RU-11679   (5-benzyl-3-furyl)methyl-trans-(+)-3-(cyclopentylilidenemethyl-2,2-dimethylcyclopropanecarboxylate)
Use: pyrethroid
D. BIOLOGICAL EFFECTS:
—Fish:

	96 hr LC_{50} (µg/l)*	
	static test	flow through test
coho salmon	0.635	0.151
steelhead trout	0.110	0.100
channel catfish	0.630	0.700

	LC_{50} (µg/l) flow through test*							
	5 d	10 d	15 d	20 d	25 d	30 d	35 d	$TILC_{50}$
coho salmon	0.176	0.160	0.109	0.093	0.089	0.069	0.064	0.0293
channel catfish	0.750	0.410	0.410	0.305	0.305	0.305		0.194

*toxicity calculated for formulation with 95% of active ingredient   (1626)
$TILC_{50}$ = time independent lethal concentration for 50% of the exposed population
= concentration at which 50% of the animals exposed would survive indefinitely

S

salicylal   *see* salicylaldehyde

salicylaldehyde   (salicylal; *o*-hydroxybenzaldehyde)

Users and formulation: analytical chemistry; perfumery (violet); synthesis of coumarins; gasoline additives; auxiliary fumigant.

A. PROPERTIES: colorless, oily liquid or dark red oil; bitter almond like oder; burning taste; m.w. 122.1; m.p. −7°C; b.p. 196°C; v.p. 1 mm at 33°C; sp.gr. 1.167 at 20/4°C; log P_{oct} 1.70/1.81

B. AIR POLLUTION FACTORS:
  —Odor threshold conc.:   detection: 0.014 mg/cu m   (840)
                               recognition: 0.38–0.41 mg/cu m   (610)

C. WATER POLLUTION FACTORS:
  —Odor threshold conc. (detection): 0.03 mg/l   (998)

D. BIOLOGICAL EFFECTS:
  —Toxicity threshold (cell multiplication inhibition test):

bacteria (*Pseudomonas putida*):	10 mg/l	(1900)
algae (*Microcystis aeruginosa*):	1.6 mg/l	(329)
green algae (*Scenedesmus quadricauda*):	4.9 mg/l	(1900)
protozoa (*Entosiphon sulcatum*):	1.4 mg/l	(1900)
protozoa (*Uronema parduczi Chatton-Lwoff*):	5.5 mg/l	(1901)

  —Amphibian:
    mexican axolotl (3–4 w after hatching): 48 hr LC_{50}: 7.0 mg/l
    clawed toad (3–4 w after hatching):   48 hr LC_{50}: 7.7 mg/l   (1823)
  —Mammals:
    ingestion: rat: single oral LD_{50}: 300–1000 mg/kg   (1546)

salicylic acid   (*o*-hydroxybenzoic acid)

A. PROPERTIES: colorless needles; m.p. 158°C; b.p. 256°C; v.p. 5 mm at 136°C, 60 mm at 182°C; sp.gr. 1.443 at 20/4°C; solub. 1.8 g/l at 20°C, 17.6 g/l at 75°C; P_{oct} 2.21/2.26

C. WATER POLLUTION FACTORS:
—Biodegradation: decomposition by a soil microflora in 2 days (176)
 adapted A.S. at 20°C—product is sole carbon source: 98.8% COD removal at 95.0 mg COD/g dry inoculum/hr (327)
—Oxidation parameters
 BOD_5: 0.95 std. dil. sew. (282, 163)
 : 0.97 std. dil. acclim. (41)
 : 41% ThOD 220)
 BOD_{20}: 1.25 (30)
 COD: 1.58 (41)
 : 100% ThOD (220)
 $KMnO_4$ value: 5.25 (30)
 : acid: 100% ThOD
 : alkaline: 90% ThOD (220)
 ThOD: 1.623 (30)
—Waste water treatment: coagulation: 17% BOD reduction with 3 lb alum per 1,000 gal (95)

D. BIOLOGICAL EFFECTS:
—Mammals:
 ingestion: rat: single oral LD_{50}: 400–800 mg/kg
—Carcinogenicity: none?
—Mutagenicity in the Salmonella test: none
 <0.02 revertant colonies/nmol
 <70 revertant colonies at 500 µg/plate (1883)

salicylic acid methylether see o-anisic acid

santicizer 711
A mixture of diheptyl-, dinonyl-, and diundecylphthalates.
C. WATER POLLUTION FACTORS:
—Biodegradation: in continuous A.S.: 51%
 in river water—1 week: 0% (1841)

sarcine see hypoxanthine

SAS see teepol 610

SBA see sec-butanol

SBP-1382 (resmethrin; (5-benzyl-3-furyl)methyl-cis-trans-(±)-2,2-dimethyl-3-(2-methylpropenyl)cyclopropanecarboxylate)
Use: a pyrethroid
D. BIOLOGICAL EFFECTS:
—Fish:

	96 hr LC_{50} ($\mu g/l$)	
	static test	flow through test
coho salmon	>150	<0.277
steelhead trout	0.450	0.275
bluegill	2.62	0.750
yellow perch	2.36	0.513 (1626)

sencorex *see* bayer 6159

dl-serine (2-amino-3-hydroxypropanoic acid)

$$\begin{array}{c} CH_2OH \\ | \\ CH-NH_2 \\ | \\ COOH \end{array}$$

A. PROPERTIES: m.w. 105.09; m.p. 246°C decomposes; solub. 50,200 mg/l at 25°C, 192,100 mg/l at 75°C

C. WATER POLLUTION FACTORS:
  —Manmade sources: excreted by man in urine: 0.35–1.4 mg/kg body wt./day
  (203)
  —Waste water treatment: A.S. after 6 hr: 8.6% ThOD (1-serine)
                                    12 hr: 21.0% ThOD
                                    24 hr: 29.0% ThOD (89)

sevin *see* carbaryl

sextone *see* cyclohexanone

shellflex
  Naphtenic and paraffinic distillates from lubricating oil manufacturing.
  Use: process oils in printing and rubber industry.
  Shellflex 310 HP: sp.gr. 0.885 at 289/289 K; p.p. 264 K; solub. 1.3 mg/l.
         Composition: 0% asphaltenes; 1.3% polar components; 29.0% aromatics; 69.7% aliphatics.
  Shellflex 212 HP: sp.gr. 0.882 at 289/289 K; p.p. <247 K; solub. 3.3 mg/l.
         Composition: <0.1% asphaltenes; 2.5% polar components; 38.5% aromatics; 58.7% aliphatics.
  Shellflex 451 NC: sp.gr. 0.905 at 289/289 K; p.p. 253 K; solub. 1.7 mg/l.
         Composition: <0.1% asphaltenes; 1.4% polar components; 33% aromatics; 65.6% aliphatics

C. WATER POLLUTION FACTORS:

	BOD_5	COD	*Carassius auratus*
Shellflex 451 NC	0.11	2.09	not toxic in saturated solution (1.7 mg/l)
Shellflex 214 BG	0.17	2.31	not toxic in saturated solution (3.3 mg/l)
Shellflex 724 BG	0.17	3.05	not toxic in saturated solution (1.5 mg/l)
Shellflex 310 HP	0.33	3.16	not toxic in saturated solution (1.3 mg/l)
Shellflex 212 HP	0.54	2.73	not toxic in saturated solution (3.3 mg/l)

(277)

Shellsol
    Shellsol T and K are high-boiling-point aliphatic solvents.
    Shellsol A and RA are high-boiling-point aromatic solvents.
    Shellsol T: sp.gr. 0.758 at 289/289 K; boiling range: 176–211°C; solub.
        3 mg/l; aromatic content <0.5%.
    Shellsol K: sp.gr. 0.789 at 289/289 K; boiling range: 193–245°C; solub.
        3 mg/l; aromatic content 3%.
    Shellsol A: sp.gr. 0.870 at 289/289 K; boiling range: 159–179°C; aromatic
        content 96%.
    Shellsol RA: sp.gr. 0.906 at 289/289 K; boiling range: 178–325°C; aromatic
        content 99.5%.
C. WATER POLLUTION FACTORS:

	Shellsol RA	A	K	T	
–BOD_5	1.99	1.81	1.75	0.11	
–COD	2.58	2.12	0.33	0.18	(277)

    –*Carassius auratus*:
      Shellsol RA: 24 hr LD_{50}: 7 mg/l
      Shellsol A: 24 hr LD_{50}: 9 mg/l
      Shellsol K: not toxic in saturated solution (3 mg/l)
      Shellsol T: not toxic in saturated solution (3 mg/l)   (277)

silicones (polyorganosiloxanes)
C. WATER POLLUTION FACTORS:
    –in Potomac sediments: avg. 1.38 ppm, range: 0.46–3.07 ppm
    in Delaware Bay sediments: avg.: 0.61 ppm, range: 0.10–1.56 ppm
    in Blue Plains waste water treatment plant:
      filter cake: avg. 36.2 ppm; range: 22.6–48.6 ppm
      sludge: avg. 96.1 ppm; range: 89.1–103.8 ppb   (1932)

silvan *see* 2-methylfuran

silvex *see* 2(2,4,5-trichlorophenoxy)propionic acid)

silvex (IOE) *see* 2(2,4,5-trichlorophenoxy)propionic acid, isooctylester

silvex (PGBE) *see* 2(2,4,5-trichlorophenoxy)propionic acid, propyleneglycolbutyl-etheresters

simazine *see* 2-chloro-4,6-*bis*(ethylamino)-1-triazine

skatole (3-methylindole)

A. PROPERTIES: leaflets; m.w. 131.17; m.p. 95°C; b.p. 266.2°C; v.p. 1 mm at 95°C,
    10 mm at 140°C, 100 mm at 197°C; solub. 500 mg/l in cold water; log P_{oct} 2.60

SODIUM DIPHENYL-4,4'-bis-AZO-2'',8''-AMINO-1''-NAPHTHOL-3'',6''-DISULFONATE

B. AIR POLLUTION FACTORS:
—Odor: characteristic: quality: excreta, pungent, irritating

odor thresholds   mg/cu m

	10^{-7}	10^{-6}	10^{-5}	10^{-4}	10^{-3}	10^{-2}	10^{-1}	1	10	10^2	10^3	10^4
detection				♦	♦♦							
recognition						♦						
not specified		♦						♦				

                                                             (10; 210; 606; 682; 710; 871)

   O.I. at 20°C: 30,000   (316)
—Control methods: activated carbon: retentivity: 25 wt % of adsorbent   (83)
    wet scrubber: water at pH 8.5: outlet: 60–100 odor units/scf
                  $KMnO_4$ at pH 8.5: outlet: 1 odor unit/scf   (115)

C. WATER POLLUTION FACTORS:
  —Oxidation parameters
     BOD_5: 1.51 std. dil. sew.   (256)
           : 59% ThOD   (274)
     COD: 98% ThOD (0.05 n Cr_2O_7)   (274)
     $KMnO_4$ value: 5.94   (30)
               : 58% ThOD (0.01 n $KMnO_4$)   (274)
     ThOD: 2.57   (30)
—Impact on biodegradation processes: 75% reduction of nitrification process in
   non acclimated activated sludge at 7 mg/l   (30)
—Reduction of amenities: T.O.C.: 0.0012 mg/l   (50)
                            : 0.019 mg/l   (98)
                     faint odor at 0.009 mg/l   (129)
                     T.O.C. (detection): 0.01 mg/l   (998)
                     taste threshold: 0.05 mg/l   (998)
—Manmade sources: content of domestic sewages: 0.00025 mg/l   (85)

sodium diphenyl-4,4'-bis-azo-2'',8''-amino-1''-naphthol-3'',6''-disulfonate *see* direct blue 6

sodium-N-methyldithiocarbamate *see* methamsodium

sorbic aldehyde *see* 2,4-hexadienal

spinacene *see* squalene

squalene (spinacene; 2,6,10,15,19,23-hexamethyl-2,6,10,14,18,22-tetracosahexane) $[(CH_3)_2C[=CHCH_2CH_2C(CH_3)]_2=CHCH_2-]_2$
   A natural raw material found in human sebum and in shark liver oil.
   Use: biochemical and pharmaceutical research.
A. PROPERTIES: m.w. 410.73; m.p. -75°C; b.p. 285° at 25 mm; sp.gr. 0.858–0.860; oil with faint odor

C. WATER POLLUTION FACTORS:
—adsorption on smectite clay particles from simulated seawater at 25°C experimental conditions: 100 µg aqualene/l water; 50 mg smectite/l adsorption: 1.14 µg/mg: 57% adsorbed (1009)

stam F-34 see N-(3,4-dichlorophenyl)propionamide

starch (amylum)
$(C_6H_{10}O_5)_n$
C. WATER POLLUTION FACTORS:
—BOD_{35}^{25}: 0.81 in seawater/inoculum: enrichment cultures of hydrocarbon oxidizing bacteria
—ThOD: 1.18 (521)

starlicide see 3-chloro-4-methylbenzenaminehydrochloride

stearamide (octadecanamide)
$CH_3(CH_2)_{16}CONH_2$
A. PROPERTIES: colorless leaflets; m.w. 283.49; m.p. 109°C; b.p. 251°C at 12 mm
C. WATER POLLUTION FACTORS:
—Waste water treatment: A.S.: after 6 hr: 0.5% ThOD
                                     12 hr: 0.7% ThOD
                                     24 hr: 1.3% ThOD (89)

stearic acid (octadecanoic acid; n-octadecylic acid)
$CH_3(CH_2)_{16}COOH$
Use: lubricants; soaps; pharmaceuticals and cosmetics; dispersing agent and softener in rubber compounds; shoe and metal polishes; coatings; food packaging
A. PROPERTIES: colorless leaflets; m.w. 284.47; m.p. 69.4°C; b.p. 383°C; sp.gr. 0.847 at 69°C; solub. 340 mg/l at 25°C, 1,000 mg/l at 37°C
B. AIR POLLUTION FACTORS:
—Ambient air quality:
 organic fraction of suspended matter:
  Bolivia at 5200 m altitude (Sept.–Dec. 1975): 1.3–1.8 µg/1000 cu m
  Belgium, residential area (Jan.–April 1976): 17–32 µg/1000 cu m (428)
 glc's in residential area (Belgium), Oct. 1976:
  in particulate sample: 35.7 ng/cu m
  in gas phase sample: 2.27 ng/cu m (1289)
 glc's in Botrange (Belgium)—woodland at 20–30 km from industrial area: June–July 1977: 6.17; 6.55 ng/cu m ($n = 2$)
 glc's in Wilrijk (Belgium)—residential area: Oct.–Dec. 1976: 26.55–33.4 ng/cu m ($n = 2$) (1233)
 glc's on Detroit freeway interchange Oct.–Nov. 1963: 113.7 µg/1,000 cu m
 in New York high traffic city location: 137.5 µg/1,000 cu m (100)
C. WATER POLLUTION FACTORS:
—Oxidation parameters:
 BOD_5: 1.44 (275)
    : 0.8 std. dil. sew. (282, 163)

1054 STEARYLALCOHOL

          : 4% ThOD (220)
COD: 30% ThOD (220)
KMnO$_4$ value: acid: 0% ThOD
               alkaline: 0% ThOD (220)
—Reduction of amenities: T.O.C. in water: 21 ppm (326)(886)
—Manmade sources: in domestic sewer effluent: 0.05 mg/l (227)
—Waste water treatment methods: A.S. after 6 hr: 0.2% ThOD
                                      12 hr: 0.6% ThOD
                                      24 hr: 1.3% ThOD (89)

D. BIOLOGICAL EFFECTS:
—Fish:
    goldfish: lethal dose: 14 mg/l (sodium salt) (522)

stearylalcohol *see* 1-octadecanol

stilbestrol *see* diethylstilbestrol

strychnine

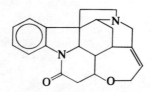

Use: in medicine as such, or as the salts; for destroying rodents and predatory animals and for trapping fur animals

A. PROPERTIES: hard, white crystals or powder; bitter taste; b.p. 270°C at 5 mm; m.w. 334.43; m.p. 284–286°C; solub. 143 ppm at room temp.

D. BIOLOGICAL EFFECTS:
—Fish:
Lepomis macrochirus: static bioassay in fresh water at 23°C, mild aeration applied after 24 hr:

material added ppm	% survival after				best fit 96 hr LC_{50} ppm
	24 hr	48 hr	72 hr	96 hr	
10	0	—	—	—	
2	100	90	70	20	
1	100	100	80	10	0.87
0.75	100	100	100	80	
0.50 (slight effects)	100	100	100	100	

Menidia beryllina: static bioassay in synthetic seawater at 23°C, mild aeration applied after 24 hr:

material added ppm	% survival after				best fit 96 hr LC_{50} ppm
	24 hr	48 hr	72 hr	96 hr	
2.0	30	30	20	0	
1.0	80	70	70	40	0.95
0.5	100	100	100	100	(352)

—Mammals:
  rats: lethal dose: 1–30 mg/kg (1855)
  man: lethal dose: 30–60 mg/kg (1854)

styphnic acid *see* 2,4,6-trinitroresorcinol

styrene (vinylbenzene; cinnamene; phenylethylene; ethenylbenzene)

Manufacturing source: organic chemical industry. (347)
Users and formulation: mfg. styrene, polystyrene; mfg. synthetic rubber; ABS plastics mfg.; mfg. resins, insulators; mfg. protective coatings (styrene-butadiene latex, alkyds). (347)

A. PROPERTIES: colorless liquid; m.w. 104.14; m.p. $-30.6°C$; b.p. 145.2°C; v.p. 5 mm at 20°C, 9.5 mm at 30°C; sp.gr. 0.9045 at 25/25°C; solub. 280 mg/l at 15°C, 300 mg/l at 20°C, 400 mg/l at 40°C; THC 1,046 kcal/mole, LHC 1,014.0 kcal/mole; sat.conc. 31 g/cu m at 20°C, 52 g/cu m at 30°C

B. AIR POLLUTION FACTORS: 1 mg/cu m = 0.23 ppm, 1 ppm = 4.33 mg/cu m
  —Odor: characteristic: if pure, sweet and pleasant, but usually containing aldehydes which have a typical penetrating smell, sharp, sweet, unpleasant

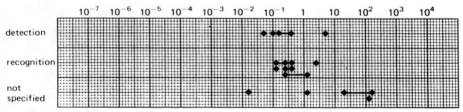

  (2; 3; 4; 19; 57; 307; 637; 643; 675; 676; 741; 743; 758; 815; 819; 866)
  O.I. (100% recogn.) 44,391 (19)
  human odor perception: non perception: 0.02 mg/cu m
                         perception: 0.036 mg/cu m
  human reflex response: no response: 0.003 mg/cu m
                         adverse response: 0.005 mg/cu m (170)
  animal chronic exposure: no effect: 0.03 mg/cu m
                           adverse effect: 0.5 mg/cu m (170)
  —Control methods:
    wet scrubber: water at pH 8.5: outlet: 2,000 odor units/scf
                  $KMnO_4$ at pH 8.5: outlet: 10 odor units/scf (115)
  —Sampling and analysis
    trapped in methanol or ethanol: U.V. spectrophotometry: det.lim.: 1 ppm
    trapped in CCl_4 or CS_2: I.R.: Det. lim.: 20 ppm
    photometry: min. full scale: 8.2 ppm (53)

1056 STYRENE

second derivative spectroscopy: det. lim.: 100 ppb (42)
4 f. abs. app.: 30 l air/30 min: UVS: det. lim.: 30 µg cu m/30 min (208)

C. WATER POLLUTION FACTORS:
—Oxidation parameters:
  BOD_5: 0.55–1.95 (21)
  65% bio. ox. (fresh dilution water) (79)
  8% bio. ox. (salt dilution water) (79)
  1.29 NEN 3235-5.4 (277)
  2.45 adapted sludge (277)
  BOD_{10}: 65% bio. ox. (fresh dilution water)
  12% bio. ox. (salt dilution water) (79)
  BOD_{15}: 78% bio. ox. (fresh dilution water)
  21% bio. ox. (salt dilution water)
  BOD_{20}: 87% bio. ox. (fresh dilution water)
  80% bio. ox. (salt dilution water) (23)
  COD: 2.88
  2.80 NEN 3235-5.3 (277)
  ThOD: 3.07
—Reduction of amenities:
  T.O.C.: 0.05 mg/l (295)
  0.036 mg/l (297)
  approx. conc. causing adverse taste in fish: 0.25 mg/l (41)
  T.O.C.: average: 0.74 mg/l; range 0.02–2.6 mg/l (97)
  organoleptic limit USSR: 0.1 mg/l (181)
  T.O.C. in water: 0.73 ppm
  0.37 ppm (326)
—Water quality: Kanawha river: 13.11.1963: raw: 18 ppb
  sand filtered: not detectable
  21.11.1963: raw: not detectable (159)
—Waste water treatment methods:
  A.C.: adsorbability: 0.028 g/g C; 88.8% reduction, infl. 180 mg/l, effl. 18 mg/l
  (32)
  Sew, D, CO_2, 20°C, 10 days observed: feed: 10 mg/l, no acclimation: 18% ThOD
  (93)

A.C.: influent ppm	carbon dosage	effluent ppm	% reduction	
200	10X	6	97	
100	10X	7	93	
20	10X	9	55	(192)

—Sampling and analysis: photometry: min. full scale: 2.2×10^{-6} mole/l (53)

D. BIOLOGICAL EFFECTS:
—Toxicity threshold (cell multiplication inhibition test):
  bacteria (*Pseudomonas putida*): 72 mg/l (1900)
  algae (*Microcystis aeruginosa*): 67 mg/l (329)
  green algae (*Scenedesmus quadricauda*): >200 mg/l (1900)
  protozoa (*Entosiphon sulcatum*): >256 mg/l (1900)
  protozoa (*Uronema parduczi Chatton-Lwoff*): 185 mg/l (1901)
—Arthropoda: brine shrimp (*artemia salina*) TLm (24 hr): 68 mg/l (158)
—Fish: fatheads: soft dil. water: TLm (24, 48, 96 hr): 56.7; 53.6; 46.4 mg/l
  : hard dil. water: TLm (24, 48, 96 hr): 62.8; 62.8; 59.3 mg/l

bluegills: soft dil. water: TLm (24, 48, 96 hr): 25.1; 25.1; 25.1 mg/l
goldfish: soft dil. water: TLm (24, 48, 96 hr): 64.7; 64.7; 64.7 mg/l
guppies: soft dil. water: TLm (24, 48, 96 hr): 74.8; 74.8; 74.8 mg/l
(158)
goldfish: LD_{50} (24 hr): 26 mg/l modified ASTM D 1345 (277)
—Mammalia: rat: acute oral LD_{50}: 1 g/kg (277)
—Man: EIR: 8.9 (49)
severe toxic effects: 1,000 ppm = 4,330 mg/cu m, 60 min
symptoms of illness: 200 ppm = 866 mg/cu m, 60 min
unsatisfactory: >100 ppm = 433 mg/cu m, 60 min (185)

styrene oxide (1-phenyl-1,2-epoxyethane; 1,2-epoxy-ethylbenzene)

A. PROPERTIES: colorless liquid; m.p. 120.2; m.p. -37°C; b.p. 194°C; v.p. 0.3 mm at 20°C; v.d. 4.14; sp.gr. 1.05; solub. 2,800 mg/l at 25°C
B. AIR POLLUTION FACTORS: 1 mg/cu m = 0.204 ppm; 1 ppm = 4.91 mg/cu m
—Odor: characteristic: quality: sweet; hedonic tone; pleasant
absolute perception limit: 0.063 ppm
50% recogn.: 0.40 ppm
100% recogn.: 0.40 ppm
O.I. (100% recogn.): 3,420 (19)
C. WATER POLLUTION FACTORS:
—Waste water treatment: A.C. adsorbability: 0.190 g/g C; 95.3% reduction, infl. 1,000 mg/l, effl. 47 mg/l (32)
D. BIOLOGICAL EFFECTS:
—Mammalia: guinea pigs, rats: single oral dose: approx. 2.0 g/kg
rat: inhalation: survived: sat. air, 2 hr
3/6: sat. air, 4 hr (211)
—Carcinogenicity: negative.
—Mutagenicity in the *Salmonella* test: positive (without liver homogenate)
0.37 revertant colonies/nmol
1292 revertant colonies at 500 µg/plate (1883)

suberic acid (octanedioic acid)
$COOH(CH_2)_6COOH$
A. PROPERTIES: colorless needles; m.w. 174.19; m.p. 140°C; b.p. 279°C at 100 mm; solub. 1,400 mg/l at 16°C; log P_{oct} 0.54/0.74 (calculated)
C. WATER POLLUTION FACTORS:
—Waste water treatment: A.S. after 6 hr: 1.2% ThOD
12 hr: 1.5% ThOD
24 hr: 8.5% ThOD (89)

succinic acid (butanedioic acid)
COOH(CH$_2$)$_2$COOH
A. PROPERTIES: colorless crystals; m.w. 118.09; m.p. 185/190°C; b.p. 235°C, decomposes; sp.gr. 1.564 at 15/4°C; solub. 68 g/l at 20°C, 1,210 g/l at 100°C; log P_{oct} −0.59
C. WATER POLLUTION FACTORS:
—Oxidation parameters

BOD$_5$:	0.21 std. dil. sp. culture	(283, 162)
	0.64 Warburg, sew.	(283, 163)
	0.419	(30)
	0.57 std. dil. acclimated	(41)
	35% ThOD	(220)
BOD$_{10}$:	0.31 std. dil. spec. culture	(283, 162)
BOD$_{20}$:	0.607	(30)
BOD$_2^{25°C}{}_{days}$:	0.45 (substrate conc.: 8.3 mg/l; inoculum: soil microorganisms)	
5 days:	0.63	
10 days:	0.86	
20 days:	0.86	(1304)
COD:	1.85	(41)
	100% ThOD	(220)
KMnO$_4$:	0.004	(30)
	acid: 2% ThOD	
	alkaline: 0% ThOD	(220)
ThOD:	1.305	(30)

—Waste water treatment methods: A.S. after 6 hr: 11.2% ThOD
                                              12 hr: 27.2% ThOD
                                              24 hr: 42.2% ThOD   (89)
D. BIOLOGICAL EFFECTS:
—Bacteria: *Pseudomonas*: no toxic effect: 125 mg/l
—Algae: *Scenedesmus*: no toxic effect: 1 mg/l
—Protozoa: *Colpoda*: LD$_0$: 125 mg/l   (30)

succinic acid, 2,2-dimethylhydrazide *see* alar

succinonitrile (butanedinitrile; ethylenecyanide)
CNCH$_2$CH$_2$CN
A. PROPERTIES: colorless waxy solid; m.p. 80.09; m.p. 54/57°C; b.p. 267°C; v.p. 6 mm at 125°C; v.d. 2.8: sp.gr. 0.985 at 63/4°C; solub. 128 g/l,
C. WATER POLLUTION FACTORS:
—Waste water treatment: A.S. after 6 hr: 1.5% ThOD
                                    12 hr: 2.4% ThOD
                                    24 hr: 3.8% ThOD   (89)
D. BIOLOGICAL EFFECTS:
—Mammalia: rat: acute oral LD$_{50}$: 450 mg/kg
            mouse: inhalation: no symptoms: sat. vap., 24 hr   (211)

sucrose

A. PROPERTIES: m.w. 342.3; m.p. 190–192°C (dec.); log P_{oct} −3.67 (calculated)
C. WATER POLLUTION FACTORS:
—BOD_{35}^{25}: 0.27 in seawater/inoculum: enrichment cultures of hydrocarbon oxidizing bacteria
  ThOD: 1.12   (521)

sugar of lead *see* leadacetate

sulfanilic acid *see* p-anilinesulfonic acid

2-sulfobenzoic acid (o-carboxybenzenesulfonic acid)

A. PROPERTIES: needles; m.w. 256.23; m.p. 141°C (anh), 105°C (3H₂O); b.p. 68/69°C; solub. 500 g/l
C. WATER POLLUTION FACTORS:
—Biodegradation: biodegradation by a soil microflora in >64 days   (176)

3-sulfobenzoic acid (m-carboxybenzenesulfonic acid)

A. PROPERTIES: deliquescent crystals; m.w. 238.21; m.p. 141°C, 98°C (anh);
C. WATER POLLUTION FACTORS:
—Biodegradation: decomposition by a soil microflora >64 days   (176)

4-sulfobenzoic acid (p-carboxybenzenesulfonic acid)

A. PROPERTIES: needles; m.w. 256.23; m.p. 259/260°C (anh), 94°C;
C. WATER POLLUTION FACTORS:
  —Biodegradation: decomposition by a soil microflora in >64 days (176)

sulfolane (thiocyclopentane-1,1-dioxyde; tetramethylenesulfone; sulphoxaline; 2,3,4,5-tetrahydrothiophene-1,1-dioxyde; dihydrobutadienesulfone; 1,1-dioxo-2,3,4,5-tetrahydrothiophene; thiolane-1,1-dioxyde; 1,1-dioxothiolane)

$$\begin{array}{c} CH_2-CH_2 \\ | \quad\quad | \\ CH_2 \quad CH_2 \\ \diagdown\diagup \\ S \\ \diagup\diagdown \\ O \quad\quad O \end{array}$$

A. PROPERTIES: m.w. 120.17; m.p. 27.4°C; b.p. 285°C; v.p. 5 mm at 118°C, 31 mm at 170°C; 85 mm at 200°C; sp.gr. 1.256 at 30/4°C
B. WATER POLLUTION FACTORS:
  —Oxidation parameters:
   BOD_5: Nil NEN 3235-5.4 (277)
   COD: 1.66 NEN 3235-5.3 (277)
D. BIOLOGICAL EFFECTS:
  —Fish: goldfish: LD_{50} (24 hr): 4800 mg/l modified ASTM 1345 (277)

sulfosalicylic acid
C. WATER POLLUTION FACTORS:
  adapted A.S. at 20°C—product is sole carbon source: 98.5% COD removal at 11.3 mg COD/g dry inoculum/hr (327)

sulfoxaline *see* sulfolane

sumithion *see* fenitrothion

sylvan *see* 2-methylfuran

sylvic acid *see* abietic acid

T

2,4,5-T *see* 2,4,5-trichlorophenoxyacetic acid

2,4,5-T, isooctylester *see* co-op brushkiller; 2,4,5-trichlorophenoxyacetic acid, isooctylester

tafazine (2-chloro-4,6-*bis*(ethylamino)-*s*-triazine)
  Use: herbicide
D. BIOLOGICAL EFFECTS:
  —Fish:

	96 hr LD_{50}	LD_{100}
Sarotherodon mossambicus	3.1 mg/l	8.0 mg/l
Puntius ticto	24.5 mg/l	64.0 mg/l
Danio spp.	12.6 mg/l	40.0 mg/l
		(1902)

  freshwater ectoprocta: no appreciable effect at 2.5 mg/l for 84 hr exposure
(1902)

talestol 0214
  Use: non-ionic surfactant; an ester of tall oil and polyethyleneglycol
D. BIOLOGICAL EFFECTS:
  —Fish:
  Northern pike yolk sac larvae: 48 hr (S) LC_{50}: 140 mg/l
  Northern pike yolk sac larvae: 48 hr (S) LC_{50}: 107 mg/l
  Northern pike-free-swimming: 48 hr (S) LC_{50}: 22 mg/l
  Northern pike—1 mo old: 48 hr (S) LC_{50}: 145 mg/l   (1113)

tallow ethoxylate
D. BIOLOGICAL EFFECTS:
  —Fish:

	48 hr LC_{50}	96 hr LC_{50}
Rasbora heteromorpha	0.8 mg/l	0.7 mg/l
Salmo trutta	0.4–0.7 mg/l	0.4 mg/l
Idus idus	2.6–2.7 mg/l	2.3–2.5 mg/l
Carassius auratus	2.1 mg/l	(1905)

tannic acid (gallotannic acid; tannin)
$C_{76}H_{52}O_{46}$
A penta(m-digalloyl)glucose. Natural substance widely found in nutgalls, tree barks, and other plant parts.
Tannins are known to be gallic acid derivatives.
Use: chemicals; alcohol denaturant; tanning; textiles; clarification agent in wine manufacture; photography.
A. PROPERTIES: m.w. 1701.23; m.p. 218
C. WATER POLLUTION FACTORS:
—Oxidation parameters: BOD_5: 0.31; 0.46 std. dil. sew.   (282; 284; 163; 167)
—Impact on biodegradation:
   NH_3 oxidation by *Nitrosomonas* sp. at 100 mg/l: 20% inhibition
                                     50 mg/l: 7% inhibition
D. BIOLOGICAL EFFECTS:
—Fish: goldfish: toxic: 100 ppm   (155)
            LD_0: 10 mg/l, longtime exposure in hard water   (245)

tannin *see* tannic acid

d*l*-tartaric acid (racemic acid)

$$\begin{array}{c} COOH \\ | \\ CHOH \\ | \\ CHOH \cdot H_2O \\ | \\ COOH \end{array}$$

A. PROPERTIES: m.w. 168.10; m.p. anh: 100°C; H_2O 204/206°C; sp.gr. 1.697; solub. 206,000 mg/l at 20°C, 92,300 mg/l at 0°C; colorless crystals; log P_{oct} -0.76/-2.02 (calculated)
C. WATER POLLUTION FACTORS:
—BOD_5: 0.350   (30)
           0.30 Warburg, sewage   (282, 163)
           34% of ThOD   (220)
—BOD_{20}: 0.460   (30)
—COD: 98% ThOD   (220)
—$KMnO_4$ value: 1.492   (30)
                acid: 73% ThOD
                alkaline: 10% ThOD   (220)
—ThOD: 0.533   (30)
—Waste water treatment: A.S.: after 6 hr: 2.6% of ThOD
                                      12 hr: 1.4% of ThOD
                                      24 hr: 0.7% of ThOD   (220)
A.S., Resp, BOD, 20°C, 1-5 days observed, feed: 720 mg/l, acclimation 1 day, 80% theor. oxidation   (93)
D. BIOLOGICAL EFFECTS:
—Protozoa: *Vorticella campanula*: perturbation level: 100 mg/l
           Paramaecium caudatum: LD_0: 250-320 mg/l   (30)

—Arthropoda: *Gammarus pulex*: perturbation level: 230 mg/l
—Fish:
  trout (*Trutta iridis*): perturbation level: 150 mg/l (30)
  goldfish: LD_0 : 200 mg/l, longtime exposure in hard water
         : 10 mg/l, longtime exposure in very soft water (245)

TBH *see* hexachlorocyclohexane

TBP *see* tri-*t*-butylphosphate

TCA *see* trichloroacetic acid

TCC *see* 3,3,4'-trichlorocarbanilide

TCDD *see* 2,3,7,8-tetrachlorodibenzo-*p*-dioxin

TCP *see* tri-*o*-cresylphosphate

TDE *see* DDD

2,4-TDI *see* toluene-2,4-diisocyanate

2,6-TDI *see* toluene-2,6-diisocyanate

TEA *see* triethanolamine

TEDP (tetraethyldithiopyrophosphate; sulfotep; bladafume; dithio; dithione)

$$\begin{array}{c} C_2H_5O \\ \diagdown \\ P{-}O{-}P \\ \diagup \\ C_2H_5O \end{array} \begin{array}{c} S \quad\quad S \\ \| \quad\quad \| \\ \\ \end{array} \begin{array}{c} OC_2H_5 \\ \diagup \\ \\ \diagdown \\ OC_2H_5 \end{array}$$

Use: insecticide; acaricide
A. PROPERTIES: m.w. 322; sp.gr. 1.19; b.p. 138°C at 2 mm; solub. 25–66 mg/l at room temp.; v.p. 1.7×10^{-4} mm at 20°C
D. BIOLOGICAL EFFECTS:
  —Crustacean:
    daphnids: 48 hr EC_{50}: 0.00023 mg/l (1591)
  —Fish:
    fathead minnow: 96 hr LC_{50}: 0.178 mg/l
    bluegill: 96 hr LC_{50}: 0.0016 mg/l
    rainbow trout: 96 hr LC_{50}: 0.018 mg/l (1591)
  —Mammals:
    acute oral LD_{50} (rat): 7–10 mg/kg (1854)

teepol 610 (*sec*-alkylsulfate; SAS)

$$\begin{array}{c} R_1 \\ \diagdown \\ CH-O-S(=O)_2-O-Na \\ \diagup \\ R_2 \end{array}$$

R_1 = long alkylchain
R_2 = short alkylchain
  Use: detergent
C. WATER POLLUTION FACTORS:
  −COD: 0.68 = 94% ThOD (277)
  Carassius auratus: 24 hr LD_{50}: 29 mg/l
  96 hr LD_{50}: 28 mg/l (277)

teepol 715 (alkylarylsulfonate; alkylbenzenesulfonate; AAS; ABS)

R—C$_6$H$_4$—S(=O)$_2$—ONa

  Use: component of detergents, degreasers, moisturisers.
  Composition: teepol 715 is a 40% aqueous solution of alkylarylsulfonate.
C. WATER POLLUTION FACTORS:
  −Biodegradability: 92.4% (OECD method) (277)

teepol GC 56
  Use: detergent.
  Composition: mixture of alkylarylsulfonate and alcoholethersulfate.
C. WATER POLLUTION FACTORS:
  −COD: 0.82 = 95% ThOD (277)
  −Biodegradability: 93.8% (OECD method) (277)
D. BIOLOGICAL EFFECTS:
  −Fish:
  Carassius auratus: 24 hr LD_{50}: 21 mg/l
  96 hr LD_{50}: 20 mg/l (277)

teepol GC 53 P
  Use: detergent.
  Composition: mixture of alkylarylsulfonate, alcoholethersulfate and alcoholethoxylate.
C. WATER POLLUTION FACTORS:
  −Biodegradability: 95% (OECD method) (277)

TEG *see* triethyleneglycol

TEL *see* tetraethyllead

telodrin (isobenzan)
  Composition: min. 82% isobornyl thiocyanoacetate
  max. 18% other active terpenes.
C. WATER POLLUTION FACTORS:
  —Aquatic reactions:
  95% disappearance from soils: 2-4 years (1815)
  persistence in river water in a sealed glass jar under sunlight and artificial fluorescent light—initial conc. = 10 µg/l:

after	1 hr	1 wk	2 wk	4 wk	8 wk	
	100	25	10	0	0	(1309)

% of original compound found

telvar *see* 3-(*p*-chlorophenyl)-1,1-dimethylurea

temephos *see* abate

tepp (tep; fosvex; hesamite; nifos; tetron; vapotone; ethylpyrophosphate; tetraethylpyrophosphate)

$$(C_2H_5O)_2 \overset{O}{\overset{\|}{P}}-O-\overset{O}{\overset{\|}{P}}(OC_2H_5)_2$$

Use: insecticide for aphids and mites, rodenticide.
A. PROPERTIES: m.w. 290.20; sp.gr. 1.20; b.p. 135-138°C at 1 mm; v.p. 1.55×10^{-4} mm at 20°C
C. WATER POLLUTION FACTORS:
  —Rapidly hydrolysed by water 1/2 t-6.8 hr at pH 7 and 25°C
D. BIOLOGICAL EFFECTS:
  —Algae:

Protococcus sp.:	1×10^5 ppb	.62 OD expt/OD control	10 day growth test
Protococcus sp.:	5×10^5 ppb	.00 OD expt/OD control	10 day growth test
Chlorella sp.:	1×10^5 ppb	.65 OD expt/OD control	10 day growth test
Chlorella sp.:	3×10^5 ppb	.27 OD expt/OD control	10 day growth test
Dunaliella euchlora:	3×10^5 ppb	.49 OD expt/OD control	10 day growth test
Phaeodactylum tricornutum:	1×10^5 ppb	.58 OD expt/OD control	10 day growth test
Monochrysis lutheri:	1×10^5 ppb	.83 OD expt/OD control	10 day growth test
Monochrysis lutheri	3×10^5 ppb	.38 OD expt/OD control	10 day growth test

  —Molluscs:
  Crassostrea virginica (American oyster), egg: 1×10^4 ppb TLm 48 hr static lab bioassay

1066 TEREPHTHALIC ACID

Crassostrea virginica (American oyster), larvae:	1×10^4 ppb	TLm	14 day static lab bioassay	(2324)
—Crustaceans:				
Gammarus lacustris:	96 hr, LC_{50}: 39 µg/l			(2124)
Gammarus fasciatus:	96 hr, LC_{50}: 210 µg/l			(2126)
—Fish:				
Pimephales promelas:	96 hr, LC_{50}: 1900 µg/l			(2123)
Lepomis macrochirus:	96 hr, LC_{50}: 1100 µg/l			(2123)
—Mammals:				
acute oral: LD_{50} (rat): 1.12 mg/kg				
acute dermal LD_{50} (male rat): 2-4 mg/kg				(1855)

terephthalic acid (1,4-benzenedicarboxylic acid; p-phthalic acid)

A. PROPERTIES: needles or amorphe; m.w. 166.13; m.p. sublimes; v.d. 5.74; sp.gr. 1.510; solub. 16 mg/l
B. AIR POLLUTION FACTORS: 1 mg/cu m = 0.15 ppm; 1 ppm = 6.91 mg/cu m
C. WATER POLLUTION FACTORS:
 —Biodegradation: decomposition by a soil microflora in 2 days (176)
 —Water quality: self purification is affected at 0.1 mg/l (181)
D. BIOLOGICAL EFFECTS:
 —Mammalia: rat: acute oral LD_{50}: >6.4 g/kg (211)

α-terpineol

Manufacturing source: extraction of essential oils, fractional distillation of pine oils, wood processing industry. (347)
Users and formulation: perfume mfg.; soap mfg.; hydrocarbon solvent; solvent for resins, cellulose esters and ethers; disinfectants; antioxidants; medicines; flavorings constituent. (347)
Natural sources (*water and air*): several essential oils; pine oil component; forest runoff. (347)

B. AIR POLLUTION FACTORS:
odor thresholds   mg/cu m

	10^{-7}	10^{-6}	10^{-5}	10^{-4}	10^{-3}	10^{-2}	10^{-1}	1	10	10^2	10^3	10^4
detection						♦	♦♦					
recognition						♦	♦♦					
not specified												

(610; 614; 682; 871; 872)

C. WATER POLLUTION FACTORS:
—Reduction of amenities: T.O.C. in water: 0.34 ppm
　　　　　　　　　　　　　　　　　　　　0.35 ppm　　　　　　　　(326; 880)
—Water quality:
　in Delaware river (U.S.A.): conc. range: winter: 0.5–4 ppb
　　　　　　　　　　　　　　　　　　　　summer: n.d.　　　　　　(1051)

D. BIOLOGICAL EFFECTS:
—Fish:
　rainbow trout: static bioassay: 96 hr TLm: 10–100 mg/l at 10°C　　(451)

terrachlor　*see* pentachloronitrobenzene

1,2,4,5-tetrabromobenzene

D. BIOLOGICAL EFFECTS:
—Fish:
　juvenile atlantic salmon: BCF at 15.5 μg/l: 1390
　　　　　　　　　　　　　　excretion half-life: 103 hr　　　　　　(1852)

2,2′,4,5′-tetrabromobiphenyl

D. BIOLOGICAL EFFECTS:
—Fish:
　Atlantic salmon: 96 hr BCF (water) (S): 474
　　　　　　　　　42 d BCF (food) (S): 0.41　　　　　　　　　　(1532)

tetrabromo-2-chlorotoluene

D. BIOLOGICAL EFFECTS:
  —Fish:
    Atlantic salmon: 96 hr BCF (water) (S): 39
                      42 d BCF (food) (F): 0.035   (1532)

1,1,2,2-tetrabromo-ethane *see* acetylenetetrabromide

tetra-*t*-butyldiphenoquinone

C. WATER POLLUTION FACTORS:
  —Water quality:
    in river water: <0.001 ppm: in river sediment: 0.2–0.5 ppm   (555)

1,2,3,4-tetrachlorobenzene

Use: component of dielectric fluids; synthesis.
Technical grade contains 30% 1,2,4,5-isomer.

A. PROPERTIES: needles; m.w. 215.90; m.p. 47.5°C; b.p. 254°C; solub. 3.5 ppm at 22°C

C. WATER POLLUTION FACTORS:
  —Impact on degradation processes
    biochemical oxygen utilization and nitrification is influenced at 0.08–0.16 mg/l   (181)
    no effect on BOD test or nitrification at 0.02 mg/l (n.s.i.)   (30)
  —Reduction of amenities: odor threshold: 0.006 mg/l
                           taste threshold: 0.0064 mg/l   (181)
    practical organoleptic threshold USSR 1970: 0.013 mg/l (n.s.i.)   (30)

T.O.C.: 0.02 mg/l
—Waste water treatment:
degradation by *Pseudomonas*: 200 mg/l at 30°C:
    parent: 33% ring disruption in 120 hr
    mutant: 74% ring disruption in 120 hr   (152)

D. BIOLOGICAL EFFECTS:
—In fat phase of the following fish from the Frierfjord (Norway), April 1975–Sept. 1976: cod, whiting, plaice, eel, sprot:
    avg.: 0.36 ppm (sum of isomers)
    range: 0.1–0.7 ppm (sum of isomers)   (1065)
guppy (*Poecilia reticulata*): 14 d LC_{50}: 0.8 ppm   (1833)
—Mammals: rats, rabbits: oral LD_{50}: 1,500 mg/l   (181)

1,2,3,5-tetrachlorobenzene

A. PROPERTIES: m.w. 215.89; m.p. 50–52°C; b.p. 246°C; solub. 2.4 ppm at 22°C
C. WATER POLLUTION FACTORS:
—Odor threshold conc. 0.4 mg/l
D. BIOLOGICAL EFFECTS:
—Fish:
guppy (*Poecilia reticulata*): 14 d LC_{50}: 0.8 ppm   (1833)
female guppies (*Poecilia reticulata*): BCF: 7.2×10^4 (1833)
BCF in bacteria (*Siderocapsa treubii*): approx. 3000   (1833)
BCF in guppies (*Poecilia reticulata*): 72,000 (on lipid content)

1,2,4,5-tetrachlorobenzene

A. PROPERTIES: needles; m.w. 215.90; m.p. 138°C; b.p. 246°C; sp.gr. 1.858 at 21/4°C; solub. 0.3 ppm at 22°C
C. WATER POLLUTION FACTORS:
—Reduction of amenities: T.O.C.: 0.13 mg/l   (225)
—Biodegradation: degradation by *Pseudomonas*: 200 mg/l at 30°C
    parent: 30% ring disruption in 120 hr
    mutant: 80% ring disruption in 120 hr   (152)
D. BIOLOGICAL EFFECTS:
—Fish:
guppy (*Poecilia reticulata*): 14 d LC_{50}: 0.30 ppm   (1833)

tetrachloro-*p*-benzoquinone *see* chloranil

tetrachlorobiphenyl
Manufacturing source: organic chemical industry. (347)
Users and formulation: mfg. electrical insulation; fire resistant heat transfer and hydraulic fluids; high temperature lubricants; elastomers; adhesives; paints; lacquers, varnishes, pigments and waxes; heat sensitive paper. (347)
Man caused sources (*water and air*): general use of PCB containing materials, e.g., paints, coatings, adhesives, paper, electrical insulation. (347)

2,3,4,5-tetrachlorobiphenyl

C. WATER POLLUTION FACTORS:
—Biodegradation:
51.6% degradation after 2 hr by *Alcaligenes* Y42 (cell number 2×10^9/ml)
38.0% degradation after 2 hr by *Acinetobacter* P6 (cell number 4.4×10^8/ml)
at 14.6 mg/l initial concentration (1086)

2,3,5,6-tetrachlorobiphenyl

C. WATER POLLUTION FACTORS:
—Biodegradation:
no degradation after 20 hours by *Alcaligenes* Y42 and *Acinetobacter* P6 at 14.6 mg/l initial concentration (1086)

2,3,2',3'-tetrachlorobiphenyl

C. WATER POLLUTION FACTORS:
—Biodegradation:
17.4% degradation after 2 hr by *Alcaligenes* Y42 (cell number 2×10^9/ml)
14.6% degradation after 2 hr by *Acinetobacter* P6 (cell number 4.4×10^8/ml)

at 14.6 mg/l initial concentration. Dichlorobenzoic acid derivatives were detected in the metabolite (1086)

2,4,2',4'-tetrachlorobiphenyl

C. WATER POLLUTION FACTORS:
  —Biodegradation:
    no degradation after 2 hr by *Alcaligenes* Y42 and *Acinetobacter* P6 at 14.6 mg/l initial concentration (1086)

2,4,3',4'-tetrachlorobiphenyl

C. WATER POLLUTION FACTORS:
  —Biodegradation:
    no degradation after 2 hr by *Alcaligenes* Y42 and *Acinetobacter* P6 at 14.6 mg/l initial concentration (1086)

2,5,2',5'-tetrachlorobiphenyl

A. PROPERTIES: solub. 16 ppb; log $P_{oct.}$ 3.91
C. WATER POLLUTION FACTORS:
  —Biodegradation:
    no metabolism in an anaerobic marine mud incubated for 6 weeks
    aerobic incubation of 3.6 mg tetrachlorobiphenyl with seawater and sandy beach sediment: from 2 to 4% metabolism was always detected within 3 days after commencement of the experiment. Levels of metabolism did not increase after periods of incubation up to several weeks. The metabolite was probably a lactone acid. Analysis by GC-MS proved unsuccessful because the compound decomposed during gas chromatography (1083)
    no degradation after 2 hr by *Alcaligenes* Y42 (cell number 2×10^9/ml)
    7% degradation after 2 hr by *Acinetobacter* P6 (cell number 4.4×10^8/ml)
      at 14.6 mg/l initial concentration (1086)

2,5,3',4'-TETRACHLOROBIPHENYL

—Aquatic reactions:

Freundlich adsorption constants:	n	k
Illite clay	0.92	21.2
Woodburn soil	1.0	48.9
humic acid	3.8	43.6

(1879)

D. BIOLOGICAL EFFECTS:

Chlorella pyrenoidosa: bioconcentration coefficient: 4800 in whole cells (1567)

—Bioaccumulation and biodegradabiltiy index (B.I.) in a laboratory model ecosystem after 33 days at 26°C:

	BCF	B.I.*
alga (*Oedogonium cardiacum*)	17,997	0.015
snail (*Physa*)	39,439	0.082
mosquito (*Culex pipiens quinquefasciatus*)	10,562	0.076
fish (*Gambusia affinis*)	11,863	0.060

$$^*B.I. = \frac{\text{ppm polar degradation product}}{\text{ppm non polar product}}$$

(1643)

2,5,3',4'-tetrachlorobiphenyl

C. WATER POLLUTION FACTORS:
—Aquatic reactions:
partition coefficient to natural sediments:

			physical-chemical characteristics of sediment			partition
sediment	TOC (%)	pH	% sand	% silt	% clay	coeff.
Oconee river	0.4	6.5	93.0	6.0	2.0	420
USDA Pond	0.8	6.4	—	—	—	580
Doe Run Pond	1.4	6.1	56.0	44.0	<1.0	990
Hickory Hill Pond	2.4	6.3	55.0	45.0	<1.0	1180

(1068)

2,6,2',6'-tetrachlorobiphenyl

C. WATER POLLUTION FACTORS:
—Biodegradation:
2.8% degradation after 2 hr by Alcaligenes Y42 (cell number 2×10^9/ml)

no degradation after 2 hr by Acinetobacter P6 (cell number 4.4 × 10^9/ml)
at 14.6 mg/l initial concentration (1086)
—Aquatic reactions:
partition coefficient to natural sediments:

sediment	TOC (%)	pH	% sand	% silt	% clay	coeff.
			physical-chemical characteristics of sediment partition			
Oconee river	0.4	6.5	93.0	6.0	2.0	510
USDA Pond	0.8	6.4	–	–	–	650
Doe Run Pond	1.4	6.1	56.0	44.0	<1.0	1080
Hickory Hill Pond	2.4	6.3	55.0	45.0	<1.0	1270

(1068)

3,4,3',4'-tetrachlorobiphenyl

C. WATER POLLUTION FACTORS:
  —Biodegradation:
    no degradation after 2 hr by *Alcaligenes* Y42 and *Acinetobacter* P6 at 14.6 mg/l
    initial concentration (1086)

2,3,7,8-tetrachlorodibenzo-p-dioxin (TCDD; *see also* dioxins)

CAVEAT: it was not always clear from literature whether reference was made to
the 2,3,7,8-isomer or to the sum of the TCDD-isomers.
Manmade sources:
Secondary reaction product in the alkaline hydrolysis of 1,2,4,5-tetrachlorobenzene
  for the production of 2,4,5-trichlorophenol.
Impurity of industrial trichlorophenol, and derivatives such as 2,4,5-T, and 2,4,5-
  TP.
Impurity of several pesticides in concentrations from 0.1 to 40 ppm.
Impurity of certain polychlorinated biphenyls such as Arochlor 1254. (583)
Is formed during pyrolysis of sodium salt of 2,4,5-T and 2,4,5-TP (2,4,5-trichloro-
  phenoxypropionic acid).
TCDD was found in 23 of 24 samples of 2,4,5-T as an impurity in concentrations
  ranging from <0.01 ppm to >10 ppm (2,3,7,8 TCDD) (1305)
A. PROPERTIES: colorless crystals; m.p. 305–306°C
C. WATER POLLUTION FACTORS:
  —Persistence:
    after weathering in soil for 1 year: 50–60% of the original concentrations (1 to
    100 ppm) remaining unchanged (585)

half-life of TCDD in a model aquatic environment was in the order of 600 days (2,3,7,8 TCDD) (1655)
photodecomposition is negligible in aqueous solutions (584)
recovery of TCDD from two soils at three concentrations over a period of 350 days:

soil	applied conc. ppm	% of TCDD recovered after				
		20 days	40 days	80 days	160 days	350 days
loamy sand	1	94	81	81	80	54
silty clay loam	1	79	77	69	83	54
loamy sand	10	80	80	80	79	57
silty clay loam	10	85	88	82	85	63
loamy sand	100	95	92	86	73	56
silty clay loam	100	107	116	92	75	71

(1305)

—Control methods:
at 700°C: 50% decomposition after 21 sec. exposure
at 800°C: complete decomposition after 21 sec. exposure (584)

D. BIOLOGICAL EFFECTS:
—Plants:
young oats and soybeans grown on a sandy loam contaminated with 60 ppb TCDD accumulated 40 ppb TCDD (585)
—Bioaccumulation:
TCDD levels in wildlife living near Seveso (Italy), 1979: TCDD level of the top 7 cm of soil varied from 0.010–12 ppb (avg. 3.5 ppb, $n = 23$):
field mouse (*Microtus arvalis*): range: 0.07–49 ng/g whole body ($n = 14$)
                                  avg.: 4.5 ng/g whole body
hare: range: 2.7–13 ng/g liver ($n = 5$)
      avg.: 7.7 ng/g liver
toad: 0.2 ng/g whole body ($n = 1$)
snake: 2.7 ng/g liver ($n = 1$)
       16 ng/g adipose tissue ($n = 1$)
earthworms: 0–12 ng/g whole body ($n = 2$) (1824)
TCDD levels in cow's milk from the contaminated area of Seveso, Italy varied between 75 and 7900 ng TCDD/l milk (July–Aug. 1976) (1865)
BCF: algae: 2000–18,600
      duckweed: 1200–5000
      snails: 1400–47,100 (1891; 1892; 1893)
BCF*: brine shrimp: 1570 (at 0.1 ppb in water)
       mosquito larvae: 5000–9222 (at 0.45–2.4 ppb in water)
*TCDD introduced into system in the form of residues on sand (1880)
mosquito fish: 7 d BCF (S): 4875 (1538)
—Toxicity:
—Birds:
American coot (*Fulcia americana*) collected in a reservoir that received runoff from a watershed treated with 2,4,5-T: no TCDD detected in any of the birds (1594)

—Mammals:
guinea pig: acute oral LD_{50}: 0.6 μg/kg (2,3,7,8-TCDD)
male rats: acute oral LD_{50}: 22 μg/kg (2,3,7,8-TCDD)
female rats: acute oral LD_{50}: 45 μg/kg (2,3,7,8-TCDD)
monkeys: acute oral LD_{50}: <70 μg/kg (2,3,7,8-TCDD)
rabbits: acute oral LD_{50}: 115 μg/kg (2,3,7,8-TCDD) (1334)
male rats: no chromosomal aberrations in the bone marrow has been observed
(584)
strong mutagenicity with Salmonella strain TA 1532 (587)
teratogenic for most animals (mouse, rat, chicken, hamster, guinea pig, rabbit, monkey) (584)

2,3',4,4'-tetrachlorodiphenylether

D. BIOLOGICAL EFFECTS:
—Fish:
Atlantic salmon: 96 hr BCF (water) (S): 2720
42 d BCF (food) (F): 0.33 (1532)

2,2',4,4'-tetrachlorodiphenyloxide

A. PROPERTIES: log P_{oct} 6.72
D. BIOLOGICAL EFFECTS:
—Fish:
rainbow trout (*Salmo gairdneri*): bioconcentration factor: 12,400 ± 2290; log bioconc. factor: 4.09 (193)

1,1,2,2-tetrachloroethane (acetylenetetrachloride;)
$CHCl_2 CHCl_2$
Manufacturing source: organic chemical industry. (347)
Users and formulation: mfg. 1,1-dichloroethylene; solvent for chlorinated rubber and other organic materials; insecticide mfg.; bleach mfg.; paint, varnish, rust remover mfg.; soil fumigant; cleansing and degreasing metals; photo films, resins and waxes; extractant of oils and fats; organic synthesis, herbicide; alcohol denaturant. (347)
A. PROPERTIES: colorless liquid; m.w. 167.86; m.p. -42.5/-43.8°C; b.p. 146.4°C; v.p. 5 mm at 20°C, 8.5 mm at 30°C; v.d. 5.79; sp.gr. 1.60 at 20/4°C; solub. 2,900 mg/l at 20°C; sat. conc. 46 g/cu m at 20°C, 75 g/cu m at 30°C

1076 1,2,2,2-TETRACHLOROETHANE

B. AIR POLLUTION FACTORS: 1 mg/cu m = 0.14 ppm; 1 ppm = 6.98 mg/cu m
—Odor threshold conc.: <3 ppm (211)
  20 mg/cu m (740)
  detection: 50 mg/cu m (643)
—Sampling and analysis: photometry: min. full scale: 5,000 ppm (53)

C. WATER POLLUTION FACTORS:
—Reduction of amenities:
  odor threshold: 5 mg/l (84)
  organoleptic limit USSR 1970: 0.2 mg/ (n.s.i.) (181)
  T.O.C. in water: 0.5 ppm (326)
—Waste water treatment: evaporation from water at 25°C of 1 ppm solution:
  50% after 56 min
  90% after less than 120 min (313)
—Sampling and analysis:
  GC-EC after n-pentane extraction: det. lim.: 0.01 μg/l (84)

D. BIOLOGICAL EFFECTS:
—Fish:
  guppy (*Poecilia reticulata*): 7 d LC_{50}: 37 ppm (1833)
—Mammalia:
  cats and rabbits: inhalation: no typical organ change: 100–160 ppm, 8 hr/d, 4 weeks
  dog: oral, dose: toxic: 0.7 g/kg
    lethal: 0.3 ml/kg (211)
—Man: severe toxic effects: 50 ppm = 350 mg/cu m, 60 min
    symptoms of illness: 20 ppm = 140 mg/cu m,
    unsatisfactory: >10 ppm = 70 mg/cu m, (n.s.i.) (185)

1,2,2,2-tetrachloroethane
CH_2ClCCl_3

A. PROPERTIES: m.w. 167.85; b.p. 138°C; sp.gr. 1.60
C. WATER POLLUTION FACTORS:
—Waste water treatment: evaporation from water at 25°C of 1 ppm solution:
  50% after 43 min
  90% after more than 120 min (313)

1,1,2,2-tetrachloroethylene (perchloroethylene, ethylenetetrachloride)
CCl_2CCl_2

Manufacturing source: organic chemical industry. (347)
Users and formulation: dry cleaning operations; metal degreasing; solvents for fats, greases, waxes, rubber, gums, caffeine from coffee; remove soot from industrial boilers; mfg. paint removers, printing inks; mfg. trichloroacetic acid; vermifuge; heat transfer medium; mfg. of fluorocarbons. (347)
A. PROPERTIES: colorless liquid; m.w. 165.83; m.p. −22.7°C; b.p. 121.4°C; v.p. 14 mm at 20°C, 24 mm at 30°C, 45 mm at 40°C; v.d. 5.83; sp.gr. 1.626 at 20°C; solub. 150 mg/l at 25°C; sat. conc. 126 g/cu m at 20°C, 210 g/cu m at 30°C; log P_{oct} 2.60 at 20°C
B. AIR POLLUTION FACTORS: 1 mg/cu m = 0.15 ppm; 1 ppm = 6.89 mg/cu m

1,1,2,2-TETRACHLOROETHYLENE

—Odor: characteristic: etheric, chlorinated solvent: recogn.: 34.6–320 mg/cu m (73)
T.O.C.: 50 ppm (298)
distinct odor: 480 mg/cu m (278)
$PIT_{50\%}$ recogn.: 4.68 ppm
$PIT_{100\%}$ recogn.: 4.68 ppm (2)
O.I. at 20°C: 370 (316)
—Ambient air quality:
  glc's in rural Washington, dec. 1974–febr. 1975: 20 ppt (315)
  glc's in Los Angeles county: 0.01–4.2 ppb (46)
  glc's in Munich (W. Germany): Fall 1974: µg/cu m:

distance from Karlsplatz	series 1			series 2		
	min.	avg.	max.	min.	avg.	max.
0 km	2	3	6	3	5	6
1 km	1	6	25	3	7	14
5 km	1	4	11	2	5.5	15
10 km	0	4	17	1	4	15

(936)

  glc's in Liverpool/Manchester: <0.1–8 µg/cu m
  glc's in forest (U.K.): 2.4 µg/cu m
  glc's in surroundings of production plant: 12–32 µg/cu m (937)
  glc's over Atlantic ocean: 0.001–0.009 µg/cu m (938)

	glc (ppb)
U.K.	
Liverpool (Childwall)	0.07–0.68
Widness (Pex Hill)	0.03
Delamere	0.03–0.14
Frodsham	<0.01–1.35
Moel Faman (N. Wales)	<0.01–0.34
Rannock Moor (Scotland)	0.04–0.14
Forest of Dean	0.4
Netherlands	
Hengelo: open country	0.03–0.08
center	0.01–0.13
Weiwerd	0.03–0.08
W. Germany	
Munich: center	0.27–0.81
1 km from center	0.14–3.38
5 km from center	0.14–1.49
10 km from center	n.d.–2.30
20 km from center	0.14
Uberackern	0.13–2.30
Ranzel	0.81–2.02
Langel	1.62–3.24
Unkendorf	<0.49
Belgium	
Brussels N.	0.27–0.41

1,1,2,2-TETRACHLOROETHYLENE

   Uccle   0.14–0.41
   Namur   0.14–1.22
  France
   Saint Auban   0.07–4.7
   Lyon: 20 km from center   2.16–4.72
    center   2.84–5.0
  Italy
   Monte Tauro (Sicily)   0.27–0.41
   Floridia (Sicily)   0.27
   Melilli (Sicily)   0.68–1.35
—Sampling and analysis: photometry: min. full scale: 270 ppm   (53)

C. **WATER POLLUTION FACTORS:**
 —Oxidation parameters
  BOD_5: 0.06   (275)
  ThOD: 0.39   (275)
 —Reduction of amenities: T.O.C.: 0.3 mg/l   (225)
              5 mg/l   (84)
 —Manmade sources: percolation water at 30–500 m from dumping ground
         contained: 35–120 ppb   (183)
 —Waste water treatment: evaporation from water at 25°C of 1 ppm solution:
  50% after 24–28 min
  90% after 72–90 min   (313)

	sample 1	sample 2	
experimental water reclamation plant:			
sand filter effluent	52 ng/l	88 ng/l	
after chlorination	21,926 ng/l	24,527 ng/l	
final water after A.C.	192 ng/l	63 ng/l	(928)

—Water and sediment quality:

		μg/l	
in River Isar (W. Germany)	at Scharnitz	0.015	
	at Wolfratshauser	0.1–1.1	
	at Stauwehr Hirschau	1.9–2.3	
	in Munich	0.5–0.7	
in Mittlerer Isarkanal		0.2–1.5	
in Lake Starnberger, Ammerland		0.2	
in Lake Lerchenauer, Munich		2.0–2.7	
in raw sewage of Munich		88	
after mechanical treatment		6.9–62	
in tapwater in Munich		0.09–2.4	(936)
rainwater in Runcorn (U.K.)		0.15	

seawater in bay of Liverpool: avg. 0.12 μg/l; max. 2.6 μg/l
well water: nil   (937)
waters from upland reservoir in N.W. England (1974):
 during dry cloudy weather: <0.1–0.1 ppb
 during prolonged heavy rain: 5–16 ppb   (933)
in Zürich Lake: at surface: 0.14 ppb; at 30 m depth: 0.42 ppb
in Zürich area: in springwater: 0.012 ppb; in groundwater: 1.85 ppb;
    in tapwater: 2.1 ppb   (513)

1,1,2,2-TETRACHLOROETHYLENE

	ppb	
U.K.		
Liverpool bay (U.K.):	0.12-2.6	
W. Germany		
River Rhine: Hoenningen:	1.5	
Luelsdorf:	2-2.5	
Wesseling:	1.75	
River Salzach: Marienberg:	0.6-1.9	
Überackern:	3.3-19.6	
River Isar: source of river:	0.01-0.02	
Munich:	0.1-1.0	
downstream of Munich:	1.9-2.5	
Lake Starnberg:	0.15-0.20	
Lake Lerchenau:	2.0-2.8	
Netherlands		
Twente, Canal Hengelo:	0.3	
Canal Delden:	<0.2	
Eems:	16.0	
Netherlands		
Oostfriese Gaatje: South (NL):	6.6	
North:	1.4	
Ranselgat:	1.7	
Huibertgat:	1.4	
France		
River Durance: Pont Craison:	<10-46	
Ste. Tulle:	<5	(511)

—Sampling and analysis: GC-EC after n-pentane extraction: detection limit = 0.01 µg/l

D. BIOLOGICAL EFFECTS:

—Fish: *Salmo gairdneri* (trout): log. bioconc. factor: 1.59 (193)

Concentrations in various organs of molluscs and fish collected from the relatively clean water of the Irish Sea, in the vicinity of Port Erin, Isle of Man (only organs with highest and lowest concentrations are given—concentrations on dry wt. basis):

			ng/g
molluscs:	*Baccinum undatum*:	digestive gland:	33
		muscle:	39
	Modiolus modiolus:	muscle:	10
		mantle:	63
	Pecten maximus:	muscle:	24
		testis:	176
fish:	*Conger conger* (eel):	muscle, gill, gut:	1-3
		liver:	43
	Gadus morhua (cod):	muscle, heart, brain:	2-3
		liver:	8
	Pollachius birens (coalfish):	muscle:	2
		liver:	6

Scylliorhinus canicula (dog fish): liver: 9
gill, brain: 12–13
Trisopterus luscus (bib): liver: 0.3
gill: 27   (1092)

Pimephales promelas Rafinesque (flow through test): ng/l: effective concentration[a]

hr	(EC) value Plotted TE_m[c]	lethal concentration (LC) value LC_{10}	LC_{50}	LC_{90}
24	14.4	15.1	23.5	36.6
		(9.1–18.5)[b]	(19.5–28.2)	(30.0–59.1)
48	14.4	13.9	19.6	27.6
		(7.8–16.7)	(15.9–22.8)	(23.6–42.7)
72	14.4	13.2	18.9	27.1
		(7.5–16.0)	(15.3–22.1)	(23.1–41.2)
96	14.4	13.2	18.4	25.6
		(7.5–16.0)	(14.8–21.3)	(22.0–38.0)

[a] the effective concentration. Concentration producing an adverse effect. In this case the effect noted was loss of equilibrium
[b] 95% confidence limits
[c] TE_m —median tolerance effect is obtained by a logarithmic plot of the data. Insufficient data was available to obtain a computer plot by Finney's probit analysis   (1054)

Pimephales promelas Rafinesque: LC_{50}, 96 hr:
flow through test: 18.4 mg/l; 95% conf. lim.: 14.8–21.3 mg/l
static test: 21.4 mg/l; 95% conf. lim.: 16.5–26.4 mg/l   (1054)
guppy (*Poecilia reticulata*): 7 d LC_{50}: 18 ppm   (1833)
fathead minnow: 96 hr LC_{50} (F): 18.4 mg/l
96 hr LC_{50} (S): 21.4 mg/l   (1524)

—Mammalia:
rats: inhalation: no patholog. effects: 70 ppm, 8 hr/day, 5 days/w, 7 months
: some patholog. changes in liver and kidneys 230 ppm, 8 hr/day, 5 days/w, 7 months   (211)
ingestion: rats: single oral LD_{50}: >5000 mg/kg
mouse: single oral LD_{50}: in range of 8000 to 11,000 mg/kg

N-(1,1,2,2-tetrachloroethylthio)cyclohex-4-ene-1,2-dicarboximide)   *see* difolitan

3,4,5,6-tetrachloroguaiacol

Chlorinated guaiacols are formed by the reaction of chlorination agents with phenolic lignins during the bleaching process in the manufacture of pulp and paper.

A. PROPERTIES: m.p. 122–124°C

D. BIOLOGICAL EFFECTS:
—Fish: juvenile rainbow trout: 96 hr LC_{50}: 0.32 mg/l (314)
—Mammals:
male Sprague–Dawley rats: single oral LD_{50}: 1690 mg/kg
male weanling rats: accumulation: 50–5000 ppm incorporated in the diet for 28 days: the accumulation in liver and kidney appears to be in a dose-related manner:

	(n.s.i.)		
conc. in diet	500 ppm	5000 ppm	
conc. in liver	0.66 ppm	6.4 ppm	
conc. in kidney	0.74 ppm	7.9 ppm	(1429)

tetrachloroisophthalonitrile *see* chlorothalonil

tetrachloromethane *see* carbon tetrachloride

tetrachloronitrobenzene
An impurity of pentachloronitrobenzene, 0.06% has been found in Terraclor®.
D. BIOLOGICAL EFFECTS:
—Residues:
in peanut butter samples: avg.: 1.1 ppb
range: 0.42–4.3 ppb ($n = 11$) (1805)

2,3,4,5-tetrachlorophenol

A. PROPERTIES: m.w. 231.89; m.p. 95–98°C
D. BIOLOGICAL EFFECTS:
—Fish:
guppy (*Poecilia reticulata*): 24 hr LC_{50}: 0.77 ppm at pH 7.3 (1833)

2,3,4,6-tetrachlorophenol

Use: fungicide.
A. PROPERTIES: needles; m.w. 231.9; m.p. 69/70°C; b.p. 164°C at 23 mm; v.p. 60 mm at 190°C, 400 mm at 250°C; sp.gr. 1.6 at 60/4°C;

2,3,5,6-TETRACHLOROPHENOL

C. WATER POLLUTION FACTORS:
 —Biodegradation: decomposition in soil suspensions: >72 days for complete disappearance (175)
 —Waste water treatment methods:
  reaction of excess hypochlorous acid with dilute aqueous solutions at initial pH's of 6.0 and 3.5:

[Reaction scheme showing compound (1) → (2) + (3) + (4) + (5) + unknown]

	(1)	(2)	(3)	(4)	(5)
yields at pH 6.0	5%	3%	31%	24%	9%
yields at pH 3.5	37%	5.5%	8%	11%	10%

(2) hexachloro-2,5-cyclohexadienone
(4); (5): octachlorophenols
(6): chloranil
(3): 2,2,3,4,6,6-hexachloro-5-hydroxy-3-cyclohexenone

at pH 3.5 4% (1091)
 —Odor threshold conc.: 0.915–47.0 mg/l (226)
  detection: 0.6 mg/l (998)
 —Taste threshold conc.: 0.001 mg/l (998)
D. BIOLOGICAL EFFECTS:
 —Fish: goldfish: 24 hr LC_{50}: 0.75 ppm
  amount found in dead fish at 0.8 ppm: 75 µg/g
  BCF (at 0.8 ppm) = 93 (1850)
 —Mammals: rats: acute oral LD_{50}: 0.14 g/kg (211)
  guinea pig: acute oral LD_{50}: 250 mg/kg (1854)

2,3,5,6-tetrachlorophenol

A. PROPERTIES: m.w. 231.89; m.p. 114–116°C
D. BIOLOGICAL EFFECTS:
 —Fish:
  guppy (*Poecilia reticulata*): 24 hr LC_{50}: 1.37 ppm at pH 7.3 (1833)

tetrachloropropane
A. PROPERTIES: m.w. 181.88; v.d. 6.28
C. WATER POLLUTION FACTORS:
 —Reduction of amenities: organoleptic limit: 0.01 mg/l (n.s.i.) (181)
 —Waste water treatment: evaporation from water at 25°C of 1 ppm solution:
  1,2,3,3'-isomer: 50% after 50 min
  90% after more than 120 min
  1,2,2',3-isomer: 50% after 47 min
  90% after more than 120 min (313)

2,3,5,6-tetrachloropyridine

Technical grade contains 2,3,4,6-tetrachloropyridine (0.5%), pentachloropyridine (0.3%), and trichloropyridine (0.2%).
A. PROPERTIES: white solid; b.p. 251°C; v.d. 7.48; sp.gr. 1.55 at 100°C; m.p. 91°C; $\log P_{oct}$ 3.32
D. BIOLOGICAL EFFECTS:
 —Mammals:
  ingestion: female rats: single oral LD_{50}: approx. 1000 mg/kg (1546)

tetrachloroquinone *see* chloranil

n-tetracosane
$CH_3(CH_2)_{22}CH_3$
A. PROPERTIES: m.w. 338.66; m.p. 49–52°C; b.p. 391°C
B. AIR POLLUTION FACTORS:
 —Ambient air quality:
  organic fraction of suspended matter:
   Bolivia at 5200 m altitude (Sept.–Dec. 1975): 0.59–1.1 µg/1000 cu m
   Belgium, residential area (Jan.–April 1976): 1.9–4.5 µg/1000 cu m (428)
  glc's: Botrange (Belgium)—woodland at 20–30 km from industrial area: June–July 1977: 0.36; 0.97 ng/cu m (n = 2)
  glc's in Wilrijk (Belgium)—residential area: Oct.–Dec. 1976: 2.99–20.46 ng/cu m (n = 2) (1233)
  glc's in residential area (Belgium) Oct. 1976:
   in particulate sample: 8.15 ng/cu m
   in gas phase sample: 4.63 ng/cu m (1289)
  glc's in the average American urban atmosphere—1963:
   480 µg/g airborne particulate or
   60 ng/g cu m air (1293)
C. WATER POLLUTION FACTORS:
 —Biodegradation:
  degradation in seawater by oil-oxidizing microorganisms: 36.2% breakdown after

21 days at 22°C in stoppered bottles containing a 1000 ppm mixture of alkanes, cycloalkanes and aromatics (1237)

n-tetracosanoic acid
$CH_3(CH_2)_{22}COOH$

B. AIR POLLUTION FACTORS:
—Ambient air quality:
glc's: Botrange (Belgium)—woodland at 20–30 km from industrial area: June–July 1977: 1.57; 3.20 ng/cu m ($n = 2$) (1233)
glc's in residential area (Belgium)—Oct. 1976: in particulate sample: 10.7 ng/cu m (1289)

n-tetradecane
$CH_3(CH_2)_{12}CH_3$

Use: organic synthesis; solvent; standardized hydrocarbon.

A. PROPERTIES: colorless liquid; m.w. 198.38; m.p. 5°C; b.p. 252°C; v.p. 1 mm at 76°C; v.d. 6.83; solub. in sea water: 0.0017 mg/l at 25°C, in distilled water: 0.0022 mg/l at 25°C

C. WATER POLLUTION FACTORS:
—Manmade sources: percolation water at 30–500 m from dumping ground contained 30 ppb (183)
—Waste water treatment methods: A.S. after 6 hr: 1.5% of ThOD
12 hr: 3.4% of ThOD
24 hr: 6.4% of ThOD (88)
rotating disk contact aerator: infl. 29.8 mg/l, effl. 1.1 mg/l; elimination: 96% or 2369 mg/m²/24 hr or 640 g/m³/24 hr (406)
first order evaporation constant of *n*-tetradecane in 3 mm No. 2 fuel oil layer in dark room at windspeed of 21 km/hr at:
5°C: 2.21×10^{-5} min^{-1}
10°C: 4.20×10^{-5} min^{-1}
20°C: 1.14×10^{-4} min^{-1}
30°C: 2.94×10^{-4} min^{-1} (438)

tetradecanoic acid *see* myristic acid

1-tetradecene (α-tetradecylene)
$CH_2CH(CH_2)_{11}CH_3$

Use: solvent in perfumes; flavors; medicines; dyes; oils; resins.

A. PROPERTIES: m.w. 196.38; sp.gr. 0.775 (20/4°C); m.p. −12°C; b.p. 256°C

C. WATER POLLUTION FACTORS:
—Waste water treatment methods:
rotating disk contact aerator: infl. 29.6 mg/l; effl. 0.8 mg/l; elimination: 97% or 2371 mg/m²/24 hr or 640 g/m³/hr (406)

α-tetradecylene *see* 1-tetradecene

tetraethyleneglycol
$(CH_3OCH_2CH_2OCH_2CH_2)_2O$

A. PROPERTIES: m.w. 194.2; m.p. $-6°C$; b.p. $327°C$; v.p. 0.001 mm at $20°C$; sp.gr. 1.12; log P_{oct} $-1.38/-2.18$ (calculated)
C. WATER POLLUTION FACTORS:
—Oxidation parameters
BOD_{10}: 0.50 std. dil. sew. (256)
—Waste water treatment:
A.C.: adsorbability: 0.116 g/g C; 58.1% reduction, infl. 1,000 mg/l, effl. 418 mg/l (32)
NFG, BOD, $20°C$, 1-10 days observed, feed: 200-1,000 mg/l, acclimation 365 + P: inhibition (93)
D. BIOLOGICAL EFFECTS:
—Mammals: ingestion: rats: single oral LD_{50}: 32.8 g/kg (1546)

tetraethyleneglycol-di(2-ethylhexanoate)
C. WATER POLLUTION FACTORS:
—Water quality:
in Delaware river (U.S.A.): conc. range winter: 1-14 ppb
summer: 1-4 ppb (1051)

tetraethyleneglycoldi(2-methylheptanoate)
C. WATER POLLUTION FACTORS:
—Water quality:
in Delaware river (U.S.A.): conc. range winter: 0.1-0.3 ppb
summer: 0.1-0.3 ppb (1051)

tetraethyldithionopyrophosphate *see* TEDP

tetraethyllead (TEL)
$(C_2H_5)_4Pb$
Composition: TEL: 61.49 wt %
ethylene dibromide: 17.86% by wt
ethylene dichloride: 18.81 % by wt
dye, stabilizer, kerosene and inerts: 1.84% by wt (1341)
Use: anti-knock compounds for gasoline; see composition above.
A. PROPERTIES: colorless liquid; m.w. 323.44; m.p. $-136°C$; decomposes at 110/$200°C$; v.p. 0.15 mm at $20°C$; v.d. 8.6; sp.gr. 1.659 at $18°C$; solub.:
in seawater: 0.1 mg/l (Pb); 2.0 mg/l (Pb) at $20°C$ (575)
in distilled water: at $20°C$: 0.8 mg/l (575)
in dilute solution in water decomposes to give triethyl salt, then diethyl salt, and finally inorganic lead: (567)

$$(C_2H_5)_4Pb \longrightarrow (C_2H_5)_3Pb^+ \longrightarrow (C_2H_5)_2Pb^{2+} \longrightarrow Pb^{2+}$$

after stirring pure TEL with seawater for 2 days, the following concentrations of lead compounds were measured (as Pb).
TEL: 0.626 mg/l; R_3PB^+: 0.039 mg/l; R_2Pb^{2+}: <0.005 mg/l;
Pb^{2+}: 0.057 mg/l ($R = (C_2H_5)$) (574)
B. AIR POLLUTION FACTORS:
—Ambient air quality:

TETRAETHYLLEAD

opposite filling station: 0.04–0.75 µg TEL + TML/cu m (expressed as Pb)
in Stockholm streets: 0.04–3.37 µg TEL + TML/cu m (expressed as Pb) (509)
at works entrance: <0.1–2.5 µg TEL/cu m (expressed as Pb) (508)

D. BIOLOGICAL EFFECTS:
  −Residues:
  in fish: n.d.–9.3 ng/g wet wt.
  in water, vegetation, weeds and sediments (taken from various lakes and rivers in Ontario (1979): n.d. (1793)
  −Bioaccumulation:
  accumulation in shrimp after 96 hr exposure at 0.02 mg/l (96 hr LC_{50}):
  $$\text{concentration factor} \left(\frac{\text{animal tissue (mg/kg wet wt.)}}{\text{water mg/l}}\right) = 650 \times \qquad (575)$$
  accumulation in mussel (*Mytilus edulis*) after 96 hr exposure at 10 mg/l (96 hr LC_{50}):
  $$\text{concentration factor} \left(\frac{\text{animal tissue (mg/kg wet wt.)}}{\text{water mg/l}}\right) = 120 \times \text{ (mean figure of 4}$$
  tissues: digestive gland, foot, gill and gonad) (575)
  accumulation in plaice after 96 hr exposure at 0.23 mg/l (96 hr LC_{50}):
  $$\text{concentration factor} \left(\frac{\text{animal tissue (mg/kg wet wt.)}}{\text{water mg/l}}\right) = 130 \times \qquad (575)$$
  −Toxicity:

	0% effect	48 hours 50% effect	100% effect
mixed coastal marine bacteria (1)	80 ppb	200 ppb	2000 ppb
algae: *Dunaliella tertiolecta* (2)	100 ppb	150 ppb	300 ppb
crustacean: 24 hr old nauplii of *Artemia salina* (3)	25 ppb	85 ppb	260 ppb
fish: 6 mm larvae of *Morone labrox* (3)	10 ppb	65 ppb	130 ppb

(996)

(1) reduction of oxygen consumption
(2) reduction of photosynthetic activity
(3) mortality

  −Algae:
  freshwater alga *Ankistrodesmus folcatus*: EC_{50}, 4 hr: <0.3 mg/l (reduction in photosynthesis) (578)
  algae: 96 hr LC_{50}: 0.1 mg/l (Pb)
  −Molluscs:
  mussel: 96 hr LC_{50}: 0.10 mg/l (Pb) (573)
  −Crustacean:
  shrimp: 96 hr LC_{50}: 0.02 mg/l (Pb) (573)
  −Fish: bluegill: TLm (24, 48, 96 hr): 2.0; 1.4; 0.2 mg/l (41)
  plaice: 96 hr LC_{50}: 0.23 mg/l (Pb) (573)
  bluegill: sunfish: 24 hr LC_{50}: 2.0 ppm (Pb)
  48 hr LC_{50}: 1.4 ppm (Pb) (571)
  bluegill: 96 hr LC_{50}: 0.02 ppm (364)
  −Mammalia: rat: single oral LD_{50}: 35 mg/kg

parenteral LD_{50}: 15 mg/kg
inhalation LC_{50}: 850 mg/cu m, 1 hr
rabbit: LD_{50}: 0.6 ml/kg (54)
rat: intravenous injection: LD_{50}: 10 mg Pb/kg body wt. (568)
ingestion: LD_{50}: 12 mg Pb/kg body wt. (569)
intraperitoneal: LD_{50}: 10 mg Pb/kg body wt. (570)

O,O,O,O-tetraethyl-S,S-methylene-*bis*-phosphorodithioate see ethion

tetraethylpyrophosphate see TEPP

1,2,3,4-tetrahydride see 1,2,3,4-tetrahydronaphthalene

1,2,3,4-tetrahydrobenzene see cyclohexene

tetrahydro-3,5-dimethyl-2H-1,3,5-thiadiazine-2-thione see dazomet

tetrahydrofuran (THF; diethyleneoxide; tetramethyleneoxide)

Use: tetrahydrofuran is a solvent for polyvinylchlorides, vinylchloride copolymers, and vinylidene chloride copolymers.
THF is stabilized with butylated hydroxytoluene to prevent the build up of peroxides. (503)
A. PROPERTIES: waterwhite liquid; m.w. 72.10; m.p. -108.5°C; b.p. 65/66°C; v.p. 0.06 atm at 0°C, 0.11 atm at 10°C, 0.173 atm at 20°C, 0.26 atm at 30°C; sp.gr. 0.888 at 20°/4°C; v.d. 2.5; sat. conc. 557 g/cu m at 20°C; solub. miscible in all proportions
B. AIR POLLUTION FACTORS: 1 mg/cu m = 0.33 ppm; 1 ppm = 3.00 mg/cu m
—Odor: characteristic: ethereal
T.O.C.: 30 ppm = 90 mg/cu m (298, 210)
0.27 mg/cu m (792)
recognition: 7.3-10.2 mg/cu m (712)
distinct odor: 180 mg/cu m (278)
O.I. at 20°C = 5800 (316)
C. WATER POLLUTION FACTORS:
—Manmade sources:
PVC-pipe cement used for joint connection caused 13 ppm, 6 months after PVC-pipe installation and 7.5 ppm after 8 months (at equilibrium condition—reached in about 48 hr) (1670)
D. BIOLOGICAL EFFECTS:
—Threshold conc. of cell multiplication inhibition of the protozoan *Uronema parduczi Chatton-Lwoff*: 858 mg/l (1901)
—Bacteria: *Pseudomonas putida*: inhibition of cell multiplication starts at 580 mg/l
(329)

—Algae: *Microcystis aeruginosa*: inhibition of cell multiplication starts at 225 mg/l
(329)
—Mammalia:
rat: inhalation: LC_{50}: 80,975 ppm, 1 hr
60,000 ppm, 2 hr
18,000–22,000 ppm, 4 hr
no clinical symptoms: 3,000 ppm, 8 hr/day, 20 months   (54)

tetrahydro-2-furancarbinol *see* tetrahydrofurfurylalcohol

tetrahydrofurfurylalcohol (tetrahydro-2-furancarbinol; THFA)

Use: solvent for many synthetic resins and leather dyes; plasticizer, synthesis of lysine; paint and varnish ingredient.
A. PROPERTIES: colorless almost odorless liquid; m.w. 102.13; m.p. $<-80°C$; b.p. 178°C; v.p. 2.3 mm at 40°C, 5 mm at 50°C; sp.gr. 1.054 at 20/20°C; THC 709 Kcal/mole
B. AIR POLLUTION FACTORS: 1 mg/cu m = 0.24 ppm; 1 ppm = 4.18 mg/cu m
C. WATER POLLUTION FACTORS:
—Biodegradation: adapted A.S. at 20°C—product as sole carbon source: 96.1% COD removal at 40.0 mg COD/g dry inoculum/hr   (327)
—Water quality:
self purification of surface water is affected at 0.5 mg/l   (181)
D. BIOLOGICAL EFFECTS:
—Mammalia: guinea pig: single oral LD_{50}: 0.8–1.6 g/kg
rat: single oral LD_{50}: 1.6–3.2 g/kg
inhalation: 2/3: 12,650 ppm, 6 hr
0/3:   655 ppm, 6 hr   (211)

2,3,4,5-tetrahydrohexanoic acid *see* mucic acid

3a,4,7,7a-tetrahydro-4,7-methanoindene *see* α-dicyclopentadiene

tetrahydromethylcyclopentadiene

D. BIOLOGICAL EFFECTS:
—Fish:
golden shiner: 96-hr LC_{50}: 0.51 mg/l (emulsion) (R)
96-hr LC_{50}: 100.0 μl/l (unemulsified) (R)
rainbow trout: 97-d BCF 9,800 (F)   (1523)

1,2,3,4-tetrahydronaphthalene (tetralin, naphthalene-1,2,3,4-tetrahydride)

Use: solvent; chemical intermediate.
A. PROPERTIES: colorless liquid; m.w. 132.20; m.p. -35/-30°C; b.p. 207°C; v.p. 0.3 mm at 20°C; 0.6 mm at 30°C, 1 mm at 38°C; v.d. 4.55; sp.gr. 0.97; sat. conc. 2.2 g/cu m at 20°C, 4.2 g/cu m at 30°C
B. AIR POLLUTION FACTORS: 1 mg/cu m = 0.18 ppm; 1 ppm = 5.50 mg/cu m
C. WATER POLLUTION FACTORS:
—Oxidation parameters
BOD_5:	0% ThOD	(274)
COD:	10% ThOD (0.05 n $K_2Cr_2O_7$)	(274)
$KMnO_4$:	0% ThOD (0.01 n $KMnO_4$)	(274)
ThOD:	3:147	(274)

—Biodegradation:
degradation in seawater by oil-oxidizing microorganisms: 31.0% breakdown after 21 days at 22°C in stoppered bottles containing a 1000 ppm mixture of alkanes, cycloalkanes, and aromatics (1237)
—Waste water treatment methods:
conventional municipal treatment + A.C.: infl. 0.057 mg/l, effl. n.d. (404)
—Odor threshold conc. (mean): 0.018 mg/l (403)

tetrahydronorbornadiene
D. BIOLOGICAL EFFECTS:
—Fish:
golden shiner: 96-hr LC_{50}: 0.61 mg/l (emulsion (R)
96-hr LC_{50}: 4,700 µl/l (unemulsified) (R)
rainbow trout: 97-d BCF: 3,900 (F) (1523)

tetrahydro-1,4-oxazine *see* morpholine

tetrahydrophthalic acid
C. WATER POLLUTION FACTORS:
—Biodegradation: adapted A.S. at 20°C—product as sole carbon source: nihil % COD removal after 20 days (327)

tetrahydrophthalimide
C. WATER POLLUTION FACTORS:
—Biodegradation: adapted A.S. at 20°C—product as sole carbon source: nihil % COD removal after 20 days (327)

2,3,4,5-tetrahydrothiophene-1,1-dioxyde *see* sulfolane

tetrahydroxybutane *see* erythritol

1,3,4,5-tetrahydroxycyclohexanecarboxylic acid *see* quinic acid

tetralin *see* 1,2,3,4-tetrahydronaphthalene

tetramethylammoniumchloride
$(CH_3)_4CCl$
  Use: chemical intermediate; catalyst; inhibitor
  A. PROPERTIES: m.w. 109.6; m.p. >300°C; a quaternary ammonium compound
  C. WATER POLLUTION FACTORS:
    —Inhibition of biodegradation processes:
      ~50% inhibition of nitrification at 2200 mg/l   (411)

1,2,4,5-tetramethylbenzene (durene)

Use: organic synthesis; plasticizers; polymers; fibers
A. PROPERTIES: m.w. 134.22; sp.gr. 0.838; m.p. 77°C; b.p. 196–197°C; camphorlike odor
B. AIR POLLUTION FACTORS:
  —Odor threshold conc. (recognition): 0.083 mg/cu m   (610)
                                        0.087 mg/cu m   (611)
C. WATER POLLUTION FACTORS:
  —Aquatic reactions:
    adsorption on smectite clay particles from simulated seawater at 25°C—experimental conditions: 100 µg durene/l; 50 mg smectite/l adsorption: 0 µg/mg
                                                                (1009)

p-(1,1,3,3-tetramethylbutyl)phenol

C. WATER POLLUTION FACTORS:
  —Water quality:
    in Delaware river (U.S.A.): conc. range: winter: 1–5 ppb
                                             summer: 0.2–2 ppb   (1051)

tetramethylenedicyanide *see* adiponitrile

tetramethyleneglycol *see* 1,4-butanediol

tetramethylene oxide *see* tetrahydrofuran

tetramethylenesulfone *see* sulfolane

tetramethylethylene *see* 2,3-dimethylbutene-2

tetramethyllead (TML)
$(CH_3)_4 Pb$
Use: anti-knock compound for gasoline; see also TML-CB
Composition of TML-CB:
TML: 50.82 wt %
ethylene dibromide: 17.86 wt %
ethylene dichloride: 18.81 wt %
dye, stabilizer, kerosene and inerts: 12.51 % (1341)
A. PROPERTIES: m.w. 267.33; m.p. $-27.5°C$; b.p. $110°C$ decomposes; v.p. 22.5 mm at $20°C$; v.d. 6.5; sp.gr. 1.99; solub in seawater 15 mg/l (Pb)
B. AIR POLLUTION FACTORS: 1 mg/cu m = 8.61 ppm as Pb
C. WATER POLLUTION FACTORS:
—Decomposes in dilute solution in water to give trimethyl salt, then diethyl salt, and finally inorganic lead:

$$(CH_3)_4 Pb \longrightarrow (CH_3)_3 Pb^+ \longrightarrow (CH_3)_2 Pb^{2+} \longrightarrow Pb^{2+}$$

D. BIOLOGICAL EFFECTS:
—Residues:
in fish taken from various lakes and rivers in Ontario: n.d.–5.2 ng/g wet wt.
in water, vegetation, weeds, and sediments taken from various lakes and rivers in Ontario (1979): n.d. (1793)
—Bioaccumulation:
accumulation in shrimp after 96 hr exposure at 96 hr LC_{50} conc. of 0.11 mg/l:

concentration factor $\left(\dfrac{\text{animal tissue (mg/kg wet wt)}}{\text{water mg/l}}\right) = 20 \text{ X}$ (575)

accumulation in mussel (*Mytilus edulis*) after 96 hr exposure at 96 hr LC_{50} conc. of 0.27 mg/l:

concentration factor $\left(\dfrac{\text{animal tissue (mg/kg wet wt)}}{\text{water mg/l}}\right) = 170 \text{ X}$ (mean (575)

figure of 4 tissues: digestive gland, foot, gill and gonad)
accumulation in plaice after 96 hr exposure at 96 hr LC_{50} conc. of 0.05 mg/l:

concentration factor $\left(\dfrac{\text{animal tissue (mg/kg wet wt)}}{\text{water mg/l}}\right) = 60 \text{ X}$ (575)

2,6,10,14-TETRAMETHYLPENTADECANE

−Toxicity:

	0% effect	48 hours 50% effect	100% effect
mixed coastal marine bacteria (1):	900 ppb	1900 ppb	4500 ppb
alga: *Dunaliella tertiolecta* (2):	450 ppb	1650 ppb	4500 ppb
crustacean: 24 hr old nauplii of *Artemia salina* (3):	180 ppb	250 ppb	670 ppb
fish: 6 mm larvae of *Morone* *labrax* (3):	45 ppb	100 ppb	250 ppb

(1): reduction of oxygen consumption
(2): reduction of photosynthetic activity
(3): mortality (996)

−Algae: 6 hr EC_{50} (photosynthetic activity): 1.3 mg/l (Pb) (573)
−Molluscs: mussel: 96 hr LC_{50}: 0.27 mg/l (Pb) (573)
−Fish:
  shrimp: 96 hr LC_{50}: 0.11 mg/l (Pb) (573)
  plaice: 96 hr LC_{50}: 0.05 mg/l (Pb) (573)

Lepomis macrochirus: static bioassay in fresh water at 23°C, mild aeration applied after 24 hr (68% TML in toluene):

material added ppm	24 hr	% survival after 48 hr	72 hr	96 hr	best fit 96 hr LC_{50} ppm
125	0	−	−	−	
90	100	90	10	0	84
79	100	100	95	70	
50	100	100	100	100	

Menidia beryllina: static bioassay in synthetic seawater at 23°C: mild aeration applied after 24 hr (68% TML in toluene):

material added ppm	24 hr	% survival after 48 hr	72 hr	96 hr	best fit 96 hr LC_{50} ppm
79	10	0	−	−	
50	0	−	−	−	13.5
25	100	100	70	30	
10	100	100	80	60	

(352)

−Mammals:
  rat: intravenous injection: LD_{50}: 80 mg Pb/kg body wt (568)
       ingestion: LD_{50}: 108 mg Pb/kg body wt (569)
       intraperitoneal: LD_{50}: 70−100 mg Pb/kg body wt (570)
       inhalation: LC_{50}: >9,000 mg/cu m, 60 min (54)
−Man: no effect level: 0.15 mg/cu m for 12−18 months (54)

2,6,10,14-tetramethylpentadecane *see* Pristane

O,O,O',O'-tetramethyl-O,O-thiodi-*p*-phenylenephosphorodithioate *see* abate

tetramethylthiourea (TMTU)

An animal carcinogen (1799)
Use: in adhesive industry; a curing compound for polychloroprene rubber.

B. AIR POLLUTION FACTORS:
 —Method and analysis: TMTU is collected from air using midget impingers containing water, or PVC or cellulose ester membrane filters which are then extracted with water. Pentacyanoamineferrate reagent is added to the filter extract or to the impinger contents to form a colored coordination complex. The absorbance of the solution is measured spectrophotometically at 590 nm. The detection limit is 3 µg/sample (1799)

tetramethylthiuramdisulfide *see* thiram

tetranitromethane
$C(NO_2)_4$
A. PROPERTIES: colorless liquid; m.w. 196.04; m.p. 13.75°C; b.p. 125.7°C; v.p. 8 mm at 20°C; 13 mm at 25°C, 15 mm at 30°C; v.d. 6.8; sp.gr. 1.650 at 13/4°C; sat. conc. 90 g/cu m at 20°C, 154 g/cu m at 30°C
B. AIR POLLUTION FACTORS: 1 mg/cu m = 0.12 ppm, 1 ppm = 8.15 mg/cu m
 —Odor: characteristic: acrid, biting
D. BIOLOGICAL EFFECTS:
 —Phytotoxicity (mg/m^3):

	EC_{50}*	(95% conf. limits)
wheat	0.68	(0-3.7)
alfalfa	0.93	(0-4.77)
tobacco	6.1	(0.14-12.2)
soybean	0.69	(0-4.31)
corn	2.1	(0.58-3.61)

*EC_{50}, concentration required to cause visible injury in 50% in the leaves of the plant exposed for a 2 hr fumigation period. Zero values indicate that statistical range includes negative values (1831)
 —Mammalia: dog: inhalation: mild symptoms: 6.35 ppm, 6 months
  rat: inhalation: 11/19: 6.35 ppm, 6 months
   20/20: 33 ppm, 10 hr
  cat: inhalation: mild irritation: 0.1-0.4 ppm, 2 X 6 hr
   5/5: 7-25 ppm, 2-5 hr (211)

tetrapropylene benzenesulphonate
D. BIOLOGICAL EFFECTS:
 —Arthropoda:
 Crangon crangon in sea water at 15°C: exposure time; LC_{50}, mg/l

exposure time	LC_{50}, mg/l
9 min	>56,000
up to 24 hr	>56,000
48 hr	42,000
72 hr	32,000
96 hr	18,000

(328)

Crangon crangon in sea water at 15°C:
exposure time; EC_{50} after recovery period in unpolluted sea water

exposure time	0 min	24 hr
27 min	>56,000 ppm	>56,000 ppm
180 min	>56,000 ppm	>56,000 ppm

(328)

—Fish:
Gobius minutus in sea water at 15°C:
exposure time; EC_{50} after recovery period in unpolluted sea water

	0 min	24 hr	
1 min	>18,000 ppm	±1,200 ppm	
3 min	>18,000 ppm	400 ppm	
9 min	±7,000 ppm	<550 ppm	(328)

Gobius minutus in sea water at 15°C: exposure time; LC_{50}, mg/l

3 min	>18,000
9 min	±7,000
27 min	150
1 hr	30
3 hr	25
6 hr	25
12 hr	25
24 hr	25
up to 96 hr	25	(328)

tetryl (2,4,5-trinitrophenylmethylnitramine; N-methyl-N-2,4,6-tetranitroaniline; methylpicrylnitramine)

A. PROPERTIES: yellow crystals; m.w. 287.15; m.p. 130°C; b.p. explodes at 187°C; v.d. 9.92; sp.gr. 1.57 at 19°C
B. AIR POLLUTION FACTORS: 1 mg/cu m = 0.08 ppm; 1 ppm = 12.0 mg/cu m
C. WATER POLLUTION FACTORS:
  —Biodegradation: biodegradation in 7 days tests: nil % degradation in original culture, neither in 1st, 2nd or 3rd subculture (87)

TGA *see* nitrilotriacetic acid

TH 6040 *see* diflubenzuron

THF *see* tetrahydrofuran

THFA *see* tetrahydrofurfurylalcohol

thallium acetate (thallous acetate)
CH_3COOTl
Use: medicine; high specific gravity solutions used to separate ore constituents by flotation.

A. PROPERTIES: white, deliquescent crystals; m.p. 131°C; sp.gr. 3.68
D. BIOLOGICAL EFFECTS:
  —Fish:
    brown shrimp: 96 hr LC_{50}: 10 ppm Tl (365)
    atlantic salmon: LD_{50}: 0.03 ppm Tl (365)
    Lepomis macrochirus: static bioassay in fresh water at 23°C, mild aeration applied after 24 hr:

material added ppm	% survival after				best fit 96 hr LC_{50} ppm
	24 hr	48 hr	72 hr	96 hr	
320	100	100	10	0	
250	100	80	70	10	
100	85	85	71	71	170
32	100	100	100	100	(132 as Tl)

Menidia beryllina: static bioassay in synthetic seawater at 23°C: mild aeration applied after 24 hr:

material added ppm	% survival after				best fit 96 hr LC_{50} ppm
	24 hr	48 hr	72 hr	96 hr	
180	0	—	—	—	
100	80	0	—	—	
75	90	40	0	—	
56	100	40	10	10	31
32	100	70	30	30	(24 as Tl)
24	100	100	90	70	
18	100	100	100	100	(352)

thallous acetate *see* thallium acetate

theine *see* caffeine

thianaphthalene
C. WATER POLLUTION FACTORS:
  —Biodegradation:
    degradation in seawater by oil oxidizing microorganisms: 13.6% breakdown after 21 days at 22°C in stoppered bottles containing a 1000 ppm mixture of alkanes, cycloalkanes, and aromatics (1237)

thianaphthene (1-benzothiophene; thionaphthene)

A. PROPERTIES: m.w. 134.20; m.p. 29–32°C; b.p. 221–222°C; sp.gr. 1.15
D. BIOLOGICAL EFFECTS:
  —Inhibition of photosynthesis of a freshwater, nonaxenic unialgal culture of *Selenastrum capricornutum*:
    at   1% saturation: 89% carbon-14 fixation (vs. controls)
        10% saturation: 82% carbon-14 fixation (vs. controls)
       100% saturation: 4% carbon-14 fixation (vs. controls) (1690)

thimet *see* phorate

thioacetamide (ethanethiomide; acetothioamide)
CH_3CSNH_2
A. PROPERTIES: yellow tablets; m.w. 75.13; m.p. 108.5°C; log P_{oct} −0.46/0.36 (calculated)
C. WATER POLLUTION FACTORS:
  —Impact on treatment processes: 75% reduction of nitrification process in non acclimated activated sludge at 0.14 mg/l (30)
  —Waste water treatment:
  A.S., Resp, BOD, 20°C, 1-5 days observed, feed: 1,000 mg/l, acclim. less than 1 day: nil % theor. oxidation; nil % removed (93)
  —Odor removal:
  median odor threshold level: 94.7 ppm in 50% aqueous ethylene glycol
                                    406 ppm in 50% aqueous ethylene glycol, after reaction with chloramine-T (1821)
D. BIOLOGICAL EFFECTS:
  —Carcinogenicity: weak
  —Mutagenicity in the *Salmonella test*: negative,
  <0.001 revertant colonies/nmol
  <70 revertant colonies at 5000 μg/plate (1883)

thiobenzamide

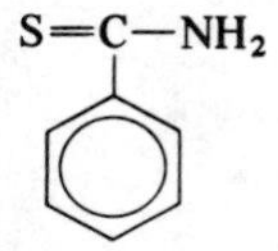

$$S=C-NH_2 \text{ (phenyl ring)}$$

A. PROPERTIES: m.w. 137.20; m.p. 116-118°C
C. WATER POLLUTION FACTORS:
  —Odor removal: median odor threshold level increases from 406 ppm to 4000 ppm in 50% aqueous ethylene glycol after reaction with chloramine-T (1821)

thiobenzoic acid:

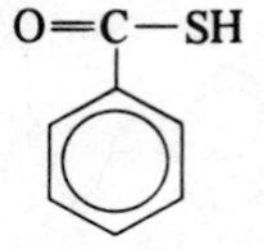

$$O=C-SH \text{ (phenyl ring)}$$

A. PROPERTIES: m.w. 138.19; m.p. 15-18°C; sp.gr. 1.17
C. WATER POLLUTION FACTORS:
  —Odor removal: median odor threshold level increases from 20.6 ppm to >125 ppm in 50% aqueous ethyleneglycol after reaction with chloramine-T (1821)

2(thiocyanomethylthio)benzothiazol *see* busan 25; busan 72; busan 74

thiocyclopentane-1,1-dioxide *see* sulfolane

thiodemeton *see* disulfoton

thioether *see* diethylsulfide

thiofuran *see* thiophene

thioglycolic acid *see* mercaptoacetic acid

thiolane-1,1-dioxide *see* sulfolane

thiometon *see* ekatin

thion
  Use: microbicide.
  Active ingredient: organosulfur compound.
D. BIOLOGICAL EFFECTS:
  —Fish:

	mg/l	24 hr	48 hr	96 hr	3m (extrap.)
harlequin fish					
(*Rasbora heteromorpha*) LC_{10} (F)	0.047	0.024	0.013		
LC_{50} (F)	0.09	0.045	0.022	0.001	
					(331)

thiophanatemethyl
D. BIOLOGICAL EFFECTS:
  —Fish: *Cyprionus carpio*: TLm: 75.0 ppm
         Salmo iridens: TLm: 8.8 ppm   (1873)
  —Birds: Japanese quail: oral LD_{50}: >5.0 g/kg   (1873)
  —Mammals:
    oral LD_{50} (mouse, rat, guinea pig, rabbit, dog): 2.27–7.50 g/kg
    dermal LD_{50} (mouse, rate, guinea pig, rabbit, dog): 10.0–>10.0 g/kg
    i.p. LD_{50} (mouse, rat): 0.79–1.64 g/kg
    s.c. LD_{50} (mouse, rat): >10.0 g/kg   (1873)

thiophene (thiofuran)

A. PROPERTIES: m.w. 84.13; m.p. -30/-38°C; b.p. 84°C; v.p. 60 mm at 20°C, 100 mm at 30°C; sp.gr. 1.06 at 20/4°C; sat. conc. 275 g/cu m at 20°C, 443 g/cu m at 30°C; solub. 3600 mg/l at 18°C; log P_{oct} 1.81
B. AIR POLLUTION FACTORS: 1 mg/cu m = 0.29 ppm, 1 ppm = 3.50 mg/cu m

THIOPHENOL

odor thresholds   mg/cu m

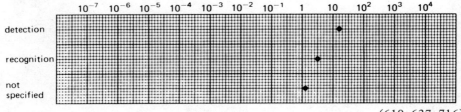

(610; 637; 716)

- Control methods:

    wet scrubber: water at pH 8.5: outlet: 4,000 odor units/scf

                $KMnO_4$ at pH 8.5: outlet: 13 odor units/scf   (115)

    catalytic combustion: platinized ceramic honeycomb catalyst: ignition temp. 335°C, inlet temp. for 90% conversion: 400–450°C   (91)

C. WATER POLLUTION FACTORS:

- Reduction of amenities: organoleptic limit USSR 1970: 2.0 mg/l   (181)
- Water quality: M.A.C. in waters class I for the prod. of drinking water: 2 mg/l   (133)

D. BIOLOGICAL EFFECTS:

- Inhibition of photosynthesis of a freshwater, nonaxenic unialgal culture of *Selenastrum capricornutum*:

    at   10 mg/l: 95% carbon-14 fixation (vs. controls)

         100 mg/l: 97% carbon-14 fixation (vs. controls)

     1000 mg/l: 68% carbon-14 fixation (vs. controls)   (1690)

thiophenol (phenylmercaptan; benzenethiol)

A. PROPERTIES: colorless liquid; m.w. 110.17; b.p. 169.5°C; v.p. 1 mm at 18.6°C, 10 mm at 56°C, 100 mm at 106.6°C; sp.gr. 1.078 at 20/4°C; solub. 470 mg/l at 15°C; log P_{oct} 2.52

B. AIR POLLUTION FACTORS:

- Odor: Characteristic: putrid, nauseating

        T.O.C.: 0.0002 ppm   (306)

                0.00026 ppm

                14 ppm   (279)

        O.I. at 20°C: 94   (316)

- Control methods:

    wet scrubber: water at pH 8.5: outlet: 1,300 odor units/scf

                $KMnO_4$ at pH 8.5: outlet: 13 odor units/scf   (115)

- Odor removal: median odor threshold level increases from 13 ppm to >125 ppm in 50% aqueous ethyleneglycol after reaction with chloramine-T   (1821)

C. WATER POLLUTION FACTORS:
—Reduction of amenities: faint odor: 0.000062 mg/l (129)
T.O.C.: average: 13.5 mg/l
range: 2.05–32.8 mg/l (294, 97)
0.001 mg/l (296)

thioranic acid *see* mercaptoacetic acid

thiosemicarbazide (aminothio-urea)
$NH_2CSNHNH_2$
Use: reagent for ketones and certain metals; photography; rodenticide.
A. PROPERTIES: m.p. 180–184°C; m.w. 91.14
C. WATER POLLUTION FACTORS:
~50% inhibition of NH_3 oxidation in *Nitrosomonas* at 0.9 mg/l (407)

thiosinamine *see* allylthiourea

thiourea (thiocarbamide)

Use: silver ion complexing agent.
A. PROPERTIES: m.w. 76.12; m.p. 182°C; b.p. decomposes; sp.gr. 1.405 at 20/4°C; solub. 91.8 g/l at 13°C; log P_{oct} −2.38/−0.95 (calculated)
C. WATER POLLUTION FACTORS:

theoretical	analytical
TOD = 2.42	reflux COD = 99.1% recovery
COD = 0.84	rapid COD = 94.3% recovery
NOD = 1.68	TKN = 102.2% recovery
	BOD_5 = 0.013
	BOD_5/COD = 0.015
	BOD_5 (acclimated) = 0.075

—Impact on conventional biological treatment systems:

	chemical conc. mg/l	effect
unacclimated system	100	no effect
	1,000	inhibitory
acclimated system	276	biodegradable

(1828)

Thiourea was found to be amenable to moderate rate biodegradation by an acclimated biomass. (1828)

75% reduction of nitrification process of non acclimated activated sludge at 0.075 mg/l (30)

D. BIOLOGICAL EFFECTS:
—Algae: *Selenastrum capricornutum*: 1 mg/l: no effect
10 mg/l: no effect
100 mg/l: inhibitory (1828)

1100 1,1-THIOXOTHIOLANE

—Crustacean: *Daphnia magna*: LC_{50}: 1.8 mg/l (1828)
—Fish: *Pimephales promelas*: LC_{50}: >100 mg/l (1828)
—Carcinogenicity: weak
—Mutagenicity in the *Salmonella* test: negative
  <0.001 revertant colonies/nmol
  <70 revertant colonies at 5000 µg/plate (1883)

1,1-thioxothiolane *see* sulfolane

threonine (2-amino-3-hydroxybutanoic acid)

$$\begin{array}{c} CH_3 \\ | \\ CHOH \\ | \\ CH-NH_2 \\ | \\ COOH \end{array}$$

An essential amino acid (*l*-isomer)
Use: nutrition and biochemical research
C. WATER POLLUTION FACTORS:
—Waste water treatment: A.S. after 6 hr: 3.9% of ThOD
                                   12 hr: 8.2% of ThOD
                                   24 hr: 16.2% of ThOD (89)
  A.S., BOD, 1–5 days observed, feed: 333 mg/l, 15 days acclim.: 91% removed
  (93)
—Manmade sources: excreted by man:
  in urine: 0.36–1.2 mg/kg body wt./day
  in feces: 3.5–5.2 mg/kg body wt./day
  in sweat: 1.7–9.1 mg/100 ml (203)
—Inhibition of biodegradation processes:
  ~50% inhibition of NH_3 oxidation in *Nitrosomonas* at 3.6 mg/l (*l*-isomer) (408)

thyme camphor *see* thymol

thymic acid *see* thymol

thymine (2,4-dioxy-5-methylpyrimidine; 5-methyluracil)

One of the pyrimidine bases of living matter.
Use: biochemical research
A. PROPERTIES: m.w. 126.12; m.p. 316–317°C; log P_{oct} −0.44 (calculated)

C. WATER POLLUTION FACTORS:
 —in primary domestic sewage plant effluent: 0.007–0.028 mg/l (517)

thymol (3-*p*-cymenol; isopropyl-*m*-cresol; thyme camphor; thymic acid)

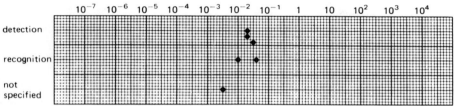

A. PROPERTIES: colorless plates; m.w. 150.21; m.p. 51.5°C; b.p. 233.5°C; v.p. 1 mm at 64.3°C, 10 mm at 107.4°C, 100 mm at 164.1°C; sp.gr. 0.969 at 20/4°C; solub. 850 mg/l at 20°C, 1,320 mg/l at 37°C; log P_{oct} 3.30
B. AIR POLLUTION FACTORS: 1 mg/cu m = 0.16 ppm; 1 ppm = 6.244 mg/cu m
 —Odor: quality: aromatic

odor thresholds   mg/cu m

(307; 606; 610; 614; 840)
 O.I. at 20°C: 155,000 (316)
C. WATER POLLUTION FACTORS:
 —Oxidation parameters: COD: 2.2
   ThOD: 2.77 (27)
 —Biodegradation: adapted A.S. at 20°C—product is sole carbon source: 94.6% COD removal at 15.6 mg COD/g dry inoculum/hr (327)
 —Odor threshold conc. (detection): 0.5 mg/l (998)

tillam (S-propylethylbutylthiocarbamate; pebulate)

$$CH_3-CH_2-CH_2-S-\underset{\underset{O}{\parallel}}{C}-N\begin{matrix}CH_2-CH_3\\CH_2-CH_2-CH_2-CH_3\end{matrix}$$

Use: herbicide
A. PROPERTIES: amber liquid; m.w. 203.35; sp.gr. 0.956 at 20°C; solub. 60 ppm, 92 ppm at 21°C
 formulations tillam 6E: an emulsifiable liquid containing 6 lbs tillam/gal.
   tillam 10G: a granular formulation containing 10% wt tillam
D. BIOLOGICAL EFFECTS:
 —Crustacean: brown shrimp (*P. aztecus*): 48 hr LC_{50}: >1 ppm (500)
 —Molluscs: commercial oyster (*C. Virginica*): EC_{50} (Shell growth inhibition) 96 hr: >1 ppm (500)

- Fish: white mullet (*M. curema*): 48 hr TLm: 6.25 ppm
  killifish (*F. similis*): 48 hr TLm: 7.78 ppm
  rainbow trout (*S. gairdneri*): 96 hr LC_{50}: 7.4 ppm
  bluegill sunfish (*L. macrochirus*): 96 hr LC_{50}: ca. 7.4 ppm (500)
- Birds: bobwhite quail: LC_{50} (7 day feed treatment): 8,400 ppm (500)
- Mammals:
  acute oral toxicity: technical tillam: rat: LD_{50}: 1,120 mg/kg
  mouse: LD_{50}: 1,652 mg/kg
  tillam 6E: rat: LD_{50}: 1,462 mg/kg
  acute dermal toxicity: rabbits: technical tillam: LD_{50}: >2,936 mg/kg
  tillam 6E: LD_{50}: >4,640 mg/kg
  acute ocular irritancy: rabbits:
  technical tillam: mild conjunctivitis clearing within 24–48 hours
  tilliam 6E: severe irritant; produces irreversible eye changes in rabbits

TMA *see* trimethylamine

TML *see* tetramethyllead

TMTU *see* tetramethylthio-urea

TNT *see* 2,4,6-trinitrotoluene

o-tolidine (3,3'-dimethylbenzidine; diaminotolyl)

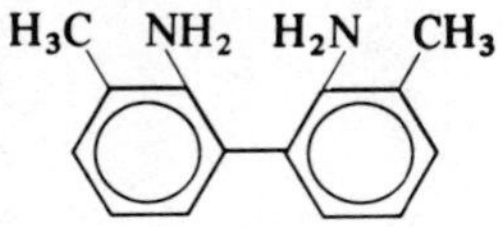

H_3C NH_2 H_2N CH_3

Use: dyes; curing agent for urethane resins.
A. PROPERTIES: m.p. 129–131°C
C. WATER POLLUTION FACTORS:
 —Waste water treatment methods:
  W, A.S. from mixed domestic/industrial treatment plant: 99–100% depletion at 20 mg/l after 6 hr at 25°C (419)

tolualdehyde

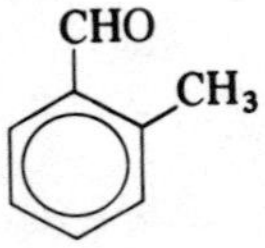

CHO
CH_3

A. PROPERTIES: m.w. 120.15
B. AIR POLLUTION FACTORS:
 —Manmade sources:
  in exhaust of gasoline engine: 1.9–7.2 vol. % of total exhaust aldehydes
  (394; 396; 397)

in exhaust of a 1970 Ford Maverick gasoline engine operated on a chassis dynamometer following the 7-mode California cycle:
from API #7 gasoline: 2.6 ppm
from API #8 gasoline: 3.7 ppm (1053)

toluene (methylbenzene; phenylmethane)

Manufacturing source: petroleum refining, coal tar distillation. (347)
Users and formulation: mfg. benzene derivatives, caprolactam, saccharin, medicines, dyes, perfumes, TNT; solvent recovery plants; component gasoline; solvent for paints and coatings, gums, resins, rubber and vinyl organosols; diluent and thinner in nitrocellulose lacquers, adhesive solvent in plastic toys and model airplanes; detergent mfg.; asphalt and naphtha constituent. (347)
Natural sources (water and air): coal tar, petroleum. (347)
Manmade sources: in gasoline (high octane number): 14.86 wt %. (387)

A. PROPERTIES: colorless liquid; m.w. 92.1; m.p. $-95.1°C$; b.p. $110.8°C$; v.p. 10 mm at $6.4°C$, 22 mm at $20°C$, 40 mm at $31.8°C$; v.d. 3.14; sp.gr. 0.867 at $20/4°C$; solub. 470 mg/l at $16°C$, 515 mg/l at $20°C$; THC 935.2 kcal/mole, LHC 903 kcal/mole; sat. conc. 110 g/cu m at $20°C$, 184 g/cu m at $30°C$; log P_{oct} 2.69 at $20°C$

B. AIR POLLUTION FACTORS: 1 mg/cu m = 0.26 ppm, 1 ppm = 3.83 mg/cu m
   —Odor: characteristic: quality: sour, burnt
                          hedonic tone: unpleasant to neutral

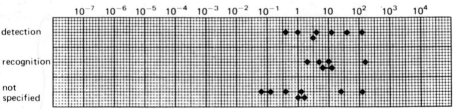

(2; 19; 57; 210; 278; 298; 307; 610; 611; 637; 642; 671; 673; 675; 676; 727; 735; 741; 749; 762; 805; 819)

odor Index: 16,609 (19)
—Atmospheric reactions: (49)
  R.C.R.: 0.98
  reactivity: HC. cons. ranking: 1
              NO ox. ranking: 0.2 (63)
  estimated lifetime under photochemical smog conditions in S.E. England: 5.8 hr
                                                                    (1699; 1707)
—Ambient air quality:
  glc's: downtown Los Angeles 1967: 10%ile: 10 ppb
                                    average: 30 ppb
                                    90%ile: 50 ppb (64)

expected glc's in USA urban air: range: 10–50 ppb (102)
glc's in the Netherlands:
  tunnel Amsterdam 1973:   avg.:   13 ppb ($n = 3$)
  tunnel Rotterdam 1974. 10.2: avg.:   38 ppb ($n = 12$)
                                    max.: 63 ppb
  The Hague 1974.10.11:   avg.:   23 ppb ($n = 12$)
                                    max.: 54 ppb
  Roelofarendsveen 1974.9.11.: avg.:   6 ppb ($n = 12$)
                                    max.:   9 ppb (1231)
glc's Los Angeles 1966: avg.:   0.037 ppm ($n = 136$)
                    highest value: 0.129 ppm (1319)

—Manmade sources:
evaporation from gasoline tank: 0.3–0.9 vol.% of total evaporated HC's
evaporation from carburetor: 0.5–2.4 vol.% of total evaporated HC's
                                                  (398; 399; 400; 401; 402)
in exhaust of gasoline engines:
  62-car survey: 3.1 vol.% of total exhaust HC's (391)
  15-fuel survey: 5 vol.% of total exhaust HC's (392)
  engine-variable study: 7.9 vol.% of total exhaust HC's (393)
in exhaust of diesel engine: 3.1% of emitted HC.s (72)
in exhaust of reciprocating gasoline engine: 6.0% of emitted HC.s
in exhaust of rotary gasoline engine: 16.3% of emitted HC.s (78)
in exhaust of gasoline engine: 0.1–7.0 ppm (partly crotonaldehyde) (195)
in gasoline: 6–7 vol% (312)

—Control methods:
activated carbon: retentivity: 30 wt % of adsorbent (83)
catalytic combustion: platinized ceramic honeycomb catalyst: ignition temp.:
  170°C; inlet temp. for 90% conversion: 250–300°C (91)
catalytic incineration over commercial Co_3O_4 (1.0–1.7 mm) granules, catalyst
  charge 13 cm^3, space velocity 45000 h^{-1} at 100 ppm inlet:
  80% decomposition at 245°C
  99% decomposition at 275°C
  80% conversion to CO_2 at 250°C
  95% conversion to CO_2 at 265°C (346)

—Comparison of catalyst performance at space velocity of 80,000/hr:

catalyst (on $\gamma - Al_2O_3$)	feed conc. ppm	reactor inlet temp °C	% odor removal ASTM	% conversion GC
0.5% Pt	1256	175	23	27
0.1% Pt	1256	195	23	30
0.5% Pd	1256	300	100	100
10% Cu	1256	497	98	100

(1221)

—Sampling and analysis:
second derivative spectroscopy: 50 ppb (42)
non dispersive I.R.: min. full scale: 125 ppm (55)
photometry: min. full scale: 185 ppm (53)
detector tubes: UNICO: det. lim.: 1 ppm (59)
              DRÄGER: det. lim.: 5 ppm (58)

AUER: det. lim.: 5 ppm (57)
efficiency of scrubbers at 325–375°F
(1): $HgSO_4/H_2SO_4$: 96% absorbed
(2): $PdSO_4/H_2SO_4$: 96% absorbed
(3): (1) + (2): 97% absorbed (198)
PMS: det. lim.: 0.7–19 ppm (200)
6 f abs. app.: 60 l air/30 min: UVS: det. lim.: 20 mg/cu m/30 min (208)

C. WATER POLLUTION FACTORS:
—BOD_5 :   0 std. dil. (27)
   0.86 (36)
   1.23 std. dil. sew., acclimated (41)
   5% of ThOD (220)
35 days: 1.47 (62)
—BOD_5^{20} : 0.86 at 10 mg/l, unadapted sew.; lag period: 1 day
—BOD_{20}^{20} : 1.02 at 10 mg/l, unadapted sew.; lag period: 1 day (554)
—BOD: 2.15 NEN 3235–5.4 (277)
—COD: 0.7; 1.41; 1.88 (27, 36, 41)
   21% of ThOD (220)
   27% of ThOD (274)
   2.52 NEN 3235–5.3 (277)
—$KMnO_4$ value: 0.039 (30)
   acid: 1% of ThOD; alkaline: 0% of ThOD (220)
—TOC: 60% of ThOD (220)
—ThOD: 3.13 (36)
—Biodegradation:
incubation with natural flora in the groundwater—in presence of the other components of high-octane gasoline (100 μl/l): biodegradation: 100% after 192 hr at 13°C (initial conc.: 2.22 μl/l) (956)
biodegradation to CO_2 :

sampling site	concentration (μg/l)	incubation month	time (hr)	degradation rate (μg/l/day)×10^3	turnover time (days)
control station	6	–	–	41 ± 8	130
near oil storage tanks	6	–	–	58 ± 15	90
near oil storage tanks	6	–	–	48 ± 7	100
Skidaway river	20	June	24	150 ± 40	130

(381)

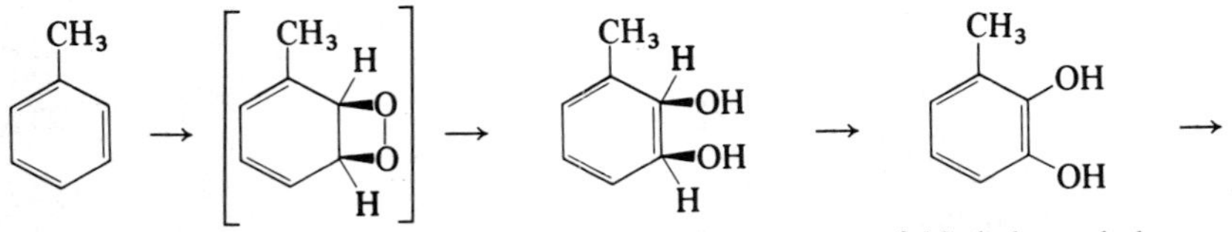

Toluene → [structure] → cis-2,3-Dihydroxy-2,3-dihydrotoluene → 3-Methyl catechol →

TOLUENE

Toluene (CH₃) → Benzyl alcohol (CH₂OH) → Benzaldehyde (CHO) → Benzoic acid (COOH) → → Catechol (OH, OH) →

initial steps in degradation by micro-organisms (1235)
- Impact on biodegradation processes:
  inhibition of degradation of glucose by *Pseudomonas fluorescens* at: 30 mg/l
  inhibition of degradation of glucose by *E. coli* at: 200 mg/l (293)
- Reduction of amenities:
  T.O.C. = 1 mg/l (295)
  approx. conc. causing adverse taste in fish: <0.25 mg/l (41)
- Water quality:
  in river Maas at Eysden (Netherlands) in 1976: median: 0.1 µg/l; range: n.d.–2.1 µg/l
  in river Maas at Keizersveer (Netherlands) in 1976: median: 0.1 µg/l; range: n.d.–0.7 µg/l (1368)
- Waste water treatment:
  A.C.: adsorbability: 0.050 g/g C; 79.2% reduction, infl.: 317 mg/l, effl.: 66 mg/l (32)
  sludge digestion is not affected yet at 440 mg/l; at 870 mg/l, gas production in the digester is reduced by 14.5% (30)
  calculated half life time based on evaporative loss for a water depth of 1 m at 25°C: 5.18 hr (330)
- Sampling and analysis:
  photometry: min. full scale: 5.7×10^{-6} mole/l (53)

D. BIOLOGICAL EFFECTS:
- Bacteria: *E. coli*: LD_0: 200 mg/l
- Toxicity threshold (cell multiplication inhibition test):
  bacteria (*Pseudomonas putida*):   29 mg/l   (1900)
  algae (*Microcystis aeruginosa*):   105 mg/l   (329)
  green algae (*Scenedesmus quadricauda*):   >400 mg/l   (1900)
  protozoa (*Entosiphon sulcatum*):   456 mg/l   (1900)
  protozoa (*Uronema parduczi Chatton-Lwoff*): >450 mg/l (1901)
- Algae:
  inhibition of photosynthesis of a freshwater, nonaxenic unialgal culture of *Selenastrum capricornutum*:
  at   1% saturation: 91% carbon-14 fixation (vs. controls)
       10% saturation: 96% carbon-14 fixation (vs. controls)
      100% saturation: 3% carbon-14 fixation (vs. controls) (1690)
  algae: kelp (*Macrocystis angustifolia*): 75% reduction in photosynthesis within 96 hr at 10 ppm
  Scenedesmus: LD_0: 120 mg/l (946)
- Protozoa: ciliate (*Tetrahymena pyriformis*): 24 hr LC_{100}: 5.97 mmole/l (1662)

—Crustaceans:
   marine isopod (*Cirolana borealis*): median effective time (ET_{50}) to narcotization for different nominal concentrations of toluene in seawater:

conc. (ppm)	ET_{50} (hr)
0	—
0.0125	—
1.25	—
5.7	400
12.5	69
25	28
125	3

(bars indicate no visible effects on any individual during 4 days of exposure)
(1443)
   grass shrimp (*Palaemonetes pugio*): 96 hr LC_{50}: 9.5 ppm   (940)
   crab larvae—stage I (*Cancer magister*): 96 hr LC_{50}: 28 ppm   (941)
   shrimp (*Crangon franciscorum*): 96 hr LC_{50}: 4.3 ppm   (942)
   Daphnia: LD_0: 60 mg/l   (30)
—Fish:
   goldfish: LD_{50}: (24 hr): 58 mg/l—modified ASTM D 1345   (277)
   fatheads: TLm (24–96 hr): 56–34 mg/l
   bluegill: TLm (24–96 hr): 24,0 mg/l
   goldfish: TLm (24–96 hr): 57.7 mg/l
   guppies: TLm (24–96 hr): 63–59 mg/l   (158)
   mosquito fish: TLm (24–96 hr): 1,340–1,280 mg/l in turbid Oklahoma   (244)
   young Coho Salmon in artificial sea water at 8°C: no significant mortality up to 10 ppm after 24 to 96 hrs
   mortality:
   9/10 at 50 ppm after 24 hrs, 10/10 at 50 ppm after 48 up to 96 hrs
   28/30 at 100 ppm after 24 hrs, 30/30 after 48 up to 96 hrs   (317)
   static test:

	96 hr TLm (ppm) at		
	4°C	8°C	12°C
pink salmon (*Oncorhynchus gorbuscha*)	6.41	7.63	8.09
shrimp (*Eualus* spp.)	21.4	20.2	14.7

   the conc. of toluene in the test containers declined with time because of evaporation losses or biodegradation. Toluene declined to nondetectable levels after 72 hr at 12°C; after 96 hr at 8°C; and to 25% of the initial conc. after 96 hr at 4°C
(1397)
   goldfish: flow through bioassay: 96 hr TLm: 23 mg/l at 17–19°C
                                     720 hr TLm: 15 mg/l at 17–19°C   (471)
   guppy (*Poecilia reticulata*): 14 d LC_{50}: 68 ppm   (1833)
   bass (*Morone saxatilis*): 96 hr LC_{50}: 7.3 ppm   (942)
   goldfish (*Carassius auratus*): 96 hr LC_{50}: 22.8 ppm   (943)
   eel: infiltration ratio: flesh/water: 0.53; eel flesh: 1.17 ng/g; water: 2.22 ng/g
(412)
   eel (*Anguilla japonica*): BCG: 13.2, 1/2 life period: 1.4 days   (1926)
—Man: EIR: 5.3
   severe toxic effects: 1,000 ppm = 3,830 mg/cu m, 60 min

symptoms of illness: 300 ppm = 1,149 mg/cu m
unsatisfactory: >100 ppm = 383 mg/cu m (185)

2,4-toluenediisocyanate (2,4-TDI; tolyene-2,4-diisocyanate)

A. PROPERTIES: white liquid; m.w. 174.16; m.p. 21°C; b.p. 251°C; v.p. 0.01 mm at 20°C, 1 mm at 80°C; v.d. 6.0; sp.gr. 1.20; solub. reacts

B. AIR POLLUTION FACTORS: 1 mg/cu m = 0.14 ppm, 1 ppm = 7.24 mg/cu m
  —Odor: characteristic: quality: medicated bandage
  hedonic tone: pungent

odor thresholds — mg/cu m (10^{-7} to 10^4)

detection, recognition, not specified

(2; 54; 57; 211; 633; 681; 869)

odor Index: 6 (316)

  —Sampling and analysis:
    4 f. abs. app.: 60 l air/30 min: VLS: det. lim.: 10 µg/cu m/30 min (n.s.i.)
    many interfering compounds (208)

D. BIOLOGICAL EFFECTS:
  —Mammalia:
    rat: single oral LD_{50}: 5,800 mg/kg
      inhalation: LC_{50}: 14 ppm, 4 hr/day, 14 days
    guinea pig: inhalation: LC_{50}: 13 ppm, 4 hr/day, 14 days
    rabbit: inhalation: LC_{50}: 11 ppm, 4 hr/day, 14 days
    mouse: inhalation: LC_{50}: 10 ppm, 4 hr/day, 14 days
    rat: inhalation: 3/20: 1 ppm, 10 × 6 hr
    rat, mouse, guinea pig: inhalation: no effect: 0.03 ppm, 6 hr/day, 5 days/w, 4 weeks (54)
    mice: RD_{50} (respiratory rate): 0.39 ppm (n.s.i.) (1330)
  —Man: eye and nose irritation: 0.1 ppm, 30 min
    no effect level: 0.01 ppm, 30 min (54)

2,6-toluenediisocyanate (2,6-TDI)

A. PROPERTIES: m.w. 174.16
B. AIR POLLUTION FACTORS: 1 mg/cu m = 0.14 ppm, 1 ppm = 7.24 mg/cu m
   —Odor:
   USSR: human odor perception: non perception: 0.15 mg/cu m
            perception: 0.020 mg/cu m
      human reflex response: no response: 0.050 mg/cu m
            adverse response: 0.1 mg/cu m
      animal chronic exposure: no effect: 0.02 mg/cu m
            adverse effect: 0.2 mg/cu m   (170)
D. BIOLOGICAL EFFECTS:

          24 hr LC_{50} 48 hr LC_{50} 96 hr LC_{50}
—Fathead minnows (*Pimephales promelas*) 195 mg/l 172 mg/l 164 mg/l
grass shrimp (*Palaemonetes pugia*): no significant mortality below 508 mg/l
                          (1904)
—Animals: similar effects as 2,4-TDI
—Man: similar effects as 2,4-TDI

p-toluene sulfonic acid (4-methylbenzenesulfonic acid)

A. PROPERTIES: leaflets; m.w. 172.19; m.p. 106°C; b.p. 140°C at 20 mm Hg
C. WATER POLLUTION FACTORS:
   —Biodegradation: decomposition by a soil microflora in 24 days (176)
    adapted A.S. at 20°C—product is sole carbon source: 98.7% COD removal at 8.4
     mg COD/g dry inoculum/hr   (327)
   —BOD_5: nil std. dil. sew. (n.s.i.)   (281, 161)
   —Waste water treatment:
    adsorption on Amberlite XAD-2: 23% retention efficiency; infl.: 9.0 ppm, effl.:
     6.9 ppm   (40)
D. BIOLOGICAL EFFECTS:
   —Algae: *Chlorella vulgaris*: 245 ppm: 50% reduction of cell numbers vs controls,
    after 1 day incubation at 20°C   (343)

toluenetrichloride *see* benzotrichloride

α-toluic acid *see* phenylacetic acid

o-toluic acid (*o*-methylbenzoic acid)

1110 *m*-TOLUIC ACID

A. PROPERTIES: colorless needles; m.w. 136.14; m.p. 103.7°C; b.p. 259.2°C; sp.gr. 1.062 at 115/4°C; solub. 1,180 mg/l at cold, 21,700 mg/l at 100°C log P_{oct} 2.54/2.83 (calculated)
C. WATER POLLUTION FACTORS:
 —Biodegradation: decomposition by a soil microflora in 16 days   (176)
D. BIOLOGICAL EFFECTS:
 —Mammalia: rat: acute oral LD_{50}: 0.4–3.2 g/kg
           rabbit: acute oral LD_{50}: 0.75 g/kg   (211)

***m*-toluic acid** (*m*-methylbenzoic acid)

A. PROPERTIES: colorless needles; m.w. 136.14; m.p. 108.75°C; b.p. 263°C; sp.gr. 1.054 at 112°C; solub. 850 mg/l at 15°C, 17,000 mg/l at 100°C; log P_{oct} 2.37
C. WATER POLLUTION FACTORS:
 —Biodegradation: decomposition by a soil microflora in 2 days   (176)
D. BIOLOGICAL EFFECTS:
 —mammalia: rat: acute oral LD_{50}: >3.2 g/kg   (211)

***p*-toluic acid** (*p*-methylbenzoic acid)

A. PROPERTIES: colorless needles; m.w. 138.14; m.p. 179.6°C; b.p. 275°C; solub. 340 mg/l at cold, 12,600 mg/l at 100°C; log P_{oct} 2.27
C. WATER POLLUTION FACTORS:
 —Biodegradation: decomposition by a soil microflora in 8 days   (176)
D. BIOLOGICAL EFFECTS:
 —Mammalia: rat: acute oral LD_{50}: 0.4–3.2 g/kg   (211)

***o*-toluidine** (*o*-aminotoluene; *o*-methylaniline)

A. PROPERTIES: colorless liquid; m.w. 107.15; m.p. α: -24.4°C, β: -16°C; b.p.

200°C; v.p. 0.1 mm at 20°C, 0.3 mm at 30°C; v.d. 3.72; sp.gr. 1.004 at 20/4°C; solub. 15,000 mg/l at 25°C; sat. conc. 0.58 g/cu m at 20°C, 1.7 g/cu m at 30°C; log P_{oct} 1.29/1.32

B. AIR POLLUTION FACTORS: 1 mg/cu m = 0.23 ppm, 1 ppm = 4.45 mg/cu m

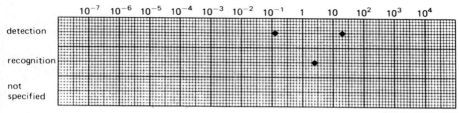

(610; 695; 829)

C. WATER POLLUTION FACTORS:
- BOD_5: 1.4 std. dil. at 1-2.5 mg/l (27)
  0.242 (30)
  0.24-1.43 std. dil. sew. (41)
- $KMnO_4$: 6.50 (30)
- ThOD: 2.54; 3.14 (30, 27)
- Biodegradation: decomposition by a soil microflora in >64 days (176)
  adapted A.S. at 20°C—product is sole carbon source: 97.7% COD removal at 15.1 mg COD/g dry inoculum/hr (327)
  degradation by *Aerobacter*: 500 mg/l at 30°C
    parent: 100% ring disruption in 64 hr
    mutant: 100% ring disruption in 6 hr (152)
- Water quality: in river Waal (The Netherlands): average in 1973: 0.75 µg/l (342)

D. BIOLOGICAL EFFECTS:
- Toxicity threshold (cell multiplication inhibition test):
  bacteria (*Pseudomonas putida*): 16 mg/l (1900)
  algae (*Microcystis putida*): 0.31 mg/l (329)
  green algae (*Scenedesmus quadricuada*): 6.3 mg/l (1900)
  protozoa (*Entosiphon sulcatum*): 76 mg/l (1900)
  protozoa (*Uronema parduczi Chatton-Lwoff*): 21 mg/l (1901)
- Bacteria:
  E. coli: no effect at 1 g/l
- Algae: *Scenedesmus*: LD_0: 8 mg/l (30)
- Arthropoda: *Daphnia*: LD_0: 8 mg/l (30)
- Fish: TLm (? hr): 100 mg/l (30)
- Man: severe toxic effects: 40 ppm = 176 mg/cu m, 60 min
  symptoms of illness: 10 ppm = 44 mg/cu m
  unsatisfactory: 5 ppm = 22 mg/cu m all isomers (185)
- Carcinogenicity: negative
- Mutagenicity in the *Salmonella* test:
  <0.0007 revertant colonies/nmol
  <70 revertant colonies at 10 µg/plate (1883)

m-toluidine (m-methylaniline)

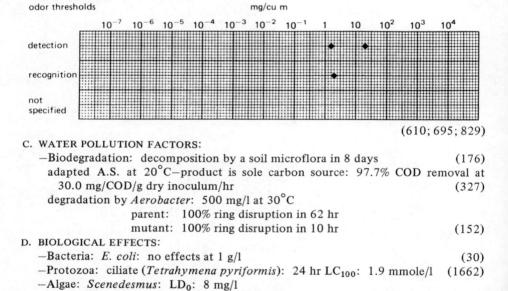

A. PROPERTIES: m.w. 107.15; m.p. -32°C; b.p. 203°C; v.d. 3.72; sp.gr. 0.99 at 20/4°C; log P_{oct} 1.40/1.43

B. AIR POLLUTION FACTORS:

odor thresholds, mg/cu m

(610; 695; 829)

C. WATER POLLUTION FACTORS:
 —Biodegradation: decomposition by a soil microflora in 8 days (176)
  adapted A.S. at 20°C—product is sole carbon source: 97.7% COD removal at 30.0 mg/COD/g dry inoculum/hr (327)
  degradation by *Aerobacter*: 500 mg/l at 30°C
    parent: 100% ring disruption in 62 hr
    mutant: 100% ring disruption in 10 hr (152)

D. BIOLOGICAL EFFECTS:
 —Bacteria: *E. coli*: no effects at 1 g/l (30)
 —Protozoa: ciliate (*Tetrahymena pyriformis*): 24 hr LC_{100}: 1.9 mmole/l (1662)
 —Algae: *Scenedesmus*: LD_0: 8 mg/l
 —Arthropoda: *Daphnia*: LD_0: 0.6 mg/l (30)

p-toluidine (p-methylaniline)

A. PROPERTIES: leaflets; m.w. 107.15; m.p. 45°C; b.p. 200.3°C; sp.gr. 1.046 at 20/4°C; solub. 7,400 mg/l at 21°C; log P_{oct} 1.39/1.41

B. AIR POLLUTION FACTORS:
odor thresholds   mg/cu m

detection
recognition
not specified

(610; 695; 829)

C. WATER POLLUTION FACTORS:
- BOD_5:   1.6 std. dil. sew. at 2.5 mg/l (27)
   1.44 (30)
   1.44–1.63 std. dil. sew. (41)
   57% of ThOD (274)
- COD: 90% of ThOD (0.05 n $K_2Cr_2O_7$) (274)
- $KMnO_4$:   6.72 (30)
   71% of ThOD (0.01 n $KMnO_4$) (274)
- ThOD: 2.54; 3.14 (30, 27)
- Reduction of amenities:
   approx. conc. causing adverse taste in fish (rudd): 20 mg/l (41)
   tainting of fish flesh: 20.0 mg/l (81)
- Biodegradation: decomposition by a soil microflora in 4 days (176)
   adapted A.S. at 20°C—product is sole carbon source: 97.7% COD removal at 20.0 mg COD/g dry inoculum/hr (327)
   degradation by *Aerobacter*: 500 mg/l at 30°C
      parent: 100% ring disruption in 48 hr
      mutant: 100% ring disruption in 3 hr (152)

D. BIOLOGICAL EFFECTS:
- Bacteria: *E. coli*: no effect at 1 g/l
- Algae: *Scenedesmus*: LD_0: 10 mg/l
   bluegreen alga *Agmenellum quadruplicatum* (strain PR 6): the addition of 50 ppb during exponential growth in a liquid culture resulted in a bending of the growth curve to a plateau within 4 hours. Cultures containing 500 ppb did not grow (1093)
- Arthropoda: Daphnia: LD_0: 0.6 mg/l (30)

α-tolunitrile   *see* benzylcyanide

tolylene-2,4,diisocyanate   *see* toluene-2,4-diisocyanate

o-tolylphosphate   *see* tri-*o*-cresylphosphate

tonka bean camphor   *see* coumarin

tordon 22 K
Aqueous solution of 240 g/l picloram as the potassium salt of 4-amino-3,5,6-trichloropicolinic acid.
D. BIOLOGICAL EFFECTS:
—Fish:

	LC$_{50}$ at 15°C		
	48 hr	96 hr	10 d
trout: flow through bioassay:	31.0 mg/l	26.0 mg/l	22.2 mg/l
zebrafish (*Brachydario rerio*): flow through bioassay:	—	35.5 mg/l	35.5 mg/l
flagfish (*Gordonella floridae*): flow through bioassay:	—	26.1 mg/l	12.3 mg/l
			(479)

tordon 101
Mixture of picloram and other pesticides
D. BIOLOGICAL EFFECTS:

	ppb		method of assessment
Chlorococcum sp.	>2 × 10^6	50% decrease in O$_2$ evolution	f
Chlorococcum sp.	1 × 10^5	50% decrease in growth	measured as ABS. (525 mu) after 10 days
Dunaliella tertiolecta	>2 × 10^6	50% decrease in O$_2$ evolution	f
Dunaliella tertiolecta	1.25 × 10^6	50% decrease in growth	measured as ABS. (525 mu) after 10 days
Isochrysis galbana	1 × 10^5	50% decrease in O$_2$ evolution	f
Isochrysis galbana	5 × 10^4	50% decrease in growth	measured as ABS. (525 mu) after 10 days
Phaeodactylum tricornutum	>72 × 10^6	50% decrease in O$_2$ evolution	f
Phaeodactylum tricornutum	1 × 10^5	50% decrease in growth	measured as ABS. (525 mu) after 10 days (2348)

toxaphene
A mixture of chlorinated bicyclic terpenes, mainly chlorinated camphene, with the approximate formula $C_{10}H_{10}Cl_8$
Use: the largest simple use is as pesticide on cotton crops. Other major uses are on cattle and swine, and on soybeans, corn, wheat, and peanuts. Substantial amounts also are used for lettuce, tomatoes, and other food crops. (1402)
Formulations: Its standard commercial formulation is a dust containing 20% toxaphene. Emulsifiable concentrates contain up to 8 pounds/gallon; oil solutions are 90% toxaphene, and wettable powders are 40% toxaphene. Toxaphene also is mixed with other insecticide chemicals for a variety of uses. (1402)
B. AIR POLLUTION FACTORS:
—Atmospheric levels in agricultural area: n.d.–701 ng/cu m (1308)
C. WATER POLLUTION FACTORS:

- Odor threshold level (detection): 0.14 mg/kg (915)
- Sediment quality:
  Chesapeake bay: suspended sediments: 280 ppt (1173)
D. BIOLOGICAL EFFECTS:
  - Molluscs:
    pelecypod (*Crassostrea virginica*): 96 hr: 3.1–3.6 µg/l: 9000–15200 X (1149)
    oyster: 2920 X
  - Crustacean: decapods:
    pink shrimp (*Penaeus duorarum*): 96 hr LC_{50}: 0.0014 mg/l;
      bioconcentration factor: 400–800
    grass shrimp (*Palaemonetes pugio*): 96 hr LC_{50}: 0.0044 mg/l
      bioconcentration factor: 800–1200
  - Fish:
    sheepshead minnow (*Cyprinodon variegatus*): 96 hr LC_{50}: 0.0011 mg/l
      bioconcentration factor: 3,100–20,600
    pinfish (*Lagodon rhomboides*): 96 hr LC_{50}: 0.00053 mg/l
      bioconcentration factor: 3800–3900
    longnose killifish: concentration factors as determined after 28 days of continuous exposure:
      embryo/fry: 13,800 to 33,000
      fry: 19,300 to 33,300
      juveniles: 23,700 to 60,000
      adults: 4,200 to 5,300
    Cyprinodon variegatus: 0.20–2.5 µg/l: 6,100–14,400 X (1476)
    Fundulus similis:

	exposure	BCF	
juveniles	0.3–1.7 µg/l: 28 days	23 700–60 000 X	
adult	0.3–0.9 µg/l: 14 days	4 200–6 800 X	
fry	0.3–1.3 µg/l: 28 days	13 300–33 300 X	(1149)

    catfish fry: bioconcentration factor: 91,000 X (1402)
    fathead minnow: 10-d LC_{50}: 4.8 µg/l (F)
      259-d MATC: 25–54 ng/l (F)
      98-d BCF: 55,000–69,000 (F)
    channel catfish: 9-d LC_{50}: 15 µg/l (F)
      240-d MATC: 129–299 ng/l (F) (1540)
      100-d BCF: 17,000–26,000 (F)
    brook trout fry: 15-d BCF: 76,000 (F)
    fathead minnow: 98-d BCF: 55 000–67 000 (F)
    channel catfish:
      fry: 90-d BCF: 27,000–50,000
      adult: 100-d BCF: 17 000–26,000 (1541)
    brook trout fry: 150-d BCF: 15,000–20,000 (F)
    fathead minnow: 150-d BCF: 70,000–107,000 (F) (1539)
  - Toxicity:
    Algae: polychlorodicyclic terpenes with chlorinated camphene 60% emulsion concentrate:
    Protococcus sp.: 40 ppb .77 O.D. expt/O.D. control Test organisms grown in

Chlorella sp.: 40 ppb .70 O.D. expt/O.D. control test media for 10 days
Dunaliella   absorbance measured
  euchlora: 70 ppb 53 O.D. expt/O.D. control at 530 m/
Phaeodactylum
  tricornutum: 10 ppb .54 O.D. expt/O.D. control
Monochrysis
  lutheri: 10 ppb .00 O.D. expt/O.D. control
—Molluscs:
  American oyster (*Crassostrea virginica*):
    96 hr EC_{50}: 0.016 mg/l (50% reduction in shell growth);
    bioconcentration factor: 9,000–15,200
  polychlorobicyclic terpenes with chlorinated camphene:
    Mercenaria mercenaria (hard clam) eggs: 48 hr static lab bioassay: TLm:
      1120 ppb
    Mercenaria mercenaria (hard clam) larvae: 12 day static lab bioassay: TLm:
      250 ppb
—Crustaceans:
  Gammarus lacustris: 96 hr LC_{50}: 26 µg/l (2124)
  Gammarus fasciatus: 96 hr LC_{50}: 6 µg/l (2126)
  Palaemonetes kadiakensis: 96 hr LC_{50}: 28 µg/l (2126)
  Simocephalus serrulatus: 48 hr LC_{50}: 10 µg/l (2137)
  Daphnia pulex: 48 hr LC_{50}: 15 µg/l (2137)
  Copepod (*Acartia tonsa*): 96 hr LC_{50} (S): 0.0072 µg/l (1129)
  Daphnia magna: 24 hr LC_{50}: 94 µg/l
                    48 hr LC_{50}: 15 µg/l (1002; 1004)
  Korean shrimp (*Palaemon macrodactylus*): 96 hr TL_{50}: static lab bioassay: 20.3
    (8.6–47.9) ppb
  blue crab (*Callinectes sapidus*), adult: 96 hr TLm: static lab bioassay: 370 (180–
    700) ppb
—Decapods:
  Penaeus duorarum: 96 hr LC_{50} (FT): 1.4 µg/l
  Palaemonetes pugio: 96 hr LC_{50} (FT): 4.4 µg/l (1145)
—Insects:
  Pteronarcella badia: 96 hr LC_{50}: 3.0 µg/l
  Claassenia sabulosa: 96 hr LC_{50}: 1.3 µg/l (2128)
  Pteronarcys: 48 hr LC_{50}: 2.3 µg/l (1003)
  Hydropsyche larvae: significant modification of net construction after 48 hr
    exposure: ± 10 µg/l (1006; 1007, 1008)
—Fish:
  Gasterosteus aculaetus, threespine stickleback: TLm: 7.8 ppb

	96 hr LC_{50}, µg/l		96 hr LC_{50}, µg/l
catfish	13	bluegill	18
bullhead	5	bass	2
goldfish	14	rainbow	11
minnow	14	brown	3
carp	4	coho	8
sunfish	13	perch	12

(1934)

Salmo gairdneri: 96 hr LC_{50}: 5.5 µg/l (1001)
Cyprinodon variegatus: 96 hr LC_{50} (FT): 1.1 µg/l
Lagodon rhomboides: 96 hr LC_{50} (FT): 0.5 µg/l (1145)
bluegill: 96 hr LC_{50}: 0.004 ppm
rainbow trout: 96 hr LC_{50}: 0.008 ppm (1878)
mosquitofish: 24 hr LC_{50} (S): 0.045 mg/l (1104)
susceptible mosquitofish: 48 hr LC_{50}: 12 ppb
resistant mosquitofish: 48 hr LC_{50}: 459 ppb (1851)
—Mammals:
  male and female mice: toxaphene has caused liver cancers in mice given the compound in a feeding study
  rats: test results of a feeding study also suggested that toxaphene caused thyroid cancers in rats (1402)

toxilic anhydride *see* maleic anhydride

2,4,5-TP *see* 2(2,4,5-trichlorophenoxy)propionic acid

n-**triacontane**
$CH_3(CH_2)_{28}CH_3$
B. AIR POLLUTION FACTORS:
  —Ambient air quality:
    glc's in Botrange (Belgium)—woodland at 20–30 km from industrial area: June–July 1977: 0.61; 0.87 ng/cu m ($n = 2$)
    glc's in Wilrijk (Belgium)—residential area: Oct.–Dec. 1976: 3.40; 6.64 ng/cu m ($n = 2$) (1233)
    glc's in residential area (Belgium) Oct. 1976:
      in particulate phase: 5.75 ng/cu m
      in gas phase: 4.87 ng/cu m (1289)
    organic fraction of suspended matter:
      Bolivia at 5200 m altitude (Sept.–Dec. 1975): 0.27–0.57 µg/1000 cu m
      Belgium, residential area (Jan.–April 1976): 2.1–2.6 µg/1000 cu m (428)

tri-N-allylamine
$(CH_2CHCH_2)_3N$
A. PROPERTIES: m.w. 137.22; m.p. $<-70°C$; b.p. $149.5°C$; v.d. 4.73; sp.gr. 0.800 at $20/4°C$; solub. 2,500 mg/l
B. AIR POLLUTION FACTORS: 1 mg/cu m = 0.178 ppm, 1 ppm = 5.61 mg/cu m
C. WATER POLLUTION FACTORS:
  —Biodegradation:
    degradation by *Aerobacter*: 200 mg/l at $30°C$
      parent: 47% degradation in 120 hr
      mutant: 100% degradation in 22 hr (152)
D. BIOLOGICAL EFFECTS:
  —Mammalia:
    rat: acute oral LD_{50}: 1.31 g/kg
      inhalation: LC_{50}: 828 ppm, 4 hr
        LC_{50}: 554 ppm, 8 hr
        deaths: 200 ppm, 50 × 7 hr
        change in kidneys and liver: 100 ppm, 50 × 7 hr (211)

2,4,6-triamino-s-triazine *see* melamine

triazotion *see* azinphosethyl

1,2,4-tribromobenzene

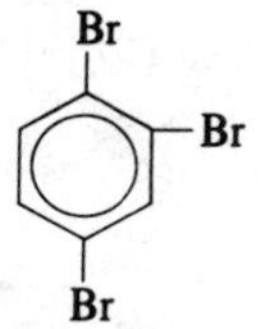

D. BIOLOGICAL EFFECTS:
—Fish:
   juvenile atlantic salmon: BCF at 4.8 µg/l: 1095
                                       excretion half-life: 104 hr   (1852)

1,3,5-tribromobenzene

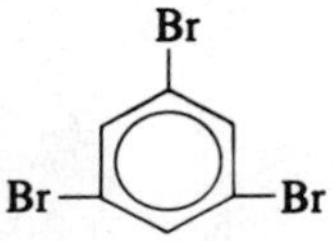

D. BIOLOGICAL EFFECTS:
—Fish:
   juvenile atlantic salmon: BCF at 2.3 µg/l: 1130
                                       excretion half-life: 95 hr   (1852)

2,4′,5-tribromobiphenyl

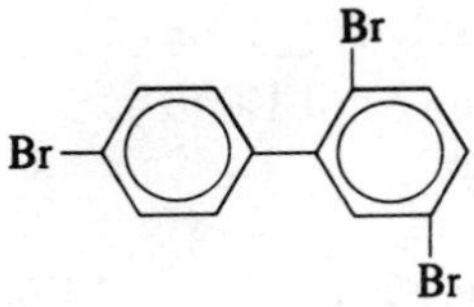

D. BIOLOGICAL EFFECTS:
—Fish:
   Atlantic salmon: 96 hr BCF (water) (S): 1996
                         42 d BCF (food) (F): 0.36

tribromomethane *see* bromoform

2,4,6-tribromophenol

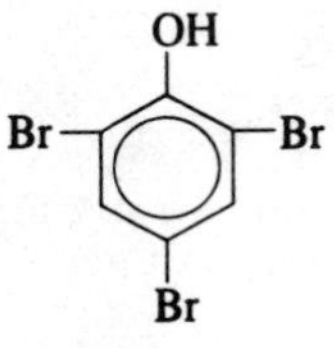

A. PROPERTIES: colorless crystals; m.w. 330.83; m.p. 96°C; b.p. sublimes; sp. gr. 2.55 at 20/20°C; solub. 70 mg/l
C. WATER POLLUTION FACTORS:
—Biodegradation:
degradation by *Pseudomonas*: 200 mg/l at 30°C
parent: 14% ring disruption in 120 hr
mutant: 92% ring disruption in 42 hr (152)
D. BIOLOGICAL EFFECTS:
—Mammalia: rat: oral LD_{50}: ±2,000 mg/kg (211)

tri(2-butoxy-ethyl)phosphate
$[CH_3(CH_2)_3O(CH_2)_2O]_3PO$
Use: primary plasticizer for resins and elastomers; floor finishes and waxes; flame retarding agent.
A. PROPERTIES: sp.gr. 1.020 at 20°C; f.p. 70°C; b.p. 215-228°C at 4 mm
C. WATER POLLUTION FACTORS:
—Water quality:
in Delaware river (U.S.A.): conc. range: winter: 0.3-3 ppb
summer: 0.4-2 ppb (1051)

tri-*n*-butylamine
$(CH_3CH_2CH_2CH_2)_3N$
Use: solvent inhibitor in hydraulic fluids; intermediate
A. PROPERTIES: delisquescent colorless liquid; m.w. 185.35; m.p. -70°C; b.p. 213°C; v.p. 0.7 mm at 20°C, 20 mm at 100°C; v.d. 6.38; sp.gr. 0.78 at 20°C; log P_{oct} 1.52 (calculated)
B. AIR POLLUTION FACTORS: 1 mg/cu m = 0.132 ppm, 1 ppm = 7.58 mg/cu m
—Sampling and analysis:
photometry: min. full scale: 12,500 ppm (53)
C. WATER POLLUTION FACTORS:
—at 100 mg/l, no inhibition of NH_3 oxidation by *Nitrosomonas* sp. (390)
D. BIOLOGICAL EFFECTS:
—Fish:
creek chub: critical range: 20-40 mg/l; 24 hr, in Detroit river water (226, 243)
gudgeon: TLm (? hr): 20-40 mg/l (30)
—Mammals:
rat: inhalation: slight lethargy; autopsy: organs normal: 29 ppm, 19 × 6 hr
(65)

tri-*n*-butylphosphate
$(CH_3CH_2CH_2CH_2)_3PO_4$
C. WATER POLLUTION FACTORS:
—Manmade sources:
percolation water at 30-500 m from waste dumping ground: 28-140 ppb (n.s.i.)
(183)
D. BIOLOGICAL EFFECTS:
—Toxicity threshold (cell multiplication inhibition test):
bacteria (*Pseudomonas putida*): >100 mg/l

green algae (*Scenedesmus quadricauda*): 3.2 mg/l
protozoa (*Entosiphon sulcatum*): 14 mg/l (1900)
protozoa (*Uronema parduczi Chatton-Lwoff*): 21 mg/l (1901)
—Fish:
rainbow trout: 96 hr LC_{50} (S): 5.0–9.0 mg/l

tri(*t*-butyl)phosphate (TBP)
$[CH_3(CH_2)_3O]_3PO$

Use: extractant for metal complexes; heat exchange medium; solvent for nitrocellulose; plasticizer; antifoam agent; dielectric.
A. PROPERTIES: m.w. 266.32; m.p. −79°C; b.p. 180–183°C at 22 mm; sp.gr. 0.979
C. WATER POLLUTION FACTORS:
—Water quality:
in Delaware river (U.S.A.): conc. range: winter: 0.4–2 ppb
summer: 0.06–0.4 ppb (1051)
in Zürich lake: at surface: 82 ppt; at 30 m depth: 54 ppt
in Zürich area: in groudwater: 10 ppt; in tapwater: 14 ppt (513)

S,S,S-tributylphosphorotrithioate (DEF)
$(CH_3CH_2CH_2CH_2S)_3PO$

Use: pesticide; cotton defoliant
A. PROPERTIES: colorless to pale yellow clear liquid; b.p. 150°C (0.3 mm); sp.gr. 1.06 at 20°C
B. AIR POLLUTION FACTORS:
—Atmospheric levels in agricultural area: n.d.–16.0 ng/cu m (1308)
D. BIOLOGICAL EFFECTS:
—Crustaceans:
Gammarus lacustris: 96 hr, LC_{50}: 100 μg/l (2124)
—Insects:
Pteronarcys californica: 96 hr, LC_{50}: 2100 μg/l (2128)
—Mammals: acute oral LD_{50} (rat): 200 mg/kg
dermal LD_{50}: >1000 mg/kg (1854)

tri-*n*-butyltinoxide
$(C_4H_9)_3SnOSn(C_4H_9)_3$

Use: bactericide, fungicide, intermediate.
A. PROPERTIES: b.p. 180°C (at 2 mm)
D. BIOLOGICAL EFFECTS:
rainbow trout: 24-hr EC_{50}: 30.8 μg/l (Median effect concentration) (R)
tilapia (*Tilapia rendalli*): 24-hr EC_{50}: 53.2 μg/l (R) (1543)

trichloroacetaldehyde *see* chloral

trichloroacetic acid (TCA)
CCl_3COOH

Users and formulations: organic synthesis; medicine; pharmacy; herbicides.
A. PROPERTIES: colorless, rhombic deliquescent crystals, sharp pungent odor; m.w.

163.4; m.p. 57°C; b.p. 197.5°C; v.p. 1 mm at 51°C; sp.gr. 1.63 at 61/4°C; log P_{oct} 0.10/1.96 (calculated); solub 13 g/kg at 25°C

B. AIR POLLUTION FACTORS:
  —Odor threshold conc. (recognition): 1.6–2.5 mg/cu m   (610)

D. BIOLOGICAL EFFECTS:
  —Toxicity threshold (cell multiplication inhibition test):

bacteria (*Pseudomonas putida*):	>1000 mg/l	(1900)
algae (*Microcystis aeruginosa*):	250 mg/l	(329)
green algae (*Scenedesmus quadricauda*):	200 mg/l	(1900)
protozoa (*Entosiphon sulcatum*):	800 mg/l	(1900)
protozoa (*Uronema parduczi Chatton Lwoff*):	435 mg/l	(1901)

  —Crustacean:
    Daphnia magna: 48 hr LC_{50}: 2000 mg/l   (1927)
  —Fish:
    fathead minnow: 96 hr LC_{50}: 2000 mg/l   (1927)
  —Mammals:
    acute oral LD_{50} (rat): 5000 mg/kg   (1854)

2,4,6-trichloroaniline

A. PROPERTIES: long needles; m.w. 196.47; m.p. 77.5°C; b.p. 262.4°C

C. WATER POLLUTION FACTORS:
  —Biodegradation:
    degradation by *Aerobacter*: 500 mg/l at 30°C
    parent: 82% ring disruption in 120 hr
    mutant: 100% ring disruption in 30 hr   (152)

D. BIOLOGICAL EFFECTS:
  —Fish:
    fathead minnows (*Pimephales promelas*): 96 hr TLm, S; 10–1.0 mg/l   (935)

1,2,3-trichlorobenzene

Manufacturing source: organic chemical industry, pesticide mfg.   (347)
Users and formulation: solvent for high melting products; coolant in electrical installations and glass tempering; mfg. 2,5-dichlorophenol; polyester dyeing; termite preparations, synthetic transformer oil; lubricants; heat transfer medium, insecticides.   (347)

1122 1,2,4-TRICHLOROBENZENE

Man caused sources (water and air): general lab use; agricultural runoff; termite control operations; use of tranformer oil. (347)
A. PROPERTIES: plates; m.w. 181.46; m.p. 52°C; b.p. 219°C; solub. 12 ppm at 22°C
B. AIR POLLUTION FACTORS: 1 mg/cu m = 0.13 ppm, 1 ppm = 7.54 mg/cu m (57)
C. WATER POLLUTION FACTORS:
 —Reduction of amenities: T.O.C. = 0.01 mg/l (225)
 —Biodegradation:
  degradation by *Pseudomonas*: 200 mg/l at 30°C
         parent: 87% ring disruption in 120 hr
         mutant: 100% ring disruption in 43 hr (152)
  in Delaware river (U.S.A.): conc. range: winter: 0.5–1 ppb
                  (n.s.i.)   summer: n.d. (1051)
  in Zürich Lake (n.s.i.): at surface: 6 ppt; at 30 m depth: 42 ppt
  in tapwater (Zürich): (n.s.i.) 4 ppt (513)
D. BIOLOGICAL EFFECTS:
 —Residues:
  in fat phase of the following fish from the Frierfjord (Norway), April 1975–Sept. 1976: cod, whiting, plaice, eel, sprot:
   average: 0.5 ppm (sum of all isomers)
   range: 0.1–1.5 ppm (sum of all isomers) (1065)
 —Bioaccumulation:
  BCF in bacteria (*Siderocapsa treubii*): approx. 200 (1833)
  BCF in guppies (*Poecilia reticulata*): 13,000 (on lipid content)
 —Toxicity:
  guppy (*Poecilia reticulata*): 14 d LC_{50}: 2.4 ppm (1833)

1,2,4-trichlorobenzene

Use: solvent in chemical manufacturing; dyes and intermediates; dielectric fluid; synthetic transformer oils; lubricants; insecticides.
A. PROPERTIES: colorless crystals; m.w. 181.46; m.p. 17°C; b.p. 213°C; sp.gr. 1.574 at 10/4°C; v.d. 6.25; clear liquid; aromatic odor; solub. 19 ppm at 22°C
B. AIR POLLUTION FACTORS: 1 mg/cu m = 0.13 ppm, 1 ppm = 7.54 mg/cu m (65)
C. WATER POLLUTION FACTORS:
 —Biodegradation at 0.1 mg/l:

	normal sewage	adapted sewage
after 24 hr	0%	22%
after 135 hr	0%	56%	(997)

 —Impact on biodegradation processes: BOD test is inhibited from 5 mg/l onwards
  (30)

effect on respiration of glucose by mixed culture derived from activated sludge:

concentration (mg/l)	increase in lag period (hr)	respiration rate (%)	
1	10	94	
10	0	71	
100	0	47	
1000	>200	0	(997)

—Reduction of amenities: T.O.C. = 0.005 mg/l (225)
—Biodegradation:
  degradation by *Pseudomonas*: 200 mg/l at 30°C
    parent: 92% ring disruption in 120 hr
    mutant: 100% ring disruption in 46 hr (152)
—Sampling and analysis:
  photometry: min. full scale: 3.0×10^{-6} mole/l (53)

D. BIOLOGICAL EFFECTS:
  —Fish:
    guppy (*Poecilia reticulata*): 14 d LC_{50}: 2.4 ppm (1833)
  —Mammals:
    rat: inhalation: no effect level: 20 ppm, 15 × 6 hr (up to 20% 1,2,3-trichlorobenzene)
    ingestion: rats: single oral LD_{50}: less than 1000 mg/kg (1546)

1,3,5-trichlorobenzene

A. PROPERTIES: long needles; m.w. 181.46; m.p. 63°C; b.p. 208.5°C; solub. 5.8 ppm at 20°C

C. WATER POLLUTION FACTORS:
  —Reduction of amenities: T.O.C. = 0.05 mg/l (225)
  —Biodegradation:
    degradation by *Pseudomonas*: 200 mg/l at 30°C
      parent: 78% ring disruption in 120 hr
      mutant: 100% ring disruption in 50 hr (152)
    biodegradation at 1.0 mg/l:

	normal sewage	adapted sewage	
after 24 hr	0%	20%	
after 135 hr	0%	47%	(997)

D. BIOLOGICAL EFFECTS:
  —Bioaccumulation:
    BCF in bacteria (*Siderocapsa treubii*): approx. 250
    BCF in guppies (*Poecilia reticulata*): 14000 (on lipid content) (1833)
  —Fish:
    guppy (*Poecilia reticulata*): 14 d LC_{50}: 3.3 ppm (1833)

2,3,6-trichlorobenzoic acid (2,3,6-TrCB acid)

C. WATER POLLUTION FACTORS:
 —30–50% of ^{36}Cl-2,3,6-TrCB acid incorporated in soil was biologically changed to ionic form after 1–5 months (1818)

D. BIOLOGICAL EFFECTS:
 —Highly phytotoxic to tomato, melon, and cucumber plants (1817)

trichlorobiphenyls

Manufacturing source: organic chemical industry. (347)
Users and formulation: mfg. electrical; insulation; fire resistant heat transfer and hydraulic fluids; high temperature lubricants; elastomers; adhesives; paints, lacquers, varnishes, pigments and waxes; heat sensitive paper. (347)
Man caused sources (water and air): general use of PCB containing materials, e.g., paints, coatings, adhesives, paper, electrical insulation, etc. (347)

2,3,4-trichlorobiphenyl

C. WATER POLLUTION FACTORS:
 —Biodegradation:
 70.2% degradation after 1 hr by *Alcaligenes* Y42 (cell number 2×10^9/ml)
 64% degradation after 1 hr by *Acinetobacter* P6 (cell number 4.4×10^8/ml)
 at 12.8 mg/l initial conc. Trichlorobenzoic acid derivatives were detected in the metabolite (1086)

2,3,4'-trichlorobiphenyl

D. BIOLOGICAL EFFECTS:
 —Amphipod (*Gammarus fasciatus*): 4 d (S) LC$_{50}$: 0.070 mg/l (1617)

2,3,6-trichlorobiphenyl

C. WATER POLLUTION FACTORS:
—Biodegradation:
no degradation after 1 hr by *Alcaligenes* Y42 and *Acinetobacter* P6 at 12.8 mg/l initial concentration (1086)

2,4,5-trichlorobiphenyl

C. WATER POLLUTION FACTORS:
—Biodegradation:
92.0% degradation after 1 hr by *Alcaligenes* Y42 (cell number 2×10^9/ml)
64.8% degradation after 1 hr by *Acinetobacter* P6 (cell number 4.4×10^8/ml) at 12.8 mg/l initial concentration. Trichlorobenzoic acid derivatives were detected in the metabolite. (1086)

2,4,6-trichlorobiphenyl

C. WATER POLLUTION FACTORS:
—Biodegradation:
6.2% degradation after 1 hr by *Alcaligenes* Y42 (cell number 2×10^9/ml)
92.0% degradation after 1 hr by *Acinetobacter* P6 (cell number 4.4×10^8/ml) at 12.8 mg/l initial concentration (1086)

2,5,2′-trichlorobiphenyl

A. PROPERTIES: solub. 16 ppb; log P_{oct} 3.89
C. WATER POLLUTION FACTORS:
—Biodegradation:
3.2% degradation after 1 hr by *Alcaligenes* Y42 (cell number 2×10^9/ml)
10.2% degradation after 1 hr by *Acinetobacter* P6 (cell number 4.4×10^8/ml) at 12.8 mg/l initial concentration. Dichlorobenzoic acid derivatives were detected in the metabolite (1086)

2,5,3'-TRICHLOROBIPHENYL

aerobic incubation with seawater produced 1.4% metabolism
proposed pathway for metabolism by mixed cultures of marine bacteria:

D. BIOLOGICAL EFFECTS:
—Bioaccumulation and biodegradability index (B.I.) in a laboratory model ecosystem, after 33 days at 26°C:

	BCF	B.I.*
alga (*Oedogonium cardiacum*)	7315	0.3
snail (*Physa*)	5795	0.17
mosquito (*Culex pipiens quinquefasciatus*)	815	0.35
fish (*Gambusia affinis*)	6400	0.60

$$^*B.I. = \frac{\text{ppm polar degradation product}}{\text{ppm nonpolar product}}$$

(1643)

2,5,3'-trichlorobiphenyl

C. WATER POLLUTION FACTORS:
—Biodegradation:
84.2% degradation after 1 hr by *Alcaligenes* Y42 (cell number 2×10^9/ml)
82.6% degradation after 1 hr by *Acinetobacter* P6 (cell number 4.4×10^8/ml)
at 12.8 mg/l initial concentration. Dichlorobenzoic acid derivatives were detected in the metabolite (1086)

2,5,4'-trichlorobiphenyl

C. WATER POLLUTION FACTORS:
—Biodegradation:
43.6% degradation after 1 hr by *Alcaligenes* Y42 (cell number 2×10^9/ml)

6.08% degradation after 1 hr by *Acinetobacter* P6 (cell number 4.4×10^8/ml) at 12.8 mg/l initial concentration (1086)

2,4,4'-trichlorobiphenyl

C. WATER POLLUTION FACTORS:
—Biodegradation:
82.6% degradation after 1 hr by *Alcaligenes* Y42 (cell number 2×10^9/ml)
80.4% degradation after 1 hr by *Acinetobacter* P6 (cell number 4.4×10^8/ml)
at 12.8 mg/l initial concentration. Dichlorobenzoic acid derivatives were detected in the metabolite (1086)

3,4,2'-trichlorobiphenyl

C. WATER POLLUTION FACTORS:
—Biodegradation:
31.2% degradation after 1 hr by *Alcaligenes* Y42 (cell number 2×10^9/ml)
77.2% degradation after 1 hr by *Acinetobacter* P6 (cell number 4.4×10^8/ml)
at 12.8 mg/l initial concentration. Dichlorobenzoic acid derivatives were detected in the metabolite (1086)

1,1,1-trichloro-2,2-bis(p-chlorophenyl)ethane *see* DDT

3,3,4'-trichlorocarbanilide (TCC)

Use: a bacteriostat in soaps and detergents, plastics
A. PROPERTIES: m.p. 250°C
C. WATER POLLUTION FACTORS:
—Biodegradation in sewage and A.S. using labelled TCC: complete biodegradation of 200 μg/l of TCC within 10 hr (1070)
D. BIOLOGICAL EFFECTS:
—Molluscs:
Mercenaria mercenaria, hard clam; egg: static lab bioassay: 48 hr TLm: 32 ppb
Mercenaria mercenaria, hard clam; larvae: static lab bioassay: 12 d TLm: 37 ppb
(2324)

2,4,4'-trichlorodiphenylether

D. BIOLOGICAL EFFECTS:
 —Fish:
  Atlantic salmon: 96 hr BCF (water) (S): 2298
  42 hr BCF (food) (F) : 0.31

trichloroethanal *see* chloral

1,1,2-trichloroethane (vinyltrichloride)
$CH_2ClCHCl_2$
 Manufacturing source: organic chemical industry. (347)
 Users and formulation: mfg. 1,1-dichloroethylene; solvent for chlorinated rubber and various organic materials (fats, oils, resins, etc.). (347)
 Man caused sources (*water and air*): general use of chlorinated rubber, lab use. (347)
A. PROPERTIES: colorless liquid; m.w. 133.41; m.p. -35/-36.7°C; b.p. 113.7°C; v.p. 19 mm at 20°C, 32 mm at 30°C, 40 mm at 35°C; v.d. 4.63; sp.gr. 1.44 at 20/4°C; solub. 4,500 mg/l at 20°C; sat. conc. 136 g/cu m at 20°C, 225 g/cu m at 30°C
B. AIR POLLUTION FACTORS: 1 mg/cu m = 0.18 ppm, 1 ppm = 5.55 mg/cu m
 —Manmade sources:
  in rural Washington Dec 74–Feb 75: glc's: <5 ppt (315)
 —Control methods: catalytic incineration over commercial Co_3O_4 (1.0–1.7 mm) granules, catalyst charge 13 cm³, space velocity 45000 h⁻¹ at 100 ppm in inlet:
  80% decomposition at 190°C
  90% decomposition at 220°C
  99% decomposition at 320°C (346)
C. WATER POLLUTION FACTORS:
 —Reduction of amenities: T.O.C. = 50 mg/l (84)
 —Waste water treatment:
  evaporation from water at 25°C of 1 ppm solution: 50% after 21 min
  90% after 102 min (313)
 —Sampling and analysis:
  GC-EC after *n*-pentane extraction: det. lim.: 0.1 µg/l (84)
D. BIOLOGICAL EFFECTS:
 —Toxicity threshold (cell multiplication inhibition test):
  bacteria (*Pseudomonas putida*): 93 mg/l (n.s.i.)
  green algae (*Scenedesmus quadricauda*): 430 mg/l (n.s.i.)
  protozoa (*Entosiphon sulcatum*): >1040 mg/l (n.s.i.) (1900)
  protozoa (*Uronema parduczi Chatton-Lwoff*): >1040 mg/l (n.s.i.) (1901)
 —Fish:
  guppy (*Poecilia reticulata*): 7 d LC_{50}: 94 ppm (1833)

—Mammals:
rat: acute lethal dose: 2,000 ppm, 4 hr inhalation
single oral LD_{50}: 0.1-0.2 g/kg
dog: single oral LD_{50}: 5 ml/kg (211)
skin absorption: rabbits: LD_{50}: >1000 mg/kg (1546)
—Carcinogenicity:
The results of this study do not provide convincing evidence for the carcinogenicity of 1,1,2-trichloroethane in Osborne-Mendel rats. Under the conditions of this bioassay 1,1,2-trichloroethane is carcinogenic in B6C3F1 mice, causing hepatocellular carcinomas and adrenal pheochromocytomas. (1761)

1,1,1-trichloroethane (methylchloroform)
CCl_3CH_3

A. PROPERTIES: colorless liquid; m.w. 133.41; m.p. -32°C; b.p. 71/81°C; v.p. 100 mm at 20°C, 155 mm at 30°C; v.d. 4.63; sp.gr. 1.35 at 20/4°C; solub. 4,400 mg/l at 20°C; sat. conc. 726 g/cu m at 20°C, 1,088 g/cu m at 30°C

B. AIR POLLUTION FACTORS: 1 mg/cu m = 0.18 ppm, 1 ppm = 5.54 mg/cu m
—Odor: characteristic: quality: sweetish

odor thresholds (mg/cu m): detection ~10^3; recognition ~10, 10^3; not specified ~10^{-3}, ~10^3

(57; 210; 278; 298; 712; 749; 804)

O.I. at 20°C = 330 (316)

—Ambient air quality:
glc's Los Angeles county: 0.01-2.3 ppb (46)
glc's in rural Washington: Dec. 74-Febr. 75: 0.1 ppb (315)
in the United Kingdom:

Liverpool (Childwall),	0.17 ppb
Widness (Pex Hill)	0.34 ppb
Delamere	0.03-0.20 ppb
Trodsham	0.05-1.01 ppb
Moel Famau (N. Wales)	0.34-0.68 ppb
Rannock Moor (Scotland)	0.17-0.25 ppb
Forest of Dean	0.47 ppb

in the Netherlands

Hengelo: open country	0.02-0.05 ppb
center	<0.02-0.13 ppb
Weiwerd	<0.02 ppb

in West Germany:

Munich: center	n.d.
1 km from center	0.17-0.67 ppb

	5 km from center	n.d.–0.50 ppb
	10 km from center	n.d.–0.34 ppb
	20 km from center	n.d.
	Ranzel	1.51–1.68 ppb
	Langel	1.51–6.55 ppb
	Unkendorf	<0.50 ppb

in Belgium:
- Bruxelles Nord   0.17–0.39 ppb
- Uccle   n.d.–<0.17 ppb
- Namur   n.d.–0.34 ppb

in France:
- Lyon: center   <0.84–2.01 ppb
- 20 km from center   <0.84–0.84 ppb

—Atmospheric reactions:
  estimated lifetime under photochemical smog conditions in S.E. England: 2470 hr   (1699; 1702)

C. WATER POLLUTION FACTORS:
  —Reduction of amenities: odor threshold: 50 mg/l; 400 ppm   (84, 279)
  —Water quality:
  in the Netherlands:
  Twente: canal Hengelo   0.07 ppb
          canal Delden   <0.1 ppb
  Eems   <0.1 ppb
  Oostfriese Gaatje: South   0.3 ppb
                     North   0.1 ppb
  Ranselgat   0.3 ppb
  Huibertgat 0.2 ppb
  in France:
  river Durance: Pont Oraison   n.d.
                 Ste. Tulle   n.d.
  waters from upland reservoir in N.W. England (1974):
    during dry cloudy weather: 12–13 ppb
    during prolonged heavy rain: 5–41 ppb   (933)

  —Waste water treatment:
  evaporation from water at 25°C of 1 ppm solution: 50% after 17–23 min
                                                    90% after 63–80 min   (313)

  —Sampling and analysis:
  GC-EC after *n*-heptane extraction: 0.1 µg/l   (84)

D. BIOLOGICAL EFFECTS:
  —Residues:
  Concentrations in various organs of fish collected from the relative clean water of the Irish sea, in the vicinity of Port Erin, Isle of Man (only organs with highest and lowest concentrations are given - concentrations on dry wt. basis):
  Conger conger (eel): gill: 2 ng/g
                         brain: 9 ng/g
  Gadus morhua (cod): liver, muscle, stomach: 5–7 ng/g
                        gill, brain: 14–16 ng/g
  Pollachius birens (coalfish): muscle: 6 ng/g   (1092)

—Toxicity:
—Fish:
fathead minnow: 96 hr LC_{50} (F): 52.8 mg/l
  96 hr LC_{50} (S): 105 mg/l
guppy (*Poecilia reticulata*): 7 d LC_{50}: 133 ppm (1833)
Pimephales promelas rafinesque, flow through test (mg/l):

hr	EC_{10}[a]	EC_{50}	EC_{90}	LC_{10}	LC_{50}	LC_{90}
24	10.5	12.1	14.1			
	(8.0-11.5)[b]	(10.9-13.5)	(12.9-18.3)			
48	10.0	11.5	13.2			
	(7.8-10.9)	(10.4-12.8)	(12.1-17.3)			
72	9.0	11.1	13.8	34.1	55.4	88.9
	(6.7-10.0)	(10.0-12.6)	(12.3-13.8)	(20.8-41.2)	(46.2-82.7)	(67.0-254.7)
96	9.0	11.1	13.8	30.8	52.8	90.8
	(6.7-10.0)	(10.0-12.6)	—	(18.8-37.6)	(43.7-77.7)	(66.4-245.9)

[a]: EC-the effective concentration. Concentration producing an adverse effect. In this case the effect noted was loss of equilibrium
[b] 95% confidence limits (1054)

Pimephales promelas rafinesque: 96 hr LC_{50}:
  flow through test: 52.8 mg/l
    95% conf. lim.: 43.7-77.7 mg/l
  static test: 105 mg/l
    95% conf. lim. 91-126 mg/l (1054)

—Mammalia:
rat: inhalation: LC_{50}: 24,000 ppm, 1 hr; 18,000 ppm, 3 hr
  14,000 ppm, 7 hr
mouse: inhalation: LC_{50}: 13,500 ppm, 10 hr
rabbit, monkey: inhalation: no response: 3,000 ppm, 7 hr/day, 5 days/w,
  2 months (211)
rat: single oral LD_{50}: 10.3-12.3 g/kg
female mice: single oral LD_{50}: 11.24 g/kg
female rabbit: single oral LD_{50}: 5.66 g/kg
male guinea pig: single oral LD_{50}: 9.47 g/kg (211)

—Man: eye irritation, headache: 500 ppm, 180 min
—Carcinogenicity:
A variety of neoplams were represented in both 1,1,1-trichloroethane-treated and matched-control rats and mice. However, each type of neoplasm has been encountered previously as a lesion in untreated rats or mice. The neoplasms observed are not believed attributable to 1,1,1-trichloroethane exposure, since no relationship was established between the dosage groups, the species, sex, type of neoplasm, or the site of occurence. Even if such a relationship were inferred, it would be inappropriate to make an assessment of carcinogenicity of 1,1,1-trichloroethane on the basis of this test, because of the abbreviated life spans of both the rats and the mice. (1735)

trichloroethylene (ethylenetrichloride)
$CCl_2=CHCl$

TRICHLOROETHYLENE

Manufacturing source: organic chemical industry. (347)

Users and formulation: dry cleaning operations and metal degreasing; solvents for fats, greases, waxes; solvents for greases and waxes from cotton, wool, etc., for caffeine from coffee; solvent for cellulose ester and ethers; solvent for dyeing; refrigerant and heat exchange liquid; organic synthesis; fumigant; anesthetic. (347)

Contaminants of technical grade:

epichlorohydrin:	0.22 % w/w	
epoxybutane:	0.20	
carbontetrachloride:	0.05	
chloroform:	0.01	
1,1,1-trichloroethane:	0.035	
diisobutylene:	0.020	
(2,2,4-trimethylpentene-1) ethylacetate:	0.052	
pentanol-2:	0.015	
butanol-2:	0.051	(1322)

A. PROPERTIES: colorless gas; m.w. 131.5; m.p. $-87°C$; b.p. $86.7°C$; v.p. 20 mm at $0°C$, 60 mm at $20°C$, 95 mm at $30°C$, v.d. 4.54; sp.gr. 1.46 at $20°C$; solub. 1.100 mg/l at $25°C$, sat. conc. 415 g/cu m at $20°C$, 643 g/cu m at $30°C$

B. AIR POLLUTION FACTORS: 1 mg/cu m = 0.18 ppm, 1 ppm = 5.46 mg/cu m
 —Odor: characteristic: soft, solventy, etheral, chloroformlike

odor thresholds   mg/cu m

[odor threshold chart with detection, recognition, and not specified rows spanning 10^{-7} to 10^4]

(2; 57; 73; 210; 278; 279; 662; 740; 741; 746; 749; 804; 861)

odor Index at $20°C = 300$ (316)

—Manmade sources:
 in municipal sewer air: 10–100 ppm   (212)

—Ambient air quality:
 in rural Washington, Dec. '74–Feb. '75: <5 ppt   (315)
 in the United Kingdom:

Liverpool (Childwall)	0.34–1.19 ppb
Widnes (Pex Hill)	1.53 ppb
Delamere	0.51–1.36 ppb
Frodsham	0.68–3.40 ppb
Moel Tamau (N. Wales)	0.17–1.53 ppb
Rannock Moore (Scotland)	0.42–1.36 ppb
Forest of Dean	0.85 ppb

 in the Netherlands:

Hengelo: open country	<0.02–0.17 ppb
center	<0.02–0.09 ppb
Weiwerd	0.09 ppb

in West Germany:
Munich: center	0.85–3.92 ppb
1 km from centre	0.17–3.59 ppb
5 km from centre	0.17–5.45 ppb
10 km from centre	0.17–3.06 ppb
20 km from centre	n.d.
Uberackern	1.36–8.5 ppb

in Belgium:
Bruxelles Nord	0.68–1.02 ppb
Uccle	n.d.–1.36 ppb
Namur	0.17–1.36 ppb

in France:
Saint Auban	0.17–8.15 ppb
Lyon: center	<0.83–4.23 ppb
20 km. from center	<0.83–2.54 ppb

—Sampling and analysis:
photometry: min. full scale: 5,530 ppb (53)

C. WATER POLLUTION FACTORS:
—Odor threshold: 10 mg/l; 0.5 mg/kg (detection) (84; 886)
—Water quality:
waters from upland reservoirs in N.W. England (1974):
during dry cloudy weather: <0.1–0.1 ppb
during prolonged heavy rain: 4–14 ppb (933)
in river Maas at Eysden (Netherlands) in 1976: median: 2.0 µg/l; range; n.d. to 14.5 µg/l
in river Maas at Keizersveer (Netherlands) in 1976: median: 0.1 µg/l; range; n.d. to 2.1 µg/l (1368)
in Zürich Lake: at surface: 38 ppt; at 30 m depth: 65 ppt
in Zürich area: in springwater: 5 ppt; in groundwater: 80 ppt; in tapwater: 105 ppt (513)

in England:
Liverpool bay	0.3–3.6 ppb

in West Germany:
river Rhine:	Hoenningen	1–1.5 ppb
	Luelsdorf	2–2.5 ppb
	Wesseling	1.5–2 ppb
river Salzach:	Marienberg	0.4–2.1 ppb
	Überackern	25–73.9 ppb
river Isar:	source of river	0.02–0.03 ppb
	Munich	0.2–0.6 ppb
	downstream of Munich	2.5–3.2
	Lake Starnberg	0.13–0.15 ppb
	Lake Lerchenau	3.2–8.5 ppb

in the Netherlands:
Twente:	canal Hengelo	0.26 ppb
	canal Delden	<0.2 ppb
Eems		11.0 ppb

Oostfriese Gaatje: South 7.5 ppb
   North 0.7 ppb
Ranselgat 0.2 ppb
Huibertgat <0.2 ppb
in France:
river Durance: Pont Oraison 6-25 ppb
   Ste. Tulle ≤3-9 ppb

—Waste water treatment:
evaporation from water at 25°C of 1 ppm solution:
50% after 19-24 min
90% after 63-80 min (313)

—Removal by aeration:

surface loading cu m/sq.m	air/water ratio	influent conc. µg/l	effluent conc. µg/l	% removal
24	22	185	48	74
16	33	185	28	85
14.5	37	185	27	85.5
10.5	51	185	14	92.5
6.8	80	185	0.1	>99

(1335)

A.C.: influent 350 µg/l; effluent <0.10 µg/l (thickness A.C. layer: 2m); filtration velocity: 5.5 mm/sec. (1335)

—Sampling and analysis:
GC-EC after n-pentane extraction: det.lim.: 0.1 µg/l (84)

D. BIOLOGICAL EFFECTS:
   —Residues:
   Concentrations in various organs of molluscs and fish collected from the relatively clean waters of the Irish sea, in the vicinity of Port Erin, Isle of Man (only organs with highest and lowest concentrations are mentioned—concentrations on dry weight basis)
   molluscs: *Baccinum undatum*: muscle: 33 ng/g
      mantle: 250 ng/g
   fish:
   Conger conger (eel): gill, gut: 29 ng/g
      brain, muscle: 62-70 ng/g
   Gadus morhua (cod): stomach, muscle: 7-8 ng/g
      brain, liver: 56-66 ng/g
   Pollachius birens (coal fish): muscle: 8 ng/g
      alimentary canal: 306 ng/g
   Scylliorhinus canicula (dogfish): muscle, gut, brain: 40-41 ng/g
      liver: 479 ng/g
   Trisopterus luscus (bib): gill: 40 ng/g
      muscle, skeletal tissue: 185-187 ng/g (1092)

   —Toxicity:
   toxicity theshold (cell multiplication inhibition test):
   bacteria (*Pseudomonas putida*): 65 mg/l (1900)
   algae (*Microcystis aeruginosa*): 63 mg/l (329)

TRICHLOROETHYLENE 1135

green algae (*Scenedesmus quadricauda*): >1000 mg/l (1900)
protozoa (*Entosiphon sulcatum*): 1200 mg/l (1900)
protozoa (*Uronema parduczi Chatton-Lwoff*): >960 mg/l (1901)
—Algae:
chlorphyl-containing algae and plants are decolored at 600 mg/l (30)
—Arthropods:
Daphnia: LD_{100}: ± 600 mg/l, 40 hr
no effect: ± 100 mg/l (30)
—Amphibian:
Mexican axolotl (3-4 wk after hatching): 48 hr LC_{50}: 48 mg/l
clawed toad (3-4 wk after hatching): 48 hr LC_{50}: 45 mg/l (1823)
—Fish:
fathead minnow: 96 hr LC_{50} (F): 40.7 mg/l
96 hr LC_{50} (S): 66.8 mg/l
guppy (*Poecilia reticulata*): 7 d LC_{50}: 55 ppm (1833)
Pimephales *promelas rafinesque*; 96 hr LC_{50}:
 flow through test: 40.7 mg/l; 95% conf. lim.: 31.4-71.8 mg/l
 static test: 66.8 mg/l: 95% conf. lim.: 59.6-74.7 mg/l
Pimephales promelas rafinesque (flow through test):

hr	$EC_{10}{}^a$	EC_{50}	EC_{90}	LC_{10}	LC_{50}	LC_{90}
24	15.2	23.0	36.2	34.7	52.4	79.1
	$(10.0-18.2)^b$	(19.8-27.4)	(30.3-51.2)	(24.4-41.4)	(44.3-65.7)	(63.7-131.6)
48	16.9	22.7	30.6	27.7	53.3	102.6
	(11.6-19.6)	(19.7-27.3)	(26.0-49.2)	(17.3-35.0)	(43.1-75.5)	(73.3-238.0)
72	15.5	22.2	31.8	20.9	39.0	72.6
	(10.0-18.2)	(18.9-27.3)	(26.2-56.0)	(11.9-26.1)	(31.8-57.5)	(51.7-109.2)
96	13.7	21.9	34.9	17.4	40.7	95.0
	(8.5-16.6)	(18.4-28.5)	(27.3-70.9)	(9.0-22.9)	(31.4-71.8)	(59.0-419.9)

^{a}EC-the effective concentration. Concentration producing an adverse effect.
 In this case the effect noted was loss of equilibrium
^{b}95% confidence limits
—Mammalia:
mouse: inhalation: LC_{50}: 49,000 ppm, 30 min
 8,450 ppm, 4 hr
rat: inhalation: LC_{50}: 8,000 ppm, 4 hr
mouse: inhalation: LC_{50}: 5,500 ppm, 10 hr
 : no effect: 1,600 ppm, 4 hr
rat: inhalation: no effect: 100 ppm, 8 hr
rabbit: inhalation: no effect: 1,200 ppm, 473 hr
ape, rabbit, rat, guinea pig: inhalation: no effect: 730 ppm, 8 hr/d, 6 weeks (54)
rat: oral toxicity: 3-5 ml/kg (211)
—Man: eye irritation: 160 ppm
 supportable during 30 min: 379-372 ppm
 full narcosis: 2,500-6,000 ppm (54)
 severe toxic effects: 2,000 ppm = 10,940 mg/cu m, 60 min
 symptoms of illness: 800 ppm = 4,376 mg/cu m
 unsatisfactory: >400 ppm = 2,188 mg/cu m (185)

trichlorofluoromethane *see* fluorotrichloromethane

trichlorofon *see* dipterex

3,4,5-trichloroguaiacol

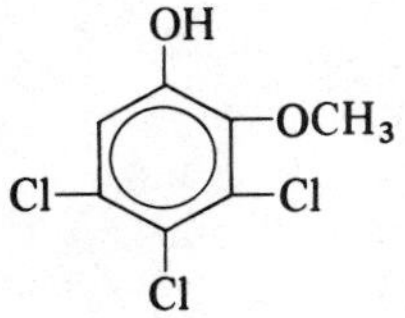

Formed by the reaction of chlorination agents with phenolic liquids during the bleaching process in the manufacture of pulp and paper.
A. PROPERTIES: white needles; m.p. 112–114°C (n.s.i.)
D. BIOLOGICAL EFFECTS:
  —Fish: rainbow trout juvenile: 96 hr LC_{50} (S): 0.75 mg/l
  —Mammals:
   male Spraque-Dawley rats: single oral LD_{50}: 3000 mg/kg
   male weanling rats: accumulation: 50–5000 ppm incorporated in the diet for 28 days: no significant level (0.2 ppm) was found in any tissue of rats, a result showing this compound was either rapidly metabolized and/or excreted (n.s.i.)

(1429)

2,4,4'-trichloro-2'-hydroxydiphenylether

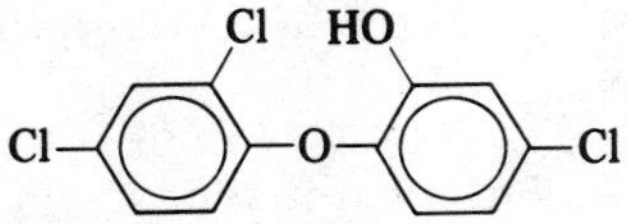

C. WATER POLLUTION FACTORS:
  —Water quality:
   in river water: 0.012–0.30 ppm; in river sediment: 1.2–5 ppm   (55)

trichloromethane *see* chloroform

***cis*-N-((trichloromethyl)thio)-4-cyclohexene-1,2-dicarboximide** *see* captan

N-trichloromethylthiotetrahydrophtalimide *see* captan

N-(trichloromethylthio)phthalimide *see* folpet

trichloronat (O-ethyl-O-(2,4,5-trichlorophenyl)ethylphosphonothioate; agritox; fenophosphon)

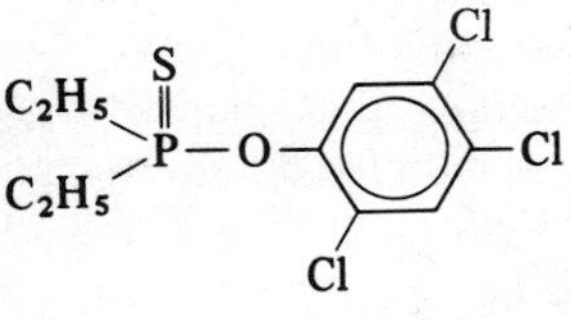

Use: insecticide
A. PROPERTIES: sp.gr. 1.365 at 20/40°C; solub. 50 ppm at 20°C
C. WATER POLLUTION FACTORS:
—Persistence in soil at 10 ppm initial concentration

	weeks incubation to	
	50% remaining	5% remaining
sterile sandy loam	>24	
sterile organic soil	>24	
non sterile sandy loam	1.5	20
non sterile organic soil	4	>24

(1433)

D. BIOLOGICAL EFFECTS:
—Mammals:
acute oral LD_{50} (rat): 35.7 mg/kg
acute dermal LD_{50} (male rat): 341 mg/kg (1854)

trichloronitromethane *see* chloropicrin

3,4,6-trichloro-2-nitrophenol

D. BIOLOGICAL EFFECTS:
—Fish: larvae of sea lamprey: LD_{100}: 5 mg/l
rainbow trout: LD_{10}: 17 mg/l
brown trout: LD_{10}: 15 mg/l (226)

2,3,5-trichlorophenol

A. PROPERTIES: long colorless needles; m.w. 197.46; m.p. 62°C; b.p. 253°C
C. WATER POLLUTION FACTORS:
—Biodegradation:
degradation by *Pseudomonas*: 200 mg/l at 30°C
parent: 100% ring disruption in 100 hr
mutant: 100% ring disruption in 52 hr (152)
—Water quality:
in Delaware river (U.S.A.): conc. range: winter: 2 ppb
(n.s.i.) summer: n.d. (1051)
D. BIOLOGICAL EFFECTS:
—Fish:
guppy (*Poecilia reticulata*): 24 hr LC_{50}: 1.6 ppm at pH 7.3 (1833)

2,3,6-trichlorophenol

C. WATER POLLUTION FACTORS:
—Taste threshold conc.: 0.0005 mg/l (998)
—Odor threshold conc.: detection: 0.3 mg (998)
D. BIOLOGICAL EFFECTS:
—Fish:
guppy (*Poecilia reticulata*): 24 hr LC_{50}: 5.1 ppm at pH 7.3 (1833)

2,4,5-trichlorophenol

Use: fungicide; bactericide.
A. PROPERTIES: m.w. 197.46; m.p. 61/63°C; b.p. 252°C; v.p. 400 mm at 225°C; solub. 1190 mg/kg at 25°C; sp.gr. 1.5 at 75°C; log P_{oct} 3.06/3.72
C. WATER POLLUTION FACTORS:
—Biodegradation:
decomposition in suspended soils: >72 days for complete disappearance (175)
—Reduction of amenities:
taste threshold conc.: 0.001 mg/l (998)
odor threshold conc. (detection): 0.2 mg/l (998)
0.011 to 0.333 mg/l (226)
D. BIOLOGICAL EFFECTS:
—Algae: *Chlorella pyrenoidosa*: toxic at 1.5 mg/l (41)
—Fish: goldfish: 24 hr LC_{50}: 1.7 ppm
amount found in dead fish at 1.8 ppm: 112 μg/g
BCF (at 1.8 ppm) = 62 (1850)
Idus idus melanotus: 48 hr LC_0: 1.0 mg/l
48 hr LC_{50}: 1.3 mg/l
48 hr LC_{100}: 1.6 mg/l (998)
—Mammals:
rats: oral LD_{50}: 0.82 g/kg (211)
rats: oral LD_{50}: 2.96 g/kg body wt
no adverse effects: at 500 mg/kg body/wt (226)

2,4,6-trichlorophenol

Manufacturing source: organic chemical industry; pesticide mfg. (347)
Users and formulation: mfg. antiseptics, bactericides, fungicides, germicides; mfg. wood and glue preservatives; used as anti-mildew agent for textiles. (347)

A. PROPERTIES: needles; m.w. 197.46; m.p. 68°C; b.p. 244.5°C; sp.gr. 1.490 at 75/4°C; solub. 800 mg/l at 25°C, 2,430 mg/l at 96°C

B. AIR POLLUTION FACTORS:
  —Odor threshold conc. (recognition): 0.0010–0.0016 mg/cu m (610)
  0.021 mg/cu m (712)

C. WATER POLLUTION FACTORS:
  —Biodegradation:
   decomposition in soil suspensions: 5 days for complete disappearance (175)
  —degradation by *Pseudomonas*: 200 mg/l at 30°C
     parent: 100% ring disruption in 120 hr
     mutant: 100% ring disruption in 50 hr (152)
  —Reduction of amenities:
   odor threshold: 0.10 > 1.0 mg/l; 0.3 mg/l (detection) (998)
   taste threshold: >1.0 mg/l; 0.002 mg/l (226)
  —Waste water treatment methods:
   ion exchange: adsorption on Amberlite XAD-4 (25°C):
   solute adsorbed: zero leakage: 12.02 lb/cu ft
        10 ppm leakage: 13.81 lb/cu ft (40)
  reaction of excess of hypochlorous acid with dilute aqueous solutions at initial pH's of 6.0 and 3.5:

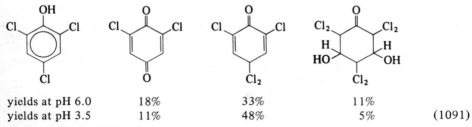

yields at pH 6.0		18%	33%	11%
yields at pH 3.5		11%	48%	5%	(1091)

D. BIOLOGICAL EFFECTS:
  —Fish:
   goldfish: 24 hr LC$_{50}$: 10.0 ppm
        amount found in dead fish at 10 ppm: 200 µg/g
        BCF (at 10 ppm) = 20 (1850)
   fathead minnows (*Pimephales promelas*): 96 hr TLm: 1.0–0.1 mg/l (static bio-assay) (935)

3,4,5-trichlorophenol

D. BIOLOGICAL EFFECTS:
  —Fish: guppy (*Poecilia reticulata*): 24 hr LC_{50}: 1.1 ppm at pH 7.3   (1833)

2,4,5-trichlorophenoxyacetic acid (2,4,5-T)

Use: 2,4,5-T salts and esters are used widely to control woody plants.
A. PROPERTIES: m.w. 255.49; m.p. 158°C; solub. 278 ppm at 25°
C. WATER POLLUTION FACTORS:
  —Biodegradation: >205 days for ring cleavage in soil suspension   (1827)
  —Aquatic reactions:
    75-100% disappearance from soils: 5 months   (1815)
  —Odor threshold conc. (detection): 2.92 mg/kg   (915)
D. BIOLOGICAL EFFECTS:
  —Residues:

plants (N.E. Finland)	conc. ppm	time of collection after herbicide spraying
lingonberry	0.07-15	2-13 weeks
wild mushroom	<0.02-1.8	2-13 weeks
birch and aspen foliage	0.1-30	13-43 weeks

    (1598)

  —Birds:
    American coot (*Fulcia americana*) collected in a reservoir that received runoff from a watershed treated with 2,4,5-T: residue conc.:
      in breast muscle: mean: 199.0 ppb
                         range: n.d.-1.338 ppb
      in fat: mean: 21.0 ppb
            range: n.d.-30.0 ppb
      in liver: mean: 39.0 ppb
             range: n.d.-118.0 ppb
      in gizzard: mean: 18.0 ppb
               range: n.d.-41.0 ppb   (1594)
  —Bioaccumulation:
    mosquitofish: 32 d BCF (S): 26 X   (1538)

—Toxicity:
  Algae:

technical acid	Chlorococcum sp.	1.5×10^5 ppb	50% decrease in O_2 evolution	f	(2348)
technical acid	Chlorococcum sp.	1.0×10^5 ppb	50% decrease in growth	Measured as ABS. (525 mu) after 10 days	(2348)
technical acid	Dunaliella tertiolecta	1.5×10^5 ppb	50% decrease in O_2 evolution	f	(2348)
technical acid	Dunaliella tertiolecta	1.25×10^5 ppb	50% decrease in growth	Measured as ABS. (525 mu) after 10 days	(2348)
technical acid	Isochrysis galbana	5×10^4 ppb	50% decrease in O_2 evolution	f	(2348)
technical acid	Isochrysis galbana	5×10^4 ppb	50% decrease in growth	Measured as ABS. (525 mu) after 10 days	(2348)
technical acid	Phaeodactylum tricornutum	7.5×10^4 ppb	50% decrease in O_2 evolution	f	(2348)
technical acid	Phaeodactylum tricornutum	5×10^4 ppb	50% decrease in growth	Measured as ABS. (525 mu) after 10 days	(2348)

  —Fish:
    rainbow trout: 96-hr LC_{50}: 0.98 mg/l (R)
    rainbow trout: 96-hr LC_{50}: 8.7 mg/l (acetone solution) (R)
                  96-hr LC_{50}: 0.15 mg/l (emulsion) (R)   (1538)
    striped bass: 96-hr LC_{50}: 14.6 mg/l (S)
    banded killifish: 96-hr LC_{50}: 17.4 mg/l (S)
    pumpkinseed: 96-hr LC_{50}: 20.0 mg/l (S)
    white perch: 96-hr LC_{50}: 16.4 mg/l (S)
    American eel: 96-hr LC_{50}: 43.7 mg/l (S)
    carp: 96-hr LC_{50}: 41.1 mg/l (S)
    guppy: 96-hr LC_{50}: 28.1 mg/l (S)   (1193)
  —Mammals:
    mice: 39% inhibition of testicular DNA synthesis at 200 mg/kg   (1325)
    ingestion: rat: single oral LD_{50}: 500 mg/kg   (1546)

2,4,6-trichlorophenoxy acetic acid

[Structure: benzene ring with OCH_2COOH at position 1, Cl at positions 2, 4, and 6]

2,4,5-TRICHLOROPHENOXYACETIC ACID, ISOOCTYLESTERS

C. WATER POLLUTION FACTORS:
—Waste water treatment method:
A.C. type BL (Pittsburgh Chem. Co.): % adsorbed by 10 mg A.C. from 10^{-4} m aqueous solution:
at pH 3.0: 65.3%
pH 7.0: 21.8%
pH 11.0: 12.5% (1313)

2,4,5-trichlorophenoxyacetic acid, isooctylesters

Use: herbicide.

A. PROPERTIES: amber liquid; v.p. 18 mm at 212°C; sp.gr. 1.223 at 68/68°F; solub. 10 mg/l

C. WATER POLLUTION FACTORS:
—Applied at rates of 1.1–4.3 kg/ha on Dykeland soil, no residues of the unchanged ester could be detected in soil 13 hours after application
—In the present study the major residue in the soil was the free phenoxyacid of which the major portion disappeared within 50 days of application (table below):

2,4,5-T residues (ppm) in soil following application of 2,4,5-T isooctyl ester:

days after treatment	1.1 kg/ha[a]		2.1 kg/ha		4.3 kg/ha	
	0–10[b] cm	10–20 cm	0–10 cm	10–20 cm	0–10 cm	10–20 cm
1	.06	—	.39	—	.26	—
14	.05	—	.28	—	1.08	—
28	.03	—	.09	—	.59	—
42	.01	.01	.02	.02	.21	.11
56	.01	.01	.02	.02	.08	.05
70	.01	.01	.02	.01	.05	.05
265	.01	.Tr	.01	Tr	.01	Tr
385	Tr	ND	Tr	ND	Tr	ND

[a] Application rate
[b] Depth
Tr = trace: <0.01 ppm
ND: <0.005 ppm (1088)

D. BIOLOGICAL EFFECTS:
—Mammals:
ingestion: rat: single oral LD_{50}: range of 500 to 1000 mg/kg (1546)
mice: inhibition of testicular DNA synthesis:
44% inhibition at 400 mg/kg ($p < 0.01$)
31%   200 mg/kg ($p < 0.05$)

10%	100 mg/kg	
1%	50 mg/kg	(1325)

2-(2,4,5-trichlorophenoxy)propionic acid (fenoprop; 2,4,5-TP; garlon; kuron; silvex)

Hormone-type weedkiller.
It should be stated which ester is present:
   propyleneglycolbutyletherester (PGBE)
   butoxyethylester   (BEE)
   isooctylester   (IOE)

A. PROPERTIES: m.p. 179–181°C; solub. 140 ppm at 25°C; 200 mg/l
C. WATER POLLUTION FACTORS:
   —Biodegradation: >205 days for ring cleavage in soil suspension   (1827)
D. BIOLOGICAL EFFECTS:
   —Potassium salt:
   —Algae:

technical acid	*Chlorococcum* sp.	2.5×10^5 ppb	50% decrease in O_2 evolution	f
technical acid	*Chlorococcum* sp.	2.5×10^4 ppb	50% decrease in growth	Measured as ABS. (525 mu) after 10 days
technical acid	*Dunaliella tertiolecta*	2×10^5 ppb	50% decrease in O_2 evolution	f
technical acid	*Dunaliella tertiolecta*	2.5×10^4 ppb	50% decrease in growth	Measured as ABS. (525 mu) after 10 days
technical acid	*Isochrysis galbana*	2.5×10^5 ppb	50% decrease in O_2 evolution	f
technical acid	*Isochrysis galbana*	5×10^3 ppb	50% decrease in growth	Measured as ABS. (525 mu) after 10 days (2348)

   —Molluscs:
    Crassostrea virginica (American oyster), egg: TLm: 5.9×10^3 ppb, 48 hr static lab bioassay
    Crassostrea virginica (American oyster), larvae: TLm: 710 ppb, 14 day static lab bioassay   (2324)
   —Fish: *Lepomis macrochirus*: 48 hr LC_{50}: 83000 µg/l   (2115)
   —Mammals:
    acute oral LD_{50} (rat): 650 mg/kg   (1854)
    500–1000 mg/kg (mixed butylesters and propyleneglycol-butyletheresters)   (1855)

2,4,5-trichlorophenoxy-ω-propionic acid

Cl—C₆H₂(Cl)(Cl)—OCH$_2$CH$_2$COOH

(structure: 2,4,5-trichlorophenyl ring with —OCH$_2$CH$_2$COOH substituent)

C. WATER POLLUTION FACTORS:
  —Biodegradation: >81 days for ring cleavage in soil suspension (1827)

2-(2,4,5-trichlorophenoxy)propionic acid, butoxyethylester
D. BIOLOGICAL EFFECTS:
  —Crustaceans:

Gammarus fasciatus:	96 hr LC$_{50}$:	250 μg/l	(2125)
Daphnia magna:	48 hr LC$_{50}$:	2100 μg/l	(2125)
Cypridopsis vidua:	48 hr LC$_{50}$:	4900 μg/l	(2125)
Asellus brevicaudus:	48 hr LC$_{50}$:	40000 μg/l	(2125)
Palaemonetes kadiakensis:	48 hr LC$_{50}$:	8000 μg/l	(2125)
Orconectes nais:	48 hr LC$_{50}$:	60000 μg/l	(2125)

  —Fish:

Lepomis macrochirus:	48 hr LC$_{50}$:	1100 μg/l	(2115)

2-(2,4,5-trichlorophenoxy)propionic acid, isooctylesters (silvex (IOE))
A. PROPERTIES: amber dark-brown liquid; b.p. 373°C; sp.gr. 1.183 at 68/68°F
D. BIOLOGICAL EFFECTS:
  —Fish:
    Lepomis macrochirus: 48 hr LC$_{50}$: 16000 μg/l   (2115)
  —Mammals:
    ingestion: rats: single oral LD$_{50}$: probably in the range of 400–800 mg/kg
                                                               (1546)

2-(2,4,5-trichlorophenoxy)propionic acid, propyleneglycolbutyletheresters (silvex (PBGE))
A. PROPERTIES: amber liquid; b.p. 327°C; v.p. 54 mm at 229°C; sp.gr. 1.22 (68/68°F)
D. BIOLOGICAL EFFECTS:
  —Crustacean:

Gammarus fasciatus:	96 hr, LC$_{50}$:	840 μg/l	(2125)
Daphnia magna:	48 hr, LC$_{50}$:	180 μg/l	(2125)
Cypridopsis vidua:	48 hr, LC$_{50}$:	200 μg/l	(2125)
Asellus brevicaudus:	48 hr, LC$_{50}$:	500 μg/l	(2125)
Palaemonetes kadiakensis:	48 hr, LC$_{50}$:	3200 μg/l	(2125)
Orconectes nais:	no effect: 100,000 μg/l, 48 hr		(2125)
Simocephalus serrulatus:	48 hr, LC$_{50}$:	2400 μg/l	(2127)
Daphnia pulex:	48 hr, LC$_{50}$:	2000 μg/l	(2127)

  —Fish:

Lepomis macrochirus:	48 hr, LC$_{50}$:	16600 μg/l	(2115)

2,4,5-trichlorophenoxy-α-valeric acid

[Structure: benzene ring with Cl at 2,4,5 positions and O-CH(COOH)-CH$_2$CH$_2$CH$_3$ substituent]

C. WATER POLLUTION FACTORS:
— Biodegradation: >81 days for ring cleavage in soil suspension (1827)

2,3,6-trichlorophenylacetic acid (chlorfenac; fenac; trifene)

[Structure: benzene ring with Cl at 2,3,6 positions and CH$_2$COOH substituent]

Use: herbicide (generally applied as the sodium salt)
A. PROPERTIES: white crystalline powder; m.p. 159–160°C; v.p. 8.5 × 10^{-3} mm at 100°C; solub. 200 ppm at 28°C
D. BIOLOGICAL EFFECTS:
— Crustaceans (sodium salt):

Gammarus lacustris:	96 hr, LC$_{50}$: 12000 μg/l	(2124)
Gammarus fasciatus:	no effect: 100,000 μg/l, 48 hr	(2125)
Daphnia pulex:	48 hr, LC$_{50}$: 4500 μg/l	(2127)
Simocephalus serrulatus:	48 hr, LC$_{50}$: 6600 μg/l	(2127)
Daphnia magna:	no effect: 100,000 μg/l, 48 hr	(2125)
Cypridopsis vidua:	no effect: 100,000 μg/l, 48 hr	(2125)
Asellus brevicaudus:	no effect: 100,000 μg/l, 48 hr	(2125)
Palaemonetes kadiakensis:	no effect: 100,000 μg/l, 48 hr	(2125)

— Insects (sodium salt):

Pteronarcys californica:	96 hr, LC$_{50}$: 55000 μg/l	(2128)

— Fish (sodium salt):

Lepomis:	48 hr, LC$_{50}$: 15000 μg/l	(2114)

— Mammals:
acute oral LD$_{50}$ (rat): 1780 mg/kg
acute dermal LD$_{50}$ (rabbit): >3160 mg/kg (1854)

2,4,6-trichlorophenyl-4'-nitrophenylether

[Structure: O$_2$N-phenyl-O-phenyl(2,4,6-trichloro)]

1146   1,2,3-TRICHLOROPROPANE

Use: herbicide for controlling various species of weeds in paddy fields.
A. PROPERTIES: solub. 0.764 ppm at 20°C; log P_{oct} 3.67 at 23°C
D. BIOLOGICAL EFFECTS:
  —Bioaccumulation:
    red snail (*Indophanorbis exustus*): 7 d/20 ppb, BCF 87
    topmouth gudgeon: 7 d/20 ppb, BCF 1109
    water sprit (*Ceratopteris thalictroides*): 7 d/20 ppb, BCF ~55   (1656)
  —Toxicity:
    daphnid: 48 hr LC_{50}: 40 ppm   (1658)
  —Fish: carp: 48 hr LC_{50}: 290 ppm   (1657)
    harlequin fish (*Rasbora heteromorpha*)

mg/l	24 hr	48 hr	96 hr	3 m (extrap.)	
LC_{10}(F)	1.4	0.66			
LC_{50}(F)	2.3	1.3	0.77	0.08	(331)

1,2,3-trichloropropane   (glyceroltrichlorohydrin; allyltrichloride; trichlorohydrin)
$CH_2ClCHClCH_2Cl$
A. PROPERTIES: colorless liquid; m.w. 147.44; m.p. -14°C; v.p. 156°C; v.p. 2 mm at 20°C, 4 mm at 30°C; v.d. 5.08; sp.gr. 1.417 at 15/4°C; sat. conc. 16 g/cu m at 20°C, 31 g/cu m at 30°C
B. AIR POLLUTION FACTORS: 1 mg/cu m = 0.16 ppm, 1 ppm = 6.13 mg/cu m
D. BIOLOGICAL EFFECTS:
  —Fish:
    guppy (*Poecilia reticulata*): 7 d LC_{50}: 42 ppm   (1833)

1,2,3-trichloropropene
$CHCl=CCl-CH_2Cl$
C. WATER POLLUTION FACTORS:
  —Waste water treatment:
    evaporation from water at 25°C of 1 ppm solution:
      50% after 49 min
      90% after >140 min   (313)

3,5,6-trichloropyridinol
D. BIOLOGICAL EFFECTS:
  —Fish:
    rainbow trout: log bioconcentration factor: 0.49   (193)

2,4-α-trichlorotoluene

D. BIOLOGICAL EFFECTS:
 —Fish:
  guppy (*Poecilia reticulata*) 14 d LC_{50}: 0.23 ppm (1833)

2,4,5-trichlorotoluene

D. BIOLOGICAL EFFECTS:
 —Fish:
  guppy (*Poecilia reticulata*) 7 d LC_{50}: 1.7 ppm (1833)

1,1,2-trichloro-1,2,2-trifluoroethane ("UCON-113; "Arklone" R-113")
$CFCl_2CF_2Cl$
A. PROPERTIES: colorless liquid; m.w. 187.38; m.p. $-35°C$; b.p. $48°C$; v.p. 270 mm at $20°C$, 400 mm at $30°C$; v.d. 6.47; sp.gr. 1.56 at $25°C$ sat. conc. 2,754 g/cu m at $20°C$, 3,945 g/cu m at $30°C$
B. AIR POLLUTION FACTORS: 1 mg/cu m = 0.13 ppm, 1 ppm = 7.79 mg/cu m
 —Odor: characteristic: quality: sweet
          hedonic tone: pleasant to unpleasant
 T.O.C.: abs. perc. lim.: 45.0 ppm
         50% recogn.: 68.0 ppm
         100% recogn.: 135 ppm
         O.I. 100% recogn.: 54
D. BIOLOGICAL EFFECTS:
 —Mammalia:
  guinea pig: inhalation: LC_{50}: >120.000 ppm, 2 hr
  rat: inhalation: 11,000 ppm, 2 hr/day, 5 days/w, 2 years
              no effect: 2,500 ppm, 7 hr/day, 30 days (54)
 —Man: no effect level: inhalation: 1,000 ppm, 6 hr/day, 5 days
                                   1,500 ppm, $2\frac{3}{4}$ hr
  10,000 ppm inhaled is expired entirely after 3 min (54)

trichlorphon *see* dimethyl-(2,2,2-trichloro-1-hydroxyethyl)phosphonate

triclosan *see* 5-chloro-2-(2,4-dichlorophenoxy)phenol

n-tricosane
$CH_3(CH_2)_{21}CH_3$
 Use: organic synthesis.
A. PROPERTIES: sp.gr. 0.779 at $48°C$; v.p. $234°C$ at 15 min; m.p. $48°C$
C. AIR POLLUTION FACTORS:
 —Ambient air quality:

organic fraction of suspended matter:
  Bolivia at 5200 m altitude (Sept.–Dec. 1975): 0.43–0.65 µg/1000 cu m
  Belgium, residential area (Jan.–April 1976):   1.1–4.5 µg/1000 cu m   (428)
in the average American urban atmosphere–1963:
  620 µg/g airborne particulates or
  77 ng/g cu m air   (1293)
glc's in Botrange (Belgium)–woodland at 20–30 km from industrial area: June–July 1977: 0.35; 0.69 ng/cu m ($n = 2$)
glc's in Wilrijk (Belgium)–residential area: Oct.–Dec. 1976: 1.59–18.05 ng/cu m ($n = 2$)   (1233)
glc's in residential area (Belgium)–Oct. 1976:
  in particulate sample: 4.75 ng/cu m
  in gas phase sample: 3.38 ng/cu m   (1289)

tricosanoic acid
$CH_3(CH_2)_{21}COOH$
A saturated fatty acid not normally found in natural fats or oils.
A. PROPERTIES: m.p. 79.1°C
C. AIR POLLUTION FACTORS:
  –Ambient air quality:
  organic fraction of suspended matter:
    Bolivia at 5200 m altitude (Sept.–Dec. 1975): 0.17–0.42 µg/1000 cu m
    Belgium, residential area (Jan.–April 1976):   1.8–5.5 µg/1000 cu m   (428)
  glc's in residential area (Belgium), Oct. 1976:
    in particulate sample: 3.23 ng/cu m
    in gas phase sample: –   ng/cu m   (1289)
  glc's Botrange (Belgium)–woodland at 20–30 km from industrial area: June–July 1977: 0.98; 0.99 ng/cu m ($n = 2$)
  glc's in Wilrijk (Belgium)–residential area: Oct.–Dec. 1976: 2.08; 5.25 ng/cu m ($n = 2$)   (1233)

tri-o-cresylphosphate (o-tolylphosphate; TCP)

Use: plasticizer for PVC; polystyrene; nitrocellulose; fire retardant for plastics; waterproofing; additive to extreme pressure lubricants; hydraulic fluid and heat exchange medium
A. PROPERTIES: colorless liquid; m.w. 368.36; m.p. -25/-30°C; b.p. 420°C; v.d. 12.7; sp.gr. 1.162 at 25/25°C

D. BIOLOGICAL EFFECTS:
  —Molluscs:
    Crassostrea virginica (American oyster), egg: TLm: 600 ppb, 48 hr static lab bioassay
    Crassostrea virginica (American oyster), larvae: TLm: 1×10^3 ppb, 14 day static lab bioassay (2324)
  —Fish:
    Lepomis macrochirus: static bioassay in fresh water at 23°C, mild aeration applied after 24 hr:

material added ppm	24 hr	% survival after 48 hr	72 hr	96 hr	best fit 96 hr LC_{50} ppm
10,000	73	44	33	33	
7,900	85	50	15	15	7,000
5,000	100	100	90	80	
3,200	100	100	100	100	

Menidia beryllina: static bioassay in synthetic seawater at 23°C, mild aeration applied after 24 hr:

material added ppm	24 hr	% survival after 48 hr	72 hr	96 hr	best fit 96 hr LC_{50} ppm
10,000	100	40	40	40	
5,600	90	90	90	90	8,700
3,200	100	100	100	100	
1,800	100	100	100	10C	(352)

tri-*p*-cresylphosphate

<chemical structure: tri-p-cresylphosphate>

A. PROPERTIES: solub. 0.074 mg/l at 24°C (practical grade) (1666)
D. BIOLOGICAL EFFECTS:
  —Fish:
    bluegill: 7 d BCF 1 589 (1545)

tricyclo(5,2,1,0)-3,8-decadiene *see* α-dicyclopentadiene

n-tridecane
$CH_3(CH_2)_{11}CH_3$
A. PROPERTIES: colorless liquid; m.p. -5.45°C; v.p. 225.5°C; sp.gr. 0.755 at 20/4°C; solub. 0.013 mg/l at 25°C

B. AIR POLLUTION FACTORS:

—Odor threshold conc.: 42 mg/cu m (737)

first order evaporation constant of n-tridecane in 3 mm layer No. 2 fuel oil, in darkened room at windspeed of 21 km/hr at

$5°C$: 1.57×10^{-4} min^{-1}
$10°C$: 1.20×10^{-4} min^{-1}
$20°C$: 2.46×10^{-4} min^{-1}
$30°C$: 5.72×10^{-4} min^{-1}

tridecanoic acid (tridecylic acid; tridecoic acid)
$CH_3(CH_2)_{11}COOH$

Use: organic synthesis; medical research.
A. PROPERTIES: m.p. 44°C; sp.gr. 0.8458 at 80/4°C; b.p. 312°C
B. AIR POLLUTION FACTORS:
—Ambient air quality:
glc's in residential area (Belgium)—Oct. 1976:
in particulate sample: — ng/cu m
in gas phase sample: 7.71 ng/cu m (1289)
C. WATER POLLUTION FACTORS:
—Odor threshold conc. (detection): 10.0 mg/kg (886)

2-tridecanone (hendecylmethylketone)
$CH_3CO(CH_2)_{10}CH_3$
A. PROPERTIES: crystals; m.w. 198.34; m.p. 28°C; b.p. 263°C; sp.gr. 0.823 at 28°C
C. WATER POLLUTION FACTORS:
—Waste water treatment:
A.S.: after 6 hr: 3.3% of ThOD
           12 hr: 8.3% of ThOD
           24 hr: 19.5% of ThOD (88)

tridecoic acid see tridecanoic acid

tridecyclic acid see tridecanoic acid

triethanolamine (2,2′,2″-trihydroxy-triethylamine; TEA)
$N(CH_2CH_2OH)_3$

Use: fatty acid soaps used in drycleaning, cosmetics, household detergents, and emulsions; wool scouring; textile antifume agent and water-repellent; corrosion inhibitor; softening agent; plasticizer; insecticide; chelating agent; rubber accelerator.
A. PROPERTIES: pale yellow liquid; m.w. 149.1; m.p. 21.1°C; b.p. 360°C; v.p. <0.01 mm at 20°C; v.d. 5.14; sp.gr. 1.12 at 20/4°C; log P_{oct} $-1.32/-1.75$ (calculated)
B. AIR POLLUTION FACTORS: 1 mg/cu m = 0.164 ppm, 1 ppm = 6.1 mg/cu m
C. WATER POLLUTION FACTORS:
—BOD_5: 0.01 std. dil. sew. (287)
         nil (255)
         nil% of ThOD (27)
10 days: 0.8% of ThOD
15 days: 2.6% of ThOD
20 days: 6.2% of ThOD (79)

—$BOD_{10}^{20°C}$: 0.8% ThOD at 2.5 mg/l in mineralized dilution water with settled
         sewage seed   (405)
—BOD_5: 0.02 NEN 3235-5.4 (TEA 85%, SHELL)   (277)
         0.17 adapted sew. (TEA 85%, SHELL)   (277)
         0.03 (TEA "commercial"-SHELL) NEN 3235-5.4   (277)
         0.90 (TEA "commercial"-SHELL) NEN 3235-5.4   (277)
—COD: 1.50 TEA "commercial" and 85%-SHELL   (277)
—BOD_{10}: nil std. dil. sew.   (256)
—ThOD: 2.04   (27)
—Impact on biodegradation processes:
   at 100 mg/l, no inhibition of NH_3 oxidation by *Nitrosomonas* sp.   (390)
—Waste water treatment:
   Activated carbon: adsorbability: 0.067 g/g C; 33.0% reduction, infl.: 1,000
      mg/l, effl.: 670 mg/l   (32)

methods	temp °C	days observed	feed mg/l	days acclim.	% removed
NFG, BOD	20	1-10	200-1,000	365+	nil
RW, BOD	20	1-10	50	28	70

                                                                               (93)

D. BIOLOGICAL EFFECTS:
   —Toxicity threshold (cell multiplication inhibition test):
     bacteria (*Pseudomonas putida*):   >10.000 mg/l   (1900)
     algae (*Microcystis aeruginosa*):   47 mg/l   (329)
     green algae (*Scenedesmus quadricauda*):   1.8 mg/l   (1900)
     protozoa (*Entosiphon sulcatum*):   56 mg/l   (1900)
     protozoa (*Uronema parduczi Chatton-Lwoff*): >10.000 mg/l   (1901)
   —Bacteria: *Pseudomonas*: LD_0: 10 g/l
   —Algae: *Scenedesmus*: LD_0: 100 mg/l
          Colpoda: LD_0: 160 mg/l
   —Arthropoda: *Daphnia*: LD_0: 2.5 g/l   (30)
   —Fish: goldfish: (TEA "85%"): LD_{50} (24 hr) = 3500 mg/l at pH 10.3
                                              = 75000 mg/l at pH 7
           (TEA "commercial"): LD_{50} (24 hr) = >5000 mg/l   (277)
   —Mammalia: rat: acute oral LD_{50}: 9.11 g/kg
              repeated oral doses: no effect: 0.08 g/kg daily   (211)

triethylamine
$(C_2H_5)_3N$
   Use: catalytic solvent in chemical synthesis; accelerator activators for rubber;
   wetting, penetrating and waterproofing agents of quaternary ammonium types;
   curing and hardening of polymers; corrosion inhibitor; propellant.
A. PROPERTIES: colorless liquid; m.w. 101.19; m.p. -115°C; b.p. 90°C; v.p. 50 mm at
   20°C; v.d. 3.48; sp.gr. 0.723 at 25/4°C; solub. 15,000 mg/l at 20°C, 19,700 mg/l
   at 65°C; log P_{oct} 1.44
B. AIR POLLUTION FACTORS: 1 mg/cu m = 0.24 ppm, 1 ppm = 4.29 mg/cu m
   —Odor: characteristic: quality: fishy, amine
                          hedonic tone: unpleasant to pleasant

T.O.C.: abs. perc. lim.: <0.09 ppm
50% recogn.:   0.28 ppm
100% recogn.: 0.28 ppm
O.I. 100% recogn.: 253,571   (19)
—Control methods:
wet scrubber: water at pH 8.5: outlet: 60–70 odor units/scf
$KMnO_4$ at pH 8.5: outlet: 50–65 odor/units/scf   (115)

C. WATER POLLUTION FACTORS:
—Impact on biodegradation processes:
inhibition of degradation of glucose by *Pseudomonas fluorescens* at: 600 mg/l
inhibition of degradation of glucose by *E. coli* at: >1000 mg/l   (293)
NH_3 oxidation by *Nitrosomonas* sp.:
at 100 mg/l: 35% inhibition
127 mg/l: ~50% inhibition   (390)
—Biodegradation:
degradation by *Aerobacter*: 200 mg/l at 30°C
parent: 100% in 28 hr
mutant: 100% in 11 hr   (152)

D. BIOLOGICAL EFFECTS:
—Bacteria: *E. coli*: no effect: 1 g/l
—Algae: *Scenedesmus*: LD_0: 1 mg/l
—Arthropoda: *Daphnia*: LD_0: 200 mg/l   (30)
—Fish: creek chub: LD_0: 50 mg/l, 24 hr in Detroit river water
LD_{100}: 80 mg/l, 24 hr in Detroit river water   (243)
—Mammalia: rat: acute oral LD_{50}: 0.46 g/kg
inhalation: 1/6: 1,000 ppm, 4 hr   (211)

triethyleneglycol (2,2(ethylenedioxy)diethanol; glycol bis(hydroxyethyl)ether; 3,6-dioxaoctane-1,8 diol; TEG)
$HOCH_2(C_2H_4O)_2CH_2OH$

Use: solvent for nitrocellulose; lacquers; organic synthesis; bactericide; humectant in printing inks; textile conditioner; fungicide.

A. PROPERTIES: colorless to pale straw colored liquid; m.w. 150.2; m.p. -4/-7°C; b.p. 287.4°C; v.p. <0.001 mm at 20°C; v.d. 5.17; sp.gr. 1.1254; log P_{oct} -2.08/ 1.32 (calculated)
B. AIR POLLUTION FACTORS: 1 ppm = 6.14 mg/cu m
C. WATER POLLUTION FACTORS:
—BOD_5: 0.03 NEN 3235-5.4   (277)
1.4% of ThOD
—BOD_{10}: 0.50 std. dil. sew.   (258)
10 days: 3.7% of ThOD
15 days: 11.5% of ThOD
20 days: 17.0% of ThOD   (79)
—COD: 1.57 NEN 3235-5.3   (277)
—Biodegradation: adapted A.S.—product as sole carbon source—97.7% COD removal at 27.5 mg/COD/g dry inoculum/hr   (327)
—Waste water treatment:

A.C.: adsorbability: 0.105 g/g C; 52.3% reduction, infl.: 1,000 mg/l, effl.: 477 mg/l (32)
A.S.: after 6 hr: 3.8% of ThOD
12 hr: 7.0% of ThOD
24 hr: 11.0% of ThOD (88)

methods	temp °C	days observed	feed mg/l	days acclim.	% removed
NFG, BOD	20	1–10	100	365 + P	6–16
NFG, BOD	20	1–10	200–600	365 + P	nil
NFG, BOD	20	1–10	800	365 + P	8
AS, BOD	11	$\frac{1}{3}$–5	333	30 +	nil

(93)

D. BIOLOGICAL EFFECTS:
 —Toxicity threshold (cell multiplication inhibition test):
  bacteria (*Pseudomonas putida*): 320 mg/l (1900)
  algae (*Microcystis aeruginosa*): 3600 mg/l (329)
  green algae (*Scenedesmus quadricauda*): >10,000 mg/l (1900)
  protozoa (*Entosiphon sulcatum*): >10,000 mg/l (1900)
  protozoa (*Uronema parduczi Chatton-Lwoff*): >10,000 mg/l (1901)
 —Fish:
  goldfish: 24 hr LD_{50} = >5000 mg/l–modified ASTM D 1345 (277)
  guppy (*Poecilia reticulata*): 7 d LC_{50}: 62,600 ppm (1833)
  Lepomis macrochirus: static bioassay in fresh water at 23°C, mild aeration applied after 24 hr:

material added ppm	% survival after				best fit 96 hr LC_{50} ppm
	24 hr	48 hr	72 hr	96 hr	
10,000	100	100	100	100	>10,000
7,900	100	100	100	100	

  Menidia beryllina: static bioassay in synthetic seawater at 23°C: mild aeration applied after 24 hr:

material added ppm	% survival after				best fit 96 hr LC_{50} ppm
	24 hr	48 hr	72 hr	96 hr	
10,000	100	100	100	100	>10,000

(352)

 —Mammalia:
  guinea pig: single oral dose: LD_{50}: 14.6 g/kg; 7.9 ml/kg
  rat: single oral dose: LD_{50}: 22.06 g/kg; 16.8 ml/kg
  mouse: single oral dose: LD_{50}: 18.7 ml/kg
  rabbit: single oral dose: LD_{50}: 8.4 ml/kg
  rat: repeated oral dose: no effect: 3–4 g/kg/day, 2 years
           5–8 g/kg/day, 30 days (211)
 —Man: very low acute and chronic toxicity (211)

triethyleneglycoldi(2-ethylhexanoate)
$C_7H_{15}COOCH_2(CH_2OCH_2)_2CH_2OOCC_7H_{15}$

A. PROPERTIES: m.w. 402.56; m.p. -58°C; b.p. 218°C at 5 mm; sp.gr. 0.9679 at 20°/20°C; v.p. 1.9 mm at 200°C; v.d. 13.9

1154 TRIETHYLENEGLYCOLMONOETHYLETHER

C. WATER POLLUTION FACTORS:
 —Water quality:
  in Delaware river (U.S.A.): conc. range: winter: 0.6–1 ppb
  summer: n.d. (1051)

triethyleneglycolmonoethylether ("trioxitol"; ethyltriglycol)
$CH_3-CH_2(O-CH_2-CH_2)_3OH$
A. PROPERTIES: m.w. 178.22; m.p. $-18.7°C$; b.p. 245–260°C; sp.gr. 1.025 at 20/20°C
C. WATER POLLUTION FACTORS:
 —BOD_5: 0.05 NEN 3235-5.4 (277)
 —COD: 1.84 NEN 3235-5.3 (277)
D. BIOLOGICAL EFFECTS:
 —Fish: goldfish: LD_{50} (24 hr): >5000 mg/l—modified ASTM D 1345 (277)

triethylleadchloride
$(CH_3CH_2)_3PbCl$
D. BIOLOGICAL EFFECTS:
 —Bioaccumulation:
  shrimp: accumulation after 96 hr exposure at 5.8 mg/l (96 hr LC_{50}):
   concentration factor $\left(\dfrac{\text{animal tissue (mg/kg wet wt)}}{\text{water mg/l}}\right)$: 2× (575)
  mussel (*Mytilus edulis*): accumulation experiment:

	control	triethylleadchloride		
exposure conc. (mg/l)	0	0.01	0.05	0.10
exposure time (days)	35	9	35	21
time to equilibrium (days)	–	–	6	5
tissue conc. (mg/kg dry wt): gill	6	20	52	60
at equilibrium: digestive gland	6	11	33	56
gonad	9	10	15	42
foot	3	6	18	35
mean	6	18	29	48
concentration factors:				
(a) tissue (wet wt)/water	–	90	66	60
(b) exposed/control	–	0.2	4.8	8.0
depuration:				
half-life (days): gill	–	–	2	–
digestive gland	–	–	12	5
gonad	–	–	4	2
foot	–	–	5	3
mean	–	–	4	3
total experiment duration (days)	62	9	62	48

(575)

  mussel (*Mytilus edulis*): accumulation after 96 hr exposure at 1.1 mg/l (96 hr LC_{50}):
   concentration factor $\left(\dfrac{\text{animal tissue (mg/kg wet wt)}}{\text{water mg/l}}\right)$: 10× (575)

—Fish:
dab (*Limanda limanda*): accumulation:

	control	triethylleadchloride	
exposure conc. (mg/l)	–	0.1	0.2
exposure time (days)	41	41	41
time to equilibrium (days): liver	–	n.d.	n.d.
muscle	–	n.d.	n.d.
tissue conc. (mg/kg dry wt): liver	1.9	9	21
after 41 days muscle	1.4	10	18
mean	1.6	10	20
concentration factors at 41 days:			
(a) tissue (wet wt)/water	–	12	13
(b) exposed/control	–	6	12
depuration:			
half-life (days): liver	–	n.d.	n.d.
muscle	–	56	n.d.
total experiment duration (days)	97	97	97

n.d. = could not be determined (575)

plaice: accumulation after 96 hr exposure at 1.7 mg/l (96 hr LC_{50}):

concentration factor: $\left(\dfrac{\text{animal tissue (mg/kg wet wt)}}{\text{water mg/l}} \right)$: 2×   (575)

—Toxicity:
Algae: 6 hr EC_{50} (photosynthetic activity): 0.1 mg/l (Pb)   (573)
Mussel: 96 hr LC_{50}: 1.1 mg/l (Pb)   (573)
Shrimp: 96 hr LC_{50}: 5.8 mg/l (Pb)   (573)
Fish: plaice: 96 hr LC_{50}: 1.7 mg/l (Pb)   (573)
       rainbow trout: 96 hr LC_{50}: 9 ppm (Pb)   (572)

—Mammals:
rat (no anion specified):
  intravenous injection: LD_{50}: 8 mg Pb/kg body wt   (568)
  intraperitoneal: LD_{50}: 5 mg Pb/kg body wt   (570)

1,1,1-trifluoro-2,6-dinitro-N,N-dipropyl-*p*-toluidine   *see* trifluralin

(trifluoromethyl)chloroaniline
C. WATER POLLUTION FACTORS:
—In Delaware river (U.S.A.): conc. range: winter: trace–2 ppb
                                                 summer: n.d.   (1051)

(trifluoromethyl)chloronitrobenzene
C. WATER POLLUTION FACTORS:
—Water quality:
in Delaware river (U.S.A.): conc. range: winter: 2–3 ppb
                                                   summer: n.d.   (1051)

3-trifluoromethyl-4-nitrophenol (α,α,α-trifluoro-4-nitro-*m*-cresol)

Use: to exterminate lampreys, especially in the Great Lakes. It is placed in tributary streams, where it kills the lamprey larvae.

A. PROPERTIES: crystals; m.p. 74–76°C
D. BIOLOGICAL EFFECTS:
  —Fish:
   sea lamprey:
     larvae: 96-hr LC_{50}: 0.35–1.30 mg/l (S)
     burrowed larvae: 96-hr LC_{50}: 1.68 mg/l (F)
     free swimming larvae: 96-hr LC_{50}: <1.48 mg/l (F)
   rainbow trout: 96-hr LC_{50}: 6.10 mg/l (F)
   brook trout: 96-hr LC_{50}: 5.95 mg/l (F) (1506)
   sea lamprey: 24-hr LC_{50}: 0.78 mg/l (S) (1201)
   rainbow trout: 24-hr LC_{50}: 3.85 mg/l (S)
   coho salmon: static bioassay:
     fingerlings: 96 hr TLm: 2.7 mg/l at 12°C
     green eggs: 96 hr TLm: 0.57 mg/l at 12°C
     green eggs: 192 hr TLm: 0.57 mg/l at 12°C
   chinook salmon: static bioassay:
     fingerlings: 96 hr TLm: 2.2 mg/l at 12°C
     green eggs: 96 hr TLm: 1.2 mg/l at 12°C
     green eggs: 192 hr TLm: 1.1 mg/l at 12°C
   lake trout: static bioassay:
     fingerlings: 96 hr TLm: 1.4 mg/l at 12°C
     green eggs: 96 hr TLm: 2.1 mg/l at 12°C
     green eggs: 192 hr TLm: 1.4 mg/l at 12°C
   brown trout (*Salmo trutta*): static bioassay:
     green eggs: 96 hr TLm: 1.52 mg/l at 12°C
             192 hr TLm: 1.39 mg/l at 12°C
   coho salmon: static bioassay:
     green eggs: 192 hr TLm: 3.65 mg/l at 12°C
   brown trout: static bioassay:
     green eggs: 192 hr TLm: >5.00 mg/l at 12°C (447)
   rainbow trout:

life stage	24 hr LC_{50} (mg/l)–hard water (96% pure TFM)
green eggs (40 hr)	19.6
eyed eggs (day 16)	>40
sac fry (day 27)	10.8
swim-up fry (day 46)	11.3

fry	19.2	
fingerlings	14.2	(1912)

rainbow trout:

life stage	24 hr LC_{50} (mg/l)–hard water (commercial TFM)	
green eggs (6 hr)	4.0	
day 10	43.0	
day 21	14.3	
day 32	11.3	(1911)

—Bioaccumulation:

Accumulation, half-life, and rate of loss of total TFM residue in several plant and animal components exposed for 24 hr to a mean measured concentration of 8.97 mg/TFM:

	BCF	half-life (h)
Pool species:		
annelid worms	50.4	5295.0
isopod		
Asellus militaris	16.8	194.4
amphipod		
Gammarus pseudolimnaeus	18.7	26.2
crayfish		
Orconectes propinquus	1.1	7.2
mayfly		
Hexagenia sp.	8.7	38.6
snail, *Physa* sp.	9.2	23.2
fingernail clam		
Pisidium sp.	13.6	7.9
green algae, *Cladophora* and *Stigeoclonium*	6.8	65.2
macrophyte		
Ceratophyllum demersum	12.2	437.5
Elodea canadensis	5.7	87.4
sediment	3.8	171.7
Riffle species:		
caddisfly,		
Glossosoma sp.	34.0	7.9
Limnephilus sp.	19.4	14.0
Brachycentrus americanus	62.2	19.5
Brachycentrus cases only	95.5	23.8
green algae, *Cladophora* and *Stigeoclonium*	11.8	25.8
moss, *Fontinalis* sp.	4.3	16.3
aufwuchs (5 cm^2) (wet wt)	0.7	24.3
leaf disks (20 mm diam)	9.6	11.0

α,α,α-**trifluoro-4-nitro-m-cresol** see 3-trifluoromethyl-4-nitrophenol

trifluralin (1,1,1-trifluoro-2,6-dinitro-N,N-dipropyl-*p*-toluidine; treflan)

$$\text{H}_7\text{C}_3-\text{N}-\text{C}_3\text{H}_7$$

[Structure: 2,6-dinitro-4-trifluoromethyl-N,N-dipropylaniline ring with O_2N and NO_2 ortho to N, and CF_3 para]

Use: herbicide, especially for cotton plant.

A. PROPERTIES: yellow orange solid; m.p. 48.5–49°C; b.p. 139–140°C at 4.2 mm; m.w. 335; v.p. 1.99×10^{-4} mm at 29.5°C; solub. 4 ppm at 27°C

C. WATER POLLUTION FACTORS:
—Persistence in soils: half lives:
    in irrigated soils: in Texas: 3 weeks   (1790)
                    in Tennessee: 5 weeks   (1791)
    in Nova Scotia (Canada): 126–>190 days in two sandy loam soils following application in May   (1789)
    under greenhouse conditions in a sandy loam soil: 50 days   (1792)
    mean overwinter loss = 38.1% in two sandy loam soils in Nova Scotia following application in November   (1789)

D. BIOLOGICAL EFFECTS:
—Bioaccumulation:
    BCF in fish exposed to 1.8 µg/l
        sauger: 5800 (half-life: 22–31 days)
        shorthead redhorse: 2800 (half-life: 17–57 days)
        golden redhorse: 1800 (half-life: 23 days)
        minnow sp.: 6000 (half-life: 3 days, in lab.)   (1914)
    BCF estimated by several methods:
        fathead minnow: 3261 (kinetic test)   (1915)
        fathead minnow: 1060 (chronic exposure)   (1916)
        mosquito fish: 1294 (predicted from solubility correlation)   (1914)
        rainbow trout: 1030 (predicted from partition correlation)   (1914)
—Toxicity:
—Algae:

technical acid	*Chlorococcum* sp.	5×10^5 ppb	50% decrease in O_2 evolution	f
technical acid	*Chlorococcum* sp.	2.5×10^3 ppb	50% decrease in growth	Measured as ABS (525 mu) after 10 days
technical acid	*Dunaliella tertiolecta*	$>5 \times 10^5$ ppb	50% decrease in O_2 evolution	f
technical acid	*Dunaliella tertiolecta*	5×10^3 ppb	50% decrease in growth	Measured as ABS (525 mu) after 10 days
technical acid	*Isochrysis galbana*	4×10^5 ppb	50% decrease in O_2 evolution	f

technical acid	*Isochrysis galbana*	2.5×10^3 ppb	50% decrease in growth	Measured as ABS (525 mu) after 10 days
technical acid	*Phaeodactylum tricornutum*	$>5 \times 10^5$ ppb	50% decrease in O_2 evolution	f
technical acid	*Phaeodactylum tricornutum*	2.5×10^3 ppb	50% decrease in growth	Measured as ABS (525 mu) after 10 days

(2348)

—Crustaceans:
Gammarus lacustris:	96 hr LC_{50}:	2200 µg/l	(2124)
Gammarus fasciatus:	96 hr LC_{50}:	1000 µg/l	(2125)
Daphnia magna:	48 hr LC_{50}:	560 µg/l	(2125)
Daphnia pulex:	48 hr LC_{50}:	240 µg/l	(2127)
Simocephalus serrulatus:	48 hr LC_{50}:	450 µg/l	(2127)
Cypridopsis vidua:	48 hr LC_{50}:	250 µg/l	(2125)
Asellus brevicaudus:	48 hr LC_{50}:	200 µg/l	(2125)
Palaemonetes kadiakensis:	48 hr LC_{50}:	1200 µg/l	(2125)
Orconectes nais:	48 hr LC_{50}:	50000 µg/l	(2125)

—Insects:
Pteronarcys californica: 96 hr LC_{50}:   3000 µg/l   (2128)

—Fish:
trifluralin (*Cyprinodon variegatus*): 96 hr LC_{50}(FT): 190 µg/l   (1472)
channel catfish: 96 hr LC_{50}(S): 417 µg/l   (1202)
fathead minnow: flow through bioassay:
   incipient TLm: 0.115 mg/l at 25°C
   MATC: 0.0019 mg/l   (444)
bluegill: 48 hr LC_{50}: 0.019 ppm
rainbow trout: 48 hr LC_{50}: 0.011 ppm   (1878)

—Mammals:
female CD-1 mice: % mortality following a single intraperitoneal injection of trifluralin in corn oil:

volume of corn oil injected ml/kg body wt	dose of trifluralin (g/kg body wt)						
	10.0	7.5	5.0	3.0	1.5	.75	.375	0
3.33	–	–	–	0	0	0	0	0
5.0	–	–	–	10	0	0	0	0
8.33	–	70	10	0	0	0	0	0
11.67	100	100	80	10	20	0	0	0
16.67	100	80	50	0	0	–	–	0
20.0	100	100	80	10	–	–	–	0
23.3	90	100	80	50	–	–	–	0
averaged % killed	97.5	90.0	60.0	11.4	4.0	0	0	0

female CD-1 mice: % mortality following a single gastric intubation of trifluralin in corn oil:

volume of corn oil intubated ml/kg body wt.	dose of trifluralin (g/kg body wt.)					
	10.0	7.5	5.0	3.0	2.5	1.5	0
<3.0	–	–	–	0	–	0	–

3.3	40	0	6.7	13.3	0	0	0
8.3	–	40	40	26.7	40	0	–
10.0	–	–	–	–	40	13.3	0
11.7	–	–	–	–	–	–	0

acute oral LD_{50} (mice): 5 g/kg
(rat): >10 g/kg
in diet: (rats) 2 years at 2000 ppm: no ill effects
(dogs) 3 years at 400 ppm: no ill effects (1855)

triforine (N,N'-(1,4-piperazinediyl-*bis*(2,2,2-trichloroethylidene)*bis*(formamide))

Use: fungicide.
A. PROPERTIES: m.p. 155°C; solub. 27–29 ppm
C. WATER POLLUTION FACTORS:
  —Chemical degradation: chemical $\frac{1}{2}$ t at 30 mg/l: 3 days

(I) = N-(1-formamido-2,2,2-trichloro-ethyl)-N'-2,2-dihydroxy-acetyl-piperazine
(II) = bis-glyoxylic piperazine dihydrate
(III) = N-glyoxylic piperazine

suggested pathway of hydrolytic breakdown of triforine in aqueous solution (30 mg/l) kept at room temperature for one year (1627)

D. BIOLOGICAL EFFECTS:
 —Acute oral LD_{50} (rat) >16 g/kg (1854)

triglycine *see* nitrilotriacetic acid

triglycollamic acid *see* nitrilotriacetic acid

γ-trihydroxybenzene *see* pyrogallol

1,3,5-trihydroxybenzene *see* phloroglucinol

3,4,5-trihydroxybenzoic acid *see* gallic acid

1,3,5-trihydroxyethylhexahydrotriazine

$$CH_2OH-CH_2-N\underset{\underset{N-CH_2CH_2OH}{}}{\overset{\overset{CH_2CH_2OH}{|}}{N}}$$

C. WATER POLLUTION FACTORS:
 —Biodegradation:
  after 3 weeks adaptation, at 70 mg/l at 22°C, product is 100% degraded under aerobic and anaerobic conditions, as sole carbon source or with synthetic sewage (512)

trihydroxypropane *see* glycerol

2,2′,2″-trihydroxytriethylamine *see* triethanolamine

triketohydrindene hydrate *see* ninhydrin

trimethylacetic acid *see* pivalic acid

trimethylamine (TMA)
$(CH_3)_3N$
 Use: organic synthesis; warning agent for natural gas; manufacture of disinfectants; flotation agent; plastics.
A. PROPERTIES: colorless gas; m.w. 59.11; m.p. -117/-124°C; b.p. 3.5°C; v.p. 760 mm at 2.9°C, 1.9 atm at 20°C, 2.6 atm at 30°C; v.d. 2.04; sp.gr. 0.662 at -5°C; log P_{oct} 0.27
B. AIR POLLUTION FACTORS: 1 mg/cu m = 0.41 ppm, 1 ppm = 2.46 mg/cu m
 —Odor: characteristic: quality: fishy, pungent

2,4,6-TRIMETHYLANILINE

[odor thresholds chart: detection ~1 mg/cu m; recognition ~10⁻⁴ and ~1–10 mg/cu m; not specified ~10⁻⁴–10⁻³ and ~10⁻¹ mg/cu m]

(2; 57; 73; 741; 801; 821; 840)

odor Index at 20°C = 493,000   (316)

—Manmade sources:
  glc's: 140 m downwind from a fish meal plant: 0.59 ppb   (50)
  in municipal sewer air: 10–50 ppb   (212)

—Control methods:
  wet scrubber: water at pH 8.5: outlet: 2,700 odor units/scf
               $KMnO_4$ at pH 8.5: outlet: 20 odor units/scf   (115)

—Sampling and analysis:
  photometry: min. full scale: 1,900 ppm   (53)

C. WATER POLLUTION FACTORS:
  —Odor threshold: average: 1.7 mg/l
                 range: 0.04–5.17 mg/l   (97)
  —Impact on biodegradation processes: 50% inhibition of nitrification at 590 mg/l
    (411)

2,4,6-trimethylaniline *see* mesidine

1,2,3-trimethylbenzene (hemimellitene; hemellitene)

[structural diagram: benzene ring with three adjacent CH_3 groups]

Manmade sources: in gasoline (high octane number): 0.73 wt %.   (387)

A. PROPERTIES: colorless liquid; m.w. 120.19; m.p. <-15°C; b.p. 176°C; sp.gr. 0.89 at 20°C

B. AIR POLLUTION FACTORS: 1 mg/cu m = 0.20 ppm, 1 ppm = 5.00 mg/cu m
  —Atmospheric reactions:
    reactivity: HC cons.: ranking: 6   (63)
    estimated lifetime under photochemical smog conditions in S.E. England: 1.5 hr
      (1699; 1707)

  —Ambient air quality:
    glc's in The Hague (The Netherlands), 1974.10.11: avg.: 2 ppb ($n = 12$)
                 max.: 5 ppb   (1231)

1,2,4-trimethylbenzene (pseudocumene)

dl-2,6,6-TRIMETHYLBICYCLO-(3,1,1)-HEPT-2-ENE

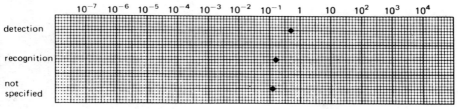

Use: manufacture of trimellitic anhydride, dyes, pharmaceuticals, and pseudocumidine.

Manmade sources: in gasoline (high octane number): 4.9 wt %. (387)

A. PROPERTIES: colorless liquid; m.w. 120.19; m.p. 61/-57°C; b.p. 169°C; v.d. 4.15; sp.gr. 0.88 at 20°C; solub. 57 mg/l at 20°C

B. AIR POLLUTION FACTORS: 1 mg/cu m = 0.20 ppm, 1 ppm = 5.00 mg/cu m

odor thresholds   mg/cu m

	10^{-7}	10^{-6}	10^{-5}	10^{-4}	10^{-3}	10^{-2}	10^{-1}	1	10	10^2	10^3	10^4
detection								◆				
recognition								◆				
not specified								◆				

(610; 611; 637)

—Atmospheric reactions:
  reactivity: HC. conc.: ranking: 6
    NO ox.: ranking: 0.6–0.7 (63)
  estimated lifetime under photochemical smog conditions in S.E. England: 0.9 hr
    (1699; 1707)

—Ambient air quality:
  glc's in The Netherlands:
    tunnel Rotterdam: 1974.10.12: avg.: 15 ppb ($n = 12$)
                                   max.: 26 ppb
    The Hague: 1974.10.11: avg.: 10 ppb ($n = 12$)
                                   max.: 29 ppb
    Roelofarendsveen: 1974. 9.11: avg.:   1.5 ppb ($n = 12$)
                                   max.:   2.0 ppb   (1231)

—Manmade sources:
  in exhaust of diesel engine: 0.4% of emitted HC.s (72)

C. WATER POLLUTION FACTORS:
  —Reduction of amenities: T.O.C. = 0.5 mg/l (295)

D. BIOLOGICAL EFFECTS:
  —Mammalia:
    rat: inhalation: initial slight eye and nose irritation, blood test normal, autopsy: organs normal: 1,000 ppm, 15 × 6 hr (65)

1,3,5-trimethylbenzene *see* mesitylene

dl-2,6,6-trimethylbicyclo-(3,1,1)-hept-2-ene *see* dl-α-pinene

2,2,3-trimethylbutane (isopropyltrimethylmethane; triptane)
$CH_3C(CH_3)_2C(CH_3)CH_3$
  Use: organic synthesis; aviation fuel
A. PROPERTIES: sp.gr. 0.691; b.p. 81.0°C; f.p. -24.96°C
C. WATER POLLUTION FACTORS:
  —Biodegradation:
    incubation with natural flora in the groundwater—in presence of the other components of high-octane gasoline (100 μl/l): biodegradation: 62% after 192 hr at 13°C (initial conc.: 0.03 μl/l) (596)

trimethylcarbinol *see* *t*-butanol

3,5,5-trimethyl-2-cyclohexene-1-one *see* isophorone

1,2,4-trimethylcyclopentane

C. WATER POLLUTION FACTORS:
  —Biodegradation:
    incubation with natural flora in the groundwater—in presence of the other components of high-octane gasoline (100 μl/l):
    biodegradation: 0% after 192 hr at 13°C (initial conc.: 0.03 μl/l) (596)

trimethylenecyanide *see* glutaronitrile

trimethylenedicyanide *see* glutaronitrile

trimethylethylene *see* 2-methyl-2-butene

trimethyllead chloride (TML, Cl)
$(CH_3)_3PbCl$
D. BIOLOGICAL EFFECTS:
  —Bioaccumulation: molluscs:
    mussel (*Mytilus edulis*): accumulation experiment:

	control	trimethyllead chloride		
exposure conc. (mg/l)	0	0.01	0.05	0.10
exposure time (days)	35	9	35	21
time to equilibrium (days)	—	—	9	7
tissue conc. (mg/kg dry wt): gill	6	10	39	99
at equilibrium: digestive gland	6	10	32	69
gonad	9	15	21	67
foot	3	5	19	39
mean	6	10	27	68
concentration factors:				
(a) tissue (wet wt)/water	—	60	60	90
(b) exposed/control	—	1.6	4.5	11

depuration half-life (days):

gill	—	—	2	1
digestive gland	—	—	2	4
gonad	—	—	4	2
foot	—	—	4	6
mean	—	—	3	3
total experiment duration (days)	62	9	62	48

(575)

mussel (*Mytilus edulis*): accumulation after 96 hr exposure at 0.5 mg/l = (96 hr LC_{50}):

concentration factor $\left(\dfrac{\text{animal tissue (mg/kg wet wt)}}{\text{water mg/l}}\right)$: 24 X

(mean figure of 4 tissues: digestive gland, foot, gill and gonad) (575)

—Crustacean:

shrimp: accumulation after 96 hr exposure at 8.8 mg/l (96 hr LC_{50}):

concentration factor $\left(\dfrac{\text{animal tissue (mg/kg wet wt)}}{\text{water mg/l}}\right)$: 1 X (575)

—Fish:

dab (*Limanda limanda*): accumulation

	control	trimethyllead chloride	
exposure conc. (mg/l)	—	1.0	2.0
exposure time (days)	41	41	41
time to equilibrium (days): liver	—	20	20
muscle	—	n.d.	n.d.
tissue conc. (mg/kg dry wt): liver	1.9	17	32
after 41 days muscle	1.4	22	33
mean	1.6	20	32
concentration factors 41 days			
(a) tissue (wet wt)/water	—	2.5	2.0
(b) exposed/control	—	12	20
depuration half-life (days): liver	—	70	62
muscle	—	n.d.	41
total experiment duration (days)	97	90	90

n.d. = could not be determined (575)

plaice: accumulation after 96 hr exposure at 24.6 mg/l (96 hr LC_{50}):

concentration factor $\left(\dfrac{\text{animal tissue (mg/kg wet wt)}}{\text{water mg/l}}\right)$: 1X (575)

—Toxicity:
—Algae: 6 hr EC_{50}: 0.8 mg/l (Pb) (photosynthetic activity)
—Crustacean: shrimps: 96 hr LC_{50}: 8.8 mg/l (Pb) (573)
—Molluscs: mussel: 96 hr LC_{50}: 0.50 mg/l (Pb) (573)
—Fish: plaice: 96 hr LC_{50}: 24.6 mg/l (Pb) (573)
          rainbow trout: 96 hr LC_{50}: 32 ppm (Pb) (572)
—Mammals:
    rat: intravenous injection: LD_{50}: 20–25 mg Pb/kg body wt (578)
         (no anion specified) intra peritoneal: LD_{50}: 17 mg Pb/kg body wt (570)

trimethylmethane see isobutane

2,3,6-trimethylnaphthalene

$$H_3C-\text{[naphthalene ring]}-CH_3, CH_3$$

A. PROPERTIES: solub in seawater at 22°C: 1.7 ± 0.6 ppm
B. AIR POLLUTION FACTORS:
 —Manmade sources:
  in coal tar pitch fumes: 13.4 wt % of which 5.8 wt % 2,3,6-isomer   (516)
D. BIOLOGICAL EFFECTS:
 —Fish:
  Neanthes arenaceodentata: 96 hr TLm in seawater at 22°C: 2.0 ppm (initial conc. in static assay)   (995)

6,10,14-trimethyl-2-pentadecanone

$$CH_3-\underset{\underset{O}{\|}}{C}-(CH_2)_3-\underset{\underset{CH_3}{|}}{CH}-(CH_2)_3-\underset{\underset{CH_3}{|}}{CH}-(CH_2)_3-\underset{\underset{CH_3}{|}}{CH}-CH_3$$

C. WATER POLLUTION FACTORS:
 —Water quality:
  in Delaware river (U.S.A.): conc. range: winter: n.d.
                                             summer: 0.8–2 ppb   (1051)

2,2,3-trimethylpentane

$$CH_3-\underset{\underset{CH_3}{|}}{\overset{\overset{CH_3}{|}}{C}}-CH-CH_2-CH_3$$

C. WATER POLLUTION FACTORS:
 —Biodegradation:
  incubation with natural flora in the groundwater in presence of the other components of high-octane gasoline (100 μl/l):
  biodegradation: 54% after 192 hr at 13°C (initial conc. 0.17 μl/l)

2,2,4-trimethylpentane see isooctane

2,3,3-trimethylpentane

$$CH_3-\underset{}{CH}-\underset{\underset{CH_3}{|}}{\overset{\overset{CH_3}{|}}{C}}-CH_2-CH_3$$

C. WATER POLLUTION FACTORS:
—Biodegradation:
incubation with natural flora in the groundwater—in presence of the other components of high-octane gasoline (100 µl/l):
biodegradation: 16% after 192 hr at 13°C (initial conc. 1.97 µl/l) (956)

2,3,4-trimethylpentane

$$CH_3-CH-CH-CH-CH_3$$
$$\quad\ \ |\ \ \ \ |\ \ \ \ |$$
$$\quad CH_3\ CH_3\ CH_3$$

C. WATER POLLUTION FACTORS:
—Biodegradation:
incubation with natural flora in the groundwater—in presence of the other components of high-octane gasoline (100 µl/l):
biodegradation: 13% after 192 hr at 13°C (initial conc. 1.89 µl/l) (956)

2,2,4-trimethyl-1,3-pentanediol-1-isobutyrate
C. WATER POLLUTION FACTORS:
—Water quality:
in Delaware river (U.S.A.): conc. range: winter: 1–6 ppb
summer: n.d. (1051)

2,2,4-trimethyl-1,3-pentanediol-3-isobutyrate
C. WATER POLLUTION FACTORS:
—Water quality:
in Delaware river (U.S.A.): conc. range: winter: 1–4 ppb
summer: n.d. (1051)

2,4,4-trimethyl-1-pentene
$CH_2C(CH_3)CH_2C(CH_3)_3$
Use: organic synthesis.
A. PROPERTIES: b.p. 101°C; v.d. 3.8; sp.gr. 0.70; m.w. 112.22
C. WATER POLLUTION FACTORS:
—BOD_5: 0.19 = 6% ThOD (n.s.i.)
—COD: 1.59 = 46% ThOD (n.s.i.) (277)
—Waste water treatment: A.S. after 6 hr: 0.5% ThOD
12 hr: 0.5% ThOD
24 hr: 1.0% ThOD (88)
D. BIOLOGICAL EFFECTS:
—Fish:
Carassius auratus: 24 hr LD_{50}: 3 mg/l (n.s.i.) (277)

2,4,4-trimethyl-2-pentene (β-diisobutylene)
$H_3CC(CH_3)CHC(CH_3)_3$
A. PROPERTIES: colorless liquid; m.p. −106.4°C; b.p. 104.55°C; sp.gr. 0.724 at 15/15°C

C. WATER POLLUTION FACTORS:
—Waste water treatment: A.S. after 6 hr: 0.4% ThOD
                                             12 hr: 0.3% ThOD
                                             24 hr: 0.7% ThOD   (88)

2,2,4-trimethyl(α-phenylisopropyl)1,2-dihydroquinoline
D. BIOLOGICAL EFFECTS:
—goldfish: approx. fatal conc.: 1.8 mg/l, 48 hr   (226)

2,4,6-trimethyl-1,3,5-trioxane *see* paraldehyde

1,3,5-trimethyl-2,4,6-*tris*(3,5-di-*t*-butyl-4-hydroxybenzyl)benzene (ionox 330)

Use: antioxidant in polyethylene, polypropylene, polystyrene, natural and synthetic rubber, lubricants, and grease etc.
A. PROPERTIES: m.w. 775.22; m.p. 517K (244°C); solub. 1.2 mg/l
C. WATER POLLUTION FACTORS:
—BOD_5: 0.92 = 31% ThOD
—COD: 0.82 = 27% ThOD   (277)
D. BIOLOGICAL EFFECTS:
—Fish:
  goldfish: no acute toxicity at 1.2 mg/l = 96 hr LC_0 (277)
—Mammals:
  rat: LD_{50} could not be determined because of too low toxicity
      1% in food during 2 years: no effect   (277)

1,3,7-trimethylxanthine *see* caffeine

trinitrin *see* nitroglycerine

trinitroglycerine *see* nitroglycerin

2,4,6-trinitrophenol *see* picric acid

2,4,6-trinitrophenylmethylnitramine *see* tetryl

2,4,6-trinitroresorcinol (styphnic acid)

A. PROPERTIES: yellow crystals; m.p. 179-180°C
C. WATER POLLUTION FACTORS:
  —Waste water treatment: adapted culture: 7% removal after 48 hr incubation, feed = 199 mg/l (292)
D. BIOLOGICAL EFFECTS:
  —Bacteria: *Pseudomonas putida*: inhibition of cell multiplication starts at >100 mg/l (329)
  —Algae:
    Microcystis aeruginosa: inhibition of cell multiplication starts at 0.32 mg/l (329)
    Selenastrum capricornutum: growth inhibition at 5 mg/l
    Microcystis aeruginosa: growth inhibition at 15 mg/l (318)
  —Fish: LC_{50}, 96 hrs: 2.58 mg/l
    EC_{50}, 96 hrs: 0.46 mg/l, behavioral response (318)

2,4,6-trinitrotoluene (TNT)

Use: explosive; intermediate in dyestuffs and photographic chemicals.
A. PROPERTIES: colorless crystals; m.w. 227.13; m.p. 80.7°C; b.p. 240°C explodes; v.d. 7.85; sp.gr. 1.654; solub. 200 mg/l at 15°C
B. AIR POLLUTION FACTORS: 1 mg/cu m = 0.11 ppm; 1 ppm = 9.44 mg/cu m
C. WATER POLLUTION FACTORS:
  —Biodegradation: self purification of surface water is affected from 0.5 mg/l (181)
  —Impact on treatment processes: biochemical oxidation is decreased from 0.5-1.0 mg/l onwards (30)
D. BIOLOGICAL EFFECTS:
  —Toxicity threshold (cell multiplication inhibition test):
    bacteria (*Pseudomonas putida*): >100 mg/l (1900)
    algae (*Microcystis aeruginosa*): 50 mg/l (329)
    green algae (*Scenedesmus quadricauda*): 1.6 mg/l (1900)
    algae (*Selenastrum capricornutum*): 9 mg/l (329)

1170 1,3,5-TRIOXANE

    protozoa (*Entosiphon sulcatum*):   1.6 mg/l   (1900)
    protozoa (*Uronema parduczi Chatton-Lwoff*):   5.9 mg/l   (1901)
—Fish: 96 hr LC_{50}: 1.60 mg/l
        96 hr EC_{50}: (behavioral response): 0.64 mg/l   (318)
        fathead minnow: flow through bioassay:
                96 hr TLm: 2.58 mg/l at 24°C (n.s.i.)
                96 hr EC_{50}: 0.46 mg/l at 24°C (n.s.i.)
                      (behavioral response)   (473)
—Man: unsatisfactory: >2 mg/cu m   (185)

1,3,5-trioxane (α-trioxymethylene)

A. PROPERTIES: needles; m.w. 90.08; m.p. 62–64°C; b.p. 115°C; v.p. 13 mm at 25°C; v.d. 3.10
D. BIOLOGICAL EFFECTS:
    —Bacteria: *Pseudomonas*: LD_0: approx. 1 mg/l
    —Algae: *Scenedesmus*: LD_0: approx. 1 mg/l
    —Protozoa: *Colpoda*: LD_0: approx. 1 mg/l
    —Arthropoda: *Daphnia*: LD_0: 5 mg/l   (30)

trioxitol *see* triethyleneglycolmonoethylether

α-trioxymethylene *see* 1,3,5-trioxane

2,6,8-trioxypurine *see* uric acid

tripalmitin (palmitin; glyceryltripalmitate)
$C_3H_5(OOCC_{15}H_{31})_3$
Use: medicine; soap; leather dressing.
A. PROPERTIES: white crystalline powder; m.p. 65.5°C; sp.gr. 0.866 (80/4°C)
C. WATER POLLUTION FACTORS:
    —BOD_{35}^{25}: 1.96 in seawater/inoculum: enrichment cultures of hydrocarbon oxidizing bacteria
    —ThOD: 2.92

triphenylene

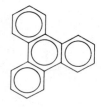

Manmade sources:
in Kuwait crude oil: 2.8 ppm
in South Louisiana crude oil: 10 ppm (1015)
in gasoline: 0.030 mg/l (1070)
A. PROPERTIES: solub. 0.038 mg/l at 26°C; m.w. 228.29; m.p. 195–198°C; b.p. 438°C
B. AIR POLLUTION FACTORS:
—Manmade sources:
in exhaust condensate of gasoline engine: 0.04–0.06 mg/l gasoline consumed
(1070)
C. WATER POLLUTION FACTORS:
—Sediment quality:
in sediment of Wilderness Lake, Colin Scott, Ontario (1976): 10 ppb (dry wt.)
(932)

triphenylphosphate

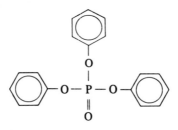

Manufacturing source: organic chemical industry. (347)
Users and formulation: mfg. plasticizers; gasoline additives; insecticides; flotation agents, stabilizers, anti-oxidants and surfactants; substitute for camphor (non-combustible); fire retardant. (347)
A. PROPERTIES: m.w. 326.29; m.p. 50°C; b.p. 245°C at 11 mm; v.p. <0.1 mm at 30°C; v.d. 11.3; solub. 0.73 mg/l at 24°C (practical grade), in buffered distilled water: 1.4–1.6 ppm, in natural waters: 0.2–0.3 ppm (1166; 1438)
C. WATER POLLUTION FACTORS:
—Aquatic reactions:
hydrolysis rates at pH 9.5: half-life: 1.3 days
at pH 8.2: half-life: 7.5 days
at natural and acid pH: too slow to reliably measure (1438)
—Water quality:
in Delaware river (U.S.A.): conc. range: winter: 0.1–0.3 ppb
summer: 0.1–0.4 ppb (1051)
D. BIOLOGICAL EFFECTS:
—Fish:
Lepomis macrochirus: static bioassay in fresh water at 23°C, mild aeration applied after 24 hr:

1172 TRIPHENYLTINHYDROXIDE

material added ppm	24 hr	% survival after 48 hr	72 hr	96 hr	best fit 96 hr LC$_{50}$ ppm
560	100	70	20	0	
420	100	90	90	80	
320	80	70	60	40	290
180	100	100	100	90	
125	100	100	100	100	

Menidia beryllina: static bioassay in synthetic seawater at 23°C: mild aeration applied after 24 hr:

material added ppm	24 hr	% survival after 48 hr	72 hr	96 hr	best fit 96 hr LC$_{50}$ ppm
560	30	0	—	—	
320	10	1	0	—	
180	100	40	20	10	95
100	100	90	70	30	
75	100	100	100	100	

(352)
(1535)

rainbow trout: 96 hr LC$_{50}$ (S): 300 µg/l

triphenyltinhydroxide (fentinhydroxide; fenolovo; Du-Ter; TPTH)

Use: a nonsystemic fungicide.
A. PROPERTIES: white odorless powder; m.p. 118–120°C
D. BIOLOGICAL EFFECTS:
  —Fish:

	mg/l	24 hr	48 hr	96 hr	3 m (extrap.)
harlequin fish					
(*Rasbora heteromorpha*)	LC$_{10}$(F)	0.038	0.024	—	
	LC$_{50}$(F)	0.062	0.042	—	—
rainbow trout (fry)	LC$_{10}$(F)	0.055	0.019	0.01	
	LC$_{50}$(F)	0.078	0.03	0.015	0.0004

(331)

  —Mammals:
    acute oral LD$_{50}$ (rats): 108–209 mg/kg
              (female guinea pigs): 27 mg/kg
              (male mouse): 245 mg/kg   (1854; 1855)

tri-*n*-propylamine
$(C_3H_7)_3N$
A. PROPERTIES: colorless liquid; m.w. 143.27; m.p. −93.5°C; b.p. 156°C; v.d. 4.9; sp.gr. 0.757 at 20/4°C; log P_{oct} 2.79
B. AIR POLLUTION FACTORS: 1 mg/cu m = 0.17 ppm, 1 ppm = 5.95 mg/cu m

C. WATER POLLUTION FACTORS:
 —Biodegradation:
  degradation by *Aerobacter*: 200 mg/l at 30°C:
           parent: 100% in 30 hr
           mutant: 100% in 10 hr (152)
D. BIOLOGICAL EFFECTS:
 —Fish: creek chub: LD_0: 30 mg/l, 24 hr in Detroit river water
                    LD_{100}: 70 mg/l, 24 hr in Detroit river water (243)

triptane *see* 2,2,3-trimethylbutane

tris(2,3-dibromopropyl)phosphate
$(CH_3-CHBr-CHBr-O)_3PO$
A. PROPERTIES: solub. 8.0 mg/l at 24°C (technical grade) (1660)
D. BIOLOGICAL EFFECTS:
 —Fish:
  rainbow trout: sac fry: 69 hr LC_{50} (S): 240 µg/l
                 fingerling: 96 hr LC_{50} (S): 1,450 µg/l (1535)

trisodiumcarboxymethyltartronate *see* carboxymethyltartronate

trisodium-2-oxa-1,1,3-propanetricarboxylate *see* carboxymethyltartronate

trithion *see* carbophenothion

tritriacontane
$C_{33}H_{68}$
B. AIR POLLUTION FACTORS:
 —Ambient air quality:
  organic fraction of suspended matter:
   Bolivia at 5200 m altitude (Sept.–Dec. 1975): 0.16–0.32 µg/100 cu m
   residential area, 10 km south of Antwerp, Belgium (Jan.–April 1976): 4.7
   µg/100 cu m (428)

N-tritylmorpholine (trifenmorph; frescon)

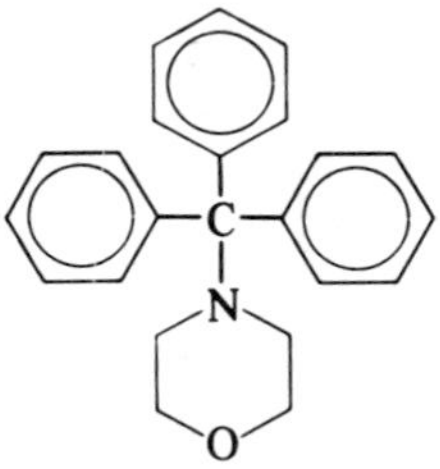

Use: molluscicide.
A. PROPERTIES: colorless crystals; m.p. 176–178°C; solub. 0.02 ppm

1174   TROPILIDENE

D. BIOLOGICAL EFFECTS:
  —Fish:
   Sarotherodon mossambicus, juvenile: 55 hr BCF (R): 1 300   (1544)
   brown trout: 48 hr LC_{50} (S): 0.083 mg/l
  —Mammals:
   acute oral LD_{50} (rat): 1200–1600 mg/kg   (1854)

tropilidene   see 1,3,5-cycloheptatriene

tryptophane   (2-amino-3-indolylpropanoic acid)

A. PROPERTIES: colorless plate; m.w. 204.22; m.p. 283–285°C;
C. WATER POLLUTION FACTORS:
  —COD: 1.780   (223)
  —Biodegradation:
   biochemical degradation reactions:
   tryptophane + 5 O_2 ⟶ indole + pyruvate + NH_3 (203)

   tryptophane + O_2 $\xrightarrow{\text{tryptophane oxygenase}}$ l-formylkynerenine   (203)

  —Waste water treatment:
   A.S.: after   6 hr: 0.6% of ThOD   (38)
                12 hr: 1.4% of ThOD
                24 hr: 4.6% of ThOD
   A.S., BOD, 20°C, 1–5 days observed; feed: 333 mg/l, 15 days acclim.: 99%
     removed (n.s.i.)   (93)
  —Manmade sources:
   excreted by man: in urine: 0.23–1.3 mg/kg body wt/day (n.s.i.)   (203)
D. BIOLOGICAL EFFECTS:
  —Carcinogenicity:
   It is concluded that under the conditions of this bioassay, l-tryptophan was not
     carcinogenic for Fischer 344 rats or B6C3F1 mice   (1738)

tubatoxin   see rotenone

dl-tyrosine   (2-amino(4-hydroxyphenyl)propanoic acid; dl-β-p-hydroxyphenylalanine)
$C_6H_4(OH)CH_2CH(NH_2)COOH$
A. PROPERTIES: short needles; m.w. 181.19; m.p. 316°C; solub. 410 mg/l at 20°C;
   log P_{oct} −2.26
C. WATER POLLUTION FACTORS:
  —Natural sources:

in soil: silty loam: 0–65 μg/kg soil
           clay soils: 10–60 μg/kg soil (174)
—Manmade sources:
   excreted by man: in urine: 0.35–1.45 mg/kg body wt/day (203)
   in primary domestic sewage plant effluent: 0.034 mg/l (n.s.i.) (517)
—Odor threshold conc. (detection): >10 mg/l (L-form) (998)
—Waste water treatment:
   A.S.: after 6 hr: 3.4% of ThOD
                 12 hr: 9.7% of ThOD
                 24 hr: 22.5% of ThOD (89)
   A.S., Resp, BOD, 20°C; 1–5 days observed, feed: 500 mg/l, acclim.: <1 day;
      13% of ThOD; 26% removed (93)

U

UDMH *see* 1,1-dimethylhydrazine

n-undecane (hendecane)
$CH_3(CH_2)_9CH_3$
A. PROPERTIES: colorless liquid; m.p. $-25.75°C$; b.p. $195.6°C$; sp.gr. 0.7402 (20/4°C); v.p. 1 mm at 32.7°C, 10 mm at 73.9°C, 100 mm at 104.4°C
B. AIR POLLUTION FACTORS:
   —Odor: T.O.C. = 0.780 mg/cu m = 0.11 ppm (307)
          recognition: 374 mg/cu m (761)
          23 mg/cu m (737)
     O.I. at 20°C = 8,400 (316)
   first order evaporation constant of n-undecane in 3 mm layer No. 2 fuel oil—in darkened room at windspeed of 21 km/hr:
     at 5°C: 4.15×10^{-4} min^{-1}
        10°C: 7.17×10^{-4} min^{-1}
        20°C: 1.31×10^{-3} min^{-1}
        30°C: 2.48×10^{-3} min^{-1} (438)
C. WATER POLLUTION FACTORS:
   —Reduction of amenities: T.O.C. = 10 mg/l (295)

undecanoic acid *see* n-undecyclic acid

1-undecanol (1-hendecanol)
$CH_3(CH_2)_9CH_2OH$
A. PROPERTIES: m.w. 172.30; m.p. 11/19°C; b.p. 131°C at 15 mm; sp.gr. 0.833 at 23/4°C
B. AIR POLLUTION FACTORS:
   —Odor threshold conc.: 46 mg/cu m (787)
C. WATER POLLUTION FACTORS:
   —Waste water treatment:
     A.S. after 6 hr: 14.5% of ThOD
               12 hr: 20.5% of ThOD
               24 hr: 21.4% of ThOD (88)

2-undecanone (2-hendecanone, methylnonylketone)
$CH_3(CH_2)_8COCH_3$

A. PROPERTIES: colorless aromatic liquid; m.w. 170.29; m.p. 15°C; b.p. 223/228°C; v.d. 5.9; sp.gr. 0.826 at 20/4°C

B. AIR POLLUTION FACTORS:

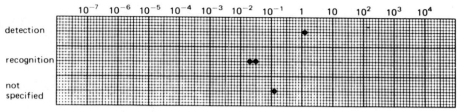

(610; 840; 842)

C. WATER POLLUTION FACTORS:
  —Waste water treatment:
    A.S. after 6 hr: 1.0% ThOD
           12 hr: 6.5% ThOD
           24 hr: 21.8% ThOD (88)
  —Odor threshold conc.: (detection): 0.007 mg/kg water (896)

10-undecenoic acid *see* undecylenic acid

undecylenic acid (10-hendecenoic acid; 10-undecenoic acid)
$CH_2CH(CH_2)_8COOH$
Use: weed killer and defoliant.
A. PROPERTIES: white crystal mass or colorless to yellowish liquid; m.w. 184.27; m.p. 24.5°C; b.p. 275°C, 295°C decomposes; v.p. 10 mm at 160°C; sp.gr. 0.907 at 25/4°C
B. AIR POLLUTION FACTORS:
  —Odor: characteristic: quality: suggestive of perspiration
C. WATER POLLUTION FACTORS:
  —Oxidation parameters: BOD_5 = 0.537 std. dil/sew. (282)
  —Waste water treatment:
    A.S. after 6 hr: 0.4% ThOD
           12 hr: 3.5% ThOD
           24 hr: 28.7% ThOD (89)
D. BIOLOGICAL EFFECTS:
  —Mammalia:
    rat: oral LD_{50}: 2.5 g/kg
          : no signs of toxicity: 0.4 g/kg daily, 6–9 months (211)

***n*-undecylic acid** (undecanoic acid; hendecanoic acid)
$CH_3(CH_2)_9COOH$
A. PROPERTIES: colorless scales; m.w. 186.29; m.p. 29.3°C; b.p. 228°C at 160 mm sp.gr. 0.891 at 30°C
C. WATER POLLUTION FACTORS:
  —Waste water treatment:
    A.S. after 6 hr: 11.6% ThOD

12 hr: 14.5% ThOD
24 hr: 20.9% ThOD (89)
—Odor threshold conc. (detection): 10.0 mg/kg water (886)

uracil (2,4-dioxypyrimidine)

A pyrimidine that is a constituent of ribonucleic acids and the coenzyme, uridine-diphosphateglucose.
Use: biochemical research.
A. PROPERTIES: crystalline needles; m.p. 335°C; log P_{oct} −1.07 (calculated)
C. WATER POLLUTION FACTORS:
 —Manmade sources:
  in primary domestic sewage plant effluent: 0.016–0.058 mg/l
  in secondary domestic sewage plant effluent: 0.016–0.030 mg/l (517)
  in sewage effluent: 0.013 mg/l (234)

uracil-6-carboxylic acid *see* orotic acid

urea (carbamide)
$NH_2 CONH_2$
A. PROPERTIES: white crystals or powder; m.w. 60.06; m.p. 132.7°C; b.p. decomposes; sp.gr. 1.335 at 20/4°C; solub. 780.000 mg/l at 5°C; 1,193,000 mg/l at 25°C; log P_{oct} −2.97/−2.26 (calculated)
C. WATER POLLUTION FACTORS:
 —Biodegradation:
  $H_2 NCONH_2 + H_2O \longrightarrow CO_2 + 2 NH_3$
  in river water at 1–15 mg/l: degradation rate is negligible below 8°C for up to 14 days, degradation at 20°C should be complete within 4–6 days (224)
 —Manmade sources:
  contents of domestic sewages 2–6 mg/l (85)
  in domestic sewer effluent: 0.020 mg/l (227)
  in primary domestic sewage plant effluent: 0.016–0.043 mg/l (517)
 —Impact on degradation processes:
  at 100 mg/l, no inhibition of NH_3 oxidation by *Nitrosomonas* sp. (390)
 —Waste water treatment: degradation rate by psychrophilic bacteria:
  at 20°C: max: 11.6 mg/l/hr
   average: 10.9 mg/l/hr
  at 2°C: max: 4.0 mg/l/hr
   average: 3.2 mg/l/hr (3)
D. BIOLOGICAL EFFECTS:
 —Toxicity threshold (cell multiplication inhibition test):
  bacteria (*Pseudomonas putida*): >10,000 mg/l

green algae (*Scenedesmus quadricauda*): >10,000 mg/l
protozoa (*Entosiphon sulcatum*): 29 mg/l (1900)
—Fish: creek chub: critical range: 16,000–30,000 mg/l in Detroit river (243)

uric acid (2,6,8 (1,3,9) purinetrione; 2,6,8-trioxypurine)
A. PROPERTIES: scales; m.w. 168.11; m.p. decomposes; sp.gr. 1.893; solub. 64.5 mg/l at 37°C, 600 mg/l in hot water; log P_{oct} −2.92
C. WATER POLLUTION FACTORS:
 —Oxidation parameters:
 BOD_5: 0.300 std. dil. sew (282)
 COD: 0.551 (223)
 $KMnO_4$ value: 0.680 (30)
 ThOD: 0.667 (30)
 —Manmade sources: contents of domestic sewages: 0.2–1.0 mg/l (85)(517)

uric oxide *see* uric acid

V

n-valeraldehyde (pentanal, n-valeric aldehyde, n-amylaldehyde)
$CH_3(CH_2)_3CHO$
   Use: flavoring; rubber accelerators:
A. PROPERTIES: m.w. 86.13; m.p. -91°C; b.p. 103°C; v.p. 50 mm at 25°C; v.d. 3.0; sp.gr. 0.818 at 11°C
B. AIR POLLUTION FACTORS: 1 mg:cu m = 0.28 ppm, 1 ppm = 3.5 mg/cu m
   —Odor threshold conc.: 0.072 mg/cu m (842)
                  recognition: 0.009–0.010 mg/cu m (610)
   —Manmade sources:
      in exhaust of gasoline engine: 0.4 vol.% of total exhaust aldehydes (395)
C. WATER POLLUTION FACTORS:
   —Reduction of amenities: T.O.C. = 0.012 mg/l (305)
                 detection: 0.100 mg/kg (900)
                            0.0606 mg/kg (874)
   —Waste water treatment: A.S. after 6 hr: 12.7% of ThOD
                                      12 hr: 16.5% of ThOD
                                      24 hr: 17.8% of ThOD (88)
     aeration by compressed air: stripping effect: 95% removal after 8 hr (30)
D. BIOLOGICAL EFFECTS:
   —Mammalia: rat: inhalation: 0/3: 1,400 ppm, 6 hr
                               3/3: 48,000 ppm, 1.2 hr (104)
              : acute oral: LD_{50}: 3.2–6.4 g/kg
              mouse: acute oral: LD_{50}: 6.4–12.8 g/kg (211)

valeramide (pentanamide)
$CH_3(CH_2)_3CONH_2$
A. PROPERTIES: m.w. 101.15; m.p. 114–116°C; sp.gr. 1.023; log P_{oct} 0.05/0.73
C. WATER POLLUTION FACTORS:
   —Waste water treatment: A.S. after 6 hr: 2.4% ThOD
                                      12 hr: 4.9% ThOD
                                      24 hr: 13.6% ThOD (89)

n-valeric acid (pentanoic acid)
C_4H_9COOH
   Use: intermediate for flavors and perfumes; ester type lubricants; plasticizers; pharmaceuticals.

n-VALERIC ACID 1181

A. PROPERTIES: colorless liquid; m.w. 102.1; m.p. $-34/-58°C$; b.p. $187°C$; v.p. 0.15 mm at $20°C$, 0.4 mm at $30°C$, 1 mm at $42°C$; v.d. 3.52; sp.gr. 0.942 at $20/4°C$; solub. 24,000 mg/l; 37,000 mg/l at $16°C$; sat. conc. 0.83 g/cu m at $20°C$, 2.2 g/cu m at $30°C$; log P_{oct} 0.99/1.69 (calculated)

B. AIR POLLUTION FACTORS: 1 mg/cu m = 0.24 ppm, 1 ppm = 4.25 mg/cu m
   —Odor: characteristics: quality: body odor
                            hedonic tone: unpleasant

odor thresholds   mg/cu m

[Chart showing detection, recognition, and not specified thresholds across 10^{-7} to 10^4 mg/cu m]

(57; 210; 307; 602; 610; 623; 646; 648; 678; 682; 683; 684; 771; 778; 781)
   odor index: USSR: animal chronic exposure: no effect: 0.006 mg/cu m
                     adverse effect: 0.14 mg/cu m   (170)
   odor index at $20°C$ = 256,300   (316)
   —Control methods: A.C.: retentivity: 7 wt% of adsorbent   (83)

C. WATER POLLUTION FACTORS:
   —Oxidation parameters:
   BOD$_5$:   1.06 std. dil. sewage   (163)
                 1.40   (30)
                 43% ThOD   (220)
   20 days: 1.90   (30)
   BOD$_{2\ days}^{25°C}$: 1.56 (substrate conc.: 4.1 mg:l; inoculum: soil microorganisms
   5 days: 1.71
   10 days: 1.83   (1304)
   COD: 1.85   (41)
   KMnO$_4$:   0.04
                 acid: 9% ThOD
                 alkaline: 3% ThOD   (220)
   TOC: 100% ThOD   (220)
   ThOD: 2.039   (30)
   —Manmade sources:
   contents of domestic sewages: 0–0.4 mg/l   (85)
   average content of secondary sewage effluents: 8.1 µg/l   (86)
   in year old leachate of artificial sanitary landfill: 5.2 g/l   (1720)
   —Odor threshold conc. (detection): 3.0 mg/kg   (886)
   —Waste water treatment:
   A.C.: adsorbability: 0.159 g/g; 79.7% reduction, infl.: 1,000 mg/l, effl.: 203 mg/l
                                                                           (32)
   A.S. after 6 hr: 13.6% ThOD
              12 hr: 18.4% ThOD
              24 hr: 22.2% ThOD   (88)

1182 VALERONE

    A.S., BOD, 20°C, 1-5 days observed, feed: 333 mg/l, acclim.: 15 days: 99% removed (93)
    powdered carbon: at 100 mg/l sodium salt (pH 7.5)—carbon dosage 1000 mg/l: 8% adsorbed (520)

D. BIOLOGICAL EFFECTS:
    —Algae: *Chlorella pyrenoidosa*: toxic: 280 mg/l (41)
    —Crustacean: *Daphnia magna*: 48 hr TLm: 45 mg/l (1295)
    —Fish: *Lepomis macrochirus*: 24 hr TLm: 5,000 mg/l (Na salt) (1294)
        fathead minnows: static bioassay in Lake Superior Water at 18-22°C: LC_{50} (1; 24; 48; 72; 96 hr): >100; >100; 77; 77; 77 mg/l (350)
    —Mammals: rat: single oral LD_{50}: >400 mg/kg (211)

valerone *see* diisobutylketone

valeronitrile *see* (pentanenitrile, *n*-butylcyanide)
$CH_3(CH_2)_3CN$
A. PROPERTIES: colorless liquid; m.w. 83.13; m.p. -96°C; b.p. 141°C; sp.gr. 0.801 at 20/4°C
C. WATER POLLUTION FACTORS:
    —Waste water treatment: A.S. after 6 hr: 2.2% of ThOD
                                      12 hr: 5.9% of ThOD
                                      24 hr: 3.8% of ThOD (89)

***dl*-valine** (*dl*-2-amino-3-methylbutanoic acid; *dl*-α-aminoisovaleric acid)
$(CH_3)_2CHCH(NH_2)COOH$
An essential amino-acid
A. PROPERTIES: leaflets; m.w. 117.15; m.p. 298°C decomposes; b.p. sublimes; solub. 74,400 mg/l at 25°C, 133,100 mg/l at 75°C; log P_{oct} -2.10 (calculated)
C. WATER POLLUTION FACTORS:
    —Natural sources: in soil: in silty loam: 3-20 µg/kg soil
                      in clay soils: 30-400 µg/kg soil (174)
    —Manmade sources:
    excreted by man in urine: 0.2-0.45 mg/kg body wt/day
    in feces: 3.6-6.2 mg/kg body wt/day
    in sweat: 1.5-4.5 mg/100 ml (203)
    —Waste water treatment: A.S. after 6 hr: 2.1% ThOD
                                      12 hr: 4.3% ThOD
                                      24 hr: 9.4% ThOD (89)
    A.S., BOD, 1-5 days observed, feed: 333 mg/l, 15 days acclim.: 99% removed (93)
    —Impact on biodegradation processes:
      ~50% inhibition of NH_3 oxidation in *Nitrosomonas* at 1.8 mg/l (*l*-isomer) (408)

vanillicaldehyde *see* vanillin

vanillin (3-methoxy-4-hydroxybenzaldehyde; vanillicaldehyde)

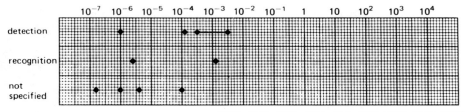

Use: perfumes; flavoring; pharmaceuticals

A. PROPERTIES: white crystalline needles; m.p. 81–85°C; b.p. 285°C; sp.gr. 1.056; v.p. 1mm at 107°C; 10 mm at 154.0°C, 100 mm at 214.5°C

B. AIR POLLUTION FACTORS:

odor thresholds

	10^{-7}	10^{-6}	10^{-5}	10^{-4}	10^{-3}	10^{-2}	10^{-1}	1	10	10^2	10^3	10^4
detection			●		●	● ●						
recognition				●			●					
not specified		● ●	●		●							

(279; 307; 607; 620; 614; 774; 775; 793; 840)

odor index at 20°C = 822,000 (316)

C. WATER POLLUTION FACTORS:
— Reduction of amenities: T.O.C. in water: 0.2 ppm (326)
　　　　　　　　　　　　4 ppm
　　　　　　　　　　detection: 0.06 mg/l; 0.2 mg/kg (995; 998)
　　　　　　　　　　recognition: 4.0 mg/kg (885)

D. BIOLOGICAL EFFECTS:
— Fish:
　fathead minnows: static bioassay in Lake Superior Water at 18–22°C:
　　LC_{50} (1; 24; 48; 72; 96 hr): >173; 131; 123; 121; 121 mg/l
　　　replicate test: 370; 125; 116; 112; 112 mg/l
　fathead minnows: static bioassay in reconstituted water at 18–22°C:
　　LC_{50} (1; 24; 48; 72; 96 hr), dissolved oxygen ≤ 4.0 mg/l: >173; 125; 116; 116; 116 mg/l (350)

VCN　*see* acrylonitrile

velsicol　*see* chlordane

ventox　*see* acrylonitrile

veratrole　*see* 1,2-dimethoxybenzene

vernolate　*see* S-propyldipropylthiocarbamate

versatic 10 (C_{10} carbonic acid)

$$R_1-\underset{\underset{R_3}{|}}{\overset{\overset{R_2}{|}}{C}}-COOH$$

$$R_1 + R_2 + R_3 = C_8$$

Composition: mixture of saturated, mainly tertiary carbon acids.
Use: manufacturing of paint-dryers, stabilizers for PVC, catalysts for polymethane foam.

C. WATER POLLUTION FACTORS:
 —BOD_5: 0.40 = 13% ThOD
  COD : 2.43 = 78% ThOD (277)
D. BIOLOGICAL EFFECTS:
 —Fish:
  Carassius auratus: 24 hr LC_{50}: 80 mg/l
        96 hr LC_{50}: 80 mg/l (277)
  rainbow trout: 96 hr LC_{50} (S): 28.0–32.0 mg/l (1500)
 —Mammals: rat: acute oral LD_{50}: 2.9 ml/kg (277)

vinegar acid *see* acetic acid

vinylacetate (ethenylethanoate)
$CH_3COOCHCH_2$
A. PROPERTIES: colorless liquid; m.w. 86.1; m.p. -100.2°C; b.p. 73°C; v.p. 83 mm at 20°C; 115 mm at 25°C; 140 mm at 30°C v.d. 2.97; sp.gr. 0.932; solub. 25,000 mg/l; sat. conc. 398 g/cu m at 20°C, 634 g/cu m at 30°C
B. AIR POLLUTION FACTORS: 1 mg/cu m = 0.28 ppm, 1 ppm = 3.58 mg/m³
 —Odor: characteristic: quality: sour, sharp
       hedonic tone: unpleasant
   odor threshold values: 1 mg/cu m = 0.28 ppm (57)
  absolute perception limit: 0.12 ppm
  50% recogn.: 0.40 ppm
  100% recogn.: 0.55 ppm
  odor index (O.I.) 100% recognition: 220,000 (19)
   USSR: human odor perception: non perception: 0.7 mg/cu m
            perception: 1.0 mg/cu m
    human reflex response: no response: 0.21 mg/cu m
            adverse response: 0.32 mg/cu m (170)
 —Sampling and analysis: I.R. spectrometry: MIRAN (Wilks): 0.006 ppm (56)
C. WATER POLLUTION FACTORS:
 —Waste water treatment: A.C.: adsorbability: 0.129 g/g C; 64.3% reduction, infl.: 1,000 mg/l, effl.: 357 mg/l (32)
  Sew D, CO_2, 26°C, 10 days observed, feed: 10 mg/l, 19 days acclim.: 42% theor. oxidation (93)

D. BIOLOGICAL EFFECTS:
  −Toxicity threshold (cell multiplication inhibition test):

bacteria (*Pseudomonas putida*):	6 mg/l	(1900)
algae (*Microcystis aeruginosa*):	35 mg/l	(329)
green algae (*Scenedesmus quadricauda*):	370 mg/l	(1900)
protozoa (*Entosiphon sulcatum*):	81 mg/l	(1900)
protozoa (*Uronema parduczi Chatton-Lwoff*):	91 mg/l	(1901)

  −Fish: fatheads: TLm (24–96 hr): 39–19 mg/l
        bluegill: TLm (24–96 hr): 18.0 mg/l
        goldfish: TLm (24–96 hr): 42.3 mg/l
        guppies: TLm (24–96 hr): 31.1 mg/l   (158)
  −Mammalia: rat: inhalation: no effect level: 100 ppm, 15 × 6 hr   (65)
           : LD_{50}: 4,000 ppm, 4 hr
           : acute oral LD_{50}: 2.9 g/kg   (286)
  −Carcinogenicity: negative
  −Mutagenicity in the *Salmonella* test: negative
     <0.0006 revertant colonies/nmol
     <70 revertant colonies at 10 μg/plate   (1883)

vinylacetic acid *see* 3-butenoic acid

vinylbenzene *see* styrene

vinylcarbinol *see* allylalcohol

vinylchloride (MVC; chloro-ethene; chloro-ethylene)
CH_2CHCl
  Use: PVC and copolymers; adhesives for plastics.
A. PROPERTIES: colorless gas; m.w. 62.5; m.p. −153/−160°C; b.p. −13.9°C; v.p. 240 mm at −40°C, 580 mm at −20°C; 2,660 mm at 25°C; v.d. 2.15; sp.gr. 0.9121 at 15/4°C; solub. 1.100 mg/l at 25°C
B. AIR POLLUTION FACTORS: 1 mg/cu m = 0.39 ppm, 1 ppm = 2.60 mg/cu m
  −Odor: characteristic; quality: mild, sweetish
                       hedonic tone: faintly pleasant at high conc.

T.O.C.: 26–52 mg/cu m	(694)
25,000 ppm	(54)
4,000 ppm slight odor	(211)
−O.I. at 20°C = 100	(316)
−Sampling and analysis: PMS: 2-11 ppm	(200)

  −Ambient air quality:
    glc's in rural Washington, Dec. '74–Feb. '75: <5 ppt   (315)
C. WATER POLLUTION FACTORS:
  −Manmade sources: in PVC bottled liquids: level of MVC: 0–0.4 ppm, depending on time of storage and quality of PVC   (206)
  −Waste water treatment: evaporation from water at 25°C of 1 ppm solution: 50% after 26 min, 90% after 96 min   (313)

D. BIOLOGICAL EFFECTS:
—Mammalia: rabbit: inhalation: liver degeneration: 200 ppm, 7 hr/day, 6 months (54)
—Man: dizziness and disorientation: 25,000 ppm, 3 min (211)
  carcinogenic: W. Germany 1977 (487)
—Carcinogenicity: positive
—Mutagenicity in the *Salmonella* test: weakly mutagenic (without liver homogenate) (1883)

vinylcyanide *see* acrylonitrile

vinylethylene *see* 1,3-butadiene

vinylidenechloride *see* 1,1-dichloroethylene

vinyltrichloride *see* 1,1,2-trichloroethane

vitavex *see* 5,6-dihydro-2-methyl-1,4-oxathün-3-carboxanilide

vondrax
  Use: herbicide.
  Active ingredients: maleic hydrazide (17.5%), chloropropham (6.7%).
D. BIOLOGICAL EFFECTS:
—Fish:

	mg/l	24 hr	48 hr	96 hr	3 m (extrap.)	
harlequin fish (*Rasbora heteromorpha*)	LC_{10} (F)	120	100	96		
	LC_{50} (F)	200	175	140	130	(331)

W

warfarin (3-(α-acetonylbenzene)-4-hydroxycoumarin; compound 42; 200 coumarin; coumafene; warf-12)

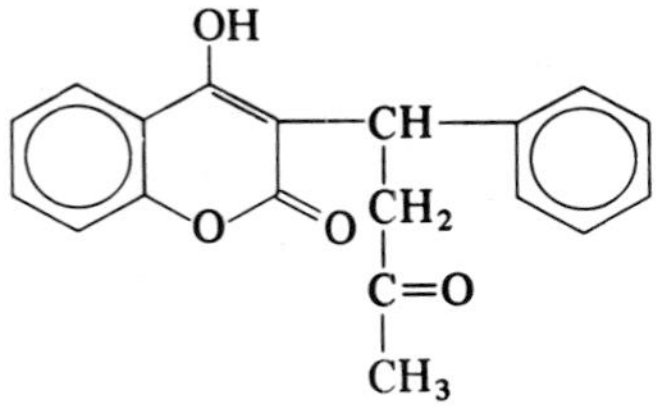

Use: a rodenticide

A. PROPERTIES: m.p. 161°C
D. BIOLOGICAL EFFECTS:
—Fish:
harlequin fish (*Rasbora heteromorpha*):
24 hr LC_{50} (S): 17 mg/l
48 hr LC_{50} (S): 14 mg/l
96 hr LC_{50} (S): 12 mg/l (331)
—Mammals:
rats are killed by 1 mg/kg for 5 days
cats are killed by 3 mg/kg for 5 days
pigs are killed by 1 mg/kg for 5 days (1855)

WSCP-busan 77 *see* polyoxyethylene(dimethylimino)ethylene(dimethylimino)ethylenedichloride

X

xanthene (dibenzopyran)

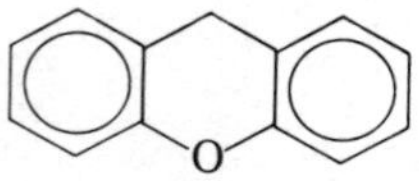

Use: organic synthesis; fungicide.
A. PROPERTIES: m.w. 182.22; m.p. 101–102°C; b.p. 310–312°C
B. AIR POLLUTION FACTORS:
　—Manmade sources:
　　in coal tar pitch fumes: 1.1 wt %　　　　　　　　　　　　　　　　　(516)

xanthine

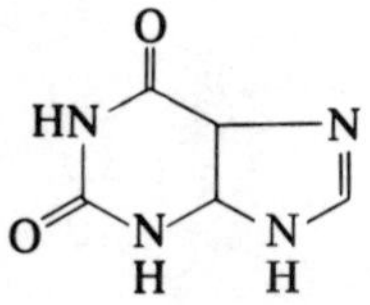

A. PROPERTIES: m.w. 152.11; m.p. >300°C; log P_{oct} −0.99 (calculated)
C. WATER POLLUTION FACTORS:
　—Manmade sources:
　　in primary domestic sewage plant effluent: 0.002–0.070 mg/l　　　(517)
　—Odor threshold conc. (detection): >10 mg/l　　　　　　　　　　　(998)

o-xylene (1,2-dimethylbenzene)

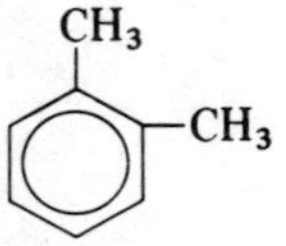

Manufacturing source: petroleum distillation, coal tar distillation, coal gas distillation, organic chemical industry.　　　　　　　　　　　　　　　　　(347)
Users and formulation: mfg. phthalic acid and anhydride; mfg. terepthalic acid for polyester; solvent recovery plants; specialty chemical manufacture; mfg. isophthalic

o-XYLENE 1189

acid, aviation gasoline, protective coatings mfg.; solvent for alkyd resins, lacquers, enamels, rubber cements; dye mfg.; insecticide mfg; pharmaceutical mfg.; asphalt and naphtha constituent. (347)
Natural sources (water and air): coal tar, petroleum. (347)
A. PROPERTIES: colorless; m.w. 106.17; m.p. -25°C; b.p. 144.4°C; v.p. 5 mm at 20°C, 9 mm at 30°C; v.d. 3.7 sp.gr. 0.88; solub. 175 mg/l at 20°C; THC 1,093.7 kcal/mole; LHC 1,051 kcal/mole; sat. conc. 29 g/cu m; at 20°C, 50 g/cu m at 30°C; $\log P_{oct}$ 2.77 (21)
B. AIR POLLUTION FACTORS: 1 mg/cu m = 0.23 ppm, 1 ppm = 4.41 mg/cu m
  —Odor: characteristic: sweet

odor thresholds   mg/cu m

	10^{-7}	10^{-6}	10^{-5}	10^{-4}	10^{-3}	10^{-2}	10^{-1}	1	10	10^2	10^3	10^4
detection							◆	◆	◆			
recognition								◆	◆			
not specified						◆—◆						

(9; 57; 73; 210; 610; 727; 829)

  USSR: human odor perception: 0.73 mg/cu m = 0.17 ppm
      human reflex response: no response: 0.5 mg/cu m
                              adverse response: 0.6 mg/cu m (170)
—Atmospheric reactions: R.C.R.: 2.34 (49)
  reactivity: HC cons.: ranking 1.5
           NO ox.:   ranking 0.4 (63)
  estimated lifetime under photochemical smog conditions in S.E. England: 2.6 hr
                                                              (1699; 1707)
—Manmade sources:
  in gasoline (high octane number): 6.27 wt % (387)
—Ambient air quality:
  Los Angeles 1966: glc's: avg.:   0.008 ppm (n = 136)
                              highest value: 0.033 ppm (1319)
  in urban air: 0.01-0.1 ppm (n.s.i.) (29)
  glc's: in carpark: 0.16 mg/cu m
         on motorway: 1,060 mg/cu m (48)
  downtown Los Angeles: 1967: 10%ile: 2 ppb
                                  average: 6.5 ppb
                                  90%ile: 11 ppb (64)
  glc's in The Netherlands:
    tunnel Amsterdam, 1973:   avg.: 3 ppb (n = 3)
    tunnel Rotterdam, 1974.10.2: avg.: 7 ppb (n = 12)
                                      max.: 11 ppb
    The Hague, 1974.10.11:   avg.: 5 ppb (n = 12)
                                max.: 14 ppb
    Roelofarendsveen, 1974.9.11: avg.: 0.5 ppb (n = 12)
                                      max.: 1 ppb (1231)

1190   o-XYLENE

  −Control methods: catalytic combustion: platinized ceramic honeycomb catalyst:
    ignition temp.: 200°C, inlet temp. for 90% conversion: 250-300°C   (91)
  −Sampling and analysis:
    second derivative spectroscopy: det. lim.: 100 ppb   (42)
    photometry: min. full scale: 150 ppm   (53)
    6 f. abs. app.: 60 l air/30 min, UVS: det. lim.: 20 mg/cu m   (208)
    PMS: 4-70 ppm (n.s.i.)   (200)
C. WATER POLLUTION FACTORS:
  −Oxidation parameters
    BOD_5: 0.64 (n.s.i.)   (275)
           nil std. dil. sew.   (26, 41)
           1.64 NEN 3235-5.4   (277)
           1.80 adapted sew. NEN 3235-5.4   (277)
    COD: 2.91 NEN 3235-5.4   (277)
    $KMnO_4$ value: 0.102   (30)
    ThOD: 3.125   (30)
  −Biodegradation:
    incubation with natural flora in the groundwater−in presence of the other com-
      ponents of high-octane gasoline (100 μl/l)−biodegradation: 100% after 192 hr
      at 13°C (initial conc. 1.62 μl/l)   (956)
    cooxidation to o-toluic acid by *Nocardio* using hexadecane and by *Pseudomonas*
      using hexane as the growth substrate   (956)
  −Reduction of amenities: taste and odor: 0.3-1.0 mg/l (n.s.i.)   (41)
    T.O.C.: average: 2.21 mg/l
            range: 0.26-4.13 mg/l (n.s.i.)   (97)
    organoleptic limit: 0.05 mg/l (n.s.i.)   (181)
    T.O.C. in water: 1.8 ppm   (325)
  −Waste water treatment:
                         carbon
    A.C.: infl. ppm   dosage   effl. ppm   % reduction
          200   10X   29   86
          100   10X   32   68   (192)
    calculated half life time based on evaporative loss for a water depth of 1 m at
      25°C: 5.61 hr   (330)
  −Sampling and analysis: photometry: min. full scale: 6.5 $\times$ 10^{-6} mole/l   (53)
D. BIOLOGICAL EFFECTS:
  −Algae: *Chlorella vulgaris*: at 55 ppm: 50% reduction of cell numbers vs. controls
    after 1 day incubation at 20°C   (343)
  −Protozoa: threshold conc. of cell multiplication inhibition of the protozoan
    Uronema parduczi Chatton-Lwoff: >160 mg/l   (1901)
  −Arthropods:
    Daphnia magna: 24 hr TLm: 100-1,000 mg/l (n.s.i.)   (26)
    grass shrimp (*Palaemonetes pugio*): 96 hr LC_{50}: 7.4 ppm (n.s.i.)   (940)
    crab larvae−stage I (*Cancer magister*): 96 hr LC_{50}: 6 ppm   (941)
    shrimp (*Crangon franciscorum*): 96 hr LC_{50}: 1.3 ppm   (942)
  −Insects:
    mayfly nymphs (*Ephemerella walkeri*): lowest observed avoidance conc.: >10
      mg/l

—Fish: goldfish: LD_{50} (24 hr): 13 mg/l modified ASTM D 1345 (277)
n.s.i.: young Coho salmon: in artificial sea water at 8°C:
no significant mortality up to 10 ppm after 24 up to 96 hrs
mortality: 30/30 at 100 ppm after 24 hrs (317)
rainbow trout (*Salmo gairdneri*): lowest observed avoidance conc.: 0.01 mg/l
(1621)
rainbow trout: 96 hr LC_{50} (S): 13.5 mg/l (n.s.i.) (1211)
eel (*Anguilla japonica*): BCF: 21.4, $\frac{1}{2}$ life period: 2.0 days (1926)
eel: infiltration ratio: flesh/water: 0.46; eel flesh: 0.80 ng/g; water: 1.74 ng/g
(412)
guppy (*Poecilia reticulata*): 7 d LC_{50}: 35 ppm (1833)
fathead minnows: static bioassay in Lake Superior Water at 18-22°C: LC_{50}
(1; 24; 48; 72; 96 hr): 46; 42; 42; 42; 42 mg/l (n.s.i.)
goldfish (*Carassius auratus*): 96 hr LC_{50}: 16.9 ppm (n.s.i.) (943)
bass (*Morone saxatilis*): 96 hr LC_{50}: 11.0 ppm (943)
—Man: EIR: 2.3
severe toxic effects: 1,000 ppm = 4,410 mg/cu m, 60 min (n.s.i.)
symptoms of illness: 300 ppm = 1,323 mg/cu m, 60 min (n.s.i.)
unsatisfactory: 100 ppm = 441 mg/cu m, 60 min (n.s.i.) (185)

m-**xylene** (1,3-dimethylbenzene)

Manmade sources: in gasoline (high octane number): 12.03 wt % (387)
A. PROPERTIES: colorless liquid; m.w. 106.16; m.p. -48/-53°C; b.p. 139°C; v.p. 6 mm at 20°C, 11 mm at 30°C; v.d. 3.66; sp.gr. 0.864 at 20°C; sat. conc. 35 g/cu m at 20°C, 61 g/cu m at 30°C; log P_{oct} 3.20
B. AIR POLLUTION FACTORS: 1 mg/cu m = 0.23 ppm; 1 ppm = 4.41 mg/cu m

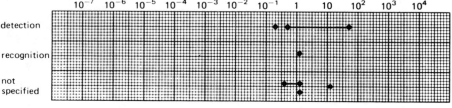

(307; 610; 642; 673; 727; 829)
O.I. at 20°C: 2100 (316)
—Atmospheric reactions:
R.C.R.: 2.67 (49)
reactivity: HC cons.: ranking 1–3 (63)
NO ox.: ranking 0.9–1 (63)
estimated lifetime under photochemical smog conditions in S.E. England: 1.5 hr
(1699; 1707)

m-XYLENE

- Ambient air quality:
  - Los Angeles 1966: glc's: avg.: 0.016 ppm ($n = 136$)
    - highest value: 0.061 ppm (1319)
  - glc's downtown Los Angeles, 1967: 10%ile: 4 ppb
    - avg.: 12 ppb
    - 90%ile: 21 ppb (64)
- Manmade sources:
  - in exhaust of gasoline engines: 62-car survey: 1.9 vol.% ($m+p$-xylene) of total exhaust HC's (391)
    - engine variable study: 2.5 vol.% ($m+p$-xylene) of total exhaust HC's (393)
  - in exhaust of diesel engine: 1.9% of emitted HC's ($m+p$-xylene) (72)
  - in exhaust of reciprocating gasoline engine: 1.3% of emitted HC's
  - in exhaust of rotary gasoline engine: 5.6% of emitted HC's ($m+p$-xylene) (78)
- Sampling and analysis: photometry: min. full scale: 200 ppm (53)

C. WATER POLLUTION FACTORS:
- Oxidation parameters: BOD: 2.53 NEN 3235-5.4 (277)
  - COD: 2.63 NEN 3235-5.3 (277)
- Biodegradation:
  - incubation with natural flora in the groundwater—in presence of the other components of high-octane gasoline (100 µl/l)—biodegradation: 100% after 192 hr at 13°C (initial conc. 3.28 µl/l) (956)

m-Xylene → (CH$_2$OH derivative) → (CHO derivative) → m-Toluic acid →→ 3-Methyl catechol →

initial steps in degradation by micro-organisms (1235)
- Reduction of amenities: T.O.C. in water: 1.0 ppm (325)
- Sampling and analysis: photometry: min. full. scale: 6.5×10^{-6} mole/l (53)

D. BIOLOGICAL EFFECTS:
- Protozoa:
  - ciliate (*Tetrahymena pyriformis*): 24 hr LC$_{100}$: 3.77 mmole/l (1662)
- Crustacean:
  - crab larvae—stage I (*Cancer magister*): 96 hr LC$_{50}$: 12 ppm (941)
  - shrimp (*Crangon franciscorum*): 96 hr LC$_{50}$: 3.7 ppm (942)
- Fish:
  - eel (*Anguilla japonica*): BCF: 23.6
    - half-life: 2.6 days ($m+p$-xylene) (1926)
  - eel: infiltration ratio: flesh/water: 0.47; eel flesh: 1.41 ng/g; water: 2.98 ng/g (412)
  - guppy (*Poecilia reticulata*): 14 d LC$_{50}$: 38 ppm (1833)
  - bass (*Morone saxatilis*): 96 hr LC$_{50}$: 9.2 ppm (942)
  - goldfish: 24 hr LD$_{50}$: 16 mg/l modified ASTM D 1345 (277)
- Man: EIR: 2.9 (49)

p-xylene (1,4-dimethylbenzene)

Manmade sources: in gasoline (high octane number): 4.68 wt % (389)

A. PROPERTIES: colorless; m.w. 106.17; m.p. 13°C; b.p. 138.4°C; v.p. 6.5 mm at 20°C, 12 mm at 30°C; v.d. 3.7; sp.gr. 0.86 at 20°C; solub. 198 mg/l at 25°C; sat. conc. 38 g/cu m at 20°C, 67 g/cu m at 30°C; log P_{oct} 3.15

B. AIR POLLUTION FACTORS: 1 mg/cu m = 0.23 ppm, 1 ppm = 4.41 mg/cu m

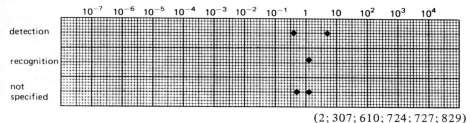

(2; 307; 610; 724; 727; 829)

O.I. at 20°C: 18,200 (316)

—Atmospheric reactions:
R.C.R.: 1.33
reactivity: HC cons.: ranking: 0.5–1.5
            NO ox.: ranking: 0.4
estimated lifetime under photochemical smog conditions in S.E. England: 2.4 hr
(1699; 1707)

—Ambient air quality:
Los Angeles 1966: glc's: avg.:   0.006 ppm (*n* = 136)
                 highest value: 0.025 ppm (1319)
glc's in carpark: 500 µg/cu m
   on motorway: 2,700 µg/cu m (48)
   downtown Los Angles 1967: 10%ile: 2 ppb
                         average: 5 ppb
                         90%ile: 10 ppb (64)

—Sampling and analysis:
photometry: min. full scale: 150 ppm (53)
scrubbers: liquid lift type: 15 ml of liquid; 4.5 l/min gas flow, 4 min scrubbing

trap method:	% removal
H_2O, 0°C	0
NH_2OH (soln), 0°C	0
H_2SO_4 (concd), 55°C	21
H_2SO_4, 4% Ag_2SO_4 (concd), 55°C	88
open tube, −80°C	87
ethanol, −80°C	89

(311)

p-XYLENE

C. WATER POLLUTION FACTORS:
- BOD_5: 0 (26)
  - 1% of ThOD (220)
  - 1.40   NEN 3235-5.4 (277)
  - 2.35 adapted sew. NEN 3235-5.4 (277)
- COD: 2.56   NEN 3235-5.3 (277)
  - 13% of ThOD (220)
- $KMnO_4$ value: 0.102 (30)
  - acid: 2% of ThOD; alkaline: 0% of ThOD
- ThOD: 3.125 (30)
- Biodegradation:
  incubation with natural flora in the groundwater—in presence of the other components of high-octane gasoline (100 μl/l)—biodegradation: 100% after 192 hr at 13°C (initial conc. 1.03 μl/l) (956)
  cooxidation to p-toluic acid by *Nocardia*—hexadecane as the growth substrate (956)

p-Xylene → (CH₂OH) → (CHO) → p-Toluic acid (COOH) → → 4-Methyl catechol (OH, OH) →

initial steps in degradation by microorganisms (1235)

- Water quality:
  in river Maas at Eysden (Netherlands) in 1976:
    median: n.d.; range: n.d.–0.1 μg/l
  in river Maas at Keizersveer (Netherlands) in 1976:
    median: n.d.; range: n.d.–0.1 μg/l (1368)
- Reduction of amenities: T.O.C. in water: 0.53 ppm (325)

D. BIOLOGICAL EFFECTS:
- Protozoa:
  ciliate (*Tetrahymena pyriformis*): 24 hr LC_{100}: 3.77 mmole/l (1662)
- Crustacean:
  shrimp (*Crangon franciscorum*); 96 hr LC_{50}: 2.0 ppm (942)
- Fish:
  guppy (*Poecilia reticulata*): 7 d LC_{50}: 35 ppm (1833)
  bass (*Morone saxatilis*): 96 hr LC_{50}: 2.0 ppm (942)
  fatheads: soft dilution water: TLm (24–96 hr): 28.8–26.7 mg/l (n.s.i.)
  fatheads: hard dilution water: TLm (24–96 hr): 28.8 mg/l (n.s.i.)
  bluegill: soft dilution water: TLm (24–96 hr): 24.0–20.9 mg/l (n.s.i.)
  goldfish: soft dilution water: TLm (24–96 hr): 36.8 mg/l (n.s.i.)
  guppies: soft dilution water: TLm (24–96 hr): 34.7 mg/l (n.s.i.) (158)
  goldfish: LD_{50} (24 hr): 18 mg/l—modified ASTM D 1345 (277)
- Man: EIR: 2.5 (49)

2,3-xylenol (2,3-dimethylphenol)

A. PROPERTIES: m.w. 122.17; m.p. 73–75°C; b.p. 217°C
C. WATER POLLUTION FACTORS:
   —BOD$_{24\ hr}^{30°C}$:   at 15 mg/l: 0% ThOD (seed water from phenol-degradation plant)
   2 days: at 15 mg/l: 40% ThOD (seed water from phenol-degradation plant)
   5 days: at 15 mg/l: 48% ThOD (seed water from phenol-degradation plant)
                                                                                  (564)
   —Biodegradation:
     adapted A.S. at 20°C—product is sole carbon source: 95.5% COD removal at 35
       mg COD/g dry inoculum/hr   (327)
   —Taste threshold conc.: 0.03 mg/l   (998)
   —Odor threshold conc.: (detection): 0.5 mg/l   (998)

2,4-xylenol (2,4-dimethylphenol)

Manufacturing source: coal tar fractionation; coal processing.   (347)
Users and formulation: intermediate in mfg. of phenolic antioxidants; pharmaceutical mfg.; plastics and resins mfg.; disinfectant (microbicide) mfg.; solvent mfg.; insecticides and fungicides, rubber chemicals, mfg. polyphenylene oxide, wetting agent, dyestuffs; cresylic acid constituent.   (347)
Natural sources (water and air): coal.   (347)
Man caused sources (water and air): asphalt and roadway runoff; general use of pharmaceuticals, fuels, plastics, pesticides; washing of dyed materials; constituent of domestic sewage.   (347)
A. PROPERTIES: colorless needles; m.w. 122.16; m.p. 26°C; b.p. 211.5°C; sp.gr. 1.036 at 20/4°C;
B. AIR POLLUTION FACTORS:
   —T.O.C.: recognition: 0.001 mg/cu m   (610)
             detection: 0.0005–0.4 mg/cu m   (829)
C. WATER POLLUTION FACTORS:
   —BOD$_5^{30°C}$: at 15 mg/l: nil (seed water from phenol-degradation plant)   (564)
   —Biodegradation: adapted A.S. at 20°C—product is sole carbon source: 94.5% COD removal at 28.2 mg COD/g dry inoculum/hr   (327)
   —Reduction of amenities:

approx. conc. causing adverse taste in fish (Rudd): 1 mg/l (41)
—Odor threshold conc.: detection: 0.4 mg/l (998)
—Taste threshold conc.: 0.5 mg/l (998)
—Waste water treatment:
  ion exchange: adsorption on Amberlite XAD-2: infl.: 0.4 ppm, effl.: 0 ppm, retention efficiency: 100% (40)
—Impact on biodegradation processes:
  inhibition of degradation of glucose by *Pseudomonas fluorescens* at: 40 mg/l
  inhibition of degradation of glucose by *E. coli* at: 500 mg/l (293)

D. BIOLOGICAL EFFECTS:
  —Bacteria: *E. coli*: LD_0: 500 mg/l
  —Algae: *Scenedesmus*: LD_0: 40 mg/l
  —Arthropoda: *Daphnia*: LD_0: 24 mg/l (30)
  —Fish: crucian carp: TLm (24 hr): 30 mg/l
        tench: TLm (24 hr): 13 mg/l
        "trout" embryo: TLm (24 hr): 28 mg/l (220)

2,5-xylenol (2,5-dimethylphenol)

A. PROPERTIES: m.w. 122.17; m.p. 71–73°C; b.p. 212°C; sp.gr. 0.971
B. AIR POLLUTION FACTORS:
  —T.O.C.: recognition: 2.0–2.3 mg/cu m (610)
          detection: 0.0005 mg/cu m (829)
C. WATER POLLUTION FACTORS:
  —$BOD_5^{30°C}$: at 15 mg/l: nil (seed water from phenol-degradation plant) (564)
  —Biodegradation:
    adapted A.S. at 20°C—product is sole carbon source: 94.5% COD removal at 10.6 mg COD/g dry inoculum/hr (327)
  —Impact on biodegradation processes:
    inhibition of degradation of glucose by *Pseudomonas fluorescens* at: 30 mg/l
    inhibition of degradation of glucose by *E. coli* at: >100 mg/l (293)
  —Odor threshold conc. (detection): 0.4 mg/l (998)
  —Taste threshold conc.: 0.5 mg/l (998)
D. BIOLOGICAL EFFECTS:
  —Fish:
    rainbow trout (*Salmo gairdneri*): 96 hr LC_{50}: 3.2–5.6 mg/l (static test) (1087)

2,6-xylenol (2,6-dimethylphenol)

A. PROPERTIES: m.w. 122.17; m.p. 44–46°C; b.p. 203°C; log P_{oct} 2.36
B. AIR POLLUTION FACTORS:
 —Odor threshold conc. (detection): 0.0002 mg/cu m (829)
C. WATER POLLUTION FACTORS:
 —$BOD_5^{30°C}$: at 15 mg/l: nihil (seed water from phenol-degradation plant) (564)
 —Biodegradation:
  adapted A.S. at 20°C–product is sole carbon source: 94.3% COD removal at 9.0 mg COD/g dry inoculum/hr (327)
 —Odor threshold conc. (detection): 0.4 mg/l (998)
D. BIOLOGICAL EFFECTS:
 —Ciliate (*Tetrahymena pyriformis*): 24 hr LC_{100}: 2.66 mmole/l (1662)

3,4-xylenol (3,4-dimethylphenol)

A. PROPERTIES: needles; m.w. 122.16; m.p. 65°C; b.p. 225°C; sp.gr. 1.023 at 17/15°C
B. AIR POLLUTION FACTORS:
 —Odor threshold conc. (detection): 0.003 mg/cu m (829)
C. WATER POLLUTION FACTORS:
 —BOD_5: 1.50 (27, 30)
 —$BOD_{24\ hr}^{30°C}$: at 15 mg/l: 1% ThOD (seed water from phenol-degradation plant)
  2 days: at 15 mg/l: 49% ThOD (seed water from phenol-degradation plant)
  5 days: at 15 mg/l: 65% ThOD (seed water from phenol-degradation plant)
  (564)
 —$KMnO_4$: 5.112 (30)
 —ThOD: 2.619 (30)
 —Biodegradation: adapted A.S. at 20°C–product is sole carbon source: 97.5% COD removal at 13.4 mg COD/g dry inoculum/hr (327)
 —Reduction of amenities:
  approx. conc. causing adverse taste in fish (carp): 5 mg/l (41)
 —Odor threshold conc. (detection): 1.2 mg/l (998)
 —Taste threshold conc.: 0.05 mg/l (998)
D. BIOLOGICAL EFFECTS:
 —Bacteria: *E. coli*: LD_0: >100 mg/l
 —Algae: *Scenedesmus*: LD_0: 40 mg/l
  Clorella pyrenoidosa: toxic: 49–81 mg/l (n.s.i.) (41)
 —Arthropoda: *Daphnia*: LD_0: 16 mg/l (30)
 —Fish:
  crucian carp: TLm (24 hr): 21 mg/l
  roach: TLm (24 hr): 16 mg/l; tench: TLm (24 hr): 18 mg/l
  "trout" embryos: TLm (24 hr): 7 mg/l (220)
  fathead minnows: static bioassay in Lake Superior Water at 18–22°C: LC_{50} (1; 24; 48; 72; 96 hr): >20; >20; 15; 14; 14 mg/l (350)

3,5-xylenol (3,5-dimethylphenol)

$$\underset{H_3C}{}\!\!\!\!\bigcirc\!\!\!\!\underset{CH_3}{}\!\!\!\text{OH}$$

A. PROPERTIES: needles; m.w. 122.16; m.p. 68°C; b.p. 219°C; log P_{oct} 2.35

B. AIR POLLUTION FACTORS:
 —Odor threshold conc. (detection): 4.10^{-5} mg/cu m (829)

C. WATER POLLUTION FACTORS:
 —BOD_5: 0.82 std. dil. sew. (30, 41)
 —$BOD_5^{30°C}$: at 15 mg/l: nil (seed water from phenol-degradation plant) (564)
 —$KMnO_4$ value: 5.112 (30)
 —ThOD: 2.619 (30)
 —Biodegradation: adapted A.S. at 20°C—product is sole carbon source: 89.3% COD removal at 11.1 mg COD/g dry inoculum/hr (327)
 —Reduction of amenities:
  approx/conc. causing adverse taste in fish (rudd): 1 mg/l (41)
 —Odor threshold conc. (detection): 5 mg/l (998)
 —Impact on biodegradation processes:
  inhibition of degradation of glucose by *Pseudomonas fluorescens* at: 25 mg/l
  inhibition of degradation of glucose by *E. coli* at: >100 mg/l (293)

D. BIOLOGICAL EFFECTS:
 —Protozoa:
  ciliate (*Tetrahymena pyriformis*): 24 hr LC_{100}: 2.3 mmole/l (1662)
 —Bacteria: *E. coli*: LD_0: >100 mg/l
 —Algae: *Scenedesmus*: LD_0: 40 mg/l
 —Arthropoda: *Daphnia*: LD_0: 16 mg/l (30)
 —Fish: crucian carp: TLm (24 hr): 53 mg/l
     tench: TLm (24 hr): 52 mg/l
     "trout" embryos: TLm (24 hr): 50 mg/l (222)

xylidines *see* dimethylanilines

Z

zectran (4-dimethylamino-3,5-xylyl-N-methylcarbamate; mexacarbate)

$(CH_3)_2N-\underset{CH_3}{\overset{CH_3}{\bigcirc}}-O-\overset{O}{\underset{}{C}}-NH-CH_3$

Use: herbicide

C. WATER POLLUTION FACTORS:
  —Persistence in riverwater in a sealed glass jar under sunlight and artificial fluorescent light - initial conc. 10 µg/l:

	% of original compound found				
after	1 hr	1 wk	2 wk	4 wk	8 wk
	100	15	0	0	0	(1309)

D. BIOLOGICAL EFFECTS:
  —Crustaceans:

Gammarus lacustris:	96 hr, LC_{50}: 46 µg/l	(2124)
Gammarus fasciatus:	96 hr, LC_{50}: 40 µg/l	(2126)
Palaemonetes kadiakensis:	96 hr, LC_{50}: 83 µg/l; 20 d LC_{50}: 25 µg/l	(2126)
Simocephalus serrulatus:	48 hr, LC_{50}: 13 µg/l	(2127)
Daphnia pulex:	48 hr, LC_{50}: 10 µg/l	(2127)

  —Insects:

Pteronarcys californica:	96 hr, LC_{50}: 10 µg/l	(2128)

  —Fish:

Pimephales promelas:	96 hr, LC_{50}: 17000 µg/l	(2121)
Lepomis macrochirus:	96 hr, LC_{50}: 11200 µg/l	(2121)
Lepomis microlophus:	96 hr, LC_{50}: 16700 µg/l	(2121)
Micropterus salmoides:	96 hr, LC_{50}: 14700 µg/l	(2121)
Salmo gairdneri:	96 hr, LC_{50}: 10200 µg/l	(2121)
Salmo trutta:	96 hr, LC_{50}: 8100 µg/l	(2121)
Oncorhynchus kisutch:	96 hr, LC_{50}: 1730 µg/l	(2121)
Perca flavescens:	96 hr, LC_{50}: 2480 µg/l	(2121)
Ictalurus punctatus:	96 hr, LC_{50}: 11400 µg/l	(2121)
Ictalurus melas:	96 hr, LC_{50}: 16700 µg/l	(2121)
bluegill: 24 hr LC_{50}: 11.2 ppm		
rainbow trout: 24 hr LC_{50}: 10.2 ppm		(1878)
rainbow trout fingerlings: 96 hr LC_{50}: 20.0 mg/l (static test)		(1101)

	96 hr LC$_{50}$, mg/l		96 hr LC$_{50}$, mg/l
catfish	11.4	bluegill	11.2
bullhead	16.7	bass	14.7
goldfish	19.1	rainbow	10.2
minnow	17.0	brown	8.1
carp	13.4	coho	1.7
sunfish	16.7	perch	2.5

(1934)

—Mammals:
acute oral LD$_{50}$ (male rat): 19 mg/kg
(dog): 15–30 mg/kg (1854)

zinc stearate
$Zn(C_{18}H_{35}O_2)_2$

Percentage of zinc may vary according to intended use, some products being more basic than others

Use: cosmetics; lacquers; plastics; powder metallurgy; dietary supplement; lubricant; medicine

D. BIOLOGICAL EFFECTS:
—Fish:
Lepomis macrochirus: chemical too insoluble in water to be toxic (1294)

zineb (zincethylene*bis*dithiocarbamate) *see also* karamate

$$CH_2-NH-\underset{\underset{S}{\|}}{C}-S\diagdown_{\displaystyle Zn}$$
$$CH_2-NH-\underset{\underset{S}{\|}}{C}-S\diagup$$

Use: fungicide (70% active ingredient).
A. PROPERTIES: solub. 10 ppm at room temp.
D. BIOLOGICAL EFFECTS:
—Fish:

	mg/l	24 hr	48 hr	96 hr	3m (extrapolated)	
harlequin fish						
(*Rasbora heteromorpha*) LC$_{10}$(F)	380	320	150			
LC$_{50}$(F)	560	400	250	100		(331)

—Mammals:
acute oral LD$_{50}$ (rat): >5200 mg/kg (1854)

zoalene *see* 3,5-dinitro-*o*-toluamide

zolone (O,O-dimethyl-S-((6-chloro-2-oxobenzoxazoline-3-yl)methyl)phosphorodithioate; phosalone)

A. PROPERTIES: m.p. 43–45°C; solub. 2.15 mg/l at 20°C; log P_{oct} 4.30
D. BIOLOGICAL EFFECTS:
 —Fish:
  zolone (35% EC): *Chingatta*: 96 hr LC_{50} (S): 0.081 mg/l (1494)
 —Mammals:
  acute oral LD_{50} (male rats): 120 mg/kg (1854)

FORMULA INDEX

H_2S	hydrogen sulfide
H_4ClN	ammoniumchloride
H_4FN	ammoniumfluoride
H_4N	ammonia
H_4N_2	hydrazine
H_6N_2O	hydrazinium hydroxide
$H_8N_2O_3S$	ammoniumsulfite
$H_8N_2O_4S$	ammoniumsulfate
$H_8N_2O_6S$	hydroxylamine sulfate
$CBrN$	cyanogenbromide
$CClN$	cyanogenchloride
CCl_2F_2	dichlorodifluoromethane
CCl_2O	phosgene
CCl_3F	fluorotrichloromethane
CCl_3NO_2	chloropicrin
CCl_4	carbon tetrachloride
CN_4O_8	tetranitromethane
COS	carbonylsulfide
CS_2	carbondisulfide
$CHBrCl_2$	bromodichloromethane
$CHBr_2Cl$	dibromochloromethane
$CHBr_3$	bromoform
$CHCl_3$	chloroform
CHN	hydrogencyanide
CH_2Cl_2	methylenechloride
CH_2O	formaldehyde
CH_2O_2	formic acid
CH_3Br	methylbromide
CH_3Cl	methylchloride
CH_3ClHg	methylmercuric chloride
CH_3ClO_2S	methanesulfonylchloride
CH_3I	methyliodide
CH_3NO	formamide
CH_3NO_2	nitromethane
$CH_3N_3O_3$	nitrourea
CH_4	methane
CH_4AsNaO	monosodiummethanearsenate
CH_4N_2O	urea
CH_4O	methanol
CH_4S	methylmercaptan

CH_5N	methylamine
CH_5N_3S	thiosemicarbazide
$C_2Cl_3F_3$	1,1,2-trichloro-1,2,2-trifluoroethane
C_2Cl_4	tetrachloroethylene
C_2Cl_6	hexachloroethane
$C_2HgN_2S_2$	mercuric thiocyanate
$C_2Na_2O_4$	oxalate, sodium
C_2HCl_3	trichloroethylene
C_2HCl_3O	chloral
$C_2HCl_3O_2$	trichloroacetic acid
$C_2H_2Cl_4$	1,1,2,2-tetrachloroethane
	1,2,2,2-tetrachloroethane
C_2HCl_5	pentachloroethane
C_2H_2	acetylene
$C_2H_2Br_4$	acetylenetetrabromide
$C_2H_2Cl_2$	1,1-dichloroethylene
	1,2-dichloroethylene
$C_2H_2Cl_2O_2$	dichloroacetic acid
$C_2H_2O_4$	oxalic acid
C_2H_3Cl	vinylchloride
$C_2H_3ClO_2$	monochloroacetic acid
$C_2H_3Cl_2NO_2$	1,1-dichloro-1-nitroethane
$C_2H_3Cl_3$	1,1,1-trichloroethane
	1,1,2-trichloroethane
$C_2H_3FO_2$	monofluoroacetic acid
C_2H_3N	acetonitrile
$C_2H_3NO_3$	peroxyacetylnitrate
C_2H_3NS	methylisothiocyanate
$C_2H_3O_2Tl$	thalliumacetate
C_2H_4	ethylene
$C_2H_4Br_2$	ethylenebromide
$C_2H_4Cl_2$	1,1-dichloroethane
	ethylenedichloride
$C_2H_4Cl_2O$	dichloromethylether
$C_2H_4NNaS_2$	metham, sodium
$C_2H_4N_2O_6$	nitroglycol
$C_2H_4N_4$	3-amino-1,2,4-triazole
C_2H_4O	acetaldehyde
	ethyleneoxide
$C_2H_4O_2$	acetic acid
	methylformate
$C_2H_4O_2S$	mercaptoacetic acid
$C_2H_4O_3$	glycolic acid
C_2H_5Cl	ethylchloride
C_2H_5ClO	2-chloroethanol
	chloromethylmethylether
$C_2H_5NaO_2$	propionate, sodium
C_2H_5N	ethyleneimine
C_2H_5NO	acetamide
$C_2H_5NO_2$	aminoacetic acid
C_2H_5NS	thioacetamide
C_2H_6	ethane

$C_2H_6N_2O$	N-dimethylnitrosamine
$C_2H_6N_2S$	ethylenethiourea
C_2H_6O	ethanol
C_2H_6OS	dimethylsulfoxide
$C_2H_6OS_2$	ethylxanthate
$C_2H_6O_2$	ethyleneglycol
$C_2H_6O_4S$	dimethylsulfate
C_2H_6S	dimethylsulfide
	ethylmercaptan
$C_2H_6S_2$	dimethyldisulfide
C_2H_7As	dimethylarsine
$C_2H_7AsO_2$	cacodylic acid
C_2H_7N	dimethylamine
	ethylamine
C_2H_7NO	ethanolamine
$C_2H_7NO_2$	ammoniumacetate
C_2H_7OS	2-mercaptoethanol
C_2H_8ClN	dimethylaminehydrochloride
$C_2H_8N_2$	1,1-dimethylhydrazine
	ethylenediamine
$C_2H_8N_2O_4$	ammoniumoxalate
$C_3H_2N_2$	malononitrile
$C_3H_3Cl_2NaO_2$	dichloropropionic acid, sodium salt
$C_3H_3Cl_3$	1,2,3-trichloropropene
C_3H_3N	acrylonitrile
C_3H_4	methylacetylene
C_3H_4Cl	2-chloro-1-propylene
$C_3H_4Cl_2$	1,3-dichloro-1-propene
	2,3-dichloro-1-propene
C_3H_4O	acrolein
	propargylalcohol
$C_3H_4O_2$	acrylic acid
$C_3H_4O_6$	malonic acid
$C_3H_5Br_2Cl$	1,2-dibromo-3-chloropropane
C_3H_5Cl	1-chloro-1-propene
	2-chloro-1-propene
	allylchloride
C_3H_5ClO	epichlorohydrine
$C_3H_5Cl_3$	1,2,3-trichloropropane
C_3H_5N	lactonitrile
C_3H_5NO	acrylamide
$C_3H_5N_3O_9$	nitroglycerine
C_3H_6	propylene
$C_3H_6ClNO_2$	1-chloro-1-nitropropane
$C_3H_6Cl_2$	1,2-dichloropropane
C_3H_6O	allylalcohol
	acetone
	propeneoxide
	propionaldehyde
$C_3H_6O_2$	ethylformate
	glycidol
	methylacetate
	propionic acid

INDEX 1205

$C_3H_6O_3$	dl-lactic acid
	glyceraldehyde
	1,3,5-trioxane
$C_3H_6O_4$	glyceric acid
C_3H_7Cl	1-chloropropane
	2-chloropropane
C_3H_7N	allylamine
C_3H_7NO	N,N-dimethylformamide
	acetoxime
$C_3H_7NO_2$	dl-alanine
$C_3H_7NO_2S$	l-cysteine
$C_3H_7NO_3$	dl-serine
C_3H_8	propane
C_3H_8O	isopropanol
	n-propanol
$C_3H_8O_2$	ethyleneglycolmonomethylether
	dimethoxymethane
	propyleneglycol
$C_3H_8O_3$	glycerol
C_3H_8S	n-propylmercaptan
C_3H_9N	isopropylamine
	n-propylamine
	trimethylamine
C_3H_9NO	isopropanolamine
	propionamide
$C_3H_{11}N_2S$	1,3-dimethylthiourea
C_4Cl_6	hexachloro-1,3-butadiene
$C_4H_2O_3$	maleic anhydride
$C_4H_4N_2$	succinonitrile
$C_4H_4N_2O_3$	uracil
C_4H_4O	furan
$C_4H_4O_4$	fumaric acid
	maleic acid
$C_4H_4O_5$	oxalacetic acid
C_4H_4S	thiophene
C_4H_5Cl	chloroprene
C_4H_5N	methacrylonitrile
C_4H_5NS	allylisothiocyanate
C_4H_6	1,3-butadiene
$C_4H_6Cl_2$	trans-1,4-dichloro-2-butene
$C_4H_6Cl_4$	tetrachloropropane
$C_4H_6Cl_2O_2$	3,4-dichloro-ω-butyric acid
$C_4H_6CuO_4$	cupric acetate
$C_4H_6MnN_2S_4$	manganeseethylene-bis-dithiocarbamate
$C_4H_6Na_2N_2S_4$	nabam
$C_4H_6N_2S_4Zn$	zineb
C_4H_6O	crotonaldehyde
	methacrolein
	methylvinylketone
$C_4H_6O_2$	3-butenoic acid
	methacrylic acid
	methylacrylate
	vinylacetate

$C_4H_6O_3$	acetic anhydride
$C_4H_6O_4$	succinic acid
$C_4H_6O_4Hg$	mercuric acetate
$C_4H_6O_4Pb$	leadacetate
$C_4H_6O_5$	*l*-malic acid
$C_4H_6O_6$	*dl*-tartaric acid
$C_4H_7Br_2Cl_2O_4P$	naled
C_4H_7Cl	methallylchloride
$C_4H_7Cl_2O_4P$	dichlorovinyldimethylphosphate
C_4H_7N	butyronitrile
C_4H_7NO	acetonecyanohydrin
	2-pyrrolidone
$C_4H_7NO_3$	N-acetylglycine
$C_4H_7NO_4$	*dl*-aspartic acid
C_4H_8	*cis*-2-butene
	trans-2-butene
	α-butylene
	isobutene
$C_4H_8Cl_2O$	β,β'-dichloroethylether
$C_4H_8Cl_3O_4P$	dimethyl(2,2,2-trichloro-1-hydroxylethyl)phosphonate
$C_4H_8N_2O_2$	dimethylglyoxime
$C_4H_8N_2O_3$	*l*-asparagine
$C_4H_8N_2S$	allythiourea
C_4H_8O	2-buten-1-ol
	1,2-butyleneoxide
	butyraldehyde
	isobutyraldehyde
	methylethylketone
	tetrahydrofuran
$C_4H_8O_2$	*n*-butyric acid
	1,4-dioxane
	ethylacetate
	3-hydroxybutanal
	isobutyric acid
	methylpropionate
$C_4H_8O_2S$	sulfolane
$C_4H_8O_3$	2-hydroxyisobutyric acid
C_4H_8S	crotylmercaptan
C_4H_9Cl	*n*-butylchloride
$C_4H_9NaO_3S$	butylsulfonate, sodium
C_4H_9NO	2-butanoneoxime
	butyramide
	N,N-dimethylacetamide
	morpholine
$C_4H_9NO_2$	N-acetylethanolamine
$C_4H_9NO_3$	homoserine
	threonine
C_4H_{10}	butane
	isobutane
$C_4H_{10}NO$	ethyldiethanolamine
$C_4H_{10}N_2O$	diethylnitrosamine
$C_4H_{10}O$	*n*-butanol
	sec-butanol

	t-butanol
	ethylether
$C_4H_{10}O_2$	1,4-butanediol
	ethyleneglycol monoethylether
$C_4H_{10}O_3$	diethyleneglycol
$C_4H_{10}O_4$	erythritol
$C_4H_{10}S$	n-butylmercaptan
	diethylsulfide
$C_4H_{10}S_2$	diethyldisulfide
$C_4H_{11}N$	n-butylamine
	sec-butylamine
	diethylamine
	isobutylamine
$C_4H_{11}NO$	N,N-diethylhydroxylamine
	N-ethylethanolamine
$C_4H_{11}NO_2$	diethanolamine
$C_4H_{11}O_3P$	diethylphosphite
$C_4H_{12}ClN$	tetramethylammoniumchloride
$C_4H_{12}Pb$	tetramethyllead
$C_4H_{13}N_3$	diethylenetriamine
$C_4H_{14}BN$	t-butylamineborane
C_5Cl_6	hexachlorocyclopentadiene
C_5HCl_4N	2,3,5,6-tetrachloropyridine
$C_5H_3Cl_3NO$	3,5,6-trichloropyridinol
$C_5H_3Na_3O_7$	carboxymethyltartronate
$C_5H_4N_2O_4$	orotic acid
$C_5H_4N_4$	xanthine
$C_5H_4N_4O$	hypoxanthine
$C_5H_4N_4O_3$	uric acid
$C_5H_4O_2$	furfural
$C_5H_4O_3$	2-furoic acid
C_5H_5N	pyridine
$C_5H_5NO_2$	methyl-2-cyanoacrylate
$C_5H_6N_2$	2-aminopyridine
	4-aminopyridine
	glutaronitrile
$C_5H_6N_2O_2$	thymine
C_5H_6O	2-methylfuran
$C_5H_6O_2$	furfurylalcohol
$C_5H_7N_3$	2,6-diaminopyridine
C_5H_8	cyclopentene
	isoprene
	2,3-pentadiene
C_5H_8O	cyclopentanone
	methylisopropenylketone
	4-pentenal
$C_5H_8O_2$	allylacetate
	ethylacrylate
	methylmethacrylate
	2,4-pentadione
$C_5H_8O_4$	dimethylmalonic acid
	glutaric acid
C_5H_9N	valeronitrile

C_5H_9NO	N-methyl-2-pyrrolidone
$C_5H_9NO_2$	*dl*-proline
$C_5H_9NO_3$	2-hydroxyproline
$C_5H_9NO_4$	*dl*-glutamic acid
C_5H_{10}	cyclopentane
	2-methyl-1-butene
	2-methyl-2-butene
	3-methyl-1-butene
	1-pentene
	2-pentene
$C_5H_{10}Cl_2$	1,5-dichloropentane
$C_5H_{10}N_2O_3$	*l*-glutamine
$C_5H_{10}N_2S_2$	dazomet
$C_5H_{10}O$	cyclopentanol
	diethylketone
	3-methylbutanal
	3-methyl-2-butanone
	2-pentanone
	n-valeraldehyde
$C_5H_{10}OS_2$	*sec*-butylxanthate
$C_5H_{10}O_2$	*n*-butylformate
	ethylpropionate
	isopropylacetate
	isovaleric acid
	pivalic acid
	n-propylacetate
	tetrahydrofurfurylalcohol
	n-valeric acid
$C_5H_{10}O_3$	ethyleneglycolmonomethyletheracetate
$C_5H_{11}Cl$	amylchloride
$C_5H_{11}NO$	N-methylmorpholine
	valeramide
$C_5H_{11}NO_2$	*dl*-valine
$C_5H_{11}NO_2S$	*dl*-methionine
C_5H_{12}	isopentane
	n-pentane
$C_5H_{12}ClO_2PS_2$	chlormephos
$C_5H_{12}NO_3PS_2$	dimethoate
$C_5H_{12}N_2O_2$	ornithine
$C_5H_{12}N_2S$	1,3-diethylthiourea
$C_5H_{12}O$	*prim*-isoamylalcohol
	sec-act-amylalcohol
	2-methyl-2-butanol
	n-pentanol
	3-pentanol
$C_5H_{12}O_2$	2-isopropoxyethanol
	1,5-pentanediol
$C_5H_{12}O_3$	diethyleneglycolmonomethylether
$C_5H_{12}O_4$	pentaerytritol
$C_5H_{12}S$	pentylmercaptan
$C_5H_{13}N$	*n*-amylamine
$C_6Cl_4O_2$	chloranil
C_6Cl_5F	pentachlorofluorobenzene

C_6Cl_5NaO	pentachlorophenate, sodium
$C_6Cl_5NO_2$	pentachloronitrobenzene
C_6Cl_6	hexachlorobenzene
$C_6HCl_4NO_2$	tetrachloronitrobenzene
C_6HCl_5	pentachlorobenzene
C_6HCl_5O	pentachlorophenol
C_6H_2	1,3-cyclohexadiene
$C_6H_2Br_4$	1,2,4,5-tetrabromobenzene
$C_6H_2Cl_3NO_3$	3,4,6-trichloro-2-nitrophenol
$C_6H_2Cl_4$	1,2,3,4-tetrachlorobenzene
	1,2,3,5-tetrachlorobenzene
	1,2,4,5-tetrachlorobenzene
$C_6H_2Cl_4O$	2,3,4,5-tetrachlorophenol
	2,3,4,6-tetrachlorophenol
	2,3,5,6-tetrachlorophenol
$C_6H_2Cl_5N$	pentachloroaniline
$C_6H_3Br_3$	1,2,4-tribromobenzene
	1,3,5-tribromobenzene
$C_6H_3Br_3O$	2,4,6-tribromophenol
$C_6H_3ClNaNO_5S$	4-nitrochlorobenzene-2-sulfonate, sodium
$C_6H_3Cl_2NO_2$	3,4-dichloronitrobenzene
$C_6H_3Cl_2NO_3$	2,5-dichloro-4-nitrophenol
$C_6H_3Cl_3$	1,2,3-trichlorobenzene
	1,2,4-trichlorobenzene
	1,3,5-trichlorobenzene
$C_6H_3Cl_3O$	2,3,5-trichlorophenol
	2,3,6-trichlorophenol
	2,4,5-trichlorophenol
	2,4,6-trichlorophenol
	3,4,5-trichlorophenol
$C_6H_3Cl_4N$	2-chloro-6-(trichloromethyl)pyridine
$C_6H_3Cl_3N_2O_2$	picloram
$C_6H_3N_3O_7$	picric acid
$C_6H_3N_3O_8$	2,4,6-trinitroresorcinol
$C_6H_4BrNO_3$	2-bromo-4-nitrophenol
	3-bromo-4-nitrophenol
$C_6H_4Br_2O$	2,4-dibromophenol
	2,5-dibromophenol
$C_6H_4ClNO_2$	o-chloronitrobenzene
	m-chloronitrobenzene
	p-chloronitrobenzene
$C_6H_4ClNO_3$	2-chloro-4-nitrophenol
	5-chloro-2-nitrophenol
$C_6H_4Cl_2$	o-dichlorobenzene
	m-dichlorobenzene
	p-dichlorobenzene
$C_6H_4Cl_2O$	2,3-dichlorophenol
	2,4-dichlorophenol
	2,5-dichlorophenol
	2,6-dichlorophenol
$C_6H_4Cl_3N$	2,4,6-trichloroaniline
$C_6H_4NNaO_5S$	m-nitrobenzene sulfonate, sodium
$C_6H_4N_2O_4$	o-dinitrobenzene

	m-dinitrobenzene
	p-dinitrobenzene
$C_6H_4N_2O_5$	2,4-dinitrophenol
	2,5-dinitrophenol
	2,6-dinitrophenol
$C_6H_4N_2O_6$	2,4-dinitroresorcinol
$C_6H_4O_2$	*p*-benzoquinone
C_6H_5Br	bromobenzene
C_6H_5BrO	*o*-bromophenol
	m-bromophenol
	p-bromophenol
C_6H_5Cl	chlorobenzene
C_6H_5ClO	*o*-chlorophenol
	m-chlorophenol
	p-chlorophenol
$C_6H_5ClO_2$	monochlorohydroquinone
$C_6H_5ClSO_2$	benzenesulfochloride
$C_6H_5ClSO_3$	*p*-chlorobenzenesulfonic acid
$C_6H_5Cl_2N$	2,4-dichloroaniline
$C_6H_5NaO_4S$	*p*-phenolsulfonate, sodium
$C_6H_5NO_2$	nitrobenzene
$C_6H_5NO_3$	citrazinic acid
	o-nitrophenol
	m-nitrophenol
	p-nitrophenol
$C_6H_5NO_5S$	*o*-nitrobenzenesulfonic acid
	m-nitrobenzenesulfonic acid
	p-nitrobenzenesulfonic acid
$C_6H_5N_3O_5$	picramic acid
C_6H_6	benzene
$C_6H_6Cl_6$	α-hexachlorocyclohexane
	β-hexachlorocyclohexane
	γ-hexachlorocyclohexane
C_6H_6ClN	*o*-chloroaniline
	m-chloroaniline
	p-chloroaniline
$C_6H_6NaO_5S$	hydroquinonemonosulfonate, sodium
$C_6H_6NaNO_5S$	4-nitrotoluene-2-sulfonate, sodium
$C_6H_6N_2O_2$	*o*-nitroaniline
	m-nitroaniline
	p-nitroaniline
$C_6H_6N_4O_7$	ammoniumpicrate
C_6H_6O	phenol
$C_6H_6O_2$	catechol
	hydroquinone
	resorcinol
$C_6H_6O_3$	phloroglucinol
	pyrogallol
$C_6H_6O_3S$	*m*-benzenesulfonic acid
$C_6H_6O_4S$	4-hydroxybenzenesulfonic acid
$C_6H_6O_6S_2$	*m*-benzenedisulfonic acid
C_6H_7N	aniline

INDEX 1211

	α-picoline
	β-picoline
C_6H_7NO	o-aminophenol
	m-aminophenol
	p-aminophenol
$C_6H_7O_3S$	aminophenolsulfonic acid
	o-anilinesulfonic acid
	m-anilinesulfonic acid
	p-anilinesulfonic acid
C_6H_7S	thiophenol
C_6H_8ClN	anilinechloride
$C_6H_8N_2$	adiponitrile
	o-phenylenediamine
	m-phenylenediamine
	p-phenylenediamine
$C_6H_8N_2O$	2,4-diaminophenol
	oxydipropionitrile
C_6H_8O	2,4-hexadienal
$C_6H_8O_7$	citric acid
C_6H_9NO	o-anisidine
	m-anisidine
	p-anisidine
$C_6H_9NO_6$	nitrilotriacetic acid
$C_6H_9N_2O_2$	dl-histidine
$C_6H_{10}Cl_2N_2O$	2,4-diaminophenolhydrochloride
C_6H_{10}	cyclohexene
$C_6H_{10}O$	cycloheanone
	mesityloxide
$C_6H_{10}O_2$	allylglycidylether
$C_6H_{10}O_4$	adipic acid
	diethyloxalate
	2,2-dimethylsuccinic acid
	glycoldiacetate
$C_6H_{10}O_6$	d-glucono-s-lactone
$C_6H_{10}O_8$	mucic acid
$C_6H_{11}ClO_2$	butylchloroacetate
$C_6H_{11}N$	di-N-allylamine
	hexanenitrile
$C_6H_{11}NO$	α-caprolactam
	cyclohexanoneoxime
$C_6H_{11}NO_2$	4-acetylmorpholine
C_6H_{12}	cyclohexane
	2,3-dimethylbutene-2
	1-hexene
	2-hexene
	methylcyclopentane
	2-methyl-1-pentene
	4-methyl-1-pentene
	4-methyl-2-pentene
C_6H_{12}	tetrahydromethylcyclopentadiene
$C_6H_{12}Cl_2O$	bis-2-chloro-1-methylethylether
	dichloroisopropylether

$C_6H_{12}N_2O_4S_2$	*l*-cystine
$C_6H_{12}N_4$	hexamethylenetetramine
$C_6H_{12}O$	butylmethylketone
	cyclohexanol
	n-hexaldehyde
	methylisobutylketone
$C_6H_{12}O_2$	*n*-butylacetate
	tert-butylacetate
	caproic acid
	1,2-cyclohexanediol
	diacetone alcohol
	dimethyldioxane
	ethylbutyrate
	isobutylacetate
	2-methylvaleric acid
	3-methylvaleric acid
	4-methylvaleric acid
$C_6H_{12}O_3$	ethyleneglycolmonoethyletheracetate
	paraldehyde
$C_6H_{12}O_6$	glucose
$C_6H_{12}O_7$	*d*-gluconic acid
$C_6H_{13}Cl$	1-chlorohexane
$C_6H_{13}N$	cyclohexylamine
	hexamethyleneimine
	l-isoleucine
$C_6H_{13}NO$	N-ethylmorpholine
$C_6H_{13}NO_2$	*l*-leucine
	dl-norleucine
C_6H_{14}	2,2-dimethylbutane
	2,3-dimethylbutane
	n-hexane
	2-methylpentane
	3-methylpentane
$C_6H_{14}N_2$	*l*-lysine
$C_6H_{14}N_4O_2$	*dl*-arginine
$C_6H_{14}O$	diisopropylether
	2-ethyl-1-butanol
	n-hexanol
	2-hexanol
	3-hexanol
	methylamylalcohol
$C_6H_{14}O_2$	butylcellosolve
	hexyleneglycol
$C_6H_{14}O_3$	diethyleneglycolmonoethylether
	dipropyleneglycol
$C_6H_{14}O_4$	triethyleneglycol
$C_6H_{14}O_6$	mannitol
$C_6H_{14}S$	di-*n*-propylsulfide
	prim-n-hexylmercaptan
$C_6H_{14}S_2$	dipropyldisulfide
$C_6H_{15}ClPb$	triethylleadchloride
$C_6H_{15}N$	diisopropylamine
	n-hexylamine

	di-N-propylamine
	triethylamine
$C_6H_{15}NO$	2-diethylaminoethanol
$C_6H_{15}NO_2$	diisopropanolamine
$C_6H_{15}NO_3$	triethanolamine
$C_6H_{15}O_2PS_3$	ekatin
$C_6H_{15}O_4PS_2$	oxydemetonmethyl
$C_6H_{16}N_2$	hexamethylenediamine
$C_6H_{16}Pb$	dimethyldiethyllead
$C_7H_3Br_4Cl$	tetrabromo-2-chlorotoluene
$C_7H_3Br_5$	pentabromotoluene
$C_7H_3Cl_2N$	dichlobenil
$C_7H_3Cl_2NO$	dichlorophenylsiocyanate
$C_7H_3Cl_2NO_4$	3-nitro-2,5-dichlorobenzoic acid
$C_7H_3Cl_3O_2$	2,3,6-trichlorobenzoic acid
$C_7H_3Cl_5O$	pentachloroanisole
	pentachlorobenzylalcohol
$C_7H_3Cl_5S$	pentachlorothioanisole
$C_7H_4ClNO_2$	6-chloropicolinic acid
$C_7H_4Cl_2O_2$	2,4-dichlorobenzoic acid
	2,5-dichlorobenzoic acid
$C_7H_4Cl_4O_2$	3,4,5,6-tetrachloroguaiacol
$C_7H_4F_2O_2$	2,6-difluorobenzoic acid
$C_7H_4F_3N_2O_3$	3-trifluoromethyl-4-nitrophenol
$C_7H_4N_2O_7$	3,5-dinitrosalicylic acid
C_7H_5ClO	benzoylchloride
	o-chlorobenzaldehyde
$C_7H_5ClO_2$	o-chlorobenzoic acid
	m-chlorobenzoic acid
	p-chlorobenzoic acid
$C_7H_5Cl_2NO_2$	3-amino-2,5-dichlorobenzoic acid
$C_7H_5Cl_3$	benzotrichloride
	2,4,5-trichlorotoluene
	2,4,α-trichlorotoluene
$C_7H_5Cl_3O_2$	3,4,5-trichloroguaiacol
$C_7H_5I_2NO_2$	2-amino-3,5-diiodobenzoic acid
C_7H_5N	benzonitrile
C_7H_5NO	phenylisocyanate
$C_7H_5NO_3$	o-nitrobenzaldehyde
	m-nitrobenzaldehyde
	p-nitrobenzaldehyde
$C_7H_5NO_4$	o-nitrobenzoic acid
	m-nitrobenzoic acid
	p-nitrobenzoic acid
	peroxybenzoylnitrate
$C_7H_5NS_2$	2-mercaptobenzothiazole
$C_7H_5N_3O_6$	2,4,6-trinitrotoluene
$C_7H_5N_5O_8$	tetryl
$C_7H_6Br_2$	2,5-dibromotoluene
$C_7H_6ClNaO_3S$	2-chlorotoluene-4-sulfonate, sodium
	2-chlorotoluene-5-sulfonate, sodium
$C_7H_6Cl_2$	2,4-dichlorotoluene
	3,4-dichlorotoluene

1214 INDEX

$C_7H_6NF_3$	aminobenztrifluoride
$C_7H_6N_2O_4$	2,3-dinitrotoluene
	2,4-dinitrotoluene
	2,6-dinitrotoluene
$C_7H_6N_2O_5$	4,6-dinitro-*o*-cresol
C_7H_6O	benzaldehyde
C_7H_6OS	thiobenzoic acid
$C_7H_6O_2$	benzoic acid
	salicylaldehyde
$C_7H_6O_3$	*m*-hydroxybenzoic acid
	p-hydroxybenzoic acid
	salicylic acid
$C_7H_6O_4$	gentisic acid
$C_7H_6O_5$	gallic acid
$C_7H_6O_5S$	2-sulfobenzoic acid
	3-sulfobenzoic acid
	4-sulfobenzoic acid
C_7H_7Cl	benzylchloride
	o-chlorotoluene
	m-chlorotoluene
	p-chlorotoluene
C_7H_7ClO	4-chloro-*o*-cresol
	p-chloro-*m*-cresol
$C_7H_7ClO_2$	5-chloro-3-methylcatechol
$C_7H_7NO_2$	*m*-aminobenzoic acid
	p-aminobenzoic acid
	anthranilic acid
	o-methylnitrobenzene
	m-methylnitrobenzene
	p-methylnitrobenzene
$C_7H_7NO_3$	1-methoxy-2-nitrobenzene
	1-methoxy-3-nitrobenzene
	1-methoxy-4-nitrobenzene
$C_7H_7NO_3$	2-nitro-*p*-cresol
	4-nitro-*m*-cresol
	6-nitro-*m*-cresol
C_7H_7NS	thiobenzamide
$C_7H_7N_2$	2-aminobenzimidazole
C_7H_8	1,3,5-cycloheptatriene
	toluene
C_7H_8ClN	4-chloro-*o*-toluidine
C_7H_8O	benzylalcohol
	o-cresol
	m-cresol
	p-cresol
	methylphenylether
$C_7H_8O_2$	2-methoxyphenol
	3-methoxyphenol
	4-methoxyphenol
$C_7H_8O_3S$	*p*-toluenesulfonic acid
C_7H_8S	4-methylbenzenethiol
$C_7H_9Cl_2N$	3-chloro-4-methylbenzenaminehydrochloride

INDEX 1215

C_7H_9N	benzylamine
	2,6-dimethylpyridine
	p-methylaminophenol
	N-methylaniline
	o-toluidine
	m-toluidine
	p-toluidine
$C_7H_9NO_3S$	4-amino-*m*-toluenesulfonic acid
$C_7H_{10}N_2$	pimelonitrile
$C_7H_{11}NO_5S$	4-methylaminophenol sulfate
C_7H_{12}	cycloheptene
$C_7H_{12}ClN_5$	2-chloro-4,6-*bis*(ethylamino)-2-triazine
$C_7H_{12}O$	2-methylcyclohexanone
	4-methylcyclohexanone
$C_7H_{12}O_2$	*n*-butylacrylate
$C_7H_{12}O_4$	2,2-dimethylglutaric acid
	3,3-dimethylglutaric acid
	pimelic acid
$C_7H_{12}O_6$	quinic acid
$C_7H_{13}N$	heptanenitrile
	mesidine
$C_7H_{13}O_6P$	mevinphos
C_7H_{14}	1,1-dimethylcyclopentane
	ethylcyclopentane
	1-heptene
	methylcyclohexane
$C_7H_{14}NO_5P$	azodrin
$C_7H_{14}N_2OS$	aldicarb
$C_7H_{14}O$	*n*-butylglycidylether
	heptanal
	2-heptanone
	3-heptanone
	4-heptanone
	4-methylcyclohexanol
	methylisoamylketone
$C_7H_{14}O_2$	*prim*-amylacetate
	2,2-dimethylvaleric acid
	4,4-dimethylvaleric acid
	enanthic acid
$C_7H_{14}O_4$	diethyleneglycolmonomethyletheracetate
C_7H_{15}	2,2,3-trimethylbutane
$C_7H_{15}Cl$	1-chloroheptane
$C_7H_{15}O_2$	4-methoxy-4-methyl-2-pentanone
C_7H_{16}	2,2-dimethylpentane
	2,3-dimethylpentane
	2,4-dimethylpentane
	3,3-dimethylpentane
	n-heptane
	2-methylhexane
	3-methylhexane
$C_7H_{16}O$	1-heptanol
$C_7H_{16}S$	*n*-heptylmercaptan

$C_7H_{17}O_2PS_3$	phorate
$C_7H_{18}Pb$	methyltriethyllead
$C_8Cl_4N_2$	chlorothalonil
$C_8H_2Cl_5NO$	pentachloromandelonitrile
$C_8H_4N_2$	phthalonitrile
$C_8H_4O_3$	o-phthalic anydride
$C_8H_5Cl_3O_2$	2,3,6-trichlorophenylacetic acid
$C_8H_5Cl_3O_3$	2,4,5-trichlorophenoxyacetic acid
	2,4,6-trichlorophenoxyacetic acid
$C_8H_5NO_2$	phthalimide
C_8H_6O	benzofuran
$C_8H_6Cl_2O_2$	2,4-dichlorophenylacetic acid
$C_8H_6Cl_2O_3$	dicamba
	3,6-dichloro-o-anisic acid
	2,4-dichlorophenoxyacetic acid
	3,4-dichlorophenoxyacetic acid
	2-methoxy-3,6-dichlorobenzoic acid
$C_8H_6O_4$	isophthalic acid
	o-phthalic acid
	terephthalic acid
C_8H_6S	thianaphthene
C_8H_7ClO	α-chloroacetophenone
$C_8H_7ClO_3$	4-chlorophenoxyacetic acid
$C_8H_7Cl_2NaO_4$	2,4-dichlorophenoxyacetic acid, sodium salt, monohydrate
C_8H_7N	benzylcyanide
	indole
C_8H_7NO	3-hydroxyindole
$C_8H_7NO_3$	p-nitroacetophenone
C_8H_8	styrene
$C_8H_8Br_2$	2,5-dibromoxylene
$C_8H_8Cl_2$	α,α'-dichloro-m-xylene
$C_8H_8Cl_3O_3PS$	ronnel
$C_8H_8HgO_2$	phenylmercuryacetate
C_8H_8O	acetophenone
	styreneoxide
	tolualdehyde
$C_8H_8O_2$	methylbenzoate
	phenylacetate
	phenylacetic acid
	o-toluic acid
	m-toluic acid
	p-toluic acid
$C_8H_8O_3$	o-anisic acid
	m-anisic acid
	p-anisic acid
	4-hydroxyphenylacetic acid
	dl-mandelic acid
	methylsalicylate
	phenoxyacetic acid
	vanillin
$C_8H_9ClNO_5PS$	dicapthon
C_8H_9NO	acetanilide

$C_8H_9NO_2$	3-nitro-o-xylene
	N-phenylglycine
$C_8H_9NO_3$	p-hydroxyphenylglycine
C_8H_{10}	ethylbenzene
	o-xylene
	m-xylene
	p-xylene
$C_8H_{10}NaN_3O_3S$	dexon
$C_8H_{10}NO_5PS$	dimethylparathion
$C_8H_{10}N_2O$	p-amino-acetanilide
$C_8H_{10}N_2O_2$	N,N-dimethyl-p-nitroaniline
$C_8H_{10}N_2O_4S$	asulam
$C_8H_{10}N_4O_2$	caffeine
$C_8H_{10}O$	o-ethylphenol
	p-ethylphenol
	α-methylbenzylalcohol
	2,3-xylenol
	2,4-xylenol
	2,5-xylenol
	2,6-xylenol
	3,4-xylenol
	3,5-xylenol
$C_8H_{10}O_2$	1,2-dimethoxybenzene
	1,3-dimethoxybenzene
	1,4-dimethoxybenzene
$C_8H_{10}O_5$	aquathol K
$C_8H_{11}N$	3,4-dimethylaniline
	3,5-dimethylaniline
	N,N-dimethylaniline
	2,3-dimethylaniline
	2,5-dimethylaniline
	2,6-dimethylaniline
	o-ethylaniline
	2-methyl-5-ethylpyridine
$C_8H_{11}NO$	4-amino-3,5-xylenol
C_8H_{12}	chrysene
	1,3-cyclooctadiene
	1,5-cyclooctadiene
$C_8H_{12}N_2$	p-aminodimethylaniline
$C_8H_{12}O_4$	diethylfumarate
C_8H_{13}	1,2,4-trimethylcyclopentane
C_8H_{14}	cyclooctene
$C_8H_{14}ClN_5$	atrazine
$C_8H_{14}O_2$	cyclohexylacetate
$C_8H_{14}O_4$	suberic acid
$C_8H_{14}O_5$	diglycoldiacetate
C_8H_{16}	cyclooctane
	1,2-dimethylcyclohexane
	ethyl-sec-amylketone
	ethylcyclohexane
	1-octene
	2-octene

1218 INDEX

	2,4,4-trimethyl-1-pentene
	2,4,4-trimethyl-2-pentene
$C_8H_{16}O$	octanal
	2-octanone
$C_8H_{16}O_2$	caprylic acid
$C_8H_{16}O_3$	butylcellosolve acetate
$C_8H_{16}O_4$	diethyleneglycolmonoethyletheracetate
$C_8H_{17}N_4OS$	4-amino-6-t-butyl-3-methylthio-1,2,4-triazin-5(4H)-one
$C_8H_{17}O$	dimethylcyclohexanol
C_8H_{18}	2,2-dimethylhexane
	2,3-dimethylhexane
	2,4-dimethylhexane
	2,5-dimethylhexane
	3,4-dimethylhexane
	isooctane
	2-methylheptane
	3-methylheptane
	4-methylheptane
	n-octane
	2,2,3-trimethylpentane
	2,3,3-trimethylpentane
	2,3,4-trimethylpentane
$C_8H_{18}O$	n-butylether
	2-ethyl-1-hexanol
	isooctanol
	1-octanol
	2-octanol
	4-octanol
$C_8H_{18}O_2$	ethyleneglycolmonohexylether
$C_8H_{18}O_3$	diethyleneglycoldiethylether
	diethyleneglycolmonobutylether
$C_8H_{18}O_4$	ethoxytriglycol
	triethyleneglycolmonoethylether
$C_8H_{18}S$	n-dibutylsulfide
	octylmercaptan
$C_8H_{19}N$	di-N-butylamine
	di-s-butylamine
	diisobutylamine
	2-ethylhexylamine
$C_8H_{19}O_2PS_3$	disulfoton
$C_8H_{19}O_3PS_2$	demeton
$C_8H_{20}Pb$	tetraethyllead
$C_8H_{20}O_5P_2S_2$	TEDP
$C_8H_{20}O_7P_2$	TEPP
$C_9H_4Cl_3NO_2S$	folpet
$C_9H_4O_3$	ninhydrin
$C_9H_6Cl_6O_3S$	endosulfan
C_9H_6N	isoquinoline
$C_9H_6N_2O_2$	2,4-toluenediisocyanate
	2,6-toluenediisocyanate
$C_9H_6O_2$	coumarin

$C_9H_7Cl_3O_3$	silvex
	2,4,5-trichlorophenoxy-ω-propionic acid
	2(2,4,5-trichlorophenoxy)propionic acid
$C_9H_7Cl_5N_2O$	imugan
C_9H_7N	quinoline
C_9H_7NO	8-hydroxyquinoline
C_9H_8	o-ethyltoluene
	m-ethyltoluene
	p-ethyltoluene
	indene
$C_9H_8Cl_2O_3$	2,4-dichlorophenoxy-α-propionic acid
	2,4-dichlorophenoxy-ω-propionic acid
$C_9H_8Cl_3NO_2S$	captan
C_9H_8O	cinnamaldehyde
$C_9H_9ClO_3$	4-chlorophenoxy-α-propionic acid
	4-chlorophenoxy-ω-propionic acid
	2-methyl-4-chlorophenoxy-acetic acid
$C_9H_9Cl_2NO$	N-(3,4-dichlorophenyl)propionamide
C_9H_9N	skatole
$C_9H_9N_3OS$	N-(2-benzothiazolyl)-N'-methylurea
$C_9H_9O_3N$	hippuric acid
$C_9H_{10}Cl_2N_2O$	3(3,4-dichlorophenyl)-1,1-dimethylurea
$C_9H_{10}Cl_2N_2O_2$	3-(3,4-dichlorophenyl)-1-methoxy-1-methylurea
$C_9H_{10}NO_3PS$	cyanox
$C_9H_{10}O_2$	phenylglycidylether
$C_9H_{10}O_3$	3(p-hydroxyphenyl)propionic acid
$C_9H_{11}ClN_2O$	3-(p-chlorophenyl)-1,1-dimethylurea
$C_9H_{11}Cl_2FN_2O_2S_2$	euparen
$C_9H_{11}Cl_3NO_3PS$	chlorpyrifos
$C_9H_{11}NO_2$	benzocaine
$C_9H_{11}NO_3$	dl-tyrosine
C_9H_{12}	isopropylbenzene
	mesitylene
	n-propylbenzene
	1,2,3-trimethylbenzene
	1,2,4-trimethylbenzene
$C_9H_{12}N_2O$	3(phenyl)-1,1-dimethylurea
$C_9H_{12}O$	2-phenyl-2-propanol
$C_9H_{12}O_2$	isopropylbenzenehydroxyperoxide
$C_9H_{13}BrN_2O_2$	bromacil
$C_9H_{13}N$	N,N-dimethyl-p-toluidine
$C_9H_{14}O$	isophorone
	phorone
$C_9H_{15}N$	tri-n-allylamine
$C_9H_{15}O_4P$	tris(2,3-dibromopropyl)phosphate
$C_9H_{17}NOS$	molinate
$C_9H_{17}NO_7$	muramic acid
$C_9H_{17}N_5S$	ametryn
C_9H_{18}	2,6-dimethyl-3-heptene
	propylene trimer

$C_9H_{18}O$	diisobutylketone
	n-nonanal
	2-nonanone
	5-nonanone
$C_9H_{18}O_2$	pelargonic acid
$C_9H_{19}NOS$	eptam
C_9H_{20}	2,2-dimethylheptane
	n-nonane
$C_9H_{20}O$	diisobutylcarbinol
	1-nonanol
$C_9H_{21}N$	tri-n-propylamine
$C_9H_{21}N_2S$	1,3-dibutylthiourea
$C_9H_{21}N_3O_3$	1,3,5-trihydroxyethylhexahydrotriazine
$C_9H_{22}O_4P_2S_4$	ethion
$C_{10}Cl_{12}$	mirex
$C_{10}H_4Cl_2O_2$	dichlone
	2,3-dichloro-1,4-naphtoquinone
	phygon
$C_{10}H_5ClN_2$	o-chlorobenzylidenemalononitrile
$C_{10}H_5Cl_7$	heptachlor
$C_{10}H_6Cl_4O_4$	dacthal
$C_{10}H_6Cl_6$	chlordene
$C_{10}H_6Cl_8$	chlordane
$C_{10}H_6O_2$	α-naphthoquinone
$C_{10}H_7Cl$	1-chloronaphthalene
$C_{10}H_7NO_2$	2-nitronaphthalene
$C_{10}H_8$	naphthalene
$C_{10}H_8O$	α-naphthol
	β-naphthol
$C_{10}H_8N_2$	2,2'-bipyridine
$C_{10}H_8O_3S$	1-naphthalenesulfonic acid
$C_{10}H_8O_4S$	1-naphthol-2-sulfonic acid
$C_{10}H_9Cl_4NO_2S$	difolitan
$C_{10}H_9Cl_7O$	heptachlorepoxide
$C_{10}H_9N$	2-methylquinoline
	6-methylquinoline
	8-methylquinoline
	α-naphthylamine
	β-naphthylamine
$C_{10}H_9NO_2$	indole-3-acetic acid
	1-phenyl-3-pyrazolidone
$C_{10}H_{10}Cl_2O_2$	bidisin
$C_{10}H_{10}Cl_2O_3$	2,4-dichlorophenoxy-ω-butyric acid
$C_{10}H_{10}Cl_6N_4O_2$	triforine
$C_{10}H_{10}O_4$	o-dimethylphthalate
	dimethylterephthalate
$C_{10}H_{11}ClO_3$	4-chlorophenoxy-ω-butyric acid
	2-(4-chlorophenoxy)-2-methylpropionic acid
$C_{10}H_{11}NO_2$	phenylalanine
$C_{10}H_{12}$	1,2,3,4-tetrahydronaphthalene
$C_{10}H_{12}ClNO_2$	isopropyl-N-(3-chlorophenyl)carbamate
$C_{10}H_{12}Cl_3O_2PS$	trichloronat

$C_{10}H_{12}N_2O_5$	2,4-dinitro-6-*sec*-butylphenol
$C_{10}H_{12}N_3O_3PS_2$	azinphosmethyl
$C_{10}H_{12}N_4O_5$	inosine
$C_{10}H_{13}ClN_2O_2$	dosanex
$C_{10}H_{13}NO_2$	isopropyl-N-phenylcarbamate
$C_{10}H_{13}ClN_2$	chlorphenamidine
$C_{10}H_{13}N_5O_5$	guanosine
$C_{10}H_{14}$	*n*-butylbenzene
	sec-butylbenzene
	tert-butylbenzene
	o-diethylbenzene
	m-diethylbenzene
	p-diethylbenzene
	isobutylbenzene
	1,2,4,5-tetramethylbenzene
$C_{10}H_{14}NO$	4-dimethylamino-3,5-xylenol
$C_{10}H_{14}NO_5PS$	parathion
$C_{10}H_{14}O$	*p-tert*-butylphenol
	menthol
	thymol
$C_{10}H_{14}O_2$	*p-tert*-butylcatechol
$C_{10}H_{15}N$	N,N-diethylaniline
$C_{10}H_{15}O_3PS_2$	baytex
$C_{10}H_{16}$	*dl*-α-pinene
$C_{10}H_{16}Cl_2NOS$	diallate
$C_{10}H_{16}N_2O_8$	ethylenediaminetetraacetic acid
$C_{10}H_{16}O$	camphor
$C_{10}H_{17}ClN$	*p*-aminodiethylanilinehydrochloride
$C_{10}H_{17}O_6N_3S$	glutathione
$C_{10}H_{18}$	decahydronaphthalene
$C_{10}H_{18}N_3O_3PS$	actellic
$C_{10}H_{18}O$	borneol
	α-terpineol
$C_{10}H_{19}ClNO_5P$	phosphamidon
$C_{10}H_{19}N_5O$	prometone
$C_{10}H_{19}O_6PS_2$	malathion
$C_{10}H_{20}$	1-decene
$C_{10}H_{20}NO_5PS_2$	mecarbam
$C_{10}H_{20}O_2$	*n*-capric acid
	2,2-dimethyloctanoic acid
$C_{10}H_{20}O_4$	butylcarbitolacetate
$C_{10}H_{21}Cl$	1-chlorodecane
$C_{10}H_{21}NOS$	pebulate
	S-propyldipropylthiocarbamate
	tillam
$C_{10}H_{22}$	*n*-decane
$C_{10}H_{22}O$	1-decanol
	isoamylether
$C_{10}H_{22}O_5$	tetraethyleneglycol
$C_{11}H_7N$	1-cyanonaphthalene
	2-cyanonaphthalene
$C_{11}H_8O_2$	naphthalenecarboxylic acid

$C_{11}H_8O_3$	2-hydroxy-3-naphthoic acid
$C_{11}H_{10}$	2-methylnaphthalene
	1-methylnaphthalene
$C_{11}H_{11}Cl_3O_3$	2,4,5-trichlorophenoxy-α-valeric acid
$C_{11}H_{11}N$	2,6-dimethylquinoline
$C_{11}H_{12}Cl_2N_2O_5$	chloramphenicol
$C_{11}H_{12}N_2O_2$	tryptophane
$C_{11}H_{12}O_3Cl_2$	2,4-dichlorophenoxy-α-valeric acid
$C_{11}H_{13}ClO_3$	4-chlorophenoxy-α-valeric acid
$C_{11}H_{14}N_2O_5$	4,6-dinitro-*o-sec*-amylphenol
$C_{11}H_{14}O_2$	*p-tert*-butylbenzoic acid
$C_{11}H_{15}NO_2S$	mesurol
$C_{11}H_{15}NO_3$	baygon
$C_{11}H_{16}$	*p-tert*-butyltoluene
$C_{11}H_{16}ClO_2PS_3$	carbophenothion
$C_{11}H_{16}N_2O$	aminocarb
$C_{11}H_{17}O_4PS_2$	fensulfothion
$C_{11}H_{18}N_2O_9$	1,3-diamino-2-propanoltetraacetic acid
$C_{11}H_{20}O_2$	undecylenic acid
$C_{11}H_{21}NOS$	ro-neet
$C_{11}H_{22}O$	2-undecanone
$C_{11}H_{22}O_2$	2-ethylhexylglycidylether
	n-undecylic acid
$C_{11}H_{23}N$	di-*n*-amylamine
$C_{11}H_{24}$	*n*-undecane
$C_{11}H_{24}O$	1-undecanol
$C_{12}H_2Cl_8$	2,2′,3,3′,4,4′,5,5′-octachlorobiphenyl
$C_{12}H_3Br_7$	2,2′,3,4,4′,5,5′-heptabromobiphenyl
$C_{12}H_4Cl_4O_2$	2,3,7,8-tetrachlorodibenzo-*p*-dioxin
$C_{12}H_5Cl_5$	2,4,5,2′,5′-pentachlorobiphenyl
$C_{12}H_5Cl_5O$	pentachlorodiphenylether
$C_{12}H_6Br_4$	2,2′,4,5′-tetrabromobiphenyl
$C_{12}H_6Cl_3NO_2$	2,4,6-trichlorophenyl-4′-nitrophenylether
$C_{12}H_6Cl_4$	2,3,2′,3′-tetrachlorobiphenyl
	2,4,2′,4′-tetrachlorobiphenyl
	2,4,3′,4′-tetrachlorobiphenyl
	2,5,2′,5′-tetrachlorobiphenyl
	2,5,3′,4′-tetrachlorobiphenyl
	2,3,4,5-tetrachlorobiphenyl
	2,3,5,6-tetrachlorobiphenyl
	2,6,2′,6′-tetrachlorobiphenyl
	3,4,3′,4′-tetrachloropbiphenyl
$C_{12}H_6Cl_4O$	2,3′,4,4′-tetrachlorodiphenylether
	2,2′,4,4′-tetrachlorodiphenyloxide
$C_{12}H_7Br_3$	2,4,5-tribromobiphenyl
$C_{12}H_7Cl_2NO_3$	2,4-dichlorophenyl-*p*-nitrophenylether
$C_{12}H_7Cl_3$	2,3,4-trichlorobiphenyl
	2,3,4′-trichlorobiphenyl
	2,3,6-trichlorobiphenyl
	2,4,5-trichlorobiphenyl
	2,4,6-trichlorobiphenyl
	2,5,2′-trichlorobiphenyl

	2,5,3'-trichlorobiphenyl
	2,5,4'-trichlorobiphenyl
	2,4,4'-trichlorobiphenyl
	3,4,2'-trichlorobiphenyl
$C_{12}H_7Cl_3O$	2,4,4'-trichlorodiphenylether
$C_{12}H_7Cl_3O_2$	5-chloro-2(2,4-dichlorophenoxy)phenol
	2,4,4'-trichloro-2'-hydroxydiphenylether
$C_{12}H_8Cl_2$	2,2'-dichlorobiphenyl
	2,3-dichlorobiphenyl
	2,4-dichlorobiphenyl
	2,4'-dichlorobiphenyl
	2,5-dichlorobiphenyl
	2,6-dichlorobiphenyl
	3,3'-dichlorobiphenyl
	3,4-dichlorobiphenyl
	3,5-dichlorobiphenyl
	4,4'-dichlorobiphenyl
$C_{12}H_8Cl_6$	aldrin
$C_{12}H_8Cl_6O$	dieldrin
	endrin
$C_{12}H_8N_2$	o-phenanthroline
$C_{12}H_9Cl$	2-chlorobiphenyl
	3-chlorobiphenyl
	4-chlorobiphenyl
$C_{12}H_9N$	carbazole
$C_{12}H_{10}$	acenaphthene
	diphenyl
$C_{12}H_{10}ClO_4S$	4-chloro-2,5-dihydroxydiphenylsulfone
$C_{12}H_{10}Cl_2$	bis(chloromethyl)naphthalene
$C_{12}H_{10}Cl_2N_2$	3,3'-dichlorobenzidine
$C_{12}H_{10}NO_2S$	5,6-dihydro-2-methyl-1,4-oxathün-3-carboxanilide
$C_{12}H_{10}N_2$	azobenzene
$C_{12}H_{10}O$	diphenylether
	o-phenylphenol
$C_{12}H_{10}O_2$	hydroquinonemonobenzylether
$C_{12}H_{10}O_4$	quinhydrone
$C_{12}H_{11}N$	N-diphenylamine
$C_{12}H_{11}NO_2$	carbaryl
$C_{12}H_{11}N_3$	p-phenylazoaniline
$C_{12}H_{12}$	1,3-dimethylnaphthalene
	2,6-dimethylnaphthalene
$C_{12}H_{12}Br_2N_2$	diquat
$C_{12}H_{12}N_2$	benzidine
$C_{12}H_{13}N$	N,N-dimethyl-1-naphthylamine
$C_{12}H_{14}ClN_2O_2PS$	chlorphoxim
$C_{12}H_{14}Cl_2N_2$	benzidine dihydrochloride
	paraquat
$C_{12}H_{14}Cl_3O_4P$	chlorfenvinphos
$C_{12}H_{14}NO_4PS_2$	imidan
$C_{12}H_{14}N_2O_6$	2-tert-butyl-4,6-dinitrophenylacetate
$C_{12}H_{14}O_4$	diethylphthalate
$C_{12}H_{15}NO_3$	carbofuran

$C_{12}H_{15}N_2O_3PS$	ekalux
$C_{12}H_{16}Cl_2N$	N-α-naphthylethylenediamine dihydrochloride
$C_{12}H_{16}Cl_2N_2O$	3-(3,4-dichlorophenyl)-1-methyl-1-n-butylurea
$C_{12}H_{16}NO_4PS_2$	zolone
$C_{12}H_{16}N_3O_3PS_2$	azinphosethyl
$C_{12}H_{17}O_4PS_2$	phenthoate
$C_{12}H_{18}$	1,5,9-cyclododecatreine
$C_{12}H_{18}N_2O_2$	mexacarbate
	zectran
$C_{12}H_{21}N_2O_3PS$	diazinon
$C_{12}H_{22}$	dicyclohexyl
$C_{12}H_{22}O_{11}$	sucrose
$C_{12}H_{24}$	1-dodecene
$C_{12}H_{24}O$	dodecanal
$C_{12}H_{24}O_2$	lauric acid
$C_{12}H_{25}Cl$	1-chlorododecane
$C_{12}H_{25}NaO_4S$	laurylsulfate, sodium
$C_{12}H_{26}$	n-dodecane
$C_{12}H_{26}O$	1-dodecanol
$C_{12}H_{26}O_6P_2S_4$	dioxathion
$C_{12}H_{27}N$	n-dodecylamine
	tri-n-butylamine
$C_{12}H_{27}OPS_3$	S,S,S-tributylphosphorotrithioate
$C_{12}H_{27}O_4P$	tri-n-butylphosphate
	tri-t-butylphosphate
$C_{12}H_{27}P$	merphos
$C_{13}H_6Cl_6O_2$	2,2'-methylene-bis(3,4,6-trichlorophenol)
$C_{13}H_7N$	1-cyanoacenaphthylene
	5-cyanoacenaphthylene
$C_{13}H_8O$	phenalen-1-one
$C_{13}H_9$	fluorene
$C_{13}H_9Cl_3N_2O$	3,3,4'-trichlorocarbanilide
$C_{13}H_9N$	acridine
	benzo(f)quinoline
	benzo(h)quinoline
	5-cyanoacenaphthene
$C_{13}H_{10}Cl_2O_2$	5,5'-dichloro-2,2'-dihydroxydiphenylmethane
$C_{13}H_{10}O$	xanthene
$C_{13}H_{10}O_2$	phenylbenzoate
$C_{13}H_{11}BrCl_2O_2PS$	leptophos
$C_{13}H_{11}NO_2$	N-phenylanthranilic acid
$C_{13}H_{11}N_3O$	2-(2'-hydroxy-5'-methylphenyl)-2H-benzotriazole
$C_{13}H_{12}$	diphenylmethane
$C_{13}H_{12}Cl_2N_2$	4,4'-methylene-bis(2-chloro-aniline)
$C_{13}H_{12}N_4O$	diphenylcarbazone
$C_{13}H_{12}N_4S$	diphenylthiocarbazone
$C_{13}H_{14}$	2,3,6-trimethylnaphthalene
$C_{13}H_{14}N_2$	4,4'-diaminodiphenylmethane
$C_{13}H_{14}N_4O$	1,5-diphenylcarbohydrazide
$C_{13}H_{16}F_3N_3O_4$	balan
	trifluralin
$C_{13}H_{16}N_2O_2$	melamine

INDEX 1225

$C_{13}H_{26}O$	2-tridecanone
$C_{13}H_{26}O_2$	tridecanoic acid
$C_{13}H_{28}$	n-tridecane
$C_{14}H_8Cl_4$	p,p'-DDE
$C_{14}H_8Cl_5$	o,p'-DDT
	DDT
$C_{14}H_8O_2$	anthraquinone
$C_{14}H_8O_5S$	anthraquinonesulfonic acid
$C_{14}H_9Cl_5O$	kelthane
$C_{14}H_9F_2N_2O_2$	diflubenzuron
$C_{14}H_{10}$	anthracene
	phenanthrene
$C_{14}H_{10}Cl_4$	DDD
$C_{14}H_{10}O_3$	o-benzoylbenzoic acid
$C_{14}H_{12}$	9,10-dihydroanthracene
	9,10-dihydrophenanthrene
$C_{14}H_{14}$	bibenzyl
$C_{14}H_{14}NO_4PS$	EPN
$C_{14}H_{14}O_4$	diallylphthalate
$C_{14}H_{15}N_3$	4-dimethylaminoazobenzene
$C_{14}H_{16}ClO_5PS$	co-ral
$C_{14}H_{16}N_2$	o-tolidine
$C_{14}H_{16}N_2O_2$	bianisidine
	o-dianisidine
$C_{14}H_{18}$	1,2,3,4,5,6,7,8-octahydroanthracene
$C_{14}H_{18}Cl_2O_4$	2,4-dichlorophenoxyacetic acid, butoxyethanolester
$C_{14}H_{18}N_4O_3$	benomyl
$C_{14}H_{18}O_4$	dipropylphthalate
$C_{14}H_{20}O_2$	2,6-di-t-butylbenzoquinone
$C_{14}H_{20}O_6P$	ciodrin
$C_{14}H_{22}O$	2,4-di-$tert$-butylphenol
	2,6-di-t-butylphenol
	p-(1,1,3,3-tetramethylbutyl)phenol
$C_{14}H_{23}O_{10}N_3$	diethylenetriaminepentacetic acid
$C_{14}H_{24}NO_4PS_2$	bensulide
$C_{14}H_{28}$	1-tetradecene
$C_{14}H_{28}O_2$	myristic acid
$C_{14}H_{30}$	n-tetradecane
$C_{15}H_{12}$	2-methylanthracene
	1-methylphenanthrene
	2-methylphenanthrene
	3-methylphenanthrene
	4-methylphenanthrene
$C_{15}H_{15}Cl$	4-chloro-4'-isopropylbiphenyl
$C_{15}H_{15}ClN_2O_2$	chloroxuron
$C_{15}H_{16}O$	p-cumylphenol
$C_{15}H_{17}NO$	p-isopropoxydiphenylamine
$C_{15}H_{18}N_2O_6$	binapacryl
$C_{15}H_{24}O$	2,6-di-$tert$-butyl-4-methylphenol
	nonylphenol
$C_{15}H_{24}O_2$	4-hydroxymethyl-2,6-di-$tert$-butylphenol
$C_{15}H_{30}O_2$	pentadecanoic acid

$C_{15}H_{32}$	n-pentadecane
$C_{15}H_{33}N_3O_2$	n-dodecylguanidineacetate
$C_{16}H_{10}$	fluoranthene
	pyrene
$C_{16}H_{12}$	2-phenylnaphthalene
$C_{16}H_{13}N$	N-phenyl-β-naphtylamine
	N-phenyl-α-naphthylamine
$C_{16}H_{15}Cl_3O_2$	methoxychlor
$C_{16}H_{17}NO$	diphenamid
$C_{16}H_{19}Cl_3O_3$	2,4,5-trichlorophenoxyacetic acid, isooctylester
$C_{16}H_{20}O_6P_2S_3$	abate
$C_{16}H_{22}O_2$	diisobutylphthalate
$C_{16}H_{22}O_4$	di-sec-butylphthalate
	di-n-butylphthalate
$C_{16}H_{26}O$	2,4-diamylphenol
$C_{16}H_{28}O_4$	butyloctylfumarate
$C_{16}H_{32}O_2$	palmitic acid
$C_{16}H_{34}$	n-hexadecane
$C_{16}H_{34}O$	1-hexadecanol
$C_{17}H_{10}O$	benzanthrone
$C_{17}H_{11}N$	benz(a)acridine
	benz(c)acridine
$C_{17}H_{12}$	benzo(b)fluorene
	benzo(a)fluorene
	benzo(c)fluorene
	1-methylpyrene
$C_{17}H_{17}ClO_6$	griseofulvin
$C_{17}H_{21}NO_2$	devrinol
$C_{17}H_{34}O_2$	heptadecanoic acid
$C_{17}H_{36}$	n-heptadecane
$C_{18}H_{10}$	cyclopenten(cd)pyrene
$C_{18}H_{12}$	benzo(c)phenanthrene
	benzo(a)anthracene
	triphenylene
$C_{18}H_{14}O$	2-diphenylphenylether
$C_{18}H_{15}O_4P$	triphenylphosphate
$C_{18}H_{16}OSn$	triphenyltinhydroxide
$C_{18}H_{20}O_2$	diethylstilbestrol
$C_{18}H_{24}N_2O_6$	dinocap
$C_{18}H_{28}O_3$	methyl-3-(3',5'-di-t-butyl-4'-hydroxyphenyl)propionate
$C_{18}H_{34}O_2$	oleic acid
$C_{18}H_{34}O_3$	9,10-epoxystearic acid
	ricinoleic acid
$C_{18}H_{36}O$	6,10,14-trimethyl-2-pentadecanone
$C_{18}H_{36}O_2$	stearic acid
$C_{18}H_{37}NO$	stearamide
$C_{18}H_{38}$	n-octadecane
$C_{18}H_{38}O$	1-octadecanol
$C_{18}H_{39}O_7P$	tri(2-butoxyethyl)phosphate
$C_{19}H_{14}$	methylbenz(a)anthracene
	2-methylchrysene
	4-methylchrysene

INDEX   1227

$C_{19}H_{16}O_4$	warfarin
$C_{19}H_{20}O_4$	butylbenzylphthalate
$C_{19}H_{26}O_2$	dimethrin
$C_{19}H_{26}O_3$	d-trans-allethrin
$C_{19}H_{30}O_5$	piperonylbutoxide
$C_{19}H_{34}O_3$	altosid-SR-10
$C_{19}H_{38}O_2$	nonadecanoic acid
$C_{19}H_{40}$	n-nonadecane
	pristane
$C_{19}H_{42}BrN$	cetyltrimethylammoniumbromide
$C_{19}H_{42}ClN$	cetyltrimethylammoniumchloride
$C_{20}H_{12}$	perylene
	benzo(a)pyrene
	benzo(e)pyrene
	benzo(j)fluoranthene
	benzo(b)fluoranthene
	benzo(ghi)fluoranthene
	benzo(k)fluoranthene
$C_{20}H_{24}O_2$	ethinyloestradiol
$C_{20}H_{28}O_2$	dehydroabietic acid
$C_{20}H_{30}O_2$	pimaric acid
	abietic acid
$C_{20}H_{40}$	1-eicosene
$C_{20}H_{40}O_2$	eicosanoic acid
$C_{20}H_{42}$	eicosane
$C_{21}H_{13}N$	dibenz(a,h)acridine
	dibenz(a,j)acridine
$C_{21}H_{21}O_4P$	tri-o-cresylphosphate
	tri-p-cresylphosphate
$C_{21}H_{23}NO_3$	fenpropanate
$C_{21}H_{22}N_2O_2$	strychnine
$C_{21}H_{24}$	n-heneicosane
$C_{21}H_{38}BrN$	cetylpyridiniumbromide
$C_{21}H_{42}O_2$	heneicosanoic acid
$C_{22}H_{13}$	3,4,9,10-dibenzopyrene
	indeno(1,2,3,cd)pyrene
$C_{22}H_{14}$	dibenz(a,c)anthracene
	dibenz(a,h)anthracene
	dibenzo(b,def)chrysene
	dibenzo(def,p)chrysene
	benzo(a)chrysene
$C_{22}H_{16}$	anthanthrene
$C_{22}H_{29}N_3O$	2-(2'-hydroxy-3',5'-di-t-amylphenyl)-2H-benzotriazole
$C_{22}H_{30}Cl_2N_{10}$	1,6-di(4-chlorophenyldiguanido)hexane
$C_{22}H_{42}O_4$	dioctyladipate
$C_{22}H_{42}O_2$	cis-erucic acid
$C_{22}H_{42}O_6$	triethyleneglycoldi(2-ethylhexanoate)
$C_{22}H_{44}O_2$	behenic acid
$C_{22}H_{44}$	1-docosene
$C_{22}H_{46}$	n-docosane
$C_{23}H_{22}O_6$	rotenone
$C_{23}H_{23}NO$	N-tritylmorpholine

$C_{23}H_{26}N_2O_4$	brucine
$C_{23}H_{46}O_2$	tricosanoic acid
$C_{23}H_{48}$	n-tricosane
$C_{24}H_{12}$	coronene
$C_{24}H_{21}ClNO_3$	fenvalerate
$C_{24}H_{38}O_4$	dioctylphthalate
$C_{24}H_{48}O_2$	n-tetracosanoic acid
$C_{24}H_{50}$	n-tetracosane
$C_{24}H_{54}OSn_2$	tri-n-butyltinoxide
$C_{25}H_{52}$	n-pentacosane
$C_{25}H_{50}O_2$	n-pentacosanoic acid
$C_{26}H_{54}$	n-hexacosane
$C_{26}H_{52}O_2$	n-hexacosanoic acid
$C_{27}H_{16}O$	cholesterol
$C_{27}H_{42}ClNO_2$	benzethonium chloride
$C_{27}H_{54}O_2$	n-heptacosanoic acid
$C_{27}H_{56}$	n-heptacosane
$C_{28}H_{40}O_2$	tetra-t-butyldiphenoquinone
$C_{28}H_{58}$	n-octacosane
$C_{28}H_{56}O_2$	octacosanoic acid
$C_{29}H_{60}$	n-nonacosane
$C_{30}H_{50}$	squalene
$C_{30}H_{50}O_4$	diundecylphthalate
$C_{30}H_{62}$	n-triacontane
$C_{31}H_{64}$	n-hentriacontane
$C_{32}H_{66}$	n-dotriacontane
$C_{33}H_{68}$	tritriacontane
$C_{35}H_{61}O_3$	octadecyl-3-(3',5'-di-t-butyl-4'-hydroxyphenyl)propionate
$C_{36}H_{70}O_4Zn$	zinc stearate
$C_{51}H_{98}O_6$	tripalmitin
$C_{54}H_{78}O_3$	1,3,5-trimethyl-2,4,6-tris(3,5-di-t-butyl-4-hydroxybenzyl)benzene
$C_{76}H_{52}O_{46}$	tannic acid

BIBLIOGRAPHY

1. NAPCA publication No. AP. 64, March 1970 "Air quality criteria for hydrocarbons."
2. Manuf. Chem. Assoc. "Research on chemical odor," Part 1, Oct. 1958.
3. A. D. Little, study for MCA, 1967.
5. USSR, "Odor perception levels."
7. Boundy-Bayer, "Styrene, its polymers and derivatives."
8. Truhaut, R., "Produits et problèmes pharmaceutiques," Sept. 1971.
9. BASF, "Arbeitsschutz im Werk" 29, 1970.
10. McCord & Whitheridge, "Odors: Physiology and control," 1949.
11. Wilby, F. V., Southern Cal. Gas Co. 1969.
13. Leithke, W., "The analysis of air pollutants."
16. Vom Wasser, XXXVII Band, Verlag Chemie, 1970.
17. Eldridge, E. F., "Industrial Waste Treatment," McGraw-Hill, 1942.
18. Herian, Ernest Dr., "Umweltforschung über nichthalogenierte Kohlenwasserstoffe." *Chemische Rundschau*, 24.10.1973.
19. Hellmann, T. M. and Small, F. H., "Characterization of odor properties of 101 petrochemicals using sensory methods," *Chem. Eng. progress*, 69, 9, Sept. 1973.
20. Fodor, "Schädliche Dämpfe," VDI Verlag, 1972.
23. Price, Kenneth S., et al., "Brine shrimp bioassay and seawater BOD of petrochemicals," *JWPCF*, 46, 1, Jan. 1974.
24. U.S. *Fed. Register* 39(81), April 25, 1974.
26. Elkins, H. F., et al., *Sewage Ind. Wastes*, 28 12, 1475, 1956.
27. Lund, Herbert F., *Industrial Pollution Control Handbook*, pp. 14-20, 14-21, table 2: Waste characteristics of some dissolved organic chemicals, McGraw-Hill, 1971.
29. Eckenfelder, W. W., "Industrial Water Pollution Control," McGraw-Hill 1966.
30. Meinck, F., Stoof, H. and Kohlschutter, H., "Les eaux résiduaires industrielles," 1970.
31. Lange Norbert Adolph, *Handbook of chemistry*, McGraw-Hill.
32. Guisti, D. M., Conway, R. A., Lawson, C. T., "Activated carbon adsorption of petrochemicals," *JWPCF* 46(5), 947-965, 1974.
33. N. Irving Sax, *Dangerous Properties of Industrial Materials*, Van Nostrand Reinhold.
34. EPA, Proposed criteria for water quality, Oct., 1973.
36. Ford, D. L., Eller, J. M., Gloyna, E. F., "Analytical parameters of petrochemical and refinery waste waters," *JWPCF*, 43(8), 1716, Aug. 1971.

Note: Refs. 4, 6, 12, 14, 15, 21, 22, 25, 28 and 35 are not cited in text.

BIBLIOGRAPHY

37. Hovious, Joseph C., Conway, Richard, A., Ganze, Charles W., "Anaerobic lagoon pretreatment of petrochemical wastes," *JWPCF* **45**(1), Jan 1973.
38. *JWPCF* **46**, July 1974.
39. Corstjens, G. H., Monnikendam, D., "Toepassing van een TOD analysator in het afvalwateronderzoek," *Water*, **6**(21), 1973.
40. Simpson, Richard M., "Progress in hazardous chemicals handling and disposal," The institute of advanced sanitation research international, Noyes Data Corporation, 1972.
41. Jones, H. R., "Environmental control in the organic and petrochemical industries," Noyes Data Corporation 1971.
42. Hager, Robert N., "Derivative spectroscopy with emphasis on trace gas analysis—LSI/Spectrometrics Inc.," *Anal. Chem.*, **45**(13), 1973.
43. Reimann, Dr. K., "Die Messung einiger Grössen des Sauerstoffhaushalts in Fliesswasser—Münchner Beitrage zur Abwasser Fischerei und Flussbiologie," Band 19, Verlag R. Oldenbourg.
44. The institute of advanced sanitation research international, "Hazardous chemicals handling and disposal," Noyes Data Corporation, 1971.
45. *Chemisch Weekblad*, 5.7.1974.
46. Simmonts, P. G., Kerrin, S. L., Lovelock, J. E., Shair, F. H., "Distribution of atmospheric halocarbons in the air over the Los Angeles basin," *Atm. Environm.* **8**, 209–216, 1974.
47. Linsky, Benjamin and Carter, Roy V., "Gaseous emissions from whiskey fermentation units," *Atm. Environm.* **8**, 57–62, 1974.
48. Perry, R. and Turbell, J. D., "A time based elution technique for the estimation of specific hydrocarbons in air," *Atm. Environm.*, **7**, p. 929, 1973.
49. Yeung, C. K. K. and Philips, C. R., "Estimation of physiological smog symptom potential from chemical reactivity of hydrocarbons," *Atm. Environm.* **7**, T. 551, 1973.
50. Okita, T., "Filter method for the determination of trace quantities of amines, mercaptans and organic sulphides in the atmosphere," *Atm. Environm.* **4**, 93, 1970.
51. Dague, Richard R., "Fundamentals of odor control," *JWPCF* **44**, 583, 1972.
52. Committee on the challenges of modern society, "Air quality criteria for photochemical oxidants and related hydrocarbons."
53. Du Pont de Nemours & Co., "Concentrations of liquids, gases and vapors measured with the Du Pont 400 photometric analyzer," Bulletin 5 A.
54. Henschler, prof. Dr. D., "Toxikologisch-Arbeitsmedizinische Begründung von MAK-werte," Verlag Chemie.
55. Hartmann & Braun, "URAS 2T nondispersive infrared gas analyzer."
56. Wilks, "Analyse des gaz par spectrometrie infrarouge: MIRAN."
57. Auer, "Auer-prüfrörchen für Auer-Toximeter und Auer-Gas tester," Auergesellschaft, 1000 Berlin 65 (west).
58. Dräger, "Gasspürgerät. 3," Drägerwerk AG, 24 Lübeck 1 Postfach 1339.
59. Matheson Scientific, "Detector tubes UNICO," Chicago, Elk Grove Village III.
60. Dräger, "Detector tube handbook."
61. Wirth, Wolfgang; Hect, Gerard; Gloxhuber, Christian, "Toxicologie Fibel," Georg Thieme Verlag, Stuttgart, 1971.
62. Gatellier, C. R., "Les facteurs limitant la biodégradation des hydrocarbures dans l'épuration des eaux," *Chimie et industrie–génie chimique*, **104**, 1971.
63. Altshuller, Aubrey P. and Bufalini, Joseph J., "Photochemical aspects of Air Pollution: A Review," *Environm. Sci. & Techn.*, **5**, Jan. 1971.
64. Altshuller, Aubrey P., Lonneman, William A., Sutterfiel, Frank D., Kopczyncki, Stanley L., "Hydrocarbon composition of the atmosphere of the Los Angeles basin 1967," *Environm. Sci. & Techn.*, **5**, 1971.
65. Cage, J. C., "The subacute inhalation toxicity of 109 industrial chemicals," *British Journal of Industrial Medicine* **27**, 1970.

66. Bürgi, Kurt, "Brandgase und ihre Toxizität. 3," *Monatzeitschrift des Gottlieb Duttweiler Instituts* 4/1971.
67. Heraeus GmbH, "Katalytische Abgasreinigung," 6450 Hanau, W. Germany.
68. Siemens, "Gaszusammensetzung, Heizwerte und Wärmetönung der Spaltgasreaktion," Wissenschaftlich technische Veröffentlichung aus dem Haus Siemens.
69. Kali-Chemie, "Engelhard Information," Hannover, W. Germany.
71. Wilby, "Variation in recognition odor threshold of a panel." *JAPCA*, Feb. 1969.
72. Weigert, Wolfgang; Roverstein, Edgard; Lakatos, Eduard, "Katalysatoren zur Reinigung von Autoabgasen," *Chemiker Zeitung*, 97, 1973.
73. Leonardos, G., et al. "Odor threshold determinations of 53 odorant chemicals," *JAPCA* 19(2), 91-95, 1969.
74. Paulet, G. "Action des hydrocarbures chlorofluorés (HCCF) utilisés dans les aerosols. Problèmes de leur rétention par l'organisme après inhalation," Cebedeau (Nov. 1970), rue Stévart 2, Liège, Belgium.
75. Du Pont Co. "450 Phenol analyzer system bulletin 450 A."
76. Murthy, B. N. and Eusebio, B. C. "Odor control research: a state of the art report and research recommendations," annual report: Research laboratory branch, div. of control systems, EPA, March 1971.
77. *Federal Register* 39(125), June 1974–subpart G: Occupational health and environmental control.
78. Schofield, Keith, "Problems with flame ionization detectors in automotive exhaust hydrocarbon measurements." *Environm. Sci. & Techn.* 8(9), Sept. 1974.
79. Lamb, C., Jenkins, G. F., "BOD of synthetic organic chemicals," Proc. 7th Ind. Waste Conf. Purdue Eng. Bull. 79–1953.
80. Brooker, Peter J. and Ellison, Michael "The determination of the water solubility of organic compounds by a rapid turbidimetric method," *Chem. and Ind.* 5 Oct. 1974.
81. Brandt, H. J. "Phenolabwässer and Abwasserphenol, ihre Entstehung, Schadwirkung und Abwasser technische Behandlung–eine monographische Studie," Wiss. Abhandl. Akademie Verlag, Berlin, 1958.
82. Yang, J. T., "Development of design parameters for biological treatment of industrial water," Dissertation, Univ. Texas, Austin, June 1968.
83. Turk, A., et al., *Am. Ind. Hyg. Assoc. Quart.* 13, 23, 1952.
84. Dietz, F., Ing. Traud J. "Bestimmung niedermolekularer Chlorkohlenwasserstoffe in Wässern und Schlämmen mittels Gaschromatographie," Ruhrverband.
85. Faust, S. J., Hunder, J. V., "Organic compounds in aquatic environment," Marcel Dekker, NY, 1971.
86. Murtaugh, J. and Bung, R. "Acidic components of sewage effluents and river water," *JWPCF* 37, 410, 1965.
87. Bunch, R. L. and Chamber C. W., "A biodegradability test for organic compounds," *JWPCF* 39(2), 181, 1967.
88. Gerhold, R. M. and Malaney, G. W., *JWPCF* 38(4), 562, 1966.
89. Malaney, G. W. and Gerhold, R. M. "Structural determinants in the oxidation of aliphatic compounds by activated sludge." *JWPCF* 41(2), part 2; R18–R33, 1969.
90. US EPA "Disposal of hazardous wastes," Report to Congress SW-115, 1974.
91. Honeycat "Air pollution control units," JMC, 74 Hatton Garden, London.
92. McKinney et al., *Sew. Ind. Wastes*, 28(4), 547, 1956.
93. Ludzack, F. J.; and Ettinger, M. B., *JWPCF* 32, 1173, 1960.
94. Sawyer, C. N., et al., *JWPCF* 32(12), 1274, 1960.
95. Masselli and Burford "A simplification of textile waste survey and treatment," New Engl. Intern. Water Pollut. Control Comm., Boston, July 1959.

Note: Ref. 70 is not cited in the text.

96. Dickerson, B. W., *Sewage Ind. Wastes*, **22**; 536, 1950.
97. American Water Works Association Inc., N.Y., 1963.
98. Carlson, D. A. and Lerser, C. P. "Soil beds for the control of sewage odors" *JWPCF* **38**(5), 829, 1966.
99. *J. Am. Water Works Assoc.*, **56**(3), 1964.
100. Hoffman, D. and Wynder, E. L., in *Air Pollution*, A. C. Stern, ed., Academic Press, 1968.
101. Cavanagh, L., *Environm. Sci. and Techn.* **3**(3), 255, March 1969.
102. Altshuller, A. P., "Atmospheric analysis by gas chromatography" U.S. Dept. of H.E.W., Public Health Services, Div. Air Pollut. 1969, and California Dept. Publ. Health, SDPH 1-SPDM-50, Aug. 1966.
103. Bloomfield, B. D., in *Air Pollution*, A. C. Stern, ed., Vol. 2, pp. 523, 1968.
104. Fassett, D. W. "Aldehydes and acetals," *Industrial Hygiene and Toxicology*, Patty, F. A., ed., Vol. II, John Wiley & Sons, 1963.
105. Katz, M. "Effects of contaminants other than sulfur dioxide, on vegetation and animals," Background papers prepared for the national conference on pollution and our environment, Canadian council of resource ministers, Montreal, Canada, Oct.–Nov. 1966.
106. Heyrith, F. F. "Hydrogen chloride" *Industrial Hygiene and Toxicology*, Patty, F. A., ed., Vol. II, John Wiley & Sons, 1963.
107. Machle, W., et al. "The effect of the inhalation of hydrogen chloride." *J. Ind. Hyg. Toxic.*, **24**, 222, 1942.
108. Cralley, L. V. "The effect of irritant gases upon the rate of ciliary activity," *J. Ind. Hyg. Toxic.*, **24**, 193, 1942.
109. Crocker, W. "Physiological effects of ethylene and other unsaturated carbon containing gases," *Growth and Plants*, chapter 4, Reinhold Publishing Co., 1948.
110. Marchesani, V. J., et al., *JAPCA*, **20**(1), 19, 1970.
111. U.S. Dept. of H.E.W. "Preliminary Air Pollution Survey of Ammonia" NAPCA, Raleigh, N.C., 1969, p. 16.
112. Eliassen, R. "Domestic and municipal sources of Air Pollution," Proceedings of the national conference on Air Pollution, Public Health Service Publ. no 654, U.S. Dept. H.E.W., 1959.
113. U.S. Dept. of H.E.W. "Preliminary Air Pollution Survey of Hydrogen Sulfide," NAPCA, Raleigh, N.C., 1969.
114. U.S. Dept. of H.E.W. "Preliminary Air Pollution Survey of Ethylene," NAPCA, Raleigh, N.C., 1969.
115. Posselt, H. S. and Reidies, A. H., "Odor abatement with potassium permanganate solutions," *Ind. Eng. Chem., Prod. Res. Developm.* **4**, 48, 1965.
117. Akamine, E. K. and Sakamoto, H. I., "Brominated charcoal to prevent fading of *Vanda* Orchid Flowers," *Amer. Orchid Soc. Bull.*, **20**; 149, 1951.
118. "California Standards for Ambient Air Quality and Motor Vehicle Exhaust, suppl. 2," California State Department of Public Health, Berkeley, Calif., 1962.
119. Davidson, O. W. "Effects of Ethylene on Orchid Flowers," *Proc. Amer. Soc. Hort. Sci.*, **53**, 440, 1949.
120. Crocker, W., Zimmerman, P. W. and Hitchcock, A. E. "Ethylene induced epinasty of leaves, and the relation of gravity of it," *Contrib. Boyce Thompson Inst.* **4**, 177, 1932.
121. Heck, W. W., Pires, E. G. and Hall, W. C. "The effects of a low ethylene concentration on the growth of cotton," *JAPCA*, **11**(12), 549, 1961.
122. Hitchcock, A. E., Crocker, W. and Zimmerman, P. W. "Effect of illuminating gas on lily, narcissus, tulip and hyacinth," *Contrib. Boyce Thompson Inst.*, **4**, 155, 1932.
123. Zimmerman, P. W., Hitchcock, A. E. and Crocker, W. "The effect of ethylene and illuminating gas on roses," *Contrib. Boyce Thompson Inst.*, **3**. 459, 1931.

Note: Ref. 116 is not cited in the text.

124. Denny, F. E. and Miller, L. P. "Production of ethylene by plant tissue as indicated by the epinastic response of leaves," *Contrib. Boyce Thompson Inst.* **7**, 97, 1935.
125. Crocker, W. "Physiological effects of ethylene and other unsaturated carbon-containing gases," in *Growth of Plants*, Reinhold Publishing Co., N.Y., 1948.
126. Pohraugh, P. W. "Measurement of small concentrations of ethylene in automobile exhaust gases and their relation to lemon storage," *Plant Physiol.*, **18**, 79, 1943.
127. Doubt, S. L. "The response of plants to illuminating gas," *Botan. Gaz.*, **63**, 209, 1917.
128. Crocker, W. and Knight, L. I., "Effect of illuminating gas and ethylene upon flowering carnations," *Botan. Gaz.*, **46**, 259, 1908.
129. Fieldnap, A. C., Sayers, R. R. et al., "Warning agents for fuel gases," U.S. Bureau of mines, Monograph No. 4, 1931.
130. *Literaturberichte über Wasser, Abwasser, Luft und feste Abfallstoffe*, **Band 15, Heft 1**.
131. idem, **Band 15, Heft 4**.
132. idem, **Band 16, Heft 1**.
133. idem, **Band 16, Heft 2**.
134. idem, **Band 16, Heft 3**.
135. idem, **Band 16, Heft 5**.
136. idem, **Band 17, Heft 2**.
137. idem, **Band 17, Heft 3**.
138. idem, **Band 17, Heft 4**.
139. idem, **Band 17, Heft 5**.
140. idem, **Band 18, Heft 2**.
141. idem, **Band 18, Heft 4**.
142. idem, **Band 19, Heft 3**.
143. idem, **Band 19, Heft 4**.
144. idem, **Band 20, Heft 1**.
145. idem, **Band 20, Heft 2**.
146. idem, **Band 20, Heft 3**.
147. idem, **Band 20, Heft 4**.
148. idem, **Band 21, Heft 1**.
149. idem, **Band 21, Heft 2**.
150. Mann, Jon B., Enos, Henry F.; Gonzales, Jorge, and Thompson, John F., "Development of sampling and analytical procedure for determining hexachlorobenzene and hexachloro-1,3-butadiene in air," *Environm. Sci. & Techn.*, **8**(6), June 1974.
151. Cormack, Douglas, Derling, Thomas A., Lynch, Bernard W. J., "Comparison of techniques for organoleptic odor-intensity assessment." *Chem. and Ind.*, 2 Nov. 1974.
152. Worne, H. E. "The activity of mutant microorganisms in the biological treatment of industrial wastes," *Tijdschrift van het BECEWA*, Liège, Belgium.
153. Dowden, B. F. and Bennett, H. J., *JWPCF* **37**(9), 1310, 1965.
154. Gauhey, P. H. Mc *Engineering Management of Water Quality*, McGraw-Hill, 1968.
155. Grindley, *Ann. Appl. Biol.*, **33**, 103, 1946.
156. Longwell and Pentelow, *J. Exp. Biol.*, **21**, 1, 1935.
157. Ellis, M. M. "Detection and measurement of stream pollution." U.S. Dept. of Commerce, Bureau of Fisheries, 1937.
158. Pickering, Q. H. and Henderson, C., *JWPCF*, **38**(9), 1419, 1966.
159. *J. Am. Water Assn.*, **57**(5), 663, 1965.
160. Ford, D. L., "Application of the Total Carbon Analyzer for Industrial Wastewater Evaluation," Proc. 23rd. Ind. Waste Conf., Purdue University, Lafayette, Ind., 1968.
161. Gellman, I., "Studies on Biochemical Oxidation of Sewage, Industrial Wastes and Organic Compounds," Ph. D. Thesis, Rutgers Univ., 1952.
162. Zobell, C. E., *Biol. Bull.* **78**, 388, 1940.
163. Meissner, B., *Wasserwirtschaft-Wassertechnik*, **4**, 166, 1954.
164. Hess, R. W., private communication.

165. Swope, H. G. and Kenna, M., *Sewage Ind. Wastes Eng.*, 21, 467, 1950.
166. Mills, E. J., Jr. and Stack, V. T., Jr., Proc. 8th Purdue Ind. Waste Conf., Lafayette, Ind., 1953.
167. Weston, R. S., "Activated Carbon in Sewage and Industrial Waste Treatment," Masters Thesis, New York Univ., 1939.
168. Staniar, W., *Plant Engineering Handbook*, McGraw-Hill, New York, 1950.
169. Masselli, J. W. and Burford, M. G., "Pollution Sources in Wood Scouring and Finishing Mills and Their Reduction through Process and Process Chemical Changes," Rept. New England Inter. Water Poll. Control Comm., 1954.
170. Stern A. C., ed., *Air Pollution*, Vol. 3, p. 661, Academic Press N.Y., London, 1968.
171. Howe, Robert H. L., "Hazardous chemicals handling and disposal," Noyes Data Corporation, 1970.
172. Howe, Robert H. L. and Kent, Robert, "Hazardous chemicals handling and disposal," Noyes Data Corporation, 1970.
173. Howe, Robert H. L., "The removal of toxic organic chemicals by a specific flotation technique," "Hazardous chemicals handling and disposal," Noyes Data Corporation, 1970.
174. Kuprevich, V. F. and Shcherbakova, T. A., "Comparative enzymatic activity in diverse types of soil," *Soil Biochemistry*, Vol. 2, Marcel Dekker, Inc., New York, 1971.
175. Woodcock, David, "Metabolism of fungicides and nematocides in soils," *Soil Biochemistry*, Vol. 2, Marcel Dekker Inc., New York, 1971.
176. Alexander, M. and Lustigman, B. K., *J. Agr. Food Chem.*, 14, 410, 1966.
177. Walker, N. and Wiltshire, G. H., I *Gen. Microbiol.* 8, 273, 1953.
178. "Technische Anleitung zur Reinhaltung der Luft," W. Germany, 1974.
179. Dhandapani, Thirumurthi, "Design criteria for waste stabilization ponds," *JWPCF*, 46(9), 1974.
180. Stern, A. C., *Air Pollution*, Vol. 3., p. 663 Academic Press, N.Y. 1968.
182. Bach, Rolf Werner; Dötsch, Erwin; Friedrichs, Hans Adolf; Marx, Lothar, "Nachweis unverbrannter Kohlenwasserstoffe in Heizölabgasen," *Staub-Reinhalt. Luft* 33, 1973.
183. Korte, F., "Percolatiewaterafvalstortplaatsen bevat gevaarlijke chemicaliën," Orientatiedag OGEM-MPC Milieutechniek, Nov. 1974.
185. Strafford, N.; Strouts, C. R. N.; Stubbings, W. V., *The Determination of Toxic Substances in Air*, Heffer, Cambridge.
186. Valland, A. and Damel, R., "L'hygiène et la sécurité dans la grande industrie chimique," Inst. nat. de recherche et de sécurité pour la prévention des accidents du travail et des maladies professionnelles, 9 avenue Montaigne, Paris 8e, 1971.
188. Sargent, J. W. and Sanks, R. L., "Light energized oxidation of organic wastes," *JWPCF*, 46(11), Nov. 1974.
190. Grupinski, L. and Staub, W., "Dosage par spectrometrie du I. R. de crésol émis par les installations d'émaillage de fils," *Staub*, May 1973.
192. Dahm, Douglas B., Pilié, Roland, J., Laforna, Joseph P., "Technology for managing spills on land and water," *Environm. Sci. and Techn.*, 8(13), Dec. 1974.
193. Neely, W. Brock; Branson, Dean R.; Blau, Gary E., "Partition coefficient to measure bioconcentration potential of organic chemicals in fish," *Environm. Sci. and Techn.* 8(13), Dec. 1974.
194. Kharasch, *Bureau of Standards Journal of Research* 2, 359 (1929).
195. Seizinger, E. Donald; and Dimitriades, "Oxygenates in exhaust from simple hydrocarbon fuels," *JAPCA*, 22(1), Jan. 72.
196. Carotti, A. A. and Kaiser, E. R. "Concentrations of twenty gaseous chemical species in the flue gases of a municipal incinerator," *JAPCA* 22(4), April 1972.

Note: Refs. 181, 184, 187, and 191 are not cited in the text.

197. Rasmussen, Reinhold A., "what do hydrocarbons from trees contribute to pollution," *JAPCA* **22**(7), July 1972.
198. Groth, Richard H. and Zaccardi, Vincent A., "Development of a high temperature substractive analyzer for hydrocarbons," *JAPCA*, Sept. 1972.
200. Driscoll, John and Warneck, Peter, "The analysis of ppm levels of gases in air by photoionisation mass spectrometry," *JAPCA*, Oct. 1973.
201. Hester, Norman E., Stephen, Edgar R.; Clifton, Taylor O., "Fluorocarbons in the Los Angeles Basin," *JAPCA*, June 1974.
203. Dugan, Patrick R., *Biochemical ecology of water pollution*, Plenum Press, New York, London, 1972.
204. Moncrieff, R. W., *Odours*, William Heinemann Medical Books, Ltd.
205. Okita, T. and Kanamori, S., "Determination of trace concentration of ammonia in the atmosphere using pyridine-pyrazolone reagert," *Atm. Environm.*, **5**, 621–627, 1971.
206. Davies, Ian W. and Perry, Roger, "Vinyl chloride monomer's effect on liquids in PVC bottles," *Environm. Poll. Managm.*, Jan.–Feb. 1975.
207. First, Nelvin W., "Control of acrylic odors in processing industries," Proceedings of the third international air congress, Düsseldorf, VDI Verlag GmbH.
208. IG-TNO, "Review of analytical methods for the determination of pollutants in indoor air, atmospheric air and in water–1970," research institute for public health engineering TNO, Schoenmakerstraat 97, Delft, The Netherlands.
209. Lauwerijs, Robert and Lavenne, F., "Précis de toxicologie industrielle et des intoxications professionnelles–1972," éditions J. Duculot, S.A., 18, rue Pierquin, 5800 Gembloux, Belgium.
210. Summer, W., "Odor pollution of air, causes and control 1971," Leonard Hill Books.
211. Patty, Frank A., "Industrial hygiene and toxicology," Vol. 2, Interscience Publishers, 1967.
212. Thisthethwayte, D. K. B. and Goleb, E. E., "The composition of sewer air," Proceedings of the 6th international conference on advances in water pollution research, Jerusalem 1972, Pergamon Press.
213. Singer, Philip C. and Zilli, William B., "Ozonation of ammonia in wastewater," *Water Research*, **9**, 127–134, Feb. 1975.
214. Croll, B. T., Arkell, G. M., Hodge, R. P. J., "Residues of acrylamide in water," *Water Res.* **8**, 989–993, Nov. 1974.
215. Croll, B. T. and Simkins, G. M., *Analyst* **97**, 1972.
216. Rump, Ole, "Phenolic acids as indicators of pollution with liquid manure. A method for their detection," *Water Res.*, **8**, 889–894, Nov. 1974.
217. Afghan, Badar K.; Leung, Ricky; Ryan, James F., "Automated fluorometric method for determination of citric acid in sewage effluents," *Water Res.* **8**, 789–785, Oct. 1974.
218. Morita, M., Nakamura, H., Mimura, S., "Phthalic acid esters in water," *Water Res.*, **8**, 781–788, Oct. 1974.
219. Oseid, Donavon M., Smith, Lloyd L., Jr., "Factors influencing acute toxicity estimates of hydrogen sulfide to freshwater invertebrates," *Water Res.* **8**, 739–746, Oct. 1974.
220. Dore, M., Brunet, N., Legube, B., "participation de différents composés organiques à la valeur des critères globaux de pollution," *La tribune du Cebedeau*, **28**(374), 3–11, Jan. 1975.
221. Duvel, William A. and Helfgott, Theodore, "Removal of wastewater organics by reverse osmosis," *JWPCF*, **47**(1), Jan. 1975.
222. European inland fishery advisory commission working party on water quality criteria for european freshwater fish, "Water quality criteria for european freshwater fish," *Water Res.*, **7**, 929–941, June 1973.

Note: Refs. 199 and 202 are not cited in the text.

223. Chudoba, J. and Dalesicky, "Chemical oxygen demand of some nitrogenous heterocyclic compounds," *Water Res.* 7, 663–668, May 1973.
224. Evans, W. H. and Patterson, Stella, "Biodegradation of urea in river waters under controlled laboratory conditions," *Water Res.*, 7, 975–985, July 1973.
225. Kölle, W., Koppe, P., Santheimer, H., "Taste and odor problems with the River Rhine."
226. McKee, Jack Edwards and Harold, W. Wolf, "Water quality criteria," The Resources Agency of California, State Water Quality Control Board, 1963.
227. Water Research Centre, Stevenage Laboratory, Elder Way, Stevenage, Herts. U.K., "Analytical methods for organic compounds in sewage effluents," 1975.
228. R. Ya. Krasnoshchekova and M. Ya. Gubergrits, "Solubility of paraffin hydrocarbons in fresh and salt water," *Petroleum Chemistry, USSR* 13; 4, 1973.
229. Arthur L. Benson, Gerard C. Coletta, Philip L. Levins, "Removal of cyanoacrylate vapor from work spaces by activated carbon," *AHIA Journal* 36(10), 741–744, Oct. 1975.
230. Chris Sutton and John A. Calder, "Solubility of higher-molecular-weight *n*-paraffins in distilled water and seawater," *Environm. Sci. and Techn.* 8(7), July 1974.
231. Groll, B. T. and Simkins, G. M., *Analyst* (London), 97, 281 1972.
232. Lowden, G. F., Saunders, C. L. and Edwards, R. W., *J. Soc. Wat. Treatm. Exam.*, 18, 275, 1969.
233. Garrison, A. W., Southeast Environmental Research Laboratory of the E.P.A., Athens, Georgia.
234. Jolley, R. L., Oak Ridge National Laboratory, Oak Ridge, Tennessee.
235. Giger, W.; Reinhard, M.; and Schaffner, C., *Vom Wasser*.
236. Rudling, L., *Wat. Res.* 5, 831, 1971.
242. McAuliffe Clayton, "Solubility in Water of paraffin, cycloparaffin, olefin, acetylene, cycloolefin, and aromatic hydrocarbons," *J. Phys. Chem.* 70(4), April 1966.
243. L. A., Gillette, D. L. Miller, and H. E. Redman, "Appraisal of a chemical waste problem by fish toxicity tests," *Sewage Ind. Wastes* 24(11), 1397–1401, 1952.
244. I. E. Wallen, W. C. Greer, and R. Lasater, "Toxicity to *Gambusia affinis* of certain pure chemicals in turbid waters," *Sewage Ind. Wastes* 29(6) 695–711, 1957.
245. M. M. Ellis, "Detection and measurement of stream pollution," U.S. Bur. Fisheries Bull. No 22, XLVIII, 365–437, U.S. Dept. Commerce, Washington, 1937.
246. Unpublished work of the Louisiana Petroleum Refiner's Waste Control Council.
247. H. Turnbull, J. H. De Mann, and R. F. Weston, "Toxicity of various refinery materials to fresh water fish," *Ind. Eng. Chem.* 46(2), 324–33, 1954.
248. J. T. Garrett and F. M. Dougherty, "Toxicity levels of hydrocyanic acid and some industrial products," *Texas J. Sci.* 3, 391–6, 1951.
249. C. A. Crandall and C. J. Goodnight, "The effect of various factors on the toxicity of sodium pentachlorophenate to fish," *Limnol. Oceanog.* 4, 53–6, 1959.
250. Thomas W. Bober and Thomas J. Dagon. "Ozonation of photographic processing wastes," *JWPCF*, 47(8), 2114–2129, 1975.
251. Eisenhauer, H. R., "The ozonation of phenolic wastes," *JWPCF*, 40, 1887, 1968.
252. C. Henderson, Q. H. Pickering, and A. E. Lemke, "The effect of some organic cyanides (nitriles) on fish," Purdue Univ. Eng. Bull. Ext. Ser. 106, 120–30, Lafayette, Ind., 1960.
253. M. W. Lammering and M. J. Burbank, Jr., "The toxicity of phenol, *o*-chlorophenol and *o*-nitrophenol to bluegill sunfish," Purdue Univ. Eng. Bull. Ext. Ser. 106, 541–55, Lafayette, Ind., 1960.
254. Private communication from Esso Research and Engineering Co.
255. Lamb, C. B. and Jenkins, G. F., "B.O.D. of Synthetic Organic Chemicals." Proc. 7th Purdue Ind. Waste Conf., p. 326, 1952.

Note: Refs. 237, 238, 239, 240 and 241 are not cited in the text.

256. Mills, E. J., Jr., and Stack V. T., Jr., "Biological Oxidation of Synthetic Organic Chemicals," Proc. 8th Purdue Ind. Waste Conf., p. 492, 1953.
257. Ettinger, M. B. Lishka, R. J. and Kroner, R. C., "Persistence of pyridine bases in polluted water," *Ind. Eng. Chem.*, **46**, 791, 1954.
258. Swope, H. G. and Kenna, M., "Effect of Organic Compounds on Biochemical Oxygen Demand," *Sew. and Ind. Wastes Eng.*, **21**, 467, 1950.
259. Hess, R. W., private communication.
260. Southwest Research Institute, "Complex water pollution problems solved by applied research," *Eng. News-Rec.*, **146**(6), 44, 1951.
261. Sawyer, C. N., Callejas, P., Moor, M. and Tom, A. Q. Y., "Primary Standards for B.O.D. Work," *Sew. and Ind. Wastes*, **22**, 26, Jan. 1950.
262. Lea, W. L. and Nichols, M. S.; "Influence of substrate on biochemical oxygen demand," *Sewage Works Jour.*, **8**, 435, 1936.
263. Lee, E. W. and Oswald, W. J., "Comparative studies of dilution and Warburg methods for determining B.O.D.," *Sew. and Ind. Wastes* **26**, 1097, 1954.
264. Moore, A. W. and Ruchhoft, C. C., "Chemical oxygen consumed and its relationship to B.O.D.," *Sew. and Ind. Wastes*, **23**, 705, 1951.
265. Symons, G. E. and Buswell, A. M., "Preparation and biochemical oxygen demand of pure sodium soaps," *Ind. Eng. Chem.* **24**, 460, 1932.
266. Heukelekian, H. and Gellman, I., "Studies of biochemical oxidation by direct methods. II. Effect of certain environmental factors on the biochemical oxidation of wastes," *Sew. ind Ind. Wastes*, **23**, 1546, 1951.
267. Gellman, I. and Heukelekian, H., "Studies of biochemical oxidation by direct methods. I. Direct method for determining B.O.D.," *Sew. and Ind. Wastes*, **23**, 1267, 1951.
268. Unpublished data from Dept. of Sanitation Files, Rutgers Univ., New Brunswick, N.J.
269. Heukelekian, H., "Use of the dilution method for determining the effect on industrial wastes of deoxygenation," *Sewage Works Jour.*, **19**, 612, 1947.
270. Gellman, I. and Heukelekian, H., "Biological oxidation of formaldehyde," *Sew. and Ind. Wastes*, **22**, 1321, 1950.
271. Gellman, I., and Heukelekian, H., "Studies on biochemical oxidation by direct methods. V. Effect of various seed materials on the rates of oxidation of industrial wastes and organic compounds," *Sew. and Ind. Wastes*.
272. C. J. M. Wolff, "COD determination of volatile compounds," *Water Res.*, **9**, 1015, 1975.
273. B. Meissner, Wasserwirtschaft-Wassertechniek, E. Germany, 1954.
274. N. Wolters, "Unterschiedliche Bestimmungsmethoden zur Erfassung der organischen Substanz in einer Verbindung," Lehrauftrag Wasserbiologie a.d. Technischen Hochschule, Darmstadt, W. Germany.
275. Joseph W. Masselli, Nicholas W. Masselli, M. Gilbert Burford, "BOD? COD? TOD? TOC?," *Textile Industries*, Sept. 1972.
277. Shell Chemie, "Shell Industrie Chemicalien gids," Shell Nederland Chemie, Afd. Industriechemicalien, Wassenaarseweg 80, 's-Gravenhage, Nederland, 1.1.1975.
278. May, J., "Geruchsschwellen von Lösemitteln zur Bewertung von Lösemittelgerüchen in der Luft," *Staub* **26**, 385-88, 1966.
279. Oelert, H. H. and Th. Florian, "Erfassung und Bewertung der Geruchsbelästigung durch Abgase von Dieselmotoren," *Staub* **32**, 400-407, 1972.
280. Gellman, I., "The Direct Oxygen Utilization Method and Its Application to Study of Oxidation Characteristics of Wastes," Master's Thesis, Rutgers Univ., 1950.
281. Gellman, I., "Studies on Biochemical Oxidation of Sewage, Industrial Wastes and Organic Compounds," Ph. D. Thesis, Rutgers Univ., 1952.

Note: Ref. 276 is not cited in the text.

282. Meissner, B., "Addition Methods of Water Analyses and Their Value in Appraising the Content of Waste Waters," *Wasserwirtschaft-Wassertechnik*, 4, 166, 1954.
283. Zobell, C. E., "Some factors which influence oxygen consumption by bacteria in lake water," *Biol. Bull.* 78, 388, 1940.
284. Weston, R. S., "Activated Carbon in Sewage and Industrial Waste Treatment," Master's Thesis, New York Univ., 1939.
285. Staniar, Wm., "B.O.D. of pure compounds," *Plant Engineering Handbook*, 1st Edition, p. 1601, McGraw-Hill Pub. Co., 1950.
286. Kaufman, S., Lederle Lab. Div. American Cyanamid Co., private communication.
287. Stafford, W. and Northrup, H. J., "The B.O.D. of textile chemicals," *Proc. Amer. Assoc. Text. Chem. & Colorists*, May 23, 1955.
288. Masselli, J. W. and Burford, M. G. "Pollution reduction in cotton finishing wastes through process chemical changes," *Sew. and Ind. Wastes*, 26, 1109, 1954.
289. Masselli, J. W. and Burford, M. G., "Pollution Sources in Wool Scouring and Finishing Mills and Their Reduction through Process and Process Chemical Changes," Rept. New England Inter. Water Poll. Control Comm., 1954.
290. Burford, N. G., Masselli, J. W., Snow, W. S., Campbell, H. and de Luise, F. J., "Industrial Waste Surveys of Two New England Cotton Finishing Mills." Rept. New England Inter. Water Poll. Control Comm., 1953.
292. Cabridenc, R., "Métabolisation du phénol et des produits phénoliques," Conférence Arts Chimiques, Paris 1965, *La Pollution des Eaux* p. 119, 1966.
293. Bringmann, G. and Kühn, R., "Vergleichende toxikologische Befunde an Wasser-Bakterien," *gwf-wasser/abwasser* 6j. 81, 337, 1960.
294. Baker, R. A. "Threshold odors of Organic Chemicals," *J.A.W.W.A.* pp. 913–916, July 1963.
295. Zoeteman, B. C. J., A. J. A. Kraayeveld, G. J. Piet, "Oil pollution and drinking water odor," H_2O 4(16), 367–371, 1971.
296. Halluta, J., "Geruchs- und Geschmacksbeeinträchtigung des Trinkwassers—Ursachen und Bekämpfung," *Gas und Wasserfach* 101(40), 1018–1023, 1960.
297. Rosen, A. A., R. T. Skell, M. B. Ettinger, "Relationship of river water odor to specific organic contaminants," *JWPCF*, 35(6), 777–782, 1963.
298. Kreisler, R., "Geruchsprobleme der Lackindustrie," *Schriftenreihe des Vereins für Wasser-, Boden- und Lufthygiene* 35, Geruchsbelästigende Stoffe, Kolloquium am 18 März 1971 in Düsseldorf, zusammengestellt von Dr. Helmut Kettner (Gustav Fischer Verlag, Stuttgart).
299. Zoeteman, B. C. J., A. J. A. Kraayeveld, G. J. Piet, unpublished data, 1971.
300. Zoeteman, B. C. J., G. J. Piet, "Drinkwater is nog geen water drinken," H_2O 6(7), 1973.
301. Koppe, P., "Identifizierung der Hauptgeruchsstoffe im Ufer filtrat des Mittel- und Niederrheins," *Vom Wasser* 32, 33–68, 1965.
302. Brady, S. O., *Proc. Div. Refining Am. Petrol. Inst.* 48, 556, 1968.
303. Gabrilevskaya, L. N., V. P. Laskina, *Prom. Zagryazneniya Vodsemov* 8, 46, 1967.
304. Burttschell, R. H., A. A. Rosen, A. M. Middleton, M. B. Ettinger, "Chlorine-Derivatives of phenol causing taste and odor," *J.A.W.W.A.* 205–214, 1950.
305. Guadaqui, D. G., R. G. Buttery, S. Okano, *J. Sci. Food. Agr.* 14, 76, 1963.
306. Drawnieks, A., "Fundamentals of odor perception: Their applicability to air pollution control programs," Atmospheric quality improvement, technical bulletin No. 54, June 1971 (National Council of the paper industry for air and stream improvement, Inc. New York, N.Y. 10016).
307. J. Stockham, A. O'Donnell, and A. Drawnieks, "Chemical species in engine exhaust and their contribution to exhaust odor," Coordinating Research Council, New York, June 1969. Report No IITRI C8150-5.

Note: Refs. 291 and 308 are not cited in the text.

BIBLIOGRAPHY 1239

309. Johnson, B. T. and Lulves, W., "Biodegradation of dibutylphthalate and di-s-ethylphthalate in freshwater hydrosoil," *J. Fish. Res. Bd Can.*, 32(3), 333–339, 1975.
310. R. A. A. Blackmann, "Toxicity of oil-sinking agents," *Marine Poll. Bull.*, 5(8), 116–118, Aug. 1974.
311. J. W. Vogh, "Nature of odor components in diesel exhaust," *J APCA* 19(10), Oct. 1969.
312. C. W. Melton, R. I. Mitchell, D. A. Trayser, "Chemical and physical characterization of automotive exhaust particulate matter in the atmosphere," EPA No. 68-02-0205, June 1973.
313. Wendell L. Dilling, Nancy B. Tefertiller, George J. Kallos, "Evaporation roles and reactivites of methylene chloride, chloroform, 1,1,1-trichloroethane, trichloroethylene, tetrachloroethylene, and other chlorinated compounds in dilute aqueous solutions," *Environm. Sci. and Techn.*, 9(9), Sept. 1975.
314. Leach, J. M. and Thakore, A. N., "Isolation and identification of constituents toxic to juvenile rainbow (*Salmo gairdneri*) in caustic extraction effluents from kraft pulp mill bleach plants," *J. Fish. Res. Bd Can.*, 32(8), 1249–1257,
315. E. P. Grimsrud and R. A. Rasmussen, "Survey and analysis of Halocarbons in the atmosphere by gas chromatography–mass spectrometry," *Atm. Envir.* 9, 1014–1017.
316. K. Verschueren, "Reuken, bronnen en oorzaken," KVIV symposium, Gent, Belgium, 11.3.1976.
317. James E. Morrow, "Effects of crude oil and some of its components on young Coho and Sockeye salmon," EPA-660/3-73-018, Jan. 1974.
318. Smock, L. A., Stoneburger, D. L. and Clark, J. R. "The toxic effect of TNT and its primary degradation products on two species of algae and the fathead minnow," *Water Res.*, 10(6), 537–543, 1976.
319. Marconi, S. and Ionescu, M. "The toxic action of acrylonitrile on fish," Studii Protectia Calitatri Apclor.
320. Daniel L. Flamm, Robert E. James, "Absorption efficiencies for source sampling of hydrogen sulfide," *Environm. Sci. and Techn.*, 10(2), Feb. 1976.
321. Dorris A. Lillard, et al., "Aqueous odor, thresholds of organic pollutants in industrial effluents," EPA-660/4-75-002, May 1975.
325. Rosen, A. A., J. B. Peter, and F. M. Middleton, "Odor thresholds of mixed organic chemicals," *JWPCF* 34, 7–14, 1962.
326. Stahl, W. H., Ed., "Compilation of odor and taste thresholds: Value data," Data Series No. 48, Am. Society for Testing and Materials, Philadelphia, Pa. 19103, 1973.
327. P. Pitter, "Determination of biological degradability of organic substances," *Water Res.* 10, 231–235, 1976.
328. Adema, D. M. M., "Acute toxiciteitstoetsen met 1,2-dichloorethaan, fenol, acrylonitrile, en alkylbenzenesulfonaat in zeewater," 1976, Central laboratory TNO, 97 Schoemakerstraat–P.O. Box 217, Delft, The Netherlands.
329. Bringmann, G., and R. Kühn, "Vergleichende Befunde der Schadwirkung wassergefährdender Stoffe gegen Bakterien (*Pseudomonas putida*) und Blaualgen (*Microcystis aeruginosa*)," *Gwf-Wasser/Abwasser*, 117(9), 1976.
330. Donald MacKay and Paul J. Leinonen, "Rate of evaporation of low solubility contaminants from water bodies to atmosphere," *Environm. Sci. and Techn.* 9(13), Dec. 1975.
331. T. E. Tooby and P. A. Hursey. "The acute toxicity of 102 pesticides and miscellaneous substances to fish," *Chemistry and Industry*, 21 June 1975.
332. V. Leoni and S. U. D'Arca. "Experimental data and critical review of the occurence of hexachlorobenzene in the Italian environment," *The Science of the Total Environment*, 5, 253–272, 1976.
333. Ockner, R. K., and R. Schmid, Nature, 189, 449, 1961.

Note: Refs. 322, 323 and 324 are not cited in the text.

334. Savitskii, I. U., *Kievsk. Med. Inst.* **1964**, 158.
335. Gurfein, L. N., and Z. K. Pavlova, *Sanit. Okhr. Vodoemov Zagryaz. Prom. Stochnymi Vodami*, 4, 117, 1960, as reported by Gehring, P. J., and MacDougall, 1971.
336. Vos, J. G., H. L. Van Der Maas, A. Musch, and E. Ram, *Toxicol. App. Pharmacol.*, 18, 944, 1971.
337. E. D. S. Corner, R. P. Harris, C. C. Kilvington, and S. C. M. O'Hara, "Petroleum compounds in the marine food web: short-term experiments on the fate of naphthalene in *Calanus*," *J. Mar. Biol. Ass. U.K.* 56, 121–133, 1976.
338. Adelman, I. R., Smith, L. L., Jr. and Sieennop, G. D., "Effect of size or age of goldfish and fathead minnows on use of pentachlorophenol as a reference toxicant," *Water Res.* 100, 685–687, 1976.
339. Metcalf, Robert L., Gary M. Booth, Carter K. Schuth, Dale J. Hansen, and Po-Yung Lu, "Uptake and fate of di-2-ethylhexylphthalate in aquatic organisms and in a model ecosystem," *Environmental Health Perspectives*, June 1973.
340. Schrödter, W., P. Studt, P. Volgtsberger, "Polycyclische aromatische Kohlenwasserstoffe im Ottomotor-Abgas," *Erdöl und Kohle-Erdgas–Petrochemie vereinigt mit Brennstoff-Chemie*, 29(4), April 1976.
341. Davies, Ian W., Roy M. Harrison, Roger Perry, Don Ratnayaka, and Roger A. Wellings, "Municipal incinerators as source of polynuclear aromatic hydrocarbons in environment," *Environm. Sci. and Techn.*, 10(5), May 1976.
342. RIWA, "De samenstelling van het Rijn en Maaswater in 1974," Secretariaat RIWA, condensatorweg 54, Pb 8169, Amsterdam-Sloterwijk.
343. Kauss, P. B., T. C. Hutchinson, "The effects of water-soluble petroleum components on the growth of *Chlorella vulgaris Beijerinck*," *Environm. Poll.* (9), 1975.
344. Anonymous, "A propos de l'accident de Seveso," *Pollution Atmosphérique* (71), July–Sept. 1976.
345. Singh, A. R., Lawrence, W. H. and Autian, J., "Teratogenicity of phthalate esters in rats," *J. Pharm. Sci.* 61, 51, 1972.
346. Popw, D., Walker, D. S., Moss, R. L., "Evaluation of cobalt oxide catalysts for the oxidation of low concentration of organic compounds in air" *Atm. Envir.* 10, 951–956.
347. EPA, "Identification of organic compounds in effluents from industrial sources," EPA-560/3-75-002, April 1975.
348. J. F. Finklea, M.D., "Metabolic precursors of a known human carcinogen, beta-naphthylamine."
349. S. E. Herbes and J. J. Beauchamp, "Toxic interaction of mixtures of two coal conversion effluent components (resorcinol and 6-methylquinoline) to *Daphnia magna*," *Bull. of Env. Cont. & Tox.* 27(1), 1977.
350. Vincent R. Mattson, John W. Arthur, Charles T. Walbridge, "Acute toxicity of selected organic compounds to fathead minnows," EPA-600/3-76-097, Oct. 1976.
351. Nissim Claude Cohen and Guy Régnier, "Solubilités des molécules déterminées à partir de leurs caractéristiques géométriques et électroniques: cas des hydrocarbures dans l'eau," *Bull. Soc. Chim. de France* (11–12), Nov.–Dec. 1976.
352. Gaynor W. Dawson, Allen L. Jennings, Daniel Drozdowski, and Eugene Rider, "The acute toxicity of 47 industrial chemicals to fresh and saltwater fishes," *J. Hazardous Materials*, 1, 1975/77.
353. G. E. Burdick and M. Lipschustz, "Toxicity of ferro and ferricyanide solutions to fish and determination of the cause of mortality," *Trans. Am. Fish. Soc.*, 78, 1948.
354. A. D. Little, Inc., "Relationship between organic chemical pollution of fresh water and health," FWQA 71632, Dec. 1970.
355. L. A. Gillette, D. L. Miller, and H. E. Redman, "Appraisal of a chemical waste problem by fish toxicity test," *Sewage and Industrial Wastes*, 24, 1397, 1975.
356. J. E. McKee and H. W. Wolf, "Water Quality Criteria," US PHS, Control Board, Pub. 3-A, April 1971.

BIBLIOGRAPHY 1241

357. Dr. Fluck, "The odor threshold of phosphine," *JAPCA*, **26**(8), Aug. 1976.
358. H. T. Kemps, R. L. Little, V. L. Holman, and R. L. Darby, "Water quality criteria data book, Vol. 5: Effects of chemicals on aquatic life," U.S. EPA, P8 234435, Sept. 1973.
359. Dow Chemical company, data provided June 1974.
360. R. Ukeles, "Growth of pure cultures of marine phytoplankton in the presence of toxicants," *Applied Microbiology*, **10**(6), 532, 1962.
361. *Oceanography International*, October 1970.
362. J. E. Portman, "The toxicity of 120 substances to marine organisms," Shellfish information leaflet, Fisheries Experimental Station, Conway, N. Wales, Ministry of Agriculture, Fisheries and Food, September 1970.
363. P. Benville, "Acute toxicity of nine solvents to rainbow trout fingerlings," (unpublished) transmitted from Tiburon Laboratory, NOAA, July 10, 1974.
364. C. G. Wilber, *The biological Effects of Water Pollution*, Charles C. Thomas, Springfield, IL, 1969.
365. V. Zitko, "Toxicity and pollution potential of thallium," *The Science of the Total Environment*, **4**, 185, 1975.
366. J. E. Portman and K. W. Wilson, "The toxicity of 140 substances to the brown shrimp and other marine animals," Shellfish information leaflet No. 22, Fisheries experiment station, Conway, N. Wales, Dec. 1971.
367. C. W. Flickinger, "The benzenediols: catechol, resorcinol and hydroquinone—A review of the industrial toxicology and current industrial exposure limits," *Am.Ind.Hyg.Assoc.J.* (37), Oct. 1976.
368. Kenneth Winters, John C. Batterton, Chase van Baalen, "Phenalen-1-one: occurence in a fuel oil and toxicity to microalgae," *Environm. Sci. & Techn.* **11**(3), March 1977.
369. Wendell L. Dilling, "Interphase transfer processes: evaporation rates of chloromethanes, ethanes, -ethylenes, -propanes and -propylenes from dilute aqueous solutions. Comparisons with theoretical predictions," *Environm. Sci. & Techn.*, **11**(4), April 1977.
370. S. F. Stepan and J. F. Smith, "Some conditions for use of macroreticular resins in the quantitative analysis of organic pollutants in water," *Water Res.* **11**, 1977.
371. F. J. Sandalls and D. B. Hatton, "Measurements of atmospheric concentrations of trichlorofluoromethane, dichlorodifluoromethane and carbontetrachloride by aircraft sampling over the British Isles" *Atm. Environm.*, **11**, 1977.
372. Grimsrud, E. P., and Rasmussen, R. A. "The analysis of chlorofluorocarbons in the troposphere by gas chromatography–mass spectrometry" *Atm. Environm.* **9**, 1010–1013, 1975.
373. Hester, N. E., Stephens, E. R. and Clifton Taylor, O., "Fluorocarbon air pollutants, measurements in lower stratosphere" *Environm. Sci. and Techn*, **9**, 875–876, 1975.
374. Lovelock, J. E., Maggs, R. J. and Wade, R. J., "Halogenated hydrocarbons in and over the Atlantic" *Nature (Lond.)* **241**, 194–196, 1973.
375. Lovelock, J. E., "Atmospheric halocarbons and stratospheric ozone," *Nature (Lond.)* **252**, 292–294, 1974.
376. Wilkniss, P. E., Swinnerton, J. W., Bressan, D. J., Lamontagne, R. A. and Larson, "CO, CCl_4, CH_4 and Rn-222 concentrations at low altitude over the Arctic Ocean in January 1974," *J. Atm. Sci*, **32**, 158–162, 1975.
377. Wilkniss, P. E., Swinnerton, J. W., Lamontagne, R. A. and Bressan, D. J., "Trichlorofluoromethane in the troposphere, distribution and increase, 1971–1974" *Science* **187**, 832–834, 1975.
378. Zafonte, L., Hester, N. E., Stephens, E. R. and Taylor, O. C., "Background and vertical atmospheric measurements of fluorocarbon-11 and fluorocarbon-12 over Southern California," *Atm. Environm.* **9**, 1007–1009, 1975.
379. G. Grimmer and H. Böhnke, "Ein Anreicherungsverfahren für die gaschromatographische Bestimmung von polyclischen aromatischen Kohlenwasserstoffen in Schmierölen," DGMK-projekt 4559, March 1975.
380. J. P. Meyer and G. Grimmer, "Optimierung und Erprobung eines Sammelverfahrens für

polycyclische aromatische Kohlenwasserstoffe im Kraftfahrzeug-Abgas unter den Bedingungen des Europäischen Fahrzyklus," BMI-DGMK-projekt 4547, July 1974.
381. Richard F. Lee "Fate of petroleum components in estuarine waters of the southeastern United States," 1977 Oil Spill Conference, U.S.A.
382. Cary T. Chiou, Virgil H. Freed, David W. Schmedding, and Rodger L. Kohnert, "Partition coefficient and bioaccumulation of selected organic chemicals," *Environm. Sci. & Techn.* 11(5), May 1977.
383. National Institute for Occupational Safety and Health, "Background information on 4,4-diaminodiphenylmethane," *AIHA J.*, May 1976.
384. William T. Roubal, Donald H. Bovee, Tracy K. Collier, Susan I. Stranahan, "Flow-through system for chronic exposure of aquatic organisms to seawater soluble hydrocarbons from crude oil: construction and applications," 1977 Oil spill conference.
385. K. Müller and J. P. Meyer, "Einfluss von Ottokraftstoffen auf die Emission von polynuklearen aromatischen Kohlenwasserstotten in Automobilabgasen im Europa-Test," DGMK-projekt 4568, June 1974.
386. Shinichi Nagata and Goro Kondo, "Photo-oxidation of crude oils," 1977 Oil spill Conference.
387. H. Schulz, N. Sedighi, H. B. Gregor, T. Din Van and S. San Min, "Gaschromatographische Analyse Olefinhaltiger Benzine," DGMK-projekt 4509/4558.
388. F. Irmann, "Eine einfache Korrelation zwischen Wasserlöslichkeit und Struktur von Kohlenwasserstoffen und Halogenkohlenwasserstoffen," *Chemie Ingenieur Technik*, 37(8), 789–872, Aug. 1965.
389. Masana Ogata, Yoshio Miyake, and Shohei Kira, "Transfer to fish of petroleum paraffins and organic sulfur compounds," *Water Res.*, 11, 333–338, 1977.
390. Melvin R. Hockenbury, C. P. Leslie Grady, Jr. "Inhibition of nitrification–effects of selected organic compounds," *JWPCF*, May 1977.
391. Papa, L. J. "Measuring exhaust hydrocarbons down to the parts per billion," pp. 43–65 in *Vehicle Emissions, part III* (selected SAE papers 1967–1970) *Progress in Technology*, Vol. 14, New York: Society of Automotive Engineers, 1971.
392. Morris, W. E., and K. T. Dishart, "Influence of vehicle emission control systems on the relationship between gasoline and vehicle exhaust hydrocarbon composition, pp. 63–101, in *Effect of Automotive Emission Requirements on Gasoline Characteristics*," symposium presented at the seventy-third annual meeting ASTM, Toronto, June 1970, ASTM Special Techn. Publ. 487.
393. Jackson, M. W. "Effects of some engine variables and control systems on composition and reactivity of exhaust emissions," *SAE Prog. Techn.*, Ser. 12, 1967.
394. Wigg, E. E., R. J. Campion, and W. L. Petersen, "The effect of fuel hydrocarbon composition on exhaust and oxygenate emissions," SAE paper No. 720251 presented at the SAE-Automotive Engineering Congress, Detroit, Michigan, Jan. 1972.
395. Oberdorfer, P. E., "The determination of aldehydes in automobile exhaust gases," *Vehicle Emissions, Part III, SAE Prog. Tech.*, Ser. 14, 32–42, 1971.
396. Wodkowski, C. S. and E. E. Weaver. "The effects of engine parameters, fuel composition and control devices on aldehyde exhaust emissions," Paper presented at West Coast Section, APCA Meeting, San Francisco, Oct 8, 1970.
397. Fracchio, M. F., F. J. Schuette, and P. K. Mueller, "A method for sampling and determination of organic carbonyl compounds in automobile exhaust," *Environm. Sci. & Techn.*, 1, 915–922, 1967.
398. Wade, D. T., "Factors influencing vehicle evaporative emissions," *Vehicle Emissions, Part III, SAE Progr. Tech.*, Ser. 14, 743–755, 1971.
399. Caplan, J. D. "Smog chemistry points the way to rational vehicle emission control," *Vehicle Emissions, Part II, SAE Progr. Techn.* Ser. 12; 20–31, 1967.
400. Muller, H. L., R. E. Ray, and T. O. Wagner, "Determining the amount and composition of

evaporation losses from automotive fuel systems," *Vehicle Emissions, Part II, SAE Progr. Tech.*, Ser. 12, 402-412, 1967.
401. Jackson, M. W., and R. L. Everett, "Effect of fuel composition on amount and reactivity of evaporative emissions," *Vehicle Emissions, Part III, SAE Prog. Techn.*, Ser. 14, 802-823, 1971.
402. McEwen, D. J., "The analysis of gasoline vapors from automobile fuel tanks," presented at the 153rd National meeting of the American Chemical Society, Miami Beach, Florida, April 1967.
403. Rosen, A. A., R. T. Skeel, and H. B. Ettinger, "Relationship of river water odor to specific organic contaminants," *JWPCF* **35**, 777-782, 1963.
404. Dostal, K. A., R. C. Pierson, D. G. Hager, and G. G. Robeck, "Carbon bed design criteria study at Nitro, W. Va.," *J. Amer. Water Works Assoc.*, **57**, 663-674, 1965.
405. Ettinger, M. B., "Biochemical oxidation: Characteristics of stream-pollutant organics," *Ind. Eng. Chem.* **48**, 256-259, 1956.
406. Bringmann, G., and R. Kuhn, "Comparative investigation into the microbial decomposition of alkanes, alkenes, cycloalkanes and cycloalkenes by the rotating disk contact aerator technique," Concawe Doc, No 4797, 1970, Den Haag, Netherlands.
407. Hooper, A. and Terry, K., "Specific inhibitors of ammonia oxidation in *Nitrosomonas*," *J. Bacteriol.* **115**, 480, 1973.
408. Clark, C., and Smidt, E. L., "Uptake and utilization of amino acids by resting cells of *Nitrosomonas europaea*," *J. Bacteriol.* **93**, 1309, 1967.
409. Lees, H., and Simpson, J. R., "The biochemistry of the nitrifying organisms—Nitrite oxidation by *Nitrobacter*," *Biochem. J.*, **65**, 297, 1957.
410. Zavarzin, G. H., "On the inducer of the second phase of nitrification, I. Concerning the participation of respiratory pigments in nitrification." *Mikrobiologiia*, **27**, 401, 1958.
411. Lees, H., and Quastel, J. H., "Biochemistry of nitrification in soils, III. Nitrification of various organic nitrogen compounds," *Biochem. J.*, **40**, 824, 1946.
412. Ogata, M. and Miyake, Y., "Compound from floating petroleum accumulating in fish," *Water Res.*, **9**, 1075-1078, 1975.
413. Ralf D. Rurainski, Hans Jürgen Theiss and Wilhelm Zimmermann, "Uber das Vorkommen von natürlichen und synthetischen Oestrogenen im Trinkwasser," *Gwf-Wasser/Abwasser* **118**(6), 1977.
414. Briggs, G. G., "A simple relationship between soil adsorption of organic chemicals and their octanol/water partition coefficients," Proceedings of the 7th British Insecticide and Fungicide Conference, 1973.
415. Davies, J. E., Barquet, A.; Freed, V. H., Harque, R., Morgade, C.; Sonneborn, R. E.; Vaclavek, C., *Arch. Env. Health*, **30**, 608, 1975.
416. Kenaga, E. E., *Res. Rev.*, **44**, 73, 1972.
417. Lu, P. Y. and Metcalf, R. L., *Env. Health Perspect.*, **10**, 269, 1975.
418. Metcalf, R. L., Sanborn, J. B., Lu, P. Y.; Nye, D., *Arch. Environm. Contam. Toxicol.* **3**, 151, 1975.
419. Rodger Baird, Luis Carmona, Richard L. Jenkins, "Behavior of benzidine and other aromatic amines in aerobic wastewater treatment," *JWPCF*, 1609-1615, July 1977.
420. Robert W. Risebrough, Brock W. De Lappe, Timothy T. Schmidt, "Bioaccumulation factors of chlorinated hydrocarbons between mussels and seawater," *Marine Poll. Bull*, 7(12), Dec. 1976.
421. J. P. Zumbrunn, "Nouvelles applications des produits peroxydés pour l'épuration des eaux résiduaires," *Extern*, **IV**, (3) 1975.
422. J. H. Canton, G. J. Van Esch, P. A. Greve, A. B. A. M. Van Hellemond, "Accumulation and elimination of hexachlorocyclohexane by the marine algae *Chlamydomonas* and *Dunaliella*," *Water Res.*, **11**, 111-115, 1977.

423. F. L. Hart and T. Helfgott, "Bio-refractory index for organics in water," *Water Research* 9, 1055-1058, 1975.
424. R. C. C. Wegman, P. A. Greve, "The microcoulometric determination of extractable organic halogen in surface water; application to surface waters of the Netherlands," *The Science of the Total Environment*, 7, 235-245.
425. David T. Gibson, "Microbial degradation of polycyclic aromatic hydrocarbons," Department of microbiology, University of Texas at Austin, Texas 78712, 1976.
426. N. A. Donaghue, M. Griffin, D. B. Norris, and P. W. Trudgill, "The microbial metabolism of cyclohexane and related compounds," Department of Biochemistry Univ. College of Wales, Aberystwyth, Dyfed, U.K., 1976.
427. R. J. Watkinson, and H. J. Somerville, "The microbial utilization of butadiene." Shell Research Limited, Sittingbourne Research Centre, Kent, U.K., 1976.
428. W. Cautreels, K. Van Cauwenberghe, L. A. Guzman, "Comparison between the organic fraction of suspended matter at a background and an urban station," *The Science of the Total Environment*, 8, 79-88, 1977.
429. M. Dong, D. Hoffman, D. C. Locke, E. Ferrand, "The occurence of caffeine in the air of New York City," *Atm. Environm.*, 11, 651-653, 1977.
430. Gregor A. Junk, Harry J. Svec, Ray D. Vick, Michael J. Avery, "Contamination of water by synthetic polymer tubes," *Environm. Sci. & Techn.*, 8(13), Dec. 1974.
431. Schoental, R., "Carcinogenic and chronic effects of 4,4'-diaminodiphenylmethane, on epoxy resin hardener," *Nature* 219, 1162, 1968.
432. Munn, A., "Occupational bladder tumors and carcinogens: Recent developments in Britain, in W. Deichmann and K. F. Lampe, *Bladder Cancer: A Symposium*, Birmingham, Alabama, Aesculapius.
433. Steinhoff, D., and E. Grundmann, "Zur Cancerogenen Wirkung von 4,4'-diaminodiphenylmethan und 2,4'-diaminodiphenylmethan," *Naturwissenschaften* 57, 247 1970.
434. Roy M. Harrison, Roger Perry, and Roger A. Wellings, "Effect of water chlorination upon levels of some polynuclear arsmatic hydrocarbons in water," *Environm. Sci. & Techn.*, 10(12), 1151-1156, Nov 1976.
435. Prairie, R. L. and Talaley, P., "Enzymatic formation of testololactone," *Biochemistry*, 2, 203-208, 1963.
436. Griffin, M., and Trudgill, P. W., "The metabolism of cyclopentanol by *Pseudomonas*," *Biochem. J.*, 129, 595-603, 1972.
437. Donald Mackay and Paul J. Leinone, "Rate of evaporation of low solubility contaminants from water bodies at atmosphere," *Environm. Sci. & Techn.*, 9(13), 1178-1180, Dec. 1975.
438. Zephyr R. Regnier and Brian F. Scott, "Evaporation rates of oil components," *Environm. Sci. & Techn.*, 9(5), 469-472, May 1975.
439. Liebmann Hans, "Handbuch der Frischwasser- und Abwasserbiologie," Oldenbourg-Munchen, 1962.
440. Gibson, D. T., Roberts, R. L., Wells, M. C. and Kobal, V. M., "Oxidation of biphenyl by a *Beijerinckia* species," *Biochem. Biophys. Res. Commun.*, 50, 211-219, 1973.
441. Leach, J. M., and Thakore, A. N., "Toxic constituent in Mechanical Pulping Effluents," *Tappi*, 59, 129, 1976.
442. McLeay, D. J., "Rapid Method for measuring acute toxicity of pulpmill effluents and other toxicants to salmonid fish at ambient room temperature," *J. Fish. Res. Bd. Can.*, 33, 1303, 1976.
443. Macek, K. J., et al., "Toxicity of four pesticides to water fleas and fathead minnows," EPA 600/3-76-099, U.S. EPA, Duluth, Minn., 1976.
444. Mauck, W. L., et al., "Toxicity of natural pyrethrins and five pyrethroids to fish," *Arch. Environm. Contam. Toxicol.*, 4, 18, 1976.
445. Schafer, E. W., Jr., and Marking, L. L., "Long-term effects of 4-aminopyridine exposures to birds and fish," *J. Wildlife Management*, 39, 807, 1975.

446. Canton, J. H., and Vanesch, G. J., "Short-term toxicity of some food additives to different freshwater organisms," *Bull. Environm. Contam. Toxicol.*, **15**, 720, 1976.
447. Olson, L. E., and Marking, L. L., "Toxicity of four toxicants to green eggs of salmonids," *Progressive Fish-Culturist*, **37**, 143, 1975.
448. Macek, K. J. et al., "Chronic toxicity of atrazine to selected aquatic invertebrates and fishes," EPA 600/3-76-047, 1976.
449. Leach, J. M., et al., "Acute toxicity to juvenile rainbow trout (*Salmo gairdneri*) of naturally occurring insect juvenile hormone analogues, *J. Fish. Res. Bd. Can.*, **32**, 2556, 1975.
450. Bender, M. E., and Westman, J. R., "Toxicity of malathion and its hydrolysis products to Eastern mudminnow, *Umbra pygmaea*," *Chesapeake Sci.*, **17**, 125, 1976.
451. Webb, M., et al., "The toxicity of various mining flotation reagents to rainbow trout," *Water Res. (G. B.)*, **10**, 303, 1976.
452. Woodward, D. F., "Toxicity of herbicides dinoseb and picloram to cutthroat (*Salmo clarki*) and lake trout (*Salvelinus namaycush*)" *J. Fish. Res. Bd. Can.*, **33**, 1671, 1976.
453. Fabacher, D. L., "Toxicity of endrin and an endrin-methylparathion formulation to large-mouth bass fingerlings," *Bull. Environm. Contam. Toxicol.* **16**, 376, 1976.
454. Biesinger, K. E., et al., "Comparative toxicity of polyelectrolytes to selected aquatic animals," *JWPCF*, **48**, 183, 1976.
455. Adelman, I. R., et al., "Chronic toxicity of guthion to fathead minnow (*Pimephales promelas Rafinesque*)," *Bull. Environm. Contam. Toxicol.*, **15**, 726, 1976.
456. Leeuwangh, P., et al., "Toxicity of hexachlorobutadiene in aquatic organisms," in *Sublethal Effects of Toxic Chemicals in Aquatic Animals*," J. H. Koeman and J. J. T. W. A. Strik (Eds.), 167, Elsevier Scientific Publ. Co., New York, 1975.
457. Canton, J. H., et al., "Toxicity, accumulation and elimination studies of α-hexachloro-cyclohexane (α-HCH) with freshwater organisms of different trophic levels," *Water Res. (G. B.)*, **9**, 1163, 1975.
458. Verma, S. R., et al., "Studies on the toxicity of lindane on *Colisa fasciatus*. Part I: TLm measurements and histopathological changes in certain tissues," *Gegenbaurs Morph. Jahrb.*, (Leipzig) **121**, 38, 1975.
459. Macek, K. J., et al., "Chronic toxicity of lindane to selected aquatic invertebrates and fishes," EPA-600/3-76-046, 1976.
460. Hobbs. E. J., et al., "Toxicity of polydimethylsiloxane in certain environmental systems," *Environm. Res.*, **10**, 397, 1975.
461. Desi, I., et al., "Toxicity of malathion to mammals, aquatic organisms and tissue culture cells," *Arch. Environm. Contam. Toxicol.*, **3**, 410, 1976.
462. Olson, L. E.; and Marking, L. L., "Toxicity of four toxicants to green eggs of salmonids," *Progressive Fish Culturist*, **37**, 143, 1975.
463. Ruesink, R. G., and Smith, L. L., "Relationship of 96 hr LC_{50} to lethal threshold concentration of hexavalent chromium, phenol and sodium pentachlorophenate for fathead minnows (*Pimephales promelas Rafinesque*)," *Trans. Amer. Fish. Soc.*, **104**, 567, 1975.
464. Cardwell, R. D., et al., "Acute toxicity of selected toxicants to six species of fish," EPA-600/3-76-008, 1976.
465. Adelman, I. R. and Smith, L. L., "Fathead minnows and goldfish as standard fish in bio-assays and their reaction to potential reference toxicants," *J. Fish. Res. Bd. Can.*, **33**, 209, 1976.
466. Crews, R. C., "Effects of propylene oxide on selected species of fishes," Rept. No. AFATL-TR-74-183, Air Force Armament Lab., Air Force Systems Command, Eglim Air Force Base, Fla., 1974.
467. Miura, T. and Takahashi, T. M., "Effects of a synthetic pyrethroid, SD 43775, on non-target organisms when utilized as a mosquito larvicide," *Mosquito News*, **36**, 322, 1976.
468. Smith, L. L., Jr. and Oseid, D. M., "Chronic effects of low levels of hydrogen sulfide on freshwater fish," *Prog. Water Technol. (G. B.)* **7**, 599, 1975.

469. Smith, L. L., et al., "Toxicity of hydrogen sulfide to various life history stages of bluegill (*Lepomis macrochirus*)." *Trans Amer. Fish. Soc.*, 105, 442, 1976.
470. Dawson, V. K., et al., "Laboratory efficacy of 3-trifluoromethyl-4-nitrophenol (TFM) on development stages of the sea lamprey," Investigations in fish control No. 64, U.S. Dept. Int. Fish & Wildlife Serv., Washington, D.C., 1975.
471. Brenniman, G., et al., "A continuous flow bioassay method to evaluate the effects of outboard motor exhausts and selected aromatic toxicants on fish," *Water Res. (G. B.)* 10, 165, 1976.
472. Mayer, F. L., et al., "Toxaphene effects on reproduction, growth and mortality of brook trout," EPA 600/3-75-013, 1975.
473. Lockhart, W. L., et al., "Chronic toxicity of a synthetic triaryl phosphate oil to fish," *Environm. Physiol. & Biochem.*, 5, 361, 1975.
474. Arnold R. Slonium, "Acute toxicity of selected hydrazines to the common guppy," *Water Res.* 11, 889-895, 1977.
475. S. Le Roux., "The toxicity of pure hydrocarbons to mussel larvae," *Rapp. P. v. Réun. Cons. Int. Explor. Mer*, 171, 189-190, 1977.
476. K. Winters, C. Van Baalen and J. A. C. Nicol., "Water soluble extractives from petroleum oils: chemical characterization and effects on microalgae and marine animals," *Rapp. P. v. Réun. Cons. Int. Explor. Mer.* 171, 166-174, 1977.
477. Richard F. Lee and M. Takahashi, "The fate and effect of petroleum in controlled ecosystem enclosures," *Rap. P. v. Réun. Cons. Int. Explor. Mer.* 171, 150-156, 1977.
478. K. J. Whittle et al., "Fate of hydrocarbons in fish," Torrey Research Station P.O. Box 31, 135 Abbey Road, Aberdeen, AB9 8DG, U.K.–*Rapp. P. v. Réun. Cons. Int. Explor. Mer* 171, 139-152, 1977.
479. A. Fagels and J. B. Sprague., "Comparative short term tolerance of zebrafish, flagfish and rainbow trout to five poisons including potential reference toxicant," *Water Res.* 11, 811-817, 1977.
480. R. F. Addison, "Diphenylether–Another marine environmental contaminant," *Marine Poll. Bull.* 8(10), Oct. 1977.
481. NIOSH, "Current NIOSH intelligence bulletin," *AIHAJ* 38(9), 1977.
482. E. M. Waters et al., "Mirex: An overview," *Environm. Research* 14, 212-222, 1977.
483. Ivie, G. W., et al., "Photodecomposition of mirex on silicagel chromatoplates exposed to natural and artificial light," *J. Agr. Food Chem.* 22(6), 933-936, 1974.
484. Hyde, K. M., et al., "Accumulation of mirex in food chains," *La. Agr. Exp. Sta. Bull.* 17(1), 10-11, 1973.
485. Hollister, T. A., et al., "Mirex and marine unicellular algae: Accumulation, population growth and oxygen evolution," *Bull. Environm. Contam. Toxicol.* 14(6), 753-759, 1975.
486. Tagatz, M. E., et al., "Seasonal effects of leached mirex on selected estuarine animals," *Arch. Environ. Contam. Toxicol.* 3(3), 371-383 1975.
488. B. Baleux and P. Caumette., "Biodégradation de quelques agents de surface cationiques," *Water Res.* 11, 833-841, 1977.
489. Harry Distler and Ernest-Heinrich Pommer, "Die Bekämpfung des Algenwachstums in Rückkühlwerken mit neuen Mikroziden," *Erdöl und Kohle-Erdgas-Petrochemie*, No. 5 May 1965.
490. R. P. Hangebrauck, J. V. von Lehmden and J. E. Meeker, "Sources of polynuclear hydrocarbons in the atmosphere," U.S. Dept. of H.E.W., Public Health Service, Publ. No 999-AP-33, 1967.
491. V. P. Puzinauskas and L. W. Corbett, "Report on emissions from asphalt hot mixes." The asphalt institute, College Park, Maryland 20740, May 1975.

Note: Ref. 487 is not cited in the text.

492. T. R. Terkelson and F. Oyen, "The toxicity of 1,3-dichloropropene as determined by repeated exposure of laboratory animals," *AIHAJ*, 38(5), 1977.
493. P. Michel, *Revue des Travaux Institut des Pêches Maritimes*, 36(1), 1972.
494. Stauffer Chemical Company, Technical Information A-10177, July 1973.
495. —— Technical Information A-10423, March 1972.
496. —— A-10167 R, Oct. 1969.
497. —— A-10104 R-71, Nov. 1970.
498. —— A-10179 R 2/72, Feb. 1972.
499. —— A-10166 R, Oct. 1969.
500. —— A-10176 R, Oct. 1969.
501. The Quaker Oats Company, Technical Bulletin No. 135.
502. —— No. 149.
503. —— No. 148.
504. —— No. 206-A.
505. Minnesota Mining and Manufacturing Company, personal communication.
506. H. J. Neumann and D. T. Kaschani, "Bestimmung und Gehalt von polycyclischen aromatischen Kohlenwasserstoffen in Bitumen 'wlb'wasser, *Luft und Betrieb*," 21 (12), 1977.
507. Nancy Isaacson Kerkvliet and Donald J. Kemeldorf, "Inhibition of tumor growth in rats by feeding a polychlorinated biphenyl, Aroclor 1254," *Bull. Environm. Contam. Toxicol.*, 18(2), 1977.
508. Hancock, S., and Slater, A, "A specific method for the determination of trace concentrations of tetramethyl- and tetraethyllead vapors in air," *Analyst* 100, 422–429, 1975.
509. Laveskog, A., "A method for the determination of TEL and TML in air," Proceedings, Second International Clean Air Congress, Washington, D.C., 1970, pp. 549–57.
510. G. Broddin, L. van Vaeck, and K. van Cauwenberghe, "On the size distribution of polycyclic aromatic hydrocarbon containing particles from a coke oven emission source," *Atm. Environm.* 11, 1061–1064, 1977.
511. Y. Correia, C. J. Martens, F. H. van Hensch and B. P. Whim, "The occurrence of trichloroethylene, tetrachloroethylene and 1,1,1-trichloroethane in Western Europe in air and water," *Atm. Environm.*, 11, 1113–1116, 1977.
512. J. P. Voets, P. Pipijn, P. van Lancker, and W. Verstraete, "Degradation of microbicides under different environmental conditions," *J. Appl. Bact.* 40, 67–72, 1976.
513. K. Grob and G. Grob, "Organic substances in potable water and its precursor–part II: Applications in the area of Zürich," *J. Chromatography*, 90 303–313, 1974.
514. John McNeill Sieburth, "Biochemical warfare among the microbes of the sea," 1962 Honors Lecture, University of Rhode Island.
515. BASF, Technical Note M 1830.
516. Douglas C. Hittle and James J. Stukel, "Particle size distribution and chemical composition of coal-tar fumes," *AIHAJ*, 199–204, April 1976.
517. W. Wilson Pitt, Jr., Robert L. Jolley, and Charles D. Scott "Determination of trace organics in Municipal sewage effluents and natural waters by high resolution ion-exchange chromatography," *Environm. Sci. & Techn.* 9(12), 1068–1073, November 1975.
518. W.S. Smith, "Atmospheric emissions from fuel oil combustion," PHS Publ. No. 999-AP-2, R.A. Taft Sanitary Engineering Center, Cincinnati, Nov. 1962.
519. M. I. Weisburd, "Air pollution control field operation Manual–A guide for inspection and enforcement," PHS Publ. No. 937, 1962.
520. U.S. Department of the Interior–FWPCA, "Summary report–Advanced waste treatment," Publication WP-20-AWTR-19, 1968.
521. Claude E. Zobell, and Joseph F. Prokop "Microbial oxidation of mineral oils in Baratoria Bay bottom deposits," *Zeitschrift für Allg. Mikrobiologie* 6(3), 143–162, 1966.
522. K. J. Bock, "Über die Wirkung der Grenzflächenaktivität auf Fische," *Münch. Beitr. Abwass.-, Fischerei- und Flussbiol.*, 9(2), 1967.

523. R. Cabridenc, Mme. Leygue and R. Raux, "Etude des possibilités de biodégradation d'un effluent de frabrication d'isoprene," I.R.Ch.A. Centre de Recherche, 91, Vert-le-Petit, France, 1968.
524. The Quaker Oats Company, "Physical data on QO Furfural—Bulletin 203-B," Merchandise Mart Plaza, Chicago, Illinois 60654.
525. The Quaker Oats Company, "General information, physical data, chemistry, uses of QO Furfuryl alcohol—bulletin 205-A."
526. R. W. Goldbach, H. van Genderen, P. Leeuwangh "Hexachlorobutadiene residues in aquatic fauna from surface water fed by the River Rhine," *The Science of the Total Environment*, 6, 31-40, 1976.
527. C. K. Wun, R. W. Walker, and W. Litsky, "The use of XAD-2 resin for the analysis of coprostanol in water," *Water Res.* 10, 955-959, 1976.
528. Borneff, J., and Kunte, H., "Carcinogenic substances in water and soil—XIV: Further investigations for PAH in soil sample," *Arch. Hyg. Bakt.* 147, 401-409, 1963.
529. Acheson, M. A., Harrison, R. M., Perry, R. and Wellings, R. A., "Factors affecting the extraction and analysis of PAH in water," *Water Res.*, 1974.
530. Borneff, J., and Kunte, H., "Carcinogenic substances in water and soil—XVI: Evidence of PAH in water samples through direct extraction," *Arch. Hyg. Bakt.* 148, 585-597, 1964.
——— XVII: "About the origin and evaluation of the PAH in water," *Arch. Hyg. Bakt.* 148, 226-243, 1965.
531. Borneff, J., and Fischer, R., "Carcinogenic substances in water and soil—IX: Investigations on filter mud of a lake water (treatment) plant for PAH," *Arch. Hyg. Bakt.* 146, 183-197, 1962.
——— XII: "PAH in surface waters," *Arch. Hyg. Bakt.*, 146, 572-585, 1963.
——— XVII: About the origin and evaluation of the PAH in water," *Arch. Hyg. Bakt.* 149, 226-243, 1965.
532. Mallet, L., Perdriau, A. and Perdriau, J., "Pollution by B(a)P type PAH of the western region of the Arctic Ocean," *C. R. Hebd. Séanc. Acad. Sci., Paris*, 256, 3487-3489, 1963.
533. Bourcart, J., and Mallet, L., "Marine pollution of the shores of the central region of the Tyrrhenian sea (Bay of Naples)," *C. R. Hebd. Séanc. Acad. Sci., Paris* 260, 3729-3734, 1965.
534. Bourcart, J., Lalou, C. and Mallet, L., "About the presence of B(a)P type hydrocarbons in the coastal muds and the beach sands along the coast of Villefranche (Alpes Maritimes)," *C. R. Hebd. Séanc. Acad. Sci., Paris* 252, 640-644, 1961.
535. Mallet, L., and Sardou, J., "Investigation on the presence of the B(a)P type PAH in the plankton environment of the region of the Bay of Villefranche," *C. R. Hebd. Séanc. Acad. Sci., Paris* 258, 5264-5267, 1964.
536. Veldre, I. A., Lakhe, L. A. and Arro, I. K., "Content of 3,4-benzpyrene in the sewage effluent of the shale-processing industry," *Hyg. Sanit.* 30, 291-294, 1965.
537. Dikun, P. P., and Makhinenko, A. I., "Detection of 3,4-benzpyrene in the schistose plant resins, in its effluents and in water basins after discharge of effluents," *Hyg. Sanit.* 28, 10-12; *C. A.* 59, 3640g 1963.
538. Fedorenko, Z. P. "The effect of biochemical treatment of waste water of a by-product coke plant on the 3,4-benzpyrene content," *Hyg. Sanit.* 29, 19-21, 1964.
539. Wedgwood, P., and Cooper, R. L., "The detection and determination of traces of polynuclear hydrocarbons in industrial effluents and sewage—II: Sewage humus and treated effluents," *Analyst. (London)* 79, 163-169, 1954.
540. Cherkinskii, S. N., Dikun, P. P. and Yakovleva, G. P., "An investigation of carcinogenic substances in the effluents of several industrial enterprises" *Hyg. Sanit.* 24(9), 11-14, 1959; *C. A.* 54, 7942f, 1960.
541. Samarlovich, L. N., and Redkin, Y. R., "3,4-Benzpyrene pollution of the River Sunzha caused by the petrochemical industry in Grozny," *Hyg. Sanit.* 33, 165-168, 1968.

542. Ershova, K. P., "Carcinogenic pollution of water by effluents from petroleum refineries," *Hyg. Sanit.*, 33, 268-270, 1968.
543. Evans, W. C., Fernley, H. N. and Griffiths, E., "Oxidative metabolism of phenanthrene and anthracene by soil *Pseudomonas*," *Biochem. J.* 95, 819-831, 1968.
544. Poglazova, M. N., Fedosceva, G. E., Khesina, A. J., Meisser, M. N. and Shabad, L. M., "Destruction of 3,4-benzpyrene by soil bacteria," *Life Sci.* 6, 1053-1063, 1967.
545. Borneff, J., and Kunte, H., "Carcinogenic substances in water and soil–XIX: The effect of sewage purification on PAH," *Arch. Hyg. Bakt.*, 151, 202-210, 1967.
546. Wedgwood, P., and Cooper, R. L., "The detection and determination of traces of polynuclear hydrocarbons in industrial effluents and sewage–IV: The quantitave examination of effluents," *Analyst (London)* 81, 42-44, 1956.
547. Graf, W., and Nothhafft, G., "Drinking water chlorination and 3,4-benzpyrene," *Arch. Hyg. Bakt.*, 147, 135-146, 1963.
548. Trakhtman, N. N., and Manita, M. D., "Effect of chlorination of water on pollution by 3,4-benzpyrene," *Hyg. Sanit.* 31, 316-319, 1966.
549. Sforzolini, G. S., Saviano, A. and Merletti, C., "The action of chlorine on some hydrocarbons, polycyclic hydrocarbons, contributing to the study of the decontamination of water from carcinogenic compounds," *Boll. Soc. Ital. Biol. Sper.* 46, 903-906, 1970.
550. Il'nitskii, A. P., Kesina, A. Y., Cherkinskii, S. N. and Shabad, L. M., "Effect of ozonation upon aromatic hydrocarbons including carcinogens," *Hyg. Sanit.* 33, 323-327, 1968.
551. R. B. Laughlin, J. R. and J. M. Neff, Y. C. Hrung, T. C. Goodwin, and C. S. Gian, "The effects of three phthalate esters on the larval development of the Grass Shrimp *Palaemonetes pugio* (Holthuis)," *Water, Air and Soil Pollution* 9, 323-336, 1978.
552. Sanders, H. O., Mayer, F. L., Jr. and Walsh, D. F., *Environm. Res.* 6, 84, 1973.
553. Sandborn, J. R., Metcalf, R. L., Yu, C. C., and Lu, P. Y.; *Arch. Environ. Contam. Toxicol* 3, 244, 1975.
554. R. H. J. Young, D. W. Ryckman, and J. C. Buzzell Jr. "An improved tool for measuring biodegradability," *JWPCF* 40, 8, August 1968.
555. Gregory A. Jungclaus, Viorica Lopex-Avila, Ronald A. Hites, "Organic compounds in an industrial wastewater: A case study of their environmental impact."
556. R. B. King, A. C. Antoine, J. S. Fordyce, H. E. Neustadter, and H. F. Leibecki, "Compounds in airborne particulates: salts and hydrocarbons" *JAPCA* 27(9), Sept. 1977.
557. "Fluorides," National Academy of Science, 1971, p. 34.
558. EPA, "Monitoring and Air Quality Trends Report, 1972," EPA 450/1-73-004, Environmental Protection Agency, 1973, p. 4.
559. N.A.P.C.A., "Air Quality Data," U.S. Dept. of Health, Education and Welfare, 1968.
560. R. J. Gordon and R. J. Bryan, "Patterns in airborne polynuclear hydrocarbons concentrations at four Los Angeles sites," *Environm. Sci. and Techn.* 7, 1050, 1973.
561. G. Chatot, R. Dangy-Caye, and R. Fontanges, "Etude de la pollution atmosphérique par les arènes polynucléaires dans la région Lyonnaise à l'aide de deux collecteurs de principe différent," *Atm. Environm.* 7, 819, 1973.
562. L. Zoccolillo, A. Liberti, and D. Brocco, "Determination of polycyclic hydrocarbons in air by gas chromatography with high efficiency packed columns," *Atm. Environm.* 6, 715, 1972.
563. M. Kertesz-Saringer and Z. Morlin, "On the occurrence of polycyclic aromatic hydrocarbons in the urban area of Budapest," *Atm. Environm.* 9, 831, 1975.
564. A. L. A. M. Bridié, "Determination of biochemical oxygen demand with continuous recording of oxygen uptake," *Water Res.*, 3, 157-165, 1969.
565. F. de Wiest, H. Della Fiorentina, "Etude de la fraction organique présente dans les particules en suspension dans l'air de la Belgique," *La Revue du CEBEDEAU-BECEWA*, 8-21, Jan. 1977.
566. J. Borneff, F. Selenka, H. Kunte, A. Maximos, "Experimental Studies on the formation of polycyclic aromatic hydrocarbons in plants," *Environm. Research*, 2, 22-29, 1968.

567. G. F. Harrison, "The cavtat incident" (The Associated Octel Co., Ltd.), presented at the International Experts Discussion Meeting on Lead—Occurrence, Fate and Pollution in the Marine Environment, Rovinj, Yugoslavia, Oct. 1977.
568. J. E. Cremer, *Ann. Occup. Hyg.* 3, 226, 1961.
569. Ethyl Corporation, private communication.
570. J. S. Buck and K. Kumro, *J. Pharm. Exp. Therap.* 38, 161, 1930.
571. H. Turnbull, J. G. De Mann, R. F. Weston, *Ind. Eng. Chem.*, 46, 324, 1954.
572. [obsolete]
573. B. G. Maddock and D. Taylor, "The acute toxicity and bioaccumulation of some lead alkyl compounds in marine animals" (Imperial Chemical Industries), presented at the International Experts Discussion Meeting on Lead—Occurrence, Fate and Pollution in the Marine Environment, Rovinj, Yugoslavia, Oct. 1977.
574. F. G. Noden, "The determination of tetraalkyllead compounds and their degradation products in natural water" (The Associated Octel Co. Ltd.), presented at the International Experts Discussion Meeting on Lead—Occurrence, Fate and Pollution in the Marine Environment, Oct. 1977.
575. J. R. Grove, "Investigation into the formation and behaviour of aqueous solutions of lead alkyls," The Associated Octel Co. Ltd., P. O. Box 17, Ellesmere Port, S. Wirral, Cheshire L65 4HF, U.K.
576. Taylor, D., "Some investigations into the effect of 1-2,dichloroethane on marine life," Brixham Laboratory report BL/B/1571, 1974.
577. Davis, J. T. and Hardcastle, W. S., "Biological assay of herbicides for fish toxicity," *Weeds* 7, 397–404, 1959.
578. Silverberg, B. A., Wong, P. T. S. and Chan, Y. K., "Effects of tetramethyllead on freshwater green algae," *Arch. Environm. Contam. Toxicol* 5, 1977.
579. Richard F. Lee, "Accumulation and turnover of petroleum hydrocarbons in marine organisms," Skidaway Institute of Oceanography, P.O. Box 13687, Georgia 31406.
580. Dunn, B. P., and H. F. Stich, "Release of the carcinogen benzo(a)pyrene from environmentally contaminated mussels," *Bull. Environm. Cont. Toxicol.* 15, 398–401, 1976.
581. Siamak Khorram and Allen W. Knight, "The toxicity of kelthane to the grass shrimp (*Crangon franciscorum*)," *Bull. Environm. Contam. Toxicol.* 18(6), 674–682, 1977.
582. Ardith A. Grote and Richard E. Kupel, "Sampling and analysis of 2,6-di-*t*-butyl-*p*-cresol (DBPC)," *Am. Ind. Hyg. Assoc. J.* 39, 78–82, Jan. 1978.
583. G. Saint-Ruf, "La dioxine," *Proceedings of TCDD Pollution*, Commission of the European Communities, 30.9.1977.
584. M. J. Mercier, "2,3,7,8-tetrachlorodibenzo-*p*-dioxin—An overview," *Proceedings of TCDD Pollution*, Commission of the European Communities, 30.9.1977.
585. Kearny, P. C., Woolson, E. A., Isensec, A. R. and Helling, C. S., *Environm. Health Perspect.* 5, 273, 1973.
586. Schwetz, B. A.; Norris, J. M.; Sparschu, G. L.; Rowe, V. K.; Gehring, P. P.; Emerson, J. L.; and Gerbig, C. G., *Environm. Health. Perspect,* 5, 87, 1973.
587. Seiler, J. P., *Experimentia* 29(5), 622–3, 1973.
588. C. J. A. van Echteld, H. L. Golterman "Toxicologische aspecten van NTA in oppervlaktewater," H_2O May 1978.
589. Nixon, G. A., *Tox. Appl. Pharm.* 18, 398, 1971.
590. J. K. Reichert, "Kanzerogene substanzen in Wasser und Boden," *Arch. Hyg.* 152(3), 277–279, 1968.
591. W. Gräf and Chr. Winter, "3,4-benzpyren im Erdöl," *Arch. Hyg.* 152(4), 289–293, 1968.
592. H. S. Majewski, J. F. Klaverkamp, and D. P. Scott, "Acute lethality and sublethal effects of acetone, ethanol and propyleneglycol on the cardiovascular and respiratory system of rainbow trout," *Water Res.* 13, 217–221, 1978.

593. Smith, A. L., "The effects of effluents from the Canadian petrochemical industry on aquatic organisms," Fish. Res. Bd. Can., Techn. Rep. 472, 1974.
594. Mohamed A. El-Dib and Osama A. Aly, "Removal of phenylamide pesticides from drinking waters—II. Adsorption on powered carbon," *Water Res.* **11**, 617–620, 1977.
595. R. J. Kociba et al., "Results of a two year chronic toxicity study with hexachlorobutadiene in rats," *AIHA J* **38**(11), 1977.
596. Gehring, P. J., and D. Mac Dougall, "Review of the toxicity of hexachlorobenzene and hexachlorobutadiene," Special release from the Dow Chemical Company, Midland, Michigan, 1971.
597. Murzakaev, F. G., "Toxicity Data for hexachlorobutadiene and its Intermediates," *Farmakol i. Toksikol* **26**, 750, 1963; *Chem. Abs.* **60**: 1377b, 1964.
598. Gulko, A. G.; N. I., Zimina; and I. G., Shroit, "Toxicological study of the insecticide hexachlorobutadiene," *Vopr. Gigieny i Sanit. Ozdorovl. Vneshn. Sredy, Kishinev*, **128**, 1964; *Chem. Abs.* **62**; 13757c, 1965.
599. Treon, J. F. and A. Edwards, "Report on the toxicity of hexachlorobutadiene vapor," Personal communication from Kettering Laboratory, University of Cincinnati, 1948.
600. Schwetz, B. A., J. M. Norris, R. J. Kociba, P. A. Keeler, R. F. Cornier, and P. J. Gehring, "Reproduction study in Japanese quail fed hexachlorobutadiene for 90 days," *Toxicol. Appl. Pharmacol.* **30**, 255, 1974.
601. Alibaev, T. S., "Hygienic standards for cyclohexane and its mixture with benzene in air," *Hyg. Sanit. USSR*, **35**(1–3), 22–28, 1970.
602. Allison, V. C. and S. H. Katz "An investigation of stenches and odors for industrial purposes," *J. Ind. Eng. Chem.*, **11**, 336–338, 1919.
603. Amdur, M. O., W. W. Melvin, and P. Drinker, "Effects of inhalation of sulphur dioxide by man," *Lancet*, **265**, 758–759, 1953.
604. Andreescheva, N. G., "Substantiation of the maximum permissible concentration of nitrobenzol in the atmospheric air" (in Russ.), *Gig. Sanit.*, **29**(8), 5–10, 1964.
605. Andreescheva, N. G., "Sanitary toxicologic assessment of certain aromatic carbohydrates in the atmosphere" (in Russ.), *Gig. Sanit.*, **33**(4), 12–16, 1968.
606. Anrooij, A. van, "Desodorisatie langs photochemischen weg," Thesis, Utrecht, the Netherlands 1931.
607. Appell, L., "Physical foundation in perfumery VIII. The minimum perceptible," *Am. Perfum. Cosmet.*, **84**(3), 45–50, 1969.
608. Armit, H. W., "The toxicology of nickel carbonyl," *J. Hyg.*, **7**, 525–551, 1907.
609. Babin, I. N., I. I. Strizhevskii, and E. I. Kordysh, "Use of the odorimeter VNIIT for comparative tests of the intensity of the odor of carbide and electrolysis acetylene," *Khim. Tekhnol. Topl. Prod. Ego Pererab.*, 152–155, 1965; *Chem. Abstr.*, **66**, 31815m, 1967.
610. Backman, E. L., "Experimentella undersökningar öfver luktsinnets fysiologi," *Upsala Laekarefoeren. Foerh.*, **22**, 319–470, 1917.
611. Backman, E. L., "The olfactologie of the methylbenzol series," *Onderzoekingen gedaan in het Physiologisch Laboratorium der Utrechtse Hogeschool*, Laboratory of Physiology, Utrecht, the Netherlands S5, **18**, 349–364, 1918.
612. Baikov, B. K., "Experimental data for substantiating the maximum permissible concentration of carbon disulfide in combination with hydrogen disulfide in the atmospheric air" (in Russ.), *Gig. Sanit.*, **28**(3), 3–8, 1963.
613. Baikov, B. K., M. Kh. Khachaturyan, E. V. Borodina, N. G. Feldman, and A. M. Tambovtseva, "Data for hygienic standardiziation of amyl alcohol in the atmosphere" (in Russ.), *Gig. Sanit.*, **38**(9), 10–14, 1973.
614. Baldus, C., "Untersuchung über Geruchsswellen," Thesis, Würzburg, W. Germany 1936.

Note: Refs. 615 and 616 are not cited in the text.

617. Belkov, A. N., "Action of small concentrations of carbon tetrachloride on the human body" *Tr., Tsent. Inst. Usoversh. Vrachei*, **135**, 90-96, 1969; *Chem. Abstr.*, **74** 2340n, 1971.
618. Berck, B., "Sorption of phosphine by cereal products," *J. Agric. Food Chem.*, **16**, 419-425, 1968.
619. Berthelot, M., "Observations sur les procédés propres à déterminer les limites de la sensibilité olfactive," *Ann. Chim. Phys.* (Serie 7), **22**, 460-464, 1901.
620. Berzins, A., "Determination of the threshold of olfactory perception of ethylenimine," *Aktual. Vop. Gig. Tr. Prof. Patol., Mater. Konf., 1st*, 47-48, 1967, *Chem. Abstr.*, **72** 35503e, 1970.
621. Bezpalková, L. E., "The threshold of reflex effects of methylacrylate vapours upon human organism" (in Czech.), *Cesk. Hyg.*, **12**, 577-582, 1967.
622. Blinova, A. E., "Industrial standards for substances emitting strong odors," *Hyg. Sanit. USSR*, **30**(1-3), 18-22, 1965.
623. Bocca, E., and M. N. Battiston "Odour perception and environment conditions," *Acta Oto-Laryngol.*, **57**, 391-400, 1964.
624. Borisova, M. K., "Experimental data for determination of the maximum allowable concentration of dichloroethane in the atmosphere" (in Russ.), *Gig. Sanit.*, **22**(3), 13-19, 1957.
625. Buchberg, H., M. H. Jones, K. G. Lindh, and K. W. Wilson, "Air pollution studies with simulated atmospheres" Report no. 61-44, California University, Los Angeles, 1961.
626. Bushtueva, K. A. "New studies of the effect of sulfur dioxide and of sulfuric acid aerosol on reflex activity of man," in: V. A. Ryazanov (Ed.), *Limits of allowable concentrations of atmospheric pollutants*, Book 5, 86-92, U.S. Department of Commerce, Office of Technical Services, Washington, D.C., 1962.
627. Cain, W. S., "Odor intensity differences in the exponent of the psychophysical function," *Percept. Psychophys.*, **6**(6A), 349-354, 1969.
628. Carpenter, C. P. "The chronic toxicity of tetrachloroethylene," *J. Ind. Hyg. Toxicol.*, **19**, 323-336, 1937.
629. Carpenter, C. P., H. F. Smyth Jr., and C. B. Shaffer, "The acute toxicity of ethylene imine to small animals" *J. Ind. Hyg. Toxicol.*, **30**, 2-6, 1948.
630. Cederlof, R., M. L. Edfors, L. Friberg, and T. Lindvall, "On the determination of odor thresholds in air pollution control. An experimental field study on flue gases from sulfate cellulose plants" *JAPCA*, **16**, 92-94, 1966.
631. Chao-Chen-Tzi, "Data for determining the standard maximum permissible concentration of methanol vapours in the atmospheric air" (in Russ.), *Gig. Sanit.*, **24**(10), 7-12, 1959.
632. Cheesman, G. H., and H. M. Kirkby, "An air dilution olfactometer suitable for group threshold measurements" *Q. J. Exp. Psychol.*, **11**, 115-123, 1959.
633. Chizhikov, V. A., "Data for substantiating the maximum permissible concentration of toluenediisocyanate in the atmospheric air" (in Russ.), *Gig. Sanit.*, **28**(6), 8-15, 1963.
634. Corbit, T. E., and T. Engen, "Facilitation of olfactory detection," *Percept. Psychophys.*, **10**, 433-436, 1971.
635. Cormack, D., T. A. Dorling, and B. W. J. Lynch, "Comparison of techniques for organoleptic odour-intensity assessment," *Chem. Ind. (London)*, (21), 857-861, 1974.
636. Davis, R. G., "Olfactory psychophysical parameters in man, rat, dog and pigeon, *J. Comp. Physiol. Psychol.*, **85**, 221-232, 1973.
637. Deadman, K. A., and J. A. Prigg, "The odour and odorization of gas," Research Communication GC59, The Gas Council, London, 1959.
638. Deese, D. E., and R. E. Joyner, "Vinyl acetate: a study of chronic human exposure," *J. Am. Ind. Hyg. Assoc.*, **30**, 449-457, 1969.
639. Dobrinsky, A. A., "Hygienic standards for certain intermediate compounds of Caprolactam production in the atmospheric air" (in Russ.), *Gig. Sanit.*, **29**(12), 8-13, 1964.
640. Doll, W., and K. Bournot, "Uber den Geruch optischer Antipoden" *Pharmazie*, **4**, 224-227, 1949.

641. Dravnieks, A., and B. K. Krotszynski, "Collection and processing of airborne chemical information II," *J. Gas Chromatogr.*, **6**, 144–149, 1968.
642. Dravnieks, A., and A. O'Donnell, "Principles and some techniques of high-resolution headspace analysis," *J. Agric. Food Chem.*, **19**, 1049–1056, 1971.
643. Dravnieks, A., "A building-block model for the characterization of odorant molecules and their odors," *Ann. N. Y. Acad. Sci.*, **237**, 144–163, 1974.
644. Duan-Fen-Djuy, "Data for determining the maximum permissible concentration of hydrogen sulfide in the atmospheric air" (in Russ.), *Gig. Sanit.*, **24**(10), 12–17, 1959.
645. Dubrovskaya, F. I., "Hygienic evaluation of pollution of atmospheric air of a large city with sulfur dioxide gas," in: V. A. Ryazanov (Ed.), *Limits of Allowable Concentrations of Atmospheric Pollutants*, Book 3, 37–51, U.S. Department of Commerce, Office of Technical Services, Washington, D.C., 1957.
646. Dubrovskaya, F. I., M. S. Katsenelenbaum, Ya. K. Yushko, G. V. Bulychev, and V. A. Koroleva, "Atmospheric air pollution with discharges from synthetic fatty acids and alcohols producing industries and their effect on the health of the population" (in Russ.), *Gig. Sanit.*, **26**(12), 3–8, 1961.
647. Dubrovskaya, F. I., "One-time maximum concentration of caproic acid in air," *Hyg. Sanit. USSR*, **34**(4–6), 331–335, 1969.
648. Dubrovskaya, F. I., and M. Kh. Khachaturyan, "Sanitary evaluation of valeric acid as an atmospheric pollutant," in: M. Y. Nuttonson (Ed.), *American Institute of Crop Ecology (AICE) survey of U.S.S.R. Air Pollution Literature*, Volume XIX, 98–103, National Technical Information Service, U.S. Department of Commerce, Springfield, PB-214 264, 1973.
649. Dunlap, M. K., J. K. Kodama, J. S. Wellington, H. H. Anderson, and C. H. Hine, "The toxicity of allyl alcohol," *AMA Arch. Ind. Health*, **18**, 303–311, 1958.
650. Eglite, M. E., "A contribution to the hygienic assessment of atmospheric ozone" *Hyg. Sanit. USSR*, **33**(1–3), 18–23, 1968.
651. Elfimolva, E. V., "Biological activity of low atmospheric concentrations of isopropylbenzene and benzene" *Vop. Gig. Planirovki San. Okhr. Atmos. Vozdukha, Moscow*, 148–158, 1966; *Chem. Abstr.*, **66**, 93947q, 1967.
652. Endo, R., T. Kohgo, and T. Oyaki, "Research on odor nuisance in Hokkaido (part 2). Chemical analysis of odors," in: *Proceedings of the 8th Annual Meeting of the Japan Society of Air Pollution, Symposium (1): Hazard nuisance pollution problems due to odors (English translation for APTIC by SCITRAN)*, 1967.
653. Eykman, H. J., "Klinische olfactometrie," Thesis, Utrecht, the Netherlands 1927.
654. Fassett, D. W., "Esters," in: F. A. Patty (Ed.), *Industrial Hygiene and Toxicology*, Volume II. *Toxicology, 1847–1934*, Interscience, New York/London, 1962.
655. Feldman, Yu. G., "Data for determining the maximum permissible concentration of acetone in the atmosphere" (in Russ.), *Gig. Sanit.*, **25**(5), 3–10, 1960.
656. Feldman, Yu. G., "The experimental determination of the maximum permissible one-time concentration of diketene in the atmosphere" *Hyg. Sanit. USSR*, **32**(1–3), 9–14, 1967.
657. Feldman, Yu. G., and T. I. Bonashevskya, "On the effect of low concentrations of formaldehyde," *Hyg. Sanit. USSR*, **36**(4–6), 174–180, 1971.
658. Fieldner, A. C., S. H. Katz, and S. P. Kinney, "Gas masks for gases met in fighting fires," U.S. Bureau of Mines, Technical Paper No. 248, 1921.
659. Filatova, V. I., "Data for substantiating the maximum permissible concentration of methylmetacrylate in the atmospheric air" (in Russ.), *Gig. Sanit.*, **27**(11), 3–8, 1962.
660. Flury, F., "Uber Kampfgasvergiftungen IX. Lokal reizende Arsenverbindungen," *Z. Gesamte Exp. Med.*, **13**, 523–578, 1921.
661. Fomin, A. P., "Biological effect of epichlorohydrine and its hygienic significance as an atmospheric contamination factor" (in Russ.), *Gig. Sanit.*, **31**(9), 7–11, 1966.
662. Frantiková, D., "Determination of the olfactory threshold to industrial poisons" (in Czech.), *Act. Nerv. Super.*, **4**, 184–185, 1962.

663. Gavaudan, P., H. Poussel, G. Brebion, and M.-P. Schutzenberger, "L'étude des conditions thermodynamiques de l'excitation olfactive et les théories de l'olfaction" *C. R. Hebd. Seances Acad. Sci.*, **226**, 1995–1396, 1948.
664. Geier, F., "Beitrag zur Bestimmung von Geruchsschwellen" Thesis, Würzburg, W. Germany, 1936.
665. Gofmekler, V. A., "Data for substantiating the maximum permissible concentration of acetates in the atmosphere" (in Russ.), *Gig. Sanit.*, **25**(4), 9–15, 1960.
666. Gofmekler, V. A., "Experimental observation of the reflex effect of acetaldehyde upon human organism" (in Czech.), *Cesk, Hyg.*, **12**, 369–375, 1967.
667. Goldenberg, D. M., "Geruchswahrnehmung und Schwellen von Duftgemischen beim Menschen," *Hals-, Nasen- und Ohrenheilkunde, Zwanglose Schriftenreihe*, Vol. 19, Barth, Leipzig, 1967.
668. Gorlova, O. E., "Hygienic assessment of isopropyl alcohol as an atmospheric pollutant" (in Russ.), *Gig. Sanit.*, **35**(8), 9–14, 1970.
669. Grigorieva, K. V., "On the atmospheric air pollution with maleic anhydride" (in Russ.), *Gig. Sanit.*, **29**(3), 8–12, 1964.
670. Grijns, G., "Messungen der Riechschärfe bei Europäern und Javanen," *Arch. Anat. Physiol., Physiol. Abt.*, **30**, 509–517, 1906.
671. Grijns, G., "Y-a-t-il une relation entre le pouvoir absorbant à l'égard de la chaleur rayonnante et le pouvoir odorant des substances," *Arch. Neerl. Physiol.*, **3**, 377–390, 1919.
672. Gundlach, R. H., and G. Kenway, "A method for the determination of olfactory thresholds in humans," *J. Exp. Psychol.*, **24**, 192–201, 1939.
673. Gusev, I. S., "Reflective effects of microconcentrations of benzene, toluene, xylene and their comparative assessment," *Hyg. Sanit. USSR*, **30**(10–12), 331–336, 1965.
674. Hamanabe, Y., S. Harima, T. Miura, T. Domon, and T. Namiki, "The sensitive measurement of bad odor using an odorless chamber," *J. Jpn Soc. Air Pollut.*, **4**(1), 115, 1969; from: Odors and air pollution: a bibliography with abstracts, Environmental Protection Agency, Office of Air Programs Publication, No. AP-113 (1972) p. 17.
675. Hellman, T. M., and F. H. Small, "Characterization of petrochemical odors," *Chem. Eng. Progr.*, **69**(9), 75–77, 1973.
676. Hellman, T. M., and F. H. Small, "Characterization of the odor properties of 101 petrochemicals using sensory methods," *JAPCA*, **24** 979–982, 1974.
677. Henderson, Y., and H. W. Haggard, "The elimination of industrial organic odors," *J. Ind. Eng. Chem.*, **14**, 548–551, 1922.
678. Henning, H., *Der Geruch*, 2nd ed., Barth, Leipzig, 1924.
679. Henning, H., "Psychologische Studien am Geruchssinn," in: *Abderhalden's Handbuch der biologischen Arbeitsmethoden* VI, A, 741–836, Urban und Schwarzenberg, Berlin/Wien, 1927.
680. Henschler, D., A. Stier, H. Beck, and W. Neumann, "Geruchsschwellen einiger wichtiger Reizgase (Schwefeldioxyd, Ozon, Stickstoffdioxyd) und Erscheinungen bei der Einwirkungen geringer Konzentrationen auf den Menschen" *Arch. Gewerbepathol Gewerbehyg.*, **17**, 547–570, 1960.
681. Henschler, D., W. Assmann, and K.-O. Meyer, "Zur Toxikologie der Toluenediisocyanate" *Arch. Toxicol.*, **19**, 364–387, 1962.
682. Hermanides, J., "Over de constanten der in de olfactometrie gebruikelijke negen standaardgeuren," Thesis, Utrecht, the Netherlands 1909.
683. Hesse, W., "Bestimmung von Geruchsschwellen in absoluten Werten," *Z. Hals-, Nasen-, Ohrenheilk.*, **16**, 359–373, 1926.
684. Hesse, W., "Ein einfacher Apparat zur Bestimmung von Geruchsschwellen in absoluten Werten," *Z. Hals-, Nasen-, Ohrenheilk.*, **19**, 348–352, 1927.
685. Heyroth, F. F., "Halogens," in: F. A. Patty (Ed.), *Industrial Hygiene and Toxicology, Volume II. Toxicology*, 831–857, Interscience, New York/London, 1962.

686. Hildenskiold, R. S., "Maximum permissible concentration of carbon bisulphide in the atmospheric air of residential districts" (in Russ.), *Gig. Sanit.*, **24**(6), 3–8, 1959.
687. Hine, C. H., H. Ungar, H. H. Anderson, J. K. Kodama, J. K. Critchlow, and N. W. Jacobson, "Toxicological studies on p-tertiary-butyltoluene" *AMA Arch. Ind. Hyg. Occup. Med.*, **9**, 227–244, 1954.
688. Hine, C. H., J. K. Kodama, R. J. Gurman, and G. S. Loquvam, "The toxicity of allylamines," *Arch. Environm. Health*, **1**, 343–352, 1960.
689. Hine, C. H., R. J. Gurman, M. K. Dunlap, R. Lima, and G. S. Loquvam, "Studies on the toxicity of glycidaldehyde," *Arch. Environm. Health*, **2**, 23–30, 1961.
690. Hofmann, F. B., and A. Kohlrausch, "Bestimmung von Geruchsschwellen," *Biochem. Z.*, **156**, 287–294, 1925.
691. Hollingsworth, R. L., V. K. Rowe, F. Oyen, H. R. Hoyle, and H. C. Spencer, "Toxicity of paradichlorobenzene," *AMA Arch. Ind. Health*, **14**, 138–147, 1956.
692. Hollingsworth, R. L., V. K. Rowe, F. Oyen, T. R. Torkelson, and E. M. Adams, "Toxicity of o-dichlorobenzene," *AMA Arch. Ind. Health*, **17**, 180–187, 1958.
693. Holmes, J. A., E. C. Franklin, and R. A. Gould, "Report of Selby Smelter Commission," Department of the Interior, Bureau of Mines, Bulletin 98, 1915.
694. Hori, M., Y. Kobayashi, and Y. Ota, "Vinyl chloride monomer odor concentration," *Plast. Ind. News*, **18**, 164–168, 1972.
695. Huijer, H., "De olfactologie van aniline en homologen," Thesis, Utrecht, the Netherlands 1924.
696. Ifeadi, N. G., "Quantitative measurement and sensory evaluation of dairy waste odor," Thesis, Columbus (Ohio State Univ.), 1972.
697. Imasheva, N. B., "The substantiation of the maximum permissible concentrations of acetophenon in the atmospheric air" (in Russ.), *Gig. Sanit.*, **28**(2), 3–8, 1963.
698. Irish, D. D., "Halogenated hydrocarbons: I. Aliphatic," in: F. A. Patty (Ed.), *Industrial Hygiene and Toxicology*, Volume II. Toxicology, 1241–1332, Interscience, New York/London, 1962.
699. Itskovish, A. A., and V. A. Vinogradova, "Norms for phenol in the atmosphere," *Okhrana Prirody Sibiri i Dal'nego Vostoka Sb.* (1), 139–145, 1962; *Chem. Abstr.*, **61**, 6256b 1964.
700. Ivanov, S. V., "Materials on toxicology and hygienic rating of ethylbenzene content in the atmosphere of industrials premises" (in Russ.), *Gig. Tr. Prof. Zabol.*, **8**(2), 9–14, 1964.
701. Jacobson, K. H., J. H. Clem, H. J. Wheelwright, W. E. Rinehart, and N. Mayes, "The acute toxicity of the vapors of some methylated hydrazine derivatives," *AMA Arch. Ind. Health*, **12**, 609–616, 1955.
702. Jacobson, K. H., E. B. Hackley, and L. Feinsilver, "The toxicity of inhaled ethylene oxide and propylene oxide vapors" *AMA Arch. Ind. Health*, **13**, 237–244, 1956.
703. Jacobson, K. H., W. E. Rinehart, H. J. Wheelwright, M. A. Ross, J. L. Papin, R. C. Daly, E. A. Greene, and W. A. Groff, "The toxicology of an aniline-furfuryl alcohol-hydrazine vapor mixture," *J. Am. Ind. Hyg. Assoc.*, **19**, 91–100, 1958.
704. Janicek, G. V. Plaska and J. Kubátová, "Olfactometric estimation of the threshold of perception of odorous substances by a flow olfactometer" (in Czech.), *Cesk. Hyg.*, **5**, 441–447, 1960.
705. Jones, F. N., "An olfactometer permitting stimulus specification in molar terms," *Am. J. Psychol.*, **67**, 147–151, 1954.
706. Jones, F. N., "The reliability of olfactory thresholds obtained by snifting," *Am. J. Psychol.*, **68**, 289–290, 1955.
707. Jones, F. N., "A comparison of the methods of olfactory stimulation: blasting vs. sniffing," *Am. J. Psychol.*, **68**, 486–488, 1955.
708. Jones, F. N., "Olfactory absolute thresholds and their implications for the nature of the receptor process," *J. Psychol.*, **40**, 223–227, 1955.
709. Jung, J., "Untersuchung über Geruchsschwellen," Thesis, Würzburg, W. Germany 1936.

710. Katz, S. H., and E. J. Talbert, "Intensities of odors and irritating effects of warning agents for inflammable and poisonous gases," U.S. Bureau of Mines, Technical Report No. 480, 1930.
711. Kay, R. E., J. T. Eichner, and D. E. Gelvin, "Quantitative studies on the olfactory potentials of *Lucilia sericata*," *Am. J. Physiol.*, **213**, 1–10, 1967.
712. Kendall, D. A., G. Leonardus, and E. R. Rubin, "Parameters affecting the determination of odor thresholds," Report C-68988, Arthur D. Little, Inc., Cambridge, Mass., U.S.A., 1968.
713. Kerka, W. F., and C. M. Humphreys, "Temperature and humidity effect on odor perception," *Heat. Piping Air Cond.*, **28**, 129–136, 1956.
714. Khachaturyan, M. Kh., M. I. Gusev, and O. E. Gorlova, "The effect of weak olfactory stimuli on the latent period of motor reaction to light in man," *Hyg. Sanit. USSR*, **33**(7–9), 274–276, 1968.
715. Khachaturyan, M. Kh., and B. K. Baikov, "On the possibility of detecting the effect of reflexes of low butanol concentrations by the incalculation of conditioned inhibition," *Hyg. Sanit. USSR*, **34**(10–12), 251–254, 1969.
716. Khikmatullaeva, Sh. S., "Maximum permissible concentration of thiophene in the atmosphere," *Hyg. Sanit. USSR*, **32**(4–6), 319–323, 1967.
717. Kimmerle, G., Persönl. Mitteilung Institut für Toxikologie der Farbenfabrikanten Bayer A. G., Wuppertal-Elberfeld, 1971; taken from: D. Henschler (Ed.), *Gesundheitschädliche Arbeitsstoffe* (2. Lieferung), Verlag Chemie, Weinheim, 1973.
718. Kincaid, J. F., E. L. Stanley, C. H. Beckworth, and F. W. Sunderman, "Nickel poisoning III," *Am. J. Clin. Pathol.*, **26**, 107–119, 1956.
719. Kinkead, E. R., U. C. Pozzani, D. L. Geary, and C. P. Carpenter, "The mammalian toxicity of ethylidenenorbornene (5-ethylidenebicyclo(2,2,1)hept-2-ene)," *Toxicol. Appl. Pharmacol.*, **20**, 250–259, 1971.
720. Kinkead, E. R., U. C. Pozzani, D. L. Geary, and C. P. Carpenter, "The mammalian toxicity of dicyclopentadiene," *Toxicol. Appl. Pharmacol.*, **20**, 552–561, 1971.
721. Kittel, G., "Die moderne Olfaktometrie und Odorimetrie," *Z. Laryngol., Rhinol., Otol. Grenzgeb.*, **12**, 893–903, 1968.
722. Kittel, G., and P. G. J. Wendelstein, "Zur Differenzierung der olfaktiven Wahrnehmungs- und Erkennungsschwelle," *Arch. Klin. Exp. Ohren-, Nasen- Kehlkopfheilk.* (Berlin), **199**, 683–687, 1971.
723. Kniebes, D. V., J. A. Chisholm, and R. C. Stubbs, "Odorizzanti del gas naturale: fattori di sensibilita' olfattiva," *Gas* (Rome), **19**, 211–217, 1969.
724. Knuth, H. W., Institut für Chemische Technologie, TU Clausthal, personal communication, 1973.
725. Korneev, Yu. E., "Effect of the combined presence of low concentrations of phenol and acetophenone in the urban atmosphere," *Hyg. Sanit. USSR*, **30**(7–9), 336–345, 1965.
726. Kosiborod, N. R., "Hygienic significance of low diethylamine concentrations in the atmosphere," *Hyg. Sanit. USSR*, **33**(1–3), 31–35, 1968.
727. Köster, E. P., "Adaptation and cross-adaptation in olfaction," Thesis, Utrecht, the Netherlands 1971.
728. Krackow, E. H., "Toxicity and health hazards of boron hydrides," *AMA Arch. Ind. Hyg. Occup. Med.*, **8**, 335–339, 1953.
729. Krasovitskaya, M. L., and L. K. Malyarova, "Small concentrations of hydrocarbons in the air of naphthachemical plants," *Biol. Deistvie i Gigien. Znachenie Atm. Zagryazenii*, 74–100, 1966; *Chem. Abstr.*, **65** 14324d, 1966.
730. Krichevskaya, I. M., "Biological effect of caprolactam and its sanitary-hygienic assessment as an atmospheric pollutant," *Hyg. Sanit. USSR*, **33**(1–3), 24–31, 1958.
731. Kristesashvili, Ts. S., "The maximum permissible concentration of pyridine in the air," *Hyg. Sanit. USSR*, **30**(10–12), 173–177, 1965.
732. Kulakov, A. E., "The effect of small concentrations of hexamethylenediamine on man" (in Russ.), *Gig. Sanit.*, **29**(4), 8–13, 1964.

733. Kulka, W., and E. Homma, "Beiträge zur Kenntnis der Laboratoriumsluft und deren schädlichen Beimengungen," *Z. Anal. Chem.*, **50**, 1-11, 1910.
734. Kurtschatowa, G., and E. Dawidkowa, "Hygienische Begründung des Kurzzeit-MIK-Wertes von 4.6-Dinitro-o-kresol in der atmosphärischen Luft," *Wiss. Z. Humboldt-Univ. Berlin, Math.-Naturwiss. Reihe*, **19**, 467-468, 1970.
735. Laffort, P., "Some new data on the physico-chemical determinants of the relative effectiveness of odorants," in: N. N. Tanyolac (Ed.), *Theories of Odor and Odor Measurement*, 247-270, Robert College Bebek, Istanbul, 1968.
736. Laffort, P., "Interactions quantitatives dans un mélange d'odeurs: niveau liminaire," *Olfactologia*, **1**, 95-104, 1968.
737. Laffort, P., and A. Dravnieks, "An approach to a physico-chemical model of olfactory stimulation in vertebrates by single compounds," *J. Theor. Biol.*, **38**, 335-345, 1973.
738. Laffort, P., Centre National de la Recherche Scientifique, Collège de France, Paris, Personal communication, 1974.
739. Laing, D. G., "A comparative study of the olfactory sensitivity of humans and rats," *Chem. Senses Flavor*, **1**, 257-269, 1975.
740. Lehmann, K. B., and L. Schmidt-Kehl, "Die 13 wichtigsten Chlorkohlwasserstoffe der Fettreihe vom Standpunkt der Gewerbehygiene," *Arch. Hyg. Bakteriol.*, **116**, 131-268, 1936.
741. Leonardus, G., D. Kendall, and N. Barnard, "Odor threshold determinations of 53 odorant chemicals," *JAPCA*, **19**, 91-95, 1969.
742. Lindvall, T., "On sensory evaluation of odorous air pollutant intensities," *Nord. Hyg. Tidskr.*, **51**, Suppl. 2, 1-181, 1970.
743. Li-Shen, "Data for substantiating the maximum permissible concentration of styrol in atmospheric air" (in Russ.), *Gig. Sanit.*, **26**(8), 11-17, 1961.
744. Loginova, R. A., "Basic principles for the determination of limits of allowable concentrations of hydrogen sulfide in atmospheric air," in: V. A. Ryazanov (Ed.), *Limits of Allowable Concentrations of Atmospheric Pollutants*, Book 3, 52-68, U.S. Department of Commerce, Office of Technical Services, Washington, D.C., 1957.
745. Makhynia, A. P., "Hygienic assessment of joint atmospheric contamination with sulphurous acid and phenol" (in Russ.), *Gig. Sanit.*, **31**(8), 103-105, 1966.
746. Malyarova, L. K., "Biological action and hygienic significance of trichloroethylene as an atmospheric pollutant," *Mater. Nauch.-Prakt. Konf. Molodykh. Inst. Gig. Sanit. Vrachei, 11th*, 56-59, 1967; *Chem. Abstr.*, **72**, 24265c; 1970.
747. Martirosyan, A. S., "Maximum permissible concentration of methyl vinyl ketone in the air of a working area," *Tr. Klin. Otd. Nauch.-Issled. Inst. Gig. Profzabol.*, no. *1*, 113-115, 1970; *Chem. Abstr.*, 77 156013g, 1972.
748. Mateson, J. F., "Olfactometry: its techniques and apparatus," *JAPCA*, **5**, 167-170, 1955.
749. May, J., "Geruchsschwellen von Lösemitteln zur Bewertung von Lösemittelgerüchen in der Luft," *Staub*, **26**, 385-389, 1966.
750. McGee, W. A., F. L. Oglesby, R. L. Raleigh, and D. W. Fassett, "The determination of a sensory response to alkyl 2-cyanoacrylate vapor in man," *J. Am. Ind. Hyg. Assoc.*, **29**, 558-561, 1968.
751. Melekhina, V. P., "The maximum permissible concentration of formaldehyde in the atmospheric air" (in Russ.), *Gig. Sanit.*, **23**(8), 10-14, 1958.
752. Minaev, A. A., "Investigation to substantiate the hygienic standards for alphamethylstyrol vapours in the atmosphere" (in Russ.), *Gig. Sanit.*, **31**(2), 3-6, 1966.
753. Mitsumoto, T., "Olfaktometrische Untersuchungen," *Z. Sinnesphysiol.*, **57**, 114-165, 317, 1926.
754. Mnatsakanyan, A. V., "Basic experimental information for the determination of the limit of allowable chloroprene concentration in atmospheric air," in: V. A. Ryazanov (Ed.), *Limits of Allowable Concentrations of Atmospheric Pollutants*, Book 5, 79-85, U.S. Department of Commerce, Office of Technical Services, Washington, D. C., 1962.

755. Morimura, S., "Untersuchung über den Geruchssinn," *Tohoku J. Exp. Med.*, 22, 417-448, 1934.
756. Moskowitz, H. R., and A. Dravnieks, "Odor intensity and pleasantness of butanol," *J. Exp. Psychol.*, 103, 216-223, 1974.
757. Moulton, D. G., and D. A. Marshall, "Odor detection in dog and man," in: *Abstracts of the First Congress of the European Chemoreception Research Organization*, Paris, 1974.
758. Mühlen, T. Zur, "Messungen von Styrol-Emissionen und Immissionen mit Hilfe der Gaschromatographie," *Zentralbl. Arbeitsmed. Arbeitsschutz*, 18, 41-43, 1968.
759. Mukhitov, B. M., "Experimental data for substantiating maximum permissible concentration of phenol in the atmospheric air" (in Russ.), *Gig. Sanit.*, 27(6), 16-24, 1962.
760. Mukhitov, B. M., and A. A. Azimbekov, "Hygienic evaluation of methyl ethyl ketone as an atmospheric pollutant," in: A. P. Filin (Ed.), *Vop. Gig. Tr. Profzabol.*, *Mater. Nauch. Konf.*, 232-234, 1971; *Chem. Abstr.* 81, 10876v, 1974.
761. Mullins, L. J., "Olfaction," *Ann. N.Y. Acad. Sci.*, 62, 247-276, 1955 (converted data taken from: P. Laffort, "Essai de standardisation des seuils olfatifs humains pour 192 corps purs," *Arch. Sci. Physiol.*, 17 75-105, 1963).
762. Nader, J. S., "An odor evaluation apparatus for field and laboratory use" *J. Am. Ind. Hyg. Assoc.*, 19, 1-7, 1958.
763. Naves, Y.-R., "Etudes sur les matières végétales volatiles XLVI. Sur le dédoublement de la d, l-α-ionone, *Helv. Chem. Acta*, 30, 769-774, 1947.
764. Naves, Y. R., and M. Delepine, "Sur les iso-α-isnones actives et racémique et leurs dérivés," *C. R. Hebd. Seances Acad. Sci.*, 237, 1167-1168, 1953.
765. Neuhaus, W., "Wahrnehmungsschwelle und Erkennungsschwelle beim Riechen des Hundes in Vergleich zu den Reichwahrnehmungen des Menschen," *Z. Vergl. Physio.*, 39, 624-633, 1957.
766. Nevers, A. D., and W. H. Oister, "Problems in the critical comparison of odor intensities," *Proc. Oper. Sect. Am. Gas. Assoc.*, 1965.
767. Nikiforov, B., "Basis for the single maximum permissible level of carbon tetrachloride in the atmosphere," *Khig. Zdraveopazvane*, 13(4), 365-370, 1970; *Chem. Abstr.*, 74, 57059h, 1971.
768. Novikov, Yu. V., "The determination of limits of allowable concentrations of benzene in atmospheric air," in: V. A. Ryazanov (Ed.), *Limits of Allowable Concentrations of Atmospheric Pollutants*, Book 3, 69 U.S. Department of Commerce, Office of technical services, Washington, D.C., 1957.
769. Odoshashvili, D. G., "Data for substantiating the level of maximum permissible concentration of dimethylformamide in the atmospheric air" (in Russ.), *Gig. Sanit.*, 27(4), 3-7, 1962.
770. Oglesby, F. L., J. H. Sterner, and B. Anderson, "Quinone vapors and their harmful effects II. Plant exposures associated with eye injuries," *J. Ind. Hyg. Toxicol.*, 29, 74-84, 1947.
771. Ohma, S., "La classification des odeurs aromatiques en sousclasses," *Arch. Neerl. Physiol.*, 6, 567-590, 1922.
772. Pangborn, R. M., H. W. Berg, E. B. Roessler, and A. D. Webb, "Influence of methodology on olfactory response," *Percept. Mot. Skills*, 18, 91-103, 1964.
773. Parker, G. H., and E. M. Stabler, "On certain distinctions between taste and smell," *Am. J. Physiol.*, 32, 230-240, 1913.
774. Passy, J., "Sur quelques minimums perceptibles d'odeurs," *C. R. Hebd. Séances Acad. Sci.*, 114, 786-788, 1892.
775. Passy, J., "Sur la perception des odeurs," *C. R. Soc. Bio.*, 44, 239-243, 1892.
776. Passy, J., "Les propriétés odorantes des alcools de la série grasse," *C. R. Hebd. Séances Acad. Sci.*, 114, 1140-1143, 1892.
777. Passy, J., "Pouvoir odorant du chloroforme, du bromoforme et de l'iodoforme," *C. R. Hebd. Séances Acad. Sci.*, 116, 769-770, 1893.
778. Passy, J., "Forme périodique du pouvoir odorant dans la serie grasse" *C. R. Hebd. Séances Acad. Sci.*, 116, 1007-1010, 1893.

779. Passy, J., "L'odeur dans la série grasse," *C. R. Soc. Biol.*, **45**, 479–481, 1893.
780. Passy, J., "Revue générale sur les sensations olfactives," *Année Psychol.*, **2**, 363–410, 1895.
781. Patty, F. A., and W. P. Yant, "Odor intensity and symptoms produced by commercial propane, butane, pentane, hexane and heptane vapor," U.S. Bureau of Mines, R.I. 2979, 1929.
782. Patty, F. A., "Alkaline materials," in: F. A. Patty (Ed.), *Industrial Hygiene and Toxicology*, *Volume II. Toxicology*, 859–869, Interscience, New York/London, 1962.
783. Patty, F. A., "Arsenic, phosphorus, selenium, sulfur and tellurium," in: F. A. Patty (Ed.), *Industrial Hygiene and Toxicology*, *Volume II. Toxicology*, 871–910, Interscience, New York/London, 1962.
784. Patty, F. A., "Inorganic compounds of oxygen, nitrogen and carbon," in: F. A. Patty (Ed.), *Industrial Hygiene and Toxicology*, *Volume II. Toxicology*, 911–940, Interscience, New York/London, 1962.
785. Piggott, J. R., and R. Harper, "Ratio scales and category scales of odour intensity," *Chem. Senses Flavor*, **1**, 307–316, 1975.
786. Pliška, V., and G. Janiček, "Olfaktometrische Studien I. Die Definition der Wahrnehmungsschwelle-Konzentration mittels Volumengraden," *Sb. Vys. Sk. Chem.-Technol. Praze, Oddil Fak. Potravin. Technol.*, **4**, pt 2, 115–127, 1960.
787. Pliška, V., "Olfaktometrische Studien IV. Fehleranalyse olfaktometrischer Messungen," *Sb. Vys. Sk. Chem.-Technol. Praze, Potraviny*, **6**, pt. 1, 37–46, 1962.
788. Pliška, V., and G. Janiček, "Die Veränderungen der Wahrnehmungsschwellen-Konzentration der Riechstoffe in einiger homologischen Serien," *Arch. Int. Pharmacodyn. Ther.*, **156**, 211–216, 1965.
789. Plotnikova, M. M., "Data on hygienic evaluation of acrolein as a pollution of the atmosphere" (in Russ.), *Gig. Sanit.*, **22**(6), 10–15, 1957.
790. Pogosyan, U. G., "The effect on man of the combined action of small concentrations of acetone and phenol in the atmosphere," *Hyg. Sanit. USSR*, **30**(7–9), 1–9, 1965.
791. Popov, I. N., E. F. Cherkasov, and O. L. Trakhtman, "Determination of sulphur dioxide odor threshold concentration" (in Russ.), *Gig. Sanit.*, **17**(5), 16–20, 1952.
792. Popov, V. A., "Hygienic evaluation of tetrahydrofuran as an atmospheric pollutant," *Hyg. Sanit. USSR*, **35**(4–6), 178–182, 1970.
793. Randebrock, E. M., "Molecular theory of odor with the α-helix as potential perceptor," in: G. Ohloff, A. F. Thomas (Eds.), *Gustation and Olfaction*, 111–125, Academic Press, London/New York, 1971.
794. Ripp, G. Kh., "Hygienic basis for the determination of the allowable concentration limit of divinyl in atmospheric air," in: *U.S.S.R. Literature on Air Pollution and Related Occupational Diseases*, **17**, 18–32 Federal Scientific and Technical Information, U.S. Department of Commerce, Springfield, 1968.
795. Rocén, E., "Contribution to the localisation of "sweet smell," *Scand. Arch. Physiol.*, **40**, 129–144, 1920.
796. Rowe, V. K., and M. A. Wolf, "Ketones," in: F. A. Patty (Ed.), *Industrial Hygiene and Toxicology*, *Volume II, Toxicology*, 1719–1770, Interscience, New York/London, 1962.
797. Rumsey, D. W., and R. P. Cesta, "Odor threshold levels for UDMH and NO_2," *J. Am. Ind. Hyg. Assoc.*, **31**, 339–342, 1970.
798. Rupp, H., and D. Henschler, "Wirkungen geringer Chlor- und Bromkonzentrationen auf den Menschen," *Int. Arch. Gewerbepathol. Gewerbehyg.*, **23**, 79–90, 1967.
799. Sadilova, M. S., "Studies in the standardization of maximum allowable hydrogen fluoride concentrations in the air of inhibited areas," in: *U.S.S.R. Literature on Air Pollution and Related Occupational Diseases*, **17**, 118–128, Federal Scientific and Technical Information, U.S. Department of Commerce, Springfield, 1968.
800. Saifutdinov, M. M., "On hygienic standard of ammonium content in the atmosphere" (in Russ.), *Gig. Sanit.*, **31**(5), 7–11, 1966.
801. Sakuma, K., T. Miura, T. Domon, K. Kanazawa, S. Harima, and T. Namiki, "Sensory measurement of odors using an odorless chamber, in: *Proceedings of the 8th Annual Meet-*

ing of the Japan Society of Air Pollution, 1967: Symposium (1): Hazard nuisance pollution problems due to odors (English translation for APTIC by SCITRAN), 1967.
802. Sales, M., "Odeur et odorisation des gaz (IGU/36-58)," 7e Congrès International de l'Industrie du Gaz, Rome, 1958.
803. Sanders, G., and R. Dechant, "Hydrogen sulphide odor threshold study," Air and Industrial Hygiene Laboratory, California, Department of Public Health, unpublished report, 1961 (cited by H. F. Droege, "A study of the variation of odor threshold concentrations reported in the literature," Proceedings 60th Annual Meeting Air Pollution Control Association, Paper No 67-117, Cleveland, 1967).
804. Scherberger, R. F., G. P. Happ, F. A. Miller, and D. W. Fassett, "A dynamic apparatus for preparing air-vapor mixtures of known concentrations," Am. Ind. Hyg. Ass. Quart., 19, 494-498, 1958.
805. Schley, O.-H., "Untersuchung über Geruchsschwellen," Thesis, Würzburg, W. Germany 1934.
806. Schneider, R. A., and S. Wolf, "Olfactory perception thresholds for citral utilizing a new type olfactorium," J. Appl. Physiol., 8, 337-342, 1955.
807. Schwarz, R., "Uber die Riechschärfe der Honigbiene," Z. Vergl. Physiol., 37, 180-210, 1955.
808. Sfiras, J., and A. Demeiliers, "Application of gas chromatography to olfactometric research," Recherches, (16), 47-57, 1967.
809. Sgibnev, A. K., "The action of low formaldehyde fumes concentrations on the human organism" (in Russ.), Gig. Tr. Prof. Zabol., 12(7), 20-25, 1968.
810. Shalamberidze, O. P., "Reflex action of a mixture of sulfur dioxide and nitrogen dioxide" (in Russ.), Gig. Sanit., 32(7), 9-13, 1967.
811. Sherrard, G. C., "The practical application of two qualitative tests for HCN in ship fumigation," Public Health Rep., 43, 1016-1022, 1928.
812. Singh, P., T. Ramasivan, and K. Krishnamurty, "A simple phosphine detector," Bull. Grain Technol., 5(1), 24-26, 1967.
813. Sinkuvene, D. S., "Hygienic evaluation of acrolein as an air pollutant," Hyg. Sanit. USSR, 35(1-3), 325-329, 1970.
814. Slavgorodskiy, L. P., "Biological effects and hygienic evaluation of air by phthalic anhydride," in: U.S.S.R. Literature on Air Pollution and Related Occupational Diseases, 17, 54-59 Federal Scientific and Technical Information, U.S. Department of Commerce, Springfield, 1968.
815. Smith, H. O., and A. D. Hochstettler, "Determination of odor thresholds in air using C14-labeled compounds to monitor concentrations," Environm. Sci. and Techn., 3, 169-170, 1969.
816. Smolczyk, E., and H. Cobler, "Chemischer Nachweis von Atemgiften und subjektive Empfindlichkeit," Gasmaske, 2, 27-33, 1930.
817. Solomin, G. I., "Data substantiating maximum permissible concentration of dinyl in atmospheric air" (in Russ.), Gig. Sanit., 26(5), 3-8, 1961.
818. Solomin, G. I., "Experimental data for hygienic substantiation of maximal permissible one-time concentrations of isopropylbenzol and isopropylbenzol hydrogen peroxide" (in Russ.), Gig. Sanit., 29(2), 3-9, 1964.
819. Stalker, W. W., "Defining the odor problem in a community," J. Am. Ind. Hyg. Assoc., 24, 600-605, 1963.
820. Steinmetz, G., G. T. Pryor, and H. Stone, "Effect of blank samples on absolute odor threshold determinations," Percept. Psychophys., 6, 142-144, 1969.
821. Stephens, E. R., "Identification of odors from cattle feed lots," Calif. Agric., 25(1), 10-11, 1976.
822. Stone, H., C. S. Ough, and R. M. Pangborn, "Determination of odor difference thresholds," J. Food Sci., 27, 197-202, 1962.

BIBLIOGRAPHY

823. Stone, H., "Techniques for odor measurement: olfactometric vs. sniffing," *J. Food Sci.*, 28, 719–725, 1963.
824. Stone, H., "Determination of odor difference thresholds for three compounds," *J. Exp. Psychol.*, 66, 466–473, 1963.
825. Stone, H., "Some factors affecting olfactory sensitivity and odor intensity," Thesis, University of California, 1963.
826. Stone, H., and J. J. Bosley, "Olfactory discrimination and Weber's law," *Percept. Mot. Skills*, 20, 657–665, 1965.
827. Stone, H., G. T. Pryor, and J. Colwell, "Olfactory detection thresholds in man under conditions of rest and exercise," *Percept. Psychophys.*, 2, 167–170, 1967.
828. Stone, H., G. T. Pryor, and G. Steinmetz, "A comparison of olfactory adaptation among seven odorants and their relationship with several physicochemical properties," *Percept. Psychophys.*, 12, 501–504, 1972.
829. Stuiver, M., "Biophysics of the sense of smell," Thesis, Groningen, the Netherlands. 1958.
830. Styazhkin, V. M., "Experimental basis for the determination of allowable concentrations of chlorine and HCl gas simultaneously present in atmospheric air," in: *U.S.S.R. Literature on Air Pollution and Related Occupational Diseases*, 8, 158–164, U.S. Department of Commerce, Office of Technical Services, Washington, D.C., 1963.
831. Styazhkin, V. M., and H. Kh. Khachaturyan, "Data for validating the maximum permissible concentration of butyric acid in atmospheric air," in: M. Y. Nuttonson (Ed.), *American Institute of Crop Ecology (AICE) Survey of U.S.S.R. Air Pollution Literature*, XIX, 104–110, National Technical Information Service, U.S. Department of Commerce, Springfield, PB-214 264, 1973.
832. Sutton, W. L., "Aliphatic and alicyclic amines," in: F. A. Patty (Ed.), *Industrial Hygiene and Toxicology Volume II. Toxicology*, 2037–2067, Interscience, New York/London, 1962.
833. Sutton, W. L., "Heterocyclic and miscellaneous nitrogen compounds," in: F. A. Patty (Ed.), *Industrial Hygiene and Toxicology, Volume II. Toxicology*, 2174–2234, Interscience, New York/London, 1962.
834. Takhiroff, M. T., "Maximum allowable concentration of chlorine in the atmosphere based on experimental data" (in Russ.), *Gig. Sanit.*, 22(1), 13–18, 1957.
835. Takhirov, M. T., "Hygienic standards for acetic and acetic anhydride in air," *Hyg. Sanit. USSR.*, 34(4-6), 122–125, 1969.
836. Takhirov, M. T., "Experimental study of the combined action of six atmospheric pollutants on the human organism" (in Russ.), *Gig. Sanit.*, 39(5), 100–102, 1974.
837. Tamman, G., and W. Oelsen, "Zur Bestimmung der Dampfdrucke von Riechstoffen," *Z. Anorg. Allg. Chem.*, 172, 407–413, 1928.
838. Tarkhova, L. P., "Materials for determining the maximum permissible concentration of chlorobenzol in atmospheric air," *Hyg. Sanit. USSR.*, 30(1-3), 327–333, 1965.
839. Taylor, E. F., and F. T. Bodurtha, "Control of dimethylamine odors," *Ind. Wastes*, 5(4), 92–94, 1960.
840. Tempelaar, H. C. G., "Over den invloed van licht op reukstoffen," Thesis, Utrecht, the Netherlands 1913.
841. Tepikine, L. A. "Biological effects of mesidine as an atmospheric pollutant," *Hyg. Sanit. USSR.*, 33(4-6), 299–302, 1968.
842. Teranishi, R., R. G. Buttery, and D. G. Guadagni, "Odor quality and chemical structure in fruit and vegetable flavors," *Ann. N.Y. Acad. Sci.*, 237, 209–216, 1974.
843. Thomas, M. D., J. O. Ivie, J. N. Abersold, and R. H. Hendricks, "Automatic apparatus for determination of small concentrations of sulfur dioxide in air," *Ind. Eng. Chem., Anal. Ed.*, 15, 287–290, 1943.
844. Tkach, N. Z., "Combined effect of acetone and acetophenone in the atmosphere," *Hyg. Sanit. USSR.*, 30(7-9), 179–185, 1965.

BIBLIOGRAPHY

845. Tkachev, P. G., "Data for substantiating the maximum permissible concentration of aniline in the atmospheric air" (in Russ.), *Gig. Sanit.*, **28**(4), 3-11, 1963.
846. Tkachev, P. G., "Monoethylamine in the atmosphere: hygienic significance and standards," *Hyg. Sanit. USSR.*, **34**(7-9), 149-153, 1969.
847. Tkachev, P. G., "Triethylamine in air: hygienic significance and hygienic standards," *Hyg. Sanit. USSR.*, **35**(10-12), 8-13, 1970.
848. "Toxicity data sheet: Mesityl oxide," Shell Chemical Corporation: Industrial Hygiene Bulletin SC: 57-106, 1957.
849. "Toxicity data sheet: Allyl chloride," Shell Chemical Corporation, Industrial Hygiene Bulletin SC: 57-80, 1958.
850. "Toxicity data sheet: Ethyl amyl ketone," Shell Chemical Corporation, Industrial Hygiene Bulletin SC: 57-99, 1958.
851. "Toxicity data sheet: Epichlorohydrin," Shell Chemical Corporation, Industrial Hygiene Bulletin SC: 57-86, 1959.
852. Treon, J. F., and F. R. Dutra, "Physiological response of experimental animals to the vapor of 2-nitropropane," *AMA Arch. Ind. Hyg. Occup. Med.*, **5**, 52-61, 1952.
853. Treon, J. F., E. P. Cleveland, and J. Cappel, "The toxicity of hexachlorocyclopentadiene," *AMA Arch. Ind. Health*, **11**, 459-472, 1955.
854. Turk, A., "Expressions of gaseous concentration and dilution ratios," *Atm. Environm.*, **7**, 967-972, 1973.
855. Ubaidullaev, R., "Data for substantiating the maximum permissible concentration of furfurol in atmospheric air" (in Russ.), *Gig. Sanit.*, **26**(7), 3-10, 1961.
856. Ubaidullaev, R., "Effect of small concentration of methanol vapours on the body of men and animals" (in Russ.), *Gig. Sanit.*, **31**(4), 9-12, 1966.
857. Valentin, G., *Lehrbuch der Physiologie des Menschen* **2**(2), 279-283, Friedrich Bieweg und Sohn, Braunschweig, 1848.
858. Valentin, G., *Lehrbuch der Physiologie des Menschen* **2**(3), 271-274, Friedrich Bieweg und Sohn, Braunschweig, 1850.
859. Vincent, G. P., J. O. MacMahon, and J. F. Synan, "The use of chlorine dioxide in water treatment," *Am. J. Public Health*, **36**, 1035-1037, 1946.
860. Weeks, M. H., "Monoethanolamine," in: *Transactions of the Symposium on Health Hazards of Military Chemicals*, 15-23, Chemical Warfare Laboratories Special Publication 2-10, 1958.
861. Weitbrecht, U., "Beurteilung der Trichloroäthylen-Gefährdung im Betrieb," *Zentralbl. Arbeitsmed. Arbeitsschutz.*, **7**, 55-58, 1957.
862. Wilby, F. V., "The odor comparator: an improved instrument for quantitative odor measurement," *Proc. Oper. Sect. Am. Gas Assoc.*, 225-231, 1964.
863. Wilby, F. V., "Variation in recognition odor threshold of a panel," *JAPCA*, **19**, 96-100, 1969.
864. Wilska, S., "Ozone. Its physiological effects and analytical determination in laboratory air," *Acta Chem. Scand.*, **5**, 1359-1367, 1951.
865. Witheridge, W. N., and C. P. Yaglou, "Ozone in ventilation—its possibilities and limitations," *Heat. Piping Air Cond.*, **11**, 648-653, 1939.
866. Wolf, M. A., V. K. Rowe, D. D. McCollister, R. I. Hollingsworth, and F. Oyen, "Toxicological studies of certain alkylated benzenes and benzene," *AMA Arch. Ind. Health*, **14**, 387-398, 1956.
867. Young, F. A., and D. F. Adams, "Comparison of olfactory thresholds obtained on trained and untrained subjects," *Proc. 74th Annu. Conv. Am. Psychol. Assoc.*, 75-76, 1966.
868. Yuldashev, T., "Maximum permissible concentration of ethylene oxide in the atmosphere," *Hyg. Sanit. USSR.*, **30**(10-12), 1-6, 1965.
869. Zapp, J. A., "Hazards of isocyanates in polyurethane foam plastic production," *Arch. Ind. Health*, **15**, 324-330, 1957.

870. Zibireva, I. A., "On comparative toxicity of microconcentrations of two chlorisocyanates (para- and meta-isomers)" (in Russ.), *Gig. Sanit.*, **32**(6), 3-9, 1967.
871. Zwaardemaker, H., "Geruch und Geschmack," in: R. Tigerstedt (Ed.), *Handbuch der Physiologischen Methodik* **III**, 1, 46-108, S. Hirzel, Leipzig [corrected by Tempelaar (1913)], 1914.
872. Zwaardemaker, H., and K. Komuro, "Ruiken bij volledige vermoeienis, resp. adaptatie voor een bepaalde geur," *Kon. Ned. Akad. Wetensch., Versl. Gewone Vergad. Afd. Natuurk.*, **30**, 1189-1195, 1921.
873. Amoore, J. E., and D. Venstrom (1966) "Sensory analysis of odor qualities in terms of the stereochemical theory," *J. Food Sci.*, **31**, 118-128, 1966.
874. Amoore, J. E., L. J. Forrester, and P. Pelosi, "Specific anosmia to isobutyraldehyde: The malty primary odor." *Chem. Senses Flavor*, **2**, 17-25, 1976.
875. Baker, R. A., "Threshold odors of organic chemicals," *J. Am. Water Works Assoc.*, **55**, 913-916, 1963.
876. Baker, R. A., and Ming-Dean Luh, "Odor effects of pyridine-montmorillonite in aqueous solution," *Sci. Total Environ.*, **2**, 13-20, 1973.
877. Boelens, M., P. J. De Valois, H. J. Wobben, and A. van der Gen, "Volatile flavor compounds from onion," *J. Agric. Food Chem.*, **19**(5), 984-991, 1971.
878. Brady, S. O., "Taste and odor components in refinery effluents," *Proc. Div. Refin. Am. Pet. Inst.*, **48**, 556, 1968.
879. Burttschell, R. H., A. A. Rosen, F. M. Middleton, and M. B. Ettinger, "Chlorine derivatives of phenol causing taste and odor," *J. Am. Water Works Assoc.*, **51**, 205-214, 1959.
880. Buttery, R. G., R. M. Seifert, D. G. Guadagni, D. R. Black, and L. C. Ling, "Characterization of some volatile constituents of carrots," *J. Agric. Food Chem.*, **16**(6), 1009-1015, 1968.
881. Buttery, R. G., R. M. Seifert, R. E. Lundin, D. G. Guadagni, and L. C. Ling, "Characterization of an important aroma component of bell peppers," *Chem. Ind. (London)*, 490-491, 1969.
882. Buttery, R. G., R. M. Seifert, D. G. Guadagni, and L. C. Ling, "Characterization of some volatile constituents of bell peppers," *J. Agric. Food Chem.*, **17**(6), 1322-1327, 1969.
883. Buttery, R. G., R. M. Seifert, D. G. Guadagni, and L. C. Ling, "Characterization of additional volatile components of tomato," *J. Agric. Food Chem.*, **19**(3), 524-529, 1971.
884. Buttery, R. G., D. G. Guadagni, and R. E. Lundin, "Some 4,5-dialkylthiazoles with potent bell pepper like aromas," *J. Agric. Food Chem.*, **24**(1), 1-3, 1976.
885. Cartwright, L. C., and P. H. Kelly, "Flavor quality and strength of propenyl guaethol as a vanilla extender or replacement," *Food Technol.* (Chicago), **6**, 372-376, 1952.
886. Cherkinski, W., "Protection of natural waterbasins against pollution with industrial wastewater," *Literature on Water Supply and Pollution Control*, Book No. 5, U.S. Department of Commerce, Office of Technical Service, Washington, D.C., TT-61-31601-5, 1961.
887. Curtis, R. F., D. G. Land, N. M. Griffiths, M. Gee, and D. Robinson, "2,3,4,6-tetrachloroanisole association with musty taint in chickens and microbiological formation," *Nature* (London), **235**, 223-224, 1972.
888. Eriksson, C. E., B. Lundgren, and K. Vallentin, "Odor detectibility of aldehydes and alcohols originating from lipid oxidation," *Chem. Senses Flavor*, **2**, 3-15, 1976.
889. Flath, R. A., D. R. Black, D. G. Guadagni, W. H. McFadden, and T. H. Schultz, "Identification and organoleptic evaluation of compounds in delicious apple essence," *J. Agric. Food Chem.*, **15**(1), 29-35, 1967.
890. Forss, D. A., E. A. Dunstone, E. H. Ramshaw, and W. Stark, "The flavor of cucumbers," *J. Food Sci.*, **27**, 90-93, 1962.
891. Gee, M. G., N. M. Griffiths, and R. G. Fenwick, "Potential metabolites of phenolic desinfectants: preparation and odour properties of some anisoles," *Pestic. Sci.*, **5**, 703-708, 1974.

892. Griffiths, N. M., and D. G. Land, "6-chloro-o-cresol taint in biscuits," *Chem. Ind.* (London), 904, 1973.
893. Griffiths, N. M., "Sensory properties of the chloro-anisoles," *Chem. Senses Flavor*, 1, 187–195, 1974.
894. Grunt, F. E., de, Unpublished data, National Institute for Water Supply, Voorburg, Netherlands, 1975.
895. Guadagni, D. G., R. G. Buttery, and S. Okana, "Odour threshold of some organic compounds associated with food flavours," *J. Sci. Food Agric.*, 14, 761–765, 1963.
896. Guadagni, D. G., R. G. Buttery, and J. Harris, "Odour intensities of hop oil components," *J. Sci. Food Agric.*, 17, 142–144, 1966.
897. Haring, D. G., F. Rijkens, H. Boelens, and A. van der Gen, "Olfactory studies on enantiomeric eremophilane sesquiterpenoids," *J. Agric. Food Chem.*, 20(5), 1018–1021, 1972.
898. Holluta, J., "Geruchs- und Geschmacksbeeinträchtigung des Trinkwassers-Ursachen und Bekämpfung," *Gas Wasserfach*, 101(4), 1018–1023, 1960.
899. Jakob, M. A., R. Hippler, and H. R. Lüthi, "Ueber das Vorkommen von hexyl-2-methylbutyrat im Apfelaroma," *Mitt. Geb. Lebensmittelunters. Hyg.*, 60, 223–229, 1969.
900. Johannson, B., B. Drake, B. Berggren, and K. Vallentin, "Detection thresholds: Effect of stimulus presentation order and addition of blanks, 1. Odor of pentanal and hexanol," *Lebensm. Wiss. Technol.*, 6, 115–122, 1973.
901. Kauffmann, M., "Ueber eigentümliche Geruchsanomalien einiger chemischer Körper," *Z. Sinnesphysiol.*, 42, 271–280, 1907.
902. Köehler, P. E., M. E. Mason, and G. V. Odell, "Odor threshold levels of pyrazine compounds and assessment of their role in the flavor of roasted foods," *J. Food Sci.*, 36, 816–818, 1971.
903. Kölle, W., K. H. Schweer, H. Güsten, and L. Stieglitz, "Identifizierung schwer abbaubaren Schadstoffen im Rhein und Rheinuferfiltrat," *Vom Wasser*, 39, 109–119, 1972.
904. Koppe, P., "Identifizierung der Hauptgeruchsstoffe im Uferfiltrat des Mittel- und Niederrheins," *Vom Wasser*, 32, 33–68, 1965.
905. Leitereg, T. J., D. J. Guadagni, J. Harris, T. R. Mon, and R. Teranishi, "Evidence for difference between the odour of the optical isomers (+)- and (–)- carvone," *Nature* (London), 230, 455–456, 1971.
906. Lillard, D. A., J. Powers, and R. G. Webb, "Aqueous odor thresholds of organic pollutants in industrial effluents," Environmental monitoring series: EPA-660/4-75-002, National Environmental Research Center, Office of Research and Development, U.S. Environmental Protection Agency, Corvallis Oregon 97330, 1975.
907. Moncrieff, R. W., "Olfactory adaptation and odor intensity," *Am. J. Psychol.*, 70(1), 1–20, 1957.
908. Mulders, E. J., "Het aroma van witbrood: Kwantitatieve samenstelling van verbindingen in de damp en berekening van aromawaarden," *Voedingsmiddelentechnologie*, 3(4), 6–11, 1972.
909. Rosen, A. A., J. B. Peter, and F. M. Middleton, "Odor thresholds of mixed organic chemicals," *JWPCF*, 7–14, 1962.
910. Rosen, A. A., R. F. Steel, and M. B. Ettinger, "Relationship of river water odor to specific organic contaminants," *JWPCF*, 35, 777–782, 1963.
911. Sega, G. M., M. J. Lewis, M. H. Woskow, "Evaluation of beer flavor compounds," *Proc. Am. Soc. Brew. Chem.*, 156–164, 1967.
912. Seifert, R. M., R. G. Buttery, D. G. Guadagni, D. R. Black, and J. G. Harris, "Synthesis of some 2-methoxy-3-alkylpyrazines with strong bell pepper-like odors," *J. Agric. Food Chem.*, 18, 246, 1970.
913. Seifert, R. M., R. G. Buttery, D. G. Guadagni, D. R. Black, and J. G. Harris, "Synthesis and odor properties of some additional compounds related to 2-isobutyl-3-methoxy pyrazine," *J. Agric. Food Chem.*, 20, 135, 1972.

914. Stahl, W. H., "Compilation of odor and taste threshold values data," ASTM Data Series DS48, American Society for Testing and Materials, 1916 Race Street, Philadelphia, Pa. 19103; personal communication from D. G. Land, 1973.
915. Sigworth, E. A., "Identification and removal of herbicides and pesticides," *J. Am. Water Works Assoc.*, 57, 1016–1022, 1964.
916. Stevens, K. L., D. G. Guadagni, and D. J. Stern, "Odour character and threshold values of nootkatone and related compounds," *J. Sci. Food Agric.*, 21, 590–593, 1970.
917. Takken, H. J., L. M. van der Linde, M. Boelens, and J. M. van Dort, "Olfactive properties of a number of polysubstituted pyrazines," *J. Agric. Food Chem.*, 23(4), 638–642, 1975.
918. Teranishi, R., R. A. Flath, D. G. Guadagni, R. E. Lundin, F. R. Mon, and K. L. Stevens, "Gas chromatographic, infrared, proton magnetic resonance, mass spectral and threshold analysis of all pentyl acetates," *J. Agric. Food Chem.*, 14(3), 253–262, 1966.
919. Teranishi, R., "High resolution gas chromatography in aroma research," *Am. Perfum. Cosmet.* 82, 43–51, 1967.
920. Teranishi, R., R. G. Buttery, and D. G. Guadagni, "Odor quality and chemical structure in fruit and vegetable flavors," *Ann. N.Y. Acad. Sci.*, 237, 209–216, 1974.
921. Teranishi, R., R. G. Buttery, and D. G. Guadagni, "Odor, thresholds and molecular structure," In: F. Drawert (Ed.), *Geruch- und Geschmackstoffe*, Verlag Hans Carl, Nürnberg, 1975.
922. Wang, L. C., B. W. Thomas, K. Warner, W. J. Wolf, and W. F. Kwolek, "Apparent odor thresholds of polyamines in water and 2% soybean flour dispersions," *J. Food Sci.*, 40, 274–276, 1975.
923. Wasserman, A. E., "Organoleptic evaluation of three phenols present in woodsmoke," *J. Food Sci.*, 31(6), 1005–1010, 1966.
924. [Reserved for future expansions.]
925. Zoeteman, B. C. J., and G. J. Piet, "Drinkwater is nog geen water drinken," H_2O, 6(7), 179–177, 1973.
926. A. W. Thomas, L. J. Bahlman, N. A. Leidel, J. C. Packer, and J. C. Millar, "Trimellitic anhydride," *AIHAJ* 39(5), 1978.
927. C. D. Becker, J. A. Lichatowich, M. J. Schneider, and J. A. Strand, "Regional survey of marine biota for bioassay standardization of oil and oil dispersant chemicals," API publication No. 4167, April 1973.
928. J. F. J. van Rensburg, J. J. van Huyssteen, and A. J. Hassett, "A semiautomated technique for the routine analysis of volatile organohalogens in water purification processes," *Water Research*, 12, 127–131, 1978.
929. James D. Navratil, Robert E. Sievers, and Harold F. Walton, "Open-pore polyurethane columns for collection and preconcentration of polynuclear aromatic hydrocarbons from water," *Anal. Chem.* 49(14), Dec. 1977.
930. Freudenberg, K., and M. Reichert, "Sulphate pulp mill odors," *Tappi*, 38, 165A–166A, 1955.
931. M. Knorr, D. Schenk, "Zur Frage der Synthese polyzyklischer Aromate durch Bakterien," *Arch. Hyg.* 152(3), 282–285, 1968.
932. R. A. Brown and P. K. Starnes, "Hydrocarbons in the water and sediment of Wilderness Lake, II," *Marine Poll. Bull.* 9, 162–165, 1978.
933. G. McConnell, "Halo-organics in water supplies," *J. of the Institution of Water Engineers and Scientists* 30(8), Nov. 1976.
934. L. M. Shabad, "On the distribution and the fate of the carcinogenic hydrocarbon benz(a)-pyrene in the soil," *Zeitschrift für Krebsforshung*, 70, 204–210, 1968.
935. MCA, "The effect of chlorination on selected organic chemicals," U.S. EPA-WPC Research series, 12020 EXG 03/72.
936. Wacker-Chemie GmbH, W. Germany, "Perchloräthylen–eine Bestandsaufnahme," Umwelt 6/76, W. Germany.

937. Pearson, C. R., and G. McConnell, "Chlorinated C_1 and C_2 hydrocarbons in the marine environment," *Proc. Roy. Soc. Lond.*, **B189**, 305-332, 1975.
938. Murray, A. I., and I. P. Riely, "Occurrence of some chlorinated aliphatic hydrocarbons in the environment," *Nature* **242**, 37-38, 1973.
939. Woodiwiss, F. S., and Fretwell, G., "The toxicities of sewage effluents, Industrial discharges and some chemical substances to brown trout in the Trent River Authority Area," *Water Poll. Control (G.B.)* **73**, 396, 1974.
940. Neff, J. M., J. W. Anderson, B. A. Cox, R. B. Laughlin, J. R. S. S. Rossi, and H. E. Tatem, "Effects of petroleum on survival, respiration and growth of marine animals," in: *Sources, Effects and Sinks of Hydrocarbons in the Aquatic Environments*, 519-539, American Institute of Biological Sciences, Washington D.C., 1976.
941. Caldwell, R. S., E. M. Caldarone, and M. H. Mallon, "Effects of a seawater-soluble fraction of cook inlet crude oil and its major aromatic components on larval stages of the Dungeness crab, *Cancer magister Dana*," in: *Proceedings, NOAA-EPA Symposium on Fate and Effects of Petroleum Hydrocarbons*, Pergamon Press, Oxford, 1977.
942. Benville, P. E., and S. Korn, "The actue toxicity of six monocyclic aromatic crude oil components to striped bass (*Morone saxatilis*) and bay shrimp (*Crangon franciscorum*)," *Calif. Fish Game* (in press).
943. Brenniman, G., R. Hartung, and W. J. Weber, Jr.. "A continuous flow bioassay method to evaluate the effects of outboard motor exhausts and selected aromatic toxicants on fish," *Water Res.* **10**, 165-169, 1976.
944. Struhsaker et al., 1974.
945. Brockson, R. W., and H. T. Bailey, "Respiratory response of juvenile chinook salmon and striped bass exposed to benzene, a water soluble component of crude oil," In: *Proceedings, Joint Conference on Prevention and Control of Oil Spills*, pp. 783-792, A.P.I. Washington, D.C., 1973.
946. Wilber, 1969.
947. Fuerstenau, M. C., et al., "Toxicity of selected sulfhydryl collectors to rainbow trout," *Trans. Soc. Mining Engr.*, **256**, 337, 1974.
948. Mukherjie, S., and Bhattacharya, S., "Effect of some industrial pollutants on fish brain cholinesterase activity," *Environm. Physiol. Biochem.* (Den.) **4**, 226, 1974.
949. Leach, J. M., and Thakore, A. N., "Isolation and identification of constituents toxic to juvenile rainbow trout (*Salmo gairdneri*) in caustic extraction effluents from kraft pulpmill bleach plants," *Jour. Fish. Res. Bd. Can.*, **32**, 1249, 1975.
950. Gutenmann, W. H., and Lisk, D. J., "Flame retardant release from fabrics during laundering and their toxicity to fish," *Bull. Environ. Contam. Toxicol*, **14**, 61, 1975.
951. Rice, S. D., and Stokes, R. M., "Acute toxicity of ammonia to several developmental stages of rainbow trout, *Salmo gairdneri*," *Fish Bull.*, **73**, 207, 1975.
952. Davis, J. C., and Hoos, R. A. W., "Use of sodium pentachlorophenate and dehydro-abietic acid as reference toxicants for salmonid bioassays," *Jour. Fish. Res. Bd. Can.*, **32**, 411, 1975.
953. Rogers, I. H., et al., "Fish toxicants in kraft effluents," *Tappi*, **58**, 136, 1975.
954. H. O. Hettche, "Die Belastung der Atmosphäre durch polyzyklische Aromaten im Grossraum eines Industriegebietes," *Schriftenreihe des Landesanstalt für Immissions und Bodennutzungsschutz des Landes Nordrhein-Westfalen in Essen* (12), 92-108, 1968.
955. J. Borneff and H. Kunte. "Kanzerogene Substanzen in Wasser und Boden," *Arch. Hyg.* **153**(3) 220-229, 1969.
956. V. W. Jamison, R. L. Raymond, and J. O. Hudson, "Biodegradation of high-octane gasoline," *Proceedings of the Third International Biodegradation Symposium*, Applied Science Publishers, 1976.
957. Richard F. Lee et al., "Fate of polycyclic aromatic hydrocarbons in controlled ecosystem enclosures," *Environm. Sci. & Techn.*, **12**(7), 832-838, July 1978.

958. Dipak K. Basu and Jilendra Saxena, "Polynuclear aromatic hydrocarbons in selected U.S. drinking waters and their raw water sources," *Environm. Sci. & Techn.* **12**(7), 795-798, July 1978.
959. Laurel E. Kane, and Yves Alarie, "Sensory irritation to formaldehyde and acrolein during single and repeated exposures in mice," *AIHAJ* **38**(10), 1977.
960. Coon, R., R. Jones, L. Jenkins and J. Siegel, "Animal inhalation studies on ammonia, ethylene glycol, formaldehyde, and dimethylamine, and ethanol," *Toxicol. Appl. Pharmacol.* **16**, 646, 1970.
961. Salem, H., and H. Cullumbine, "Inhalation toxicities of some aldehydes," *Toxicol. Appl. Pharmacol.* **2**, 183, 1960.
962. Horton, A., R. Tye and K. Stemmar, "Experimental carcinogenesis of the lung. Inhalation of gaseous formaldehyde or an aerosol of coal tar by C3H mice," *J. Nat. Cancer. Inst.*, **30**, 31, 1963.
963. Iwanoff, N., "Experimentelle studien uber den einflus technisch und hygienisch wichtiger gase under dampfe auf den organisms," *Arch. Hyg.* **73**, 307, 1911. (Translation by Mr. William Van Stone, Dept. of Germanic Languages and Literature, University of Pittsburgh).
964. Skog, E., "A toxicological investigation of lower aliphatic aldehydes," *Acta Pharmacol. Toxicol.* **6**, 299, 1950.
965. Dalhamn, T., and A. Rosengren, "Effect of different aldehydes on tracheal mucosa," *Arch. Otolaryngol.* **93**, 496, 1971.
966. Cralley, L., "The effect of irritant gases upon the rate of ciliary activity," *J. Ind. Hyg. Toxicol.* **24**, 193, 1942.
967-969. [Reserved for future expansion.]
970. Amdur, M., "The response of guinea pigs to inhalation of formaldehyde and formic acid alone and with a sodium chloride aerosol," *Int. J. Air Pollut.* **3**(4), 201, 1960.
971. Murphy, S., and C. Ulrich, "Multi-animal test system for measuring effects of irritant gases and vapors on respiratory function of guinea pigs," *Am. Ind. Hyg. Assoc. J.*, **25**, 28, 1964.
972. Davis, T., S. Battista, and C. Kensler, "Mechanism of respiratory effects during exposure of guinea pigs to irritants," *Arch. Environ. Health* **15**, 412, 1967.
973. Kulle, T., and G. Cooper, "Effects of formaldehyde and ozone on the trigeminal nasal sensory system," *Arch. Environ. Health* **30**, 237, 1975.
974. Barnes, E., and H. Speicher, "The determination of formaldehyde in air," *J. Ind. Hyg. Toxicol.* **24**, 10, 1942.
975. Pattle, R., and H. Cullumbine, "Toxicity of some atmospheric pollutants," *Brit. Med. J.* **2**, 913, 1956.
976. Sim, V. M., and R. E. Pattle, "Effect of possible smog irritants on human subjects," *J. Amer. Med. Assoc.* **165**, 1908, 1957.
977. Schuck, E., and N. Renzetti, "Eye irritants formed during photo-oxidation of hydrocarbons in the presence of oxides and nitrogen," *JAPCA* **10**, 389, 1960.
978. Stephens, E., E. Darley, O. Taylor, and C. Scott, "Photochemical reaction products in air pollution," *J. Air & Water Pollut.* **4**, 79, 1961.
979. Hendrick, D., and D. Lane, "Formalin asthma in hospital staff," *Brig. Med. J.* **1**, 607, 1975.
980. Morrill, E., "Formaldehyde exposure from paper process solved by air sampling and current studies." *Air Cond. Heat. Vent.*, **58**, 94, 1961.
981. Bourne, H., and S. Seferian, "Formaldehyde in wrinkle-proof apparel processes–tears for milady," *Ind. Med. Surg.* **28**, 232, 1959.
982. Kerfoot, E., and T. Mooney, "Formaldehyde and paraformaldehyde study in funeral homes," *Am. Ind. Hyg. Assoc. J.* **36**, 533, 1975.
983. Schoenber, J., and C. Mitchell, "Airway disease caused by phenolic (phenol-formaldehyde) resin exposure," *Arch. Environ. Health* **30**, 575, 1975.
984. Smith, H., "Improved Communication–Hygienic Standards for Daily Inhalation," *Am. Ind. Hyg. Assoc. Q.* **17**, 129, 1956.

985. Lyon, J., L. Jenkins, R. Jones, R. Coon, and J. Siegel, "Repeated and continuous exposure of laboratory animals to acrolein," *Toxicol. Appl. Pharmacol.* **17**, 726, 1970.
986. Carpenter, C., H. Smyth, and U. Pozzani, "The assay of acute vapor toxicity, and the grading and interpretation of results on 96 chemical compounds," *J. Ind. Hyg.* **31**, 343, 1949.
987. Murphy, S., D. Klingshirn, and C. Ulrich, "Respiratory response of guinea pigs during acrolein inhalation and its modification by drugs," *J. Pharmacol. Exp. Ther.* **141**, 79, 1963.
988. Murphy, S., "A review of effects on animals of exposure to auto exhaust and some of its components," *JAPCA* **14**, 303, 1964.
989. Bouley, G., A. Dubreuil, J. Godin, M. Boisset, and Cl. Boudene, "Phenomenon of adaptation in rats continuously exposed to low concentrations of acrolein," *Ann. Occup. Hyg.* **19**, 27, 1976.
990. Yant, W. B., H. Schrenk, F. Patty, and R. Sayers, "Acrolein as a warning agent for detecting leakage of methyl chloride from refrigerators," *U.S. Bur. Mines Rep. of Inves.* #3027, 1930.
991. Darley, E., J. Middleton, and M. Garber, "Plant damage and eye irritation from ozone-hydrocarbon reactions," *J. Agr. Food Chem.* **8**, 483, 1960.
992. Schuck, E., and G. Doyle, "Photo-oxidation of hydrocarbons in mixtures containing oxides of nitrogen and sulfur dioxide," Report No. 29 *Air Pollut. Found.* San Marino, California, 1959.
993. R. C. Lao, R. S. Thomas, and J. L. Monkman, "Computerized gas chromatographic-mass spectrometric analysis of polycyclic aromatic hydrocarbons in environmental samples," *J. Chromatogr.*, **112**, 681–700, 1975.
994. E. A. Smith and C. I. Mayfield, "Paraquat: Determination, degradation and mobility in soil," *Water, Air and Soil Pollut.*, **9**, 439–452, 1978.
995. S. S. Rossi and J. M. Neff, "Toxicity of polynuclear aromatic hydrocarbons to the polychaete *Neanthes arenaceodentata*," *Marine Poll. Bull.* **9**, 220–223, 1978.
996. R. Marchetti, "Acute toxicity of alkylleads to some marine organisms," *Marine Poll. Bull.* **9**, 206–207, 1978.
997. P. E. Gaffney, "Carpet and rug industry case study, II: Biological effects," *JWPCF* **48**(12), 2731–2737, Dec. 1976.
998. F. Dietz and J. Traud, "Geruchs- und Geschmacks-Schwellen-Konzentrationen von Phenolkörper," *Gwf-Wasser/Abwasser* **119**(6), 1978.
999. Morris Katz, Takeo Sakuma, and Andrew Ho, "Chromatographic and spectral analysis of polynuclear aromatic hydrocarbons–Quantitative distribution in air of Ontario cities," *Environm. Sci. & Techn.*, **12**(8), Aug. 1978.
1000. Mitchell D. Erickson, Larry C. Michael, Ruth A. Zweidinger, and Edo D. Pellizari, "Development of methods of sampling and analysis of polychlorinated naphthalenes in ambient air," *Environm. Sci. & Techn.*, **12**(8), Aug. 1978.
1001. Macek, K. J., and W. A. McAllister, "Insecticide susceptibility of some common fish family representatives," *Trans. Am. Fish Soc.* **99**, 20–27, 1970.
1002. Sanders, H. O. and O. B. Cope, "Toxicities of several pesticides to two species of cladocerans," *Trans. Am. Fish Soc.* **95**, 165–169, 1966.
1003. Sanders, H. O., and O. B. Cope, "The relative toxicities of several pesticides to naiads of three species of stoneflies," *Limnol. Oceanogr.* **13**, 112–117, 1968.
1004. Frear, D. E. H., and J. E. Boyd. "Use of *Daphnia magna* for microbioassay of pesticides, I," *J. Econom. Entomol.* **60**, 1228–1238, 1967.
1005. Schoettger, R. A., "Toxicology of thiodan in several fish and aquatic invertebrates," *Res. Publ. U.S. FWS* **35**, 1–31, 1970.
1006. Décamps, H., "Untersuchungen zur verwendung der Larven der Gattung *Hydropsyche* (Trichoptera, Insecta) zu Standard-Toxiztätstests. Teil II," Unpublished information from Landesanstalt für Umweltschutz Baden-Württemberg, W. Germany, 1972.
1007. Décamps, H., W. K. Besch, and H. Vobis, "Influence de produits toxiques sur la construc-

tion du filet des larves d'*Hydropsyche* (Insecta, Trichoptera)," *C. R. Acad. Sci. Paris* **276**, Ser. D. 375–378, 1973.
1008. Wulf Besch, "Studien zum Gewässerschutz–II. Bioteste in der limnischen Toxikologie," Landesanstalt für Umweltschutz Baden-Württemberg, Institut für Wasser- und Abfallwirtschaft, Karlsruhe, W. Germany, 1977.
1009. Philip A. Meyers, and Terence G. Oas, "Comparison of associations of different hydrocarbons with clay particles in simulated seawater," *Environm. Sci. & Techn.* **12**(8), Aug. 1978.
1010. NIOSH current intelligence bulletin No. 25, "Ethylene dichloride (1,2-dichloro-ethane)," *AIHAJ* **39**(9), 1978.
1011. John Palassis, "The sampling and determination of azelaic acid in air," *AIHAJ* **39**(9), 1978.
1012. National Academic of Sciences, "Particulate polycyclic organic matter," Washington, D.C., 1972.
1013. Ralph A. Brown and Fred T. Weiss, "Fate and effects of polynuclear aromatic hydrocarbons in the aquatic environment," API Publication No. 4297, Oct. 1978.
1014. Suess, Michael J., "The environmental load and cycle of polycyclic aromatic hydrocarbons," *The Science of the Total Environment* **6**, 239–250, 1976.
1015. Pancirov, R. J., and Brown, R. A., "Analytical methods for polynuclear aromatic hydrocarbons in crude oils, heating oils and marine tissues," *Proceedings of Joint Conference on Prevention and Control of Oil Pollution (San Francisco, 1973)*, API, pp. 103–115.
1016. Greffard, J., and Meury, J., "Carcinogenic hydrocarbon pollution in Toulon harbor," *Cah. Oceanogr.* **19** 457–468, 1967.
1017. Pancirov, R. J., and Brown, R. A., "Polynuclear Aromatic Hydrocarbons in Marine Tissues." *Environm. Sci. & Techn.* **11**(10), 989–992, 1977.
1018. Dunn, B. P., and Stich, J. F., "The use of mussels in estimating benzo(a)pyrene contamination of the marine environment," *Proc. Soc. Exptl. Biol. and Med.*, **150**, 49–51, 1975.
1019. Cahnmann, H. J., and Kuratsune, M., "Determination of PAH in oysters collected in polluted waters," *Anal. Chem.* **29**, 1312–1317, 1957.
1020. Mallet, L., "Marine pollution by BP-type PH of the North and West coasts of France and their incidence in biological media, especially in plankton," *Cah. Oceanogr.* **19**, 237–243, 1967.
1021. Mallet, L., Perdriau, A. and Perdriau, J., "Pollution by BP-type PH of the Western Region of the Arctic Ocean," *C. R. Acad. Sci. Paris*, **256**, 3487–3489, 1963.
1022. Brown, R. A., and Pancirov, R. J., unpublished work.
1023. Lenges, J. and Luks, D., "Determination of 3,4-benzopyrene in smoked meat and fish products. I. Quantitative determination of a 3,4-benzopyrene in smoked meat products," *Revue des Alimentations et des Industries Alimentairs*, **29**(5), 134–139, 1974.
1024. Fretheim, K., "Carcinogenic polycyclic aromatic hydrocarbons in Norwegian smoked meat sausages," *J. Agric. Food Chem.*, **24**(5), 976–979, 1976.
1025. Howard, J. W., and Fazio, T., "A review of polycyclic aromatic hydrocarbons in foods," *J. Agric. Food Chem.*, **17**, 527–531, 1969.
1026. Malanoski, A. J., Greenwood, E. L., Barnes, C. J., Worthington, J. M. and Joe, F. L., Jr., "A survey of polycyclic aromatic hydrocarbons in smoked foods," *J AOAC*, **51**(1), 114–121, 1968.
1027. Lijinski, W. and Shubik, B., "Benzo(a)pyrene and other polynuclear hydrocarbons in charcoal-broiled meat," *Science*, **145**(3267), 53–55, 1964.
1028. Fritz, W., "Formation of carcinogenic hydrocarbons during heat treatment of foods. V. Studies of contamination during grilling over charcoal," *Deutsche Lebensmittel Rundschau*, **69**(3), 119–122, 1973.
1029. Steinig, J., and Meyer, V., "3,4-Benzpyrene contents of smoked fish," *Lebensmittel-Wissenschaft und Technologie*, **9**(4), 215–217, 1976.
1030. Graff, W., and Diehl, H., "The natural normal levels of carcinogenic polycyclic aromatic hydrocarbons and the reasons therefore," *Arch. Hyg. Bakt.* **150**(1–2), 49–59, 1966.

1270 BIBLIOGRAPHY

1031. Grimmer, G., "Cancerogene Kohlenwasserstoffe in der Umgebung des Menschen," *Erdöl und Kohle, Erdgas Petrochemie* 19(8), 578–583, 1966.
1032. Siegfried, R., "Effects of garbage compost on the 3,4-benzpyrene content of carrots and lettuce," *Naturenwissenschaften* 62, 300, 1975.
1033. Engst, R., and Fritz, W., "Contamination of food with carcinogenic hydrocarbons of environmental origin," *Zakl. Hig. Roez. Parstov. Zakl. Hig.*, 20(1), 113–118, 1975.
1034. Shiraishi, Y., Shirotori, T., and Takabatake, E., "Polycyclic aromatic hydrocarbons in foods. III. 3,4-Benzopyrene in vegetables," *J. Food, Hygienic Soc. Japan*, 15(1), 18–21, 1974.
1035. Sokolowska, R., "Analysis of some food products for presence of benzopyrene," *Roczniki Pantswoweg Zaklader Higieny*, 27(3), 253–259, 1976.
1036. Endo, F., Hirokado, H., Handa, Y., Ishii, K., Usami, H. and Jujii, T., "Survey of polynuclear aromatic hydrocarbons (3,4-benzopyrene) in food additives. I. Citric acid," *Annual Report of Tokyo Metropolitan Research Laboratory of Public Health*, 25 271–277, 1975.
1037. Shiraishi, Y.; Shirotori, T.; and Takabatake, E., "Determination of polycyclic aromatic hydrocarbons in foods. V. 3,4-Benzopyrene in fruits," *J. Food Hygienic Soc. Japan*, 16(3), 187–188, 1975.
1038. Siegfried, R., "3,4-Benzopyrene in oils and fats," *Naturwissenschaften*, 62(12) 576, 1975.
1039. Prokhorova, L. T., and Mironova, A. N., "Presence of 3,4-benzopyrene in margarine," *Trudy Vsesoyuznogo Nauchno-Isseldovatel'skogo Instituta Zhirov*, 30 36–38, 1973.
1040. Grimmer, G. von, and Hildebrandt, A., "Hydrocarbons in the human environment. VI. Levels of polycyclic hydrocarbons in crude vegetable oils," *Arch. Hyg. Bakt.*, 152, 255–259, 1968.
1041. Berner, G., and Biernoth, G., "The content of polycyclic aromatic hydrocarbons in heated oils and fats. Investigation of an industrial fried fat," *Zeitschrift für Lebensmittel-Untersuchung und -Forshung*, 140(6), 330–331, 1969.
1042. Fritz, W., "Solution characteristics of polyaromatic compounds during cooking of coffee substitutes and real coffee," *Deutsche Lebensmittel-Rundschau*, 65(3), 83–85, 1969.
1043. Soos, K., and Foezy, I., "The content of poly aromatic hydrocarbon carcinogens in various coffee types," *Edesipar*. 25(1), 7–10, 37–40, 65–69, 1974.
1044. Strobel, R. J. K., "The determination of 3,4-benzpyrene in coffee products," *6th Inter. Colloq. on the Chemistry of Coffee, Bogota, June 4-9, 1973*, pp. 128–134, 1974.
1045. Shiraishi, Y.; Shirotori, T.; and Sakagami, Y., "Determination of polycyclic aromatic hydrocarbons in foods. I. 3,4-Benzpyrene in Japanese teas," *J. Food Hygienic Soc. Japan*, 13(1), 41–46, 1972.
1046. Dikun, P. P.; Kalinina, I. A.; Lepaie, Y. A. and Pani, E. A., "Influence of drying grain in drum-type and shaft-type dryers on their content of 3,4-benzopyrene," *Gigiena i Sanitarija* (4), 94–96, 1976.
1047. Soos, K., "Content of carcinogenic polynuclear hydrocarbons in Hungarian grain," *Zeitschrift für Lebensmittel-Untersuchung and -Forschung*, 156(6), 344–346, 1974.
1048. Neff, J. M. and Anderson, J. W., "Accumulation, release and distribution of benzo(a)pyrene C_{14} in the clam *Rangia cuneata*," *Proceedings of Joint Conference on Prevention and Control of Oil Pollution (San Francisco, 1975)*, API, pp. 469–471.
1049. Gibson, D. T., "Microbial degradation of carcinogenic hydrocarbons and related compounds," *Symposium on Sources, Effects and Sinks of Hydrocarbons in the Aquatic Environment*, pp. 224–238. American Institute of Biological Sciences, August 1976.
1050. George R. Southworth, John J. Beauchamp, and Patricia K. Schmieder, "Bioaccumulation potential and acute toxicity of synthetic fuels effluents in freshwater biota: Azaarenes," *Environm. Sci. & Techn.* 12(9), 1062–1066, Sept. 1978.
1051. Linda S. Sheldon and Ronald A. Hites. "Organic compounds in the Delaware River," *Environm. Sci. & Techn.* 12(10), 1118–1194, Oct. 1978.
1052. R. A. Brown et al., "Rapid methods of analysis for trace quantities of polynuclear aromatic

hydrocarbons and phenols in automobile exhaust, gasoline and crankcase oil," CRC-APRAC Project CAPE-12-68, Dec. 1971.
1053. Bartlesville Petroleum Research Center, "Oxygenates in automotive exhaust gas: Part III: Carbonyls and noncarbonyls in exhausts from simple hydrocarbon fuels," Bureau of Mines, U.S. Dept. of the Interior, Bartlesville, Oklahoma 74003, Nov. 1970.
1054. H. C. Alexander, W. M. McCarty, and E. A. Bartlett. "Toxicity of perchloroethylene trichloroethylene, 1,1,1-trichloroethane, and methylene chloride to fathead minnows," *Bull. Environm. Contam. Toxicol.* 20, 344-352, 1978.
1055. Jukka Särkkä, Hattula, Jorma Janatuinen, and Jaakko Paasivirta, "Chlorinated hydrocarbons and Mercury in aquatic vascular plants of Lake Paijänne, Finland," *Bull. Environm. Contam. Toxicol.* 20, 361-368, 1978.
1056. G. R. B. Webster, L. P. Sarna, and S. R. Macdonald, "Nonbiological degradation of the herbicide metribuzin in Manitoba soils," *Bull. Environm. Contam. Toxicol.* 20, 401-408, 1978.
1057. W. B. Wilson, C. S. Giam, T. E. Goodwin, A. Aldrich, V. Carpenter, and Y. C. Hrung, "The toxicity of phthalates to the marine dinoflagellate *Gymnodinium breve*," *Bull. Environm. Contam. Toxicol.* 20, 149-154, 1978.
1058. Dan H. Martin, Roger Lewis, and F. Donald Tibbitts, "Teratogenicity of the fungicides captan and folpet in the chick embryo," *Bull. Environm. Contam. Toxicol.* 20, 155-158, 1978.
1059. A. B. McKague, and R. B. Pridmore, "Toxicity of altosid and dimilin to juvenile rainbow trout and coho salmon," *Bull. Environm. Contam. Toxicol.* 20, 167-169, 1978.
1060. David T. Williams, Frank Benoit, and Karel Muzika, "The determination of N-nitrosodiethanolamine in cutting fluids," *Bull. Environm. Contam. Toxicol.* 20, 206-211, 1978.
1061. Fan, T. Y., J. Morrison, D. P. Raunbehler, R. Ross, D. H. Fine, W. Miles, and N. P. Sen., *Science* 196, 70, 1977.
1062. W. Y. Lee, and J. A. C. Nicol, "The effect of naphthalene on survival and activity of the amphipod *Parhyale*," *Bull. Environm. Contam. Toxicol.* 20, 233-240, 1978.
1063. Nicoletta Pacces Zaffaroni, Elio Arias, Giorgio Capodanno, and Teresa Zavanella, "The toxicity of manganese ethylene*bis*dithiocarbamate to the adult newt, *Triturius cristatus*," *Bull. Environm. Contam. Toxicol.* 20, 261-267, 1978.
1064. Philip J. Conklin, and K. Rango Rao, "Toxicity of sodium pentachlorophenate (Na-PCP) to the grass shrimp, *Palaemonetes pugio*, at different stages of the molt cycle," *Bull. Environm. Contam. Toxicol.* 20, 275-279, 1978.
1065. Elizabeth Baumann Ofstad, Gulbrand Lunde and Kari Martinsen, "Chlorinated aromatic hydrocarbons in fish from an area polluted by industrial effluents," *The Science of the Total Environment* 10, 219-230, 1978.
1066. D. Quaghebeur and E. De Wulf, "Polynuclear atomatic hydrocarbons in the main Belgian aquifers," *The Science of the Total Environment* 10, 231-237, 1978.
1067. Akio Nakamura, and Takashi Kashimoto, "Quantitation of sulfur containing oil compounds and polychlorinated biphenyls (PCB) in marine samples," *Bull. Environm. Contam. Toxicol.* 20, 248-254, 1978.
1068. W. C. Steen, D. F. Paris, and G. L. Baughman, "Partitioning of selected polychlorinated biphenyls to natural sediments," *Water Res.* 12, 655-657, 1978.
1069. G. Grimmer, "Analysis of automobile exhaust condensates from air pollution and cancer in man," IARC Scientific Publications No. 16, 1977.
1070. G. Grimmer, H. Böhnke, and A. Glaser, "Investigation on the carcinogenic burden by air pollution in man. XV. Polycyclic aromatic hydrocarbons in automobile exhaust gas–An inventory," *Zbl. Bakt. Hyg. I. Abt. Orig. B* 164, 218-234, 1977.
1071. N. L. Wolfe, R. G. Zepp, and D. F. Paris, "Carbaryl, propham, and chlorpropham: A comparison of the rates of hydrolysis and photolysis with the rate of biolysis," *Water Res.* 12, 565-571, 1978.

1072. William E. Gledhill, "Microbial degradation of a new detergent builder carboxymethyl-tartronate (CMT) in laboratory activated sludge systems," *Water Res.* 12, 591–597, 1978.
1073. J. H. Canton, R. C. C. Wegman, Theresa J. A. Vulto, C. H. Verhoef, and G. J. van Esch, "Toxicity, accumulation, and elimination studies of α-hexachlorocyclohexane (α-HCH) with saltwater organisms of different trophic levels," *Water Res.* 12, 687–690, 1978.
1074. Zepp, R. G., and Cline, D. M., "Rates of direct photolysis in aquatic environment," *Environm. Sci. and Techn.* 11, 359–366, 1977.
1075. Moe, P. G., "Kinetics of the microbial decomposition of the herbicides IPC and CIPC," *Environm. Sci. and Techn.* 4, 429–431, 1970.
1076. Gledhill, W. E., "Biodegradation of 3,3,4′-trichlorocarbanilide, TCC, in sewage and activated sludge," *Water Res.* 9, 649, 1975.
1077. Haider, K., and Jagnow, G., "Degradation of ^{14}C, ^{3}H and ^{36}Cl labelled γ-hexachlorocyclohexane by anaerobic soil microorganisms," *Arch. Mikrobiol.* (Ger.) 104, 113, 1975.
1078. Tucker, E. S., et al., "Activated sludge primary biodegradation of polychlorinated biphenyls." *Bull. Environm. Contam. Toxicol.* 14, 705, 1975.
1079. American Cyanamid Company–Cyanamid International, Wayne, New Jersey, USA 07470.
1080. U.S. Dept. of H. E. W., National Institute of Health, National Cancer Institute, "Report on carcinogenesis bioassay of 1,2-dichloroethane (EDC)," *Am. Ind. Hyg. Assoc. J.* (39), Nov. 1978.
1081. Sidney L. Beck, "Concentration and toxicity of trifluralin in CD-1 mice, presented intragastrically or intraperitoneally," *Bull. Environm. Contam. Toxicol.* 20, 554–560, 1978.
1082. Talaat I. Rihan, Hanaa T. Mustafa, George Caldwell Jr., Leroy F. "Chlorinated pesticides and heavy metals in streams and lakes of northern Mississippi Water," *Bull. Environm. Contam. Toxicol.* 20, 568–572, 1978.
1083. Anne A. Carey, and George R. Harvey, "Metabolism of polychlorinated biphenyls by marine bacteria," *Bull. Environm. Contam. Toxicol.* 20, 527–534, 1978.
1084. Robert W. Moore, John V. O'Connor, and Steven D. Aust, "Identification of a major component of polybrominated biphenyls as 2,2′,3,4,4′,5,5′-heptabromobiphenyl," *Bull. Environm. Contam. Toxicol.* 20, 478–483, 1978.
1085. J. Aubert, L. Petit, D. Puel, "Etude de la degradabilité des nitrosamines en milieu marin," *Rev. Int. Océanogr. Méd.*, Vol. L1–L11, 1978.
1086. Kensuke Furukawa, Kenzo Tonomura, and Akira Kamibayashi, "Effects of chlorine substitution on the biodegradability of polychlorinated biphenyls," *Applied and Environmental Microbiology*, 223–227, Feb. 1978.
1087. Madelyn Webb, H. Ruber, and G. Leduc, "The toxicity of various mining flotation reagents to rainbow trout (*Salmo gairdneri*)," *Water Res.*, 10, 303–306, 1976.
1088. D. K. R. Stewart and Sonia O. Gaul, "Persistence of 2,4-D, 2,4,5-T and dicamba in a dykeland soil," *Bull. Environm. Contam. Toxicol.* 18(2), 1977.
1089. Pierre Fusey, M. F. Lampin, and Jean Dudot, "Recherches sur l'élimination des hydrocarbures par voie biologique," *Material und Organismen*, 10(2), 1975.
1090. P. Maggi, *Revue des Travaux Institut des Pêches maritimes* 36(1), 121–124, 1972.
1091. James G. Smith, Scow-Fong Lee, and Aharon Netzer, "Model studies in aqueous chlorination: The chlorination of phenols in dilute aqueous solutions," *Water Res.* 10, 985–990, 1976.
1092. A. G. Dickson, and J. P. Riley, "The distribution of short-chain halogenated aliphatic hydrocarbons in some marine organisms," *Marine Poll. Bull.* 7(9), Sept. 1976.
1093. J. Batterton, K. Winters, C. van Baalen, "Anilines: Selective toxicity to blue-green algae," *Science*, 199, 10 March 1978.
1094. G. P. Fitzgerald, G. C. Gerloff, F. Skoog, *Sewage Ind. Wastes* 24, 888, 1952.
1095. A. N. Smirnova et al., *Vodosnabzh. Kanaliz. Gidrotekh. Sooruzh*, 5, 17, 1967.
1096. Helen D. Haller, "Degradation of mono-substituted benzoates and phenols by wastewater," *JWPCF*, 2771–2777, Dec. 1978.

1097. D. J. Robertson, R. H. Groth, A. G. Glastris, "HCN content of turbine engine exhaust," *JAPCA*, **29**(1), Jan, 1979.
1098. Jun Kanazawa, "Bioconcentration ratio of diazinon by freshwater fish and snail." *Bull. Environm. Contam. Toxicol.* **20**, 613-617, 1978.
1099. Cope, O. B., "Sport fishery investigation," in: *The Effect of Pesticides on Fish and Wildlife*, 51-64, U.S. Fish Wildl. Serv. Cir. 226, 1965.
1100. Bills, T. D., "Toxicity of formalin, malachite green, and the mixture to four life stages of rainbow trout," Master Science Thesis, Univ. Wisconsin, La Cross, WI, 1974.
1101. Marking, L. L., and Mauck, W. L., "Toxicity of paired mixtures of candidate forest insecticides to rainbow trout," *Bull. Environm. Contam. Toxicol.*, **13**, 518, 1975.
1102. Skea, J. C. et al., "Toxicity of two synthetic pyrethrins to brown trout," *N. Y. Fish & Game Jour.*, **22**, 62, 1975.
1103. Toor, H. S. and Kaur, K., "Toxicity of pesticides to the fish, *Cyprinus carpio communis* Linn," *Indian Jour. Exp. Biol.*, **12**, 334, 1974.
1104. Chaiyarch, S. et al., "Acute toxicity of insecticides toxaphene and carbaryl and herbicides propanil and molinate to four species of aquatic organisms," *Bull. Environm. Contam. Toxicol.*, **14**, 281, 1975.
1105. Verma, S. R., et al., "Studies on the toxicity of DDT on fresh water teleost fishes. Part I. TLm measurements and accumulation of DDT in liver and intestine of *Colisa fasciatus* and *Notopterus notopterus*," *Gegenbaurs Morphol. Jahrb.* (Leipzig), **120**, 439, 1974.
1106. Bhattacharya, S., et al., "Toxic effects of endrin on hepatopancreas of the teleost fish, *Clarias batrachus* (Linn.)," *Indian Jour. Exp. Biol.*, **13**, 185, 1975.
1107. Olson, L. E., et al., "Dinitramine: residues in the toxicity to freshwater fish," *J. Agric. Food Chem.*, **23**, 437, 1975.
1108. Ingham, B., and Gallo, M. A., "Effects of asulam in wildlife species: Acute toxicity to birds and fish," *Bull. Environm. Contam. Toxicol.*, **13**, 194, 1975.
1109. Anderson, A. C., et al., "Acute toxicity of MSMA to black bass (*Micropterus dolomieu*), crayfish (*Procambarua* sp.) and Channel catfish (*Ictalurus lacustris*)," *Bull. Environm. Contam. Toxicol.*, **14**, 330, 1975.
1110. Sun, L. T., and Gorman, M. L., "The toxicity to Fish of herbicides recommended for weed control in the Rewa," *Fiji Agric. J.*, **35**, 31, 1973.
1111. Marking, L. L., and Olson, L. E., "Toxicity of the lampricide 3-trifluoromethyl-4-nitrophenol (TFM) to nontarget fish in static tests," Investigations in Fish Control No. 60, Fish & Wildlife Serv., USDI, Washington, D.C., 1975.
1112. Kimura, S., et al., "Acute toxicity and accumulation of PCB (KC 300) in freshwater fish." *Bull. Freshwater Fish. Res. Lab.* (Jap.), **23**, 115, 1973.
1113. Woodiwiss, F. S., and Fretwell, G., "The toxicities of sewage effluents, industrial discharges and some chemical substances to brown trout (*Salmo trutta*) in the Trent River Authority area," *Water Poll. Control* (G. B.), **73**, 396, 1974.
1114. Davis, J. C., and Hoos, R. A. W., "Use of sodium pentachlorophenate and dehydroabietic acid as reference toxicants for salmonid bioassays," *J. Fish. Res. Bd. Can.*, **32**, 411, 1975.
1115. Fuerstenau, M. C., et al., "Toxicity of selected sulfhydryl collectors to rainbow trout," *Trans. Soc. Mining Engr.*, **256**, 337, 1974.
1116. Hakkila, K., and Niemi, A., "Effects of oil and emulsifiers on eggs and larvae of Northern pike (*Esox lucius*) in brackish water," *Aqua Fennica*, p. 44, 1973.
1117. Mann, H., and Stache, H., "Effect of a Mixture of succinic acid mono-sulfo-ester and olefin sulfonate on fish," *Arch. Fisch. Wiss.* (Ger.), **25**, 53, 1974.
1118. Doster, R. C., et al., "Acute intraperitoneal toxicity of ochratoxin A and B derivatives in rainbow trout (*Salmo gairdneri*)," *Food Cosmet. Toxicol.* (Ger.), **12**, 499, 1974.
1119. Mukherjee, S., and Bhattacharya, S., "Effect of some industrial pollutants on fish brain cholinesterase activity," *Environm. Physiol. Biochem.* (Den.), **4**, 226, 1974.
1120. Leach, J. M., and Thakore, A. N., "Isolation and identification of constituents toxic to

juvenile rainbow trout (*Salmo gairdneri*) in caustic extraction effluents from kraft pulpmill bleach plants," *J. Fish. Res. Bd. Can.*, **32**, 1249, 1975.

1121. Sturm, R. N., et al., "Fluorescent whitening agents: Acute fish toxicity and accumulation studies," *Water Res.* (G. B.), **9**, 211, 1975.

1122. Tovell, P. W. A., et al., "Effect of water hardness on the toxicity of a nonionic detergent to fish," *Water Res.* (G. B.), **9**, 31, 1975.

1123. Ramusino, M. C., and Rossaro, B., "Comparative toxicity of two groups of commercial detergents on *Carassius auratus* L.," *La Riv. Ital. Sostanze Grasse*, **52**, 7, 1975.

1124. Gafa, S., "Studies on relationship between acute toxicity to fish and surface activity of anionic surfactants," *La Riv. Ital. Sostanze Crasse*, **51**, 183, 1974.

1125. Macek, K. J., and Krzeminski, S. F., "Susceptibility of bluegill sunfish (*Lepomis macrochirus*) to nonionic surfactants," *Bull. Environm. Contam. Toxicol.*, **13**, 377, 1975.

1126. Gutenmann, W. H., and Lisk, D. J., "Flame-retardant release from fabrics during laundering and their toxicity to fish," *Bull. Environm. Contam. Toxicol.*, **14**, 61, 1975.

1127. Rice, S. D., and Stokes, R. M., "Acute toxicity of ammonia to several developmental stages of rainbow trout, *Salmo gairdneri*," *Fish. Bull.*, **73**, 207, 1975.

1128. Gradom, L., et al., "Toxicity of altosid to the crustacean *Gammarus aequicauda*," *Mosquito News*, **36**, 294, 1976.

1129. Khattat, F. H., and Farley, S., "Acute toxicity of certain pesticides to *Acartiatonsa dana*," Environmental Protection Agency Final Report, EPA-68/01/0151, 1976.

1130. Zitko, V., et al., "Toxicity of alkyldinitrophenols to some aquatic organisms," *Bull. Environm. Contam. Toxicol.*, **16**, 508, 1976.

1131. Parrish, P. K., et al., "Chronic toxicity of methoxychlor, malathion, and carbofuran to sheepshead minnows (*Cyprinodon variegatus*)," Environmental Protection Agency *Ecol. Res. Series*, EPA-600/3-77-059, 1977.

1132. Parrish, P. R., et al., "Chlorodane: Effects on several estuarine organisms," *J. Toxicol. Environm. Health*, **1**, 485, 1976.

1133. Thirugnanam, M., and Forgash, A. J., "Environmental impact of mosquito pesticides: Toxicity and anticholinesterase activity of chlorpyrifos to fish in a salt marsh habitat," *Arch. Environm. Contam. Toxicol.*, **5**, 415, 1977.

1134. Hanks, K. S., "Toxicity of some chemical therapeutics to the commercial shrimp, *Penaeus californiensis*," *Aquaculture*, **7**, 293, 1976.

1135. Crawford, R. B., and Guarino, A. M., "Effects of DDT in *Fundulus*: Studies on toxicity, fate and reproduction," *Arch. Environm. Contam. Toxicol.*, **4**, 334, 1976.

1136. Hansen, D. J., et al., "Endrin: Effects on the entire life cycle of a saltwater fish, *Cyprinodon variegatus*," *Jour. Toxicol. Environm. Health*, **3**, 721, 1977.

1137. Schimmel, S. C., et al., "Acute toxicity to and bioconcentration of endosulfan by estuarine animals," In *Aquatic Toxicology and Hazard Evaluation*, F. L. Mayer and J. L. Hamelink (Eds.), Amer. Soc. Testing Materials, 241, 1977.

1138. Schimmel, S. C., et al., "Heptachlor: Toxicity and uptake by several estuarine organisms," *J. Toxicol. Environm. Health*, **1**, 955, 1976.

1139. Walsh, G. E., et al., "The toxicity and uptake of Kepone® in marine unicellular algae," *Chesapeake Sci.*, **18**, 222, 1977.

1140. Schimmel, S. C., and Wilson, A. J. Jr., "Acute toxicity of Kepone® to four estuarine animals," *Chesapeake Sci.*, **18**, 224, 1977.

1141. Hansen, D. J., et al., "Kepone®: Chronic effects on embryo, fry, juvenile and adult sheepshead minnows (*Cyprinodon variegatus*)," *Chesapeake Sci.*, **18**, 227, 1977.

1142. Abrahams, D., and Brown, W. D., "Evaluation of fungicides for *Haliphthoros milfordensis* and their toxicity to juvenile European lobsters," *Aquaculture*, **12**, 31, 1977.

1143. Armstrong, D. V., et al., "Toxicity of the insecticide methoxychlor to the Dungeness crab— *Cancer magister*," *Marine Biol.* (W. Ger.), **38**, 239, 1976.

1144. Zitto, V., et al., "Toxicity of pyrethoids to juvenile Atlantic salmon," *Bull. Environm. Contam. Toxicol.*, **18**, 35, 1977.

BIBLIOGRAPHY 1275

1145. Schimmel, S. C., et al., "Uptake and toxicity of toxaphene in several estuarine organisms," *Arch. Environm. Contam. Toxicol.*, 5, 353, 1977.
1146. Weis, J. S., and Weis, P., "Optical malformations induced by insecticides in embryos of the Atlantic silverside, *Menidia menidia*," *U.S. Natl. Marine Fish Service Fish. Bull.*, 74, 208, 1976.
1147. Peterson, R. H., "Temperature selection of juvenile Atlantic salmon (*Salmo salar*) as influenced bv various toxic substances," *J. Fish. Res. Bd. Can.*, 33, 1722, 1976.
1148. Ogilvie, D. M., and Miller, D. L., "Duration of a DDT-influenced shift in the selected temperature of Atlantic salmon (*Salmo salar*)," *Bull Environm. Contam. Toxicol.*, 16, 86, 1976.
1149. Bookout, C. G., and Costlow, J. D., Jr., "Effects of mirex, methoxychlor, and malathion on development of 'crabs," *Env. Prot. Agency Ecol. Res. Series*, EPA-600/3-76-007, 1976.
1150. Bookout, C. G., et al., "Effects of methoxychlor on larval development of mud crabs and blue crabs," *Water Air Soil Poll.*, 5, 349, 1976.
1151. Ward, D. V., and Busch, D. V., "Effects of temefos, an organophosphorus insecticide, on survival and escape behavior of the marsh fiddler crab, *Uca pugnax*," *Oikos* (Den.), 27, 331, 1976.
1152. Price, N. R., "The effect of two insecticides on the Ca^{2+} and Mg^{2+}-activated ATPase of the sarcoplasmic reticulum of the flounder, *Platichthys flesus*," *Comp. Biochem. Physiol.*, 55, 91, 1976.
1153. Khorram, S., and Knight, A W., "Effects of temperature and kelthane on grass shrimp," *J. Environm. Eng. Div.*, EE5. 1043, 1976.
1154. Tagatz, M. E., "Effects of mirex on predator-prey interaction in an experimental estuarine ecosystem," *Trans. Amer. Fish. Soc.*, 105, 546, 1976.
1155. Fisher, W. S., et al., "Toxicity of malachite green to cultured American lobster larvae," *Aquaculture*, 8, 151, 1976.
1156. Schoor, W. P., and New man, S. M., "The effect of mirex on the burrowing activity of the lugworm (*Arenicola cristata*)," *Trans. Amer. Fish. Soc.*, 105, 700, 1976.
1157. Payen, G. G., and Costlow, J. D., "Effects of a juvenile hormone mimic on male and female gametogenesis of the mud-crab, *Rhithropanopeus harrisii* (Gould) (Brachyura: Xanthidae)," *Biol. Bull.*, 152, 199, 1977.
1158. Bresch, H., and Arendt, U., "Influence of different organochlorine pesticides on the development of the sea urchin embryo," *Environm. Research*, 13, 121, 1977.
1159. Couch, J. A., et al., "Kepone-induced scoliosis and its histological consequences in fish," *Science*, 147, 585, 1977.
1160. Weis, P., and Weis. J. S., "Abnormal locomotion associated with skeletal malformations in the sheepshead minnow, *Cyprinodon variegatus*, Exposed to malathion," *Environm. Research*, 12, 196, 1976.
1161. Overnell, J., "Inhibition of marine algal photosynthesis by heavy metals," *Marine Biol.* (W. Ger.), 38, 335, 1976.
1162. Mason, J. W., and Rowe, D. R., "The accumulation and loss of dieldrin and endrin in the Eastern oyster," *Arch. Environm. Contam. Toxicol.*, 4, 349, 1976.
1163. Harding, G. C. H., and Vass, W. P., "Uptake from sea water and clearance of ^{14}C-p, p'-DDT by the marine copepod *Calanus finmarchicus*," *J. Fish. Res. Bd. Can.*, 34, 177, 1977.
1164. Warlen, S. M., et al., "Accumulation and retention of dietary ^{14}C-DDT by Atlantic menhaden," *Trans. Amer. Fish Soc.*, 106, 95, 1976.
1165. Schimmel, S. C., et al., "Heptachlor: Uptake, depuration, retention and metabolism by spot, *Leiostomus xanthurus*." *J. Toxicol. Environm. Health*, 2, 169, 1976.
1166. Bahner, L. H., et al., "Kepone® bioconcentration accumulation, loss, and transfer through estuarine food chains," *Chesapeake Sci.*, 18, 299, 1977.
1167. Cripe, C. R., and Livingston, R. J., "Dynamics of mirex and its prinicpal photoproducts in a simulated march system," *Arch. Environm. Contam. Toxicol.*, 5, 295, 1977.
1168. Stadler, D., and Ziebarth, U., "$p,$-p'-DDT, dieldrin and polychlorinated biphenyls (PCBs) in

the surface water of the Western Baltic, 1974," *Deutsche Hydrographische Zeitschrift* (W. Ger.), **29**, 25, 1976.
1169. Jonas, R. B., and Pfaender, F. K., "Chlorinated hydrocarbon pesticides in Western North Atlantic Ocean," *Environm. Sci. and Techn.*, **10**, 770, 1976.
1170. Walker, W. W., "Pesticides and their residues in the Mississippi coastal zone," *J. Mississippi Acad. Sci.*, **21**, 148, 1977.
1171. Dawson, R., and Riley, J. P., "Chlorine-containing pesticides and polychlorinated biphenyls in British coastal waters," *Estuarine Coastal Marine Sci.*, **5**, 55, 1977.
1172. Eder, G., "Polychlorinated biphenyls and compounds of the DDT group in sediments of the Central North Sea and the Norwegian Depression," *Chemosphere*, **5**, 101, 1976.
1173. Munson, T. O., "A Note on toxaphene in environmental samples from the Chesapeake Bay region," *Bull. Environm. Contam. Toxicol.*, **16**, 491, 1976.
1174. Tanita, R., et al., "Organochlorine pesticides in the Hawaii Kai Marina, 1970-1974," *Pesticides Monitoring J.*, **10**, 24, 1976.
1175. Choi, W. W., and Chen, K. Y., "Association of chlorinated hydrocarbons with fine particles and humic substances in near-shore surficial sediments," *Environm. Sci. and Techn.*, **10**, 782, 1976.
1176. Ahr, W. M., "Chlorinated hydrocarbons in Bay Sediments," in: *Shell Dredging and Its Influence on Gulf Coast Environments*," A. H. Bouma (Ed.), Houston, Texas, 151, 1976.
1177. MacGregor, J. S., "DDT and its metabolites in the sediments off Southern California," *U.S. Natl. Marine Fish. Serv. Fish Bull.*, **74**, 27, 1976.
1178. Young, D. R., and McDermott-Ehrlich, D., "Sediments as sources of DDT and PCB," in: *Southern California Water Research Project Annual Report*," 49, 1976.
1179. Young, D. R., et al., "DDT in sediments and organisms around Southern California outfalls," *JWPCF*, **48**, 1919, 1976.
1180. Clausen, J., and Berg, O., "The content of polychlorinated hydrocarbons in Arctic ecosystems," *Pure Appl. Chem.*, **42**, 223, 1976.
1181. Jensen, S., and Jansson, B., "Anthropogenic substances in seal from the Baltic: Methyl sulfon metabolites of PCB and DDE," *Ambio*, **5**, 257, 1976.
1182. Sarat, W. F., et al., "Pesticides in people," *Pesticides Monitoring Jour.*, **11**, 1, 1977.
1183. Risebrough, R. H., et al., "Bioaccumulation factors of chlorinated hydrocarbons between mussels and seawater," *Marine Poll. Bull.*, **7**, 225, 1976.
1184. Riley, J. P., and Wahby, S., "Concentrations of PCB's, dieldrin and DDT residues in marine animals from Liverpool Bay," *Marine Poll. Bull.*, **8**, 9, 1977.
1185. Cubit, D. A., et al., "Determination of organochlorine pesticides in the tissues of the black mullet (*Mugil cephalus*) and the silver mullet (*Mugil curema*)," *Amer. Inc. Hyg. Assn. Jour.*, **37**, 8, 1976.
1186. Lock, J. W., and Solly, S. R. B., "Organochlorine residues in New Zealand birds and mammals. I. Pesticides," *New Zealand Jour. Sci.*, **19**, 43, 1976.
1187. Palmer, H. D., et al., "Transport of chlorinated hydrocarbons in sediments of the Upper Chesapeake Bay," Westinghouse Ocean Research Lab., Annapolis Md., Office of Water Research and Technology, Final Report W76-10607, OWRT-C-5160 (4204) (1), 1976.
1188. Deubert, K. H., and Gray, R. S., "Parathion residues in environmental samples from untreated areas," *Bull. Environm. Contam. Toxicol.*, **15**, 613, 1976.
1189. Dean, H. J., et al., "Toxicity of methoxychlor and naled to several life stages of land-locked Atlantic salmon," *N.Y. Fish & Game*, **24**, 144, 1977.
1190. Dolan, J. M., and Hendricks, A. C., "The lethality of an intact and degraded linear alkylbenzene sulfonate mixture to bluegill sunfish and a snail," *JWPCF*, **48**, 2570, 1976.
1191. Kimerle, R. A., and Swisher, R. D., "Reduction of aquatic toxicity of linear alkylbenzene sulfonate (LAS) by biodegradation," *Water Res.* (G. B.), **11**, 31, 1977.
1192. Bills, T. D., et al. "Formalin: Its toxicity to nontarget aquatic organisms, persistence, and counteraction," Investigations in Fish Control No. 73, U.S. Dept. Int. Fish & Wildlife Ser., Washington, D.C., 1977.

1193. Rehwoldt, R. E., et al., "Investigations into acute toxicity and some chronic effects of selected herbicides and pesticides on several freshwater fish species," *Bull. Environm. Contam. Toxicol.*, **18**, 361, 1977.
1194. Jarvinen, A. W., et al., "Long-term toxic effects of DDT food and water exposure on fatheat minnows (*Pimephales promelas*)," *J. Fish Res. Bd. Can.*, **34**, 2089, 1977.
1195. Jarvinen, A. W., et al., "Toxicity of DDT food and water exposure to fathead minnows," EPA-600/3-76-114, U.S. Environmental Protection Agency, Duluth, Minn., 1976.
1196. Zitko, V., et al., "Toxicity of pyrethroids to juvenile Atlantic salmon," *Bull. Environm. Contam. Toxicol.*, **18**, 35, 1977.
1197. Johnson, W. W., and Sanders, H. O., "Chemical forest fire retardants: Acute toxicity to five freshwater fishes and a scud," Tech. Rept. No. 91, U.S. Dept. Int., Fish & Wildlife Serv., Washington, D.C., 1977.
1198. Marking, L. L., and Bills, T. D., "Toxicity of rotenone to fish in standardized laboratory tests," Investigations in Fish Control No. 72, U.S. Dept. Int., Fish & Wildlife Serv., Washington, D.C., 1976.
1199. Huda, J., and Svobodova, Z., "Acute toxicity of the actellic biocide to fish," *Vodnany* (Czech.), **12**, 24, 1976; *Aquatic Sci. & Fish. Abs.*, 7, 7Q10524, 1977.
1200. Paflitschek, J., "Investigations on toxic effects of Bayer 73 (Bayluscid Wp) on eggs and yolk-sac larvae of *Tilapia leucosticta* (Cichlidae)," *Experientia* (Switz.), **32**, 1537, 1976.
1201. Rye, R. P., Jr., and King, E. L., Jr., "Acute toxic effects of two lampricides to twenty-one freshwater invertebrates," *Trans. Amer. Fish. Soc.*, 105, 322, 1976.
1202. McCorkle, F. M., et al., "Acute toxicities of selected herbicides to fingerling Channel catfish, *Ictalurus punctuatus*," *Bull. Environm. Contam. Toxicol.*, **18**, 267, 1977.
1203. Davey, R. B., et al., "Toxicity of five rice-field pesticides to the mosquitofish, *Gambusia affinis*, and green sunfish, *Lepomis cyanellus*, under laboratory and field conditions in Arkansas," *Environm. Entomology*, 5, 1053, 1976.
1204. Cardwell, R. D., et al., "Acute and chronic toxicity of chlordane to fish and invertebrates," EPA-600/3-77-019, U.S. Environmental Protection Agency, Duluth, Minn., 1977.
1205. Marking, L. L., et al., "Toxicity of furanace to fish, aquatic invertebrates, and frog eggs and larvae," Investigations in Fish Control No. 76, U.S. Dept. Int. Fish & Wildlife Serv., Washington, D.C., 1977.
1206. Spehar, R. L., et al., "A rapid assessment of the toxicity of three chlorinated cyclodiene insecticide intermediates to fathead minnows," EPA-600/3-77-099, U.S. Environmental Protection Agency, Duluth, Minn., 1977.
1207. Julin, A. M., and Sanders, H. O., "Toxicity and accumulation of insecticide imidan in freshwater invertebrates and fishes," *Trans. Amer. Fish. Soc.*, 106, 386, 1977.
1208. Abel, P. D., "Toxic action of several lethal concentrations of an anionic detergent on gills of brown trout (*Salmo trutta* L.)," *Jour. Fish Biol.*, 9, 441, 1976.
1209. Bills, T. D., et al., "Malachite green: Its toxicity to aquatic organisms, persistence, and removal with activated carbon," Investigations in Fish Control No. 75, U.S. Dept. Int., Fish & Wildlife Serv., Washington, D.C., 1977.
1210. Dawson, G. W., et al., "Acute toxicity of 47 industrial chemicals to fresh and saltwater fishes," *J. Hazard. Materials*, 1, 303, 1977.
1211. Walsh, D. F., et al., "Residues of emulsified xylene in aquatic weed control and their impact on rainbow trout, *Salmo gairdneri*," Rept. No. REC-ERC-77-11, Engineering & Res. Center, Bureau of Reclamation, Denver, Colo., 1977.
1212. F. Geike and C. D. Parasher. "Effect of hexachlorobenzene on photosynthetic oxygen evolution and respiration of *Chlorella pyrenoidosa*," *Bull. Environm. Contam. Toxicol.* 20, 647–651, 1978.
1213. P. A. Gerakis, D. S. Veresoglou, and E. Sfakiotakis, "Ethylene pollution from wheat stubble burning," *Bull. Environm. Contam. Toxicol.* 20, 657–661, 1978.
1214. Darley, E. F., et al., *J.A.P.C.A.* **16**, 685, 1966.
1215. M. Th. H. Tulp et al., "Environmental Chemistry of PCB-replacement compounds. III. The

metabolism of 4-chloro-4'-isopropylbiphenyl and 2,5-dichlor-4'-isopropylbiphenyl in the rat," *Chemosphere* 6, 109, 1977.
1216. H. Appleton et al. "Fate of 3,3'-dichlorobenzidine in the aquatic environment," in *Aquatic Pollutants: Tranformation and Biological Effects*, Pergamon Press, 1978.
1217. Ryckman, Edgerley, Tomlinson and Associates, Inc., "Laboratory test methods to assess the effects of chemicals on terrestrial animal species," Report No. EPA 560/5-75-004, April 1975.
1218. Sujit Banerjee, Harish C. Sikka, Richard Gray, and Christine M. Kelly, "Photodegradation of 3,3'-dichlorobenzidine," *Environm. Sci. & Techn.* 12(13), Dec. 1978.
1219. David T. Gibson, "Microbial transformation of aromatic pollutants," in *Aquatic Pollutants: Transformation and Biological Effects*, Pergamon Press, 1978.
1220. German Society for Petroleum Sciences and Coal Chemistry, "Influence of engine/driving-conditions and operating time of engine oil on the exhaust emission of polycyclic aromatic hydrocarbons from gasoline passenger cars. Part I. Emissions," BMI-DGMK Project 110, 2000 Hamburg 1, Nordkanalstrasse 28, W. Germany, August 1977.
1221. P. Leinster, R. Perry, and R. J. Young, "Ethylenebromide in urban air," *Atmosph. Environm.* 12, 2383-2398, 1978.
1222. Adel K. Wasfi et al., "Evaluation of catalysts for vapor phase oxidation of odorous organic compounds," *Atm. Environm.* 12, 2389-2398, 1978.
1223. W. D. Saunders, "F-11 and N_2O in the North American troposphere and lower stratosphere," *Water, Air and Soil Poll.*, 10, 421-439, 1978.
1224. S. A. Penkett, F. J. Sandalls, and B. M. R. Jones, "PAN measurements in England—Analytical methods and results," in: VDI-Berichte 270, *Ozon und Begleitsubstanzen im Photochemischen Smog*," 47-56, 1977.
1225. Heuss, J. M., and W. A. Glasson, "Hydrocarbon reactivity and eye irritation," *Environm. Sci. & Techn.* 2, 1109-1116, 1968.
1226. G. M. Meyer and H. Nieboer. "Determination of peroxybenzoylnitrate (PBzN) in ambient air," in: VDI-Berichte No. 270, *Ozon und Begleitsubstanzen im Photochemischem Smog*, 55-57, 1977.
1227. A. Kruysse and V. J. Feron, "Acute and sub-acute inhalation toxicity of peroxyacetylnitrate and ozone in rats," in: VDI-Berichte No. 270, *Ozon und Begleitsubstanzen in photochemischen Smog*, 101-109, 1977.
1228. Campbell, K. I., G. L. Clarke, L. O. Emik, and R. L. Plata, "The atmospheric contaminant peroxyacetylnitrate. Acute inhalation toxicity in mice," *Arch. Environm. Health* 15, 739-744, 1967.
1229. Dungworth, D. L., G. L. Clarke, and R. L. Plata. "Pulmonary lesions produced in A-strain mice by long-term exposure to peroxyacetylnitrate," *Am. Rev. Resp. Dis.* 99, 565-574, 1969.
1230. J. S. Jacobson, Yonkers, N.Y., "The effect of photochemical oxidants on vegetation," in: VDI-Berichte No. 270, *Ozon und Begleitsubstanzen im photochemischen Smog*, 1977.
1231. R. Jeltes, "Messung organischer Luftverunreinigungen in den Niederlanden," in: VDI-Berichte No. 270, *Ozon und Begleitsubstanzen im photochemischen Smog*, 1977.
1232. L. J. Brasser, R. Guicherit, and C. Huygen, "The occurrence of photochemical smog formation in Western Europe," VDI-Berichte No. 270, *Ozon und Begleitsubstanzen im photochemischen Smog*, 1977.
1233. L. van Vaeck, G. Broddin, W. Cautreels and K. van Cauwenberghe, "Aerosol collection by cascade impaction and filtration; influence of different sampling systems on the measured organic pollutant levels," *The Science of the Total Environment*, 11, 41-52, 1979.
1234. J. D. Butler and P. Crossley, "An appraisal of relative airborne suburban concentrations of polycyclic aromatic hydrocarbons monitored indoors and outdoors," *The Science of the Total Environment* 11, 53-58, 1979.
1235. I. J. Higgins and P. D. Gilbert, "The biodegradation of hydrocarbons," in: *The Oil Industry*

and *Microbial Systems*, edited by K. W. A. Chater and H. J. Somerville, Van Heyden and Son Ltd., 1978.
1236. A. Björseth and G. Lunde, "Long-range transport of polycyclic aromatic hydrocarbons," *Atm. Environm.* **13**, 45–53, 1979.
1237. P. McKenzie, and D. E. Hughes, "Microbial degradation of oil and petrochemicals in the sea," in: *Microbiology in Agriculture, Fisheries and Food*, edited by F. A. Skinner and J. G. Carr, Academic Press, 1976.
1238. D. J. Hopper, In: *Development in Biodegradation of Hydrocarbons*, edited by R. J. Watkinson, Applied Science, London, 1977.
1239. J. I, Davies and W. C. Evans, *Biochem. J.* **91**, 251, 1964.
1240. E. A. Barnsley, *Biochem. Biophys. Res. Commun.* **72**, 1116, 1976.
1241. W. C. Evans, H. N. Fernley, and E. Griffiths, *Biochem. J.* **95**, 819, 1965.
1242. D. Lunt, and W. C. Evans, *Biochem. J.* **118**, 54, 1970.
1243. D. Catelani, A. Colombi, C. Sorlini, and V. Trecanni, *Biochem. J.* **134**, 1063, 1973.
1244. L. A. Stirling, R. J. Watkinson, and I. J. Higgins, *J. Gen. Microbiol.* **99**, 119, 1977.
1245. D. B. Norris, and P. W. Trudgill, *Biochem. J.* **121**, 363, 1971.
1246. E. Heuniger et al., "Recommendations concerning the problem of the reduction of air pollution caused by carcinogenic substances contained in exhaust gases of road vehicles," Published by VEB Wissenschaftlich-Technisches Zentrum Automobilbau (WTZ) Abgasprüfstelle der DDR, DDR-1199 Berlin, Rudower Chaussee 6 and Oncologic Scientific Center (ONZ) of the Academy of Medical Sciences of the USSR, Moscow, Kaschirskoje Chaussee.
1247. Schramm, T. and Bierwolf, D.; "Ursachen des Krebses—Neue Erkenntnisse und Tendenzen," *Wissenschaft und Fortschritt* **27**(2), 61–65, 1977.
1248. Schramm, T.; Vortrag, Akademie für Ärztliche Fortbildung der DDR, Berlin, 7.9.77.
1249. Schramm, T. and Teichmann, B.; "Zur Problematik von Grenzwerten für chemische Kanzerogene," *Deutsches Gesundheitswesen*, **32**(20), 940–944, 1977.
1250. Doll, R.. Ztiologij raka legkih uspehi v izuchenii raka, 1957, 3, 11-66-M (ler 5 angl)
1251. Pott, P., *Chirurgical Observations Relative to the Cataract, the Polupus of the Nose, the Cancer of the Scrotum, the Different Kinds of Ruptures and the Mortification of the Toes and Feet*, London, Carnegy, 1775.
1252. Kennaway, E. and Hieger, I., "Carcinogenic substances and their fluorescence spectra," *Brit. Med. Journ.*, **1**, 1044–1046, 1930.
1253. Hartwell, J., "Survey of compounds which have been tested for carcinogenic activity," 2nd ed., National Cancer Institute U.S. Public Health Service, Washington D.C., 1951.
1254. Shubik, D. S. and Hartwell, K. L., "Survey of compounds which have been tested for carcinogenic activity," Suppl. 2, National Cancer Inst. U.S. Public Health Service, Washington D.C., 1969.
1255. Doll, R.; 1971. Profilaktika raka na osnove dannyh ėpidemiologii. M[oscow?]. Translated from English.
1256. Shabad, L. Sh.[sic]; 1973. O cirkuljacii kancerogenov v okruzhajushchei srede. Moscow, 367 pp.
1257. Janyshcheva, N. Ja.; Chernichenko, I. A.; Balenko, N. V.; Kireeva, I. S.; 1977. Kancerogennye veshchestva i ih gigienicheskoe normirovanie v okruzhajushchei srede. Kiev. 134 pp.
1258. Shabad, L. M., "On the so-called MAC (maximal allowable concentration) for carcinogenic hydrocarbons," *Neoplasma* **22**(5), 459–468, 1975.
1259. Shabad, L. Sh.[sic]; 1977, 1978. Mezhdunarodnyi simpozium "Zagrjaznenie vozduha i rak" Stockholm. 8-II/III/1977 gig. i san., 1978, 2 121–122.
1260. Kertesz-Saringer, M. and Morlin, Z., "On the occurence of polycyclic aromatic hydrocarbons in the urban area of Budapest," *Atm. Environm.* **9**(9), 831–834, 1975.
1261. Gimmer, G., Hildebrant A. and Böhnke, H., "Probenahme und Analytik polycylischer aromatischer Kohlenwasserstoffe in Kraftfahrzeugabgasen," *Erdöl und Kohle* **25**(8), 442–447, 1972; **25**(9), 531–536, 1972.

1262. O. C. Taylor, "Importance of peroxyacetylnitrate (PAN) as a phytotoxic air pollutant," *JAPCA*, **19**(5), May 1969.
1263. Joseph M. Colucci and Charles R. Begeman, "Carcinogenic air pollutants in relation to automotive traffic in New York." *Environm. Sci. and Techn.* **5**(2), 145-150, Feb. 1971.
1264. Colucci, J. H.; Begeman, C. R., *JAPCA* **15**, 113-22, 1965.
1265. Sawicki, E.; Elbert, W. C.; et al., *A.I.H.A.J.* **21**, 443-451, 1960.
1266. De Maio, L.; Corn, M., *JAPCA* **16**, 67-71, 1966.
1267. Hettche, H. O., *Int. J. Water Pollut.* **8**, 185-191, 1964.
1268. Waller, R. E., *Brit. J. Cancer* **6**, 8-21, 1952.
1269. Waller, R. E.; Cummins, B. T.; Lawther, P. J., *Brit. J. Ind. Med.* **22**, 128-138, 1965.
1270. D'Ambrosio, A.; Pavelka, F.; et al., Centro Provincial Per lo Studio Sugli Inquinamenti Atmosferici, 1958.
1271. Campbell, J. M.; Clemmesen, J., *Dan. Med. Bull.* **3**, 205, 1956.
1272. Skramovsky, V., *Acta Un. Int. Cancr.* **19**, 733-736, 1963.
1273. Saringer, M., *Egiszigtudomany* **7**, 25-32, 1963.
1274. Louw, G. W., *Amer. Ind. Hyg. Ass. J.* **26**, 520-526, 1965.
1275. Watanabe, H.; Tomita, K., *Proc. Int. Clean Air Congress Part. I*, 226-228, 1966.
1276. Cleary, G. L.; Sullivan, J. L., *Med. J. Aust.* **1**, 758-763, 1965.
1277. P. J. Blokzijl, and R. Guicherit, "The occurrence of polynuclear aromatic hydrocarbons in outdoor air," *TNO-Nieuws*, Nov. 1972.
1278. *Brit. J. of Cancer.* **10**, 4201, Sept. 1956.
1279. *Danish Med. Bull.* **3**, 205, Nov. 1956.
1280. *Roy. Soc. Hlth. J.*, **76**, 677, 1956.
1281. *Arch. Belg. Med. Soc.*, **9**(10), 578, 1963.
1282. *Int. J. of Air Poll.*, **1**, 14, 1958.
1283. *Int. J. of Air and Wat. Poll.*, **8**, 185, 1964.
1284. *Air Poll. Aspects of Organic Carc.*, Litton Syst. Inc. Prep. for the APCA, 1969.
1285. Centro per Lo Studio Sugli Ing. Attm., 1958.
1286. *Revue Poll. Atmosph.*, **7**, 316, 1956.
1287. Badger, G. M., "Mode of formation of carcinogens in human environment," *Natl. Cancer Inst. Monograph.* **9**, 1, 1962.
1288. Badger, G. M., J. K. Donnelly, and T. M. Spotswood, "The formation of aromatic hydrocarbons at high temperature," *Australian J. Chem.* **19**, 1023, 1965.
1289. W. Cautreels, and K. Van Cauwenberghe, "Experiments on the distribution of organic pollutants between airborne particulate matter and the corresponding gas phase," *Atm. Environm.* **12**, 1133-1141, 1978.
1290. Kirby I. Campbell et al., "The atmospheric contaminant peroxyacetylnitrate-Acute inhalation toxicity in mice," *Arch. Environm. Health*, **15**, Dec. 1967.
1291. A. Candeli, V. Mastrandrea, G. Morozzi, and S. Toccaceli, "Carcinogenic air pollutants in the exhaust from an European car operating on various fuels," *Atm. Environm.* **8**, 693-705, 1974.
1292. D. J. Von Lehmden et al., "Polynuclear hydrocarbon emissions from selected industrial processes," *JAPCA* **15**(7), July 1965.
1293. E. Sawicki et al., "Quantitative composition of the urban atmosphere in terms of polynuclear aza heterocyclic compounds and aliphatic and polynuclear aromatic hydrocarbons," *Air, Water Poll. J.*, 515-524, Sept. 1965.
1294. Dowden, B. F., "Cumulative toxicities of certain inorganic salts to *Daphnia magna* as determined by median tolerance limits," *Proc. La. Acad. Sci.*, **23**, 77, 1960.
1295. Freeman, L., "A standardized method for determining toxicity of pure compounds to fish," *Sewage and Industrial Wastes*, **25**(7), 845, 1953.
1296. Robert K. Hinderer, "Toxicity studies of methylcyclopentadienyl manganese tricarbonyl (MMT)," *AIHAJ* **40**(2), 164-167, 1979.

1297. Katherine W. Wilson, "Survey of eye irritation and lachrymation in relation to air pollution," Coordinating Research Council, Inc., 30 Rockefeller Plaza, New York, N.Y. 10020, Contract No. CAPM-17-71, Jan. 1973.
1298. Wynder and Hoffman, *Tobacco and Tobacco Smoke*, Academic Press, New York, 1967.
1299. Edwin A. Woolson, and Philip C. Kearney, "Persistence and reactions of ^{14}C-cacodylic acid in soils," *Environ. Sci. & Techn.* 7(1), 47–50, Jan. 1973.
1300. Allan R. Isensee et al., "Distribution of alkyl arsenicals in model ecosystem," *Environm. Sci. & Techn.*, 7(9), 841–845, Sept. 1973.
1301. Donald R. Buhler, M. E. Rasmusson, and H. S. Nakaue, "Occurrence of hexachlorophene and pentachlorophenol in sewage and water," *Environm. Sci. & Techn.* 7(10), 929–934, Oct. 1973.
1302. William D. Youngs, Walter H. Gulenmann, and Donald J. Lisk, "Residues of DDT in lake trout as a function of age," *Environm. Sci. and Techn.* 6(5), 451–452, May 1972.
1303. Francis R. Boucher, and G. Fred Lee, "Adsorption of lindane and dieldrin pesticides on unconsolidated aquifer sands," *Environm. Sci. & Techn.* 6(6), 539–543, June 1972.
1304. Max W. Hammond and Martin Alexander, "Effect of chemical structure on microbial degradation of methyl-substituted aliphatic acids," *Environm. Sci. & Techn.* 6(8), 732–735, 1972.
1305. Philip C. Kearney, Edwin A. Wodson, and Charles P. Ellington Jr., "Persistence and metabolism of chlorodioxins in soils," *Environm. Sci. & Techn.*, 6(12), 1017–1019, Nov. 1972.
1306. Michael A. Poirrier, Billy Ray Bordelon, and John L. Laseter, "Adsorption and concentration of dissolved carbon-14 DDT by coloring colloids in surface waters," *Environm. Sci. & Techn.*, 6(12), 1033–1035, Nov. 1972.
1307. Kishore P. Goswami, and Richard E. Green. "Microbial degradation of the herbicide atrazine and its 2-hydroxy analog in submerged soils," *Environm. Sci. & Techn.*, 5(5(, 426–429, May 1971.
1308. Charles W. Stanley et al., "Measurement of atmospheric levels of pesticides." *Environm. Sci. & Techn.*, 5(5), 430–435, May 1971.
1309. James W. Eichelberger, and James J. Lichtenberg, "Persistence of pesticides in river water," *Environm. Sci. & Techn.*, 5(6), 541–544, June 1971.
1310. Nagalaxmi Sridharan and G. Fred Lee, "Coprecipitation of organic compounds from lake water by iron salts," *Environm. Sci. & Techn.*, 6(12), 1031–1033, Nov. 1972.
1311. Krishna C. Patil et al., "Metabolic transformation of DDT, dieldrin, aldrin and endrin by marine micro-organisms," *Environm. Sci. & Techn.*, 6(7), 629–632, July 1972.
1312. Robert L. Metcalf et al., "Model ecosystem for the evaluation of pesticide biodegradability and ecological magnification." *Environm. Sci. & Techn.*, 5(8), 709–713, Aug. 1971.
1313. Thomas M. Ward, and Forrest W. Getzen, "Influence of pH on the adsorption of aromatic acids on activated carbon," *Environm. Sci. & Techn.*, 4(1), 64–67, Jan. 1970.
1314. U.S. Dept. H.E.W., N.I.H., N.C.I., "Report on carcinogenesis bioassay of 1,2-dibromo-ethane (EDB)," *AIHA J* 40(2), A-31-A-35, 1979.
1315. Don E. Henley et al., "Isolation and identification of an odor compound produced by a selected aquatic actinomycete," *Environm. Sci. & Techn.*, 3(3), 268–271, March 1969.
1316. Lloyd L. Medsker, David Jenkins, and Jerome F. Thomas, "Odorous compounds in natural waters, 2-*exo*-hydroxy-2-methylbornane, the major odorous compound produced by several actinomycetes," *Environm. Sci. & Techn.*, 3(5), 476–477, May 1969.
1317. Bernhard H. Pfeil, and G. Fred Lee, "Biodegradation of nitrilotriacetic acid in aerobic systems," *Environm. Sci. & Techn.*, 2(7), 543–546, July 1968.
1318. C. E. Castro, and N. O. Belser, "Biodehalogenation. Reductive dehalogenation of the boicides ethylene dibromide, 1,2-dibromo-3-chloropropane, and 2,3-dibromobutane in soil," *Environm. Sci. & Techn.*, 2(10), 779–783, Oct. 1968.
1319. W. A. Lonneman, T. A. Bellar, and A. P. Altshuller, "Aromatic hydrocarbons in the atmosphere of the Los Angeles basin," *Environm. Sci. & Techn.*, 2(11), 1017–1020, Nov. 1968.

1320. Alexander J. Fatiadi, "Effects of temperature and of ultraviolet radiation on pyrene adsorbed on garden soil," *Environm. Sci. & Techn.*, **1**(7), 570–572, July 1967.
1321. Moira J. MacKenzie, and Joseph V. Hunter, "Sources and fates of aromatic compounds in urban stormwater runoff," *Environm. Sci. & Techn.*, **13**(2), 179–183, Feb. 1979.
1322. D. Henschler, E. Eder, T. Neudecker, and M. Metzler, "Carcinogenicity of trichloroethylene: Fact or artifact?" *Arch. Toxicol.* **37**, 233–236, 1977.
1323. Roger Bluzat, and Jacqueline Seuge, "Effects de trois insecticides (lindane, fenthion et carbaryl): Toxicité aiguë sur quatre espèces d'invertébrés limniques; toxicité chronique chez le mollusque pulmoné *Lymnea*," *Environm. Pollut.* **18**, 51–70, 1979.
1324. Masako Veji and Jun Kanazawa, "Degradation of *o-sec*-butylphenyl-N-methylcarbamate (BPMC) in soil," *Bull. Environm. Contam. Toxicol.* **21**, 29–36, 1979.
1325. J. P. Seiler, "Phenoxyacids as inhibitors of testicular DNA synthesis in male mice," *Bull. Environm. Contam. Toxicol.* **21**, 89–92, 1979.
1326. Seiler, J. P., *Mutation Res.* **46**, 305, 1977.
1327. David T. Williams, and Frank M. Benoit, "The determination of polychlorinated biphenyls in selected household products," *Bull. Environm. Contam. Toxicol.* **21**, 179–184, 1979.
1328. *Science*, **203**, 559, 1979.
1329. M. H. Weeks et al., "The toxicity of hexachloroethane in laboratory animals," *AIHAJ* **40**(3), 1979.
1330. Laurel E. Kane et al., "A short-term test to predict acceptable levels of exposure to airborne sensory irritants," *AIHAJ* **40**(3), 1979.
1331. Merrill D. Jackson, "Volatilization of methylparathion from fields treated with microencapsulated and emulsifiable concentrate formulations," *Bull. Environm. Contam. Toxicol.* **21**, 202–205, 1979.
1332. C. H. Schaefer et al., "The accumulation and elimination of diflubenzuron by fish," *Bull. Environm. Contam. Toxicol.* **21**, 249–254, 1979.
1333. M. Shariat et al., "Screening of common bacteria capable of demethylation of methylmercuric chloride," *Bull. Environm. Contam. Toxicol.* **21**, 255–261, 1979.
1334. Rebecca L. Rawls, "Dow finds support, doubt for dioxin ideas," *Chem. Eng. News.*, Febr. 12, 1979.
1335. J. van der Laan, "Onderzoek naar de verwijdering van trichlooretheen uit water door intensieve beluchting en door filtratie over geactiveerde kool," H_2O **12**(7), 141–145, 1979.
1336. R. C. Gupta, and B. S. Paul, "'No-effect-level' of malathion (*o,o'*-dimethyldithiophosphate of diethyl mercaptosuccinate) in *Bubalus bubalis*," *Bull. Environm. Contam. Toxicol.* **20**, 819–825, 1978.
1337. FAO/WHO, "Evaluation of the toxicity of pesticides residues in Food," FAO Meeting Rep. No. PL/1965/10, WHO/Food Add/26, 1965.
1338. R. Dean Blevins, "Organochlorine pesticides in gamebirds of eastern Tennessee," *Water, Air and Soil Poll.* **11**, 1979.
1339. Roales, R. R., and Perlmutter, A., "Toxicity of methylmercury and copper, applied singly and jointly to the blue gourami, *Trichogaster trichopterus*," *Bull. Environm. Contam. Toxicol.* **12**, 633, 1974.
1340. Bhattacharya, S., et al., "Toxic effects of endrin on hepatopancreas of the teleost fish, *Clarius Batrachus* (Linn.)," *Indian J. Exp. Biol.*, **13**, 185, 1975.
1342. Environmental Protection Agency, Draft report for Congress, "Preliminary assessment of suspected carcinogens in drinking water," Office of Toxic Substances, Washington, D.C., October 17, 1975.
1343. Environmental Protection Agency, "Sampling and analysis of selected toxic substances, Task II–Ethylene dibromide," Final report, Office of Toxic Substances, EPA, Washington, D.C., Sept. 1975.
1344. Anon., "Ethylene dibromide 'ubiquitous' in air, EPA report says, *Toxic Mater. News*, **3**, 12, 1976.

Note: Ref. 1341 is not cited in the text.

1345. Anon., "EDB presents close to 100% cancer risk for citrus fumigators, EPA finds," *Pesticide Chem. News*, 5(46), 3-4, 1977.
1346. M. J. Prival et al., "Tris(2,3-dibromopropyl)phosphate: Mutagenicity of a widely used flame retardant," *Science*, 195, 76-78, 1977.
1347. A. Blum, and B. N. Ames, "Flame-retardant additives as possible cancer hazards," *Science*, 195, 17-23, 1977.
1348. C. Maltoni, "Up to date conclusions and comments on the long-term carcinogenicity bioassays on dichloroethane, performed at the Tumor Center and Institute of Oncology of Bologna, Italy," Report to the EEC, Bologna, May 8, 1978.
1349. R. E. Albert et al., "Mortality Patterns among workers exposed to chloromethyl ethers," *Environm. Health Persp.* 11, 209-214, 1975.
1350. W. G. Figueroa, R. Raszkomski, and W. Weiss, "Lung cancer in chloromethylmethylether workers," *New Engl. J. Med.* 288, 1094-1096, 1973.
1351. Josef Seifert, Ctibor Blattny, Helena Henzlerova, and Jiri Davidek, "Persistence of fungicide euparen on strawberry and/or in some canned products of strawberry," *Bull. Environm. Contam. Toxicol.* 20, 702-706, 1978.
1352. S. Kar, and P. K. Singh, "Toxicity of carbofuran to blue-green alga *Nostoc uniscorum*," *Bull. Environm. Contam. Toxicol.* 20, 707-714, 1978.
1353. Samuel S. Epstein, "Kepone—Hazard evaluation," *The Science of the Total Environment* 9, 1-62, 1978.
1354. P. A. Butler, "Pesticide-wildlife studies: A review of fish and wildlife service investigations during 1961 and 1962," Fish and Wildlife Service, Circ. 169, 1963, pp. 11-25.
1356. Bionomics Inc., Laboratory for Environmental Research Wareham, MA, "Acute toxicity of kepone to bluegill and rainbow trout," unpublished, Allied Contract Report, Feb. 1974.
1357. Bionomics Inc., Marine Research Laboratory, Pensacola, FL, "Acute toxicity of kepone to fiddler crabs," unpublished, Allied Contract Report, April 1975.
1358. J. B. de Witt, and J. L. George, U.S. Dept. Interior, Bur. Sport Fisheries and Wildlife Circ. 84, Pesticide-Wildlife Review, 1959 (pub. 1960).
1359. M. Sherman, and E. Ross, *Toxicol. Appl. Pharmacol.*, 3, 521, 1961.
1360. J. B. de Witt et al., "Effects on wildlife," in U.S. Dept. of Interior, Bur. Sport Fisheries and Wildlife Circ. 143, "Effects of pesticides on fish and wildlife in 1960, 1962."
1361. United States Testing Co., Inc., Report of Test Number 4299, Feb. 28, 1949.
1362. Environment Protection Agency (EPA), The Carcinogen Assessment Group," Analysis of kepone," July 1976.
1363. W. A. Knapp, Allied Chemical, memorandum to I. Swisher, "Depone biodegradability," May 1970.
1364. J. C. Tou, L. B. Westover, and L. F. Sonnabend, *J. Phys. Chem.*, 78, 1096, 1974.
1365. Lawrence Fishbein, "Potential halogenated industrial carcinogenic and mutagenic chemicals. III. Alkane halides, alkanols and ethers," *The Science of the Total Environment* 11, 223-257, 1979.
1366. L. Collier, "Determination of BH-chloromethylether at the ppb level in air samples by H164 resolution mass spectrocopy," *Environm. Sci. and Technol.* 6, 930, 1972.
1367. M. Alverez, and R. T. Rosen, "Formation and decomposition of *bis*(chloroethyl)ether in aqueous solution," *Int. J. Environm. Anal. Chem.*, 4, 241-246, 1976.
1368. Rijncommissie Waterleidingsbedrijven, "De samenstelling van het Maaswater in 1976," Sekretariaat KIWA, Pb. 8169, 1005 AD Amsterdam, April 1979.
1369. James T. Stevens et al., "The acute inhalation toxicity of the technical grade organo-arsenical herbicides, cacodylic acid and disodium methanearsonic acid: A route comparison," *Bull. Environm. Contam. Toxicol.* 21, 304-311, 1979.
1370. Wilhelm P. Schoor, "Distribution of mirex in an experimental estuarine ecosystem," *Bull. Environm. Contam. Toxicol.* 21, 315-321, 1979.
1371. Wilhelm P. Schoor, *"Bull. Environm. Contam. Toxicol.* 12, 136, 1974.

Note: Ref. 1355 is not cited in the text.

1372a. Tagatz, M. E., et al., *Arch. Environm. Contam. Toxicol.* **3**, 371, 1975.
1372b. C. Ramakrishna et al., "Effect of benomyl and its hydrolysis products, MCB and AB, on nitrification in a flooded soil," *Bull. Environm. Contam. Toxicol.* **21**, 328-333, 1979.
1373. Helweg A., *Tidsskr. f. Planteavl.* **77**, 232, 1973.
1374. Kilgore, W. W., and E. R. White, *Bull. Environm. Contam. Toxicol.* **6**, 1, 1970.
1375. S. C. Quinlivan et al., "Sources, characteristics and treatment and disposal of industrial wastes containing hexachlorobenzene," *J. Hazardous Mater.* **1**, 343-359, 1975-77.
1376. Jos Mes, and David J. Davies, "Presence of polychlorinated biphenyl and organochlorine pesticide residues and the absence of polychlorinated terphenyls in Canadian human milk samples," *Bull. Environm. Contam. Toxicol.* **21**, 381-387, 1979.
1377. Ware, G. W., and E. E. Good, *Toxicol. Appl. Pharmacol.* **10**, 54, 1967.
1378. J. L. Wolfe et al., "Lethal and reproductive effects of dietary mirex and DDT on Oldfield mice, *Peromyscus polionotus*," *Bull. Environm. Contam. Toxicol.* **21**, 397-402, 1979.
1379. Asphalt Institute, "Asphalt hot-mix emission study," Report 75-1 (RR-75-1), March 1975.
1380. Wellcave et al., *Toxicol. Appl. Pharmacol*, **18**, 41, Jan. 1971.
1381. Smith W. M., *Yearbook Am. Iron and Steel Institute*, 163-180, 1970.
1382. Hangebranck et al., "Sources of polynuclear hydrocarbons in the atmosphere," U.S. Public Health Service, Publication No. 999-AP-33, 1967.
1383. John L. Egle, Jr., and Bethe J. Gochberg, "Respiratory retention and acute toxicity of furan," *Am. Ind. Hyg. Assoc. J.* **40**, April 1979.
1384. Newsome, J. R., V. Norman, and C. H. Keith, "Vapor phase analysis of tobacco smoke," *Tox. Sci.* **9**, 1965.
1385. D. F. Adams et al., "Preliminary measurements of biogenic sulfur-containing gas emissions from soils," *J. A.P.C.A.*, **29**(4), 380-383, April 1979.
1386. George J. Nebel, "Benzene in auto exhaust," *JAPCA*, **29**(4), 391-392, April 1979.
1387. Donavon M. Oseid, and Lloyd L. Smith Jr., "The effect of hydrogen cyanide on *Asellus communis* and *Gammarus pseudolimnaeus* and changes in their competitive response when exposed simultaneously," *Bull. Environm. Contam. Toxicol.* **21**, 439-447, 1979.
1388. D. L. Elder, S. W. Fowler, and G. G. Polikarpov, "Remobilization of sediment-associated PCB's by the worm *Nereis diversicolor*," *Bull. Environm. Contam. Toxicol.* **21**, 448-452, 1979.
1389. K. Kuwabara et al., "Levels of polychlorinated biphenyls in blood of breast-fed children whose mothers are non-occupationally exposed to PCB's," *Bull. Environm. Contam. Toxicol.* **21**, 458-462, 1979.
1390. Marja Liisa Hattula, "Toxicity of 4-chloro-o-cresol to rat: I. Light microscopy and chemical observations," *Bull. Environm. Contam. Toxicol.* **21**, 492-497, 1979.
1391. George R. Soutworth, "The role of volatilization in removing polycyclic aromatic hydrocarbons from aquatic environments," *Bull. Environm. Contam. Toxicol.* **21**, 507-514, 1979.
1392. Valanne Glooschenko et al., "Bioconcentration of chlordane by the green alga *Scenedesmus quadricauda*," *Bull. Environm. Contam. Toxicol.* **21**, 515-520, 1979.
1393. National Research Council, Canada. Rept. No. 14094, 1974.
1394. Velsicol Chem. Corp., Technical Bulletin, Aug. 1971, "Standard for technical chlordane," Velsicol Chem. Corp., 341 East Ohio Ave., Chicago, IL.
1395. Glooschenko, V., and J. N. A. Lott, *Can. J. Bot.* **55**, 2866, 1977.
1396. Sanborn, J. R., et al., *Environm. Entomol.* **5**, 533, 1976.
1397. Sid Korn, D. Adam Moles, and Stanley D. Rice, "Effects of temperature on the median tolerance limit of pink salmon and shrimp exposed to toluene, naphthalene and Cook inlet crude oil," *Bull. Environm. Contam. Toxicol.* **21**, 521-525, 1979.
1398. R. L. Spehar et al., "Toxicity and bioaccumulation of hexachlorocyclopentadiene, hexachloronorbornadiene, and heptachloronorbornene in larval and early juvenile fathead minnows, *Pimephales promelas*," *Bull. Environm. Contam. Toxicol.* **21**, 576-583, 1979.
1399. Brooks G. T., *Chlorinated Insecticides, Vol. I. Technology and Application*, CRC Press, Inc., Cleveland, Ohio, 1974.

1400. W. W. Walker et al., "Acute toxicity of 3-chloro-4-methylbenzenamine hydrochloride to shrimp and crabs," *Bull. Environm. Contam. Toxicol.* 21, 643–651, 1979.
1401. D. Alan Hansen, "Survey of benzene occurrence in ambient air at various locations throughout the United States," API publication 4305, Dec. 1978.
1402. U.S. Dept. of H. E. W., National Institutes of Health, National Cancer Institute, "Report on carcinogenesis bioassay of toxaphene," *AIHAJ* 40(5), A26–A23, 1979.
1403. H. P. Stein, N. A. Leidel, and J. M. Lane, "Glycidylethers," *AIHAJ* 40, A36–A51, May 1979.
1404. U.S. Dept. of H. E. W., Public Health Service, Center for Disease Control, NIOSH, "NIOSH criteria for a recommenaeu standard: Occupational exposure to glycidyl ethers," 1978.
1405. *Standard Industrial Classification Manual 1967*, Executive office of the President, Bureau of the Budget, Washington, 1967.
1406. Kodama, J. K., et al., "Some effects of epoxy compounds on the blood," *Arch. Environm. Health* 2, 56–57, 1961.
1407. Anderson, H. H., et al., "Chronic vapor toxicity of *n*-butylglycidyl ether," Report to Shell Development Company, Emeryville, California from Dept. of Pharmacology and Experimental Therapeutics, Univ. California School of Medicine, San Francisco, U. C. report No. 270, 1957.
1408. Hine, C. H., et al., "Effects of diglycidyl ether on blood of animals," *Arch. Environm. Health* 2, 37–50, 1961.
1409. Terrill, J. B., and H. J. Trochimowicz, "A two generation reproduction and mutagenic study in rats," E. I. Du Pont de Nemours and Co., Haskell Laboratory for Toxicology and Industrial Medecine, Haskell Laboratory Report No. 163-75, Wilmington, Delaware, April 1975.
1410. Shimkin, M. B., et al., "Bioassay of 29 alkylating chemicals by the pulmonary tumor response in strain A mice," *J. Nat. Cancer Inst.* 36, 915–935, 1966.
1411. Hendry, J. A., et al., "Cytotoxic agents. II. *bis*-Epoxides and related compounds," *Brit. J. Pharmacol.* 6, 235–255, 1951.
1412. OSHA, "Occupational safety and health general industry standards," U.S. Dept. Labor, OSHA Administration, OSHA Publication 2206 (29 CFR 1910,1000), Washington, D.C., 1976.
1413. W. F. Randall et al., "Acute toxicity of dechlorinated DDT, chlordane and lindane to bluegill (*Lepomis macrochirus*) and *Daphnia magna*," *Bull. Environm. Contam. Toxicol.* 21, 849–854, 1979.
1414. H. R. A. Scorgie, and A. S. Cooke, "Effects of the triazine herbicide cyanatryn on aquatic animals," *Bull. Environm. Contam. Toxicol.* 22, 135–142, 1979.
1415. Haddow, B. C., et al., *Proc. British Weed Control Conf., 12th, Brighton, England*, 239, 1974.
1416. D. Rose, and J. M. Lane, "Epichlorohydrin," NIOSH Current Intelligence Bulletin No. 30, *AIHAJ* 40, June 1979.
1417. U.S. Dept. of H. E. W., Public Health Service, Center for Disease Control, NIOSH, "Criteria for a recommended standard:. Occupational exposure to epichlorohydrin," Washington, D.C., 1976.
1418. James N. Dumont et al., "Toxicity and teratogenicity of aromatic amines to *Xenopus laevis*," *Bull. Environm. Contam. Toxicol.* 22, 159–166, 1979.
1419. Thomas F. Gale, "Toxic effects of cadmium and amaranth on the developing hamster embryo," *Bull. Environm. Contam. Toxicol.* 22, 175–181, 1979.
1420. W. J. Vincent et al., "Monitoring personal exposure to ethylenediamine in the occupational environment," *AIHAJ* 40, June 1979.
1421. Frank W. Judd, "Acute toxicity and effects of sublethal dietary exposure of monosodium methanearsonate herbicide to *Peromyscus leucopus* (Rodentia: Cricetidae)," *Bull. Environm. Contam. Toxicol.* 22, 143–150, 1979.
1422. Dickinson, J. D., *Am. J. Vet. Res.* 33, 1889, 1972.
1423. Exon, J. H., et al., *Nutr. Rpts. Internat.* 9, 351, 1974.
1424. S. Ramamoorthy, and D. R. Miller, "Removal of mercury and methylmercury from waste waters by sorption," *Bull. Environm. Contam. Toxicol.* 22, 196–201, 1979.

1425. Melvin Dwaine Reuber, "Carcinogenicity of endrin," *The Science of the Total Environment* **12**, 101-135, 1979.
1426. T. P. Nicholls, R. Perry, and J. N. Lester, "The influence of heat treatment on the metallic and polycyclic aromatic hydrocarbon content of sewage sludge," *The Science of the Total Environment* **12**, 137-150, 1979.
1427. NIOSH, "Recommended exposure limits for ketones," Dept. HEW (NIOSH) publication No. 78-173, June 1978.
1428. NIOSH, "Occupational exposure to Vinyl acetate," Dept. HEW (NIOSH) publication No. 78-205, Sept. 1978.
1429. Ih Chu et al., "Toxicity studies on chlorinated guaiacols in the rat," *Bull. Environm. Contam. Toxicol.* **22**, 293-296, 1979.
1430. Elio Arias, and Teresa Zavanella, "Tetratogenic effects of manganese ethylene*bis*dithiocarbamate (Maneb) on forelimb regeneration in the adult newt *Triturus cristatus carnifex*," *Bull. Environm. Contam. Toxicol.* **22**, 297-304, 1979.
1431. Paul R. Walsh, and Ronald A. Hites, "Dicofol solubility and hydrolysis in water," *Bull. Environm. Contam. Toxicol.* **22**, 305-311, 1979.
1432. Johnson R. E., *Pestic. Rev.* **61**, 1, 1976.
1433. J. R. W. Miles et al., "Persistence of eight organophosphorus insecticides in sterile and nonsterile mineral and organic soils," *Bull. Environm. Contam. Toxicol.* **22**, 312-318, 1979.
1434. Jack C. Skea et al., "Bioaccumulation of Aroclor 1016 in Hudson River fish," *Bull. Environm. Contam. Toxicol.* **22**, 332-336, 1979.
1435. Mayer F. C. et al., *Arch. Environm. Contam. and Toxicol.* **5**, 501, 1977.
1436. Nebeker, A. A., et al., *Trans. Am. Fish. Soc.* **103**, 562, 1974.
1437. Bills, T. D., et al., *Prog. Fish. Cult.* **39**, 150, 1977.
1438. Philip H. Howard, and Padmaker G. Doe, "Degradation of arylphosphates in aquatic environments," *Bull. Environm. Contam. Toxicol.* **22**, 337-344, 1979.
1439. Arrigo Collina, and Paolo Maini, "Analysis of methylisothiocyanate derived from the soil fumigant metham-sodium in workroom air," *Bull. Environm. Contam. Toxicol.* **22**, 400-404, 1979.
1440. Helmut Goerke et al., "Patterns of organochlorine residues in animals of different trophic levels from the Weser Estuary," *Marine Poll. Bull.* **10**, 127-133, 1979.
1441. ICES, "The ICES coordinated monitoring programmes, 1975 and 1976," Cooperative Research Report No. 72, International Council for the Exploration of the Sea, Charlottenlund, Denmark, 1977.
1442. ICES, "The ICES coordinated monitoring programmes in the North Sea, 1974," Cooperative Research Report No. 58, International Council for the Exploration of the Sea, Charlottenlund, Denmark, 1977.
1443. Torgeir Bakke, and Hein Rune Skjoldal, "Effects of toluene on the survival, respiration and adenylate system of a marine isopod," *Marine Poll. Bull.* **10**(4), April 1979.
1444. WHO-EHE, "Polychlorinated biphenyls and terphenyls," Environmental Health Criteria 2, UNEP, WHO, Geneva, 1976.
1445. Olsson, M., Jensen, S. and Renberg, L., "PCB in coastal areas of the Baltic," in: *PCB Conference II*, National Swedish Environmental Protection Board, Stockholm, Sweden, 1973.
1446. CEC, "Study of the contamination of continental fauna by organochlorine pesticides and PCB's," Environment and quality of Life, Health and Safety Directorate, Commission of the European Communities, Luxembourg, 1978.
1447. NAS, "Drinking water and health," National Academy of Science, Washington, D.C. 1977.
1448. Klimmer, O. R., "Nutritional physiology, analytical and toxicological studies with the fungicide triphenyltin acetate," *Zentralbl. Veterinärmed.*, **11**, 29-37, 1964.
1449. NIOSH, "Criteria for a recommended standard: Occupational exposure to organotin compounds," National Institute for Occupational Safety and Health, Washington, D.C., 1976.
1450. Lewis, P. J., Sr. and Tatken, R. L., "Registry of toxic effects of chemical substances," NIOSH, 1978.

1451. Pate, B. D., and Hays, R. L., "Histological studies of testes in rats treated with certain insect chemosterilants," *J. Econ. Entomol.* **61**, 1968.
1452. Wolfe, N. L. et al., "Kinetic investigation of malathion degradation in water," *Bull. Environm. Contam. Toxicol.* **13**, 707–713, 1975.
1453. Bourquin, A. W., "Microbial malathion interaction in artificial salt-march ecosystems," EPA-600/3-75-035, US EPA, office of Research and Development, Cornvallis, Oregon, 1975.
1454. Cook, G. H. and Moore, J. C., "Determination of malathion, malaoxon and mono- and dicarboxylic acids of malathion in fish, oyster and shrimp tissue," *J. Agric. Food Chem.*, **24**, 631–634, 1976.
1455. A. Seidell, *Solubility of Organic Compounds*, Vol. II, 34d ed., Van Nostrand, Princeton, N.J., 1941.
1456. F. Helmer, K. Kiehs, and C. Hansch, *Biochemistry* **7**, 2858, 1968.
1457. Albert Leo, Corwin Hansch, and David Elkins, "Partition coefficients and their uses," *Chemical Review* **71**(6), Dec. 1971.
1458. W. Strumm, and D. Schwarzenbach, "Die Schadstoffe in unserer Umwelt und ihre Auswirkungen auf Ökologie, Mensch und Tier," in: *Arbeitstagung 1979*, IAWR, Postfach 8169 NL-1005 AD Amsterdam, The Netherlands.
1459. Davies, J. E., et al., *Arch. Environ. Health*, **30**, 608, 1975.
1460. Marja Liisa Hattula et al., "Toxicity of 5-chloro-3-methyl-catechol to rat: Chemical observations and light microscopy of the tissue," *Bull. Environm. Contam. Toxicol.* **22**, 457–461, 1979.
1461. Ih Chu, David D. Villeneuve, Viateur Secours, and André Viau, "Absorption, distribution and metabolism of epoxystearic acid in the rat," *Bull. Environm. Contam. Toxicol.* **22**, 462–466, 1979.
1462. Christina Rosenberg and Hilkka Siltanen, "Residues of mancozeb and ethylenethiourea in grain samples," *Bull. Environm. Contam. Toxicol.* **22**, 475–478, 1979.
1463. Marja Liisa Hattula et al., "Toxicity of 4-chloro-*o*-cresol to fish. Light microscopy and chemical analysis of the tissue," *Bull. Environm. Contam. Toxicol.* **22**, 508–511, 1979.
1464. Environmental Protection Agency, "Toxic substances control—Discussion of premanufacture—Testing policy and technical issues; request for comment," *Federal Register*, 16251-16292 March 16, 1979.
1465. Räsänen, L., et al., *Bull. Environm. Contam. Toxicol.* **18**, 5, 1977.
1466. K. S. Khera et al., "Teratogenicity studies on pesticidal formulations of dimethoate, diuron and lindane in rats," *Bull. Environm. Contam. Toxicol.* **22**, 522-529, 1979.
1467. Edward D. John, Michael Cooke, and Graham Nickless, "Polycyclic aromatic hydrocarbon in sediments taken from the Severn Estuary drainage system," *Bull. Environm. Contam. Toxicol.* **22**, 653–659, 1979.
1468. R. Engst et al., "Interim results of studies of microbial isomerization of gamma-hexachlorocyclohexane," *Bull. Environm. Contam. Toxicol.* **22**, 699–707, 1979.
1469. Mark Hite et al., "Acute toxicity of methylfluorosulfonate (Magic methyl)," *AIHAJ* **40**, July 1979.
1470. Tsai, S. C., "Control of chironomids in milk-fish (*Chanos chanos*) ponds with abate (*Temephos*) insecticide," *Trans. Amer. Fish Soc.*, **107**, 493, 1978.
1471. Schimmel, S. C., et al., "Toxicity and bioconcentration of BHC and lindane in selected estuarine animals," *Arch. Environm. Contam. Toxicol.*, **6**, 355, 1977.
1472. Parrish, P. R., et al., "Chronic toxicity of chlordane, trifluralin, and pentachlorophenol to sheepshead minnows (*Cyprinodon variegatus*)," U.S. EPA Eco. Res. Ser., EPA-600/9-78-010, 121, 1978.
1473. Borthwick, P. W., and Schimmel, S. C., "Toxicity of pentachlorophenol and related compounds to early life stages of selected estuarine animals," in: *Pentachlorophenol*, K. R. Rao (Ed.), Plenum Press, 1978.
1474. Green, F. A., and Neff, J. M., "Toxicity, accumulation, and release of three polychlorinated

napthalenes (Halowax 1000, 1013, and 1099) in postlarval and adult grass shrimp, *Palaemonetes pugio*," *Bull. Environm. Contam. Toxicol.*, 17, 399, 1977.
1475. Shimmel, S. C., et al., "Effects of sodium pentachlorophenate on several estuarine animals: Toxicity, uptake, and depuration," in: *Pentachlorophenol*, K. R. Rao (Ed.), Plenum Press, 147, 1978.
1476. Goodman, L. R., et al., "Effects of heptachlor and toxaphene on laboratory-reared embryos and fry of the sheepshead minnows," *Proc. 30th Annual Conf. Southeast. Assoc. Game Fish Comm.*, 192, 1976.
1477. Ernst, W., "Determination of the bioconcentration potential of marine organisms—A steady state approach," *Chemosphere*, 6, 731, 1977.
1478. Emanuelsen, M., et al., "The residue uptake and histology of American oysters (*Crassostrea virginica*, Gemlin) exposed to dieldrin," *Bull. Environm. Contam. Toxicol.*, 19, 121, 1978.
1479. Zitko, V., "Nonachlor and chlordane in aquatic fauna," *Chemosphere*, 1, 3, 1978.
1480. Fowler, S. W., and Elder, D. L., "PCB and DDT residues in a Mediterranean pelagic food chain," *Bull. Environ. Contam. Toxicol.*, 19, 244, 1978.
1481. Wharfe, J. R., and Van Den Broek, W. L. F., "Chlorinated hydrocarbons in macroinvertebrates and fish from the lower Medway Estuary, Kent," *Marine Poll. Bull.* (G. B.), 9, 76, 1978.
1482. Drescher, H. E., et al., "Organochlorines and heavy metals in the harbour seal *Phoca vitulina* from the German North Sea coast," *Marine Biol.* (W. Ger.), 41, 99, 1977.
1483. Cheng, L., and Bidleman, T. F., "Chlorinated hydrocarbons in marine insects," *Estuarine Coastal Marine Sci.*, 5, 289, 1977.
1484. Sims, G. G., et al., "Organochlorine residues in fish and fishery products from the Northwest Atlantic," *Bull. Environm. Contam. Toxicol.*, 18, 697, 1977.
1485. Giam, C. S., et al., "Phthalate ester plasticizers: A new class of marine pollutant," *Science*, 199, 419, 1978.
1486. Lipka, E., and Doboszynska, B., "Studies on DDT appearance in Baltic Sea fish. Part I. DDT level in fresh fish," *Bromat. Chem. Toksykol.* (Pol.), 11, 167, 1978.
1487. McDermott-Ehrlich, D. J., et al., "Chlorinated hydrocarbons in Dover sole, *Microstomus pacificus*: Local migrations and fin erosion," *U.S. Natl. Marine Service Fish. Bull.*, 25, 513, 1977.
1488. Addison, R. F., and Brodie, P. F., "Organochlorine residues in maternal blubber, milk, and pup blubber from grey seals (*Halichoerus grypus*) from Sable Island, Nova Scotia," *J. Fish. Res. Bd. Can.*, 34, 937, 1977.
1489. Lipka, E., et al., "Studies on DDT appearance in Baltic Sea fish. Part II. Determination of DDT and its metabolites in cod liver and cod liver products," *Bromat Chem. Toksykol.* (Pol.), 11, 171, 1978.
1490. Marchetti, R., "Acute toxicity of alkyl leads to some marine organisms," *Marine Poll. Bull.* (G. B.), 9, 206, 1978.
1491. Wilson, W. B., et al., "The toxicity of phthalates to the marine dinoflagellate *Gymnodinium breve*," *Bull. Environm. Contam. Toxicol.*, 20, 149, 1978.
1492. Cunningham, P. A., and Grosch, D. S., "A comparative study of the effects of mercuric chloride and methyl mercury chloride on reproductive performance in the brine shrimp, *Artemia salina*," *Environm. Poll.*, 15, 83, 1978.
1493. Inman, C. B. E., and Lockwood, A. P. M., "Some effects of methylmercury and lindane on sodium regulation in the amphipod *Gammarus duebeni* during changes in the salinity of its medium," *Comp. Biochem. Physiol.*, 58C, 67, 1977.
1494. Verma, S. R., et al., "Biocides in relation to water pollution. Part 2. Bioassay studies of few biocides to a fresh water fish, *Channa gachua*," *Acta Hydrochim. Hydrobiol.* (Ind.), 6, 137, 1978.
1495. Leach, J. M., and Thakore, A. N., "Compounds toxic to fish in pulp mill waste streams," *Progr. Water Technol.* (G. B.), 9, 787, 1978.

1496. Majewski, H. S., et al., "Acute lethality and sublethal effects of acetone, ethanol, and propylene glycol on cardiovascular and respiratory systems of rainbow trout *(Salmo gairdneri),*" *Water Res.* (G. B.), **12**, 217, 1978.
1497. Slonim, A. R., "Acute toxicity of Selected Hydrazines to common guppy," *Water Res.* (G. B.), **11**, 889, 1977.
1498. Llewellyn, G. C., et al., "Aflatoxin B(1) induced toxicity and teratogenicity in Japanese medaka eggs *(Oryzias latipes),*" *Toxicon*, **15**, 582, 1977.
1499. Stephenson, G. A., et al., "Toxic and teratogenic effects of aflatoxin B(1) in Japanese medaka eggs," *Va. Jour. Sci.*, **28**, 67, 1977; *Sport Fish. Abs.*, **23**, 78-001157, 1978.
1500. Dave, G., and Lidman, U., "Biological and toxicological effects of solvent extraction chemicals: Range finding acute toxicity in rainbow trout *(Salmo gairdneri* Rich) and in rat *(Rattus norwegicus* L.)," *Hydrometallurgy* (Neth.), **3**, 201, 1978.
1501. McKague, A. B., and Pridmore, R. B., "Toxicity of altisod and dimilin to juvenile rainbow trout and coho salmon," *Bull. Environm. Contam. Toxicol.*, **20**, 167, 1978.
1502. Mauck, W. L., et al., "Effects of water quality on deactivation and toxicity of mexacarbate (zectran) to fish," *Arch. Environm. Contam. Toxicol.*, **6**, 385, 1977.
1503. Rubin, A. J., and Elmaraghy, G. A., "Studies on toxicity of ammonia, nitrate, and their mixtures to guppy fry," *Water Res.* (G. B.), **11**, 927, 1977.
1504. Thurston, R. V., et al., "Acute toxicity of ammonia and nitrate to cutthroat trout fry," *Trans. Amer. Fish. Soc.*, **107**, 361, 1978.
1505. Buckley, J. A., "Acute toxicity of un-ionized ammonia to fingerling coho salmon," *Progressive Fish-Culturist*, **40**, 30, 1978.
1506. Dawson, V. K., et al., "Efficacy of 3-trifluoromethyl-4-nitrophenol (TFM), 2',5-dichlor-4'-nitrosalicylanilide (Bayer 73), and a 98:2 mixture as lampricides in laboratory studies," Investigations in Fish Control No. 77, U.S. Dept. Int., Fish & Wildlife Serv., Washington, D.C., 1977.
1507. Kühnhold, W. W., and Busch, F., "Uptake of three different types of hydrocarbons by salmon eggs *(Salmo salar* L.)," *Meeresforsch.* (Ger.), **26**, 50, 1978.
1508. Saxena, P. K., and Garg, M., "Effects of insecticidal pollution on ovarian recrudescence in fresh water teleost *Channa punctatus* (B1)," *Indian J. Exp. Biol*, **16**, 689, 1978.
1509. Verma, S. R., et al., "Toxicity of selected organic pesticides to a freshwater teleost fish, *Saccobranchus fossilis,* and its application in controlling water pollution," *Arch. Environm. Contam. Toxicol.*, **7**, 317, 1978.
1510. Julin, A. M., and Sanders, H. O., "Toxicity of IGR, diflubenzuron to freshwater invertebrates and fishes," *Mosquito News*, **38**, 256, 1978.
1511. Krzeminski, S., et al., "A pharmacokinetic model for predicting pesticides residues in fish," *Arch. Environm. Contam. Toxicol.*, **5**, 157, 1977.
1512. Smith, L. L., et al., "Acute toxicity of hydrogen cyanide to freshwater fishes," *Arch. Environm. Contam. Toxicol.*, **7**, 325, 1978.
1513. Kam-Wing, L., and Furtado, J. I., "The chemical control of *Salvinia molesta* (Mitchell) and some related toxicological studies," *Hydrobiologia* (Den.), **56**, 49, 1977.
1514. Mulla, M. S., et al., "Toxicity of mosquito larvicidal pyrethroids to four species of freshwater fishes," *Environm. Entomol.*, **7**, 428, 1978.
1515. Allison, D. T., and Hermanutz, R. O., "Toxicity of diazinon to brook trout and fathead minnows," EPA-600/3-77-060, U.S. Environmental Protection Agency, Duluth, Minn., 1977.
1516. Chambers, H., et al., "Hydrolytic activation and detoxication of 2,4-dichlorophenoxyacetic acid esters in mosquitofish," *Pesticide Biochem. Physiol.*, **7**, 297, 1977.
1517. Grieco, M. P., et al., "Carcinogenicity and acute toxicity of dimethylnitrosamine in rainbow trout *(Salmo gairdneri),*" *J. Natl. Cancer Inst.*, **60**, 1127, 1978.
1518. Hermanutz, R. O., "Endrin and malathion toxicity to flagfish *(Jordanella floridae),*" *Arch. Environm. Contam. Toxicol.*, **7**, 159, 1978.

1519. Dalela, R. C., et al., "Biocides in Relation to water pollution. Part 1. Bioassay studies on the effects of a few biocides on fresh water fishes." (in preparation).
1520. Wohlgemuth, E., "Toxicity of endrin to Some species of aquatic vertebrates," *Prirodoved Pr. Ustavu Cesk. Akad. Ved Brne* (Czech.), **11**, 1, 1977.
1521. Nevins, M. J., and Johnson, W. W., "Acute toxicity of phosphate ester mixtures to invertebrates and fish," *Bull. Environm. Contam. Toxicol.*, **19**, 250, 1978.
1522. Hattula, M. L., et al., "The toxicity of MCPA to fish, light and electron microscopy and the chemical analysis of the tissue," *Bull. Environm. Contam. Toxicol.*, **19**, 465, 1978.
1523. Jenkins, D., et al., "Fish toxicity of jet fuels I–The toxicity of the synthetic fuel JP-9 and its components," *Water Res.* (G. B.), **11**, 1059, 1977.
1524. Alexander, H. C., et al., "Toxicity of perchloroethylene, trichloroethylene, 1,1,1-trichloroethane, and methylene chloride to fathead minnows," *Bull. Environm. Contam. Toxicol.*, **20**, 344, 1978.
1525. Chan, K. K., "Chronic effects of methylmercury on the reproduction of the teleost fish, *Oryzias latipes*," *Dissertation Abs.*, **39**, 1978.
1526. Melancon, M. J., Jr., and Lech, J. J., "Distribution and elimination of naphthalene and 2-methylnaphthalene in rainbow trout during short- and long-term exposures," *Arch. Environm. Contam. Toxicol.*, **7**, 207, 1978.
1527. Kawatsu, H., "Studies on the anemia of fish VIII–Hemorrhagic anemia of carp caused by a herbicide, molinate," *Bull. Jap. Soc. Sci. Fish.*, **43**, 905, 1977.
1528. Gingerich, W. H., and Dalich, G. M., "Evaluation of liver toxicity in rainbow trout following treatment with monochlorobenzene," *Proc. West. Pharmacol. Soc.*, **21**, 475, 1978.
1529. Strange, R. J., and Schreck, C. B., "Anesthetic and handling stress on survival and cortisol concentration in yearling Chinook salmon (*Oncorhynchus tshawytscha*)," *J. Fish. Res. Bd. Can.*, **35**, 345, 1978.
1530. DeFoe, D. L., et al., "Effects of Aroclor 1248 and 1260 on fathead minnow (*Pimephales promelas*)," *J. Fish. Res. Bd. Can.*, **35**, 997, 1978.
1531. Mauck, W. L., et al., "Effects of Polychlorinated biphenyl Aroclor 1254 on growth, survival, and bone development in brook trout (*Salvelinus fontinalis*)," *J. Fish. Res. Bd. Can.*, **35**, 1084, 1978.
1532. Zitko, V., and Carson, W. G., "Uptake and excretion of chlorinated biphenyl ethers and brominated toluenes by fish," *Chemosphere*, **6**, 293, 1977.
1533. Cairns, J., Jr., et al., "Effects of temperature on aquatic organism sensitivity to selected chemicals," Bulletin No. 106, Virginia Water Resources Res. Ctr., Blacksburg, Va., 1978.
1534. Jolly, A. L., et al., "Effects of a new insecticide on aquatic animals," *Louisiana Agric.*, **21**, 1978.
1535. Sitthichaikasem, S., "Some toxicological effects of phosphate esters on rainbow trout and bluegill," *Dissertation Abs.*, **39**, 7813246, 1978.
1536. Broderius, S. L., et al., "Relative toxicity of free cyanide and dissolved sulfide forms to the fathead minnow (*Pimephales promelas*)," *J. Fish. Res. Bd. Can.*, **34**, 2323, 1977.
1537. Hawkes, C. L., and Norris, L. A., "Chronic oral toxicity of 2,3,7,8-tetrachlorodibenzo-*p*-dioxin (TCDD) to rainbow trout," *Trans. Amer. Fish. Soc.*, **106**, 641, 1977.
1538. Yockim, R. S., et al., "Distribution and toxicity of TCDD and 2,4,5-T in an aquatic model ecosystem," *Chemosphere*, **7**, 215, 1978.
1539. Mehrle, P. M., and Mayer, F. L., "Bone development and growth of fish as affected by toxaphene," in: *Fate of Pollutants in Air and Water Environments–Part 2*, I. H. Suffet (Ed.), 301, Wiley-Interscience, New York, N.Y., 1977.
1540. Mayer, F. L., Jr., et al., "Toxaphene: Chronic toxicity to fathead minnows and Channel catfish," EPA-600/3-77-069, U.S. EPA, 1977.
1541. Mayer, F. L., and Mehrle, P. M., "Toxicological aspects of toxaphene in fish: A summary," in: *Transactions of the 42nd North American Wildlife and Natural Resources Conference*, 365, Wildlife Management Institute, Washington, D.C., 1977.

1542. Yokote, M., et al., "Effects of some herbicides applied in the forest to the freshwater fishes and other aquatic organisms. IV–Experiments on the assessment of acute and subacute toxicites of 2,4,5-T to the rainbow trout," *Bull. Freshwater fish. Res. Lab.* (Jap.), **26**, 85, 1977.
1543. Chliamovitch, Y. -P., and Kuhn, C., "Behavioural, haematological and histological studies on acute toxicity of BIS (tri-n-butyltin) oxide on *Salmo gairdneri* Richardson and *Tilapia rendalli* Boulenger," *J. Fish Biol.*, **10**, 575, 1977.
1544. Matthiessen, P., "Uptake metabolism and excretion of the molluscicide N-tritylmorpholine by the tropical food fish (*Sarotherodon mossambicus-Peters*)." *J. Fish. Biol.*, **11**, 497, 1977.
1545. M. E. Stewart et al., "By-products of oxidative biocides: Toxicity to oyster larvae," *Marine Poll. Bull.* **10**, 166–169.
1546. Dow Chemical, "Material Safety Data Sheet," 1978.
1547. H. Lizuka, and T. Masuda, "Residual fate of chlorphenamidine in rice plant and paddy soil," *Bull. Environm. Contam. Toxicol.* **22**, 745–749, 1979.
1548. Ehrhardt, D. A., and C. O. Knowles, *J. Econ. Entomol.* **63**, 1306, 1970.
1549. C. O. Knowles, *J. Agr. Food Chem.* **18**, 1038, 1970.
1550. S. Gupta A. K., and C. O. Knowles, *J. Agr. Food Chem.* **17**, 593, 1969.
1551. W. H. Dennis et al., "Catalytic dechlorination of organochlorine compounds. V. Polychlorinated biphenyls–Aroclor 1254," *Bull. Environm. Contam. Toxicol.* **22**, 750–753, 1979.
1552. Berg, O. W., P. L. Drosody, and G. A. V. Pees, *Bull. Environm. Contam. Toxicol* **7**, 338, 1972.
1553. Lapierre, R. B., EPA Report No. 60013-77-018, Jan. 1977.
1554. D. R. Nimmo et al., "Effect of diflubenzuron on an estuarine crustacean," *Bull. Environm. Contam. Toxicol.* **22**, 767–770, 1979.
1555. Miura, T., and R. M. Takahashi, *Environm. Entomol.* **3**, 631, 1974.
1556. Cunningham, P. A., *Environm. Entomol.* **5**, 701, 1976.
1557. D. W. McLeese, and C. D. Metcalfe, "Toxicity of creosote to larval and adult lobster and crangon and its accumulation in lobster hepatopancreas," *Bull. Environm. Contam. Toxicol.* **22**, 796–799, 1979.
1558. D. W. McLeese, V. Zitko, and D. B. Sergeant, "Uptake and excretion of fenitrothion by clams and mussels," *Bull. Environm. Contam. Toxicol.* **22**, 800–806, 1979.
1559. J. R. W. Miles, and P. May, "Degradation of endosulfan and its metabolites by a mixed culture of soil micro-organisms," *Bull. Environm. Contam. Toxicol.* **23**, 613–619, 1979.
1560. Archer, T. E., et al., *J. Agric. Food Chem.* **20**, 954, 1972.
1561. Martens, R., *Appl. and Environ. Microbiol.* **31**, 853, 1976.
1562. Martens, R., *Bull. Environm. Contam. Toxicol.* **17**, 438, 1977.
1563. A. J. Niimi, "Hexachlorobenzene (HCB) levels in Lake Ontario salmonids," *Bull. Environm. Contam. Toxicol.* **23**, 20–24, 1979.
1564. Norström, R. J., et al., *J. Fish Res. Board Can.* **35**, 1401, 1978.
1565. Laska, A. L., et al., *Bull. Environm. Contam. Toxicol.* **15**, 535, 1976.
1566. M. A. Cole, P. B: Reichart, and D. K. Button, "Reversible bioconcentration of monochlorobiphenyls by *Rhodotorula rubra*: Correlations with aqueous solubility of substrate," *Bull. Environm. Contam. Toxicol.* **23**, 44–50, 1979.
1567. Urey, J. C., J. C. Kricher, and J. M. Boylan, *Bull. Environm. Contam. Toxicol.* **16**, 81, 1976.
1568. Cox, J. L., *Bull. Environm. Contam. Toxicol.* **5**, 218, 1970.
1569. M. Feroz, and M. A. Q. Khan, "Fate of ^{14}C-*cis*-chlordane in goldfish, *Carassius auratus* (L.)," *Bull. Environm. Contam. Toxicol.* **23**, 064–069, 1979.
1570. G. R. Southworth, J. J. Beauchamp, and P. K. Schmieder, "Bioaccumulation of carbazoles: A potential effluent from synthetic fuels," *Bull. Environm. Contam. Toxicol.* **23**, 073–078, 1979.

1571. Russell J. Hall et al., "Organochlorine residues in eggs of the endangered American crocodile (*Crocodylus acutus*)," *Bull. Environm. Contam. Toxicol.* 23, 087–090, 1979.
1572. Ogden J. C. et al., South Florida Environ. Proj., Ecol. Rept. No. D1-SFEP-74-16, 1973.
1573. Ralph G. Stahl Jr., "Effect of a PCB (Aroclor 1254) on the striped hermit crab, *Clibanarius vittatus* (Anomura: Diogenidea) in static bioassays," *Bull. Environm. Contam. Toxicol.* 23, 091–094, 1979.
1574. Tagatz, M. E., et al., *Arch. Environ. Contam. Toxicol.* 4, 435, 1977.
1575. Eisler, R., *Crustaceana* 16, 302, 1969.
1576. S. K. Sharma et al., "Acute endrin toxicity on oxidases of *Ophiocephalus punctatus* (Bloch)," *Bull. Environm. Contam. Toxicol.* 23, 153–157, 1979.
1577. J. R. Coats, and N. L. O'Donnell-Jeffery, "Toxicity of four synthetic pyrethroid insecticides to rainbow trout," *Bull. Environm. Contam. Toxicol.* 23, 250–255, 1979.
1578. D. F. Berard, and D. P. Rainey, "Dissipation of ^{14}C-N-nitroso-di-N-propylamine from field soil and residue determinations in field-grown soybeans," *Bull. Environm. Contam. Toxicol.* 23, 141–144, 1979.
1579. Ross R. E., et al., *J. Agric. Food Chem.* 25, 1416, 1977.
1580. West, S. D., and E. W. Day Jr., *Abstr. Am. Chem. Soc. Nat. Meeting*, March 1978, Anaheim, Calif.
1581. Jensen S. et al., *Nature* 240, 358, 1972.
1582. Linko R. R., et al., *Bull. Environm. Contam. Toxicol.* 12, 733, 1974.
1583. R. R. Linko, et al., "Polychlorinated biphenyls in plankton from the Turku Archipelago (Finland)," *Bull. Environm. Contam. Toxicol.* 23, 145–152, 1979.
1584. Reisebrough, R. W., et al., *Bull. Environm. Contam. Toxicol.* 8, 345, 1972.
1585. Williams, R., and A. V. Holden, *Mar. Poll. Bull.* 4, 109, 1973.
1586. Holden, A. V., *PCB Conference II*, Stockholm 1972, National Swedish Environmental Protection Board Publications 4E, 23, 1973.
1587. Clayton, J. R., Jr., et al., *Environm. Sci. Techn.* 11, 676, 1977.
1588. Särkkä et al., *Environ. Pollut.* 16, 41, 1978.
1589. *Biochem. Biophys. Res. Commun.* 49, 364, 1972.
1590. Gessner G. Hauley, *The Condensed Chemical Dictionary*, Van Nostrand Reinhold Company, 8th edition.
1591. E. P. Meier, et al., "Sulfotepp, a toxic impurity in formulations of diazinon," *Bull. Environm. Contam. Toxicol.* 23, 158–164, 1979.
1592. Meier, E. P., et al., "Chemical degradation of military standard formulations of organophosphate and carbamate pesticides. I. Chemical hydrolysis of diazinon," U.S. Army Medical Bio-engineering Laboratory, Technical Report 7611, AD No. A036051, Fort Detrick, MD, Nov. 1976.
1593. Faust, S. D., and H. M. Gomma, *Environ. Letters* 3, 171, 1972.
1594. John D. Garcia, and Mark Rhodes, "Residues of 2,4,5-T in the American coot (*fulcia americana*)," *Bull. Environm. Contam. Toxicol.* 23, 231–235, 1979.
1595. R. P. Moody, et al., "The fate of fenitrothion in an aquatic ecosystem," *Bull. Environm. Contam. Toxicol.* 19(1), 8–14, 1978.
1596. Cope O., *Pesticide Wildlife Studies*, U.S. Dept. Int., Fish and Wildl. Serv. Circ. 199, 1963.
1597. Stephen E. Herbes, and George F. Risi. "Metabolic alteration and excretion of anthracene by *Daphnia pulex*," *Bull. Environm. Contam. Toxicol.* 19(2), 147–155, 1978.
1598. Hilkka Siltanen, and Christina Rosenberg, "Analysis of 2,4-D and 2,4,5-T in lingonberries, wild mushrooms, birch and aspen foliage," *Bull. Environm. Contam. Toxicol.* 19(2), 177–182, 1978.
1599. Harvey, G. R., and W. G. Steinhauer, "Biogeochemistry of PCB and DDT in the North Atlantic," in: *Environmental Biogeochemistry* (J. O. Nriagu, ed.) Ann Arbor Science Pub., 1976.
1600. H. G. Langer, and R. P. Brady, "Formation of dioxins and other condensation pro-

ducts from chlorinated phenols by differential thermal analysis," in: *Thermal Analysis, Vol. 2, Proceedings Fourth ICTA, Budapest 1974*, Akademiai Kiado, Budapest, 1975.
1601. Melvin D. Reuber, "Carcinogenicity of lindane," *Environm. Research* 19, 460-481, 1979.
1602. Scott W. Fowler, and D. L. Elder, "PCB and DDT residues in a Mediterranean pelagic food chain," *Bull. Environm. Contam. Toxicol.* 19(2), 244-249, 1978.
1603. Michael J. Nevins, and W. Waynon Johnson, "Acute toxicity of phosphate ester mixtures to invertebrates and fish," *Bull. Environm. Contam. Toxicol.* 19(2), 250-255, 1978.
1604. C. S. Wiese, and D. A. Griffin, "The solubility of Aroclor 1254 in seawater," *Bull. Environm. Contam. Toxicol.* 19(4), 403-411, 1978.
1605. Kuang Yang Lue, and Armando A. de la Cruz, "Mirex incorporation in the environment: Toxicity in *Hydra*," *Bull. Environm. Contam. Toxicol.* 19(4), 412-415, 1978.
1606. Marja Liisa Hattula, et al., "The toxicity of MCPA to fish. Light and electron microscopy and the chemical analysis of the tissue," *Bull. Environm. Contam. Toxicol.* 19(4), 465-470, 1978.
1607. Dennis Y. Takada, et al., "Alteration of O,O-dimethyl-S-[α-(carboethoxy)benzyl] phosphorodithioate(penthoate) in citrus, water and upon exposure to air and sunlight," *Arch. Environm. Contam. Toxicol.* 5, 63-86, 1976.
1608. Pelegrini, G., and R. Santi, "The potentiation of toxicity of organophosphorus compounds containing carboxylic ester functions toward warm-blooded animals by some organophosphorus impurities," *J. Agric. Food Chem.* 20, 944, 1972.
1609. D. I. Chkanikov, et al., "Variety of 2,4-D metabolic pathways in plants, its significance in developing analytical methods for herbicides residues," *Arch. Environm. Contam. Toxicol.* 5, 97-103, 1976.
1610. R. W. Meikle, et al., "Measurement and prediction of the disappearance rates from soil of 6-chloropicolinic acid," *Arch. Environm. Contam. Toxicol.* 5, 105-117, 1976.
1611. J. R. Coppedge, D. L. Bull, and R. L. Ridgway, "Movement and persistence of aldicarb in certain soils," *Arch. Environm. Contam. Toxicol.* 5(2), 129-141, 1977.
1612. D. L. Grant, W. E. J. Phillips, and G. V. Hatina, "Effect of hexachlorobenzene on reproduction in the rat," *Arch. Environm. Contam. Toxicol.* 5(2), 207-216, 1977.
1613. Joseph G. Zinkl, "Skin and liver lesions in rats fed a polychlorinated biphenyl mixture," *Arch. Environm. Contam. Toxicol.*, 5(2), 217-228, 1977.
1614. Steven C. Schimmel, James M. Patrick, Jr., and Jerrold Forester, "Uptake and toxicity of toxaphene in several estuarine organisms," *Arch. Environm. Contam. Toxicol.*, 5, 353-367, 1977.
1615. M. Thirugnanam, and A. J. Forgash, "Environmental impact of mosquito pesticides: Toxicity and anticholinesterase activity of chlorpyrifos to fish in a salt marsh habitat," *Arch. Environm. Contam. Toxicol.* 5, 415-425, 1977.
1616. B. Das, and P. K. Singh, "The effect of 2,4-dichlorophenoxyacetic acid on growth and nitrogen-fixation of blue-green alga *Anabaenopsis raciborskii*," *Arch. Environm. Contam. Toxicol.* 5, 437-445, 1977.
1617. Foster L. Mayer, et al., "Residue dynamics and biological effects of polychlorinated biphenyls in aquatic organisms," *Arch. Environm. Contam. Toxicol.* 5, 501-511, 1977.
1618. William T. Roubal, Tracy K. Collier, and Donald C. Malins, "Accumulation and metabolism of carbon-14 labeled benzene, naphthalene, and anthracene by young coho salmon (*Oncorhynchus kisuth*)," *Arch. Environm. Contam. Toxicol.* 5, 513-529, 1977.
1619. Peter E. Berteau, and Wallace A. Deen, "A comparison of oral and inhalation toxicities of four insecticides to mice and rats," *Bull. Environm. Contam. Toxicol.* 19(1), 113-120, 1978.
1620. Mads Emanuelsen, et al., "The residue uptake and histology of American oysters (*Crassostrea virginica* Gmelin) exposed to dieldrin," *Bull. Environm. Contam. Toxicol.* 19(1), 1978.

1621. L. C. Folmar, "Avoidance chamber responses of mayfly nymphs exposed to eight herbicides," *Bull. Environm. Contam. Toxicol.* **19**(3), 312–318, 1978.
1622. J. S. Lee, et al., "Identification and quantitation of 3-hydroxy-N-nitrosopyrrolidine in fried bacon," *Bull. Environm. Contam. Toxicol.* **19**(5), 511–517, 1978.
1623. Kenji Isshiki, et al., "Residual piperonyl butoxide in agricultural products," *Bull. Environm. Contam. Toxicol.* **19**(5), 518–523, 1978.
1624. R. W. Meikle, "The hydrolysis rate of diamidafos in dilute aqueous solution," *Bull. Environm. Contam. Toxicol.* **19**(5), 589–599, 1978.
1625. J. P. E. Anderson, and K. H. Domsch. "Microbial degradation of the thiolcarbamate herbicide, diallate, in soils and by pure cultures of soil micro-organisms," *Arch. Environm. Contam. Toxicol.* **4**, 1–7, 1976.
1626. Wilbur L. Mauck, and Lee E. Olson, "Toxicity of natural pyrethrins and five pyrethroids to fish," *Arch. Environm. Contam. Toxicol.* **4**, 18–29, 1976.
1627. A. Fuchs, and W. Ost, "Translocation, distribution and metabolism of triforine in plants," *Arch. Environm. Contam. Toxicol.* **4**, 30–43, 1976.
1628. K. Knoevenagel, and R. Himmelreich, "Degradation of compounds containing carbon atoms by photo-oxidation in the presence of water," *Arch. Environm. Contam. Toxicol.* **4**, 324–333, 1976.
1629. J. W. Mason, and D. R. Rowe, "The accumulation and loss of dieldrin and endrin in the eastern oyster," *Arch. Environm. Contam. Toxicol.* **4**, 349–360, 1976.
1630. M. E. Tagatz, P. W. Borthwick, J. M. Ivery, and J. Knight, "Effects of leached mirex on experimental communities of estuarine animals," *Arch. Environm. Contam. Toxicol.* **4**(4), 435–442, 1976.
1631. NIOSH "Current Intelligence Bulletin 22–Ethylene thiourea"–April 1978, U.S. Dept. H. E. W., NIOSH.
1632. Stula, E. F., and Krauss, W. C., "Embryotoxicity in rats and rabbits from cutaneous application on amide-type solvents and substituted ureas," *Toxicol. Applied Pharmacol.* **41**, 35–55, 1977.
1633. Khera, K. S., and Tryphonas, L., "Ethylenethiourea-induced hydrocephalus: Pre- and postnatal pathogenesis in offspring from rats given a single oral dose during pregnancy," *Toxicol. Applied Pharmacol.*, **42**, 85–97, 1977.
1634. Ulland, B. M., et al., "Thyroid cancer in rats from ethylene thiourea intake," *J. Nat. Cancer Inst.*, **49**, 583–584, 1972.
1635. *IARC Monographs on the Evaluation of Carcinogenic Risk of Chemicals to Man*, **7**, 45–52, 1974.
1636. Innes, J. R. M., et al., "Bioassay of pesticides and industrial chemicals for tumorigenicity in mice: A preliminary note," *J. Nat. Cancer Inst.*, **42**, 1101–1114, 1969.
1637. Graham, S. L., et al., "Effects of prolonged ethylene thiourea ingestion on the thyroid of the rat," *Food and Cosmetic Toxicol.*, **13**, 493–499, 1975.
1638. Kenneth W. Moilanen and Donald G. Crosby, "The photodecomposition of Bromacil," *Arch. Environm. Contam. Toxicol.*, **2**(1), 3–8, 1974.
1639. R. B. Baird, C. L. Kuo, J. S. Shapiro, and W. A. Yanko, "The fate of phenolics in wastewater determination by direct-injection GLC and Warburg respirometry," *Arch. Environm. Contam. Toxicol.* **2**(2), 165–178, 1974.
1640. Leland E. Dannals, et al., "Dissipation and degradation of Alar® in soils under greenhouse conditions," *Arch. Environm. Contam. Toxicol.* **2**(3), 213–221, 1974.
1641. H. M. Khan and M. A. Q. Khan, "Biological magnification of photodieldrin by food chain organisms," *Arch. Environm. Contam. Toxicol.* **2**(4), 289–301, 1974.
1642. Robert I. Starr and Donald J. Cunningham, "Leaching and degradation of 4-aminopyridine-^{14}C in several soil systems," *Arch. Environm. Contam. Toxicol.* **3**(1), 72–83, 1975.
1643. Robert L. Metcalf et al., "Laboratory model ecosystem studies of the degradation and fate

of radiolabeled tri-, tetra-, and pentachlorobiphenyl compared with DDE," *Arch. Environm. Contam. Toxicol.* 3(2), 151-165, 1975.
1644. James R. Sanborn et al., "Plasticizers in the environment: the fate of di-N-octylphthalate (DOP) in two model ecosystems and uptake and metabolism of DOP by aquatic organisms," *Arch. Environm. Contam. Toxicol.* 3(2), 244-255, 1975.
1645. I. Dési et al., "Toxicity of malathion to mammals aquatic organisms and tissue culture cells," *Arch. Environm. Contam. Toxicol.* 3, 410-425, 1975/76.
1646. Maier-Bode, H., *Pflanzenschutzmittel-Rückstände*, Verlag E. Ulmer, Stuttgart, 1955.
1647. Bordás, S., *Veszélyes Növényvédöszerek*, 6ed., Mezögard Kiad, Budapest, 1967.
1648. Perkow, V., *Die Insektizide*, 2 ed., A. Hüthig Verlag, Heidelberg, 1968.
1649. Anonymous, *Pesticides in the Modern World*, p. 19, Newgate, London, 1972.
1650. J. H. Caro, A. W. Taylor, and H. P. Freeman, "Comparative behavior of dieldrin and carbofuran in the field," *Arch. Environm. Contam. Toxicol.* 3, 437-447, 1975/76.
1651. W. S. G. Maass, O. Hutzinger, and S. Safe, "Metabolism of 4-chlorobiphenyl by lichens," *Arch. Environm. Contam. Toxicol.* 3, 470-478, 1975/76.
1652. Richard W. Meikle and Charles R. Youngson, "The hydrolysis rate of chlorpyrifos, O,O-diethyl-O-(3,5,6-trichloro-2-pyridyl)phosphorothioate and its dimethyl analog, chlorpyrifos-methyl, in dilute aqueous solution," *Arch. Environm. Contam. Toxicol.* 7, 13-22, 1978.
1653. R. C. Muirhead-Thomson, "Relative susceptibility of stream macroinvertebrates to temephos and chlorpyrifos, determined in laboratory continuous-flow systems," *Arch. Environm. Contam. Toxicol.* 7, 129-137, 1978.
1654. Richard W. Meikle, et al., "The hydrolysis and photolysis rates of nitrapyrin in dilute aqueous solution," *Arch. Environm. Contam. Toxicol.* 7, 149-158, 1978.
1655. Claudia T. Ward and F. Matsumura, "Fate of 2,3,7,8-tetrachlorodibenzo-*p*-dioxin (TCDD) in a model aquatic environment," *Arch. Environm. Contam. Toxicol.* 7, 349-357, 1978.
1656. Jun Kanazawa, and Chojiro Tomizawa, "Intake and excretion of 2,4,6-trichlorophenyl-4'-nitrophenylether by topmouth gudgeon, *Pseudorasbora parva*," *Arch. Environm. Contam. Toxicol.* 7, 397-407, 1978.
1657. Toyama, T., and Y. Takazawa, "Herbicide, MO®," *Noyaku Seisan Gijutsu*, (23), 1, 1971.
1658. Nishiuchi, Y., and Y. Hashimoto, "Toxicity of pesticide ingredients to some fresh water organisms." *Botyu-Kagaku* 32, 5, 1967.
1659. William T. Roubal et al., "The accumulation of low molecular weight aromatic hydrocarbons of crude oil by coho salmon (*Oncorhynchus kisuth*) and starry flounder (*Platichthys stellatus*)," *Arch. Environm. Contam. Toxicol.* 7, 237-244, 1978.
1660. Alfred W. Jarvinen, and Robert M. Tyo. "Toxicity to fathead minnows of endrin in food and water," *Arch. Environm. Contam. Toxicol.* 7, 409-421, 1978.
1661. Thomas R. Goes, Eldon P. Savage, and William L. Boyd, "In vitro inhibition of oral viridans streptococci by chlordane," *Arch. Environm. Contam. Toxicol.* 7, 449-456, 1978.
1662. T. Wayne Schultz, Lola M. Kyte, and James N. Dumont, "Structure-toxicity correlations of organic contaminants in aqueous coal-conversion effluents," *Arch. Environm. Contam. Toxicol.* 7, 457-463, 1978.
1663. N. Ahmad, D. D. Walgenbach, and G. R. Sutter, "Degradation rates of technical carbofuran and a granular formulation in four soils with known insecticide use history," *Bull. Environm. Contam. Toxicol.* 23, 572-574, 1979.
1664. Mayo, Z. B., et al., *Insecticide Acaricides Test.* 2, 87, 1977.
1665. Sechriest R. E., C. Johnson, and R. Foster, *Insecticide Acaricides Test.* 2, 89, 1977.
1666. Henry C. Hollifield, "Rapid nephelometric estimate of water solubility of highly insoluble organic chemicals of environmental interest," *Bull. Environm. Contam. Toxicol.* 23, 579-586, 1979.
1667. Charles J. Spillner, Victor M. Thomas, and Jack R. Debaun, "Effect of fenitrothion on

micro-organisms which degrade leaf-litter and cellulose in forest soils," *Bull. Environm. Contam. Toxicol.* 23, 601–606, 1979.
1668. M. R. Riskallah, M. M. El-Sayed, and S. A. Hindi, "Study on the stability of leptophos in water under laboratory conditions," *Bull. Environm. Contam. Toxicol.* 23, 607–614, 1979.
1669. M. R. Riskallah, E. G. Esaac, and M. M. El-Sayed, "Photodegradation of leptophos," *Bull. Environm. Contam. Toxicol.* 23, 636–641, 1979.
1670. T. C. Wang and J. L. Bricker, "2-butanone and tetrahydrofuran contamination in the water supply," *Bull. Environm. Contam. Toxicol.* 23, 620–623, 1979.
1671. Michael C. Mix and Randy L. Schaffer, "Benzo(a)pyrene concentrations in mussels (*Mytilus edulis*) from Yaquina Bay, Oregon, during June 1976–May 1978," *Bull. Environm. Contam. Toxicol.* 23, 677–684, 1979.
1672. Tomohiro Miyazaki et al., "Identification of *trans*-nonachlor in goby-fish from Tokyo Bay," *Bull. Environm. Contam. Toxicol.* 23, 631–635, 1979.
1673. Sovocol, G. W., et al., *Anal. Chem.* 49, 734, 1977.
1674. Cochrane et al., *J. Assoc. Offic. Anal. Chem.* 53, 769, 1970.
1675. Dennis Stainken and Janice Rollwagen, "PCB residues in bivalves and sediments of Raritan Bay," *Bull. Environm. Contam. Toxicol.* 23, 690–697, 1979.
1676. S. H. Sandifer et al., "Spermatogenesis in agricultural workers exposed to dibromochloropropane (DBCP)," *Bull. Environm. Contam. Toxicol.* 23, 703–710, 1979.
1677. G. K. Iwama and G. L. Greer. "Toxicity of sodium pentachlorophenate to juvenile Chinook salmon under conditions of high loading density and continuous-flow exposure," *Bull. Environm. Contam. Toxicol.* 23, 711–716, 1979.
1678. S. C. Fang, "Uptake and biotransformation of phenylmercuric acetate by aquatic organisms," *Arch. Environm. Contam. Toxicol.* 1(1), 18–26, 1973.
1679. Matsumura, F., Y. Gotoh, and G. M. Boush, "Phenylmercuric acetate: metabolic conversion by micro-organisms," *Science* 173, 49, 1971.
1680. J. W. Bedford and M. J. Zabik, "Bioactive compounds in the aquatic environment: Uptake and loss of DDT and dieldrin by freshwater mussels," *Arch. Environm. Contam. Toxicol.* 1(2), 1973.
1681. M.A.Q. Khan et al., "Toxicity–metabolism relationship of the photoisomers of certain chlorinated cyclodiene insecticide chemicals," *Arch. Environm. Contam. Toxicol.* 1(2), 159–169, 1973.
1682. A. A. de la Cruz, and S. M. Naqvi, "Mirex incorporation in the environment: Uptake in aquatic organisms and effects on the rates of photosynthesis and respiration," *Arch. Environm. Contam. Toxicol.* 1(3), 255–264, 1973.
1683. W. Winterlin and G. Walker, "Carbaryl residues in bees, honey and bee-bread following exposure to carbaryl via the food supply," *Arch. Environm. Contam. Toxicol.* 1(4), 362–380, 1973.
1684. Georgacakis, E., and M. A. Q. Khan, "Toxicity of the photo-isomers of cyclodiene insecticides to freshwater animals," *Nature* 233, 120, 1971.
1685. Maitra, N., H. M. Khan, and M. A. Q. Khan, "Biological concentration of the photo-isomers of cyclodiene insecticides and their metabolites," *Bull. Environm. Contam. Toxicol.* 1, 1973.
1686. Raul Morales, Stephen M. Rappaport, and Robert G. Hermes, "Air sampling and analytical procedures for benzidine, 3,3'-dichlorobenzidine and their salts," *AIHA Journal* 40, 970–978, Nov. 1979.
1687. C. H. Hine, et al., "Three month inhalation exposure study with methane sulfonylfluoride," *AIHAJ* 40, 986–992, Nov. 1979.
1688. F. J. Murray, et al., "Embryo toxicity of inhaled benzene in mice and rabbits," *AIHAJ* 40, 993–998, Nov. 1979.
1689. Ronad J. Young et al., "Phorate intoxication at an insecticide formulating plant," *AIHAJ* 40, 1013–1016, Nov. 1979.

1690. Jeffrey M. Giddings, "Acute toxicity to *Selenastrum capricornutum* of aromatic compounds from coal conversion," *Bull. Environm. Contam. Toxicol.* **23**, 360-364, 1979.
1691. Lynn H. Zakitis, "Extraction and gas-liquid chromatographic analysis of chlorphoxim in water and fish," *Bull. Environm. Contam. Toxicol.* **23**, 391-397, 1979.
1692. Le Berre R. et al., "Control of *Simulium damnosum*, the vector of human onchocerciasis in West Africa. II. Test by classic application of new insecticides and new formulations," 70 Oncho du 8 Mai Orstom-Occge, BP171, Bobo-Dioulasso, Upper Volta, 1972.
1693. Day, A., and H. Crosby, *J. Econ. Entomol.* **65**, 1164, 1972.
1694. Harris, C. R., and H. Svic. *J. Econ. Entomol.* **63**, 605, 1970.
1695. Dale, W. E., J. W. Miles, and F. C. Churchill, II, *J. Assoc. Offic. Anal. Chem.* **59**, 1088, 1976.
1696. Kevser Taymaz, David T. Williams, and Frank M. Benoit, "Chlorine dioxide oxidation of aromatic hydrocarbons commonly found in water," *Bull. Environm. Contam. Toxicol.* **23**, 398-404, 1979.
1697. M. M. Hanumante, and S. S. Kulkarni, "Acute toxicity of two molluscicides, mercuric chloride and pentachlorophenol, to a freshwater fish (*Channa gachua*)," *Bull. Environm. Contam. Toxicol.* **23**, 725-727, 1979.
1698. Frank M. Benoit, Guy L. Lebel, and David T. Williams, "Polycyclic aromatic hydrocarbon levels in Eastern Ontario drinking waters," *Bull. Environm. Contam. Toxicol.* **23**, 774-778, 1979.
1699. K. A. Brice, and R. G. Derwent. "Emissions inventory for hydrocarbons in the United Kingdom," *Atm. Environm.* **12**, 2045-2054, 1978.
1700. Darnall, K. R., Lloyd, A. C., Winer, A. M. and Pitts, J. N., "Reactivity scale for atmospheric hydrocarbons based on reaction with hydroxyl radical," *Environm. Sci. and Technol.* **10**, 692-696, 1976.
1701. Cox, R. A., Derwent, R. G. and Holt, P. M., "Relative rate constants for the reactions of OH radicals with H_2, CH_4, CO, NO and HONO at atmospheric pressure and 296 K," *J. Chem. Soc. Faraday Trans. I*, **72**, 2031-2043, 1976.
1702. Howard, C. J., and Evenson, K. M., "Rate constants for the reactions of OH with ethane and some halogen substituted ethanes at 296 K," *J. Chem. Phys.* **64**, 4303-4306, 1976.
1703. Greiner, N. R., "Hydroxyl radical kinetics by kinetic spectroscopy. VI. Reactions with alkanes in the range 300-500 K," *J. Chem. Phys.* **53**, 1010-1076, 1970.
1704. Greiner, N. R., "Hydroxyl-radical kinetics by kinetic spectroscopy. II. Reactions with C_2H_6, C_3H_8 and iso-C_4H_{10} at 300 K," *J. Chem. Phys.* **46**, 3389-3392, 1967.
1705. Stuhl, F., "Rate constant for the reaction of OH with n-C_4H_{10}," *Z. Naturforsch* **28a**, 1383-1384, 1973.
1706. Smith, I. W. M., and Zellner, R. D., "Rate measurements of reactions of OH by resonance absorption. Part 2. Reactions of OH with CO, C_2H_4 and C_2H_2," *J. Chem. Soc. Faraday Trans. II*, **69**, 1617-1627, 1973.
1707. Perry, R. A., Atkinson, R. and Pitts, J. N., "Kinetics and mechanism of the gas phase reaction of OH radicals with aromatic hydrocarbons over the temperature range 296-473 K," *J. Phys. Chem.* **81**, 296-304, 1977.
1708. P. H. Howard, J. Saxina, P. R. Durkin and L.-T. Ou, "Review and evaluation of available techniques for determining persistence and routes of degradation of chemical substances in the environment," EPA-560/5-75-006, May 1975.
1709. Russell J. Hall, and Douglas Swineford, "Uptake of methoxychlor from food and water by the American toad (*Bufo americanus*)," *Bull. Environm. Contam. Toxicol.*, **23**, 335-337, 1979.
1710. NIOSH, "Current Intelligence Bulletin 23," U.S. Dept. H. E. W., Public Health Service, Center for Disease Control, NIOSH Publication No. 78-145, 1978.
1711. *IARC Monographs on the Evaluation of Carcinogenic Risk of Chemicals in Man*, **12**, 85-95, 1976.

1712. NIOSH, "Current intelligence Bulletin 24.," U.S. Dept. H. E. W., NIOSH Publication No. 78 - 148, 1978.
1713. NIOSH, "Current Intelligence Bulletin 25," U.S. Dept. H. E. W., NIOSH Publication No. 78 - 149, 1978.
1714. Arne Helweg, "Influence of temperature, humidity and inoculation on the degradation of ^{14}C-labeled 2-aminobenzimidazole in soil," *Water, Air and Soil Poll.* **12**, 275–281, 1979.
1715. G. R. Southworth, B. R. Parkhurst, and J. J. Beauchamp, "Accumulation of acridine from water, food and sediment by the fathead minnow, *Pimephales promelas*," *Water, Air and Soil Poll.* **12**, 331–341, 1979.
1716. J. F. de Kreuk, and A. O. Hansveit, "Assessment of biodegradation," in *The Oil Industry and Microbial Ecosystems*, edited by K. W. A. Chater and H. J. Somerville, Heyden & Son, 1978.
1717. Wen Yuh Lee, "Some laboratory cultured crustaceans for marine pollution studies," *Marine Poll. Bull.*, **8**(11), 258–259, 1977.
1718. Bayer, "Eulan–Okotoxikologisches Verhalten," Bayer ecotoxicological data sheet.
1720. W. Dickinson Burrows and Robert S. Rowe, "Ether soluble constituents of landfill leachate," *JWPCF* **47**(5), May 1975.
1721. Tomoyuki Miyazaki, et al., "Identification of chlordanes and related compounds in gobyfish from Tokyo Bay," *Bull. Environm. Contam. Toxicol.* **24**, 1–8, 1980.
1722. A. C. Anderson, and A. A. Abdelghani, "Toxicity of selected arsenical compounds in short term bacterial bioassays," *Bull. Environm. Contam. Toxicol.* **24**, 124–127, 1980.
1723. S. Krishnan, et al., "Cyanoarenes in soot: synthesis and mutagenicity of cyanoacenaphthylenes," *Environm. Sci. & Techn.* **13**(12), 1532–1534, Dec. 1979.
1724. C. S. Gram, E. Athos, H. S. Chan, and G. S. Neff, "Phthalate esters, PCB and DDT residues in the gulf of Mexico atmosphere," *Atm. Environm.* **14**, 65–69, 1980.
1725. Bidleman, T. F., Rice, C. P. and Olney, C. E., "High molecular weight chlorinated hydrocarbons in the air and sea: Rates and mechanisms of air/sea transfer," in: *Marine Pollutant Transfer* (edited by H. L. Windom, and R. A. Duce), Lexington Books, Mass.
1726. Kamiyama, K., Takai, T. and Yamanaka, Y., "Correlation between volatile substances released from plants and meteorological conditions," *Proc. Int. Clean Air Conf., Brisbane, Australia*, 365–372, 1978.
1727. Tingey, D. T., Manning, M., Ratsch, H. C., Burns, W. F., Grothaus, L. C. and Field, R. W., "Monoterpene emission rates from slash pine," U.S. EPA, Rept. EPA-904/9-78-013, 1978.
1728. Arnts, R. R., et al., "Measurements of α-pinene fluxes from a loblolly pine forest," *Proc. 4th Jt. Conf. Sensing Environ. Pollut.*, 829–833, Am. Chem. Soc., Washington, D.C., 1978.
1729. Tingey, D. T., et al., "Isoprene emission rates from live oak," U.S. EPA, Rept. EPA-904/9-78-004, 1978.
1730. R. Panter and R. D. Penzhorn, "Alkyl sulfonic acids in the atmosphere," *Atm. Environm.* **14**, 149–151, 1980.
1731. Dept. HEW, "Bioassay of hydrazobenzene," TRS No. 92, Dept. HEW publication No. (NIH) 78-1342, 1978.
1732. Dept. HEW, "Bioassay of chloropicrin," TRS No. 65, Dept. HEW publication No. (NIH) 78-1315, 1978.
1733. Dept. HEW, "Bioassay of estradiol mustard," TRS No. 59, Dept. HEW publication No. (NIH) 78-1309, 1978.
1734. Dept. HEW, "Bioassay of diarylanilide yellow," TRS No. 30, Dept. HEW publication No. (NIH) 78-830, 1978.
1735. Dept. HEW, "Bioassay of 1,1,1-trichloroethane," TRS No. 3, Dept. HEW publication No. (NIH) 77-803, 1978.
1736. Dept. HEW, "Bioassay of 2,4-diaminoanisole sulfate," TRS No. 84, Dept. HEW publication No. (NIH) 78-1334, 1978.

Note: Ref. 1719 is not cited in the text.

1737. Dept. HEW, "Bioassay of dichlorvos," TRS No. 10, Dept. HEW publication No. (NIH) 77-810, 1977.
1738. Dept. HEW, "Bioassay of *l*-tryptophan," TRS No. 71, Dept. HEW publication No. (NIH) 78-1321, 1978.
1739. Dept. HEW, "Bioassay of 1H-benzotriazole," TRS No. 88, Dept. HEW publication No. (NIH) 78-1338, 1978.
1740. Dept. HEW, "Bioassay of dimethoate," TRS No. 4, Dept. HEW publication No. (NIH) 77-804, 1977.
1741. Dept. HEW, "Bioassay of 2-chloro-*p*-phenylenediaminesulfate," TRS No. 113, Dept. HEW publication No. (NIH) 78-1368, 1978.
1742. Dept. HEW, "Bioassay of dioxathion," TRS No. 125, Dept. HEW publication No. (NIH) 78-1380, 1978.
1743. Dept. HEW, "Bioassay of 3-amino-4-ethoxyacetanilide," TRS No. 112, Dept. HEW publication No. (NIH) 78-1367, 1978.
1744. Dept. HEW, "Bioassay of ICRF-159," TRS No. 78, Dept. HEW publication No. (NIH) 78-1328, 1978.
1745. Dept. HEW, "Bioassay of acronycine," TRS No. 49, Dept. HEW publication No. (NIH) 78-849, 1978.
1746. Dept. HEW, "Bioassay of 5-nitro-*p*-toluidine," TRS No. 107, Dept. HEW publication No. (NIH) 78-1357, 1978.
1747. Dept. HEW, "Bioassay of iodoform," TRS No. 110, Dept. HEW publication No. (NIH) 78-1365, 1978.
1748. Dept. HEW, "Bioassay of trichlorofluoromethane," TRS No. 106, Dept. HEW publication No. (NIH) 78-1356, 1978.
1749. Dept. HEW, "Bioassay of *m*-cresidine," TRS No. 105, Dept. HEW publication No. (NIH) 78-1355, 1978.
1750. Dept. HEW, "Bioassay of 3-sulfolane," TRS No. 102, Dept. HEW publication No. (NIH) 78-1352, 1978.
1751. Dept. HEW, "Bioassay of phenazopyridine, hydrochloride," TRS No. 99, Dept. HEW publication No. (NIH) 78-1349, 1978.
1752. Dept. HEW, "Bioassay of methoxychlor," TRS No. 35, Dept. HEW publication No. (NIH) 78-835, 1978.
1753. Dept. HEW, "Bioassay of dicofol," TRS No. 90, Dept. HEW publication No. (NIH) 78-1340, 1978.
1754. Dept. HEW, "Bioassay of *o*-anisidine hydrochloride," TRS No. 89, Dept. HEW publication No. (NIH) 78-1339, 1978.
1755. Dept. HEW, "Bioassay of 4-chloro-*m*-phenylenediamine," TRS No. 85, Dept. HEW publication No. (NIH) 78-1335, 1978.
1756. Dept. HEW, "Bioassay of trimethylphosphate," TRS No. 81, Dept. HEW publication No. (NIH) 78-1331, 1978.
1757. Dept. HEW, "Bioassay of 1,4-dioxane," TRS No. 80, Dept. HEW publication No. (NIH) 78-1330, 1978.
1758. Dept. HEW, "Bioassay of N-phenyl-*p*-phenylenediamine," TRS No. 82, Dept. HEW publication No. (NIH) 78-1332, 1978.
1759. Dept. HEW, "Bioassay of pyrimethamine," TRS No. 77, Dept. HEW publication No. (NIH) 78-1327, 1978.
1760. Dept. HEW, "Bioassay of hexachloroethane," TRS No. 68, Dept. HEW publication No. (NIH) 78-1318, 1978.
1761. Dept. HEW, "Bioassay of 1,1,2-trichloroethane," TRS No. 74, Dept. HEW publication No. (NIH) 78-1324, 1978.
1762. Dept. HEW, "Bioassay of aroclor 1254," TRS No. 38, Dept. HEW publication No. 78-83, 1978.

1763. Dept. HEW, "Bioassay of 1-nitronaphthalene," TRS No. 64, Dept. HEW publication No. 78-1314, 1978.
1764. Dept. HEW, "Bioassay of endosulfan," TRS No. 62, Dept. HEW publication No. (NIH) 78-1312, 1978.
1765. Dept. HEW, "Bioassay of pentachloronitrobenzene," TRS No. 61, Dept. HEW publication No. (NIH) 78-1311, 1978.
1766. Dept. HEW, "Bioassay of phenesterin," TRS No. 60, Dept. HEW publication No. (NIH) 78-1310, 1978.
1767. Dept. HEW, "Bioassay of thio-tepa," TRS No. 58, Dept. HEW publication No. (NIH) 78-1308, 1978.
1768. Dept. HEW, "Bioassay of β-TGdR," TRS No. 57, Dept. HEW publication No. (NIH) 78-1363, 1978.
1769. Dept. HEW, "Bioassay of ethionamide," TRS No. 46, Dept. HEW publication No. (NIH) 78-846, 1978.
1770. Dept. HEW, "Bioassay of chlorothalonil," TRS No. 41, Dept. HEW publication No. (NIH) 78-841, 1978.
1771. Dept. HEW, "Bioassay of trisodium ethylenediaminetetraacetate trihydrate (EDTA)," TRS No. 11, Dept. HEW publication No. (NIH) 77-811, 1978.
1772. Dept. HEW, "Bioassay of malathion," TRS No. 24, Dept. HEW publication No. (NIH) 78-824, 1978.
1773. Dept. HEW, "Bioassay of proflavine," TRS No. 5, Dept. HEW publication No. (NIH) 77-805, 1977.
1774. Vincenzo Amico et al., "Levels of chlorinated hydrocarbons in marine animals from the central mediterranean," *Marine Poll. Bull.*, 10, 282-284, 1979.
1775. V. Contardi et al., "PCB's and chlorinated pesticides in organisms from the liqurian Sea," *Marine Poll. Bull.* 10, 307-311, 1979.
1776. R. Dubrawski and J. Falandysz, "Chlorinated hydrocarbons in fish-eating birds from the Gdańsk Bay, Baltic Sea," *Marine Poll. Bull.* 11, 15-18, 1980.
1777. G. Puccetti, and V. Leoni, "PCB and HCB in the sediments and waters of the Tiber Estuary," *Marine Poll. Bull.*, 11, 22-25, 1980.
1778. Prestt, I., and Jefferies, D. J., "Winter numbers, breeding success and organochlorine residues in the great crested grebe in Britain," *Bird Study*, 16, 168-185, 1969.
1779. Bourne, W. R. P., and Bogan, J. A., "Seabirds and pollution," in: *Marine Pollution*, R. Johnston (ed.), 482-493, Academic Press, New York, 1976.
1780. Särkkä, J., et al., "Chlorinated hydrocarbons and mercury in birds of Lake Päijänne, Finland—1972-74," *Pest. Monit. J.* 12, 26-35, 1978.
1781. Lester L. Lamparski, Rudolph H. Stehl, and Robert L. Johnson, "Photolysis of pentachlorophenol-treated wood. Chlorinated Dibenzo-p-dioxin formation," *Environm. Sci. & Techn.* 14(2), 196-200, Feb. 1980.
1782. Mieure, J. P.; Hicks, O.; Kaley, R. G.; Michael, P. R., *J. Chromatogr. Sci.* 15, 275, 1977.
1783. Williams, P. M., and Robertson, K. Y., "Chlorinated hydrocarbons in seasurface and subsurface waters at nearshore stations and in the North Central Pacific Gyre," *Fish. Bull. O. S.* 73, 445, 1975.
1784. Harvey, G. R., et al., "Polychlorinated biphenyls in North Atlantic Ocean Waters," *Science*, 180, 643, 1973.
1785. Harvey, G. R., et al., "Atmospheric transport of polychlorinated biphenyls to the North Atlantic," *Atm. Environm.* 8, 777, 1974.
1786. Duce, R. A., et al., "Enrichment of heavy metals and organic compounds in the surface microlayer of Narragansett Bay, Rhode Island," *Science*, 176, 161, 1972.
1787. Elder, D., "PCB's in N.W. Mediterranean coastal waters," *Marine Poll. Bull.*, 7, 65.
1788. Leroy C. Folmar, "Effects of short-term field applications of acrolein and 2,4-D (DMA)

on flavor of the flesh of rainbow trout," *Bull. Environm. Contam. Toxicol.* **24**, 217-224, 1980.
1789. K. I. N. Jensen, and E. R. Kimball, "Persistence of dinitramine and trifluralin in Nova Scotia, Canada," *Bull. Environm. Contam. Toxicol.* **24**, 238-243, 1980.
1790. Menges, R. M., and J. L. Hubbard, *Weed Sci.* **18**, 247, 1980.
1791. Duseja, D. R., and E. E. Holmes, *Soil Sci.* **125**, 41, 1978.
1792. Savage, K. E., *Weed Sci.* **26**, 465, 1978.
1793. Y. K. Chau et al., "Occurrence of tetraalkyllead compounds in the aquatic environment," *Bull. Environm. Contam. Toxicol.* **24**, 265-269, 1980.
1794. R. L. Escalona, M. T. L. Rosales, and E. F. Mandelli, "On the presence of fecal steroids in sediments from two Mexican harbors," *Bull. Environm. Contam. Toxicol.* **24**, 289-295, 1980.
1795. Fruton, J. S., and S. Simmonds, *General Biochemistry*, 2nd ed. Wiley International, New York, 1958.
1796. Eneroth, P., et al., *J. Lipid Res.* **5**, 245, 1964.
1797. Goodfellow, R. M., et al., *Marine Poll. Bull.* **8**, 272, 1977.
1798. Hatcher, P. G., and P. A. McGillivary, "Sewage contamination in the marine environment of the New York Bight: Fecal steroids as indicators," Miami, Fla., National Oceanic and Atmospheric Administration, Atlantic Oceanographic and Meteorological Laboratories, 1978.
1799. John Palassis, "Sampling and analytical determination of airborne tetramethyl and ethylene thiourea," *AIHAJ* **41**, Feb. 1980.
1800. Frederic W. Yeager et al., "Dialkylnitrosamines in elastomers," *AIHAJ* **41**, 148-150, 1980.
1801. A. L. Goldblatt, *Aflatoxin: Scientific background, Control and Implications*, Academic Press, New York, 1969; *J.A.O.A.C.*, **53**, 92-96, 1970.
1802. "Suggested guide for the use of insecticides to control insects affecting crops, livestock, households, stored products, forests, and forest products—1968," U.S. Dept. Agr. Handbook 331, 1968.
1803. M. A. Lindenberg, *Compt. Rend.*, **243**, 2057, 1956.
1804. P. J. McCall et al., "Estimation of chemical mobility in soil from liquid chromatographic retention times," *Bull. Environm. Contam. Toxicol.* **24**, 190-195, 1980.
1805. David L. Heikes, "Residues of pentachloronitrobenzene and related compounds in peanut butter," *Bull. Environm. Contam. Toxicol.*, **24**, 338-343, 1980.
1806. Manske, D. D., and R. D. Johnson, *Pest. Monit. J.* **10**, 134, 1977.
1807. Dejonckheere, W., et al., *Meded. Fac. Landbouwet.* **40**, 1187, 1975.
1808. Nakanishi, T., and H. Oku, *Nippon Shokubutsu Byor Gakkaiho* **35**, 339, 1969.
1809. J. J. T. W. A. Strik et al., "Toxicity of photomirex with special reference to porphyria, hepatic P-450 and glutathione levels, serum enzymes, histology and residues in the quail and rat," *Bull. Environm. Contam. Toxicol.* **24**, 350-355, 1980.
1810. Hallett, D. J., et al., *J. Agr. Food Chem.* **26**, 388, 1978.
1811. Thibodeaux, L. J., and D. G. Parker, "Desorption of selected gases and liquids from aerated basins," Presented at 76th National Meeting American Institute Chemical Engineers, Tulsa, Oklahoma, March 1974, Paper No. 30d.
1812. Thibodeaux, L. J., "A test method for volatile component stripping of waste water," EPA-660/2-74-044, May 1974.
1813. Fumis Matsumura, *Environmental Toxicology of Pesticides*, Academic Press, N.Y., 1972.
1814. Hidetsugu Ishikura, "Impact of pesticide use on the Japanese environment," in: *Environmental Toxicology of Pesticides*, 25, Academic Press, N.Y., 1972.
1815. Edwards, C. A., *Residue Rev.* **13**, 83, 1966.
1816. Kearney, P. C., et al., *Encycl. Chem. Techn.* **18**, 515, 1965.

1817. Mitsuo Ishida, "Phytotoxic metabolites of pentachlorobenzylalchol." in: *Environmental Toxicology of Pesticides*, ed. by F. Matsumura, Academic Press, 1972.
1818. Dewey, O. R., et al., *Nature*, **195**, 1232, 1962.
1820. J. E. Drifmeyer, C. L. Rosenberg, and M. A. Heywood, "Chlordecone (kepone) accumulation on estuarine plant detritus," *Bull. Environm. Contam. Toxicol.* **24**, 364-368. 1980.
1821. Norman M. Trieff et al., "Chloramine-T as a potential scrubbing agent: Removal of odorous sulfur-containing environmental pollutants," *Bull. Environm. Contam. Toxicol.* **24**, 383-388, 1980.
1822. Guy L. Lebel and David T. Williams, "Determination of low ng/l levels of polychlorinated biphenyls in drinking water by extraction with macroreticular resin and analysis using a capillary column," *Bull. Environm. Contam. Toxicol.* **24**, 397-403, 1980.
1823. W. Slooff and R. Baerselman, "Comparison of the usefulness of the Mexican axolotl (*Ambystoma mexicanum*) and the clawed toad (*Xenopus laevis*) in toxicological bioassays," *Bull. Environm. Contam. Toxicol.* **24**, 439-443, 1980.
1824. R. Fanelli et al., "Presence of 2,3,7,8-tetrachlorodibenzo-*p*-dioxin in wildlife living near Seveso, Italy: A preliminary study," *Bull. Environm. Contam. Toxicol.* **24**, 460-462, 1980.
1825. George I. Wingfield, "Effect of asulam on cellulose decomposition in three soils," *Bull. Environm. Contam. Toxicol.* **24**, 473-476, 1980.
1826. Ishida, M., H. Sumi, and H. Oku, *Residue Rev.* **25**, 139, 1969.
1827. M. Alexander, "Microbial degradation of pesticides," in: *Environmental Toxicology of Pesticides*, ed. by F. Matsumura, Academic Press, N.Y., 1972.
1828. National Association of Photographic Manufacturers, Inc. in cooperation with hydroscience, Inc., *Environmental Effect of Photoprocessing Chemicals*," Vols. I & II, National Association of Photographic Manufacturers, Inc., 600 Mamaroneck Ave., Harrison, N.Y., 10528, 1974.
1829. Ian J. Tinsley, *Chemical Concepts in Pollutant Behavior*," J. Wiley & Sons, New York, 1979.
1830. William E. Gledhill et al., "An environmental safety assessment of butylbenzylphthalate," *Environm. Sci. & Techn.* **14**(3), 301-305, March 1980.
1831. C. Ray Thompson, Gerrit Kats, and Robert W. Lennox, "Phytotoxicity of air pollutants by high explosive production," *Environm. Sci. & Techn.* **13**(10), 1263-1268, Oct. 1979.
1832. Kopperman, H. L., R. M. Carlson, and R. Caple: "Aqueous chlorination and ozonation studies. I. Structure-toxicity correlation of phenolic compounds to *Daphnia magna*," *Chem. Biol. Interact.* **9**, 245, 1974.
1833. Könemann, W. H., "Quantitative structure–activity relationships for kinetics and toxicity of aquatic pollutants and their mixtures in fish," Univ. Utrecht, Netherlands, 1979.
1834. Junshi Miyamoto et al., "Metabolism of organophosphorus insecticides in aquatic organisms, with special emphasis on fenitrothion," in: *Pesticide and Xenobiotic Metabolism in Aquatic Organisms*, American Chemical Society Series 99, 1979.
1835. J. Miyamoto, *Botyu-Kagaku (Scientific Pest Control)* **36**, 189, 1971.
1836. Takimoto, Y.; Kagoshima, M.; and Miyamoto, J., unpublished observation, 1978.
1837. Hutzinger, O., Safe, S.; and Zitko, V., in: *The Chemistry of PCB's*, CRC Press, Inc., Cleveland, Ohio, 1974.
1838. "Toxic pollutant effluent standards: Standards for polychlorinated biphenyls (PCB's); Final decision," *Fed. Reg.* **42**, 6532, 1977.
1839. M. A. Q. Khan, M. Feroz, and P. Sudershan, "Metabolism of cyclodiene insecticides by fish," in: *Pesticide and Xenobiotic Metabolism in Aquatic Organisms*, American Chemical Society, Symposium Series 99, 1979.
1840. Graham, P. R., "Phthalate ester plasticizers: Why and how they are used," *Environm. Health Perspect.* **4**, 116-128, 1973.
1841. Johnson, B. T., and Lulves, W., "Biodegradation of di-*n*-butylphthalate and di-2-ethylhexylphthalate in freshwater hydrosoil," **32**, 333-339, 1975.

Note: Ref. 1819 is not cited in the text.

1842. Saeger, V. W., and Tucker, E. S., "Biodegradation of phthalic acid esters in river water and activated sludge," 31, 29–34, 1976.
1843. Mark J. Melancon, "Metabolism of phthalate esters in aquatic species," in: *Pesticide and Xenobiotic Metabolism in Aquatic Organisms*, American Chemical Society, Symposium Series 99, 1979.
1844. Ming-Muh Lay et al., "Metabolism of the thiocarbamate herbicide molinate (ordram) in Japanese carp," in: *Pesticide and Xenobiotic Metabolism in Aquatic Organisms*, American Chemical Society, Symposium Series 99, 1979.
1845. *Herbicide Handbook of the Weed Science Society of America*, 3rd ed., p. 252–254, 1974.
1846. Chaiyarach, S., V. Ratananun, and R. C. Harrel, *Bull. Environm. Contam. Toxicol.* 14, 281–284, 1975.
1847. Kawatsu, H., *Bull. Japanese Soc. Sci. Fisheries* 43, 905–912, 1977.
1848. Skryabin, G. K., et al., *Dokl. Akad. Nauk SSSR* 239(3), 717–720, 1978.
1849. Golovleva, L. A., et al., Abstract V-610, Fourth International Congress of Pesticide Chemistry, Zurich, Switzerland, 1978.
1850. Kunio Kobayashi, "Metabolism of pentachlorophenol in fish," in: *Pesticide and Xenobiotic Metabolism in Aquatic Organisms*, American Chemical Society, Symposium Series 99, 1979.
1851. James D. Yarbrough and Janice E. Chambers, "The disposition and biotransformation of organochlorine insecticides in insecticide-resistant and -susceptible mosquitofish," in: *Pesticide and Xenobiotic Metabolism in Aquatic Organisms*, American Chemical Society, Symposium Series 99, 1979.
1852. V. Zitko, "The fate of highly brominated aromatic hydrocarbons in fish," in: *Pesticide and Xenobiotic Metabolism in Aquatic Organisms*, American Chemical Society, Symposium Series 99, 1979.
1853. J. F. Estenik and W. J. Collins, "In Vivo and in vitro studies of mixed function oxidase in an aquatic insect, *Chironomus riparius*," in: *Pesticide and Xenobiotic Metabolism in Aquatic Organisms*, American Chemical Society, Symposium Series 99, 1979.
1854. *Pesticide Dictionary 1976*, published by Farm Chemicals, Meister Publishing Company, 37841 Euclid Avenue, Willoughby, Ohio 44094.
1855. Hubert Martin, *Pesticide Manual*, British Crop Protection Council, Clacks Farm, Boreley, Ombersley, Droitwich, Worcester, U.K., 1968.
1856. Jukes, T. H., and Shaeffer, C. B., *Science N.Y.* 132, 296, 1960.
1857. Astwood, E. B., *J. Am. Med. Ass.*, 172, 1319, 1960.
1858. G. E. Parris et al., "Waterborne methylene*bis*(2-chloroaniline) and 2-chloroaniline contamination around Adrian, Michigan," *Bull. Environm. Contam. Toxicol.* 24, 497–503, 1980.
1859. Stula E. F. et al., *J. Environ. Pathol. Toxicol.* 1, 31, 1977.
1860. Thomas Cairns and Charles H. Parfitt, "Persistence and metabolism of TDE in California clear lake fish," *Bull. Environm. Contam. Toxicol.* 24, 504–510, 1980.
1861. Dora R. May Passino and Janet Matsumoto Kramer, "Toxicity of arsenic and PCB's to fry of deepwater ciscoes (*Coregonus*)," *Bull. Environm. Contam. Toxicol.* 24, 527–534, 1980.
1862. Robert D. Rogers, James C. McFarlane, and Amy J. Cross, "Adsorption and desorption of benzene in two soils and montmorillonite clay," *Environm. Sci. & Techn.* 14(4), 457–460, April 1980.
1863. Douglas W. Kuehl and Ralph C. Dougherty, "Pentachlorophenol in the environment. Evidence for its origin from commercial pentachlorophenol by negative chemical ionization mass spectrometry," *Environm. Sci. & Techn.* 14(4), 447–449, April 1980.
1863a. T. N. Shaver, and D. L. Bull. "Environmental fate of methyl eugenol," *Bull. Environm. Contam. Toxicol.* 24, 619–626, 1980.
1864. D. W. McLeese et al., "Uptake and excretion of aminocarb, nonylphenol and pesticide diluent 585 by mussels (*Mytilus edulis*)," *Bull. Environm. Contam. Toxicol.* 24, 575–581, 1980.
1865. R. Fanelli et al., "2,3,7,8-tetrachlorodibenzo-*p*-dioxin levels in cow's milk from the contaminated area of Seveso, Italy," *Bull. Environm. Contam. Toxicol.* 24, 634–639, 1980.

1866. S. A. Peoples et al., "A study of samples of well water collected from selected areas in California to determine the presence of DBCP and certain other pesticide residues," *Bull. Environm. Contam. Toxicol.* **24**, 611-618, 1980.

1867. J. D. Isbister, R. S. Shippen, and J. Caplan, "A new method for monitoring cellulose and starch degradation in soils," *Bull. Environm. Contam. Toxicol.* **24**, 570-574, 1980.

1868. C. G. Wright and R. B. Leidy, "Insecticide residue in the air of buildings and pest control vehicles," *Bull. Environm. Contam. Toxicol.* **24**, 582-589, 1980.

1869. T. L. Batchelder, H. C. Alexander, and W. M. McCarty, "Acute fish toxicity of the Versene family of chelating agents," *Bull. Environm. Contam. Toxicol.* **24**, 543-549, 1980.

1870. Calvin M. Menzie, "Effects of pesticides on fish and wildlife," in: *Environmental Toxicology of Pesticides*, edited by Fumio Matsumura, et al., Academic Press, N.Y. and London, 1972.

1871. Tsutamu Nakatsugawa, and Peter A. Nelson, "Studies of insecticide detoxication in invertebrates: An enzymological approach to the problem of biological magnification," in: *Environmental Toxicology of Pesticides*, edited by Fumio Matsumura et al., Academic Press, N.Y. and London, 1972.

1872. Fumio Matsumura, "Biological effects of toxic pesticidal contaminants and terminal residues," in: *Environmental Toxicology of Pesticides*, edited by Fumia Matsumura et al., Academic Press, N.Y. and London, 1972.

1873. Teruhisa Noguchi, "Environmental evaluation of systemic fungicides," in: *Environmental Toxicology of Pesticides*, edited by Fumio Matsumura et al., Academic Press, N.Y. and London, 1972.

1874. Katherine Alben, "Coal tar coatings of storage tanks. A source of contamination of the potable water supply," *Environm. Sci. & Techn.* **14**(4), 469-470, April 1980.

1875. Robert S. Boethling and Martin Alexander, "Microbial degradation of organic compounds at trace levels," *Environm. Sci. & Techn.* **13**(8), 989-991, Aug. 1979.

1976. Eric H. Snider and F. C. Alley, "Kinetics of the chlorination of biphenyl under conditions of waste treatment processes," *Environm. Sci. & Techn.* **13**(10), 1244-1248, Oct. 1979.

1877. Nariyoshi Kawabata, and Kozo Ohira, "Removal and recovery of organic pollutants from aquatic environment. 1. Vinylpyridine-divinylbenzene copolymer as a polymer adsorbent for removal and recovery of phenol from aqueous solution," *Environm. Sci. & Techn.* **13**(11), 1396-1402, Nov. 1979.

1878. Edwards, C. A., "Nature and origins of pollution of aquatic systems by pesticides," in: *Pesticides in Aquatic Environments*, ed. by M. A. Q. Khan, Plenum Press, N.Y., 1977.

1879. Hague, R., et al., "Dynamics of pesticides in aquatic environments," in: *Pesticides in Aquatic Environments*, ed. by M. A. Q. Khan, Plenum Press, N.Y., 1977.

1880. Matsumura, F., "Absorption, accumulation and elimination of pesticides by aquatic organisms," in: *Pesticides in Aquatic Environments*, ed. by M. A. Q. Khan, Plenum Press, N.Y., 1977.

1881. Metcalf, R. L., and J. R. Sanborn., *Bull. Illinois Natural History Survey* **31**, Art. 9, 381-436, 1975.

1882. "Environmental pathways," EPA-600/7-78-074, May 1978.

1883. Joyce McCann et al., "Detection of carcinogens as mutagens in the *Salmonella*/microsome test: Assay of 300 chemicals," *Proc. Nat. Acad. Sci. USA*, **72**(12), pp. 5135-5139 Medical Sciences, Dec. 1975.

1884. Ames B. N. in: *Chemical Mutagens: Principles and Methods for Their Detection*, ed. A. Hollaender, Vol. 1, 267-282, Plenum Press, New York, 1971.

1885. Ames, B. N.; Durstan, W. E.; Yamasaki, E.; and Lee, F. D., *Proc. Nat. Acad. Sci. USA* **70**, 2281-2285, 1973.

1886. Ames, B. N.; Lee, F. D.; and Durston, W. E., *Proc. Nat. Acad. Sci. USA* **70**, 782-786, 1973.

1887. McCann, J.; Spingarn, N.; Kobori, J.; and Ames, B. N., *Proc. Nat. Acad. Sci. USA* **72**, 979-983, 1975.

1888. Miller, E. C., and Miller, J. A., in: *Chemical Mutagens: Principles and Methods for Their Detection*, ed. A. Hollaender, Vol. 1, 83-119, Plenum Press, New York, 1971.
1889. McCann, J., and Ames, B. N., *Ann. N.Y. Acad. Sci.*, 1975.
1890. Ames, B. N.; McCann, J.; and Yamasaki, E., *Mutat. Res.*, 1975.
1891. Isnee, 1976.
1892. Isnee and Jones, 1975.
1893. Kanazawa et al., 1975.
1894. Allison et al., 1964.
1895. Ferguson et al., 1966.
1896. Mount and Putnicki, 1966.
1897. Hansen, 1966.
1898. Cope, 1966.
1899. Sport Fisheries and Wildlife, 1969.
1900. G. Bringmann, and R. Kühn, "Comparison of the toxicity thresholds of water pollutants to bacteria, algae and protozoa in the cell multiplication inhibition test," *Water Research* 14, 231-241, 1980.
1901. G. Bringmann, and R. Kühn, "Bestimmung der biologischen Schadwirkung wassergefahrdender Stoffe gegen Protozoen. II. Bakterienfressende Ciliaten," *Z. Wasser/Abwasser Forsch.* (1), 26-31, 1980.
1902. K. S. Rao and N. K. Dad, "Studies of herbicide toxicity in some freshwater fishes and ectoprocta," *J. Fish. Biol.* 14, 517-522, 1979.
1903. Peter Biro, "Acute effects of the sodium salt of 2,4-D on the early development stages of bleak, *Alburnus alburnus*," *J. Fish. Biol.* 14, 101-109, 1979.
1904. M. W. Curtis, T. L. Copeland, and C. H. Ward, "Acute toxicity of 12 industrial chemicals to freshwater and saltwater organisms," *Water Res.*, 13, 137-141, 1979.
1905. B. Reiff et al., "The acute toxicity of eleven detergents to fish: results of an interlaboratory excercise," *Water Res.* 13, 207-210, 1979.
1906. William D. Stanbro and Wendy D. Smith, "Kinetics and mechanism of the decomposition of N-chloroalanine in aqueous solution," *Environm. Sci. & Techn.* 13(4), 446-451, April 1979.
1907. Henry T. Appleton, and Harish C. Sikka. "Accumulation, elimination and metabolism of dichlorobenzidine in the bluegill sunfish," *Environm. Sci. & Techn.* 14(1), 50-54, Jan. 1980.
1908. Mohsen Moussavi, "Effect of polar substituents on autoxidation of phenols" *Water Res.* 13, 1125-1128, 1979.
1909. Michael C. Lee et al., "Solubility of polychlorinated biphenyls and capacitor fluid in water," *Water Res.* 13, 1249-1258, 1979.
1910. J. Seuge and R. Bluzat, "Toxicité chronique du carbaryl et du lindane chez le mollusque d'eau douce *Lymnea stagnalis* L," *Water Res.* 13, 285-293, 1979.
1911. P. D. Niblett and B. A. McKeown, "Effect of the lamprey larvicide TFM (3-trifluoromethyl-4-nitrophenol) on embryonic development of the rainbow trout (*Salmo gairdneri*, Richardson)," *Water Res.* 14, 515-519, 1980.
1912. Olson, L. E., and Marking, L. L. "Toxicity of TFM (lampricide) to six early life stages of rainbow trout (*Salmo gairdneri*)," *J. Fish. Res. Bd. Can.* 30, 1047-1052, 1973.
1913. F. S. H. Abram, and P. Wilson, "The acute toxicity of CS to rainbow trout" *Water Res.* 13, 631-635, 1979.
1914. Anne Spacie and Jerry L. Hamelink, "Dynamics of trifluralin accumulation in river fishes," *Environm. Sci. & Techn.* 13(7), 817-822, 1979.
1915. Spacie, A., Ph. D. Thesis, Purdue University, Lafayette, Ind., 1975.
1916. Macek, K., et al., "Toxicity of four pesticides to water fleas and fathead minnows," U.S. EPA-60013-76-099, 1976.

1917. Wilbur L. Mauck et al., "Effects of the polychlorinated biphenyl Aroclor 1254 on growth, survival and bone development in brook trout (*Salvelinus fontinalis*)," *J. Fish. Res. Bd. Can.* 35, 1084-1088, 1978.
1918. Maxwell B. Eldridge and Tina Echevarria, "Fate of ^{14}C-benzene in eggs and larvae of pacific herring," *J. Fish. Res. Bd. Can.* 35, 861-865, 1978.
1919. D. L. De Foe, G. D. Veith, and R. W. Carlson, "Effects of Aroclor 1248 and 1260 on the fathead minnow (*Pimephales promelas*)," *J. Fish. Res. Bd. Can.* 35, 997-1102, 1978.
1920. Alan W. Maki and Howard E. Johnson, "Kinetics of lampricide (TFM, 3-trifluoromethyl-4-nitrophenol) residues in model stream communities," *J. Fish Res. Bd. Can.* 34, 1977.
1921. Wilson, 1966.
1922. Ludke et al., 1971.
1923. Butler, 1964.
1924. Wilson, 1965.
1925. Miller et al., 1966.
1926. Masana Ogata and Yoshio Miyake, "Disappearance of aromatic hydrocarbons and organic sulfur compounds from fish flesh reared in crude oil suspension," *Water Res.* 13, 75-78, 1979.
1927. William H. Dennis, et al., "Degradation of diazinon by sodium hypochlorite. Chemistry and aquatic toxicity," *Environm. Sci. & Techn.* 13(5), 594-598, May 1979.
1928. Yasushi Hiraizumi, et al., "Adsorption of polychlorinated biphenyl onto sea bed sediment, marine plankton, and other adsorbing agents," *Environm. Sci. & Techn.* 13(5), 580-584, 1979.
1929. Hiraizumi, Y., et al., *La Mer*, 13, 163, 1975.
1930. Lawrence, J.; Tosine, H., *Environm. Sci. & Techn.* 10, 381, 1976.
1931. Haque, R.; Schmedding, D.; Freed, V., *Environm. Sci. & Techn.* 8, 139, 1974.
1932. R. Pellenbarg, "Environmental poly(organosiloxanes) (silicones)," *Environm. Sci. & Techn.* 13(5), 565-569, May 1979.
1933. Meinrat O. Andreae and David Klumpp, "Biosynthesis and release of organoarsenic compounds by marine algae," *Environm. Sci. & Techn.* 13(6), 738-741, June 1979.
1934. Macek, K. J., and McAllister, W. A., "Insecticide susceptibility of some common fish family representatives," *Trans. Amer. Fish. Soc.* 99(1), 20-27, 1970.
1935. Lawson, C. T. and Hovious, J. C., "Realistic performance criteria for activated carbon treatment of waste-waters from the manufacture of organic chemicals and plastics," Union Carbide Corporation, Research and Development Dept., South Charleston, W. Va., Technical Center, Feb. 14, 1977.
1936. Dobbs, Richard A. and Cohen, Jesse M., "Carbon adsorption isotherms for toxic organics," EPA 600/8-80-023, April 1980.
1937. Kapoor, I. P., R. L. Metcalf, A. S. Hirwe, J. R. Coats, and M. S. Khalsa, *J. Agr. Food Chem.* 21, 310, 1973.
1938. Kapoor, I. P., R. L. Metcalf, A. S. Hirwe, Po-Yung Lu, J. R. Coats, and R. F. Nystrom, *J. Agr. Food Chem.* 20, 1, 1972.
1939. Kapoor, I. P., R. L. Metcalf, R. F. Nystrom, and G. K. Sangla, *J. Agr. Food Chem.* 18, 1145, 1970.
1940. Metcalf, R. L., G. K. Sangha, and I. P. Kapoor, *Environm. Sci. & Technol.* 5, 709, 1971.
1941. Hirwe, A. S., R. L. Metcalf, Po-Yung Lu, and Li-Chun Chio, *Pesticide Biochem. Physiol.* 5, 65, 1975.
1942. Rumiantsev, G. I., and S. M. Novikov, "Predicting skin-adsorptive properties of new chemicals," *Gig. Sanit.*, 4, 9-95, 1975.
1943. MacKay, D., and P. J. Leinonen, "Rate of evaporation of low-solubility contaminants from water bodies to atomosphere," *Environm. Sci. and Techn.* 9, 1178-80, 1975.
1944. Dilling, W. L. "Interphase transfer processes. II. Evaporation rates of chloromethanes,

ethanes, ethylenes, propanes and propylenes from dilute aqueous solutions. Comparisons with theoretical predictions." *Environm. Sci. and Techn.* 11, 405–9, 1977.
1945. Smith, J. H., et al. Environmental Pathways of Selected Chemicals in Freshwater Systems. Part I Background and Experimental Procedures, EPA-600/7-77-113, 1977.
1946. Smith, J. H., et al. Environmental Pathways of Selected Chemicals in Freshwater Systems. Part II. Laboratory Studies, EPA-600/7-78-074, 1978.
1947. Wolfe, N. L., et al. Chemical and Photochemical Transformations of Selected Pesticides in Aquatic Systems. EPA-600/3-76-067, 1976.
1948. Zepp, R. G., and D. M. Cline. "Rates of direct photolysis in aquatic environment," *Environm. Sci. & Tech.* 1, 359, 1977.
1949. Zepp, R. G. "Quantum yields for reactions of pollutants in dilute aqueous solution," *Environ. Sci. & Techn.* 12, 327, 1978.
1950. Zepp, R. G., et al. "Photochemical transformation of the DDT and methoxychlor degradation products, DDE and DMDE, by sunlight," *Arch. Environm. Contam. Toxicol.* 6, 305, 1977.
1951. Zepp, R. G., et al. "Light-induced transformations of methoxychlor in aquatic systems," *J. Agric. Food Chem.*, 24, 727, 1976.
1952. Zepp, R. G., et al. "Dynamics of 2,4-D esters in surface waters. Hydrolysis, photolysis, and vaporization," *Environm. Sci. and Techn.*, 9, 1145, 1975.
1953. Wolfe, N. L., et al. "Kinetics of chemical degradation of malathion in water," *Environm. Sci. & Techn.* 9, 88, 1977.
1954. Mabey, W., and T. Mill. "Critical review of hydrolysis of organic compounds in water under environmental conditions," *J. Phys. and Chem. Ref. Data* 7, 383, 1975.
1955–2105. [Reserved for further expansion]
2106. Bond, C. E., R. H. Lewis, and J. L. Fryer, "Toxicity of various herbicidal materials to fish," *Second Seminar on Biological Problems in Water Pollution*, R. A. Taft San. Eng. Cen. Tech. Rept. W603, pp 96–101, 1960.
2107. Bridges, W. R., *Biological Problems in Water Pollution, Third Seminar (1961)*, U.S.P.H.S. Pub. No. 999-WP-25, pp. 247–249, 1961.
2108. Burdick, G. E., H. J. Dean, and E. J. Harris, "Toxicity of aqualin to fingerling brown trout and bluegills," *N.Y. Fish Game J.* 11(2), 106–114, 1964.
2109. Cairns, J., Jr., and A. Scheier, "The effect upon the pumpkinseed sunfish *Lepomis gibbosus* (Linn.) of chronic exposure to lethal and sublethal concentrations of dieldrin," *Notulae Natur.* (Philadelphia) No. 370, 1–10, 1964.
2110. Carlson, C. A., "Effects of three organophosphorus insecticides on immature *Hexagenia* and *Hydropsyche* of the upper Mississippi River," *Trans. Amer. Fish. Soc.* 95(1): 1–5, 1966.
2111. Eaton, J. G., "Chronic malathion toxicity of the bluegill (*Lepomis macrochirus* Rafinsque)." *Water Res.* 4, 673–684, 1971.
2112. Gilderhus, P. A., "Effects of diquat on bluegills and their food organisms," *Progr. Fish-Cult.* 29(2), 67–74, 1967.
2113. Henderson, C., Q. H. Pickering, and C. M. Tarzwell, "Relative toxicity of ten chlorinated hydrocarbon insecticides to four species of fish," *Trans. Amer. Fish. Soc.* 88(1), 23–32, 1959.
2114. Hughes, J. S., and J. T. Davis, "Comparative toxicity to bluegill sunfish of granular and liquid herbicides," *Proceedings Sixteenth Annual Conference Southeast Game and Fish Commissioners*, 319–323, 1962.
2115. Hughes, J. S., and J. T. Davis, "Variations in toxicity to bluegill sunfish of phenoxy herbicides," *Weeds* 11(1), 50–53, 1963.
2116. Hughes, J. S., and J. T. Davis, "Effects of selected herbicides on bluegill and sunfish," *Proceedings Eighteenth Annual Conference, Southeastern Assoc. Game & Fish Commissioners*, Oct. 18–21, 1964, 480–482, 1964.

2117. Jensen, L. D., and A. R. Gaufin, "Long-term effects of organic insecticides on two species of stonefly naiads," *Trans. Amer. Fish. Soc.* 93(4), 357-363, 1964.
2118. Jensen, L. D., and A. R. Gaufin, "Acute and long-term effects of organic insecticides on two species of stonefly naiads," *JWPCF* 38(8), 1273-1286, 1966.
2119. Katz, M., "Acute toxicity of some organic insecticides to three species of salmonids and to the threespine stickleback," *Trans. Amer. Fish. Soc.* 90(3), 264-268, 1961.
2120. Lane, C. E., and R. E. Livingston, "Some acute and chronic effects of dieldrin on the sailfin molly, *Poecilia latipinna*," *Trans. Amer. Fish. Soc.* 99(3), 489-495, 1970.
2121. Macek, K. J., and W. A. McAllister, "Insecticide susceptibility of some common fish family representatives," *Trans. Amer. Fish. Soc.* 99(1), 20-27, 1970.
2122. Mount, D. I., and C. E. Stephen, "A method of establishing acceptable toxicant limits for fish—malathion and the butoxyethanol ester of 2,4-D," *Transactions American Fisheries Society*, 96(2), 185-193, 1967.
2123. Pickering, Q. H., C. Henderson, and A. E. Lemke, "The toxicity of organic phosphorus insecticides to different species of warmwater fishes," *Trans. Amer. Fish. Soc.* 91(2), 175-184, 1962.
2124. Sanders, H. O., *Toxicity of Pesticides to the Crustacean, Gammarus lacustris*, Bureau of Sport Fisheries and Wildlife technical paper 25, Government Printing Office, Washington, D.C., 1969.
2125. Sanders, H. O., "Toxicities of some herbicides to six species of freshwater crustaceans," *JWPCF* 42(8, part 1), 1544-1550, 1970.
2126. Sanders, H. O., The toxicities of some insecticides to four species of malocostracan crustacea," Fish Pesticide Res. Lab. Columbia, Mo., Bureau of Sport Fish and Wildlife," 1972.
2127. Sanders, H. O., and O. B. COpe, Toxicities of several pesticides to two species of cladocerans," *Trans. Amer. Fish. Soc.* 95(2), 165-169, 1966.
2128. Sanders, H. O., and O. B. Cope, "The relative toxicities of several pesticides to naiads of three species of stoneflies," *Limnol. Oceanogr.* 13(1), 112-117, 1968.
2129. Schoettger, R. A., *Toxicology of Thiodan in Several Fish and Aquatic Invertebrates*, Bureau of Sport Fisheries and Wildlife Investigation in Fish Control 35, Government Printing Office, Washington, D.C., 1970.
2130. Solon, J. M., and J. H. Nair, "The effect of a sublethal concentration of LAS on the acute toxicity of various phosphate pesticides to the Fathead Minnow *Pimephales promelas* Rafinesque," *Bull. Environm. Contam. Toxicol.* 5(5), 408-413, 1970.
2131. Surber, E. W., and Q. H. Pickering, "Acute toxicity of endothal, diquat, hyamine, dalapon, and silvex to fish," *Progr. Fish-Cult.* 24(4), 164-171, 1962.
2132. Walker, C. R., "Toxicological effects of herbicides on the fish environment," *Water & Sewerage Works*, 111(3), 113-116, 1964.
2133. Wilson, D. C., and C. E. Bond, "The effects of the herbicides diquat and dichlobenil (casoron) on pond invertebrates. Part I. Acute toxicity," *Transactions Am. Fishery. Soc.* 98(3), 438-443, 1969.
2134. Bell, H. L., unpublished data, National Water Quality Laboratory, Duluth, Minnesota, 1971.
2135. Biesinger, K. E., unpublished data, National Water Quality Laboratory, Duluth, Minnesota, 1971.
2136. Carlson, C. A., unpublished data, National Water Quality Laboratory, Duluth, Minnesota, 1971.
2137. FPRL, unpublished data, Fish Pesticide Res. Lab. Annual Rept. Bur. Sport Fish. and Wildlife. Columbia, Mo., 1971.
2138. Merna, J. W., unpublished data, Institute for Fisheries Research, Michigan Department of Natural Resources, Ann Arbor, Michigan, E.P.A. Grant No 18050-DLO, 1971.
2139-2320. [Reserved for further expansion.]

2321. Buchanan, D. V., R. E. Milleman, and N. E. Stewart, "Effects of the insecticide Sevin® on survival and growth of the Dungeness crab *Cancer magister*," *J. Fish. Res. Bd. Can.* **26**, 1969.
2322. Butler, P. A., R. E. Milleman, and N. E. Stewart, "Effects of insecticide Sevin on survival and growth of the cockle clam *Clinocardium nuttalli*," *J. Fish. Res. Bd. Can.* **25**, 1631-1635, 1968.
2323. Chin, E., and D. M. Allen, "Toxicity of an insecticide to two species of shrimp, *Penaeus aztecus* and *Penaeus setiferus*," *Texas J. Sci.* **9**(3), 270-278, 1957.
2324. Davis, H. C., and H. Hidu, "Effects of pesticides on embryonic development of clams and oysters and on survival and growth of the larvae," *Fish. Bull.* **67**(2), 383-404, 1969.
2325. Derby, S. B. (Sleeper), and E. Ruber, "Primary production: Depression of oxygen evolution in algal cultures by organophosphorus insecticides," *Bull. Environm. Contam. Toxicol.* **5**(6), 553-558, 1971.
2326. Eisler, R., "Effects of apholate, an insect sterilant, on an estuarine fish, shrimp, and gastropod," *Progr. Fish-Cult.* **28**(2), 154-158, 1966.
2327. Eisler, R., "Acute toxicities of insecticides to marine decapod crustaceans," *Crustaceana* **16**(3), 302-310, 1969.
2328. Eisler, R., Factors Affecting Pesticide-Induced Toxicity in an Estuarine Fish, Bureau of Sport Fisheries and Wildlife Technical Paper 45, Government Printing Office, Washington, D.C., 1970.
2329. Eisler, R., Acute Toxicities of Organochlorine and Organophosphorus Insecticides to Estuarine Fishes, Bureau of Sport Fisheries and Wildlife Technical Paper 46, Government Printing Office, Washington, D.C., 1970.
2330. Eisler, R., "Latent effects of insecticide intoxication to marine molluscs," *Hydrobiologia* **36**(3/4), 345-352, 1970.
2331. Erickson, S. J., T. E. Maloney, and J. H. Gentile, "Effect of nitrilotriacetic acid on the growth and metabolism of estuarine phytoplankton," *JWPCF* **42**(8 part 2), R329-R335, 1970.
2332. Hansen, D. J., P. R. Parish, J. I. Lowe, A. J. Wilson, Jr., and P. D. Wilson, "Chronic toxicity, uptake, and retention of a polychlorinated biphenyl (Aroclor 1254) in two estuarine fishes," *Bull. Environm. Contam. Toxicol.* **6**(2), 113-119, 1971.
2333. Katz, M., "Acute toxicity of some organic insecticides to three species of salmonids and to the three-spine stickleback," *Trans. Amer. Fish. Soc.* **90**(3), 264-268, 1961.
2334. Katz, M., and G. G. Chadwick, "Toxicity of endrin to some Pacific Northwest fishes," *Trans. Amer. Fish. Soc.* **90**(4), 394-397, 1961.
2335. Lane, C. E., and R. J. Livingston, "Some acute and chronic effects of dieldrin on the sailfin molly, *Poecilia latipinna*," *Trans. Amer. Fish. Soc.* **99**(3), 489-495, 1970.
2336. Lane, C. E., and E. D. Scura, "Effects of dieldrin on glutamic oxaloacetic transaminase in *Poecilia latipinna*," *J. Fish. Res. Board Can.* **27**(10), 1869-1871, 1970.
2337. Litchfield, J. T., and F. Wilcoxon, "A simplified method of evaluating dose-effect experiments," *J. Pharmacol. Exp. Ther.*, **96**, 99-113, 1947.
2338. Lowe, J. I., "Some effects of endrin on estuarine fishes," *Proc. Southeast Ass. Game Fish Commissioners* **19**, 271-276, 1965.
2339. Lowe, J. I., "Effects of prolonged exposure to sevin on an estuarine fish, *Leiostomus xanthurus* Lacepede," *Bull. Environm. Contam. Toxicol.* **2**(3), 147-155, 1967.
2340. Lowe, J. I., P. R. Parrish, A. J. Wilson, Jr., P. D. Wilson, and T. W. Duke, "Effects of mirex on selected estuarine organisms," in *Transactions of the 36th North American Wildlife and Natural Resources Conference*, J. B. Trefethen, ed., Wildlife Management Institute, Washington, D.C. Vol. 36, pp. 171-186, 1971.
2341. Lowe, J. I., P. D. Wilson, A. J. Rick, and J. Wilson, Jr., "Chronic exposure of oysters to DDT, toxaphene and parathion," *Proc. Nat. Shellfish Ass.* **61**, 71-79, 1971.

2342. Mahood, R. K., M. D. McKenzie, D. P. Middaugh, S. J. Bollar, J. R. Davis, and D. Spitsbergen, "*A Report on the Cooperative Blue Crab Study—South Atlantic States*, U.S. Department of the Interior, Bureau of Commercial Fisheries, 1970.
2343. Millemann, R. E., "Effects of dursban on shiner perch," in: *Effects of Pesticides on Estuarine Organisms*, Progress Report, Res. Grant 5 Ro1 CC 00303, U.S. Public Health Service, National Communicable Disease Center, pp. 63–76, 1969.
2344. National Marine Water Quality Laboratory, "An evaluation of the toxicity of nitrilotriacetic acid to marine organisms," Progress report F.W.Q.A. Project 18080 GJ4, 1970.
2345. Nimmo, D. R., A. J. Wilson, Jr., and R. R. Blackman, "Localization of DDT in the body organs of pink and white shrimp," *Bull. Environm. Contam. and Toxicol.*, 5(4), 333–341, 1970.
2346. Stewart, N. E., R. E. Millemann, and W. P. Breese, "Acute toxicity of the insecticide sevin and its hydrolytic product 1-naphtol to some marine organisms," *Trans. Amer. Fish. Soc.* 96(1), 25–30, 1967.
2347. Ukeles, R., "Growth of pure cultures of marine phytoplankton in the presence of toxicants," *Appl. Microbiol.* 10(6), 532–537, 1962.
2348. Walsh, G. E., "Effects of herbicides on photosynthesis and growth of marine unicellular algae," *Hyacinth Control J.* 10, 45–48, 1972.
2349. Walsh, G. E., and T. E. Grow, "Depression of carbohydrate in marine algae by urea herbicides," *Weed Sci.* 19(5):568–570, 1971.
2350. Cooley, N. R., and J. Keltner, unpublished data from Gulf Breeze Laboratory, Environmental Protection Agency, Gulf Breeze, Florida.
2351. Cooley, N. R., J. Keltner, and J. Forester, unpublished data from Gulf Breeze Laboratory, Environmental Protection Agency, Gulf Breeze, Florida.
2352. Coppage, D. L., unpublished, "Organophosphate Pesticides: Specific level of brain AChE inhibition related to death in Sheepshead minnows," submitted to *Trans. Amer. Fish Soc.*
2353. Earnest, R., unpublished data, "Effects of pesticides on aquatic animals in the estuarine and marine environment," in *Annual Progress Report 1970*, Fish-Pesticide Research Laboratory, Bur. Sport Fish. Wildl. U.S. Dept. Interior, Columbia, Mo., 1976.
2354. Earnest, R. D., and P. Benville, unpublished, "Acute toxicity of four organochlorine insecticides to two species of surf perch," Fish-Pesticide Research Laboratory, Bureau Sport Fisheries and Wildlife, U.S. D. I., Columbia, Missouri.
2355. Nimmo, D. R., R. R. Blackman, A. J. Wilson, Jr., and J. Forester, unpublished data, "Toxicity and distribution of Aroclor® 1254 in pink shrimp (*Penaeus duorarum*)," Gulf Breeze Laboratory, Environmental Protection Agency, Gulf Breeze, Florida.

Ref TD 196 .073 V47 1983
Verschueren, Karel.
Handbook of environmental
 data on organic chemicals